BIOLOGICAL SCIENCE

WILLIAM T. KEETON

CORNELL UNIVERSITY

BIOLOGICAL SCIENCE

THIRD EDITION

ILLUSTRATED BY PAULA DI SANTO BENSADOUN

W · W · NORTON & COMPANY · NEW YORK · LONDON

COVER PAINTING: GEORGIA O'KEEFFE, *Shell I, 1928*
PHOTOGRAPHED BY MALCOLM VARON, N.Y.C.

Copyright © 1976 by Georgia O'Keeffe
Oil on canvas, 7 × 7 in.
From the artist's private collection
Used by permission of the artist

BOOK DESIGN: ANTONINA KRASS

THIRD EDITION

Copyright © 1980, 1979, 1978, 1972, 1967 by W. W. Norton & Company, Inc.
Published simultaneously in Canada by George J. McLeod Limited, Toronto
Printed in the United States of America
All rights reserved

LIBRARY OF CONGRESS CATALOGING IN PUBLICATION DATA

Keeton, William T
Biological science.
Includes bibliographies and index.
1. Biology. I. Title.
QH308.2.K44 1980   574   79–24387

ISBN 0–393–95021–2
  3 4 5 6 7 8 9 0

CONTENTS

PREFACE

Publication of this third edition of *Biological Science* comes eight years after the appearance of the second edition—an unconscionable lapse of time from the point of view of many users, who have been obliged to cope with a dated treatment of some very important topics. Two factors must be blamed for much of the delay. First, like so many other areas of biology, my own field of research—the orientation and navigation of birds—has been enjoying a period of unprecedented advances. I'll confess that the excitement of it often lured me away from strict attendance to book revision. Second, once I started work on the book in earnest, I saw that updating by itself—let alone a mere cosmetic once-over—would not be adequate. What was called for, it became clear, was a thorough revision in presentation as well as content—even at the cost of another year's work. My hope is that readers will find the long delay justified by the results.

Let me single out some of the major changes in this edition to give some idea of its character:

1. In Chapter 2, on chemistry, a new emphasis on weak bonds and their role in biological processes lays the groundwork for a fuller explanation of those special properties of water that make it fundamental to life, and for a more thorough discussion of the conformation of proteins and the action and regulation of enzymes. These subjects are not introduced merely to be ignored later; subsequent chapters draw on this earlier examination to help explain many important phenomena—among them, the way osmotically active particles affect the free energy of water, the mechanism of active transport across membranes, the oxygen-carrying ability of hemoglobin, control mechanisms in cellular metabolism, and antigen-antibody interactions.

2. Chapter 3 includes a detailed examination of the fluid-mosaic model of the cell membrane. This model later illuminates a variety of other subjects, e.g. the chemiosmotic hypothesis of oxidative phosphorylation (Chapter 4) and the mode of action of the sodium-potassium pump—the mechanism basic to both kidney function (Chapter 8) and nervous conduction (Chapter 10).

3. Completely overhauled, Chapter 4 is a fresh attempt at clarifying energy transformations in cells—particularly the complicated processes of photophosphorylation, glycolysis, and oxidative phosphorylation. The subject of C_4 photosynthesis is new to this edition; I have tried to relate it to what students may more readily identify as a "biological" feature—the Kranz anatomy that usually characterizes C_4 plants.

4. The reorganization of Chapter 11, on effectors, reflects our improved knowledge of microfilaments and microtubules, which has made possible a unified interpretation of the most diverse types of movement—including muscle contraction, the beating of cilia, amoeboid movement, and the migration of chromosomes during mitosis.

5. Chapter 12, on animal behavior, introduces the analysis of "evolutionary strategies" as a method of interpreting differences in feeding behavior and in social systems. I have avoided detailed descriptions of the social behavior of such species as marmots, lions, and baboons, appealing as such

descriptions can be; instead, I have focused on some of the chief concepts of interest to present-day sociobiology (e.g. parental investment, inclusive fitness, kin selection, reciprocal altruism) and on the application of these concepts to the development of diverse social systems and (in Chapter 18) to the evolution of altruism. It is this evolutionary approach to the study of social behavior, I believe—and not the description of social systems per se—that constitutes the principal contribution of modern sociobiology.

6. The book discusses cancer in many contexts. Chapter 15, for example, examines the relationship of mutagenicity to carcinogenicity, and the use of such tools as the Ames test in evaluating the health hazards present in the increasing number of chemicals to which we are exposed. Chapter 16 deals with the malignant process as a failure in normal cellular control.

7. The burgeoning field of developmental biology, covered in a single chapter in the second edition, requires two in the third (Chapters 16 and 17). Both chapters introduce much new material—e.g. gene amplification, post-transcriptional control, the somatic cell-fusion technique, pattern formation. In both, the presentation of topics previously included—e.g. control of gene transcription, development of immunity, the embryonic development of a vascular plant, the role of cell polarity in differentiation—has been thoroughly revised.

8. Completely reorganized, Chapter 19, on ecology, has a more statistical orientation and deals more fundamentally with many key ecological questions—including population growth and regulation, community stability, and the distribution of species.

9. An unfailingly fascinating subject—the history of life on earth—may gain added depth and interest from consideration of the Berkner-Marshall hypothesis (Chapter 20) and continental drift (Chapter 19).

10. The third edition adopts the now widely accepted five-kingdom classification championed by my colleague at Cornell, Robert H. Whittaker. Thus what were previously three chapters on the groups of organisms are now five (Chapters 21–25). They include brief descriptions of three newly discovered groups (viroids, Prochlorophyta, Placozoa) that may provide exciting new insights into the evolutionary relationships of other organisms.

This sketch of some of the major changes in the third edition should not be taken to imply that other parts of the book have been left untouched. Far from it. No part has escaped careful scrutiny—and change where change seemed warranted. Some examples: the description of the respiratory system of birds in Chapter 6, which was utterly wrong in the second edition; the model for active transport in Chapter 8; the treatment of the mechanism of hormonal action in Chapter 9, which demanded extensive updating. But any attempt to list all such changes here would mean incorporating the whole book in the preface, a procedure not favored by my publisher.

Despite my best efforts, some parts of this book will be out of date before students come to them, either because I failed to catch up with the most recent advances or because further progress occurred between the time of writing and the time of reading. This is not all bad: Students should understand the growth of information and the renewal of hypotheses as aspects of the dynamic, searching, inquiring character of science. As a way of reinforcing that understanding, I have noted time and again in the course of discussion that more research is needed before some of the ideas presented can be reliably evaluated. Often I have cited evidence both for and against two or three divergent hypotheses, and have made it clear that a final decision is not yet possible. This approach, which encourages students to evaluate the evidence for themselves, can bring it home to them that much remains to be done and that the effort will be exciting.

As in previous editions, I have assumed that if a topic is to be included at all it must be given sufficient depth to be made meaningful. My experience in teaching introductory biology courses at Cornell has convinced me that it is often more difficult for students to understand an oversimplified "elementary" presentation than one rigorous enough to give them insight into the meaning and relevance of the information. To be sure, I do not expect them to commit the entire contents of the book to memory—I haven't done that myself! Much of the material is meant as background, to provide a richer context for the ideas that really should be remembered.

After long consideration I chose to retain my original sequence of chapters, even though some teachers suggested that molecular genetics and development be moved forward to Part I. I did, however, introduce some changes (e.g. a description in Chapter 2 of the molecular structure of nucleic acids) that should make it easier to go directly from Part I to Part III.

A new feature in this edition is the inclusion of short discussions of special topics in boxes, designated *Exploring further* in the table of contents. Some of these topics (e.g. the chemiosmotic hypothesis, the three-dimensional arrangement of thylakoids, the genetic basis of the Rh factor, evolution of pine cones) may be difficult for many students; since the discussions function as asides, it will be easy for teachers to include or omit them from assignments as they wish. Others of these special topics (e.g. the digestion of lactose by human adults, thromboembolic and hypertensive disease in human beings, trisomy in humans) will interest students because of their human and social relevance; they add greater depth to the discussion, but are not essential.

As before, all illustrations have been designed or selected to work intimately with the text; in this edition the full-color illustrations are not relegated to separate sections, but appear near the pertinent text discussion. Every figure is expressly referenced in the text. The number of illustrations is significantly greater than in previous editions. Where a single diagram may have illustrated a structure or process in the second edition, several diagrams and photographs often help clarify that same structure or process in the third edition (note, for example, the illustrations for xylem vessels and for the circulatory system in Chapter 7, for contractile vacuoles in Chapter 8, for muscular contraction in Chapter 11, and for succession in Chapter 19).

Paula DiSanto Bensadoun, whose work has been an essential part of the earlier editions, has revised the majority of her drawings and diagrams and contributed a large number of new ones, many of them based on her own examination of source materials (e.g. the arrangement of thylakoids, the gills of fish). Both she and I are grateful to Clark Carroll for invaluable technical assistance in preparing the figures for the printer.

Ruth Mandel, as picture editor, located literally thousands of photographs for my examination. I express my debt to her as well as to the many people who contributed the photographs, both color and black and white. The legends identify the contributors of the pictures we finally selected.

Few things are so infuriating as to be unable to locate in the index some item that one knows is discussed and even illustrated in the text. A defective index undermines the reference value of a book, and it is a hindrance to study. From the beginning my publisher and I have attached prime importance to the index. We have been fortunate enough, in this edition, to enlist the help of Ellis H. Whitaker of Southeastern Massachusetts University, who has devoted many months, and applied his high ideals of accuracy and order, to the compilation of an index that is at once exceptionally comprehensive and very easy to use.

In the past I resisted all suggestions that I authorize preparation of a study guide and instructor's manual to go with *Biological Science*, because of my low (indeed unprintable) opinion of most of the guides I had seen. But my colleague at Cornell Carol Hardy McFadden, who has had extensive experience teaching introductory biology and other undergraduate courses, has happily obliged me to revise my opinion. I am convinced that the study guide and instructor's manual she has prepared will be valuable supplements to the book.

Many biologists reviewed parts of the previous edition and offered valuable advice for this revision. Others kindly sent me information from their own research that they knew would be helpful to me in my writing. And still others were patient with me when I took up hours of their valuable time asking for clarification of points I found confusing in their fields of specialty. I can't list everyone who helped me, but I do want to acknowledge the following: Michael W. Berns (University of California, Irvine), William L. Brown (Cornell), Hal Caswell (University of Connecticut), Peter J. Davies (Cornell), Sidney W. Fox (University of Miami), Joseph Frankel (University of Iowa), John C. Gerhart (University of California, Berkeley), Karl G. Grell (University of Tübingen), Jack P. Hailman (University of Wisconsin), Walter Hewitson (Bridgewater State College, Mass.), Peter C. Hinkle (Cornell), Andre Jagendorf (Cornell), Lewis J. Kleinsmith (University of Michigan), Nancy Maizels (University of California, Berkeley), Carol Newlon (University of Iowa), Gerard A. O'Donovan (Texas A & M University), William Overlease (West Chester State College, Pa.), Dominick J. Paolillo (Cornell), Efraim Racker (Cornell), Richard B. Root (Cornell), David G. Shappirio (University of Michigan), Virginia Utermohlen (Cornell), Steven Vogel (Duke University), Robert H. Whittaker (Cornell), Stanley Zahler (Cornell).

Melvin Kreithen and Douglas Quine of Cornell aided me throughout the preparation of this third edition by searching references, running down obscure facts, and participating with me in the enormous task of sorting through photographs and arriving at final selections. I owe both of them a great debt.

Bertha Blaker assisted me in many ways during preparation of the book. It is a pleasure to acknowledge her help.

I appreciate more than I can say the constant help and encouragement of Neil Patterson and Joseph B. Janson of W. W. Norton. Even more important, I value and appreciate my friendship with two such fine people.

It gives me great pleasure finally to acknowledge publicly my debt to Kenneth B. Demaree, the man who convinced me to write the first edition of this book and shepherded it and the second edition through the publishing process. He would never permit me to thank him in earlier prefaces, but now that he has retired he no longer has the power to say me nay.

Once again, as with former editions, my closest colleague and most stalwart aid has been my editor Ruth Thalheimer. Without her sure hand and constant good judgment, this book would have suffered greatly.

Finally, I thank the many students throughout the world who have taken the time to write to me expressing appreciation for my book. Many have been undergraduates just awakening to the excitement of biological studies; others have been graduate students preparing for their Ph.D. comprehensive examinations. Their kind comments have made the long hours of labor worthwhile. After all, the book is for them.

W. T. K.

Ithaca, New York
October 1979

BIOLOGICAL SCIENCE

Chapter 1

INTRODUCTION

The subject of this book is biological science—the study of life. But what is life? We all have some idea. After all, we ourselves are manifestations of life, and endlessly varied forms of it abound around us. But when asked to say explicitly what life is, we find that a satisfying answer eludes us. One dictionary defines life as "the condition which distinguishes animals and plants from inorganic objects and dead organisms," and defines dead as "deprived of life"; in other words, life is what animals and plants have when they are not dead, and dead is what those same organisms are when they lack life. Clearly, such circular definitions get us nowhere. Here is a word for something at the very center of our concern, and yet we are at a loss to offer a meaningful definition of it.

In lieu of a definition, many biology texts give a list of attributes that, taken together, are said to characterize life. The list usually includes metabolism, reproduction, growth, responsiveness, and movement. And there can be no doubt that investigation of these attributes does help toward an understanding of what life is. But merely setting down one-paragraph descriptions of each of these properties with the assertion that they are peculiar to life seems a singularly inadequate way to describe something so rich and complex, so varied and changeful. This entire book is about life; yet when you have finished it, chances are that you still won't be able to define life convincingly or describe it satisfactorily, although you will perhaps have a deeper appreciation and understanding of what it means to be alive.

BIOLOGY AS A SCIENCE

We live in a science-conscious age. Radio and television advertise every product as the latest result of "scientific" research. Commercials are full of "scientists" in starched white lab coats. The news media deal every day with

1

some fresh development in space science or medical science or agricultural science. The President and Congress have scientific advisers, and millions of dollars are spent every year by both the national and the state governments on scientific research and development. Yet there is a surprising lack of understanding of what science is. Some view it as akin to magic or as a cold, relentless mathematical game. And while there are those who feel that science should be able to provide definitive solutions to all the problems of modern life—from pollution of water and air, to the shortage of energy, to the ravages of cancer—there are many others who regard science with anxious suspicion as threatening to a humane existence.

The advent of moon probes and space stations has changed the public image of at least some scientists. But the new picture of scientists as glamorous adventurers is as erroneous as the older one of blinkered recluses in their ivory towers. Most scientists are just people who have a deep curiosity about the world they live in. They want to know more about the earth and stars, rocks and rivers, atoms and molecules, plants and animals. For many scientists, the need to know is in itself sufficient motivation. For others, the possibility that what they discover may benefit humanity is additional motivation. But whatever moves them, they are usually neither naïve recluses nor glamorous adventurers. And they are no strange breed of modern magicians waving a wand called "scientific method."

THE SCIENTIFIC METHOD

So much has been said about the powers of the scientific method that many suspect it involves some formula too complicated for ordinary people to understand. It doesn't. It is used to some extent by almost everyone every day. Its power in the hands of a scientist stems from the rigor of its application. Let us briefly examine the elements of this method.

Science is concerned with the material universe, seeking to discover facts about it and to fit those facts into conceptual schemes, called theories or laws, that will clarify the relations among them. Science must therefore begin with observations of objects or events in the physical universe. The objects or events may occur naturally, or they may be the products of planned experiments, but the important point is that they must be observed, either directly or indirectly. Science cannot deal with anything that cannot be observed.

Science rests on the assumption (justified by its pragmatic success) that all events of the universe can be described by physical theories and laws, and that we get the data with which to formulate those theories and laws through our senses. (Notice that the laws with which we are concerned are descriptive, not prescriptive. They do not say how things should be; they say how things are and probably will be.) Scientists readily acknowledge the imperfection of sensory perception, which is itself a subject of scientific study. They know that it is variable and unreliable. And they know that there is always an unavoidable interaction between the observer and the phenomenon observed. As George Gamow has said, "There is no such thing as a physical phenomenon *per se*—the observer and his instruments become an integral part of the phenomenon under investigation." But to recognize the imperfection of sensory perception is not to suggest that we may get information by other means (except, of course, by genetic transmission, which is also a physical process). No other means are open to us as scientists.

Once scientists have formulated the questions they want to ask, and have made careful observations in an attempt to answer them, they must do something with their data. It is not enough simply to amass data; the data must be analyzed and fitted into some sort of coherent pattern or generalization. The step from isolated observations to generalization can be taken with confidence only if enough observations have been made to give a firm basis for generalization and only if the individual observations have been reliably made.

But the making of general statements is not the end of the process. Scientists must devise new ways of testing their generalizations, perhaps by basing simple predictions on them and checking if the predictions are accurate. If they find a discrepancy between their generalizations and observable fact, they must change their generalizations. Eventually, if all the evidence continues to support the new idea, it may become widely accepted as probably true and be dignified by the appellation *theory*.[1] But the testing never completely stops. No theory in science is ever absolutely and finally proved. All good scientists must be ready to alter or even abandon their most cherished generalizations when new facts contradict them. They must always remember that their theories, even their physical laws, are dependent on observable facts and not vice versa.

In actual practice, the steps followed in a scientific investigation may be far from clear. For example, hypotheses may not flow obviously from previous generalizations; they may often depend more on educated guesses or hunches than on strict deduction. Indeed, insight, the ability to make good guesses, is often the distinguishing characteristic of the great scientist. But in the final analysis the basic rules are the same. The hypothesis must be testable—otherwise it is of no value—and as testing proceeds it must be altered when necessary to conform with the evidence. The scientist's query must always be: "What is the evidence?"

Though no aspect of the scientific method is unique to science, nowhere else is the method applied with such good results, for nowhere else is it consistently applied with such rigor. A telling comparison between the scientific method and the same procedures used less rigorously was made by T. H. Huxley: "Science is . . . nothing but trained and organized common sense, differing from the latter only as a veteran may differ from a raw recruit; and its methods differ from those of common sense only so far as the guardsman's cut and thrust differ from the manner in which a savage wields his club."

---

[1] Notice that scientists do not use the word "theory" as the general public does. To many people a theory is a highly tentative statement, a poor makeshift for fact. But when scientists dignify a statement by the name of theory, they imply that it has a very high degree of probability and that they have great confidence in it. A theory is a hypothesis that has been repeatedly and extensively tested and always found to be true. It is supported by the facts and helps order and explain those facts. Many scientific theories, such as the cell theory and the theory of evolution, are so well supported by all known facts that they themselves are "facts" in the nonscientific application of that term.

Limitations of the scientific method   The insistence on testability in science limits its range of applications. For example, the idea that there is a God working through the natural laws of the universe is not testable, and hence cannot be evaluated by science. Science can neither confirm nor refute it. Yet throughout history human beings have made statements in the name of God—that the earth is at the center of the universe, for example—and have asserted that denial of their statements is a denial of their God. Those who make the existence of a deity stand or fall on some supposed fact about the universe risk having science destroy that deity.

Science cannot make value judgments; it cannot say that a painting or a sunset is beautiful. And science cannot make moral judgments; it cannot say that war is immoral. It cannot even say that a river should not be polluted. Science can, however, analyze response to a painting; it can analyze the biological, social, and cultural implications of war; and it can demonstrate the consequences of pollution. It can, in short, try to predict what people will consider beautiful or moral, and it can provide them with information that may help them make value or moral judgments about war or pollution. But the act of making the judgment itself is not science.

THE RISE OF MODERN BIOLOGICAL SCIENCE

The scientific approach based on observation and testing is of relatively recent origin. It may seem obvious to you that the way to learn about nature is to look at nature, to seek evidence by observation, but for many centuries such a procedure was far from obvious. True, major scientific discoveries were made in the ancient civilizations of Egypt, Babylonia, and Greece, but this early scientific activity began a steady decline more than a century before the birth of Christ. Between about 200 and 1200 A.D., there were almost no important scientific advances, and much of what the ancients had known was forgotten. Greater reliance was placed on religious dogma and on the writings of a few venerated ancient scholars than on observation of the universe itself. Haggling over the exact meaning of a sentence in one of Aristotle's treatises on plants or animals took the place of studying the plants or animals themselves; the possibility that Aristotle might sometimes have erred seems not to have been entertained. No real distinction was made between science and theology, and the one was questioned as little as the other.

Then came a period of intellectual reawakening in Europe. Two very influential theologians and philosophers, Albertus Magnus (1193–1280) and Thomas Aquinas (1225–1274), both of whom taught at the University of Paris, accepted the idea of a distinction between natural truth and revealed truth. This rationalistic approach, separating large segments of human knowledge and speculation from theology, prepared the way for a relatively independent development of science. At about the same time, Roger Bacon (ca. 1210–1293) at Oxford University was calling for an end to unthinking acceptance of traditionally authoritative writings such as Aristotle's: "Cease to be ruled by dogmas and authorities; look at the world!" Three centuries later, Francis Bacon (1561–1626) vigorously championed the experimental approach to knowledge, urging that no statements be trusted without verification, that all things be tested with the utmost rigor.

The new era in the physical sciences   The intellectual climate was changing; it was no longer beneath the dignity of an educated person to look at the material objects of nature. Thus Nicolaus Copernicus (1473–1543), a Polish astronomer, analyzed the movements of the heavenly bodies and announced that the earth moves around the sun rather than the sun around the earth. Now, the intellectual climate may have been changing and becoming more friendly to science, but it was not yet ready for such a proposition as this. After all, if the earth moved around the sun, then the earth was not the center of the universe and—outrageous suggestion!—humanity did not stand at the center of creation. When the great Galileo Galilei (1564–1642) embraced Copernicus' theory, he was forced to recant publicly under threat of excommunication. H. G. Wells[2] describes this early clash between science and dogma:

> the church . . . decided that to believe that the earth was smaller and inferior to the sun made man and Christianity of no account . . . so Galileo, under threats of dire punishment, when he was an old man of sixty-nine, was made to recant this view and put the earth back in its place as the immovable centre of the universe. He knelt before ten cardinals in scarlet, an assembly august enough to overawe truth itself, while he amended the creation he had disarranged. The story has it that as he arose from his knees, after repeating his recantation, he muttered, *"Eppur si muove"*—"it moves nevertheless."

Galileo may have publicly recanted, but the ideas of Copernicus, Galileo, and the other great scientists who followed them could not be suppressed. The conviction grew that the physical universe could be understood in terms of orderly relationships, of universal laws, that physical events have comprehensible impersonal causes, that the whims of gods and magicians and evil spirits need no longer be invoked to explain a physical event. No one else stands out so prominently during this period as Isaac Newton (1642–1727), who was born in the year of Galileo's death. His discovery of the Law of Gravitation and his explanation in 1685 of the movements of the planets caused a revolution in thought and carried physical science into a new era. In a very real sense, the work of Newton marks the birth of modern physics.

The science of the new era was restricted largely to physics, astronomy, and chemistry, i.e. to the physical sciences. If many were now ready to give up their place at the center of the universe and admit that the earth circles the sun, they were nevertheless not yet ready to admit that they themselves and other living creatures could be understood in terms of impersonal forces. Life seemed too full of purpose, of design, to be studied in the same way as chemicals and moving particles. If Copernicus was a threat to human dignity and pride, how much greater a threat lay in an explanation of life processes in mechanistic terms! It was not until nearly two hundred years after Newton that biological science experienced its own revolution and entered its modern era.

The new era in the biological sciences   The mid-nineteenth century witnessed a number of major biological breakthroughs. The cell theory, one of the fundamental generalizations of biology, had been given essentially its modern form by the year 1858; and in 1868 Louis Pasteur conclusively

[2] In *The Outline of History* (3rd ed.; Macmillan, 1922), pp. 732–733.

disproved the idea of spontaneous generation of life. The birth of the modern era of biological science, however, may be placed—somewhat arbitrarily—in 1859, because it was in that year that Charles Darwin (1809–1882), the great English naturalist, published *The Origin of Species*, in which he proposed his theory of evolution by natural selection.[3] We do not wish to imply that no important biological work was done in the centuries before Darwin—that would be nonsense. But important as the work of earlier investigators was, it did not spark the sort of explosive growth of biological science that Newton's work had stimulated in the physical sciences.

An explosion is precisely what Darwin's book did cause. Even those who had reconciled themselves to living on a small planet far from the center of the universe were not prepared to accept the ultimate indignity—that they had descended from some lowly form of life in the distant past and shared common ancestors with monkeys and apes and even worms. Again, the old cry of heresy was raised. Again, some felt that the very basis of all they held dear had been challenged, that if Darwin's views should prevail religion would be destroyed. The outcry was loud and anguished and has not fully subsided yet, although most major Western religions and denominations now accept the theory of evolution and no longer consider it a threat to their existence.

Copernicus' theory was not generally accepted until long after his death, as we have seen. But delayed recognition was not to be the fate of Darwin's theory. And despite violent denunciations from some quarters, Darwin never had to recant like Galileo. The times were ripe for Darwin. A major part of the scientific community welcomed and promptly championed his views. In fact, he himself seldom debated the merits of his theory in public. He didn't need to. Some of his greatest contemporaries eagerly acted as his defenders. His spark had ignited an excitement in the scientific community that was not to be extinguished. Biology was never again to be the same. Almost immediately interest in biological research began to grow rapidly. Whether to prove or to disprove Darwin, there was an increasing willingness to investigate the phenomena so long considered beyond the scope of science. The dynamic growth of biological research, which started more than a century ago, has never slackened; in fact, the rate of growth is greater now than ever before. And to this day the theory of evolution by natural selection remains one of the most important unifying principles in all biology. We shall have cause to refer to it in every chapter of this book.

DARWIN'S THEORY

The theory of evolution, as modified in the years since Darwin, will be treated at some length in Chapter 18. But since we shall be referring to it in

---

[3] Actually, the theory was first announced in a short paper read before the Linnaean Society of London in 1858. A paper by Alfred Russel Wallace (1823–1913) containing essentially the same conclusions was also presented on that occasion. Darwin had conceived the theory first and had labored for many years to provide convincing proof for it. For this reason, and because what he published the following year was far more complete than anything Wallace ever published on evolution, the theory is usually credited to him. It is, however, a good indication that the times were ripe for the theory that two men advanced it independently at almost the same time. In fact, earlier in the nineteenth century several other workers had published comments that foreshadowed the ideas of Darwin and Wallace.

interpreting much of the material covered in earlier chapters, let us briefly examine the central concepts of Darwin's theory here.

The theory consists of two major parts: the concept of evolutionary change and the concept of natural selection. First, Darwin rejected the notion that living creatures are the immutable products of a sudden creation, that they exist now in precisely the form in which they have always existed. He maintained that, on the contrary, change is the rule, that the organisms living today have descended by gradual changes from ancient ancestors quite unlike themselves. Second, Darwin declared that it is **natural selection** that determines the course of the change, and that this guiding factor can be understood in completely mechanistic terms, without reference to conscious purpose or design. Let us examine the two parts of Darwin's theory separately.

THE CONCEPT OF EVOLUTIONARY CHANGE

In the mid-twentieth century the idea that lineages of organisms change with time seems far from revolutionary. We are used to change. We see change on every hand. We should probably be surprised to find anything that remained the same for any very long period of time. But in Darwin's day things moved more slowly. The idea of a world in constant flux had only few supporters. The vast majority accepted without question the proposition that the universe was created a few thousand years before the birth of Christ, and that all the species of plants and animals were put on the earth at that time and had perpetuated themselves without change ever since. What sorts of evidence could Darwin bring forward to combat this view of a static universe?

First, he could point to the fossils. During the latter part of the eighteenth and the first half of the nineteenth century, geologists had unearthed many fossils and realized that most of them represented species no longer living on the earth and, conversely, that few living species were represented in the fossil record. In other words, forms of life different from those known today inhabited the earth in past ages. This is a point now familiar to grade-school children, who are aware that dinosaurs once roamed the earth in vast numbers, but are not encountered nowadays. For they can go to museums and see dioramas of ancient seas filled with strange fish and shelled creatures unlike anything living today, or they can see reconstructions of weird coal-age forests with plants that became extinct millions of years ago. They may study about the early cave dwellers, who, though clearly human, were also clearly different in many ways from modern human beings. To us the existence of fossils seems convincing evidence that the history of life on earth has been marked by change. But when it was first suggested that the extinct creatures whose remains are preserved in the rocks as fossils represented the ancestors from which the organisms living today are descended, it was urged instead that these extinct species indicated the occurrence of catastrophic extinctions at various times in the history of the earth, followed by new episodes of divine creation. According to this view, each species would have remained unchanged from the time of its creation until the time of its extinction; there would have been no evolution.

Soon, however, the fossil record itself made this interpretation untenable. As more and more fossils were discovered and studied, it became evident that gradual shifts in characters (physical traits) could be traced through time. If an investigator studied the fossils in one rock layer and then studied the fossils in a slightly more recent layer, he would often find that those in the more recent layer, though very similar to the older ones, showed slight differences. If he then studied a third layer slightly more recent than the second, he would again find that slight changes in the characters of the fossil species could be detected. In this way, by studying a series of successive rock layers, he could reconstruct the sequence of changes through which a given lineage had passed. He could even predict what the fossils in some interme-diate layer not yet studied would be like and then test his hypothesis by locating and studying such a layer. The notion of catastrophic extinctions and repeated creations hardly seemed adequate to explain such fossil se-quences. It was far more likely that the changes seen in the fossils were the result of accumulation of many small alterations as the generations passed.

Second, Darwin could point to resemblances between living species. If one looks at the forelimbs of a variety of different mammals, for example, one will find essentially the same bones arranged in the same order (Fig. 1.1). The basic bone structure of a human arm, a dog's front leg, or a seal's flipper is the same; the same bones are present even in a bird's wing. True, the size and shape of the individual bones vary from species to species, and some bones may be missing entirely in one or another species, but the basic construction is unmistakably the same. To Darwin the resemblance sug-gested that all these species had descended from a common ancestor from which each had inherited its forelimb, with distinctive modifications. The fact that some species possess in reduced and nonfunctional form structures that in other species have important functions further convinced Darwin of the validity of his theory. Why would the Creator have given pigs, which walk on only two toes per foot, two other toes that dangle uselessly well above the ground? Why would he have given human embryos gill pouches and well-developed tails only to make them disappear before the time of birth? It seemed much simpler to assume that such structures were inherited vestiges of structures that functioned in ancestral forms and that still func-tion in other species descended from the same ancestor.

Third, and particularly convincing, Darwin could point to changes pro-duced in domesticated plants and animals. How could anyone confronted with the historical evidence of the changes in domesticated forms doubt that great changes can occur in organisms with time? Where were French poo-dles and Mexican Chihuahuas two thousand years ago? Where were Guern-sey cattle and Leghorn chickens? Where were the modern strains of toma-toes and corn and roses? None of them existed. Their ancestors existed, but those ancestors bore little resemblance to poodles or Chihuahuas or Guern-seys or Leghorns or garden tomatoes, corn, and roses. Obviously, radical changes have occurred in a few thousand, or even a few hundred, years. The ancestors of the poodles and Chihuahuas were wolves. Those of modern corn were small wild plants with ears less than an inch long. Let these facts be explained by anyone who would still insist that species cannot change.

THE CONCEPT OF NATURAL SELECTION

It was easy for Darwin to see that evolutionary change had occurred. But it took him many years to figure out what caused the changes. His first clue

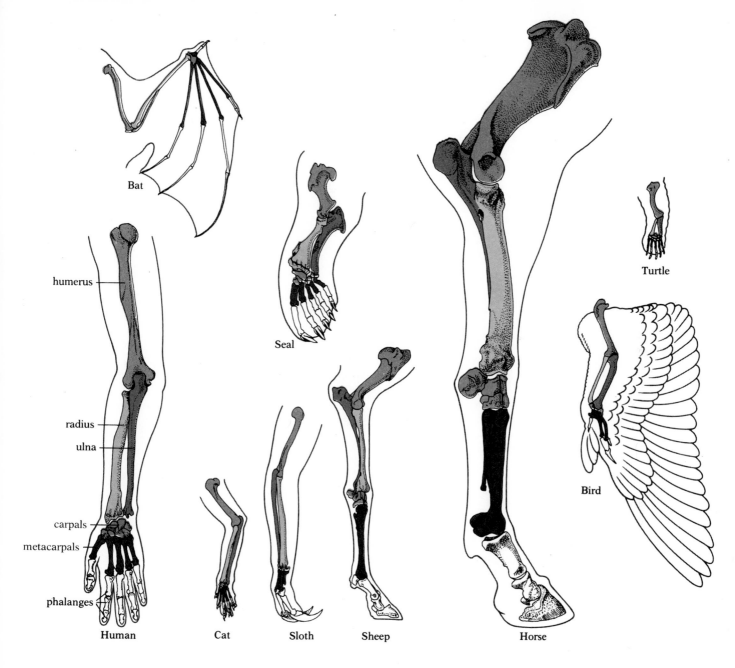

Bat

humerus

radius

ulna

carpals

metacarpals

phalanges

Human   Cat   Sloth   Sheep   Horse

Seal

Turtle

Bird

1.1 The bones in some vertebrate forelimbs compared

The labeled and color-coded bones of the human arm at left will permit identification of the same bones in the other forelimbs depicted. In the bat the metacarpals (hand bones) and phalanges (finger bones) are elongated as supports for the membranous wing. In the seal the bones are shortened and thickened in the flipper. The cat walks on its toes, the metacarpals having come to form a part of the leg. The sloth normally hangs upside down from tree limbs—hence its recurved claws. The horse walks on the tip of one toe, which is covered by a hoof (specialized claw), and the sheep walks on the hoofed tips of two toes (i.e. it is cloven-hoofed, though only one hoof can be seen in this side view). Note that the carpals (wrist bones) of both the horse and the sheep are elevated far off the ground, because the much-elongated metacarpals (hand bones) have become a section of the leg. Small splintlike bones that are vestiges of other ancestral metacarpals can be seen on the back of the upper portion of the functional metacarpals of both horse and sheep. All the animals mentioned so far—human, bat, seal, cat, sloth, sheep, and horse—are mammals, but the same bones can also be seen in the leg of a turtle or the wing of a bird. (All limbs are drawn to the same scale except that of the turtle, which is enlarged.)

9

1.2 Breeds of pigeons
The ancestral rock dove is shown in the center. The domestic breeds (clockwise from upper left) are: fantail, Saxon swallow, jacobin, scandaroon, Norwich cropper, and satinette oriental frill. [Based on photographs in W. M. Levi, *The pigeon*, 1963, and *Encyclopedia of pigeon breeds*, 1965, Levi Publishing Co., Inc.]

came from the breeding of domesticated plants and animals. When pigeon breeders, for example, are developing a new strain, they exploit the variation always seen among individuals by selecting the ones best endowed with the characteristics they want to propagate and using them as the parents for the next generation. The same procedure is followed in each successive generation; those individuals most nearly approximating the desired type are selected as breeders, and individuals that deviate markedly from the desired type are eliminated. After many generations of such selection, the pigeons will be very different from the ones with which the breeders began (Fig. 1.2). Essentially the same procedure is used in developing a new breed of dog or horse or wheat or chrysanthemum. Since individual variation occurs in all populations of wild organisms, just as it does in populations of domesticated ones, Darwin reasoned that evolutionary change in these populations must be caused by some sort of natural selection for individuals with certain characteristics and elimination of individuals with other characteristics. But what sort of selective force might be at work in nature? The answer eluded Darwin for several years.

Then in 1838 he read a work entitled *An Essay on the Principle of Population*, written by Thomas R. Malthus in 1798. It suggested to him how he could account for the selection he felt sure must be operating in nature. Consider for a moment a population of gray squirrels. If this population is to be perpetuated at a stable level, each pair of squirrels must leave enough offspring to replace itself—two, if we assume that all the offspring survive to reproduce. If the average number of progeny per pair were more than two, then the population density would rise; if the average number were less than two, then the population density would fall. Now, even a casual survey of actual populations will reveal that the average number of offspring per pair is always more than two, usually far more. A single female frog may lay many thousands of eggs each year; a single pair of robins usually has two clutches per year of four to five eggs each; and a pair of gray squirrels usually has two litters per year containing two to four young each. A single oak tree may produce millions of seeds during its lifetime. Very large reproductive potentials are, in short, the rule in all types of organisms. Yet natural populations usually remain relatively stable over long stretches of time; they may fluctuate noticeably, but they never even approach the level that would be expected if all their progeny survived to reproduce. It is obvious, therefore, that a very high percentage of the young of any species fail to survive and reproduce.

Once Darwin recognized that in nature the majority of the offspring of any species die before they reproduce, and that survival and reproduction are never totally random, he had the clue he needed to explain natural selection. If each individual in a large population had exactly the same chance of surviving and reproducing as every other individual, then there would probably be no significant evolutionary change in the population. But some individuals are born with such gross defects that they stand almost no chance of surviving to reproduce. And even among individuals not so severely afflicted, differences in the ability to escape predators or obtain nutrients or withstand the rigors of the climate or find a mate, etc., ensure that survival will not be totally random. The individuals with characteristics that weaken their capacity to escape predators or obtain nutrients or withstand the rigors of the climate, etc., will have a poorer chance of surviving and reproducing than individuals with characteristics enhancing these capacities. In each generation, therefore, a slightly higher percentage of the well-adapted individuals will leave progeny. If the characteristics are inherited, those favorable to survival will slowly become more common as the generations pass and those unfavorable to it will become less common. Given enough time, these slow shifts can produce major evolutionary changes.

Now let us compare the propagation of favorable characteristics in nature, as outlined above, with their propagation in domesticated organisms. In each case far more offspring are born than will survive and reproduce; i.e. in each case there is differential reproduction, or selection. In the breeding of domesticated plants and animals, the selection (differential reproduction) results from the deliberate choice of the breeder. In nature the selection (differential reproduction) results simply from the fact that individuals with different inherited characteristics have unequal chances of surviving and reproducing. Both sorts of selection, artificial and natural, cause some inherited characteristics to become more prominent in the population and others to become less so as the generations pass. Notice that individuals, once born, are not changed by selection. An individual cannot evolve. The change is in the makeup of the population.

One difference between evolutionary change in nature and the change produced by breeders should be noted. That is the rate of the change. Breeders can practice very rigorous selection, eliminating all undesirable individuals in every generation and allowing only a few of the most desirable to reproduce. They can thus bring about very rapid change, as we all know. Natural selection, which allows considerable play to chance, is usually much less rigorous: Some poorly adapted individuals manage to survive and reproduce, and some well-adapted individuals are eliminated. Hence evolutionary change is usually rather slow; major changes may take many thousands or even millions of years. Fortunately for Darwin and his theory, the geologists of his day, particularly his friend, the great Charles Lyell, had provided evidence that the earth could not possibly have been created in 4004 B.C., as many churchmen insisted, but that it must be several billion years older. Without the geologists' gift of immense spans of time, Darwin's theory of natural selection could not adequately have explained evolutionary change.

In summary, we see that Darwin's explanation of evolutionary change in terms of natural selection depends on five basic assumptions:

1. Many more individuals are born in each generation than will survive and reproduce.
2. There is variation among individuals; they are not identical in all their characteristics.
3. Individuals with certain characteristics have a better chance of surviving and reproducing than individuals with other characteristics.
4. At least some of the characteristics resulting in differential reproduction are heritable.
5. Enormous spans of time are available for slow, gradual change.

All the known evidence supports the validity of these five assumptions.

THE VARIETY OF ORGANISMS

According to the fossil record, the most primitive organisms known—the bacteria and blue-green algae—date back over 3 billion years, the first land plants and insects over 400 million years, the first birds and mammals over 180 million years. Since the simplest forms of life arose, innumerable different kinds of organisms, increasingly complex and adapted to widely varying environments, have evolved. The process by which the different groups of organisms have come into being—one to be treated at length in a later chapter—generally begins with the geographic separation of a population that, before its separation, could interbreed.

Natural selection, as we have seen, can bring about significant changes in the characteristics of a population as its members breed together generation after generation. Now suppose that some members of that population emigrate, or that a segment of the population is cut off from the rest by some change in the terrain. The new population will inevitably find itself in a slightly different environment, and natural selection may favor other characteristics than in the original population; in other words, the most adaptive characteristics in the original population may not be the most adaptive in the new population. Eventually, as natural selection acts on individuals in each of the two populations, the differences between them will become such that they can no longer interbreed, even if the geographical barrier between them should disappear. It is essentially such a process, repeated over billions of years, under the varied and, in the course of time, profoundly altered conditions of terrestrial life, that has given rise to the radically different groups of organisms that inhabit the earth today.

Living things are classified on the basis of the evolutionary relationships thought to exist among the various groups. Since the relationships are often far from clear, a number of different systems of classification are possible. The one used in this book recognizes five broad categories, or kingdoms: Monera, Protista, Plantae, Fungi, and Animalia.

The survey that follows will serve as a brief introduction to the five kingdoms and their most important subdivisions. We shall refer to these major groups again and again in the course of this text, as we discuss the basic problems faced by all living things and the different adaptations these have evolved in response to them. The groups mentioned here, and many others, will be described in far greater detail in Part V.

MONERA

Members of the Monera differ from those of the four other kingdoms in that their cells lack a membrane-enclosed nucleus, as well as other subcellular membranous structures present in the cells of all other types of organisms.

Bacteria   The bacteria are single-celled organisms visible only under magnification (Fig. 1.3); it was not until the advent of the electron microscope that their internal structure could be elucidated. With very few exceptions, bacteria lack the green pigment chlorophyll, which enables plants to carry out photosynthesis—the process by which they manufacture sugar from

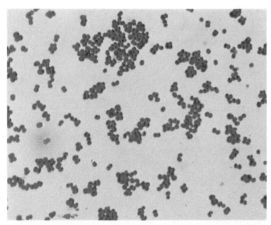

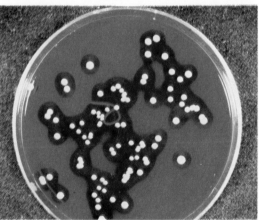

1.3 The bacterium *Staphylococcus aureus*
Top: High magnification (here × 1,500) is needed—and the staining is helpful—in making this bacterium detectable by the eye. It is an agent for many infections, including boils and abscesses. Bottom: The colonies *S. aureus* forms on a blood agar medium each contain many thousands of cells. [Courtesy Department of Microbiology, The Mount Sinai Hospital, New York (top); Elliot Scientific Corp. (bottom).]

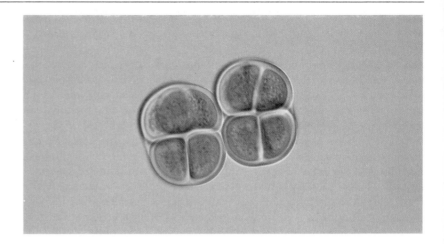

1.4   A blue-green alga *(Chroococcus)* **from the pine barrens of New Jersey**
Small groups of cells are enclosed within common gelatinous sheaths. × 1,600.   [Courtesy T. E. Adams.]

1.5   A diatom, *Craspedodiscus coscinodiscus*
The diatoms are single-celled organisms common both in freshwater habitats and in the oceans. They characteristically have elaborately ornamented glasslike cell walls. × 525.   [Courtesy Turtox/Cambosco, Macmillan Science Co., Inc.]

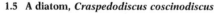

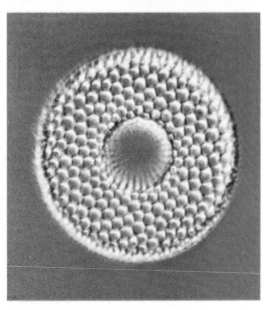

carbon dioxide and water with the help of light energy; hence bacteria must generally obtain preformed nutrients from the environment. Most bacteria have strong cell walls, which protect them from damage. The chemical components in the walls differ according to the kinds of bacteria; these differences are the basis for many of the diagnostic techniques used to identify them.

Many bacteria are disease-producing agents that attack human beings, other animals, and plants. Many other bacteria, however, are beneficial, acting as scavengers that destroy the dead bodies and bodily wastes of other organisms and, in the process, recycle important compounds from these organisms to environmental reservoirs, where they become available to other living things. Some bacteria are also used in industrial and manufacturing processes. Finally, study of bacteria has enabled biologists to learn many important principles about life that have proved applicable to other organisms, including ourselves.

Blue-green algae   All blue-green algae (Fig. 1.4) contain chlorophyll and are photosynthetic. They occur in an amazing variety of habitats, in moist places on land as well as in both freshwater and saltwater.

PROTISTA

The kingdom Protista includes a variety of groups whose members are predominantly unicellular, sometimes colonial, only rarely (and on a very rudimentary level) multicellular. Like true plants and animals, and unlike the unicellular Monera, protists have cells that contain a membrane-bounded nucleus and other intracellular membranous structures.

The plantlike protists   The members of some protistan groups—the euglenoids, dinoflagellates, diatoms (Fig. 1.5), yellow-green algae, and golden-brown algae—have chlorophyll and are photosynthetic. In older classifications they are identified as primitive plants. Because they are indeed plantlike in many respects, it is still common practice (a practice we shall often follow) to use the term "plant" to include both the photosynthetic protists and the true plants.

The animal-like protists   Many protistan groups lack chlorophyll and are not photosynthetic. Some of these are funguslike, but others, known as the *Protozoa,* have traditionally been viewed as unicellular animals, and we shall occasionally use the term "animal" to include both these groups and the true animals. Protozoa are often highly specialized, their single cell exhibiting a complexity and separation of function analogous to those observable in multicellular animals.

Protozoa are generally much larger than bacteria. They are very mobile, swimming rapidly in the water in which they live or crawling along the bottom or on submerged objects. Some propel themselves by the whiplike motion of long hairlike structures known as flagella. Others, like *Paramecium* (Fig. 1.6), bear many shorter hairlike structures called cilia, which often function in both locomotion and feeding. Still others, like *Amoeba* and *Pelomyxa* (Fig. 1.7), have neither flagella nor cilia, but move by a complex flowing

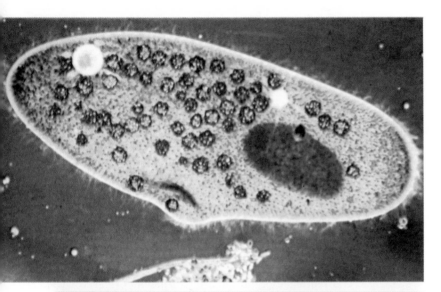

1.6 A ciliate protozoan, *Paramecium caudatum*
The colors seen here, which were produced by optical staining with the interference microscope, help differentiate the component parts of this living specimen; they are not the organism's natural colors. Ciliate protozoans are unusual in having two kinds of nuclei—as seen here, a large brown structure with a small darker one overlapping with it. The yellow cilia, which seem to edge the cell, actually cover all of it. The numerous small green circles are food vacuoles, chambers here containing yeast on which the protozoan has been feeding. The two white circles with radiating canals are contractile vacuoles, whose main function is eliminating water from the cell. × 520. [Courtesy E. V. Gravé.]

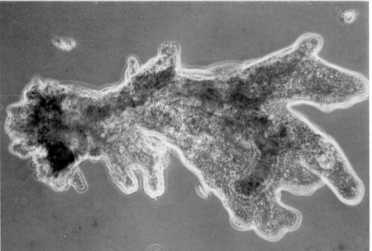

1.7 An amoeboid protozoan, *Pelomyxa carolinensis*
"Amoeboid"—from the Greek word for change—describes a cell that can alter its shape as it thrusts out or withdraws many armlike extensions. × 150.   [Courtesy E. V. Gravé.]

1.8 A brown alga, *Taonia atomaria*
Brownish pigments mask the green pigment
chlorophyll also present in this seaweed. The flattened
leaflike blades are supported by the water in which the
plant is growing; when uncovered at low tide, the
blades lie flat on the substrate. [Courtesy Hervé
Chaumeton.]

of the cell contents as the cell, constantly changing shape, sends out extensions into which the rest of the cell contents flows. We shall refer to representative protozoans many times as we examine their fascinating evolutionary adaptations.

PLANTAE

Common to all plant groups are cells with rigid walls and subcellular membrane-bounded structures containing chlorophyll, the green pigment essential for photosynthesis. Thus plants, unlike fungi and animals, can themselves synthesize the high-energy compounds they need for maintenance and growth.

Brown algae and red algae   Both the brown algae and the red algae are primarily marine and are commonly known as seaweeds. They are especially prevalent in the intertidal zone along rocky coasts, where they can easily be observed at low tide. Their color is due to brown and red pigments, which often mask the chlorophyll these algae also contain (Figs. 1.8, 1.9). They are always multicellular, but show relatively little of that differentiation into distinct tissues—associations of cells usually similar in structure and function—found in the higher land plants. Even though some brown algae do show tissue differentiation and many other characteristics recalling the higher land plants, it is believed that they and the higher land plants evolved independently of each other and are not closely related. Indeed, even the brown and red algae are thought not to be closely related, despite many superficial similarities.

1.9 A red alga, *Scinaia furcellata*
Like the brown algae, the red algae are mostly
seaweeds, but they generally grow at greater depths
than the brown algae. [Courtesy Hervé Chaumeton.]

Green algae   Green algae are relatively simple plants that live only in the water or in very moist environments on land. Some are unicellular, others multicellular. In multicellular green algae, most of the cells are similar to one another and are not highly specialized; they form what might be considered a single continuous tissue, which is the plant body (Fig. 1.10). The plants are green because their chlorophyll is not masked by accessory pigments. In this and many other biochemical characteristics, the green algae are similar to the higher land plants, and most botanists agree that it was probably from ancestral flagellated green algae that the land-plant groups arose.

Mosses, liverworts, and their relatives   Although these plants live on land, they are not entirely independent of the ancestral aquatic environment; thus they occur only in very moist habitats (Fig. 1.11). One reason is that parts of their reproductive cycle are dependent on abundant moisture. Another is that, unlike the vascular plants, they have not evolved an efficient internal transport system through which water from the soil can be carried to all parts of the plant.

Vascular plants   Of all members of the plant kingdom, the vascular plants show the greatest internal specialization into tissues and organs—roots, stems, leaves, and reproductive organs (cones, flowers). Because they possess vascular tissue, which provides conduits in which water and dissolved substances can move from one part of the plant body to another, they are less dependent than the other plant groups on water in the surrounding environment. They are the dominant plant group on land today.

The vascular-plant group is usually subdivided into several sections, three of which are doubtless familiar to you: the ferns, the conifers and their allies (gymnosperms), and the flowering plants (angiosperms). Of these three groups, the ferns are the most primitive; they appeared on the ancient earth before the other two groups, dominated the land for a long time, but eventually gave way to the other groups and are now largely overshadowed by them.

1.10   A marine green alga, _Ulva lactuca_
This multicellular green alga, also known as sea lettuce, is one of the most advanced green algae—a group that contains many unicellular and colonial forms. Its leaflike blades contrast with the filamentous bodies of most multicellular green algae.   [Courtesy Hervé Chaumeton.]

1.11   Mosses and liverworts
Growing in a forest in New Zealand, these mosses form a nearly continuous carpet on the forest floor. In their midst are liverworts, whose flattened bodies bear numerous cuplike structures (gemmae cups) in which cells for asexual reproduction are produced.   [Courtesy G. R. Roberts.]

1.12 *Iris cristata,* **a monocot (left), and** *Rosa rubrifolia,* **a dicot (right)**
Left: The irises have flowers with three petals and three sepals, which in these plants have the same color as the petals. The leaves have parallel veins. Right: The wild rose has five petals and leaves with a network of veins. [E. R. Degginger (left) and Jane Burton (right), Bruce Coleman Inc.]

The gymnosperms and flowering plants are known collectively as the ***seed plants;*** they are more highly specialized for a terrestrial existence than the ferns. Some common gymnosperms are pine, cedar, spruce, fir, and hemlock; all of these bear cones and have needlelike leaves, though not all gymnosperms do.

The angiosperms, or flowering plants, are the most advanced of the three groups. The majority of the land plants familiar to you belong to this group, which includes plants of every shape and size from grasses to cacti, and from tiny herbs and wildflowers to large oaks and maple trees. This huge and very diverse group is customarily divided into two subgroups: the ***dicotyledons*** (dicots, for short), which include beans, buttercups, privets, dandelions, oak trees, maple trees, roses, potatoes, and a great variety of other plants; and the ***monocotyledons*** (monocots, for short), which include the grasses and grasslike plants such as corn, lilies, irises, and palm trees. The two subgroups differ in many characteristics, including two easily observable ones: (1) The leaves of dicots usually show a network of veins; those of monocots usually show parallel veins. (2) The petals of dicot flowers occur in fours or fives (or multiples of these); those of monocots occur in threes or multiples of three (Fig. 1.12).

FUNGI

It has been customary to include the fungi in the plant kingdom, because they are predominantly sedentary and because their cells, like plant cells, have walls. But unlike true plants, they lack chlorophyll and therefore cannot manufacture their own food. Like animals, they must obtain the complex high-energy nutrients they need in an already synthesized form; unlike animals, however, they cannot ingest particulate food, and depend entirely on absorption of nutrient molecules; hence they must live in or on their nutrient sources, which are either other living organisms or the dead remains of other organisms. Since fungi differ from both plants and animals in so many ways, most recent classifications assign them to a kingdom of their own.

Some fungi are unicellular (e.g. yeast), but most are multicellular (e.g. bread mold, fruit molds, mushrooms, toadstools, bracket fungi) (Fig. 1.13). The latter, however, are not multicellular in the sense that plants and animals are, for the membranous partitions between adjacent fungal cells, unlike those between plant or animal cells, are usually absent or incomplete; thus the cell contents are essentially continuous within the long tubular body. Rather than call these organisms multicellular, it might be more appropriate to describe them as multinucleate—as having many nuclei.

ANIMALIA

Of the many characteristics that distinguish animals from the two other main categories of multicellular organisms—the plants and the fungi—we shall here mention but two: First, animal cells differ from plant and fungal cells in lacking rigid cell walls. Second, the principal mode of nutrition in animals is ingestion of particulate food; most plants, by contrast, depend on photosynthesis, and most fungi on absorption.

1.13 Two representative fungi
Left: The black bread mold *Rhizopus* is made up of numerous threadlike filaments hung with what look like black and white balls. These are spore-producing reproductive structures; the black ones are ripe and ready to release their spores. × 100. Right: The velvet-stemmed collybias growing among lichens on a tree trunk are composed of filaments so densely packed that the unaided eye cannot make them out individually. [Left: W. H. Amos, Bruce Coleman Inc. Right: Courtesy D. G. Allen.]

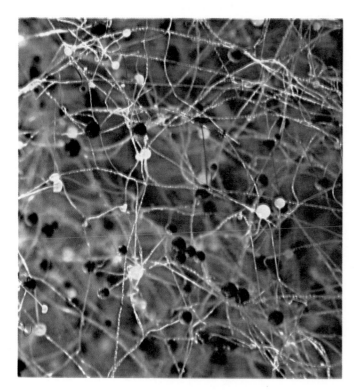

1.14 A sea anemone, *Epiactis*
A veritable thicket of tentacles surrounds the animal's mouth. Many newly budded young anemones are attached to the stalk portion of its body. [R. N. Mariscal, Bruce Coleman Inc.]

1.15 Hydra, feeding
The green hydra *Chlorohydra* owes its color to the cells of a green alga that it incorporates into its own cells, thus benefiting from the alga's photosynthetic activity. Note the buds with tentacles on the lower portion of the stalks; hydras reproduce asexually by budding. Left: The hydra is manipulating a small crustacean with its tentacles. Right: The food has been taken into the digestive cavity, where it makes a prominent bulge. × 2½. [Oxford Scientific Films.]

1.16 A planarian, one of the free-living flatworms
The animal is naturally grayish; this specimen has been stained to bring out the details of its anatomy. Note the eyespots, which allow a primitive sort of vision, the much-branched digestive cavity (reddish brown), and the tubular pharynx in the center, which is extruded during feeding. × 17½.   [Courtesy E. V. Gravé]

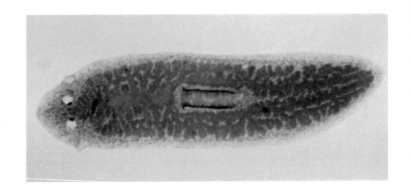

Coelenterates The coelenterates constitute a large group of primitive aquatic animals whose body plan, or symmetry, is radial rather than bilateral, a reflection of the relatively sedentary existence of many coelenterates during at least part of their life cycle. Their saclike bodies are composed of two distinct tissue layers with a much less distinct third layer between them. They have a digestive cavity, but it has only one opening, which must serve as both mouth and anus. Tentacles are often present around the mouth and are used in capturing prey. The nerves and muscles of these simple animals are of an exceedingly primitive type, and no circulatory system is present. The group includes jellyfish, sea anemones (Fig. 1.14), and corals. A freshwater form to which we shall often refer as representative is the hydra (Fig. 1.15).

Flatworms The flatworms are more complex than the coelenterates in some ways, but they, too, have a digestive tract with only one opening. The body is composed of three primary tissue layers, and the symmetry is bilateral. Many flatworms, such as flukes and tapeworms, are parasites and show numerous interesting specializations for this mode of existence. Others, like planarians (Fig. 1.16), small animals to which we shall make frequent reference, are free-living aquatic organisms.

Molluscs Fairly complex animals, most of which have shells, the molluscs include snails (Fig. 1.17), clams, oysters, and scallops, as well as octopuses and squids, which do not have obvious shells. These animals are particularly abundant in the oceans, as anyone who has collected their shells along the seashore will know. They are also common in freshwater. Some snails have evolved lungs and become fully terrestrial.

1.17 Florida tree snail
The majority of molluscs live in the water, but many snails, such as this one, are adapted for life on land. The snail moves by means of a muscular "foot" extruded from the shell. Note the eyes located on the ends of long retractable stalks. [Courtesy J. H. Carmichael, Jr.]

1.18 An earthworm, *Lumbricus terrestris*
The animal's body is divided into numerous ringlike segments. Its front end, with the mouth, is at upper left. The prominent girdlelike structure is the clitellum—a group of glandular segments that secrete mucus during mating and produce a cocoon for the eggs.   [Oxford Scientific Films, Bruce Coleman Inc.]

1.19 *Nereis*, a marine annelid worm
This worm has a pair of large flaplike appendages on each of its body segments, which function in respiration and in locomotion.   [S. C. Bisserot, Bruce Coleman Inc.]

Annelids   The annelids are often called the segmented worms. As this term implies, the bodies of these highly evolved organisms are divided into a series of units or segments, which are often clearly visible externally. Though most annelids are aquatic, some, like the earthworm, occur on land, but always in moist places. We shall generally take the earthworm as representative of this group (Fig. 1.18); occasionally, however, we shall mention *Nereis*, a marine worm with large lobes growing from each side of the body segments (Fig. 1.19).

Arthropods   The arthropods constitute an immense group of very advanced animals that includes more different species than all other animal groups combined. All arthropods have jointed legs and a hard outer skeleton. Spiders, scorpions, crabs, lobsters, crayfish, centipeds, millipeds, and insects all belong to this major group (Fig. 1.20); of these, the insects are by far the largest subgroup and are among the most successful of all land animals, rivaled only by the mammals and particularly humans.

Echinoderms   All echinoderms are strictly marine; they have apparently never been able to invade either the freshwater or the terrestrial habitats. Though their symmetry is radial, they are fairly advanced in many ways. The group includes sea stars (starfish), sand dollars, sea urchins, sea cucumbers, and a variety of other forms (Fig. 1.21). Though their appearance hardly suggests it, the echinoderms are regarded by most biologists as the group of animals most closely related to the next group, the chordates, which includes humans.

Chordates   This very important group contains a major subgroup called the *vertebrates,* which comprises all animals possessing an internal bony skeleton, particularly a backbone. Fish, amphibians (e.g. frogs, salamanders), reptiles (e.g. snakes, lizards, turtles, alligators), birds, and mammals (including humans) belong to this group. We shall pay special attention to the vertebrates throughout this book.

1.20   Two representative arthropods
Left: The centiped has a segmented body with a pair of jointed legs on each segment. Two hornlike antennae curve outward at lower right.   Right: Like all insects, the lubber grasshopper *(Romalea microptera)* has only three pairs of legs, the abdominal segments being legless.   [David Hughes (left) and R. L. Dunne (right), Bruce Coleman Inc.]

1.21   The blue sea star *Linkia laevigata*
The animal rests on a piece of yellow coral. Only four of its five arms are visible.   [Allan Power, Bruce Coleman Inc.]

PART I

THE CHEMICAL AND
CELLULAR BASIS OF LIFE

Chapter 2

SOME SIMPLE CHEMISTRY

One of the fundamental maxims of modern biological science is that life processes obey the laws of chemistry and physics. In the behavior of atoms and molecules lies the key to such crucial phenomena as the trapping and storing of energy by green plants; the extraction by living things of usable energy from nutrients; the growth and development of organisms; impulse transmission along nerve pathways; immunological response to substances foreign to the organism; genetic inheritance and the regulation by genes of the activities of living cells. In recent years biochemistry has greatly broadened our insight into the nature of life, and profoundly altered the course of biological research. Clearly, there is no understanding modern biology without at least some knowledge of chemistry, especially the chemistry of the elements and classes of compounds that form living material.

THE ELEMENTS

All the matter of the universe is composed of a limited number of basic substances called elements, which cannot be decomposed into simpler substances by chemical reactions. There are 92 naturally occurring elements; many additional, synthetic elements have been manufactured in the laboratory, raising the current total for both natural and artificial elements to well over a hundred.

Each element is designated by one or two letters that stand for its English or Latin name. Thus H is the symbol for hydrogen, O for oxygen, C for carbon, Cl for chlorine, Mg for magnesium, K for potassium (Latin, *kalium*), Na for sodium (Latin, *natrium*), etc.

Matter is not continuous; i.e. it cannot be subdivided without limit. Progressive subdivision leads ultimately to units indivisible by ordinary chemi-

TABLE 2.1 *Elements important in living material**

Element	Symbol	Atomic number	Approximate atomic weight	Approximate percentage of earth's crust by weight	Approximate percentage of human body by weight	Approximate percentage of human body by number of atoms
Hydrogen	H	1	1	0.14	9.5	63
Boron	B	5	11	Trace	Trace	Trace
Carbon	C	6	12	0.03	18.5	9.5
Nitrogen	N	7	14	Trace	3.3	1.4
Oxygen	O	8	16	46.6	65.0	25.5
Fluorine	F	9	19	0.07	Trace	Trace
Sodium	Na	11	23	2.8	0.2	0.03
Magnesium	Mg	12	24	2.1	0.1	0.01
Silicon	Si	14	28	27.7	Trace	Trace
Phosphorus	P	15	31	0.07	1.0	0.22
Sulfur	S	16	32	0.03	0.3	0.05
Chlorine	Cl	17	35	0.01	0.2	0.03
Potassium	K	19	39	2.6	0.4	0.06
Calcium	Ca	20	40	3.6	1.5	0.31
Vanadium	V	23	51	0.01	Trace	Trace
Chromium	Cr	24	52	0.01	Trace	Trace
Manganese	Mn	25	55	0.1	Trace	Trace
Iron	Fe	26	56	5.0	Trace	Trace
Cobalt	Co	27	59	Trace	Trace	Trace
Copper	Cu	29	64	0.01	Trace	Trace
Zinc	Zn	30	65	Trace	Trace	Trace
Selenium	Se	34	79	Trace	Trace	Trace
Molybdenum	Mo	42	96	Trace	Trace	Trace
Tin	Sn	50	119	Trace	Trace	Trace
Iodine	I	53	127	Trace	Trace	Trace

* All the elements listed have been shown essential for one or more species, but not all are essential for every species.

cal means. These units are called *atoms.* The atoms of a particular element are alike in many essential characteristics and differ in many measurable ways from the atoms of all other elements. A single atom is customarily represented by the chemical symbol of the element concerned; e.g. N stands for a single atom of nitrogen.

ELEMENTS IMPORTANT IN BIOLOGY

Not all 92 naturally occurring elements are of significance in the bodies of living organisms. Table 2.1 lists the important ones. Of these, six play such a prominent role that they demand special mention. They are hydrogen, carbon, oxygen, nitrogen, phosphorus, and sulfur. They occur in all living creatures and are indispensable to life as we know it on this planet. We shall have countless occasions to refer to them in our examination of life processes.

These six, however, are not the only elements essential for life. Others such as calcium, sodium, potassium, magnesium, and iron are of great importance in most organisms. Still others, commonly called trace elements, are present in minute amounts in many cells and, despite the extremely small quantities in which they occur, may be indispensable for the maintenance of life. Table 2.1 indicates the relative quantities of the various elements in the human body.

ATOMIC STRUCTURE

Although atoms can be considered the basic chemical units of matter, they themselves are composed of still smaller particles. Many of these particles belong to the world of subatomic physics and are of little immediate concern to biologists, but three of them—the so-called primary particles—play such a central role in determining the properties of an element that we cannot hope to understand its chemical behavior without first becoming familiar with those particles.

The atomic nucleus All the positive charge and almost all the mass of an atom are concentrated in its center, or nucleus, which contains two kinds of primary particles, *protons* and *neutrons.* Each proton carries a charge of $+1$ electronic charge unit.[1] The neutrons, as their name implies, have no charge. Protons and neutrons have roughly the same mass, which is close to one dalton (see Table 2.2).

The number of protons in the nucleus is unique for each element. This number, called the *atomic number,* is sometimes written as a subscript immediately before the chemical symbol. Thus $_1$H indicates that the atomic number of hydrogen is one; i.e. its nucleus contains only one proton. Similarly, $_8$O indicates that all oxygen nuclei contain eight protons.

It is often desirable to indicate the total number of protons and neutrons in a nucleus; this number is called the *mass number,* because it approximates the total mass of the nucleus. The mass number is commonly written as a superscript immediately following (or sometimes preceding) the chemical symbol. For example, most atoms of oxygen contain eight protons and

[1] For definitions of units of measurement see Glossary.

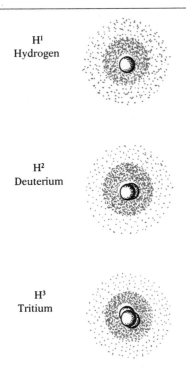

H¹
Hydrogen

H²
Deuterium

H³
Tritium

2.1 The three principal isotopes of hydrogen
Each of the three isotopes has one proton in its nucleus
and one electron. They differ in that ordinary hydrogen
(H¹) has no neutrons in its nucleus, deuterium (H²) has
one, and tritium (H³), which is radioactive, has two.
The single electron is not represented, but the volume
within which it moves (a sphere) is indicated by color
stippling; the denser the stippling, the greater the
likelihood that at any given moment the electron will
be found in that portion of the sphere.

eight neutrons; the mass number is therefore 16, and the nucleus can be symbolized as O^{16} or, if we wish to show both the atomic number and the mass number, as $_8O^{16}$.

Although the number of protons is the same for all atoms of the same element, the number of neutrons is not always the same, and neither, consequently, is the mass number. For example, most oxygen atoms, as we have seen, contain eight protons and eight neutrons and have a mass number of 16; some, however, contain nine neutrons and thus have a mass number of 17 (symbolized as O^{17}), and still others have ten neutrons and a mass number of 18 (symbolized as O^{18}). Atoms of the same element that differ in mass, because they contain different numbers of neutrons, are called *isotopes.* Thus O^{16}, O^{17}, and O^{18} are three isotopes of oxygen. Some elements have as many as 20 naturally occurring isotopes; others have as few as two.

Apparently the number of neutrons in the nucleus does not affect the chemical properties of an atom, for all isotopes of the same element have essentially the same chemical characteristics. However, the number of neutrons does affect the physical properties of an atom. The nuclei of some isotopes are unstable and tend to break down spontaneously to more stable forms, emitting high-energy radiation in the process. Figure 2.1 illustrates three different isotopes of hydrogen: $_1H^1$, which is the usual form of the element; $_1H^2$, a nonradioactive isotope generally called deuterium; and $_1H^3$, a radioactive isotope called tritium. Both deuterium and tritium have been used extensively in tracing the movements of hydrogen in biochemical reactions. As we shall note again and again, isotopes of various other elements as well are invaluable research tools for biologists.

TABLE 2.2 *Fundamental particles*

Particle	Mass (daltons)	Charge (electronic charge units)
Electron	0.00055	−1
Proton	1.00728	+1
Neutron	1.00867	0

The electrons The portion of the atom outside the nucleus contains the third kind of primary particle—the electrons. Electrons have very little mass (see Table 2.2). Almost the total mass of the atom is therefore contributed by the protons and neutrons in the nucleus, even though the extranuclear region constitutes most of the volume of the atom. Each electron carries a charge of −1 electronic charge unit; i.e. its charge is exactly the opposite of a proton's.

In a normal neutral atom, the number of electrons around the nucleus is exactly the same as the number of protons in the nucleus. The positive charges of the protons and the negative charges of the electrons cancel each other, making the total atom neutral. Consequently, in a neutral atom, the atomic number represents both the number of protons inside the nucleus and the number of electrons outside the nucleus. If, then, we see the symbol $_{17}Cl^{35}$, we can tell that a neutral atom of this isotope of chlorine has 17 protons, 18 neutrons, and 17 electrons. Similarly, the symbol $_{19}K^{39}$ means that this isotope of potassium contains 19 protons, 20 neutrons, and 19 electrons.

The electrons are not in fixed positions outside the nucleus. Each is in constant motion, and it is impossible to know exactly where a given electron is at any particular moment. For this reason some illustrations of atoms, such as Fig. 2.1, do not show the electron itself, but indicate the region where the electron is likely to be by stippling.

The distance of an electron from the nucleus is a function of its energy; the higher its energy, the greater its probable distance from the nucleus. Thus, while we can never say precisely where an electron is—at any moment it might be anywhere from right at the nucleus to an infinite distance from it—we can, on the basis of its energy level, make statistical estimates of where it is likely to be. For example, knowing that the single electron of a hydrogen atom is at a low energy level, we can predict that it will most often be close to the nucleus.

The volume within which an electron can be found 90 percent of the time is known as the **orbital** of the electron. In illustrations of atoms the orbital is often represented by a circle, the electrons sometimes being shown as round spots on the circle (Fig. 2.2). Such representations, of course, should not be taken to imply that the electrons are orbiting the nucleus at a precise distance from it, as some older atomic models envisioned; the circle is merely a convention for indicating the border of the volume that probably contains the electrons.

The energy level of the hydrogen electron, which is the level nearest the nucleus, is often referred to as the K level (or K shell), and the orbital of the electron, which is spherical, is designated the $1s$ orbital. The K level can contain only two electrons. What happens, then, when an atom contains more than two electrons, such as the oxygen atom, with its eight electrons? Two of these can be accommodated at the K level, but the other six must move at higher energy levels, farther from the nucleus. It was discovered many years ago that only certain energy levels are possible for electrons in atoms, for the energy of electrons is quantized; if an electron gains or loses energy, it must do so by certain discrete amounts. The next possible energy level beyond the K level is called the L level; it can contain a maximum of eight electrons. Since the most stable configuration for an atom is one in

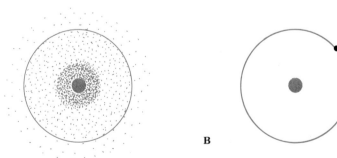

A   B

2.2   Two ways of representing the hydrogen atom
(A) The nucleus is shown here as a central gray area, with the "cloud" around it representing the region where the electron is likely to be. The circle encloses the orbital of the electron—the volume, a sphere, within which the electron will be found 90 percent of the time (see also Fig. 2.1).   (B) Sometimes only the circle indicating the circumference of the orbital is shown; the electron may be designated by a small ball on the circle.

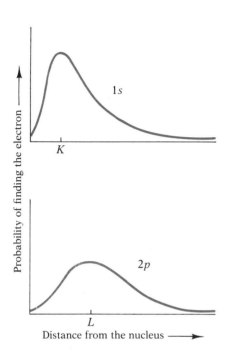

which its electrons have minimum energy, the six electrons outside the *K* level in an oxygen atom are all at the *L* energy level. Thus an oxygen atom has two *K* electrons and six *L* electrons.

The most likely distance from the nucleus of each of the six *L* electrons of an oxygen atom is roughly the same; as shown in Fig. 2.3, it is somewhat greater than the most likely distance of the *K* electrons. However, the orbitals of the *L* electrons are not all of the same shape. Two of these electrons have a spherical orbital (called the 2*s* orbital), which is like the 1*s* orbital of the *K* electrons except that it extends farther from the nucleus (Fig. 2.4). But the other *L* electrons have dumbbell-shaped orbitals (symbolized by 2*p*). The *L* energy level can contain three of these *p* orbitals, each oriented at right angles to the other two, i.e. each aligned along a different one of the three dimensions of space (Fig. 2.5). The combination of the three 2*p* orbitals, which can contain a maximum of two electrons each, and of the 2*s* orbital, which can also contain two electrons, accounts for the overall maximum of eight electrons for the *L* energy level. Obviously, not all these possible orbitals are filled in an oxygen atom (Fig. 2.6).

When elements have more than 10 electrons (two at the *K* level and eight at the *L*), the additional electrons are accommodated at energy levels beyond *L*. Like the first two levels, each of these levels is limited as to the number of electrons it can contain. Thus the third level (*M*) can contain a maximum of 18 electrons; the fourth (*N*) can contain 32; and so forth. In addition to *s* orbitals (spherical) and *p* orbitals (dumbbell-shaped), orbitals of other shapes may occur at these outer energy levels.

Although the third and successive levels can hold more than eight electrons, they are in a particularly stable configuration when they contain only

2.3 Graphs of the probability of finding electrons at various distances from the nucleus
Note that at no distance is the probability zero; i.e. the electrons could be at any distance, from right at the nucleus to infinity. But the most likely distance, *K*, for the 1*s* electron, which is at the first energy level, is less than the most likely distance, *L*, for the 2*p* electron, which is at the second energy level.

2.4 Representations of electron orbitals
The orbitals of *s* electrons are approximately spherical, those of *p* electrons roughly dumbbell-shaped. The numerals before *s* and *p* indicate the energy level. Thus the 1*s* electron is at the first energy level (*K*), that nearest the nucleus; the 2*s* and 2*p* electrons are at the *L* level, a higher energy level, and hence at a greater average distance from the nucleus than the 1*s* electron. Note that, despite the very different shapes of their orbitals, the 2*s* and 2*p* electrons are at the same energy level; i.e. their most probable distances from the nucleus are nearly the same.

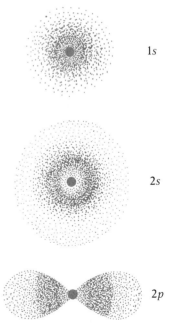

1*s*

2*s*

2*p*

2.5 The three 2*p* electron orbitals
Each of the dumbbell-shaped orbitals is oriented in a different dimension of space, at right angles to the other two. (Imaginary planes and arrow inserted to aid visualization.)

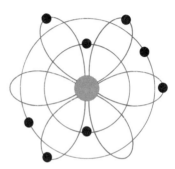

A

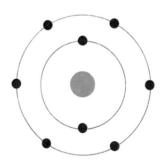

B

2.6 Two ways of showing distribution of the eight electrons of an oxygen atom
(A) Circles indicate the 1*s* and 2*s* orbitals (each containing two electrons), and dumbbells the three 2*p* orbitals (only one of which contains the maximum of two electrons). (B) Because the scheme in (A) is cumbersome, it is more customary to show each major energy level (*K*, *L*, *M*, etc.) as a single circle carrying all the electrons on that level, no attempt being made to show the actual shapes of the orbitals.

eight. For our purposes, then, the first level can be considered complete when it holds two electrons, and every other level when it holds eight electrons.

Electron distribution and the chemical properties of elements   When the elements are arranged in sequence according to their atomic numbers—beginning with hydrogen, which has the number one, and proceeding to uranium, the last of the natural elements, which has the number 92—it can be seen that elements with very similar properties occur at regular intervals in the list. For example, fluorine, number 9, is more like chlorine, number 17, bromine, number 35, and iodine, number 53, than like oxygen, number 8, or neon, number 10, the two elements immediately adjacent to it in the list. This tendency for chemical properties to recur periodically throughout the sequence of elements is called the Periodic Law.

The explanation for this periodicity is that the chemical properties of elements are largely determined by the number of electrons in their outermost shell (i.e. at their outermost energy level). If that shell is complete, as in helium (atomic number, 2), neon (10), or argon (18), the element has very little tendency to react chemically with other atoms (Fig. 2.7). If the outermost shell has one electron less than the full complement, the element has certain characteristic chemical properties; if it lacks two electrons, the element has somewhat different properties; if it lacks seven electrons, the element has very different properties.

We have said that fluorine is chemically similar to chlorine, bromine, and iodine. The critical characteristic shared by these four elements is that each

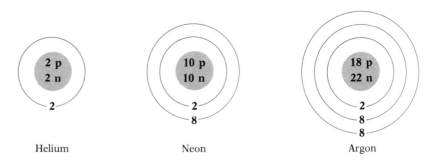

2.7 Atomic structure of the so-called inert gases
Each of these elements has a complete outer shell (two electrons for helium and eight electrons for each of the other two). Hence they show very little chemical activity compared to most other elements (though they are not totally inactive, as was once thought).

Helium   Neon   Argon

has seven electrons in its outer shell (Fig. 2.8). Oxygen (Fig. 2.6B), the element immediately preceding fluorine in the periodic table, has six electrons in its outer shell and hence is chemically quite different from fluorine. Neon (Fig. 2.7), the element just after fluorine, has a full eight electrons in its outer shell and therefore lacks the reactivity of fluorine.

A convenient way to represent the electronic configuration of the outer shell is to symbolize each electron by a dot placed near the chemical symbol for the element under consideration. Thus fluorine and chlorine, which, as we have said, have seven electrons at their outer energy level, would have the electronic symbols

$$\overset{..}{:}\text{F}\cdot\quad\quad\overset{..}{:}\text{Cl}\cdot$$

Similarly, hydrogen with one electron in its shell, carbon with four in its outer shell, nitrogen with five, and oxygen with six would be shown as follows:

$$\text{H}\cdot\quad\cdot\overset{.}{\text{C}}\cdot\quad\cdot\overset{..}{\text{N}}\cdot\quad:\overset{..}{\text{O}}\cdot$$

It goes without saying that the placement of the dots has no significance and in no way indicates the actual positions of the electrons concerned.

2.8 Atomic structure of halogens
The four elements have similar chemical properties, because each has seven electrons in its outermost shell.

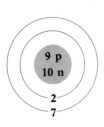
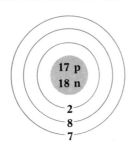
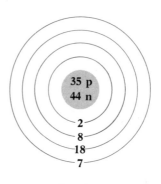
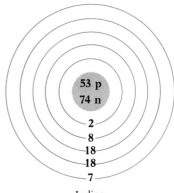

Fluorine   Chlorine   Bromine   Iodine

CHEMICAL BONDS

The atoms of most elements possess the property of binding to other atoms to form new and more complex aggregates. When two or more atoms are bound together in this fashion, the force of attraction that holds them together is called a chemical bond. The atoms of a particular element characteristically can form only a certain precise and limited number of such bonds; atoms of some other element may be capable of forming a different but equally precise and limited number of bonds. Our previous examination of atoms already suggests how atoms bond together, and why each element has its own characteristic bonding capacity.

IONIC BONDS

We have said that atoms are in a particularly stable configuration when their outer electron shell is complete, i.e., in most cases, when it contains eight electrons. There is consequently a general tendency for atoms to form complete outer shells by reacting with other atoms. These reactions are the stuff of chemistry; the tendency of atoms to gain complete outer shells forms the basis upon which all chemistry is built.

Consider, for example, an atom of sodium (atomic number, 11). This atom has two electrons in its first shell, eight in the second, and only one in the third. One way sodium might gain a complete outer shell would be to acquire seven more electrons from some other atom or atoms. But the sodium would then have an enormous excess of negative charge, and, since like charges repel each other, the electrons would tend to push each other away from the sodium. In point of fact, sodium cannot obtain a full outer shell by appropriating seven additional electrons. An alternative way, and the way actually followed in nature, is for the sodium atom to give up the lone electron in its third shell to some electron acceptor, leaving the complete second shell as the new outer shell (Fig. 2.9).

Next, consider an atom of chlorine (atomic number, 17). This atom has two electrons in its first shell, eight in its second shell, and seven in its third shell. In other words, its outer shell is almost complete, lacking only a single electron. It cannot lose the seven electrons in its outer shell for reasons similar to those preventing sodium from gaining seven electrons; the electrons are too strongly attracted to the protons of the chlorine nucleus for seven to be removed, an operation that would leave an excess charge of +7 in the atom. It is by gaining an extra electron from some electron donor that chlorine can acquire a complete outer shell.

If a strong electron donor like sodium (i.e. an atom with a strong tendency to get rid of an electron) and a strong electron acceptor like chlorine (i.e. an atom with a strong tendency to acquire an extra electron) come into contact, an electron may be completely transferred from the donor to the acceptor. The result, in the present example, is a sodium atom with one less electron than normal and a chlorine atom with one more electron than normal. Once it has lost an electron, the sodium is left with one more proton than it has electrons, and it therefore has a net charge of +1. Similarly, the chlorine atom that gained an electron has one more electron than it has protons and

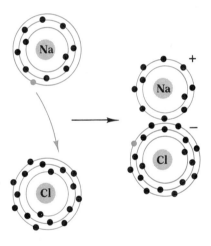

2.9 Ionic bonding of sodium and chlorine
Sodium has only one electron in its outer shell, while chlorine has seven. Sodium acts as an electron donor, giving up the one electron in its outer shell, whereupon the complete second shell functions as its new outer shell. Chlorine acts as an electron acceptor, picking up an additional electron to complete its outer shell. But after sodium has donated an electron to chlorine, the sodium, left with one more proton than it has electrons, has a positive charge. Conversely, the chlorine, with one more electron than it has protons, has a negative charge. The two charged atoms, called ions, are attracted to each other by their unlike charges. The result is sodium chloride (NaCl).

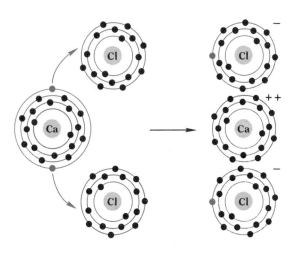

2.10   Ionic bonding of calcium and chlorine
Calcium has two electrons in its outer shell. It donates
one to each of two chlorine atoms, and the two
negatively charged chloride ions thus formed are
attracted to the positively charged calcium ion to form
calcium chloride ($CaCl_2$).

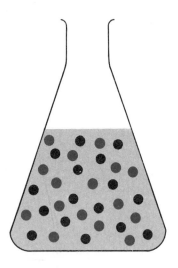

2.11   Ionization of sodium chloride
When in solution, the NaCl dissociates into separate
Na^+ (black) and Cl^- (color) ions.

has a net charge of -1. Such charged atoms (or charged aggregates of
atoms) are called *ions,* and are symbolized by the appropriate chemical
symbol followed by a superscript indicating the charge. Sodium and chlorine ions are written Na^+ and Cl^-.

A sodium ion with its positive charge and a chlorine ion (usually called a
chloride ion) with its negative charge tend to attract each other, since opposite charges attract. Consequently the two ions are held together by an
electrical attraction and form the compound we know as table salt, or sodium chloride, NaCl (Fig. 2.9). Such a bond, involving the complete transfer
of an electron from one atom to another and the mutual attraction of the
two ions thus formed, is termed an *ionic bond.*

Ionic bonding may entail the transfer of more than one electron, as in
calcium chloride, another common salt. Calcium (atomic number, 20) has
two electrons in its outermost shell, and it loses both to form the calcium ion
Ca^{++} (Fig. 2.10). Chlorine, however, need gain only one electron to complete
an octet in its outer shell, as we have already seen. Hence it takes two
chlorine atoms to act as acceptors for the two electrons from a single calcium atom, and the calcium chloride formed involves the bonding together
of a total of three ions, symbolized as $CaCl_2$, where the subscript 2 indicates
that there are two chlorine atoms for each calcium atom in this compound.
We can say, then, that calcium has a bonding capacity, or valence, of $+2$,
while sodium has a valence of $+1$ and chlorine a valence of -1.

Ionic bonds occur between strong electron donors (configurations with
only a few electrons at their outer energy level) and strong electron acceptors (configurations with nearly eight electrons at their outer energy level).
Ionic bonding is not common between configurations that have intermediate numbers of electrons in the outer shells, or between units both of which
are strong electron donors or electron acceptors.

In many instances *ionization* (i.e. the transfer of one or more electrons
from one atom to another to form ions) occurs without true molecular
formation. A *molecule* is generally defined as an electrically neutral aggregate of atoms bonded together strongly enough to be regarded as a single
entity. Substances like sodium chloride (NaCl) or calcium chloride ($CaCl_2$),
in which the bonds are almost exclusively ionic, have a pronounced tendency to dissociate into separate ions when in solution. When they are
ionized in this manner, they do not exist as molecules. In solution, NaCl
forms two separate entities, a Na^+ ion and a Cl^- ion (Fig. 2.11). Similarly,
$CaCl_2$ in solution forms three separate entities, a Ca^{++} ion and two Cl^- ions.
Even in the solid state, ionic compounds often do not form discrete molecules in the usual sense. In solid sodium chloride, many sodium and chlorine
atoms are bound together into a large crystal (Fig. 2.12); there are no separate molecules composed of one sodium atom bonded to one chlorine atom,
as the molecular symbol NaCl might seem to indicate. In a sense, the entire
crystal is a single molecule, although in practice the term "molecule" is not
used in such cases.

Ions, being charged particles, behave differently from neutral atoms or
molecules in living systems, and substances wholly or partly ionized in water
play many important roles in the functioning of biological systems. In later
chapters we shall be concerned with the effects of charge on the movements
of materials through the membranes of living cells, and with the partitioning

of positive and negative ions that gives rise to the electric potential differences essential for nerve and muscle activity.

Acids and bases   There are a number of possible definitions of these two highly important classes of ionic compounds. But for our purposes an **acid** can be characterized simply as a substance that increases the concentration of hydrogen ions (H⁺) in water, a **base** as a substance that decreases the concentration of hydrogen ions (which in water is equivalent to increasing the concentration of hydroxyl ions, OH⁻).

The degree of acidity or basicity (usually called alkalinity) of a solution is commonly measured in terms of a value known as **pH,** which is the negative logarithm of the concentration of hydrogen ions:

$$\mathrm{pH} = \frac{1}{\log [\mathrm{H^+}]} = -\log [\mathrm{H^+}]$$

On the pH scale, which ranges from 0 on the acidic end to 14 on the alkaline end, a solution is neutral, neither acidic nor alkaline (i.e. it contains equal concentrations of H⁺ ions and OH⁻ ions), if its pH is exactly 7. Substances with a pH of less than 7 are acidic (i.e. contain a higher concentration of H⁺ ions than of OH⁻ ions); the lower the pH, the more acidic the substance. Conversely, substances with a pH higher than 7 are alkaline (i.e. contain a higher concentration of OH⁻ ions than of H⁺ ions); the higher the pH, the more alkaline the substance. Since the pH scale is a log scale, you should realize that a change of one pH unit means a tenfold change in the concentration of hydrogen ions. Thus the concentration of H⁺ ions in the solution of a very strong acid may be as much as 100,000,000,000,000 (i.e. 10^{14}) times greater than in the solution of a very strong base.

Living matter is extraordinarily sensitive to pH, functioning best (with a few notable exceptions, e.g. in the lumen of parts of the digestive tract) when conditions are nearly neutral. Thus most of the interior material of living cells (excluding the nucleus and some vacuoles) has a pH of about 6.8. The blood plasma and other fluids that bathe the cells in our own bodies have a pH of 7.2–7.3. Numerous special mechanisms aid in stabilizing these fluids, so that cells will not be subject to appreciable fluctuations in pH. Foremost among these mechanisms are certain chemical substances known as **buffers,** which have the capacity to bond to H⁺ ions, thereby removing them from solution whenever their concentration begins to rise, and, conversely, to release H⁺ ions into solution whenever their concentration begins to fall. Buffers thus help minimize fluctuations in pH, which would otherwise be considerable, since many of the biochemical reactions normally occurring in living organisms either release or use up H⁺ ions.

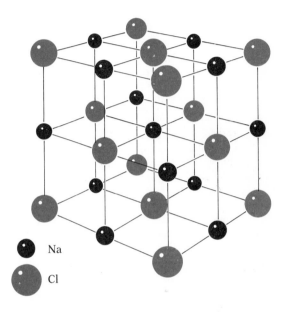

2.12   The arrangement of ions in crystalline table salt (sodium chloride)

The imaginary lattice indicates the spatial arrangement of the Na⁺ and Cl⁻ ions.

COVALENT BONDS

Ionic bonds, as we have seen, involve complete transfer of electrons from one atom to another. But in many, indeed most, cases bonding occurs without complete transfer, by a sharing of electrons between the atoms involved. Bonds of this sort, based on shared electrons, are called **covalent bonds.**

Consider the first element, hydrogen. An atom of hydrogen has only one electron. A complete first shell would contain two electrons. If the hydrogen

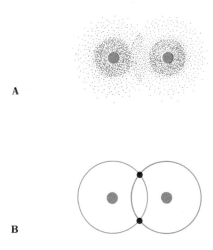

A

B

2.13 Covalent bonding of two hydrogen atoms
(A) The sharing of electrons is indicated by overlapping electron clouds.   (B) Alternatively, the sharing may be indicated by interlocking orbital rings.

gained an electron, it would have a full shell, but there would be twice as much negative charge as positive charge in the atom (one proton and two electrons). Hydrogen does not, in fact, ionize in this manner. It tends to do the reverse; it loses its single electron, forming H^+ ions, which are simply isolated protons, since the hydrogen nucleus contains no neutrons. But suppose there is no strong electron acceptor available and the hydrogen cannot ionize. One possible reaction is for two atoms of hydrogen to bond with each other to form what is called molecular hydrogen, H_2

$$H \cdot + H \cdot \longrightarrow H : H$$

In this molecule, each atom shares its electron with the other atom, so that each hydrogen has, in a sense, two electrons (Fig. 2.13).

Clearly, when two atoms of the same element bond together, as in this case, one will not have more attraction for the two electrons than the other; instead, the electrons will be shared equally by the two atoms. A bond in which the negatively charged electrons are likely to be no closer to one atom than to the other is said to be a **nonpolar** covalent bond.

Suppose now that, instead of being bonded to each other, two hydrogen atoms are covalently bonded to an oxygen atom, forming water, H_2O

$$H \cdot + H \cdot + : \overset{..}{\underset{.}{O}} \cdot \longrightarrow H : \overset{..}{\underset{..}{O}} : H$$

Oxygen has six electrons in its outer shell (Fig. 2.6) and needs two more. By sharing electrons with two hydrogen atoms, the oxygen atom can obtain a full outer octet, while at the same time each hydrogen obtains a complete first shell of two electrons. A covalent bond between a hydrogen atom and an oxygen atom is somewhat different from one between two hydrogen atoms or between two oxygen atoms, however. No two elements have exactly the same affinity for electrons. Consequently, when a covalent bond forms between two different elements, the shared electrons tend to be pulled closer to the one element than to the other. Such a bond is called a **polar** covalent bond; i.e. the charge is distributed asymmetrically within the bond. In water, the electrons are closer to the oxygen than to the hydrogen.

Since many covalent bonds are polar, and since there are all degrees of polarity, from bonds where the electrons are much closer to one atom than to the other to bonds where they are only slightly closer to one than to the other, there is actually no sharp distinction between ionic bonds and covalent bonds. Ionic bonds are simply one extreme, where the electrons are pulled completely from one atom to the other, and nonpolar covalent bonds are the other extreme, where the electrons are pulled with equal force by two atoms. Polar covalent bonds represent the middle ground between these two extremes; the electrons are pulled closer to one atom than to the other, but not all the way. Since electrons do not remain in one position but are constantly moving, a bond may be essentially ionic one instant and covalent the next. Thus even a primarily covalent compound may be very slightly ionized.

The phenomenon of polarity helps explain many of the properties of various molecules in living systems. Whole molecules can be polar as a result of the polarity of bonds within them. If we return to the example of water, we can see (Fig. 2.14) that, even though the two hydrogen-oxygen bonds are polar, the atoms in the molecule might be aligned so that the charge would be distributed symmetrically within the molecule, which would then be nonpolar. But this is not the actual arrangement. The three atoms, instead of being in a straight line, form a bent-chain or V-shaped structure, with the oxygen forming the apex of the V and the two hydrogen atoms forming the arms. Since the electrons are closer to the oxygen than to the hydrogen, there is a concentration of negative charge near the oxygen end of the molecule (making it electronegative) and a concentration of positive charge near the hydrogen end (making it electropositive). Therefore the water molecule is polar. The polarity of molecules often has important biological implications, e.g. in cell membranes, where the polarity of the molecular components is a major factor in determining how they are arranged.

The atoms of some elements are capable of forming more than one or two covalent bonds. For example, carbon (atomic number, 6), which has only four electrons in its outer shell (L) and needs four more if it is to have a complete octet, forms four covalent bonds with other atoms. The gas methane, CH_4, illustrates this fourfold bonding:

$$\begin{array}{c} H \\ \cdot\cdot \\ H:C:H \\ \cdot\cdot \\ H \end{array}$$

Covalent bonds are not limited to single bonds, i.e. to the sharing of one electron pair between two atoms. Sometimes two atoms share two or three electron pairs and form double or triple bonds. When two atoms of oxygen bond together, they form a double bond (remember that an oxygen atom needs two electrons to complete its outer shell), and when two atoms of nitrogen (atomic number, 7) bond together, they form a triple bond, because each nitrogen atom needs three additional electrons to fill its outer shell:

$$:O::O: \qquad :N:::N:$$

A covalent bond may be represented simply by a line between two atoms, rather than by a pair of dots; the other electrons in the outer shells are then ignored. Shown in this manner, H_2, H_2O, O_2, N_2, and CH_4, the molecules we have discussed, appear as follows:

$$H-H \qquad H-O-H \qquad O=O \qquad N\equiv N \qquad \begin{array}{c} H \\ | \\ H-C-H \\ | \\ H \end{array}$$

We have seen that hydrogen atoms tend to form only one bond, oxygen two bonds, nitrogen three bonds, and carbon four bonds. In other words, hydrogen has a covalent bonding capacity, or valence, of 1; oxygen a valence of 2; nitrogen usually a valence of 3; and carbon a valence of 4. Notice that a plus or minus does not precede the valence in covalent bonding, as it does in ionic bonding. The reason, of course, is that electrons are not completely captured or lost, but are simply shared.

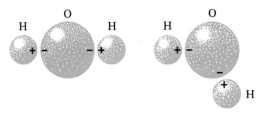

2.14 Molecular structure of water
Two hydrogen atoms are bonded covalently to one oxygen atom, but the shared electrons are pulled closer to the oxygen than to the hydrogen. The bonds are thus polar, with the oxygen side more negative. If the three atoms were arranged linearly, as at left, the charge distribution within the whole molecule would be symmetrical and the molecule would be nonpolar. But in reality the atoms are arranged at an angle of 104.5°, as shown at right; the charge distribution is therefore asymmetrical, and the molecule is polar.

While the four elements we have used as illustrations of covalent bonding—hydrogen, carbon, nitrogen, and oxygen—are not the most common in the earth's crust, they are the most common in the bodies of living organisms (see Table 2.1). What is the explanation? Among elements that form covalent bonds, the lighter the element, the stronger the bond it forms. Since hydrogen, carbon, nitrogen, and oxygen are the lightest of these elements (excluding boron, which is relatively rare), they are especially well suited to forming the complex stable molecules fundamental to life.

BIOLOGICALLY IMPORTANT WEAK BONDS

When two atoms are bonded together, they can be broken apart only through the expenditure of energy. A way of determining the strength of a chemical bond is to measure the energy necessary to break it. Covalent bonds are strong bonds, their bond energies usually ranging between 50 and 110 kilocalories per mole;[2] they are therefore stable and show little tendency to rupture spontaneously. By contrast, ionic bonds in aqueous solutions are relatively weak, averaging about 5 kcal/mole. (They are much stronger in the solid state, e.g. in dry crystals of salt, but since virtually all biological reactions take place in water, we shall deal only with the aqueous condition.)

In addition to ionic bonds, three other kinds of weak bonds, or "interactions" as they are sometimes called, play important roles in biological systems. These are hydrogen bonds, van der Waals interactions, and hydrophobic interactions.

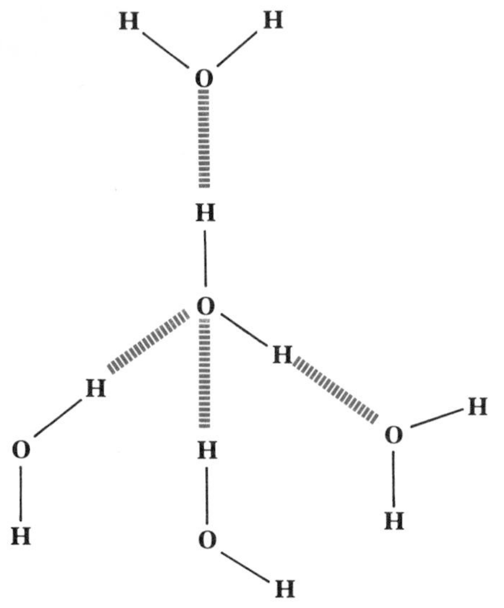

Hydrogen bonds arise when a single hydrogen atom is shared between two electronegative atoms, usually oxygen or nitrogen. The hydrogen is covalently bonded to one oxygen or nitrogen atom, but because of the polarity of the bond—the electrons being closer to the other atom than to the hydrogen—the hydrogen is electropositive. It is therefore attracted by an electronegative atom in another nearby molecule; the bond energy of this attraction is usually 4 to 5 kcal/mole. Water molecules, which have a pronounced tendency to form hydrogen bonds with each other, provide a good example of this type of interaction. As we have already seen, water molecules are polar (Fig. 2.14). Since the oxygen end is negative and the hydrogen end positive, each of the hydrogens, while remaining covalently bonded to the oxygen atom of its own molecule, can form a weak attachment with the oxygen of another water molecule; similarly, the oxygen can form a weak attachment with two external hydrogens. Thus each water molecule has the potential for being simultaneously linked by hydrogen bonds to four other water molecules (Fig. 2.15). In a sense, then, a volume of water is a continuous chemical entity, because of the hydrogen bonding between the individual water molecules.

Much weaker than ionic or hydrogen bonds are the linkages known as *van der Waals interactions*, which have bonding energies of only 1 to 2 kcal/mole. These linkages occur between electrically neutral molecules (or parts of molecules) when they are very close to each other. They are due to

2.15 Hydrogen bonding between water molecules
Like the central H_2O molecule shown here, each water molecule can form hydrogen bonds (striped bands) with four other water molecules.

---

[2] One calorie (spelled with a small *c*) is defined as the quantity of energy, in the form of heat, required to raise the temperature of one gram of pure water one degree, from 14.5°C to 15.5°C. One kilocalorie (kcal) is 1,000 calories. Nutritionists use a different scale to measure energy; their Calorie (spelled with a capital *C*) is equal to one kilocalorie on the standard scale.

very slight perturbations of the electron distributions within the molecules as a consequence of the interactions of their respective electron clouds.

Hydrophobic interactions are also very weak, usually ranging between 1 and 3 kcal/mole. As the name implies, they occur between groups that are insoluble in water. Such groups, which are nonpolar, tend to clump together in the presence of water, thus partially masking one another and minimizing their direct exposure to the water (Fig. 2.16).

Weak bonds play a crucial role in stabilizing the shape of many of the large molecules found in living matter, and they also often hold groups of such molecules together in orderly arrays. But the various kinds of weak bonds—ionic, hydrogen, van der Waals, hydrophobic, and a few others not discussed here—all resemble one another in that they are readily broken under normal physiological conditions. It is thus relatively easy to alter the molecular configurations of the molecules whose shapes they stabilize and to disassemble and rearrange the arrays they hold together. Many essential life processes depend on these sorts of changes.

SOME IMPORTANT INORGANIC MOLECULES

Chemists have traditionally referred to complex molecules containing the element carbon as *organic* compounds. All other compounds are called *inorganic,* a designation that should not mislead you into assuming that these compounds play no role in life processes. Many inorganic substances are, in fact, basic to the chemistry of life. We shall examine a few of the most important below, for without some knowledge of them it is hardly possible to understand the more complex organic compounds to which we will soon turn our attention.

WATER

In attempting to answer the intriguing question whether there is life on Mars and other planets besides earth, one of the first facts scientists try to establish—notably with the help of explorer satellites—is whether or not water is present. The reason is that life as we know it on earth is totally dependent on water. A high percentage of living things, both plant and animal, are found in aquatic habitats. Not only is earthly life thought to have arisen in an aquatic medium, but the bodies of living organisms themselves are composed largely of water; roughly 70 to 90 percent of living material is water. Moreover, the chemical reactions that characterize life all take place in a water medium. Besides facilitating reactions, water is itself often a reactant or a product of chemical reactions. In short, the chemistry of life is water chemistry. It is perhaps conceivable that life elsewhere in the universe is based on some substance other than water, but such life would surely be vastly different from anything in our experience—so different, in fact, that we might not recognize it as life even if we should stumble on it.

Water as a solvent One of the main reasons why water is so well adapted as the medium for life is that it is a superb solvent for many important classes of chemicals. It is a better solvent than most common liquids—first, because of the marked polarity of the water molecule; second, because of the great

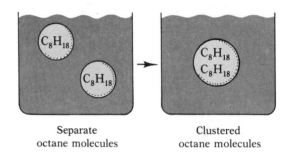

Separate
octane molecules
　
Clustered
octane molecules

2.16 Hydrophobic interactions
Left: Two molecules of octane, a strongly hydrophobic substance, are separately suspended in a volume of water. Each occupies its own cavity in the water and is directly exposed to it on all sides.   Right: The octane molecules have clustered together and occupy a common cavity in the water. Because each octane has part of its surface in contact with the other, less total octane surface is exposed to water—an energetically more favorable arrangement.

2.17   Hydration spheres of Na⁺ and Cl⁻

When dissolved in water, each of the Na⁺ and Cl⁻ ions is surrounded by water molecules electrostatically attracted to it. Note that it is the oxygen of the water molecules that is attracted to the positively charged Na⁺, and that it is the hydrogen of the water molecules that is attracted to the negatively charged Cl⁻. Water molecules in a hydration sphere are called bound water. With a spatial arrangement different from that of pure water, they exhibit corresponding differences in their properties.

propensity of water to form hydrogen bonds. Thanks to these two properties, both ionic substances and substances that are nonionic but polar are soluble in water. Let us consider each of these properties in turn.

We have mentioned that the ionic bonds linking together the atoms within a dry crystal of salt, such as NaCl, are comparatively strong, but that those same bonds are relatively weak when the salt is in an aqueous medium. Why the difference? Within the dry crystal the electrostatic attractions between the alternating positive sodium ions and negative chloride ions are very strong, and much energy would be required to pull these ions away from each other. But when the crystal is put in water, the attraction of the electronegative oxygen end of the water molecules for the positively charged sodium ions, and the similar attraction of the electropositive hydrogen ends of the water molecules for the negatively charged chloride ions, are greater than the mutual attraction between the Na⁺ and Cl⁻ ions. Thus, in water, the ionic bonds are broken with extreme ease, because of the competitive attraction of the water. Consequently the Na⁺ and Cl⁻ ions dissociate (i.e. the NaCl ionizes), and each of the ions becomes surrounded by spheres of regularly arranged water molecules that are electrostatically attracted to it (Fig. 2.17). Such an ion, surrounded by water loosely bound to it, is said to be hydrated; the process of formation of the spheres of water is called **hydration.**

From the point of view of some biological processes, an ion and its hydration spheres must be regarded as a single entity—as though the whole were a true molecule. For example, if the question is whether or not a given type of ion can move through tiny pores in cell membranes, then it is the size of the hydrated ion that must be compared with the dimensions of the pores.

As we have said, water is also an excellent solvent for nonionic molecules if they are polar. The reason, clearly, is an electrostatic attraction between the charged parts of the solute molecules and the oppositely charged parts of the water molecules. This is especially true when, as in many biologically important compounds, the solute molecule has an oxygen together with a hydrogen (—OH) attached to it. As in water, the hydrogen in such a grouping is electropositive and is therefore attracted by the electronegative oxygen of the surrounding water, with the result that a hydrogen bond is formed. The solute molecules and the water molecules thus become weakly linked with each other.

Substances that do not dissolve in water are electrically neutral and nonpolar. They thus show no tendency to interact electrostatically with water and, indeed, are repulsed by it. A hydrophobic substance (e.g. oil) stirred into water will soon begin to separate out, because the water molecules tend slowly to re-establish the hydrogen bonds broken by the physical intrusion of the insoluble material and thus to push out that material. Since, moreover, the molecules of the insoluble material are subject to hydrophobic interactions of the sort discussed above, they tend to coalesce to form droplets, which in general eventually fuse and form a separate layer outside the water.

Special physical properties of water   We have seen that substances dissolve in water if their molecules can form weak bonds or interactions with the water molecules. Such interactions have important implications for the

2.18 A water strider on the surface of the water
Water striders can move rapidly across the surface of still water, where they often congregate in large numbers. Note the dimples in the water surface where each foot rests.   [W. H. Amos, Bruce Coleman, Inc.]

water molecules themselves. As Fig. 2.17 shows, water molecules in the hydration spheres of ions are arranged in orderly arrays; such water is referred to as ***bound water,*** because it is essentially immobilized. The same is true of water molecules around polar groups of nonionic compounds. But the orderly arrays of bound water are very different from those of pure water (Fig. 2.15). The physical properties of bound water are consequently different from those of free water; e.g. the greater the proportion of bound water in a given volume, the lower the freezing point, the higher the boiling point, and the lower the vapor pressure of that volume of water. Since much of the water inside living cells is bound water, it follows that the physical properties of the cell contents are very different from those of pure water, even though water is the principal constituent of the cells.

Some of the special physical properties of water attributable to the strong tendency of its molecules to form hydrogen bonds with each other—and the consequent ordering of the molecules to which this tendency gives rise— have important implications for life processes. For example, water has a high ***surface tension;*** i.e. the surface of a volume of water is not easily broken. You witness the effects of surface tension when you watch a water strider or other insect walk on the surface of a pond without breaking the surface (Fig. 2.18), or when you fill a glass with water to a point slightly above the top without spilling; the water has so much cohesion that the extra water resists breaking away. The reason why water has a high surface tension is that hydrogen bonds link the molecules at the surface to each other and to the molecules below them. Before the legs of the water strider (or any other object, for that matter) can penetrate the water's surface, they must break some of these hydrogen bonds and deform the orderly array of water molecules. Similarly, the cohesion that prevents the extra water in an overfilled glass from spilling is due to the hydrogen bonds that bind the extra water molecules to the other water molecules below them.

Just as water molecules are attracted electrostatically to areas of charge on dissolved molecules, so they are also attracted to charged groups on a hydrophilic surface. Consequently such surfaces are wettable (i.e. water

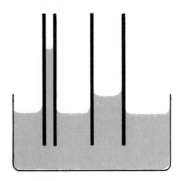

2.19 Capillarity in glass tubes
Water rises higher in a tube of small bore (left) than in one of large bore (right), because in the smaller tube a higher percentage of the water molecules are in direct contact with the glass and can form hydrogen bonds with charged groups on the glass.

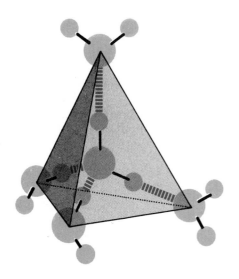

2.20 The tetrahedron configuration of hydrogen-bonded water molecules
When a water molecule in ice is simultaneously linked by hydrogen bonds (striped bands) to four other water molecules, each of the four is positioned at maximal divergence from the other three, thereby forming one of the corners of an imaginary tetrahedron (here drawn in to aid visualization). (Large circles represent oxygen atoms, small circles hydrogen.)

spreads over them and binds loosely to them). By contrast, hydrophobic surfaces (which lack surface charge), such as those of most plastics and waxes, are not wettable; water on them will form isolated droplets, but will not spread out over the surface.

The propensity of water to bind to hydrophilic surfaces explains the phenomenon of *capillarity*—the tendency of aqueous liquids to rise in narrow tubes. Thus, if the end of a narrow glass tube is inserted below the surface of a volume of water, water will rise in the tube to a level well above the water level outside (Fig. 2.19). The reason is that glass is very hydrophilic, having numerous charged groups on its surface. The water molecules, electrostatically attracted to the glass, tend to creep along the surface of the tube. But as the ring of water molecules in contact with the inner surface creeps upward, it pulls along other water molecules to which it is linked by hydrogen bonds, especially those inside the ring (think of a doughnut moving up in the tube and pulling along with it enough material to keep the hole filled). Clearly, the smaller the diameter of the tube, the higher the water will rise. In a very large tube such a small percentage of the water molecules would be in direct contact with the glass that they would be unable to lift the others. Even though the relatively few molecules in contact with the glass would have a tendency to creep upward, they would be held back because of their cohesion with the rest of the water in the tube, which they could not lift with them.

We have said that each water molecule has the potential for forming hydrogen bonds to four other water molecules. In the liquid state this potential is not fully realized, but as water is cooled the extent of hydrogen bonding increases. Hydrogen bonding reaches its full potential when the water has frozen into ice. Now, hydrogen bonds are highly directional, and when all four possible bonds have formed, each is oriented in space with maximal divergence from the other three. Consequently the bonds are directed toward the four corners of a tetrahedron (Fig. 2.20). The resulting three-dimensional lattice of water molecules in ice is a rather open one (Fig. 2.21); i.e. the packing of the molecules is not so tight as would be possible if the molecules were not so rigidly tetrahedral.

When ice is warmed to the melting point, a few of the hydrogen bonds rupture, and those that remain become less rigidly oriented. The resulting deformation of the lattice and tighter packing of the molecules make the water denser. The density reaches its maximum when the water is at 4°C; above this temperature, the further packing that might be expected as a consequence of increasing disruption of the lattice is more than compensated by the tendency, shown by virtually all substances, to expand owing to the increased molecular motion of heat energy. In summary, we see that, unlike most other substances that continue to become denser as the temperature falls, water first becomes denser and then begins to expand again below 4°C. This means that ice, being less dense than cold water, floats and, further, that ponds and streams freeze from the top down rather than from the bottom up. The crust of ice that forms at the surface insulates the water below it from the cold air above and thereby often prevents the pond or stream from freezing solid, even in very cold winter weather. The importance of this special property of water for organisms living in ponds or streams is obvious.

The role of water in regulating environmental temperature Not only do the hydrogen bonds in water produce high internal cohesion, which gives it an unusually high melting point, boiling point, and heat of vaporization, but they also enable water to absorb much heat energy without a very large increase in temperature (and, conversely, to release much heat energy without a great drop in temperature). When most substances absorb heat, their molecules move more rapidly in relation to one another; it is the amount of such molecular motion that is expressed as temperature. In water, by contrast, much of the absorbed energy is dissipated by more rapid vibration of shared hydrogens back and forth along the bond axis between the two oxygen atoms to which they are linked. As a result, relatively little of the added heat energy is expressed as translational motion of whole water molecules, and the temperature increase is therefore modest. The high *heat capacity* of water (the amount of energy that must be added or subtracted to change the temperature by one degree), together with its high *heat of vaporization* (which means that evaporation of water requires much heat energy and thus has a marked cooling effect), permits it to act as an effective buffer against extreme temperature fluctuations in the environment. In this way water helps stabilize the earth's temperatures within the range favorable to life.

Besides damping fluctuations in temperature, water plays an important role in determining the absolute temperature at the earth's surface, because the water vapor in the atmosphere exerts what has been called a "greenhouse effect." The vapor absorbs much of the sunlight striking it from above and also much of the radiation re-emitted by the earth. The absorbed radiation warms the atmosphere, which, in turn, warms the earth's surface.

CARBON DIOXIDE

As we have seen, carbon has only four electrons in its outer electron shell and, as a result, has a covalent bonding capacity of 4. Carbon dioxide (CO_2) is the compound formed when two atoms of oxygen bond to one atom of carbon. Though this substance contains carbon, it is generally thought of as inorganic, because it is simpler than all but a few of the compounds classified as organic.

Only a very small fraction of the atmosphere, roughly 0.033 percent, is CO_2;[3] yet atmospheric carbon dioxide is the principal inorganic source of carbon, and carbon is the principal structural element of living matter. Before CO_2 can take part in chemical reactions, it must usually first dissolve in water, which it does very readily, and then react with the water to form carbonic acid, H_2CO_3

$$CO_2 + H_2O \longrightarrow H_2CO_3$$

This reaction involves so little energy change that it is easily reversible, and CO_2 can readily be released from water solution when conditions are appropriate:

$$H_2CO_3 \longrightarrow CO_2 + H_2O$$

[3] The value of 0.033 percent is based on data for 1965. Prior to 1900, the figure was 0.029 percent. The increase is due primarily to the burning of fossil fuels.

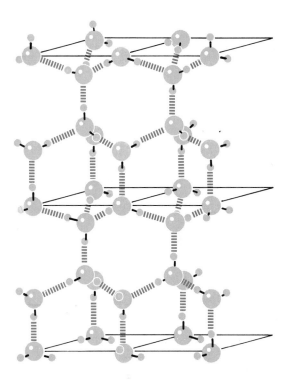

2.21 The lattice formed by water molecules in ice
Because of the tetrahedral arrangement around each water molecule, the lattice is an open one, with considerable space between molecules. In liquid water the arrangement is not quite so rigid, and the packing is therefore slightly denser, but the general lattice arrangement is nonetheless largely preserved. (Planes added to help show the three-dimensional disposition of the molecules.)

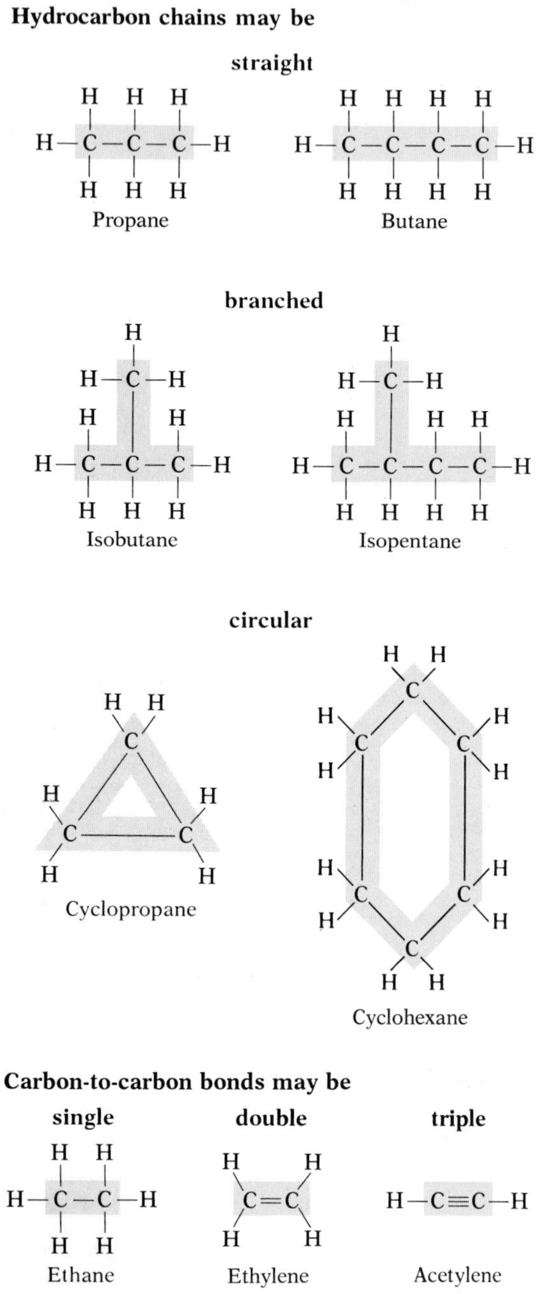

Hydrocarbon chains may be

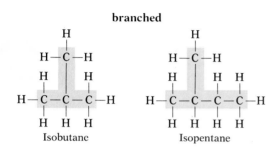

Propane   Butane

Isobutane   Isopentane

Cyclopropane

Cyclohexane

Carbon-to-carbon bonds may be

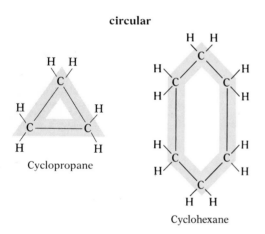

single   double   triple

Ethane   Ethylene   Acetylene

2.22   Examples of hydrocarbons
The molecules appear flat in these conventionally drawn structural diagrams, though they are, in fact, three-dimensional. The bonds around a carbon atom that forms only single bonds (all the carbons seen here except those in the last two molecules) are oriented toward the four corners of a tetrahedron.

Carbon dioxide and water are the raw materials from which green plants manufacture many complex organic compounds essential to life, as we shall see in detail in Chapter 4. And when these complex compounds have run their course in the life system, they are broken down again to carbon dioxide and water, and the carbon dioxide is eventually released into the atmosphere. The simple compound carbon dioxide, then, is the beginning and the end of the immensely complex carbon cycle in nature.

OXYGEN

Molecular oxygen (O_2) constitutes approximately 21 percent of the atmosphere. It is a necessary material for maintenance of life in most organisms, although a few can live without it. It can be utilized directly, without change, by both plants and animals in the process of extracting usable energy from nutrient molecules. Although oxygen is not very soluble in water, enough dissolves to supply the needs of aquatic organisms, provided (1) that the water is not too hot and (2) that its surface is exposed to the air or, alternatively, that green plants are growing in it, thus constantly releasing oxygen into it by the process of photosynthesis.

SOME SIMPLE ORGANIC CHEMISTRY

Organic compounds are based on the element carbon (atomic number, 6), which has a covalent bonding capacity of 4. This element is present in an enormous number of known chemical compounds; of all the elements, only hydrogen has more known compounds. Although carbon can and does bond to a variety of different elements, it is most commonly bonded to hydrogen, oxygen, nitrogen, or more carbon.

Of central importance in organic chemistry are the compounds containing carbon and hydrogen; the number of different compounds of this kind, called *hydrocarbons,* is immense. A basic reason for the great variety of hydrocarbons is the readiness with which carbon-to-carbon bonds can form, producing chains of varying lengths and shapes (Fig. 2.22). These chains may be simple, as in propane, butane, and other still longer substances; or they may be branched, as in isobutane and isopentane; or they may form circles of varying numbers of carbons, as in cyclopropane and cyclohexane. Obviously, the more atoms there are in a molecule, the more different arrangements of those atoms are possible. Compounds of the same atomic content and molecular formula but differing structures and hence differing properties are called *isomers* (Fig. 2.23). Very large organic molecules may have hundreds of different isomers.

Another factor adding variety to hydrocarbon compounds is the capacity of adjacent carbon atoms to form single, double, or triple bonds (Fig. 2.22). And, of course, substitution of other elements or groups of elements for hydrogen atoms makes possible an almost infinite number of derivative hydrocarbons. The total number of hydrocarbons and derivatives has been estimated at more than half a million, and this may be a conservative estimate.

Because carbon-to-hydrogen bonds are nonpolar, simple hydrocarbon molecules are nonpolar and hence not soluble in water, as anyone who has tried unsuccessfully to mix a gasoline such as octane (see Fig. 2.16) with water will know.

Obviously, we cannot even attempt in these few pages to cover the enormous field of organic chemistry, but as the word "organic" suggests, organisms are composed of organic substances, and it is impossible to understand many of the most important attributes of life without a few simple facts of organic chemistry. We shall mention only four major classes of organic compounds: carbohydrates, lipids, proteins, and nucleic acids.

CARBOHYDRATES

Carbohydrates are compounds composed of carbon, hydrogen, and oxygen.[4] In simple carbohydrates the hydrogen and oxygen are characteristically present in the same proportions as in water; i.e. there are two hydrogen atoms and one oxygen atom for each carbon atom. Consequently the grouping CH_2O, diagrammed H—C—OH, recurs frequently in carbohydrate molecules.

Some carbohydrates, such as starch and cellulose, are very large and complex molecules. Fortunately for our grasp of them, however, these molecules, like most very large organic molecules, are composed of many simpler "building-block" compounds bonded together. An understanding of the constituent or building-block compounds greatly facilitates comprehension of the more complex substances.

Simple sugars The basic carbohydrate molecules are simple sugars, or *monosaccharides.* All sugars, when in straight-chain form, contain a C=O group (Fig. 2.23). If the double-bonded O is attached to the terminal C of a chain, the combination is called an aldehyde group; if it is attached to a nonterminal C, the combination is called a ketone group. The OH (hydroxyl) groups attached to all the carbons except those with a double-bonded oxygen are polar. Hence sugars readily form hydrogen bonds with water and are soluble.

The carbon chain that forms the backbone of the sugar can be of different lengths. Some monosaccharides contain as few as three carbons (trioses); others contain five carbons (pentoses), six carbons (hexoses), or more. Though both trioses and pentoses play important biological roles and will be mentioned in later chapters, it is the hexoses, six-carbon sugars, that are the most important as building-block compounds for more complex carbohydrates.

There are many six-carbon sugars, of which glucose and fructose are two of the most important. Since they all have the proportions of oxygen and hydrogen typical of carbohydrates, all have the same molecular formula, $C_6H_{12}O_6$, and are thus isomers of each other. Glucose and fructose, an aldehyde and a ketone sugar respectively, are structural isomers (Fig. 2.23).

In addition to structural isomerism, which is readily understandable, there is another, more subtle, kind of isomerism called stereoisomerism. In a given pair of stereoisomers, identical groups are attached to the carbon atoms, but the spatial arrangements of the attached groups are different.

[4] Derivative carbohydrates may contain other elements in addition to these three.

2.23 Three isomeric hexoses
Each of these six-carbon sugars has the same molecular formula, $C_6H_{12}O_6$; hence each is an isomer of the others. Fructose, which is a ketone sugar (a sugar with the double-bonded oxygen attached to an internal carbon), is a structural isomer of the other two. Glucose and galactose, which are both aldehyde sugars (with the double-bonded oxygen attached to a terminal carbon), are optical isomers of each other.

STRUCTURAL ISOMERS

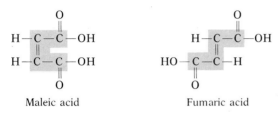

Ethyl alcohol   Dimethyl ether

GEOMETRIC ISOMERS

Maleic acid   Fumaric acid

OPTICAL ISOMERS

$H_3C-\overset{\overset{\textstyle COOH}{|}}{C}-H$ $H-\overset{\overset{\textstyle COOH}{|}}{C}-CH_3$

            $\overset{|}{OH}$ $\overset{|}{OH}$

l-Lactic acid   *d*-Lactic acid

2.24   Three types of isomerism
The two structural isomers differ in the basic grouping of their constituent atoms, one being an alcohol (characterized by an OH group) and the other an ether (characterized by an oxygen bonded between two carbons).   The two geometric isomers are fixed in different spatial arrangements by their inability to rotate around the double bond between the middle two carbons.   The two optical isomers are mirror images of each other; like the two asymmetrical molecules shown in Fig. 2.25, they cannot be superimposed.

2.25   Models of two optical isomers
No matter which way you turn these models, you cannot superimpose them so as to get all atoms in the same positions. Note that the four bonds around each carbon are so oriented as to determine a tetrahedron.

The two middle compounds shown in Fig. 2.24 are geometric stereoisomers; if all the carbon-to-carbon bonds in these two molecules were single bonds, the two would in fact be the same compound, because free rotation is possible around a single bond, though not around a double bond.

When considering optical stereoisomers (Fig. 2.24, bottom), keep in mind that molecules are not flat, even though they are often drawn that way. In a carbon atom the four valences form the apexes of a tetrahedron; if the groups attached to each of the four valences are different, some alternative arrangements of these groups will not be superimposable (Fig. 2.25). Glucose and galactose are optical isomers of each other because the position of one OH group is different in the two molecules (Fig. 2.23). Although the optical isomers of a compound may appear very similar, their biological properties are usually completely different.

Glucose does not always exist as a straight-chain compound; indeed, this is probably its least common form. Sometimes it exists in ring form—most often, as a ring composed of five carbons and one oxygen (Fig. 2.26).

Glucose plays a unique role in the chemistry of life. It is in a very real sense the crossroads of the chemical pathways in the bodies of plants and animals. Other six-carbon monosaccharides, among them fructose and galactose, are constantly being converted into glucose or synthesized from glucose. The more complex carbohydrates such as disaccharides and polysaccharides are composed of monosaccharides bonded together in sequence. And even such classes of compounds as fats and proteins can be converted into glucose or synthesized from glucose in the living body.

In addition to ordinary monosaccharides composed only of carbon, oxygen, and hydrogen, there are a variety of derivative monosaccharides containing other elements. For example, some have a phosphate group attached to one of the carbons, and others an ***amino*** group (a nitrogen with two hydrogens, NH_2) (Fig. 2.27).

Disaccharides   Disaccharides, or double sugars, are compound sugars composed of two simple sugars bonded together through a reaction that involves the removal of a molecule of water. This kind of reaction is called a ***condensation reaction*** or a dehydration reaction.

Let us first examine the double sugar ***maltose,*** or malt sugar. This compound is synthesized by a condensation reaction between two molecules of glucose. The reaction can be described by the following equation:

$$2C_6H_{12}O_6 \longrightarrow C_{12}H_{22}O_{11} + H_2O$$

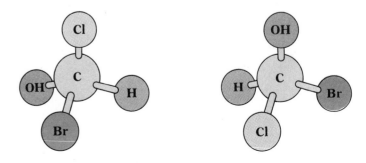

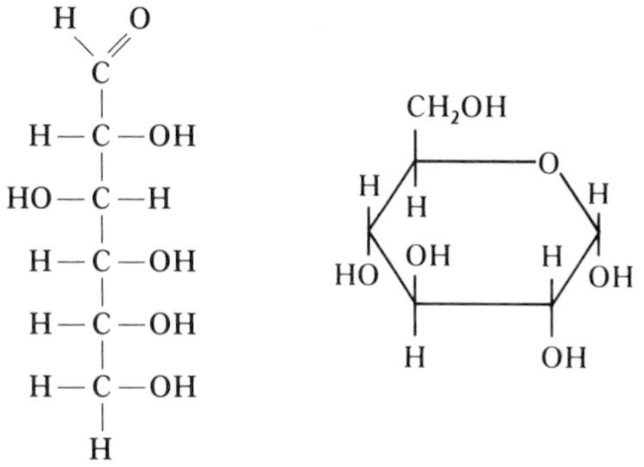

2.26 Two forms of glucose
Glucose may exist in the straight-chain aldehyde form shown at left or as a ring structure, as shown at right. The ring structure (of which there is an alternative form, not shown here) is the more common. (By convention, the unmarked corners of a hexagon signify carbon atoms.)

This equation tells us very little, however. Any simple six-carbon sugar has the formula $C_6H_{12}O_6$, and any double sugar synthesized from such building blocks will have the formula $C_{12}H_{22}O_{11}$. Consequently the above equation can describe many different reactions involving a variety of reactants and products.

To understand specifically what is involved in the condensation synthesis of maltose, it is helpful to look at a diagram indicating the structures of the molecules. Figure 2.28 shows that the hydrogen atom from a hydroxyl group (OH) of one molecule of glucose combines with a complete hydroxyl group from the other molecule of glucose to form water.[5] The oxygen valence vacated by removal of hydrogen and the carbon valence vacated by removal of OH are filled by the bonding together of the oxygen of the one glucose molecule with the carbon of the other glucose molecule. As a result, the two glucose units are connected by way of an oxygen atom shared between them. The product is the double sugar maltose.

Sucrose, our common table sugar, is also a disaccharide. It is synthesized by a condensation reaction between a molecule of glucose and a molecule of fructose. *Lactose,* or milk sugar, is a double sugar composed of glucose and galactose.

Synthesized by condensation reactions, disaccharides can be broken

[5] This is not exactly the way sugars are combined in the bodies of living organisms. The actual reaction is considerably more complex and involves the temporary formation of some intermediate compounds, but the end result is just what it would be if water had been directly removed.

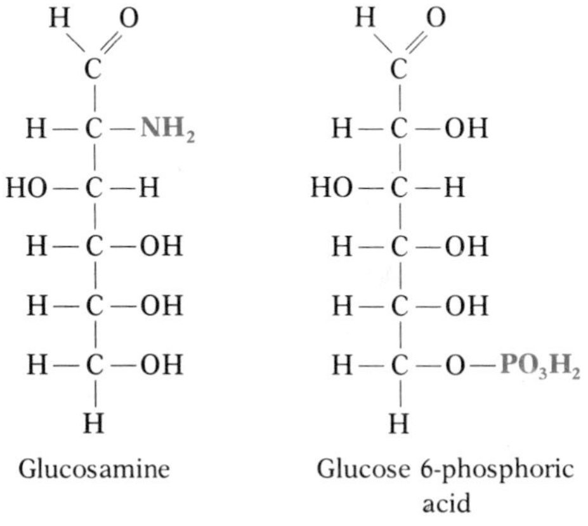

Glucosamine   Glucose 6-phosphoric
                     acid

2.27 Two examples of derivative monosaccharides
Glucosamine is merely a glucose molecule with an amino group (NH_2) substituted for an OH group (cf. Fig. 2.26). Similarly glucose 6-phosphoric acid is glucose with a substituted phosphate group.

2.28 Synthesis of a double sugar
Removal of a molecule of water between the two molecules of glucose results in formation of a bond (color) between the two. The double sugar produced is maltose.

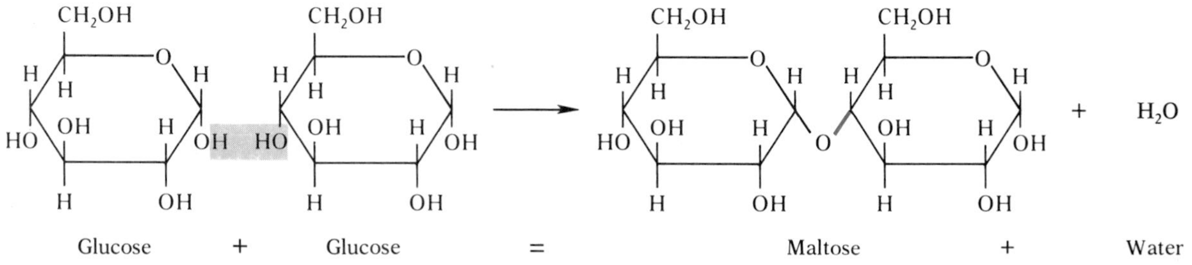

Glucose   +   Glucose   =   Maltose   +   Water

down to their constituent simple sugars by the reverse process. This reaction, called **hydrolysis**, involves addition of a water molecule:

$$C_{12}H_{22}O_{11} + H_2O \longrightarrow 2C_6H_{12}O_6$$

Hydrolysis reactions are of particular importance in digestion, as we shall see in a later chapter.

We are now in a position to define a simple sugar, or monosaccharide, more precisely than we have so far done. By contrast with compound sugars, a monosaccharide is a sugar that cannot be hydrolyzed into smaller carbon-containing substances.

Polysaccharides   The prefix *poly-* means many, and polysaccharides are complex carbohydrates composed of many simple-sugar building blocks bonded together in long chains (Fig. 2.29). They are synthesized by exactly the same kind of condensation reaction as the disaccharides and, like them, can be broken down to their constituent sugars by hydrolysis.

A number of complex polysaccharides are of great importance in biology. **Starches**, for example, are the principal carbohydrate storage products of higher plants. They are composed of many hundreds of glucose units bonded together. In some forms of starch the chain of sugars is unbranched, and in others it is branched; both types are common in plant material. **Glycogen** is the principal carbohydrate storage product in animals and is sometimes called "animal starch." Its molecules are much like those of starch; they have the same type of bond between adjacent glucose units, but the chains are more extensively branched. **Cellulose** is a highly insoluble unbranched polysaccharide common in plants, where it is a major supporting material. The bonds between its glucose units are aligned differently from those of starch and glycogen, and most animals are unable to hydrolyze these bonds.

Reactions like those that form polysaccharides—i.e. reactions in which small molecules bond together to form long chains—are called polymerization reactions. The products formed are called **polymers.** From our exami-

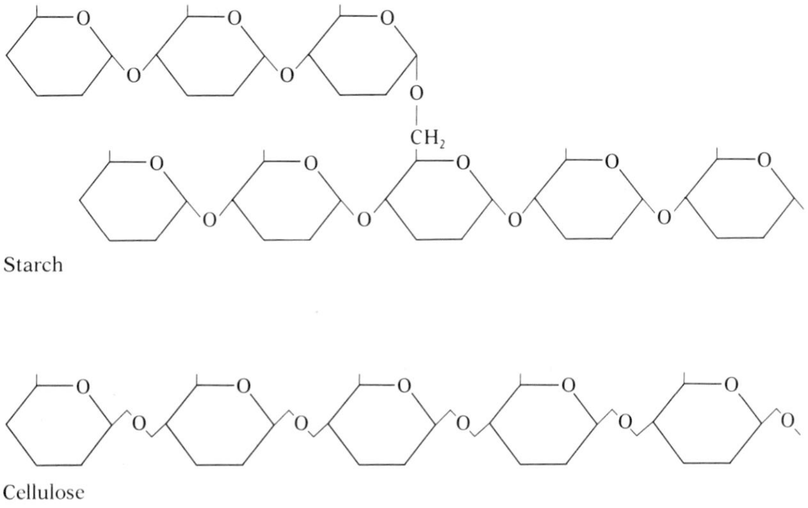

2.29   Two polysaccharides
Shown here are small segments of a molecule of starch and of a molecule of cellulose. Although the starch is shown branching, some forms of starch are unbranched. Cellulose molecules are always unbranched, but one molecule is usually tightly linked to another by multiple hydrogen bonds. Both starch and cellulose are polymers of glucose, but the bonds between adjacent glucose units are not aligned in the same way; the differences in the bonds are correlated with the different functions fulfilled by these two compounds within living organisms.

Starch

Cellulose

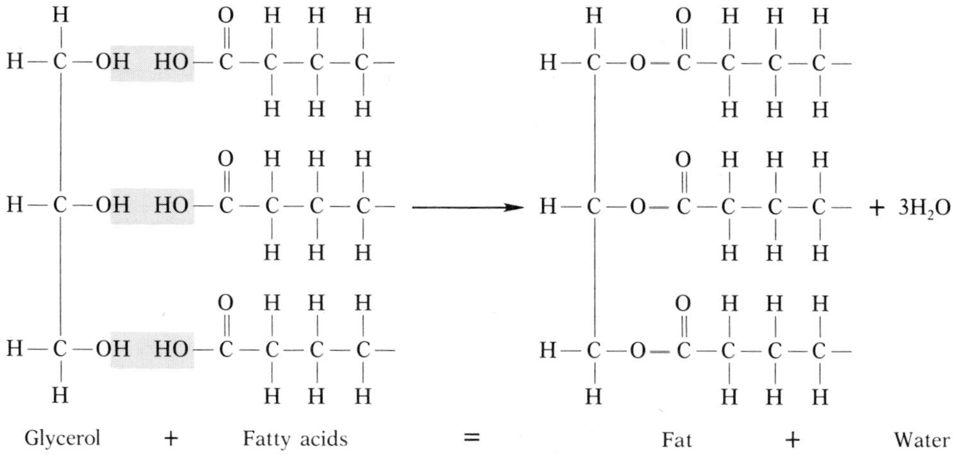

Glycerol + Fatty acids = Fat + Water

2.30 Synthesis of a fat
Removal of three molecules of water results in the bonding of three molecules of fatty acid to a single molecule of glycerol. The carbon chains of the fatty acids are usually longer than shown here.

nation of polysaccharides, it is plain that if polymerization between molecules A, B, C, D, etc. is to occur, two requirements must be met: One end of molecule A must be capable of interacting with one end of molecule B so as to split out some small group such as water; and each molecule must contain two functional groups, so that after the first two combine, the free ends can continue to react and extend the polymer. Polymers of several types play a critical role in biology, as we shall see.

LIPIDS

Like carbohydrates, lipids, a second major group of biological compounds, are composed principally of carbon, hydrogen, and oxygen, but they may also contain other elements, particularly phosphorus and nitrogen. They differ from carbohydrates in that they contain a much smaller proportion of oxygen. Unsubstituted lipids, being nonpolar, are relatively insoluble in water, but soluble in organic solvents such as ether.

Fats Among the best-known lipids are the neutral fats. Important as energy-storage molecules in living organisms, the fats also provide insulation, cushioning, and protection for various parts of the body. Each molecule of fat is composed of two different types of building-block compounds: an alcohol called glycerol and fatty acids.

Glycerol (also sometimes called glycerin) has a backbone of three carbon atoms, each carrying a hydroxyl (OH) group (Fig. 2.30). (It is the presence of an OH group attached to a carbon atom that characterizes a compound as an alcohol; hence the hydroxyl group is sometimes termed the alcoholic functional group.)

Fatty acids, like all organic acids, contain a COOH group (called a carboxyl group):

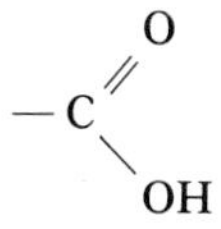

$$-C\overset{\displaystyle O}{\underset{\displaystyle OH}{\big\Vert}}$$

HO   O
  \\ //
   C
   |
   CH₂
   |
   CH₂
   |
   CH₂
   |
   CH₂
   |
   CH₂
   |
   CH₂
   |
   CH₂
   |
   CH₂
   |
   CH₂
   |
   CH₂
   |
   CH₂
   |
   CH₂
   |
   CH₂
   |
   CH₂
   |
   CH₃

Palmitic acid

HO   O
  \\ //
   C
   |
   CH₂
   |
   CH₂
   |
   CH₂
   |
   CH₂
   |
   CH₂
   |
   CH₂
   |
   CH₂
   |
   CH
   ‖
   CH
   |
   CH₂
   |
   CH
   ‖
   CH
   |
   CH₂
   |
   CH₂
   |
   CH₂
   |
   CH₂
   |
   CH₃

Linoleic acid

2.31 Examples of saturated and unsaturated fatty acids

Palmitic acid is saturated with hydrogen; i.e. it contains the maximum number of hydrogens possible. By contrast, linoleic acid, with its two carbon-to-carbon double bonds, contains four fewer than the maximum number of hydrogens.

(Notice that when both a double-bonded oxygen and an OH group are attached to the same carbon atom, the OH does not make the compound an alcohol. In this case the entire COOH unit acts as a single functional group, making the compound an acid.)

There are many different fatty acids, varying in carbon-chain length, in the number of carbon-to-carbon bonds that are single or double, and in other characteristics. The fatty acids in edible fats and oils contain an even number of carbon atoms, and most of them have relatively long carbon backbones, usually from 4 to 24 carbons, or more; three of the most common are stearic acid (18 carbons), palmitic acid (16 carbons) (Fig. 2.31), and oleic acid (18 carbons).

Organic acids and alcohols have a tendency to combine through condensation reactions. Since glycerol has three alcoholic groups, it can combine with three molecules of fatty acid to form a molecule of fat (Fig. 2.30). Hence fats are sometimes also called triglycerides.

The various fats differ in the specific fatty acids, or types of fatty acids, of which they are composed. You have doubtless read of the controversy in medical and nutritional circles concerning saturated and unsaturated fats. Saturated fats are simply those incorporating fatty acids with the maximum possible number of hydrogen atoms attached to each carbon, and hence no carbon-to-carbon double bonds (Fig. 2.31). The fatty acids in unsaturated fats (or perhaps we should say oils, since they are usually liquid at room temperature) have at least one carbon-to-carbon double bond; i.e. they are not completely saturated with hydrogen. Regarding health effects, there is evidence that an elevated intake of saturated fats is one of many factors that predispose human beings to atherosclerosis—a disease of the arteries in which fatty deposits in the arterial walls cause partial obstruction and thus interfere with blood flow.

Since fats are synthesized by condensation reactions (removal of water), they, like complex carbohydrates, can be broken down to their building-block compounds by hydrolysis, as happens in digestion.

Phospholipids A variety of substituted lipids contain a phosphate group. Among the most common phospholipids are those composed of one unit of glycerol, two units of fatty acid, and a phosphate group often linked with a nitrogen-containing group (Fig. 2.32). The phosphate group is bonded to the glycerol at the point where the third fatty acid would be in a fat. Because the phosphate group has a marked tendency to lose a hydrogen ion, one of the oxygens becomes negatively charged; similarly, the nitrogen tends to attract a hydrogen ion and thus to become positively charged. In short, the end of the phospholipid molecule with the phosphate and nitrogenous groups is strongly polar and hence soluble in water, whereas the other end, composed of the two long hydrocarbon tails of the fatty acids, is nonpolar and insoluble. This curious, as it were schizoid, property of solubility at one end but not at the other makes phospholipids especially well suited to function as major constituents of cell membranes, as we shall see in Chapter 3.

Steroids Though commonly classified as lipids because their solubility characteristics are similar to those of fats, oils, waxes, and phospholipids, the steroids differ markedly in structure from the other lipids we have discussed (Fig. 2.33). They are not based upon a bonding together of fatty acids and an alcohol. Instead, they are complex molecules composed of four interlocking rings of carbon atoms, with various side groups attached to the rings. Steroids are very important biologically. Some of the vitamins and hormones are steroids, and steroids often occur as structural elements in living cells, particularly in cellular membranes.

PROTEINS

Far more complex than either carbohydrates or lipids, proteins are fundamental to both the structure and function of living material. Like carbohydrates and lipids, proteins are composed of relatively few, simple building-block compounds.

The building blocks and primary structure of proteins All proteins contain four essential elements: carbon, hydrogen, oxygen, and nitrogen; most proteins likewise contain some sulfur. These elements are bonded together to form compounds called **amino acids,** which, being organic acids, contain the COOH (carboxyl) group. In addition, they each have an amino group, NH_2. Both the COOH and the NH_2 group are attached to the same carbon atom—designated as the α (alpha) carbon—in all amino acids found in proteins; accordingly, these are known as α-amino acids. Finally, each amino acid has a side chain, designated as R:

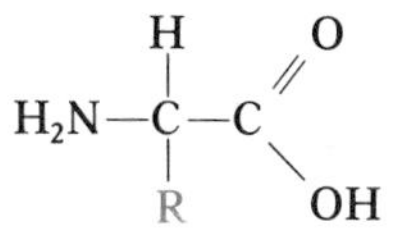

$$H_2N-\underset{\underset{R}{|}}{\overset{\overset{H}{|}}{C}}-C\overset{\displaystyle O}{\underset{\displaystyle OH}{\big<}}$$

In the normally very slightly acidic pH range prevailing within cells, a high percentage of the amino acid molecules are ionized; the carboxyl group,

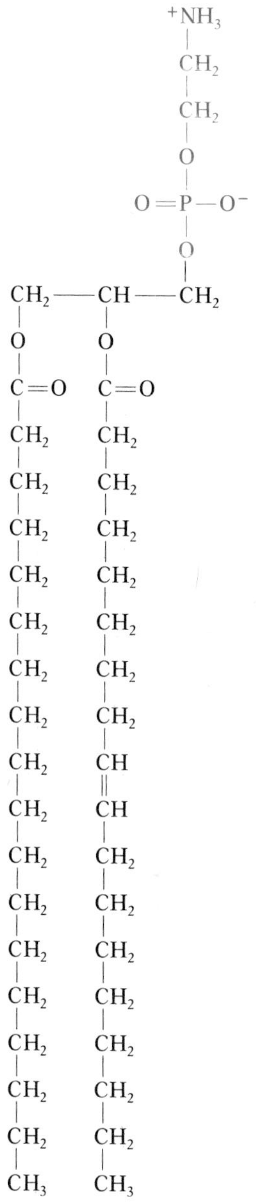

2.32   A phospholipid

The portion of the molecule with the phosphate and nitrogenous groups (color) is soluble in water, whereas the two hydrocarbon chains are not. This particular phospholipid, ethanolamine phosphoglyceride, is one of the two most abundant in higher plants and animals.

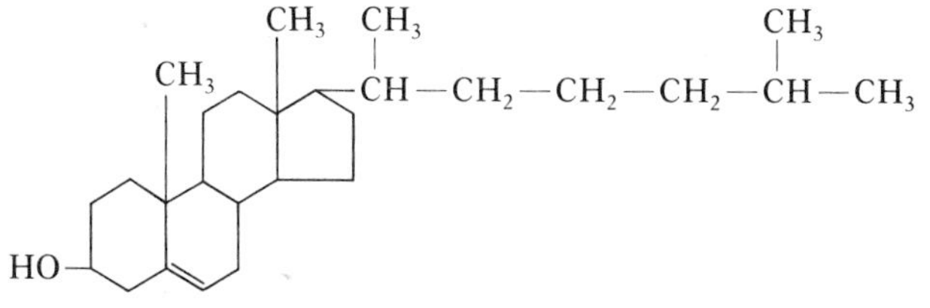

2.33   A steroid

All steroids have the same basic unit of four interlocking rings, but differ in their side groups. This particular steroid is cholesterol. (By convention, a hexagon signifies a six-carbon ring with its valences completed by hydrogens; see cyclohexane in Fig. 2.22. A pentagon signifies a five-carbon ring, also with hydrogens attached to the carbons.)

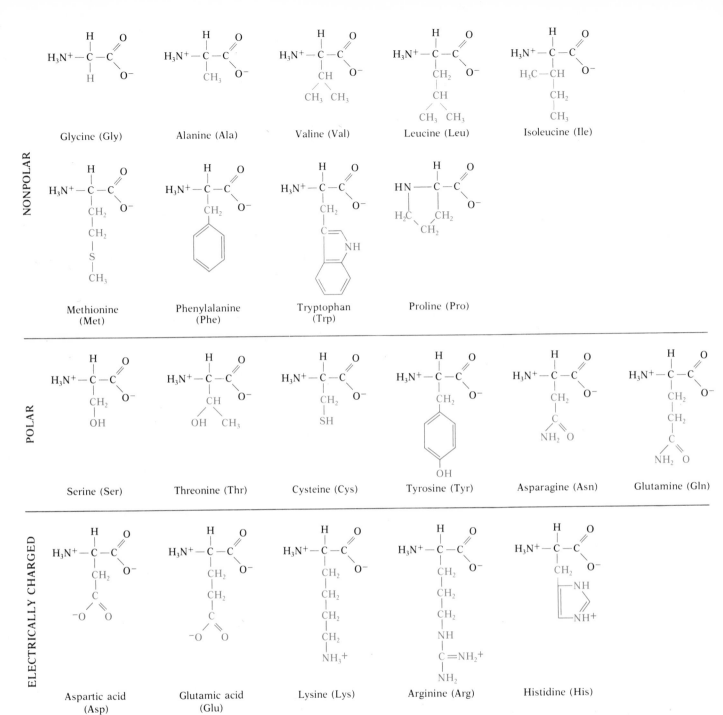

2.34 Structural formulas of the 20 amino acids common in proteins

The amino acids are shown in their ionized form. All have the same arrangement of a carboxyl group and an amino group attached to the α carbon; they differ in their R groups (shown in color).   Rows 1 and 2: The nine amino acids here have nonpolar R groups and are insoluble. (Glycine is an exception. Although its R group, a single hydrogen atom, is nonpolar, it is too small to outweigh the charge of the amino and carboxyl groups. The molecule therefore behaves more like a polar amino acid and is water-soluble.) Row 3: The six amino acids have polar R groups and are soluble.   Row 4: The five amino acids, with R groups ionized at intracellular pH levels, are electrically charged and thus water-soluble; the first two, being negatively charged, are acidic, whereas the last three, with a positive charge, are alkaline.

having lost a hydrogen ion, is negatively charged, and the amino group, having attracted an extra hydrogen ion, is positively charged:

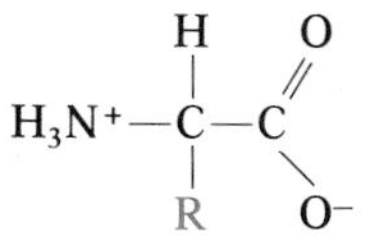

$$H_3N^+ - \underset{\underset{R}{|}}{\overset{\overset{H}{|}}{C}} - \overset{\overset{O}{\parallel}}{C} \diagdown O^-$$

The various α-amino acids differ in their side chains, or R groups. R may be very simple, as in glycine, where it is only a hydrogen atom, or it may be very complex, as in tryptophan, where it includes two ring structures. Twenty-odd different amino acids are commonly found in proteins; their structural formulas are shown in Fig. 2.34. The various R groups give each of the amino acids different characteristics, which, in turn, greatly influence the properties of the proteins incorporating them. For example, some amino acids are relatively insoluble in water, owing to R groups that are nonpolar at pH 6.5–7 (Fig. 2.34, top two rows), whereas other amino acids are water-soluble, because their R groups are polar (third row) or electrically charged (bottom row).

Proteins are long and complex polymers of the twenty-odd amino acids. The amino acid building blocks bond together by condensation reactions between the COOH groups and the NH_2 groups (Fig. 2.35). Such bonds are called **peptide bonds,** and the chains they produce are called **polypeptide chains.** The number of amino acids in a single polypeptide chain within a protein molecule is usually between 40 and 500, although shorter and longer chains sometimes occur. It must be emphasized that the variation is between polypeptides of different kinds; for any given kind of polypeptide, the chain length is constant.

Protein molecules often consist of more than one polypeptide chain. The chains may be held together by numerous weak bonds, especially hydrogen bonds; e.g. a single molecule of hemoglobin, the red oxygen-carrying protein in blood, is composed of four polypeptide chains linked by hydrogen bonds. Insulin, an important hormone secreted by the pancreas of vertebrates, exemplifies a protein whose polypeptide chains are held together by both

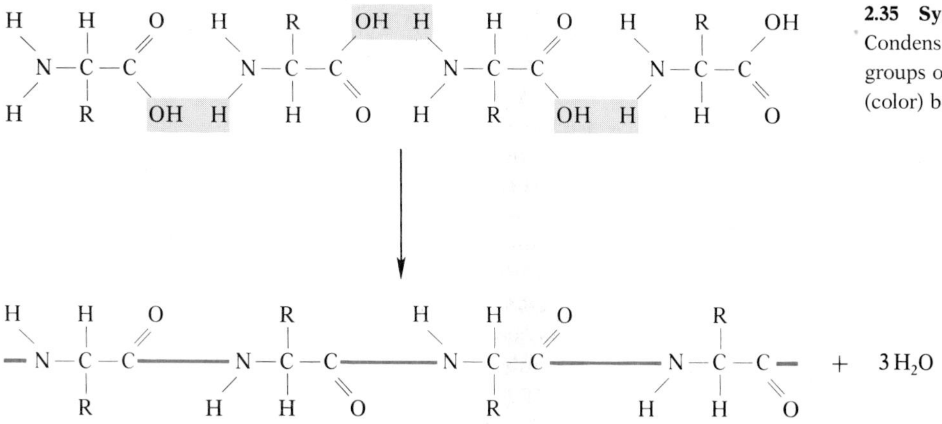

2.35 Synthesis of a polypeptide chain
Condensation reactions between the COOH and NH_2 groups of adjacent amino acids result in peptide bonds (color) between the acids.

2.36 The structural formula of cystine
A molecule of cystine, formed by linkage of two molecules of cysteine via a disulfide bond, is a sort of molecular twin. On either end is the characteristic α-amino acid arrangement of a carboxyl group and an amino group attached to the α carbon. Each end of the molecule can be independently incorporated into a polypeptide chain.

hydrogen and covalent bonds. The covalent bonds, called *disulfide bonds,* are between the sulfur atoms of two units of the amino acid cysteine (Fig. 2.34); two cysteines readily react with each other to form a symmetrical molecule called cystine (Fig. 2.36). As Fig. 2.37 shows, disulfide bonds can also link two parts of a single polypeptide chain, maintaining it in a bent or folded shape. Clearly, disulfide bonds play a very important role in linking the constituent polypeptide chains of some complex proteins, whose folding patterns they thus help determine.

We have so far discussed the so-called *primary structure* of protein molecules—the number of polypeptide chains and the number, type, and sequence of amino acids in each. Since all these aspects of primary structure can vary among proteins of different kinds, the potential number of different proteins is enormous, indeed nearly infinite. Here lies the answer to the question, how the many millions of different species of organisms can have proteins peculiar to each of them. In fact, in many cases individual organisms have their own unique proteins—a property we shall examine in a later chapter in connection with immunologic reactions.

Determining the primary structure   For a better understanding of the primary structure of proteins, it may be instructive to trace some of the steps by which Frederick Sanger and his colleagues at Cambridge University determined the structural formula for insulin, the first protein for which a structural formula could be written (Fig. 2.37). Insulin was well suited to be the object of pioneering work on protein structure; it can easily be obtained in pure form, and it is one of the smallest proteins known. Nevertheless, about ten years of exhaustive work were required before Sanger, in 1954, could feel confident that he finally knew the full sequence of amino acids in insulin. His achievement, a milestone in the history of biochemical knowledge, won him the Nobel Prize in 1958.

Sanger's method of determining the sequence of amino acids in the two polypeptide chains of insulin involved breaking the chains into fragments and then trying to establish how the pieces fitted together. If a polypeptide chain is heated in an acid solution for about 24 hours, all its peptide bonds are hydrolyzed, i.e. broken by the addition of a water molecule at the site of each bond, the amino acids being uncoupled in the process. The results can then be analyzed.

The technique used by Sanger for analyzing the amino acids was chromatography, a process utilizing the different affinities of the unknown substances for two other materials. From this procedure he could find out which

Chromatography is one of the most valuable techniques available for separating substances in a complex liquid or gaseous mixture, e.g. the amino acids in the mixture produced by completely hydrolyzing a polypeptide chain, or the various proteins in a sample of blood. There are many kinds of chromatography, but all share the same fundamental mechanism—exposure of the mixture simultaneously to two different substances, such as two immiscible solvents or a solvent and an adsorbent solid. (An adsorbent is a substance to the surface of which molecules of a gas, dissolved solute, or liquid tend to adhere.) Each solute in the mixture will be partitioned between the two substances in concentrations proportional to the relative affinities of the solute for those two substances. For example, if solute *A*, a material highly soluble in water but only minimally soluble in phenol, is shaken in a jar with water and phenol, the molecules of *A* will be partitioned between the two solvents in such fashion that most will be in the water phase and only a few in the phenol phase.

Paper chromatography is one of the simplest kinds of chromatography. Several drops of the mixture in some solvent other than water are placed near the bottom edge of a piece of moist filter paper. As the solvent (called the flowing phase) migrates up the paper by capillary action, those materials in the mixture that have a much higher affinity for the solvent than for the moist filter paper travel freely up the paper with the solvent. By contrast, materials with a much higher affinity for the moist filter paper (called the stationary phase) will quickly transfer from the flowing solvent to the stationary film of water and hence will not travel far up the paper. Materials with intermediate relative affinities for the flowing phase and for the stationary phase will travel intermediate distances. At the end of a measured time interval, the materials originally in the mixture will be at different places along the paper.

Column chromatography works in a similar fashion, except that, instead of filter paper, a glass column packed with some hydrated adsorbent material such as starch or silica gel is used. The test mixture, in a nonaqueous solvent, is poured into the top of the column and allowed to filter downward (or it is forced down by a pump). Depending on the relative affinities of the various components in the test mixture for the flowing solvent phase and the stationary moist adsorbent phase, each component will tend to move at its own rate. If enough solvent is poured into the column to keep the separated materials moving, they will emerge from the bottom of the column at different times and can be collected in different containers, ready for analysis.

More sophisticated chromatographic techniques, making possible very precise and very rapid separation and analysis of materials, often rely on tubes filled with electrically charged resins and on carefully manipulated pH values to encourage formation of ionic bonds between the test materials and the stationary phase (the test materials exchange places with other materials loosely bonded to the resins). Though these more elaborate techniques were not available to Sanger when he performed his classic study of insulin, they greatly facilitate analysis of proteins and other complex substances by present-day researchers.

For the amino acid content of proteins, the entire process of analysis has been automated, and analyses that would have demanded years of work in Sanger's day can be performed in two or three hours. The mixture of amino acids obtained by complete hydrolysis of a polypeptide chain is forced under pressure through ion-exchange columns, and each component is caught by an automatic fraction collector as it emerges from the column. A special reagent (ninhydrin) that reacts with amino acids to produce a blue color is added automatically to each container, and the intensity of color is then measured by a photometer attached to a recorder. The graph produced by the recorder tells, by the position of the peaks on the graph (which are determined by times of emergence), what amino acids were present in the original test mixture, and it tells the quantities of each by the heights of the peaks (which are determined by the intensity of the blue color, which, in turn, is proportional to the amount of amino acid in the container).

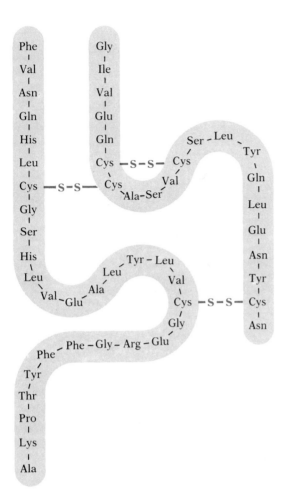

2.37 The structure of beef insulin

The molecule consists of two polypeptide chains joined by two disulfide linkages. There is also one disulfide linkage within the shorter chain (right). Hydrogen bonds (not shown) between the chains and between segments of the same chain are also present.

amino acids insulin contained and in what amounts. To learn the amino acid sequence, he had to use other approaches.

By less drastic treatment of insulin with hydrolyzing agents, he could preserve some of the peptide bonds and thus obtain many protein fragments consisting of two, three, four, five, or more amino acids. He analyzed these fragments for their amino acid content, utilizing particularly a technique that enabled him to determine which amino acid was on the end with the free amino group. After analyzing vast numbers of such pieces, he attempted to fit them together in proper sequence, by looking for fragments with regions of apparent overlap. For example, he found the following two fragments that seemed to have overlapping sequences at their ends:

Leu–Val–Cys–Gly–Glu–Arg–Gly–Phe–Phe
Gly–Phe–Phe–Tyr–Thr–Pro–Lys

He reasonably concluded that one part of the insulin molecule contained the sequence

Leu–Val–Cys–Gly–Glu–Arg–Gly–Phe–Phe–Tyr–Thr–Pro–Lys

He then hunted for other fragments that overlapped this sequence, so that he could extend it. After long and laborious investigations of this sort, he finally determined the entire amino acid sequence of the two polypeptide chains. Later, by other techniques, he determined the positions of the three disulfide bonds.

Since Sanger's discovery of the structural formula of insulin—after more than a century of effort by scientists to learn the composition and structure of proteins—a number of other proteins have been similarly elucidated. The more recent work has been aided, as you would expect, by many technological advances, especially by rapid methods for determining the amino acid sequence in the relatively short chain fragments produced by partial hydrolysis. But the basic approach—complete hydrolysis and amino acid analysis, followed by partial hydrolysis and fragment matching—remains the same.

The spatial conformation of proteins Proteins are not laid out simply as two-dimensional chains of amino acids, as our discussion so far might seem to imply. Instead, they are coiled and folded into very complex spatial patterns, called conformations, which play a crucial role in determining the distinctive biological properties of each protein.

Let us first consider the simplest level of conformation, namely the spatial relationship of each amino acid to its immediate neighbors in the polypeptide chain. In 1951 Linus Pauling and Robert B. Corey of the California Institute of Technology found that, although in principle there can be unlimited rotation around the bonds on both sides of the α carbon in each amino acid, in fact only a very few angles are assumed by these bonds—those angles leading to energetically favorable and therefore stable arrangements of the molecule. Because of the tendency for the same bond angles to be repeated in successive amino acids, proteins are characterized by a limited number of recurrent structural patterns, called *secondary structure.*

One set of stable recurrent angles within the amino acids of a polypeptide chain causes the chain to be twisted into an *alpha helix* (a helix may be

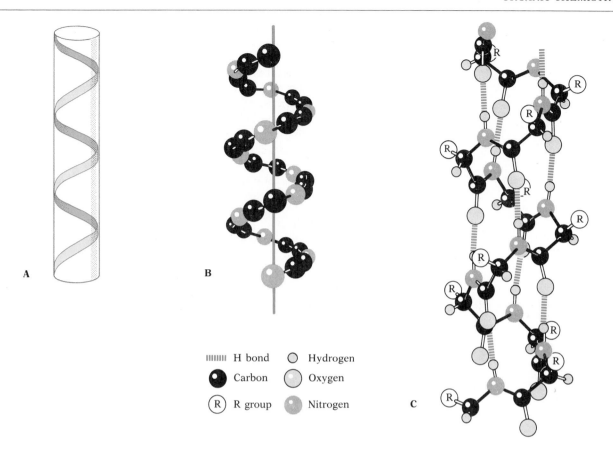

⦀⦀⦀ H bond	◯	Hydrogen
⬤ Carbon	◯	Oxygen
Ⓡ R group	◯	Nitrogen

visualized as a chain wound around a regular cylinder). In protein each complete turn of the helix takes up approximately 3.6 amino acid units of the polypeptide chain (Fig. 2.38A). The chain is held in this helical shape by hydrogen bonds formed between the nitrogen of one amino acid and the oxygen of the third amino acid beyond it in the polypeptide chain—which is the amino acid next to it in the axial direction of the helix (Fig. 2.38C).

The helical pattern is seen at its simplest in some *fibrous proteins.* One category of these insoluble proteins, extensively used in studies of protein structure, comprises the *keratins,* which provide the structural elements for many of the specialized derivatives of skin cells. Some keratins with α-helical secondary structure, such as nails, hooves, and horns, are hard and brittle. Others, such as hair and wool, are soft and flexible and can easily be stretched (especially when moistened and warmed). The stretching is possible because the intrachain hydrogen bonds are easily broken, and the polypeptide chains can then be pulled out of their compact helical shapes into a more extended form. The chains tend to contract to their normal length, with re-formation of the hydrogen-bonded α helix, when the tension on them is released (or when they are dried and cooled).

Another possible set of stable recurrent angles within a polypeptide chain gives rise to the second major type of secondary structure. This type, desig-

2.38   Alpha-helical secondary structure of some proteins

(A) The helix may be visualized as a chain wrapped around a regular cylinder.   (B) The backbone of a polypeptide chain (i.e. the repeating sequence of N—C—C—N—C—C—N—C—C—) is shown coiled in a helix (all other atoms and R groups are omitted). It can be seen that it takes approximately 3.6 amino acid units (N—C—C) to form one complete turn of the helix. The vertical rod is meant to aid visualization. (C) A ball-and-stick model of the α helix of a protein, showing some of the intrachain hydrogen bonds that help stabilize the helix; the hydrogen bonds shown extend between the nitrogen of one amino acid and the oxygen of the third amino acid beyond it in the polypeptide chain.   [C, redrawn from A. L. Lehninger, *Biochemistry,* 2nd ed., Worth Publishers, New York, 1975, p. 129. Based on G. H. Haggis *et al., Introduction to molecular biology,* Wiley, 1964.]

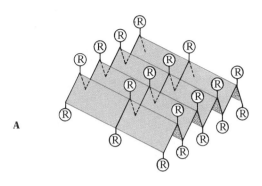

A

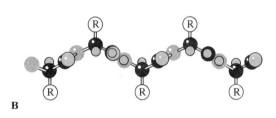

B

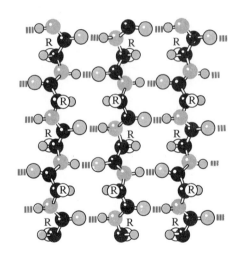

C

⦀⦀ H bond	⚪ Hydrogen	
⚫ Carbon	⚪ Oxygen	
Ⓡ R group	⬤ Nitrogen	

nated beta (β) structure, is often called the **pleated sheet.** In this pattern, also seen at its simplest among the keratins, many side-by-side polypeptide chains are cross-linked by interchain hydrogen bonds (Fig. 2.39). The resulting arrangement is flexible and strong, but resists stretching, because the polypeptide chains are already almost fully extended. Probably the best-studied β-keratin is silk, but other examples include spider webs and the scales, claws, and beaks of reptiles and birds.

In addition to the keratins, another kind of fibrous protein, with its own distinctive secondary structure, should be mentioned here, because it is the most abundant protein in higher vertebrates. This is **collagen,** which may constitute one third or more of all the body protein; it is especially abundant in skin, tendons, ligaments, and bones, and in the cornea of the eye. A molecule of collagen is composed of three polypeptide chains, each first helically coiled and then wound around the other two to form a triple helix (Fig. 2.40). What facilitates the intertwining of the three chains is that every third amino acid in the chains is glycine, whose R group, being only a single hydrogen atom (Fig. 2.34), takes up very little room. The chains are held together by hydrogen bonds. Collagen fibers are exceedingly strong and very resistant to stretching.

Far more complex in spatial conformation than the fibrous proteins are the **globular proteins,** whose polypeptide chains are folded into complicated spherical or globular shapes (Fig. 2.41). Typically, these soluble proteins—which include enzymes, proteinaceous hormones, antibodies, and most blood proteins—are made up of sections of α helix interspersed with nonhelical regions. The protein myoglobin, for example, shown in Fig. 2.41, consists of eight major sections of α helix connected by short regions of irregular coiling. At each nonhelical region, the three-dimensional orientation of the polypeptide changes, giving rise to the protein's characteristic folding pattern. This three-dimensional folding pattern, which is superimposed on the secondary structure, is called **tertiary·structure.**

When a globular protein is composed of two or more independently folded subunits—separate polypeptide chains loosely held together, usually by weak bonds—the manner in which the already folded subunits fit together is called **quaternary structure** (Fig. 2.42).

Several aspects of a protein's primary structure (i.e. its amino acid content and sequence) contribute to producing its tertiary and quaternary structure. If, for example, a polypeptide chain contains two cysteine units, the intrachain disulfide bond joining them introduces a fold in the chain (Fig. 2.37).

2.39 Pleated-sheet secondary structure of some proteins
(A) Diagrammatic representation of three parallel polypeptide chains in β conformation, with the imaginary pleated sheet between them shown in color.   (B) Edge and (C) top views of a ball-and-stick model of polypeptide chains in β conformation. [B and C, redrawn from H. D. Springall, *The structural chemistry of proteins,* Butterworths & Co. and Academic Press, 1954.]

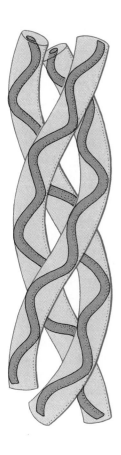

2.40 Model of a portion of a molecule of collagen
Three polypeptide chains, each helically coiled, are wound around each other to form a triple helix. The "sheaths" here and in Figs. 2.41 and 2.42 are intended as a reminder that the molecules consist not merely of a backbone, but also of R groups, which give the molecules volume.

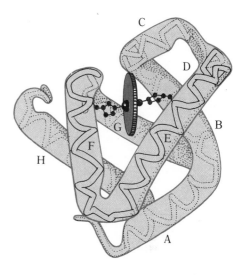

2.41 Spatial conformation of a molecule of myoglobin
Myoglobin, a globular protein related to hemoglobin and, like hemoglobin, characterized by a strong affinity for molecular oxygen, is a single complexly folded polypeptide chain of 151 amino acid units; attached to the chain is a nonproteinaceous prosthetic group called heme. The polypeptide chain consists of eight sections of α helix (labeled A through H) with nonhelical regions between them. These nonhelical regions are a major factor in determining the tertiary structure of the molecule—i.e. the way the helical sections are folded together. (Section D cannot be seen in this drawing, because it is oriented perpendicular to the plane of the page.)

2.42 Quaternary structure of hemoglobin
A single molecule of hemoglobin is composed of four independent polypeptide chains, each of which has a globular conformation and its own prosthetic group. The spatial relationship between these four subunits —i.e. the way the four fit together—is called the quaternary structure of the protein.

2.40, 2.41, 2.42 Adapted by permission from *The structure and action of proteins* by Richard E. Dickerson and Irving Geis, W. A. Benjamin, Inc., Menlo Park, Calif., Publisher. Copyright 1969 by Dickerson and Geis.

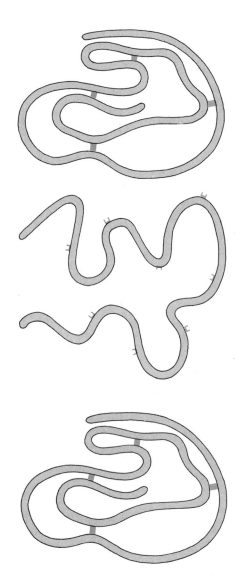

2.43 Denaturation and renaturation of ribonuclease
When ribonuclease, a normally globular protein (top), is denatured, with both its weak bonds and its four intrachain disulfide bonds (color) broken, it unfolds into an irregularly coiled state (middle). In this denatured condition the ribonuclease lacks its characteristic enzymatic activity. When the denaturing agents are removed and favorable conditions restored, the protein spontaneously refolds into its native conformation (bottom); even the four disulfide bonds are correctly re-formed.

Or if the chain contains proline, a kink or bend occurs, because the structure of proline is such that it cannot conform to the geometry of an α helix; proline, though one of the building-block units of protein, is not technically a true amino acid, having an R group that circles around and links with its amino group (Fig. 2.34).

The distinctive properties of the various R groups of the amino acids also impose constraints on the shape of the protein. For example, hydrophobic groups tend to be close to each other in the interior of the folded chains—as far away as possible from the surrounding water—whereas hydrophilic groups tend to be on the outside, in contact with the water. Polar R groups, such as that of tyrosine (Fig. 2.34), tend to assume positions where they can form hydrogen bonds with other such groups; similarly, electrically charged groups tend to be in positions where they can form ionic bonds with oppositely charged groups (e.g. aspartic acid with lysine). Thus the various kinds of weak bonds we discussed earlier all play important roles in stabilizing the tertiary structure of proteins.

As this discussion suggests, there are compelling reasons to believe that the primary structure of a protein determines its spatial conformation. More specifically, it appears that the primary structure determines the energetically most favorable, and therefore most stable, possible arrangement of the polypeptide chains. Hence the question, long perplexing to biochemists, how conformation is specified when a protein is being synthesized becomes synonymous with the question how amino acid content and sequence are specified—a question no longer so perplexing to scientists, as we shall see in a later chapter.

Further support for the idea that primary structure determines conformation comes from studies of *denatured* proteins—proteins that have lost most of their secondary, tertiary, and quaternary structure, and with it their normal biological activity, through exposure to high temperature or extreme pH (Fig. 2.43). That a denatured protein should lack the characteristic biological activity of the natural protein is an indication that its conformation is functionally essential. Since conformation is dependent in large part on weak bonds (which are very sensitive to temperature and pH), it is easily disrupted by anything that breaks or alters those bonds; it is stable only within a limited range of temperature and pH. Even brief exposure to high temperatures (usually above 60°C) or to extremes of pH will cause denaturation of a globular protein. But it has been discovered that under favorable test-tube conditions some denatured proteins can spontaneously regain their native three-dimensional conformation; they can refold, and recover their normal biological activity. Since only the primary structure is available to dictate the folding pattern in such cases, it alone must be sufficient to determine all other aspects of the protein's structure.

Conjugated proteins Attached to some proteins are nonproteinaceous groups called *prosthetic groups;* an example is the heme group of myoglobin, a ringlike structure with an iron atom in its center (Fig. 2.41). Prosthetic groups may be as simple as a single metal ion bonded to the polypeptide chain; they may be sugars or other carbohydrate entities; or they may be of lipid form. Whatever the nature of the prosthetic group, its presence alters

X-ray crystallography has been an invaluable tool in discovering the three-dimensional conformation of proteins. The first step is to crystallize the protein. (That proteins can be crystallized indicates that all the molecules have the same shape and are ordered in a regular array.) Then X rays are sent through the crystal while it is being turned in various directions. Now, X rays passing through a crystal do not travel in a straight line, but are scattered, or diffracted, by the electrons of the atoms in the crystal. The pattern of the scattering is a function of the angle at which the X rays strike the planes of the crystal—themselves a function of the arrangement of the atoms in the crystal. This pattern can be recorded if the X rays, as they emerge from the crystal, are made to impinge on a photographic plate. In crystallography, then, the crystal being studied is rotated in a carefully planned way in an X-ray camera, so that its sets of planes come into correct reflecting position in an orderly sequence. The successive scattering patterns thus produced make their imprint on the photographic plates.

Since crystals of different substances have different atomic (and hence electronic) arrangements, each substance produces its own characteristic diffraction patterns. For a single protein, tens of thousands of diffraction patterns are recorded, and their details fed into high-speed computers for analysis. From the results, investigators try to deduce the position of each atom in the three-dimensional molecule. The process is so complex that it is practical only if some preliminary models of the molecules are already available from other evidence, so that the diffraction results can be compared with predictions based on the models. So far, the detailed conformation of only 25 proteins has been worked out.

the properties of the protein in important ways; e.g. without their heme groups, myoglobin and hemoglobin lose their normal high affinity for molecular oxygen.

NUCLEIC ACIDS

Nucleic acids constitute a fourth major class of organic compounds crucial to all life. They are the materials of which genes, the units of heredity, are composed. They are also the messenger substances that convey information from the genes in the nucleus to the rest of the cell, information that not only determines the structural attributes of the cell but also regulates its ongoing functional activities.

Like polysaccharides and proteins, nucleic acid molecules are long polymers of smaller building-block units. In this case the building blocks, called **nucleotides,** are themselves composed of still smaller constituent parts: a five-carbon sugar, a phosphate group, and an organic nitrogen-containing base. Both the phosphate group and the nitrogenous base are covalently bonded to the sugar (Fig. 2.44).

Deoxyribonucleic acid   This nucleic acid, commonly called **DNA** for short, is the one genes are made of. Four different kinds of nucleotide building blocks occur in DNA. All have, as the name suggests, deoxyribose as their sugar component, but they differ in their nitrogenous bases, which may be one of four different substances. Two of these, **adenine** and **guanine,** are

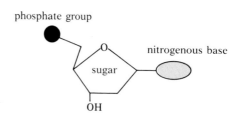

2.44   Diagram of a nucleotide
A phosphate group and a nitrogenous base are attached to a five-carbon sugar.

63

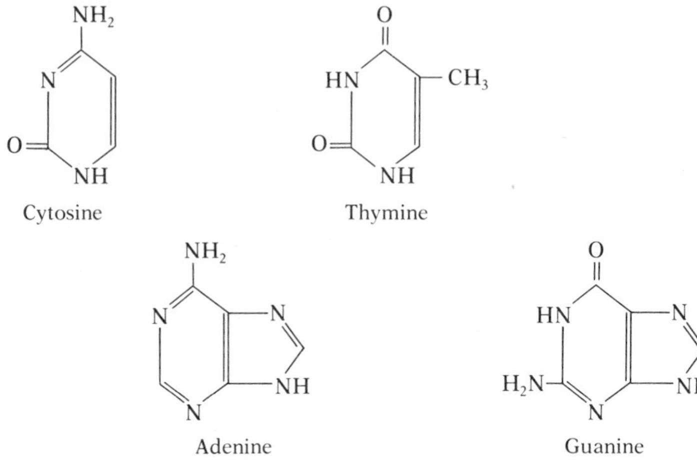

2.45 The four nitrogenous bases in DNA
The two single-ring bases, cytosine and thymine, are pyrimidines; the two double-ring bases, adenine and guanine, are purines.

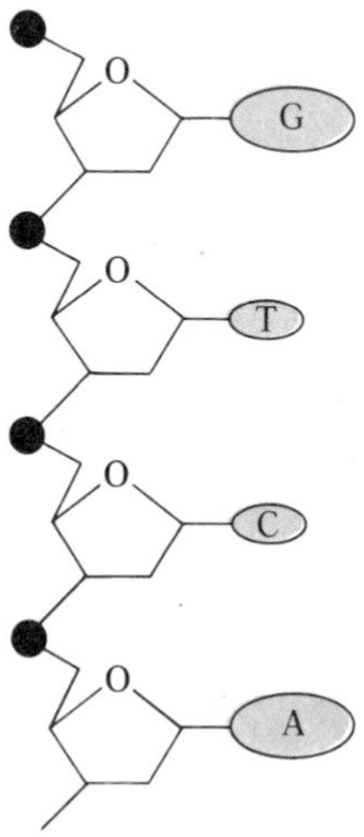

2.46 Portion of a single chain of DNA
Nucleotides are hooked together by bonds between their sugar and phosphate groups. The nitrogenous bases (G, guanine; T, thymine; C, cytosine; A, adenine) are side groups.

double-ring structures of a class known as purines; the other two, *cytosine* and *thymine,* are single-ring structures known as pyrimidines (Fig. 2.45).

The nucleotides within a DNA molecule are bonded together in such a way that the sugar of one nucleotide is always attached to the phosphate group of the next nucleotide in the sequence (Fig. 2.46). Thus a long chain of alternating sugar and phosphate groups is established, with the nitrogenous bases oriented as side groups off this chain. The sequence in which the four different nucleotides occur is essentially constant in DNA molecules of the same kind, but differs in different kinds of DNA molecules. It is this sequence that determines the specificity of each type of DNA. It is, in fact, the sequence of the nucleotides in DNA that encodes hereditary information, which is expressed through control of protein synthesis. More particularly, the sequence of nucleotides in DNA determines the sequencing of amino acids (i.e. the primary structure) in proteins; we shall examine this process in considerable detail in Chapter 15.

DNA molecules do not ordinarily exist in the single-chain form shown in Fig. 2.46. Instead, two such chains, oriented in opposite directions, are arranged side by side like the uprights of a ladder, with their nitrogenous bases constituting the cross rungs of the ladder (Fig. 2.47). The two chains are held together by hydrogen bonds between adjacent bases. Finally, the entire double-chain molecule is coiled into a double helix (Fig. 2.48).

The regular helical coiling and hydrogen bonding between bases impose two extremely important constraints on how the cross rungs of the ladder-like DNA molecule can be constructed. First, each rung must be composed of a purine (double ring) and a pyrimidine (single ring); only in this way will all cross rungs be of the same length and permit formation of a regular helix. Second, if the purine is adenine, then the pyrimidine must be thymine, and, similarly, if the purine is guanine the pyrimidine must be cytosine; only these two pairs are capable of forming the required hydrogen bonds (Fig. 2.47). Since it does not matter, however, in which order the members of a pair appear (A–T or T–A; G–C or C–G), the double-chain molecule can have four different kinds of cross rungs, as shown in Fig. 2.47.

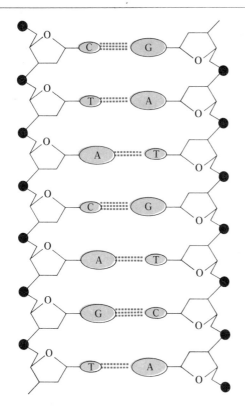

2.47 Portion of a DNA molecule uncoiled
The molecule has a ladderlike structure, with the two uprights composed of alternating sugar and phosphate groups and the cross rungs composed of paired nitrogenous bases. Each cross rung has one purine base (large oval) and one pyrimidine base (small oval). When the purine is guanine (G), then the pyrimidine with which it is paired is always cytosine (C); when the purine is adenine (A), then the pyrimidine is thymine (T). Adenine and thymine are linked by two hydrogen bonds (striped bands), guanine and cytosine by three. Note that the two chains run in opposite directions, i.e. that the free phosphate is at the upper end of the left chain and at the lower end of the right chain.

Ribonucleic acid A second important category of nucleic acids comprises the ribonucleic acids, or *RNA* for short. There are several types of RNA, each with a different role in protein synthesis. Some act as messengers carrying instructions from the DNA genes to the sites of protein synthesis in the cell. Others are structural components of cytoplasmic organelles, called ribosomes, on which the process of protein synthesis takes place. Still others transport amino acids to the ribosomes, so that they may be incorporated into proteins. We shall discuss each of these types of RNA in much more detail in Chapter 15. Here let us note only that all types of RNA differ from DNA in three principal ways: (1) The sugar in RNA is ribose, whereas that in DNA is deoxyribose (it is the sugars that give the two nucleic acids their different first initials). (2) Instead of thymine, one of the four nitrogenous bases of DNA, RNA contains a very similar base called *uracil.* (3) RNA is ordinarily single-stranded, whereas DNA is usually double-stranded.

CHEMICAL REACTIONS

In previous sections we have mentioned several types of chemical reactions that take place within organisms: condensation reactions between simple sugars to form polysaccharides, and hydrolysis of polysaccharides back to simple sugars; condensation reactions of fatty acids and glycerol to form fats, and the reverse hydrolysis; condensation of amino acids to form polypeptide chains and proteins, and the reverse hydrolysis. But we have said nothing about the conditions under which these reactions will take place. It is now time for a brief examination of those conditions.

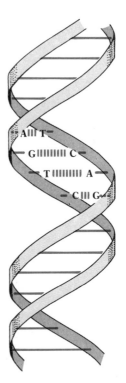

2.48 A model of the DNA molecule
The double-chained structure is coiled in a helix. As shown in detail in the second segment, it consists of two polynucleotide chains held together by hydrogen bonds (striped bands) between their adjacent bases.

Let us examine further the free-energy change in a chemical reaction, using G as the symbol for free energy and Δ (the Greek letter delta) as the symbol for change. An exergonic reaction is one in which $\Delta G < 0$ (read: ΔG is less than zero), i.e. in which the change is negative; the products of the reaction contain less free energy than the reactants. Conversely, an endergonic reaction is one in which $\Delta G > 0$ (ΔG is greater than zero), i.e. in which the change is positive; the products contain more free energy than the reactants.

The free energy lost in an exergonic reaction may be transformed into heat or dissipated in increased disorder in the system. If the heat content of the system, **enthalpy**, is symbolized by H and the amount of disorder, **entropy**, by S, the changes in free energy in the system (ΔG) can be related to the changes in heat content (ΔH) and in the amount of disorder (ΔS) by the following equation:

$$\Delta G = \Delta H - T\,\Delta S$$

Here T is the absolute temperature in degrees Kelvin. (For reference: 0°K, often called absolute zero, equals -273.16°C, and 294.15°K equals 21°C, a normal room temperature.) The ΔS term must be multiplied by T because disorder increases with increasing temperature.

What accounts for the negative ΔG in exergonic reactions, which, as we have said, proceed downhill energetically and can therefore occur spontaneously? Clearly, the right side of the above equation takes a negative value if much heat energy is lost from the system (ΔH has a large negative value) or if the disorder of the system increases greatly ($T\,\Delta S$ has a large positive value), or both.

The great majority of exergonic biochemical reactions are driven by loss of enthalpy (negative ΔH); i.e. they give off heat. Such reactions are sometimes designated as exothermic, while reactions that absorb heat are called endothermic. There are, however, a few exergonic and therefore spontaneous reactions, such as the dissolving of NaCl in water and the melting of ice, that absorb heat, i.e. that are endothermic. In these cases there is so much increase in entropy, or disorder, that the large $T\,\Delta S$ term outweighs the positive ΔH term and yields a negative ΔG.

ENERGETIC CONDITIONS FOR CHEMICAL REACTIONS

Instead of talking at this point about condensation reactions or hydrolysis reactions or any other specific type of reaction, let us consider some hypothetical generalized reactions. Suppose, for example, that two substances, A and B, can react with each other to form a new compound AB:

$$A + B \longrightarrow AB$$

What determines whether this reaction will tend to take place spontaneously?

The answer to this question turns on a concept of physics—**energy,** which is defined as the capacity to do work. **Free energy** (as the term is used in chemistry and biology) is the energy available in a system for doing work under conditions of constant temperature and pressure. Reactions that release free energy are said to be **exergonic;** reactions that require the addition of free energy from an external source are said to be **endergonic.** All spontaneous reactions are exergonic, proceeding downhill, so to speak, as far as free energy is concerned. This rule is a particular expression of the principle that all systems, of whatever kind, tend to minimize their free-energy content.

The reaction of A and B to form AB will proceed spontaneously only if it is exergonic, i.e. only if it releases free energy. But if exergonic reactions are the only ones that can take place spontaneously, how can organisms carry out endergonic reactions, such as synthesis of the many complex energy-rich molecules necessary for life? Endergonic reactions are made possible within living organisms by close coupling with exergonic reactions; some of the energy released by the exergonic reaction of a coupled pair is used to drive the endergonic reaction:

Coupled reactions
$$\begin{cases} XY \longrightarrow X + Y + energy & \text{Exergonic} \\ K + L + energy \longrightarrow M + N & \text{Endergonic} \end{cases}$$

Reaction coupling does not contradict the rule that systems tend to minimize their free-energy content, for only a portion of the energy released by the exergonic reaction is stored in the endergonic reaction. If the two reactions are considered together, the net change is still energetically downhill.

FACTORS THAT INFLUENCE THE RATE OF CHEMICAL REACTIONS

That exergonic reactions can proceed spontaneously does not necessarily mean that they proceed rapidly. Indeed, at physiological temperatures and pressures, most exergonic biochemical reactions are so slow as to be practically negligible. For a molecule to react with another substance, some bonds must be broken, but even the weakest covalent bonds are too strong to break easily.

The need for activation energy For covalent bonds to be broken, they must temporarily acquire extra energy; the bonded atoms must be pushed farther apart. This extra energy necessary to initiate a reaction—known as *activation energy* (E_a)—makes the bonds less stable and thus more reactive. The molecules, once activated, proceed to react, releasing, in an exergonic reaction, both the extra energy they briefly possessed during activation and the free energy of reaction (ΔG) characteristic of that reaction (Fig. 2.49). The need for activation energy acts as a barrier to rapid spontaneous and uncontrolled reactions between covalently bonded molecules that must coexist within living cells. Without such barriers, the complex high-energy molecules (e.g. carbohydrates, lipids, proteins, nucleic acids) on which life depends would be unstable and would promptly break down.

Beneficial as activation-energy barriers may be in conferring stability on living systems, the barriers must be overcome in some way if organisms are to carry out the chemistry of life. For example, how does the body hydrolyze starch to sugar? We all know that simple mixing of starch and water is not sufficient to produce significant hydrolysis, even though the reaction is an exergonic one. How can organisms make normally exceedingly slow reactions take place with the extreme rapidity often necessary in a functioning cell?

The effect of heating In the laboratory a standard way of speeding up chemical reactions—i.e. of supplying activation energy—is to apply heat. Indeed, this procedure is used in everyday life, e.g. in the application of extra

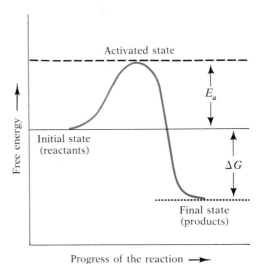

2.49 The energy changes in an exergonic reaction
Although the reactants are at a higher energy level than the products, the reaction cannot begin until the reactants have been raised from their initial energy state to an activated state by the addition of activation energy (E_a). It is the need for activation energy that ordinarily prevents high-energy substances from breaking down, and hence makes them stable; the higher the activation-energy barrier, the slower the reaction and thus the more stable the substance. When activation energy is available, the reactants form a temporary and unstable activated complex, which breaks down to yield the end products of the reaction; in the process both the activation energy and the free energy of reaction (ΔG) are released.

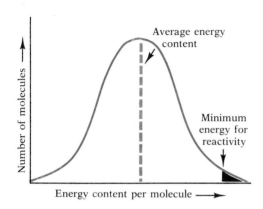

2.50   Distribution of energy among molecules at ordinary physiological temperatures
The vast majority of the molecules possess too little energy for reactivity. Only a few (black area at right) are above the minimum energy level for reactivity.

heat from a match or some other hot object to a piece of paper (or coal, or wood, or gasoline) to make it burn (i.e. combine with oxygen). Once ignited, the material continues to burn, because some of the heat energy released in this exergonic reaction by the first molecules that react supplies the activation energy for other molecules of the material to undergo the same reaction.

Extra heat, then, can supply activation energy. Now, all forms of energy involve motion of some sort. For heat energy, the motion is that of the atoms and molecules of which the substance in question is composed. At all temperatures above absolute zero (−273°C), all atomic and molecular particles possess a vibratory randomly directed type of motion, often called thermal agitation. Let us suppose the molecules of two substances, A and B, are moving in this fashion within the same space. Collisions between them will sometimes occur, and they will have an opportunity to react. They will not react, however, unless they collide with enough force to form an activated complex—in short, unless the collision provides the necessary activation energy. At physiological temperatures, only a very tiny percentage of the molecules possess sufficient energy of thermal agitation to collide with enough force to form activated complexes (Fig. 2.50). But thermal agitation increases as the temperature of the system rises. At higher temperatures, therefore, more molecules will collide per unit time; moreover, a greater percentage of the collisions will occur with sufficient force to produce activation. This is the reason why the application of heat is so effective in speeding up chemical reactions.

Heating, a fine way to accelerate reactions in the chemist's test tube or when something is to be burned, clearly cannot be the principal method of speeding up reactions inside living organisms. The amount of heat that would be necessary for the high rates of biochemical reactions would destroy the organisms. The chemistry of life is "cold" chemistry, and we must search further for an explanation.

The effect of concentration   As Fig. 2.50 shows, even at ordinary temperatures a few molecules possess sufficient energy to be reactive. It follows that anything that could increase the probability of collisions between them would increase the rate of reaction. An obvious way to increase the probability of collisions would be to increase the concentrations of the reactants, and hence of the small proportion of energized molecules; in other words, the more energized molecules per unit volume, the more collisions between them per unit time. This concept is sometimes formalized as the Law of Mass Action, which says that, all other conditions being constant, the rate of reaction is proportional to the concentrations of the reactants.

Let us now suppose that the reaction

$$A + B \longrightarrow C + D$$

has been going on for some time and that high concentrations of the products, C and D, are accumulating, while the concentrations of A and B are dwindling. This means that the probability of collisions between C and D is increasing, while that for A and B is decreasing. When the molecular concentrations of C and D become greater than those of A and B, there may actually be more collisions per unit time between product molecules than between reactant molecules. What happens then? Does the direction of the reaction become reversed? The answer is: Perhaps. In principle, all chemical reactions are reversible. It is possible, therefore, that as the concentrations of the products (C and D) start rising, these will begin to react to form A and B. At first, there will be very little C and D and much A and B. According to the Law of Mass Action, the forward reaction (written with an arrow pointing toward the right) would proceed at a much higher rate than the back reaction (written with an arrow pointing toward the left). The relative rates can be symbolized by the lengths of the arrows:

$$A + B \rightleftharpoons C + D$$

But later, when the concentrations of the substances on the two sides of the equation are approximately equal, the forward and back reactions, obeying the Law of Mass Action, would proceed at the same rate:

$$A + B \rightleftharpoons C + D$$

When such a point is reached, how can the concentrations of C and D increase further? Isn't the system now in equilibrium? If no energy considerations were involved, the answer would be yes. If the energy content of C and D were exactly equal to that of A and B, equilibrium would be reached when the concentrations of C and D were equal to the concentrations of A and B. But suppose the forward reaction is slightly exergonic and releases a small amount of free energy. Under these conditions, the equilibrium point is pushed to the right; the loss of energy means that the concentrations of C and D must be higher than the concentrations of A and B before the rate of the back reaction equals that of the forward reaction. If the reaction is highly exergonic, the equilibrium point is pushed so far to the right that, for all practical purposes, there will be no back reaction; such essentially irreversible reactions are common in biological systems. If we consider energy one of the products of the reaction, such a result agrees with the Law of Mass Action. Because one of the products (i.e. energy) is being lost from the system, the concentration of products can never be sufficient to drive a back reaction. A similar effect occurs when one or more of the material products precipitates out of solution as a relatively insoluble solid or when a product escapes as a gas.

Although changes in the concentrations of reactants sometimes play an important role in determining the rates at which reactions occur in living organisms, they cannot explain the extreme rapidity of most biochemical reactions, whose rates often far exceed anything that could be produced merely by increasing concentrations. The explanation obviously lies elsewhere.

The effect of catalysts   A simple mixture of hydrogen gas (H_2) and oxygen gas (O_2) shows no appreciable reaction. If, however, a small quantity of finely divided platinum is added, an explosive reaction takes place, and water is formed. At the end of the reaction, the platinum is still present, unchanged.

A substance that, like the platinum, speeds up a reaction but is itself unchanged when the reaction is over (it may have been temporarily changed during the reaction) is known as a *catalyst.* It must be emphasized that a

2.51   Reduction of necessary activation energy by catalysts

The activation energy (E_a) necessary to initiate the reaction is much less in the presence of a catalyst than in its absence. It is this lowering of the activation-energy barrier by enzyme catalysts that makes possible most of the chemical reactions of life. Note that the free energy of reaction (ΔG) is unchanged by the catalyst; i.e. it is the same for both the catalyzed and the uncatalyzed reaction; only the activation energy is changed.

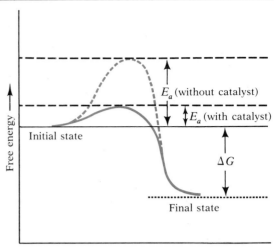

catalyst affects only the rate of reaction; it simply speeds up reactions that are thermodynamically possible to begin with. No catalyst can alter the direction of a reaction, its final equilibrium, or the reaction energy involved.

In terms of our discussion, catalysts decrease the activation energy needed for reactions to take place. *The principle of reduction of activation energy by catalysts is fundamental to life.* Here is the explanation for the "cold" chemistry in living things. The vast majority of chemical reactions within organisms are catalyzed by special catalysts called enzymes, which are globular proteins.

ENZYMES

Enzyme specificity and the concept of active site   Most of the inorganic catalysts used in the chemical laboratory are relatively nonspecific. For example, platinum, often used to catalyze the formation of water from hydrogen gas and oxygen gas, will catalyze almost any reaction in which H_2 is one of the reactants, and reactions of other materials as well. By contrast, enzymes are characteristically very specific, in two ways: (1) Most will interact with only one set of reactants (customarily called ***substrates***), or occasionally with a few closely related ones. (2) They speed up only one of the various possible reactions of these substrates; i.e. they steer the substrates into a particular reaction pathway and thus play a fundamental role in guiding the chemistry of life. It is enzymes, moreover, that make possible the precise coupling of exergonic reactions to the endergonic reactions resulting in the numerous complex energy-rich compounds on which life is based.

Like all catalysts, enzymes reduce the activation energy necessary for a reaction to begin. More specifically, by providing an alternative pathway for the reaction, they lower the activation-energy barrier to a point where it is easily overcome by the thermal motion of molecules at ordinary physiological temperatures (Fig. 2.51). Again like all catalysts, enzymes cannot speed up a reaction that could not take place on its own. They are potentially capable of catalyzing both the forward and the reverse reactions; the enzymes themselves do not determine where the equilibrium point between these two reactions lies.

Biochemists have long held that the key to enzyme function—as to the operation of simpler inorganic catalysts—is surface activity. Enzymes, as we have said, are globular proteins, of which each kind has its own distinctive surface geometry; they are enormously complex molecules with intricate three-dimensional contours. It seems reasonable, then, that a given enzyme would interact only with substrates whose molecular configurations "fit" that enzyme's surface. Thus the specificity of enzymes can be viewed as depending on their molecular conformation, especially their tertiary and quaternary structure.

The conclusion that the action of enzymes depends on their three-dimensional shape is consistent with the observation that when proteins are denatured—i.e. when their three-dimensional conformation is disrupted—they lose their characteristic biological activity; their enzymatic properties vanish (Fig. 2.52). Also consistent with this conclusion is the fact that most enzymes are highly sensitive to changes in pH, being active only within a limited pH range (Fig. 2.53). Apparently, change in pH results in the breakage of many of the weak bonds that help stabilize the conformation of proteins and, at the same time, leads to formation of new bonds, with consequent changes in the shape of the protein.

Enzyme and substrate have traditionally been visualized as fitting together like a lock and key, or like pieces of a puzzle. And it is true that the two must be roughly complementary if they are to combine temporarily; i.e. at least the reactive portion of the substrate molecule and the portion of the enzyme known as the **active site** must fit together in space intimately enough to become temporarily bonded and form a transient enzyme-substrate complex:

$$E + S \longrightarrow ES \longrightarrow E + P$$

(E stands for enzyme, S for substrate, and P for product.) But E and S probably do not always have to fit together exactly before ES can form; according to the recently advanced **induced-fit hypothesis,** E sometimes

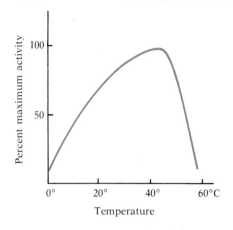

2.52   Enzyme activity as a function of temperature
Although temperature sensitivity varies somewhat from one enzyme to another, the curve shown here may be taken as applying to an "average" enzyme. Its activity rises steadily with temperature (approximately doubling for each 10°C increase) until thermal denaturation causes a sudden sharp decline, beginning between 40° and 45°. The enzyme becomes completely inactivated at temperatures above 60°, presumably because its three-dimensional conformation has been severely disrupted.

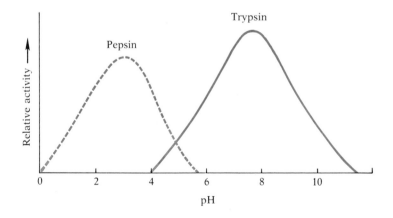

2.53   Enzyme activity as a function of pH
Most enzymes are very sensitive to pH, but they differ markedly in their pH optima. Pepsin and trypsin are both enzymes that digest protein, but the pH ranges within which they are active overlap only slightly. Pepsin is most active under strongly acidic conditions, trypsin under neutral and slightly alkaline conditions.

SUBSTRATE

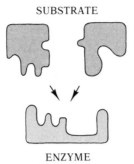

ENZYME

Enzyme-substrate complex

PRODUCT

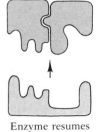

Enzyme resumes
original conformation

2.54   Induced-fit model of enzyme-substrate interaction

The enzyme molecule is thought to have an active site onto which the substrate molecules can fit (left), forming an enzyme-substrate complex (middle). The binding of the substrate is believed to induce conformational changes in the enzyme that maximize the fit and force the complex into a more reactive state. The enzyme molecule reverts to its original conformation when the product is released (right).

undergoes conformational changes in the course of bonding, which improve the fit and make ES more reactive (Fig. 2.54).

By itself spatial complementarity does not fully explain enzyme-substrate interactions. It is evident that ES can form only if E and S are chemically compatible and capable of bonding with each other. In other words, the concept of enzyme-substrate fit should include not only spatial fit but also bonding fit.

Although enzyme and substrate are sometimes held together by covalent bonds, the bonds are more often weak ones of the same types—ionic, hydrogen, hydrophobic, and van der Waals—that stabilize protein conformation. They are bonds that can be made and broken rapidly as a result of collisions due to random thermal motion at normal temperatures. The type of sub-

2.55   Model of an active site of an enzyme

Shown here in schematic form is the cleft where the active site of carboxypeptidase is located. Part of a substrate molecule is shown in the cleft, linked to the enzyme by five weak bonds (striped bands). Seven of the amino acids of the active site are indicated by their abbreviated names; the numbers beside the names refer to the positions of the amino acids in the enzyme polypeptide chain.   [Modified from A. L. Lehninger, *Biochemistry*, 2nd ed., Worth Publishers, New York, 1975, p. 233.]

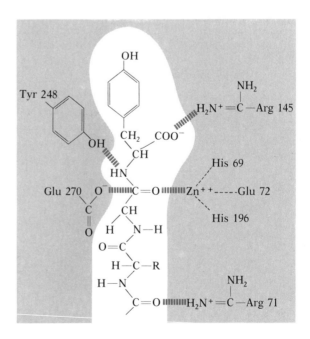

strate to which a given enzyme molecule can become bonded depends on the amino acids constituting its active site—more specifically, on the exposed R groups of these amino acids and the details of their arrangement relative to one another. Suppose the active site of a particular enzyme is a curving groove into which the reactive portion of the substrate must fit. Suppose, further, that the exposed R groups in this groove are mostly electrically charged ones. It is obvious that the reactive portion of the substrate must be complementarily charged or polar; a nonpolar substrate molecule could not react with the active site of this enzyme, even if it could, by chance, fit into the groove. Conversely, only a hydrophobic substrate could interact with an active site made up largely of hydrophobic R groups; both electrically charged and polar molecules would be incompatible with such a site.

Figure 2.55 offers a model of what is currently known about the active site of one actual enzyme (carboxypeptidase, which catalyzes the removal of the terminal amino acid from one end of a polypeptide chain). As can be seen, the site is visualized as a cleft into which the end of the substrate molecule can fit. The substrate is thought to form several weak bonds (five are shown here) with the R groups of amino acids that constitute part of the active site. But note that the critical amino acids in the active site (nos. 71, 196, 72, 69, 145, 248, and 270) are not adjacent to each other in the polypeptide chain, which means that the complex folding of the protein—its tertiary structure—has brought amino acids from several regions of the protein into close spatial proximity to form the active site. This is the usual pattern; active sites nearly always include some nonadjacent amino acids (Fig. 2.56 illustrates this principle in a different way). We can now understand more fully why an elevated temperature or a major pH change may greatly reduce enzyme activity; anything that changes the folding pattern of the polypeptide chains is likely to alter the critically important arrangement of amino acids in the active site.

Figure 2.55 illustrates another important point. Many enzymes contain a prosthetic group essential to their activity. A metal atom is often part of the prosthetic group; in carboxypeptidase, shown here, the metal is zinc.

Some enzymes that do not have a prosthetic group require a cofactor to which they bond only briefly and loosely during the reactions they catalyze. The cofactors may be metal ions, or they may be nonproteinaceous organic molecules; the latter are called *coenzymes.* Coenzyme molecules are much smaller and less complex than protein molecules. Like enzymes, they are not used up or permanently altered by the reactions in which they participate, and hence can be used over and over again. Only very tiny amounts are needed, therefore, but if the supply falls below normal, the health or even the life of the organism may be endangered. This is why vitamins, which act as parts of essential coenzymes, are so necessary in the diet.

The mechanism of enzyme catalysis   Assuming that enzymes (or enzyme-coenzyme complexes) form temporary unstable complexes with their substrates, how does this association reduce the amount of activation energy needed for the reaction being catalyzed? There is no single simple answer to this question, because a variety of plausible mechanisms have been proposed, which may apply to some enzymes, but not to others. We shall here mention a few of the more promising hypotheses.

2.56   Location of the active site in a polypeptide chain
The folding of the chain brings together in space nonadjacent amino acids of the active site (black).

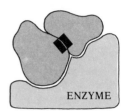

2.57 Orientation of substrate for maximal reactivity
When substrate molecules collide in the absence of an enzyme (left), their orientation often fails to bring together their reactive groups (black). When the substrate molecules are bound to an enzyme, they are oriented for maximal interaction of their reactive groups.

Enzyme-substrate complex

Competitive inhibitor bound to enzyme

Noncompetitive inhibitor bound to enzyme

2.58 Competitive and noncompetitive inhibition of an enzyme
Top: The substrate is bound to the catalytic site of the enzyme.   Middle: The binding of a competitive-inhibitor molecule to the catalytic site prevents the substrate from binding.   Bottom: A noncompetitive inhibitor bound to a different site on the enzyme blocks access to the catalytic site.

One of the oldest proposals is that enzymes, by binding substrate molecules on their active sites, greatly increase the effective concentrations of the substrates in certain regions and thus facilitate the contact between substrate molecules required for their interaction. In short, the frequency of collision of bound substrate molecules is far higher than that of molecules dispersed in a fluid medium.

The elevation of collision frequency is surely not the only advantage derived from formation of the enzyme-substrate complex. It is thought that substrate molecules are oriented in a very precise way when bound to the enzyme (Fig. 2.54). Very likely, their respective reactive groups, as well as the catalytic group of the enzyme itself, are so aligned as to favor formation of an activated complex. No longer is great force of collision necessary for interaction, because collisions on enzyme surfaces are oriented for maximal effect (Fig. 2.57).

When substrate molecules interact with the active site of an enzyme, forming weak bonds or, at times, covalent bonds, the distribution of electrons within the substrate molecules must of necessity undergo some alteration. The susceptible bonds in the substrate molecules may be weakened as a result, becoming more breakable and thus more reactive. In other words, enzymes may not only steer substrate molecules so that their reactive groups are properly aligned on collision, but they may also enhance the reactivity of those groups.

There is another way in which enzymes may enhance the reactivity of substrates. X-ray diffraction studies suggest that in at least some cases the formation of the enzyme-substrate complex induces conformational changes in the enzyme itself (Fig. 2.54). Such changes in the shape of the enzyme could impose strains or distortions on the susceptible bonds in the substrate molecules, thereby making them easier to break.

Finally, it should be kept in mind that a reaction on the active site of an enzyme takes place in a chemical environment very different from that of a noncatalyzed reaction. For example, the microenvironment of the active site may be more acid or more alkaline than the surrounding medium, or it may be nonpolar while the medium is polar. Presumably the active site of each type of enzyme has a microenvironment especially suited to heightening the reactivity of the particular substrates for which that enzyme is specific.

Control of enzyme activity   Since enzymes control the myriad chemical reactions within living organisms, it is not surprising that a variety of mechanisms should have evolved for controlling the activity of the enzymes themselves. These mechanisms depend not only on physical parameters such as temperature, pH, and substrate or enzyme concentration, but also on chemical agents, which mask, block, or alter the active sites of the enzymes they help regulate.

One common form of enzyme control, called *competitive inhibition,* involves an inhibitor substance sufficiently similar to the normal substrate of the enzyme to bind reversibly to its active site, but differing from the substrate in not being chemically changed in the process. By binding with the active site, the inhibitor (I) masks the site and prevents the normal substrate molecules from gaining access to it (Fig. 2.58). Thus the reaction

$$E + I \longrightarrow EI$$

competes with the reaction

$$E + S \longrightarrow ES \longrightarrow E + P$$

because both involve the same enzyme, which is present in only very small quantities. Which of the two reactions will predominate depends on their relative energetics and, even more, on the relative concentrations of I and S. If there is much I and a low concentration of S, then a high percentage of the enzyme will be bound as EI and therefore unavailable; on the other hand, if there is much S and only a small concentration of I, then most of the enzyme molecules will be free to catalyze the reaction of S to form P.

A second category of reversible inhibition, called *noncompetitive inhibition,* depends on the operation of two kinds of binding sites in the same enzyme molecules—the usual active sites to which substrate can bind and other sites to which inhibitors can bind. In simple noncompetitive inhibition, the bound inhibitor molecules are thought to interfere physically with the active sites, perhaps by partly or wholly blocking access to them (Fig. 2.58).

A special type of noncompetitive inhibition, called *allosteric inhibition,* deserves special mention. An allosteric enzyme is one that can exist in two distinct spatial conformations, which usually reflect alterations of quaternary structure (allosteric enzymes are composed of two or more independently folded subunits). Often when the molecule is in one conformation the enzyme is active, and when it is in the other conformation it is inactive (or less active), because the substrate-binding site has been disrupted. In allosteric inhibition, the binding of inhibitor molecules—usually called negative *modulators* (or sometimes negative effectors)—stabilizes the enzyme in its inactive conformation (Fig. 2.59).

Other types of allosteric enzymes have binding sites for positive modulators, which induce conformational changes enhancing enzyme reactivity.

Instead of having different kinds of binding sites, some allosteric enzymes have two or more sites of a single kind; i.e. they can bind substrate at two or more locations simultaneously. The binding of substrate at one active site causes conformational changes that make the remaining sites more reactive. This phenomenon, called *cooperativity,* is exemplified in hemoglobin (the red oxygen-carrying protein in vertebrate blood, which, though not

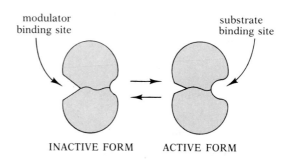

modulator binding site   substrate binding site

INACTIVE FORM   ACTIVE FORM

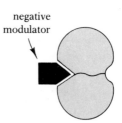

negative modulator

2.59 An allosteric enzyme with negative modulation
Top: The enzyme can exist in two conformations, one inactive and the other active; the two are interconvertible.   Bottom: When a negative modulator binds to the inactive form, it locks the enzyme in the inactive conformation and maintains the binding site for substrate (shown here partly closed) in a nonreactive condition.   An alternative model supposes the unbound enzyme to be present in only one form (say, the inactive), which is converted to the other form by binding a modulator molecule.

technically an enzyme, is very enzymelike). A single molecule of hemoglobin is capable of carrying four oxygen molecules. The binding of the first oxygen molecule induces changes in the quaternary structure of the hemoglobin molecule that give the other three binding sites a higher affinity for oxygen. Thus, in cooperativity, the first substrate molecule functions as the modulator for stabilizing the allosteric protein in one of its possible conformations—in the case of hemoglobin, in its most reactive conformation.

Because of the different effects that competitive inhibitors, noncompetitive inhibitors, and allosteric modulators have on the active sites of enzymes, biochemists find them exceedingly valuable in probing the nature of enzymes. For example, to gain a clearer idea of the active site—its physical shape and its reactive R groups—they may synthesize a series of slight variants of a competitive inhibitor and study the effectiveness of each in binding to the active site. Or, to investigate the roles of the various parts of an enzyme molecule, they may make use of irreversible inhibitors—reagents that act as enzyme poisons by forming permanent covalent bonds with the functional groups necessary for catalysis.

SUGGESTED READING

BAKER, J. J. W., and G. E. ALLEN, 1974. *Matter, energy, and life.* 3rd ed. Addison-Wesley, Reading, Mass. *Easy-to-read treatment of chemical topics important to an understanding of biology.**

FRIEDEN, E., 1972. "The chemical elements of life," *Scientific American*, July. *On procedures for determining whether an element is essential to life, with particular emphasis on four elements (fluorine, silicon, tin, and vanadium) recently found to be important.*

GREEN, D. E., 1960. "The synthesis of fat," *Scientific American*, February. (Offprint 67) *A demonstration that the process is not merely the reverse of hydrolysis.*

KENDREW, J. C., 1961. "The three-dimensional structure of a protein molecule," *Scientific American*, December. (Offprint 121) *How the complete folding pattern of myoglobin—the first protein whose conformation was determined—was worked out.*

KOSHLAND, D. E., 1973. "Protein shape and biological control," *Scientific American*, October. (Offprint 1280) *On the importance of protein conformation in determining enzymatic activity; how substances that cause changes in the shape of a protein can regulate its activity.*

SPEAKMAN, J. C., 1966. *Molecules.* McGraw-Hill, New York. *Helpful summary of basic chemical principles.**

THOMPSON, E. O. P., 1955. "The insulin molecule," *Scientific American*, May. (Offprint 42) *On the first determination of the primary structure of a protein—by Frederick Sanger, who labored for ten years before he worked out the amino acid sequence of insulin in 1954.*

* Available in paperback.

Chapter 3

CELLS: UNITS OF STRUCTURE AND FUNCTION

We saw in the last chapter that organisms are composed of a great variety of chemicals, some simple and some complex. But these chemicals do not of themselves possess the properties we recognize as life—properties observable only in association with the organization superimposed on the chemicals in living systems. For the chemicals in living systems are not simply dispersed in random fashion in an aqueous medium; they are highly ordered, forming elaborate structural and functional complexes. In this chapter we shall examine in some detail the fundamental unit of structure and function in living things: the cell.

THE CELL THEORY

The discovery of cells and of their structure is linked with the development of magnifying lenses, particularly the microscope. Although some of the optical properties of curved surfaces had been known since 300 B.C., it was not until the seventeenth century that Antoni van Leeuwenhoek (1632–1723) and his contemporaries refined the production of lenses sufficiently to construct microscopes satisfactory for simple observations. Thus in 1665 Robert Hooke was able to report to the Royal Society of London on "the first microscopical pores I ever saw, and perhaps, that were ever seen," in a piece of cork; "these pores, or cells, were not very deep, but consisted of a great many little Boxes, separated out of one continued long pore, by certain Diaphragms." Hooke's microscopic examination of cork marks the beginning of the study of cells. Intensive work on cells was not pursued, however, until the early nineteenth century.

The idea that all living things are composed of cells—the **cell theory**—is commonly credited to two German investigators, the botanist Matthias

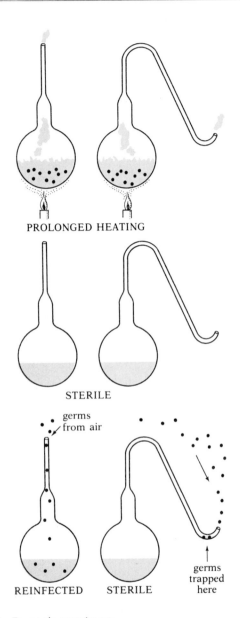

PROLONGED HEATING

STERILE

germs
from air

REINFECTED   STERILE

germs
trapped
here

3.1 Pasteur's experiment
Nutrient broths in two kinds of flask, one with a
straight neck, the other with a bent neck, were boiled
to kill any germs (here vastly exaggerated in size) they
might contain (top). The sterile broths were then
allowed to sit in their open-mouthed containers for
several weeks (middle). Microorganisms entering the
straight-necked flask contaminated the broth, but those
entering the bent neck of the other flask were trapped
in films of moisture in the curves of the neck and did
not contaminate the broth (bottom).

Jakob Schleiden and the zoologist Theodor Schwann, who published their
conclusions in 1838 and 1839 respectively. One of the most important generalizations of modern biology, this idea had already been expressed earlier—
among others, by the French naturalist Jean Baptiste de Lamarck, who
wrote in 1809: "no body can have life if its constituent parts are not cellular
tissue or are not formed by cellular tissue." But Schleiden and Schwann
stated the principle with particular clarity, and they helped it gain general
acceptance.

An important extension of the cell theory, proposed in 1858 by the German
physician Rudolf Virchow, was that all living cells arise from pre-existing
living cells *("omnis cellula e cellula")*, that there is no spontaneous creation
of cells from nonliving matter. The theory of **biogenesis,** life from life, con-
tradicted the belief in spontaneous generation, then widely held not only by
the general public but by scientists as well. It was Louis Pasteur in France
who, a few years later (1862), supplied proof for Virchow's theory in a series
of now classic experiments.

Pasteur's first step was to place various nutrient broths into long-necked
flasks and then bend the necks of the flasks into curves (Fig. 3.1). Next, he
boiled the broths in the flasks to kill any microorganisms (germs) that might
be in them. While the flasks were left standing, germ-laden dust particles in
the air moving into the flasks were trapped in the films of moisture on the
humid curves of the neck; the curved neck acted as a filter. Though the
broths might be left standing in their swan-neck containers for months or
even a year or more, no life appeared in them. Control broths boiled in flasks
with straight necks did not remain free of microorganisms and were soon
teeming with life. Similarly, if the swan-neck was broken off, the experimental broth rapidly developed colonies of molds and bacteria. Thus Pasteur showed that the source of the microorganisms that fermented or pu-
trefied such substances as milk, wine, or sugar-beet juice was the air. The
organisms did not arise spontaneously from the nutrient media.

The principle of biogenesis has needed some modification in recent years,
as we shall see in a later chapter. Current theory holds that spontaneous
generation of life from nonliving matter, while it does not occur under
present conditions, probably did occur under the conditions existing on the
primitive earth when life first arose.

The two components of the cell theory—that all living things are composed
of cells and that all cells arise from other cells—give us the basis for a
working definition of living things: Living things are chemical organizations
composed of cells and capable of reproducing themselves. Notice we have
said "a working definition." In fact, as we shall see later, any attempt to draw
a sharp line between the nonliving and the living is essentially arbitrary.
Viruses, for example, have some of the central attributes of living things,
even though they are not composed of cells in the usual sense.

SUBCELLULAR ORGANIZATION AND FUNCTION

In the traditional view, cells consist of a substance called **protoplasm,** which
is subdivided into two major areas: the **nucleus,** which is the control center
of the cell, and the **cytoplasm,** which includes all the rest of the cell. What
sort of substance is protoplasm? The standard definition, "living substance,"

conveys remarkably little. There is, in fact, no satisfactory short definition of protoplasm, and we shall generally avoid using the term because of its indefiniteness. An adequate description of protoplasm would actually be a description of cells—the materials of which they are composed and their intricate organization.

THE USE OF MICROSCOPES

Much of our knowledge of subcellular organization has been made possible by the development of better and more powerful microscopes. In the detailed analysis of subcellular structure, three attributes of microscopes are of particular importance: magnification, resolution, and contrast. Magnification is a means of increasing the apparent size of the object being viewed until it provides an adequate stimulus to our eyes. Resolution is the capacity to separate adjacent forms or objects, i.e. to see them as distinct. Contrast is important in distinguishing one part of a cell from another.

Although ordinary light microscopes can be endowed with very high magnification, their resolving power is limited. It is about 500 times better than that of the unaided human eye, but this is still not enough for viewing some of the smaller subcellular structures. Contrast is often obtained in microscopy by fixing and staining the material being studied. Since different parts of the material often differ in their affinities for various dyes, it is possible to stain these parts different colors and make them stand out from each other.

The electron microscope opened up whole new vistas in the study of cells (Fig. 3.2). This microscope, as its name implies, uses a beam of electrons instead of light as its source of illumination. The electrons pass through the specimen and fall on a photographic plate, where they produce an image of the specimen. Because resolution improves as the wavelength of the illumination becomes shorter, and because electron beams have much shorter wavelengths than visible light, electron microscopes are capable of resolving objects about 10,000 times better than the unaided human eye. Many of the details of cellular structure discussed in this chapter would not be known but for the electron microscope.

The standard transmission electron microscope poses a problem that has been solved with the recent development of the scanning electron microscope. Because the transmission EM requires that the specimen be sliced into ultrathin sections, information about the three-dimensional shape of a structure can be obtained only through laborious examination of countless separate pictures of different sections. In the scanning EM the probe electron beam does not pass through the specimen, but instead causes secondary electrons to be emitted from its surface. If the specimen has a contoured surface, the intensity of emission of these secondary electrons varies with the angle at which the probe beam strikes each point on the specimen. Thus a point by point recording of the emission provides a picture of three-dimensional effect that conveys much information about the shape of the specimen (see Fig. 3.40, for example). The specimen need not be sectioned, but is usually frozen. The resolution of a scanning EM is not as great as that of a transmission EM, but its much greater depth of field is an inestimable advantage for some types of studies.

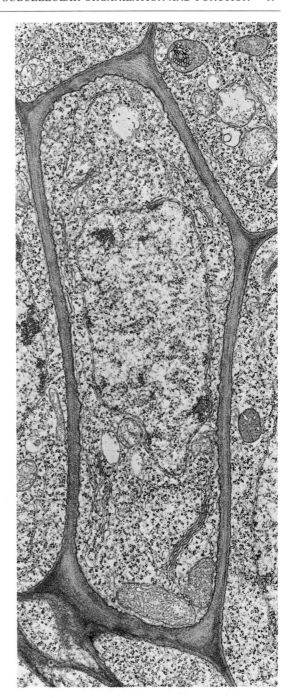

3.2 Electron micrograph of a differentiating cell from the root tip of a bean plant
Many of the membranous or particulate intracellular organelles shown here could have been seen only faintly, if at all, under the light microscope. × 15,600. [Courtesy E. H. Newcomb, University of Wisconsin.]

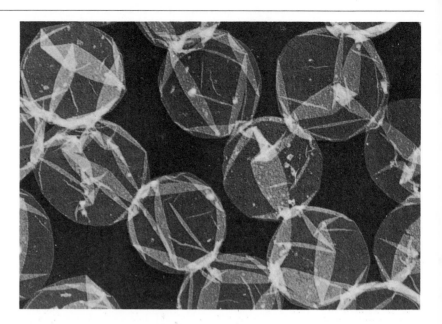

3.3   Ghosts of human red blood cells
The "ghosts" are now known to be cell membranes. The whitish areas are places where the membrane is folded. × 4,500.   [Courtesy J. F. Hoffman, Yale University.]

CELL SIZE

Most cells are very small and can be distinguished only with a microscope. Some, however, such as birds' egg cells, can be seen with the naked eye. Others, like nerve cells, may be very small in some of their dimensions, but extremely long; a single human nerve cell may be over a meter in length, and that of an elephant can be even longer. To say that cells are generally small is not saying much, however, because even among microscopic cells there is a wide range in size. The diameter of a human red blood cell is about 35 times greater than that of some very tiny microorganisms, while that of a human egg cell is about 14 times greater than that of a red blood cell. The diameter of an ostrich egg cell, in turn, is about 1,500 times greater than that of a human egg cell. Most cells, however, have a diameter ranging between 0.5 and 40 μm (micrometers; formerly called microns).[1]

One probable reason why cell size has remained restricted throughout the course of evolution is that the ratio of surface area to volume strongly affects the functioning of cells. Cells obtain necessary materials such as oxygen and nutrients from the area surrounding them. These materials must enter across the surface of the cell, and waste products must leave by the same route. As cell size increases, the volume increases much more rapidly than the surface area (volume increases as the cube of the cell radius, surface area as the square of the cell radius). Thus increasing size entails the problem of adequate exchange surface for support of the greatly increased volume. Cells very active in carrying out chemical reactions tend to be smaller than cells with lower metabolic rates, for the problem of exchange of materials is much greater for them. This whole question of the ratio of surface area to volume is not limited to single cells; we shall encounter it again and again in reference to whole organisms.

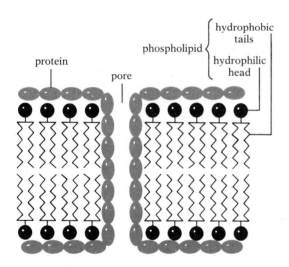

3.4   The Davson-Danielli model of the cell membrane
Two layers of lipids are sandwiched between two layers of protein. The phospholipids are oriented with their hydrophilic polar heads near the surfaces and their nonpolar hydrophobic tails projecting into the interior of the structure, at right angles to the surfaces. Pores are thought to penetrate the membrane at some points.

[1] For units of measurement see Glossary.

Another factor limiting cell size is the ability of the nucleus to exert control over the rest of the cell. As a cell increases in size, more and more of its parts must be located far from the control center and proper interaction becomes progressively more difficult.

THE STRUCTURE OF THE CELL MEMBRANE

Despite the widespread, almost routine acceptance of the idea that cells are bounded by a ***plasma membrane,*** it is only in the last three decades that direct proof of its existence has been obtained. Most of the earlier conceptions of the membrane were deduced from the characteristics of cells, for the membrane is usually not visible even under the most powerful light microscopes. Although something believed to be the membrane could be isolated from red blood cells, there was no conclusive proof that these red-cell "ghosts" (Fig. 3.3) were really cell membranes and not artifacts of the procedures used to obtain them.

The Davson-Danielli and unit-membrane models   Permeability studies had long shown that lipids and many substances soluble in lipids move with relative ease between the cell and the surrounding medium. From this fact it was deduced that the outer boundary of the cell, the cell membrane, must contain lipids, and that fat-soluble substances could move across the membrane by being dissolved in it. It was also observed that many small water-soluble molecules move quite freely between the inner portion of the cell and its external environment; it was therefore postulated that the cell membrane is a kind of sieve, containing pores or nonlipid patches. But still another observation had to be accounted for. Small water-soluble ions move through the cell membrane less freely than uncharged particles of roughly the same size; moreover, different ions do not all exhibit the same facility for crossing the cell boundary, some moving rather freely, others only very slightly. It was therefore assumed that the cell membrane itself possesses charge, a property that tends to interfere with the movement of charged particles. Finally, the physical properties of the cell boundary, especially its wettability and elasticity, seemed to indicate the presence of protein in the membrane.

These various ideas about the cell membrane—its lipid and protein components and pores—all stemming primarily from permeability studies, were put together in the late 1930s by J. F. Danielli of Princeton University and H. Davson of University College, London. They concluded that the membrane consists of at least two layers of phospholipids, oriented with their polar (hydrophilic) ends near the surface and their nonpolar (hydrophobic) hydrocarbon chains in the interior of the membrane, as far away as possible from the surrounding water (Fig. 3.4). Danielli and Davson envisioned a membrane about 8 nm (nanometers) thick, with layers of protein on both the inner and the outer surface of a bimolecular lipid core.

In later years, both electron microscopy and X-ray diffraction studies provided more direct information about the membrane. Electron micrographs made in the 1950s by J. David Robertson, then of the Harvard Medical School, suggested that the cell membrane is composed of two electron-dense layers separated by a somewhat wider lighter area (Fig. 3.5). His

3.5   Electron micrograph showing cell membrane of sectioned human red blood cell

The cytoplasm of the cell is in the upper right half of the picture. The membrane consists of two dark lines—probably protein—separated by a lighter area, which is probably lipid. × 280,000. [Courtesy J. David Robertson, Duke University.]

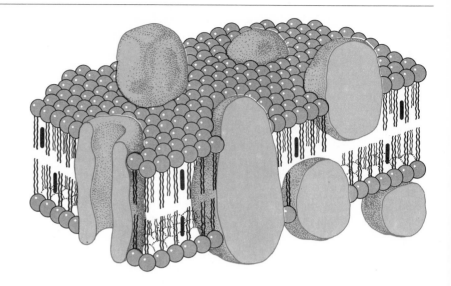

3.6   The fluid-mosaic model of the cell membrane
A double layer of lipids forms the main continuous
part of the membrane; the lipids are mostly
phospholipids, but in plasma membranes of higher
organisms cholesterol (solid bars) is also present.
Proteins (shown here as blocks, but keep in mind that
they are really coiled polypeptide chains) occur in
various arrangements. Some, called extrinsic proteins,
are entirely on the surface of the membrane. Others,
called intrinsic proteins, are partly embedded in the
lipid layers; some of these may penetrate all the way
through the membrane. There is also evidence that at
least some of the pores in the membrane are entirely
protein-lined (lower left).

measurements of the total thickness of the membrane were very close to the
8 nm hypothesized by Danielli and Davson. X-ray diffraction studies indicated that the molecules of the material in the lighter middle portion of the
cell membrane are oriented parallel to the radius of the cell. Again, these
observations supported the Davson-Danielli hypothesis of the arrangement
of the phospholipid molecules (see Fig. 3.4).

Robertson, impressed with the close agreement between his electron-
microscope studies and the Davson-Danielli lipoprotein-sandwich model (a
bimolecular lipid layer between two layers of protein), argued that this
model, which he termed the ***unit membrane,*** must be the basic structure of
all cellular membranes, including both the plasma membrane and the various intracellular membranes, to be discussed later in this chapter. More
recent research, however, has cast doubt on the idea of a uniform structure
for all membranes. Although most membranes appear to be composed of
lipids and proteins, they vary considerably in other ways: Thicknesses range
from about 5 to 10 nm; lipid content ranges from about 30 to 80 percent;
cholesterol is ordinarily abundant in the lipid portion of plasma membranes
of higher organisms, but is absent from intracellular membranes and from
the plasma membranes of bacteria; the ratios of the various other types of
lipids differ from one membrane to another; and so forth. It seems likely
that these differences reflect corresponding differences in structural detail,
and a variety of alternatives to, and modifications of, the unit-membrane
model have been suggested.

The fluid-mosaic model   Of the various alternatives to the Davson-Danielli
model of the plasma membrane (as modified in Robertson's unit-membrane
version), the one that enjoys the widest acceptance today is the fluid-mosaic
model, postulated in 1972 by S. J. Singer of the University of California, San
Diego, and G. L. Nicolson of the Salk Institute. This model, like the unit-
membrane model, envisions a bilayer of phospholipids (plus cholesterol for

the plasma membranes of higher organisms) oriented with their hydrophilic heads toward the surfaces of the membrane and their hydrophobic tails toward the interior. According to this model, however, the proteins are not confined to the surfaces of the membrane, nor do they form continuous layers; instead, they are distributed in a mosaic pattern, both on the surfaces and in the interior of the membrane (Fig. 3.6).

Of the proteins confined to the surfaces (extrinsic proteins), those on the inner surface may differ markedly from those on the outer surface; some membranes may have no extrinsic proteins at all. The proteins located wholly or partly within the lipid bilayer (intrinsic proteins) may exhibit a variety of arrangements: Some are entirely buried within the bilayer, whereas others have parts that project through the surface; some are confined to the outer half of the lipid core, and others to the inner half; some may extend entirely through the bilayer, projecting into the watery medium on both sides. As would be expected, hydrophilic amino acids (i.e. those with polar or electrically charged R groups) predominate in the portions of the protein molecules that project out of the lipid bilayer into the water, whereas hydrophobic (i.e. nonpolar) amino acids are abundant in the portions buried in the lipid bilayer (Fig. 3.7).

According to the fluid-mosaic model, the structure of the membrane is not static. The individual lipid molecules (which are linked to one another only by weak bonds, not by covalent bonds) can move laterally, so that a particular molecule found in one position at a given moment may be in an entirely different position several hours later. Mobility of the lipids is greatest in membranes that contain no cholesterol.

The proteins, too, can move laterally to some extent, but much less than the lipids. Complete freedom of movement would be incompatible with the specific functional demands placed on the membrane proteins. Certain proteins in the membrane of nerve cells, for example, are essential to the transmission of nerve impulses from one cell to another; they are found only at points where one nerve cell is in close proximity to another—not in other positions, where they could not fulfill their function. Similarly, proteins responsible for pumping sodium ions out of the cells that line the intestine are located in the membrane on only one side of the cells (the side away from the intestinal cavity). In short, at least some of the membrane proteins are anchored in place; there is a limit to the fluidity of the membrane. In some cases the anchoring is probably due to tight associations between two or more intrinsic proteins and the resulting formation of structural and functional complexes too large to move easily. In other cases there may be complexing between intrinsic and extrinsic proteins. Owing to association with proteins, even the lipid molecules may not all be free to move; there is evidence that the lipids immediately adjacent to intrinsic proteins may be loosely bound to the proteins and thus immobilized.

In the most recent versions of the fluid-mosaic model, the pores in the membrane are no longer envisioned as lipid-lined. Instead, they are thought to be bounded by protein molecules (Fig. 3.6), or even in some cases to be channels through a single protein molecule. The distinctive properties of the various R groups of the amino acids in the proteins give the pores some selectivity; not all ions or molecules small enough to fit in the pores can actually move through them.

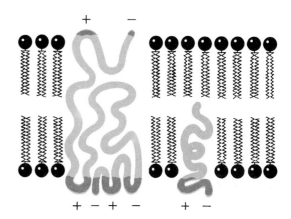

3.7 Orientation of proteins within membranes
The parts of the polypeptide chain containing most of the hydrophilic amino acids (polar or charged R groups) (dark color) tend to project into the watery medium outside the lipid layers, whereas the parts of the chain with hydrophobic amino acids (light color) tend to be folded into the inner, lipid portion of the membrane.

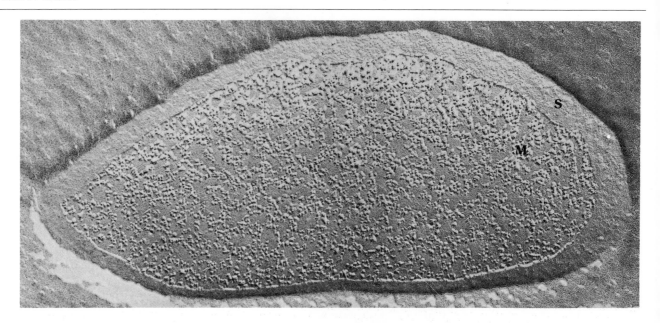

3.8 Electron micrograph of freeze-etched plasma membrane of red blood cell

In freeze-etching, the specimen is first frozen and then fractured; here the plasma membrane is fractured along the plane between the two layers of lipids, i.e. along the middle of the bimolecular lipid core (see sketch). Removal of some of the ice by sublimation etches the surface, which is next coated with carbon and a metal (usually platinum) applied at an angle so as to shadow any irregularities. Once the original specimen is removed, the detailed coating, or surface replica, can be studied. Electron micrographs of replicas thus produced have a three-dimensional appearance. The numerous spherical particles visible in the micrograph are interpreted as protein (see colored entities in sketch). They appear where the unit-membrane model predicts only lipid, but their presence is explained satisfactorily by the fluid-mosaic model. S: outer surface of membrane; M: interior of fractured membrane. × 40,000. [Courtesy Daniel Branton, Harvard University.]

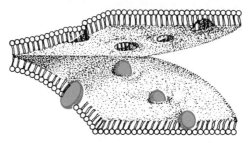

The fluid-mosaic model is a powerful one. Besides accounting for the permeability and wettability properties of membranes, it accommodates many findings that the unit-membrane model cannot. Thus it views the widely varying protein content of different membranes—a variation hard to reconcile with the continuous protein layers postulated by the earlier model—as due to differences in the number of protein molecules per unit area of membrane. It attributes the varying thicknesses of membranes to differences in the proteins and the degree to which they protrude from the lipid bilayer. It accounts for the fairly rapid changes sometimes observed in the location of functional groups on the membrane. And, in marked contrast to the unit-membrane model, it can explain the presence, revealed by freeze-etch microscopy, of proteins in the center of the lipid portion of the membrane (Fig. 3.8).

Here we should pause for a moment and consider the value of model construction in science. A good conceptual model organizes what is known about a subject and thereby helps define the problems yet to be tackled. The model may eventually be abandoned for one more consonant with new evidence, but it will have fulfilled its function if it has stimulated research leading to a persuasive new conception. The Davson-Danielli-Robertson model of the cell membrane accomplished just such a purpose; it remained dominant for some thirty years, and though it is now giving way to the fluid-mosaic model, it is due every honor for the discoveries it inspired. Should the fluid-mosaic model, in its turn, be displaced someday, it will deserve high praise if it provides but half the impetus for research that was provided by its predecessor.

FUNCTION OF THE CELL MEMBRANE

The cell membrane not only serves as an envelope that gives mechanical strength and shape and some protection to the cell. It is also an active component of the living cell, preventing some substances from entering it and others from leaking out. It regulates the traffic in materials between the precisely ordered interior of the cell and the essentially unfavorable and potentially disruptive outer environment. All substances moving between the cell's environment and the cellular interior in either direction must pass through a membrane barrier.

Diffusion and osmosis   Before examining the role of the membrane in more detail, we must consider the movement of materials in general. In the previous chapter we saw that thermal agitation—the motion with which small particles are endowed by heat energy—affects the rates of chemical reactions. Thermal agitation is also a major factor in the tendency of particles to move from one place to another.

Let us consider a small rectangular box containing 20 marbles, all placed in a tight cluster near one end (Fig. 3.9A). The box is shaken, and the marbles are scattered almost evenly over the bottom of the box (Fig. 3.9B). Obvious as this result might seem, it is worth a closer look.

First, it must be observed that, among all the possible directions in which a given marble might move, more lead away from the center of the cluster than toward it. Hence random movement will tend to disrupt the cluster rather than maintain it. Or, to put it another way, in the absence of any counteracting external influence, a dynamic system will tend to move toward the more probable disorganized state rather than toward the less probable organized state.

Another factor favors dispersion of the marbles. Movement toward the cluster has a high probability of resulting in a collision of two or more marbles, which will then be deflected. Movement away from the cluster carries much less probability of collision; a marble has a good chance of continuing on an uninterrupted path to the outer portions of the box. On the average, then, more marbles move away from the center of concentration than toward it.

Notice that the above arguments are both statistical. It is possible that, as a result of random motion, 20 scattered marbles will all come to form a tight cluster at one end of the box. This result has a finite possibility, but one so slight that it can justifiably be disregarded. The kind of reasoning used here is typical of most scientific reasoning. The facts and laws of science are statistical rather than absolute. They describe nature in terms of probable occurrence.

We can now make a generalization based on our example of the marbles in the box and on others like it: All other factors being equal, *the net move-*

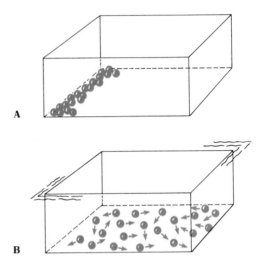

3.9 Mechanical model for diffusion

(A) All 20 marbles are placed in a cluster at one end of a rectangular box. (B) When the box is shaken to make the marbles move randomly, they become distributed throughout the box in nearly uniform density. For fuller explanation, see text.

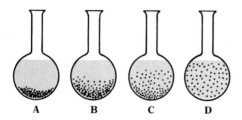

3.10 Diffusion in a liquid

Particles of solute are at the bottom of a flask of water in (A). The particles slowly diffuse away from the cluster until in (D) they are distributed with nearly uniform density through the water. If there are no convection currents, and all movement is by simple diffusion, it may take years to reach the condition shown in (D).

ment of the particles of a particular substance is from regions of higher to regions of lower concentration of that substance. Note that we said the *net* movement. There will always be some particles moving in the opposite direction, but, overall, the movement will be away from the centers of concentration. An obvious result is that the particles of a given substance tend to become distributed with relatively uniform density within any available space. When this uniform density is reached, the system is in equilibrium; the particles continue to move, but there is little net change in the system.

Movement of molecular-sized particles from one place to another in the manner we have been discussing is called **diffusion.** Diffusion is fastest by far in gases, where there is much space between the molecular particles and hence relatively little chance of collision. Diffusion in a liquid is much slower (Fig. 3.10); in the absence of convection currents, it takes a very long time, years in fact, for a substance to move in appreciable quantity only a few feet through water. Diffusion in solids is, of course, much slower still; there is very little space between the molecular particles of a solid, and collisions occur almost before the particles get going. In all these instances, however, regardless of the rate of diffusion, the net effect is movement from the regions of higher concentration, as long as all regions are at the same temperature and pressure.

Thus far we have discussed diffusion in terms of concentration gradient, but diffusion is not strictly a function of concentration. More accurately, it is a function of a gradient in the free energy of the particles. Now, the free energy of a system declines if heat energy is lost from the system or if the disorder (entropy) of the system increases (see p. 66). We have seen previously that it is usually loss of heat energy that drives exergonic chemical reactions. By contrast, it is increase in entropy that drives diffusion. An orderly, improbable arrangement is, of course, more likely to undergo net change than a disorderly, more probable arrangement. Look again at Figs. 3.9 and 3.10. The marbles in Fig. 3.9A and the particles of solute in Fig. 3.10A are arranged in tight clusters—arrangements most unlikely to arise purely by chance. Because the tight clusters are orderly, improbable arrangements, they possess less entropy and hence more free energy than the more disorganized arrangements shown in Figs. 3.9B and 3.10D. Thus we can rephrase our earlier generalization as follows: *The net movement of the particles of a particular substance is from regions of greater to regions of less free energy of that substance.*

Free energy is not only the more correct basis for understanding diffusion but also the more broadly applicable. Consider a situation where there is a slight concentration gradient from point Y to point Z and a pronounced temperature gradient in the reverse direction. If concentration alone were a factor, there would be net diffusion from Y, the region of higher concentration, to Z, the region of lower concentration. But temperature and pressure also play a role. In this case, Z has a higher temperature than Y. Now, the higher the temperature in a given system, the greater the heat motion of the particles in that system; and the greater the heat motion, the greater the free-energy content. Because the difference in free energy associated with the temperature gradient from Z to Y may outweigh the difference in free energy associated with the concentration gradient from Y to Z, net diffusion may be from Z, the region of lower concentration but higher temperature, to

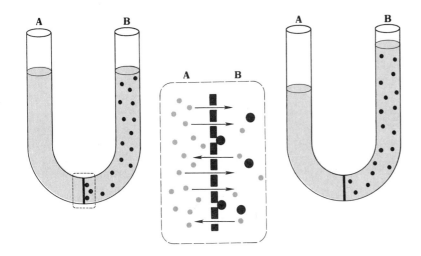

3.11 U tube divided by a semipermeable membrane
The membrane at the base of the U tube is permeable to water, but not to sugar molecules (black balls). Left: Side A contains only water; side B contains a sugar solution. Initially, the quantity of fluid in the two sides is the same. Center: A larger number of water molecules (colored balls) bump into the membrane per unit time on side A than on side B. Right: Because more water molecules move from A to B than from B to A, the level of fluid on side A falls while that on side B rises. For fuller explanation, see text.

Y, the region of higher concentration but lower temperature. Thus it is clear that the crucial factor in determining the movement of substances is not concentration (or temperature or pressure) but free energy.

Now let us consider another situation, a chamber divided into two halves by a membrane partition. Let us assume, further, that particles of some substances can pass through the membrane while particles of other substances cannot. Such a membrane is said to be **differentially permeable** (or selectively permeable). How will the membrane affect the diffusion of materials between the two halves of the chamber? Suppose the chamber is a U tube divided in half by a differentially permeable membrane (Fig. 3.11). Suppose side A contains pure water and side B an equal quantity of sugar solution (i.e. sugar dissolved in water), both sides being subject to the same initial temperature and pressure. If the membrane is permeable to water but not to sugar, water molecules will be able to pass in both directions, from A to B and from B to A. Such a membrane, permeable to water but not to solute, is said to be **semipermeable,** and movement of water through it is called **osmosis.**[2] Unlike our hypothetical membrane in the U tube, biological membranes are not, strictly speaking, semipermeable, because they are not completely impermeable to solutes; they are sufficiently like semipermeable membranes, however, for the movement of water through them to be routinely termed osmosis.

---

[2] Actually, osmosis is the movement of the principal solvent of a system—whether water or another substance—through a semipermeable membrane, but since the principal solvent in living systems is water, this book will always use the term with reference to water.

It should also be noted that osmosis is more than a process of simple diffusion through a semipermeable membrane. Even though both the direction of movement and the equilibrium conditions can be understood in terms of simple diffusion, the movement of water in a membrane is far too rapid to be explained by the random molecular motions on which diffusion depends. Apparently there is bulk flow of water in the membrane. The mechanism may be the following:

Much of the movement of water through a membrane is through the pores. There is an extremely steep concentration (free-energy) gradient at one end of each pore, which results in such rapid diffusion at that point that the density of the water within the pore decreases. The consequent hydrostatic-pressure gradient along the length of the pore causes bulk flow of water through it.

But let us return to our U-tube model. Since water is already present on both sides of the membrane, it might at first be supposed that the net effect of the movement of water molecules across the membrane would be zero, but such a supposition would be wrong. Consider the differences between the pure water and the sugar solution more carefully. On side A, all the molecules that bump into the membrane during a given interval are water molecules and, because the membrane is permeable to water, many of these molecules will pass through the membrane from A to B. By contrast, on side B, some of the molecules bumping into the membrane during the same interval will be water molecules, which may pass through, and some will be sugar molecules, which cannot pass, because the membrane is impermeable to them. At any given instant, then, part of the membrane surface on side B is in contact with sugar molecules and part is in contact with water, whereas on side A all the membrane surface is in contact with water. Hence more water molecules will move across the membrane from side A to side B per unit time than in the opposite direction; the net osmosis will be from A to B.

We may think of the matter in another way. The arrangement of water molecules in pure water is orderly, in that every molecular location is occupied by a water molecule, whereas the arrangement in the sugar solution is disorderly, in the sense that any given molecular location may be occupied either by a water molecule or by a sugar molecule. Now, we have said that, all other factors being equal, an orderly system possesses more free energy than a disorderly one. It follows that the orderly water molecules in the pure water (side A) have more free energy than the disorderly water molecules in the sugar solution (side B). There is a free-energy gradient for water from side A to side B, and, according to our generalization concerning diffusion, there will be a net movement of water down this gradient, from A to B.

We are now in a position to make some additional generalizations. *The free energy of water molecules is always decreased if osmotically active substances* (i.e. dissolved or colloidally suspended particles) *are present in the water.* (Colloidal particles are generally larger than the separated individual molecules of a dissolved substance, yet small enough so that—unlike the still larger particles of a true suspension—they do not settle out at an appreciable rate, but remain dispersed within the fluid medium.) The reason for the diminution in free energy is that the osmotically active particles to some degree disrupt the orderly three-dimensional array of the water molecules (see Fig. 2.21, p. 45). An important consideration here is the **osmotic concentration,** the number of osmotically active particles per unit volume. *The decrease in the free energy of water molecules is proportional to the osmotic concentration.* From these generalizations, and from our earlier generalization that net movement is always from the region of greater to the region of less free energy, it follows logically that if, under conditions of constant temperature and pressure, two different solutions are separated by a membrane permeable only to water, *the net movement of water will be from the solution with the lower to the solution with the higher osmotic concentration.* The steeper the osmotic-concentration gradient, the more rapid the movement.

Another question arises in connection with the U-tube example of Fig. 3.11. If, as we have seen, the net movement of water is from side A to side B,

The pressure that must be exerted on a solution to keep it in equilibrium with pure water when the two are separated by a semipermeable membrane—in our U-tube example, the hydrostatic pressure exerted by the sugar solution at equilibrium—is known as *osmotic pressure*. Clearly, then, *the osmotic pressure of a solution is a measure of the tendency of water to move by osmosis into it.* The more dissolved particles in a solution, the greater the tendency of water to move into it, and the higher the osmotic pressure of the solution. Thus, under constant temperature and pressure, water will move from the solution with the lower osmotic pressure to the solution with the higher osmotic pressure when the two solutions are separated by a semipermeable membrane.

Because the concept of osmotic pressure is not an easy one to grasp intuitively, the inverse concept, *osmotic potential*, is sometimes applied instead. Pure water is assigned the osmotic-potential value of zero. Since the osmotic potential decreases as the osmotic concentration increases, all solutions have values of less than zero. Under constant temperature and pressure, water will move from the solution with the higher osmotic potential to the solution with the lower osmotic potential when the two solutions are separated by a semipermeable membrane.

While the terms "osmotic pressure" or "osmotic potential" are regularly used by physiologists studying animals, plant physiologists more often refer to *water potential*, which is essentially the same as the free energy of water. At a pressure of one atmosphere, pure water is assigned a water potential of zero. Since the water potential decreases as the osmotic concentration increases, all solutions have values of less than zero. In this sense, water potential is like osmotic potential. But unlike osmotic potential, which is a function of solute concentration alone, water potential (like free energy) is also a function of temperature and pressure. When two solutions are separated by a semipermeable membrane, water will move from the solution with the higher water potential to the solution with the lower water potential.

Familiarity with all three of these terms is useful, because they are all common in the biological literature. In this book, however, we shall ordinarily use the term "osmotic pressure."

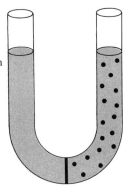

Lower osmotic concentration

Higher free energy of water

Higher osmotic potential

Higher water potential

Lower osmotic pressure

Higher osmotic concentration

Lower free energy of water

Lower osmotic potential

Lower water potential

Higher osmotic pressure

the volume of fluid will increase on side B and decrease on side A. How long can this process continue? Will an equilibrium point be reached?

Clearly, conditions on the two sides of the membrane will never be equal, no matter how many water molecules move from A to B, because the fluid in B will remain a sugar solution, though an increasingly weak one, and the fluid in A will remain pure water, if the membrane is completely impermeable to sugar molecules. The net movement of water from A to B might be expected to continue indefinitely. However, this is not in fact what happens. Under normal conditions, the fluid level in B will rise to a certain point and then cease to rise further. Why? The column of fluid is, of course, being pulled downward by gravity; i.e. it has weight. As the column rises, therefore, it exerts increasing downward hydrostatic pressure. Now, as the pressure increases, the free energy of the water in the sugar solution rises, because free energy is proportional to pressure. Eventually the column of sugar solution becomes so high, and the pressure and free energy so great, that water molecules begin to move across the membrane from B to A as fast as they move into B from A. When this point is reached—when water is passing through the membrane in opposite directions at the same rate—the system is in dynamic equilibrium. Under given conditions of temperature and pressure, the equilibrium point is determined by the difference in osmotic concentration on the two sides of the membrane; the greater the difference, the higher the column will rise before equilibrium is reached.

It should be emphasized that osmotic concentration is not concentration by weight, but rather molecular or ionic concentration, i.e. the total number of solute particles per unit volume. If there are several kinds of solutes in the same solution, then the osmotic concentration of that solution is determined by the total (per unit volume) of *all* the particles of all kinds. If a dissolved substance ionizes, then each ion functions osmotically as a separate particle; e.g. one mole of sodium chloride (NaCl) dissolved in water produces two moles of particles—one of Na^+ ions and one of Cl^- ions. Colloidal particles may also contribute to the total osmotic concentration.

By now you doubtless realize that we have discussed diffusion and osmosis and the role played by semipermeable membranes at such length because the cell membrane is semipermeable and the processes of diffusion and osmosis are fundamental to cell life. Although the membranes of different types of cells vary widely in their permeability characteristics—e.g. the membrane of a human red blood cell is over one hundred times more permeable to water than the membrane of *Amoeba*, a single-celled organism—a few rough generalizations can be made: Cell membranes are relatively permeable to water and to certain simple sugars, amino acids, and lipid-soluble substances. They are relatively impermeable to polysaccharides, proteins, and other very large molecules. Their permeability to small inorganic ions differs greatly depending on the particular ion, but in general negatively charged ions can cross cellular membranes more rapidly than positively charged ions, though neither can do so as readily as uncharged particles.

The role of carriers in the membrane So far we have treated the cell membrane as though it were an inert partition through which substances move by simple diffusion. And, indeed, some of the movement through the membrane is the result of free-energy gradients (usually due to concentration gradients) between the cell contents and the external medium.[3] But simple diffusion is far from the whole story. The membrane is a part of the living cell, and as such it is an active participant in the exchange of materials through it.

It is an indication of the dynamic role played by the cell membrane that simple sugars and polar amino acids regularly move through it, even though they are insoluble in lipid and far too large to move through pores in the membrane. Clearly, there must be a special mechanism whereby some substances not soluble in the lipid bilayer can nonetheless be transported through it.

It is easy to demonstrate that the transport system is highly selective. If a molecule that can readily enter a cell is slightly altered, but not in such a way as to change its solubility properties, it often loses its capacity to move through the membrane. Such a degree of selectivity suggests that the transport agents, or "carriers," are enzymelike proteins—a hypothesis supported by a variety of experiments. It has been shown, for example, that the movement of some substances through the membrane can be competitively inhibited by structurally related ones. Inhibition would not occur if both substances were moving by simple diffusion; apparently the two substances compete for access to specific binding sites on enzymelike carrier molecules in the membrane. Other experiments show that the movement of substances can sometimes be blocked altogether by reagents that seem to act as irreversible inhibitors, i.e. that combine permanently with the carrier molecules and inactivate them.

We have seen that the number and kinds of protein molecules in the cell membrane vary among different types of cells. At least some of those proteins must function as carriers—or **permeases,** as they are often called. Since different types of cells differ in their permeability characteristics, each type must have in its membrane permeases specific to the particular materials it takes in or releases. That the concept of permeases in the membranes is correct has been confirmed in recent years by the isolation of several such substances and the characterization of their binding properties.

The mechanism whereby permeases transport their substrates through membranes remains uncertain. It is assumed that the permease is oriented in such a way that its active site is exposed at the membrane surface, and that the molecule to be transported binds to this site (probably by weak bonds). But how transport is achieved after formation of the permease-substrate complex (P-S) is still not known.

According to one long-popular model, the complex, envisioned as lipid-soluble, would diffuse from the surface where it was formed to the other side of the membrane, where the substrate would be released. The free P would then diffuse back through the membrane, from the release side to the pickup

[3] In some cases passive diffusion results from free-energy gradients attributable to electric-potential gradients. Suppose, for example, that a cell contains a high concentration of large negatively charged molecules to which the membrane is impermeable. Suppose, further, that the membrane is permeable to a small positively charged ion that is in slightly greater concentration inside the cell than out. The positively charged ions may exhibit a net movement into the cell, against their concentration gradient, because of their attraction to the negatively charged molecules in the cell.

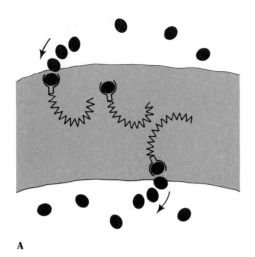

A

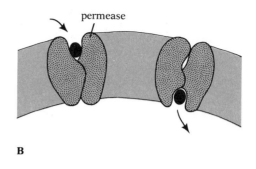

permease

B

3.12   Two models of active transport across a membrane
(A) Anchored in the membrane, the permease picks up a molecule of substrate at one surface of the membrane; it then undergoes a conformational change that causes its loaded end to move to the other side of the membrane, where the substrate is released.   (B) As the permease (which is probably composed of at least two subunits) picks up the substrate at a first binding site, it undergoes a conformational change that channels the substrate to a second binding site near the other side of the membrane, where it is released. The second model is the one best supported by current evidence.

side, where it would be available for reloading. In short, the permease would keep shuttling back and forth across the membrane, moving as part of the P-S complex in one direction and as free P in the other. Unfortunately the weight of recent evidence seems to be against this shuttle model; no membrane proteins have been found capable of such transmembrane movement.[4]   Another model that seems more in agreement with current evidence depends on a conformational change in the permease. According to this model, the binding of substrate causes the permease to undergo a change in its three-dimensional folding that might have one or another effect: It might move the active site with its substrate load to the other side of the membrane (Fig. 3.12A), or it might shift the substrate along the permease molecule (perhaps through a channel in the molecule) from the first

[4] Some relatively small, nonproteinaceous substances (called ionophores) that can transport ions across membranes may function in shuttle fashion. However, all known ionophores are antibiotic poisons produced by microorganisms, and none have thus far been found as normal components of membranes. Two examples of antibiotic ionophores are valinomycin and nigericin.

SIMPLE DIFFUSION

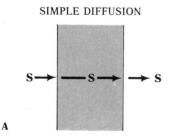

A

FACILITATED DIFFUSION

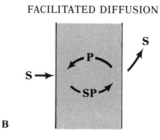

B

ACTIVE TRANSPORT

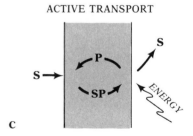

C

3.13   A comparison of simple diffusion, facilitated diffusion, and active transport
(A) In simple diffusion the substrate, S, is transported through the membrane without the aid of permease and without expenditure of energy by the cell.   (B) In facilitated diffusion a permease, P, transports the substrate through the membrane, but no energy is expended.   (C) In active transport a permease is involved, and energy must be expended to drive the process.

binding site near the pickup side of the membrane to a second binding site near the release side (Fig. 3.12B).[5]

Whatever the actual mechanism, two categories of carrier-mediated transport can be distinguished. In one case the permease merely accelerates movement of a substance through the membrane by enabling it to penetrate the membrane more easily, but the direction of movement is determined by the concentration gradient of the substance itself. Because the movement is with a favorable concentration gradient, and hence makes no energetic demands on the cell, the process is called passive transport, or *facilitated diffusion* (Fig. 3.13B).

In *active transport,* on the other hand, the substance being transported is moved against its concentration gradient, i.e. from a region where it is in lower concentration to one where it is in higher concentration. Because the movement is up a free-energy gradient, it is possible only through the expenditure of energy by the cell; i.e. the cell must actively perform work (Fig. 3.13C). For example, many cells perform active transport to maintain a concentration of sodium that is lower inside the cell than in the surrounding fluid. Since the membrane is somewhat permeable to sodium, sodium diffuses passively into the cell, but as fast as it does so, the cell actively pumps it out again, against the concentration gradient. Conversely, many of these same cells actively transport potassium into the cell against a concentration gradient and thereby maintain a much higher concentration of potassium inside than outside.

Another method by which the cell may play an active role in determining what substances will move into it, and in what quantities, depends on the cellular contents rather than the membrane. As a given substance moves into the cell (by simple or facilitated diffusion), some other molecule may bind with it; formation of the compound prevents a buildup in concentration of the unbound substance and maintains a concentration gradient that favors movement of more of the substance into the cell (Fig. 3.14). Alternatively, the substance may be precipitated as soon as it enters the cell and thus prevented from diffusing.

Endocytosis and exocytosis   Substances may enter a cell without actually moving through the cell membrane. By an active process called *endocytosis,* the cell encloses the substance in a membrane-bounded vesicle pinched off from the cell membrane. Two types of endocytosis are recognized:

1. When the material engulfed is in the form of large particles or chunks of matter, the process is called *phagocytosis* (Fig. 3.15). Usually, armlike processes of the cell, called *pseudopodia,* flow around the material, enclosing it within a vesicle, which then becomes detached from the plasma membrane and migrates into the interior of the cell.

2. When the engulfed material is liquid or consists of very small particles, the process is termed *pinocytosis.* The material first becomes adsorbed on the cell membrane, probably at selective binding sites. Then the loaded

---

[5] It has repeatedly been suggested that the conformational change induced by the binding of substrate causes the entire permease molecule to rotate, so that the side with substrate is turned to the opposite surface of the membrane. Since research on membrane proteins has not thus far uncovered any proteins that can rotate in this fashion, this "rotating-door" mechanism, though it cannot be completely ruled out as this writing (1979), must be regarded as unlikely.

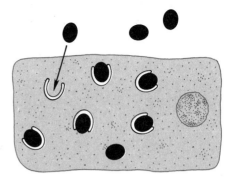

3.14   Formation of a complex inside the cell
As fast as S molecules (black ovals) enter the cell, they combine with molecules of X (white) already in the cell to form a new substance, XS. The concentration of free S inside the cell therefore remains low, and S continues to diffuse inward.

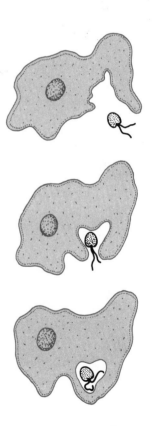

3.15   Phagocytosis of food by *Amoeba*
Pseudopodia flow around the prey until it is entirely enclosed within a vacuole.

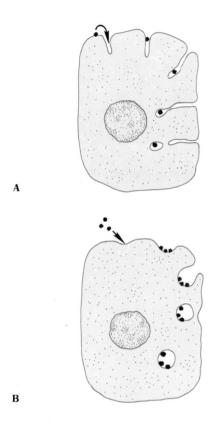

A

B

3.16 Pinocytosis
(A) Membrane with an adsorbed particle flows into a deep narrow channel, at the end of which the particle is enclosed in a vesicle that detaches and moves into the interior of the cell.   (B) Alternatively, adsorbed particles may be enclosed in vesicles detached directly from the cell surface.

membrane either flows inward into a deep narrow channel, at the end of which vesicles are formed, or the small vesicles are simply detached directly from the membrane at the cell surface (Fig. 3.16).

When material is enclosed within endocytotic vesicles, it has not yet entered the cell in the fullest sense. It is still separated from the cellular substance by a membrane, and it must eventually cross that membrane (or the membrane must disintegrate) if it is to become incorporated into the cell. Often the material is first acted on by enzymes in the vesicles and broken down into smaller, simpler substances that can move more easily across the vesicular membranes. In this manner substances that cannot penetrate the membrane can enter the cell in the form of their breakdown products. But sometimes a substance that could not pass through the cell membrane can pass through the vesicular membrane, even though that membrane originated as a part of the cell membrane. For example, the single-celled organism *Amoeba* has a cell membrane impermeable to glucose; yet when this sugar is enclosed in an endocytotic vesicle, it rapidly moves through the vesicular membrane into the cytoplasm of the cell. Apparently the membrane undergoes important functional changes as it is converted from cell membrane into vesicular membrane. Here is but one of many examples demonstrating that living membranes are dynamic and changing structures.

In a process essentially the reverse of endocytosis, called *exocytosis,* materials contained in membranous vesicles are conveyed to the periphery of the cell, where the vesicular membrane fuses with the cell membrane and then bursts, releasing the materials to the exterior (Fig. 3.17). Many secretory materials are released from cells in this way; e.g. the hormone insulin is released by exocytosis from the pancreatic cells that synthesize it. Exocytosis also functions in release of waste products from cells (e.g. from digestive vacuoles of *Amoeba*). And in some cases a coupling of endocytosis and exocytosis moves substances entirely through a cellular barrier, such as the wall of a blood vessel; the substance is picked up by endocytosis on one side of the cell, and the vesicle then moves through the cell to the other side, where the substance is released by exocytosis (see p. 288).

The osmotic effects of the fluids bathing cells   It would be a mistake to suppose that the plasma membrane can completely regulate the exchange of materials between the cell and the surrounding medium and always maintain optimum conditions within the cell. Some poisons can apparently move freely across the plasma membrane, much to the detriment of the cell. And some beneficial substances are lost to the cell because the membrane cannot prevent them from diffusing out.

Furthermore, the great permeability of the membrane to water can be harmful or even fatal to the cell. When a cell is in a medium that is *hypertonic* relative to it (i.e. in a medium to which it loses water by osmosis, usually because the medium contains a higher concentration of osmotically active particles), the cell tends to shrink (Fig. 3.18); and if the process goes too far, it may die. Conversely, when a cell is in a medium *hypotonic* to it (i.e. in a medium from which it gains water, usually because the medium contains a lower concentration of osmotically active particles), the cell tends to swell; and unless it has special mechanisms for expelling the excess water, or special structures that prevent excessive swelling (as most plant cells do), it may burst. A cell in an *isotonic* medium (i.e. one with which the cell is in osmotic balance, usually because it contains the same concentration of osmotically active particles) neither loses nor gains appreciable quantities of water by osmosis.

Obviously, the osmotic relationship between the cell and the medium surrounding it is a critical factor in the life of the cell. Some cells are normally bathed by an isotonic fluid and therefore have no serious osmotic problems. Human red blood cells are an example; they are normally bathed by blood plasma, with which they are in relatively close osmotic balance. Most simpler oceanic plants and animals also exemplify cells in an isotonic medium; their cellular contents have an osmotic concentration close to that of seawater. All cells, however, have a higher osmotic concentration than freshwater. Freshwater organisms thus live in a hypotonic medium and face the problem of accumulating excessive water within their cells by osmosis. Their very existence has depended on the evolution of ways of preventing their cells from becoming so turgid (i.e. so distended by their fluid contents) that they would burst. A variety of evolutionary solutions to this problem will be examined in a later chapter. But let us mention one example here. We stated earlier that the membranes of *Amoeba* are less permeable to water than those of human red blood cells. Now, *Amoeba* lives in freshwater and has therefore been exposed to selection pressures favoring evolution of more impermeable membranes, whereas red blood cells, bathed in an isotonic medium, have not been subjected to any significant selection for such membranes.

CELL WALLS AND COATS

For as long as biologists have been examining cells under the microscope, they have been aware that plant cells are encased in a conspicuous cell wall. This wall, which is located outside the plasma membrane, is composed primarily of carbohydrate. Biologists have also long known that the cells of fungi and most bacteria have strong, thick walls containing much carbohydrate. But only in recent years have they come to realize that most animal cells, too, have a carbohydrate coat on the outer surface of their plasma membranes, and that this coat plays an important role in determining certain properties of the cells. Thus the presence of carbohydrate materials on their outer surfaces appears to be a general property of cells. Nonetheless, the conspicuous, thick, relatively rigid walls of plant, fungal, and bacterial cells, on the one hand, and the inconspicuous, thin, nonrigid coats of animal cells, on the other, remain among the most striking differences between these groups.

Cell walls   Located outside the cell membrane, the plant cell wall is generally not considered part of the cellular protoplasm, although it is a product of the cell. The principal structural component of the cell wall is the complex polysaccharide *cellulose,* which is generally present in the form of long

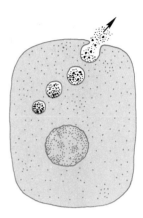

3.17   Exocytosis
A membranous vesicle moves to the periphery of the cell, where it bursts, releasing its contents to the exterior.

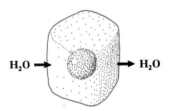

ISOTONIC MEDIUM

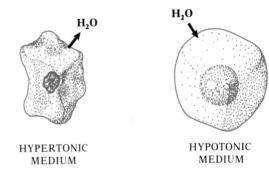

HYPERTONIC   HYPOTONIC
MEDIUM   MEDIUM

3.18   Osmotic relationships of a cell
In an isotonic medium water gain and water loss are equal; hence the cell neither shrinks nor swells. In a hypertonic medium there is a net loss of water from the cell, which therefore shrinks. In a hypotonic medium the cell has a net gain of water and swells.

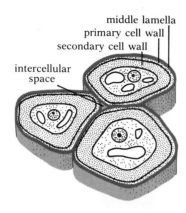

middle lamella
primary cell wall
secondary cell wall
intercellular
space

3.19 Cell walls and middle lamella of three adjacent plant cells

3.20 Electron micrograph of cellulose microfibrils from the cell wall of a green alga
The microfibrils are laid out in parallel lines in two directions; each is about 20 nm wide. The specimen, from the green alga *Chaetomorpha melagonium*, was shadowed with palladium-gold. × 14,000. [Courtesy Eva Frei and R. D. Preston, University of Leeds.]

threadlike structures called fibrils. The cellulose fibrils are cemented together by a matrix of other organic compounds, including the carbohydrates pectin and hemicellulose (a material not structurally related to cellulose). The spaces between the fibrils are not entirely filled with matrix, however; hence they generally allow water, air, and dissolved materials to pass freely through the cell wall. The wall does not usually exert a discriminating effect as to which materials can enter the cell and which cannot. This function is mostly reserved to the plasma membrane located below the cell wall.

The first portion of the cell wall laid down by a young growing cell is the **primary wall.** As long as the cell continues to grow, this wall, which is stretchable, owing to the action of certain plant hormones, is the only one formed. Where the walls of two cells abut, an intercellular layer between them, known as the **middle lamella,** binds them together. **Pectin,** a complex polysaccharide generally present in the form of calcium pectate, is one of the principal constituents of the middle lamella. If the pectin is dissolved away, the cells become less tightly bound to each other. That is what happens, for example, when fruits ripen. The calcium pectate is partly converted into other more soluble forms, the cells become looser, and the fruit becomes softer. Many of the bacteria and fungi that produce soft rots of the tissues of higher plants do so by first dissolving the pectin, reducing the tissue to a soft pulp on which they can feed.

Cells of the soft tissues of the plant have only primary walls and intercellular middle lamellae. After ceasing to grow, the cells that eventually form the harder, more woody portions of the plant add further layers to the cell wall, forming what is known as the **secondary wall.** Since this wall, like the primary wall, is deposited by the protoplasm of the cell, it is located internal to the earlier-formed primary wall, lying between it and the membrane (Fig. 3.19). The secondary wall is often much thicker than the primary wall and is composed of a succession of compact layers, or lamellae. The cellulose fibrils in each lamella lie parallel to each other and are generally oriented at angles of 60–90 degrees to the fibrils of the adjacent lamellae (Fig. 3.20). This arrangement gives added strength to the cell wall. In addition to cellulose, secondary walls usually contain other materials such as **lignin,** which make them stiffer. Once deposition of the secondary wall is completed, many cells die, leaving the hard tube formed by their walls to function in mechanical support and internal transport for the body of the plant.

The cellulose of plant cell walls is commercially important as the main component of paper, cotton, flax, hemp, rayon, celluloid, and, obviously, wood itself. Lignin extracted from wood is sometimes used in the manufacture of synthetic rubber, adhesives, pigments, synthetic resins, and vanillin.

Plant cell walls generally do not form completely uninterrupted boundaries around the cells. There are often tiny holes in the walls through which delicate cytoplasmic connections between adjacent cells may run. These connections are called **plasmodesmata** (Fig. 3.44). They are often located in areas, called pits, where the cell wall is very thin (see Fig. 7.9, p. 267). Thus the cytoplasm of an individual cell in a multicellular plant body is not isolated, but is in contact and communication with the cytoplasm of other cells by way of the plasmodesmata. The cytoplasms of cells interconnected by plasmodesmata constitute a continuous system called the **symplast.** A large portion of the intercellular exchange of such materials as sugars and amino acids probably takes place through the plasmodesmata of the symplast.

The cell walls of both fungi and bacteria differ from those of plant cells. In fungi the main structural component of the wall is not cellulose but **chitin**, a polymer that is a derivative of the amino sugar glucosamine (Fig. 2.27, p. 49). In bacteria the cell walls contain several kinds of organic substances, which vary from subgroup to subgroup. The distinctive responses of these organic substances to diagnostic stains are a regular means of identifying bacteria in the laboratory. Structurally, however, the cell walls of all bacterial groups are alike. The entire wall has a rigid framework of polysaccharide chains covalently cross-linked by short chains of amino acids; the resulting structure can be regarded as a single enormous molecule, often called **murein.**

The presence of cell walls means that the cells of plants, fungi, and bacteria can withstand very dilute external media without bursting. In such media the cells are, of course, in a condition of turgor (distention). Water tends to move into them by osmosis, as a result of the high osmotic concentration of the cell contents. The cell swells, building up **turgor pressure** against the cell walls. The walls exert an equal opposing pressure against the swollen cell (from physics you may recall Newton's third law: To every action there is an equal and opposite reaction). The cell wall of a mature cell can usually be stretched only by a minute amount. Equilibrium is reached when the resistance of the wall is so great that no further increase in the size of the cell is possible and, consequently, no more water can enter the cell. Thus the cells of plants, fungi, and bacteria are not as sensitive as animal cells to the difference in osmotic concentration between the cellular material and the surrounding medium. Because of their walls, such cells can withstand much wider fluctuations in the osmotic makeup of the surrounding medium than animal cells.

Animal cell coats   In plants, fungi, and bacteria, the cell wall is entirely separate from the plasma membrane; if the cell shrinks in a hypertonic medium, the membrane separates from the much more rigid wall (see Fig. 8.2, p. 310). By contrast, the cell coat of an animal cell is not an independent entity. The carbohydrates (short chains of sugars called oligosaccharides) of which it is composed are covalently bonded to protein or lipid molecules in the plasma membrane (Fig. 3.21). The resulting complex molecules, made up of carbohydrate and protein or carbohydrate and lipid components, are termed glycoproteins or glycolipids, and the cell coat itself is often called the **glycocalyx.**

According to recent research, it is the glycocalyx that provides the recognition sites on the surface of the cell enabling it to interact with other cells. For example, if individual liver and kidney cells are mixed in a culture medium, the liver cells will recognize one another and reassociate; similarly, the kidney cells will seek out their own kind and reassociate. Apparently some property of the glycocalyx enables the cells to distinguish liver from kidney cells in such a situation. Cell recognition in the process of embryonic development doubtless also depends, at least in part, on the glycocalyx. And the same is probably true for the control of cell growth. When normal cells grown in tissue culture touch each other, they cease moving and their

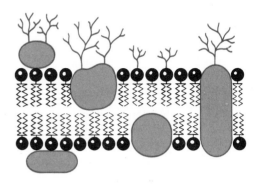

3.21   Plasma membrane with glycocalyx
The glycocalyx of an animal cell is composed of oligosaccharides (branching structures) attached to some of the protein and lipid molecules of the outer surface of the membrane.

growth slows down or stops altogether. This phenomenon of **contact inhibition** appears to be absent in cancer cells, which continue growing without restraint, most likely because they have an abnormal glycocalyx that does not permit them to interact properly.

The glycocalyx also provides recognition sites for interaction between the cell and important molecules in the extracellular medium. For example, the antigenic sites that make cells foreign to the organism recognizable by antibody molecules are probably provided by the oligosaccharides sticking outward from the glycocalyx of the foreign cells; it has already been shown that oligosaccharides provide the sites responsible for the ABO blood groups in human red blood cells (see p. 601). The recognition of host cells by invading viruses probably also depends on the oligosaccharides of the glycocalyx.

THE NUCLEUS

Within the cells of most organisms (though not of bacteria and blue-green algae), the largest and one of the most conspicuous structural areas is the membrane-bounded nucleus (Fig. 3.22). The nucleus plays the central role in cellular reproduction, the process by which a single cell divides and forms two new cells. It also plays a crucial part, in conjunction with the environment, in determining what sort of differentiation a cell will undergo and what form it will exhibit at maturity. And the nucleus directs the metabolic

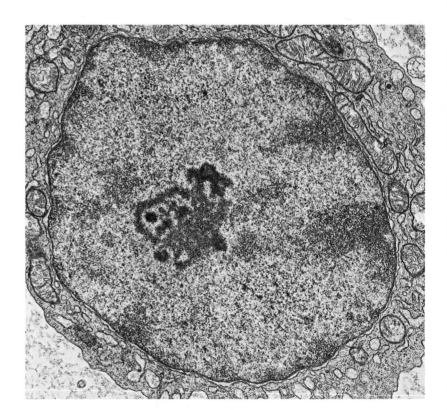

3.22 Electron micrograph of an ovarian-cell nucleus from a rat
The dark area near the center of the nucleus is a nucleolus. $\times$ 13,000. [Courtesy N. B. Gilula, Rockefeller University.]

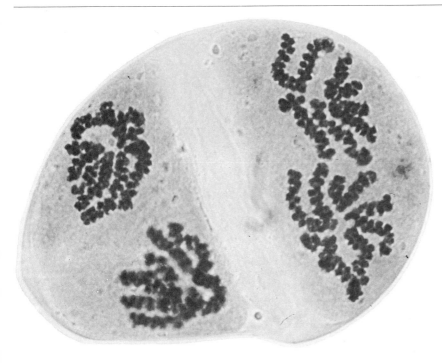

3.23 Chromosomes in a dividing cell of the reproductive organ of *Trillium*

The separate chromosomes in the two nuclei can easily be distinguished. × 1,400.   [Courtesy A. H. Sparrow and R. F. Smith, Brookhaven National Laboratory.]

activities of the living cell. In short, it is from the nucleus that the "instructions" emanate that guide the life processes of the cell as long as it lives.

We have said that the cells of bacteria and blue-green algae differ from those of all other kinds of organisms in lacking a membrane-bounded nucleus (though they do possess genetic material that controls the cell's activities). Similarly, these two groups lack many of the other cellular structures found in other organisms. These differences are so fundamental that bacteria and blue-green algae are classified in a kingdom of their own (the Monera), and their cells are designated as *procaryotic* ("having a primitive nucleus"), whereas the cells of all other organisms are designated as *eucaryotic* ("having a true nucleus"). The following discussion of the nucleus and other cellular organelles will be concerned only with eucaryotic cells; the characteristics of procaryotic cells will be discussed in a later section.

Within the eucaryotic nucleus are two relatively distinct types of structures, the chromosomes and the nucleoli. They are embedded in a mass of rather amorphous, granular-appearing nucleoplasm. The entire nucleus is bounded by a nuclear membrane.

The *chromosomes* (Fig. 3.23) are elongate, threadlike bodies clearly visible only when the cell is undergoing division. They are composed of DNA and protein. The chromosomes bear, apparently in linear arrangement, the basic units of heredity, called *genes,* which are composed of DNA. It is the genes that, passed from generation to generation, determine the characteristics of cells and act as the units of control in the day-to-day activities of living cells. They are the code units, if you will, for the transmission of information from parent to offspring.

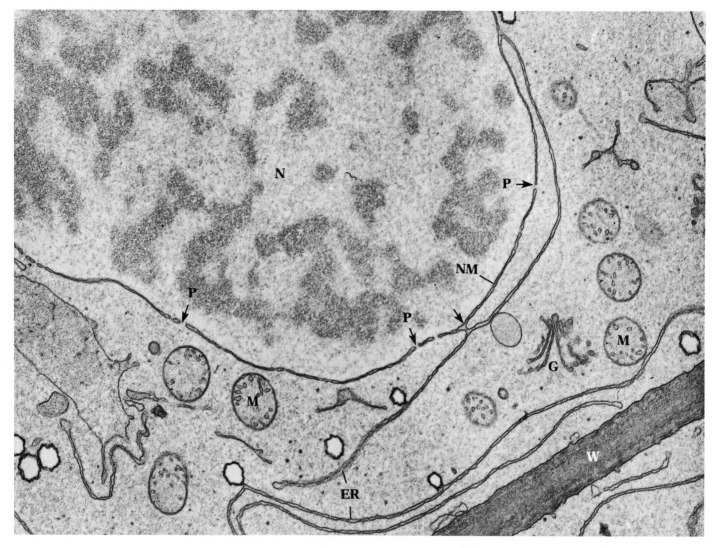

3.24 Electron micrograph showing nuclear membrane of a cell from corn root
The large structure filling the upper left quarter of the picture is the nucleus. The unlabeled arrow indicates a point where the endoplasmic reticulum and the double nuclear membrane interconnect. ER, endoplasmic reticulum; G, Golgi apparatus; M, mitochondrion; N, nucleus; NM, nuclear membrane; P, pore in nuclear membrane; W, cell wall. × 16,700. [Courtesy W. G. Whaley, H. H. Mollenhauer, and J. H. Leech, *Am. J. Botany*, vol. 47, 1960.]

The hereditary information is written in the sequence of the nucleotide building blocks of the DNA molecules. Since the DNA genes remain in the nucleus while most of the processes the genes control take place in the cytoplasm, some mechanism must exist for conveying the information outside the nucleus. The first step is transcription of the nucleotide sequence in the DNA into a corresponding nucleotide sequence in so-called messenger RNA (mRNA). The mRNA can leave the nucleus and move to the sites of protein synthesis in the cytoplasm. There amino acids are ordered in a sequence corresponding to the sequence of nucleotides in the mRNA and are then linked by peptide bonds to form proteins (including enzymes). As we saw in Chapter 2, the sequence of amino acids—the primary structure of a protein—determines the three-dimensional conformation of the protein and the biological activity that this conformation bestows. Thus the genes are at the very hub of life; they encode all the information necessary for the synthesis of the enzymes regulating the myriad interdependent chemical reactions that determine the characteristics of cells and organisms. A later chapter will deal in more detail with the question of how genes work, and with some of the experiments that underlie the current conception of the process.

The other prominent structures in the nucleus, besides the chromosomes, are the ***nucleoli,*** dark-staining, generally oval bodies clearly visible as a rule within the nuclei of nondividing cells. There may be one or more nucleoli per nucleus, depending on the species of organism. Nucleoli form in association with particular regions of specific chromosomes; they are, in fact, simply specialized parts of the chromosome and, like the rest of the chromosome, are composed of DNA and protein. The DNA of the nucleoli includes multiple copies of the genes from which a type of RNA called ribosomal RNA (rRNA) is transcribed. After this rRNA is synthesized, it combines with proteins, and the resulting complex detaches from the nucleolus, leaves the nucleus, and enters the cytoplasm, where it becomes a part of the protein-synthesizing organelles called ribosomes (to be discussed in more detail below). Thus the nucleoli are responsible for manufacturing and exporting to the cytoplasm the precursors of the particles on which proteins will be synthesized. They are small or absent in cells that carry out little protein synthesis.

The presence of a ***nuclear membrane*** surrounding the nucleus permits maintenance within the nucleus of an environment different from that in the surrounding cytoplasm. Unlike the plasma membrane, the complete nuclear envelope consists of two membranes; i.e. it is double (Figs. 3.22 and 3.24). A distinct space is enclosed between the inner and the outer membrane.

Electron-microscope studies indicate that the double-membrane envelope is interrupted at intervals by fairly large pores at points where the outer and inner membrane are continuous (Figs. 3.24, 3.25, 3.27). Nevertheless, the membrane is highly selective. According to permeability experiments, some substances that can cross the cell membrane into the cytoplasm apparently cannot readily cross the nuclear membrane into the nucleus and are consequently restricted to the cytoplasm. For example, when isotopically labeled albumin[6] is injected into rats and enters certain cells of the liver, microscopic studies of these cells show that the labeled albumin is in the cytoplasm only. Such experiments with molecules much smaller than the pores show clearly that simple unrestricted movement through them is definitely not possible. Yet some macromolecules readily pass through the pores. These seem to be primarily substances produced on the genes (such as mRNA) that are moving out of the nucleus, proteins moving into the nucleus to be incorporated into nuclear structures or to catalyze chemical reactions in the nucleus, and various substances from the cytoplasm that

---

[6] An isotopically labeled substance is one containing a radioactive isotope. Such an isotope acts as a tag, enabling an investigator to locate the substance after it has been incorporated by other material, e.g. living cells.

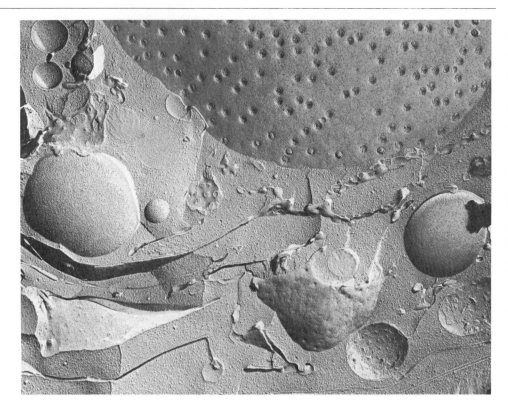

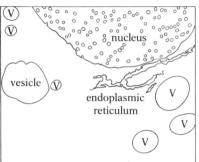

3.25 Freeze-etch electron micrograph of an onion root-tip cell

At upper right is the surface of the nuclear membrane, with numerous pores. A variety of vesicles can be seen in the cytoplasm. × 18,000. [Courtesy Daniel Branton, Harvard University.]

move into the nucleus and help regulate gene activity. There is thus a carefully controlled and highly selective two-way exchange of materials between the nucleus and the cytoplasm through the pores.

The electron microscope has revealed another particularly interesting fact about the nuclear membrane—the continuity, at some points, of this double membrane with an extensive cytoplasmic membrane system called the endoplasmic reticulum (Fig. 3.24).

THE ENDOPLASMIC RETICULUM AND RIBOSOMES

For many years cell chemists have successfully used the technique of differential centrifugation to study the various fractions of cells. The first step is to grind up the tissue in a liquid medium of proper osmotic concentration to form a homogenate, which is then put in a centrifuge and spun. Initially it is spun at a relatively slow rate to separate out the larger, heavier parts of the homogenate, such as any remaining whole cells and the nuclei. When these fractions have been removed, the remaining material is again spun, this time at a higher speed, and cellular components of intermediate size, such as mitochondria, plastids, etc. (parts of the cell we shall discuss later), precipitate out and are removed. The supernatant liquid can then be spun at still higher speeds and smaller, lighter cellular fractions precipitated. By centrifuging the homogenate at increasing speeds in this manner, a series of fractions, segregated according to their particular size and density, can be obtained. These fractions can then be analyzed both morphologically and chemically to determine their characteristics and probable functions.

In 1938 Albert Claude of the Rockefeller Institute isolated by centrifugation at very high speeds some essentially submicroscopic components of the cytoplasm, which were later termed microsomes (small bodies). This fraction made up as much as 15 to 20 percent of the total cell mass and could be isolated from almost any kind of cell—plant or animal. Chemical analysis showed that the microsomes had a very high nucleic acid content; in fact, they contained almost all the cytoplasmic nucleic acid. They also contained a high percentage of the cytoplasmic phospholipids. However, since the microsomes were not visible under the light microscope, there was much argument whether they were actually discrete portions of the living cells or simply artifacts produced by the breaking up and centrifugation of the cells.

Biologists had long wondered if the cytoplasm had some sort of invisible structural organization; faint traces of such a cytoplasmic "skeleton" had been reported at various times. Now, the cytoplasm contains much dissolved protein, and protein molecules are so large that, even when they are truly in solution, they behave physically like colloidal particles. Depending on such factors as temperature, pH, and salt concentration, colloidal particles (and dissolved protein molecules) may be dispersed randomly in the medium, forming a sol; or they may be arranged in a more orderly three-dimensional network, forming a gel. Hence the reported cytoplasmic organization might have been due merely to the colloidal properties of the cytoplasm, some regions being in the sol state and others in the gel state. That the organization of the cytoplasm is not attributable solely to sol-gel transitions was finally demonstrated in 1945 when Keith R. Porter, then of the Rockefeller Institute, using a phase microscope, which has much greater resolution than the usual light microscope, described a complex system of membranes forming a network in the cytoplasm. This system, called the *endoplasmic reticulum* by Porter, has since been extensively studied with the electron microscope. It has been shown to be present in all nucleated cells (though it is often not as well developed in Protozoa as it is in higher animals and plants).

Although the endoplasmic reticulum (ER) varies greatly in appearance in different cells—its components may look like long tubules or round or oblong vesicles, or they may form stacks of flattened sacs (Fig. 3.26)—it always forms a system of membrane-enclosed fluid-filled spaces. In many cells, though not all, these spaces are interconnected, forming a true reticulum. Sometimes the membranes of the ER are lined on their outer surfaces by small particles called *ribosomes,* in which case the ER is spoken of as "rough"; when no ribosomes line the membranes, the ER is described as "smooth" (Fig. 3.27). Just as the ER may exist without associated ribosomes, ribosomes may occur independently of the ER.

It is now known that the microsomal fraction obtained by differential centrifugation is composed of ribosomes and fragments of the ER. It has been shown that rRNA is one of the principal constituents of ribosomes; thus the high concentration of nucleic acid in the microsome is explained. Similarly, the phospholipids of the ER membranes account for the high concentrations of these materials in the microsome.

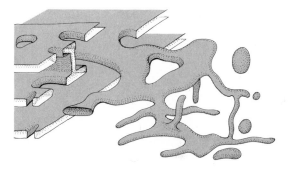

3.26 Drawing of three-dimensional aspect of the endoplasmic reticulum

Flattened cisternal elements are shown at left, tubular and vesicular elements at right.

There is now general agreement that ribosomes, which are present in all actively metabolizing cells, are the sites of protein synthesis. When radioactive amino acids are injected into cells engaged in protein synthesis, a high percentage of the radioactivity is localized on the ribosomes within 10 or 15 minutes; soon thereafter the radioactive amino acids are found incorporated into proteins. Much exciting recent work (to be examined in a later chapter) concerns the mechanism of protein synthesis on the ribosomes.

But what of the endoplasmic reticulum? What functions might it have? Early in the electron-microscopic examinations it was noticed that the ER was almost always closely associated with the nuclear membrane. Electron micrographs often show connections between the outer portion of the double nuclear membrane and adjacent elements of the ER (Fig. 3.24), an indication that the spaces between the two components of the nuclear membrane are continuous with the membrane-enclosed spaces (cisternae) and channels of the ER. It seems, therefore, that the nuclear membrane is only a specialized part of the general cell-membrane system, a part of the endoplasmic reticulum.

One possible function of the ER is immediately apparent. Its channels could serve as routes for transport of materials between various regions of the cytoplasm or between the various parts of the cytoplasm and the nucleus, forming a communications network, as it were, between the nuclear control center and the rest of the cell. There is considerable evidence, in fact, that materials do move within the ER. Some of it comes from the investigations of P. Siekevitz and G. E. Palade, of the Rockefeller Institute, on the movement of zymogen, an enzyme precursor synthesized in large quantities by certain cells in the pancreas of guinea pigs. Siekevitz and Palade demonstrated that the zymogen, which is doubtless synthesized on ribosomes attached to the ER, soon crosses the membranes of the ER into the cisternae and then moves within the ER to other parts of the cell, particularly the Golgi apparatus (an organelle of the cytoplasm, discussed below). It is probable that most, and perhaps all, proteins to be transported in the ER channels are synthesized on ribosomes attached to rough ER; indeed, such association of the ribosomes with the ER membrane may be essential to the penetration of the protein through the membrane. Proteins synthesized on free ribosomes are apparently not destined for export or for incorporation into membranes; it seems that they are released to function as enzymes in the *cytosol* (the more fluid part of the cytoplasm).

It would be surprising if the ER functioned only as a passageway for intracellular transport. The ER membrane has a large protein content, and proteins, you will remember, may act both as structural elements in cells and as enzymes catalyzing chemical reactions. There is now abundant evidence that at least some of the many protein molecules of which the ER membranes are composed act as enzymes and that the ER functions as a cytoplasmic framework providing catalytic surfaces for some of the biochemical activity of the cell; its complex folding provides an enormous surface for such activity. Some of the evidence has come from studies of the smooth ER found prominently in cells in which large quantities of lipids are synthesized; apparently the enzymes involved in lipid synthesis are structural components of the ER membranes, for all efforts to separate them from the membranes have failed.

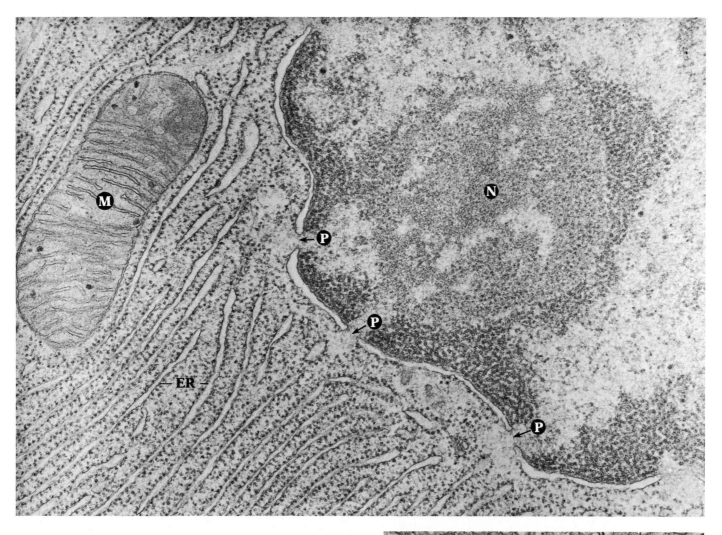

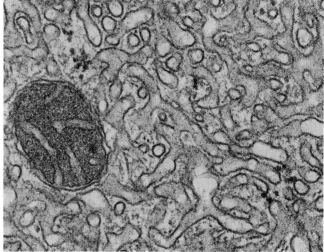

3.27 Electron micrographs of endoplasmic reticulum
Top: Thin section of a pancreatic cell from a bat, showing many flattened cisternae of rough ER; the ribosomes lining the ER membranes can be clearly distinguished. In the upper right portion of the micrograph is part of the nucleus (N); note the very prominent pores (P) in the double nuclear membrane. A mitochondrion (M) is at upper left. × 40,000. Bottom: Smooth ER from an interstitial cell of boar testis. Note the absence of ribosomes from the surfaces of the numerous membranes. The dark structure at left is a mitochondrion. × 61,000. [Top: Courtesy K. R. Porter, University of Colorado. Bottom: Courtesy D. S. Friend, University of California, San Francisco.]

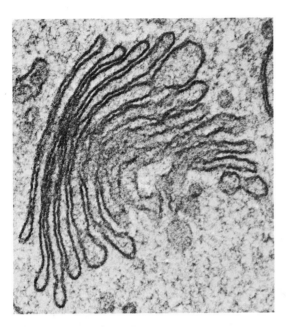

3.28   Electron micrograph of Golgi apparatus from the green alga *Chlamydomonas*
Note the stack of membranous cisternae. The ends of the cisternae become enlarged and are pinched off as secretory vesicles. × 78,000.   [Courtesy M. C. Ledbetter, Brookhaven National Laboratory.]

In the liver of vertebrates, smooth ER has an additional function. It plays a crucial role in detoxifying many poisons and drugs, including barbiturates, amphetamines, morphine, and codeine.

THE GOLGI APPARATUS

In 1898 Camillo Golgi in Italy first described a new "reticular apparatus" in certain cells of the vertebrate brain. This "apparatus" was characterized by its reactions with certain chemicals (silver nitrate or osmium tetroxide), which became impregnated in it and made it visible under the light microscope. Similar cytoplasmic regions were subsequently found by numerous workers in a great variety of animal and plant cells. Although there was variation in their form and several different names were at first applied to them, all eventually came to be called Golgi apparatus (or Golgi bodies). A long controversy developed as to whether they represented actual structures of the living cell or whether they were merely artifacts produced by the fixation process.

With the advent of the electron microscope and its extensive use in the examination of cells, the controversy was at last settled. It was shown that subcellular elements identified as Golgi apparatus have a characteristic fine structure, regardless of the type of cell being studied. They consist of a system of membrane-delimited vesicles arranged approximately parallel to each other (Figs. 3.24 and 3.28). There is evidence that the smooth membranes of the Golgi apparatus often have connections (probably transient) with the membranes of the endoplasmic reticulum and therefore constitute another portion of the complex cellular membrane system.

Many years ago it was noticed that the Golgi apparatus was particularly prominent in cells thought to be involved in the secretion of various chemical products and that, as the level of secretory activity of these cells changed, corresponding changes occurred in the morphology of the organelle. The inference could be drawn that the Golgi played some part in the secretory process. In the preceding section we mentioned that in certain cells of the pancreas of guinea pigs, zymogen, presumably synthesized on the ribosomes, moves into the channels of the endoplasmic reticulum and through these to the Golgi apparatus. With the help of the electron microscope, it has been established that the zymogen is concentrated and stored there, until it is eventually released from the cell via secretory vesicles that are produced by the outer portion of the organelle and that move to the cell surface. Thus the role of the Golgi apparatus in secretion is confirmed; its functions are seen to include storage, modification (e.g. removal of water or emulsification of lipids), and packaging of secretory products. According to the prevailing evidence, no protein synthesis occurs in the Golgi apparatus. However, in some cases, polysaccharides may be synthesized from simple sugars in the Golgi, and these polysaccharides may then be attached to proteins and lipids to form glycoproteins and glycolipids, some of which, in animal cells, will later be incorporated into the glycocalyx.

It has been suggested that the secretory vesicles produced by the Golgi apparatus may also play an important role in adding surface area to the plasma membrane. When a secretory vesicle moves from the Golgi apparatus to the cell surface, it becomes attached to the plasma membrane and then ruptures, releasing its contents to the exterior in the process of exocytosis. The membrane of the ruptured vesicle may remain as a permanent addition to the plasma membrane. Biochemical and morphological studies of the membranes show that the inner portion of the Golgi apparatus resembles the membranes of the nuclear envelope and endoplasmic reticulum, and that there is a progressive change within the Golgi apparatus until the outer portion, where secretory vesicles are produced, resembles the plasma membrane. It has been suggested, therefore, that there is a dynamic relationship between the different parts of the cellular membrane system, with the following directional flow of membrane: nuclear envelope ⟶ rough ER ⟶ smooth ER ⟶ Golgi apparatus ⟶ secretory vesicles (and lysosomes) ⟶ plasma membrane (Fig. 3.29).

MITOCHONDRIA

During the latter part of the nineteenth century, a number of scientists carried out intensive investigations of those granules and formed inclusions of the cytoplasm that could be detected with the light microscopes of that time. Among these subcellular components were some small bodies of variable shape (granules, rods, filaments), but with numerous properties in common (particularly an affinity for certain stains), that seemed to be a constant feature of cytoplasm; they were found in almost all types of eucaryotic cells from most kinds of organisms (Figs. 3.24, 3.27, 3.30). These bodies were given the name "mitochondria" (singular: mitochondrion).

Speculation concerning the function of mitochondria ranged far and wide, until finally, in 1934, a fraction containing the mitochondria was sepa-

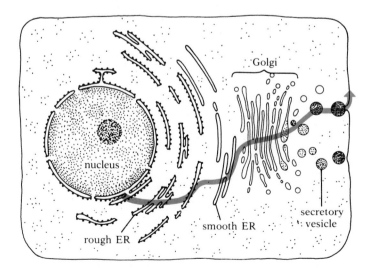

3.29 Membrane flow within the cellular membrane system
New membrane may be formed in the nuclear envelope, whence it may flow to the rough ER, then to the smooth ER, then to the Golgi apparatus, then to secretory vesicles derived from the Golgi, and finally to the plasma membrane.

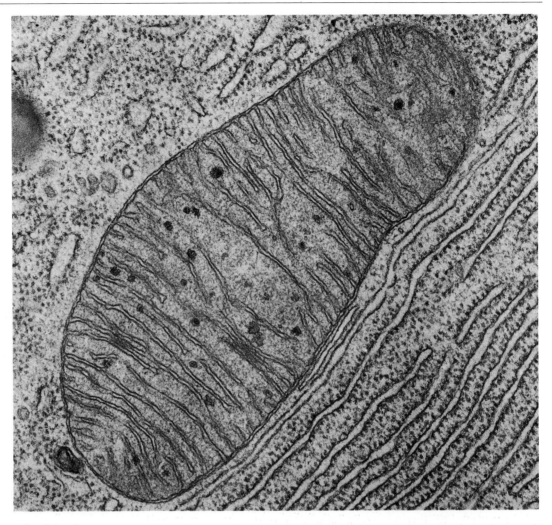

3.30　Electron micrograph of a mitochondrion from the pancreas of a bat

Note the double outer membrane and the numerous cristae, which can be seen to arise as folds of the inner membrane. The dark granules in the matrix of the mitochondrion are thought to be sites where divalent cations, particularly Mg^{++} and Ca^{++}, are bound. $\times$ 50,000.　[Courtesy K. R. Porter, University of Colorado.]

rated from the other cellular components by differential centrifugation.[7] From this time forth, an enormous amount of research was done on the chemical properties of the mitochondrial fraction. It was found that these organelles are the site of chemical reactions that extract energy from food-stuffs and make it available to the cell for its innumerable energy-demanding activities. These chemical reactions, which are fundamental to life, will be one of the main subjects of the next chapter.

Mitochondria are too small for the light microscope to reveal much about their morphology, but they are much larger than many other cytoplasmic bodies, and since they can easily be studied in detail with the electron microscope, knowledge of their structure has made notable progress during

[7] A mitochondrial fraction had already been prepared in the 1920s by Otto Heinrich Warburg, and some of its chemical properties had been determined; but, as has happened so often in the history of science, the importance of this work was not generally recognized, and it made little impression on other scientists.

the last decades. Each mitochondrion is a double-walled vessel; the outer wall is a smooth membrane, and the inner wall is a membrane with many inwardly directed folds (Fig. 3.30; see also Fig. 4.28, p. 176). These folds, called *cristae,* extend into a semifluid amorphous matrix.

Some substances apparently cannot pass from the cytoplasmic matrix into the mitochondria, while others do so fairly easily. Clearly, the mitochondrial membranes, which are composed of a mosaic of protein and lipid similar to that in the membranes already discussed, are differentially permeable. Moreover, their permeability is subject to alteration. It is affected by changes in the concentration of certain chemicals, particularly phosphate ions, in the surrounding cytoplasm. It is also thought to be affected by the degree of stretching the membranes presumably undergo as the mitochondria swell in a hypotonic medium or shrink in a hypertonic one. Insofar as changes in the permeability of the membranes influence what goes into and out of the mitochondria, they would also influence the rates at which chemical reactions take place within the mitochondria. Thus factors that alter the permeability of the membranes may be important in regulating mitochondrial function. The mitochondria illustrate particularly well how the numerous membrane-bounded compartments within a cell permit not only the physical separation of competing reactions, but also the maintenance of distinctive chemical environments for different biochemical pathways.

LYSOSOMES

Lysosomes were first described in the 1950s by Christian de Duve of the Catholic University of Louvain in Belgium. Slightly smaller than mitochondria, with which they were often confused in the past, they are membrane-enclosed bodies that apparently function as storage vesicles for many powerful digestive (hydrolytic) enzymes (Fig. 3.31). The lysosome membrane, which is single (unlike the double-membrane envelope of mitochondria), is both impermeable to the outward movement of these enzymes and capable of resisting their digestive action. Packaged in the lysosomes, the enzymes are prevented from digesting the material of the cell itself. If the lysosome membrane is ruptured, the enzymes are released into the surrounding cytoplasm and immediately begin to break it down.

It is now thought that lysosomes act, in a sense, as the digestive system of the cell, enabling it to process some of the bulk materials taken in by phagocytosis or pinocytosis (see Fig. 5.18, p. 215). They may also play a role in some developmental processes. We shall discuss their functions in more detail in later chapters.

Lysosomes are probably produced by the Golgi apparatus. First discovered in rat liver cells, they have since been found in many kinds of animal cells, and may occur in all of them. They have also been found in some fungi, but it is not yet clear how widely they occur in plant cells.

PEROXISOMES

Improved cell-fractionation methods have revealed in some cells (particularly those of kidney and liver) a type of membrane-bounded organelle

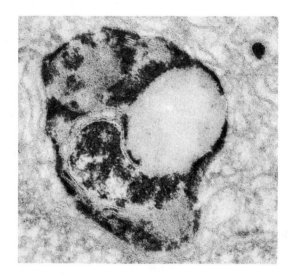

3.31   Lysosomes in a connective-tissue cell from the vas deferens of a rat
The small dark body at upper right is a primary lysosome. The much larger body at left is a secondary lysosome (digestive vacuole) formed by fusion of a primary lysosome with a phagocytic or pinocytic vesicle. (The dark appearance of the lysosomes is due to staining for acid phosphatase, a digestive enzyme whose presence is used as the definitive test for these organelles.) × 100,000. [Courtesy D. S. Friend, University of California, San Francisco.]

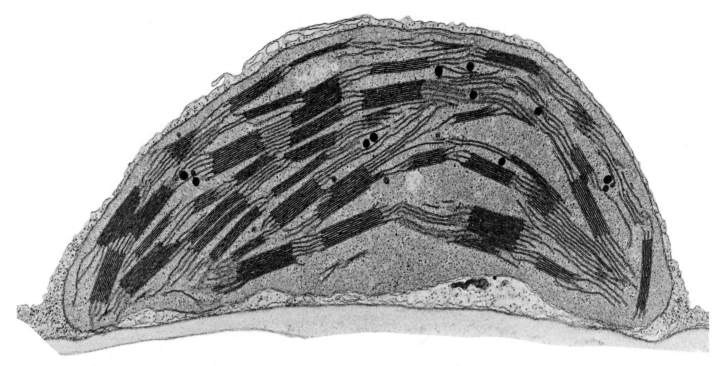

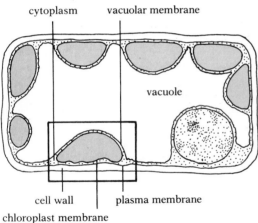

cytoplasm   vacuolar membrane

vacuole

cell wall   plasma membrane

chloroplast membrane

3.32   Electron micrograph of a chloroplast from a corn leaf
The stacks of disclike lamellae are the grana; the less tightly packed lamellae running between the grana are the stroma lamellae. The sketch identifies the structures surrounding the chloroplast. × 29,000. [Micrograph by W. P. Wergin. Courtesy E. H. Newcomb, University of Wisconsin.]

previously confused with lysosomes and now known as peroxisomes. Like lysosomes, these organelles contain an assortment of powerful enzymes.[8] These are oxidative rather than digestive enzymes, catalyzing, for example, the oxidative removal of amino groups from amino acids and the oxidative decomposition of hydrogen peroxide. In the leaves of green plants peroxisomes are the site of photorespiration (to be discussed in Chapter 4).

[8] A variety of organelles similar to peroxisomes in structure and appearance, but containing enzymes of other kinds, have been reported in recent years from one or another group of organisms. Some plant cells, for example, especially in seeds with large fat reserves, possess organelles called glyoxysomes that contain enzymes for converting fat into carbohydrate. It seems likely that the packaging of enzymes in membranous vesicles of this sort will prove to be a common phenomenon. So far there is little agreement how such vesicles should be designated. Some investigators use a different name for each type, depending on the enzymes they contain, whereas others prefer to group them all under the more neutral term "microbodies."

PLASTIDS

Plastids are large cytoplasmic organelles found in the cells of most plants, but not in fungal or animal cells. They can easily be observed with an ordinary light microscope. There are two principal categories of plastids: *chromoplasts* (colored plastids) and *leucoplasts* (white or colorless plastids).[9]

Chloroplasts, which are chromoplasts containing the green pigment *chlorophyll*, are extremely important to all life. Energy from sunlight is trapped in them by chlorophyll and used in the manufacture of complex organic molecules (particularly sugar) from simple inorganic raw materials. Chloroplasts contain, in addition to chlorophyll, various yellow or orange pigments called *carotenoids.*

The electron microscope reveals that the typical chloroplast is bounded by two concentric membranes and has, in addition, a complex internal membranous organization (Fig. 3.32). The fairly homogeneous internal proteinaceous matrix is called the stroma. Numerous double-membrane lamellae are embedded in it. In most higher plants these lamellae are differentiated into two varieties: separate lamellae that run through the stroma and stacks of platelike lamellae forming regions known as *grana.* In some other plants, the brown algae for example, no grana occur; all the lamellae are stroma lamellae. The chlorophyll and carotenoids are located in the lamellae, apparently bound to the proteins and lipids of the membranes. The arrangement of the protein, lipid, and pigment components of the lamellae is evidently a very precise one without which complete photosynthesis cannot take place. We shall consider the internal organization of chloroplasts in more detail in the next chapter when we examine the photosynthetic process.

Chromoplasts lacking chlorophyll are usually yellow or orange (occasionally red) because of the carotenoids they contain. It is these kinds of chromoplasts that give many flowers, ripe fruits, and autumn leaves their characteristic yellow or orange color. Some of these chromoplasts have never contained chlorophyll, while others are formed from chloroplasts whose chlorophyll has been lost. The latter are particularly common in ripe fruits and autumn leaves, structures that were once green.

The colorless plastids, or leucoplasts, are primarily organelles in which materials such as starch, oils, and protein granules are stored. Plastids filled with starch (amyloplasts) are particularly common in storage roots and stems (e.g. carrots, potatoes) and in seeds, although they also occur in the cells of many other parts of the plant. The starch is deposited as a grain or group of grains in the plastid (Fig. 3.33); no starch is found in other parts of the cell.

All types of plastids form from small colorless bodies called proplastids. Once formed, many kinds of plastids can be converted into other types under appropriate conditions. It can be demonstrated, for example, that synthesis of chlorophyll is dependent on light and that under certain conditions leucoplasts exposed to light develop chlorophyll. When a leucoplast is

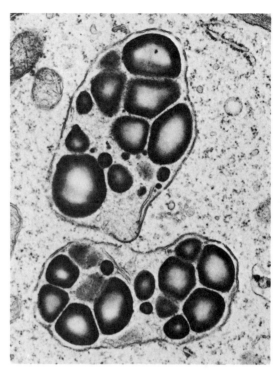

3.33   Electron micrograph of leucoplasts from the root tip of *Arabidopsis*
Because they contain numerous prominent starch grains, these leucoplasts from the small desert plant *Arabidopsis* are called amyloplasts. × 24,500. [Courtesy M. C. Ledbetter, Brookhaven National Laboratory.]

[9] Some classifications of plastids recognize three principal categories: chloroplasts, chromoplasts, and leucoplasts. However, chloroplasts, or green plastids, can appropriately be considered a subcategory of chromoplasts.

converted into a chloroplast, the internal membranous lamellae characteristic of a chloroplast develop from invaginations of the inner boundary membrane of the plastid.

VACUOLES

Membrane-enclosed, fluid-filled spaces called vacuoles are found in both animal and plant cells, though they have their greatest development in plant cells. There are various kinds of vacuoles with a corresponding variety of functions. In some Protozoa specialized vacuoles, called contractile vacuoles, play an important role in expelling excess water and some wastes from the cell; we shall discuss them in greater detail in a later chapter. Many Protozoa also possess food vacuoles, chambers that contain food particles. Similar vacuoles, or vesicles, are formed by many kinds of cells when they take in material by endocytosis.

In most mature plant cells, a large vacuole occupies much of the volume of the cell. The immature cell usually contains many small vacuoles (Fig. 3.34). As the cell matures, the vacuoles take in more water and become larger, eventually fusing to form the very large definitive vacuole of the mature cell. This process pushes the cytoplasm to the periphery of the cell, where it forms a relatively thin layer.

The plant vacuole contains a liquid called cell sap—primarily water, with a variety of substances dissolved in it. Since the cell sap is generally hypertonic relative to the external medium, the vacuole tends to take in water by osmosis. As the vacuole swells, the vacuolar membrane (or tonoplast, as it is often called) pushes outward against the cytoplasm, which, being essentially fluid, resists compression and transmits the pressure to the cell wall. The wall is strong enough to limit the swelling and prevent the cell from bursting, but the outward push of the vacuolar membrane is sufficient to maintain cell turgidity.

Many substances of importance in the life of the plant cell are stored in the vacuoles, e.g. high concentrations of soluble organic nitrogen compounds, including amino acids; sugars; various organic acids; and some proteins.

The vacuoles also function as dumping sites for noxious wastes. Enzymes secreted into them degrade some of these wastes into simpler substances that can be reabsorbed into the cytosol and reused. The presence of poisonous wastes and powerful precipitating or denaturing agents in the vacuoles has greatly complicated studies of plant biochemistry, for when entire cells are disrupted the mixing of these substances with the cytosol often results in the alteration or destruction of the compounds under investigation.

As might be expected, many of the substances accumulated in the vacuoles are selectively prevented from leaving by the vacuolar membrane, which must have its own distinctive permeability characteristics and must be capable of regulating the direction of movement of substances across it. If living beet cells, whose red pigment is "a ***betacyanin***" in the cell sap, are placed in distilled water, the pigment does not diffuse out, even though it is in much higher concentration inside the vacuoles than outside. As soon as the beet cells die, the vacuolar membranes lose their selectivity and the ***betacyanin*** diffuses out.

3.34 Development of the plant-cell vacuole
The immature cell (top) has many small vacuoles. As the cell grows, these vacuoles fuse and eventually form a single large vacuole, which occupies most of the volume of the mature cell, the cytoplasm having been pushed to the periphery (bottom).

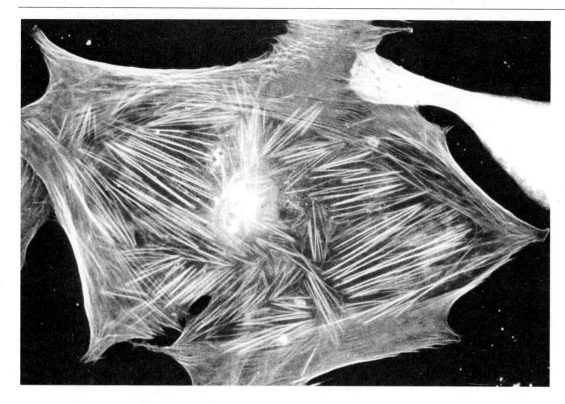

3.35 Immunofluorescent visualization of actin microfilaments in a rat embryo cell

× 300. [Courtesy Elias Lazarides, California Institute of Technology.]

Anthocyanins, another group of red pigments in the cell sap, are responsible for many of the purples, blues, and dark reds commonly seen in flowers, fruits, and autumn leaves (we have already seen that the carotenoids in the plastids are responsible for orange, yellow, and sometimes light red in these same structures). The relative amounts of anthocyanins and carotenoids in autumn leaves differ for different species of plants and also differ under different conditions for the same species. A high accumulation of sugars, low temperatures, and adequate light favors anthocyanin formation.

MICROFILAMENTS

Examined under very powerful electron microscopes, cells often reveal long threadlike microfilaments of extreme thinness. Such filaments were first seen clearly in skeletal-muscle cells, where they are lined up parallel to each other and interdigitated with thicker filaments to give the cell a striated appearance (see Fig. 11.12, p. 476); it is the mechanochemical interaction of the very thin filaments, composed of a protein called actin, with the thicker filaments, composed of another protein, myosin, that enables the muscle cell to move. Eventually, thin microfilaments were also found in many other kinds of cells where, because they are often arranged in less orderly fashion (Fig. 3.35), they are not as conspicuous as in skeletal-muscle cells.

Wherever microfilaments composed of actin or an actinlike protein occur, they appear to be associated with some sort of cellular movement. Thus they are found near the advancing edge of pseudopodia like those of *Amoeba*

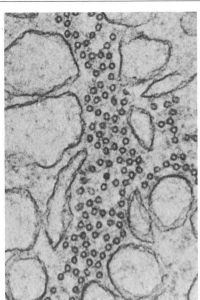

3.36 Electron micrographs of microtubules from a hamster spermatid
Left: Longitudinal section. Right: Cross section. The larger structures near the microtubules are immature mitochondria. × 60,000 (est.). [Courtesy D. W. Fawcett, Harvard Medical School.]

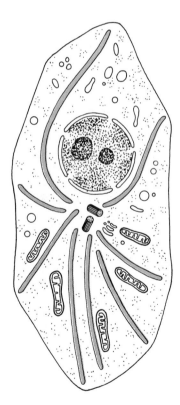

3.37 The microtubular cytoskeleton
It has been proposed that the microtubules (color), which often radiate outward from the region of the centrioles, help hold the cell in its characteristic shape.

(Fig. 3.15) and in some plant cells when cytoplasm streams from one region of the cell to another. When developing cells change shape, constricting in one region or putting out protuberances or forming invaginations, microfilaments are found in the region of movement that brings about the change (see Fig. 17.4, p. 700). When endocytotic vesicles are formed and move into a cell, or when vesicles containing secretory products move toward the plasma membrane and discharge their contents in the process of exocytosis, actinlike microfilaments appear to play a role. All these types of movements, as well as some others not mentioned here, cease if the cell is treated with cytochalasin B, a reagent that alters the properties of actin and is therefore regularly used to reveal cellular processes in which actin or actinlike microfilaments may be involved.

Since some microfilaments in the cytoplasm are not sensitive to cytochalasin B, they are presumably not composed of actinlike protein. These microfilaments are probably not involved in producing motion, but may, instead, help support and strengthen the cell.

MICROTUBULES

Long, hollow, cylindrical structures, the microtubules (Fig. 3.36) seem to play a role in intracellular movement and support, just like the microfilaments. Though not recognized as tubules at the time, they were probably first seen in the spindle of dividing cells. (The spindle, a basketlike arrangement of tubules formed during the process of cell division, is instrumental in moving chromosomes into new nuclei; see Fig. 13.10, p. 566). Later, microtubules were detected elsewhere in the cell, and evidence accumulated that they may help guide the movement of a variety of materials within the cell.

For example, they may delimit the pathways followed by secretory vesicles as they move from their sites of formation to the points on the plasma membrane where they will release their contents by exocytosis.

There is evidence that microtubules also provide a supportive cytoskeleton that helps maintain the cell's shape (Fig. 3.37). If the cell is treated with colchicine, a substance that causes breakdown of microtubules and prevents assembly of new ones, the distinctive shape of the cell is quickly lost. Just as cytochalasin B does with microfilaments, colchicine produces functional disruptions in the microtubules that help reveal their possible physiological roles.

Not all microtubules occur free in the general cytoplasm. Some are incorporated as essential components in several cellular organelles, including centrioles, cilia, and flagella, which we shall discuss below.

The microtubules of the spindle appear at one stage in the process of cell division and disappear at another stage. Microtubules in other parts of the cell are sometimes similarly transitory. In short, except when incorporated in other organelles such as centrioles, cilia, and flagella, microtubules are not stable features of the cell, but may be assembled and disassembled at various times. They are made of a globular protein called tubulin, each molecule of which consists of two independent subunits (polypeptide chains). Assembly of a microtubule involves the stacking, in helical fashion, of tubulin molecules to form the wall of the tube (Fig. 3.38). It is this process that colchicine blocks when it binds to tubulin.

3.38 Structure of portion of a microtubule
The molecules of tubulin protein, each of which consists of two subunits (one shown colored, the other white), are helically stacked to form the wall of the tubule, which is usually 13 subunits in circumference.

CENTRIOLES

Centrioles are small dark bodies located just outside the nucleus of most animal cells in a region of specialized cytoplasm. As electron-microscope studies have revealed, they are cylindrical structures, and two of them normally lie close together, oriented at right angles to each other; in cross section, each is seen to contain nine groups of microtubules, with three tubules in each group (Fig. 3.39).

Centrioles are absent from most of the cells of higher plants, although they do occur in some algae and fungi and in a few reproductive cells of higher plants.

CILIA AND FLAGELLA

Some cells of both plants and animals have one or more movable hairlike structures projecting from their free surfaces. If there are only a few of these appendages and they are relatively long in proportion to the size of the cell, they are called *flagella.* If there are many and they are short, they are called *cilia* (Fig. 3.40). Actually, the basic structure of flagella and cilia is the same, and the terms are often used interchangeably. Both usually function either in moving the cell or in moving liquids or small particles across the surface of the cell. They occur commonly on unicellular and small multicellular organisms and on the male reproductive cells of most animals and many plants, in both of which they may be the principal means of locomotion. They are also common on the cells lining many internal passageways and

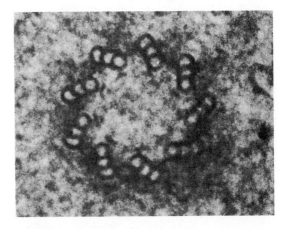

3.39 Electron micrograph of cross section of centriole
Note the nine triplet microtubules. × 180,000. [Courtesy Jean André, Université de Paris-Sud.]

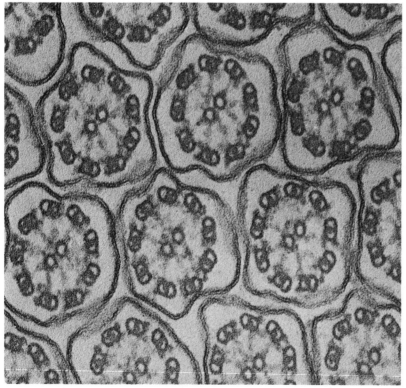

3.40 Scanning electron micrograph of ciliated surface of the trachea (windpipe) of a hamster
Coordinated movements of the many cilia function to sweep dust particles and other foreign material from the respiratory surfaces to the mouth. × 3,000. [Courtesy E. R. Dirksen, University of California, Los Angeles.]

3.41 Electron micrograph of cross sections of cilia
Note that each cilium has nine peripheral and two central sets of microtubules. The peripheral sets are doublets, the central ones singlets. × 140,000. [Courtesy Peter Satir, Albert Einstein College of Medicine.]

ducts in animals, where their beating aids in moving materials through the passageways.

Electron-microscope studies of the flagella and cilia of eucaryotic cells have revealed a remarkable uniformity in their internal structure, regardless of the organism to which they belong, whether plant or animal, simple or complex. The slender cylindrical stalk, an extension of the cell membrane, contains a cytoplasmic matrix, with eleven groups of microtubules embedded in the matrix. Invariably, nine of these groups are arranged around the periphery of the cylinder and the other two are in the center (Fig. 3.41). At the base of the stalk, within the main portion of the cell is a ***basal body,*** whose structure is the same as that of a centriole. The basal body is essential to the function of the cilium or flagellum, and it is the part that gives rise to the stalk. Sometimes fibrous rootlets project from the basal body into the more internal cytoplasm of the cell (see Fig. 3.53). We shall discuss the mechanism of movement of cilia and flagella, with special attention to the role played by the microtubules inside them, in a later chapter.

EUCARYOTIC VS. PROCARYOTIC CELLS

In spite of the extreme variety among cells, we have seen that two fundamental categories are recognized: eucaryotic cells, found in the great majority of organisms, and procaryotic cells, found in bacteria and blue-green algae. The preceding discussion has been almost wholly concerned with eucaryotic cells. Here we offer a mainly pictorial summary of the main features of eucaryotic cells, before turning to the contrasting characteristics of procaryotic cells.

"TYPICAL" EUCARYOTIC CELLS

How complex cells are was not truly appreciated until the advent of the electron microscope and modern biochemical techniques, which combined to change our whole picture of the cell. Three or four decades ago, biology books still regularly included a diagram of a so-called typical cell that showed only five or six simple internal components, a diagram easily memorized by students (Fig. 3.42). No simple diagram of this sort can be given today. In the first place, there is no such thing as a typical cell, not even a typical eucaryotic cell. Not only do plant and animal cells differ in many important ways, but the various cells of the body of any one plant or animal are often strikingly different from one another in shape, size, and function. This much, of course, has been known for a long time. But now that the number of known cellular components has grown so large and their great variability has been so well demonstrated, it becomes even more obvious that no single diagram, or even series of diagrams, can really portray a "typical" cell. Nevertheless, in an effort to help you visualize the arrangement of the organelles discussed in the preceding pages, two such diagrams (Figs. 3.43, 3.44) are given here. As you examine them, keep in mind that not all the components shown always occur together in any one real cell.

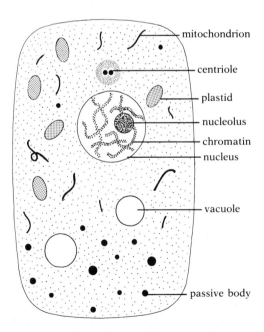

3.42 A "typical" cell as visualized about 1925
[Modified from E. B. Wilson, *The cell in development and heredity,* Macmillan, 1925.]

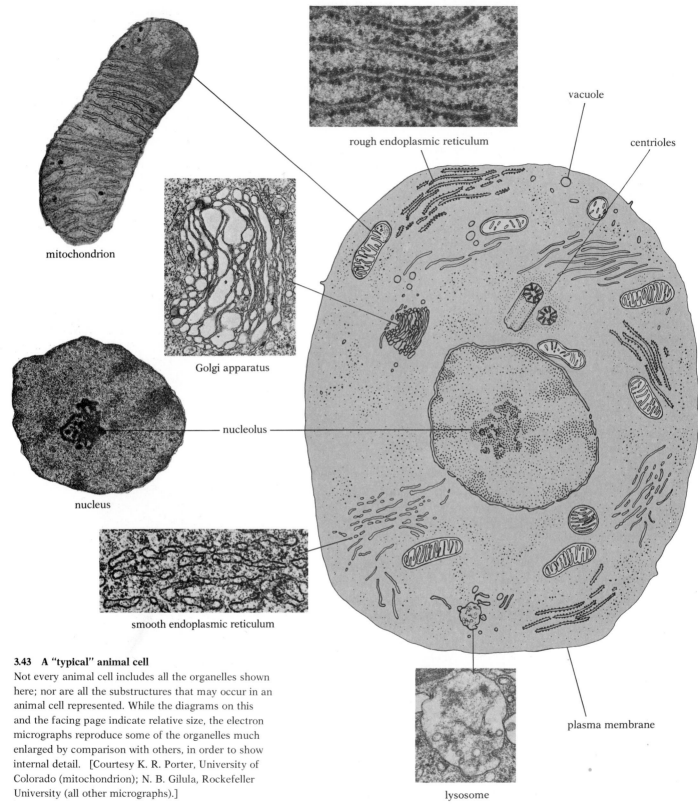

rough endoplasmic reticulum

vacuole

centrioles

mitochondrion

Golgi apparatus

nucleolus

nucleus

smooth endoplasmic reticulum

plasma membrane

lysosome

3.43   A "typical" animal cell

Not every animal cell includes all the organelles shown here; nor are all the substructures that may occur in an animal cell represented. While the diagrams on this and the facing page indicate relative size, the electron micrographs reproduce some of the organelles much enlarged by comparison with others, in order to show internal detail.   [Courtesy K. R. Porter, University of Colorado (mitochondrion); N. B. Gilula, Rockefeller University (all other micrographs).]

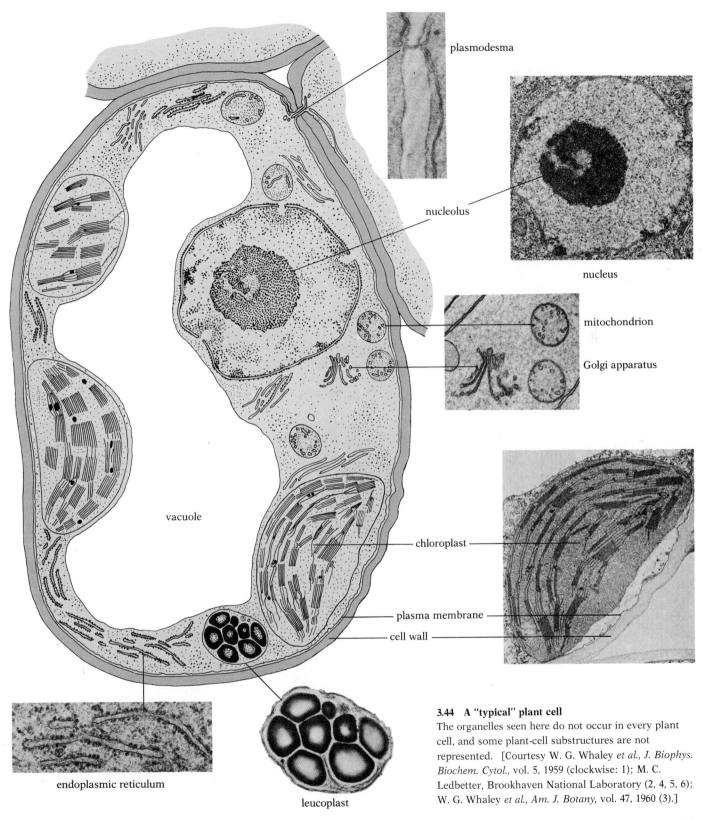

plasmodesma

nucleolus

nucleus

mitochondrion

Golgi apparatus

chloroplast

plasma membrane

cell wall

vacuole

endoplasmic reticulum

leucoplast

3.44 A "typical" plant cell
The organelles seen here do not occur in every plant cell, and some plant-cell substructures are not represented. [Courtesy W. G. Whaley *et al., J. Biophys. Biochem. Cytol.,* vol. 5, 1959 (clockwise: 1); M. C. Ledbetter, Brookhaven National Laboratory (2, 4, 5, 6); W. G. Whaley *et al., Am. J. Botany,* vol. 47, 1960 (3).]

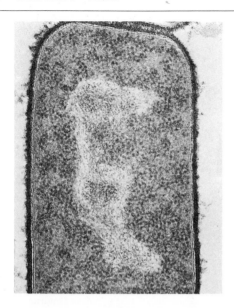

3.45   Electron micrograph of part of a bacterial cell
The light area in the center of the cell, called the
nucleoid, contains DNA, but is not bounded by a
membrane. Note the prominent cell wall, with the
plasma membrane visible just inside it. × 68,000.
[Courtesy A. Ryter, Institut Pasteur, Paris.]

PROCARYOTIC CELLS

Procaryotic cells lack most of the cytoplasmic organelles present in eucaryotic cells. We have already mentioned that they have no nuclear membrane; they also lack other membranous structures such as an endoplasmic
reticulum, a Golgi apparatus, lysosomes, peroxisomes, and mitochondria
(many of the functions of mitochondria are carried out by the inner surface
of the plasma membrane). There is an exception, however, to the general
rule that procaryotic cells lack intracellular membranous structures. All
blue-green algae and the few photosynthetic bacteria contain chlorophyll,
which is associated with membranous vesicles or lamellae, but the lamellae
are not contained in membrane-bounded plastids.

For a long time it was thought that procaryotic cells had no chromosomes.
With the advent of the electron microscope, however, it became possible to
detect in each procaryotic cell a nuclear region, or ***nucleoid*** (Fig. 3.45),
containing a single large DNA molecule, which, though not tightly associated with proteins as DNA is in eucaryotic cells, may nonetheless be considered a chromosome. Unlike eucaryotic chromosomes, which are usually
linear, the procaryotic chromosome is ordinarily circular (Fig. 3.46).

Like eucaryotic chromosomes, the procaryotic chromosome bears, in linear array, the genes that control both the hereditary traits of the cell and its
ordinary activities. The DNA functions by directing protein synthesis on
ribosomes via mRNA, in the way already described for eucaryotic cells. Note
that ribosomes are the most prominent cytoplasmic organelles to occur in

TABLE 3.1   *A comparison of procaryotic and eucaryotic cells*

Characteristic	Procaryotic cells	Eucaryotic cells
Nuclear membrane	Absent	Present
Chromosomes	Single, composed only of nucleic acid	Multiple, composed of nucleic acid and protein
Golgi apparatus, endoplasmic reticulum, lysosomes, peroxisomes	Absent	Present
Mitochondria	Absent, but same function performed by plasma membrane	Present
Photosynthetic apparatus	May contain chlorophyll, but not in chloroplasts	Chlorophyll, when present, contained in chloroplasts
Ribosomes	Small	Large
Microtubules	Absent*	Present
Flagella	Lack 9–2 tubular structure	Have 9–2 tubular structure
Cytoplasmic streaming or amoeboid movement	Does not occur	May occur
Cell wall	Contains murein	When present, does not contain murein

* According to a report published in 1978, there are microtubules in certain spirochete bacteria that inhabit the digestive tract of termites.

both eucaryotic and procaryotic cells. Those of procaryotic cells, however, are structurally different from those of eucaryotic cells, and they are considerably smaller.

Some bacterial cells possess hairlike organelles used in swimming, and these have traditionally been called flagella. But since these organelles do not have microtubules, which do not exist in most procaryotic cells, their internal structure is completely different from that of eucaryotic flagella and the mechanism of their movement must therefore also be different.

We have already (p. 97) contrasted the procaryotic cell walls with those of true plants and fungi. Suffice it to say here that their remarkable murein structure is a distinctive procaryotic trait.

Table 3.1 gives a summary of some of the most important differences between procaryotic and eucaryotic cells.

MULTICELLULAR ORGANIZATION

Some animals and some plants are unicellular, or single-celled, but most are multicellular, or many-celled. Ordinarily the bodies of multicellular organisms, particularly animals, are organized on the basis of tissues, organs, and systems. A *tissue* is composed of many cells, usually similar in both structure and function, that are bound together by intercellular material. An *organ,* in turn, is composed of various tissues (not necessarily similar) grouped together into a structural and functional unit. Similarly, a *system* is a group of interacting organs that "cooperate" as a functional complex in the life of the organism.

The following sections will introduce you briefly to some of the basic plant and animal tissues, organs, and systems. We shall refer to them repeatedly and examine them in more detail in later chapters.

PLANT TISSUES

Plant tissues have been classified in a variety of ways by different botanists. The system used here is not necessarily better than other possible systems; it is simply one of several acceptable ones. The lack of full agreement on any one classification springs from characteristics of the plant cells themselves. The different cell types intergrade, and a given cell may even change from one type to another during the course of its life. Consequently the tissues formed from such cells also intergrade and may share structural and functional characteristics. Furthermore, plant tissues may contain cells of only one type, or they may be complex, containing a variety of cell types. In short, plant tissues cannot be fully characterized or distinguished on the basis of any single criterion such as structure, function, location, or mode of origin.

All plant tissues can be divided into two major categories: meristematic tissue and permanent tissue. *Meristematic tissues* are composed of immature cells and are regions of active cell division; *permanent tissues* are composed of more mature, differentiated cells. This distinction is not an absolute one, however, for some permanent tissues may revert to meristematic activity under certain conditions.

The permanent tissues fall into three subcategories: surface tissues, fundamental tissues, and vascular tissues. Each of these, in turn, contains sev-

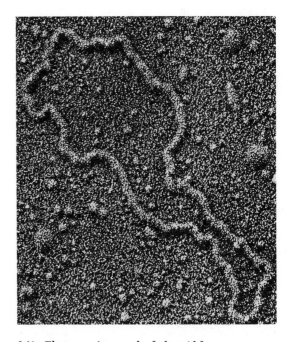

3.46   Electron micrograph of plasmid from *Escherichia coli*

A plasmid is a molecule of DNA that functions as a small accessory chromosome; like the main chromosome of most procaryotic cells, it is generally circular. This plasmid from the common bacterium *E. coli* carries genetic information for conferring resistance to the antibiotic tetracycline. (The plasmid was shadowed with platinum-palladium.) × 106,600. [Courtesy S. N. Cohen, Stanford University.]

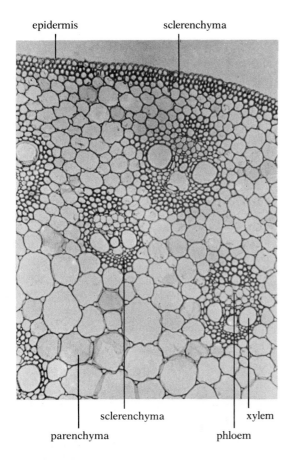

epidermis   sclerenchyma

sclerenchyma   xylem

parenchyma   phloem

3.47 Photograph of portion of cross section of corn stem
× 20. [Courtesy K. D. Laser, State University of New York, Stony Brook.]

eral different tissue types. The classification used here may be summarized as follows:

I. Meristematic tissue
II. Permanent tissue
   A. Surface tissue
      1. Epidermis
      2. Periderm
   B. Fundamental tissue
      1. Parenchyma
      2. Collenchyma
      3. Sclerenchyma
      4. Endodermis
   C. Vascular tissue
      1. Xylem
      2. Phloem

It must be emphasized that this classification is based on the higher land plants—the vascular plants. It has little relevance for other plants, where multiple tissue types seldom occur.

MERISTEMATIC TISSUE

Meristematic tissues are composed of embryonic, undifferentiated cells capable of active cell division. Cell division occurs throughout the very early embryo, but as the young plant develops, many regions become specialized for other functions and cease playing a primary role in the production of new cells. Consequently cell division becomes restricted largely to certain undifferentiated tissues in localized regions; these tissues are the meristems.

It is hard to make any general statement on the characteristics of meristematic cells, because they show much variation. There is no such thing, in fact, as a typical meristematic cell. Nevertheless, we can say that these cells tend to be small, to have thin walls, and to be rich in cytoplasm (i.e. to have only small vacuoles), and that meristematic tissues tend to lack intercellular spaces. New cells produced by a meristem are initially like those of the meristem itself, but, as they grow and mature, their characteristics slowly change and they become differentiated as components of other tissues.

There are regions of meristematic tissue at the growing tips of roots and stems. These apical meristems are responsible for increase in length of the plant body. In many plants, there are also meristematic areas toward the periphery of the roots and stems, and these lateral meristems are responsible for increase in girth.

SURFACE TISSUE

As the term implies, surface tissues form the protective outer covering of the plant body. In young plants and herbaceous adult plants that lack active lateral meristems, the principal surface tissue of roots and stems is the *epidermis* (Fig. 3.47); epidermis is also the surface tissue of all leaves (Fig. 3.48). Often the epidermis is only one cell thick, though it may be thicker, as it is in some plants living in very dry habitats, where protection against water loss is critical. Most epidermal cells are relatively flat. They generally have a very large vacuole and only a thin layer of cytoplasm. Often their outer and side walls are thicker than the inner wall. Epidermal cells on the aerial parts of the plant often secrete a waxy, water-resistant *cuticle* on their outer surface; this, combined with the thick outer wall, which is often impregnated with cutin, aids in protection against loss of water, mechanical injury, and invasion by parasitic fungi. What makes the barrier even more complete is the fact that the irregularly shaped cells usually interlock tightly like pieces of a puzzle, leaving no intercellular space.

Epidermal tissues of the aerial parts sometimes give rise to unicellular or multicellular hairs, spines, or glands. Some epidermal cells, particularly of the leaves, are specialized as guard cells and regulate the size of small holes in the epidermis through which gases can move into or out of the leaf (Fig. 3.48). Epidermal cells of the roots, which have no cuticle and function in water absorption, commonly bear long hairlike processes that greatly increase the total absorptive surface area (see Fig. 5.5, p. 197).

As the stems and roots of plants with active lateral meristems increase in diameter, the epidermis is slowly replaced by another surface tissue, the *periderm* (Fig. 7.7, p. 265). This tissue constitutes the corky outer bark so characteristic of old trees. Functional cork cells are dead; it is their cell walls, waterproofed with suberin, that function as the protective outer covering of the plant.

3.48 Photograph of portion of cross section of privet leaf
Note the numerous chloroplasts (bubblelike structures) in the parenchyma cells. The very dark spots are nuclei. E, epidermis; G, guard cell; P, parenchyma. × 510. [Courtesy Thomas Eisner, Cornell University.]

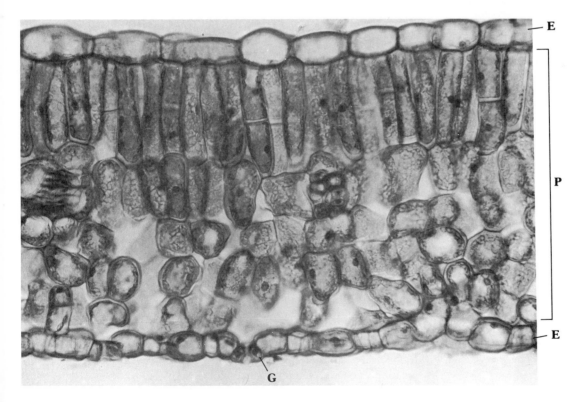

FUNDAMENTAL TISSUE

Most of the fundamental tissues are simple tissues; i.e. each is usually composed of a single type of cell. Often these same types of cells also occur as components of the complex vascular tissues. The various fundamental tissues are not necessarily structurally similar, although they form from the same embryonic regions. They are often defined simply as those tissues that are neither surface tissues nor vascular tissues.

Parenchyma   Parenchyma tissue occurs in roots, stems, and leaves. The parenchyma cells are relatively unspecialized vegetative cells, like those that make up almost the whole body of the lower plants. They have not lost the capacity for cell division, and in some circumstances they take on meristematic activity; they also sometimes undergo further specialization, forming other cell types. Parenchyma cells usually have thin primary walls and no secondary walls. They generally have a large vacuole surrounded by a peripheral layer of cytoplasm. The cells are ordinarily loosely packed; consequently intercellular spaces are abundant in parenchyma tissue (Figs. 3.47, 3.48). Most of the chloroplasts of leaves are in the cells of parenchyma tissue, and it is largely here that photosynthesis occurs. Parenchyma of stems and roots functions in storage of nutrients and water. When turgid, parenchyma is important in giving support and shape to the plant.

Collenchyma   Like parenchyma, collenchyma is a simple tissue whose cells remain alive during most of their functional existence. Though collenchyma cells are characteristically more elongate, they are structurally similar to parenchyma cells, except that their walls are irregularly thickened. The thickened areas are usually most prominent at the edges (the "corners" when viewed in cross section) (Fig. 3.49). Collenchyma functions as an important supporting tissue in young plants, in the stems of nonwoody older plants, and in leaves.

Sclerenchyma   Sclerenchyma is a type of simple fundamental tissue that, like collenchyma, functions in support. However, sclerenchyma cells are far more specialized than collenchyma cells; at functional maturity, most are dead, and their uniformly very thick, heavily lignified secondary walls give strength to the plant body. Often these walls are so thick that the lumen (internal space) of the cell has been nearly obliterated.

Sclerenchyma cells are customarily divided into two categories: fibers and sclereids (Fig. 3.50). *Fibers* are very elongate cells with tapered ends. They are tough and strong, but flexible; commercial flax and hemp are derived from strands of sclerenchyma fibers. *Sclereids* are of variable, often irregular, shape. The simpler, unbranched sclereids are frequently called stone cells; they are common in nutshells and the hard parts of seeds, and are scattered in the flesh of hard fruits (the gritty texture of pears is due to small clusters of stone cells; Fig. 3.50).

Endodermis   Endodermis is a type of tissue difficult to place in any classification. It occurs as a layer surrounding the vascular-tissue core of roots

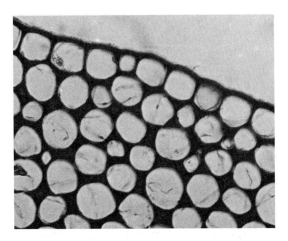

3.49   Collenchyma cells from petiole of beet leaf
Notice the particularly thick walls at the corners of the cells. [Used by permission from J. D. Dodd, *Course book in general botany*, © 1977 by The Iowa State University Press.]

and, less frequently, of stems (Fig. 5.4, p. 196). Young endodermal cells are much like elongate parenchyma cells, except that a band of chemically distinctive thickening runs around the cell on its radial (side) and end walls. This lignified and suberized ("waterproofed") band is called the ***Casparian strip.*** In older endodermal cells, the walls may become secondarily so thickened (sometimes almost obliterating the lumen) that the Casparian strip is obscured, but it can be detected by chemical tests. The cells of endodermal tissue occur in a single layer and are compactly arranged without intercellular spaces. Possible functions of the endodermis will be discussed in Chapter 5.

VASCULAR TISSUE

Vascular, or conductive, tissue is a distinctive feature of the higher plants, one that has made possible their extensive exploitation of the terrestrial environment. It incorporates cells that function as tubes or ducts through which water and numerous substances in solution move from one part of the plant body to another. There are two principal types of vascular tissue: xylem and phloem. Both of these are complex tissues; i.e. they consist of more than one kind of cell.

Xylem   Xylem is a vascular tissue that functions in the transport of water and dissolved substances upward in the plant body. It forms a continuous pathway running through the roots, the stem, and appendages of the stem such as leaves. In its most advanced form (i.e. in the form least like the ancestral one)—in the flowering plants—it commonly includes two types of cells unique to xylem: ***tracheids*** and ***vessel elements.*** It also includes numerous parenchyma and sclerenchyma cells (especially thick-walled fibers, but some sclereids as well). The parenchyma cells are the only living cells in mature functioning xylem, inasmuch as both the cytoplasm and the nucleus

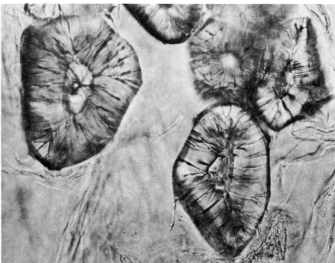

3.50 Sclerenchyma
Left: Cross section of fibers from corn stem. Right: Stone cells of pear fruit. [Used by permission from J. D. Dodd, *Course book in general botany*, © 1977 by The Iowa State University Press.]

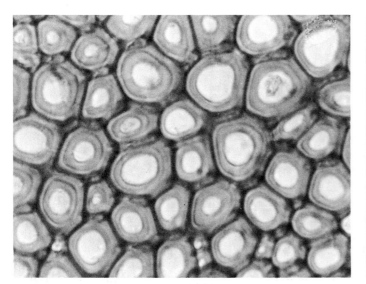

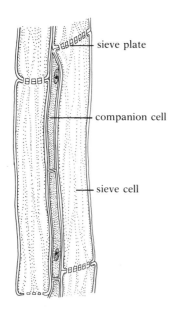

3.51 Longitudinal section of sieve cells and companion cells
The sieve cells lose their nuclei at maturity, but retain their cytoplasm.

sieve plate

companion cell

sieve cell

of tracheids, vessel elements, and sclerenchyma cells disintegrate at maturity, leaving the thick cell walls as the functional structures. In tracheids and vessels, the walls form passages or tubes in which vertical movement of materials can take place (Figs. 7.8, 7.10, 7.11, pp. 266–268).

Transport is not the only function of xylem. Another is support, particularly of the aerial parts of the plant. The numerous fibers in the xylem function almost exclusively in this way, and the thick-walled tracheids are also important as supportive elements. As a reminder of the enormous strength characteristic of xylem, keep in mind that its common name is wood.

Phloem The second vascular tissue, phloem, is unlike xylem in that materials can move both up and down in it. Phloem functions particularly in the transport of organic materials such as carbohydrates and amino acids. For example, newly synthesized organic molecules are moved in the phloem from the leaves to the stem and roots for storage or to the growing points of the plant for immediate use. Like xylem, phloem is a complex tissue and contains both parenchyma and sclerenchyma cells in addition to those found only in phloem: sieve cells and companion cells. The *sieve cells* (Fig. 3.51) are the vertical transport units of phloem; at maturity, their nuclei disintegrate, but their cytoplasm remains. *Companion cells,* which retain both their nuclei and their cytoplasm at maturity, are closely associated with the sieve cells in the most advanced plants; their possible function will be discussed in Chapter 7, where both xylem and phloem will be described in greater detail.

PLANT ORGANS

The body of the higher land plants is customarily divided into two major parts: the *root* and the *shoot* (Fig. 3.52). These two fundamental organ systems can be distinguished on the basis of numerous morphological characteristics, especially the arrangement and mode of origin of the vascular tissue, the ways in which lateral roots and branch stems are formed, and the presence of leaves on the shoot but not on the root. Yet note that many of the tissues are essentially continuous throughout the entire axis, both root and shoot. For example, the vascular tissue of root and shoot, despite a somewhat different arrangement in each, forms an uninterrupted transport system.

The roots of a plant function mainly in procurement of inorganic nutrients such as minerals and water, in transport, in nutrient storage, and in anchoring the plant to the substrate.

The structurally somewhat more complex shoot consists of the stem and the appendages of the stem, particularly the foliage leaves and the reproductive organs. The stem, of course, functions in support and in internal transport, while the foliage leaves are organs in which the critical process of photosynthesis takes place.

You will notice that the number of distinct organs mentioned here—root, stem, foliage leaf, and reproductive organs—is much smaller than it would be for higher animals. In general, the plant body is simply not as clearly subdivided into readily distinguishable functional components, or organs, as the animal body; the parts of the plant grade more imperceptibly into each other and, in some ways, form a more continuous whole. The arrangement of the tissues within these organs will be described in some detail in later chapters.

ANIMAL TISSUES

It is traditional to divide all animal tissues into four categories: epithelium, connective tissue, muscle, and nerve. Each of these, particularly the second, is a diverse assemblage containing numerous subtypes. It should be emphasized that the subtype classification is based primarily on vertebrate animals, especially human beings, and that its application to other animals, particularly the lower invertebrates, is difficult and sometimes even meaningless.

The classification of animal tissues used here can be summarized as follows:

    I. Epithelium
        A. Simple epithelium
           1. Squamous
           2. Cuboidal
           3. Columnar
        B. Stratified epithelium
           1. Stratified squamous
           2. Stratified cuboidal
           3. Stratified columnar
    II. Connective tissue
        A. Vascular tissue
           1. Blood
           2. Lymph
        B. Connective tissue proper
           1. Loose connective tissue
           2. Dense connective tissue
        C. Cartilage
        D. Bone
   III. Muscle
        A. Skeletal muscle
        B. Smooth muscle
        C. Cardiac muscle
   IV. Nerve

EPITHELIUM

Epithelial tissue forms the covering or lining of all free body surfaces, both external and internal. The outer portion of the skin, for example, is epithelium, as are the linings of the digestive tract, the lungs, the blood vessels, the various ducts, the body cavity, etc. Epithelial cells are packed tightly together, with only a small amount of cementing material between them and almost no intercellular spaces. Thus they provide a continuous barrier protecting the underlying cells from the external medium. Because anything

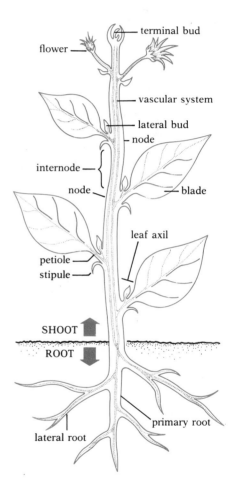

3.52 Diagram of flowering plant body
Note the difference between root and shoot in the arrangement of vascular tissue.

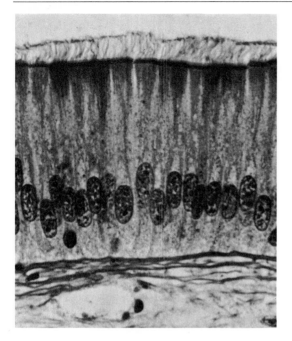

3.53 Micrograph of columnar ciliated epithelium from digestive tract of a freshwater mussel
The dark basal bodies of the cilia and the long cones of fibrous rootlets are plainly visible. Mitochondria are prominent in the basal portion of the cells. Note the dark basement membrane on which the epithelial cells rest. × 960. [Courtesy D. W. Fawcett, Harvard Medical School.]

entering or leaving the body must cross at least one layer of epithelium, the permeability characteristics of the cells of the various epithelia play an exceedingly important role in regulating the exchange of materials between different parts of the body and between the body and the external environment.

Since one surface of an epithelium is generally exposed to air or fluid and the opposite surface rests upon other cell layers, and since the epithelium plays a crucial part in the directional passage of materials, it is no surprise that epithelial cells should show significant differences between their free ends and their attached ends. Often highly specialized, the free ends commonly bear cilia, hairs, or short fingerlike processes; they may also have deep depressions and are sometimes covered with waxy or mucous secretions. Inside the cells, pigments and such organelles as mitochondria are often more abundant at one end than at the other (Fig. 3.53). Recent studies have shown that, as might be expected, the plasma membrane of an epithelial cell is not uniform in its permeability characteristics; the portion of the membrane on the outer surface of the cell, that exposed to the extracellular environment, is quite different from the portions of the membrane adjacent to other epithelial cells. When a form of sodium fluorescent in ultraviolet light is injected into a single epithelial cell, it can be detected within a few minutes in neighboring cells, but it does not appear in the external medium.

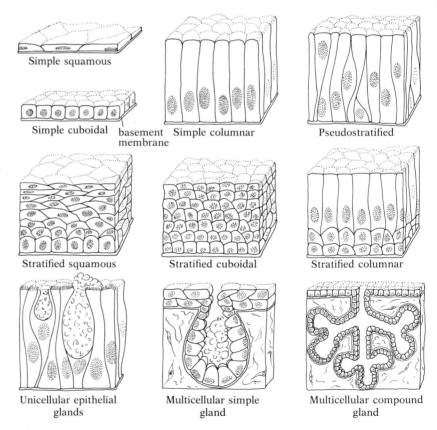

3.54 Epithelial tissues

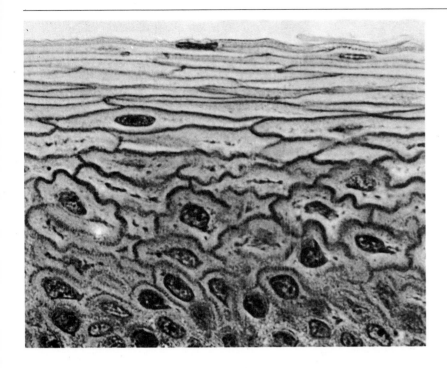

3.55 Stratified squamous epithelium from the gum of a cat
Beneath the layers of squamous cells, which give the epithelium its name, are several layers of cuboidal cells. × 1,300. [Courtesy D. W. Fawcett, Harvard Medical School.]

Apparently the sodium can move across the plasma membrane at the junctions between epithelial cells, but cannot penetrate the plasma membrane where it is in contact with the external medium. Clearly, the chemical and/or physical properties of the membrane are not the same at those two locations.

It is customary to group epithelial cells into three categories: squamous, cuboidal, and columnar. *Squamous cells* are much broader than they are thick and have the appearance of thin flat plates (Fig. 3.54). *Cuboidal cells* are roughly as thick as they are wide and, as their name implies, have a rather square or cuboidal shape when viewed in a section perpendicular to the tissue surface; in surface view, however, they look like polygons, often with six sides. *Columnar cells* are much thicker than they are wide and, in vertical section, look like rectangles set on end.

Epithelial tissue may be only one cell thick, in which case it is called *simple epithelium,* or it may be two or more cells thick and is then known as *stratified epithelium.* There is, in addition, a third category, called *pseudo-stratified epithelium,* in which the tissue looks stratified, but actually is not; whereas in true stratified epithelium only the cells in the lowest layer are in contact with the underlying membrane, in pseudostratified epithelium all the cells are in contact with it. The various types of epithelia are named on the basis of cell type and number of cell layers; it is the cells in the outermost layer of stratified epithelia that determine the name: simple squamous epithelium, simple cuboidal epithelium, simple columnar epithelium, stratified squamous epithelium (Fig. 3.55), stratified cuboidal epithelium, stratified columnar epithelium, etc. Epithelium, regardless of type, is usually

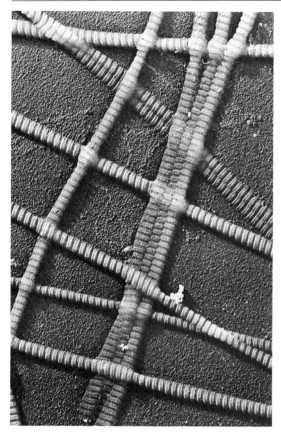

3.56 Electron micrograph of collagen fibrils from calf skin

The preparation was shadowed with chromium. × 30,000. [Courtesy Jerome Gross, Harvard Medical School.]

separated from the underlying tissue by an extracellular fibrous **basement membrane** (Fig. 3.54).

Epithelial cells often become specialized as gland cells, secreting substances at the epithelial surface (Fig. 3.54). Sometimes a portion of the epithelial tissue becomes invaginated, and a multicellular gland is formed.

CONNECTIVE TISSUE

In connective tissue the cells are always embedded in an extensive intercellular matrix. Much of the total volume of connective tissue is matrix, the cells themselves often being widely separated. The matrix may be liquid, semisolid, or solid. Connective tissue is often divided into four main types: (1) blood and lymph; (2) connective tissue proper; (3) cartilage; (4) bone. The last three are sometimes collectively described as supporting tissues.

Blood and lymph   Blood and lymph are rather atypical connective tissues with liquid matrixes. They will be discussed in some detail in Chapter 7.

Connective tissue proper   Connective tissue proper is very variable, but its intercellular matrix always contains numerous fibers. These fibers are of three types:

Collagenous fibers (or white fibers), which are very common, are composed of numerous fine fibrils of collagen, a protein that constitutes a very high percentage of the total protein in the animal body (Fig. 3.56; see also Fig. 2.40, p. 61). Such fibers are flexible, but resist stretching and confer considerable strength on the tissues containing them.

Elastic fibers (or yellow fibers) can, as their name implies, easily be stretched. When the stretching force ceases, the fibers return to their former length. Elastic fibers are often much thinner than collagenous fibers. They are composed of the protein elastin.

Reticular fibers, as the term "reticular" indicates, branch and interlace to form complex networks. They are important at points where connective tissues and other tissues join; e.g. there are many in the basement membrane between epithelium and connective tissue.

Several kinds of cells are generally found in connective tissue proper (Fig. 3.57). They perform a variety of functions:

1. Fibroblasts secrete the proteins from which fibers form.

2. Macrophages, irregularly shaped cells particularly common near blood vessels, become mobilized when there is an inflammation. They can move by amoeboid motion and actively engulf particles such as dead red blood cells and such foreign material as bacteria.

3. Mast cells produce a substance (heparin) that tends to prevent blood clotting and another substance (histamine) that increases the permeability of blood capillaries.

4. Fat cells are cells highly specialized for fat storage. When they are very numerous in a region of connective tissue, the tissue is often called adipose tissue.

5. The various kinds of white blood cells help fight infection. Some can move fairly easily between the blood or lymph and the connective tissue proper—a clear demonstration of the close interrelationship between these tissues.

Both cells and fibers are embedded in a rather amorphous ground substance, which is a mixture of water, proteins, carbohydrates, and lipids. Associated with the ground substance is the *tissue fluid,* a liquid derived from the blood.

Connective tissue proper is customarily subdivided into two basic types—loose connective tissue and dense connective tissue—though there is no rigid separation between them, and intermediate types sometimes occur.

Loose connective tissue is characterized by the loose, irregular arrangement of its fibers, the large amount of ground substance, and the presence of numerous cells of a variety of types (Fig. 3.57). It is very widely distributed in the animal body, no microscopic section of which is free of it. It has been said that if all other tissues were destroyed, the loose connective tissue alone would still show the exact contours of the body and the detailed shape of most internal organs. Much of the framework of the lymph glands, bone marrow, and liver is loose connective tissue; and loose connective tissue supports, surrounds, and connects the elements of all other tissues. For

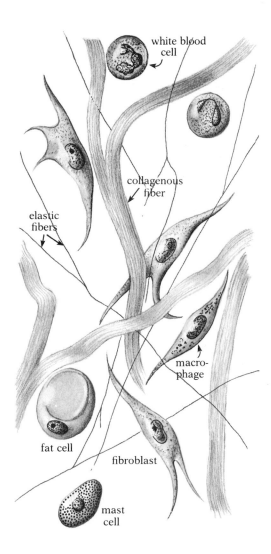

white blood cell

collagenous fiber

elastic fibers

macro-phage

fat cell

fibroblast

mast cell

3.57 Loose connective tissue
The several varieties of cells are embedded in an extensive extracellular matrix of fibers and ground substance.

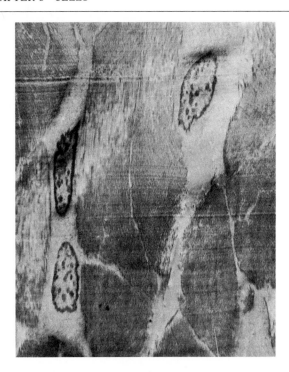 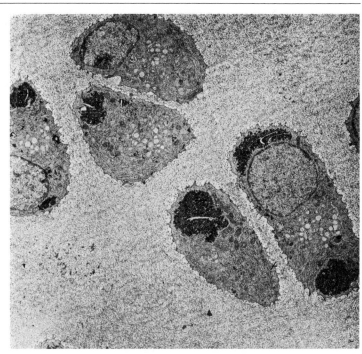

3.58   Photographs of tendon and cartilage
Left: Longitudinal section of tendon from the tail of a
young rat, showing three cells surrounded by bundles
of collagen fibrils. × 3,500.   Right: Section of cartilage
from a newborn mouse, showing five cells embedded in
an extracellular matrix; the matrix looks almost
homogeneous, but actually contains a dense network of
very thin fibrils. × 2,750.   [Left: Courtesy J. Dlugosz
and L. J. Gathercole, University of Bristol. Right: A.
Jurand, University of Edinburgh.]

example, it binds muscle fibers together; attaches the skin to underlying
tissues; forms the membranes that line the heart and abdominal cavities;
forms the membranes, called mesenteries, that suspend the internal organs
in their proper position and the membranes that bind together the parts of
internal organs or that bind various organs together; functions as packing
material in the spaces between organs; and forms a thin sheath around
blood vessels, consequently penetrating with them into the interior of most
organs. Because of its flexibility, loose connective tissue allows movement
between the units it binds or connects.

   Dense connective tissue is characterized by the compact arrangement of
its many fibers, the limited amount of ground substance, and the relatively
small number of cells. The fibers may be irregularly arranged into an interlacing network, as in the dermis of the skin or the sheaths (periostea) of
bone; or they may be arranged in a definite pattern—usually parallel bundles
oriented to withstand tension from one direction, as in tendons connecting
muscle to bone (Fig. 3.58, left) or ligaments connecting bone to bone.

Cartilage   Cartilage (gristle) is a specialized form of dense fibrous connective tissue in which the intercellular matrix has a rubbery consistency (Fig.
3.58, right). The relatively few cells are located in cavities in the matrix.
Cartilage can support great weight; yet it is often flexible and somewhat
elastic. It varies in texture, color, and elasticity, and several major types are
recognized.

   Cartilage is found in the human body in such places as the nose and ears
(where it forms pliable supports), the larynx and trachea (you can feel the

rings of cartilage in the front of your throat), intervertebral discs, surfaces of skeletal joints, and ends of ribs. Most of the skeleton of the early vertebrate embryo is composed of cartilage; the developing bones follow this model and slowly replace it. Some vertebrate groups, the sharks for example, retain a cartilaginous skeleton even in the adult.

Bone   Bone has a hard, relatively rigid matrix. This matrix, which contains numerous collagenous fibers and a surprising amount of water, is impregnated with inorganic salts such as calcium carbonate and calcium phosphate. This inorganic material may constitute as much as 65 percent of the dry weight of an adult bone. The few bone cells are widely separated and are located in spaces in the matrix (see Fig. 11.5, p. 469). We shall discuss the histology of bone in more detail in Chapter 11.

MUSCLE

The cells of muscle have greater capacity for contraction than most other cells, although all protoplasm probably possesses this capacity to some extent. Muscles are responsible for most movement in higher animals. The individual muscle cells are usually elongate and are bound together into sheets or bundles by connective tissue. Three principal types of muscle tissue (see Fig. 11.7, p. 471) are recognized in vertebrates: (1) skeletal or striated muscle, which is responsible for most voluntary movement; (2) smooth muscle, which is involved in most involuntary movements of internal organs; and (3) cardiac muscle, the tissue of which the heart is composed. This classification does not hold for many invertebrates. In these animals one or more of the three types may be missing entirely, and the distribution of the types within the body is nearly always different from that outlined here for vertebrates.

NERVE

To some extent, all protoplasm possesses the property of irritability—the ability to respond to stimuli—but nerve tissue is highly specialized for such response. Nerve cells are easily stimulated and can transmit impulses very rapidly. Each cell is composed of a cell body, containing the nucleus, and one or more long thin extensions called fibers (Fig. 3.59). Nerve cells are thus admirably suited to serve as conductors of messages over long distances. An individual nerve cell may be a meter long, or even longer; no other kind of cell even approaches such length. Many nerve fibers bound together by connective tissue constitute a nerve.

The functional combination of nerve and muscle tissue is fundamental to all multicellular animals except sponges. It is these tissues that confer on animals their characteristic ability to move rapidly in response to stimuli.

ANIMAL ORGANS

The bodies of some of the simpler multicellular animals generally show few clearly distinct organs, but most larger, more advanced animals characteristically have numerous organs, which, in turn, are organized into func-

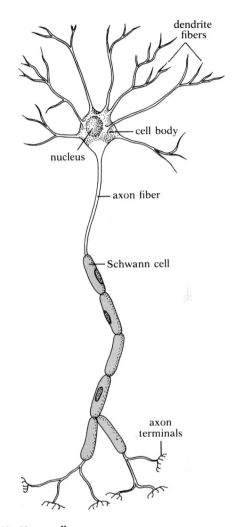

3.59   Nerve cell
The dendrites carry impulses toward the cell body; the axon carries impulses away from the cell body. The axons of vertebrates are often sheathed along much of their length by Schwann cells.

tional complexes called organ systems. Structural organization at the organ and organ-system levels is far more advanced in the higher animals than in the higher plants. This fundamental difference is doubtless correlated with the different modes of life pursued by these two major groups of organisms, particularly their different feeding methods and their different adaptations for a sedentary existence on the one hand (plants) and active locomotion on the other (animals).

An example of the complex integration of different types of cells and tissues to form an animal organ is human skin (Fig. 3.60). It is clear from the illustration that skin is far more complex than a first impression might convey. It contains elements of all four primary animal tissue types: epithelium, connective tissue, muscle, and nerve. Portions of these tissues, in turn, are organized into relatively complex structures like glands, ducts, hairs, blood vessels, and sensory devices. All these structural elements are integrated to form the functional organ. In this case the organ functions in many ways: as a protective covering for the body that resists penetration by many harmful substances and disease-producing organisms, and helps prevent excessive water loss from the body; as a mechanism of excretion for several different waste materials; as an aid in regulating the temperature of the body; as an instrument whose sensory nerve endings receive impulses from the outside environment; and as a depot in which reserve nutrients are stored.

The numerous organs, in addition to skin, that together constitute the body of a higher animal will be discussed in later chapters. They are commonly grouped into organ systems, of which the following will be of particular concern to us:

1. The digestive system, which functions in procuring and processing nutrients;
2. The respiratory system, functioning in the gas-exchange process by which oxygen is taken into the body and waste carbon dioxide released;
3. The circulatory system, the internal transport system of animals;
4. The excretory system, which not only functions in the release of certain metabolic wastes from the body, but acts as a critical regulator of the chemical makeup of the body fluids;
5. The endocrine system, whose glands, and the hormones they produce, play an important role in internal control;
6. The nervous system, a control system essential in coordinating the myriad functions of a complex multicellular animal;
7. The skeletal system, which provides support and determines shape in some animals;
8. The muscular system, of fundamental importance in the movement of animals;
9. The reproductive system, which functions in the production of new individuals.

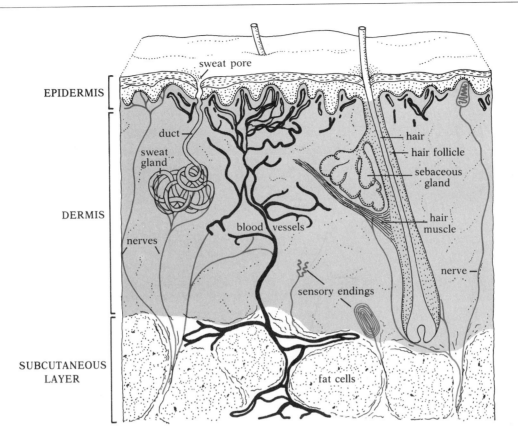

EPIDERMIS

DERMIS

SUBCUTANEOUS
LAYER

sweat pore

duct

sweat
gland

nerves

blood vessels

sensory endings

hair

hair follicle

sebaceous
gland

hair
muscle

nerve

fat cells

3.60 Section of human skin
The outer portion of the skin, the epidermis, is composed of stratified squamous epithelial tissue. Its outermost layer (*stratum corneum*) consists of hardened dead cells that are constantly being sloughed off. Active cell division in the deeper layers of the epidermis produces new cells that are pushed outward and take the place of those that are lost. Beneath the epidermis is a layer, called the dermis, composed chiefly of connective tissue. Blood vessels penetrate into the dermis but not into the epidermis. Sweat glands are embedded in the deeper layers of the dermis, and their ducts push outward through both dermis and epidermis to open onto the surface of the skin through sweat pores. Both the glands and their ducts are

derived from the epidermis; they form initially as invaginations of the epidermis that push downward into the connective tissue of the dermis.

Hairs, and the inner layers of the hair follicles in which they are encased, are also derived from the epidermis and also develop as invaginations into the dermis. When the hair follicle is fully developed, the bulbous base of the follicle and the hair root lie deep in the dermis; the shaft of the hair extends at a slant from the root to the surface of the skin and beyond. A small muscle runs diagonally from the upper portion of the dermis to the hair follicle near its lower end; when this muscle contracts, it pulls the hair erect. One or more sebaceous (oil) glands empty into the hair follicle.

Numerous nerves penetrate into the dermis, and a few even penetrate into the epidermis. Among them are nerves to the hair muscles, sweat glands, and blood vessels, and also nerves terminating in the sensory structures for touch, temperature, and pain.

Beneath the dermis, and not sharply delimited from it, is a subcutaneous layer, which is not considered a part of the skin itself. This is a layer of very loose connective tissue, usually with abundant fat cells. It is this layer that binds the skin to the body. The extent and form of its development determine the amount of possible skin movement.

SUGGESTED READING

ALBERSHEIM, P., 1975. "The walls of growing plant cells," *Scientific American,* April. *Some insights into the special properties of cell walls, derived from study of the arrangements of the various polysaccharides they contain.*

CAPALDI, R. A., 1974. "A dynamic model of cell membranes," *Scientific American,* March. (Offprint 1292) *Good discussion of the fluid-mosaic model of membrane structure.*

DEDUVE, C., 1963. "The lysosome," *Scientific American,* May. (Offprint 156) *The discoverer of lysosomes on his own pioneering work.*

GROSS, J., 1961. "Collagen," *Scientific American,* May. (Offprint 88) *On the molecular structure of the most abundant protein in the human body.*

JENSEN, W. A., 1970. *The plant cell,* 2nd ed. Wadsworth, Belmont, Calif. *Short and easy to read.**

LOEWY, A. G., and P. SIEKEVITZ, 1970. *Cell structure and function,* 2nd ed. Holt, Rinehart & Winston, New York. *Good summary of cell biology.**

LURIA, S. E., 1975. "Colicins and the energetics of cell membranes," *Scientific American,* December. *How antibiotics made by bacteria can be used to study active transport by cell membranes.*

NEUTRA, M., and C. P. LEBLOND, 1969. "The Golgi apparatus," *Scientific American,* February. (Offprint 1134) *How radiography was used in working out the function of the Golgi apparatus.*

ROTHMAN, J. E., and J. LENARD, 1977. "Membrane asymmetry," *Science,* vol. 195, pp. 743–753. *On the functional implications of membrane asymmetry, i.e. the presence of different proteins on the outer and inner surfaces of cell membranes.*

SATIR, B., 1975. "The final steps in secretion," *Scientific American,* October. *How the membrane of a secretory vesicle interacts with the plasma membrane in the process of exocytosis.*

SHARON, N., 1974. "Glycoproteins," *Scientific American,* May. (Offprint 1295) *Good summary of what is currently known about glycoproteins, whose extreme importance to the life of the cell has only recently been recognized.*

SINGER, S. J., and G. NICOLSON, 1972. "The fluid mosaic model of the structure of cell membranes," *Science,* vol. 175, pp. 720–731. *The original publication of Singer's model.*

SWANSON, C. P., 1977. *The cell,* 4th ed. Prentice-Hall, Englewood-Cliffs, N.J. *Readable up-to-date summary of basic cell biology.**

WESSELLS, N. K., 1971. "How living cells change shape," *Scientific American,* October. (Offprint 1233) *The roles of microtubules and microfilaments in cell movement.*

* Available in paperback.

Chapter 4

ENERGY TRANSFORMATIONS

A basic principle of physics, as we saw in Chapter 2, is that all systems have a natural tendency toward increasing disorder. The more orderly any arrangement of matter, the less probable it is, and the less likely to be maintained if energy is not expended to counteract the tendency toward greater disorder. For example, the organization of wood, bricks, and other materials in the form of a house is one of an infinitely large number of possible arrangements of those materials, but it is one so orderly that there is almost no likelihood it would occur spontaneously. Similarly, if all the characters on this page were shaken together and dropped on the floor, one possible arrangement in which they could come to rest is the one you see before you, but this orderly arrangement of letters into words and sentences is so very unlikely to occur by chance that you doubtless feel confident it could not be achieved except by some expenditure of energy. The living cell, whose complex structure we examined in the last chapter, is also an inherently unstable and improbable organization. Only by constant use of energy can it maintain itself and keep from falling into the more stable and more probable random and disorganized state. The acquisition of energy in usable form is thus a necessity for every living cell.

CHARACTERISTICS OF ENERGY

Energy, defined as the capacity to do work, may exist in active (or kinetic) form or in stored (or potential) form. For work actually to be done, the energy must become kinetic, which is to say that there must be motion of some sort. Thus a car parked on a steep hill has potential energy by virtue of its position; when its brake is released, the car rolls down the hill, converting its potential energy into the kinetic energy of motion. Similarly, a lump of coal contains potential energy, which is released as kinetic energy when the coal burns.

Energy can occur in many different forms. Light is energy, and so is electricity; there is heat energy, mechanical energy, chemical energy, etc. All these involve motion, whether motion of photons, as in light, motion of electrons, as in electricity, or motion of atoms and molecules, as in heat. All forms of energy are interconvertible, at least partly. According to the *First Law of Thermodynamics* (also called the Law of Conservation of Energy), when energy is converted from one form into another no energy is either gained or destroyed (for this law to be strictly valid, matter must be considered a form of energy).

As you know, heat, in the form of heat engines, can be used to do work, but heat is not usable in this way by cells (even though temperature influences the rates of all cellular reactions). The reason is that, at constant pressure, work can be done only if there is a temperature gradient—a flow of heat from a region of higher temperature to one of lower temperature. But each cell is essentially isothermal; i.e. at any given moment the entire cell is at the same temperature, and there can be no heat flow. Rather than rely on heat energy, living cells draw primarily on chemical energy; they do work by utilizing the energy stored in complex organic molecules. Every such molecule represents an amount of potential energy equal to the amount of energy necessary to synthesize it originally. Living cells, then, are transducers that turn other forms of energy into chemical energy—or the reverse—or that move energy from one chemical compound to another.

A living organism is a storehouse of potential chemical energy, which can be used, when necessary, to do work. But as this stored energy is converted into other forms, less and less remains in reserve. A source of usable energy outside the organism must be available to replenish its supply. For many organisms, that outside energy source is other organisms; i.e. one living thing obtains new supplies of energy-rich molecules by eating other living things. Since all these energy conversions, according to the First Law of Thermodynamics, are accomplished without reduction of the total amount of energy, it might seem at first glance that the same energy could be passed continuously from organism to organism and that no source of energy outside the system composed of all living things would be required. But a little further thought shows that this is not true, because energy is constantly passed from organisms to nonliving matter, as when you throw a rock or move a pencil or when heat from your body warms the air; such energy is lost to the life system. Furthermore, the molecules of substances that leave the body retain some energy, and this energy, too, may be lost. But there is another basic reason why life on earth would run down if there were no nonliving source of energy to be tapped. The *Second Law of Thermodynamics* says that every energy transformation results in a reduction in the usable, or free, energy of the system; or, to put it another way, the amount of energy that cannot be used to do work increases steadily as a consequence of increasing entropy. We saw in Chapter 2 that entropy is a measure of the randomness of a system. Thus the Second Law is essentially a more formal statement of the idea, expressed earlier, that there is a general tendency toward greater disorder.

PHOTOSYNTHESIS

As you doubtless know, the ultimate energy souce for most living things is sunlight, and the organisms that transform the light energy into chemical energy are primarily the green plants. The process by which this transformation is carried out is called photosynthesis. There is still much about the process that is not known, but in the last decades enormous strides have been made in analyzing the chemical pathways involved. It is not our purpose here to discuss in detail all the chemistry of photosynthesis, or to mention all reactions and compounds now thought to be involved in it; but photosynthesis is so fundamental to life that you ought to understand at least the broad outlines of the process and some of the principles of energy transformation in living systems exemplified by it.

HISTORICAL DEVELOPMENT

As early as 1772, an English clergyman and chemist, Joseph Priestley, demonstrated that green plants affect air in such a way as to reverse the effects of burning or of breathing. In his words:

> I flatter myself that I have accidentally hit upon a method of restoring air which has been injured by the burning of candles, and that I have discovered at least one of the restoratives which nature employs for this purpose. It is vegetation. In what manner this process in nature operates, to produce so remarkable an effect, I do not pretend to have discovered; but a number of facts declare in favour of this hypothesis. I shall introduce my account of them, by reciting some of the observations which I made on the growing of plants in confined air, which led to this discovery.
>
> One might have imagined that, since common air is necessary to vegetable, as well as to animal life, both plants and animals had affected it in the same manner, and I own I had that expectation, when I first put a sprig of mint into a glass-jar, standing inverted in a vessel of water; but when it had continued growing there for some months, I found that the air would neither extinguish a candle, nor was it at all inconvenient to a mouse, which I put into it.
>
> Finding that candles burn very well in air in which plants had grown a long time, and having had some reason to think, that there was something attending vegetation, which restored air that had been injured by respiration, I thought it was possible that the same process might also restore the air that had been injured by the burning of candles.
>
> Accordingly, on the 17th of August, 1771, I put a sprig of mint into a quantity of air, in which a wax candle had burned out, and found that, on the 27th of the same month, another candle burned perfectly well in it. This experiment I repeated, without the least variation in the event, not less than eight or ten times in the remainder of the summer. Several times I divided the quantity of air in which the candle had burned out, into two parts, and putting the plant into one of them, left the other in the same exposure, contained, also, in a glass vessel immersed in water, but without any plant; and never failed to find, that a candle would burn in the former, but not in the latter. I generally found that five or six days were sufficient to restore this air, when the plant was in its vigour; whereas I have kept this kind of air in glass vessels, immersed in water many months without being able to perceive that the least alteration had been made in it.

Priestley's important experiments were the first demonstration that plants produce oxygen, though he himself did not realize that this was what was happening; nor did he realize that light was essential for the process he observed. But his findings stimulated interest in photosynthesis, as we now call it, and led to further work. Only seven years later, the Dutch physician Jan Ingen-Housz demonstrated the necessity of sunlight for oxygen production (though, like Priestley, he knew nothing about oxygen at that time and explained his results·in other terms), and he also showed that only the green parts of the plant could photosynthesize. He reported his results in a book with the richly descriptive title *Experiments upon Vegetables, Discovering Their Great Power of Purifying the Common Air in the Sun-Shine, and of Injuring It in the Shade and at Night.* In 1782 a Swiss pastor and part-time scientist, Jean Senebier, showed that the process depended on a particular kind of gas, which he called "fixed air" (and we call carbon dioxide). Finally, in 1804, another Swiss worker, Nicolas Théodore de Saussure, found that water is necessary for the photosynthetic production of organic materials.

Thus, early in the nineteenth century, all the important components of the photosynthetic process were at least vaguely known, and could be summarized by the following equation:

$$\text{carbon dioxide} + \text{water} \xrightarrow[\text{light}]{\text{green plants}} \text{organic material} + \text{oxygen}$$

Later, scientists came to believe that light energy splits carbon dioxide, CO_2, and that the carbon is then combined with water, H_2O, to form the unit (CH_2O),[1] the atomic grouping on which carbohydrates are based. According to this view, the oxygen released by the plant during photosynthesis comes from CO_2. This idea received a severe blow when, about 1930, C. B. van Niel of Stanford University showed that some photosynthetic bacteria, which use hydrogen sulfide (H_2S) instead of water as a raw material for photosynthesis, give off sulfur instead of oxygen as a by-product. Now, H_2S and H_2O have obvious chemical similarities, and if the sulfur produced by the bacteria during photosynthesis came from H_2S, it seemed reasonable to suppose that the oxygen produced by higher plants during photosynthesis might come from H_2O and not from CO_2. This was at last conclusively shown by use of a heavy isotope of oxygen (O^{18} instead of the usual O^{16}). If photosynthesizing plants are given normal carbon dioxide plus water composed of heavy oxygen, the heavy isotope appears as molecular oxygen:

$$CO_2 + 2H_2O^{18} \longrightarrow O_2^{18} + (CH_2O) + H_2O$$

It is now known that **chlorophyll,** the green pigment of plants, traps the energy required to split the water. The currently accepted equation for green-plant photosynthesis is given in more detail below; the dashed lines indicate the fates of all the atoms involved:

$$CO_2 + 2H_2O + \text{light} \xrightarrow{\text{chlorophyll}} O_2 + (CH_2O) + H_2O$$

[1] The parentheses indicate that this combination of atoms does not represent an actual mole-

Multiplying this summary equation by 6 has the advantage of showing that glucose, a six-carbon simple sugar, is often the end product:

$$6CO_2 + 12H_2O + light \xrightarrow{\text{chlorophyll}} 6O_2 + C_6H_{12}O_6 + 6H_2O$$

It may seem curious that water should appear on both sides of the equation. The reason is that the water produced by the photosynthetic process is new water; it is not the water used as a raw material.

Although the above equation is a convenient summary of photosynthetic carbohydrate synthesis, it tells us nothing about how the synthesis is actually achieved. The process is certainly not one gross chemical reaction, as the summary equation might imply. Many reactions are involved, some that require light, others that can take place without it—so-called "dark" reactions.

THE CHEMICAL NATURE OF PHOTOSYNTHESIS

Carbon dioxide is an exceedingly energy-poor compound; sugar is energy-rich. Photosynthesis, then, is a process that converts light energy into chemical energy and stores it by synthesizing energy-rich sugar from energy-poor carbon dioxide. In chemical terms, the energy is stored by the reduction of carbon dioxide.

Basically, *reduction* means the addition of an electron (e^-), and its converse, *oxidation,* means the removal of an electron. The addition of an electron, reduction, stores energy in the reduced compound. The removal of an electron, oxidation, liberates energy from the oxidized compound.

Since an electron added to one molecule must have been removed from some other molecule, it follows that whenever one substance is reduced another is oxidized:

$$\underset{\substack{\text{electron} \\ \text{donor}}}{A^{e^-}} + \underset{\substack{\text{electron} \\ \text{acceptor}}}{B} \longrightarrow \underset{\substack{\text{has been} \\ \text{oxidized} \\ \text{(lost energy)}}}{A} + \underset{\substack{\text{has been} \\ \text{reduced} \\ \text{(gained energy)}}}{B^{e^-}}$$

Among the many ways a compound may be reduced, two of the most common involve removal of oxygen or addition of hydrogen, both of which have the effect of adding an electron; similarly, oxidation often involves addition of oxygen or removal of hydrogen:

$$\underset{\substack{\text{electron} \\ \text{donor}}}{A} + \underset{\substack{\text{electron} \\ \text{acceptor}}}{BO} \longrightarrow \underset{\substack{\text{has been} \\ \text{oxidized}}}{AO} + \underset{\substack{\text{has been} \\ \text{reduced}}}{B}$$

or:

$$\underset{\substack{\text{electron} \\ \text{donor}}}{AH} + \underset{\substack{\text{electron} \\ \text{acceptor}}}{B} \longrightarrow \underset{\substack{\text{has been} \\ \text{oxidized}}}{A} + \underset{\substack{\text{has been} \\ \text{reduced}}}{BH}$$

In biological systems, removal or addition of an electron derived from hydrogen constitutes the most frequent mechanism of oxidation-reduction reactions.

It now becomes clear why the synthesis of sugar from carbon dioxide constitutes reduction of the CO_2. Hydrogen obtained by splitting water

cule, but only a grouping within some larger compound; thus it takes six (CH_2O) groupings to make a simple six-carbon sugar, $C_6H_{12}O_6$.

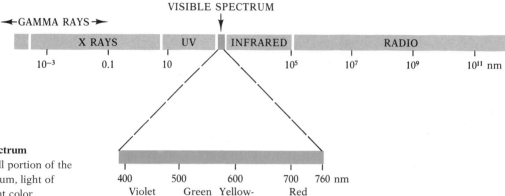

4.1   Portion of the electromagnetic spectrum
Visible light constitutes only a very small portion of the total spectrum. Within the visible spectrum, light of different wavelengths stimulates different color sensations in us. Not only vision and photosynthesis, but also other radiation-dependent biological processes, rely on this same small portion of the electromagnetic spectrum (sometimes extended very slightly into the ultraviolet or infrared). The reason is that the shorter wavelengths have so much energy they tend to destroy living material, whereas the longer wavelengths have too little energy to drive biological reactions. (Wavelengths in nanometers.)

molecules is added to the CO_2 to form compounds based on (CH_2O) units, and energy from light is stored in the process. Our attention must therefore be focused on two key points: the mechanism of trapping and handling energy and the mechanism of transferring hydrogen from H_2O to CO_2.

LIGHT AND CHLOROPHYLL

Light waves constitute one small region of the spectrum of electromagnetic radiations (Fig. 4.1). Each radiation in this spectrum has a characteristic wavelength and energy content. These two characteristics are inversely related; i.e. the longer the wavelength, the smaller the energy content. Within the narrow band visible to human beings, the shortest light waves stimulate the sensation of violet and the longest stimulate the sensation of red. Radiations such as ultraviolet, X rays, and gamma rays, which are of shorter wavelength than violet, are invisible to us, as are infrared and radio-TV radiations, which are of longer wavelength than red.

Is all light, regardless of wavelength, equally effective for photosynthesis? To answer this question, we must turn to the all-important green pigment chlorophyll, which, we have said, traps light energy and helps convert it into chemical energy. Of the several slightly different kinds of chlorophyll, the most widespread is chlorophyll *a*, and it will be to this compound that our discussion will primarily refer. Figure 4.2 shows the rather complex structure of a molecule of chlorophyll *a*.

Light falling on an object may pass through the object (be transmitted), be absorbed by it, or be reflected from it (Fig. 4.3). We can see transmitted or reflected light, but, obviously, we cannot see light that has been absorbed. Now, if chlorophyll is the material that traps the incident light, and if it appears green to our eyes, several facts should immediately be plain. First, chlorophyll cannot be absorbing much radiation of the wavelengths that stimulate in us the sensation of green, or we wouldn't see green color. Second, chlorophyll must be absorbing radiation of some wavelengths within the visible part of the spectrum, or the light transmitted or reflected to us would appear white (when all visible wavelengths are combined, they stimulate the sensation of white). At this point we have already partly answered the question whether all light is equally effective for photosynthesis: Green light is not as effective as light of some other colors, since it is not absorbed as readily by chlorophyll.

More precise information can be obtained if chlorophyll is extracted from the leaf and exposed to light of varying wavelengths to determine the amount of absorption at each wavelength. The absorption spectrum of chlorophyll *a* thus obtained (Fig. 4.4) shows that it is primarily light in the violet and red regions that is absorbed, while green, yellow, and orange light is absorbed only very slightly. It must be noted that the action spectrum of photosynthesis—a measure of the effectiveness of light of various wavelengths in driving photosynthesis—is somewhat different from the absorption spectrum of chlorophyll *a*. As shown by Fig. 4.5, there is relatively high activity in parts of the spectrum where the chlorophyll absorbs very little light—an indication that some wavelengths not readily absorbed by chlorophyll *a* are nonetheless effective in driving photosynthesis. Apparently other pigments, principally the yellow and orange carotenoids and other forms of chlorophyll, absorb light in these regions of the spectrum and then pass the energy to the chlorophyll *a*. Accessory pigments like the carotenoids thus enable the plants to use light of more different wavelengths than could be trapped by chlorophyll *a* alone.

What happens when light of a proper wavelength strikes a chlorophyll molecule? The exact answer is not known, but many aspects of the process have been elucidated.

In the chloroplasts of a living photosynthesizing cell, the chlorophyll and accessory pigments are organized into functional groups called ***photosynthetic units.*** Each unit contains some 300 pigment molecules, including chlorophyll *a*, chlorophyll *b*, and carotenoid. One pigment molecule in each

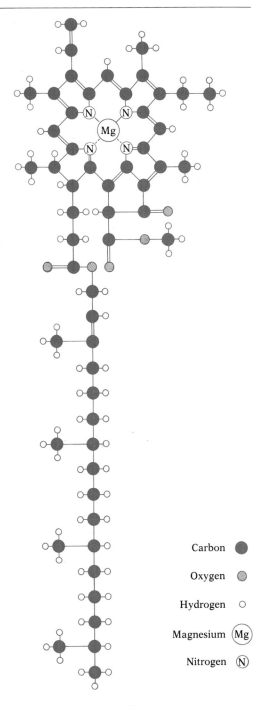

Carbon ●
Oxygen ◉
Hydrogen ○
Magnesium (Mg)
Nitrogen (N)

4.2 Molecular structure of chlorophyll *a*

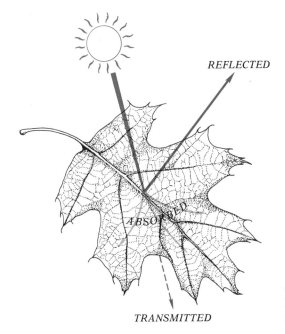

REFLECTED

ABSORBED

TRANSMITTED

4.3 Light striking a leaf
Light striking an object, such as a leaf, may be reflected, absorbed, or transmitted.

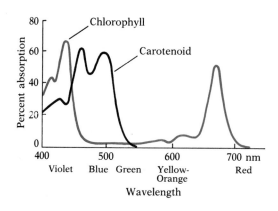

4.4 Absorption spectrum of chlorophyll *a* and carotenoid in alcohol solution
A note of caution: There is evidence that absorption of green light by chlorophyll in the intact leaf—as it occurs in photosynthesis—is substantially higher than it is in an alcohol solution.

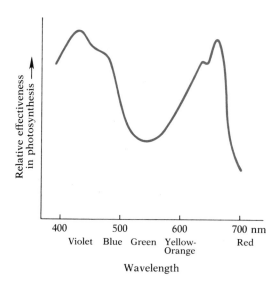

4.5 Action spectrum of photosynthesis
Light of intermediate wavelengths is more effective in driving photosynthesis than would be predicted on the basis of the absorption spectrum of chlorophyll *a* alone. Apparently other pigments, such as carotenoids, absorb these intermediate wavelengths to some extent and pass the energy to chlorophyll *a* for photosynthesis.

unit is distinct from all the rest; it is a specialized form of chlorophyll *a*, which acts as a reaction center. The other pigment molecules function something like antennas responsive to light energy.

Light energy comes in discrete packets, or units, called photons. When a photon strikes a chlorophyll (or carotenoid) molecule and is absorbed, its energy is apparently transferred in some manner to an electron of the pigment molecule. This electron is raised from its normal stable energy level to a higher, relatively unstable energy level (Fig. 4.6); accordingly it will show a marked tendency to fall back to its normal level, giving up the absorbed energy. Isolated chlorophyll in a test tube promptly loses the energy it captures by re-emitting it as visible-light fluorescence; evidently the pigment molecules alone, dissociated from the photosynthetic units, are incapable of converting light energy into chemical energy.

In the functioning chloroplast, once light energy has raised an electron in an "antenna molecule" to a high-energy state, the energized electron is passed from pigment molecule to pigment molecule and may eventually reach the reaction-center molecule, which traps it (Fig. 4.7). Once trapped, the energized electron does not simply fall back to its normal lower energy level. Instead, it is passed to an acceptor molecule with a high affinity for electrons and enters a series of enzyme-catalyzed reactions, which convert the energy into a form more readily usable by the cell.

CYCLIC PHOTOPHOSPHORYLATION

Two types of photosynthetic units occur in most plants; each set of units constitutes a different *photosystem.* Photosystem I will be considered here, Photosystem II in the next section.

Electron transport We have said that the specialized chlorophyll molecule that serves as the reaction center of a photosynthetic unit is capable of passing the energized electron to an acceptor molecule. In Photosystem I the reaction-center molecule is designated P700, and the acceptor molecule,

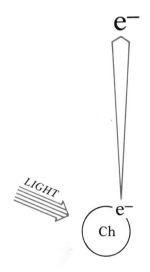

4.6 Effect of light on chlorophyll
Light striking a chlorophyll molecule (Ch) causes an electron (e^-) of the chlorophyll to be raised to a higher energy level.

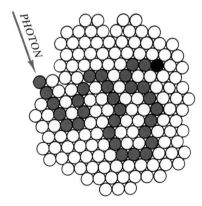

4.7 Flow of an energized electron within a photosynthetic unit
A photon of light strikes one of the antenna pigments (open circles) and raises an electron in the pigment to a higher energy level. The excited electron is then passed from pigment molecule to pigment molecule in a random sequence (colored pathway) until it eventually reaches the reaction-center molecule (black circle), which traps it.

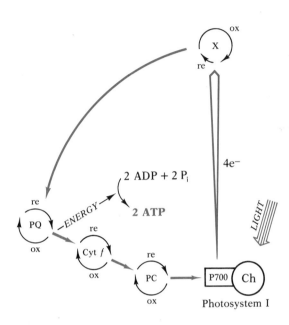

4.8 Cyclic photophosphorylation

Photons of light strike pigment molecules in Photosystem I. The electrons thus energized eventually reach a molecule of P700, a specialized form of chlorophyll *a*. The electrons are then passed (one at a time, though a total of four are shown here in order to balance the chemical equations) from P700 to X, a very strong acceptor molecule, which passes them to a series of further acceptor molecules (circles), each at a slightly lower energy level. As each acceptor molecule receives an electron, it becomes reduced (re), and as it releases the electron it becomes oxidized (ox). Eventually the electrons return to the chlorophyll from which they started. Some of the energy released as the electrons are eased step by step down the energy gradient is used in the synthesis of ATP from ADP and inorganic phosphate (P_i). (PQ, plastoquinone; Cyt *f*, cytochrome *f*; PC, plastocyanin.)

which has not yet been fully characterized chemically, is called X. In the mid-1950s Daniel I. Arnon and his co-workers at the University of California, Berkeley, discovered in test-tube experiments that X, in turn, often passes the energized electron to a second electron-acceptor molecule, which passes it to still another acceptor molecule (Fig. 4.8). As each acceptor in the series receives the electron, it becomes reduced, and as it gives up the electron it becomes oxidized. Each of the oxidation-reduction reactions in the chain is catalyzed by a different enzyme; such a series of enzyme-catalyzed reactions is sometimes termed a multienzyme system.

When a light-energized electron leaves a chlorophyll molecule, it is energy-rich; when it finally returns, it is energy-poor. The transition is a gradual one. As the electron is passed from transport molecule to transport molecule in the chain, it releases some of its extra energy with each transfer; in other words, it is eased down the energy gradient step by step from the excited state to the normal state. Thus when the electron finally falls back into the chlorophyll, it has discharged all its extra energy, but it has not discharged it all at once, as would happen with isolated chlorophyll in a test tube. Instead, the energy has been released in a series of small portions of manageable size.

Many everyday situations provide ready analogies to the step-by-step energy release by the excited chlorophyll electrons. Consider, for example, an automobile with its tank of gasoline. Locked in the chemical bonds of the gasoline is much energy. One way to release that energy would be to put a match to the entire supply. The energy would promptly be released, but it wouldn't do the automobile much good! A far more effective way of ensuring that the released energy will be harnessed to propel the car is to burn the gasoline bit by bit. Or consider a waterfall where the water is not allowed to descend unobstructed, but is made to strike the paddles of a mill wheel, to which it imparts energy that can be used to do work. In each case—excited electron, gasoline, and waterfall—the total amount of energy released is unaffected by whether the change in energy levels is abrupt or gradual, but the results of the release are considerably affected. Step-by-step reactions of this kind are a general property of biological chemical systems. In living cells, almost all major chemical conversions are accomplished by a series of smaller conversions, each catalyzed by its own enzyme.

Synthesis of ATP When the energy of the excited chlorophyll electron is released in a series of small bursts, some of the energy, we have implied, can be harnessed to do useful work. How? At some point along the transport chain, part of the released energy is captured and used in the synthesis of a compound named adenosine triphosphate, generally abbreviated ***ATP***. This compound, found in every living thing, is one of the essential substances of life. It plays the key role in most biological energy transformations.

The ATP molecule (Fig. 4.9) is composed of a nitrogen-containing compound (adenosine) plus three phosphate groups bonded in sequence:

$$\text{adenosine} - \textcircled{P} \sim \textcircled{P} \sim \textcircled{P}$$

According to convention, $\textcircled{P}$ stands for the entire phosphate group,[2] and the

---

[2] At pH 7 the phosphate (actually phosphoryl) group is in ionized form: $-PO_3^{--}$; so, too, is inorganic phosphate (orthophosphate), HPO_4^{--}, symbolized by P_i.

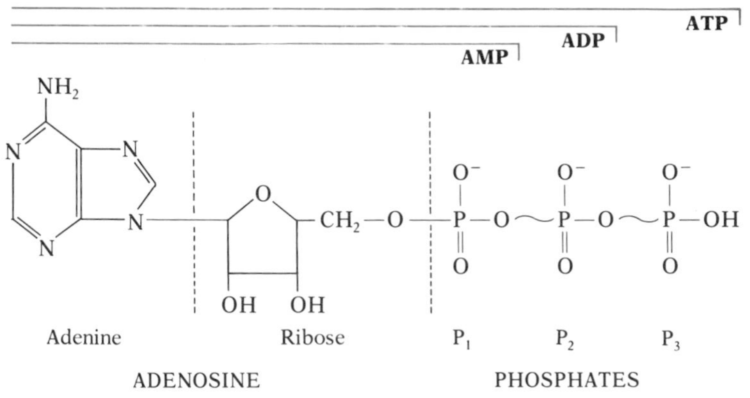

$$NH_2$$

Adenine · Ribose · P_1 · P_2 · P_3

ADENOSINE · PHOSPHATES

AMP · ADP · ATP

4.9 The ATP molecule
As its full name, adenosine triphosphate, implies, this molecule is composed of an adenosine unit (a complex of adenine and ribose sugar) and three phosphate groups arranged in sequence. The last two phosphates are attached by so-called high-energy bonds (wavy lines). When either of these bonds is hydrolyzed, the result is a release of energy; i.e. the free energy of the products is much less than that of the reactants. Removal of one phosphate group leaves ADP; removal of two leaves AMP. (Note that AMP is one of the nucleotides of which RNA is composed.)

wavy lines between the first and second and the second and third phosphate groups represent so-called high-energy bonds; if they are broken by hydrolysis, far more free energy is released than if the other bonds in the ATP molecule are broken.

Actually, it is often only the terminal phosphate bond of ATP that is involved in energy conversions. The exergonic reaction by which this bond is hydrolyzed and the terminal phosphate group removed leaves a compound called adenosine diphosphate, or **ADP** (adenosine plus two phosphate groups), and inorganic phosphate (symbolyzed by P_i):

$$ATP + H_2O \xrightarrow{\text{enzyme}} ADP + P_i + \text{energy}$$

If both the second and third phosphate groups are removed from ATP, the resulting compound is called adenosine monophosphate, or **AMP.**

New ATP can be synthesized from ADP and inorganic phosphate if adequate energy is available to force a third phosphate group onto the ADP. Addition of phosphate is termed **phosphorylation:**

$$ADP + P_i + \text{energy} \xrightarrow{\text{enzyme}} ATP + H_2O$$

ATP is often called the universal energy currency of living things, and this characterization is justified. Although some other compounds can supply energy to drive endergonic reactions, ATP is the one most often used by cells in the various kinds of work they perform: synthesis of more complex compounds, muscular contraction, nerve conduction, active transport across cell membranes, light production, etc. The energy price of the work is paid in ATPs; i.e. the energy-releasing hydrolysis of ATP to ADP is coupled with the energy-demanding reactions the cell must carry out.

ATP molecules can be synthesized from ADP in a variety of ways in living organisms, as will be seen later in this chapter. But although neither synthesis of ATP nor step-by-step electron transport is unique to photosynthesis, the light-induced production of the high-energy electron that starts the process is. Any organism can manufacture energy currency by using up other energy-rich compounds, but only photosynthetic organisms (and a few chemosynthetic bacteria) can manufacture such currency from energy-poor inorganic raw materials. In short, it is the unique capacity of chlorophyll to absorb light energy and act as a donor of high-energy electrons that is critical for photosynthesis.[3]

In the photosynthetic phosphorylation process so far described, chlorophyll acts both as electron donor and as the ultimate electron acceptor, donating excited electrons and eventually accepting the electrons in a low-energy state (Fig. 4.8). Because electrons can be carried round and round the system and no outside source of electrons is involved, this method of synthesizing ATP is called *cyclic photophosphorylation.*

NONCYCLIC PHOTOPHOSPHORYLATION

It is probable that, when chlorophyll molecules first appeared in the evolution of life on earth, the sole energy-storing process driven by light energy in the primitive organisms was cyclic photophosphorylation, and cyclic photophosphorylation is still the only such process in the photosynthetic bacteria. Under certain conditions (as when carbon dioxide is not available), it seems likely that modern green plants also use light for this particular synthesis. But most present-day photosynthetic organisms usually employ light energy to drive another, more complicated process that not only synthesizes ATP, but also transfers hydrogen electrons from water to an intermediate compound, which can then act as an electron donor in the reduction of CO_2. Much remains to be learned about this aspect of photosynthesis, and there is disagreement among authorities concerning the details of the process. The overall scheme, however, has become clearer in the past few years.

Like the process of cyclic photophosphorylation already described, this process begins when photons of light strike molecules of chlorophyll in photosynthetic units of Photosystem I and raise electrons to an excited state (Fig. 4.10). Again, the excited electrons are led away from the P700 molecule by a strong electron acceptor, probably the one we have called X. But here the similarity to cyclic photophosphorylation ends. Instead of passing the electrons to the transport chain of cyclic photophosphorylation, X donates the electrons to a different series of electron-transport molecules, beginning with an iron-containing compound called ferredoxin (Fd). The ferredoxin, in turn, passes the electrons, via an intermediate compound, to an extremely important substance called nicotinamide adenine dinucleotide phosphate, or *NADP,* as it is usually abbreviated.[4] NADP, unlike the reduced electron-acceptor molecules in cyclic photophosphorylation, does not promptly pass the electrons along to another acceptor molecule. Instead, the NADP retains a pair of energized electrons, and the now energy-rich compound can eventually act as an electron donor in the reduction of CO_2 to carbohydrate. Thus the electrons move from the chlorophyll to X to NADP to carbohydrate (via other intermediate compounds).

By acquiring the two extra electrons, the NADP also attracts a hydrogen

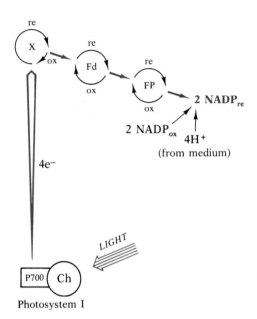

4.10 Electron transport from Photosystem I to NADP After the light-energized electrons captured by P700 reach X, they may either follow the cyclic photophosphorylation pathway (see Fig. 4.8), or they may be passed by a different series of acceptor molecules to NADP$_{ox}$ to form NADP$_{re}$. The second route is believed to be the more common. (Fd, ferredoxin; FP, flavoprotein.)

---

[3] Although chlorophyll is the light-absorbing pigment that drives photosynthetic production of ATP in most photosynthetic organisms, it has recently been discovered that certain bacteria *(Halobacterium)* that live in very salty media can make ATP with energy from light trapped by a pigment much like rhodopsin, the light-sensitive pigment in the rod cells of the vertebrate eye.

[4] NADP was formerly known as TPN, and NAD (a compound to be discussed later in this chapter) was known as DPN.

proton (H^+), and an additional proton is released into the medium. Thus the reduced NADP may be written $NADPH + H^+$, but in this chapter we shall generally use the designation $NADP_{re}$; oxidized NADP will be designated $NADP_{ox}$.

Now, if the energized electrons from the chlorophyll of Photosystem I are retained by $NADP_{re}$ and eventually incorporated into carbohydrate, it follows that Photosystem I is left short of electrons, that it is left with "electron holes." These electron holes are filled, indirectly, by electrons derived from water through a process we shall now examine.

At first, it was thought that the electrons from the water passed, via a few transport molecules, directly to Photosystem I, and hence that there was only one light-driven event in photosynthesis—the one that initially energized the chlorophyll electrons of Photosystem I. Later, however, it became apparent that there are two light-driven events, the one already discussed and a second one more intimately related to the splitting of water. This second light event involves photosynthetic units of Photosystem II, each of which is composed of about 200 molecules of chlorophyll a, about 200 molecules of chlorophyll b, c, or d, depending on the species of plant,[5] and one molecule of a specialized form of chlorophyll a called P680, which acts as the reaction center for the photosynthetic unit.

When light of the proper wavelength strikes a pigment molecule of Photosystem II, the energy, in the form of excited electrons, is passed around within the photosynthetic unit until it finally reaches the molecule of P680. This molecule, in turn, donates the high-energy electrons to a strong electron acceptor designated Q (Fig. 4.11). Substance Q then passes the electrons to a chain of acceptor molecules, which lead the electrons to the electron holes in Photosystem I. As the electrons move along the chain of transport molecules, they are eased step by step down an energy gradient. Some of the energy thus released is used in the synthesis of ATP from ADP and inorganic phosphate.

Thus the electron holes created in Photosystem I by the first light event are refilled by electrons moved from Photosystem II by the second light event. But this process alone would leave electron holes in Photosystem II; the electron deficit would simply have been shifted from Photosystem I to Photosystem II. It is at this point that the electrons from water, mentioned earlier, play their role. As Fig. 4.11 indicates, it is thought that P680 pulls replacement electrons away from water[6] (perhaps via some intermediate compound), leaving behind free protons and molecular oxygen:

$$2H_2O \longrightarrow 4e^- + 4H^+ + O_2$$
$$\downarrow$$

to P680

The protons, as we saw previously, become associated with $NADP_{re}$. The oxygen is released as a gaseous by-product (note that two molecules of H_2O must be split to yield one molecule of O_2).

[5] Chlorophyll b is found in green algae, bryophytes, and vascular plants. Chlorophyll c occurs in brown algae, and chlorophyll d in red algae.

[6] Some investigators think that additional ATP is synthesized as electrons move down the energy gradient from water to P680.

The electrons involved in the second light event move from water to Photosystem II to Q to the transport chain to Photosystem I (Fig. 4.11). If we combine these steps with the electron movement associated with the first light event, as traced above, we obtain the following abbreviated sequence showing the overall electron movement:

$$H_2O \longrightarrow \text{Photosystem II} \longrightarrow Q \longrightarrow \text{transport chain} \longrightarrow \text{Photosystem I} \longrightarrow$$
$$X \longrightarrow \text{transport chain} \longrightarrow \text{NADP}_{re} \longrightarrow \text{carbohydrate}$$

This sequence reinforces our earlier statement that the electrons necessary to reduce carbon dioxide to carbohydrate come from water, but it shows that the movement of electrons from water to carbohydrate is an indirect and complex process.

Since electrons are not passed in a circular chain in this process, some leaving the system via $NADP_{re}$ and others entering the system from water as replacements, this series of reactions is termed ***noncyclic photophosphorylation.*** The whole process results in the formation of both ATP and $NADP_{re}$ and in the release of molecular oxygen. The reactions of photophosphorylation have traditionally been known as the "light" reactions of photosynthesis, but, as we have seen, only two steps are directly light-dependent.

We should point out that the details given above apply only to photosynthesis by green plants. As we have already indicated, some bacteria possess a form of chlorophyll and can use light energy in synthesis of ATP and $NADP_{re}$, but they do not use water as the source of electrons. Some of these use hydrogen sulfide (H_2S), which is very much like water, and they give off sulfur instead of oxygen. Others use inorganic compounds not much like water, but the basic processes are essentially the same, although oxygen is not a by-product. Green plants themselves can be experimentally induced to use a source of electrons other than water. If oxidation of water is blocked by chemical inhibitors, and then a strong electron donor is provided as a substitute, noncyclic photophosphorylation can continue without production of oxygen; in other words, green-plant photosynthesis has been experimentally converted into an essentially bacterial type of photosynthesis. Water is therefore only one of many possible electron sources for photosynthesis. It is not surprising, however, that the most successful photosynthetic organisms, the green plants, evolved photosynthesis based on water, that ubiquitous agent in the chemistry of life.

CARBOHYDRATE SYNTHESIS BY THE CALVIN CYCLE

Early in this chapter we said that in studying photosynthesis attention should be focused on two key points: the mechanism of trapping and handling energy and the mechanism of transferring hydrogen from water to carbon dioxide. Both of these key mechanisms have now been explained in terms of ATP and $NADP_{re}$. What we must still do is indicate how these two energy-rich compounds make possible the synthesis of carbohydrates from CO_2.

Actually, we have already discussed the reactions unique to photosynthetic organisms and at the heart of photosynthesis: the synthesis of ATP and $NADP_{re}$ by means of light energy. The utilization of these two key compounds in the synthesis of carbohydrate can be carried out in the dark; the dark reactions of carbohydrate synthesis are fully separable from the light-driven synthesis of ATP and $NADP_{re}$ on which they depend. One way to separate the light and dark phases of photosynthesis experimentally in both time and space is to expose a suspension of chloroplasts to light with a plentiful supply of ADP, inorganic phosphate, and $NADP_{ox}$, but no CO_2. Under such conditions, the chloroplasts will synthesize large quantities of ATP and $NADP_{re}$. If the chloroplasts are then fractionated and the solid green portion in which photophosphorylation takes place is discarded, and

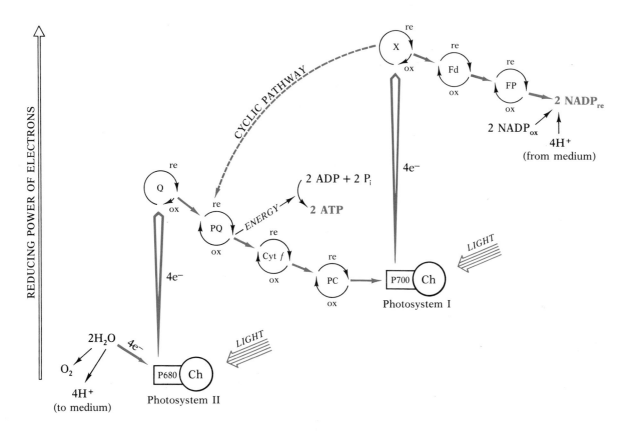

4.11 Noncyclic photophosphorylation
Photons of light strike pigment molecules in Photosystem II (lower left), and the electrons thus energized are passed from P680 (a special form of chlorophyll *a*) to the strong electron acceptor Q. The electrons lost from the chlorophyll are replaced by electrons from water (perhaps via an intermediate compound); the splitting of water also produces molecular oxygen (O_2) and releases protons (H^+) into the medium. Q passes the energized electrons to a series of transport molecules that lower the electrons step by

step down an energy gradient until they reach the pigment molecules of Photosystem I; some of the energy released by this series of exergonic reactions is used in the synthesis of ATP from ADP and inorganic phosphate. The electrons in Photosystem I are energized by another light event and passed to X, whence they may move to NADP via a second series of transport molecules. In summary, electrons move (colored arrows) from H_2O to P680, to Q, along a transport chain to P700, to X, and along a second transport chain to NADP. As the dashed

arrow indicates, the electrons may follow an alternative pathway, that of cyclic photophosphorylation (the relationship between cyclic and noncyclic photophosphorylation can clearly be seen in this diagram, which incorporates the pathways shown in Figs. 4.8 and 4.10). The two important products of noncyclic electron transport are ATP and $NADP_{re}$, which make possible the synthesis of carbohydrate from CO_2. Molecular oxygen is released as a by-product.

THE HILL REACTION

In the late 1930s R. Hill of Cambridge University, England, in the course of experiments he conducted with chloroplasts removed from cells, came to the conclusion that these could perform only a part of photosynthesis. While such chloroplasts could produce oxygen from water in the light if they were provided with a very strong electron acceptor, they could not, Hill thought, assimilate carbon dioxide.

Hill's technique of providing illuminated isolated chloroplasts, or fragments of chloroplasts, with a strong electron acceptor (such as a ferric salt or a reducible organic dye) and then studying factors such as temperature, pH, etc. that influence oxygen production from water by the chloroplasts became an important research approach. If we designate the artificially provided, nonphysiological electron acceptor as X, we can summarize the so-called Hill reaction by the following equation:

$$2 X + 2H_2O + \text{light} \xrightarrow{\text{chloroplast}} 2 H_2X + O_2$$

As long as X is continuously supplied (and other conditions are properly controlled), chloroplasts will produce oxygen from water in the light at nearly the same rate as a whole green cell.

How is it, in view of the evidence accumulated since the mid-1950s that isolated chloroplasts can perform complete photosynthesis, that hundreds of research workers using variants of the Hill reaction for over fifteen years remained convinced that isolated chloroplasts could not assimilate CO_2? The answer apparently lies in the procedure used. It is a good example of how minor, seemingly irrelevant details of technique may radically alter experimental results. The enzymes that catalyze CO_2 assimilation are very soluble, and they were simply lost from the chloroplasts in the process of separating the chloroplasts from the cells. If the chloroplasts are isolated by more careful procedures that leave the enzyme systems of the stroma intact, a single chloroplast can act as a complete photosynthetic factory.

This was a very important discovery, because it demonstrated clearly that any ATP used in CO_2 assimilation is synthesized in the chloroplasts themselves and is not, as was once believed, supplied to them by the mitochondria. Recognition of this fact led, in turn, to the discovery of photophosphorylation and to the current understanding of the "light" reactions of photosynthesis.

if CO_2 is added to the remaining mixture in the dark, the CO_2 will be assimilated into carbohydrate at the expense of the earlier-synthesized ATP and $NADP_{re}$. In other words, the production of ATP and $NADP_{re}$, on the one hand, and the assimilation of CO_2, on the other, may occur at different times, under different conditions, and even in different parts of the chloroplasts. It can be shown, furthermore, that almost any cell, whether it contains chlorophyll or not, and whether it is from a plant or an animal, can synthesize carbohydrate from CO_2 if it is furnished with the necessary ATP and $NADP_{re}$, which function as free-energy donor and reducing agent respectively.

Now, carbohydrate contains much chemical energy, while CO_2 contains very little. As you might predict from what you have already learned about the chemistry of living cells, the reduction of CO_2 to form glucose proceeds by many steps, each catalyzed by an enzyme. In effect, CO_2 is pushed up an energy gradient through a series of intermediate compounds, some of them unstable, until the stable carbohydrate end product is formed. An analogy would be a man moving a large and very heavy chest up a flight of stairs from the first floor of his home to the second floor. The man might be able to lift the chest just high enough to get it up one step at a time, balancing it on each step long enough to marshal his strength before the next heave. If he let go (i.e. stopped applying energy) at any point between the stable level of the first floor and the stable but higher energy level of the second floor, the chest would come crashing to the bottom. The steps, then, make it possible to move the chest up an energy gradient, but they themselves are unstable intermediate levels. In this case, the energy necessary to move the chest through the series of unstable intermediate levels to the stable high energy level at the top is supplied by the man. In the case of synthesis of carbohydrate from CO_2, the energy comes from light via ATP and $NADP_{re}$ (Fig. 4.12).

If there are many sequential steps in the reduction of CO_2 to carbohydrate, and if many of the intermediate compounds occur also in other processes leading to different end products, you may well wonder how the exact sequence of steps could be discovered. The tool that made such discoveries possible was a radioactive isotope of carbon, designated C^{14}. Samuel Ruben and Martin D. Kamen at the University of California at Berkeley, who discovered this isotope about 1940, immediately recognized its potential as a tool in research on photosynthesis. They showed that plants exposed to carbon dioxide containing the radioactive isotope ($C^{14}O_2$ instead of the normal $C^{12}O_2$) incorporated the isotope into a variety of compounds. Later, in 1946, Melvin Calvin and his associates, likewise at Berkeley, began an intensive long-term investigation of carbon dioxide fixation in photosynthesis, utilizing C^{14} as their principal tool. They exposed algal cells to light in an atmosphere of $C^{14}O_2$ for a few seconds and then killed the cells by immersing them in alcohol. The alcohol not only killed the cells but also inactivated the enzymes that catalyze the reactions of photosynthesis. With the enzymes inactivated, whatever amount of each intermediate compound existed in the cell at the moment of inactivation was, in effect, locked in. Calvin and his co-workers could then determine which of these locked-in intermediate compounds contained C^{14}. How long the algal cells were exposed to the $C^{14}O_2$ before being killed determined the number of compounds in which C^{14} was detected; if the time was very short, the C^{14} would reach only the first few compounds in the synthetic sequence, while, if the time was longer, the isotope would have moved through more steps in the sequence and would appear in a great variety of compounds. After years of painstaking research, Calvin, who in 1961 was awarded the Nobel Prize for his critically important investigations, worked out a sequence of reactions now called the *Calvin cycle,* which we shall outline in very abbreviated form below (Fig. 4.13).

According to Calvin, the CO_2 first combines with a five-carbon sugar called *ribulose diphosphate* (abbreviated RuDP) to form a hypothetical six-carbon compound, which is promptly broken into two three-carbon molecules called phosphoglyceric acid, or PGA. Each molecule of PGA is then phosphorylated by ATP and reduced by hydrogen from $NADP_{re}$. The resulting energy-rich three-carbon compound is called phosphoglyceraldehyde, or *PGAL* for short. PGAL is a true sugar and, in a sense, is the stable end product of photosynthesis. Because PGAL is a three-carbon compound, as are the intermediate compounds leading to its formation, the Calvin cycle is often called *C_3 photosynthesis.*

Most (five of every six) of the molecules of PGAL are used in the formation of new RuDP (by a complicated series of reactions powered by ATP), with which more CO_2 can be processed. But some (one molecule out of six) of the

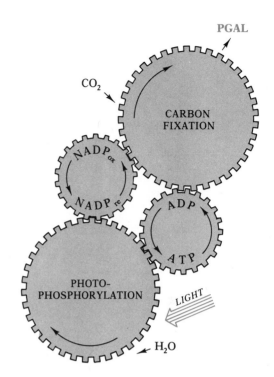

4.12   The relationship between photophosphorylation and carbon fixation

The entire photosynthetic process can be visualized as a series of interlocking gears. Energy from light turns the photophosphorylation gear. The turning of this gear causes both the ATP synthesis and the $NADP_{re}$ synthesis gears to turn. These two gears cause the carbon-fixation gear to turn, with resultant production of carbohydrate (PGAL) from CO_2.

4.13 Synthesis of carbohydrate by the Calvin cycle
Each CO_2 molecule combines with a molecule of
ribulose diphosphate (RuDP), a five-carbon sugar, to
form a hypothetical unstable six-carbon intermediate,
which promptly splits into two molecules of a
three-carbon compound called PGA. Each PGA is
phosphorylated by ATP and then reduced by $NADP_{re}$ to
form PGAL, a three-carbon sugar. Thus each turn of
the cycle produces two molecules of PGAL. Five of
every six new PGAL molecules formed are used in
synthesis of more RuDP by a complicated series of
reactions (not shown separately here) driven by ATP.
The sixth new PGAL molecule may be used in the
synthesis of glucose. The path of carbon from CO_2 to
glucose is here traced by colored arrows. Since it takes
three turns of the cycle to yield one PGAL for glucose
synthesis, the diagram begins with three molecules of
CO_2; it would require a total of six turns to produce
one molecule of glucose, a six-carbon sugar. Note that
the cycle is driven by energy from ATP and $NADP_{re}$,
formed by the light reactions of photophosphorylation.

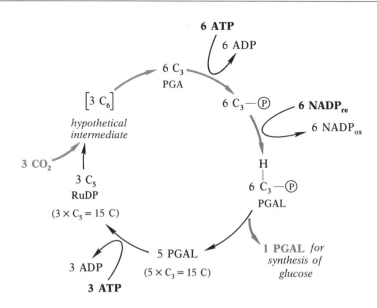

PGAL is combined and rearranged in a series of steps to form the six-carbon
sugar glucose, traditionally regarded as the end product of photosynthesis.
Once synthesized, the glucose molecules may be broken down to yield energy for doing work; or they may serve as raw material for synthesis of other
classes of compounds such as fats; or they may be bonded together to form
more complex carbohydrates such as sucrose, starch, or cellulose.[7]

Carbohydrate is most often transported in the vascular tissue of plants in
the form of the double sugar sucrose, and it is generally stored in higher
plants in the form of starch. One of the most important advantages of
storing carbohydrate as starch rather than sugar is that starch, which is
insoluble, has much less osmotic activity than sugar, which dissolves in the
cellular cytoplasm. An excessive accumulation of sugar would raise the
osmotic concentration of the cytoplasm to a very high level and severely
upset the osmotic balance between the cells and the surrounding fluid, with
the result that too much water would be taken in by the cells.

Photorespiration Before leaving the Calvin cycle, we should mention one
perplexing process associated with it. It appears that the enzyme that catalyzes the carboxylation of ribulose diphosphate (the addition of CO_2 to
RuDP) at the start of the Calvin cycle can also catalyze the oxidation of
RuDP by molecular oxygen (the addition of O_2 to RuDP). In other words,
CO_2 and O_2 are alternative substrates for the enzyme; some investigators
think they are competitive inhibitors, i.e. that they compete for the very
same binding sites on the enzyme and that each inhibits the binding of the
other. When the concentration of CO_2 is high and that of O_2 low, carboxylation is favored and carbohydrate synthesis by the Calvin pathway proceeds.

[7] Although glucose has traditionally been considered the end product of photosynthesis, free
glucose is not present in significant amounts in most higher plants. Glucose is present primarily as
building-block units in double sugars, starch, cellulose, and other polysaccharides.

But when the reverse conditions prevail—when the concentration of CO_2 is low and that of O_2 high—oxidation is favored. The oxidation alternative is also favored by high temperatures.

The details of the oxidation process are still uncertain, and several conflicting hypotheses have been put forward; but it appears certain that one substance produced by the oxidation is a two-carbon compound called glycolic acid, which is then further broken down to CO_2. Under certain conditions, then, high-energy molecules, such as RuDP, that function as intermediates in photosynthesis and are themselves produced by photosynthesis are destroyed by a series of reactions initiated by the very same enzyme that under more favorable conditions would facilitate photosynthesis. This breakdown of photosynthetic intermediates to CO_2 is called **photorespiration.** Since it does not result in synthesis of ATP, as other types of respiration do, it would appear to be a wasteful process, which short-circuits the Calvin cycle. Its biological function, if any, is still a mystery.

Because photorespiration predominates over photosynthesis at low concentrations of CO_2, plants that depend exclusively on the Calvin cycle for CO_2 fixation cannot synthesize carbohydrate unless the CO_2 concentration in the air is above a critical level (commonly about 50 parts per million); even at normal levels much of the production of photosynthesis is undercut by concurrent photorespiration. At atmospheric CO_2 concentrations, net photosynthesis by such plants could be increased by as much as 50 percent if oxygen inhibition of photosynthesis and the associated photorespiration could be stopped. We shall return to this idea later in this chapter.

THE LEAF AS AN ORGAN OF PHOTOSYNTHESIS

The morphology and anatomy of leaves Although photosynthesis can occur in all green parts of the plant, in most higher vascular plants it is the leaves that expose the greatest area of green tissue to the light and are therefore the principal organs of photosynthesis.

Figure 4.14 shows leaves of a variety of familiar land plants. Most dicot leaves consist of a stalk, or **petiole,** and a flattened **blade** (some leaves, particularly those of monocots, lack petioles, the blade being sessile on the stem). In addition, some leaves bear small appendages, called stipules, at their bases. The blade is usually broad and thin and contains a complex system of veins. Because of the flatness of the blade, the leaf exposes to the light an area that is very large in relation to its volume.

If a transverse section of a leaf is examined microscopically (Fig. 4.15), it can be seen that the outer surfaces are formed by layers of epidermis, usually only one cell thick, but sometimes two, three, or more cells thick. A waxy layer, the **cuticle,** usually covers the outer surfaces of both the upper and lower epidermis; it is generally thicker on the former. The chief function of the epidermis is protection of the internal tissues of the leaf from excessive water loss, from invasion by fungi, and from mechanical injury. Most epidermal cells do not contain chloroplasts.

The entire region between the upper and lower epidermis contains parenchyma cells, which constitute the **mesophyll** portion of the leaf. The mesophyll is commonly (but by no means always) divided into two fairly distinct parts: an upper palisade mesophyll, consisting of cylindrical cells arranged

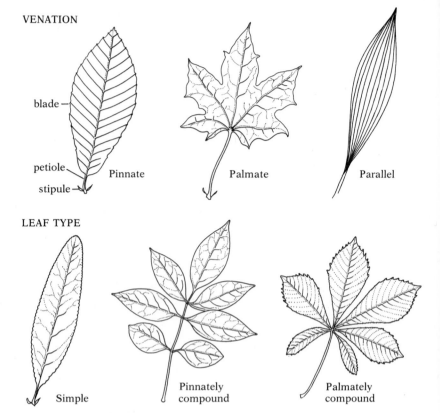

VENATION

blade

petiole

stipule

Pinnate   Palmate   Parallel

LEAF TYPE

Simple   Pinnately   Palmately
                compound   compound

4.14 Leaf types
A leaf usually consists of a blade and of a petiole, which sometimes has stipules at its base. Veins run from the petiole into the blade. The main veins may branch in succession off the midvein (pinnate venation); or they may all branch from the base of the blade (palmate venation); or they may be parallel. The blade may be simple; or it may be compound, i.e. divided into leaflets that may be pinnately or palmately arranged.

vertically, and a lower spongy mesophyll, composed of irregularly shaped cells. The cells of both parts of the mesophyll are very loosely packed and have many intercellular spaces between them. These spaces are interconnected and communicate with the atmosphere outside the leaf by way of holes in the epidermis called *stomata.* The size of the stomatal openings is regulated by a pair of modified epidermal cells called *guard cells.*

A conspicuous system of veins (also called vascular bundles) branches into the leaf blade from the petiole (Fig. 4.14). The veins form a structural framework for the blade and also act as transport pathways, which are connected with the transport system of the rest of the plant. Each vein contains cells of the two principal vascular tissues, xylem and phloem; and each is usually surrounded by a *bundle sheath* composed of parenchymatous cells packed so tightly together that there are few spaces between them. In most cases the branching of the veins is such that no mesophyll cell is far removed from a veinlet; in one study the veins were found to attain a combined length of 102 cm per square centimeter of leaf blade.

Chloroplasts, the organelles of photosynthesis   As we saw in Chapter 3, in eucaryotic cells photosynthesis takes place within membrane-bounded chloroplasts. The molecules of chlorophyll and carotenoid pigment are not simply scattered randomly within the chloroplasts, but are arranged in precise and orderly fashion within the membranes of flattened sacs called *thylakoids,* which are embedded in a colorless protein matrix called the *stroma* (see Fig. 3.32, p. 110). In many parts of the chloroplasts of higher plants, numerous disc-shaped thylakoids lie very close to each other, forming structures called *grana;* a granum looks like a stack of coins (Fig. 4.16). The "light" reactions of photosynthesis (i.e. those of cyclic and noncyclic photophosphorylation) take place in the thylakoids, particularly those of the grana. The "dark" reactions of photosynthesis (i.e. those of carbon dioxide assimilation) take place in the more fluid stroma. As we mentioned earlier, the stroma can continue to perform its carbon-assimilation function even after it has been physically separated from the thylakoids by disruption of the chloroplasts; it is only necessary for the stroma to have an adequate supply of CO_2, ATP, and $NADP_{re}$.

Although the general structure and appearance of the thylakoid membranes are like those of the membranes of other parts of the cell, chemical analyses indicate that most of the lipids, and probably the proteins, in them are of types found only there. Thus the chemical composition of the thylakoid membranes mirrors the uniqueness of the photosynthetic function of these structures. According to electron-microscope studies of freeze-etched thylakoid membranes, the inner surface of the membrane is not smooth, but shows a regular pattern of repeating structures with a granular, often

4.15 The anatomy of C_3 and C_4 leaves
In a C_3 leaf the palisade mesophyll cells typically form a layer in the upper part of the leaf; the corresponding mesophyll cells in a C_4 leaf are usually arranged in a ring around the bundle sheath. While the bundle-sheath cells of C_4 leaves have chloroplasts (color), those of C_3 leaves usually lack them.

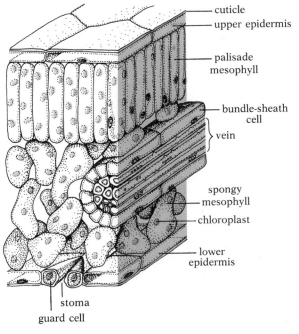

A C_3 leaf

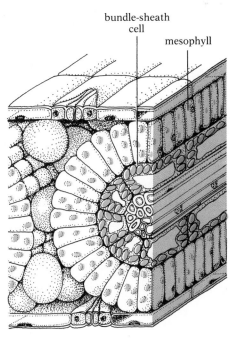

B C_4 leaf

THE THREE-DIMENSIONAL
ARRANGEMENT OF THYLAKOIDS

The diagram of Fig. 4.16 shows the usual way of portraying the three-dimensional arrangement of granal and stromal thylakoids. Recent evidence, however, suggests that the actual arrangement is much more complicated—as the sketch below attempts to show. It presents parts of five grana (stacks of disc-shaped thylakoid sacs) with stromal thylakoids running between them.

In the granum at left, all but one of the stromal thylakoids have been cut away; the frets that connected them to the granal thylakoids can be seen as boxlike extensions from these. As indicated by the one example remaining, each stromal thylakoid usually winds upward and to the right in helical fashion around a granum, attaching by a single fret to each granal sac in succession.

The granum in the center shows clearly that each granal sac has eight frets, each of which attaches to a different stromal thylakoid.

The granum on the right has been sectioned in such fashion as to reveal its internal structure approximately as it appears in normal electron micrographs. Note the parallel arrangement of flattened sacs and, in places where the cut goes through the center of a fret, the continuation of the sac membranes beyond the granum to form the stromal thylakoids (cf. Fig. 4.16).

The difficulties that have beset the determination, and accurate representation, of the three-dimensional form of thylakoids are typical of the difficulties presented by microscopic structures generally, for such structures are usually viewed only in two-dimensional photographs of sections. It is important to realize that with respect to such structures photographs and drawings cannot fully represent reality, but are merely aids to understanding.

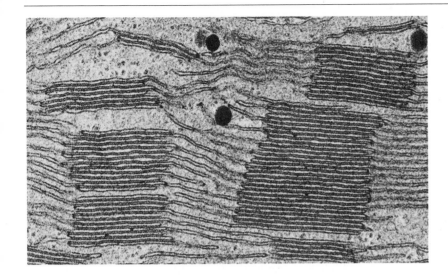

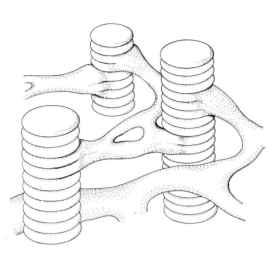

spherical, appearance. Some workers believe these granular structures are the site of the "light" reactions of photosynthesis and, as such, contain photosynthetic units of Photosystems I and II together with a complete set of the electron-transport molecules that functionally unite the two photosystems. Although this hypothesis has not been demonstrated conclusively, it is an appealing one, because the concept that structure and function are two sides of the same coin is not only intellectually satisfying, but an experimentally productive principle of biology.

Leaves with Kranz anatomy   As early as 1904 German plant anatomists, notably B. Haberlandt, observed that the leaf anatomy of some angiosperm plants of tropical origin—plants associated with xeric (dry) habitats—showed a combination of features generally not found in plants native to the temperate zones. Haberlandt called this complex of features Kranz anatomy.[8] The bundle-sheath cells of plants with Kranz anatomy contain numerous chloroplasts, whereas those of other plants often do not. In plants with Kranz anatomy the mesophyll cells that correspond to the palisade layer tend to be clustered in a ringlike arrangement around the veins, just outside the bundle sheaths (Fig. 4.15B). These mesophyll cells contain numerous chloroplasts, but the spongy mesophyll cells outside the rings often have reduced numbers of chloroplasts, or even none at all. In Kranz plants the chloroplasts of the bundle-sheath cells and mesophyll cells usually differ in a number of ways (Fig. 4.17). In the bundle-sheath cells the chloroplasts are bigger, they accumulate large amounts of starch in the light, and the grana are few and poorly developed; in the mesophyll cells the chloroplasts are smaller, they usually do not accumulate much starch in the light, and they have numerous large grana.[9]

4.16   Granal and stromal thylakoids in a chloroplast
Left: Electron micrograph showing several grana from a chloroplast of timothy grass; note the continuity between granal and stromal thylakoids. × 64,000. Right: Traditional interpretation of the three-dimensional arrangement of granal and stromal thylakoids. [Micrograph by W. P. Wergin. Courtesy E. H. Newcomb, University of Wisconsin.]

[8] *Kranz* means wreath in German. Here it refers to the ringlike or wreathlike arrangement of photosynthetic cells around the leaf veins.

[9] From the frequent use of "usually," "often," and other qualifiers, you will have gathered, correctly, that there is considerable variation among species exhibiting Kranz anatomy, and that the various Kranz characteristics do not always occur together.

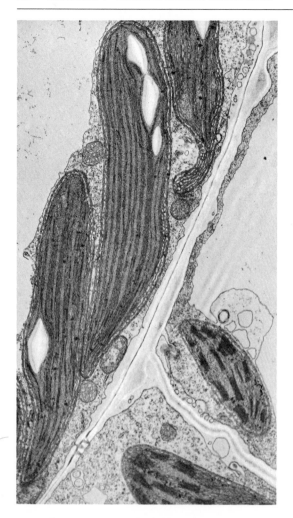

4.17   Electron micrograph of dimorphic chloroplasts of corn

At right and bottom are parts of two mesophyll cells; their chloroplasts contain numerous grana, but no starch grains. At left is part of a bundle-sheath cell; its chloroplasts are larger, have no grana, and contain starch grains (light areas). × 12,000.   [Courtesy Raymond Chollet, University of Nebraska.]

Most suggestions concerning the functional significance of Kranz anatomy put forward between 1900 and 1965 turned either on enhanced conservation of water under xeric conditions or on unusually rapid transport of photosynthetic products away from the sites of synthesis. Although Kranz anatomy may well serve both these functions, investigations of the late 1960s and early 1970s suggest that it is also associated with special biochemical adaptations that enhance the ability of the plants to carry out photosynthesis under conditions of high temperature, intense light, low moisture, low CO_2 and high O_2 concentrations—all conditions far from optimal for plants that depend entirely on the Calvin cycle for CO_2 fixation. We shall next examine the special photosynthetic pathways found in Kranz plants.

C_4 PHOTOSYNTHESIS

As Fig. 4.18A shows, corn, which originated in the tropics and has Kranz anatomy, can carry out photosynthesis at very low concentrations of CO_2, whereas bean, a plant native to the temperate zone, cannot, because of photorespiration. When the CO_2 concentration falls below about 50 parts per million (in 21 percent O_2 at 25°C), bean becomes incapable of CO_2 fixation; and at CO_2 concentrations of 200 or 300 ppm, where corn approaches its maximum photosynthetic capacity, bean performs at a mere fraction of its potential capacity. Again because of photorespiration, a far lower concentration of O_2 inhibits photosynthesis in bean than it does in corn (Fig. 4.18B). As was first realized in 1965, these contrasts between corn and bean are characteristic of the differences in photosynthetic capacity between Kranz plants and other plants.

Now, consider a plant exposed to great heat, dryness, and brilliant light, as in a desert or on a dry savanna. Under such conditions, when the moist membranes of the mesophyll cells risk losing an excess of water by evaporation, the guard cells close the stomata almost completely. Water loss is thus reduced, but now gases can no longer move freely between the atmosphere and the air spaces inside the leaf. As CO_2 is used up in photosynthesis, the nearly closed stomata prevent the supply inside the leaf from being fully replenished, and the CO_2 concentration in the air spaces around the mesophyll cells falls. At the same time the concentration of O_2, a by-product of photosynthesis, rises. Under such conditions, as we have seen, a non-Kranz plant like bean will carry out so much photorespiration that it will be unable to synthesize carbohydrate from CO_2. By contrast, corn and other Kranz plants *can* synthesize carbohydrate under xeric conditions; they are thus able to survive in climates that would be fatal to other plants. Even under less extreme conditions they can often carry out photosynthesis at a higher rate than other plants.

Surely it is not the distinctive anatomy of Kranz plants, by and of itself, that enables them to carry out photosynthesis under conditions inhospitable to other plants. It must be, rather, some special biochemical capability correlated with their anatomy—perhaps an ability to avoid photorespiration, which we saw earlier limits the photosynthetic ability of many plants under conditions of low CO_2 and high O_2. A finding in 1965 that plants with Kranz anatomy do not leak CO_2 to the environment in light seemed to indicate that they are indeed capable of avoiding photorespiration. But how? Here a third piece of the structure-function puzzle must be added to the other two. This third piece is a special way of fixing CO_2 initially.

In 1954 Hugo Kortschak of Hawaii, using the C^{14} tracer technique much as Calvin did, found that one of the main early products of photosynthesis in sugar cane is not one of the three-carbon intermediates (C_3) of the Calvin cycle, but a four-carbon compound (C_4) instead. It was not until the late 1960s, however, that M. D. Hatch and C. R. Slack of Queensland, Australia, worked out the biochemical pathway responsible for this C_4 type of photosynthesis. They found that in the mesophyll cells of sugar cane and other Kranz plants (where, you will recall, the mesophyll cells richest in chloroplasts are arranged in rings around the veins) CO_2 is combined, not with ribulose diphosphate as in the Calvin cycle, but rather with a three-carbon compound called PEP (for phosphoenolpyruvate) to form a four-carbon compound. The enzyme that catalyzes this carboxylation of PEP, unlike the one that catalyzes the carboxylation of RuDP in the Calvin cycle, does not have O_2 as an alternative substrate and is not inhibited by high O_2 concentrations. Thus it enables Kranz plants (which we shall henceforth call C_4 plants) to fix CO_2 under conditions when photorespiration would predominate over photosynthesis in C_3 plants (plants using only the Calvin cycle).

Curiously enough, the C_4 compound formed in the mesophyll cells is not used for growth or nutrition by the plant. Instead, it is passed (in reduced form) into the bundle-sheath cells (which in Kranz plants, you will remem-

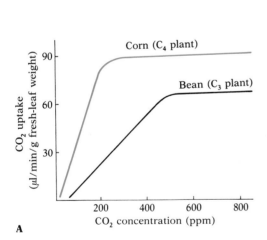

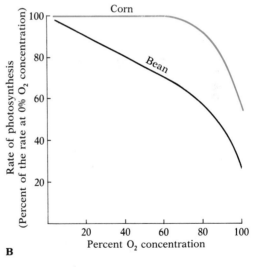

4.18 Rate of photosynthetic CO_2 fixation in corn and bean as a function (A) of CO_2 concentration and (B) of O_2 concentration

(A) Corn can fix CO_2 at CO_2 concentrations as low as one part per million, and it carries out photosynthesis at a very high rate at concentrations of 200–300 ppm (a normal concentration of CO_2 in the atmosphere in about 330 ppm). By contrast, bean performs no net CO_2 fixation at CO_2 concentrations below about 50 ppm, and its rate of photosynthesis at concentrations of 200–300 ppm is not very high.   (B) Photosynthesis in corn shows no inhibition at all at O_2 concentrations below 65 percent, whereas the photosynthetic rate of bean falls steadily as the O_2 concentration rises (a normal concentration of O_2 in the atmosphere is about 21 percent). The steep fall of both curves above about 65 percent O_2 is due, not to competitive inhibition, but to irreversible oxidative damage.   Both (A) and (B) are for temperatures of 20°C and light intensity of 2,000 foot-candles (ca. 2.15 lamberts).

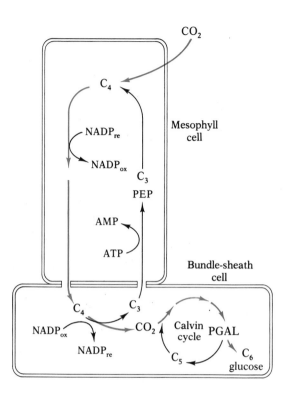

4.19 The Hatch-Slack pathway of C₄ photosynthesis
The path of carbon is here traced by color arrows. A mesophyll cell absorbs CO_2 from the intercellular air spaces (top). The CO_2 combines with PEP, a three-carbon compound to form a four-carbon compound. After reduction by $NADP_{re}$, the C_4 substance moves (probably via plasmodesmata) into an adjacent bundle-sheath cell. There it is oxidized by $NADP_{ox}$ and split into a C_3 substance and CO_2. The C_3 compound moves back to the mesophyll cell and is converted (by a reaction probably driven by energy from ATP) into PEP. The CO_2 is fed into the Calvin cycle in the bundle-sheath cell and is incorporated into carbohydrate.

ber, are very well developed and contain chloroplasts), where it is decarboxylated; i.e. the C_4 compound is broken down to CO_2 and a C_3 compound (Fig. 4.19). The C_3 residue moves back to the mesophyll cells, where it is reconverted into PEP and starts the C_4 cycle over again. But the CO_2 remains in the bundle-sheath cells, where it is picked up by the chloroplasts and incorporated into carbohydrate via the Calvin cycle.

Note, then, that in both C_3 and C_4 plants the ultimate assimilation of CO_2 into carbohydrate is by the Calvin cycle. The difference is that in C_3 plants the Calvin cycle is the only pathway of CO_2 fixation, whereas in C_4 plants there is another, preliminary fixation pathway. It may at first seem strange that a plant would have evolved a special mechanism for fixing CO_2 as C_4 in the mesophyll, only to combine it with a mechanism for promptly breaking off the CO_2 again and refixing it as carbohydrate in the bundle-sheath cells. But remember that the C_4 fixation in the mesophyll cells, because it is insensitive to O_2 concentration, cannot be short-circuited by photorespiration and can, therefore, operate under conditions where C_3 plants could not carry out net photosynthesis. The mesophyll cells can then pump enough CO_2 (via the C_4 intermediate) into the bundle-sheath cells to maintain an artificially high CO_2 concentration in which the Calvin cycle is able to function. Moreover, since the bundle sheath is surrounded by mesophyll cells, any CO_2 lost from the sheath cells as a result of photorespiration may be reclaimed by the mesophyll cells.

In summary, C_4 plants have an advantage over C_3 plants under conditions of high temperature and intense light, when stomatal closure results in low CO_2 and high O_2 in the air spaces inside the leaf. Under such conditions, C_3 plants have a net loss of CO_2 through photorespiration, but C_4 plants can fix CO_2, because the mesophyll cells, acting as CO_2 pumps, can elevate the CO_2 concentration in the bundle-sheath cells to a level where carboxylation of ribulose diphosphate (leading into the Calvin cycle) exceeds its oxidation. The Kranz anatomy, with its concentric rings of mesophyll and bundle-sheath cells, facilitates the compartmentalization on which the process of CO_2 pumping depends.

The combination of Kranz anatomy and C_4 photosynthesis has evolved independently in a variety of unrelated plants, including both monocots (e.g. corn, sugar cane, sorghum, crabgrass) and dicots (e.g. saltbush, Portulaca). It is therefore an especially impressive illustration of the intimate relationship between structure and function in living systems.

CHEMOSYNTHESIS

Some bacteria are able to synthesize high-energy organic compounds from inorganic raw materials without benefit of light energy. Such bacteria oxidize inorganic compounds and trap the energy thus released. One type oxidizes ammonia (NH_4^+):

$$2NH_4^+ + 3O_2 \longrightarrow 2NO_2^- + 2H_2O + 4H^+ + energy$$

Another bacterium oxidizes nitrite (NO_2^-) to nitrate (NO_3^-):

$$2NO_2^- + O_2 \longrightarrow 2NO_3^- + energy$$

Still another oxidizes sulfur to sulfate:

$$2S + 3O_2 + 2H_2O \longrightarrow 2SO_4^{--} + 4H^+ + energy$$

Many other examples of bacteria that oxidize various nitrogen and sulfur compounds, and even molecular hydrogen, could be cited. They share an ability to trap the small quantities of energy released by their oxidation of inorganic raw materials and to use this energy in synthesis of carbohydrate. Such organisms are called **chemosynthetic.** They are fascinating subjects for study, but their importance in the overall energy economy of nature is very small compared with that of the photosynthetic organisms. Some of the chemosynthetic bacteria that act on nitrogen compounds do, however, play an extremely important role in the movement of nitrogen within the life system, as we shall see in a later chapter.

CELLULAR METABOLISM

"Metabolism" is a general term embracing all the myriad enzyme-mediated reactions of a living cell. These reactions may be viewed as belonging to one or the other of two phases—**anabolism,** the biosynthetic building-up phase, and **catabolism,** the degradative breaking-down phase.

Photosynthesis is an anabolic process that binds energy from solar radiation into complex organic compounds such as glucose and compounds synthesized from glucose. Before this potential energy can be used in doing work, either by the green plant itself or by some organism that has eaten the green plant, the large energy-rich molecules must be broken down chemically and the energy transferred to ATP. This catabolic process must occur in every living thing and is thus a fundamental characteristic of life.

As you might anticipate from our examination of photosynthesis, the breakdown does not occur as a single gross reaction, but rather as a series of smaller step-by-step reactions, each catalyzed by its own enzyme; and the release of packets of energy is coupled with phosphorylation reactions synthesizing ATP from ADP and inorganic phosphate.

The complete degradation of an energy-rich compound such as glucose to carbon dioxide and water depends on a multienzyme system in which numerous interrelated series of reactions play their part. Only the more important steps in the breakdown will be examined here, although others will be indicated in the diagrams to give you a better notion of the extreme complexity of the process. Our discussion will deal, first, with a chain of reactions that can take place whether oxygen is present or not (anaerobic metabolism), second, with a chain of reactions that are dependent on oxygen (aerobic metabolism).

ANAEROBIC METABOLISM

Glycolysis   We shall begin our examination of cellular metabolism with the anaerobic breakdown of carbohydrate—specifically, glucose to pyruvic acid, a process known as glycolysis.

Glucose, as we have already seen, is a stable compound; i.e. it has little tendency to break down to simpler products. If the energy locked in its molecular configuration is to be released, the glucose must first be made

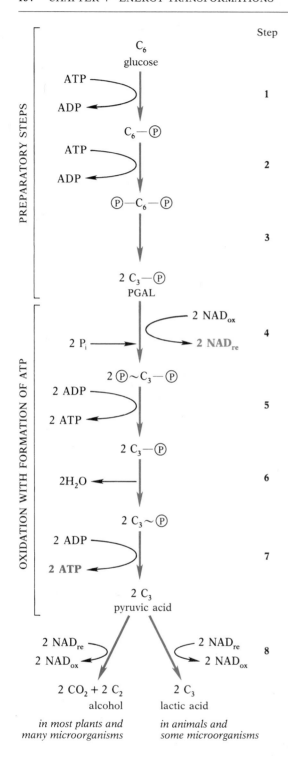

Step

more reactive; a small amount of energy must be invested by the cell to initiate the glycolytic process.

As you might expect, it is ATP, described previously as the energy currency of life, that provides the energy for initiating glycolysis (Fig. 4.20). This initial reaction, like the succeeding ones, is facilitated by its own specific enzyme. A molecule of ATP donates its terminal phosphate group to the glucose:

(1) $\underset{\text{glucose}}{C-C-C-C-C-C} + ATP \xrightarrow{\text{enzyme}} \underset{\text{glucose phosphate}}{C-C-C-C-C-C-\text{\textcircled{P}}} + ADP$

(The simplified summary equations given here show only the carbon skeleton, and you should bear in mind that oxygen and hydrogen are attached to the carbons; see Fig. 2.26, p. 49. Similarly \textcircled{P} here represents an entire phosphate group.) When the phosphate group is transferred to glucose, some energy is lost as heat, and the phosphorylated glucose formed is not especially energy-rich; i.e. its free energy of hydrolysis is not as great as that of ATP (which is why the bond attaching the \textcircled{P} to the glucose is symbolized by a straight and not a wavy line).

After a reaction that converts the glucose phosphate into fructose phosphate, a related six-carbon compound, another molecule of ATP donates its terminal phosphate group to the other end of the six-carbon chain, again with the loss of some energy as heat:

(2) $\underset{\text{fructose phosphate}}{C-C-C-C-C-C-\text{\textcircled{P}}} + ATP \xrightarrow{\text{enzyme}}$

$\underset{\text{fructose diphosphate}}{\text{\textcircled{P}}-C-C-C-C-C-C-\text{\textcircled{P}}} + ADP$

The enzyme that catalyzes this reaction is allosteric, and its activity is regulated by several modulators, both positive and negative. By regulation of this enzyme, the entire glycolysis pathway is effectively regulated (see p. 170, box).

4.20   Glycolysis and fermentation

Energy to initiate the breakdown of glucose is supplied by two molecules of ATP (Steps 1–2). The resulting compound is then split into two molecules of PGAL (3). This completes the preparatory reactions. Next, the PGAL is oxidized by removal of hydrogen, which is picked up by NAD_{ox} to form two molecules of NAD_{re}; in the same reaction inorganic phosphate is added to each of the three-carbon molecules (4). A series of reactions then results in synthesis of four new molecules of ATP, for a net gain of two (5–7). The pyruvic acid produced by this anaerobic breakdown can be further oxidized if O_2 is present (by reactions not shown here). But in the absence of sufficient O_2, the pyruvic acid may accept hydrogen from NAD_{re} to form CO_2 and ethyl alcohol in some kinds of organisms or lactic acid in others (8). *Note:* Some of the steps indicated in the figure summarize several reactions.

Next, the diphosphorylated six-carbon compound (fructose diphosphate) is split between the third and fourth carbons, two essentially similar three-carbon molecules being formed; one is PGAL, and the other is immediately converted into PGAL. We can summarize these reactions:

(3) $\text{P}-\text{C}-\text{C}-\text{C}-\text{C}-\text{C}-\text{C}-\text{P} \xrightarrow{\text{enzyme}} 2 \text{ C}-\text{C}-\text{C}-\text{P}$
        fructose diphosphate   PGAL

PGAL, you will recall, is a major product of photosynthesis; the photosynthetic process involved synthesis of glucose from PGAL. So far, then, glycolysis looks like photosynthesis in reverse. And so far, instead of releasing energy from the carbohydrate and forming new ATP molecules, glycolysis has actually resulted in the loss of two molecules of ATP. All the reactions up to this point have been preparatory.

The next reaction, a rather complicated one that begins the changes leading to production of new ATP, is really two reactions in one. The PGAL is both oxidized (by NAD, an electron acceptor similar to the NADP already discussed) and phosphorylated (by inorganic phosphate):

(4) $2 \text{ P}-\text{C}-\text{C}-\text{C} + 2 \text{ NAD}_{ox} + 2 \text{ P}_i \xrightarrow{\text{enzyme}}$

                    $2 \text{ P}-\text{C}-\text{C}-\text{C} \sim \text{P} + 2 \text{ NAD}_{re}$

Now, the oxidation phase of this reaction, taken alone, is strongly exergonic, whereas the phosphorylation phase is strongly endergonic. Since the two processes occur together, the energy that would have been released by the oxidation is conserved by the formation of a high-energy phosphorylated derivative of PGAL, a process exemplifying substrate-level phosphorylation.[10] Indeed, the free energy of hydrolysis of the newly formed phosphate bonds (shown above as a wavy line) is considerably greater than that of the terminal phosphate bonds of ATP. Hence the next reaction in the sequence can transfer the high-energy phosphate groups to ADP exergonically, to form two new molecules of ATP:

(5) $2 \text{ P}-\text{C}-\text{C}-\text{C} \sim \text{P} + 2 \text{ ADP} \xrightarrow{\text{enzyme}} 2 \text{ P}-\text{C}-\text{C}-\text{C} + 2 \text{ ATP}$

At this point, then, the cell regains the two ATP molecules used in phosphorylating glucose (Steps 1 and 2) to start glycolysis. The initial energy investment has been repaid.

Next come several reactions that result in converting the remaining phosphate groups into energy-rich phosphate:

(6) $2 \text{ P}-\text{C}-\text{C}-\text{C} \xrightarrow{\text{enzyme}} 2 \text{ P} \sim \text{C}-\text{C}-\text{C} + \text{H}_2\text{O}$

The transfer of these phosphate groups to ADP results in the formation of two more ATP molecules plus two molecules of a three-carbon compound named *pyruvic acid:*[11]

(7) $2 \text{ P} \sim \text{C}-\text{C}-\text{C} + 2 \text{ ADP} \xrightarrow{\text{enzyme}} 2 \text{ C}-\text{C}-\text{C} + 2 \text{ ATP}$
                                                              pyruvic acid

[10] Substrate-level phosphorylation is the addition of inorganic phosphate to a substance being oxidized and the subsequent formation of an energy-rich phosphate group in the oxidized product. The energy-rich phosphate group may then be transferred to ADP to form ATP.

[11] Since pyruvic acid is ionized at typical cellular pH levels, it is often referred to as pyruvate in this context. Similarly, lactic, citric, and other acids to be discussed below may be referred to as lactate, citrate, etc.

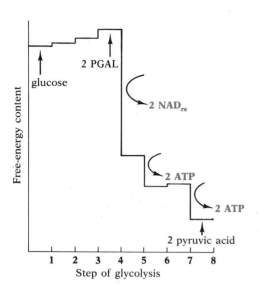

4.21 Changes in free-energy content at successive steps in glycolysis
In the three preparatory steps, which convert glucose into PGAL, the free-energy content of the substances is slightly increased owing to the investment of two molecules of ATP. In Step 4 there is a sharp drop in free energy associated with the formation of two molecules of NAD_{re}. There are also major drops in Steps 5 and 7, each associated with the formation of two molecules of ATP.

Because the two ATP molecules used up in Steps 1 and 2 have already been regained (in Step 5), the two additional ATP molecules formed here represent a net gain in ATP for the cell.

We can now summarize the most important features of glycolysis:

1. Each molecule of glucose (a six-carbon compound) is broken down to two molecules of pyruvic acid (a three-carbon compound).
2. Two molecules of ATP are used to initiate the process. Later, four new molecules of ATP are synthesized as a result of substrate-level phosphorylation, for a *net* gain of two molecules of ATP from each glucose molecule broken down. The energy stored in the new ATP molecules represents only about 2 percent of the energy initially present in the glucose molecule.
3. Two molecules of NAD_{re} are formed.
4. Because no molecular oxygen is used, glycolysis can occur whether or not O_2 is present. It is a process encountered in all living cells, whatever their mode of life.

Figure 4.21 shows graphically the changes in free-energy content at each successive step in glycolysis, from glucose to pyruvic acid.

Fermentation We have seen that in glycolysis two molecules of NAD_{ox} are reduced to NAD_{re}. Like the very similar NADP so important in photosynthesis, NAD functions in the cell as an electron-transport compound, shuttling electrons between one substance and another. Thus NAD is only a temporary acceptor of electrons, which promptly passes the extra electrons to some other compound and then goes back for another load. The cell has only a limited supply of NAD molecules, which must be used over and over. If the NAD_{re} molecules formed in glycolysis could not quickly unload electrons (i.e. be reconverted into NAD_{ox}), all the cell's NAD would soon be tied up (i.e. all would be in the NAD_{re} form); with no NAD_{ox} available, Step 4 of glycolysis could not take place and the whole glycolysis process would come to a stop.

In most cells, if molecular oxygen is available, it becomes the ultimate acceptor of electrons from NAD_{re}, by a process we shall examine later. But under anaerobic conditions, with no O_2 to accept the electrons, it is the pyruvic acid formed by glycolysis that accepts electrons from NAD_{re} (Fig. 4.21). This reduction of pyruvic acid results in the formation of *lactic acid* (another three-carbon compound) in animal cells and in some unicellular organisms or in the formation of *ethyl alcohol* (a two-carbon compound), plus CO_2, in plant cells and in many unicellular organisms:

(8) $$\text{pyruvic acid} + NAD_{re} \xrightarrow{\text{enzyme}} \text{lactic acid} + NAD_{ox}$$

or

$$\text{pyruvic acid} + NAD_{re} \xrightarrow{\text{enzyme}} \text{alcohol} + CO_2 + NAD_{ox}$$

Thus, under anaerobic conditions, NAD shuttles back and forth between Steps 4 and 8, picking up electrons (i.e. becoming NAD_{re}) in Step 4 and giving up the electrons (i.e. becoming NAD_{ox}) in Step 8.

The process, of which the glycolytic pathway is the first stage, whereby carbohydrate is transformed into alcohol or lactic acid is called *fermenta-*

tion.[12] We can thus speak of alcoholic fermentation or lactic acid fermentation, depending on the end product of the process. (In a few organisms, fermentation leads to products other than alcohol or lactic acid, but these are of less general importance and will not be discussed here.)

Whatever the end product, fermentation enables a cell to continue synthesizing ATP by breakdown of nutrients under anaerobic conditions. But this process extracts only a very small portion of the energy present in the original glucose; the end products of fermentation still contain much of that original energy.

Fermentation by yeast cells and other microorganisms is, of course, the basis for the extensive and economically vital industry that produces both commercial alcohol and alcoholic beverages. Microbial fermentations are also essential to the production of most cheeses, yogurt, and a variety of other dairy products.

AEROBIC METABOLISM (RESPIRATION)

Next, let us consider what happens when metabolism takes place in the presence of abundant molecular oxygen. Under aerobic as under anaerobic conditions, the metabolic breakdown of glucose initially follows the glycolytic pathway to pyruvic acid. But in the presence of O_2, which can act as the ultimate acceptor of electrons from NAD_{re}, pyruvic acid need not act as an electron acceptor and become converted into lactic acid or alcohol. Instead, it can be further broken down and yield energy for synthesis of still more new ATP. In other words, under aerobic conditions, ATP synthesis does not end with the pyruvic acid step. Indeed, if lactic acid has already been formed, it may be reconverted into pyruvic acid, with accompanying resynthesis of NAD_{re}, when sufficient oxygen becomes available; this pyruvic acid, too, may then be further oxidized.

The process of aerobic breakdown of nutrients with accompanying synthesis of ATP is called *cellular respiration.* It comprises the second and third stages in the metabolic breakdown of glucose—namely, the oxidation of pyruvic acid to acetyl-CoA and the reactions of the Krebs citric acid cycle; the first stage, glycolysis, is, as we saw, anaerobic. All three stages in the breakdown yield the energy-rich compound NAD_{re}. It is the transfer of electrons from this compound to O_2 via the respiratory electron-transport chain that produces most of the ATP generated by the metabolic breakdown of one molecule of glucose.

Oxidation of pyruvic acid to acetyl-CoA The aerobic oxidation of pyruvic acid begins with a complicated set of reactions whose net effect is to break

[12] The term "fermentation" has been used in countless ways in the scientific literature. It is often restricted to the breakdown of glucose to alcohol. It is also applied to the production of either alcohol or lactic acid by microorganisms—lactic acid production in animal cells being called glycolysis. Both these uses lead to confusion between the terms "fermentation" and "glycolysis," and both tend to obscure the general occurrence of the same basic fermentation process in all living cells. Accordingly, "fermentation" is here applied to the process of anaerobic production of alcohol or lactic acid (or other products of the reduction of pyruvic acid), whether by plant, animal, or microorganism, and the glycolytic pathway to pyruvic acid is taken both as a preparatory reaction sequence leading to the Krebs citric acid cycle when sufficient oxygen is present and as the initial portion of fermentation in the absence of sufficient oxygen (a few microorganisms carry out fermentation in the presence of abundant oxygen).

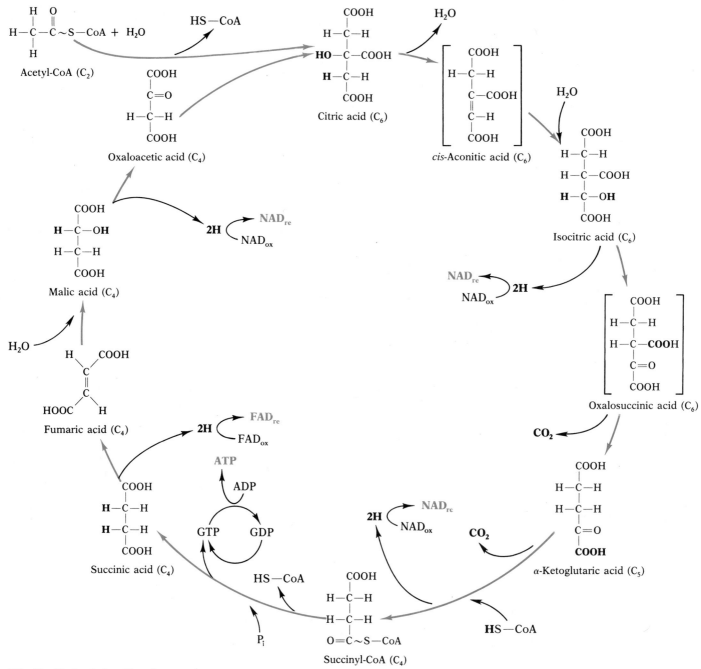

4.22 The Krebs citric acid cycle

The complete cycle is shown here only to help you appreciate the characteristic complexity of metabolic pathways; there is no point in your trying to learn all the reactions involved. As shown at upper left, the acetyl group (2 carbons) from acetyl-CoA enters the cycle by combining with oxaloacetic acid (4 carbons) to form citric acid (6 carbons). During subsequent reactions two of the carbons are removed as CO_2 (between oxalosuccinic acid and α-ketoglutaric acid, and between α-ketoglutaric acid and succinyl-CoA), and a total of eight hydrogens are removed. These hydrogens are picked up by NAD (or by the related acceptor molecule FAD). One molecule of ATP is synthesized at substrate level (bottom). Finally, oxaloacetic acid is regenerated and can combine with a new acetyl group to start the cycle over again. The cycle is completed twice for each molecule of glucose oxidized. (The atoms removed at each step are shown in boldface in the structural formulas. The two substances in brackets—cis-aconitic acid and oxalosuccinic acid—are enzyme-bound intermediates that seldom exist as free compounds.)

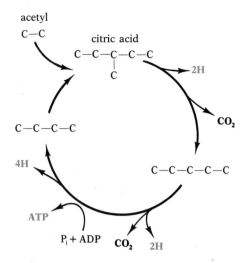

4.23 Simplified version of the Krebs citric acid cycle
The two carbons of the acetyl group combine with a four-carbon compound to form citric acid, a six-carbon compound. Removal of one carbon as CO_2 leaves a five-carbon compound. And removal of a second carbon as CO_2 leaves a four-carbon compound, which can combine with another acetyl group and start the cycle over again. In the course of the cycle, one molecule of ATP is synthesized and eight hydrogens are released (they are picked up by carrier compounds not shown here). Since one molecule of glucose gives rise to two acetyl units, two turns of the cycle occur for each molecule of glucose oxidized, with production of four molecules of CO_2, two molecules of ATP, and 16 hydrogens.

down the three-carbon pyruvic acid to CO_2 and an activated form of the two-carbon compound acetic acid. The acetic acid is said to be activated because it is not present as free acetic acid, but is bonded to a coenzyme called coenzyme A, or CoA for short; the complete compound is called *acetyl-CoA.* When a molecule of pyruvic acid is oxidized to acetyl-CoA and CO_2, hydrogen is removed and a molecule of NAD_{re} is formed. Since two molecules of pyruvic acid were formed from each glucose molecule, two molecules of NAD_{re} are formed here. This complicated series of reactions can be summarized by the following equation:

$$2 \text{ pyruvic acid} + 2 \text{ CoA} + 2 \text{ NAD}_{ox} \longrightarrow 2 \text{ acetyl-CoA} + 2CO_2 + 2 \text{ NAD}_{re}$$

Note that, at this stage, two of the six carbons present in the original glucose have been released as CO_2. Note also that the newly formed NAD_{re} must be oxidized if the breakdown process is to continue; we shall return to this problem shortly.

The Krebs citric acid cycle   The acetyl-CoA is next fed into a complex circular series of reactions called the Krebs citric acid cycle (after the British scientist Sir Hans Krebs, who was awarded a Nobel Prize for his elucidation of this system).[13] The cycle is shown in some detail in Fig. 4.22; its essential features are outlined in Fig. 4.23. Briefly, each of the two two-carbon acetyl-CoA molecules formed from one molecule of glucose is combined with a four-carbon compound (oxaloacetic acid) already present in the cell, to form a new six-carbon compound called *citric acid.* Each of the citric acid molecules is then oxidized to a five-carbon compound plus CO_2. The five-carbon unit, in turn, is oxidized to a four-carbon compound plus CO_2. This four-carbon compound is then converted into the same four-carbon compound as the one to which acetyl-CoA was originally attached; it can now pick up more acetyl-CoA, forming new citric acid and beginning the cycle again.

[13] The Krebs cycle is also frequently called the tricarboxylic acid circle.

REGULATION OF GLUCOSE BREAKDOWN

There are several major points at which the interconnected multienzyme systems shown in Fig. 4.24 can be regulated. One of the most important is at the step in glycolysis where fructose diphosphate is produced. The allosteric enzyme that catalyzes this reaction is positively modulated (made more reactive) by ADP and AMP; it is negatively modulated (made less reactive) by ATP and by citric acid. Thus the enzyme is most active, and glucose breakdown is greatest, when there is a shortage of ATP and a buildup of ADP (and AMP). The enzyme is least active, and glucose breakdown is slowed, when there is an accumulation of ATP and citric acid. Regulation of the enzyme for fructose diphosphate by citric acid is a good example of *feedback inhibition*—the inhibition of an allosteric enzyme by one of the products of a later reaction in the biochemical pathway.

Other important regulatory enzymes are to be found at Step 1 of glycolysis (where the first phosphate is added to glucose) (see Fig. 4.20), at Step 7 of glycolysis (where pyruvic acid is formed), at the reaction that breaks down pyruvic acid to acetyl-CoA, and at the reaction in the Krebs cycle where isocitric acid is converted into α-ketoglutaric acid (see Fig. 4.22). The enzymes at all these points are inhibited by ATP, and most are made more reactive by ADP. In other words, all stages of glucose breakdown tend to be turned off when there is an excess of ATP and to be turned on again when the ATP level falls and more glucose has to be broken down to provide energy for ATP synthesis.

We see, then, that two carbons are fed into the Krebs cycle as the acetyl group and that the same number is released as two molecules of CO_2. Since each glucose molecule being oxidized yields two molecules of acetyl-CoA, two turns of the cycle are required, and a total of four carbons are released as CO_2 during this stage of glucose breakdown. With the two other carbons already released as CO_2 during the oxidation of pyruvic acid to acetyl-CoA, all six carbons of the original glucose are accounted for.

The oxidative breakdown of each molecule of acetyl-CoA via the Krebs citric acid cycle also involves the removal of eight hydrogens, which are picked up by NAD_{ox} (or by a related electron-carrier compound called FAD_{ox});[14] four units of reduced carrier are thus formed (Fig. 4.22). Since the breakdown of one molecule of glucose leads to two turns of the Krebs cycle, a total of eight molecules of reduced carrier (6 NAD_{re} and 2 FAD_{re}) are formed during this stage of the breakdown of glucose. Two molecules of ATP are also synthesized (as a result of substrate-level phosphorylation) in the Krebs cycle.

Figure 4.24 summarizes the yield of ATP, NAD_{re} (or FAD_{re}), and CO_2 from the three stages of breakdown of glucose.

The respiratory electron-transport chain We began this discussion of catabolism by saying that glucose is energy-rich, and that its breakdown enables the cell to synthesize new ATP, the cell's energy currency. But in our examination of the three stages of catabolism—glycolysis, conversion of pyruvic acid into acetyl-CoA, and the Krebs cycle—we have so far seen a net

---

[14] FAD is an abbreviation for flavin adenine dinucleotide.

gain of only four new ATP molecules (two in glycolysis and two in the Krebs cycle). These represent but a small fraction of the energy originally available in the glucose. Where has the rest of the energy gone?

To answer this question, let us first consider the three kinds of atoms (C, O, and H) present in the original glucose. The carbon and oxygen released as CO_2, a low-energy compound, clearly do not contain the energy we are seeking. But what about the hydrogen from the glucose? It has been picked up by NAD_{ox} (and other carrier substances) to form NAD_{re} (which, you will recall, stands for $NADH + H^+$). Now look again at Fig. 4.21, where the free-energy changes in the various steps of glycolysis are shown. Notice that the single biggest energy drop occurs in the step where NAD_{re} is formed. This drop, considerably bigger than those associated with ATP formation, immediately suggests that much energy must go into the formation of NAD_{re} (i.e. that synthesis of NAD_{re} is strongly endergonic). Indeed, both NAD_{re} and FAD_{re} are energy-rich substances. And 12 molecules of these are synthesized in the process of breakdown of a molecule of glucose: 2 NAD_{re} in glycolysis, 2 NAD_{re} in breakdown of pyruvic acid to acetyl-CoA, and 6 NAD_{re} plus 2 FAD_{re} in the Krebs cycle (Fig. 4.24). It is these reduced carrier molecules that harbor much of the energy from the original glucose.

How is this energy used to synthesize ATP? Here we must return to a statement we have already made several times without explanation: Under aerobic conditions the regeneration of NAD_{ox} from NAD_{re} is achieved by

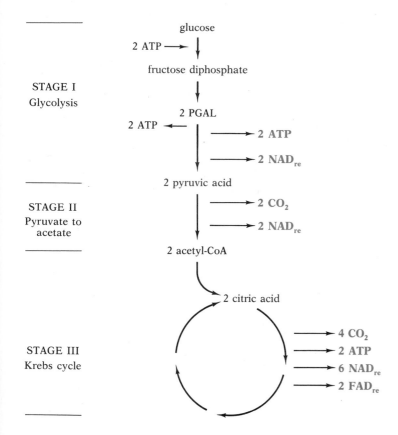

4.24 Summary of the most important products of Stages I, II, and III in the complete breakdown of one molecule of glucose

Stage I (glycolysis) begins with expenditure of two molecules of ATP to produce fructose diphosphate, which is broken down to two molecules of PGAL. After these preparatory steps, the two PGAL molecules are broken down to two molecules of pyruvic acid, a process that first pays back the two ATP molecules originally invested and then yields two molecules each of ATP and NAD_{re} (color). Stage II (the breakdown of two molecules of pyruvic acid to two molecules of acetyl-CoA) yields two molecules each of CO_2 and NAD_{re}. Stage III, in which the two molecules of acetyl-CoA are fed into the Krebs cycle and further broken down, yields four CO_2 molecules, two ATP molecules, six NAD_{re} molecules, and two FAD_{re} molecules.

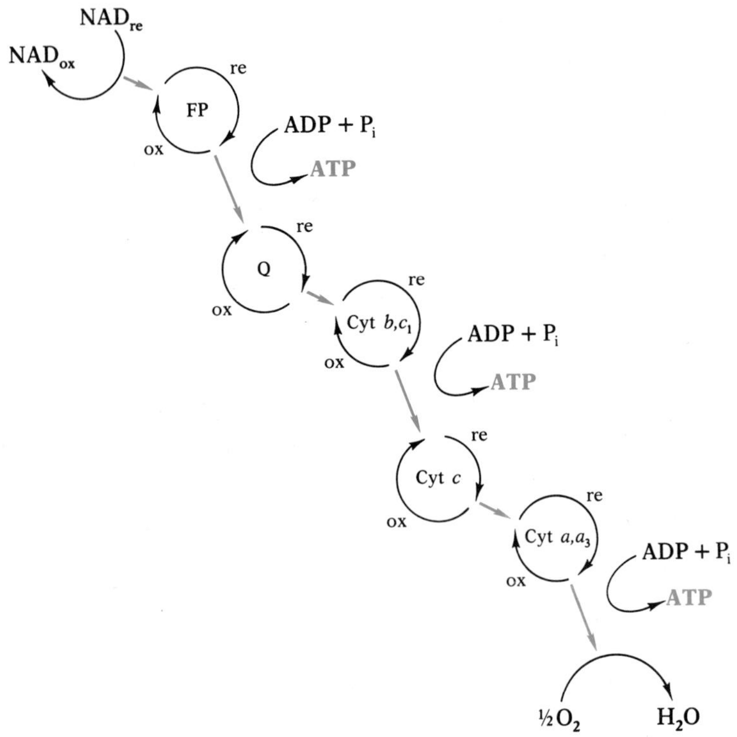

4.25 The respiratory electron-transport chain
The pair of electrons donated to the chain by NAD_{re} are passed from one acceptor substance to the next, step by step down an énergy gradient from their initial high energy level in NAD_{re} to their final low energy level in H_2O. Each successive acceptor molecule is cyclically reduced when it receives the electrons and then oxidized when it passes them on to the next acceptor molecule. At three sites along the chain, some of the free energy released is used in the synthesis of ATP. (The electron acceptor substances are flavoprotein, FP; coenzyme Q; cytochromes *b* and c_1; cytochrome *c*; cytochromes *a* and a_3.)

passage of the hydrogen (especially hydrogen electrons) from NAD_{re} to O_2; in other words, O_2 acts as the ultimate acceptor of hydrogen, and water is formed:

$$O_2 + 2\ NADH + 2H^+ \longrightarrow 2H_2O + 2\ NAD_{ox}$$

The NAD_{re} does not, however, pass its hydrogen directly to the oxygen, as this summary equation might seem to indicate. What happens, instead, is that the hydrogen electrons are passed down a "respiratory chain" of electron-transport compounds, many of which are iron-containing pigments called cytochromes[15] (Fig. 4.25). In this manner, the electrons are lowered step by step from their high energy level in NAD_{re} to a low energy level in H_2O.

As the electrons are lowered down the energy gradient step by step through the respiratory electron-transport chain, energy is released, and some of this energy is used in the synthesis of ATP from ADP and inorganic phosphate. This process, often called *oxidative phosphorylation,* is very similar to the electron transport and ATP synthesis of photophosphorylation, already discussed. In both photo- and oxidative phosphorylation, some of the energy released during the flow of electrons from a high-energy donor to an electron acceptor is used in ATP synthesis. But there is a basic difference between the two processes. The overall effect of photophosphorylation is an increase in the amount of stored energy within the organism, the additional energy coming from light, whereas the overall effect of oxidative

[15] The respiratory electron-transport chain is sometimes called the cytochrome system.

phosphorylation is a decrease in the amount of stored energy within the organism, because the process cannot be thermodynamically 100 percent efficient.

It has been demonstrated that, if a molecule of NAD_{re} can donate its extra pair of electrons directly to the electron-transport chain, three new ATP molecules are synthesized as the electrons move through the chain to H_2O (Fig. 4.25). As Fig. 4.26 shows, this is the case for the two NAD_{re} molecules produced by breakdown of pyruvic acid to acetyl-CoA and for the six NAD_{re}

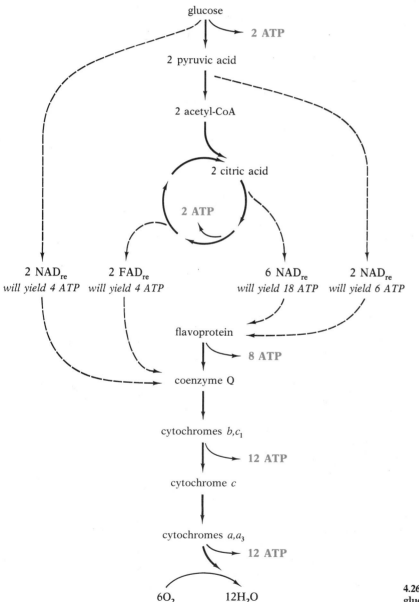

4.26 **The ATP yield from complete breakdown of glucose to carbon dioxide and water**

molecules produced in the Krebs cycle; the electrons from each of these give rise to three molecules of ATP as they move through the respiratory electron-transport chain (a total of 24 new ATP molecules from the 8 NAD_{re}). But as Fig. 4.26 also shows, the two FAD_{re} molecules from the Krebs cycle feed their electrons into the respiratory chain at a point beyond the first phosphorylation site; hence these electrons yield only two molecules of ATP per molecule of FAD_{re} (a total of 4 ATP from the 2 FAD_{re}). Since the two NAD_{re} molecules produced in glycolysis often also feed their electrons into the chain beyond the first phosphorylation site, they, too, yield only two molecules of ATP per molecule of NAD_{re} (a total of 4 ATP).[16] In summary, of the 12 molecules of reduced carrier produced by complete breakdown of one molecule of glucose, eight donate their extra electron pairs to the respiratory chain in such fashion as to get the maximum ATP yield (3 ATP per pair of electrons), and the other four donate their electrons farther along in the chain and get only two molecules of ATP per pair of electrons. The total yield from transport of the 12 pairs of electrons down the chain is 32 molecules of ATP.

SUMMARY OF THE ATP YIELD OF THE RESPIRATION OF GLUCOSE

We can summarize the yield of ATP molecules from the complete metabolic breakdown of one molecule of glucose to carbon dioxide and water as follows (Fig. 4.26):

1. A grand total of 36 new ATP molecules are formed. This means that about 38 percent of the free energy initially present in the glucose has been trapped; the rest has been lost owing to heat release and increased entropy.[17]

2. Only two of the 36 ATP molecules (about 5 percent) are synthesized anaerobically; the other 34 (about 95 percent) are the product of aerobic respiration. We now see why oxygen is essential to the life of human beings

---

[16] The enzymes of the Krebs cycle and the electron-transport chain are located inside mitochondria, whereas the enzymes of glycolysis are in the cytosol outside the mitochondria. This means that the two molecules of NAD_{re} produced during glycolysis are separated from the chain by the mitochondrial membrane, which is impermeable to NAD_{re}. Hence the two NAD_{re} produced during glycolysis cannot donate electrons directly to the electron-transport chain. They can, however, pass electrons to shuttle molecules that readily penetrate the mitochondrial membrane, and these shuttle molecules can then donate the electrons to the chain.

Of the shuttle molecules now known, one, glycerol phosphate, enters the system *after* the first site of ATP synthesis, and therefore allows production of only two ATP molecules. A second, newly discovered shuttle molecule, malate, enters the chain *before* the first site of ATP synthesis, and therefore allows production of three ATP molecules. It is not yet certain which shuttle system predominates in normal living tissue; the present text assumes that it is glycerol phosphate, and hence that each NAD_{re} from glycolysis results in the production of only two ATP molecules.

[17] The efficiency figures given by different texts vary considerably, ranging from a low of 34 percent to a high of 60 percent. The reason is that there is disagreement both about the exact energy content of ATP and about the number of ATP molecules yielded by the complete breakdown of one molecule of glucose; some workers believe this number to be 38–40 rather than 36. The 38 percent efficiency figure is calculated as follows:

The free-energy content of a mole of glucose is about 686 kcal. According to the best recent evidence, the free energy of hydrolysis of the terminal phosphate of a mole of ATP is 7.3 kcal. If 36 moles of ATP are formed per mole of glucose, then $(36 \times 7.3)/686$, or approximately 38 percent, of the free energy initially present in the glucose has been trapped in the high-energy phosphate groups of the ATP. (If 38 moles of ATP are formed, as when the malate shuttle is operating, then the energy yield is about 40 percent.)

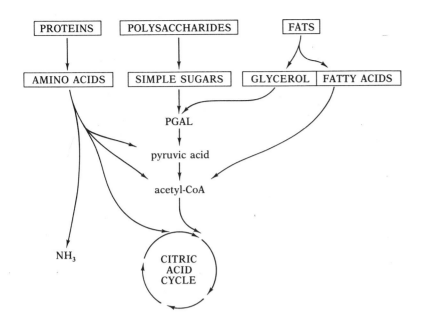

4.27 The relationships of the metabolism of proteins and fats to the metabolism of carbohydrates

and most other organisms; if such energy-demanding organisms could derive only two molecules of ATP from each glucose molecule, they would require far more food than it would be possible for them to obtain and process. In fact, in the absence of O_2 such organisms soon die of asphyxiation, because their cells run out of energy. The importance of the anaerobic processes of glycolysis and fermentation should not be overlooked, however. Many microorganisms rely exclusively on fermentation for their energy. Our own tissues, particularly our muscles during violent activity, often need so much energy so fast that the oxygen supplied from breathing is insufficient. In such cases glycolysis provides the needed energy. Later, the oxygen debt is paid back by deep breathing or panting, and the lactic acid, which accumulated in the muscles as a result of glycolysis and fermentation, and which in excess contributes to muscle fatigue, is removed.

3. Of the 34 aerobically synthesized ATP molecules, 32, or 94 percent, result from electron transport via NAD (or FAD) and the respiratory electron-transport chain. This proportion points up the critical importance of the chain, and makes it easy to understand why cyanide and certain other poisons that block the chain are lethal; a cell unable to utilize O_2 because its electron-transport chain has been poisoned can derive so little energy from its nutrients that it soon dies.

RESPIRATION OF FATS AND PROTEINS

Cells can extract energy in the form of ATP not only from carbohydrates, on which we have focused so far, but also from the two other major categories of nutrients: fats and proteins. As Fig. 4.27 shows, early breakdown products of these substances are fed into the multienzyme systems we have already discussed.

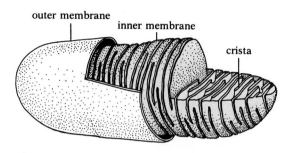

outer membrane   inner membrane   crista

4.28   Structure of a mitochondrion

Much of the outer membrane has been cut away, and
the interior has been sectioned to show how the inner
membrane folds into cristae. The mitochondria of
metabolically very active cells have more cristae than
those of less active cells.

Metabolism of fats begins with their hydrolysis to glycerol and fatty acids.
The glycerol (a three-carbon compound) is then converted into PGAL and
fed into the glycolytic pathway at the point where PGAL normally appears.
The fatty acids are broken down to a number of two-carbon fragments,
which are converted into acetyl-CoA and fed into the respiratory pathway at
the appropriate point. Since fats are more completely reduced compounds
than carbohydrates, their full oxidation yields more energy per unit weight;
one gram of fat yields slightly more than twice as much energy as one gram
of carbohydrate.

The amino acids produced by hydrolysis of proteins are metabolized in a
variety of ways. After removal of the amino group (deamination) in the form
of ammonia (NH_3), some amino acids are converted into pyruvic acid, some
into acetyl-CoA, and some into one or another compound of the citric acid
cycle. Complete oxidation of a gram of protein yields roughly the same
amount of energy as one gram of carbohydrate.

Such compounds as pyruvic acid, acetyl-CoA, and the compounds of the
citric acid cycle, which are common to the metabolism of several different
types of substances, not only play a crucial role in the oxidation of energy-
rich compounds to carbon dioxide and water, but they also function as
important intermediates in interconversions between various types of compounds, serving as carbon skeletons from which a variety of substances can
be synthesized. For example, carbohydrates can be converted into fat, and
many amino acids into carbohydrate, via the common intermediates in their
metabolic pathways. However, most higher animals lack enzymes for converting fatty acids into carbohydrate.

The inability of animals to make carbohydrate from fatty acids, despite
the convergence of the metabolic pathways for these two classes of compounds, points up the fact that synthesis pathways are seldom simply the
reverse of degradative pathways. For example, while it is technically true
that the same enzymes can catalyze the back reactions as well as the forward
reactions outlined in Fig. 4.20, several of the reactions have equilibrium
points so far forward that for all practical purposes they are irreversible in
the living cell. Thus the enzymes that catalyze the preparatory steps of
glycolysis (those that convert a molecule of glucose into two molecules of
PGAL) are not necessarily the same enzymes that catalyze the synthesis of
glucose from PGAL in photosynthesis. Similarly, some of the reactions in the
citric acid cycle are essentially irreversible, but the effect of the back reactions may be achieved by alternative pathways. It is the irreversibility of the
reaction from pyruvic acid to acetyl-CoA, as well as the absence of an alternative pathway, that makes higher animals unable to convert fatty acids into
carbohydrate.

MITOCHONDRIA AS ORGANELLES OF CELLULAR RESPIRATION

Aerobic respiration is confined to the mitochondria. We have already discussed these subcellular structures in Chapter 3; here we shall examine them
in somewhat greater detail, attempting to relate their function to their
structure.

The electron microscope, you will recall, shows that each mitochondrion
is bounded by an outer membrane and by an inner membrane, which has a

series of folds, or cristae, projecting into the matrix (Fig. 4.28); the inner surface of the inner membrane bears numerous tiny spherical particles (Fig. 4.29). Since mitochondria have two membranes and two chambers, the mitochondrial enzymes may be located in four different regions: (1) They may be built into the outer membrane; (2) they may be free as soluble proteins in the compartment between the outer and the inner membrane (O compartment); (3) they may be built into the inner membrane; or (4) they may be free in the matrix of the inner compartment (M compartment). Each of these four parts of the mitochondrion has its own distinctive functions.

Let us first consider the enzymes of the Krebs cycle. Since most of these can easily be isolated in soluble form from disrupted mitochondria, they are probably not parts of membranes. And since they are not set free by removal of the outer mitochondrial membrane as long as the inner membrane remains intact, it is likely that they are located in the matrix of the inner compartment. A few enzymes of the Krebs cycle that do not conform to this description are apparently built into the inner mitochondrial membrane.

The great precision with which electrons are passed from one acceptor molecule to the next in the respiratory electron-transport chain suggests regular spatial ordering. And the difficulty of isolating the acceptor molecules suggests that they are integral structural components of the inner membrane. Indeed, studies of fragmented membrane indicate that the various acceptor molecules are arranged in clusters, each of which is a self-contained functional unit known as a ***respiratory assembly.***

The fact that 32 of the 36 new ATP molecules synthesized as a result of the complete oxidation of a molecule of glucose come from electron transport implies that one determinant of a cell's capacity to synthesize ATP is the number of its respiratory assemblies. As would be expected, the cells of tissues requiring large amounts of energy have more mitochondria than cells with low energy demands. Furthermore, the individual mitochondria of cells with high energy requirements tend to have more cristae, and hence more respiratory assemblies.

The respiratory assemblies are apparently distributed with relatively constant density in the plane of the inner mitochondrial membrane; if, therefore, a mitochondrion is to have more respiratory assemblies, it must have more inner-membrane surface area, which means more cristae. In rats the surface area of inner membrane is about 40 square meters per gram of liver mitochondria and about 200–250 square meters per gram of heart mitochondria; it has been calculated that in vertebrates each liver mitochondrion contains about 5,000 respiratory assemblies, whereas each heart mitochondrion contains about 20,000. The flight muscles of blowflies have more than 400 square meters of inner membrane per gram of mitochondria. It is clear that the surface area of inner membrane, and the number of respiratory assemblies it determines, are correlated with the respiratory demands of the tissues.

The effort to determine where in the inner membrane the various elements of the electron-transport chain are located has been greatly aided by the generation of submitochondrial vesicles. These vesicles, derived from the inner membrane, are formed when mitochondria are fragmented by treatment with appropriate detergents or by exposure to high-frequency sonic irradiation (Fig. 4.29). The vesicles are "inside out"; i.e. the side of the

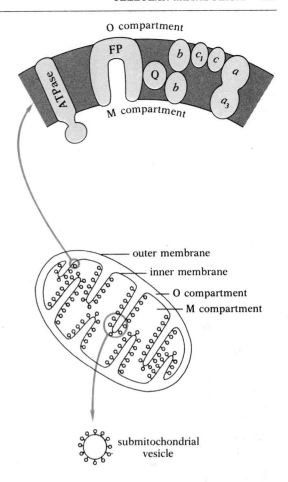

4.29 Location of the respiratory chain in the mitochondrion

Top: A small portion of the inner mitochondrial membrane is enlarged to show the hypothesized arrangement of the electron-transport molecules in the membrane. The flavoprotein (FP), the cytochromes *(b, c_1, c, a, a_3)*, and the ATPase are all proteins, and their positions in the membrane are probably relatively fixed. Coenzyme Q, by contrast, is a small lipid-soluble molecule that can move easily; at times it is probably at the surface of the membrane and at other times more in the interior, as shown here. (Q is drawn disproportionately large; drawn in proper scale, it would hardly show up.)   Bottom: Submitochondrial vesicles, which are formed from a piece of the inner mitochondrial membrane, have been very valuable in studies of the respiratory chain, because the M side of the membrane (distinguished by the spherical particles of ATPase) is exposed on the outer surface of the vesicle.   [Modified from E. Racker, *A new look at mechanisms in bioenergetics*, Academic Press, 1976.]

THE CHEMIOSMOTIC HYPOTHESIS OF
OXIDATIVE PHOSPHORYLATION

Among the various hypotheses advanced to explain how electron transport in the inner mitochondrial membrane is coupled to ATP synthesis, one has gained much favor in recent years. It was proposed by P. Mitchell, of the Glynn Research Laboratories in England, who was awarded the Nobel Prize in 1978 for his contribution.

According to his model, the electron-transport chain functions in translocating hydrogen protons (H$^+$) through the inner mitochondrial membrane, from the M compartment to the O compartment, and thus helps establish a steep pH gradient (and also an electrical-charge gradient) across the membrane. In other words, the energy released as pairs of electrons are transported down the energy gradient from NAD$_{re}$ to H$_2$O is converted into the energy of an electrochemical gradient of H$^+$ ions across the membrane. At special sites in the membrane (marked by the spherical particles on the M surface), where enzymes capable of catalyzing ATP synthesis (ATPase) are located, the protons move back from the O compartment to the M compartment. Since they are moving down an energy gradient, energy is released, and it is this energy that ATPase uses in catalyzing ATP synthesis (figure A). The process is much like the reverse of active transport across a membrane.

Evidence from a variety of experiments supports Mitchell's chemiosmotic hypothesis:

1. The model predicts that the membrane can establish a proton gradient by the translocation of H$^+$ ions from the M side to the O side, against

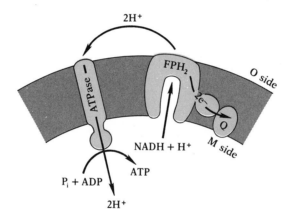

A ATP synthesis and the first site of H$^+$ translocation by the respiratory electron-transport chain

NADH + H$^+$ (NAD$_{re}$) on the M side of the membrane donates a pair of electrons and two H$^+$ ions to FP, forming FPH$_2$. The FP must pass the electrons along to Q, via an iron-containing enzyme (not a cytochrome) entirely embedded in the interior of the membrane. But this enzyme will not accept the H$^+$ ions; hence when FP is oxidized by donating electrons to the enzyme, it releases the H$^+$ ions into the medium on the O side (FP can pick up H$^+$ on one side and release it on the other, because the FP molecule penetrates all the way through the membrane). When Q receives the electron pair, it will pick up another two H$^+$ ions from the medium on the M side (the pickup is not shown here), and the process of moving H$^+$ ions across the membrane to the second site of release will begin. The pumping of H$^+$ ions from the M side to the O side establishes an H$^+$ gradient across the membrane. Some H$^+$ ions can move back across the membrane at sites where ATPase (a protein complex with a spherical particle on the M side) is located. This movement of H$^+$ ions down a concentration gradient releases energy, which enables the ATPase to catalyze synthesis of ATP from ADP and inorganic phosphate. [Redrawn from E. Racker, *A new look at mechanisms in bioenergetics*, Academic Press, 1976.]

their concentration gradient. That it can actually do so has been confirmed with submitochondrial vesicles, which accumulate H^+ ions in their interior.

2. The model predicts that electron transfer can drive ATP synthesis only if the membrane is intact, so that protons can be partitioned between the inner and the outer compartments of the mitochondrion. This has been confirmed experimentally.

3. The model predicts that the membrane must be impermeable to protons, except at the pumping sites where they are moved from the M to the O side and at the return sites where ATPase is located (otherwise leakage of protons through the membrane would short-circuit the system). This, too, has been confirmed experimentally.

4. The model predicts that some reagents that uncouple electron transport from ATP synthesis do so by making the membrane permeable to protons, and thus destroying the proton gradient generated by electron transport. This has been confirmed.

5. The model predicts that a proton gradient is capable of driving ATP synthesis. That it actually can has been verified in chloroplasts (the chemiosmotic hypothesis likewise undertakes to explain how ATP synthesis is driven by photosynthetic electron transport). If the pH of the interior of a chloroplast is lowered to 4 by holding the chloroplast in the dark in an acid bath, the H^+ (proton) concentration in the chloroplast will greatly increase. If the medium is then suddenly made alkaline (pH 8.5), the H^+ will tend to move out of the chloroplast, across the membranes, and this movement will drive ATP synthesis (figure B).

If electron transport does indeed give rise to a proton gradient, which then drives ATP synthesis, how is that proton gradient established? According to the model, the process depends on two characteristics of the electron-transport molecules. The first is the different locations of the transport molecules in the inner mitochondrial membrane, some being exposed on the O side, some on the M side, and some on both sides. The second is their different affinities for H^+ ions; the model assumes that when the transport molecules are reduced by the addition of electrons, some pick up H^+ ions as well (e.g. FP_{re} may also be written FPH_2), but others accept only the electrons. Suppose that one reduced link in the respiratory chain has picked up H^+ ions, but that the next link in the chain—the one to which the electron pair must be passed—is one that does not pick up H^+ ions. Under these conditions, the first link will pass the electrons to the second, but, unable to pass along the H^+ ions, will release them into the surrounding medium (figure A). If the electron-transport molecules are so arranged in the membrane that pickup of H^+ ions will occur on the M side but release will occur on the O side, then the H^+ ions will be distributed in such a way as to establish a gradient across the membrane.

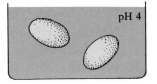

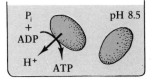

B ATP synthesis in the dark by an artificially induced pH gradient across chloroplast membranes
Top: Spinach chloroplasts are held in an acid bath until their contents come into equilibrium with the medium (pH 4). Bottom: When the medium is then made alkaline (pH 8.5) and ADP and inorganic phosphate are added, movement of H^+ ions out of the chloroplasts, down their concentration gradient, drives synthesis of ATP. [Based on experiments by A. Jagendorf, Cornell University.]

membrane bearing spherical particles (the M side), which in the intact mitochondrion borders the inner compartment, is on the outside of the vesicle. Thus membranous components in the interior of the intact mitochondrion, where they are protected from many test reagents by the outer portion of the membrane, are on the surface of the vesicles, where they can be more readily studied.

Most electron-transport molecules are proteins. Thus one technique for determining their position in the inner membrane relies on a radioactive substance that will bind to protein but cannot penetrate the membrane. By applying the radioactive substance to submitochondrial vesicles, where it can contact only the M side of the membrane (which is on the outside of the inside-out vesicles), and then determining to which membrane proteins it has become bound, the biochemist can find out which of the proteins are exposed on the M side of the normal membrane. The same procedure, but with application of the radioactive substance to intact mitochondria, where it can contact only the O side of the inner membrane, reveals which membrane proteins are exposed on the O side. Membrane proteins labeled in both experiments are probably transmembranous.

Another technique uses antigen-antibody reactions. Each membrane protein is isolated by chemical methods; then antibodies against it are manufactured, usually by injecting the protein into a rabbit, which responds to the foreign protein by synthesizing antibodies specific to that particular protein. Application of the antibodies to submitochondrial vesicles (with exposed M side of the membrane) and to intact mitochondria (with exposed O side), followed by assays to determine to which membrane proteins the antibodies have attached, gives an indication of the location of the proteins in the membrane.

While the positions in the membrane of some components of the respiratory chain are not yet certain, it is already apparent that the spatial sequence of the components of the respiratory chain corresponds to their functional sequence in electron transport (Fig. 4.29)—a beautiful example of the close relationship between structure and function in biological systems. Moreover, the way in which the various components are inserted in the membrane—whether they are exposed only on the M side of the membrane or only on the O side, or whether they span the full thickness of the membrane and are thus exposed on both sides—appears to be related to the manner in which electron transport drives ATP synthesis. According to the so-called chemiosmotic hypothesis, the asymmetrically placed components of the respiratory chain help convert the energy released by the transported electrons into the energy of an electrochemical gradient of hydrogen ions (H^+) across the membrane, and this energy, in turn, is converted into the energy of the terminal phosphate of ATP.

BODY TEMPERATURE AND METABOLIC RATE

We have seen that cellular metabolism captures some of the energy released by the oxidation of carbohydrates, fats, and proteins and converts it into the energy of high-energy phosphate groups in ATP. But this process fails to capture roughly 60 percent of the energy, most of which is released in the form of heat. The vast majority of animals, as well as all plants, promptly lose most of this heat energy to their environment; thus more than half the energy from the food they metabolize is lost without appreciably benefiting them. Such animals are known as cold-blooded; a more accurate term is **poikilothermic,** i.e. of variable temperature, for their body temperature fluctuates with the environmental temperature; when they are at rest, it is nearly the same as that of the surrounding medium, particularly if the medium is water.

The metabolism of an organism is closely tied to temperature. Within the narrow range of temperatures to which the active organism is tolerant, the metabolic rate[18] increases with increasing temperature and decreases with decreasing temperature in a very regular fashion. The relationship between metabolic rate and temperature is often expressed in terms of a value called the Q_{10}. This value is a measure of the rate increase for each 10°C rise in temperature. Thus if the rate doubles for each 10° rise in temperature, the Q_{10} is said to be 2; if the rate triples for each 10° rise, the Q_{10} is said to be 3; and so forth. Metabolic rates frequently have a Q_{10} of about 2. Let us denote the metabolic rate of a given animal at 0°C by X. If its metabolic rate has a Q_{10} of 2, then at 10°C the rate will be $2X$, at 20° $4X$, at 30° $8X$, and at 40° $16X$. Notice that the rate increases more and more rapidly as the temperature increases; when represented graphically, as in Fig. 4.30, this type of exponential increase produces a curve that becomes steeper and steeper as the temperature rises.

As would be expected, the activity of poikilothermic animals is radically affected by temperature changes in their environment. As the temperature rises (within narrow limits), they become more active; as the temperature falls, they become sluggish and lethargic. Such animals, then, are restricted as to the habitats they can effectively occupy, because they are at the mercy of the temperatures in those habitats.

A few animals, notably mammals and birds, can make use of the heat produced during the exergonic reactions of their metabolism, because they have evolved mechanisms—often including insulation by fat, hair, feathers, etc.—whereby heat loss to the environment is retarded. Such animals are commonly called warm-blooded; biologists use the term **homeothermic,** i.e. of uniform temperature.[19] Their body temperature is fairly high—usually higher than the environmental one—and relatively constant even when the environmental temperature fluctuates widely. Their metabolic rate can accordingly be maintained at a uniformly high level, and they remain very active. They are thus less dependent on environmental temperatures than poikilotherms and are freed for successful exploitation of more varied habitats.

It is not surprising that homeotherms generally have body temperatures considerably higher than the average temperature of their environment. Not only does a high temperature produce a high metabolic rate and make possible a high activity level, but a constant temperature higher than that of the surroundings is much easier to maintain than one lower than that of the surroundings. The animal can have very effective insulation, such as the

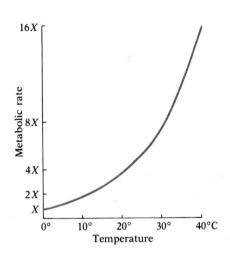

4.30   Graph of exponential rise in metabolic rate with increasing temperature
The hypothetical organism has a Q_{10} of 2. See text for fuller explanation.

---

[18] The basal metabolic rate (BMR), i.e. the metabolic rate of the organism at rest under standardized conditions, is usually determined by measuring the rate of oxygen comsumption or the rate of carbon dioxide production, or both.

[19] In recent years the term "endothermic"—emphasizing the importance of internally produced heat—has often been used instead of "homeothermic." Similarly, "ectothermic" has been used in place of "poikilothermic," to emphasize the importance of heat from external sources.

thick fur of polar bears or the massive fat layers of seals and whales; it can shunt blood away from blood vessels near the body surface; and it can speed up heat production by such responses as shivering, which is intensive muscle activity that rapidly uses up ATP energy and thus stimulates more cellular respiration and more heat production. In these ways an arctic mammal like the Eskimo dog, with a body temperature of 38.3°C, may be comfortable at −30°C or less, a temperature over 60° below its body temperature.

But no mammal or bird could live more than a very short time in an environment 60° hotter than its body temperature. In fact, few homeotherms can long withstand environmental temperatures more than a few degrees above their body temperature. Their cooling mechanisms are simply not effective enough. Their metabolism, with its unavoidable heat production, is by its very nature a furnace, not a refrigeration unit; it can easily be speeded up to counteract environmental cold, but it cannot be made into a cooling device. The external heat will tend, in fact, to do just the reverse of what is needed: speed up the metabolic rate according to the relationship we have already discussed. The animal can, of course, avoid the heat to some extent by seeking out shady spots or retiring to cool burrows. It may also have mechanisms for shunting much blood into surface capillaries, where heat loss is greatest, and for evaporative cooling such as sweating in humans or panting in dogs or evaporation from licked parts in cats. But these methods are effective only in temperatures below, at, or just above body temperature; they cannot long counteract very high temperatures. In short, the most effective thermal regulatory devices available to animals are based largely on heat production and conservation, not on heat loss and cooling. Homeothermy, then, involves mechanisms that can produce relative metabolic stability at a body temperature above that of the environment, but not at one much below that of the environment.

Under certain conditions many poikilotherms can, like homeotherms, take advantage of metabolic heat energy. For example, on a cold day butterflies and moths are apt to vibrate their wings for several minutes before launching into the air. The heat produced by cellular respiration in the vibrating muscles may increase the muscle temperature by as much as 15°C in five or six minutes, the muscles thus getting into a condition where they can contract fast enough to produce normal flight. Obvious analogies are the warming-up of an automobile or airplane engine and the warming-up exercises of athletes. Similarly, any rapidly moving poikilotherm produces metabolic heat faster than that heat can radiate away into the surrounding medium, and the animal's body temperature consequently rises well above that of the environment. Honeybees provide a particularly interesting example of thermal regulation by poikilotherms. When the temperature in their hive falls below a critical value, the bees become very active, releasing enough body heat to raise the hive temperature and maintain it at a level well above that of the outside environment. Conversely, when the hive temperature starts rising too high in summer, the bees create ventilating drafts with their wings.

In both homeothermic and poikilothermic animals and in plants, the normal metabolic rate is inversely related to body size; the smaller the organism, the higher the relative metabolic rate (Fig. 4.31). This is easily understood in the case of homeotherms, for the smaller the animal, the greater its surface-to-volume ratio and, consequently, the greater its relative

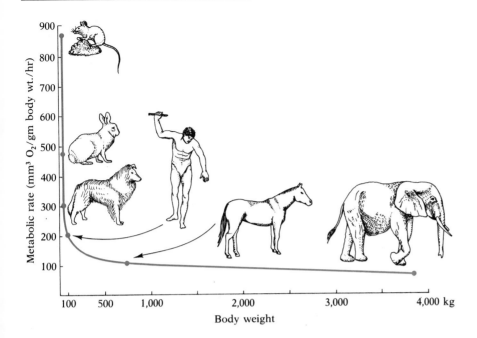

4.31 Graph showing inverse relationship of relative metabolic rate and body size in mammals

heat loss to the environment per unit time. To maintain a constant high body temperature despite rapid heat loss across its body surface, a small animal must oxidize food materials at a very fast rate. Because the amount of food intake, digestion, respiration, etc. per unit time must rise with decreasing size, there is a lower limit on the size of homeotherms. The smallest living mammals are shrews weighing only about 4 grams. They must eat nearly their own body weight of food every day, and can starve to death in a few hours if deprived of food.

It is more difficult to explain the inverse relationship between size and metabolic rate in poikilothermic animals and in plants. Since these organisms promptly lose their metabolic heat to the environment and do not normally respond to heat loss by increased metabolism, it would be expected that any effect the surface-to-volume ratio might have on metabolic rate would be the reverse of what is actually found, because larger size and smaller surface-to-volume ratio should retard heat loss slightly and the conserved heat should automatically speed up the metabolism. Why, in fact, larger size should be correlated with lower metabolic rate has never been fully explained. One probable factor is that increasing size generally involves a disproportionate increase of skeletal and other connective tissues in animals and of supportive fibers and mature xylem in plants; since these tissues are not particularly active metabolically, the average metabolic rate per unit weight for all tissues would fall as the proportion of these less active tissues rises. This is seen in the course of the development of an embryo; the early embryo is composed almost exclusively of metabolically very active cells and has a high metabolic rate, while later embryos develop a higher proportion of the less active types of tissues and their metabolic rate falls.

The relationship between metabolic rate and body size in homeotherms has serious implications for small animals during the cold seasons of the year. Not only does the rate of heat loss rise at such times, but the food supply is generally low. Small mammals belonging to three groups—the insectivores, bats, and rodents—have evolved a mechanism that enables them partly to evade this problem. When winter comes, they **hibernate;** i.e. their body temperature falls far below its normal level, and their metabolism, heart rate, respiration, etc. are greatly depressed. The animal passes the winter in this dormant state, using up its energy reserves very slowly. Bats are of particular interest, because they not only hibernate in winter in cold climates, but go into a similar dormant state during the daylight hours every day, thus conserving the energy derived from the food they eat during the night. The reverse occurs in hummingbirds (which are very near the lower size limit for homeotherms); they are active during the day and become torpid during the night. With a few exceptions, however, most birds do not hibernate during the winter; they either remain active, spending much of their time feeding, or avoid the cold by migrating to a warmer region.

Large mammals like bears are good heat conservers and, as we have seen, have a relatively low metabolic rate. Although they do not hibernate during the winter, they become relatively inactive and spend much of their time sleeping, while using up their extensive fat reserves; but their body temperature decreases by only a few degrees and they are not truly dormant.

SUGGESTED READING

BASSHAM, J. A., 1962. "The path of carbon in photosynthesis," *Scientific American*, June. (Offprint 122) *How the Calvin cycle was worked out: an older, but informative article by one of those who did the work.*

BJÖRKMAN, O., and J. BERRY, 1973. "High-efficiency photosynthesis," *Scientific American*, October. (Offprint 1281) *The photosynthetic pathway and leaf anatomy of a group of C$_4$ plants.*

GOVINDJEE, and R. GOVINDJEE, 1974. "The primary events of photosynthesis," *Scientific American*, December. (Offprint 1310) *Summary account of photosynthesis; includes some intermediate steps for which evidence is scant.*

HINKLE, P. C., and R. E. MCCARTY, 1978. "How cells make ATP," *Scientific American*, March. *An attempt at explaining oxidative and photosynthetic phosphorylation in terms of the chemiosmotic model. Difficult, but rewarding.*

LEHNINGER, A. L., 1971. *Bioenergetics*, 2nd ed. Benjamin, New York. *A briefer, more elementary treatment of energy transformations in living organisms than in the same author's* Biochemistry.

LEVINE, R. P., 1969. "The mechanism of photosynthesis," *Scientific American*, December. (Offprint 1163) *The light events of photosynthesis clearly explained.*

RACKER, E., 1968. "The membrane of the mitochondrion," *Scientific American*, February. (Offprint 1101) *One of the leading authorities in the field on the role of membranes in the handling of energy.*

STOECKENIUS, W., 1976. "The purple membrane of salt-loving bacteria," *Scientific American*, June. (Offprint 1340) *On a unique photosynthetic mechanism based, not on chlorophyll, but on a pigment very similar to the visual pigment of animals.*

PART II

THE BIOLOGY OF ORGANISMS

Chapter 5

NUTRIENT PROCUREMENT
AND PROCESSING

Both the synthesis of new protoplasm and the respiratory oxidation of high-energy organic compounds—the process by which most organisms obtain energy for carrying out life functions—demand the procurement of two main categories of molecules from the environment: (1) already synthesized high-energy compounds, or else the raw materials from which they and new protoplasm can be synthesized; and (2) the oxygen used in cellular respiration. This chapter will deal with the procurement and processing of the first of these two categories of materials, the *nutrients.*

Organisms can be divided into two classes on the basis of their methods of nutrition. Fully *autotrophic* organisms can subsist in an exclusively inorganic environment because they can manufacture their own organic compounds from inorganic raw materials taken from the surrounding media. Since the molecules of these raw materials are small enough and soluble enough to pass through cell membranes, autotrophic organisms do not need to pretreat, or digest, their nutrients before taking them into their cells. As you would guess, most autotrophs are photosynthetic, although a few are chemosynthetic. The green plants are by far the most important of the earth's autotrophic organisms.

Heterotrophic organisms (most bacteria, fungi, and animals) are incapable of manufacturing their own complex organic compounds from simple inorganic nutrients. Hence they must obtain prefabricated organic molecules from the environment. Many of the organic molecules found in nature are too large to be absorbed unaltered through cell membranes, and they must first be broken down into smaller, more easily absorbable molecular units; i.e. they must be digested.

It is clear, then, that autotrophic and heterotrophic organisms differ both in their nutrient requirements and in the problems associated with nutrient procurement. And it is not surprising that they should have evolved radically different adaptations in response to the different selection pressures acting upon them. We shall therefore discuss these two great groups of organisms separately in this chapter, but shall indicate similarities between them where appropriate.

NUTRIENT REQUIREMENTS OF GREEN PLANTS

RAW MATERIALS FOR PHOTOSYNTHESIS

The raw materials most obviously needed by higher photosynthetic organisms are carbon dioxide and water. These two compounds supply the carbon, oxygen, and hydrogen that are the predominant elements in organic molecules. Carbon dioxide, one of the constituent gases of the earth's atmosphere, is obtained directly from the air by the leaves of terrestrial plants; submerged aquatic plants absorb the dissolved gas from the surrounding water. Terrestrial plants obtain the other raw material, water, from the substrate in which they grow; most higher plants absorb water from the soil by roots.

A very high percentage of the total dry body weight of a large tree is carbohydrate, and much of the rest of it was synthesized from carbohydrate. This fact has some rather startling implications. The formula for glucose, which may be taken as the central carbohydrate in protoplasm, is $C_6H_{12}O_6$, which means, you will recall, that one molecule of glucose contains six atoms of carbon, twelve of hydrogen, and six of oxygen. The atomic weight of carbon is 12, that of hydrogen is 1, and that of oxygen is 16; consequently the molecular weight of glucose is 180 (really 180.162 if isotopes are considered). Now, you will remember that all the carbon and oxygen incorporated into glucose by photosynthesis comes from carbon dioxide, which in turn comes from the air. Since the combined weight of the six atoms of carbon and six atoms of oxygen in glucose is 168, which is about 93 percent of the total weight of glucose, it follows that about 93 percent of the weight of a large, immensely heavy tree comes initially from the air. The hydrogen in glucose comes from water, and hydrogen constitutes roughly 7 percent of the weight of glucose; hence about 7 percent of the dry weight of the tree comes initially from water. It is the modern elucidation of the process of photosynthesis that has established the amazing fact that most of the mass of a plant's body comes from air, not from the solid earth in which it grows.

For many centuries it was assumed that the structural material of the plant body comes from the soil. Then, about 1450, Cardinal Nicolai de Cusa suggested that the weight gained by a growing plant comes from water, not earth. He was sure that if one were to put a carefully weighed quantity of earth in a pot, plant some seeds in it, wait until the seeds had germinated and the plants grown large and heavy, remove the plants, and again weigh the earth, very little loss of weight by the earth would be found. This, he thought, would prove that water had contributed the bulk of the weight of the plants. There is no evidence that the Cardinal ever performed the experiment he suggested; he lived in a day when the gathering of empirical data by experiments was not as commonplace a procedure as it is today.

The experiment suggested by Nicolai de Cusa was finally performed by Jean-Baptiste van Helmont, a Flemish physician; the results were published in 1648. Van Helmont describes his experiment as follows:

> . . . I took an earthenware vessel, placed in it 200 pounds of soil dried in an oven, soaked this with rainwater, and planted in it a willow branch weighing 5 pounds. At the end of five years, the tree grown from it weighed 169 pounds and about 3 ounces. Now, the earthenware vessel was always moistened (when necessary) only with rainwater or distilled water, and it was large enough and embedded in the ground, and, lest dust flying about be mixed with the soil, an iron plate coated with tin and pierced by many holes covered the rim of the vessel. I did not compute the weight of the fallen leaves of the four autumns. Finally, I dried the soil in the vessel again, and the same 200 pounds were found, less about 2 ounces. Therefore 164 pounds of wood, bark, and root had arisen from water only.[1]

Though van Helmont clearly demonstrated that most of the material of a plant's body does not come from the soil, he was not prepared to consider that it might have come from so weightless a thing as air. Finally, in 1727, Stephen Hales, an Englishman, suggested that plants get at least part of their nourishment from the air, but the extent of such nourishment was not understood for a long time. And only with the advent of isotopic tracer techniques in the present century could it be conclusively demonstrated, as we saw in the last chapter, that CO_2 gas contributes the oxygen as well as the carbon for photosynthesis and that water contributes only the hydrogen.

MINERAL NUTRITION

Clearly, carbon dioxide and water cannot be the only nutrient materials needed by a green plant. These two compounds provide only three elements: carbon, oxygen, and hydrogen; yet we know that other elements, too, enter into the composition of the plant. Nitrogen, for example, is always present in amino acids, the building-block units of proteins, which are essential components of protoplasm; two very important amino acids also contain sulfur. Phosphorus is present in ATP, nucleic acids, and many other critically important compounds. Chlorophyll, the essential mediator of photosynthesis, contains magnesium (see Fig. 4.2, p. 143), and the cytochromes, so important in electron transport, contain iron. Where does the green plant obtain the nitrogen, sulfur, phosphorus, magnesium, iron, and other elements it needs? Obviously not from carbon dioxide or water. Here, finally, we see the role of the soil itself as a source of plant nutrients. It is from the soil that the plant derives the minerals[2] essential to its life. Perhaps those 2 ounces of weight lost from the soil in van Helmont's experiment were more important than he knew.

During the nineteenth century, there was much interest in Europe in determining the mineral needs of crop plants and in devising ways of supplementing the amounts of essential mineral elements in the soil. By 1900,

[1] Translated from the original Latin.

[2] "Mineral" is a term applied to naturally occurring inorganic substances. As used here, it refers to an element in inorganic ionic form. Potassium and nitrogen, for example, are often available to a plant in soil in the form of potassium nitrate ($K^+NO_3^-$).

TABLE 5.1 *Essential minerals for higher plants.**

Element	Approx. number of pounds needed to grow 100 bushels of corn	Function
MACRONUTRIENTS		
Nitrogen (N)	160	Structural component of amino acids, many hormones and coenzymes, etc.
Phosphorus (P)	40	Structural component of nucleic acids, phospholipids, ATP, coenzymes, etc.
Potassium (K)	125	Plays a role in the ionic balance of cells; cofactor for enzymes involved in protein synthesis and carbohydrate metabolism
Sulfur (S)	75	Structural component of the two amino acids cysteine and methionine, and of several vitamins
Magnesium (Mg)	50	Structural component of chlorophyll; cofactor for many enzymes involved in carbohydrate metabolism, nucleic acid synthesis, and the coupling of ATP with reactants
Calcium (Ca)	50	Influences permeability of membranes; component of pectic salts in middle lamellae and necessary for wall formation; activator for several enzymes
Iron (Fe)	2	Structural component of iron-porphyrins (hemes), which are incorporated in cytochromes, peroxidases, catalases, and some other enzymes; plays a role in the synthesis of chlorophyll
MICRONUTRIENTS		
Manganese (Mn)	0.3	Cofactor of many enzymes involved in cellular respiration, photosynthesis, and nitrogen metabolism
Boron (B)	0.06	Function unknown; may play a role in translocation of sugar; perhaps necessary for utilization of calcium in wall formation
Chlorine (Cl)	0.06	Plays an essential role in photosynthesis
Zinc (Zn)	Trace	Necessary for synthesis of tryptophan (a precursor of auxin); activator of many dehydrogenase enzymes; may play a role in protein synthesis
Copper (Cu)	Trace	Structural component of many enzymes that catalyze oxidation reactions, and of plastocyanin, which is important in electron transport in chloroplasts
Molybdenum (Mo)	Trace	Structural component of the enzyme that reduces nitrate to nitrite; essential for fixation of N_2 by nitrogen-fixing bacteria

* An element must meet three criteria to be regarded as essential: (1) The element is needed for normal growth and reproduction in several different plants; (2) it is not replaceable by other elements; (3) its function is a direct one (i.e. it is not needed only to correct a deficiency, or a toxic condition, due to some other substance).

seven of these were known: nitrogen, phosphorus, sulfur, potassium, calcium, magnesium, and iron. Three of them—nitrogen, phosphorus, potassium—were stressed particularly, as they are to this day in the manufacture of fertilizer. Modern commercial fertilizers are often designated by their N-P-K percentages; e.g. the widely used garden fertilizer called 5-10-5 contains 5 percent nitrogen, 10 percent phosphoric acid, and 5 percent soluble potash by weight. These three are the elements most rapidly removed from the soil; consequently it is essential to replenish them if crops are to continue to flourish. Many modern fertilizers are also fortified by small amounts of some of the other essential minerals.

Much of the important research on the mineral requirements of plants was done by growing plants in distilled water to which measured amounts of minerals were added. This water-culture technique allowed a degree of control and a precision of measurement unattainable with plants growing in soil. Nevertheless, it was not until about 1920, after more than fifty years of water-culture research, that it became apparent that other elements, in addition to the seven already known, were essential to plants. These additional minerals (manganese, boron, chlorine, zinc, copper, molybdenum) are required in such small amounts that the traces present as contaminants in the water or salts used in the early experiments were sufficient to meet the needs of the plants. Only with very elaborate purification procedures could their presence be controlled and their effects determined. Such elements, essential in minute amounts but sometimes toxic in excess, are now called trace elements or micronutrients.

Table 5.1 lists all the essential nutrients known at the present time and gives some indication of the relative amounts needed and of their known functions. Two other elements, vanadium and cobalt, are under investigation and may someday be added to the list. Silicon and aluminum, two of the commonest elements in the earth's crust, often occur in quantity in plants, but they appear to be dispensable; however, the same thing was said of chlorine only a few years ago, and of other essential elements before that, which makes it seem advisable to reserve judgment.

The functions listed in Table 5.1 make it clear why the trace elements are needed in only minute amounts. Most of them are components of enzymes or coenzymes. You will remember that enzymes can be used over and over and that a very small quantity of each is sufficient. Only a small amount of mineral is required, therefore, to synthesize the enzyme or coenzyme initially, and to replenish the supply as the enzyme molecules are slowly broken down.

NUTRIENT PROCUREMENT BY GREEN PLANTS

We have seen that three classes of nutrients are needed by green plants: carbon dioxide, water, and minerals. Carbon dioxide is absorbed by the leaves (and occasionally by stems that are green and carry out photosynthesis). This gas, together with the other components of air, moves into the internal spaces of the leaf through openings in the epidermis called stomata. Inside the leaf, the air circulates throughout the numerous intercellular

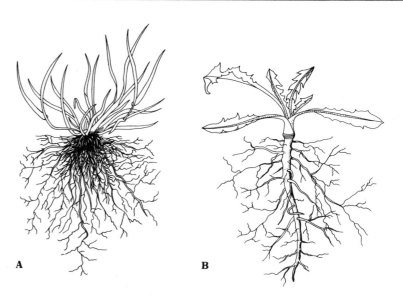

5.1   Two types of root systems
(A) Fibrous root system of grass.   (B) Taproot system of dandelion

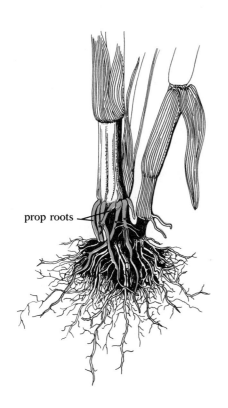

5.2   The prop roots of corn
These roots arise from a portion of the stem and are thus adventitious roots.

spaces (see Fig. 6.3, p. 240). Carbon dioxide dissolves in the film of water on the surfaces of the leaf cells and diffuses into the cells, where it is used as a raw material for photosynthesis. Details of leaf structure and adaptations for gas exchange, both photosynthetic and respiratory, will be discussed in more detail in the next chapter.

This chapter will be more concerned with the procurement of water and minerals. Aquatic plants such as algae absorb both of these from the surrounding medium. These plants usually lack specialized procurement organs and absorb their nutrients through the general body surface. Some nonaquatic plants such as Spanish moss (really a flowering plant, not a moss), which grow as epiphytes on the branches of large trees, absorb all their nutrients through their leaves. Foliar feeding has been found an effective method of providing nutrients even for plants that normally obtain minerals through their roots, such as most house and garden plants; minerals in a water solution are sprayed on the leaves, where they are absorbed (minerals can also be lost from leaves by the leaching action of rainwater). But most higher land plants take in water and minerals primarily through their roots.

ROOTS AS ORGANS OF PROCUREMENT

Root structure   The first root formed by the young seedling is called the **primary root.** Later, **secondary roots** branch from the primary root, and a root system is formed. If the branching results in a system of numerous slender roots, with no single root predominating, as in grass or clover, the plant is said to have a **fibrous root system** (Fig. 5.1A). If, however, the primary root remains dominant, with smaller secondary roots branching from it, the arrangement is called a **taproot system** (Fig. 5.1B). Dandelions, beets, and carrots, among others, are plants with taproots. As these examples suggest, taproots are frequently specialized as storage organs for the products of photosynthesis. Storage is a function of all roots, but particularly of taproots.

Obviously, procurement of water and minerals and storage of high-energy organic compounds are not the only functions of roots; they also serve to anchor the plant to the substrate. Usually the substrate is soil, but climbing vines commonly have, in addition to their normal root system in the ground, short specialized roots arising from the stem, which fasten the plant to a vertical surface such as a tree trunk or the side of a building. These aerial roots of vines are examples of *adventitious roots;* the term "adventitious" is applied to any root that arises after the embryo stage from a structure that is not a part of the root system. The prop roots of corn are also adventitious roots; they arise from the lower portion of the stem (Fig. 5.2), penetrate the soil, and become important components of the root system.

The root system of a plant is normally very extensive, far more extensive than is ordinarily realized. When we pull up a plant, we seldom get anything even approaching the entire root system, since most of the smaller roots are so firmly embedded in the soil that they break off and are lost. The amplitude of the system is, of course, important both in anchoring the plant and in providing sufficient absorptive surface. When we discussed possible limitations on potential cell size in an earlier chapter, we mentioned the problem of surface-to-volume ratio—that as a cell or an organism gets bigger, its volume increases much faster than its surface area. A large multicellular organism thus faces a serious problem; it must have an absorptive surface extensive enough to enable it to obtain the nutrients it needs to support its large volume, particularly if most of the absorption is restricted to a limited region of the body, in this case the roots. As an adaptation toward solving this problem, many organisms have evolved extensively subdivided absorptive surfaces, far greater in total area than those of an undivided system of the same volume. The manifold branching of a typical root system is an example of this kind of adaptation; it was found that a rye plant less than one meter tall had some 14 million branch roots of a combined length of over 600 kilometers.

Roots have evolved yet another adaptation that increases their absorptive capacity. Just behind the growing tip of each rootlet, there is usually an area bearing a dense cluster of tiny hairlike extensions of the epidermal cells (Fig. 5.3). The zone of these *root hairs* on each rootlet may be anywhere from a centimeter long in some species to over a meter in length. It is in this region that most absorption of water and minerals takes place. Although the root-hair zone on any one rootlet is not very long, the number of root hairs on all the many rootlets is so vast that the total absorptive surface they provide is enormous. The rye plant cited above may have had as many as 14 billion root hairs with a total surface area of over 400 square meters.

When viewed in cross section, the root of a young dicot plant can be seen to consist of a series of different tissue layers (Fig. 5.4). On its outer surface is a layer of *epidermis* one cell thick. Unlike the epidermis of the aerial parts of the plant, that of the root usually has no waxy cuticle on its surface. The explanation is obvious. The epidermis of a root functions in water absorption, that of the aerial parts as a barrier against diffusion of water. A single

5.3 Root of radish seedling with many prominent root hairs
[Courtesy W. E. Loomis, Iowa State University.]

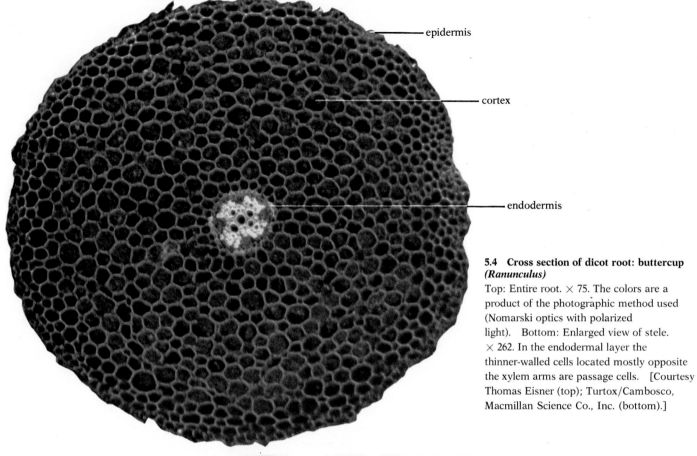

epidermis

cortex

endodermis

5.4 Cross section of dicot root: buttercup *(Ranunculus)*

Top: Entire root. × 75. The colors are a product of the photographic method used (Nomarski optics with polarized light). Bottom: Enlarged view of stele. × 262. In the endodermal layer the thinner-walled cells located mostly opposite the xylem arms are passage cells. [Courtesy Thomas Eisner (top); Turtox/Cambosco, Macmillan Science Co., Inc. (bottom).]

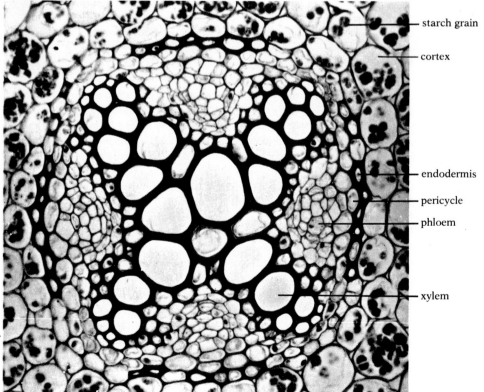

starch grain

cortex

endodermis

pericycle

phloem

xylem

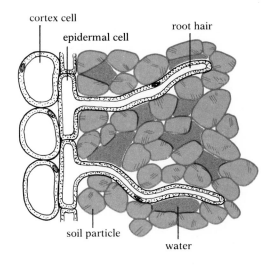

5.5 Root hairs penetrating soil
Each root hair, which is an extension of a single epidermal cell, is in contact with many soil particles (color) and with soil spaces, some of which contain air, some water (gray).

epidermal cell gives rise to each of the root hairs in the area just back of the growing tip of each rootlet (Fig. 5.5).

Beneath the epidermis is the *cortex,* a wide area composed primarily of parenchyma tissue, with numerous intercellular spaces. Sometimes the cortex also contains sclerenchyma cells; these are commonly in the outer portion of the cortex adjacent to the epidermis. Large quantities of starch are often stored in the cells of the cortex. This tissue, so prominent and important in young roots, is frequently much reduced or even lost in older roots, where both cortex and epidermis may be replaced by a corky periderm.

Next interior to the cortex is the *endodermis,* a layer one cell thick (Fig. 5.4). You will recall that endodermal cells are characterized by a waterproof band, the *Casparian strip,* which runs through their radial (side) and end walls (Fig. 5.6). The walls of mature endodermal cells are often very thick and lignified. (In some plants, however, a few of the endodermal cells, called passage cells, have relatively thin walls, though they still have Casparian strips.) A well-differentiated endodermis is always present in roots, but occurs less regularly in stems.

The endodermis forms the outer boundary of a central core of the root that contains the vascular cylinder. This core is called the *stele.* Just inside the endodermis is a layer, often only one cell thick, of thin-walled parenchymatous cells. The cells of this layer, called the *pericycle,* readily take on meristematic activity and may give rise to lateral roots (see Figs. 17.32, 17.33, pp. 724–725).

The central portion of the dicot stele, surrounded by endodermis and pericycle, is filled with the two vascular tissues, *xylem* and *phloem.* The thick-walled xylem cells often form a cross- or star-shaped figure (Fig. 5.4). Bundles of phloem cells are located between the arms of the xylem. Thus, instead of forming a continuous cylinder like the epidermis, cortex, endodermis, and pericycle, xylem and phloem alternate in this portion of the stele.

Large roots of monocots commonly have a region of parenchyma tissue, called *pith,* located at the very center of the stele (Fig. 5.7). The xylem

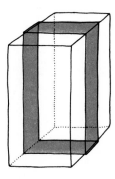

5.6 Endodermal cell with Casparian strip
The Casparian strip (color) is located in the radial and end walls of the cell.

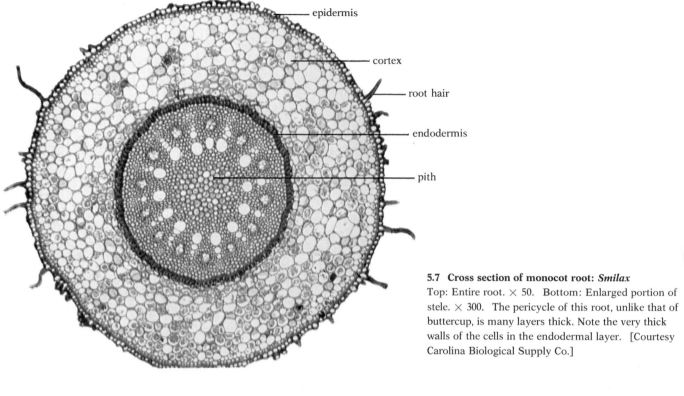

epidermis

cortex

root hair

endodermis

pith

5.7 Cross section of monocot root: *Smilax*
Top: Entire root. × 50.   Bottom: Enlarged portion of stele. × 300.   The pericycle of this root, unlike that of buttercup, is many layers thick. Note the very thick walls of the cells in the endodermal layer.   [Courtesy Carolina Biological Supply Co.]

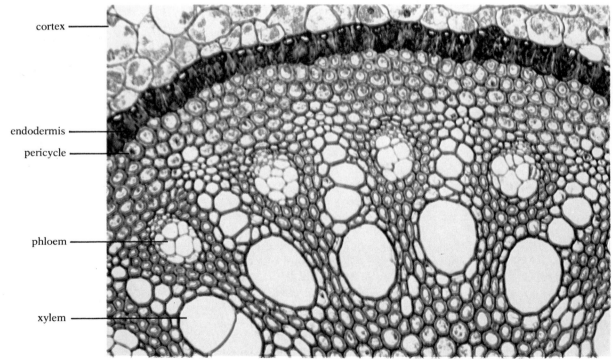

cortex

endodermis

pericycle

phloem

xylem

therefore does not form the star-shaped figure described as characteristic of dicots, but even in such roots the bundles of xylem and phloem alternate.

Absorption of nutrients Rainwater generally becomes available in the soil to plants as a loose film of water around soil particles, known as capillary water. The roots, and particularly the root hairs, are in contact with this water. Since the root epidermis lacks a cuticle, the capillary water can easily move into the root—either by osmosis or by flowing along the continuous path provided by the cell walls and intercellular spaces (Fig. 5.8).

The concentration of dissolved substances such as sugar and other organic compounds in an epidermal cell is normally higher than the concentration of dissolved substances in the soil water. With the osmotic concentration higher inside the cell than outside, a simple osmotic system is established and water can move across the membrane into the cell.

Once water has entered an epidermal cell, it dilutes the contents of that cell. The concentration of dissolved substances in it becomes lower than in the adjacent cell of the cortex, and water can move from it to the cortex cell by osmosis. But now a new concentration gradient is established, as water moving into the outermost cell of the cortex dilutes the contents of that cell and lowers its osmotic concentration to a point below that of the next cell of the cortex. As a result, water moves from the first cortex cell to the second cortex cell, following the concentration gradient. Again, dilution of the recipient cell occurs, a new gradient is established, and water moves on to the next cell. In this way, water can move fairly easily from the capillary films of the soil into the epidermis and thence across the cortex to the stele. Once

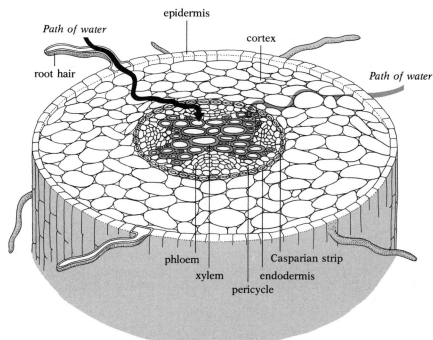

epidermis

Path of water

cortex

root hair

Path of water

phloem

xylem

Casparian strip

endodermis

pericycle

5.8 Movement of water from soil to xylem in root
Some water (black arrow) is absorbed by the epidermal cells, particularly the root hairs, and moves from cell to cell either by osmosis or by diffusion through plasmodesmata. More of the water (color arrow) flows along cell walls and does not cross membranes of living cells until it reaches the endodermis. The Casparian strip (color) of the endodermal cells prevents flow of water along their radial and end walls; hence all water entering the stele must move through the living cells of the endodermis.

inside the xylem of the vascular cylinder, the water can rise to other parts of the plant body. Removal of water from the center of the root via the xylem maintains the concentration gradient from the epidermis to the xylem and allows the process of water absorption to continue.

Water can probably move from an epidermal cell to a cell in the cortex, and from one cortex cell to the next, by means other than osmosis through cell membranes. You will recall that the cytoplasms of adjacent plant cells often form a *symplast,* an association in which plasmodesmata interconnect the contents of the cells. Once water or some other substance has entered an epidermal cell, it can thus move to other cells through the plasmodesmata, without having to cross any additional cell membranes.

Analogous to the symplast, in which the contents of adjacent cells are interconnected, is the *apoplast,* a network composed of the cell walls and intercellular spaces. The apoplast provides another means, besides osmosis, by which water can enter the root. Because cell walls are hydrophilic—their main component, cellulose, has a strong tendency to imbibe water—water from the soil can move in the apoplast across the epidermis and the entire cortex of a root without ever actually penetrating a membrane or entering a cell (Fig. 5.8). Indeed, recent experiments suggest that, in the normal uptake of water by roots, movement of water in the apoplast is much more important than movement in the symplast. But the water cannot flow across the endodermis in this manner, because the Casparian strip acts as a barrier; being hydrophobic, it interrupts the apoplast. Consequently all water entering the vascular cylinder must cross through the living cells of the endodermis. This gives the plant an opportunity to control the movement into the stele of substances dissolved in the water.

Minerals are usually absorbed in ionic form; e.g. nitrogen is absorbed as nitrate (NO_3^-) or ammonium (NH_4^+) ions; phosphorus as dihydrogen phosphate ($H_2PO_4^-$) or monohydrogen phosphate (HPO_4^{--}) ions; sulfur as sulfate ions (SO_4^{--}); and potassium, calcium, magnesium, and iron as their simple ions (K^+, Ca^{++}, Mg^{++}, Fe^{++} or Fe^{+++}). Note that, whether inorganic or organic fertilizer is applied to crops, the minerals are absorbed by the plants primarily as inorganic ions. The two kinds of fertilizers, which have very different physical and chemical effects on the soil, may, however, influence plant growth differently in other ways; e.g. organic material may possibly provide vitaminlike growth stimulants, but at times it may also contain substances toxic to plants.

The ions available to plants for absorption are in solution in the soil water, their concentration varying according to the fertility and the acidity of the soil and other factors. When the soil minerals are not in solution, but are bound by ionic bonds to soil particles, they are not available to plants. Agricultural soil management often involves changing the soil acidity to free more such bound minerals for absorption by roots. For example, the addition of lime to very acid soil in order to raise the pH may increase the availability of phosphorus, potassium, and molybdenum, but an excess of lime may decrease the available iron, copper, manganese, and zinc (Fig. 5.9). Obviously, a careful balance, appropriate to the particular crop to be grown, must be achieved for maximum yield.

The rate of absorption of each mineral by roots is essentially independent of the rates of absorption of water and of other minerals. Each nutrient

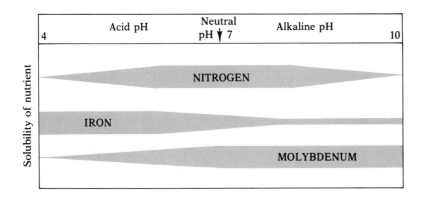

5.9 Solubility of three mineral nutrients as a function of pH
The changing width of each band indicates the relative solubility of the mineral between pH 4 and 10. Nitrogen is most soluble, and hence most available to plants, in the neutral pH range; iron is most soluble under acid conditions, and molybdenum under alkaline conditions. Other minerals have their own distinctive solubility curves. Since it is obviously impossible for soil to have a pH at which all minerals will be maximally available to plants, farmers or gardeners must adjust the soil pH according to the nutrient requirements of the particular plants they wish to grow.

moves into the root at a rate determined by such factors as its concentration both inside and outside the root, the ease with which it can passively penetrate cell membranes, and the extent to which carrier molecules are involved. Although some of the inward movement of minerals, like that of water, is a result of passive diffusion along a concentration gradient—the mineral being in higher concentration in the soil solution than in the cells— numerous experiments have shown that simple diffusion alone cannot account for all the absorption of mineral nutrients by roots. Even when the concentration gradient favors inward movement, the rate of absorption is often greater than would be possible by passive diffusion alone—which means that facilitated diffusion is taking place. Moreover, plants can often take in a mineral that is in higher concentration inside the root cells than in the soil solution and that should, therefore, move the other way, if simple diffusion alone were involved. Active transport is clearly a factor; the plant expends energy in the process of procuring the mineral nutrients essential to its continued existence. As will be evident throughout this book, active transport is the rule rather than the exception in most kinds of organisms, whether plant or animal, when substances are moving across the membranes of living cells.

Two factors help minerals enter roots by preventing their concentration in the root cells from becoming too high: As fast as a mineral enters a root, it may be removed and transported to some other part of the plant; or it may be rapidly utilized in the synthesis of a different compound. For example, nitrogen, absorbed as nitrate, is quickly reduced and built into organic compounds such as amino acids, amides, and other nitrogen-rich compounds, which are then transported and stored. Much of the storage is in cell vacuoles, where the concentration of the nitrogen compounds is often much higher than in the cytoplasm itself—a clear indication that the vacuolar membrane, or tonoplast, exerts a selective activity, secreting the compounds into the vacuole but preventing their escape from it.

INSECTIVOROUS GREEN PLANTS

A few photosynthetic plants supplement their inorganic diet with organic compounds obtained by trapping and digesting insects and other small animals. Such plants can survive without capturing any prey, but when they

5.10 Pitcher plant (*Sarracenia purpurea*) from a bog in Michigan
[R. P. Carr, Bruce Coleman Inc.]

do capture prey the nutrients thus obtained stimulate more rapid growth. Apparently it is the nitrogenous compounds of the animal's body that are of most benefit to insectivorous plants, which often grow in nitrogen-poor soils, particularly acid bogs and heavy volcanic clays, and whose root systems are not extensive. Their prey-capturing adaptations are interesting and worth examining here as examples of highly specialized leaves.

Pitcher plants (Fig. 5.10) have leaves modified into a tube or sac, which is partly filled with water. The end of the leaf is further modified to form a hood, which partly covers the open mouth of the pitcher. Insects that fall into the sac are prevented from climbing out by numerous stiff downward-pointing hairs. The proteins of the trapped insects are digested by enzymes secreted into the water, and the products of this digestion are absorbed by the inner surface of the leaf.

The leaf of the Venus's-flytrap (Fig. 5.11) is composed of two lobes with a midrib between them. There is a row of long stiff teeth along the margin of each lobe. When an insect touches small sensitive hairs on the surface of the leaf, the lobes quickly change shape and come together with their marginal teeth interlocked. The trapped animal is then slowly digested by enzymes secreted from glands on the leaf surface, and the resulting amino acids are absorbed. The rapid movement exhibited by the leaves of the Venus's-flytrap is not yet fully understood; it probably involves changes in the relative turgor pressure of the cells of the inner and outer epidermal layers.

The leaves of sundews (Fig. 5.12) show still another type of modification for carnivorous activity. They bear numerous hairlike tentacles, each with a gland at its tip. The gland secretes a sticky fluid in which small insects become trapped. The stimulus from a struggling trapped insect causes nearby tentacles to bend over the animal, further entangling it. As in the pitcher plants and Venus's-flytraps, the proteins of the insect are digested by enzymes and the amino acids are then absorbed.

NUTRIENT REQUIREMENTS OF HETEROTROPHIC ORGANISMS

Heterotrophic organisms cannot manufacture their own high-energy compounds from low-energy inorganic raw materials. Yet they, like all living things, must extract from complex molecules the energy necessary for both maintenance and growth. They must therefore obtain prefabricated high-energy organic nutrients.

There are three main groups of heterotrophic organisms: the bacteria, the fungi, and the animals. (In addition, some green plants may be at least partly heterotrophic, as we have just seen.) The first two groups—the bacteria and the fungi—lack internal digestive systems and hence depend mainly on absorption as their mode of feeding. They are usually either *saprophytic* (living and feeding on dead organic matter) or *parasitic* (living on or in other organisms and feeding on them). By contrast, the principal mode of feeding in the animal kingdom is ingestion—the taking in of particulate or bulk food, which must then be digested. Animals may be *herbivores* and eat green plants, thereby obtaining high-energy compounds directly from the organisms that first made them. Or they may be *carnivores* and eat the animals that ate the plants. Or they may be *omnivores,* eating both plant and animal

5.11 Leaf of Venus's-flytrap (Dionaea muscipula) from North Carolina
One of the fly's legs is about to touch a trigger hair on the leaf surface. The two halves of the leaf will then quickly move together, interlocking their marginal teeth and thus trapping the fly. [Oxford Scientific Films, Bruce Coleman Inc.]

material. Whether a heterotrophic organism is saprophytic or parasitic, or herbivorous, carnivorous, or omnivorous, it is clear that its energy-yielding nutrients came originally from green plants, which used radiant energy from the sun to make them.

In many ways, the absorptive and the ingestive organisms are as different from each other as each is from the green plants, with their photosynthetic mode of nutrition. Indeed, much of the diversity among living things can be viewed as reflecting alternative adaptations within three principal evolutionary trends based on the three major nutritive modes—photosynthetic, absorptive, and ingestive.

NUTRIENTS REQUIRED IN BULK

Carbohydrates, fats, and proteins are the main classes of compounds serving as energy sources for heterotrophic organisms. Of these, carbohydrates alone would suffice if organic nutrients functioned only as an energy source. But these nutrients perform another very important function—providing the carbon skeletons and functional groups necessary for the synthesis of new organic compounds. Assuming that enough inorganic minerals are included in the diet, can carbohydrates alone fulfill this second function?

For some heterotrophs, the answer is yes. Many bacteria and fungi can flourish on a diet consisting solely of carbohydrate and minerals. They need no protein in their diets, because they, like green plants, can combine inorganic nitrogen with carbon skeletons from carbohydrates to make amino acids. In similar fashion, they can synthesize for themselves all the other classes of compounds necessary for life.

But for many other heterotrophs, such a limited diet cannot sustain life. Some bacteria and fungi, having apparently lost the ability to synthesize certain organic compounds they need, must obtain these, in addition to

5.12 Leaf of sundew (Drosera rotundifolia)
There is sticky fluid on the ends of the glandular hairs. [W. Harstrick, Bruce Coleman Inc.]

carbohydrate, in their diets. Animals are especially deficient in synthetic ability. Among their extensive dietary requirements are proteins or the amino acids of which they are composed; a diet restricted to carbohydrates is soon fatal. It is true that many animals and protozoans, if not all, can use the inorganic compound ammonia (NH_3) to a limited extent, but this capacity falls far short of their requirements; organic nitrogen is indispensable.[3]

Suppose an animal were fed a diet containing only one kind of amino acid. Would this single source of organic nitrogen suffice? For most animals, the answer is no. A mixture of amino acids is necessary. Most animals have apparently lost the ability to synthesize certain amino acids and must get them in their diets. These are called the *essential amino acids*—a somewhat misleading term, since it seems to imply that the other amino acids commonly occurring in proteins are not essential; what is meant, of course, is that the designated amino acids are essential *in the diet,* whereas the others, which are also necessary for life, can be synthesized by the organism itself from other amino acids or organic nitrogen compounds. Although the essential amino acids vary for different species of animals, and even for different stages in the life history of the same species, the basic pattern is similar for all. Nine amino acids (histidine, isoleucine, leucine, lysine, methionine, phenylalanine, threonine, tryptophan, valine) are essential for almost all animals; the others may be essential for some, but not for all.

For all the essential amino acids to be included in the diet, several different proteins should be eaten, because a single protein may not contain them all. For example, zein, the main protein in corn, is deficient in tryptophan, lysine, and cysteine. Someone who depended exclusively on a "poor quality" protein such as zein would suffer from a deficiency not only of these three amino acids but of other essential amino acids as well. The reason is that effective utilization of amino acids in protein synthesis requires that all the essential amino acids be present simultaneously in the correct relative amounts. If one is not present in sufficient quantity, then utilization of the others is reduced proportionately, and since they cannot be stored they will be lost through excretion.

One way to avoid a deficiency of an essential amino acid is to include a variety of different proteins in the diet, since it is unlikely that all will be deficient in the same amino acids. It has been recommended that an average adult human being include in his daily diet at least 70 grams (about 1.85 ounces) of protein, of which at least half should be of animal origin. The proportions of the various amino acids in plant proteins are often quite different from those in animal proteins; hence plant proteins are less reliable than animal proteins as a source of essential amino acids for human beings. *Kwashiorkor,* a protein-deficiency disease characterized by degeneration of the liver, severe anemia, and inflammation of the skin, is particularly common among children in countries where the diet consists primarily of a single plant material—as in Indonesia, where rice forms much of the diet, or parts of Africa where corn is the principal staple.

In an entirely vegetarian diet, care should be devoted to selecting a combination of plant proteins that will complement one another, making up for one another's deficiencies. For example, the proteins in beans are deficient

---

[3] A few heterotrophic flagellated Protozoa can survive with ammonia as their only nitrogen source.

in methionine, whereas those in wheat are deficient in lysine; if both beans and wheat are eaten at the same meal, they will complement each other and there will be no deficiency of either methionine or lysine. It should be emphasized that complementary proteins must be eaten *at the same meal;* because amino acids cannot be stored in the body, it would be futile to eat beans at one meal and wheat at the next.

Some animals can survive and grow with little or no fat in their diets; because they can interconvert carbohydrate and fat, they may deposit much fat as a storage product in their bodies even when they do not eat any. But some animals (including rats and humans) cannot synthesize enough linoleic acid (a common fatty acid) for their needs, no matter how many other organic compounds are available to them. Severe disease symptoms or even death may result if such animals do not eat enough fat to provide presynthesized linoleic acid. For these animals, linoleic acid may be designated an *essential fatty acid,* i.e. one essential in their diets.

VITAMINS

Vitamins are organic compounds necessary in small quantities to given organisms that cannot synthesize them and must therefore obtain them prefabricated in the diet. Note that a compound may be a vitamin for species A and not for species B, because B can synthesize it. Vitamins are necessary only in very small quantities, because they ordinarily function as coenzymes or as parts of coenzymes; you will recall that enzymes and coenzymes are catalysts that can be reused many times and hence are not needed in large amounts.

That certain diseases are connected with dietary deficiencies, now identified as deprivation of vitamins, was recognized long ago. By 1752 it was known that fresh fruits help prevent *scurvy,* a painful disease, common among sailors long at sea, the symptoms of which are bleeding gums and loosening of teeth, anemia, delayed healing of wounds, and painful and swollen joints. Shortly before 1800, in an effort to control this debilitating disease, lime and lemon juice were made standard parts of rations for British sailors—hence their nickname "limeys." But the full significance of the effect of citrus fruit (now known to contain vitamin C) on a disease was not realized.

Almost a century later, the Dutch government became concerned over the high incidence among troops in the East Indies of a crippling disease called *beriberi,* which is characterized by muscle atrophy, paralysis, mental confusion, and sometimes congestive heart failure. A team of investigators was sent to the East Indies to study the problem. Aware of the discoveries of Pasteur, Koch, and others on microorganisms as causative agents in disease, they bent their efforts toward finding such an agent for beriberi; however, two years of work met with no success. Then Christiaan Eijkman, one of the members of the team, discovered that chickens fed primarily on polished rice dropped in the kitchen and dining area of the military quarters developed symptoms similar to those of human beriberi; because it was so cheap, such rice was the main food supplied the troops. Eijkman found that if unpolished rice was added to the diet neither chickens nor men developed the disease. He showed later that the antiberiberi factor was water-soluble and could be extracted from rice polishings.

In 1906 F. G. Hopkins of Cambridge University demonstrated that, in addition to carbohydrates, fats, proteins, minerals, and water, normal foods contain minute traces of other substances essential to health. Hopkins called such substances accessory factors. Then in 1911 Casimir Funk isolated and crystallized the antiberiberi factor, which, because it was chemically an amine, he proposed calling "vitamine" (conveying that the amine was vital). When it was found later that many accessory factors are not amines, some workers wanted to abandon Funk's term; a compromise resulted in abandoning only the final *e*, and the name became "vitamin." The antiberiberi factor is now called vitamin B_1 or thiamine.

It is often hard to demonstrate conclusively that a particular chemical compound is a vitamin, because a diet supposedly free of a given compound might contain trace amounts that would be sufficient to prevent symptoms from developing in an experimental animal. Even the most elaborate purification techniques are not always successful. A list of compounds that may be vitamins for human beings is under investigation, but it may be years before any certainty about them is reached. It is even more difficult to establish reliable minimum daily requirements for known vitamins. Those supposedly established are still very much open to question. Little is known, for example, about how requirements alter with age or with changing health. Although much research remains to be done, one thing can be asserted with reasonable confidence: Healthy persons who eat a varied diet including meats, fruits, and vegetables will probably get all the vitamins they need, numerous advertisements to the contrary notwithstanding.

The relation between the pathological symptoms of vitamin deficiency and the actual biochemical function of the vitamin is often obscure. For example, the symptoms of beriberi, a vitamin B_1–deficiency disease, give no indication that vitamin B_1 functions in the conversion of pyruvic acid into acetic acid and carbon dioxide. In fact, the exact biochemical roles of many vitamins are still unknown, despite extensive clinical information on the symptoms that a lack of them will cause.

Water-soluble vitamins   Vitamins can be classified as water-soluble or as fat-soluble. The water-soluble vitamins include vitamin C and the vitamins of the so-called B complex. They function as coenzymes in metabolic reactions that take place in almost all animal cells. Some animals can synthesize one or more of these coenzymes, and for them, of course, such coenzymes are not vitamins and are not required in the diet.

Vitamin C, or ascorbic acid, is the previously mentioned factor in fruit that prevents scurvy. One of its most important known functions is its role in the formation of collagen fibers, which you will recall are the chief components of connective tissue. When the diet is severely deficient in ascorbic acid, collagen formation ceases and the gravest symptoms of scurvy result. Severe scurvy is rare among adults in this country, but occurs occasionally in infants; very mild cases, which are difficult to recognize, are more frequent. Since inclusion of citrus fruits or tomatoes in the diet provides an ample supply of ascorbic acid, supplements are usually necessary only for infants, pregnant women, and the seriously ill.

The B complex includes a large number of compounds, unrelated chemically, but somewhat similar in function and tending to occur together. Several of them are components of coenzymes functioning in cellular respira-

5.13 Folic acid–deficient chick
The two birds are both four weeks old. The one on the right was fed a folic acid–deficient diet, while the one on the left received a plentiful supply of the vitamin.   [From the *Vitamin manual,* courtesy The Upjohn Company.]

tion. As already mentioned, thiamine (vitamin B_1) is a part of the coenzyme that catalyzes the oxidation of pyruvic acid. Pantothenic acid is a component of coenzyme A, which, as we saw in the last chapter, plays an essential role in carrying the acetyl group into the Krebs cycle. Riboflavin (vitamin B_2) is one of the carrier compounds in the respiratory electron-transport system. Pyridoxine (vitamin B_6) is a component of a coenzyme involved in transaminations—reactions transferring amino groups from one compound to another. Nicotinamide, another B vitamin, is a major component of both NAD and NADP (commercial vitamin preparations often contain niacin, which is converted into nicotinamide in the body); the nicotinamide-deficiency disease *pellagra* is a severe problem in many poor areas, but it can also be caused by chronic alcoholism. It is clear why the B vitamins, as indispensable catalysts of the energy-releasing reactions of cellular respiration, are of prime importance in the diets of heterotrophs such as humans.

Some of the B vitamins—particularly B_{12} (cobalamin), a very important vitamin containing the element cobalt—seem to be involved in the formation of red blood cells. Vitamin B_{12} deficiency results in *pernicious anemia,* a chronic disease most common in older people. This vitamin, like several others (e.g. vitamin E, niacin, pantothenic acid, and folic acid), is usually synthesized in mammals by microorganisms in the digestive tract, and may be absorbed from this source without having been present as such in the diet. When human beings develop pernicious anemia, the problem often is not insufficient vitamin B_{12} in the intestine, but rather an inability to absorb it or an inability to convert it into an active form once absorbed.

Folic acid, another of the B vitamins, is apparently also involved in red blood cell formation. Its primary role, however, is in the synthesis of some of the nucleotides that are building blocks for nucleic acids, and it is thus essential for cell division (Fig. 5.13).

5.14   Night blindness
Left: Normal individual's view of road.   Right:
Approximate view of road of vitamin A–deficient
individual under same lighting conditions. That person
cannot see the road sign at all.   [From the *Vitamin
manual,* courtesy The Upjohn Company.]

Fat-soluble vitamins   The compounds collectively known as the fat-soluble vitamins are vitamins only for vertebrate animals (vitamin E, which is also required by a few invertebrates, is an exception). Though the same compounds, or very similar ones, occur in a great variety of other organisms, they apparently function differently in such organisms and are not dietary essentials. It is a common evolutionary pattern for compounds present in ancestral organisms to take on new functions in the course of evolution. In other words, natural selection, instead of giving rise to completely new compounds, frequently acts upon already existing compounds, giving them new functions.

The four principal fat-soluble vitamins are A, D, E, and K.

Symptoms of vitamin A deficiency include retarded growth, excessive keratinization of epithelia (hardening by deposition of keratin, the chief component of claws, nails, and horns), and degeneration of columnar and cuboidal epithelia into stratified squamous epithelia. But the most serious result of vitamin A deficiency is *xerophthalmia*—a keratinization of tissues of the eye that can lead to permanent blindness. Indeed, xerophthalmia is the commonest cause of childhood blindness in many underdeveloped countries. A less extreme manifestation of vitamin A deficiency is night blindness, a marked impairment of vision in dim light (Fig. 5.14). The explanation for night blindness is that vitamin A is the chief component of the light-sensitive pigment in the rod cells of the eye. Deficiency of this vitamin is not common in the United States, because it can be synthesized in the animal body from the yellow pigment carotene, which, you will recall, is present in green and yellow vegetables and fruit; although carotene can be synthesized only by plants, this precursor of vitamin A is also abundant in such animal products as butter, cheese, milk, and egg yolk. An excess of vitamin A, as of some other vitamins, is decidedly harmful, and the indiscriminate use of vitamin preparations should be avoided.

Vitamin D is involved in calcium absorption and metabolism, and a deficiency in children results in the condition known as *rickets,* in which the skeleton is deformed, because the bones, lacking sufficient calcium, are very soft. Exposure to sunlight is the best preventive for rickets, because the ultraviolet radiation in sunlight acts on sterols in the skin to produce vitamin D. Rickets is therefore confined to the temperate zones, where people spend much time indoors and wear much clothing while outside; it is almost unknown in the tropics. The vitamin is present in egg yolk, milk, and fish oils.

Vitamin E is important in maintaining good muscle and nerve condition, normal liver function, and male fertility, and in preventing red blood cell rupture. A deficiency of this vitamin occurs only rarely in human beings.

Vitamin K is essential for the formation of one of the chemicals necessary for blood clotting. A deficiency results in slow blood clotting and, sometimes, in hemorrhages. In human beings enough vitamin K is normally synthesized by bacteria in the digestive tract, but a deficiency may develop if anything interferes with the absorption of fats and fat-soluble materials. At times, a deficiency of this and other vitamins synthesized by intestinal bacteria is a side effect of taking large doses of antibiotics: The drugs kill not only disease-producing bacteria but the desirable intestinal bacteria as well.

Table 5.2 lists the main vitamins for human beings and indicates deficiency symptoms and important food sources.

Evolutionary and genetic studies of loss of the ability to synthesize particular vitamins have been very rewarding in recent years. Current theories concerning the mode of action of genes hold that the genes exert their control over cellular functions by controlling the synthesis of enzymes. It has been shown that mutation of a single gene may result in loss of an enzyme necessary for the synthesis of a particular coenzyme. An organism showing such a mutant trait must obtain the coenzyme in its diet; for such an organism, the coenzyme in question has become a vitamin. Similarly, it has been shown that a mutation may result in loss of ability to synthesize a particular amino acid. For an organism showing this mutant characteristic, the amino acid in question has become an essential amino acid and must be included in the diet. Experimental organisms like bread mold can be exposed to mutation-producing X rays in the laboratory. The irradiated organisms can then be placed on a variety of nutrient media, each deficient in a different compound, and those organisms that have undergone mutations altering their nutritional requirements can be identified by determining on which media they can grow and on which they cannot. By such experiments much has been learned about the genetic basis for the evolution of the kinds of nutritional requirements that characterize modern heterotrophs.

MINERAL NUTRITION

Like the green plants, heterotrophs require certain minerals, which are usually absorbed as ions. Some they need in relatively large quantity, e.g. sodium, chlorine, potassium, phosphorus, magnesium, and calcium; in human beings the minimum daily requirement for these varies from about 0.35 gram for magnesium to nearly 3 grams for sodium chloride. Other minerals are needed in much smaller amounts, e.g. iron, manganese, and iodine. And still others, though essential to life, are needed only in trace amounts, e.g. copper, zinc, molybdenum, selenium, and cobalt. Some elements, like vanadium, barium, tin, silicon, and nickel, are necessary in some species of animals, but have not been proved essential for human beings.

The function of some of the minerals is obvious. Calcium is a major

TABLE 5.2 *Some vitamins needed by human beings*

Vitamin	Some deficiency symptoms	Important sources
FAT-SOLUBLE		
Vitamin A (retinol)	Dry, brittle epithelia of skin, respiratory system, and urogenital tract; xerophthalmia and night blindness	Green and yellow vegetables and fruit, dairy products, egg yolk, fish-liver oil
Vitamin D (calciferol)	Rickets or osteomalacia (very low blood-calcium level, soft bones, distorted skeleton, poor muscular development)	Egg yolk, milk, fish oils
Vitamin E (tocopherol)	Malfunction of muscular and nervous systems; anemia (from rupture of red blood cells); male sterility	Widely distributed in both plant and animal food, e.g. meat, egg yolk, green vegetables, seed oils
Vitamin K (phylloquinone, etc.)	Slow blood clotting and hemorrhage	Green vegetables
WATER-SOLUBLE		
Thiamine (B_1)	Beriberi (muscle atrophy, paralysis, mental confusion, congestive heart failure)	Whole-grain cereals, yeast, nuts, liver, pork
Riboflavin (B_2)	Vascularization of the cornea, conjunctivitis, and disturbances of vision; sores on the lips and tongue; disorders of liver and nerves in experimental animals	Milk, cheese, eggs, yeast, liver, wheat germ, leafy vegetables
Pyridoxine (B_6)	Convulsions, dermatitis, impairment of antibody synthesis	Whole grains, fresh meat, eggs, liver, fresh vegetables
Pantothenic acid	Impairment of adrenal cortex function, numbness and pain in toes and feet, impairment of antibody synthesis	Present in almost all foods, especially fresh vegetables and meat, whole grains, eggs
Biotin	Clinical symptoms (dermatitis, conjunctivitis) extremely rare in humans, but can be produced by great excess of raw egg white in diet	Present in many foods, including liver, yeast, fresh vegetables
Nicotinamide	Pellagra (dermatitis, diarrhea, irritability, abdominal pain, numbness, mental disturbance)	Meat, yeast, whole wheat
Folic acid	Anemia, impairment of antibody synthesis, stunted growth in young animals	Leafy vegetables, liver
Cobalamin (B_{12})	Pernicious anemia	Liver and other meats
Ascorbic acid (C)	Scurvy (bleeding gums, loose teeth, anemia, painful and swollen joints, delayed healing of wounds, emaciation)	Citrus fruits, tomatoes

constituent of bones and teeth in vertebrates and plays a variety of other roles in most organisms. Phosphorus is a component of many high-energy organic compounds of critical importance. Iron is a constituent of the cytochromes and of hemoglobin. Sodium, chlorine, and potassium are important components of the body fluids, playing a role in osmotic phenomena and in such processes as nerve and muscle action. Iodine is a component of the hormones produced by the thyroid gland. But the function of some of the minerals, particularly those needed only in trace amounts, is less obvious. Apparently most of them act as components of coenzymes, or perhaps as cofactors that help catalyze reactions without being actually incorporated into enzymes or coenzymes. Such minerals are comparable to vitamins, functioning in the same way. The only distinction, and an arbitrary one at that, is that vitamins are organic compounds and minerals are inorganic.

NUTRIENT PROCUREMENT BY FUNGI

The fungi constitute a large and diverse group of sedentary heterotrophic organisms that live on or in their food supply. Bread mold is an example (see Fig. 1.13, p. 19). The bread on which it grows is composed mostly of starch, a rich source of energy. But starch is a polysaccharide, whose very large and insoluble molecules cannot move across the cell membranes of the mold. Before absorption can take place, the starch must be broken down to its constituent building-block compounds, the simple sugars; in short, the starch must be digested. *Digestion* is nothing more than enzymatic hydrolysis, which, you will recall, involves the addition of water (see p. 50). In bread mold the hydrolysis takes place outside the cells, a process called *extracellular digestion.* Digestive enzymes synthesized inside the cells of the mold are released from the cells onto the bread and hydrolyze the starch. The simple sugars that are the products of this digestion are then absorbed, often by rootlike structures called *rhizoids* (Fig. 5.15A).

Mold living on bread exemplifies a saprophytic way of life, but many fungi are parasitic. Indeed, bread mold itself is not restricted to saprophytic nutrition; it is one of the commonest destructive fungi on fresh fruits and vegetables. The various parasitic fungi differ in their relationships to their plant or animal hosts. Some small fungi live entirely within a single host cell. Other more filamentous fungi grow between the cells of their host, but send rootlike food-absorbing structures called *haustoria* into the host cells (Fig. 5.15B). Still other filamentous types occupy many host cells simultaneously, penetrating through the host-cell membranes (and walls in the case of plant hosts) that divide one cell from the next. Whatever the details of their relationship to their hosts, the parasitic fungi employ basically the same mode of nutrition as saprophytes such as mold on bread. Enzymes are secreted into the food supply on (or in) which the fungus lives; digestion takes place extracellularly; and the products of digestion are then absorbed by the fungus. Note that fungi, unlike most animals, have no internal cavity where bulk food can be digested; they simply release digestive enzymes into their surroundings and absorb organic nutrients across the body surface, much as plant roots absorb inorganic nutrients.

A few fungi, departing somewhat from the usual pattern of feeding on their substrate, supplement their diets by trapping small animals such as

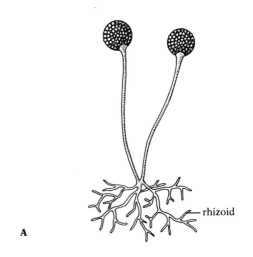

A

rhizoid

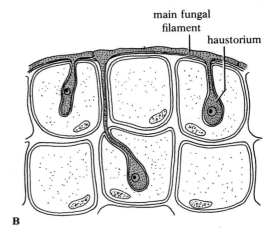

main fungal
filament

haustorium

B

5.15 Nutrient-procurement structures of fungi
(A) The rootlike rhizoids of a saprophytic fungus.
(B) The haustoria of a fungus parasitic on a multicellular plant. The body of the fungus (color) is filamentous and can grow between the cells of the plant host. The haustoria penetrate the cell wall and make deep invaginations in the cytoplasm of host cells, from which they absorb nutrients. Note that the haustoria are not in direct contact with the cytoplasm of the host cells, because the haustorial invaginations are lined with host-cell membrane.

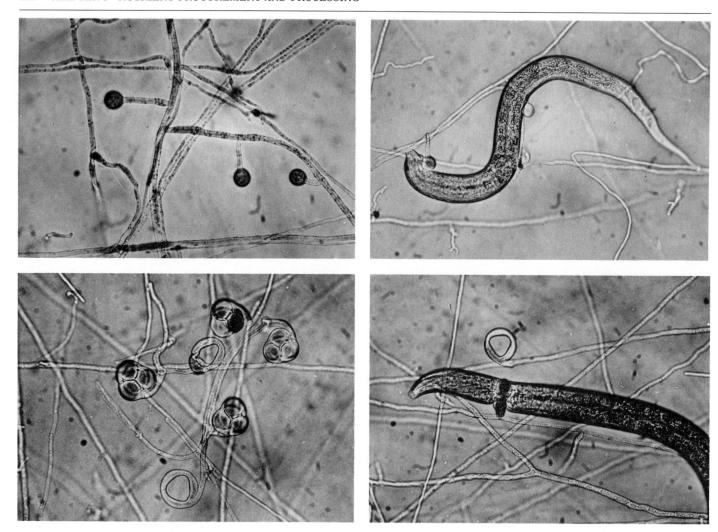

5.16   Fungi that trap nematode worms
Top: *Dactylella drechsleri* has sticky knobs (left), which hold a worm that contacts them (right).   Bottom: *Arthrobotrys dactyloides* has rings formed of three cells (left); the three cells can be seen plainly in the closed rings. When a worm enters the ring, the cells swell and constrict the ring, trapping the worm (right). [Courtesy David Pramer, *Science*, vol. 144, 1964. Copyright 1964 by the American Association for the Advancement of Science.]

nematode worms. Some of the predatory fungi, such as *Dactyllela*, produce spherical knobs with sticky surfaces; nematode worms coming in contact with the knobs adhere to them and are trapped (Fig. 5.16, top). Other predatory species, such as *Arthrobotrys*, form rings composed of three cells; when a worm enters the ring, the cells swell rapidly, constricting the ring and trapping the worm (Fig. 5.16, bottom). Branches of the fungus then penetrate the worm's body and release digestive enzymes; extracellular digestion takes place, and the products are absorbed.

Arthrobotrys provides an excellent illustration of the nutritional flexibility some fungi exhibit. When isolated from the soil, it can easily be grown on a variety of organic culture media. In such a culture it seldom forms traps. However, if live nematodes or extracts of nematode tissue (or even water in which nematodes have been living) are added to the culture, the fungus responds by developing traps. Apparently some chemical produced by the nematodes can induce the fungus to form traps; the prey stimulates the development of prey-capturing structures in the predator.

NUTRIENT PROCUREMENT BY ANIMALS AND PROTOZOANS

Nutrient procurement by animals usually involves much more activity than it does in plants. Animals must often resort to elaborate methods of locating and trapping their food. Their incredibly varied feeding habits may be classified in any number of ways. We have already mentioned one possible classification, which distinguishes carnivores, herbivores, and omnivores, depending on whether the diet consists primarily of animals, plants, or both. Another possible criterion for classification is the size of the food. Thus we can recognize microphagous feeders, which strain microscopic organic materials from the surrounding water by an array of cilia, bristles, legs, nets, etc. And we can recognize macrophagous feeders, which subdivide larger masses of food by teeth, jaws, pincers, or gizzards, or solely by the action of enzymes. Smaller groups would include the sucking animals, adapted to extract fluid from plants or from animal prey, and those parasitic animals that are bathed in the nutrients of the host and absorb them directly through the body surface.

Like the fungi already examined, most animals must digest their food before it can cross the membranes of their cells. Only seldom can they obtain as food such comparatively simple and diffusible compounds as glucose, glycerol, fatty acids, and amino acids; usually food material is in the form of polysaccharides, fats, proteins, etc., and must be hydrolyzed. But animals rarely simply secrete digestive enzymes directly onto their food, in the manner of fungi. The vast majority ingest particles or lumps of food into some sort of digestive structure, in which enzymatic digestion takes place. Often the lumen of the structure is extracellular; hence digestion in it is extracellular. In other instances, the food is ingested directly into a cell by phagocytosis, or some similar process, and then digested in a food vacuole. Although this process is classified as *intracellular digestion,* it should be noted that the food material is separated from the rest of the cellular material by a membrane that it cannot cross until after digestion has occurred. Thus extracellular ingestion and intracellular ingestion are alike in that digestion always precedes the actual absorption of complex foods across a membrane.

Although both the nutritional requirements and the basic processes of digestion are essentially alike in protozoans and all types of animals, from worms to human beings, the body plans of these oganisms vary so greatly that the structures involved in food processing and the details of that processing are often very different. In the following sections we shall briefly examine the digestive mechanisms of protozoans and a variety of animals.

NUTRIENT PROCUREMENT BY PROTOZOANS

Since Protozoa, as single-celled organisms, have a body plan obviously very different from that of multicellular animals, we would expect their adaptations for food procurement to be likewise markedly different. And the differences are, in fact, considerable. But a more interesting point, one with important biological implications, is that the similarities are often more striking than the differences.

Let us look first at an amoeboid protozoan, which constantly changes shape as its protoplasm flows along, pushing out new armlike pseudopodia and withdrawing others (Fig. 1.7, p. 15). When an amoeba is stimulated by nearby food, some of the pseudopodia may flow around the food until they have completely surrounded it. This is the process known as phagocytosis. The food is completely engulfed by the cytoplasm and is enclosed in a *food vacuole,* where it will be digested (see Fig. 3.15, p. 93). Amoebas exemplify protozoans without specialized permanent digestive structures, though their transitory food vacuoles are functional analogues of the digestive systems of higher animals.

The ciliates, another important group of Protozoa, of which *Paramecium* is an example, are characterized by innumerable cilia covering the surface of their bodies (Fig. 1.6). Like all Protozoa, they are commonly regarded as unicellular. But though lacking actual subdivision into recognizable cellular units, the more complex ciliates show much of the internal specialization usually associated with multicellularity. For this reason many biologists prefer to regard them, not as single cells, but as acellular (i.e. as organisms whose bodies are not built of cells in the usual sense). Unlike the amoeba, *Paramecium* has a permanent structure, an organelle, that functions in feeding. Food particles are swept into an *oral groove,* a ciliated channel located on one side of the cell (Fig. 5.17A), by water currents produced by the beating of the cilia, and are carried down the groove into a *cytopharynx.* As food accumulates at the lower end of the cytopharynx, a food vacuole forms around it (Fig. 5.17B). Eventually the vacuole breaks off and begins to move toward the anterior end of the cell. Digestive enzymes are secreted into the vacuole and digestion begins. As digestion proceeds, the products (simple sugars, amino acids, etc.) diffuse across the membrane of the vacuole into

5.17 *Paramecium*

(A) Drawing showing major structures.   (B) Food vacuole forms at lower end of cytopharynx, then breaks off and moves toward anterior end of cell while enzymes are secreted into it; digestion takes place, and the products of digestion are absorbed into the general cytoplasm. Vacuole then moves posteriorly, attaches to anal pore, and expels digestive wastes. The vacuole undergoes several changes in size and appearance as it moves.

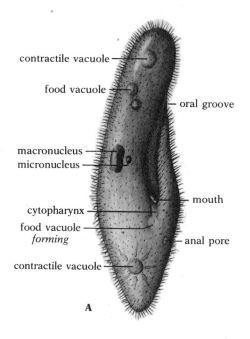

contractile vacuole

food vacuole

oral groove

macronucleus

micronucleus

mouth

cytopharynx

food vacuole *forming*

anal pore

contractile vacuole

A

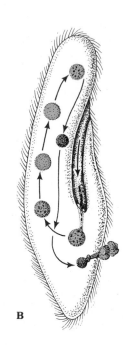

B

the cytoplasm, and the vacuole begins to move back toward the posterior end of the cell. When the vacuole reaches a tiny specialized region of the cell surface called the anal pore, it becomes attached there and ruptures, expelling by exocytosis any remaining bits of indigestible material. Not only does the food vacuole function as a digestive chamber, but by its movement it helps distribute the products of digestion to all parts of the cell.

We have said that digestive enzymes are secreted into the food vacuoles of both the amoeba and *Paramecium*. But if these powerful enzymes are capable of hydrolyzing such compounds as polysaccharides, fats, proteins, and nucleic acids, and if the cell itself is composed of these kinds of compounds, how can the cell contain the digestive enzymes without being destroyed by them? A partial answer was given in Chapter 3. Digestive enzymes are packaged in lysosomes, vesicles whose membranes are apparently both impermeable to the enzymes and capable of resisting their hydrolytic action. The digestive enzymes are presumably synthesized on the ribosomes, move through the endoplasmic reticulum to the Golgi apparatus, and there become surrounded by a membrane to form the lysosome. When a food vacuole, sometimes also called a phagosome, is formed, a lysosome soon fuses with it (Fig. 5.18). Food materials and the digestive enzymes are mixed in the resulting digestive vacuole. As already described, this vacuole circulates in the cytoplasm, the products of digestion are absorbed, and indigestible materials are eventually expelled from the cell by exocytosis.

Although this description of lysosome activity pertains to digestion in Protozoa, it applies equally well to intracellular digestion in any animal cell. Lysosomes, you will recall, were, in fact, first discovered in rat liver cells. Here is but one example of the similarity between the basic processes of digestion in Protozoa and in higher animals.

NUTRIENT PROCUREMENT BY COELENTERATES

With the evolution of multicellularity came a corresponding evolution of cellular specialization resulting in a division of labor among cells. The coelenterates provide a comparatively simple example of this phenomenon. These radially symmetrical animals have a saclike body composed of two principal layers of cells (Fig. 5.19), with a jellylike layer, called mesoglea, between them. The cells of the outer layer function as a protective and sensory epithelium, while those of the inner layer, or gastrodermis, act as a nutritive epithelium. Some cells of both layers are specialized as muscle fibrils, and others as nerves. The central cavity of this saclike body functions as a digestive cavity. It has only one opening to the outside, which is surrounded by mobile tentacles. A digestive cavity of this sort, with a single opening that functions as both mouth and anus, is called a *gastrovascular cavity.*

Coelenterates are strictly carnivorous (see Fig. 1.15, p. 20). Embedded in their tentacles are numerous stinging structures called *nematocysts.* Each nematocyst consists of a slender thread coiled within a capsule, with a tiny hairlike trigger penetrating to the outside. When appropriate prey comes in contact with the trigger, the nematocyst fires, the thread turns inside out, spines on its surface unfold, and it either penetrates the body of the prey or entangles it in sticky loops. The nematocysts also eject poisons, which have a

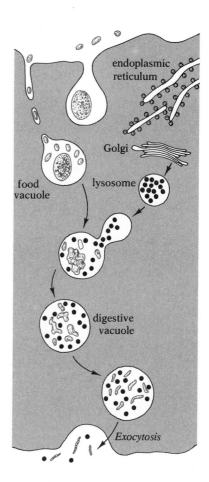

5.18   The role of lysosomes in intracellular digestion
Food material (color) that the cell takes in by phagocytosis is enclosed in a food vacuole, which fuses with a lysosome containing digestive enzymes. Digestion takes place within the composite structure thus formed (digestive vacuole), and the products of digestion are absorbed across the vacuolar membrane. The vacuole eventually fuses with the cell membrane and then ruptures, expelling indigestible wastes to the outside.

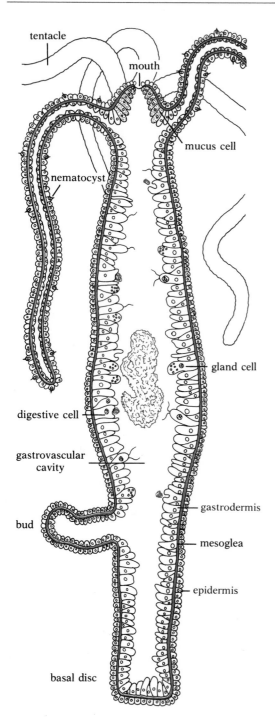

5.19 Hydra, showing gastrovascular cavity
The cavity contains food material (color). The
mesoglea layer in the body wall is much more
extensively developed in some other coelenterates.

paralyzing action on the prey. The tentacle then grasps the prey, and, if it continues to struggle, neighboring tentacles may also become involved. The tentacles draw the prey toward the mouth, which opens wide to receive it. Once the food is inside the gastrovascular cavity, digestive enzymes are secreted into the cavity by the gastrodermal cells, and extracellular digestion begins. This extracellular digestion, largely limited to proteins in coelenterates, does not break down these substances completely to their constituent amino acids. As soon as the food has been reduced to small fragments, the gastrodermal cells engulf them by phagocytosis, and digestion is completed intracellularly in food vacuoles. Indigestible remains of the food are expelled from the gastrovascular cavity via the mouth.

If phagocytosis and intracellular digestion are going to take place anyway, we can ask what adaptive advantage the evolution of the additional process of extracellular digestion might have. Why should not coelenterates rely exclusively on intracellular digestion as the Protozoa do? The answer is obvious. Intracellular digestion severely limits the size of the food the organism can handle. Extracellular digestion enables it to utilize much larger pieces of food; even whole multicellular animals become possible prey. Extracellular digestion is the rule rather than the exception in multicellular animals.

We have seen, then, that coelenterates exhibit a variety of interesting evolutionary adaptations for prey capture and digestion. As a result of cellular specialization and division of labor, certain cells, those of the gastrodermis, carry out digestion for the whole organism. The products of digestion can be distributed from the gastrodermal digestive cells to cells specialized for other functions such as protection or movement or stimulus reception. Since the bodies of coelenterates are relatively small, and no cells are far removed from the gastrodermal layer, this distribution can be effected without any specialized transport system.

NUTRIENT PROCUREMENT BY FLATWORMS

Unlike the radially symmetrical coelenterates, the flatworms are bilaterally symmetrical; they have distinct anterior (front) and posterior (rear) ends, and also distinct dorsal (upper) and ventral (lower) surfaces. Their bodies are composed of three well-formed tissue layers. Many flatworms are parasitic on other animals, but some are free-living, and it is to these we shall turn first, using planaria as an example (Figs. 5.20 and 1.16, p. 20).

The mouth of planaria is located on the ventral surface near the middle of the animal. It opens into a muscular tubular *pharynx,* which can be protruded through the mouth directly onto prey. The pharynx leads into a gastrovascular cavity (i.e. a cavity with only one opening to the outside). This cavity, though functionally similar to that of the coelenterates, is far more branched, ramifying throughout the animal's body. Literally a gastrovascular cavity (*gastro-* refers to the stomach and *vascular* to a circulatory vessel), it functions in both digestion and transport of food to all parts of the body. The extensive branching has another important function: It greatly increases the total absorptive surface of the cavity. We saw with plants that as organisms increase in size, and particularly as their volume increases, the problem of sufficient absorptive surface becomes more acute. Many organ-

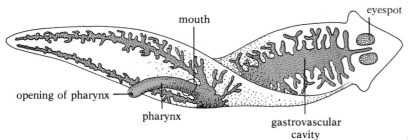

5.20 Planaria, showing much-branched gastrovascular cavity and extruded pharynx

isms have evolved greatly subdivided absorptive surfaces, thereby compacting much total surface area into relatively little space. The root hairs of plants were one example, and the branched gastrovascular cavity of planaria is another; we shall encounter many more in this and later chapters.

Some extracellular digestion occurs in the gastrovascular cavity of planaria, but most of the food particles are engulfed by gastrodermal cells and digested intracellularly.

The members of one class of flatworms, the tapeworms, have become so highly specialized as parasites living in the digestive tracts of other animals that in the course of their evolution they have lost their own digestive systems. They are constantly bathed by the products of the host's digestion and can absorb them without having to carry out any digestion themselves. Evolutionary adaptation can involve the loss of structures as well as their acquisition.

ANIMALS WITH COMPLETE DIGESTIVE TRACTS

Animals above the level of coelenterates and flatworms have a **complete digestive tract**, i.e. one with two openings, a **mouth** and an **anus.** The advantages of such a system over a gastrovascular cavity are obvious. No longer must incoming food material and outgoing wastes pass through the same opening. Instead, food can be passed in one direction through a tubular system, which can be divided into a series of distinct sections or chambers, each specialized for a different function. As the food passes along this assembly line, it is acted upon in a different way in each section. The sections may be variously specialized for mechanical breakup of bulk food, temporary storage, enzymatic digestion, absorption of the products of digestion, reabsorption of water, storage of wastes, etc. The overall result is a much more efficient digestive system, as well as a potential for special evolutionary modifications fitting different animals for different modes of existence.

The digestive system of an earthworm is a good example of subdivision into specialized compartments (Fig. 5.21). Food, in the form of decaying organic matter mixed with soil, is drawn into the mouth by the sucking actions of a muscular chamber called the **pharynx.** It passes from the mouth through a short passageway into the pharynx and then through a connecting passage called the **esophagus,** after which it enters a relatively thin-walled **crop** that functions as a storage chamber. Next, it enters a compartment with thick muscular walls, the *gizzard,* where it is ground up by a churning action; the grinding is often facilitated by small stones in the gizzard. The pulverized food, suspended in water, now passes into the long *intestine,* where enzymatic digestion and absorption take place. Finally, in the rear of the intestine, some of the water involved in the digestive process is reabsorbed, and the indigestible residue is eliminated from the body through the anus.

Notice that earthworms use extracellular digestion. Glandular cells in the epithelial lining of the intestine secrete hydrolytic enzymes into the intestinal cavity, and the end products of digestion—the simple building-block

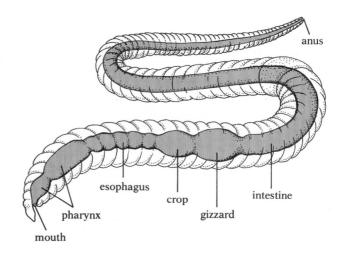

5.21   Digestive system of an earthworm

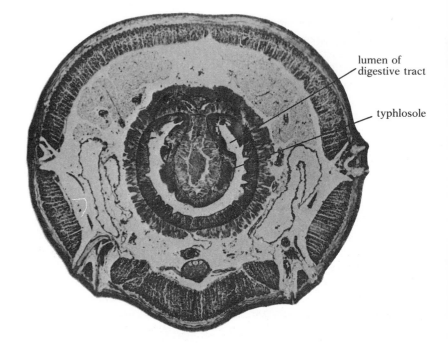

5.22   Photograph of cross section in the intestinal region of an earthworm
The typhlosole, which projects into the cavity of the intestine, greatly increases the surface area available for absorption of food.   [Courtesy Turtox/Cambosco, Macmillan Science Co., Inc.]

compounds—are absorbed. Here the problem of surface area arises again. The total interior surface presented by a plain tubular digestive tract would be inadequate in relation to the total volume of an animal the size of an earthworm. In point of fact, the earthworm's intestine is not a plain round tube internally (Fig. 5.22). A large dorsal fold, the typhlosole, projects downward into the digestive cavity, greatly increasing the total absorptive area exposed to the food, without making the outer dimensions of the intestine prohibitively large. The typhlosole is therefore functionally analogous to the root hairs of higher plants and to the branching of the gastrovascular cavity in planaria.

We have already seen that extracellular digestion is an adaptation for eating sizable pieces of food; the gizzard is obviously another such adaptation in earthworms. Mechanical breakup of bulk food is common among animals, and a variety of structures that serve this function have evolved. In our own case, food is torn and ground by the teeth. Many snails have a hard toothed pharyngeal plate, the radula, with which they rasp off small particles from larger pieces of food. Cockroaches and many other insects that feed on solid food have a chamber (the proventriculus) similar to the earthworm's gizzard except that its inner wall often bears several very hard ridges and teeth. Note that the grinding or chewing device need not be in the first section of the digestive tract, as in our own case; in both earthworms and cockroaches, the grinding chamber comes after the crop, which corresponds functionally to our stomach, and mechanical breakup thus follows temporary storage instead of preceding it. The same arrangement exists even in some vertebrates; birds have a muscular gizzard, posterior to the less specialized stomach (Fig. 5.23), in which hard food is ground with rocks and pebbles (often called grit). Why might a gizzard be adaptively superior to teeth in a flying animal like a bird?

It might easily be concluded that all except the tiniest animals eat large pieces of food and have special masticating devices, but that is not so. Some animals, such as bloodsucking or sap-sucking insects, have liquid diets. Other animals are *filter feeders,* straining small particles of organic matter from water. Clams and many other molluscs filter water through tiny pores in their gills; microscopic food particles are trapped in streams of mucus that flow along the gills and enter the mouth, kept moving by beating cilia. In such molluscs digestion is largely intracellular, as might be expected in animals that eat microscopic food.

The larvae of mosquitoes are also filter feeders. They eat bacteria and other small particles of organic matter in the water where they live. Two small hair-covered brushes (Fig. 5.24) near the mouth of the larva beat in a circular scooping motion, setting up water currents toward the mouth. The particles and water pass through the mouth and into the pharynx. Now, if the larva swallowed all the water, the salt and water balance of its body fluids would be seriously disturbed. The pharynx is specially adapted for eliminating water while filtering out food particles. Muscles in the wall of the pharynx contract, expelling the water through two small canals. Tiny combs in the canals strain out the food while the water passes through. The larva then swallows the clump of food that remains.

Clams and mosquito larvae are only two of many possible examples of filter feeders. Current theory holds that the earliest vertebrates fed in this way. Even some of the largest present-day vertebrates—certain species of

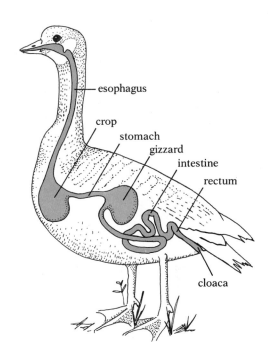

5.23   Digestive system of a bird
The chamber for mechanical breakup (gizzard) is located posterior to the stomach.

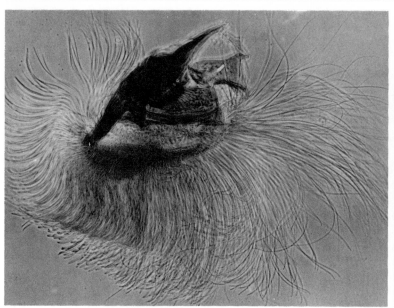

5.24   Food brush and comb of mosquito larva
Left: Isolated food brush. Its motion causes water to
flow into the mouth.   Right: Much-enlarged portion of
comb from the pharyngeal filter. It strains food
particles from water.   [Courtesy Thomas Eisner]

whales—are filter feeders, straining small planktonic plants and animals
from the vast quantities of water they take into their mouths.

But let us return to the earthworm and examine the implications of another of the specialized compartments of its digestive tract, the crop, which,
we have said, serves in food storage. The functional significance of a storage
chamber should be clear after a moment's thought. It enables the animal to
take in large amounts of food in a short time, when it is available, and then to
utilize this food over a considerable period of time. Such discontinuous
feeding makes it possible for the animal to devote much of its time to
activities other than feeding, such as searching for a mate, mating, egg
laying, and, in some animals, caring for young. Our own stomachs function
as storage organs analogous to the earthworm's crop; they enable us to live
well on only three or four meals a day and to devote the rest of our time to
other pursuits. A man can survive if his stomach is removed surgically, but
he is unable to eat more than a few bites at a time and must therefore eat
very frequently. It is not surprising that the vast majority of higher animals
have evolved adaptations for discontinuous feeding, thereby gaining time
for a behaviorally more varied existence.

Discontinuous feeding is frequently also adaptively advantageous in the
feeding process itself. If an animal had to eat constantly to maintain its
metabolic activity, it would be unable to spend time in searching for a new
food supply, which might be at some distance, or in capturing more prey
when its original supply had been depleted. In short, it would have to live in
an essentially unlimited and continuous source of food; otherwise it would
soon die or become inactive.

This is actually the case with tapeworms, nematode worms, and some
other animals that lack storage ability; they must be almost constantly in
contact with food. It would be a mistake, however, to assume that such

animals are unsuccessful or poorly adapted. Their long evolutionary history and their large numbers today testify to the contrary. They are simply adapted—successfully adapted—for a different way of life. Biological success is not measured by structural complexity or by the possession of any particular organ. The earthworm with its crop and the nematode worm without are both successful from a biological point of view, which equates success with survival.

Different kinds of food-storage organs occur in different species of animals. In many birds an expanded region of the esophagus anterior to the stomach forms a thin-walled chamber, the crop, which is functionally analogous to the earthworm's crop (Fig. 5.23). Some birds also use the crop in carrying food to their young; they fill it with seeds, berries, fish, or whatever their food may be, and then fly to the nest, where they disgorge the food for their young. In many animals storage organs take the form of blind sacs, or diverticula, branching off the digestive tract. A good example is seen in adult female mosquitoes, which have a very large diverticulum (Fig. 5.25) that opens off the anterior portion of the digestive tract and runs posteriorly, occupying much of the abdominal cavity. The female mosquito locates a suitable animal, pierces its skin with long needlelike mouthparts, and sucks blood until this diverticulum is filled. A single large blood meal may suffice to carry the female through the entire process of locating an egg-laying site and laying her eggs—a matter of four or five days.

THE DIGESTIVE SYSTEM OF HUMANS AND OTHER VERTEBRATES

Although an examination of the structure of the human digestive tract reveals little in the way of general principles that could not as easily be seen in an earthworm, natural interest in ourselves and our own species prompts a more detailed examination of human systems.

The oral cavity   The first chamber of the digestive tract is, of course, the oral cavity. Located here are the teeth, which function in the mechanical breakup of food by both biting and chewing. The internal structure of a human tooth is shown in Fig. 5.26. Human teeth are of several different

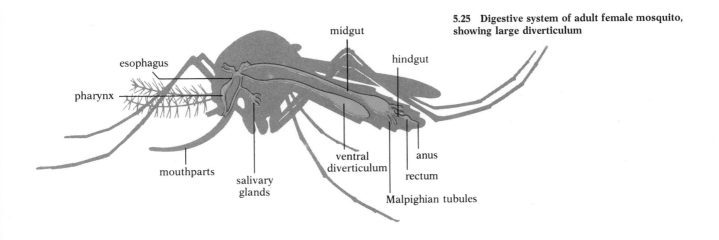

5.25   Digestive system of adult female mosquito, showing large diverticulum

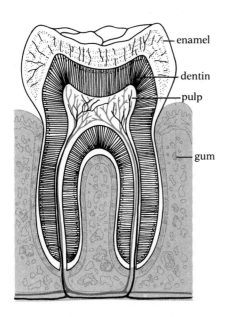

5.26 Internal structure of a human tooth
Blood vessels (color) and nerves penetrate into the pulp, but not into the outer harder layers.

5.27 Human teeth
(A) Lower jaw of adult. (B) Upper jaw of adult.
(C) Lower jaw of child, showing permanent teeth (color) in gums below milk teeth. [After Frank H. Netter, M.D., *The Ciba collection of medical illustrations*, vol. 3, 1959.]

types, each adapted to a different function (Fig. 5.27). In front are the chisel-shaped *incisors,* four in the upper jaw and four in the lower, which are used for biting. Then come the more pointed *canine* teeth, one on each side in each jaw, which are specialized for tearing food. Behind each canine are two *premolars* and three *molars* in adults; these have flattened, ridged surfaces, and function in grinding, pounding, and crushing food. A child's first set of teeth does not include all those mentioned here; the first (or milk) teeth are lost as the child gets older and are replaced with the permanent teeth that have been growing in the gums (Fig. 5.27C).

The teeth of different species of vertebrates are specialized in a variety of ways and may be quite unlike human teeth in number, structure, arrangement, and function. For example, the teeth of snakes are very thin and sharp (Fig. 5.28A) and usually curve backward; they function in capturing prey, but not in mechanical breakup, for snakes do not chew their food, but swallow it whole. The teeth of carnivorous mammals, such as cats and dogs, are more pointed than human teeth (Fig. 5.28C); the canines are long, and the premolars lack flat grinding surfaces, being more adapted to cutting and shearing (often the more posterior molars are lost). On the other hand, such herbivores as cows and horses have very large flat premolars and molars with complex ridges and cusps; the canines are often absent in such animals.

Notice that sharp pointed teeth poorly adapted for chewing seem to characterize meat eaters like snakes, dogs, and cats, whereas broad flat teeth, well adapted for chewing, seem to characterize vegetarians. How can this difference be explained? Remember that plant cells are enclosed in a cellulose cell wall. Very few animals can digest cellulose; most of them must break up the cell walls of the plant they eat if the cell contents are to be exposed to the action of digestive enzymes. Animal cells, like those in meat, do not have any such nondigestible armor and can be acted upon directly by digestive enzymes. Therefore chewing is not as necessary for carnivores as for herbivores. You have doubtless seen how dogs gulp down their food, while cows and horses spend much time chewing. But carnivores have other problems. They must capture and kill their prey, and for this, sharp teeth capable of piercing, cutting, and tearing are well adapted. Humans, being omnivores, have teeth that belong, functionally and structurally, somewhere in between the extremes of specialization attained by the teeth of carnivores and herbivores.

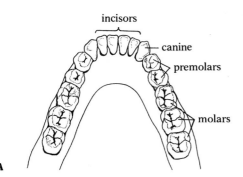

A

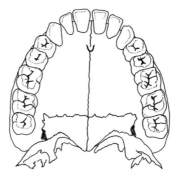

B

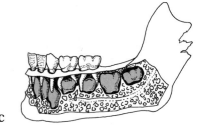

C

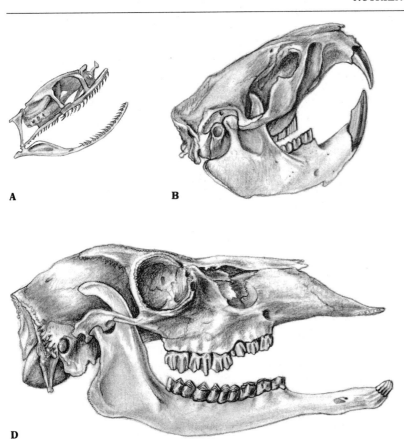

A   B   C

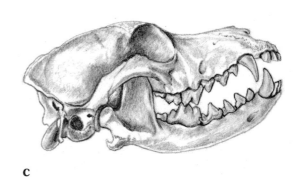

D

5.28   Structure and arrangement of teeth in different animals
(A) Snake: thin, sharp, backward-curved teeth that have no chewing function (the snake skull is here shown disproportionately large in relation to the other three).   (B) Beaver (gnawing herbivore): few but very large incisors, no canines, premolars and molars with flat grinding surfaces.   (C) Dog (carnivore): large canines, premolars and molars adapted for cutting and shearing.   (D) Deer (grazing and browsing herbivore): six lower incisors (three on each side), but no upper incisors (these are functionally replaced by a horny gum); premolars and molars with very large grinding surfaces. Notice the large gap between the incisors and premolars.

The oral cavity has other functions besides those associated with the teeth. Here food is tasted and smelled, activities of great importance in food selection, and the food is mixed with saliva secreted by several sets of salivary glands. The saliva dissolves some of the food and acts as a lubricant, facilitating passage through the next portions of the digestive tract. Human saliva contains a starch-digesting enzyme, which initiates the process of enzymatic hydrolysis. It also contains an antimicrobial agent, the thiocyanate ion, together with a special enzyme that facilitates entry of the ion into microbial cells; these substances help prevent infection by potentially harmful microbes, which are regularly introduced into the mouth.

The muscular tongue manipulates the food during chewing and forms it into a mass, called a bolus, in preparation for swallowing, then pushes the bolus backward through a cavity called the *pharynx* and into the *esophagus* (Fig. 5.29; see also Fig. 6.17, p. 251). The pharynx functions also as part of the respiratory passageway; the air and food passages cross here, in fact. Swallowing, therefore, involves a complex set of reflexes that close off the opening into the nasal passages and trachea (windpipe), thereby forcing the food to move into the esophagus. As you know, these reflexes occasionally fail to occur in proper sequence, and the food, entering the wrong passageway, makes you choke.

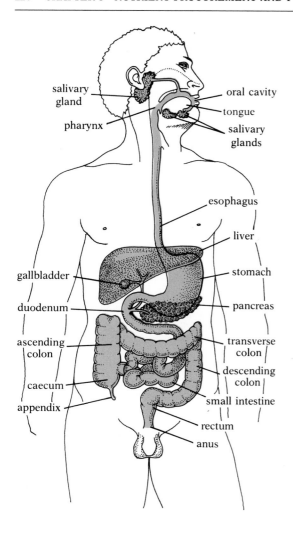

salivary gland

pharynx

oral cavity

tongue

salivary glands

esophagus

liver

gallbladder

stomach

duodenum

pancreas

ascending colon

transverse colon

descending colon

caecum

small intestine

appendix

rectum

anus

5.29 Human digestive system
The organs are slightly displaced, and the small intestine is greatly shortened.

The esophagus and the stomach The esophagus is a long tube running downward through the throat and thorax and connecting to the stomach in the upper portion of the abdominal cavity (Fig. 5.29). Food moves quickly through the esophagus, pushed along by waves of muscular contraction in a process called *peristalsis.* Circular muscles in the wall of the esophagus just behind the food bolus contract, squeezing the food forward (Fig. 5.30). As the food moves, the muscles it passes also contract, so that a region of contraction follows the bolus and constantly pushes it forward, much as though you were to keep a ball moving through a soft rubber tube by giving the tube a series of squeezes, with your hand always just behind the ball.

At the junction between the esophagus and the stomach is a special ring of muscle called a *sphincter,* which, when it is contracted, closes the entrance to the stomach. It is normally closed, thus preventing the contents of the stomach from moving back into the esophagus when the stomach moves during digestion. It opens when a wave of peristaltic contraction coming down the esophagus reaches it.

The stomach lies slightly to the left side in the upper portion of the abdomen, just below the lower ribs. It is a large muscular sac, which, as we have already seen, functions as a storage organ, making possible discontinuous feeding. It has other functions, too. Its thick walls are composed of three layers: an inner mucous membrane composed of connective tissue and columnar epithelium with many glands, a thick middle layer of smooth muscle, and an outer layer of connective tissue. The muscle layer contains fibers running around the stomach, others running longitudinally, and still others oriented diagonally. Thus the stomach is capable of a great variety of movements. When it contains food, it is swept by powerful waves of contraction, which churn the food, mixing it and breaking the larger pieces. In this manner, the stomach supplements the action of the teeth in mechanical breakup of food. The glands of the stomach lining are of several types. Some secrete mucus, which covers the stomach lining—hence the name *mucosa* or mucous membrane for the inner layer of the stomach wall; others secrete *gastric[4] juice,* a mixture of hydrochloric acid and digestive enzymes. Enzymatic digestion, then, is a third important function of the human stomach.

The small intestine The food leaves the stomach as a soupy mixture. It passes through the *pyloric sphincter* into the small intestine, which is the portion of the digestive tract where most of the digestion and absorption takes place. The first section of the small intestine, attached to the stomach, is called the *duodenum* (Fig. 5.29). It leads into a very long coiled section lying lower in the abdominal cavity. The entire small intestine of an adult man is about 23 feet long and an inch in diameter.

The length of the small intestine shows interesting variations in different animals. The intestine is usually very long and much coiled in herbivores, much shorter in carnivores, and of medium length in omnivores like humans. These differences, like those of the teeth, are correlated with the difficulty of digesting plant material because of the cellulose cell walls. Even if the cellulose has been well broken up, it remains mixed with the digestible portions of the cells and tends to mask them from the digestive enzymes.

[4] The adjective "gastric" and the prefix "gastro-" always refer to the stomach.

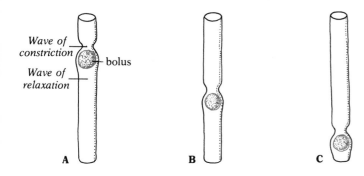

Wave of constriction

Wave of relaxation

bolus

A   B   C

5.30 Peristalsis
The wave of muscular contraction pushes the bolus of food ahead of it.

This interference makes digestion and absorption of plant material much less efficient than the processing of animal material, with the result that a longer intestine is an adaptive advantage in extracting a maximum amount of nutrients from a herbivorous diet. A striking example of adaptation of the small intestine is seen in frogs, where the immature stage, or tadpole, is herbivorous and has a long coiled small intestine, while the adult is carnivorous and has a relatively much shorter one (Fig. 5.31).

Since the small intestine is the place where absorption of the products of digestion occurs, we would expect it to have special structural adaptations that increase its absorptive surface area. Clearly, its great length plays a role here. But examination of its internal surface reveals other modifications that vastly increase its surface area over that of a smooth-walled tube of equal length and girth. First, the mucosa lining the intestine is thrown into numerous folds and ridges (Figs. 5.32 and 5.33). Second, small fingerlike outgrowths, called *villi,* cover the surface of the mucosa. And third, as the electron microscope reveals, the individual epithelial cells covering the folds and villi have what is called, for obvious reasons, a brush border, consisting of countless closely packed cylindrical processes, the *microvilli* (Fig. 5.34). Thus the total internal surface of the small intestine, including folds, villi, and microvilli, is incredibly large.

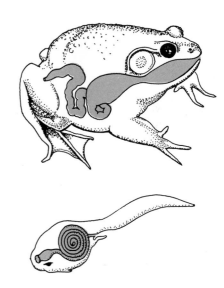

5.31 Intestines of adult frog and tadpole
The much-coiled intestine of the tadpole is far longer relative to the size of the animal than the intestine of the adult frog.

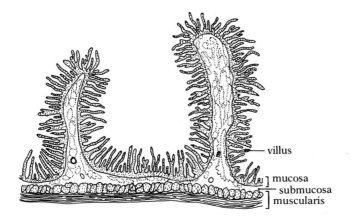

villus

] mucosa
submucosa
] muscularis

5.32 Drawing of cross section through the wall of the human intestine
Two folds of the lining of the intestine are shown, each bearing numerous villi.

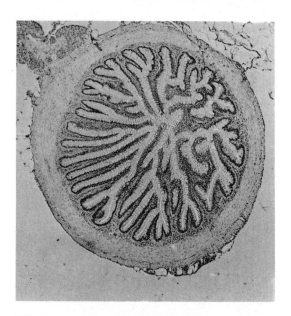

5.33   Cross section of intestine of calico bass, showing extensive folding
[Courtesy Warren Andrew, *Textbook of comparative histology*, Oxford University Press, 1959.]

Some vertebrates show other adaptations, besides those seen in humans, for increasing absorptive surface area. For instance, it is not unusual for special blind sacs, called *caeca*, to branch from the anterior end of the small intestine; in many fish such caeca are present in the pyloric region. Another example is the spiral valve of many primitive fish and of sharks. The spiral valve is an epithelial fold extending the length of the intestine. Like a carpenter's bit tightly enclosed in a tube, it forms a spiral within the intestine, to whose wall its base is attached (Fig. 5.35); hence food cannot move in a straight path, but must follow the spiral of the valve and thus contact much more epithelial surface than it could by moving straight through a tubular intestine of the same length.

The large intestine   In humans the junction between the small intestine and the large intestine (or colon) that follows it is usually in the lower right portion of the abdominal cavity. A blind sac, the *caecum*, projects from the large intestine near the point of juncture (Fig. 5.29). (Notice that most blind diverticula of the digestive tract are called caeca, even though their location and function may vary greatly. Thus the pyloric caeca of fish, mentioned above, are diverticula of the small intestine and often function in absorption, while the human caecum is a diverticulum of the large intestine and does not function in that way.) In humans there is a small fingerlike process, the

5.34   Electron micrograph of microvilli on an epithelial cell of the intestinal lining of a cat
Notice the prominent glycocalyx covering the ends of the microvilli. A bundle of microfilaments (composed of actin) runs into each microvillus. $\times$ 55,000.   [Courtesy Susumu Ito, Harvard Medical School.]

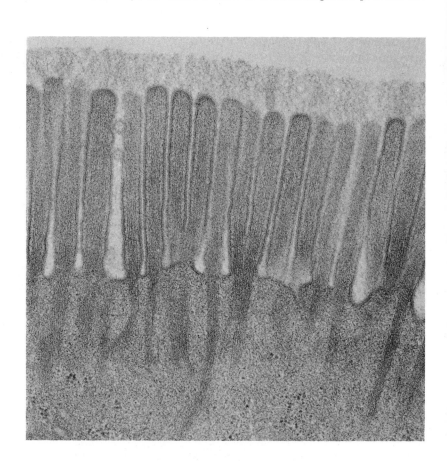

SPECIAL JUNCTIONS BETWEEN EPITHELIAL CELLS

The epithelial tissue lining the digestive tract, which acts as a protective and regulatory barrier between the tissue fluid and blood on one side and the lumen of the digestive tract on the other, determines which substances will be absorbed and which will not. But effective discrimination is possible only if there is no leakage between adjacent cells, only if all substances moving across the tissue are obliged to pass through cell membranes. This condition is met in the epithelium by means of special regions of contact between adjacent cells called *tight junctions,* which function as seals between the cells. Electron microscopy reveals that these are places where the outer layers of the adjacent membranes have fused, and thus obliterated the intercellular space. Tight junctions are not confined to digestive epithelia; they are also characteristic of tissues lining other cavities (e.g. glands and the urinary bladder).

As food passes through the digestive tract, its bulk and the peristaltic contractions cause innumerable transient changes in the shape of the digestive chambers. The strain on the epithelial tissues might tend to pull the cells apart, except for the presence of specially reinforced intercellular junctions called *desmosomes,* which act something like rivets or spot welds between the epithelial cells. There are numerous microfilaments in the junctional region, both in the cytoplasm of the adjoining cells and in the space between them. Desmosomes are likewise found in other parts of the body where tissues are subject to considerable stretching (e.g. skin and heart muscle).

We saw in Chapter 3 that some substances that cannot readily cross the outer or inner membranes of the cells move easily from one epithelial cell to another. This differential is due, in part, to the frequent presence between epithelial cells of special regions called *gap junctions,* where tiny channels run from the cytoplasm of one cell to the cytoplasm of the adjacent cell. Though much smaller and less permanent than the plasmodesmata that often interconnect plant cells, gap junctions nevertheless permit direct movement of ions and some small molecules from cell to cell; they thus facilitate rapid communication between epithelial cells. Gap junctions also occur in some smooth muscles and in certain invertebrate nervous tissue where electrical signaling between cells takes place.

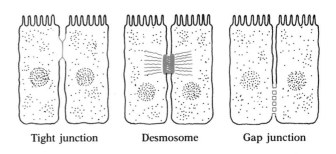

Tight junction Desmosome Gap junction

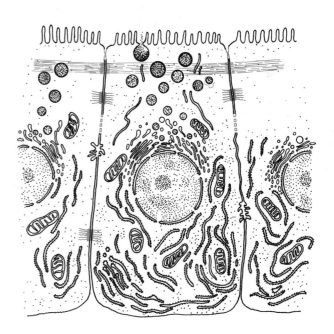

Junctions between epithelial cells
In the three epithelial cells pictured at bottom, all three types of junction (color) can be seen in their normal positions; a band of microfilaments that usually encircles epithelial cells just below the tight junctions is also shown.

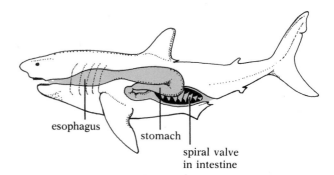

5.35 Digestive system of a shark
In the spiral valve of the intestine, food material must follow a winding course and is thus exposed to more surface area.

esophagus

stomach

spiral valve
in intestine

appendix, at the tip of the caecum. As you know, the appendix often becomes infected and must be surgically removed.

In humans the caecum is small and functionally unimportant, but in some mammals, particularly herbivorous ones, it is large and contains many microorganisms (bacteria and Protozoa) capable of digesting cellulose. Since the mammal cannot itself digest cellulose, it benefits from the action of the microbes. The caecum, however, is not located where the mammal can derive maximal benefit from the microbial action. It is too far back on the digestive tract, behind the small intestine, where most digestion and absorption take place. Thus, even though horses have an enormous caecum, much coarse undigested plant material remains in their feces. A compensating adaptation has evolved in rabbits, which form two types of feces; one of these is material from the caecum, which they promptly re-eat and expose to further digestion and absorption.

Ruminants, such as the cow, also employ microbial digestion, but the microorganisms are not held in a posterior caecum. Instead, such animals have four different stomachlike chambers (Fig. 5.36), of which the first three are thought to be expanded sections of the esophagus. Vast numbers of bacteria and Protozoa live in the *rumen,* which is the largest of the four chambers, and in the reticulum. Swallowed food enters the rumen and reticulum, where the microbes begin digesting and fermenting it, breaking down not only protein, polysaccharides, and fats, but cellulose as well. The larger, coarser material is periodically regurgitated for further chewing; i.e. the animal "chews its cud." This rechewed material is again swallowed and mixed with the fermenting material in the rumen. Slowly the products of the microbial action and the microbes themselves move on into the true stomach (the abomasum) and intestine, where the more usual type of digestion and absorption takes place. Thus, by using microbial digestion in the anterior portion of their digestive tracts rather than in a posterior caecum, ruminants derive maximal benefit from the microbial action.[5] Much less

[5] It is worth noting—though the point has no direct bearing on digestion—that the microorganisms in the rumen can use ammonia and such comparatively simple organic compounds as urea as sources of nitrogen for synthesizing amino acids. On digesting the microorganisms, the ruminant obtains the amino acids, thus benefiting from the nitrogen metabolism of the microbes. Modern agriculture often takes advantage of this microbial metabolism by supplying nitrogen in the form of cheap ammonium salts in cattle feed, rather than in the more expensive form of protein.

undigested plant material remains in their feces than in the feces of horses.

Digestion of cellulose by symbiotic microorganisms is not limited to mammals. A variety of insects, notably the termites, feed on wood, which they could not use as food were it not for intestinal microbes that can ferment the cellulose. A few species of wood-eating beetles do, however, themselves secrete cellulase, an enzyme that digests cellulose, and such beetles do not have to rely on intestinal microbes.

But let us return to the large intestine in humans. From the caecum, the large intestine ascends on the right side to the mid-region of the abdominal cavity, then crosses to the left side, and descends again (Fig. 5.29). The three sections thus formed are frequently termed the ascending, transverse, and descending colons. One of the chief functions of the colon is reabsorption of much of the water used in the digestive process. If all the water in which enzymes are secreted into the digestive tract were lost with the feces, there would be a severe problem of desiccation. Occasionally the intestine becomes irritated, and peristalsis moves material through it too fast for enough water to be reabsorbed; this condition is known as diarrhea. Conversely, if material moves too slowly, too much water is reabsorbed and constipation results. A proper amount of roughage (indigestible material, primarily cellulose) in the diet provides the bulk needed to stimulate enough peristalsis in the large intestine to prevent constipation. A second function of the colon is the excretion of certain salts, such as those of calcium and iron, when their concentration in the blood is too high. The salts are excreted into the colon and are eliminated from the body in the feces. The large intestine also contains large numbers of bacteria, which live on the undigested food that reaches the colon. The significance of these bacteria in the life of a healthy person is not clearly understood. Approximately half the dry weight of the feces is made up of masses of these bacteria.

The last portion of the large intestine, the **rectum,** functions as a storage chamber for the feces until defecation. The feces are eliminated from the rectum through an opening called the anus.

ENZYMATIC DIGESTION IN HUMANS

Digestion by saliva Having traced the human digestive tract from mouth to anus, let us next consider the chemical changes that occur in a meal as it passes through this complex tubular system. We have said that enzymatic digestion starts in the mouth. The saliva contains an enzyme called **amylase**[6] (also known as ptyalin), which begins but does not complete the hydrolysis of starch to glucose. Although amylase produces some glucose, it yields primarily the double sugar maltose (Fig. 5.37) and, in lesser amounts, fragments three or four glucose units long, which must be further digested in the intestine.

Since the food remains in the mouth only a short time, the amylase has little opportunity to work there. Much of its action occurs inside each bolus after it is swallowed into the stomach. The acid of the stomach soon inac-

[6] Note that the names of most enzymes end with the suffix *-ase,* which designates enzymes by international agreement. The first part of the name usually indicates the substrate upon which the enzyme acts; thus *amyl-* (from *amylum,* the Latin for starch) indicates that amylase acts upon starch.

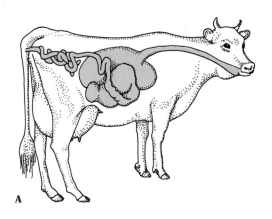

A

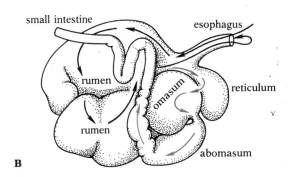

small intestine

esophagus

rumen

omasum

reticulum

rumen

abomasum

B

5.36 The digestive system of a ruminant
(A) A cow with the various chambers in approximately their correct locations.   (B) Detailed path of food as it passes through the four "stomachs." Food moves initially (black arrows) into the rumen and reticulum (the drawing shows it going only into the rumen), where it is fermented by microorganisms. Then, as the cud, it is regurgitated for more chewing and later reswallowed. In this second phase (color arrows), fluids and finely divided particles move through the reticulum into the omasum and then into the abomasum, which is the true stomach.

tivates the enzyme, however, and salivary amylase actually digests only a small percentage of the starch in the food. In fact, the saliva of many mammals contains no amylase at all; dogs are an example. Can you think of reasons why there would have been little selection pressure for the evolution of such an enzyme in dogs and their relatives?

Digestion in the stomach   Once in the stomach, food is exposed to the action of gastric juice secreted by numerous gastric glands of the mucosa of the stomach wall. This juice contains much hydrochloric acid and several enzymes. The acid makes the contents of the stomach very acidic (pH of about 1.5–2.5). Note that, advertisements for many patent medicines to the contrary, an acid stomach is both normal and necessary for proper function.

The principal enzyme of the gastric juice is ***pepsin,*** which digests protein. Unlike most protein-digesting enzymes, it will function only in a strongly acid medium. Pepsin is a characteristic enzyme of vertebrates; most invertebrates do not have any proteolytic enzymes active in strongly acid solutions. It has been hypothesized that the evolution of pepsin is correlated with feeding on animals with bones, since bones disintegrate more easily in acid.

Pepsin does not hydrolyze protein all the way to its amino acid components. It splits the peptide bonds adjacent to only a few amino acids, particularly tyrosine and phenylalanine (Fig. 5.38). The specificity of proteolytic enzymes is readily understandable. Since proteins are composed of a variety of building-block compounds, not just of one, the structural configuration around the various peptide bonds varies, depending on which two amino acids the bond joins; some of the bonds may fit in the active site of a particular enzyme and others may not. Pepsin, for example, seems to have an active site complementary to peptide bonds at the amino end of amino acids whose R groups include a six-carbon ring (see Fig. 2.34, p. 54).

Any discussion of protein digestion immediately raises the question: Why isn't the wall of the digestive tract digested by the proteolytic enzymes? After

5.37   Digestion of starch
Amylase in the saliva and in the pancreatic juice hydrolyzes some of the bonds between glucose units, producing small amounts of free glucose, but much larger quantities of the double sugar maltose. The maltose is then digested to glucose by maltase secreted by intestinal glands.

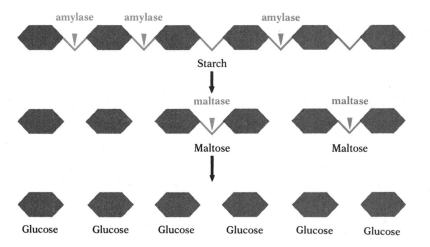

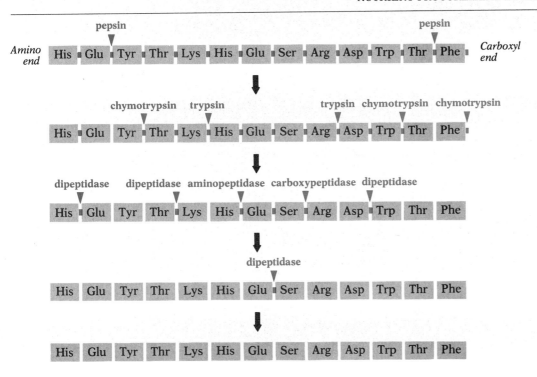

all, the walls of both stomach and intestine are composed primarily of protein. Two reasons can be given.

First, the wall of the digestive tract is covered with mucus, which apparently shields it from enzymes. Sometimes, however, this defense breaks down, and the digestive enzymes begin to eat away a small portion of the lining; the resulting sore is known as an ulcer. Occasionally, an ulcer is so severe that a hole develops in the wall of the digestive tract, and the contents of the tract spill into the abdominal cavity. Why and how ulcers first develop is not clear; nor is it fully understood why the wall is not normally digested by the proteolytic enzymes. The presence of sufficient mucus may not be the whole story; much research remains to be done.

Second, the gastric glands secrete, not active pepsin, but pepsinogen, an inactive enzyme precursor known as a *zymogen.* Pepsinogen has no proteolytic activity and, so long as it is stored in the glands of the stomach wall, poses no threat to the wall. It is changed into active pepsin only after exposure to free pepsin under the very acid conditions in the lumen of the stomach. Activation results from the splitting off of 42 amino acids from one end of the pepsinogen molecule; the shorter polypeptide chain that remains is pepsin.[7]

Digestion in the small intestine   It is in the next section of the digestive tract, the small intestine, that by far the most digestion takes place. When partially digested food passes from the stomach into the duodenum, its

5.38   Digestion of protein
Pepsin in the stomach hydrolyzes peptide bonds at the amino end of tyrosine (Tyr) and phenylalanine (Phe). Then the food moves into the intestine, where trypsin and chymotrypsin from the pancreas hydrolyze bonds at the carboxyl end of lysine and arginine and of tyrosine, tryptophan, and phenylalanine, respectively (chymotrypsin also hydrolyzes bonds adjacent to leucine and methionine when they are present). Pepsin, trypsin, and chymotrypsin hydrolyze only internal bonds, not bonds attaching terminal amino acids to the chains. Terminal bonds at the amino end of chains may be split by aminopeptidase and those at the carboxyl end by carboxypeptidase. Bonds between pairs of amino acids are split by dipeptidases, whereupon digestion is completed.

---

[7] In addition to pepsin, the gastric juice of suckling ruminants contains another enzyme, rennin, which clumps milk proteins together. Many texts imply erroneously that rennin occurs in humans.

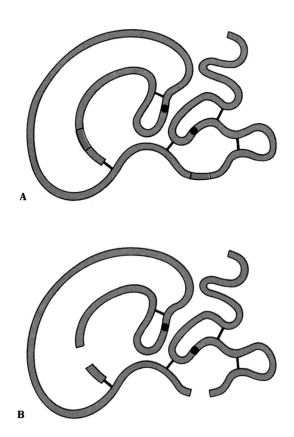

A

B

5.39 Conversion of chymotrypsinogen into chymotrypsin

(A) The single polypeptide chain constituting a molecule of the enzyme precursor chymotrypsinogen is stabilized by disulfide bonds (black bars). The two portions of the polypeptide chain removed to produce the active enzyme chymotrypsin are shown in color. (B) The disulfide bonds hold together the three polypeptide chains constituting a molecule of chymotrypsin. The black circles indicate important parts of the catalytic site of the enzyme.

acidity stimulates the release of a large number of different digestive enzymes into the lumen of the intestine. These enzymes are secreted from two principal sources, the *pancreas* and the *intestinal glands.* The pancreas is a large glandular organ, lying just below the stomach (Fig. 5.29), which originates in the embryo as an outgrowth of the digestive tract; it retains a connection to the duodenum called the *pancreatic duct.* When food enters the duodenum, the pancreas secretes a mixture of enzymes that flows through the pancreatic duct into the duodenum. Included in this mixture are enzymes that digest all three principal classes of foods—carbohydrates, fats, and proteins—as well as some that digest nucleic acids.

One of the pancreatic enzymes is *pancreatic amylase* (sometimes also called diastase or amylopsin), which, as its name implies, acts like salivary amylase, splitting starch into the double sugar maltose. It is far more important than salivary amylase, for it carries out most of the starch digestion.

Lipase, secreted by the pancreas, is the body's principal fat-digesting enzyme, but it completely hydrolyzes only a relatively small percentage of the fat to glycerol and fatty acids. Some of the fat is partly digested (e.g. by removal of only one of the three fatty acids), and some is not digested at all. But since fats, and the products of the partial digestion of fats, are lipid-soluble, they can be absorbed across cell membranes.

Like pepsin, *trypsin* and *chymotrypsin,* two of the proteolytic enzymes of the pancreas, cleave only the peptide linkages adjacent to certain specific amino acids (Fig. 5.38). Trypsin splits the peptide bonds adjacent to lysine and arginine (both of which have positively charged R groups; see Fig. 2.34); chymotrypsin splits those adjacent to tyrosine, phenylalanine, and tryptophan (which have six-carbon rings in their R groups), and, to a lesser extent, methionine and leucine. Notice that chymotrypsin resembles pepsin in hydrolyzing bonds adjacent to tyrosine and phenylalanine. But this resemblance is superficial: The two enzymes do not split the same bonds. Pepsin cleaves the bonds on the amino side of tyrosine and phenylalanine, while chymotrypsin cleaves those on the carboxyl side. This specificity, far from being unusual, is typical of most enzymes. Actually, the digestive enzymes are less specific than many others in that they will catalyze the reactions of several different substrates.

Again like pepsin, both trypsin and chymotrypsin are secreted in inactive (zymogen) forms, known as trypsinogen and chymotrypsinogen respectively. Trypsinogen is converted into active trypsin (through the removal of the terminal six amino acids from its polypeptide chain) in the intestine by enterokinase, an enzyme secreted by intestinal glands. Chymotrypsinogen is converted into active chymotrypsin by the trypsin thus formed through removal of two small internal pieces of the chymotrypsinogen polypeptide chain; the remaining molecule (chymotrypsin) consists of three separate polypeptide chains held together by disulfide bonds (Fig. 5.39). Note that all three zymogens we have discussed—pepsinogen, trypsinogen, and chymotrypsinogen—have polypeptide chains longer than those of the active enzymes; in each case activation simply involves cutting off one or more pieces of the chain.

Recent studies of the chemical structure of trypsin and chymotrypsin have helped clarify both their probable evolution and the basis for their catalytic specificity. These two enzymes are sufficiently alike—in amino acid sequence, in conformational folding pattern, and even in their catalytic sites—to make it reasonable to believe that they evolved from the same ancestral proteolytic enzyme. There is, however, a slight, but functionally crucial difference between them, which pertains to a pocketlike portion of the binding site that holds the R group of the substrate amino acid. In trypsin one of the amino acids that form this pocket is aspartic acid (which has a negatively charged R group); in chymotrypsin the same position is occupied by serine (which has an uncharged R group). This seemingly small variation accounts for the affinity of the two enzymes for different amino acids. That it should do so is a dramatic illustration of the critical importance of amino acid sequence (primary structure) in determining the functional properties of enzymes.

In summary, then, the action of pepsin in the stomach and of trypsin and chymotrypsin from the pancreas results in a splitting of proteins into fragments of varying lengths, but does not produce many free amino acids. These three enzymes are known as *endopeptidases,* i.e. enzymes that hydrolyze peptide bonds between amino acids located within the protein, not bonds linking terminal amino acids to the chain. Another class of enzymes, called *exopeptidases,* hydrolyze off the terminal amino acids, thereby completing the digestive process. There is a great variety of exopeptidases, each highly specific in its action. One, for example (carboxypeptidase), hydrolyzes the linkage binding the terminal amino acid at the free carboxyl end of the chain. Another (aminopeptidase) hydrolyzes the linkage of the terminal amino acid at the free amino end of the chain. Still others (dipeptidases) break apart pairs of amino acids; one breaks only the bond of a fragment consisting of glycine linked to leucine, another only the bond of a fragment consisting of two molecules of glycine linked together, and so on. It is these exopeptidases that complete the job begun by the endopeptidases. Most of them are secreted by intestinal glands, but carboxypeptidase is produced in the pancreas.

The chemical action of the various proteolytic enzymes has been described at some length, not because it is important for you to remember in detail what bonds each enzyme hydrolyzes, but because the proteolytic mechanism provides a good example of enzyme specificity and of the way enzymes often work in teams. If we had simply said that proteins are digested by pepsin, trypsin, and a variety of other peptidases, you would have had no indication of the elaborate interplay that characterizes these seemingly commonplace processes.

Just as certain enzymes from the intestinal glands complete the digestion of protein, other intestinal enzymes complete the digestion of carbohydrate begun by salivary and pancreatic amylase. These enzymes split double sugars into simple sugars. For example, *maltase* splits maltose (Fig. 5.37), *sucrase* splits sucrose, and *lactase* splits lactose.

Absorption of the products of digestion involves active transport, as you might expect. Recent research on the "pumps" for amino acids and simple sugars has shown that both depend on a steep sodium ion (Na^+) gradient across the cell membrane, with a high concentration of Na^+ outside the cell and a low concentration inside. These active-transport pumps thus appear to be coupled in some way to the sodium-potassium pump, which we shall discuss at length in Chapter 8.

Digestive capabilities vary not only among species (we have already mentioned the absence of salivary amylase in dogs), but also within a species. The digestion of lactose, a sugar found only in milk, provides a striking example.

Secretion of milk by the mammary glands of female mammals evolved as a way of feeding the young. The only food provided very young mammals, milk is a nearly complete food, containing, in most species, carbohydrate (in the form of lactose),* fat, and protein, as well as important minerals. But, except for humans, adult mammals do not use milk as a food. It is not surprising, then, that secretion of lactase, the lactose-digesting enzyme, usually greatly diminishes or even ceases altogether once an animal is past the age of weaning.

It has only recently been realized, however, that this pattern applies also to most human beings; in most parts of the world, humans more than four years old secrete little or no lactase. Indeed, of the various peoples studied to date, only those of European ancestry and those belonging to a few pastoral tribes in Africa have been found to secrete enough of the enzyme to be able to digest the lactose in large quantities of milk (see graph). When people of other ancestries drink more than modest quantities they will often become ill, getting a bloated feeling, cramps, and diarrhea. One reason is that the undigested sugar in the intestine upsets the normal osmotic balance to the point where an excessive amount of water moves into the intestinal lumen from the cells; another is that fermentation of the lactose by bacteria in the large intestine produces large quantities of acids and carbon dioxide. The lactose tolerance (i.e. continued production of lactase in adults) of Europeans and pastoral Africans must have evolved during the roughly 10,000 years since the milking of domestic animals began.

How widely peoples living near one another may differ is shown by the major tribes of Nigeria. The Ibo and Yoruba live in the southern part of the country, where conditions are unfavorable for cattle; milk has not traditionally been a part of their diet after weaning, and they cannot tolerate lactose. By contrast, the nomadic Fulani in north-

ern Nigeria have been raising milk cattle for thousands of years, and they are lactose-tolerant (see graph).

Most American blacks are descended from nonpastoral tribes of western Africa, and they are relatively intolerant of lactose, though not so much as native Africans. Their somewhat greater tolerance may be due, in part, to evolutionary change during the generations they have lived in dairying regions and, in part, to admixture of European genes.

In view of the widespread lactose intolerance in most underdeveloped countries, it has become clear that the former large-scale distribution of powdered milk in these countries was ill-advised. If milk is sent, it should be as powder from which the lactose has been removed or as milk products such as yogurt or cheese in which the lactose has been broken down by microbial action.

*There is no lactose in the milk of seals and their close relatives (the Pinnipedia).

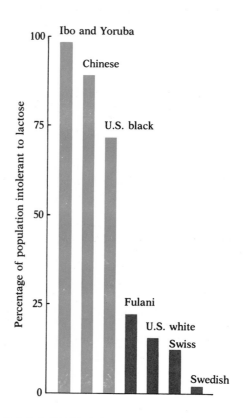

Bile One more secretion should be mentioned in this discussion of human digestion. The *liver,* a critically important organ about which much will be said in later chapters, produces a fluid called bile, which aids in fat digestion. The liver is a very large organ occupying much of the space in the upper part of the abdomen. On its surface is a small storage organ, the *gallbladder* (Fig. 5.29). Bile, produced throughout the liver, is collected by a series of branching ducts and emptied into the gallbladder. When food enters the duodenum, the muscular wall of the gallbladder is stimulated to contract, and the bile is forced down the bile duct into the duodenum.

Bile is not a digestive enzyme; it is not even a protein. It is a complex solution of bile salts, bile pigments, and cholesterol. The bile salts act as emulsifying agents, causing large fat droplets to be broken up into many tiny droplets suspended in water. This action is much like that of a good detergent. The many small fat droplets expose much more surface area to the digestive action of lipase than a few large droplets would. Bile salts apparently aid also in the absorption of fats. When insufficient bile salts are present in the intestine, both fat digestion and absorption are seriously impaired. The bile salts are reabsorbed by the large intestine, transported back to the liver, and used again.

The bile pigments and cholesterol play no perceptible role in digestion. The pigments are produced through the destruction of red blood cells in the liver; it is they that give the characteristic brown color to feces. The cholesterol, a relatively insoluble compound, sometimes causes trouble by becoming concentrated into hard gallstones, which may block the bile duct and interfere with the flow of bile.

SUGGESTED READING

CARLSON, A. J., V. JOHNSON, and H. M. CAVERT, 1961. *The machinery of the body,* 5th ed. University of Chicago Press, Chicago. *Very clearly written text on human physiology (see esp. Chapters 7–8).*

DAVENPORT, H. W., 1972. "Why the stomach does not digest itself," *Scientific American,* January. (Offprint 1240)

DEDUVE, C., 1963. "The lysosome," *Scientific American,* May. (Offprint 156) *Report by the discoverer of lysosomes, who has since been awarded the Nobel Prize for his work in cell biology.*

EPSTEIN, E., 1973. "Roots," *Scientific American,* May. (Offprint 1271) *The mechanisms by which roots take up nutrients from the soil.*

GALSTON, A. W., 1964. *The life of the green plant,* 2nd ed. Prentice-Hall, Englewood Cliffs, N.J. *Good short book on plant physiology (see esp. Chapter 3).**

HARPSTEAD, D. D., 1971. "High-lysine corn," *Scientific American,* August. (Offprint 1229) *On the fascinating attempt, still under way, to breed a variety of corn richer in essential amino acids, which would be of enormous benefit wherever corn is the principal staple in the diet.*

HESLOP-HARRISON, Y., 1978. "Carnivorous plants," *Scientific American,* February.

JARNICK, J., C. H. NOLLER, and C. I. RHYKERD, 1976. "The cycles of plant and animal nutrition," *Scientific American,* September. *The movement of important nutrients through plants, animals, and the nonliving environment; some agricultural practices that manipulate this movement.*

*Available in paperback.

KRETCHMER, N., 1972. "Lactose and lactase," *Scientific American,* October. (Offprint 1259) *The differences in human tolerance of milk sugar.*

PRAMER, D., 1964. "Nematode-trapping fungi," *Science,* vol. 144, pp. 382–388.

RAY, P. M., 1972. *The living plant,* 2nd ed. Holt, Rinehart & Winston, New York. *Short elementary text on plant physiology (see esp. Chapter 6).**

SCHMIDT-NIELSEN, K., 1970. *Animal physiology,* 3rd ed. Prentice-Hall, Englewood Cliffs, N.J. *Well-written elementary text on animal physiology, stressing the comparative approach (see esp. Chapter 1).**

SCRIMSHAW, N. S., and V. R. YOUNG, 1976. "The requirements of human nutrition," *Scientific American,* September. *A useful tabulation, but contains a number of items for which the evidence is weak.*

STAEHELIN, L. A., and B. E. HULL, 1978. "Junctions between living cells," *Scientific American,* May. (Offprint 1388) *On tight junctions, desmosomes, and gap junctions.*

STEWARD, F. C., 1964. *Plants at work.* Addison-Wesley, Reading, Mass *Clearly written elementary treatment of plant physiology. Especially good section on plant nutrition (see Chapters 4, 9).**

YOUNG, V. R., and N. S. SCRIMSHAW, 1971. "The physiology of starvation," *Scientific American,* October. (Offprint 1232) *On the remarkable biochemical responses of a starving person's body, which permit relatively long-continued minimal nutrition of the most important organs, especially the brain.*

*Available in paperback.

Chapter 6

GAS EXCHANGE

We have seen that when nutrient compounds are broken down in the process of respiration, over 90 percent of the energy yield depends on the presence of oxygen, which makes possible the complete oxidation of the compounds to carbon dioxide and water. Thus a basic problem for the great majority of living organisms is the procurement of oxygen and the elimination of carbon dioxide.

It is true that a few unicellular organisms—notably the anaerobic microbes mentioned in Chapter 4—can subsist indefinitely in the total absence of oxygen, exposure to it even being fatal to some. Other organisms can survive for limited periods under anaerobic conditions. But in these cases the respiratory degradation of nutrients stops far short of completion, the end products usually being lactic acid or ethyl alcohol, relatively large molecules that still contain much chemical energy. Anaerobic respiration thus appears very wasteful in comparison with aerobic respiration. Nevertheless, its importance in some cases should not be overlooked. By resorting to anaerobic metabolism, an organism (or part of an organism) may survive short periods of oxygen deprivation. For organisms capable of living anaerobically indefinitely, environments otherwise totally uninhabitable are open for colonization; many bacteria, for example, are anaerobic and live in habitats where no other type of organism could survive. Nonetheless, aerobic rather than anaerobic respiration is the chief method of respiration in both plants and animals.

It is a common misconception that oxygen procurement is a problem faced only by animals, and that gas exchange in green plants consists exclusively of intake of carbon dioxide and release of oxygen. The latter is the exchange that takes place in association with photosynthesis, but the carbohydrate products of photosynthesis are of little value to the plant if they cannot be respired to provide usable metabolic energy. Thus plants, like animals, are constantly taking in oxygen and releasing carbon dioxide as they carry out the process of cellular respiration. Both photosynthetic and respiratory gas exchange are usually taking place when a green plant is exposed to bright light; since the rate of photosynthesis then greatly exceeds the rate of respiration, the *net* effect is one of uptake of carbon dioxide and release of oxygen. The reverse is true, of course, when the green plant is in the dark or when it has no leaves in winter. Thus respiratory gas exchange is a requirement for plants as much as it is for animals.

THE PROBLEM

Gas exchange between a living cell and its environment always takes place by diffusion across a moist cell membrane. The gases must be in solution if they are to move across the membrane. In unicellular organisms and many small multicellular ones, particularly those that are aquatic, this requirement poses no serious problem, because each cell is either in direct contact with the surrounding medium or only a few cells removed from that medium. Thus these organisms have usually not evolved special respiratory devices.

When large body size is predominantly in two dimensions, gas exchange can still take place chiefly by direct diffusion between the individual cells and the surrounding medium. Some brown algae, the kelps, may grow to a length of 60–90 m, but the blades of even the longest kelps remain very thin (Fig. 6.1). As a result, no cell is far from the surface, and the total gas-exchange area is fairly large in relation to the volume of the plant. The thicker stipe has numerous intercellular spaces filled with water that is continuous with the external medium. In these plants, then, which are large in only two dimensions, no special gas-exchange mechanism is needed.

When increase in body size involves three dimensions, as it generally does, the maintenance of a respiratory surface of adequate dimensions relative to the volume becomes a problem, because area (a square function) increases much more slowly than volume (a cube function). The problem is most acute for the more active animals, whose rapid utilization of energy demands a large amount of oxygen per unit of body volume per unit time.

An additional complicating factor is that many organisms have evolved relatively impermeable outer body coverings. Coverings such as the waxy epidermis of the leaves of terrestrial plants, or animal skin with its derivative scales, feathers, and hair, function as protective barriers between the fragile internal tissues and organs and the often hostile outer environment, but their presence, which demands a gas-exchange surface confined to a restricted region of the body, makes the problem of adequate exchange area even more critical.

Another complication brought on by large three-dimensional size in animals is that many cells are deep within the body of the organism, far from the gas-exchange surface. Diffusion alone is incapable of moving gases in adequate concentrations across the immense number of cells that may intervene between these more distant cells and the exchange surface. In general, simple diffusion suffices for movement of substances through aqueous media only when the distances are less than one millimeter. Thus some other

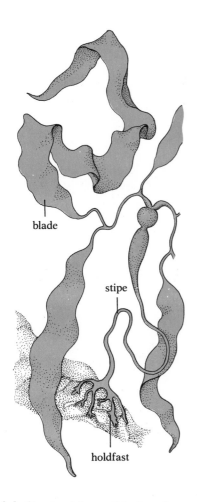

6.1   A kelp *(Laminaria),* one of the large brown algae

blade

stipe

holdfast

mechanism for conveying gases to the individual cells of the organism becomes essential.

The need for direct contact between the moist membranes across which gas exchange occurs and the environmental medium also poses serious difficulties, especially for terrestrial organisms. The moist membranes must be exposed to the environment, but they must be exposed in such a way as to minimize the chances of desiccation. Since a large, thin, moist surface is often fragile and easily suffers mechanical damage, the tendency has been toward the evolution of protective devices, particularly when the respiratory surfaces are evaginated ones.

In general, specialized respiratory surfaces may be grouped in two categories: inward-oriented and outward-oriented extensions of the body surface (see Fig. 6.8). Each category embraces a diversity of form and detail, but the diversities become less bewildering if one bears in mind that each type of respiratory system represents merely one way of meeting the basic needs discussed above: (1) a respiratory surface of adequate dimensions; (2) for many organisms, methods of transporting gases between the area of exchange with the environment and the more internal cells; (3) means of protecting the fragile respiratory surface from mechanical injury; and (4) means of keeping the surface moist.

SOLUTIONS IN TERRESTRIAL PLANTS

As indicated earlier, most primitively aquatic plants, notably the algae, carry out gas exchange across almost the entire body surface; but in response to the problems of desiccation and large three-dimensional size, most terrestrial plants have evolved more elaborate mechanisms.

LEAVES

Gas exchange associated with both photosynthesis and cellular respiration takes place at a particularly high rate in green leaves, organs strikingly adapted for this process.

Recall that most of the visible outer surface of a leaf, covered as it is by a waxy cuticle, is more or less dry and impermeable, and hence ill suited for diffusion of gases. Exchange must therefore take place elsewhere. You will remember that the mesophyll parenchyma in a leaf is so arranged as to leave large intercellular spaces (Figs. 6.2, 6.3). A high percentage of the total surface of each mesophyll cell is exposed to the air in these spaces, which are interconnected and continuous with the external atmosphere by way of openings in the epidermis: the *stomata.* Gases can thus move easily between the surrounding atmosphere and the internal spaces of the leaf. The actual gas exchange—the diffusion of gases into and out of living cells—takes place across the thin moist membranes of the cells inside the leaf.

Let us briefly consider how the structures of the leaf help it meet the four previously stated requirements for respiratory systems.

1. The surface area available for gas exchange in the leaf is very large. By comparison with the outer area of the leaf, the total area of cell membrane exposed to the intercellular spaces is enormous. The principle involved is a very elementary one: A chamber irregularly shaped and greatly subdivided

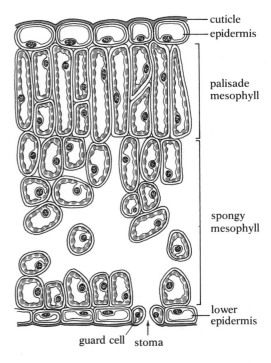

cuticle
epidermis

palisade mesophyll

spongy mesophyll

lower epidermis

guard cell   stoma

6.2   Cross section of part of a leaf of privet hedge

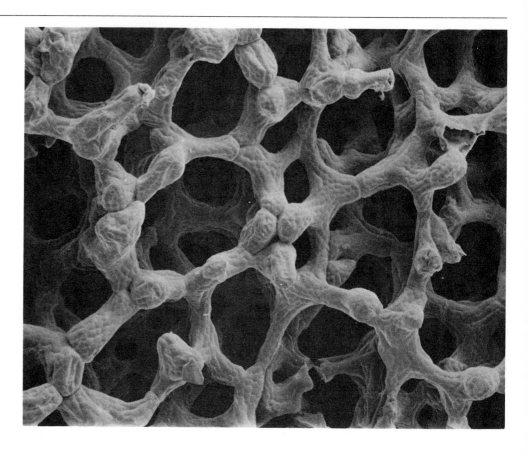

6.3 Scanning electron micrograph of spongy mesophyll in a bean leaf
The micrograph gives an especially clear view of the extensive system of interconnecting air spaces. × 375. [Courtesy J. H. Troughton, Department of Scientific and Industrial Research, Lower Hutt, New Zealand.]

by partial partitions will have far more wall space than a round or square one of equal volume.

2. Internal transport of gases occurs in leaves without any special adaptations. Gases can reach each individual cell directly via the intercellular spaces.

3. The danger of mechanical injury is relatively minor for an internal exchange surface. The epidermis with its various hairs, spines, and other derivatives functions as a protective covering for the entire leaf.

4. The exchange surfaces remain moist because they are exposed to air only in intercellular spaces. With the humidity within those spaces nearly 100 percent, the membranes of the mesophyll cells always retain a thin film of water on their surfaces. Gases dissolve in this water before moving into the cells. The protective epidermal tissues and the layers of waxy cuticle on their outer surfaces act as barriers between the dry outside air and the moist inside air.

But these barriers are not complete; if they were, movement of gases between the outside and inside could not take place. Thus, although openings are essential, the stomata in a sense constitute weak links in the protective armor of the leaf. Here we see the sort of compromise that has been a constant feature of evolutionary adaptation. Few characters, however beneficial, are without possible deleterious effects. What determines the evolu-

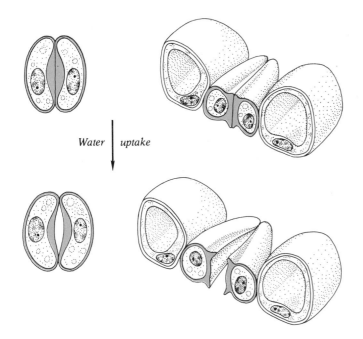

Water | uptake

6.4 Guard cells
Left: View from below. Right: Median transverse
section. With increasing osmotic concentration, the
guard cells take up water; the outer wall, which is
thinner than the inner one, starts to bulge, the cells pull
apart, and the stoma opens.

tionary fate of a character is not whether it is exclusively beneficial or
harmful, but whether or not the beneficial effects outweigh the harmful
ones. In this case, stomata present advantages that outweigh the danger of
desiccation; moreover, other adaptations minimize the danger.

Each opening, or stoma, in the epidermis is bounded by two highly specialized epidermal cells called *guard cells,* which, unlike most other epidermal cells, contain chloroplasts (Fig. 6.2). These bean-shaped cells have walls
of unequal thickness; the walls next to the stoma are considerably thicker
than those on the side away from the stoma (Fig. 6.4). When the guard cells
contain much fluid and are turgid (usually in the light), the thin outer wall of
each cell buckles outward, pulling the rest of the cell with it and opening the
stoma (Fig. 6.5). Gas exchange can now take place, and the leaf can obtain
carbon dioxide for photosynthesis. In the dark the reverse usually occurs:
The cells lose water, and, as they become flaccid, their thick inner walls close
the stoma.

That the stomata should be open during the day, when the air is usually
drier, and ordinarily closed at night, when the air is often damper, may seem
an inefficient way to combat desiccation. The fact is that the requirements of
water conservation conflict with those of photosynthesis. Photosynthesis,
without which the plant would perish, requires both light and carbon dioxide. For the plant to perform photosynthesis at a high rate, it must take in
CO_2 during the day, and for this purpose the stomata must be open. Imperfect as the regulation of the stomata may be from the point of view of water
conservation, it is in fact remarkably good; the rate of water loss through the
stomata is much lower than the rate of CO_2 uptake.

Although most plants lose large quantities of water by evaporation
through the stomata every day in the process called *transpiration,* the humidity in the intercellular spaces of the leaf probably does not often drop

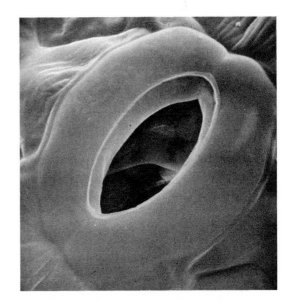

**6.5 Scanning electron micrograph of a stoma in a
cucumber leaf**
The stoma is open, because the two large
crescent-shaped guard cells have pulled away from
each other. Mesophyll cells in the interior of the leaf
can be glimpsed through the stoma. × 2,800.
[Courtesy J. H. Troughton, Department of Scientific
and Industrial Research, Lower Hutt, New Zealand.]

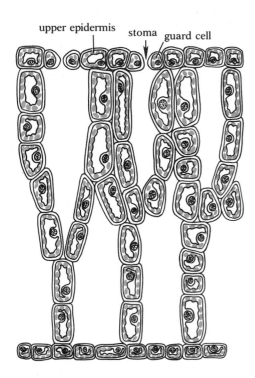

6.6   Cross section of portion of a floating leaf of pondweed (Potamogeton)
The stomata are in the upper surface, and the mesophyll has very large intercellular air spaces. There are chloroplasts in the epidermal cells.

appreciably, because the lost water is steadily replaced by water drawn up through the stem and distributed throughout the leaf by the many small veins. Indeed, the evolution of an effective transport system has involved the harnessing of the process of transpiration. Transpiration also serves as a cooling mechanism, preventing excessive buildup of heat inside the leaves even when they are exposed to the direct rays of the sun. Besides, there is a safety valve in the system. If the loss of water by transpiration is too rapid and exceeds the capacity of the plant to supply more to the leaves, the leaf cells, including the guard cells, may become flaccid (i.e. wilted), whereupon the stomata close, preventing further water loss. This closing of the stomata as a result of wilt has the effect, however, of limiting the supply of CO_2 to that produced by cellular respiration; the rate of photosynthesis is therefore reduced until it is approximately equal to the rate of respiration. Respiration, in turn, is limited to the amount of oxygen released as a by-product of photosynthesis.

How the rapid changes in guard-cell turgidity are effected is still a matter of dispute. One hypothesis takes its cue from the unusual starch metabolism of guard cells. By contrast with most cells, which convert starch into sugar mainly in the dark, guard cells do so in the light. It is suggested that the enzyme responsible for converting starch into sugar in the guard cells becomes more active in response to the decreased acidity due to consumption of CO_2 in photosynthesis. Thus conversion of stored starch into sugar would supplement photosynthetic production of sugar and bring about a rapid increase in the osmotic concentration of the guard-cell cytoplasm. Water would then tend to move into the guard cells by osmosis, causing them to swell and open the stoma.

Although the pH changes, the starch-sugar conversions, and the changes in sugar concentration unquestionably occur, and no doubt play a role in guard-cell movement, this hypothesis does not fully explain the extremely rapid movements of guard cells often observed. A more recent hypothesis posits an additional factor that causes the guard cells to take up water: an ATP-driven pump that alters the osmotic balance between the cells and the surrounding fluid by moving potassium ions (K^+) into the guard cells. It has been demonstrated that there is indeed more K^+ in the guard cells of open than of closed stomata, and that rapid guard-cell movement can be inhibited by chemicals known to poison processes dependent on ATP. There is also evidence, at least under experimental conditions, that abscisic acid, a plant hormone, can induce stomatal closing. Obviously, much more research is needed into the surprisingly complex process by which a plant regulates the size of the stomatal openings.

Before leaving the subject of stomata, let us note several other evolutionary adaptations associated with them. In most plants the stomata are located primarily, if not exclusively, in the lower epidermis—the side of the leaf usually turned away from the sun's rays, i.e. the side where the drying tendency is less severe. Further, the lower epidermis of many plants is covered with short hairs, which function in reducing direct air currents across the stomatal openings. Plant species adapted for life in particularly dry or particularly wet habitats often exhibit special adaptations of the stomata. In the oleander (Nerium), which lives in a very dry habitat, the stomata are located in deep hair-lined depressions in the lower epidermis, an

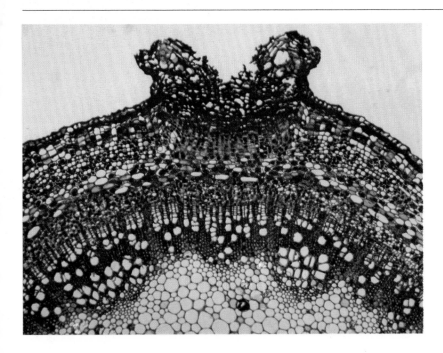

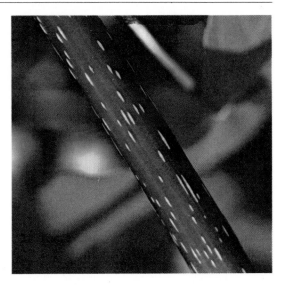

6.7 Lenticels
Left: The waterproofed outer bark (layer of dark cells on the surface of the stem) in this section of elderberry *(Sambucus)* stem is interrupted at the center of the lenticel. Thus the more loosely arranged cell layers beneath, with their numerous intercellular air spaces, are exposed to the atmosphere. × 54. Right: The individual lenticels can be seen as white areas on the surface of a young sycamore maple *(Acer)* stem. [Courtesy Thomas Eisner, Cornell University (left); G. R. Roberts (right).]

adaptation that eliminates convection currents across the stomata. In the pondweed *Potamogeton*, the stomata of the floating leaves are located in the very thin upper epidermis and not in the lower epidermis (Fig. 6.6). The adaptive significance of this arrangement is obvious.

STEMS AND ROOTS

Do not infer from this discussion that leaves constitute the only gas-exchange areas in plants. We have simply chosen to examine these beautifully adapted organs in some detail in order to illustrate the types of problems and solutions characteristic of gas-exchange mechanisms.

The relatively impervious layer of bark on many old stems would effectively cut off most of their oxygen supply were it not for the development of numerous small areas of loosely arranged cells with many intercellular air spaces between them through which gases can move freely. Each such loose group of cells is called a ***lenticel*** (Fig. 6.7). The spongy streaks always seen in cork (an outer bark) are traces of the lenticels of the cork oak. Since most of the cells in the inner layers of large stems are dead, there is little need for oxygen in the intercellular air spaces to penetrate deep into the stem.

Roots also carry out gas exchange, though they usually possess no special structures for this function. Gases can diffuse readily across the moist membranes of root hairs and other epidermal cells. For roots to obtain enough oxygen, however, the soil in which they grow must be well aerated. Different types of soils vary in their aeration characteristics, which depend on the amount of pore space, the affinity of the soil particles for water, and many other factors. Soils with very high percentages of clay particles, for example, have many pore spaces, but the tiny clay particles are so hydrophilic and absorb so much water during wet periods that the air spaces

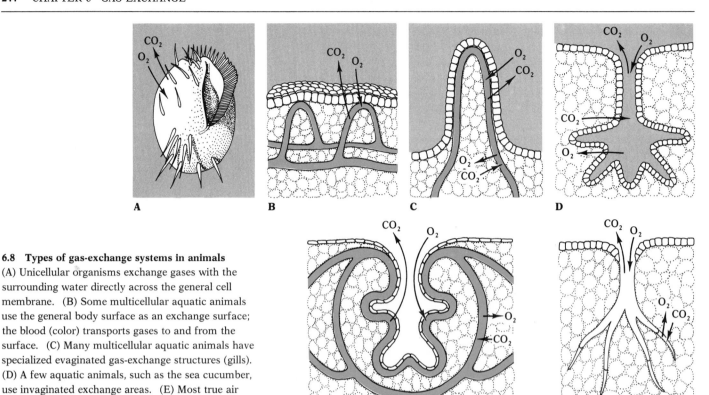

6.8   Types of gas-exchange systems in animals
(A) Unicellular organisms exchange gases with the surrounding water directly across the general cell membrane.   (B) Some multicellular aquatic animals use the general body surface as an exchange surface; the blood (color) transports gases to and from the surface.   (C) Many multicellular aquatic animals have specialized evaginated gas-exchange structures (gills).   (D) A few aquatic animals, such as the sea cucumber, use invaginated exchange areas.   (E) Most true air breathers have lungs, specialized invaginated areas that depend on a blood transport system.   (F) Land arthropods have tracheal systems, invaginated tubes that carry air directly to the tissues without the intervention of a blood transport system.

become filled with water, a condition known as waterlogging. Frequent waterlogging may reduce the air content of the soil to the point where many species of plants are stunted or cannot grow at all. In some poorly aerated soils, the circulation of gases between the pore spaces and the atmosphere above the soil is so slow that the air in the pore spaces becomes deficient in oxygen and often contains a deleteriously high concentration of carbon dioxide. One of the benefits of hoeing, raking, plowing, or otherwise cultivating the soil is the increased air circulation they make possible.

Plants, unlike animals, do not seem to need any special gas-transporting mechanisms. Most of the intercellular spaces in the tissues of land plants are filled with air, by contrast with those in animal tissues, which are filled with fluid. These air-filled spaces are interconnected to form an intercellular air-space system that opens to the outside through stomata and lenticels and penetrates to the innermost parts of the plant body. Thus incoming gases can move in gaseous form directly to the internal parts of the plant from the environmental atmosphere without having to cross membranous barriers, and they do not have to diffuse long distances through water or cell fluids, because they do not go into solution until they reach the film of water on the surfaces of the individual cells. Since oxygen can diffuse some 300,000 times faster through air than through fluids, the intercellular air-space system ensures that all cells, even the more internal ones, are adequately supplied. If the oxygen had to diffuse through fluid from the surface of a plant organ, it would penetrate less than one millimeter, and all the more internal cells would be deprived of oxygen and could not respire. Experiments show, in fact, that if the air-space system is blocked, as by waterlogging, the innermost cells soon start dying.

SOLUTIONS IN AQUATIC ANIMALS

As we have already indicated, unicellular animals have no special gas-exchange devices; simple diffusion across their cell membranes is sufficient (Fig. 6.8A). Some of the smaller and simpler multicellular animals like jellyfish, hydra, and planaria show little further development, although their multipurpose gastrovascular cavities do facilitate the exposure of the more internal cells to the environmental water (containing dissolved oxygen) they draw in through the mouth; no cell in these animals is far from the water medium. A few larger aquatic animals, particularly some of the marine segmented worms, lack special respiratory systems and use the skin of the general body surface, which is usually richly supplied with blood vessels (Fig. 6.8B). Most larger multicellular animals, however, have evolved true respiratory systems.

GILLS

With a few exceptions, the respiratory systems of multicellular aquatic animals involve evaginated exchange surfaces, usually known as gills. Gills vary in complexity all the way from the simple bumplike skin gills of some sea stars (Fig. 6.9), the flaplike parapodia of many segmented marine worms (Fig. 6.10), or the mantle-protected gills of squids (Fig. 6.11), to the minutely subdivided gills of fish (Fig. 6.12). Such diverse animals as clams and lobsters and salamanders possess gills, these having evolved independently countless times in the history of animal life on earth.

Most gills, particularly those of very active animals, have such finely dissected surfaces that a few small gills may expose an immense total exchange surface to the water. Thus, although the gas-exchange surface takes up a very limited part of the animal body, most of which can thus be

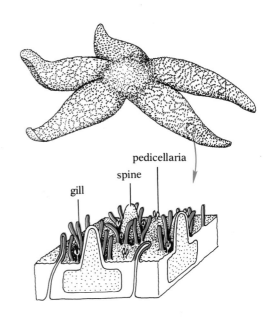

6.9 Sea star
The tiny skin gills are protected from damage by the spines and by the pincerlike pedicellariae, which repel small animals that might otherwise settle on the surface of the sea star.

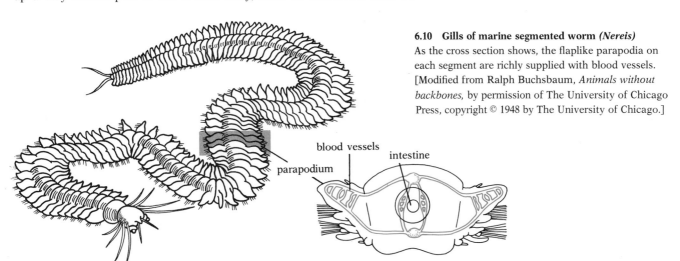

6.10 Gills of marine segmented worm *(Nereis)*
As the cross section shows, the flaplike parapodia on each segment are richly supplied with blood vessels. [Modified from Ralph Buchsbaum, *Animals without backbones*, by permission of The University of Chicago Press, copyright © 1948 by The University of Chicago.]

protected by relatively impermeable coverings, the surface-to-volume ratio of the exchange surface remains high.

Another characteristic of most gills is that they contain a rich supply of blood vessels. Often the blood in these vessels is separated from the external water by only two cells: the single cell of the wall of the vessel and a cell of the gill surface. Sometimes even the vessel wall is eliminated, only one cell remaining between the blood and the water. Oxygen moves by diffusion from the water, across the intervening cells, and into the blood, where it is ordinarily picked up by a carrier pigment (transport by the blood will be discussed in the next chapter). The blood then distributes the oxygen throughout the body to the individual cells. Carbon dioxide produced by cellular metabolism moves in the opposite direction, being transported to the gills and discharged into the surrounding water.

One intriguing feature of the arrangement for exchange of oxygen and carbon dioxide between water and blood in fish gills deserves special mention. The water flows over the surface of a gill lamella in a direction opposite to the flow of blood in the vessels of the lamella (Fig. 6.13). Consequently blood just about to leave the gill, and already almost fully loaded with O_2, encounters water that has not yet given up any of its O_2, because it is just reaching the gill. The O_2 gradient therefore favors pickup of more O_2 by the blood. At the other end of the lamella, oxygen-poor blood entering the gill encounters water that has already lost much of its O_2, but because even this water contains more O_2 than blood that has not yet picked up any, the gradient here too favors diffusion of O_2 from the water to blood. In short, this ***countercurrent exchange system,*** thanks to a favorable gradient between blood and water at every point along the lamella, maximizes the amount of O_2 the blood can pick up from the water. This would not be the case if the two fluids had the same direction of flow.

The fragile gills are easily damaged, and a variety of protective devices have evolved for them. Frequently these devices are coverings, such as the carapace of lobsters or the operculum of fish (Fig. 6.12A), but sometimes they take other forms, as in the spines and pedicellariae that surround the skin gills of sea stars (Fig. 6.9).

Obtaining enough oxygen is a greater problem for aquatic animals than for air breathers, for two reasons: First, O_2 has a low solubility in water, constituting only about 0.004 percent of seawater (the percentage is usually slightly higher in freshwater, but far more variable) as compared with approximately 21 percent in air. Second, the diffusion of O_2 is many thousands of times slower in water than in air. Most aquatic animals must therefore move water across the exchange surfaces. If the water remained still, the O_2 in the vicinity of the exchange surfaces would soon be depleted and would not be renewed by diffusion fast enough to sustain the animal. Most fish actively pump water into the mouth, across the gill filaments, and out behind the operculum. Many fast-swimming fish keep their mouths open as they swim, so that their forward motion forces water across the gills; some species are so dependent on this method of ventilation that they will die for lack of O_2 if prevented from moving. Because water is far more viscous than air and therefore much harder to move, some aquatic organisms use up as much as 20 percent of their metabolic energy in moving water across their gills.

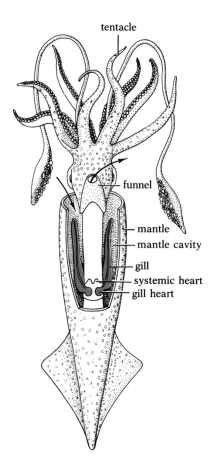

6.11   The gills of a squid
Part of the protective mantle has been cut away to expose the gills within the mantle cavity. Water (left-hand arrow) flows into the mantle cavity when the mantle is relaxed. When the mantle contracts, its collar seals the opening and the water is forced out of the funnel. The jetlike expulsion of water when the mantle is contracted propels the animal backward with great force.

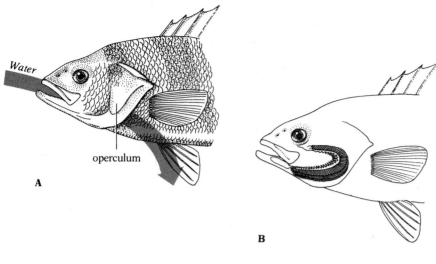

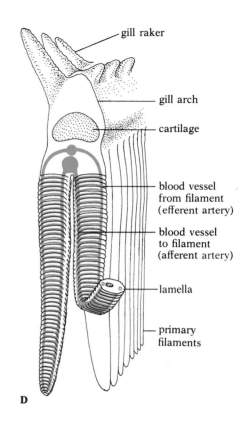

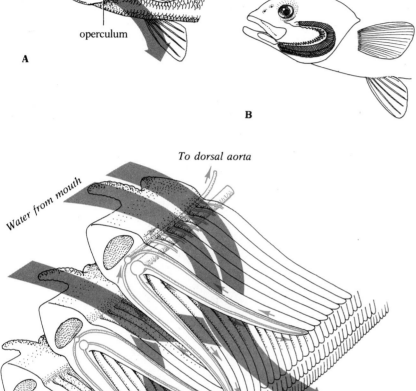

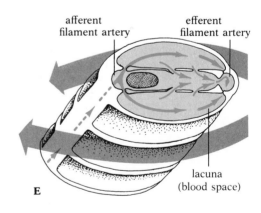

6.12 Gills of fish

(A) Head with operculum covering gills. Water carrying O_2 is drawn in through the mouth, flows across the gills, and exits behind the operculum. (B) Head with the operculum cut away and the gills exposed. (C) Portions of three adjacent gill arches. Each arch bears two rows of primary filaments. The main paths of blood-flow to and from the filaments are shown in color; the gray arrows trace the path of water across the gills (D) Each primary filament bears many disclike lamellae, which contain capillaries that run from the afferent artery to the efferent artery. The end of one filament has been cut off to show a lamella more clearly. (E) The lamella is magnified to show how the blood (color) in the lamellae and the water (gray) flowing between the lamellae move in opposite directions (countercurrent). The lamellae are the actual sites of gas exchange.

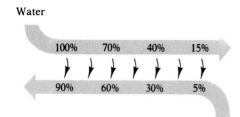

Water

Blood

6.13   Countercurrent exchange system in lamella of fish gill
Figures indicate oxygen tension for both water and blood. Because of the counterflow, the O_2 gradient between water and blood always favors diffusion of O_2 (vertical arrows) from water to blood, and the blood can extract a high percentage of the O_2 from the water. If flow were parallel, the blood could extract much less O_2 and would leave the gills far from fully loaded.

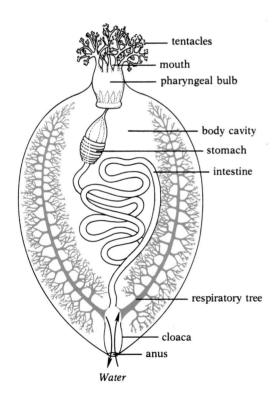

6.14   Respiratory tree of sea cucumber

SOLUTIONS OTHER THAN GILLS

Biology is a science of exceptions. Not all aquatic animals with special respiratory systems use evaginated gills. For example, sea cucumbers—relatives of sea stars—have specialized invaginated systems called respiratory trees (Fig. 6.14). These long branched tubes are diverticula of the cloaca; water is drawn into and expelled from the system by cloacal contractions. This system is an elaboration of a simpler one used by a few animals, in which gas exchange takes place as water is alternately drawn in and expelled through the enlarged thin-walled posterior portion of the digestive tract.

Many insects that live in water are not fully aquatic because they must periodically come to the surface to breathe air. Of these, some exhibit highly interesting adaptations for obtaining oxygen. For example, there are beetles (*Dytiscus*) that store air under their hard shell-like forewings when they surface, then dive with this air bubble and breathe from it. And there are spiders that construct an underwater web in which they store a large bubble of air (Fig. 6.15). You might think the O_2 in the bubble would very soon be depleted, but this is not the case. Remember that air contains high concentrations of other gases besides O_2. Since these gases are not used up, a gas bubble remains, and as the partial pressure[1] of O_2 in the bubble falls, O_2 from the surrounding water tends to diffuse into the bubble, renewing the supply. In short, the bubble acts as a gill. In a few insects this mechanism is so refined that their store of gases is permanent and they do not have to surface to renew it. These insects have thick layers of nonwettable hairs covering parts of their bodies; the hairs hold the water away and maintain a film of air over the body surface.

SOLUTIONS IN TERRESTRIAL ANIMALS

A few land animals have evolved highly modified gill-like respiratory structures that function in air (e.g. the book lungs of spiders). But the hazards of desiccation for most such evaginated surfaces are considerable, and major structural problems are associated with an array of filaments or a branched structure both sufficiently strong to maintain its shape against surface tension and gravity and sufficiently thin-walled to allow easy passage of gases. It is not surprising, therefore, that most terrestrial animals have evolved invaginated respiratory systems. These invaginated systems are of two principal types, **lungs** and **tracheae**. In both, the air inside the system is kept moist, and the cells of the exchange surface are covered by a film of water in which gases can dissolve. Thus the process of gas exchange has remained essentially aquatic in land animals, as it has in the leaf.

LUNGS

Lungs, which are invaginated gas-exchange organs limited to a particular region of the animal and dependent on a blood transport system, are most

---

[1]The partial pressure of a gas is the total pressure of the mixture of gases in which it occurs multiplied by the percentage of the total volume that it occupies. Thus, if the total pressure of all the atmospheric gases is about 760 mm Hg, and if O_2 is about 20 percent of this mixture, then the partial pressure of O_2 in the atmosphere is equal to 760×0.20, or 152 mm Hg.

6.15 The water spider *Argyoneta aquatica*
The male spider is beneath his underwater web, which holds a large bubble of air and to which he returns periodically to breathe. He also retains air among hairs on his abdomen (silvery area). [Jane Burton, Bruce Coleman Inc.]

typical of two unrelated animal groups, the land snails and the higher vertebrates, including some fish, most amphibians, and all reptiles, birds, and mammals. In their simplest form, lungs are little more than chambers with slightly increased vascularization in their walls and with some sort of passageway leading to the outside. This simple type of lung is found, for example, in some snails that inhabit the lower levels of the ocean beach, where the necessity for air breathing is seldom pressing, because oxygen is available from the water. From such a rudimentary beginning, the evolution of the lung has tended toward a greatly increased surface area, by subdivision of its inner surface into many small pockets or folds, and toward increased vascularization of its exchange surface. The latter tendency can be observed in almost diagrammatic simplicity in four closely related snails, called periwinkles, which inhabit successively higher levels of the ocean beach; the increase in vascularization of the lung (or mantle cavity) in these snails is precisely correlated with increased distance from the ocean and the concomitant increased necessity for air breathing.

It is not surprising that terrestrial vertebrates have lungs, but it may surprise you that some presently living relict species (species surviving from very ancient times) of fish have them also. Indeed, many biologists are now convinced that the ancestral fish from which both modern fish and the land vertebrates evolved had lungs that enabled them to live in stagnant, poorly aerated water for long periods of time when necessary. These primitive lungs were simple sacs that arose as ventral evaginations of the digestive

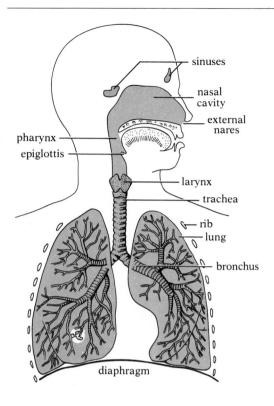

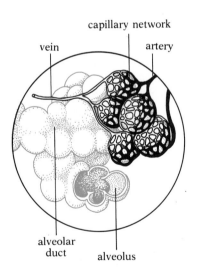

6.16   The human respiratory system
The enlargement shows a few sectioned alveoli and
other alveoli surrounded by blood vessels. Actually, all
alveoli, including those lying along alveolar ducts, are
surrounded by networks of capillaries.

tract in the pharyngeal region behind the gills. A few salamanders still have
such simple lungs, but in most terrestrial vertebrates the inner surface has
become increasingly folded and subdivided, providing a far higher surface-
to-volume ratio for the exchange process. This evolutionary tendency
reaches its culmination in mammals and birds, the two warm-blooded
classes, which must expend vast amounts of metabolic energy to maintain
their stable high body temperatures, and hence have exceedingly high oxygen demands.

Let us look at the human respiratory system (Fig. 6.16) in some detail, as
an example of the mammalian type. Air is drawn in through the *external
nares,* or nostrils, and enters the *nasal cavities,* which function in warming
and moistening the air, filtering out dust particles, and smelling. Bony ridges
in the cavities, causing eddies in the air stream, help expose the air to these
processes. The mucous layer on the epithelium of the nasal passages, and the
cilia of many of the epithelial cells, increase the efficiency with which these
processes occur; in addition, the mucus may have some bactericidal
properties. Curiously enough, of the various functions mentioned, smelling
is the primitive one; the external nares and the nasal passages originally had
nothing to do with respiration, but evolved as smelling devices. Fish have
nostrils, but they do not breathe through them; they take in water for respiratory purposes through the mouth. Since smelling and feeding are so intimately related, it is not surprising that in fish, amphibians, and many
reptiles there is little or no separation between the nasal cavity and the
mouth cavity. It is in mammals that the separation has been furthest developed, a new "roof of the mouth" having evolved that consists of an anterior
bony palate and a posterior soft palate (Fig. 6.17).

Even in mammals, however, the air and food passages ultimately join in a
region known as the *pharynx.* During inhalation air leaves the pharynx via a
ventral opening, the *glottis,* which leads into the larynx. (We are here using
the terms "dorsal" and "ventral" as though the human were standing on
four legs like other mammals.) Since air enters the pharynx dorsally and
exits ventrally, and since food enters ventrally and exits dorsally into the
esophagus, it follows that the air and food passages not only join but actually cross in the pharynx. (This rather inefficient arrangement is the price we
pay because natural selection modified the already existing smelling apparatus into respiratory passages instead of starting from scratch and building
a totally new system. But this is typical of much evolution: The new is built
from the old.) Elaborate mechanisms help ensure that when food is forced
back into the pharynx it will not enter the nasal cavity or the larynx, but
must be swallowed into the esophagus. Without attempting to describe the
whole complex action in detail, we can point out that the internal nares,
which connect the nasal cavities with the pharynx, are closed by the soft
palate and that the glottis is closed by a flap of tissue called the *epiglottis*
when the larynx is raised against it during swallowing.

After leaving the pharynx through the glottis, the air enters the *larynx,* a
chamber surrounded by a complex of cartilages (commonly called the
Adam's apple). In many animals, including humans, the larynx functions as
a voice box. It contains a pair of vocal cords—elastic ridges stretched across
the laryngeal cavity that vibrate when air currents pass between them;
changes in the tension of the cords result in changes in the pitch of the
sounds emitted.

The *trachea* is an air duct leading from the larynx into the thoracic cavity. Its epithelial lining is ciliated; the cilia beat in waves that carry foreign particles and mucus up the trachea away from the lungs. A series of C-shaped rings of cartilage are embedded in the walls of the trachea and prevent it from collapsing upon inhalation. At its lower end, it divides into two *bronchi*, tubes that lead toward the two lungs. (It is at the lower end of the trachea, where the bronchi branch away, that the voice box of birds, the syrinx, is located. Birds have no vocal cords in the larynx.) Each bronchus branches and rebranches, and the bronchioles thus formed branch repeatedly in their turn, forming smaller and smaller ducts that ultimately terminate in tiny air pockets, each of which has a series of small chamberlike bulges in its walls termed *alveoli.* The total alveolar surface is enormous: about 100 square meters—many times greater than the total area of the skin.

The walls of the alveoli are exceedingly thin, being usually only one cell thick, and each alveolus is surrounded by a dense bed of blood capillaries. The alveoli are the site of the actual gas exchange and may therefore be regarded as the primary functional units of the lungs. Oxygen entering an alveolus dissolves in the film of water on its wall and then moves by diffusion across the intervening cells to the blood. Experiments have demonstrated that both this movement and the reverse movement of CO_2 are cases of simple diffusion; no active transport across the cell barriers is involved. O_2 is in higher concentration in the air of the alveolus than in the blood, and CO_2 is in higher concentration in the blood than in the alveolus. Each gas simply moves from the region of its higher concentration to the region of its lower concentration, in accordance with the principles of diffusion that you have already learned. There are actually very few instances where diffusion takes place within the body totally unaided by active transport, but it has been demonstrated that the body is incapable of active transport of O_2 in the lungs. As a result, when the partial pressure of O_2 in the atmosphere falls below normal, as at high altitudes, symptoms of O_2 deprivation rapidly develop, because passive transport is insufficient to meet O_2 demands.

Air is drawn into and expelled from the lungs by the mechanical process called *breathing.* In mammals this process generally involves muscular contractions of two regions, the *rib cage* and the *diaphragm.* The latter is a muscular partition separating the thoracic and abdominal cavities (Fig. 6.18). Inhalation (or inspiration) occurs whenever the volume of the thoracic cavity, in which the lungs lie, is increased; such an increase reduces the air pressure within the chest below the atmospheric pressure and draws air into the lungs. The increase in thoracic volume is accomplished by contractions of the rib muscles that draw the rib cage up and out, and by contraction, or downward pull, of the normally upward-arched diaphragm; the first mechanism is popularly called chest breathing, while the second is called abdominal breathing. Normal exhalation (or expiration) is a passive process; the muscles relax, allowing the rib cage to fall back to its resting position and the diaphragm to arch upward. This reduction of thoracic volume, combined with the elastic recoil of the lungs themselves, causes a rise in the pressure inside the lungs to a level above that of the outside atmosphere and drives out the air.

The air moved by a single normal breath—the tidal air—represents only a small fraction of the total capacity of the lungs. Additional air (complementary air) can be forcibly inhaled, and, similarly, forcible exhalation can expel

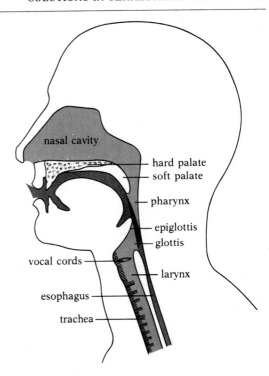

6.17   Detail of upper portion of human respiratory and digestive systems
Path of air in color; path of food, gray. The two paths cross in the pharynx.

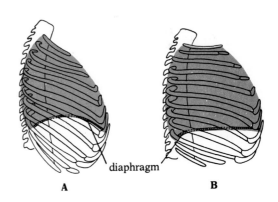

6.18   The mechanics of human breathing
(A) Resting position.   (B) Inhalation. The rib cage is raised up and out, and the diaphragm is pulled downward. Both these motions increase the volume (indicated by color shading) of the thoracic cavity. The consequently reduced air pressure in the cavity causes more air to be drawn into the lungs.

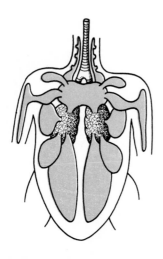

6.19 Respiratory system of a bird
Attached to the lungs are many air sacs (color), some of which even penetrate into the marrow cavities of the wing bones.

additional air (reserve air). The total breathing capacity, or *vital capacity,* is thus the tidal air plus the complementary and reserve air; though it varies greatly from person to person, 4 liters is probably a rough approximation of the average. Trained athletes usually develop larger lungs, and, as would be expected, their vital capacity is usually substantially greater than normal.

The pattern of air flow in the respiratory system of birds differs fundamentally from that of mammals. In addition to paired lungs, birds possess several (most commonly eight or nine) thin-walled air sacs that occupy much of the body cavity and even penetrate into the interior of some of the bones (Fig. 6.19). The air sacs are poorly supplied with blood vessels and do not themselves absorb O_2 or release CO_2. Their arrangement and bellows-like action, however, make possible continuous unidirectional flow of air through the lungs. Let us look at this process in more detail.

Like mammals, birds suck in air by increasing the volume of the body cavity. As Fig. 6.20A shows, most of the air drawn in during inhalation does not go directly to the lungs, but flows through the bronchus to the posterior air sacs; simultaneously air already in the lungs moves forward into the anterior air sacs via connecting passages called recurrent bronchi. During exhalation air from the posterior sacs moves into the lungs, primarily via recurrent bronchi, while air from the anterior sacs empties to the outside (Fig. 6.20B). Thus air moves forward through the lungs during both inhalation and exhalation. Instead of alveoli, dead-end chambers incompatible with unidirectional air flow, bird lungs have tiny air ducts (parabronchi) running through the lung tissue, and it is across their walls that gas exchange takes place (Fig. 6.21).

Birds are far more efficient than mammals in extracting O_2 from air, both because of the continuous unidirectional flow of air through their lungs, and because blood in the capillaries associated with the parabronchi flows at an angle to the flow of air and thereby provides some of the same benefits as the countercurrent exchange system of fish gills. This superior efficiency enables

6.20 Respiratory cycle of a bird
(A) During inhalation, new air (color) is drawn into the posterior air sacs; a small amount also enters the posterior portion of the lungs. Air already in the system (gray) is simultaneously moved forward through the lungs and into the anterior air sacs. (B) As air is exhaled from the anterior air sacs and air from the posterior air sacs moves forward into the lungs, the air sacs decrease in volume. The bolus of air (color) inhaled in (A) will be exhaled during the following respiratory cycle. Note that during both inhalation and exhalation oxygen-rich air is moving unidirectionally through the lungs.

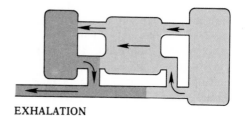

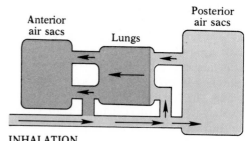

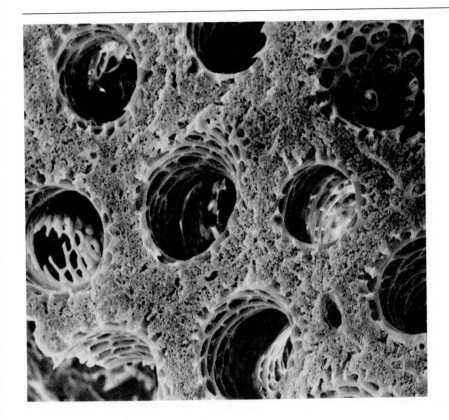

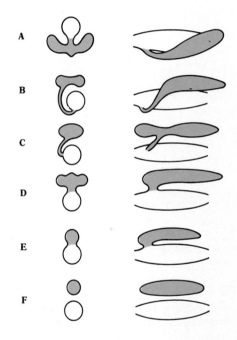

6.21 Scanning electron micrograph of cross section of parabronchi in a two-week-old chick
It is through the parabronchial tubes that the unidirectional flow of air in the bird lung occurs. Exchange of gases takes place across the walls of these tubes, each of which is less than 0.5 mm in diameter. × 56. [Courtesy H.-R. Duncker, Justus Liebig University, Giessen, West Germany.]

birds to fly actively at high altitudes, where the partial pressure of O_2 is low. Vance Tucker of Duke University, who experimentally exposed sparrows and mice to an atmosphere simulating that at 6,000 m altitude (350 mm Hg), found that while the sparrows could fly vigorously the mice were unable to stand up and could barely crawl.

The mammalian and avian method of breathing is known as **negative-pressure breathing,** by contrast with **positive-pressure breathing,** where air is forced into the lungs rather than drawn in. Both these methods are used by adult frogs. With the mouth closed and nostrils open, the frog lowers the floor of the mouth, thereby sucking air into the mouth cavity (negative-pressure method). Then it closes the nostrils and raises the mouth floor; this reduction in the volume of the mouth cavity exerts pressure on the imprisoned air and forces it into the lungs (positive-pressure method). (We should note in passing that a frog is an excellent example of an animal that uses a variety of gas-exchange mechanisms. The lungs are only occasionally filled, much exchange surface being provided by the thin membrane of the mouth cavity and by the soft moist skin.)

The swim bladder of fishes   Let us return to the hypothesis that the ancestral fish had lungs. The fate of the ancient lung is biologically a very interesting question, well worth pursuing here. A widely held view is that the primitively ventral lung evolved into the swim bladder of modern fish by gradually shifting to a dorsal position (Fig. 6.22). Its attachment to the pharynx also gradually shifted to a dorsal position, where it still is in some

6.22 Evolution of swim bladder
Swim bladder colored, pharynx uncolored. Each stage is shown in cross section (left) and longitudinal section (right). According to one hypothesis, the swim bladder of modern fishes evolved from a primitive ventral lung (A). The lung may first have moved to a dorsal position while retaining a ventral attachment to the pharynx, as is still true in some living lungfish (B). The attachment to the pharynx may then have moved laterally (C) and finally become dorsal, as in the typical modern swim bladder (D and E). In some modern fishes the connection to the pharynx has been lost entirely (F). [Modified from A. S. Romer, *The vertebrate body,* Saunders, 1949.]

6.23 Tracheae and spiracle of an insect
At top center is a spiracle (brown), from which numerous branching tubes—the tracheae—can be seen running to many parts of the insect's body. The spiracles are usually located on the sides of the body segments, the number varying in different kinds of insects. [Courtesy Roman Vishniac.]

species, but in other modern species the connecting duct has been lost and there no longer is any direct entrance or exit to the swim bladder.

The swim bladder enables the fish to remain at a given level in the water without sinking by adjusting its density to that of the surrounding water (as you might expect, bottom-dwelling fish seldom have swim bladders). The volume of gas in the bladder changes as the fish shifts to a water level with a different pressure. If the fish swims upward, the swim bladder expands in response to the reduced pressure, and gases must be removed from it to restore it to its normal size; if the fish swims downward, gases must be added to the swim bladder. The addition and removal of gases are accomplished by a region of specialized glandular cells (the gas gland), associated with many blood vessels, in the walls of the swim bladder.

Analysis of the gases in the swim bladder reveals a surprisingly high concentration of O_2—a concentration often far above that in the surrounding water. Still more surprising is a high concentration of molecular nitrogen (which is chemically inactive in the bodies of animals) and sometimes even of such inert gases as argon. Normally, the accumulation of substances against a concentration gradient can be attributed to active transport. But active transport of O_2—a feat the lungs cannot perform—would be very unusual. Moreover, active transport requires that the transported substances enter into chemical reactions, which such chemically inactive gases as nitrogen and argon would be very unlikely to do. How, then, might gases be accumulated in the swim bladders of fishes? A partial answer has come from the investigations of Johan B. Steen of the University of Oslo, Norway. Steen succeeded in taking minute blood samples from the tiny capillaries that enter and leave the gas gland in eels. His analyses indicate that gases are released from the blood, despite the apparent adverse concentration gradients, because lactic acid secreted into the blood by the gas gland greatly reduces the blood's capacity for carrying the gases. In short, no active transport is involved; the process of gas secretion is more physical than chemical.

TRACHEAL SYSTEMS

The second principal type of invaginated respiratory system evolved for air breathing is the tracheal system. It is typical of most terrestrial arthropods, in which it has evolved independently many times. Here we find no localized respiratory organ and little or no significant transport of gases by the blood. Instead, the system is composed of many small tubes, called *tracheae,* that ramify throughout the body (Fig. 6.23). The tracheae and the smaller tracheoles into which they branch carry air directly to the individual cells, where diffusion across the cell membranes takes place (Fig. 6.24).

Air enters the tracheae by way of the *spiracles,* apertures in the body wall that can usually be opened and closed by valves (Fig. 6.25). Some of the larger insects actively ventilate their tracheal systems by muscular contraction, but most small insects and some fairly large ones apparently do not. Calculations have shown that the rate of diffusion of oxygen in air is rapid

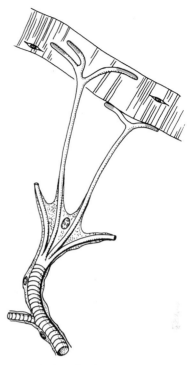

6.24 Terminal tracheoles
Unlike the larger tracheae (segments at bottom), the tracheoles do not have thickened supportive rings in their walls. The branching ends of the tracheoles lie on the surface of the target cell, to which they bring oxygen. The ends usually contain a small amount of liquid (color); when the oxygen requirements of the target cell rise, some of the liquid is withdrawn from the tracheoles, so that more air comes into direct contact with the cell.

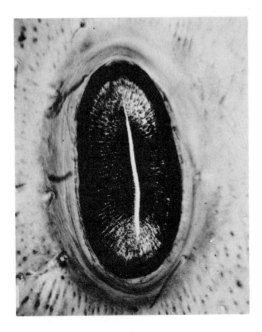

6.25 Spiracle of a grasshopper, showing valves
The spiracle is slightly open; the black areas are the valves. [Courtesy Warren Andrew, *Textbook of comparative histology,* Oxford University Press, 1959.]

enough to maintain at the tracheal endings an O_2 concentration only slightly below that of the external atmosphere. This type of respiratory system, however, has doubtless been a factor in limiting the size attainable by insects.

The aquatic larvae of some insects, such as damselflies and mayflies, have tracheal gills—platelike or feathery structures richly supplied with tracheae. Whereas in the other gills we have discussed absorbed O_2 is transported in the blood, the O_2 absorbed by tracheal gills comes out of solution and moves in gaseous form into the general tracheal system.

SUGGESTED READING

AVERY, M. E., N. S. WANG, and H. W. TAEUSCH, 1973. "The lung of the newborn infant," *Scientific American*, April. *Fascinating account of the changes, at the time of birth, that prepare the infant lung to start functioning.*

SCHMIDT-NIELSEN, K., 1971. "How birds breathe," *Scientific American*, December. (Offprint 1238) *On the recently discovered mechanism of unidirectional air flow in the avian respiratory system.*

SCHMIDT-NIELSEN, K., 1972. *How animals work.* Cambridge University Press, New York. *A small, clearly written book that pays particular attention to gas exchange in various animals.**

* Available in paperback.

Chapter 7

INTERNAL TRANSPORT

Every living cell, whether it exists alone as a single-celled organism or is a component of a multicellular one, must perform its own metabolic activities. It must synthesize its own ATP by cellular respiration (and/or photosynthesis) and carry out for itself those activities necessary for its growth and maintenance. It follows, then, that every cell must obtain the necessary raw materials to support its metabolism. It must obtain nutrients, and, if it uses aerobic respiration, it must obtain oxygen. At the same time, it must rid itself of metabolic wastes such as carbon dioxide and, in animals, nitrogenous compounds. In short, every cell must be exposed to a medium from which it can extract raw materials and into which it can dump wastes. In unicellular organisms and some of the structurally simpler multicellular ones, each cell is either in direct contact with the environmental medium or only a short distance from it. But in the larger and structurally more complex multicellular plants and animals, the more internal cells are far from the body surface and from the general environmental medium. We have already seen that in such organisms nutrient procurement, gas exchange, and waste expulsion take place in certain restricted regions of the body specialized for those functions. Obviously, some mechanism is needed for transporting substances between the specialized systems of procurement, synthesis, or elimination and the individual living cells throughout the body.

ORGANISMS WITHOUT SPECIAL TRANSPORT SYSTEMS

In bacteria, Protozoa, and unicellular algae (and within single living cells in general), diffusion plays an important role in the movement of materials. Since the individual particles of all substances within the cell exhibit random thermal agitation, they tend to become distributed throughout the cell

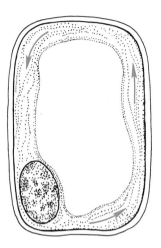

7.1 Cytoplasmic streaming in a plant cell
The cytoplasm flows around the large central vacuole.

if they are not prevented from doing so by specialized intracellular membranes. Diffusion is also important in movement of materials from cell to cell within the body of a multicellular organism; we have examined its part in the procurement of water by plant roots, for example. Such intercellular diffusion may be facilitated in plants by the plasmodesmata, which interconnect the cytoplasmic contents of adjacent cells. We have also seen the importance in plants of diffusion of water along cell walls and of diffusion of gases through intercellular spaces.

But diffusion is a very slow process. If only diffusion were involved, it would take a long time for a substance to move from one cell to another, or even from one end of a single large cell to the other end. It is not surprising, therefore, that even in unicellular and small multicellular organisms diffusion is supplemented by other transport mechanisms. We have seen, for example, that food vacuoles commonly move along a fairly precise path within the cell, thereby distributing the products of digestion to all parts of the cytoplasm. We have also seen that the endoplasmic reticulum may provide a pathway for intracellular movement of some substances. And the cytoplasm itself is seldom motionless; it frequently exhibits rapid massive flow within the cell. The flowing cytoplasm of an active amoeba is an example. The cytoplasm of many plant cells undergoes a characteristic movement called *cytoplasmic streaming* (or cyclosis), in which the cytoplasm flows in definite currents along the surface of the cell vacuole (Fig. 7.1). Sometimes the streaming is restricted to local regions of the cell; at other times most of the cytoplasm becomes involved, and a general circulation results. Such mass flow can transport substances from one part of a cell to another many times faster than simple diffusion.

Among multicellular plants, it is not only the very tiny ones that lack a specialized internal-transport system. Many algae, particularly the brown and red algae, have large multicellular bodies, yet usually lack vascular tissue.[1] As we have already seen, the cells of such plants are seldom far from the surrounding water or from water in intercellular spaces continuous with the external medium. Furthermore, nutrient and gas procurement is not limited to specialized restricted regions of the body, and photosynthesis is seldom localized in specific structures. Consequently each cell gets ample supplies locally, and long-distance transport is rarely necessary. Nevertheless, there is evidence that certain substances, particularly hormones, sometimes move over long distances surprisingly fast in some large nonvascular plants. How this happens is not clear.

Unlike plants, animals are usually adapted for active locomotion. The more rapid metabolism required for an active life makes them less able to rely on such a slow process as diffusion, even when it is supplemented by the other intracellular processes mentioned above. Furthermore, because of their way of life, animals are much less likely than plants to have bodies large in one or two dimensions but flat and thin in the third. Accordingly, very few even moderately large animals lack a circulatory system. There are exceptions to this rule, however. Some relatively large animals live an essentially sedentary life and can, like some large aquatic plants, keep one dimension thin. Large sea fans, a curious group of coelenterates, are an

---

[1] A very few brown algae do possess limited conducting tissue.

example of such plantlike animals. Tapeworms provide another example; they may be 20 m long or more, but they are always flat and thin, and no cell is far from the food supply that bathes them in the host's digestive tract.

In general, however, only very small animals lack a circulatory system; and even these often exhibit some adaptations for transport. Consider hydra (Fig. 5.19, p. 216). Its body wall is basically only two cells thick, but even the cells of the inner layer are exposed directly to water containing dissolved oxygen, because such water is drawn into the gastrovascular cavity. It might seem at first glance that the cells in the tentacles would be far removed from the food supply, but closer examination shows that branches of the gastrovascular cavity penetrate into each tentacle and that food particles can be absorbed directly from this cavity by the tentacle cells. Planaria (Fig. 5.20) is another example. We have already seen that its gastrovascular cavity branches profusely, ramifying into all parts of the body and functioning as a primitive transport system. In short, though animals like hydra and planaria lack a true blood circulatory system, they do have compensatory adaptations that free them from complete dependence on diffusion and intracellular transport.

VASCULAR PLANTS

The vascular plants, as their name indicates, incorporate the two principal types of plant vascular tissue, xylem and phloem. Thanks to these specialized internal-transport tissues, they have been free to evolve bodies large in all dimensions and to develop far greater specialization of parts and more complete integration of function than other plant groups. Thus water and mineral uptake can be restricted primarily to the roots, while photosynthesis can be restricted largely to the leaves. In some very tall forest trees, the distance between the roots and the leaves may be enormous; yet the xylem and phloem form continuous pathways between them, and they can exchange materials with relative ease. Clearly, the successful exploitation of the land environment by plants was dependent on the evolution of such a transport system.

STRUCTURE OF STEMS

Stems of plants serve many functions. Some contain chlorophyll and carry out photosynthesis. Others are highly specialized as storage organs; potato tubers, which are underground stems, are an example. Here, however, we shall concentrate on stems as organs of transport and support, and examine in some detail the structural adaptations associated with these two functions. Keep in mind that, although our discussion of transport is based on stems, the vascular tissue of the stem is continuous with that in the roots and leaves, and that internal transport is no less important in those organs than it is in stems.

Gross anatomy   Let us first examine in cross section the stem of a herbaceous dicot like alfalfa or buttercup (Fig. 7.2). ("Herbaceous" usually refers to plants whose stems remain soft and succulent; the contrasting term is "woody.") The outer tissue layer of a herbaceous stem is epidermis. Next

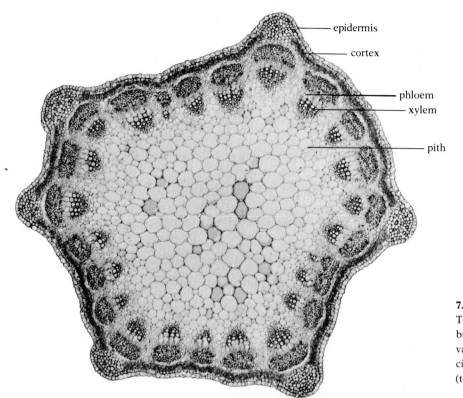

epidermis

cortex

phloem

xylem

pith

7.2 Cross sections of herbaceous dicot stems
Top: Whole stem of alfalfa. × 52.   Bottom: Vascular
bundle of buttercup *(Ranunculus)*. × 330.   The
vascular bundles in dicot stems are arranged in a
circle.   [Courtesy Thomas Eisner, Cornell University
(top); Roman Vishniac (bottom).]

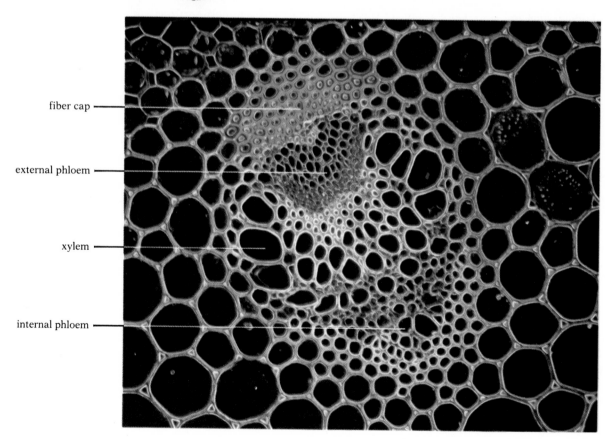

fiber cap

external phloem

xylem

internal phloem

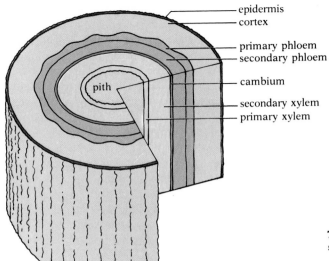

epidermis
cortex
primary phloem
secondary phloem
cambium
secondary xylem
primary xylem
pith

7.3 The tissue layers of a dicot stem after one year of secondary growth

comes the cortex, which is often divided into an area of collenchyma just beneath the epidermis and an area of parenchyma more internally. In some stems an indistinct layer of endodermis lies just internal to the cortex, but endodermis is absent in most stems, being primarily a tissue of the roots. Internal to the cortex (or to the endodermis, when present) lies the vascular tissue, which may be arranged in a continuous hollow cylinder or in a series of discrete bundles, as in alfalfa. In either case, the phloem lies outside the xylem, with a layer of meristematic tissue, called *vascular cambium,* between them. The center of the stem is filled with pith, which is parenchyma tissue and functions as a storage area.

Notice that the arrangement of the tissues in the stele of a dicot stem differs in two obvious ways from that in a typical young dicot root. (1) The phloem and xylem form concentric rings in the stem, whereas the two tissues alternate in the young root. (2) Dicot stems characteristically have pith, whereas most dicot roots do not.

The cambium of many herbaceous dicots never becomes active and never produces additional phloem or xylem cells. In such plants all the vascular tissue is said to be *primary tissue*—tissue derived originally from the apical meristem as the stem (or root) grew in length. The apical meristem of a stem is, of course, in the bud, while that of the root is near the root tip. The new cells produced in the apical meristem soon begin to differentiate, some forming epidermis, some forming the fundamental tissues of the cortex and pith, and some forming the primary phloem, primary xylem, and cambium.

In some species of herbaceous dicots, the cambium does become active, however. As the cambial cells divide, they give rise to new cells both to the inside and to the outside. Though the ratio varies, six to eight new cells are usually produced to the inside for every one produced to the outside. The new cells formed on the outer side of the cambium differentiate as *secondary phloem;* those formed on the inner side of the cambium differentiate as *secondary xylem* (Fig. 7.3). Secondary vascular tissue, then, is tissue derived from the lateral meristem, the cambium, and is a result of growth in diameter rather than growth in length. As secondary phloem is produced by the cambium, it pushes the older, primary phloem farther and farther away from the cambium toward the outside of the stem. Similarly, as secondary xylem is produced, the cambium becomes increasingly distant from the primary xylem, which is left in the inner portion of the vascular cylinder. In a stem that has undergone secondary growth, therefore, the sequence of tissues in the stele (moving from the outside toward the center) is: primary phloem, secondary phloem, cambium, secondary xylem, primary xylem, pith.

Stems of monocots are similar to those of herbaceous dicots. Their vascular tissue, however, always forms discrete bundles, never a continuous cylindrical shell, and the bundles are usually not arranged in a definite circle, as in dicots, but tend to be scattered through the stem (Fig. 7.4). This means that there is no clear distinction between cortex and pith in such stems. Most monocots do not have cambium (some plants in the lily family are exceptions).

A cross section of a "woody" stem made early in its first year would not appear very different from that of a herbaceous stem; in both, the primary vascular tissue is arranged in a continuous ring in some species and in discrete bundles in others. Secondary growth, however, soon makes the rings continuous in all woody stems, and as this growth continues, the stem looks less and less like a herbaceous one (Fig. 7.5). The secondary xylem becomes thicker and thicker until almost the entire stem of an older plant is xylem tissue—commonly called wood.

Since new xylem cells produced early in the growing season, when conditions are best, grow larger than cells produced later in the season, a series of concentric annual rings, clearly visible in cross sections of the stem, are formed. Each ring is made up of an inner area of spring wood with large cells and an outer area of summer wood with smaller cells (Fig. 7.6). A fairly accurate estimate of the age of a tree can be made by counting the annual rings. The width of the rings may vary depending on such factors as the vigor of the tree and the climatic conditions during the growing season. Study of the rings of a large sample of very old trees can give a clue to the climate of an area in past ages. Since trees frequently live for many centuries, and some, like the huge redwoods and giant sequoias of California, may be as much as 3,000–4,000 years old, they can provide a valuable historical record. Tree-ring dating has even been used in archaeological studies; by matching the rings in wooden artifacts found among ancient ruins with cross sections of very old trees from the area, a reliable estimate of the age of the artifacts can be made.

In older trees a variety of chemical and physical changes begin to occur in the older rings of xylem toward the center of the stem. The conducting cells become plugged, the parenchyma cells die, and pigments, resins, tannins, and gums are deposited. As these changes take place, the older xylem ceases to function in transport, but remains important as a strong supportive component of the tree. The rings in which these changes have occurred are known as *heartwood,* while the newer outer rings, which are still functional in transport, constitute the *sapwood.* A tree can continue to live after its

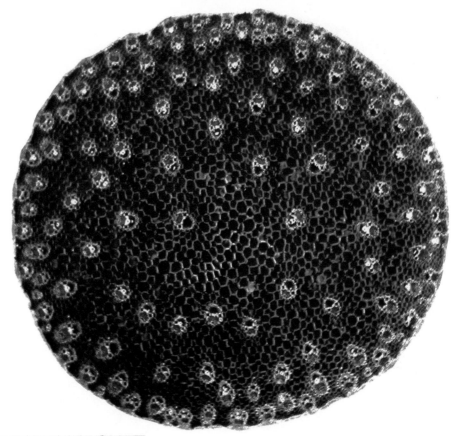

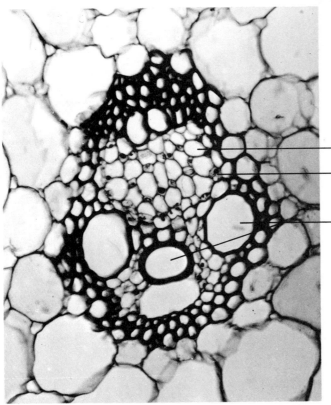

— sieve cell of phloem

— companion cell of phloem

— xylem vessels

7.4 Cross section of monocot stem: corn (Zea)
Top: Entire stem. × 27½. In monocots the vascular
bundles tend to be scattered throughout the
stem.   Bottom: Vascular bundle. × 295.   [Courtesy
Thomas Eisner, Cornell University.]

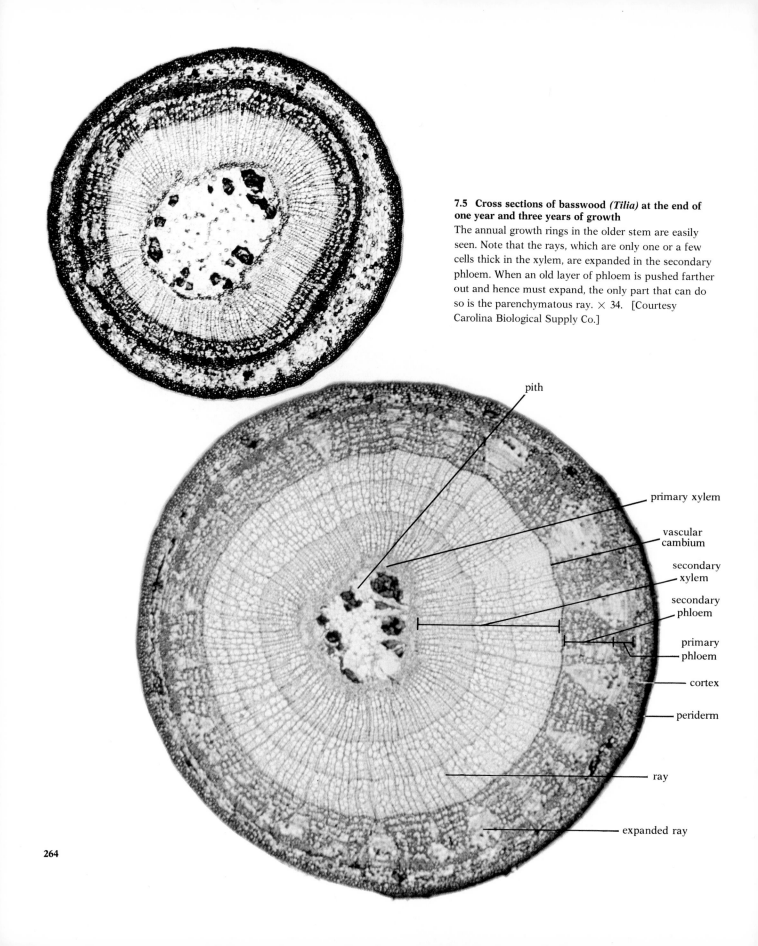

7.5 Cross sections of basswood *(Tilia)* at the end of one year and three years of growth
The annual growth rings in the older stem are easily seen. Note that the rays, which are only one or a few cells thick in the xylem, are expanded in the secondary phloem. When an old layer of phloem is pushed farther out and hence must expand, the only part that can do so is the parenchymatous ray. × 34. [Courtesy Carolina Biological Supply Co.]

pith

primary xylem

vascular cambium

secondary xylem

secondary phloem

primary phloem

cortex

periderm

ray

expanded ray

heartwood has burned or rotted away, but it is much weakened and cannot withstand strong winds.

We have discussed changes in the aging woody stem internal to the cambium; now let us examine the portions of the stem outside the cambium. As woody stems (or roots) grow in diameter, a layer of cells outside the phloem takes on meristematic activity and becomes the *cork cambium.* As growth continues, the original epidermis and cortex flake off and are replaced by *cork cells* produced by cell division in the cork cambium; the cork cambium and the cell layers derived from it are collectively termed the *periderm* (Fig. 7.7). The walls of most cork cells usually develop layers of a fatty substance called suberin and often also layers of waxes; they thus provide a waterproof coating for the plant. As more and more layers of cork cells are produced, the cells in the older, outermost layers usually die and may begin to flake off. Because the breaking of the cork tends to be patchy, the outer bark of some species of trees is very rough and uneven.

The layers of cork cells constitute the outer bark of the older stem or root. The inner bark is the phloem tissue. Annual growth rings are very difficult, if not impossible, to detect in it, since the phloem layer never becomes thick like the xylem layer. There are a number of reasons: Fewer phloem than xylem cells are produced by the vascular cambium; the phloem cells have thinner walls and are easily crushed; and with the new layers of cork cambium often forming internal to the older layers of phloem, these are pushed to the outside and periodically sloughed off. Unlike the xylem, therefore, the

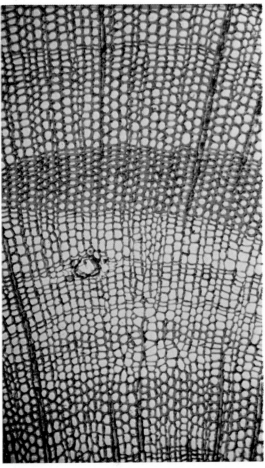

7.6 Cross section of part of secondary xylem (wood) of pine
The smaller, thick-walled cells are summer wood and the larger, thinner-walled cells are spring wood. A large resin duct can be seen in the spring wood. × 136.
[Courtesy Thomas Eisner, Cornell University.]

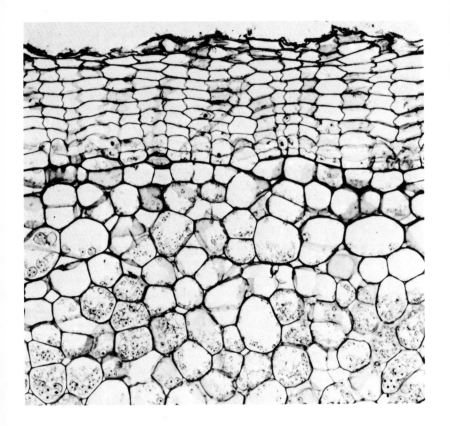

7.7 Periderm on root of sweet potato
The periderm layer, here consisting of 10–12 rows of flattened cork cells overlying the parenchyma tissue of the root, provides a protective outer layer on roots and stems that undergo secondary growth. × 129.
[Courtesy L. L. Morris, University of California, Davis.]

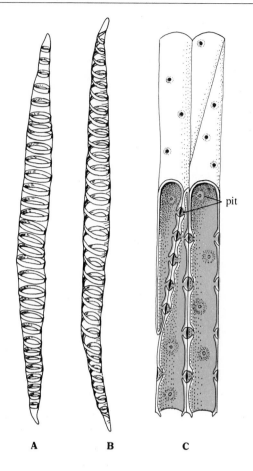

7.8   Tracheids
(A)   Primary tracheid with annular secondary walls.
(B)   Primary tracheid with spiral secondary walls.
(C)   Secondary tracheids. Parts of four cells are shown, with one wall cut away from portions of three of the cells to expose their lumina and give a clearer view of the junction between cells. Notice the pits, which are particularly abundant along the tapering ends of the cells.   [Modified from V. A. Greulach and J. E. Adams, *Plants: An introduction to modern botany,* Wiley, 1962.]

phloem of an older woody plant does not function as an important supportive tissue, but its role in transport is very important, as we shall see.

In summary, then, the old woody stem of a tree has no epidermis or cortex. Its surface is covered by an outer bark of cork tissue. Beneath the cork cambium is the thin layer of phloem, or inner bark, and beneath this is the vascular cambium, which is usually only one cell thick. The rest of the stem is mostly secondary xylem, or wood, of which only the outer annual rings, or sapwood, still function in transport.

The xylem   Xylem is a complex tissue containing several types of cells. Two of these, tracheids and vessel cells, are important as conductive elements after they have matured. Actually, the cell walls are all that remains of a tracheid or vessel cell functioning in transport; the cellular contents, both cytoplasm and nucleus, have disintegrated. The main transport in the xylem occurs, then, in tubular remnants of cells, not in the living cells themselves.

Tracheids are elongate, tapering cells with heavily lignified secondary cell walls; the walls are particularly thick in summer wood and are important as supportive elements. Tracheids of the first-matured primary xylem are stretched during their development, and their secondary walls are usually in the form of rings or spirals (Fig. 7.8A–B). Those of secondary xylem arise after all lengthwise growth has ceased, and they are not stretched during their development; their secondary walls are more continuous, being interrupted only by numerous *pits* (Fig. 7.8C). The pits may occur anywhere on the cell wall, but they are often particularly numerous on the tapered ends of the cell, where it abuts on the next cell beyond it. Water and dissolved substances move from tracheid to tracheid through the pits.

Tracheids have *bordered pits,* which are of rather intricate structure (Fig. 7.9B–D). At these pits, the secondary walls of two adjacent cells are interrupted and overhang the pit chamber, forming the pit borders. The primary walls and middle lamella are continuous through the pit and constitute the pit membrane, which, generally very thin and highly permeable to water and dissolved substances, offers little resistance to the movement of materials from one cell to the next. The bordered pits of conifers and a few other plants are particularly interesting in that the pit membrane, though very thin toward its edges, is thickened centrally to form a buttonlike *torus.* If the pressure in one of the cells becomes much greater than in the adjacent cell—perhaps because air entering the first cell has formed a bubble—the torus, forced against the pit borders of the adjacent cell, blocks the pit aperture and obstructs the flow of materials. In such plants the pit membrane with its torus functions as a valve between the cells.

Vessel cells are more highly specialized conductive elements than tracheids, from which they probably evolved. They are characteristic of the flowering plants and do not occur in most gymnosperms (e.g. conifers). In general, vessel cells are shorter and wider than tracheids. They have bordered pits along their sides, through which some lateral movement of substances may take place, but materials move chiefly through their ends, which are extensively perforated or may even be entirely open (Figs. 7.10, 7.11). Since the perforations lack the membrane (primary wall plus middle lamella) found in pits, material moving from one vessel cell to the next in a vertical sequence forms a continuous column. A vertical series of vessel cells is called a *vessel.*

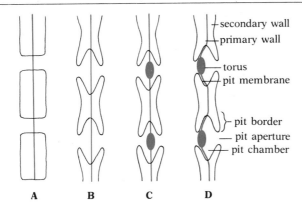

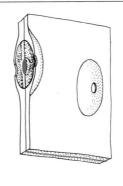

secondary wall
primary wall

torus
pit membrane

pit border
pit aperture
pit chamber

A   B   C   D

E

In addition to tracheids and vessel cells, xylem contains fiber cells and parenchyma cells. The fibers are elongate, very thick-walled cells that function as supportive elements. They apparently evolved from tracheids; numerous intermediate cell types still exist in some species.

Some of the parenchyma cells of the xylem are scattered among the other cells, but many of them are grouped together to form **rays** that run through the xylem in a radial direction (Fig. 7.5) and function as pathways for lateral movement of materials and as storage areas. Rays may be small (Fig. 7.12) or large, depending on the species of plant. The rays in oaks are typically macroscopic and greatly affect the appearance of the wood.

The number, form, and distribution of tracheids, vessels, fibers, and parenchyma cells vary from species to species, and cause the woods of different species to differ in appearance and properties. The wood of pine, a conifer, lacks vessels (Fig. 7.6) and is thus very different from that of oak, which has vessels (Fig. 7.13); oak wood, with its relatively few vessels, is, in turn, different from elm wood or tulip-tree wood, both of which are very porous and have numerous vessels. These differences have both aesthetic and mechanical implications when the wood is used in construction.

7.9   Diagrams of pit structure
(A) Simple pit pairs. The secondary walls are interrupted but do not overhang the pits. The primary walls of any two adjacent cells, and the middle lamella between them, are continuous through the pit and constitute the pit membrane.   (B) Bordered pit pairs without torus. The secondary walls overhang the pit chamber.   (C–D) Bordered pit pairs of pine with torus. In (D) the torus of each pit pair has been pushed against the pit borders on one side; thus movement of materials through the pit is impeded.   (E) Three-dimensional representation, showing one bordered pit in section and another in surface view.

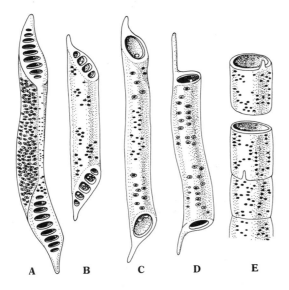

A   B   C   D   E

7.10   Vessel cells
Five different types of vessel cells are shown—those thought to be the more primitive on the left, those thought to be the more advanced on the right. The last example (E) shows a single vessel cell on top, three cells linked in sequence to form a vessel below. The evolutionary trend seems to have been toward shorter and wider cells, larger perforations in the end walls until no end walls remained, and less oblique, more nearly horizontal ends.

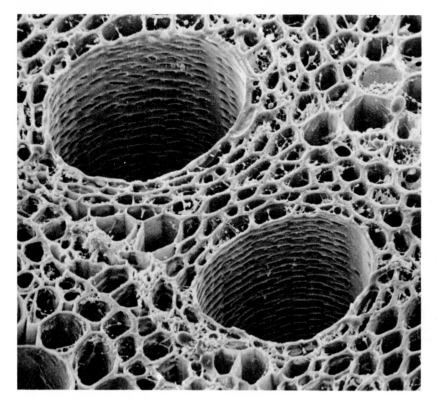

A

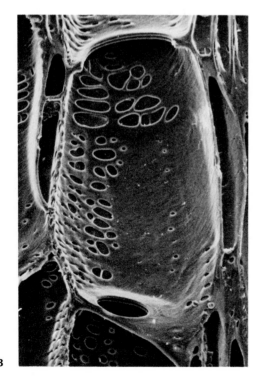

B

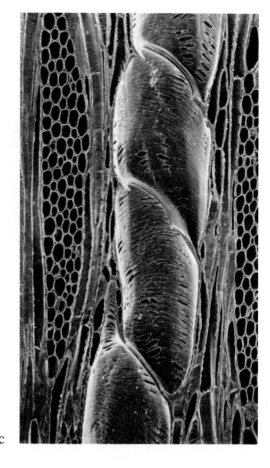

C

7.11 Scanning electron micrographs of xylem vessels
(A) Portion of cross section of corn root showing two large xylem vessels. $\times$ 600. (B) Inner surface of a single vessel cell. Numerous pits can be seen in the side walls. The large opening at the lower end is the perforation through which water moves from one cell to the next in vertical sequence. $\times$680. (C) Lateral view of four vessel cells stacked one above the other. $\times$ 360. [(A) Courtesy J. H. Troughton, Department of Scientific and Industrial Research, New Zealand. (B, C) Courtesy B. G. Butterfield, Canterbury University, and B. A. Meylan, Department of Scientific and Industrial Research, New Zealand.]

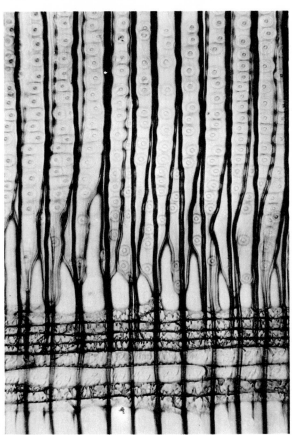

7.12 Radial section of wood of pine
Note the oblique junctions between successive tracheids. Bordered pits are clearly visible. Part of a ray about eight cells high is shown in the lower quarter of the photograph. × 150. [Courtesy U.S. Forest Products Laboratory, Madison, Wis.]

7.13 Cross sections of woods
Left: White oak. Middle: Tulip tree. Right: American elm. Note the differences in the number and size of vessels. [Left and middle: Courtesy Thomas Eisner, Cornell University; right: U.S. Forest Products Laboratory, Madison, Wis.]

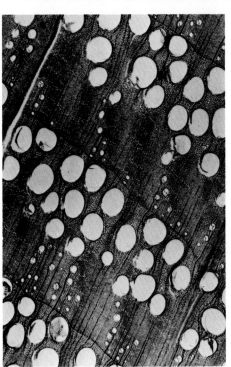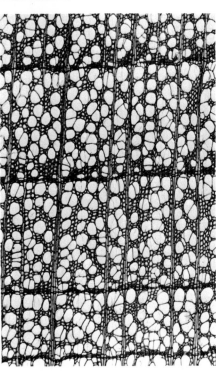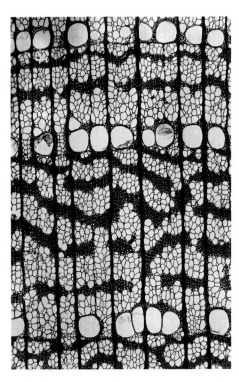

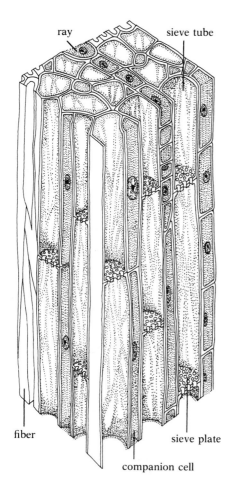

7.14   Phloem
[Adapted from S. and O. Biddulph, "The circulatory system of plants," *Sci. Am.*, February 1959. Copyright © 1959 by Scientific American, Inc. All rights reserved.]

The phloem   Like xylem, phloem is a complex tissue. It contains supportive fibers and also parenchyma; the phloem rays are continuous with the xylem rays. The principal vertical conductive elements in phloem are the ***sieve elements,*** which, in vertical series, form a ***sieve tube.*** In their most advanced form, sieve elements are elongate cells with specialized areas on their end walls, called ***sieve plates*** (Fig. 7.14). As the name implies, a sieve plate is an area with numerous perforations or pores; through these, strands of proto-plasm connect the contents of one cell with those of the next. Unlike the tracheids and vessels of xylem, the sieve elements retain their cytoplasm at maturity, but their nucleus disintegrates.

Closely associated with the sieve elements of most flowering plants are usually one or more specialized, elongate, parenchymatous cells. These ***companion cells*** are derived from the same original cell as the associated sieve element. Mature companion cells retain both their cytoplasm and their nucleus. Some biologists have suggested that the nucleus of the companion cell controls both its own cytoplasm and the cytoplasm of the adjoining sieve element after the nucleus of the latter has disintegrated. Such an association, they think, would help explain why the mature sieve element can continue to carry out many of the normal activities of a living cell even though it has no nucleus of its own. What might seem a refutation of this hypothesis is that the sieve cells of conifers and of a few primitive flowering plants lack companion cells and yet are fully active. There is evidence, however, that certain parenchyma cells are closely associated with the sieve cells in conifers, and they may function in the same way as companion cells. The most that can be said at this point is that an intimate association clearly exists between sieve elements and companion cells, but that its exact nature and a full explanation for the activity of the sieve elements remain to be discovered.

THE ASCENT OF SAP

Sap—the water and dissolved elements absorbed by the roots—moves up-ward through the plant body in the mature tracheids and vessels of the xylem. That the upward movement is primarily in the xylem can easily be demonstrated by ringing experiments; if the cork, phloem, and cambium are removed from a ring around the trunk of a tree, the leaves will still remain turgid, even though they are connected to the roots only by xylem. It can be shown, moreover, that the movement does not depend on the few cells that remain alive in mature xylem; if these cells are killed by heat or poisons, the rise of sap continues unabated.

Any general explanation for the ascent of sap in xylem must identify the forces capable of raising water to the tops of the tallest trees, which may be 90–120 m high. A pressure of one atmosphere can support a column of water 10.4 m high at sea level (less than 10.4 m at higher altitudes). It follows that a pressure of about 12 atm would be needed to support a column 120 m high. But the column must be more than supported; the water must be moved upward at a rate that may sometimes be as fast as one meter or more per minute, and this movement must take place in a system that offers frictional resistance to it. It has been calculated that an additional pressure of at least 18 atm is necessary to achieve the observed rate. Therefore a total force of at least 30 atm is necessary in the tallest trees, and any general theory of water movement in the xylem must account for forces of this magnitude. For example, the common idea that capillarity (the tendency of water to rise in a thin wettable tube, because of its tendency to flow along the walls of the tube; see p. 44) is solely responsible for the ascent of sap must be rejected: The forces involved are far too small (though, as we shall see later, capillarity probably does play an important role in helping support columns of sap in the xylem).

The driving force might be at the base of the plant and push the water upward; or it might be at the top of the plant and pull the water up; or a combination of the two might be involved. Each of these possibilities has had its advocates in the past.

Root pressure   Let us first consider the possibility of a force applied as a push from below. When the stems of certain species of plants are cut, sap flows from the surface of the stump for some time, and if a tube is attached to the stump, a column of water a meter high or more may rise in it. Similarly, when conditions are optimal for water absorption by the roots, but the humidity is so high that little water is lost by transpiration, water under pressure may be forced out at the ends of the leaf veins, forming droplets along the edges of the leaves (Fig. 7.15). This process of water secretion is called *guttation.* When the water in the xylem is under pressure, as in these instances of bleeding and guttation, the pushing force involved is apparently in the roots, and is called *root pressure.*

How is root pressure built up? We have already seen that water moves from the soil, through the epidermis, cortex, endodermis, and pericycle of the root to the xylem, in which it then flows upward to the rest of the plant. But a tall column of water under positive pressure in the xylem would exert a strong downward hydrostatic force by virtue of its weight, and this force might be expected to drive water out of the xylem in the roots. Yet water is not only held in the stele of the root, but also continues to move into the stele in sufficient quantity to build up a force capable of pushing the column upward. However, if the roots are killed, all root pressure disappears; or if the roots are simply deprived of oxygen, the root pressure ceases—an indication that respiratory production of ATP is necessary to provide the energy for development of root pressure.

Apparently the energy from ATP drives active absorption of ions from the apoplast by cells of the cortex. The ions then move from cell to cell in the cortex through plasmodesmata, until they have crossed the endodermal layer of the root and entered the stele. Now, the cells of the stele, being metabolically less active than those of the cortex, apparently allow the ions to leak out of their symplast into the surrounding apoplast. In this way an ion-concentration gradient may be built up across the root, from a low level in the apoplast of the cortex (where the ions are being removed from the apoplast by the cells) to a high level in the apoplast of the stele (where ions are being released into the apoplast). This gradient produces a corresponding osmotic gradient, with the result that water moves by osmosis into the stele. In short, energy is expended in active pumping of ions into the stele, and water follows passively in such quantity as to build up root pressure in the xylem.

7.15 Guttation by strawberry leaves
[Laura Riley, Bruce Coleman Inc.]

The question must now be asked whether root pressure can reach the magnitude we have said is required. The answer appears to be no. Some plants, particularly the conifers and their relatives, are incapable of developing any root pressure to speak of. Attempts to measure the root pressure in species in which it does occur have rarely yielded values exceeding 1 or 2 atm; however, it has been shown that isolated root tips of tomatoes growing in a tissue-culture medium can develop root pressures as high as 6–10 atm.

The low values for root pressure found by most investigators are not the only reason for doubting that it is the principal motive force for the ascent of sap, at least in tall trees. When a puncture is made in a xylem vessel during the summer, it is uncommon to find water under pressure; i.e. water is seldom forced out of the wound. Instead, one can often hear a short hissing sound as air is drawn into the vessel. Yet it is in summer that much of the upward movement of water occurs. Water under pressure is sometimes found in the xylem when the soil is very moist and the transpiration rate low, but under such conditions as these the water moves only very slowly.

In short, root pressure is not the explanation we are seeking, though it may be involved in the ascent of sap in some plants, particularly very young plants, some of the time, especially in early spring.

The cohesion theory   What about the alternative possibility, that the water is pulled from above? The presumed mechanism has been explained as follows: As water in the walls of parenchyma cells in the leaves (or other parts of the shoot) is lost by transpiration, it is replaced by water from the cell contents. With the removal of water, the osmotic concentration of the leaf cells rises, and they take up water from adjoining cells—which, in turn, withdraw water from cells adjacent to them. In this way a gradient extends to the xylem in the veins of the leaf, and the parenchyma cells next to the xylem withdraw water from the column in the xylem, in the process pulling the column upward. Notice that this is not a matter of pulling the column up by air pressure or vacuum; the mechanism is not analogous to sucking up a liquid through a straw. Air pressure could raise water only 10.4 m, but we are dealing with a mechanism presumed to move a water column that may be over 100 m high. What is assumed here is a continuity between the water on the evaporating surfaces of the cell wall and the water in the xylem, and a continuity between the water at the top of the xylem and that in the roots. If this continuity of water all the way from leaf cell to root were broken by the entrance of air into the system, that particular xylem pathway would cease to function.

This idea of pull from above as a result primarily of transpiration was stated in tentative form early in the eighteenth century by the English clergyman and pioneering botanist Stephen Hales. It was given a more complete formulation in 1894 by the Irish botanist H. H. Dixon and his physicist collaborator J. Joly. Its validity hinges on the existence of great cohesion between individual water molecules. Water molecules moving out of the top of the xylem must pull other water molecules behind them; there can be no break, no separation between the molecules.

As we have already seen, water molecules do indeed exhibit great cohesion, owing to hydrogen bonding. Thus the theoretical cohesive strength of the entire column of water—a continuous system of hydrogen-bonded molecules from roots to transpiration surfaces in the leaves—is as high as 15,000 atm. Actual experimental values are lower, but may reach 300 atm or more. If the ascent of sap requires a pressure of 30 atm or more, as we said previously, these values are compatible with Dixon and Joly's thesis—commonly called the cohesion theory (or transpiration theory). Although the pull of transpiration, transmitted throughout the entire column by the hydrogen bonds between water molecules, is thought to be the principal force in the ascent of sap, capillarity—the electrostatic attraction of the water molecules to the hydrophilic (i.e. charged) walls of the thin tubes formed by tracheids and vessels—no doubt provides additional support for the column.

Although the cohesion theory has not been tested under conditions duplicating those in very tall trees, some interesting experiments have been made on a smaller scale. In the 1890s Josef Böhm in Austria and E. Askenasy in Germany showed that if water (previously boiled to remove all dissolved air) is evaporated from the top of a thin tube to which water molecules can adhere and whose lower end is immersed in mercury, a column of mercury can be pulled up the tube to a far greater height than it could be pulled by a vacuum (Fig. 7.16). If the base of a cut branch is inserted tightly in the upper end of the tube and the leaves of the branch become the site of evaporation, similar results are obtained; the tension developed pulls the mercury to a height above the barometric height. The water molecules have adhered to each other (and to the walls of the tube) tightly enough to lift a heavy column of mercury.

Support for the cohesion theory comes from other types of experiments too. In 1935 the German botanist Bruno Huber inserted small electric elements into the xylem and heated the sap. He then measured the time it took for the warmed sap to pass a thermocouple placed a short distance higher on the tree. He found that water begins to move in the upper parts of the tree earlier in the morning than it does in the lower parts of the trunk—an indication that the upward movement of the sap is initiated at the top of the tree, not at the bottom.

More evidence comes from measurements of the girth of tree trunks at different times of day. With the aid of a dendrometer, it can be shown that during the daylight hours, when the xylem should be under maximum tension because of high rates of transpiration, the trunk diminishes in girth—shrinking, just as rubber tubing does when its contents are under lower than atmospheric pressure.

Though the cohesion theory has gained wide acceptance among plant physiologists, it leaves a number of problems unresolved. The theory requires the maintenance of continuity within the water column in the xylem; yet breaks in the column occur rather frequently. For example, during times of drought some of the gases dissolved in the sap may form gas bubbles; and when sap freezes in winter the dissolved gases are forced out, forming bubbles that break the water column. It is not entirely clear how these difficulties are met. Perhaps broken columns rejoin at night when the tension is relaxed; and perhaps the gases forced out of solution during freezing redissolve under more favorable environmental conditions. At any rate, the breaks do not appear to affect water movement seriously; but they remain a challenge to further research.

Another puzzle is how the water column that enables sap to move upward

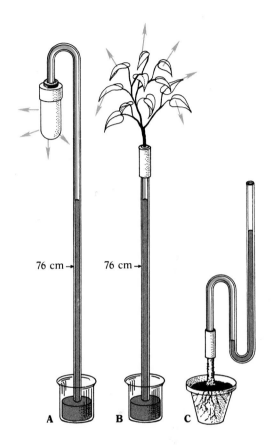

7.16 Demonstrations of rise of water by pull from above (A, B) and root-pressure push from below (C)
(A) Water is evaporated from a clay pot attached to the top of a thin tube whose lower end is in a beaker of mercury. The water in the tube rises and pulls a column of mercury to a point well above the 76 cm to which a vacuum could pull it.   (B) The same results are obtained when transpiration from the leaves of a shoot is substituted for evaporation from a clay pot. (C) In some plants root pressure can raise a column of mercury.

in the xylem is originally established, how it gets up there in the first place. The answer seems to be that the water, in a manner of speaking, grows there. The cambium produces a new layer of potential xylem cells every year. This layer lies just outside the previous year's xylem. The as yet undifferentiated and still living cells draw water laterally from the older xylem as they grow. These cells may, in fact, absorb so much water from the older xylem that their contents are under considerable positive pressure early in the growing season, and if they are punctured liquid may exude from them. As these cells mature, their end walls become more permeable and their contents can move more freely; a conductive system has been established. As the new spring leaves grow larger and the transpiration rate begins to rise, the liquid contents of the newly formed xylem elements are pulled upward. Tension develops in the column of liquid, and the positive pressure of the early part of the growing season is lost.

Notice that if the water in the xylem is under tension, the question how water moves into the xylem in the roots presents less difficulty. We said earlier that a water column under positive pressure (as it would be when there is root pressure) would exert a downward hydrostatic force opposing the entrance of more water into the stele. But if the water in the xylem is under tension, it will exert no such downward force, but will, instead, exert a pull tending to draw water from the soil, across the root tissues, and into the xylem.

TRANSLOCATION OF SOLUTES

Translocation of organic solutes   Two principal classes of solutes are transported—or translocated, as plant physiologists generally call it—within the plant body: organic solutes and inorganic solutes. Let us consider the organic solutes first. We can conveniently divide these into two principal types: carbohydrates (usually transported as sucrose) and organic nitrogen compounds.

The classical picture of the translocation of solutes was that all upward movement was through the xylem and all downward movement through the phloem. About 1920, however, this view had to be revised. It became apparent that most of the movement of carbohydrates, whether up or down, was through the phloem. Most of the early work on the path of movement involved ringing experiments. It could be demonstrated that if a ring of bark (which includes, of course, the phloem) was removed from the trunk of a tree, the supply of carbohydrates to all parts of the plant below the ring was cut off, and those parts eventually died when they had depleted their stored reserves. Downward movement of carbohydrates was clearly through the phloem, not through the xylem, which had been left intact in these experiments. But it could also be demonstrated that if a branch was ringed a short distance behind the growing bud, the supply of carbohydrates moving to the bud was cut off. Again, the movement must have been in the phloem, but in this case the movement was upward. From numerous such experiments as these, the majority of botanists came to regard it as a valid generalization that most carbohydrate movement is through the phloem.

But not all botanists accepted this view. Some objected that far too much carbohydrate moves within the plant body, and moves too rapidly, for the phloem to be the exclusive channel for this movement. After all, the total amount of functional phloem tissue in the trunk of a large tree is rather small. Surely, the argument ran, it is physically inconceivable that so much material should pass through so few sieve tubes, particularly since these tubes are not open pathways like xylem vessels. Numerous workers, however, most notably T. G. Mason and E. J. Maskell of the Cotton Research Station, Trinidad, showed by careful ringing experiments that, hard to conceive as it is, the phloem is indeed the principal pathway of sugar movement, and that this movement is amazingly rapid. Another kind of demonstration was provided by Susann and Orlin Biddulph of the State University of Washington. They grew plants in an atmosphere containing carbon dioxide made from radioactive carbon. When a thin section was cut from the stem of one of these plants and placed in contact with a photographic film, the resulting exposure showed that the radioactivity was restricted to the phloem; the sugar synthesized from the radioactive carbon had clearly traveled only in the phloem.

The situation is less clear with organic nitrogen compounds. It was formerly thought that nitrogen, absorbed by the roots primarily as nitrate, was carried upward in inorganic form through the xylem to the leaves, there to be used in synthesis of organic compounds, which were then transported through the phloem. This sequence probably holds true for some plants. But there is now good evidence that many species promptly incorporate incoming nitrogen into organic compounds such as amides and amino acids in the roots. In some species, especially herbaceous ones, these organic nitrogen compounds move upward in the phloem, but in other species, especially trees, they move in the xylem. We see, then, that not *all* upward transport of organic compounds is in the phloem. But most of it appears to be, and virtually all downward transport of organic compounds seems to be likewise in the phloem.

Translocation of inorganic solutes   Inorganic ions such as calcium, sulfur, and phosphorus ions are translocated upward from the roots to the leaves primarily through the xylem. Use of radioactive forms of these minerals indicates, however, that some are quite mobile in the plant, traveling rapidly back down the plant in the phloem, or moving out of the older leaves through the phloem and being transferred to the newer, more actively growing leaves. Phosphorus, for example, easily moves upward in the xylem and downward in the phloem, often circulating rapidly throughout the plant in this manner. If a plant is grown for a short time in a solution containing radioactive phosphorus, and the plant is then placed against a photographic plate, the younger leaves will be found to contain the greatest concentrations of radioactive phosphorus (Fig. 7.17). If the plant is then moved into a normal solution (one without the radioactive tracer), allowed to grow for a day or so, and again placed against a photographic plate, the resulting pictures will show that the radioactive phosphorus has moved from the leaves in which it was first concentrated to the new leaves just beginning to develop. Calcium, on the other hand, is not mobile in the phloem, and thus cannot move from old leaves to newer ones. Consequently plants must obtain a steady supply of new calcium from the soil, whereas they can easily survive with only intermittent feedings of phosphorus, since this element

7.17 Movement of radioactive phosphorus in a growing plant

The plant was grown for one hour in a nutrient solution containing P^{32}. It was then removed to a nonradioactive solution. At the end of 6 hours (left), the P^{32} was particularly concentrated in the youngest leaves. By the end of 96 hours (right), much of the P^{32} had moved from the leaves in which it was formerly most concentrated to new leaves that had developed above them. (The darker the area, the more P^{32} it contains.) [Courtesy O. Biddulph *et al.*, *Plant Physiol.*, vol. 31, 1958.]

can shift from place to place within the plant and be reused many times. Well-designed fertilization programs take into account such differences in the properties of the different mineral nutrients.

Hypotheses of phloem function   We have seen that most transport of organic solutes, both up and down, is through the phloem; most downward transport of minerals is also through the phloem. How phloem functions in transport is a problem that has been under investigation for a very long time. A number of hypotheses have been put forward, but it must be admitted that none is fully convincing.

There are several facts that any hypothesis about the transport mechanism in phloem must account for: (1) The movement is often rapid, much more so than simple diffusion alone could make it. In fact, it has been estimated that sugar moves through the phloem of a cotton plant more than 40,000 times faster than it diffuses in a liquid! (2) The speed of movement through the sieve tubes differs for different substances. (3) The direction of movement may be reversed periodically within a given sieve tube. (4) The movement in neighboring sieve tubes may be in opposite directions. (5) The movement takes place through sieve-tube cells that, unlike xylem, retain their cytoplasm (though the cytoplasm is not exactly like that of most other cells). (6) Unlike the ends of xylem vessels, the ends of the individual sieve-tube cells are not broadly open, but are penetrated only by the tiny pores of the sieve plates. Clearly, we are here dealing with transport through active cells, not merely with movement through dead tubes by purely mechanical processes, as was the case with mature xylem.

One hypothesis is that materials are carried the length of each sieve cell by cytoplasmic streaming. The suggestion is that materials diffusing into one end of a sieve cell through the sieve plate are picked up by the streaming cytoplasm and carried to the other end of the cell; there they diffuse across the sieve plates at that end and, on entering the next cell in the tube, are again picked up by streaming cytoplasm. In this way, by alternately streaming within cells and diffusing between cells, the materials would move long distances through the sieve tubes of the phloem. The diffusion across the sieve plates might, of course, involve active transport. One argument against this hypothesis is that there is little, if any, evidence that cytoplasmic streaming occurs in mature sieve-tube cells. Another objection is that measurements of the velocities of streaming in other cells, where the process does occur, yield values much lower than the known rates of solute movement through sieve tubes. At the present time, not many botanists accept the streaming hypothesis.

A second hypothesis turns on the fact that substances which lower the surface tension at interfaces spread rapidly along these interfaces. According to this hypothesis, substances move through the sieve cells by flowing along the intracellular membranous interfaces, or perhaps along proteinaceous strands that run longitudinally within the sieve cells. Current evidence, however, makes it doubtful that there is sufficient interface surface area in the sieve cells to account for the quantities of material known to be transported.

A third hypothesis, probably the one most widely supported by botanists today, invokes *pressure flow* or mass flow—namely, the mass flow of water and solutes through the sieve tubes along a turgor-pressure gradient. Cells like those of the leaf contain high concentrations of such osmotically active substances as sugar. Much water therefore tends to diffuse into them, and their turgor pressure rises. This pressure impinges on the next cell and tends to force substances from the first cell into the second. Thus, under hydrostatic pressure, substances are forced en masse from cell to cell along the sieve tubes. At the same time, in storage organs or actively growing tissues, where sugars are being used up and removed from the sieve tubes, the osmotic concentration in the tubes falls. The tubes therefore tend to lose water, and their turgor pressure drops.

What we have, then, is a situation where the contents of the sieve tubes in one portion of the plant (the "source") are under considerable turgor pressure and the contents in another portion of the plant (the "sink") are under lower turgor pressure. The result is a mass flow of the contents of the sieve tubes from the region under high pressure (usually the leaves, but sometimes storage organs when reserves are being mobilized for use, as in early spring) to the region under lower pressure (usually actively growing regions or storage depots). The whole process depends on massive uptake of water by cells at the source end, because of their high osmotic concentrations, and massive loss of water by cells at the sink end, because their osmotic concentrations are lowered by their loss of sugar (Fig. 7.18).

The mass-flow hypothesis is open to an obvious objection. It assumes that material can flow with relative freedom from one sieve-tube cell to the next. But the openings in the sieve plates between successive sieve-tube cells are very tiny indeed. Furthermore, the cytoplasm of sieve-tube cells, particularly in the vicinity of the sieve plates, seems to be rather viscous and might be expected to offer great resistance to mass flow. It has been argued, however,

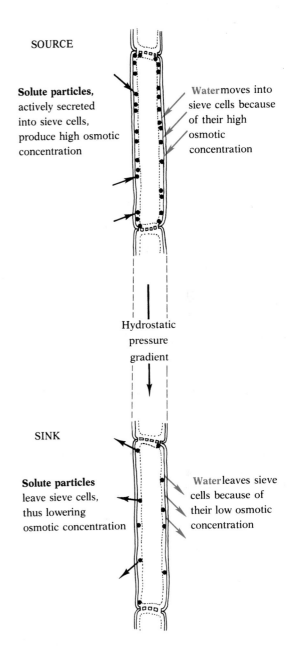

7.18 Pressure-flow model of phloem transport

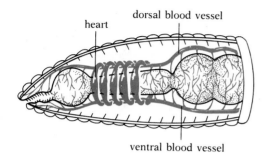

7.19   The circulatory system of an earthworm
Ten hearts, five on each side, pump blood through the longitudinal vessels, which pulsate themselves and help move the blood.

that the high viscosity results from the attempt to examine the cytoplasm; it is known that when sieve cells are damaged the pores in the sieve plates quickly become plugged. It has been further suggested that movement through the sieve plates may be accelerated by some sort of electrical phenomenon.

Future research may supply proof for one of the hypotheses outlined here. Or it may produce new and different hypotheses. Or it may show that more than one mechanism is involved; perhaps some substances move in one manner and other substances move in a different manner. The most that can be said at present is that the mass-flow hypothesis is considered by many botanists to be the best so far proposed, but that there are major weaknesses in it.

It is interesting to note that some very large brown algae, the kelps, have evolved tissue closely resembling the phloem of the true vascular plants. Considering the respective functions of xylem and phloem, can you suggest reasons why the selection pressure for evolution of phloem in these large aquatic plants should have been greater than for evolution of xylem?

CIRCULATION IN HIGHER ANIMALS

Upward transport in the xylem of plants, as we have seen, depends on loss of water from the upper end of the system. Similarly, transport in the phloem depends on loss of large quantities of materials from one end. Neither the xylem nor the phloem is therefore a circulatory system in the strict sense, although substances may move from the xylem to the phloem, or the reverse, and circulate in this manner. Most higher animals, however, have a true circulatory system; blood is moved round and round through the body along a fairly definite path.

Animal circulatory systems usually include some sort of pumping device called a **heart.** There may be only one heart, as in our own case, or a number of separate hearts, as in earthworms, where five blood vessels on each side of the animal pulsate, pumping blood from the main dorsal longitudinal vessel into the main ventral longitudinal vessel (Fig. 7.19). Many insects have both a large general heart and a series of smaller accessory hearts at the bases of their legs and wings.

The one-way pumping action of the heart, usually combined with a system of one-way valves, moves the blood in a regular fashion through the circuit. This circuit may be rigidly encompassed in well-defined channels or vessels, in which case it is called a **closed circulatory system.** Or the circuit may have some sections where definite vessels are absent and the blood flows through large open spaces known as sinuses; such a system is called an **open circulatory system.** Closed circulatory systems are characteristic of a great variety of animals, including our old friends the earthworms and all vertebrates. Open circulatory systems are characteristic of most molluscs (snails, oysters, clams, etc.) and all arthropods (insects, spiders, crabs, crayfish, etc.).

THE INSECT CIRCULATORY SYSTEM

Since movement of the blood through an open system is not as fast, orderly, or efficient as through a closed system, it may seem surprising that such active animals as insects, which must have relatively high metabolic rates and precise internal regulation, should have open circulatory systems. But you will recall that insects do not rely on the blood to carry oxygen to their tissues, that function being fulfilled by the much-branched tracheal system; consequently it is not vital for insects that their blood flow very fast and in a precise pathway. This is a good example of the way the various systems of a living creature are interrelated in adaptation to its needs.

The circulatory systems of insects are even more reduced than those of most other arthropods. Ordinarily the only definite blood vessel in an insect is a longitudinal vessel, often designated as the heart, which runs through the dorsal portion of the animal's thorax and abdomen (Fig. 7.20). The posterior portion of the heart is pierced by a series of openings, or ostia, each regulated by a valve that will allow movement of blood only into the vessel. When the heart contracts it forces blood out of its open anterior end into the head region. When it relaxes again, blood is drawn in through the ostia. Once outside the heart, the blood is no longer in vessels; there are no veins, capillaries, or arteries, other than the heart itself with its valve segments and the short so-called artery that forms its anterior end. The blood simply fills the spaces between the internal organs of the insect, which are thus bathed directly by blood.

The action of the heart causes the blood to move sluggishly through the body spaces from the anterior end where it was released to the posterior end where it will again enter the heart. The movement of the blood is accelerated by the stirring and mixing action of the muscles of the body wall and gut during activity. Thus, when the animal is most active, as in running or flying, and its organs are in most need of rapid delivery of nutrients and removal of wastes, the blood moves relatively fast because of the activity itself.

THE VERTEBRATE CIRCULATORY SYSTEM

All vertebrates have a closed circulatory system, which consists basically of a heart and numerous arteries, capillaries, and veins. An *artery* is a blood vessel carrying blood away from the heart, while a *vein* is a vessel carrying blood back toward the heart. Note that, contrary to a common impression, the definitions of these two types of vessels are not based on the condition of the blood carried. Although it is true that the majority of arteries carry oxygenated blood and the majority of veins carry deoxygenated blood, oxygen content is not always a reliable way to distinguish them. *Capillaries* are tiny blood vessels that interconnect the arteries with the veins; they usually run from very small arteries, called arterioles, to very small veins, called venules. It is across the thin walls of the capillaries that most of the exchange of materials between the blood and the other tissues takes place.

The idea that the blood circulates may seem a perfectly obvious one to you. After all, the constant beating of the heart is one of the most conspicuous of body functions; and, in humans, blood vessels are clearly visible through the wrist or the back of the hand, and the pulse in these vessels can hardly be missed. Yet for centuries this seemingly obvious idea was anything but obvious even to the well-informed. The pumping action of the heart went unrecognized, and even after the notion of the heart as a pump prevailed, the idea of circulation remained alien; it was thought that blood ebbs and flows in the veins until it seeps into the tissues.

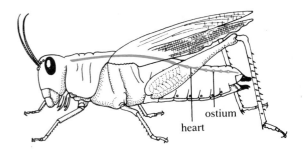

7.20 The dorsal heart of a grasshopper
Blood enters the heart through the ostia and is pumped forward and out at its open anterior end.

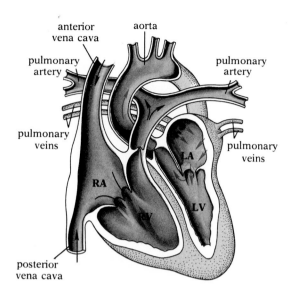

anterior
vena cava

aorta

pulmonary
artery

pulmonary
artery

pulmonary
veins

pulmonary
veins

LA

RA

LV

RV

posterior
vena cava

7.21 The human heart
The arrows show direction of blood flow (color arrows:
oxygenated blood; black arrows: deoxygenated blood).
RA, right atrium; RV, right ventricle; LA, left atrium;
LV, left ventricle. [Modified from N. D. Millard and
B. G. King, *Human anatomy and physiology,* Saunders,
1951.]

A major turning point in understanding the functioning of the human body must be credited to the great English biologist William Harvey. In 1628 he published a short work founded on his extensive examinations of many different species of animals, from worms and insects to human beings, in which he clearly enunciated the idea of circulation of blood. Though he had never actually seen a capillary, he succeeded in outlining the basic components of the circulatory system as we know them today. Harvey's work not only improved knowledge of the circulatory system; it marked the beginning of the modern science of physiology, the attempt to understand bodily processes in terms of physics and chemistry.

The circuit in humans Let us trace the movement of blood through the human circulatory system, beginning with that returning to the heart from the legs or arms. Such blood enters the upper right chamber of the heart, called the *right atrium* (or auricle) (Figs. 7.21 and 7.22). This chamber then contracts, forcing the blood through a valve (the tricuspid valve) into the *right ventricle,* the lower right chamber of the heart. Now, this blood, having just returned to the heart from its circulation through tissues, contains little oxygen and much carbon dioxide. It would be of little value to the body simply to pump this deoxygenated blood back out to the general body tissues. Instead, contraction of the right ventricle sends the blood through a valve (the pulmonary semilunar valve) into the *pulmonary artery,* which soon divides into two branches, one going to each lung. In the lungs the pulmonary arteries branch repeatedly, and each terminal branch connects with dense beds of capillaries lying in the walls of the alveoli. Here gas exchange takes place, carbon dioxide being discharged from the blood into the air in the alveoli and oxygen being picked up by the hemoglobin in the red cells of the blood. From the capillaries, the blood passes into small veins, which soon join to form large *pulmonary veins* running back toward the heart from the lungs. The four pulmonary veins (two from each lung) empty into the upper left chamber of the heart, called the *left atrium* (or auricle). When the left atrium contracts, it forces the blood through a valve (the bicuspid or mitral valve) into the *left ventricle,* which is the lower left chamber of the heart. The left ventricle, then, is a pump for recently oxygenated blood. When it contracts, it pushes the blood through a valve (the aortic semilunar valve) into a very large artery called the *aorta.*

After the aorta emerges from the anterior portion of the heart (the upper portion, in humans standing erect), it forms a prominent arch and runs posteriorly along the middorsal wall of the thorax and abdomen (Fig. 7.22). Numerous branch arteries arise from the aorta along its length, and these arteries carry blood to all parts of the body. For example, the first branch of the aorta is the coronary artery, which carries blood to the muscular wall of the heart itself. Other early branches of the aorta, which arise in the region of the aortic arch, are the arteries that supply the head, neck, and arms. As the aorta runs posteriorly, arteries to the body wall, stomach, intestines, liver, pancreas, spleen, kidneys, legs, etc. arise from it. Each of these arteries, in turn, branches into smaller arteries, until eventually the smallest arterioles connect with the numerous tiny capillaries embedded in the tissues. Here oxygen, nutrients, hormones, and other substances move out of the blood into the tissues; such waste products as carbon dioxide and nitrogenous wastes are picked up by the blood, and substances to be transported, such as hormones secreted by the tissues, or nutrients from the intestine and liver, are also picked up. The blood then runs from the capillary bed into tiny veins, which fuse to form larger and larger veins, until eventually one or more large veins exit from the organ in question. These veins, in turn, empty into one of two very large veins that empty into the right atrium of the heart: the *anterior vena cava* (sometimes called the superior vena cava), which drains the head, neck, and arms, and the *posterior vena cava* (or inferior vena cava), which drains the rest of the body.

Very little, if any, exchange of materials between the blood and the other tissues occurs across the walls of the arteries or veins. The walls of these vessels are apparently impermeable to the substances in the blood and tissue fluid. Walls of arteries are composed of three layers: (1) an outer connective-tissue layer with numerous fibers, which give the vessels their characteristic elasticity; (2) a middle layer of smooth muscle, which can change the size of the vessels (some of the largest arteries also have many fibers in this middle layer); and (3) an inner layer of connective tissue lined with simple squamous endothelium (Fig. 7.23). The two outer layers and the connective-tissue portion of the inner layer terminate at the ends of the

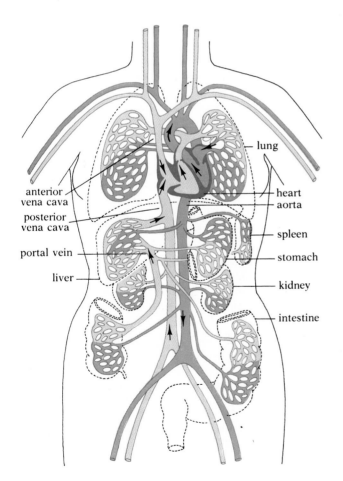

7.22 Diagram of the human circulatory system
Dark vessels contain oxygenated blood, light vessels deoxygenated blood. Only a very few of the vast number of arteries that branch off the aorta are shown here.

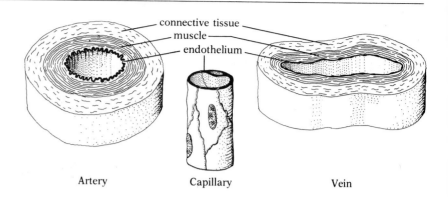

7.23   Walls of artery, vein, and capillary compared
Arteries and veins have the same three layers in their walls, but the walls of veins are much less rigid and they readily change shape when muscles press against them. Capillaries have walls composed only of a thin endothelium.

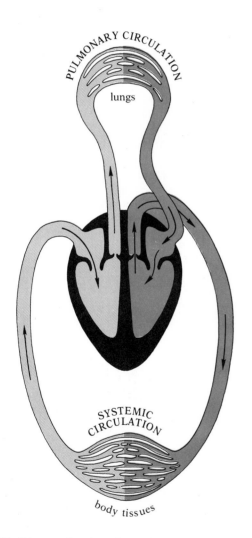

7.24   Diagram showing relationship between pulmonary and systemic circulations in mammals
Dark vessels contain oxygenated blood, light vessels deoxygenated blood.

arterioles, leaving the capillaries with walls composed only of the one-cell-thick endothelium. It is across these very thin walls of the capillaries that the exchange of materials takes place.

Let us now retrace the complete circuit traveled by blood. First, it entered the right side of the heart and was pumped to the lungs, where it picked up oxygen and gave up carbon dioxide, and then it returned to the left side of the heart. This portion of the circulatory system is called the **pulmonary circulation** (Fig. 7.24). Note that in the pulmonary circuit the arteries carry deoxygenated blood and the veins carry oxygenated blood. From the left side of the heart, the blood was pumped into the aorta and its numerous branches, from which it moved into capillaries, and then into veins, and finally back in the anterior or posterior vena cava to the right side of the heart. This portion of the circulatory system is called the **systemic circulation.** The arteries of the systemic circulation carry oxygenated blood and the veins carry deoxygenated blood—a reversal of their roles in the pulmonary circulation.

The circuit in other vertebrates   We have seen that the human heart, which lies in the thoracic cavity just beneath the sternum (breastbone), is, in effect, two hearts in one, since blood in the left side of a normal heart is completely separated from blood in the right side. This type of heart—four-chambered, with complete separation of sides—is characteristic of mammals and birds, the two groups of vertebrates commonly termed warm-blooded. These animals, which maintain relatively constant high body temperatures, regardless of fluctuations in the environmental temperature, have high metabolic rates and very precise internal control mechanisms. Constant perfusion of the tissues with blood rich in oxygen is clearly essential to them. It would be highly disadvantageous to such animals if the oxygen-rich blood from the pulmonary circulation were mixed with the oxygen-poor blood returning from the systemic circulation.

Most of the so-called cold-blooded vertebrates, whose body temperature and metabolic rate fluctuate with the environmental temperature, do not have completely separate left and right hearts. Indeed, the hearts of primitive vertebrates apparently had only one atrium and one ventricle. Modern fish retain this type of heart (Fig. 7.25), but no mixing of oxygenated and deoxygenated blood occurs, because blood aerated in the capillaries of the gills goes straight from the gills to the systemic circulation without first returning to the heart; in other words, most fish have only one basic circulation, with the blood going from heart, to gills, to body tissues, and back to the heart.

In amphibians and reptiles, as in mammals and birds, there is a clear division into pulmonary and systemic circulations, with return to the heart after each. In these animals the atrium is divided into left and right chambers, but the ventricle is only partly divided; i.e. the septum between the left and right sides is not complete (except in crocodilians). One might therefore expect mixing of oxygenated and deoxygenated blood in the ventricle to be unavoidable, but in fact very little mixing occurs. Apparently the muscular ridges that partially divide the ventricle are sufficient to keep the two streams of blood separate under most conditions.

The heart Even though the human heart is double, the two halves beat essentially in unison. The beating is inherent in the heart itself and not dependent on stimulation from the central nervous system. If all nerve connections to the heart are cut, the heart will continue to beat in a normal manner, although the rate of beat may change slightly. As you probably know, the heart of a frog or turtle can continue to beat even after its complete removal from the animal's body, if it is placed in a solution with the proper osmotic concentration. Even though the initiation of the beat and the beat itself are intrinsic properties of the heart, the rate of beat is partly regulated by stimulation from two sets of nerves of the involuntary nervous system; one set tends to accelerate the heartbeat and the other to decelerate it (p. 415).

The initiation of the heartbeat normally comes from the sino-atrial node, or *S-A node,* often called the pacemaker of the heart; it is a small mass of *nodal tissue* on the wall of the right atrium near the point where the anterior vena cava empties into it. Nodal tissue is unique to the heart; it has the contractile properties of muscle and can transmit impulses like nerve. A second mass of nodal tissue, the atrio-ventricular node or *A-V node,* is located in the lower part of the partition between the two atria. A bundle of nodal-tissue fibers (the bundle of His) runs from the A-V node into the walls of the two ventricles, branching into all parts of the ventricular musculature.

At regular intervals, a wave of contraction spreads from the S-A node across the walls of the atria. When this wave of contraction reaches the A-V node, the node is stimulated and excitatory impulses are rapidly transmitted from it to all parts of the ventricles via the fibers of the bundle of His. These impulses stimulate the ventricles to contract.

The heart rate—the alternation of systole (contraction) and diastole (relaxation)—is inversely related to body size and thus varies from species to species. In the Asiatic elephant, for example, a normal rate is 30 beats per minute; in the tiny masked shrew the reported average is 780 beats per minute; in a normal human being at rest it is about 70 times per minute, but there is much individual variation.

In the course of the beat, the heart emits several characteristic sounds, which can be heard easily through a stethoscope placed against the chest. First, there is a long, low-pitched sound produced by the closing of the valves between the atria and the ventricles and by the contraction of the ventricles. Then there is a shorter, louder, higher-pitched sound produced by

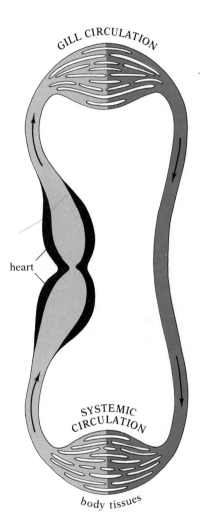

7.25 Diagram of fish circulatory system
Blood from the two-chambered heart goes to the gills, where it is aerated, and then goes straight to the body tissues of the systemic circulation before returning to the heart.

the closing of the valves between the ventricles and the arteries leading from them. Changes in these sounds often indicate to a physician that the heart is defective. A normal heart valve opens when the pressure in front of it is greater than the pressure behind it. For example, when the atria start contracting, they put pressure on the blood they contain, and as soon as this pressure is greater than the pressure in the ventricles, the tricuspid and bicuspid valves are forced open and the blood can flow into the ventricles. As soon as the atria begin to relax and the pressure in them falls below that in the ventricles, the valves snap shut. Similarly, when the ventricles contract and the pressure in them exceeds the pressure in the arteries leading from them, the semilunar valves open and the blood is forced into the arteries; as soon as the ventricles start to relax, the valves snap shut and prevent the blood in the arteries from flowing backward into the ventricles. The normal heart sounds are an indication that all these valves are functioning properly. If, however, a valve has been damaged and cannot shut completely, a hissing or murmuring sound can be heard as blood leaks backward through the damaged valve; this condition is called a diastolic heart murmur. Sometimes a damaged valve partly obstructs blood flow during systole, and the resulting sound (due to the turbulence in the blood flow) is called a systolic murmur. Heart murmurs are a common result of rheumatic fever and some other diseases. The more extensive the damage to the valve, the less efficient the heart action and the greater the strain placed on the heart.

In the course of contraction, the heart muscle undergoes a series of electrical changes. These changes can be detected by electrodes attached to the skin and can be graphed by a device called an electrocardiograph. Abnormalities in heart action alter the pattern of the graph, or electrocardiogram. These alterations aid a trained physician in diagnosing the abnormality.

The heart of a resting human adult pumps about 5 liters of blood every minute, which is approximately equal to the total amount of blood in the body. This does not mean, of course, that each individual drop of blood passes through the heart every minute; blood that happens to flow into one of the shorter circuits, such as one of those supplying the neck or chest, may return to the heart quickly and make several rounds per minute, while blood going to more distant parts of the body, such as the legs, may take several minutes to return to the heart of a resting person. During exercise both the rate of contraction and the amount of blood pumped per beat (the stroke volume) increase greatly. The combination of elevated heart rate and increased stroke volume may raise cardiac output (total amount of blood pumped per minute) to a level four to seven times the resting level. Under such conditions a given drop of blood may pass through the heart many times every minute.

Blood pressure and rate of flow   When the left ventricle contracts, it forces blood under high pressure into the aorta, and blood surges forward in each of the arteries. The walls of the arteries are elastic, and the pulse wave stretches them. During diastole, the relaxation phase of the heart cycle, the heart is not exerting pressure on the blood in the arteries and the pressure in them falls, but elastic recoil of the previously stretched artery walls maintains some pressure on the blood. There is thus a regular cycle of pressure in

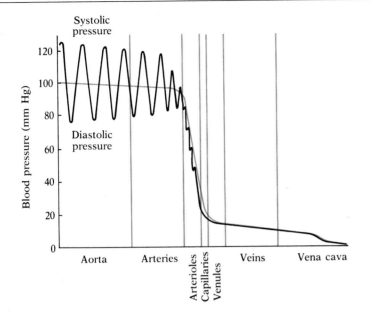

The color curve traces the mean-pressure values. As shown by the black curve, there is considerable fluctuation between the systolic and the diastolic pressure in the arteries. This fluctuation diminishes in the arterioles and no longer occurs in the capillaries and veins, because the elasticity of the vessel walls has effectively damped out the oscillations. The most rapid fall in pressure is in the arterioles and capillaries.

the larger arteries, the pressure reaching its high point during systole and its low point during diastole.

In humans arterial blood pressure in the systemic circuit is usually measured in the upper arm, where systolic values of about 120 mm Hg and diastolic values of about 80 mm are normal in young adult males at rest. Note that these pressures apply to the upper arm only; the values would not be the same for the lower arm or for the leg or for any other part of the body. The blood pressure decreases continuously as the blood moves farther and farther away from the heart. Greatest in the part of the aorta close to the heart, it falls off steadily in the more distant parts of the aorta and its branches, falls even more rapidly in the arterioles and capillaries, then declines more slowly in the veins, reaching its lowest point in the veins nearest the heart (Fig. 7.26). The decline of the blood pressure in successive parts of the circuit is the result of friction between the flowing blood and the walls of the vessels. Such a gradient of pressure is essential, of course, if the blood is to continue to flow; the fluid can only move from a region of higher pressure toward a region of lower pressure.

Several other changes occur along the route of blood flow. First, the difference between systolic and diastolic pressures diminishes, until it disappears in the capillaries and veins. This means that the cyclic, surging type of flow characteristic in the arteries is replaced by a constant rate of flow in the capillaries and veins; the elasticity of the artery walls tends to damp out the fluctuations in blood pressure in the capillaries and veins. Second, the rate of flow tends to fall as the blood moves through the branching arteries and arterioles; the rate is lowest in the capillaries and increases again in the venules and veins. These changes in rate of flow result from changes in the total cross section of the vessel system. Linear rate of flow is inversely proportional to cross-sectional area. In other words, if a fluid is flowing

Both the systolic pressure and the diastolic pressure are important diagnostic indicators to the physician, as you know. Ordinarily these pressures are measured in the artery of the arm with a sphygmomanometer. This instrument has a rubber cuff, which is attached around the upper arm and exerts more and more pressure as air is pumped into it, until finally all blood flow through the artery is blocked. A graduated column of mercury (a manometer) attached to the cuff indicates the pressure the cuff is exerting on the arm. Next, the bell of a stethoscope is placed against the artery just beyond the cuff, and the air in the cuff is gradually released, with consequent reduction of pressure on the arm. When the pressure of the cuff has fallen to a value slightly lower than the maximum systolic pressure in the artery, a small stream of blood can squirt through the artery for an instant during each pulse, producing vibrations that can be heard through the stethoscope. The value shown on the column of mercury the instant before this sound is first detected is taken as the systolic pressure. As the pressure in the cuff is gradually lowered beneath this value, the sound produced as more and more blood surges through the artery at each systole becomes louder and louder. Eventually the pressure in the cuff drops to a value only slightly above that in the artery at diastole, and at this point the sound becomes muffled; the flow of blood through the artery is now continuous. The value shown on the column of mercury at this point is taken as the diastolic pressure. The systolic and diastolic pressures are frequently written together as a fraction, e.g. 120/80.

through a tube that has a smaller cross section in some regions than in others, the fluid will move faster in the regions with the smaller cross sections and more slowly in the regions with the larger cross sections. The same rule applies if the tube is divided into many branches; the greater the effective cross section—i.e. the greater the total cross-sectional area of all the branches in any given region—the slower the flow. As the arteries break up into arterioles, and as the arterioles break up into capillaries, the total cross-sectional area increases, and the rate of flow becomes slower. As the capillaries unite to form venules and these join to form veins, the total cross-sectional area diminishes again, and the rate of flow increases (Fig. 7.27).

The hydrostatic pressure is so low by the time the blood reaches the veins that some other mechanisms besides pressure from the beating heart must be at work in moving the blood. The walls of veins have the same three layers as arteries, but because the muscle layer is much less developed and there is more connective tissue, the walls are easily collapsible. When nearby muscles contract as the body moves, they put pressure on the veins, compressing their walls and forcing the fluid in them forward. The fluid can move only toward the heart, because the numerous one-way valves with which veins are equipped prevent it from flowing backward into the section from which it just came. When standing very still for a long period, you may have noticed your feet beginning to swell and a sudden onset of fatigue. The reason is that there is not enough muscle action in your legs to push the fluids upward against the pull of gravity. If you can manage, while standing, to keep moving your feet and legs, or regularly contracting and relaxing the leg muscles, the unpleasant symptoms will not be so pronounced. Similarly, to prevent pooling of blood in the lower extremities, modern medical practice encourages hospitalized patients to begin walking as soon as they can, often only a day or two after surgery; for such pooling of blood may lead to formation of a clot in a leg vein (thrombophlebitis)—a potentially fatal development all too common in bedridden patients.

The motions of the chest during breathing also aid in moving blood in the veins. When the chest expands during inhalation, the pressure in the thorax falls. There is thus a pressure gradient from other parts of the body to the thorax, and blood tends to be drawn into the large vessels of the thorax and into the heart.

Capillary function   The capillaries are so numerous that they penetrate into all parts of every tissue; no cell is far removed from at least one capillary. It is estimated that in muscle tissue there are as many as 60,000 capillaries per square centimeter of cross section. The diameters of the capillaries are very small, being seldom much larger than those of the blood cells that must pass through them (Fig. 7.28). The extensive branching and small diameters of individual capillaries are functionally important in several respects. They ensure not only that all portions of the tissues will be supplied with capillaries, but also that a very great capillary surface area will be available for the exchange process. It has been estimated that every cubic centimeter of blood contacts nearly one square meter of capillary surface each time it passes through a capillary bed! As we have seen, the branching also increases the total cross-sectional area of the system and thus makes blood flow more slowly in the capillaries than in the arteries or veins (Fig. 7.27). This slower flow allows more time for the exchange process. Furthermore, the very small bore of the capillaries makes for high frictional resistance to blood flow; the resulting considerable drop in blood pressure in the capillary bed plays an important role in the exchange process.

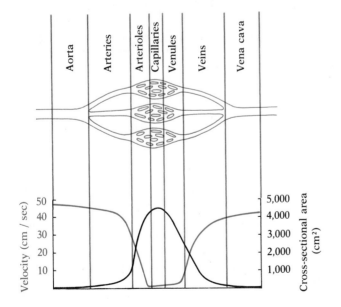

7.27 Changes in the velocity of blood flow in the various parts of a circulatory pathway
As the color curve shows, the velocity falls precipitously as the blood flows through the arterioles, where the total cross-sectional area (black curve) rises. The velocity remains low in the capillaries and venules, but rises again in the veins as the total cross-sectional area falls.

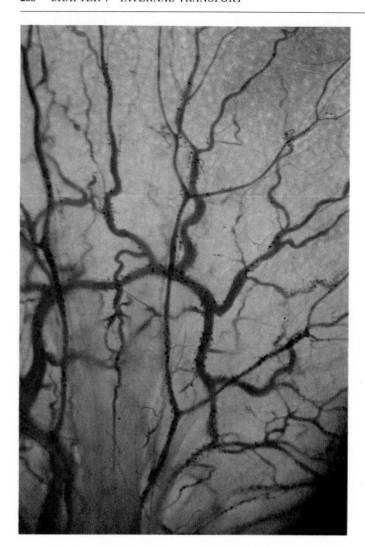

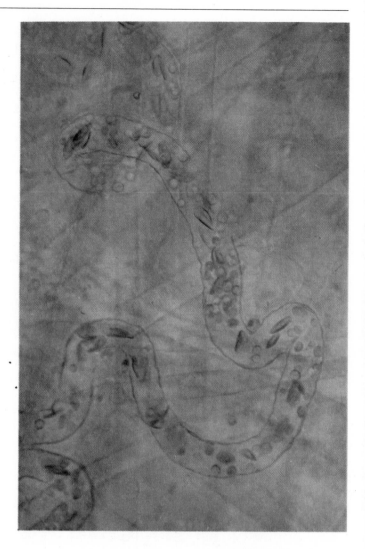

7.28 Blood vessels in the web of a living frog's foot
Left: The vessels branch repeatedly, the smallest
capillaries (seen here as very thin, faintly red lines)
reaching all cells. × 10.   Right: Individual erythrocytes
and leukocytes can be seen in this more highly
magnified capillary. × 400.   [Courtesy Thomas Eisner,
Cornell University.]

Exchange of materials between the blood in the capillaries and the tissue
fluids outside the capillaries can occur in at least three ways: (1) The
materials may move entirely by diffusion through the membrane of an
endothelial cell in the wall of the capillary, across the cytoplasm of the cell to
the other side, and out through the cell membrane on that side. (2) Electron-microscope studies of the endothelial cells of capillaries have revealed
large numbers of vesicles that apparently pick up materials by endocytosis
on one side of the cell, move across the cell, and then expel the materials by
exocytosis on the other side (Fig. 7.29). (3) Electron-microscope studies have
also shown that in the capillaries of most parts of the body (the central
nervous system is an exception) there are clefts between adjacent endothelial cells wide enough to permit filtration of water and most dissolved molecules, but not proteins (Fig. 7.29). The fact that the second and third mechanisms—transport in vesicles and filtration between cells—do not require
movement through cell membranes explains why the rates at which dis-

THROMBOEMBOLIC AND HYPERTENSIVE DISEASE IN HUMAN BEINGS

Thrombosis is the formation of a solid mass or plug of blood constituents in a blood vessel. The mass, or thrombus, may block (wholly or only in part) the vessel in which it forms (see photograph), or it may become dislodged and be carried to some other location in the circulatory system, in which case it is called an embolus. Thromboembolisms are among the leading causes of serious illness and death in Western civilizations.

Many factors can predispose a person to formation of a thrombus. These include irritation or infection of the lining of a blood vessel and a reduced rate of blood flow through a vessel, which may result from disease or merely from long periods of inactivity. For example, formation of a thrombus in a vein, especially of the leg, a condition known as thrombophlebitis, is particularly common in postoperative patients who remain immobilized in bed; it is also common in the elderly, whose leg muscles have lost much of their pumping action. The thrombus often leads to an inflammatory reaction and pain.

A thrombus that becomes detached from its site of formation and moves in the bloodstream as an embolus is extremely dangerous, because it may become lodged in a vessel of some essential organ such as a lung, the heart, or the brain and cut off its blood supply. Such emboli often lodge in the lung (pulmonary embolism) and cause death (necrosis) of a portion of the lung tissue. After pneumonia, pulmonary embolism is the most common acute disease of the lungs seen in hospitalized patients in the United States.

When an embolus (or a locally formed thrombus) blocks a blood vessel in the brain and causes necrosis of the surrounding neural tissue (owing to lack of oxygen), the condition is known as a stroke (or cerebral infarction). The symptoms of stroke vary depending on the part of the brain that has been damaged. When the infarction is in the cerebral hemispheres, there is usually some loss of muscular control in part of the body, sometimes sensory impairment, and often some loss of language ability, which may be a difficulty of expression (expressive dysphasia) or a difficulty of understanding (receptive dysphasia), or both.

Blockage of a blood vessel in the heart by an embolus (or by a locally formed thrombus) causes necrosis of a portion of the heart muscle, a condi-

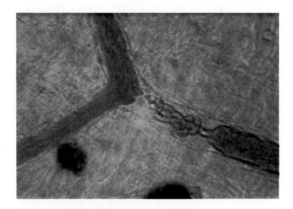

A thrombus in a small blood vessel
The thrombus (tangled red mass) has blocked blood flow near a point where the vessel branches. The blood has pulled away from the left end of the thrombus and is beginning to pull away from the right end also. [Courtesy Roman Vishniac.]

tion familiarly known as a heart attack (more technically, as a myocardial infarction). A high percentage of the deaths that occur in the first few hours after a heart attack result from disruptions of the control system of the heart, with accompanying arrhythmias, especially ventricular fibrillation.

In the vast majority of cases of cerebral or myocardial infarction, the patient already suffers from atherosclerosis, a condition in which fatty deposits in the arteries and thickening of their walls diminish the size of the lumen; the reduced blood flow makes it easier for an embolus to become lodged in the vessel. Indeed, conditions caused wholly or in part by atherosclerosis are the leading cause of death in the United States; they are reponsible for more deaths than the next two leading causes, cancer and accidents, combined.

Atherosclerosis is not the only major condition predisposing a person to a heart attack. Another extremely common precondition is damage to the lining of the arteries due to prolonged high blood pressure (hypertension). Hypertension can also lead eventually to weakening of the heart muscle (which has become thickened owing to the continuing strain imposed on it) and to declining efficiency of its pumping action. Blood may then back up in the heart and lungs, an often fatal condition called congestive heart failure.

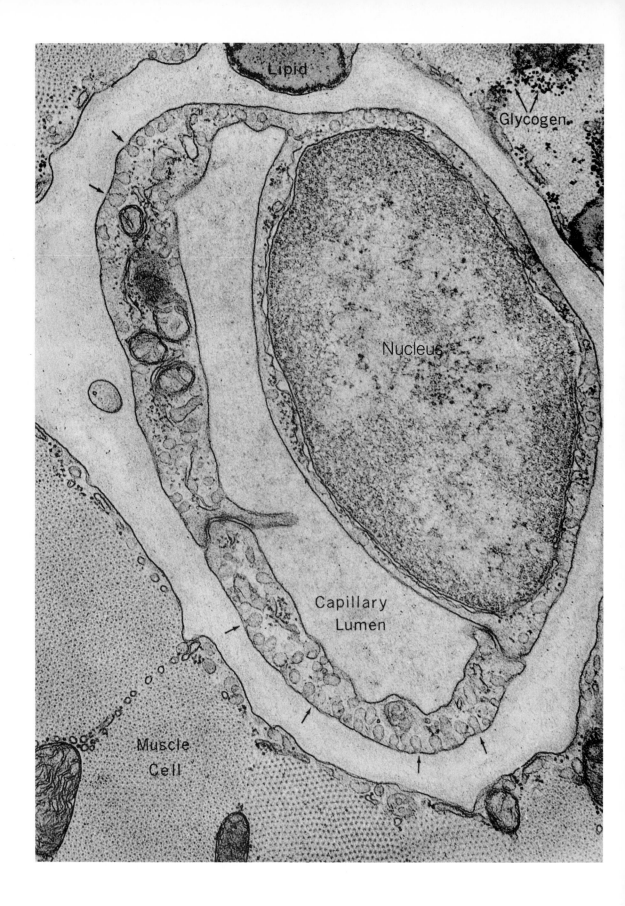

Lipid

Glycogen

Nucleus

Capillary
Lumen

Muscle
Cell

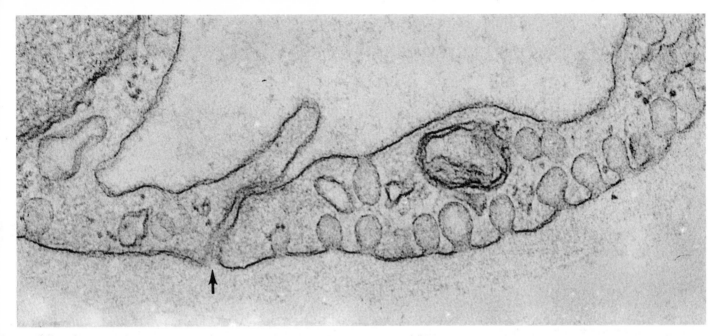

7.29 Electron micrographs of cross section of capillary
Left: Two endothelial cells (the section shows the large nucleus of one of them) make up the capillary wall. Note the clefts at each of the two junctions between the cells and the numerous pinocytic vesicles (arrows) in the cytoplasm. These vesicles may transport materials from outside the capillary, across the endothelial cell, into the lumen of the capillary, or they may take the reverse route.   Above: Enlarged view of wall of a capillary, showing the cleft where two endothelial cells join (arrow) and numerous pinocytic vesicles opening on the outer face of the right-hand cell.   [Courtesy D. W. Fawcett, Harvard Medical School.]

solved substances move between blood and tissue fluid often bear little relation to the rates predicted on the basis of their individual diffusion characteristics.

Let us examine the third mechanism in more detail as it operates in human beings. At the arteriole end of a representative capillary bed, the hydrostatic blood pressure is, on the average, about 36 mm Hg higher than the hydrostatic pressure of the tissue fluid outside the capillaries (Fig. 7.30). The pressure differential has fallen to about 15 by the time the blood reaches the venule end of the capillary bed. The hydrostatic blood pressure tends to force materials out of the capillaries into the surrounding tissue fluid. If this were the only force involved, there would be a steady loss from the blood by filtration of both water and those dissolved substances that can readily be carried by the water through the clefts in the capillary walls. It can be demonstrated, however, that normally there is relatively little net loss of water from the blood in the capillaries. Clearly, some other force must act in opposition to the hydrostatic force.

This other force derives from the difference in osmotic concentration between the blood and the tissue fluid. The blood of mammals contains a relatively high concentration of proteins, and these large molecules cannot easily pass through the capillary walls. The same kinds of proteins occur in the tissue fluids, but in much lower concentration. Because of the difference in protein concentration on the two sides of the capillary wall, the blood and

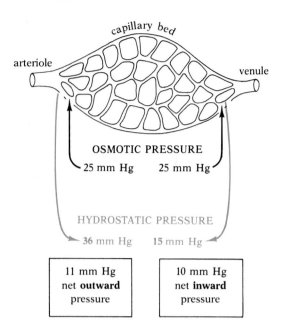

7.30   Diagram of forces involved in the filtration of materials across capillary walls
The blood in the capillaries has both a greater hydrostatic and a greater osmotic pressure than the surrounding tissue fluid. The hydrostatic pressure differential tends to force water and dissolved materials out of the capillaries into the tissue fluid; the osmotic pressure differential has the opposite effect, causing the capillaries to take up water and dissolved materials from the tissue fluid. At the arteriole end of a characteristic capillary bed, the hydrostatic pressure differential of 36 mm is greater than the osmotic pressure differential of 25 mm; the difference of 11 mm favors the outflow of materials. At the venule end, the osmotic pressure differential, which remains 25 mm, is greater than the hydrostatic pressure differential, which has dropped to 15 mm; here the difference of 10 mm favors the pickup of materials from the tissue fluid.

tissue fluids will have different osmotic pressures. Normally, the osmotic pressure of the blood is about 25 mm Hg higher than that of the tissue fluid, with the result that water tends to move into the capillaries from the tissue fluid by osmosis.

We have, then, a system in which hydrostatic pressure developed by the heart tends to force water out of the capillaries and osmotic pressure reflecting differences in protein concentration tends to force water into the capillaries. Obviously, the net movement of water will be determined by the relative magnitudes of these two opposing forces. Notice that at the arteriole end of our representative capillary bed the hydrostatic pressure differential is 36 and the osmotic pressure differential is 25. Subtracting one from the other, we find that there is a net pressure of 11 tending to force water out of the capillaries. At the venule end of the capillary bed, the hydrostatic pressure differential has fallen to 15, while the osmotic pressure differential has not changed greatly.[2] Therefore, there is now a net pressure of at least 10 tending to force water into the capillaries. In summary, the balance between hydrostatic blood pressure and osmotic pressure is such that water is forced out of the capillaries at the ateriole end and into the capillaries at the venule end. The net effect, in a capillary bed such as the one we have just described, is that nearly all (about 99 percent) of the water filtered out of the capillaries at the arteriole end is reabsorbed at the venule end. Since the water carries with it molecules of many dissolved substances, we can say that the blood in the capillaries first unloads materials for the tissues at the arteriole end and then picks up materials for transport at the venule end. In the process, there is normally only a slight net loss of water from the blood.

The balance of hydrostatic and osmotic pressures in the capillaries is a very delicate one. Since it plays such an important role in the exchange of materials between the blood and the tissue fluid, any disturbance of it may have profound effects on the condition of the organism. For example, an increase in blood pressure in a capillary would tend to increase loss of fluid from the blood, while a decrease in blood pressure would have the opposite effect (Fig. 7.31A, B). Such changes in blood pressure could be produced by any one or more of a variety of factors, such as changes in rate or strength of heart action, increase or decrease in total blood volume, changes in the elasticity of the walls of the arteries, or increased dilation or constriction of arterioles and capillary sphincters.

The last-mentioned factor—degree of dilation or constriction of capillary sphincters—is also important in determining the extent of the blood supply to any given tissue at a given time. The capillaries of the body are never fully open all at the same time; many capillaries are usually closed by constriction of rings of sphincter muscle at their bases. In a resting muscle, for example, only certain "thoroughfare" capillaries (also called metarterioles) are generally open; but once the muscle becomes active, and its needs for oxygen and nutrients increase, the numerous smaller branch capillaries become dilated, and the local blood supply is greatly increased (Fig. 7.32). Similarly, the capillaries in the wall of the intestine are extensively dilated following a large meal, when a major portion of the blood supply is channeled into this

[2] Loss of water from the blood has, of course, slightly increased the concentration of protein in the blood and raised the osmotic pressure accordingly, but this change is relatively slight and can be ignored for our purposes here.

region where much absorption of digestive products is occurring. Increased dilation of skin capillaries often gives the skin a reddish hue, seen in blushing, while constriction of these same capillaries gives the skin a bleached, whitish look. A major form of heat loss from the body is by radiation from the blood in the superficial capillaries of the skin; changes in dilation of these capillaries, by altering the amount of blood flow, are an important factor in helping to regulate heat loss.

Clearly, simultaneous dilation of a high percentage of the body's capillaries, such as those of the muscles, intestine, and skin, tends to lower the blood pressure in any given capillary, because the same amount of blood is now distributed in a greater total space. If all the capillaries were fully open at the same time, which never actually happens, they would contain all the blood in the body, and arterial blood pressure would fall to zero or below. When blood pressure falls because of extensive vasodilation (a condition known as vascular shock) or because of loss of blood by hemorrhage, the consequent increased absorption of tissue fluid by the capillaries increases the total blood volume (though not the total number of blood cells) and tends to compensate partially for the deficiency. At such times, the supply of circulating blood is also augmented by reserve blood previously stored in the **spleen.** This organ stores blood in large cavities connected to its capillaries. Contractions of the smooth muscles in the walls of the spleen can expel this blood into the general circulation.

Changes in the relative concentrations of proteins in the blood and in the tissue fluids can also severely alter the balance of forces operating in the capillaries (Fig. 7.31C). Numerous experiments have been performed in which the protein concentration in the blood supply to a limb of a frog, cat, or dog was artificially regulated. As was predicted, increasing the protein

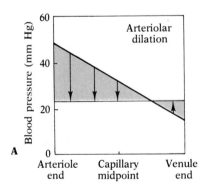

A

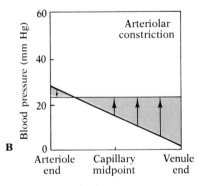

B

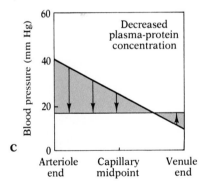

C

7.31 Conditions that alter the balance between blood pressure and osmotic pressure in the capillaries
(A) When the arterioles are dilated, and much blood is flowing into the capillary bed, the hydrostatic pressure differential (diagonal line) at the arteriole end of the capillary is much greater than the osmotic pressure

differential (horizontal line); hence flow of water out of the capillary (left color area) much exceeds reabsorption near the venule end (right color area). (B) When the arterioles are constricted, and little blood is entering the capillary bed, the hydrostatic blood pressure in the capillary is low; hence

reabsorption of water into the blood much exceeds outward filtration. (C) When the plasma-protein concentration is low, the osmotic pressure of the blood falls (in this case lowering the differential—horizontal line—to below 20 mm Hg); hence the balance is shifted toward outward filtration of water.

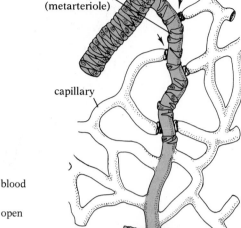
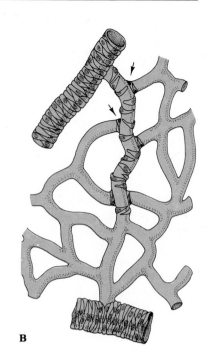

A   venule   B

7.32   The vessels of a capillary bed
When the precapillary sphincters are closed (A), blood flows only through the capillaries known as thoroughfare channels. When the sphincters are open (B), blood flows through all the capillaries.

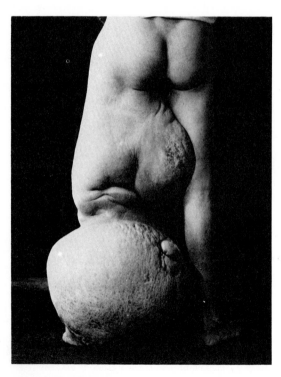

7.33   Elephantiasis
Elephantiasis is a condition of extreme edema that occurs when lymph vessels become blocked by filarial worms. Here the left leg is swollen with the fluids accumulated in the tissues as a result of the blockage.

concentration in the blood decreases loss of fluids from the blood and increases absorption from the tissue fluid. Conversely, decreasing the protein concentration in the blood increases loss of fluids from the blood and decreases reabsorption of fluid from the tissues, the result being an abnormal accumulation of fluid in the tissues that causes swelling, a condition known as edema.

One unexplained peculiarity may be mentioned here, namely that birds normally have very low levels of plasma protein. Since most birds have high blood pressures (e.g. 135/105 in pigeons, 180/130 in starlings), one might expect heavy loss of fluid from the capillaries, but this does not occur. Further research may uncover the reason.

The lymphatic system   We have seen that approximately 99 percent of the water that leaves the capillaries at the arteriole end is normally reabsorbed at the venule end. But what about the remaining one percent? Can this fluid return to the blood by any other means than direct reabsorption into the blood capillaries? The answer is yes. Vertebrates have a special system of vessels that function in returning materials from the tissues to the blood. These vessels are called lymph vessels, and together they constitute the lymphatic system, which includes lymph veins and lymph capillaries, but no arteries. The lymph capillaries, which like the blood capillaries are distributed throughout most of the body, are closed at one end. Their walls, like those of blood capillaries, are composed of a single layer of endothelium. Tissue fluid is absorbed into the lymph capillaries (whereupon it is called *lymph*) and slowly flows through the capillaries into small lymph veins, which unite to form larger and larger veins until finally two very large lymph ducts empty into veins of the blood circulatory system in the upper portion of the thorax near the heart.

Although the walls of lymph capillaries are structurally similar to those of blood capillaries, their permeability characteristics are different. Lymph often contains a small concentration of proteins of the same types as those in the blood; these proteins can apparently move easily across the walls of the lymph capillaries. Since the lymphatic system steadily carries small quantities of proteins from the tissue fluid to the blood, but the protein concentration in the tissue fluid does not normally decrease, there must ordinarily be some slight leakage of proteins from the blood capillaries, even though the walls of these capillaries are highly impermeable to proteins. The lymph vessels, whose walls are very permeable to proteins, return such proteins to the blood. The importance of this process in maintaining the normal osmotic balance between the blood and the tissue fluid is very great. Under certain conditions major lymph vessels may become blocked; the protein concentration in the tissue fluid then steadily rises, and the difference in osmotic concentration between it and the blood steadily diminishes, which means that less and less fluid is reabsorbed by the blood capillaries. The result is severe edema (Fig. 7.33).

The lymphatic system performs many other functions besides returning excess tissue fluid and proteins to the blood. For example, much of the fat absorbed from the intestine is picked up by lymph vessels rather than by blood capillaries. Absorption of fats thus differs from the absorption of sugars and amino acids, which are picked up by blood capillaries.

Lymph nodes are present in the lymphatic systems of mammals and some birds, but are absent in most lower vertebrates. Located along major lymph vessels and composed of a meshwork of connective tissue harboring many phagocytic cells, they act as filters and are sites of formation of certain types of white blood cells. As the lymph trickles through the nodes, it is filtered and such particles as dead cells, cell fragments, and invading bacteria are destroyed by the phagocytic cells. Nondigestible particles such as dust and soot, which the phagocytic cells cannot destroy, are stored in the nodes. Since the nodes are particularly active during an infection, they often become swollen and sore, as the lymph nodes at the base of the jaw are apt to be during a throat infection.

Since the lymphatic system is not connected to the arterial portion of the blood circulatory system, it is obvious that lymph is not moved by pressure developed by the heart. Its movement, like that of blood in the veins, is due to the contractions of skeletal muscles that press on the lymph vessels and push the lymph forward past one-way valves (Fig. 7.34). Though this mechanism of movement holds for all mammals, including humans, many other vertebrates have lymphatic hearts, which are pumping devices located along major lymph vessels. Most animals that have lymph hearts lack valves in their lymph vessels.

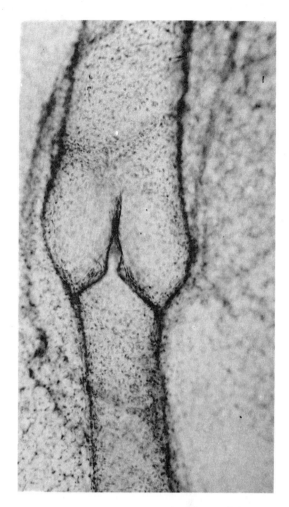

7.34 Photograph of a valve in a lymph vessel
[Courtesy Turtox/Cambosco, Macmillan Science Co., Inc.]

BLOOD

We have so far discussed the routes followed by the circulating blood, the mechanism of circulation, and the process of exchange of certain materials with the tissue fluid. We should now examine the blood itself in more detail. It is one of the most important and unusual tissues in the animal body.

Composition of the blood   In Chapter 3 we classified blood as a type of connective tissue with a liquid matrix. The extracellular liquid matrix of blood is called *plasma.* Suspended in the plasma are the formed elements, which are of three major types in vertebrates: (1) the red blood cells or *erythrocytes;* (2) the white blood cells or *leukocytes;* and (3) the *platelets,* which are small disc-shaped bodies that arise as cell fragments.[3] If whole blood, treated to prevent clotting, is left standing in a test tube, the formed elements will settle slowly to the bottom, leaving the fluid plasma above. The specific gravity of the formed elements does not greatly exceed that of the plasma, however, and the agitation associated with normal circulation is sufficient to prevent separation within the circulatory system. Normally, the formed elements constitute about 40–50 percent of the volume of whole blood, while the plasma constitutes the other 50–60 percent.

The basic solvent of the plasma is, of course, water, which constitutes roughly 90 percent of the plasma. A great variety of substances are dissolved in the water; the relative concentrations of these vary with the condition of the organism and with the portion of the system under examination. For convenience, let us divide these solutes into six categories: (1) inorganic ions; (2) plasma proteins; (3) organic nutrients; (4) nitrogenous waste products; (5) special products being transported; and (6) dissolved gases.

1. The principal inorganic cations (positively charged ions) in the plasma are sodium (Na^+), calcium (Ca^{++}), potassium (K^+), and magnesium (Mg^{++}). The chief inorganic anions (negatively charged ions) are chloride (Cl^-), bicarbonate (HCO_3^-), phosphate (HPO_4^{--} and $H_2PO_4^-$), and sulfate (SO_4^{--}); of these, chloride and bicarbonate are by far the most abundant. Together, the inorganic ions and salts make up about 0.9 percent of the plasma of mammals by weight; more than two thirds of this amount is sodium chloride, ordinary table salt.

The concentrations of the individual ions remain relatively stable, regulated as they are by a variety of agencies, particularly the kidneys and other excretory organs, as well as a number of hormones. This stability, this maintenance of equilibrium—the technical term is *homeostasis*—is essential to the normal function of the organism. Any appreciable shift in the concentrations of sodium chloride (NaCl) and sodium bicarbonate ($NaHCO_3$), for example, would cause severe disturbance, and even death, to the cells, for these compounds (together with plasma proteins) help determine the osmotic balance between the plasma and the fluids bathing the cells. Even if the total concentration of dissolved substances remains the same, shifts in the concentrations of particular ions can create serious disturbances. Nerves and muscles, for example, are highly sensitive to changes in the concentrations of K^+ and Ca^{++}. Similarly, the integrity of cell membranes depends on proper balance of Ca^{++}, Mg^{++}, K^+, and Na^+ in the extracellular medium. The concentrations of certain ions are also very important in determining the pH of the body fluids, and even the slightest changes in pH (normally slightly alkaline in plasma) may kill the organism.

2. The plasma proteins constitute 7–9 percent by weight of the plasma. Most of these proteins, which are of three types—fibrinogen, albumins, and

---

[3] True platelets are found only in mammals. The blood of most other vertebrates contains cells called thrombocytes, which function in blood clotting in a manner similar to platelets.

globulins—are synthesized in the liver, though some of the globulins are synthesized in lymphoid tissue.

We have already discussed the great importance of proteins in determining the osmotic pressure of the plasma and the influence they consequently have on the exchange of materials in the capillary beds and on the general water balance of the body. In a later section, we shall examine their role in stabilizing the pH of the plasma. These proteins are also instrumental in determining the viscosity of the plasma (viscosity is a measure of the internal friction of a fluid, i.e. the friction between molecules as they slide past each other). The heart can maintain normal blood pressure only if the viscosity of the blood is nearly normal. Injection of an isotonic salt solution into the circulatory system as an emergency measure following extensive hemorrhage can restore the blood volume to normal and thus raise the blood pressure, but it cannot raise the pressure to the normal level, because the saline solution has too low a viscosity.

The plasma proteins have other functions as well. By binding to certain hormones, fatty acids and other lipids, some vitamins, and various minerals, they greatly facilitate the transport of such substances by the blood. The functions of some particular proteins—e.g. those of fibrinogen and certain globulins in blood clotting and of gamma globulin in antibody reactions to infections—will be discussed later.

3. Organic nutrients in the blood include glucose, fats, phospholipids, amino acids, and lactic acid. Some of these are picked up from the intestine; others enter the blood from storage areas such as the liver or the fat depots. Lactic acid is a product of glycolysis, especially in muscles; it is transported by the blood to the liver, where some of it may be used in resynthesis of carbohydrates and some may be further oxidized to carbon dioxide and water.

Another substance found in the plasma is cholesterol. It is metabolized to some extent as a source of energy, but plays its major role as the precursor of most other important steroids, such as the bile acids and the steroid hormones.

4. The plasma also carries nitrogenous waste products from their sites of formation to such organs of excretion as the kidneys. In mammals this waste is primarily in the form of urea. Small amounts of ammonia and uric acid are usually also included.

5. Among the special products carried by the plasma, the hormones are of particular significance. These substances, synthesized by the endocrine tissues, are very important regulatory chemicals; much attention will be devoted to them in a later chapter.

6. Three principal gases are found dissolved in the plasma. One of these, nitrogen, which diffuses into the blood in the lungs, seems to be physiologically inert and can be disregarded here. The other two, oxygen and carbon dioxide, are of critical importance, and the details of their transport will be discussed in a later section. Actually, in vertebrates most of the oxygen and much of the carbon dioxide are transported in the red blood cells rather than in the plasma, though small quantities are also present in the plasma.

Blood clotting   Normally, the plasma of the circulating blood remains a liquid with its colloidal protein in the sol state. Under certain conditions, however, when a blood vessel has been ruptured or otherwise damaged, or when certain kinds of foreign substances have gained entrance into the circulating blood, or when the blood has been removed from the body, one of the plasma proteins, ***fibrinogen,*** comes out of solution and forms a hard lump or clot. In this way a small hole in a vessel may be plugged, or a weakened place on the vessel wall may be strengthened. Clearly, blood clotting is an evolutionary adaptation for emergency repair of the circulatory system and for preventing excessive blood loss. Clotting occurs in all vertebrates and in some invertebrates. An alternative adaptation serving the same basic function is seen in some invertebrates with powerful muscles that contract and close off the hole or damaged area.

It can be shown that clotting does not, as might seem plausible, result from exposure of the blood to air or from interference with its flow. If blood is carefully removed from a vessel without allowing it in any way to contact the damaged portion of the vessel, and if this blood is then put in an open dish lined with a nonwettable plastic, it will remain liquid for many hours. In this experiment the blood is simultaneously exposed to air and prevented from flowing; yet it does not clot. Nor does it clot when it is held stationary in a portion of a blood vessel that has been tied off. Evidently the factor or factors responsible for clotting must be sought elsewhere.

Since blood clotting occurs when a vessel is damaged, and since it occurs at the site of the damage, might the damaged tissue release some chemical that initiates the clotting process? If we prepare an extract from cells and add this to the liquid blood in the plastic-lined dish of the previous experiment, a clot will promptly start to form. We seem to be on the right track. But when we remove the blood from the vessel in the same careful way, preventing it from contacting damaged tissue, and put it in a glass dish instead of one lined with nonwettable plastic, the blood promptly clots even without the addition of tissue extract. Where does this leave us? Might there be some other source of the clot-initiating chemical we have postulated? If we allow fresh blood to contact glass while we watch it under a high-powered microscope, we can see that blood platelets tend to disintegrate on contacting the glass. Perhaps the disintegrating platelets release a clot-initiating chemical similar to that released by damaged tissues. Platelets seem to disintegrate much more readily on contacting wettable foreign surfaces like glass than nonwettable surfaces; this would explain why clotting did not occur in the dish lined with nonwettable plastic. It can be shown, finally, that platelets also disintegrate when they contact damaged tissue.

The hypotheses suggested by the various experiments outlined here have been corroborated. It has been shown that both damaged tissues and disintegrating platelets release a group of substances called ***thromboplastins*** that initiate blood clotting. For reasons not well understood, thromboplastins are not effective, however, unless calcium ions are present. Some of the anticoagulants used in storing blood for later use in transfusions are chemicals that remove calcium ions from the blood and thus prevent clotting.

Why platelets disintegrate so readily on contact with damaged tissue or wettable foreign surfaces has not been satisfactorily explained. In fact, the platelets themselves, which appear to arise as fragments from certain large cells called megakaryocytes in the red bone marrow, are not well understood. They occur in all mammals, but are absent from the blood of other vertebrates, which, however, have certain spindle-shaped cells that seem to play a similar role in blood clotting.

We have now mentioned three substances essential for normal blood clotting: fibrinogen, thromboplastin, and calcium ions. But if we mix these three substances in a dish, no clotting occurs. Clearly, something else must be involved. That something else seems to be one of the globulin proteins of the plasma, known as **prothrombin.** If prothrombin is added to the mixture of fibrinogen, thromboplastin, and calcium, a clot will form. It can be demonstrated, however, that prothrombin itself has no effect on clotting; it must first be converted into **thrombin,** which is the substance that converts fibrinogen into its gel form, called **fibrin,** of which the clot is made.

Now we have identified the main ingredients in the clotting process. Thromboplastin, produced by disintegrating platelets or damaged tissue, converts the plasma protein prothrombin into thrombin; it is in this reaction that calcium ions participate. The thrombin then converts another plasma protein, fibrinogen, into fibrin. The fibrin fibers form a meshwork, which begins to shrink; finally, a fluid called serum is squeezed out, and a hardened clot is left in place. The reactions can be summarized as follows:

$$(1) \qquad \text{prothrombin} \xrightarrow[\text{Ca}^{++}]{\text{thromboplastin}} \text{thrombin}$$

$$(2) \qquad \text{fibrinogen} \xrightarrow{\text{thrombin}} \text{fibrin}$$

These simplified summary equations show the relations between the essential substances. Actually, numerous other substances play roles in the clotting process, as accelerators, inhibitors, etc.[4]

It would perhaps be well to clarify here the meanings of the three terms "whole blood," "blood plasma," and "blood serum," which occur frequently in references to medical procedures, particularly blood transfusions. Whole blood is blood just as it exists in the circulatory system, i.e. with none of its constituents removed. Blood plasma is whole blood minus the formed elements. Blood serum is plasma minus fibrinogen.

Erythrocytes and their function Human erythrocytes, or red blood cells, are small, biconcave, disc-shaped cells (Fig. 7.35) that lack nuclei. Normally, there are roughly five million of them per cubic millimeter of blood. Though the number of red cells remains amazingly constant from day to day, there is continual destruction of some cells and formation of new ones; the normal survival time of an erythrocyte is 120 days. More than two million erythrocytes are destroyed every second, chiefly in the liver and the spleen, where they are engulfed by large phagocytic cells. There are also phagocytic cells in the lymph nodes, which destroy any erythrocytes that escape from the blood and get into the lymph. The common supposition that the phagocytic cells of

---

[4] A total of 12 or more different factors have been identified as playing a role in blood clotting. The details of the clotting process are believed to differ, moreover, depending on whether initiation is by damaged tissue or only by platelets. The two-step scheme given above, sometimes called the classical model, is a simplified version of what is thought to happen when clotting is initiated by damaged tissue; several important factors have been omitted (a Factor VII must activate the thromboplastin, which then activates a Factor X, which, together with Factor V, converts prothrombin into thrombin).

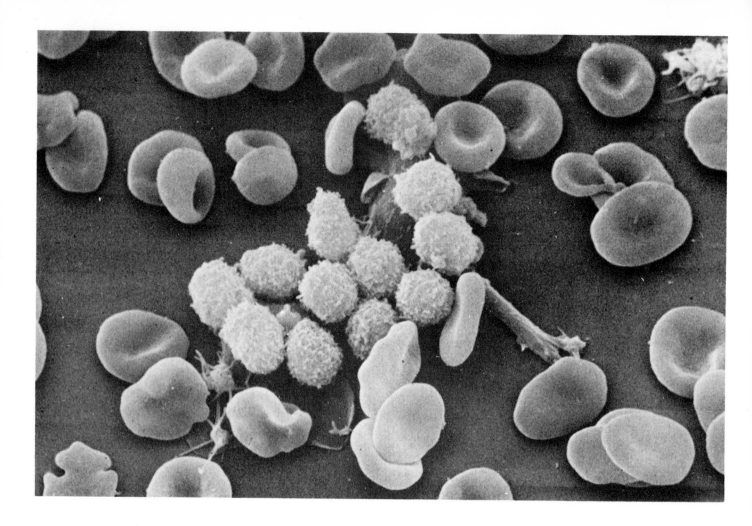

7.35 Scanning electron micrograph of blood cells
The erythrocytes look like biconcave discs. In the center is a cluster of lymphocytes, a type of white blood cells important in immunological defense against diseases. A large phagocytic macrophage can be seen under the lymphocytes. × 3,250. [Courtesy K. R. Porter and Ginny Fonte, University of Colorado.]

the liver and spleen destroy only old, worn-out red cells has not been proved.

The erythrocytes of adults are formed in the red bone marrow, which fills the interior of the upper ends of the long bones and the shafts of flat bones like those of the skull, ribs, and pelvis. Though the mature erythrocytes of mammals are devoid of nuclei, mitochondria, Golgi apparatus, etc., and therefore lack many of the characteristics of living cells, they arise from normally nucleated, rapidly dividing connective-tissue cells of the bone marrow. Toward the end of their development, they lose their nuclei and acquire the red oxygen-carrying pigment **hemoglobin,** a protein with iron-containing prosthetic groups. They then enter the circulating blood. The developmental sequence in vertebrates other than mammals is somewhat different in that their mature erythrocytes retain nuclei. It has been suggested that the evolutionary loss of the nuclei in mammals has the adaptive advantage of leaving room for more hemoglobin in each cell. Extra hemoglobin might, in turn, be correlated with the high metabolic rates—and therefore high oxygen demands—of the tissues of homeothermic animals. (However, since birds are also homeotherms but retain nucleated erythrocytes, this explanation is not fully convincing.) Analysis of the hemoglobin content of the red cells of a variety of vertebrates reveals that the red cells of mammals, though smaller than those of many of the lower vetebrates, do, in fact, contain more hemoglobin.

Not all invertebrates have hemoglobin. In those that have it, it is sometimes in cells, as in vertebrates, but in many it is simply dissolved in the plasma. Though a hemoglobin molecule can function just as well in the plasma as in erythrocytes, its location in cells has a decided adaptive advantage in animals with high metabolic rates; more pigment molecules can then be carried per unit volume of blood, and the oxygen-transporting capacity is correspondingly increased. A single human erythrocyte usually contains about 280 million molecules of hemoglobin. If all this hemoglobin were in the plasma, the concentration of plasma protein would be about three times higher, with profound effects on the osmotic balance between the blood and the tissue fluid. Erythrocytes, then, are a convenient method of packaging large amounts of hemoglobin with relatively little disturbance of the osmotic concentration of the blood, and of providing a chemical environment especially favorable for hemoglobin function.

Many of the invertebrates that lack hemoglobin have different oxygen-transporting pigments, which, like hemoglobin, combine a metal with protein. For example, many molluscs and arthropods have a pigment called hemocyanin, which contains copper instead of iron; when oxygenated, it is blue instead of red. Hemocyanin never occurs in cells, but is dissolved in the plasma.

Hemoglobin and its role in the transport of oxygen   The hemoglobin molecule is a globulin protein composed of four independent polypeptide chains (see Fig. 2.41, p. 61). Each of the four chains enfolds a complex prosthetic group called heme, which has an iron atom at its center (Fig. 7.36).

Human hemoglobin was one of the proteins used in developing methods for analysis of protein structure. The sequence of amino acids in the four chains (two pairs, one of type α, the other of type β) was determined by Gerhardt Braunitzer of the Max Planck Institute for Biochemistry in Munich, Germany, and by William H. Konigsberg and Robert J. Hill of the Rockefeller Institute in New York. Similar determinations for the hemoglobin of other vertebrates have demonstrated that they differ from one another in the number and sequence of amino acids. Nevertheless, as shown by Nobel Prize winner M. F. Perutz of Cambridge University through X-ray crystallographic studies, all normal hemoglobins have essentially the same three-dimensional structure. In some hereditary blood diseases, however, which involve changes in only one or two amino acids, conformational alterations do occur—alterations that severely impair the oxygen-transporting capability of the hemoglobin molecule.

As would be expected, the more closely related two animals are, the more similar their hemoglobins tend to be; e.g. those of humans and apes are much more alike than those of humans and fish. Sometimes hemoglobins differ within one and the same species. Thus the hemoglobin of human embryos, which is replaced by adult hemoglobin shortly after birth, is slightly different from that of adults.

Each of the four iron atoms in a hemoglobin molecule can, by virtue of its structural relationships within the molecule, combine loosely with one mol-

7.36   Structure of the heme group
A single molecule of hemoglobin has four of these iron-containing prosthetic groups.

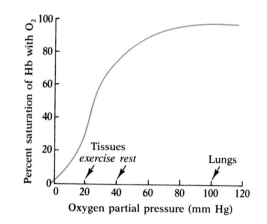

7.37 Dissociation curve of adult human hemoglobin

ecule of oxygen. The compound formed by the union of one molecule of hemoglobin (Hb for short) with four molecules of oxygen (Hb + $4O_2$) is called oxyhemoglobin.

Hemoglobin, as we mentioned on p. 75, exemplifies cooperativity in an allosteric molecule. When it binds with the first O_2 molecule, the conformational changes that occur enhance its affinity for oxygen and greatly facilitate the binding of the other three O_2 molecules. X-ray analysis shows that the important changes pertain, not to the tertiary structure of the four polypeptide chains, but to the way the chains fit together (the quaternary structure); thus the heme groups of the two β chains move farther apart, while those of the α chains move closer together. Just how these conformational changes influence the affinity of hemoglobin for oxygen is not yet fully clear. Presumably the altered positions of the heme groups relative to the R groups of neighboring amino acids permit easier access of O_2 to the iron atoms. An incidental effect of the formation of oxyhemoglobin, and the concomitant conformational changes, is to make the blood redder; it is on account of the oxyhemoglobin that arterial blood in the systemic circulation is more crimson than venous blood.

We have said that the combination of hemoglobin with O_2 is a loose one. Under certain conditions the combination will form, and under other conditions it will break down. Clearly, conditions in the lungs must favor formation of oxyhemoglobin, and conditions in the capillary beds of the systemic circulation must favor release of O_2 and re-formation of hemoglobin. The critical condition in determining whether hemoglobin will load or unload O_2 is the partial pressure[5] of O_2 in the medium to which the hemoglobin is exposed. When the partial pressure of O_2 is high, the hemoglobin picks up O_2; when the partial pressure of O_2 is low, the hemoglobin releases O_2. This is simply another way of saying that hemoglobin loads O_2 when there is a relatively high percentage of O_2 in the surrounding medium, and that it unloads O_2 when the percentage is relatively low. There is, of course, a relatively high partial pressure of O_2 in the air in the alveoli of the lungs and a relatively low partial pressure of O_2 in the tissues serviced by the systemic circulation, where the O_2 is being consumed in cellular respiration. Consequently hemoglobin tends to pick up O_2 in the capillaries of the lungs and to release O_2 in the capillaries of the systemic circulation.

Figure 7.37 shows a graph of the percentage of O_2 saturation of human hemoglobin at different partial pressures of O_2; the lower the partial pressure, the greater the tendency for oxyhemoglobin to dissociate into hemoglobin and O_2—hence the name "dissociation curve" for this graph. As you can see, the hemoglobin is about 98 percent saturated with O_2 at the partial pressure of O_2 typical of the lungs (100 mm), while it is only about 68 percent saturated at the partial pressure of O_2 typical of the tissues at rest (40 mm). The difference (30 percent) represents the approximate percentage of the O_2 carried by hemoglobin that is actually released to the tissues at rest. We see, then, that less than half the oxyhemoglobin releases its O_2 to the tissues and that venous blood still contains much O_2. During exercise the oxyhemoglobin releases more of its O_2 to the muscle tissues, so that the saturation of

[5] The partial pressure of a gas dissolved in a liquid refers to the pressure in the gas phase that would produce the observed concentration of gas molecules in the liquid.

venous blood may fall as low as 25 percent. This extra release of O_2 results from its more rapid utilization by the tissues during exercise, which causes a drop in its partial pressure in the tissues to a level of 20 mm Hg or even lower.

The S shape of the dissociation curve of hemoglobin (Fig. 7.37) is due to the cooperativity factor in the binding of O_2. Thus at very low O_2 pressures the curve rises slowly, because the binding of the first O_2 molecules to hemoglobin is difficult. But after those first molecules are bound, the binding of additional O_2 is easy, and the curve rises steeply until it finally levels off when the hemoglobin is nearly saturated with O_2. That hemoglobin saturation should vary with O_2 partial pressure according to an S-shaped curve has important implications for the release of O_2 to the tissues. Notice that the steepest portion of the curve is in the range of O_2 partial pressures prevalent in tissue fluid. Only a slight drop in the pressure here, as it occurs in the transition from rest to exercise, results in a very sizable increase in the amount of O_2 released from the oxyhemoglobin (e.g. the drop in pressure from 40 to 20 mm discussed in the preceding paragraph means an increase in the total percentage of O_2 released from about 30 to about 73 percent).

The affinity of hemoglobin for O_2 is markedly influenced by pH, because H^+ ions act as negative allosteric modulators for the hemoglobin. CO_2, produced as metabolic waste by the cells, is abundant in the tissue fluid of most parts of the body; its reaction with water results in increased acidity, i.e. increased concentration of H^+. Hence hemoglobin in the capillaries of the systemic circulation is exposed to an environment where its affinity for O_2 is reduced, which means that it will readily unload its O_2. Conversely, in the pulmonary capillaries, where CO_2 is released and the acidity is consequently lower, the hemoglobin is in an environment where its affinity for O_2 is high and where it therefore readily loads O_2. We see, then, that the waste product CO_2 plays an important regulatory role in shifting the condition of the hemoglobin back and forth between a propensity for loading O_2 in the lungs and for unloading it in the other tissues.

Temperature, as well as pH, affects the O_2 affinity of blood pigments. As shown by Fig. 7.38, hemoglobin releases O_2 more easily at higher temperatures, such as occur in muscles during strenuous exercise, when extra O_2 is needed.

Species-specific differences appear in the hemoglobin dissociation curves of mammals, the curves for smaller animals generally being to the right of those for larger animals (Fig. 7.39). It is, of course, advantageous to the smaller animals, which have a higher metabolic rate and whose tissues therefore require more O_2 per unit time, that their oxyhemoglobin should dissociate more readily.

The curves for birds, which have very high metabolic rates, as would be expected in such active animals, tend to be to the right of those for mammals. Those for cold-blooded animals such as fish and amphibians are generally to the left of those for warm-blooded animals; on account of their lower metabolic rates, the rate of demand for O_2 by their tissues is lower, and they have less need for oxyhemoglobin that dissociates very easily.

We have already mentioned that human fetal hemoglobin is slightly different chemically from adult hemoglobin. Figure 7.40 shows that fetal he-

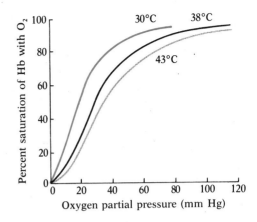

7.38 The effect of temperature on the affinity of human hemoglobin for oxygen

At higher temperatures the dissociation curve of hemoglobin is displaced to the right. In physiological terms, the hemoglobin has a lower affinity for O_2 and will therefore unload it at higher partial pressures of O_2, thus making it available to the tissues more readily.

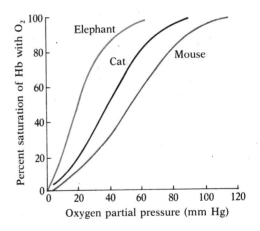

7.39 Dissociation curves for the hemoglobin of different animals

In general, the smaller the animal, the farther to the right will its curve be located. This means that small animals with high metabolic rates and correspondingly high O_2 requirements have hemoglobin that tends to unload more readily.

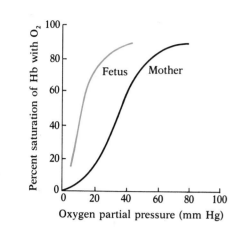

7.40 Dissociation curves for fetal and maternal hemoglobin in the cow
Because fetal hemoglobin has a higher affinity for O_2 than maternal hemoglobin, it can take O_2 from the maternal hemoglobin.

TABLE 7.1 *Percent of blood by volume occupied by red cells**

Animal	Altitude	Cells
Human	Sea level	46.0%
	5,360 m	59.9
Sheep	Sea level	35.3
	4,700 m	50.2
Dog	Sea level	34.6
	4,540 m	50.0
Rabbit	Sea level	35
	5,340 m	57
Vicuña	Sea level	29.8
	4,700 m	31.9

* Data from C. L. Prosser and F. A. Brown, *Comparative animal physiology*, Saunders, 1961.

moglobin has a higher affinity for O_2 than adult hemoglobin.[6] The adaptive significance of this difference is readily apparent. The fetus gets its O_2 from the mother's blood, not directly from the air. If the hemoglobin of the fetus is to be able to take O_2 from the hemoglobin of the mother, it must have the greater affinity for O_2; in other words, it must be able to compete successfully with the mother's hemoglobin.

The human fetus exemplifies an organism that must get its O_2 from a medium in which the partial pressure of O_2 is lower than in the atmosphere we ordinarily breathe. Animals like the South American llama and vicuña live at very high altitudes in the mountains, where the partial pressure of O_2 is appreciably lower than it is nearer sea level (as you will know if you have ever experienced shortness of breath at high altitudes). It is not surprising, therefore, to find that these animals, like the human fetus, have hemoglobin with a higher affinity for O_2 than the average mammal; i.e. their dissociation curves are to the left of the average, meaning that their hemoglobin loads more easily.

The llama and the vicuña have evolved a genetically determined type of hemoglobin that adapts them for life at high altitudes. But what of an animal such as a human, a sheep, a dog, or a rabbit that is moved to a high altitude? We know that eventually such an animal will become better acclimatized to its new environment and will no longer experience the severe shortness of breath it experienced at first. Does its hemoglobin change? The answer is no. An individual animal's genes, and the hemoglobin they determine, are not changed simply by changing the environment; acclimatization involves other processes. Table 7.1 shows that acclimatized humans, sheep, dogs, and rabbits at high altitudes have more erythrocytes per unit volume of blood than they normally have at sea level. This is apparently the result of at least two different reactions of the body to the decreased amount of O_2 reaching the tissues. First, erythrocytes stored in such areas as the spleen and skin capillaries are released into the general circulation. Second, in response to a hormonelike substance produced in the kidneys, the red bone marrow becomes more active and produces erythrocytes at a faster rate. Thus, through an increase in red cells, these animals partly compensate for the reduced percentage of saturation of the hemoglobin resulting from the lower partial pressure of O_2 at high altitudes. Notice, however, that the vicuña, which has hemoglobin adapted to high altitudes, shows little difference in red-cell count between sea level and high altitudes.

There are other gases besides O_2 that will also bind loosely to hemoglobin. One that binds even more readily than O_2 is carbon monoxide (CO). This gas, common in coal gas used for heating and cooking, in the exhaust from automobiles, and in tobacco smoke, is a dangerous poison because, even when its partial pressure in the air is relatively low, such a high percentage of the hemoglobin may bind with it that not enough is left to carry sufficient O_2 to the tissues. Severe symptoms of asphyxiation (impairment of vision, hearing, and thought) or even death may thus result from exposure to carbon monoxide.

[6] Isolated adult and fetal hemoglobin have the same affinity for O_2. But in the living animal adult hemoglobin has a much greater tendency to bind a substance called 2,3-diphosphoglycerate, and, as a result, a lesser tendency to bind O_2.

The role of hemoglobin in transport of carbon dioxide   The blood not only transports O_2 from the lungs to the tissues, but it also has the very important function of transporting CO_2 in the reverse direction, from the tissues to the lungs. Some of this CO_2 is carried as gas dissolved in the plasma and some in loose combination with hemoglobin in the red cells, but most of it is carried as bicarbonate ions in the red cells and plasma.

Relatively little of the CO_2 released from the tissue cells remains in the form of dissolved gas. Instead, it tends to combine with water to form carbonic acid (H_2CO_3):

$$CO_2 + H_2O \longrightarrow H_2CO_3$$

This reaction takes place particularly fast within the erythrocytes because these cells contain an enzyme that accelerates the reaction. But the blood must transport much CO_2, and if all of it were converted into carbonic acid and transported in this form, the pH of the blood would drop considerably. Now, cells are very sensitive to pH changes and can live only within a very narrow pH range. Any major drop in pH would clearly be very harmful to the organism.

Let us look at the problem of CO_2 transport more carefully. The difficulty arises because carbonic acid in the blood tends to dissociate into H^+ ions and HCO_3^- (bicarbonate) ions:

$$H_2CO_3 \longrightarrow H^+ + HCO_3^-$$

It is the increase of free H^+ ions that increases the acidity of the solution. If the H^+ ions could somehow be combined tightly with something else, instead of being left in the solution as free ions, then the acidity would not increase so much. This is exactly what happens in the blood—and hemoglobin and other proteins play the critical role in the process. Much of the hemoglobin (Hb) is normally present as an almost completely ionized potassium salt (K^+Hb^-). Carbonic acid reacts with the potassium hemoglobin to form acid hemoglobin (HHb) and potassium bicarbonate ($KHCO_3$), which is usually completely ionized ($K^+HCO_3^-$):

$$H^+ + HCO_3^- + K^+ + Hb^- \longrightarrow K^+ + HCO_3^- + HHb$$

You might well question the advantage of exchanging carbonic acid for acid hemoglobin; after all, both are acids. But there is a big difference between them: Acid hemoglobin is a much weaker acid than carbonic acid. Within the pH range of blood, very little of the acid hemoglobin is ionized. In other words, the reaction

$$HHb \longrightarrow H^+ + Hb^-$$

which would release free H^+ ions, does not occur at a significant rate. The reverse reaction

$$H^+ + Hb^- \longrightarrow HHb$$

is much more likely. This means that the formation of acid hemoglobin removes free H^+ ions from solution and prevents them from affecting the pH to any great extent. This is an example of the buffering action of proteins. Hemoglobin is the principal buffer substance within the erythrocytes; it even indirectly buffers the plasma and increases its bicarbonate transport-

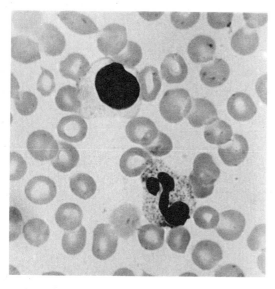

7.41 Photograph of human blood
The cells without nuclei are erythrocytes. Two types of leukocytes can be seen: The cell with the very large round nucleus is a lymphocyte, and the cell with the segmented nucleus is a polymorphonuclear neutrophil. × 1,750. [Courtesy Dorothea Zucker-Franklin, New York University Medical Center.]

ing capacity. The plasma proteins also act as buffers in the plasma. Transport of lactic acid, for example, involves buffering action by plasma proteins.

In summary, CO_2 released by the tissues and picked up by the blood reacts with water to form carbonic acid, most of which ionizes into H^+ ions and bicarbonate ions. It is in the form of bicarbonate that most of the transport takes place, the excess H^+ ions being bound to hemoglobin. In the lungs the situation is reversed. Here the CO_2 pressure is less than in the blood, and the gradient therefore favors release of the CO_2 by a reversal of the chemical reactions outlined above.

Leukocytes and their functions Human leukocytes, or white blood cells, are bigger than the erythrocytes and have large, often irregularly shaped nuclei (Fig. 7.41). At least five different types can be distinguished on the basis of the shape of the nucleus and the density of granules in the cytoplasm. The granular types are formed in the red bone marrow, while the agranular or clear types are formed in lymphoid tissues such as those of the lymph nodes, spleen, tonsils, adenoids, and the thymus. Leukocytes are not restricted to blood, being even more abundant in the lymphatic system. And they are also found wandering free in loose connective tissues and occasionally in other tissues. They are capable of amoeboid movement, and can escape from the blood and lymph vessels by squeezing through the vessel walls at the points of contact between the endothelial cells. In essence, then, all connective tissues, including blood and lymph, form one continuous system so far as the leukocytes are concerned.

The leukocytes play a variety of very important roles in the body's defenses against disease and infection. Apparently both damaged tissues and invading bacteria release chemicals that attract leukocytes. Some kinds of leukocytes act as phagocytes, engulfing and destroying bacteria and remnants of damaged tissue cells. Other kinds produce powerful enzymes that help detoxify foreign proteins and other potentially dangerous substances. In a severe infection the leukocyte count in the blood and lymph increases enormously, and vast numbers may invade the infection area. The resulting mixture of dead tissue, bacterial cells, and living and dead leukocytes that accumulates at the site of the infection is commonly known as pus. Abnormal dilation of the blood vessels in the infected area produces a local increase in temperature and reddening known as an inflammation. Local swelling occurs, because the blood vessels are often more permeable than normal and lose fluid to the tissues. Both the vasodilation and the increased capillary permeability are stimulated by a substance called *histamine,* which is secreted by one kind of leukocyte and also by some other cells in the infected area.

Although the leukocytes of other vertebrates are often not morphologically identical with those of human beings, and the corresponding cells of invertebrates differ even more, most animals have blood cells that serve similar phagocytic, detoxifying, and histamine-secreting functions.

Leukocytes have still another way of fighting disease and infection. Certain nonphagocytic leukocytes known as *lymphocytes* give rise to specialized cells that play a central role in immunologic reactions. These cells respond to the presence of certain kinds of foreign substances, called *antigens,* by making *antibodies* that destroy or inactivate the antigens. Each type of antibody is usually very specific and will inactivate only the antigen that stimulated its synthesis. The antibodies are globulin proteins in the blood plasma. Whereas phagocytosis and detoxification are particularly important as first defenses against acute infections, production of antibodies confers a degree of active immunity against the infection and is critical in fighting long-term chronic infections and in building resistance against further infection. We shall discuss the topic of immunologic competency at greater length in a later chapter.

SUGGESTED READING

ADOLPH, E. F., 1967. "The heart's pacemaker," *Scientific American,* March. (Offprint 1067) *How nodal tissue regulates the fundamental rhythm of the heart.*

BIDDULPH, O., and S. BIDDULPH, 1959. "The circulatory system of plants," *Scientific American,* February. (Offprint 53) *Excellent discussion of the physiology of phloem by two investigators who carried out some of the early radioactive-tracer studies of mineral transport.*

KILGOUR, F. G., 1952. "William Harvey," *Scientific American,* June. *The story of Harvey's classic work on the circulatory system.*

MAYERSON, H. S., 1963. "The lymphatic system," *Scientific American,* June. (Offprint 158)

PERUTZ, M. F., 1964. "The hemoglobin molecule," *Scientific American,* November. (Offprint 196) *The research that led to the discovery of the structure of hemoglobin—an account by one of the principals in that discovery.*

PERUTZ, M. F., 1978. "Hemoglobin structure and respiratory transport," *Scientific American,* December. (Offprint 1413)

RAY, P. M., 1972. *The living plant,* 2nd ed. Holt, Rinehart & Winston, New York. *Short elementary text on plant physiology (see esp. Chapters 5, 7).**

WOOD, J. E., 1968. "The venous system," *Scientific American,* January. (Offprint 1093) *How veins, by constricting or dilating, help determine the distribution of blood in the human body.*

ZIMMERMANN, M. H. 1963. "How sap moves in trees," *Scientific American,* March. (Offprint 154)

ZWEIFACH, B. W., 1959. "The microcirculation of the blood," *Scientific American,* January. (Offprint 64) *On capillaries, metarterioles, and venules.*

* Available in paperback.

Chapter **8**

REGULATION OF BODY FLUIDS

Evidence of many types has led biologists to the conclusion that life had its origin in the ancient seas. Of the major environmental media of the earth—seawater, freshwater, air—seawater exhibits by far the greatest stability. In such crucial characteristics as temperature, acidity, and salt concentration, the seas fluctuate remarkably little over immense spans of time, their vast bulk making any change very gradual and slow.

We have already seen that a living cell interacts constantly with its surrounding environmental medium. Such critical functions as nutrient procurement, gas exchange, metabolism—indeed, life itself—are closely dependent on the properties of the surrounding medium. It is not surprising, therefore, that the protoplasm of the early cells had many characteristics in common with the seawater that bathed them, and that the life processes evolved a close dependence on the stable conditions existing in seawater. Nor is it surprising that the evolution of complex multicellular marine animals involved the development of body fluids—tissue fluid, blood, etc.—that could provide even the innermost body cells with a relatively nonfluctuating aquatic environment, and that the internal body fluids of those primitive marine animals resembled in many important ways the seawater that had been the cradle of life.

As the ages passed and evolution continued, the body fluids of different organisms evolved in different ways, just as their other characteristics did. Present-day marine animals, for example, differ noticeably in the chemical makeup of their body fluids (Table 8.1), though these fluids are more similar to one another and to seawater than they are to the body fluids of freshwater or terrestrial organisms, not to speak of those of plants. Nonetheless, all these fluids have much in common, and, as Ernest Baldwin of Cambridge University has said, "The conditions under which cell life is possible are very restricted indeed and have not changed substantially since life first began."

All living things must maintain within themselves a fluid environment favorable to the continued life of their cells. The reason was stated succinctly by the great nineteenth-century physiologist Claude Bernard: *"La fixité du milieu intérieur est la condition de la vie libre"*; or, to paraphrase: A constant internal fluid environment is the prerequisite to survival under varying external conditions. Thus the evolutionary development of immense diversity among living organisms has necessarily involved the concomitant evolution of diverse mechanisms for maintaining homeostasis in their body fluids.

THE EXTRACELLULAR FLUIDS OF PLANTS

Multicellular marine algae differ markedly from multicellular marine animals in the sort of fluid environment to which their cells are exposed. Roughly 50 percent of the water in the body of a complex animal is extracellular, being in the form of tissue fluid, lymph, or blood plasma. This extracellular fluid, which bathes most of the cells, is separated from the environmental water by cellular barriers and has a characteristic composition, differing both from that of the intracellular fluid and from that of the surrounding water. By contrast, most of the fluid content of a multicellular alga is intracellular. The fluid filling its intercellular spaces is essentially continuous with the environmental water and cannot be regarded as separate or distinct. The alga thus has no fluid fully analogous to the tissue fluid and blood of an animal. Hence, unlike the animal, which must regulate the composition of both intracellular and extracellular fluids, the alga must regulate the composition of its intracellular fluids alone.

TABLE 8.1   *Concentrations of ions in seawater and in body fluids* (millimoles/liter)*

	Na^+	K^+	Ca^{++}	Mg^{++}	Cl^-
Seawater	470.2	9.9	10.2	53.6	548.3
Marine invertebrates					
Jellyfish (*Aurelia*)	454.0	10.2	9.7	51.0	554.0
Sea urchin (*Echinus*)	444.0	9.6	9.9	50.2	522.0
Lobster (*Homarus*)	472.0	10.0	15.6	6.8	470.0
Crab (*Carcinus*)	468.0	12.1	17.5	23.6	524.0
Freshwater invertebrates					
Clam (*Anodonta*)	13.9	0.3	11.0	0.3	12.0
Crayfish (*Cambarus*)	146.0	3.9	8.1	4.3	139.0
Terrestrial animals					
Cockroach (*Periplaneta*)	161.0	7.9	4.0	5.6	144.0
Honeybee (*Apis*)	11.0	31.0	18.0	21.0	?
Japanese beetle (*Popillia*)	20.0	10.0	16.0	39.0	19.0
Chicken	154.0	6.0	5.6	2.3	122.0
Human	145.0	5.1	2.5	1.2	103.0

* This table is based on a larger one in C. L. Prosser and F. A. Brown, *Comparative animal physiology*, Saunders, 1961.

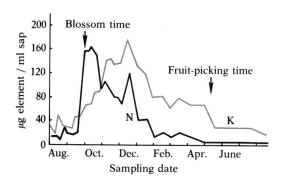

8.1 Yearly fluctuations in nitrogen and potassium concentrations in the xylem sap of apple trees in New Zealand
The fluctuations of both substances are very great—far greater than could be tolerated by animal cells. [Modified from E. G. Bollard, *J. Exp. Botany*, vol. 4, 1953.]

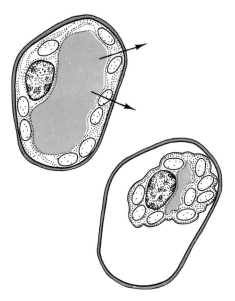

8.2 Plasmolysis
A plant cell in a hypotonic medium will lose so much water (left) that, as it shrinks, it will pull away from its more rigid wall (right).

A similar contrast appears between an animal and a large vascular land plant. Such a plant obviously contains much extracellular fluid in the form of xylem sap and the water imbibed in its cell walls. But this fluid is not as fully distinct from the environmental water as the tissue fluid and blood of an animal. You will recall that water can penetrate far into the cortex of a root by flowing along cell walls without having to cross any membranous barrier. Thus, much of the fluid that directly bathes the plant cells, even those far inside the plant body, is essentially continuous with the environmental water and is therefore not fully analogous to animal tissue fluid, which is separated from the environmental medium by a membranous barrier. This means, of course, that the composition of much of the extracellular fluid of the plant cannot be so well regulated as the tissue fluid and blood of animals. Even the composition of the xylem sap, which is separated from external water by the membranous barrier of the endodermis, fluctuates widely, being dependent on such factors as the environmental conditions, the health of the plant, and the season of the year (Fig. 8.1).

It is easy to understand why the inability of marine algal cells to regulate the composition of the fluid that bathes them poses no serious problem for the life of the cells; that fluid, after all, is essentially the same as seawater, which we have already said is the nonfluctuating medium in which life arose. (It is true that modern seawater has a composition rather different from that of the ancient seas, but the change was so very gradual that the organisms living in the seas had ample time to evolve with their evolving environment.) But what about a plant living in freshwater or on land? The fluids to which the internal cells of these plants are exposed will fluctuate much more than the tissue fluids of animals, and even their normal composition is one that would be quickly fatal to the internal cells of most higher animals.

The reasons why plant cells seem able to withstand much greater fluctuations in the makeup of the fluids bathing them are complex and not yet well understood. But a partial explanation can be given. Animal cells are seriously affected by changes in the osmotic concentration of the extracellular body fluids, which, under normal conditions, are approximately isotonic with the cells; osmotic shifts often severely alter the physiology of the cells or even kill them. Unlike the animal cell, the cell of a land or freshwater plant almost always exists in a medium that is much more dilute than the cell's contents. In other words, the plant cell is decidedly hypertonic relative to the fluid that bathes it. In such a situation an animal cell would take in so much water by osmosis that it would burst, unless it had some special mechanism for expelling the excess water. But the plant cell is surrounded by its cell wall, and as the cell takes in more water and becomes more turgid, the wall resists further expansion. Eventually the resistance of the wall will be as great as the tendency of water to enter the cell by osmosis, and no further net gain of water by the cell is possible.

We see, then, that the plant cell can withstand rather pronounced changes in the osmotic concentration of the surrounding fluids as long as those fluids remain more dilute than the cell's contents, i.e. as long as the fluids remain appreciably hypotonic relative to the cell. If the external fluids become decidedly hypertonic relative to the cell, the cell may lose so much water and shrink so grievously that it pulls away from its more rigid wall; such a cell is said to be plasmolyzed, and the phenomenon is called **plasmolysis** (Fig. 8.2).

The presence of the cell wall in plants and its absence in animals thus make the problem of salt and water balance quite different in the two.

To say that plants can tolerate much greater changes in the osmotic concentration of the fluids that bathe their cells is not to say that they are unaffected by changes in the concentration of individual ions in the surrounding medium. Such changes sometimes have pronounced effects on their health and growth. But the effects are usually attributable to an alteration in the chemical makeup of the plant, rather than to a disruption of its osmotic balance.

Land plants, like land animals, are often exposed to conditions that may cause excessive water loss by evaporation. We have already examined some of the adaptations whereby plants resist desiccation—cuticles on their exposed surfaces, regulation of stomatal openings by guard cells, stomata in deep hair-lined pits in some plants living in very dry regions, etc. Some plants show little tolerance of drought and soon die when the soil moisture becomes deficient; many plants that grow only in the shade are in this category. Some plants, including many mosses, lichens, and ferns, are drought-tolerant because their cells can be dehydrated without permanent injury. Other plants can survive through long periods of drought because they store very large quantities of water, and because they lose little by evaporation, thanks to very thick cuticles, few stomata, and a small surface-to-volume ratio; cacti and other succulent desert plants are good examples. Other plants are only moderately well equipped to withstand droughts; with a limited ability to endure dehydration, they frequently combine some adaptations for preventing water loss with large deep-penetrating root systems that increase absorptive capacity. Study of adaptations that enable plants to withstand drought is an interesting field in its own right.

THE VERTEBRATE LIVER

As a first example of the problems attendant on keeping the internal fluids of a complex vertebrate animal, such as a human, relatively constant in composition, consider the blood leaving the intestinal capillaries shortly after a meal. Digestion is taking place in the small intestine, and the products of digestion are moving in large quantities into the capillaries of the intestinal villi. This means that the blood leaving these capillaries contains high concentrations of such compounds as simple sugars and amino acids—concentrations considerably greater than those normally found in the blood in most parts of the circulatory system. But wholesale addition of these materials to the blood, if not controlled, would drastically alter the composition of the blood and other body fluids and make impossible the maintenance of a relatively nonfluctuating fluid environment for the cells.

Vertebrates meet this difficulty with the help of a very important homeostatic organ, the liver. Blood from the intestine and stomach is collected in the *portal vein,* which does not empty into the vena cava as might be expected, but goes to the liver, where it breaks up into a network of capillaries (or, more precisely, sinuses) in the liver tissue (Fig. 8.3). The liver is one of only three places in the human body where blood passes through a second set of capillaries before returning to the heart;[1] other blood circuits involve only a single capillary bed.

[1] The others, to be discussed later, are the kidney and the pituitary gland.

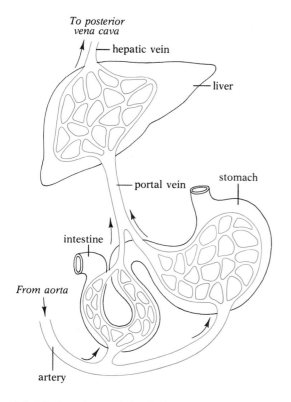

*To posterior
vena cava*

— hepatic vein

— liver

— portal vein

stomach

intestine

From aorta

artery

8.3 The hepatic portal circulation
Blood from the capillaries of the digestive tract and
stomach is carried by the portal vein to a second bed
of capillaries in the liver. It then flows via the hepatic
vein into the posterior vena cava, which takes it to the
heart.

The liver's role in regulation of the blood-sugar level   After a meal the
blood coming to the liver via the portal vein has a higher than normal
concentration of glucose. Under these conditions the liver removes most of
the excess, converting it into the insoluble polysaccharide glycogen, which is
the principal storage form of carbohydrate in animal cells. Thus the blood
leaving the liver via the **hepatic vein** (a vein that leads into the posterior vena
cava) contains a concentration of glucose only slightly higher than that
normally found in the arteries. After this blood is mixed in the vena cava
with blood from other parts of the body—blood that contains a lower concentration of sugar because it has given up glucose to the tissues through
which it has passed—the blood entering the heart has a glucose concentration within the normal tolerance range.

If the incoming supply of glucose from the intestine exceeds all the body's
immediate needs, and the liver has stored its full capacity of glycogen, the
liver begins converting glucose into fat, which can then be stored in the
various regions of adipose tissue throughout the body. Thus, in spite of the
great quantities of sugar absorbed by the intestine, the blood-sugar level in
most of the circulatory system is not greatly raised and homeostasis is
maintained.

The whole process is reversed if some time has elapsed since the last meal
and no sugar is being absorbed from the intestine. At such times, blood in
the intestinal capillaries gives up glucose to the intestinal cells, just as blood
in any capillary bed will give up nourishment to the surrounding tissues.
This means that the blood reaching the liver via the portal vein is poor in
glucose. Under these conditions the liver converts some of its stored glycogen into glucose and adds it to the blood. The result is that blood leaving
the liver in the hepatic vein has its normal glucose concentration, and that
the blood entering the heart likewise has the normal blood-sugar level. We
see, then, that the liver functions in helping maintain the blood-sugar concentration in a steady state.

The human liver is capable of storing enough glycogen to supply glucose
to the blood for a period of about 24 hours. What happens if at the end of
that period no new glucose has come to the liver from the intestine? A drop
in the blood-sugar concentration to a level much below normal would soon
be fatal; the brain cells are particularly sensitive to such a drop, because
they cannot store adequate amounts of glucose themselves or use fats or
amino acids as energy sources, and are thus wholly dependent on a regular
supply of glucose from the blood. Under such conditions the liver begins
converting other substances, such as amino acids, into glucose, and in this
way maintains the normal blood-sugar level. Some other tissues of the body,
particularly muscle, can store glucose as glycogen, but muscle glycogen
apparently serves as a local fuel deposit only and is not generally available
for maintaining the blood-glucose level.

The liver's activity in carbohydrate metabolism is regulated in a complex
fashion by several hormones, as will be described in Chapter 9. An abnormal
balance of these hormones may result in an unusually high blood-sugar level
or an unusually low one. Either condition can be dangerous.

The liver's role in the metabolism of lipids and amino acids   The liver is
not only responsible for carbohydrate conversions essential to the mainte-

nance of a relatively constant blood-sugar level; it is also important in the processing and modification of fatty acids and other lipid materials. For example, much of the fatty acid mobilized from adipose tissues during periods of starvation is first incorporated into lipoprotein by the liver before it is used as a source of metabolic energy by other tissues of the body. (Note, however, that most animals are unable to convert fatty acids into glucose. Hence the brain, which cannot metabolize fatty acids, does not benefit directly from the mobilization of fat reserves.)

Like the glucose, the amino acids absorbed by the villi of the intestine pass into the portal vein and thence to the liver. The liver removes many of these amino acids from the blood, temporarily storing small quantities and later gradually returning them to the blood, which carries them to other tissues for use in the synthesis of enzymes, hormones, or new protoplasm. But the usual diet contains far more amino acid than can be used in such syntheses. Unlike the plant, the animal body is capable of very little long-term storage of amino acids, proteins, or other nitrogenous compounds; those used as an energy source when supplies of carbohydrates and fats are exhausted are not stored products, but the actual structural material of living cells. Hence the excess amino acids must be converted into other substances such as glucose, glycogen, or fat. Such conversions take place in the liver.

You will recall that amino acids differ from most carbohydrates and fats in containing nitrogen in the form of an amino group ($-NH_2$). It is not surprising, therefore, that the first step in converting amino acids into these other substances is *deamination,* or the removal of the amino group. In the deamination reaction the amino group is converted into *ammonia* (NH_3).[2] In some animals the liver simply releases this ammonia, which is a waste material, into the blood, and it is soon removed from the blood and from the body of the organisms by excretory mechanisms. In many other animals, including humans, the liver first combines the ammonia with carbon dioxide to form a more complex but less toxic nitrogenous compound called *urea* (Fig. 8.4) and then releases the urea into the blood. In still other animals the liver converts the waste ammonia into a compound more complex than urea, called *uric acid,* which it releases into the blood. In short, whether the nitrogenous waste product is ammonia, urea, uric acid, or some other compound, the liver dumps it into the blood, and it becomes necessary for another system of the body to prevent the wastes from reaching too high a concentration in the body fluids.

Notice that animals differ markedly from green plants in being unable to reuse much of the nitrogen from the amino acids they metabolize. Green plants can shift nitrogen-containing groups from one organic compound to another more freely than animals, and they can also use inorganic nitrogen to synthesize organic nitrogen-containing compounds. Hence excretion of nitrogenous wastes is essentially an animal activity.

A summary of the liver's functions Over and over, as we have examined the physiology of vertebrate animals, we have mentioned some important role played by the liver. Before we leave the present discussion of this vital and versatile organ, let us compile a list of its functions. Though probably far

[2] At physiological pH, most of the ammonia is rapidly converted into the ammonium ion (NH_4^+).

8.4 **Three important nitrogenous waste compounds**

from complete, the list will convey how extensive and varied a part the liver plays in the maintenance of life.

1. This organ removes excess glucose from the blood and stores it as glycogen, and reconverts glycogen into glucose to maintain the blood-sugar level when the incoming supply is insufficient.
2. It resynthesizes glycogen from some of the lactic acid produced by muscles during glycolysis.
3. It plays a major role in the interconversion of various nutrients, e.g. the conversion of carbohydrates into fats, of incoming fats into fats more typical of the organism's own body, of amino acids into carbohydrates or fats.
4. It deaminates amino acids, may convert the ammonia thus obtained into urea, uric acid, or some other compound, and releases the nitrogenous wastes into the blood.
5. It detoxifies a great variety of injurious chemical compounds and is thus one of the body's most important defenses against poisons.
6. It manufactures many of the plasma proteins, including fibrinogen, prothrombin, albumin, and some globulin.
7. It manufactures some plasma lipids, including cholesterol, and plays an important role in processing and modifying lipids mobilized from adipose tissues.
8. It stores various important substances such as vitamins and iron.
9. It forms erythrocytes in the embryo.
10. It destroys red blood cells.
11. It excretes bile pigments.
12. It synthesizes bile salts.

THE PROBLEM OF EXCRETION AND SALT AND WATER BALANCE IN ANIMALS

We have seen that animals need mechanisms for ridding their bodies of metabolic wastes—particularly nitrogenous ones, but many others as well. The process of releasing such useless, even poisonous, substances is called excretion. It should not be confused with elimination (defecation). Whereas excretion is release of wastes that have been inside the cells, tissue fluids, or blood of the organism, elimination is release of unabsorbed wastes from the digestive tract.

In general, excretory mechanisms also serve a second very important function: They help regulate the water and salt balance of the organism. Our examination of excretion will focus on both these aspects, which are in most instances inextricably intertwined.

THE PROBLEM IN AQUATIC ANIMALS

As we saw, the first nitrogenous waste formed by deamination of amino acids is ammonia. Now, ammonia is an exceedingly poisonous compound, and no organism can survive if its concentration in the body fluids gets very high. But the small, highly soluble molecules of ammonia readily diffuse across cell membranes, and there is no great difficulty in getting rid of them if an adequate supply of water is available. The water keeps the solution dilute while the ammonia is in the body, acts as a vehicle for its expulsion from the body, and flushes it rapidly away from the vicinity of the animal. In view of the plentiful supply of water available to aquatic animals, it is not surprising that for many of these the characteristic nitrogenous excretory product is ammonia.

Marine invertebrates Many marine invertebrates lack special excretory systems, relying instead on release of wastes across the general surface membranes. Such organisms seldom have any problem with water balance, because they are essentially isotonic with the surrounding seawater, and hence neither take in much excess water nor lose too much. A variety of these organisms supplement the excretory process by phagocytic excretion; i.e. certain cells pick up solid particles of waste material by phagocytosis and then move to the outer body surface or to the surface of the digestive cavity, where the materials are released.

Maintenance of the proper nonfluctuating internal fluid environment is relatively simple for marine invertebrates as long as they remain in the sea; it is quite a different matter when they move into hypotonic media such as the brackish water of estuaries or the fresh water of rivers and lakes. Many marine animals are incapable of moving into such habitats, because their body fluids always lose salts until they have about the same salinity and osmotic concentration as the external fluids. Since their cells generally cannot tolerate much change in the makeup of the fluids bathing them, these animals soon die when they are put into brackish or fresh water. An example is the spider crab *(Maia)* (Fig. 8.5).

Some marine animals, however, have evolved adaptations that enable them to move into hypotonic media. The adaptations may be of an evasive character, as in oysters and clams, which simply close their shells and thereby exclude the external water during those parts of the tidal cycle when the water in the estuaries is very dilute. But by far the most important adaptations for survival in dilute media—and the ones that have played the principal role in the evolutionary movement of animals into freshwater—are those that enable animals to regulate the osmotic concentrations of their body fluids and keep them constant despite fluctuations in the external medium. Such organisms are said to have the power of **osmoregulation.**

The shore crab *(Carcinus)* is an example of a marine invertebrate that has evolved a degree of osmoregulation enabling it to live in both seawater and brackish water (Fig. 8.5). In seawater the crab's body fluids are in osmotic equilibrium, but in brackish water they are hypertonic relative to the surrounding medium. To maintain the internal fluids near their normal concentration in brackish water, cells on the crab's gills remove salt from the surrounding water and actively secrete it into the blood, while the excretory organs eliminate the excess water that constantly pours in.

Freshwater animals Once the ancestors of the modern freshwater animals had made the transition to the freshwater environment, presumably by way of the estuaries, there was no longer any great advantage to their descendants in continuing to maintain body fluids as concentrated as seawater, as long as they remained in their new environment. Such excessively hypertonic internal conditions simply aggravated the problems of obtaining enough salt and bailing out excess water. Thus it is understandable that

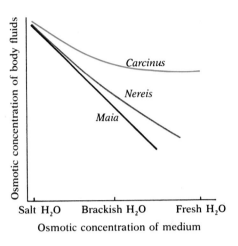

8.5 Variation of internal osmotic concentration with external osmotic concentration in three marine invertebrates

The spider crab *(Maia)* has no osmoregulatory capacity, and the concentration of its body fluids falls in direct proportion to the fall in the concentration of the external medium. The clam worm *(Nereis)* has very slight osmoregulatory capacity; the concentration of its body fluids does not bear a straight-line relationship to the concentration of the external medium. The shore crab *(Carcinus)* has considerable osmoregulatory capability and can maintain relatively concentrated body fluids even in a very dilute external medium. [Modified from E. Baldwin, *An introduction to comparative biochemistry*, Cambridge University Press, 1948.]

natural selection should have favored a reduction of the osmotic concentration of the body fluids within the bounds possible for the continuance of the life of the tissues, and that modern freshwater animals, both invertebrate and vertebrate, have osmotic concentrations decidedly lower than seawater (Table 8.2). It seems incompatible with cellular existence, however, for the body fluids to be as dilute as freshwater, for no organisms are actually isotonic with their freshwater medium; the body fluids of freshwater animals are typically hypotonic relative to seawater, but hypertonic relative to freshwater (Table 8.2).

Now, if freshwater animals are hypertonic relative to the surrounding environmental medium, there will be a strong tendency for water to move into the organism and for salts to be lost from the organism to the surrounding water. At first glance, the obvious evolutionary solution to this problem might seem to be the development of completely impermeable membranes covering the entire body, but further thought shows that this solution would have been impracticable, since a truly aquatic organism must maintain some permeable membranes exposed to the water for gas exchange. Because mammals that live in the water breathe air and hence need never expose permeable respiratory membranes to the water, they can maintain an impermeable barrier between their body fluids and the water in which they live. But fully aquatic freshwater animals cannot use the "method of evasion" exclusively. They must also be able to carry out active osmoregulation, which usually involves excretory organs that can pump out the water as fast as it floods in—preferably through the production of urine more dilute than the body fluids—and/or special secretory cells somewhere on the body that can absorb salts from the environment and release them into the blood. Both corrective measures—production of dilute urine and absorption of salts—entail movement of materials against concentration gradients and therefore necessitate expenditure of energy.

An examination of the water and salt regulation typical of modern freshwater bony fishes will provide a good example of the above-mentioned processes. The blood and tissue fluids of the fish are more concentrated than the environmental water (Table 8.2). The method of evasion is used to the extent that much of the body is covered by relatively impermeable skin and

TABLE 8.2 *Concentrations of ions in the blood of freshwater animals compared with seawater* * *(millimoles/liter)*

	Na$^+$	K$^+$	Ca^{++}	Mg^{++}	Cl$^-$
Seawater	470.2	9.9	10.2	53.6	548.3
Brown trout (*Salmo*)	144.0	6.0	5.3	?	151.0
Crayfish (*Cambarus*)	146.0	3.9	8.1	4.3	139.0
Clam (*Anodonta*)	13.9	0.3	11.0	0.3	12.0
Freshwater†	0.65	0.01	2.0	0.2	0.5

* This table is based on a larger one in C. L. Prosser and F. A. Brown, *Comparative animal physiology*, Saunders, 1961.

† The values for freshwater are representative only; actual values vary greatly, as the water ranges from "soft" (low concentrations of dissolved minerals, e.g. 0.22 Ca^{++}) to "hard" (high concentrations of dissolved minerals, e.g. 5.0 Ca^{++}).

scales, and that the fish almost never drink. There is, however, a constant osmotic intake of water across the membranes of the gills and of the mouth, and a constant loss of salts across the same membranes. The method of correction is used in two ways: The excess water is eliminated in the form of very dilute and copious urine produced by the kidneys, and salts are actively absorbed by specialized cells in the gills (Fig. 8.6).

Marine vertebrates   Curiously enough, bony fishes living in the sea have the reverse problem: They live in water, yet they steadily lose water to their environment and are in constant danger of dehydration. The explanation is that the ancestors of the bony fishes apparently lived in freshwater, not in the sea, and that when some of their descendants moved to the marine environment they retained their dilute body fluids. Thus marine bony fishes are hypotonic relative to the surrounding water, and they have the problem of excessive water loss and excessive salt intake. Besides benefiting from the evasive adaptation of relatively impermeable skin and scales, they use two corrective measures: They drink almost continuously to replace the water they are constantly losing, and, by means of specialized cells in the gills, they actively excrete the salts they unavoidably take in with the water (Fig. 8.6). Most of the nitrogenous wastes are excreted as ammonia[3] through the gills; hence only a small quantity of urine is produced by the kidneys, and little water need be lost in this manner. Apparently fish kidneys have not evolved the capacity to produce concentrated urine, and they are consequently of no help in salt elimination.

The marine elasmobranch fishes (sharks and their relatives) probably also evolved from freshwater ancestors, but they solved the osmotic problem in a very different way. Their blood contains about the same concentrations of salts as the blood of marine bony fishes, but their blood also contains high concentrations of urea, to which they are much more tolerant than most vertebrates. By conserving urea instead of excreting it, the marine elasmobranchs maintain a total osmotic concentration in their blood slightly greater than that of seawater. They therefore have no problem of water loss. Excess salt is excreted by special glandular cells in the rectum.

THE PROBLEM IN TERRESTRIAL ANIMALS

We have already seen that one of the conditions of animal life in freshwater is a relatively impermeable covering for all but certain portions of the body surface, as an aid in preventing excessive absorption of water. For this reason freshwater animals had an important preadaptive advantage over primitively marine animals in colonizing the terrestrial environment. The evidence strongly supports the view that the movement to land was by way of freshwater, not directly from the sea.

On land the greatest threat to life is desiccation. Water is lost by evaporation from the respiratory surfaces (lungs, tracheae, etc.), by evaporation from the general body surface, by elimination in the feces, and by excretion in the urine. The lost water must obviously be replaced if life is to continue. It is replaced by drinking, by eating foods containing water, and by the

---

[3] Fish also excrete some nitrogen (usually less than 10 percent) as urea.

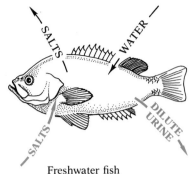

Freshwater fish
(hypertonic relative to medium)

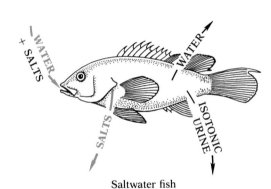

Saltwater fish
(hypotonic relative to medium)

8.6 Osmoregulation in bony fishes
Freshwater fishes tend to take in excessive amounts of water and to lose too much salt. They compensate by seldom drinking, by actively absorbing salts through specialized cells on their gills, and by excreting copious dilute urine. Saltwater fishes tend to lose too much water and to take in too much salt. They compensate by drinking constantly and by actively excreting salts across their gills. They cannot produce hypertonic urine; hence the kidneys are of little aid to them in osmoregulation.

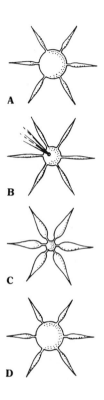

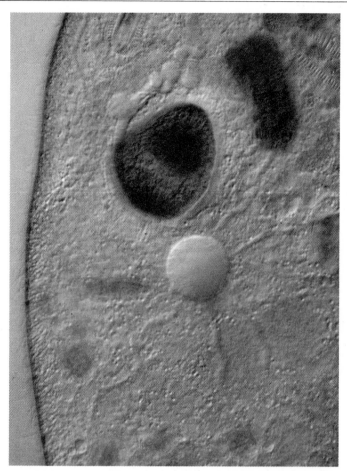

A

8.7 Contractile vacuole of *Paramecium caudatum*
The sequence shown diagrammatically above can also
be observed in the three photographs of a live
organism taken in the course of the expansion-
contraction cycle of its contractile vacuoles
(only one vacuole is seen here). The large green
object and the brown ones are remains of other
microorganisms ingested by the *Paramecium*. (A) The
vacuole is full. As shown in the diagram, a system of
radiating canals brings fluid from the cytoplasm to the
vacuole. (B) The vacuole is in the process of expelling
its contents; in the photograph the opening to the
outside can be seen as a small circle on the surface of
the vacuole. (C) The vacuole is nearly empty, but the
radiating canals are collecting more fluid from the
cytoplasm and fill the reservoir again (D). Photos
taken by the Nomarski process. × 400. [Courtesy
Thomas Eisner, Cornell University.]

oxidation of nutrients (remember that water is one of the products of cellular respiration).

We saw that ammonia is a satisfactory nitrogenous excretory product for
aquatic animals. It is far from satisfactory for terrestrial ones, because of the
difficulty of getting rid of this highly toxic substance on land, where an
unlimited water supply is not available. Amphibians and mammals, therefore, rapidly convert ammonia into urea, a compound that, though very
soluble, is relatively nontoxic. Urea can remain in the body for some time
before being excreted, and we can regard its production as an adaptation to
the conditions of water shortage characteristic of terrestrial existence.

Although urea is a far more satisfactory excretory product than ammonia
for land animals, it has the disadvantage of draining away some of the
critically needed water, for, being highly soluble, it must be released in an
aqueous solution. If, however, uric acid, a very insoluble compound, is
excreted instead of urea, almost no water need be lost. It is not surprising,
therefore, that many terrestrial animals—most reptiles, birds, insects, and
land snails—excrete uric acid or its salts. The excretion of this substance not
only allows them to conserve water, but has another advantage, which may
have been even more important in the evolution of uric acid metabolism. All

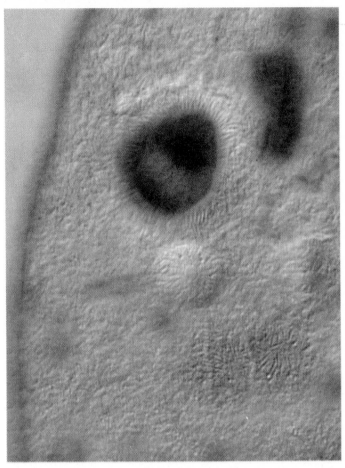

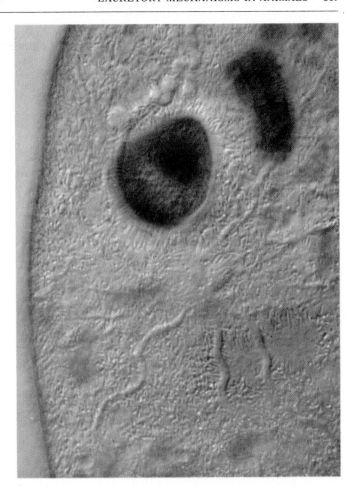

B

C

these animals lay eggs enclosed within a relatively impermeable shell or membrane. If the embryos excreted ammonia, they would rapidly be poisoned, and if they produced urea, the concentration in the egg by the latter part of development would become decidedly harmful. Uric acid, on the other hand, is so insoluble that it can be precipitated in almost solid form and stored in the egg without exerting harmful toxic or osmotic effects. In the nitrogen metabolism of fully terrestrial animals, uric acid excretion is correlated with egg laying, while urea excretion is correlated with viviparity (giving birth to living young).

EXCRETORY MECHANISMS IN ANIMALS

CONTRACTILE VACUOLES

Special excretory structures are absent in many unicellular and simple multicellular animals. Nitrogenous wastes are simply excreted across the general cell membranes into the surrounding water. Some Protozoa do, however, have a special excretory organelle, the contractile vacuole (Fig. 8.7). Each vacuole goes through a regular cycle consisting of a stage in which it

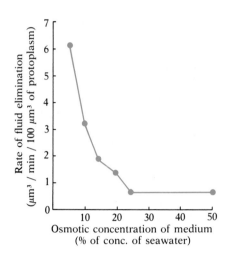

8.8   Rate of fluid elimination by contractile vacuole of *Amoeba lacerata* **as a function of the osmotic concentration of the medium**
Contractile-vacuole activity falls precipitously as the concentration of the medium goes up.   [After D. L. Hopkins, *Biol. Bull.*, vol. 90, 1946.]

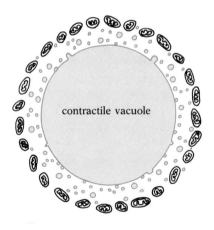

8.9   Contractile vacuole of *Amoeba proteus*
The numerous tiny vesicles around the vacuole fill with fluid, then fuse with the vacuole and empty their contents into it. The ring of mitochondria just outside the vesicle layer presumably provides the ATP necessary to drive the secretory process.   [After E. H. Mercer, *Proc. Roy. Soc. Lond. B*, vol. 150, 1959.]

fills with liquid and becomes larger and larger, followed by a contraction stage in which the contents of the vacuole are ejected from the cell. Although there is now evidence that contractile vacuoles excrete some nitrogenous wastes, it seems clear that their primary function is elimination of excess water. As might be predicted, they are much more common in freshwater Protozoa than in marine forms, and their rate of contraction becomes slower as the osmotic concentration of the environmental medium increases (Fig. 8.8).

If the principal function of contractile vacuoles is expulsion of water from the cell, then the fluid in the vacuoles should have a lower osmotic concentration than the cytosol. That it actually does, during all stages of the cycle, has been confirmed by means of samples of the fluid, withdrawn from the vacuoles with micropipettes. But how can the vacuole take in fluid when, according to the rules of osmosis, water should move out of it, not in? Since water itself cannot be moved by active transport, there must be some other explanation.

In most Protozoa the vacuole is surrounded by a layer of tiny vesicles, and these, in turn, are surrounded by a layer of mitochondria (Fig. 8.9). It has been proposed that the vesicles initially contain a fluid isotonic with the cytosol, but that they later actively pump out ions, using energy from ATP manufactured in the mitochondria. Eventually, when the osmotic concentration of the vesicular fluid has fallen to about one third that of the cytosol, the vesicles move to the contractile vacuole and fuse with it. The contractile vacuole grows larger as more and more vesicles merge with it and empty their fluid into it. The membrane of the vacuole itself must be nearly impermeable to water, which is held in the vacuole until sudden contraction expels it from the cell.

In a few ciliate Protozoa fluid enters the contractile vacuole from radially oriented feeder canals rather than from vesicles (Fig. 8.7). The feeder canals, in turn, appear to collect fluid from a network of tiny tubules probably derived from the endoplasmic reticulum. Despite these anatomical differences, the mechanism of production of dilute fluid is probably similar to the one already described; i.e. ions are pumped out before the fluid reaches the contractile vacuole itself.

FLAME-CELL SYSTEMS

The beginnings of a tubular excretory system can be seen in the flatworms (planaria, flukes, tapeworms, etc.). These animals are relatively small and lack a functional body cavity; i.e. there is no major break in the tissue mass between the outer epithelium of the body and the gastrovascular cavity. No circulatory system exists in flatworms.

Flatworm excretory systems usually consist of two or more longitudinal branching tubules running the length of the body (Fig. 8.10). In planaria and its relatives, the tubules open to the body surface through a number of tiny pores. In some other flatworms, such as the flukes, the tubules unite to form an enlarged bladder that opens to the outside. The critical portions of the systems are many small bulblike structures located at the ends of side branches of the tubules. Each bulb has a hollow center into which a tuft of long cilia projects. The hollow centers of the bulbs are continuous with the cavities of the tubules. Water and some waste materials move from the tissue fluids into the bulbs. The constant undulating movement of the cilia creates a current that moves the collected liquid through the tubules to the excretory pores, where it leaves the body. The motion of the tuft of cilia resembles the flickering of a flame, and for this reason this type of excretory system is often called a flame-cell system.

Like the contractile vacuoles discussed earlier, flame-cell systems seem to function primarily in the regulation of water balance; most metabolic wastes of flatworms are excreted from the tissues into the gastrovascular cavity and eliminated from the body through the mouth.

NEPHRIDIA OF EARTHWORMS

Note that flame-cell systems, because they function in animals without circulatory systems, pick up substances only from the tissue fluids. In animals that have evolved closed circulatory systems, the blood vessels have become intimately associated with the excretory organs, making possible direct exchange of materials between the blood and the excretory system.

The critical role of the circulatory system in excretion can be observed in the earthworm. The earthworm's body is composed of a series of segments internally partitioned from each other by membranes. In general, each of the compartments thus formed has its own pair of excretory organs, called nephridia, which open independently to the outside. A typical nephridium (Fig. 8.11) consists of an open ciliated funnel, or nephrostome (which corresponds functionally to the bulb of a flame-cell system), a coiled tubule running from the nephrostome, an enlarged bladder into which the tubule empties, and a nephridiopore through which materials are expelled from the bladder to the outside. Blood capillaries form a network around the coiled tubule. Materials move from the body fluids into the nephridium through the open nephrostome, but some materials are also picked up by the coiled tubule directly from the blood in the capillaries. There is probably also some reabsorption of materials from the tubule into the blood capillaries. The principal advance of this type of excretory system over the flame cell, then, is the association of blood vessels with the coiled tubule.

THE VERTEBRATE KIDNEY

Structure of the kidney   Like the nephridial system of earthworms, the excretory systems of vertebrates are closely associated with the closed circulatory system. When an efficient circulatory system can bring wastes to the excretory organs, the functional excretory units no longer have to be scattered throughout the body tissues, as in planaria. And the absence of internal segmentation of the body obviates the need for a series of individual excretory organs, as in earthworms. Higher vertebrates have typically evolved compact discrete organs, the kidneys, in which the functional units are massed. In humans the kidneys are located in the back of the abdominal cavity.

The functional units of the kidneys of higher vertebrates are called **nephrons.** Each nephron consists of a closed bulb called a ***Bowman's capsule*** (or renal capsule) and a fairly long coiled tubule. The tubules of the

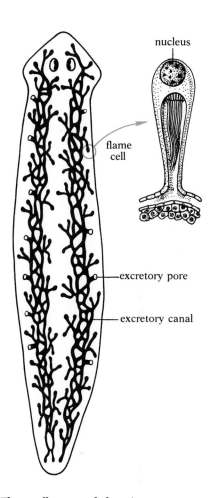

8.10   Flame-cell system of planaria
Each of the two excretory canals consists of a longitudinal network of tubules, of which some end in flame cells (one is shown enlarged at right) and others in excretory pores. The cilia in the flame cells create currents that move water and waste materials through the canals and out through the pores.   [Modified from Ralph Buchsbaum, *Animals without backbones,* by permission of The University of Chicago Press, copyright © 1948 by The University of Chicago.]

various nephrons empty into collecting tubules, which in turn empty into the central cavity of the kidney, the pelvis. From the pelvis, a large duct leaves each kidney and runs posteriorly. In some animals (frogs, birds, etc.), these ducts empty into the **cloaca,** which is a common chamber through which pass materials from the digestive, excretory, and reproductive systems. In mammals, which have no cloaca, the ducts, called **ureters,** empty into the **urinary bladder.** This storage organ drains to the outside via another duct, the **urethra** (Fig. 8.12).

Blood capillaries and the capsules and tubules of the nephrons are intimately associated in the modern vertebrate kidney. No longer are materials picked up from the general body fluids; exchange of substances takes place almost exclusively between blood capillaries and nephrons. Blood reaches each kidney via a **renal artery,** a short vessel leading directly from the aorta to the kidney (Fig. 8.13). The renal artery enters the kidney at its median depression and then breaks up into many smaller branches that run through the inner portion of the kidney (the medulla) into the outer kidney layer (the cortex), where each of the many tiny branch arterioles penetrates into a cuplike depression in the wall of a Bowman's capsule. Within each capsule, the arteriole breaks up into a tuft of capillaries called the **glomerulus** (Fig. 8.14). Blood leaves the glomerulus via an arteriole formed by the rejoining of the glomerular capillaries. After emerging from the capsule, the arteriole promptly divides again into many small capillaries that form a second dense network around the tubule of the nephron. Finally, these capillaries unite once more to form a small vein. The veins from the many nephrons then fuse to form the **renal vein,** which leads from the hilum of the kidney to the posterior vena cava. The kidney, you will note, is the second place in the mammalian body where blood circuits incorporate two sets of capillaries.

8.11   Nephridia of an earthworm

Each segment of the worm's body contains a pair of nephridia, one on each side. The open nephrostome of each nephridium is located in the segment ahead of the one containing the rest of the nephridium. The tubule from the nephrostome penetrates through the membranous partition between the segments and is then thrown into a series of coils, with which a network of blood capillaries is closely associated (the capillaries are shown here only on the nephridia of one segment). The coiled tubule empties into a storage bladder that opens to the outside through a nephridiopore (the bladder and nephridiopore can be seen in the photograph of a cross section of an earthworm, Fig. 5.22, p. 218).

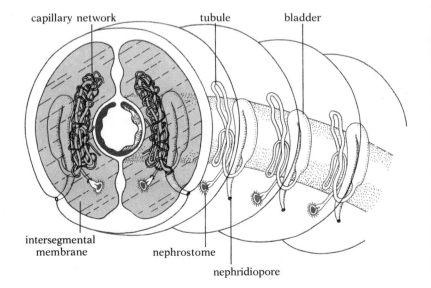

capillary network   tubule   bladder

intersegmental membrane

nephrostome

nephridiopore

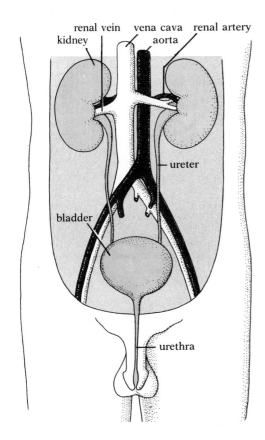

8.12   The human excretory system
The organs and vessels are shown larger relative to the body than they actually are.

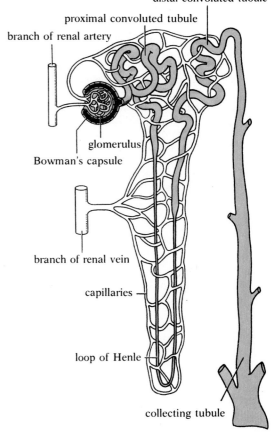

8.14   The human nephron
For description see text.   [Modified from H. W. Smith, *The kidney,* Oxford University Press, 1951.]

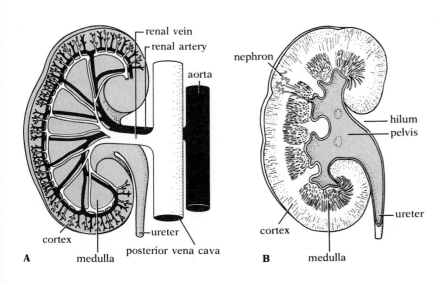

8.13   Sections of the human kidney
(A) The blood circulation of the kidney.   (B) The cortex and the medulla, and the large renal pelvis into which the collecting tubules of the nephrons empty. One nephron is shown (color); note that the glomerulus and convoluted tubules are in the cortex, but that the loop of Henle runs down into the medulla.

It has long been known that high blood pressure (or hypertension, as it is known technically) is commonly associated with kidney malfunction. But it was not until 1934 that H. Goldblatt and his associates at Western Reserve University showed that constriction of the renal arteries consistently causes pronounced and permanent hypertension. They fastened tiny screw clamps on the renal arteries of dogs. By means of the screw adjustment, they could restrict the blood flow to the kidneys to any desired amount. The dogs consistently developed hypertension to a degree correlated with the restriction in blood flow.

The result of restricting blood flow to the kidney is an oxygen deficiency. It was later shown that the kidney responds to the deficiency by secreting an enzyme (renin), which reacts in the blood with a protein secreted by the liver to form the vasoconstrictor **angiotensin.** Angiotensin stimulates the smooth muscles in the walls of small blood vessels to contract. The resulting constriction of the vessels causes the blood pressure to rise, both because the constricted vessels offer more resistance to flow and because the heart compensates for the lessened flow by increased output. The higher blood pressure can force more blood through the partly blocked renal arteries into the glomeruli of the kidneys.

Thus the kidneys have a way of compensating for the reduced blood flow caused by constrictions or other obstructions in their arteries, but that very compensation may prove to be a critical causal element in the onset of hypertension, a very common and dangerous pathologic condition in our society.

The formation of urine   With the structural relationships in mind, we are now in a position to examine the mechanism of urine formation in the human kidney. We must consider three processes: ultrafiltration, reabsorption, and tubular excretion.

In 1844 the German physiologist Carl Ludwig suggested that a glomerulus acts as a simple mechanical filter—that molecules small enough to pass through the capillary walls and through the thin membranous walls of the capsule filter from the blood into the nephron as a result of the high hydrostatic pressure in the glomerulus. The advent of modern techniques of microscopy has suggested how such filtration might take place. Pores can be detected in the walls of both the glomerular capillaries (Fig. 8.15) and the Bowman's capsule (Fig. 8.16). The blood pressure presumably forces small molecules through the pores from the capillaries into the lumen of the capsule.

If the filter explanation is correct, the liquid entering the lumen of the nephron should have basically the same percentage composition as blood, lacking only the formed elements and the plasma proteins, both of which are too large to filter through the membranes to any appreciable extent. Collection of capsular urine to verify this conclusion is a very difficult undertaking, one that, at best, can yield only minute quantities for analysis. In spite of the technical difficulties, A. N. Richards of the University of Pennsylvania was able to insert microscopic glass pipettes directly into a Bowman's capsule and to draw off small samples of glomerular filtrate. Analysis of these samples showed that the filtrate has essentially the same concentration of dissolved substances (glucose, urea, salts, amino acids, etc.) as blood plasma,

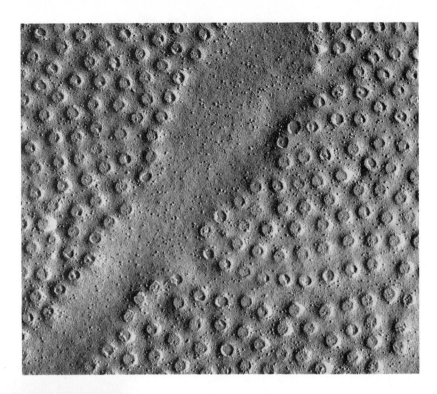

8.15 Freeze-etch preparation of the membrane of an endothelial cell from a glomerular capillary of a rat
The membrane is perforated by many pores of uniform size. It is presumably through these pores that substances filter out of the glomerulus, driven by the high blood pressure built up by the beating of the heart. × 50,000. [Courtesy D. A. Goodenough, Harvard Medical School.]

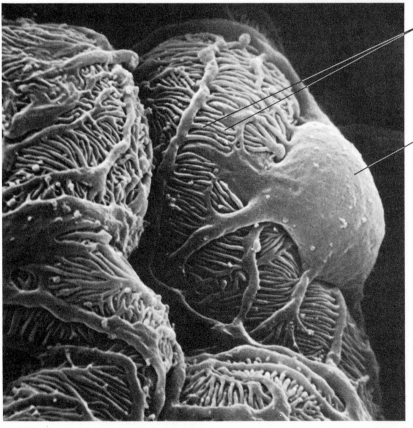

interdigitating cell processes with slit pores between them

cell body

8.16 Scanning electron micrograph showing epithelial cells of Bowman's capsule on two glomerular capillaries of a rat
The cells that form the cup of the capsule become so highly modified that they bear little resemblance to any other epithelial cells. Called podocytes, these cells consist of a cell body and numerous interdigitating processes that embrace the glomerular capillaries. The spaces, called slit pores, between the interdigitating processes provide an enormous area for filtration. × 6,600. [Courtesy F. Spinelli, Ciba-Geigy Research Laboratory, Basel, Switzerland.]

just as had been anticipated. Furthermore, it has been demonstrated that if the hydrostatic pressure in the glomerular capillaries is increased, the volume of the filtrate is increased proportionately, and that if the hydrostatic pressure is decreased, the filtrate volume declines proportionately. The experimental evidence also indicates that changes in filtrate volume are not accompanied by such changes in the kidney's oxygen consumption as would occur if the kidney were performing work in moving materials from the blood to the capsule. All the evidence, therefore, supports Ludwig's thesis that the cells of the glomerular capillaries and of the Bowman's capsules do not carry out active transport in the movement of materials from the glomeruli into the capsules, but that the work involved is performed by the beating heart as it drives the blood under high hydrostatic pressure into the glomeruli.

If the filtrate were expelled from the body without modification, many very valuable, indeed essential, substances would be lost and the process would be wasteful in the extreme. If we consider water alone, it is overpowering to contemplate the drinking that would be necessary to replace the 170 liters of filtrate formed every day in the average person's kidneys! Selective reabsorption of most of the water and many of the dissolved materials is one of the functions of the tubules of the nephrons. In humans the filtrate passes first through the *proximal convoluted tubule,* then through the long *loop of Henle,* then through the *distal convoluted tubule,* and finally into the *collecting tubule* (Fig. 8.14). The Bowman's capsule and the proximal and distal convoluted tubules are in the kidney cortex layer, while most of the loop of Henle and the collecting tubule are in the kidney medulla. As the filtrate moves through the tubules, as much as 99 percent of the water may be reabsorbed by the cells of the tubule walls and returned to the blood in the capillary network. Thus human kidneys (and the kidneys of other mammals and of birds) can produce concentrated urine, i.e. urine that is hypertonic relative to the blood plasma even though the initial filtrate was nearly isotonic.

To understand by what process the urine becomes concentrated, we must examine the structure of the nephron more closely. The fact that the concentration of sodium ions (Na^+) in the distal convoluted tubule is only slightly lower than that in the proximal tubule (Fig. 8.17) might well make one wonder about the function of the loop of Henle. Why wouldn't the system work just as well without the loop? The answer is that the loop of Henle functions to establish and maintain an Na^+ concentration gradient in the tissue fluid of the kidney medulla. This gradient makes possible the passive diffusion of water out of the collecting tubule, so that the urine ultimately produced can attain a much greater concentration than the original filtrate.

There is no evidence that any cells have ever been able to evolve a mechanism for active transport of water itself. When water must be moved, the usual mechanism is active pumping of ions with the water following passively. In this case the ions are Na^+ and the pumps are in the ascending limb of the loop of Henle. The cells in the wall of the ascending limb actively pump Na^+ ions out of the tubule into the surrounding tissue fluid. Some, but not all, of this Na^+ passively diffuses back into the descending limb, so that there is, in effect, a cycling of some of the Na^+ from ascending limb to tissue fluid to descending limb to ascending limb to tissue fluid, etc. The result is that an Na⁺ concentration gradient is maintained in the tissue fluid along the loop, with the concentration lowest in the outer part of the kidney cortex, where the convoluted tubules are located, and highest in the medulla, where the tips of the loops of Henle are located. The walls of the ascending limb of the loop must be impermeable to water since water does not diffuse out of the tubule as the Na⁺ is pumped out. Consequently the net effect of passage of the filtrate through the loop of Henle is the removal from it of some Na⁺ but very little water. Hence the urine in the distal convoluted tubule is actually less, instead of more, concentrated than the initial filtrate. But now the urine flows into the collecting tubule, which runs from the cortex through the medulla to the renal pelvis, i.e. through a region of increasing Na⁺ concentration in the tissue fluid. Since the walls of the collecting tubule are permeable to water, water moves by passive osmosis from the dilute urine in the tubule to the surrounding hypertonic tissue fluid, until the final urine may become essentially isotonic with the highly concentrated tissue fluid in the inner region of the medulla. Whether the collecting tubule finally releases dilute or concentrated urine depends on whether there is a deficiency or an excess of water in the body at the moment; the pituitary gland responds to changes in the amount of water in the body by releasing a hormone, vasopressin, that regulates the permeability of the collecting tubule to water.

Water is not the only substance reabsorbed by the tubules of the nephrons. In a normal healthy person, all the glucose, almost all the amino

8.17 Countercurrent-multiplier model for production of concentrated urine by the nephron

The nephron acts as a countercurrent-multiplier system, i.e. a system with a hairpin-loop structure that multiplies the effect of active transport, bringing about higher concentration differences than might otherwise be attainable. The hairpin structure in the nephron is the loop of Henle. The ascending limb of the loop of Henle actively pumps out Na⁺ ions (solid color arrows), some of which diffuse back passively into the descending limb (dashed color arrows). This pumping action is relatively weak; at any given point the concentration of the filtrate in the ascending limb and that of the surrounding tissue fluid differ by only two units. Yet because the Na⁺ pumps act upon a filtrate passed along a hairpin loop, they can establish and maintain, in the tissue fluid between the cortex and the inner portion of the medulla, a concentration gradient of roughly nine units. As a result of the gradient, the filtrate passing down the collecting tubule is at every point hypotonic relative to the surrounding tissue fluid. Thus the urine, dilute when it enters the collecting tubule, can steadily lose water (dashed black arrows) by osmosis through the permeable walls of the tubule and become concentrated.

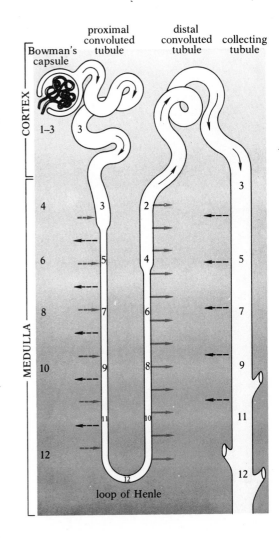

Active transport of Na⁺
Passive transport of Na⁺
Passive transport of H₂O
Flow of filtrate or urine

Figures indicate solute concentrations in hundreds of milliosmols per liter

acid, and much of the salt are also reabsorbed and returned to the blood. Much of this reabsorption involves active transport, and thus energy expenditure by the tubule cells.

In spite of the extensive reabsorption that may take place, urine is more than a concentrated solution of urea and some uric acid (in humans, apes, and Dalmatian dogs, a small amount of uric acid is formed by breakdown of nucleic acids). Most substances have what is called a kidney threshold level. If the concentration of such a substance in the blood exceeds its kidney threshold level, the excess is not reabsorbed from the filtrate by the tubules but instead appears in the urine. Glucose is an example of a substance with a high threshold value; ordinarily all glucose in the filtrate is reabsorbed, because the threshold level for glucose is higher than the normal blood-glucose level. If, however, the blood-sugar level is abnormally high, as in diabetes, sugar appears in the urine. This elimination of excess sugar by the kidneys points up once again that excretory organs do far more than just remove nitrogenous wastes; they play a critical role in maintaining the relatively nonfluctuating internal fluid environment of the organism. In this case, when the liver and/or the peripheral tissues are not functioning properly and the blood-sugar level rises, the kidneys act as a second line of defense. The kidneys likewise help regulate the composition of the blood by keeping the relative concentrations of such inorganic ions as sodium, potassium, and chloride in the blood plasma at a nearly constant level. Whenever the concentration of an ion in the blood, and hence in the glomerular filtrate, exceeds its kidney threshold value, the excess in the filtrate is not reabsorbed but is released in the urine. The remarkably steady level of ionic concentration in the blood and the considerable variation noticeable in the urine suggest the extent of the regulation.

The movement of materials between the tubules and the capillaries surrounding them is not completely one-way. Some chemicals are actively removed from the blood by the tubules and deposited in the urine. This tubular excretion supplements glomerular excretion and increases the efficiency of the overall excretory regulation of blood composition.

The glomerular kidney, which probably arose first in ancestral marine vertebrates, doubtless played an important role in enabling the ancestors of the modern bony fishes to enter freshwater, a hypotonic medium relative to their body fluids. We have seen that freshwater fishes are constantly being flooded by osmotic water. The glomerular kidney is particularly well suited to pumping out excess water. Modern marine fishes, on the other hand, have a problem of water conservation; it is not surprising, therefore, that the development and activity of their glomeruli have declined. Terrestrial vertebrates have no need to excrete large quantities of water, either. Thus in reptiles and birds, as in the marine fishes, the glomeruli have declined, and much of the excretion of uric acid is tubular rather than glomerular. In mammals, by contrast, evolution has tended, not to reduced glomeruli, but to longer loops of Henle and more efficient water reabsorption; as might be expected, the longest loops occur in species that inhabit very dry environments.

Special excretory adaptations of vertebrates The kidneys of vertebrates vary considerably in their capacity to produce concentrated urine. You will recall that marine fishes are incapable of producing urine more concentrated than their blood and that they must therefore excrete excess salt by another method, special cells on the gills. Similarly lacking very efficient kidneys, marine birds (albatrosses, penguins, etc.) and sea turtles must excrete the excess salt in the water they drink by some other mechanism; these animals have special glands in their heads, near the eye, that are capable of excreting salt in very concentrated solution.

Seals and some whales seldom drink; they get their water from the body fluids of the fish they eat, and thus benefit from the fishes' ability to excrete salt through the gills. A fish diet means much protein, however, and much urea to excrete. These animals have kidneys capable of excreting urine with a high urea concentration. Some whales eat marine invertebrates instead of fish, and accordingly take in much excess salt; apparently these species can produce urine with a high salt concentration.

Kangaroo rats living in deserts almost never drink; nor do they get water by eating succulent food. Most of their water is metabolic water obtained during the respiratory breakdown of the dry grains they eat. They must, of course, conserve water extremely well: They are not active during the heat of the day; they do not sweat; they eliminate very dry feces; and they have extraordinarily efficient kidneys capable of producing extremely concentrated urine.

The human kidney is incapable of producing urine with a very high concentration of either salt or urea. And humans have no alternative excretory mechanism like those of marine fishes, birds, and turtles. Adrift at sea, humans are in serious danger, as you know. Drinking seawater aggravates their condition because they will excrete more water than they drink, in the process of removing salt from their bodies. If they try to get their water by eating fish, as seals do, they excrete much water in the process of removing urea. Human kidneys are simply not adapted to life at sea or to life in very dry habitats.

MALPIGHIAN TUBULES

Proceeding from flame-cell systems to earthworm nephridia to vertebrate kidneys, we have noted an increasingly close interrelationship between the excretory structures and closed circulatory systems. In the flame-cell system, no circulatory system is involved. In the earthworm nephridium, blood capillaries are associated with the tubule, but not with the nephrostome. Finally, in the advanced vertebrate kidney, there are both tubule capillaries and glomerular capillaries, the glomerulus and renal capsule forming a compact interacting unit. We do not mean to imply that this sequence represents a true evolutionary progression; indeed, the evidence indicates that the evolution of earthworms and the evolution of vertebrates had little to do with each other, and that the excretory systems of the two animal groups almost certainly evolved independently. Nonetheless, these systems illustrate the trend, seen in many animal groups, of increasing dependence of the excretory process on blood-capillary beds.

But we have said before, and must say again, that in biology almost all generalizations have exceptions. Insects are a case in point here. The phylum to which this immense class of animals belongs probably evolved from an ancestral form similar to the ancestor of segmented worms like the earthworm. The evidence indicates that this ancestor had nephridia. Yet

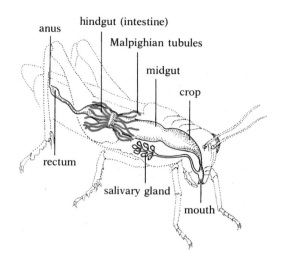

8.18 Malpighian tubules of an insect
These excretory organs arise as diverticula of the
digestive system at the junction between the midgut
and the hindgut.

insects do not have nephridia, nor have their excretory organs evolved from
nephridia. The evolution of an open circulatory system in insects (the ancestral system was most likely a closed one) and the consequent lack of
blood capillaries probably account for the evolutionary loss of nephridia,
which are dependent on capillaries. Consequently insects and many of their
relatives evolved an entirely new excretory system, one that functions well in
association with an open circulatory system.

The excretory organs of insects are called Malpighian tubules. They are
diverticula of the digestive tract located at the junction between the midgut
and the hindgut (Fig. 8.18). These blind sacs, variable in number, are bathed
directly by the blood in the open sinuses of the animal's body. Fluid is
absorbed from the blood into the blind distal end of the Malpighian tubules.
As the fluid moves through the proximal portion of the tubules, the nitrogenous material is precipitated as uric acid and much of the water and
various salts are reabsorbed. The concentrated, but still fluid, urine next
passes into the hindgut and then into the rectum. The rectum has very
powerful water-reabsorptive capacities, and the urine and feces leave the
rectum as very dry material.

The highly effective role played in water conservation by the insect rectum
is similar to that of the cloaca of birds and some other vertebrates. In those
animals the urine must move through the posterior portion of the digestive
tract, where it is subjected to the powerful reabsorptive action of the rectum; the uric acid is therefore eliminated as a nearly dry powder or hard
mass.

THE CELLULAR BASIS OF ACTIVE TRANSPORT OF SALT

In our discussion of excretion and osmoregulation so far, we have paid little
attention to events at the cellular level. We have indicated that active transport of some substances, particularly salts, is carried out by the cells in the
walls of the kidney tubules and Malpighian tubules, by the osmoregulatory
cells in the nasal glands of marine birds and turtles, and by the salt-secreting
cells of the gills of bony fishes and of the rectal glands of elasmobranchs.
What is known about the active-transport process?

The sodium-potassium pump   Our repeated assertion that animals tend to
maintain a nonfluctuating fluid environment for their cells, and that this
fluid has approximately the same osmotic concentration as the cells, should
not be taken to imply that the extracellular and intracellular fluids have the
same ionic composition. On the contrary, their compositions are very different. All cells, plant and animal, tend to accumulate certain ions in much
higher concentrations than are found in the surrounding fluids and to stabilize intracellular concentrations of other ions at levels far below those in the
extracellular fluids. For example, the vast majority of cells maintain an
internal concentration of sodium ions (Na^+) far below that in the fluids
bathing them, while at the same time accumulating potassium ions (K^+) to a
concentration many times that in the extracellular fluid.

It can be demonstrated that the unequal concentrations of most small
ions, such as Na^+ and K^+, on the two sides of the cellular membrane cannot
be maintained under conditions that prevent the cell from carrying out
energy-yielding chemical reactions such as cellular respiration. The cell

must do work to accumulate some ions and to expel others; it must carry out active transport, for which it needs an abundant supply of ATP. Active transport of ions is not a phenomenon restricted to the cells of excretory and osmoregulatory organs, but is a general property of cells. Excretory and osmoregulatory cells have simply become highly specialized for this activity, regulating not only their own intracellular composition, but also the composition of the extracellular fluids for the whole animal. Such cells have been found to use roughly 70 percent of their total ATP output for pumping Na^+ and K^+.

The so-called sodium-potassium pump has been the subject of intensive research in recent years. Although much about it is still unknown, enough facts have been garnered for a tentative model to be proposed. A protein complex in the cell membrane apparently acts both as the permease for transport of Na^+ and K^+ and as the enzyme for breaking down ATP to ADP and inorganic phosphate. The protein has been designated ***Na⁺K⁺–ATPase*** to indicate that it is an ATPase active only in the presence of Na^+ and K^+ ions, for which it functions as a transmembrane pump.

According to the model, the Na^+K^+–ATPase extends through the entire thickness of the membrane; it has two kinds of binding sites, one for Na^+ and one for K^+; and it exists in two conformations, E_1, in which its binding sites are turned to the interior of the cell (Fig. 8.19A), and E_2, in which its binding sites are turned to the exterior of the cell (Fig. 8.19C). The binding sites for Na^+ tend to be highly reactive in the E_1 conformation and less reactive in the E_2 conformation; they therefore pick up Na^+ inside the cell

8.19 A model for the action of the sodium-potassium pump in the plasma membrane.
(A) The transmembrane enzyme, Na^+K^+– ATPase (color), is in its E_1 conformation, with its binding sites for Na^+, which are very reactive, exposed to the interior of the cell.   (B) The binding of Na^+ triggers phosphorylation of the enzyme by ATP.   (C) The phosphorylation favors conversion of the enzyme to the E_2 conformation, with the binding sites for Na^+ and K^+ exposed to the extracellular medium; in E_2 the binding sites for Na^+ are less reactive and the binding sites for K^+ are more reactive; hence the enzyme releases Na^+ and picks up K^+ (actually $3Na^+$ and $2K^+$).   (D) The binding of K^+ triggers dephosphorylation.   (E) The dephosphorylation favors reconversion of the enzyme to the E_1 conformation, which, because of its lower affinity for K^+, releases the K^+ into the interior of the cell.   The net effect of the whole process is transport of $3Na^+$ out of the cell and $2K^+$ into the cell, with breakdown of one molecule of ATP to ADP and P_i. (The conformational differences between E_1 and E_2 are greatly exaggerated in this diagram, for ease of visualization. The actual differences may be slight; a shift of a few atoms by only 0.2 nm would probably suffice.)

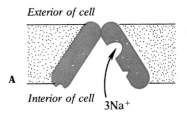

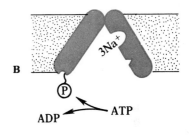

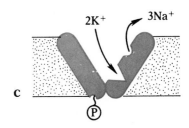

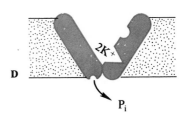

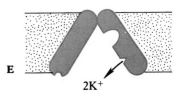

and release it outside. The reverse is true for the K^+ binding sites. They tend to be highly reactive in the E_2 conformation, picking up K^+ outside the cell, and to be less reactive in the E_1 conformation, releasing the ion into the cell.

How is the shift from E_1 to E_2, and back again, brought about? According to the model, it is powered by the hydrolysis of ATP to ADP and inorganic phosphate. The binding of the Na^+ by the enzyme when it is in the E_1 conformation triggers the transfer of a phosphate group from ATP to the enzyme (Fig. 8.19B), which, once phosphorylated, assumes the E_2 conformation. The binding of K^+ in the E_2 conformation triggers dephosphorylation of the enzyme; inorganic phosphate is released into the interior of the cell (Fig. 8.19D), and the enzyme resumes the E_1 conformation (Fig. 8.19E).

Recent measurements indicate that each complete activity cycle pumps out three Na^+ ions and pumps in two K^+ ions, with expenditure of only one molecule of ATP. The chemical changes occurring in the course of one cycle can be summarized as follows:

$$3Na^+_{inside} + 2K^+_{outside} + ATP + H_2O$$
$$\longrightarrow 3Na^+_{outside} + 2K^+_{inside} + ADP + P_i$$

The whole process depends on conformational changes in the Na^+K^+–ATPase that permit it to act as a channel through the membrane. Once again, then, we see the importance of the three-dimensional structure of proteins and of the changes in that structure which may be induced by the binding of substrate.

Let us now return to the epithelial cells involved in active transport in excretory and osmoregulatory organs. These cells appear to use the same basic pump described above as characteristic of cells in general, but the task they perform is a particularly complex one. Consider the osmoregulatory cells in the gills of a marine fish: These cells must remove Na^+ from the blood and tissue fluid and actively secrete it into the surrounding seawater (chloride will follow passively to balance the electric charge). Or consider the cells in the walls of the ascending limb of the loop of Henle in a mammalian kidney: These cells must remove Na^+ from the urine and actively secrete it into the tissue fluid surrounding the nephron. In both instances the cells are doing more than simply expelling Na^+ ions from their cytoplasm: They are picking up Na^+ ions on one side and expelling them on the other side. In other words, Na^+ ions are being moved completely across the cell barrier that separates the tissue fluids of the fish from the seawater or that separates the contents of the mammalian nephron from the tissue fluid. Apparently the membranes on the two sides of the cell function differently.

It has been hypothesized that Na^+ ions diffuse passively into the cell on one side and are then actively expelled from the cell on the other side. How would this hypothesis apply to the osmoregulatory cells in the gill of a freshwater fish? These cells must take in salt from the surrounding water and extrude it into the blood-derived tissue fluid. The pump would be active only at the tissue-fluid side of the cell, pumping Na^+ ions from the cell contents into the tissue fluid. This removal of Na^+ would lower the sodium concentration in the cell to a point below that in the environmental water, and if the membrane on the environmental side of the cell were permeable to Na^+, these ions would tend to diffuse passively into the cell from the environmental water. In other words, Na^+ would diffuse passively into the cell from the medium on one side and then be actively secreted by the pump from the cell into the tissue fluid on the other side. In a marine fish the situation would be reversed: The pump would be in the membrane on the side of the cell exposed to the seawater, and Na^+ would diffuse passively into the cell on the tissue-fluid side.

SUGGESTED READING

RAMSAY, J. A., 1968. *Physiological approach to the lower animals*, 2nd ed. Cambridge University Press, New York. *Stimulating little book with a broadly comparative approach. Chapter 4, on excretion, is short, but very much to the point.**

SCHMIDT-NIELSEN, K., 1959. "Salt glands," *Scientific American*, January. *The salt-secreting glands of marine birds and turtles.*

SCHMIDT-NIELSEN, K., 1959. "The physiology of the camel," *Scientific American*, December. (Offprint 1096) *Fascinating discussion of the special adaptations that enable camels to survive and prosper in the desert.*

SCHMIDT-NIELSEN, K., and B. SCHMIDT-NIELSEN, 1953. "The desert rat," *Scientific American*, July. (Offprint 1050) *The extraordinary osmoregulatory abilities that allow the desert rat to survive without ever drinking.*

SMITH, H. W., 1953. "The kidney," *Scientific American*, January. (Offprint 37)

SMITH, H. W., 1953. *From fish to philosopher.* Little, Brown, Boston. (Paperback edition by Doubleday Anchor Books, 1961.) *A little classic on vertebrate evolution in terms of osmoregulation and excretion, enlivened by liberal doses of personal philosophy.*

*Available in paperback.

Chapter 9

CHEMICAL CONTROL

How complex the life functions are, and how intricately interwoven, has emerged clearly from our study of various aspects of the biology of organisms. Coordination of the myriad processes they carry out depends on special control mechanisms, of which we can recognize two principal types: chemical control mechanisms, which are found in all organisms, and nervous control mechanisms, which, in the strict sense, are found only in multicellular animals. This chapter will be concerned with the first of these mechanisms—chemical control.

We have already seen that within any living cell vast numbers of different chemical reactions are occurring at any given instant. As some substances are synthesized and others are destroyed or removed, the chemical environment in every part of the cell is constantly changing slightly. But any change in the cellular environment will affect subsequent chemical reactions and thereby exert some control over them. Whether the effect of one chemical reaction on another is slight or whether it is pronounced, each reaction will in some way influence all other reactions. This simple relationship is the form in which much of the chemical regulation of all living cells still takes place; it also constitutes the raw material for the evolution of more complex chemical regulatory mechanisms.

Suppose that within a certain cell the following two reactions are occurring:

$$(1) \quad A + B \xrightarrow{\text{enzyme}} C + D$$

$$(2) \quad X + Y \xrightarrow{\text{enzyme}} Z$$

Suppose, further, that Z, the product of the second reaction, exerts a significant effect on the rate of the first reaction. If this effect is advantageous, natural selection might lead in time to cells with genetic systems that enhance the role of reaction 2 as a control mechanism for reaction 1.

Now suppose that, instead of occurring in the same cell, reactions 1 and 2 occur in different cells in the body of a multicellular organism—that Z is synthesized with particular ease by one type of cell and that reaction 1 proceeds most readily in another type of cell. If Z is to play a major role in regulating reaction 1 in such an organism, it must be secreted (released) by the cells where it is produced and must reach the cells in which reaction 1 is occurring. If this happens, intercellular control has been established; the product of one cell is regulating reactions in other so-called target cells.

There is ample evidence that such intercellular interactions occur whenever two or more cells are in either direct or indirect contact. It is significant that the characteristics of a single cell growing in a culture in the laboratory are different from those of a cell from the same source grown next to another cell. True, some intercellular interaction involves mechanical, electrical, or other physical influences. But chemical influences are also at work; it can be shown, for example, that cells separated by a thin sheet of agar influence each other via chemicals that diffuse through the agar. Release of chemicals that affect other cells is characteristic of all cells.

The multicellular organism, as we have noted repeatedly in other chapters, is characterized by division of labor among its parts. It is not unexpected, therefore, that evolution should have led recurrently in such organisms to specialization of certain cells or tissues as producers of control chemicals. These chemicals often have important control functions in parts of the body far removed from their sites of synthesis. Transport between these sites and the sites of action is often effected by the vascular system in higher plants and the blood circulatory system in higher animals. Control chemicals produced in a regular fashion by tissues or organs specialized for that function and exerting their highly specific effects on other tissues of the body are usually called *hormones.* They are effective in very low concentrations, an indication that they probably influence the synthesis or the activity of enzymes.

PLANT HORMONES

Plant hormones, at least those so far known, are produced most abundantly in the actively growing parts of the plant body, such as the apical meristems of the shoot and the root, young growing leaves, or developing seeds or fruits. The tissues in which these hormones are produced, frequently the meristematic tissues themselves, are specialized for hormone production, but they are not so highly specialized as to be concerned with little else, as is often the case with the most highly specialized hormone-producing tissues in animals. There are no separate hormone-producing organs in plants analogous to the endocrine glands of higher animals. Furthermore, plant hormones are predominantly involved in regulating growth and development (they are often called growth regulators), while animal hormones mediate a great variety of functions in addition to growth.

Work on plant hormones is one of the most active areas of modern botanical research, and knowledge of this subject is growing rapidly. The following sections do not pretend to be complete in their coverage; they are

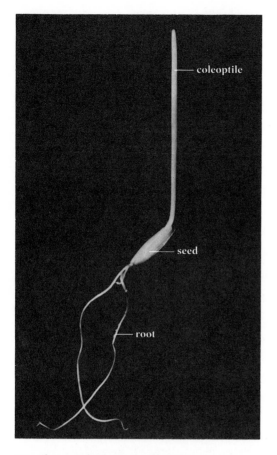

9.1   Photograph of an oat coleoptile
The first leaves are rolled up inside the protective
coleoptile sheath.   [From *Botany: An ecological
approach* by W. A. Jensen and F. B. Salisbury. © 1972
by Wadsworth Publishing Company, Inc., Belmont,
Calif. Reprinted by permission of the publisher.]

intended, rather, as an introduction providing the background necessary for
understanding advances as they occur.

AUXINS

One of the earliest-investigated and best-known groups of plant hormones,
or growth regulators, includes those hormones collectively known as auxins.
Of the amazing variety of effects they have on different parts of the plant,
one of the most important and most extensively studied is their control of
cell elongation in stems. Let us begin with an examination of this function of
auxins.

Auxins and phototropism of shoots   Everyone is familiar with the strong
tendency of many plants to turn toward the light. A potted plant in the living
room bends toward a window; you turn the plant so that it will look nicer to
people in the room, but discover that in a disconcertingly short time the
shoot is again oriented toward the light of the window. This phenomenon of
responding to light by turning is called phototropism, from the Greek words
for light and turning. (Other tropisms involve turning responses to other
stimuli. Geotropism is a turning response to gravity, hydrotropism a turning
response to water.) In plant shoots the phototropism is positive, a turning
toward the stimulus; roots, on the other hand, exhibit negative phototropism, a turning away from a light stimulus.

Among the first to investigate the phototropism of plants was the wide-
ranging Charles Darwin, who worked on the problem with his son Francis,
about 1880. They, like many who followed them, performed their experiments on the cylindrical sheath that encloses the first leaves of seedlings of
grasses and their relatives (Fig. 9.1). This sheath, called the *coleoptile,* grows
principally by cell elongation, and it exhibits a very strong positive phototropic response. The Darwins showed that if the tip of the coleoptile is
covered by a tiny black cap, it fails to bend toward light coming to it from

9.2   The Darwins' experiments on phototropism
(A) A coleoptile of canary grass bends toward the
light.   (B–C) The coleoptile does not bend if its tip is
removed or is covered by an opaque cap.   (D) The
coleoptile does bend if its tip is covered by a
transparent cap.   (E) It also bends if its base is
covered by an opaque tube.

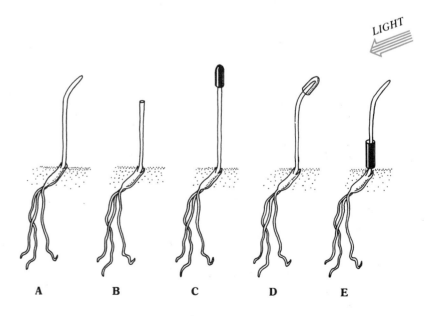

one side, while control coleoptiles with their tips exposed or covered with transparent caps bend, as expected, toward the light (Fig. 9.2). They observed that a black tube placed over the base of the coleoptile, but not covering the tip, fails to prevent bending. It seemed to be the tip of the coleoptile, therefore, that played the key role in the phototropic response. This was confirmed by experiments in which the Darwins cut off the tip and found that the coleoptile failed to bend, even though control coleoptiles damaged in other ways, but with their tips intact, bent normally. Clearly, it was the absence of the tip and not a reaction to wounding that blocked the phototropic response. The Darwins concluded that light is detected by the tip of the coleoptile and that "some influence is transmitted from the upper to the lower part, causing the latter to bend."

About thirty years later P. Boysen-Jensen in Denmark obtained the first clear evidence that the "influence" postulated by the Darwins was probably chemical rather than electrical or nervous. He removed the tips of oat coleoptiles (which made the coleoptiles stop growing), placed a thin layer of gelatin on the cut end of the stump, and then placed the tip on the gelatin. Thus the tip was separated from the rest of the coleoptile by a thin layer of galatin (Fig. 9.3). The coleoptiles resumed growing. If the tip was then illuminated with a light from the side, the coleoptile base bent toward the light. The tip had received the light stimulus, and a message from the tip had moved across the gelatin barrier and induced bending in the base. Although this experiment did not completely rule out an electrical or nervous message, it made such a possibility appear highly unlikely and strongly indicated that a diffusible chemical was involved.

That the tip could cause the base of the coleoptile to bend even in the dark was demonstrated by A. Paál in Hungary in 1918. He cut off the tip and then replaced it off center on the stump (Fig. 9.4). If he put the tip on the right side

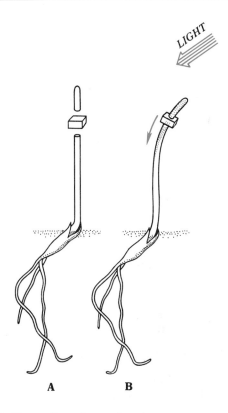

9.3 Boysen-Jensen's experiment
When the tip of an oat coleoptile is cut off, a layer of gelatin put on the end of the stump, and the tip replaced (A), the coleoptile will grow and turn toward the light (B). Presumably a chemical (color) moves from the tip, through the gelatin, into the base, and stimulates it to turn.

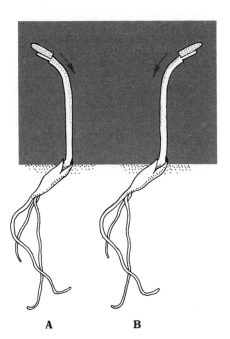

9.4 Paál's experiment
This experiment is performed in the dark. If the tip of a coleoptile is cut off and then replaced right of center, the coleoptile will bend to the left (A); if the tip is placed left of center, it bends to the right (B). (Note that only the tip of the coleoptile was cut off; the rolled up leaf inside was left intact.)

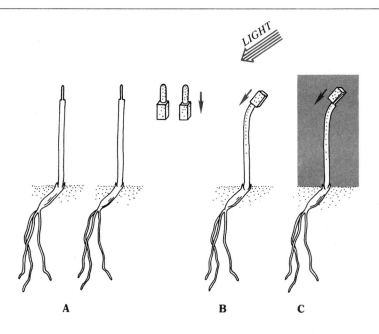

9.5 Went's experiment

When the tips of coleoptiles are cut off and placed on blocks of agar for about an hour (A), and one of the blocks alone is then put on a stump (B), the stump will resume growing and will respond to light. If a block is placed off center on a stump in the dark (C), the stump will grow and will bend away from the side on which the block rests. Apparently a hormone has diffused from the tips into the blocks, and this hormone can then diffuse from the blocks into the stumps.

of the stump in the dark, the coleoptile bent to the left; if he put the tip on the left side, the coleoptile bent to the right. Apparently that part of the coleoptile directly under the replaced tip grew much faster than the part not under the tip. This asymmetric elongation of the coleoptile caused it to bend away from the side undergoing the greatest elongation.

Experiments conclusively demonstrating that the growth stimulus moving downward from the tip is a chemical were reported in 1926 by Frits Went, then in Holland. He removed the tips from coleoptiles and placed these isolated tips, base down, on blocks of agar for about an hour (Fig. 9.5). (Agar is a gelatinlike material, made from seaweeds, often used as the base for laboratory culture media.) He then put the blocks of agar, minus the tips, on the cut ends of the coleoptile stumps. The stumps behaved as though their tips had been replaced; they resumed growth, responded to lateral light by bending toward it, and, if the agar blocks were put on off center, could be made to bend even in darkness. Plain agar blocks used as controls produced none of these effects. Apparently a growth-stimulating substance had diffused out of the tips and into the blocks of agar while the tips were sitting on the blocks. When the blocks containing the chemical were placed on the stumps, the chemical moved down into the stumps and stimulated elongation. This experiment ruled out the possibility that the stimulus was electrical or nervous, because these types of stimuli cannot be stored in agar blocks. Went named the diffusible hormone presumably involved "auxin" (from a Greek word meaning to grow). To this day, the identification of auxins is based on Went's experiment: If an agar block containing the substance in question causes a decapitated oat coleoptile to bend in the dark when the block is placed on one side of the cut end, the substance is classed as an auxin.

Many chemicals, some of them found naturally in plants and some synthesized only in the laboratory, have passed Went's test and are commonly called auxins. The one most thoroughly investigated is ***indoleacetic acid*** (Fig. 9.6), which has been isolated from numerous natural sources. There is evidence that indoleacetic acid is the principal (perhaps only) directly active natural auxin. The apparent auxin activity of other natural compounds may be due to their being converted into indoleacetic acid by the plant. Current ideas about the way this hormone brings about cell elongation will be discussed in Chapter 17 (see pp. 722–723).

The experiments we have discussed have shown that the tip of the coleoptile releases auxin, which moves downward and stimulates cell elongation in the coleoptile. As the results obtained by Paál and Went suggest, there is normally little lateral movement of the auxin after it has been released from the tip; the hormone reaches and stimulates only those cells directly under the point of release. But what about phototropism? It was this phenomenon that concerned us initially when we examined Darwin's experiments. Extensions of those experiments by Boysen-Jensen, Paál, and Went revealed much about the more general problem of hormonal control of growth, but what do these experiments say about the more specific problem of phototropism? The superficial explanation of the phototropic response is fairly obvious. When light strikes the tip of the plant from one side, it must reduce the auxin supply on that side. Consequently the illuminated side of the plant grows more slowly than the shaded side, and this asymmetrical growth produces bending toward the slower-growing illuminated side.

We have called this a superficial explanation, because plant physiologists are still not certain how the tip detects the light; some think β-carotene (one of the carotenoids) is the detection pigment, others that it is riboflavin, and still others that both pigments are involved. Nor do they know how detection of light is coupled to asymmetric auxin distribution. The evidence indicates that there is active lateral transport of auxin in the tip from the illuminated to the shaded side (Fig. 9.7), but by what mechanism this transport is effected remains unclear. Only further research can fully explain the intriguing phenomenon of plant phototropism, which captured Darwin's interest so many years ago.

Auxins and geotropism of shoots If you lay a potted plant on its side and leave it for a few hours, you will find that the shoot has begun to bend upward (Fig. 9.8). This is a negative geotropic response; the shoot turns away from the pull of gravity. (How could you prove that the shoot is responding primarily to gravity and not to some other stimulus such as light?)

It was shown by Herman Dolk in Holland that the concentration of auxin in the lower side of a horizontally placed shoot increases while the concentration in the upper side decreases. This unequal distribution of auxin stimulates the cells in the lower side to elongate faster than the cells in the upper side, and the shoot thus turns upward as it grows. Again, the external stimulus, in this case gravity, is apparently detected by the meristematic tissue in the shoot tip. It has been suggested that the meristematic cells sense the pull of gravity by its effect on the distribution of cellular inclusions, such as starch grains and other small bodies; as the inclusions respond to the pull of gravity, they tend to accumulate in the lower parts of the cell and bring about—by what mechanism is unknown—an asymmetric auxin distribution. Cellular inclusions that are believed to affect the position of a part or organ by changes in their own position are known as statoliths.

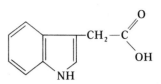

9.6   Indoleacetic acid

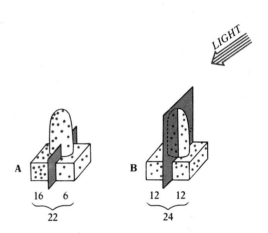

9.7   An experiment demonstrating lateral movement of auxin

The tips of two coleoptiles were placed on agar blocks partitioned by an auxin-impermeable barrier.   (A) If the barrier extended only slightly into the base of the coleoptile tip, more auxin was later found in the side of the block away from the light (16) than in the side toward the light (6). (The numbers refer to the degrees of bending the blocks induce in decapitated coleoptiles in the dark; the amount of bending indicates the amount of auxin in the blocks.)   (B) Approximately the same total amount of auxin was found in the agar block under a coleoptile tip that was completely partitioned, but the amounts in the two sides of the block were the same.   This experiment showed that light produces asymmetric auxin distribution by causing lateral movement of auxin, not by destroying auxin on the lighted side.

9.8 The geotropism of shoot and root
When a growing plant is left lying on its side, the shoot will bend upward and the roots will bend downward.

Auxins and the geotropism of roots Roots, unlike shoots, exhibit positive geotropism; they turn toward the pull of gravity (Fig. 9.8). The gravitational stimulus is apparently detected by cells in the root cap, where, as in the shoot tip, starch grains may act as statoliths.

The previously discussed effects of auxin on shoots are consistent with each other; in both phototropic and geotropic turning by shoots, the side receiving the more auxin elongates faster. But something different seems to happen in roots. Examination of the auxin distribution in the roots of a plant placed on its side reveals a greater quantity of auxin in their lower than in their upper sides; in other words, the asymmetry of auxin distribution in the roots seems to duplicate that in the shoot (though the total amount of auxin is much less). Why, then, does the root turn toward the pull of gravity while the shoot turns away from it? An obvious possibility is that auxin stimulates elongation in the shoot, but inhibits it in the root—in other words, that the same chemical induces diametrically opposite responses in stem cells and root cells.

Further experiments have suggested another hypothesis. It has been shown that auxins do not always inhibit root elongation, but, in very low concentrations, stimulate it. Perhaps roots normally maintain an optimal amount of auxin, and any increase, such as occurs in the lower side of a horizontal root, raises the concentration to an inhibiting level.

It can be demonstrated that elongation of stems, too, can be inhibited if the concentration of auxin in them is raised to a very high level. The basic difference, then, between the reactions of root and shoot cells to auxin seems to be one of sensitivity. Root cells must be far more sensitive to auxin, being stimulated to elongate at very low concentrations of the hormone and inhi-

9.9 Graph showing different sensitivities of roots, buds, and stems to auxin
Roots are much more sensitive than stems; a concentration of auxin that produces maximal growth of roots is not sufficient to maintain even normal growth of stems. Buds exhibit a sensitivity to auxin intermediate between roots and stems. (The line at 0 represents normal growth; values above the line indicate increases and values below the line decreases.) [Redrawn from L. J. Andus, *Plant growth substances*, Leonard Hill, 1959.]

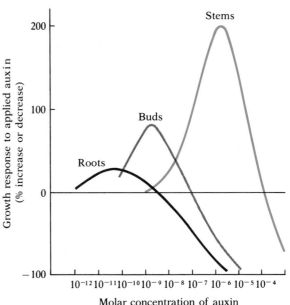

bited at higher concentrations (Fig. 9.9). Stem cells must be less sensitive, being stimulated by moderate concentrations of auxins and inhibited only by very high concentrations. An increase in auxin concentration above the normal moderate level thus inhibits root elongation, but stimulates shoot elongation (provided the increase is not too great).

Auxins and inhibition of lateral buds   Once it is recognized that the response curves to auxins of different plant organs may differ, i.e. that a stimulating concentration for one organ may be an inhibiting concentration for another, it becomes easy to understand another important function of auxin—its inhibiting effect on lateral buds in many plants. Auxin produced in the terminal bud moves downward in the shoot and inhibits development of the lateral buds, while at the same time stimulating elongation of the main stem. An examination of the graph in Fig. 9.9 shows how these two apparently opposite effects can occur at the same time: An auxin concentration that stimulates stem elongation is high enough to inhibit the more sensitive buds. The terminal bud thus exerts *apical dominance* over the rest of the shoot, ensuring that the plant's energy for growth will be funneled into the main stem and produce a tall plant with relatively short lateral branches. Longer branches usually develop only from buds far enough below the terminal bud to be partly free of the apical dominance. If the terminal bud is removed, however, apical dominance is temporarily destroyed, and several of the upper lateral buds will begin to grow, producing branches whose terminal buds soon exert dominance over any buds below them (Fig. 9.10). Flower and shrub growers frequently pinch out the terminal buds of their plants one or more times each season in order to produce bushy well-branched plants with many flowering points instead of tall spindly ones with sparse flowers. Pinching buds will not work for some plants, however, in which it is the young leaves, not the terminal buds, that exert control over the lateral buds.

Notice in Fig. 9.10 that, once two or more branches have begun to develop, neither inhibits the other. Auxin secreted by the terminal bud of one branch does not reach the terminal bud of the other branch in any significant quantity, because it moves mostly downward in stems (Fig. 9.11). This movement in one direction, for which no satisfactory explanation has been found, indicates that the transport cells, like many (perhaps all) other cells, are physiologically polarized; i.e. their ends differ from each other in some way.

The action of auxins in inhibiting growth of lateral buds has sometimes been applied commercially to prevent sprouting of stored potatoes (which are stems, not roots). In former years the buds (commonly called eyes) of potatoes usually began to grow during storage, often producing numerous long sprouts. The sprouting drained nutrients out of the potato tuber itself, often leaving only a shriveled remnant. Treatment with auxin made it possible to store potatoes for long periods, sometimes as long as three years, with relatively little loss. More recently this use of auxins has declined, and another compound, maleic hydrazide, which is not an auxin, has been employed instead. This compound blocks cell division in plants and thus stops growth.

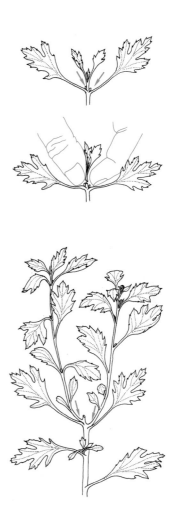

9.10 Inhibition of lateral buds by the terminal bud in chrysanthemum

As long as the terminal bud is intact, the lateral buds marked by arrows will grow very little, if at all. But if the terminal bud is removed, those buds are released from inhibition and grow rapidly, forming the new leaders of the plant.

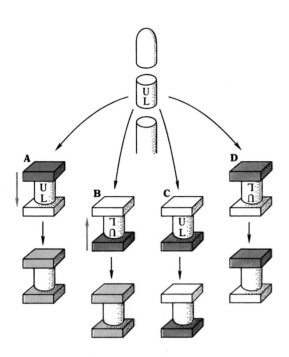

9.11 Experiment demonstrating polarity of auxin movement

A segment is cut from a coleoptile (top).   (A) An agar block containing auxin (dark color) is placed on the upper end of the segment and a block without auxin (white) is placed on the lower end. Some auxin moves from the one block, through the coleoptile segment, into the other block.   (B) The same thing happens even if the whole preparation is inverted, an indication that the movement is not a response to gravity. (C) When the agar block containing auxin is put on the lower end of the coleoptile segment and the block without auxin is put on the upper end, virtually no movement of auxin occurs, and inverting the group (D) makes no difference.   Conclusion: Auxin moves primarily in only one direction through a coleoptile, and that direction is determined by properties of the cells of the coleoptile itself, not by the pull of gravity.

Auxins and fruit development   We have seen that auxins in certain concentrations will stimulate cell elongation in stems, buds, and roots, while the same hormones in higher concentrations will inhibit such growth. Whether the stimulatory or the inhibitory effect will be operative in the plant's development depends for each organ on the particular concentration range to which that organ responds.

One organ whose normal development depends on the stimulatory effect of auxins is the fruit, which develops from the ovary or from the flower receptacle of the plant. In the absence of fertilization, fruit usually does not develop; instead, a weak layer of thin-walled cells, called an *abscission layer,* forms at the base of the flower stalk. This layer soon breaks, under any slight strain, and the withered flower with its ovary falls to the ground. If, on the other hand, fertilization does occur, no abscission layer forms, and the ovary at once begins to grow rapidly. It has been shown that this period of rapid growth by the ovary (and by the receptacle in some plants), as it develops into the fruit, is initiated by auxin released from the pollen grains that bring about the fertilization, and that the continued growth and development of the fruit depend on stimulation by auxins produced by the seeds contained within it. It is these same auxins that prevent formation of an abscission layer at the base of the flower stalk.

Once the role of auxins in stimulating fruit development and in inhibiting abscission-layer formation was recognized, production of seedless fruit became a possibility. It was known that a few plants sometimes produce seedless fruit naturally, and it seemed likely that in these cases tissues of the ovary (or associated structures) themselves produced so much auxin that fertilization and the resulting development of seed became unnecessary for fruit growth. It was reasoned that if fertilization of plants that normally produce seeds, such as tomatoes, cucumbers, squash, and figs, could be prevented, and if auxins could be artificially supplied to take the place of those normally produced by the seeds, seedless fruits should develop. Experiments along these lines were tried, and in 1934 S. Yasuda in Japan produced seedless cucumbers and in 1936 F. G. Gustafson at the University of Michigan produced seedless tomatoes. Since that time seedless fruits of many other plants have been produced. In some plants, however—notably most single-seeded fruits such as plums, cherries, and peaches—all attempts to produce seedless fruits have failed, for reasons as yet unknown. In other plants, such as strawberries and blackberries, seedless fruits have been produced, but the hard cases (derived from the ovaries) that normally cover the seeds have remained, so that a person eating the berry can't tell whether it is seedless or not. The failure of the breeders has its reassuring side: Who wants a strawberry without any crunch and without any seeds to stick between the teeth!

Treatment of fruit crops with auxins has other commercial applications. Often used to supplement normal pollination in the setting of fruit,[1] it ensures a larger crop. In some cases the size of the individual fruits can also be increased by auxin sprays. It has become common practice with many fruits to apply auxin sprays to orchards as the time of ripening approaches,

[1] Fruit-set is the development of the ovary of a flower into fruit. It usually begins after pollination of the flower, but may occur without pollination—by the application of auxins, for example.

thereby inhibiting abscission-layer formation and reducing preharvest fruit drop (fruit that drops to the ground early is largely useless commercially).

Auxins and leaf abscission We have said that unfertilized flowers drop off the plant because a special layer of cells, the abscission layer, forms at the base of the flower stalk in the absence of high auxin production in the floral organs. And we have indicated that ripe-fruit drop is also a result of abscission-layer formation as auxin production by the mature seeds declines. Similarly, the shedding of leaves in autumn (or of diseased leaves at any time of year) by deciduous trees and shrubs usually (though not always) involves abscission-layer formation at the base of the petiole (Fig. 9.12), in part as a result of declining auxin production by the leaf blade. In short, auxin acts as an inhibitor of abscission; another hormone, ethylene (to be discussed below), is the principal promoter of abscission. The actual break in the abscission layer can be initiated by any slight strain, e.g. a gentle wind, because the cell walls have been weakened by an increasing concentration of cellulase in the abscission layer; in addition, in some plants, the middle lamella between the cells may become soft and gelatinous, and thus less able to cement the cells firmly together. Notice that, unlike the previously discussed functions of auxins, their function in leaf and fruit abscission apparently does not involve regulation of cell enlargement.

Sprays of chemicals that are auxin antagonists are commonly applied to the leaves of cotton just before harvest. The anti-auxins induce formation of abscission layers at the base of the leaves, causing the leaves to fall prematurely and thus making it easier for mechanical pickers to move through the fields and harvest the bolls.

Auxins and cell division There is good reason to believe that, besides playing a part in cell elongation and abscission-layer formation, auxins are involved in cell division. Apparently it is auxin, moving downward from the buds in early spring, that stimulates renewed activity in the cambium, leading to production of new vascular tissue. As autumn approaches, auxin production by the buds and leaves declines, with the result that cambial activity also declines.

Auxins probably also initiate formation of lateral roots. Such roots usually have their origin in the layer of relatively undifferentiated cells called the pericycle (see Fig. 17.33, p. 725), which is located just internal to the endodermis. Most of the time the cells of the pericycle show no meristematic activity. At intervals, however, a small group of cells in the pericycle changes into actively dividing meristematic tissue, giving rise to a new lateral root that bursts through the outer tissues of the main root and enters the soil. There is evidence that the stimulus initiating this meristematic activity in the pericycle comes from auxins. Auxins can, in fact, be applied to the roots of plants to induce lateral branching, but it is uncertain whether the auxin stimulates cell division directly or whether it does so indirectly—by eliciting increased production of another hormone (ethylene) that acts as the direct stimulant. Unfortunately the concentrations of auxin necessary to initiate formation of new lateral roots are high enough to inhibit root elongation; the result of auxin application, therefore, is a short bushy root system rather than a longer, less bushy one.

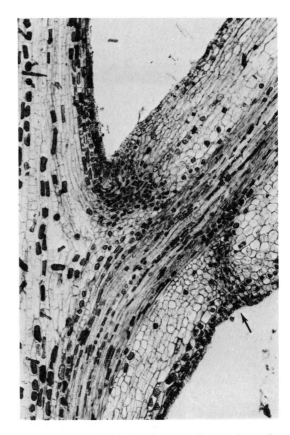

9.12 Photograph showing abscission layer at base of petiole of *Coleus* leaf
The arrow indicates the small cells of the abscission layer. [Courtesy Carolina Biological Supply Co.]

9.13   Gibberellic acid

It is in the development of adventitious roots from cuttings of such organs as stems or leaves that auxins (generally called "rooting hormones" in this context) have found a particularly important commercial use. Cuttings from some plants, such as geraniums and willows, will readily root in water or soil without application of hormone, but many plants cannot be propagated in this manner. Application of auxins will often induce formation of roots in these cases, making it possible to propagate vegetatively many valuable strains of plants that might otherwise be lost.

Chemical weed control   Two widely used modern weed killers, or herbicides, are 2,4-dichlorophenoxyacetic acid, usually abbreviated 2,4-D, and 2,4,5-trichlorophenoxyacetic acid, abbreviated 2,4,5-T. Both 2,4-D and 2,4,5-T have many of the properties of auxins, though they do not meet all the auxin tests. They have been used in vast quantities since the 1940s for control of broad-leaved weeds. Because they are selective in their action and, when used in proper concentrations, will not kill grasses or related monocots, they have been of enormous commercial value in sprays applied to kill broad-leaved weeds in lawns and pastures and in fields of corn, wheat, oats, or rice. Apparently these chemicals, which in low concentrations would have effects like those already described for auxins, kill plants when applied in higher concentrations by stimulating rapid, uncoordinated, and distorted growth of some body parts while seriously inhibiting the function of other parts. The exact manner in which these effects are produced is not well understood. Even less understood, unfortunately, is the basis of the selective action of these herbicides; future research may explain why broad-leaved plants are so much more susceptible than grasses.

In recent years a number of new herbicides even more selective than 2,4-D or 2,4,5-T have been developed and put into use; most of them are not auxins. The goal, of course, is the eventual discovery of herbicides so selective that any given species of weed can be killed with minimum disturbance to the other plants growing around it.

GIBBERELLINS

The Japanese have long been familiar with a disease of rice that they call "foolish-seedling disease." Afflicted plants grow unusually tall, but seldom live to maturity. In 1926 a Japanese botanist, E. Kurosawa, found that all such plants are infected with a fungus named *Gibberella fujikuroi.* He showed that when the fungus was moved to healthy seedlings they developed the typical disease symptom of rapid stem elongation. He could also produce the symptoms with an extract made from the fungus, and even with an extract made from culture media on which the fungus had grown. Clearly, some chemical was involved.

Several Japanese scientists, working on the problem of foolish-seedling disease during the 1930s, succeeded in isolating and crystallizing a substance from *Gibberella,* now known as **gibberellin,** that produced typical disease symptoms when applied to rice plants. Western scientists paid little attention to the Japanese investigations until about 1950, but since that time work on gibberellins has become widespread. More than 50 different substances that can be classed as gibberellins have been isolated from fungi and from higher plants; more will doubtless be discovered in years to come. The gibberellin most often used in experimental work is called **_gibberellic acid_** (Fig. 9.13).

The most dramatic effect of gibberellins is their stimulation of rapid stem elongation in dwarf plants and other plants that normally undergo little stem elongation (Fig. 9.14). They have much less effect on most normally tall plants. An attractive hypothesis is that the dwarf varieties are genetically incapable of producing sufficient gibberellin, and that administration of extra quantities of the hormone simply makes up for the deficiency and allows the plants to grow more normally.

Though both gibberellin and auxin stimulate stem elongation, they cannot substitute for each other in controlling plant growth. For one thing, the stages of development at which the plant is most sensitive to these two hormones often differ; in wheat coleoptiles, for example, responsiveness to gibberellin appears earlier than responsiveness to auxin. For another thing, gibberellin, which can move freely in the plant body through both the xylem and the phloem, does not exhibit the polarity of transport so characteristic of auxin; hence it exerts systemic rather than directional regulatory influences and cannot produce the bending movements that mark auxin-induced tropistic responses. The usual tests for gibberellins are based on their ability to stimulate growth in dwarf plants that have a very low natural gibberellin content (e.g. dwarf corn, dwarf pea); in such a test auxins would yield negative results, because they have virtually no effect on elongation when applied to intact plants.

Gibberellins play a role in a host of other developmental processes besides stem elongation. They can (1) often break seed and bud dormancy; (2) induce the embryos in germinating seeds to produce enzymes that hydrolyze starch reserves in the seeds; (3) stimulate some biennials to flower during their first year of growth; (4) induce some long-day plants to bolt and flower when the day length is too short for them to flower normally; and (5) stimulate fruit-set in some species.

Possible practical applications for gibberellins in agriculture are being extensively investigated. An obvious application would be to induce growth to greater height, e.g. in hay crops. Unfortunately the tallness induced by gibberellins is usually offset by poorer leaf formation and overall spindliness, so that the total weight yield of the crop is increased little, if at all. Gibberellins do, however, have a beneficial effect on celery by inducing rapid growth, which produces tenderer stalks. They have also proved useful in accelerating seed germination in some plants in spring and in producing larger clusters of seedless grapes. Other important uses for these hormones will doubtless be found in the future.

CYTOKININS

The technique of **_tissue culture_**—the growing of cells or bits of tissue on sterile nutrient media in the laboratory—has greatly facilitated research in both plant and animal developmental biology. It was thanks to this technique that a new class of hormones was discovered in the 1950s by Folke Skoog, Carlos O. Miller, and their associates, at the University of Wisconsin.

These workers developed methods of growing parenchyma tissue from

9.14 Effect of gibberellic acid on cabbage
The plant at left is normal. The one at right was treated with gibberellic acid.   [Courtesy S. H. Wittwer, Michigan State University.]

$$NH-CH_2-CH=C\begin{array}{c}CH_3\\CH_2OH\end{array}$$

9.15 Zeatin

tobacco plants on tissue-culture media. The cells formed a tumorlike mass of tissue called a callus, in which the constituent cells often grew to huge size. They did not, however, undergo complete cell division (sometimes the nuclei divided, but new cell walls did not form). Skoog and his colleagues found that extracts made from old nucleic acids would cause the cells in the callus to divide, even though these cells possessed fully developed vacuoles—an indication that they were mature differentiated cells bearing little resemblance to normal meristematic cells. The compound responsible for providing the stimulus was eventually isolated; it is a degradation product of nucleic acids and can easily be produced in the laboratory.

The substance isolated by Skoog and his associates was not a naturally occurring compound, but in 1964 D. S. Letham and his associates in New Zealand isolated from corn seeds a compound called *zeatin* (Fig. 9.15), which is the most active known naturally occurring representative of a class of hormones promoting cell division: the *cytokinins*.

The action of cytokinins on a callus in tissue culture is enhanced when auxins are also present. Indeed, the ratio of cytokinin to auxin appears to be of fundamental importance in determining the differentiation of the new cells (see Fig. 17.47, p. 735). Thus, when there is more cytokinin than auxin, buds are formed by the callus; when there is more auxin, root growth is initiated. In the normal growing plant cytokinins and auxins act synergistically in some situations and antagonistically in others; e.g. they act synergistically in promoting DNA replication, but antagonistically in influencing the growth of lateral buds, cytokinins here being stimulatory and auxins inhibitory. Both hormones influence cell growth, but auxin stimulates elongation whereas cytokinin promotes transverse swelling, at the same time tending to inhibit the auxin-induced elongation.

Among its numerous other functions, cytokinin (1) stimulates conversion of proplastids into functional chloroplasts; (2) helps break dormancy in some seeds; (3) enhances flowering in some plants; (4) promotes fruit development in some species; and (5) helps retard the onset of senescence (aging), especially in leaves, by maintaining protein and nucleic acid synthesis and helping preserve membrane integrity. As this list, and the similar ones for auxin and gibberellin, suggest, all the major plant hormones seem to participate in some fashion in nearly all aspects of plant growth and development.

INHIBITORS

Relatively little is known about growth inhibitors, which have effects opposite to those of auxins, gibberellins, and cytokinins. A few have been isolated and identified, but the existence of many others has simply been inferred.

The role of inhibitors in maintaining dormancy has attracted particular interest. It is believed that inhibitors block the activity of some buds and seeds in autumn, thus making sure that they will not begin to grow during a few warm days, only to be killed by the rigors of winter. Dormancy is broken —and the buds and seeds are set free to become active in the next growing season—when gradual breakdown over time, prolonged exposure to cold, or the leaching action of water has helped destroy the inhibitors. In addition, there may be a rise in another hormone (especially gibberellin) that opposes the inhibitors and helps break dormancy.

Inhibitors that must be leached out by water before the seeds can germinate constitute an important evolutionary adaptation in some desert plants. The seeds that fall to the ground will germinate only after long hard rains. Light showers, which might provide enough moisture for germination by seeds not adapted for life in the desert, do not leach out enough inhibitor to allow germination to begin in the desert-adapted seeds; thus no tender young seedlings remain to be killed by the dry conditions that would soon follow.

The most important known inhibitor is the hormone ***abscisic acid.*** It not only helps induce dormancy in buds and seeds, but, when applied to an actively growing twig, induces the complex of other changes (reduced cell division, production of protective scales instead of foliage leaves, deposition of waterproofing substances, etc.) that prepare the plant for overwintering. As its name implies, abscisic acid promotes leaf abscission in some plants, though this action is far from universal. The hormone also participates in the control of flowering in some species.

Most of the other known growth inhibitors are not true hormones. So-called ***secondary plant substances,*** they include quinones, phenolic acid and its derivatives, and a variety of other compounds. They are usually effective only in very high concentrations and do not appear to play any direct role in metabolism. Being often toxic to insects and other herbivorous animals, they may be of special importance to plants as a means of chemical defense.

ETHYLENE

"One rotten apple spoils the lot." That piece of folk wisdom rests on a familiar fact: When one apple in a barrel goes bad, most of the other apples in that barrel soon go bad too. We now know that the bad apple affects other fruit by the production of a very volatile compound called ethylene (C_2H_4). (see Fig. 2.22, p. 46). Only in recent years, however, have plant physiologists realized that ethylene has an importance beyond its association with plant disease, that it is, in fact, a hormone that plays a variety of roles in the life of a normal plant.

One of the best-studied effects of ethylene is stimulation of fruit ripening. Once the fruit has attained its maximum size, a host of chemical changes begin that cause it to ripen. The ripening process starts with a sudden sharp increase in carbon dioxide output, followed shortly by a sharp decline. It can be shown that this burst of metabolic activity, called the climacteric, is triggered by an approximately hundredfold increase in the concentration of ethylene. Inhibition of ethylene production, or removal of the ethylene as fast as the fruit produces it, prevents the climacteric, and no ripening occurs.

Ethylene has also been shown to contribute to leaf abscission and to various other changes characterizing senescence in a plant or parts of a plant. In addition, it can sometimes stimulate radial growth of stems and roots; it can aid in breaking dormancy in the buds and seeds of some species; and it can help initiate flowering in some plants (e.g. pineapple). Moreover, some effects customarily attributed to auxin (e.g. lateral bud inhibition, root geotropism) may in some cases be due to an auxin-induced increase in ethylene.

Let us summarize the conclusions that can be drawn from our discussion of chemical control of plant growth so far:

1. Cell division, the first phase of growth, is stimulated by cytokinins and auxins, and by other factors that enhance their activity.
2. Cell division is inhibited by a variety of substances, which are but poorly known.
3. The balance between the cytokinins and the inhibitors determines whether a cell will divide or not.
4. Control of cell enlargement, the second phase of growth, involves substances such as auxins and gibberellins, which promote elongation, and cytokinins (sometimes also ethylene), which stimulate lateral swelling.
5. The various plant hormones, by their mutual interactions and their differential effects on various parts of the plant body, help integrate and coordinate the development of form and function.
6. The aging process, leading to death, is brought on by various senescence-inducing substances, of which ethylene is one of the most important.

CONTROL OF FLOWERING

As you are doubtless aware, flowering is not a random process. Some plants flower early in the spring, others flower in midsummer, and still others, like chrysanthemums, flower in the fall. These simple facts have been known for centuries. But only since 1920 has anything been known of the control mechanisms involved, and some aspects are still not well understood.

Photoperiodism and flowering The intense modern interest in the flowering process dates from the investigations of W. W. Garner and H. A. Allard of the U.S. Department of Agriculture. Working in Beltsville, Maryland, these two men found that a new mutant variety of tobacco, called Maryland Mammoth, grew unusually large (as much as 10 feet tall), but would not flower, thus making it unusable in breeding experiments. They propagated the new variety by cuttings and discovered that it would flower in the greenhouse in winter. Though flowering was not the subject Garner and Allard were originally investigating, they became interested in the question why Maryland Mammoth would flower in the greenhouse in winter, but not in the fields in summer. Accordingly they began a series of experiments that were to open the way to a whole new area of botanical research. This is but one of many examples that could be cited of important leads uncovered almost accidentally through research originally devoted to a different subject. It is the mark of good scientists that they do not overlook such leads, but recognize and pursue them, even though they might have to change the course of their research.

Garner and Allard realized that winter greenhouses and summer fields differ in temperature, moisture, light intensity, day length, etc. They began experiments that painstakingly eliminated one after another of these environmental factors, until only one was left as the probable controlling factor in flowering—day length. They concluded that the short days of late autumn and early winter induced flowering in Maryland Mammoth tobacco. They could get the plants to bloom in summer if they shielded them from the light for a part of each day. Conversely, they could prevent blooming in the greenhouse in wmter by extending the day length with electric lights.

Garner and Allard also experimented with Biloxi soybeans. They planted soybeans at two-week intervals from early May through July and found that all the plants flowered at the same time in September, even though their growing periods had differed by as much as 60 days. It was as though they were waiting for some signal from the environment. Garner and Allard were sure that the signal was short days.

Experiments with other species revealed that most plants can be placed in one of three groups: (1) short-day plants, which flower when the day length is below some critical value, usually in spring or fall (examples are chrysanthemum, poinsettia, dahlia, aster, cocklebur, goldenrod, ragweed, Maryland Mammoth tobacco, and Biloxi soybean); (2) long-day plants, which bloom when the day length exceeds some critical value, usually in summer (beet, clover, petunia, larkspur, black-eyed Susan); and (3) day-neutral plants, which are independent of day length and can bloom under conditions of either long or short days (dandelion, sunflower, carnation, pansy, tomato, corn, string bean). Garner and Allard called the response by an organism to the duration and timing of the light and dark conditions *photoperiodism.*

If the critical element in the photoperiodism of flowering is day length, as the terms "long day" and "short day" imply, then we should be able to prevent a long-day plant from flowering at the proper season by shielding it from light for an hour or so during the middle of the day. But if this is done, nothing happens; the plant flowers normally. If, however, a short-day plant is illuminated by a bright light for a few minutes, or even seconds, in the middle of the night during the normal flowering season it will not bloom (Fig. 9.16H). The same sort of experiment will induce flowering at the wrong season by a long-day plant. It is clear, then, that the critical element of the photoperiod is actually the length of the night, not the length of the day. Instead of speaking of long-day and short-day plants, it would be more accurate to speak of short-night and long-night plants.

It has been established that the difference between long-day and short-day plants does not hinge on the absolute length of the night at the time of flowering. The difference is that a long-day (short-night) plant will flower only when the night is *shorter than* a critical value, whereas a short-day (long-night) plant will flower only when the night is *longer than* a critical value (Fig. 9.16). The critical night length is thus a maximum value for flowering by long-day plants and a minimum value for flowering by short-day plants.

Flowering hormone The question now becomes: How does photoperiod exert its effect on flowering? It was suggested long ago that hormones might be involved, but the first convincing evidence for a floral hormone did not come until 1936, from the experiments of M. H. Chailakhian in Russia. Chailakhian removed the leaves from the upper half of chrysanthemums (which are short-day plants), but left the leaves on the lower half (Fig. 9.17). He then exposed the lower half to short days while simultaneously exposing the defoliated upper half to long days; the plants flowered. Next, he reversed the procedure, exposing the lower half to long days and the defoliated upper half to short days; the plants did not flower. He concluded that day length does not exert its effect directly on the flower buds, but causes the leaves to manufacture a hormone that moves from the leaves to the buds and induces flowering. This hypothetical hormone has been named *florigen.*

Further evidence for the existence of a moving stimulus, probably a hor-

9.16   Comparison of long-day and short-day plants
White bars indicate days and gray bars nights. The hypothetical long-day (short-night) plant shown here has a rather long critical night length of 13 hours, and the hypothetical short-day (long-night) plant has a rather short critical night length of $8\frac{1}{2}$ hours. In other words, in this example the critical night length for the short-night plant is actually longer than that for the long-night plant. The difference is that the critical night length is a *maximum* value for the short-night plant and a *minimum* value for the long-night plant. Thus the short-night plant will flower when the night length is slightly *below* the critical value (A) or when it is much below the critical value (B), but will not flower when it is above the critical value (C); the plant will flower, however, if given a long night interrupted by a flash of light that reduces the period of continuous dark below the critical value (D). Conversely, the long-night plant will flower when the night length is slightly *above* the critical value (E) or when it is much above the critical value (F), but will not flower when it is below the critical value (G); the plant will not flower if given a long night interrupted by a flash of light that reduces the period of continuous dark below the critical value (H).

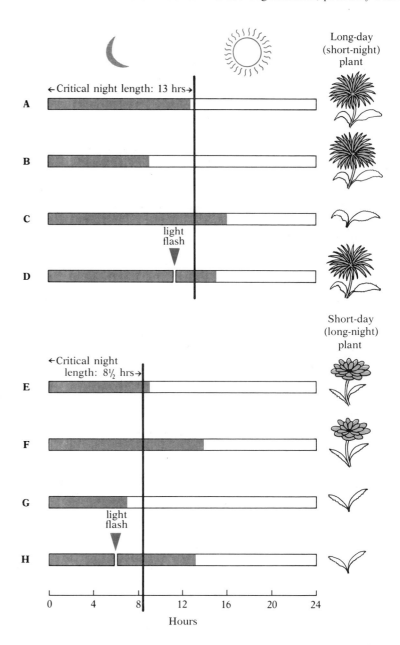

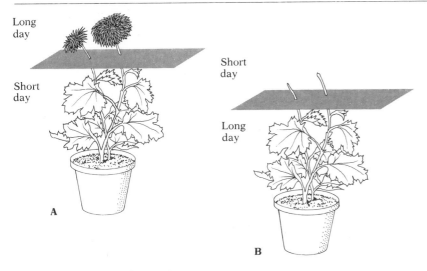

9.17 Chailakhian's experiment
(A) Chailakhian removed the leaves from the top half of a chrysanthemum (a short-day plant) and then exposed the top half of the plant to long days and the bottom half to short days. The plant flowered.
(B) When he did the reverse experiment, the plant did not flower.

mone, has come from grafting experiments with cockleburs (which are short-day plants) (Fig. 9.18). If one plant is grafted onto another through a light-tight partition, and if the first plant is exposed to an inducing photoperiod (short days/long nights) while the other is exposed to a noninducing photoperiod (long days/short nights), the plant exposed to short days will flower, and soon thereafter the plant exposed to long days will also flower. Presumably a stimulus from the first plant moves through the graft and induces flowering in the second plant, even though the second plant is exposed to the wrong photoperiod. The same results are obtained if only a single leaf is left on the plant in the inducing photoperiod; apparently the one leaf can produce enough hormone to cause flowering in both plants.

There is evidence that the flower-inducing hormone is the same in both short-day and long-day plants, because in many cases, if a long-day and a short-day plant are grafted together and then exposed to short-day conditions, both will flower. Apparently hormone produced by the induced short-day plant is able to move through the graft and make the noninduced long-day plant flower. Indeed, cross induction can also be obtained in grafts between long-day and day-neutral plants or between short-day and day-neutral plants. Ringing experiments show that the stimulus is transported in the phloem.

In some plants there is an added complication, namely that leaves exposed to a noninducing photoperiod actively inhibit flowering. Thus, if a light-tight barrier is placed across a cocklebur leaf and the basal half of the leaf is exposed to a flower-inducing photoperiod while the distal half is

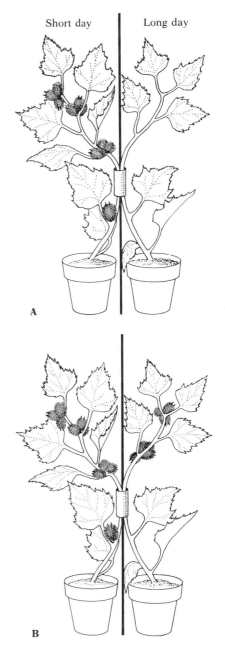

9.18 Grafting experiment with cockleburs
The two plants are separated by a light-tight barrier, but are connected by a graft. The plant exposed to an inducing photoperiod (short days) flowers (A), and shortly thereafter the other plant begins to flower (B).

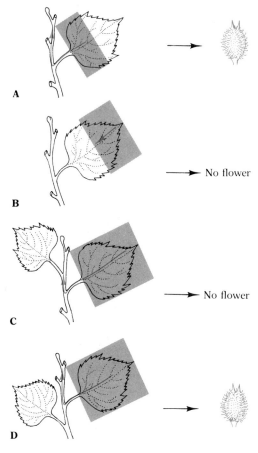

9.19   Some experiments illustrating the inhibitory effect of long days on flowering in the cocklebur
(A) When the tip of the uppermost leaf is exposed to continuous illumination, but the base is covered with black paper to give it a short day, the plant flowers, presumably because some inducing substance (florigen?) synthesized in the shaded part of the leaf has moved (arrow) to the flower bud.   (B) The reverse proceeding, shading the leaf tip and illuminating the leaf base, does not result in flowering, presumably because the inducing substance is destroyed as it passes through the illuminated part of the leaf. (C) When one leaf is shaded, but a leaf above it on the stem is illuminated, the plant does not flower; perhaps something at the base of the illuminated leaf destroys the inducing substance.   (D) However, if the shaded leaf is located above the illuminated leaf, no such destruction occurs presumably, and the plant flowers.   [After *Botany: An ecological approach* by W. A. Jensen and F. B. Salisbury. © 1972 by Wadsworth Publishing Company, Inc., Belmont, Calif. By permission of the publisher.]

exposed to a noninducing photoperiod, a nearby bud will flower (Fig. 9.19A). If the reverse experiment is run, with the basal half of the leaf exposed to a noninducing photoperiod and the distal half to an inducing photoperiod, the bud will either not flower or flower only weakly; perhaps hormone produced under inducing conditions in the distal half of the leaf is destroyed as it passes through the noninduced basal half. The inhibiting effect can also be seen if one leaf is exposed to an inducing photoperiod and another nearby leaf is in a noninducing photoperiod, but only if the noninduced leaf is located between the induced leaf and the bud (Fig. 9.19C, D). In cocklebur and other species the inhibition may be local and not transmissible, but in some species (e.g. strawberries) there is evidence for a transmissible inhibitor.

The explanation of flowering that emerges from the sum total of the above experiments is that an inducing photoperiod causes the leaves to produce florigen, which moves to the buds and stimulates development of the flower. A noninducing photoperiod causes the leaves of many, but not all, plants to inhibit florigen in some way. Under natural conditions flower induction is triggered when, as the season changes, the photoperiod passes a critical value and production of florigen by the leaves exceeds the inhibiting tendency.

This is a tidy explanation, but it still awaits proof. All attempts to isolate a flower-inducing hormone have failed. It has been demonstrated that gibberellins will induce flowering in some plants even in the absence of an inducing photoperiod, but that they have no effect on flowering in many species, and actually inhibit flowering in still other species. The evidence suggests that gibberellins probably do play an important role in the flowering of some species, especially long-day plants that undergo great stem elongation during the transition from the vegetative to the reproductive state, but that they do not qualify as florigen. The other major plant hormones—auxins, cytokinins, abscisic acid, and ethylene—have all been shown to promote flowering in one or more species, but they too fall far short of exhibiting all the properties ascribed to florigen.

Some scientists hold that there is no such thing as florigen, that interactions among the known hormones, including the many gibberellins (each of which has its own distinctive biological activity), might be the basis for control of flowering. Although it seems difficult to reconcile this hypothesis with all the results from grafting experiments, and although most botanists continue to favor some version of the florigen hypothesis, the fact remains that it will not rest on firm foundations until a flower-inducing hormone is actually isolated and identified.

Detection of the photoperiod   In our discussion so far, we have avoided asking one fundamental question: How do plants detect and measure the photoperiod? Since interrupting the dark period prevents flowering by a short-day (long-night) plant and induces flowering by a long-day (short-night) plant, the light itself must be detected by the plant. What wavelengths of light are involved? H. A. Borthwick, S. B. Hendricks, and their associates of the U.S. Department of Agriculture, Beltsville, Maryland, began investigating this question in 1944. They exposed Biloxi soybeans to light of different wavelengths, and found that red light (wavelength about 660 nm) is by far the most effective in inhibiting flowering in these short-day plants; the

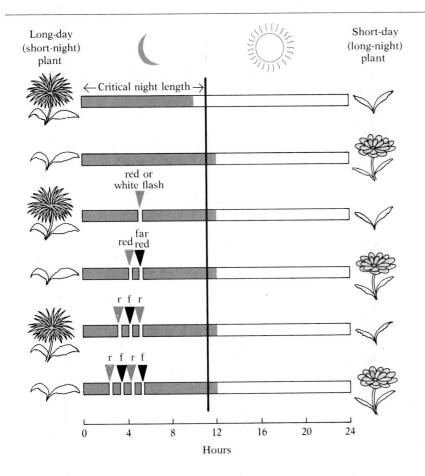

9.20 Reactions of long-day and short-day plants under a variety of light regimes

White bars are days; gray bars are nights. Long-day (short-night) plants flower when the night is shorter than the critical value, or when a longer night is interrupted by an intense red or white flash or by a series of flashes of which the last is red or white. Short-day (long-night) plants give the reverse responses.

same red light is very effective in inducing flowering by long-day plants. Later, it was found that far-red light (wavelength about 730 nm), which is invisible to the human eye, has effects exactly contrary to those of red light; it induces flowering in short-day plants and inhibits flowering in long-day plants. Not only do red and far-red light have opposite effects, but each reverses the effect of prior exposure to the other (Fig. 9.20). A short-day (long-night) plant will not flower if its long night is interrupted by a bright flash of red light; if, however, the red flash is followed immediately by a far-red flash, the plant flowers normally. Almost any number of successive flashes can be used, the final effect depending solely on whether the last flash was red or far-red.

The fact that red and far-red light can reverse each other led Borthwick and Hendricks to conclude that a single receptor pigment, which they called **phytochrome,** is involved, and that this pigment exists in two forms: one that absorbs red light (P_r) and one that absorbs far-red light (P_{fr}). It has since been identified as a protein with a prosthetic group that gives it the properties of a pigment—notably the ability to absorb light of certain wavelengths. When P_r absorbs red light, it is rapidly converted into P_{fr}. Conversely, absorption of far-red light by P_{fr} rapidly converts it into P_r. The P_r form is apparently the more stable of the two; in darkness P_{fr} may revert to P_r over the course of several hours in some plants. In addition, P_{fr} may be enzymatically destroyed. We can summarize these conversions as follows:

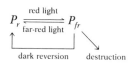

When phytochrome is exposed to red and far-red light simultaneously, the red light dominates and most of the pigment is converted into P_{fr}. Sunlight or light from ordinary electric lamps contains both red and far-red wavelengths; hence, during the day, the phytochrome exists predominantly as P_{fr}. During the night the P_{fr} supply dwindles as a result of both reversion and destruction. The pigment thus gives the plant a way of detecting whether it is day or night.

We began this part of our discussion with the question: How does the plant detect and measure the dark period? We have answered the first half of the question; the plant possesses a sensitive pigment, phytochrome, that responds to presence or absence of light. But what about the crucial second half of the question? It is, after all, the measuring of the dark period that is fundamental to control of flowering.

The most obvious hypothesis would be that the metabolic conversion of P_{fr} into P_r in the dark proceeds so slowly that the amount of conversion occurring between two light periods provides a measure of the length of the intervening dark period. In other words, the system would work like an hourglass. Light would convert all the pigment into P_{fr}. Then during the following dark period the amount of P_{fr} converted into P_r before the next light period would tell the plant how long the dark period had lasted. The evidence is unfortunately against this hypothesis. Although the rate of dark conversion is strongly temperature-dependent, it can be shown that the plant's measure of time is little influenced by temperature. Moreover, it appears that the supply of P_{fr} declines much too fast to be the basis for the time measurement.

The mechanism by which the plant measures the length of the dark period is apparently tied to a phenomenon—now believed to occur in all living cells—involving persistent and regular rhythms in function, rhythms that must be dependent on some internal time-measuring system, or "internal clock." Biological time measurement is an exciting area of modern research, and we shall discuss it at greater length in a later chapter. For the moment, let us say just this: The phytochrome enables the plant to sense whether it is in light or in darkness, but the actual measuring of the time lapse between the moment the plant senses onset of darkness and the moment it senses the next exposure to light must depend on an internal clock.

Once the phytochrome mechanism and the internal-clock mechanism have together indicated to the plant that the photoperiod is appropriate to flowering, the leaves must begin synthesizing florigen, which is then transported in the phloem to the buds. How phytochrome is actually coupled to florigen synthesis is unknown. Apparently it is P_{fr} that is the active form of the pigment; it must help regulate processes that alter the developmental state of the leaf cells, causing them to become more inductive.

The phytochrome molecules, which are located in the cell membrane and occur in only very tiny concentrations, can powerfully influence a variety of cell activities besides those associated with flowering. For example, germination of some types of seeds requires exposure to red light, which is sensed by phytochrome. Gibberellin-controlled cell elongation, expansion of new leaves, breaking of dormancy in spring, and formation of plastids in cells also involve this pigment.

Recent evidence suggests that one of the major functions of phytochrome in plants growing under natural conditions lies in detecting the extent of shading. Because foliage tends to absorb or reflect wavelengths below 700 nm, but to transmit wavelengths in the 700–800 nm range (i.e. the far red), the ratio of red to far-red light in the sunlight reaching a leaf is an indication of the amount of shading by other leaves. The plant responds with modifications in stem elongation, amount of branching, leaf pigmentation, and, of course, flowering.

HORMONES IN INVERTEBRATE ANIMALS

Much interest has developed in the hormones of invertebrate animals in recent years. Hormonal mechanisms have been found in a variety of invertebrates, including arthropods, annelid worms, molluscs, and echinoderms. It seems likely that hormonal control is a general phenomenon in both plants and animals, and that the list of animals in which such control is demonstrated will become steadily longer. At present, however, knowledge of the hormones of most invertebrates is extremely rudimentary. The arthropods, particularly the insects, have been most extensively studied, and our discussion here will be limited to these animals. We shall not attempt to mention all the insect hormones now known. Our intent is simply to give you a quick introduction to the subject of hormonal control in invertebrates, for it sheds light on developmental processes in general.

Although most known hormones of insects, including the three discussed below, regulate growth and development, it has become clear that insect hormones are not restricted to regulation of these processes. In 1963 a hormone was found in cockroaches that regulates the blood-sugar level—an important discovery, because it indicated that hormones are involved in homeostatic regulation of body fluids in insects just as they are in vertebrates. Other hormones of this type, regulating such conditions as salt and water balance and protein metabolism, have since been discovered.

Much of the early work on insect hormones was done in the 1930s in England by V. B. Wigglesworth, who was studying the metamorphosis of insects. Now, insects show a pattern of growth different from that of vertebrates. Their body is encased in a hard outer covering, or exoskeleton, that severely limits size increase. The insect's tissues grow until they exert considerable pressure against the inner surface of the exoskeleton; further growth is impossible unless the exoskeleton is shed. This is exactly what happens. The insect periodically molts its old exoskeleton and develops a new larger one in its place. Wigglesworth was interested in the mechanisms that control this molting.

Most of his experiments were performed on a bloodsucking bug from South America named *Rhodnius*. This bug goes through five immature or nymphal stages, each separated by a molt, before it becomes an adult.

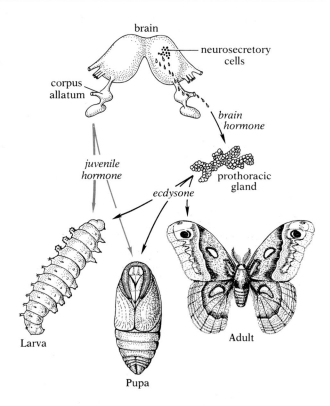

9.21   Diagram showing interactions of juvenile hormone, brain hormone, and molting hormone (ecdysone) in *Cecropia* silkworm
If much juvenile hormone is present when the insect molts, it will molt into another larval stage. If a low concentration of juvenile hormone is present, the larva will molt into a pupa. If no juvenile hormone is present, the pupa will molt into an adult.   [Modified from H. A. Schneiderman and L. I. Gilbert, *Science*, vol. 143, 1964. Copyright 1964 by the American Association for the Advancement of Science.]

During each nymphal stage it must obtain a blood meal, which engorges and stretches the abdomen. As Wigglesworth demonstrated, this repletion apparently stimulates release of hormones that cause molting at the end of a definite time interval following the meal.

Ordinarily, the last molt (from fifth nymphal stage to adult) occurs about 28 days after the blood meal. Wigglesworth showed that if *Rhodnius* is decapitated during the first few days after this meal, molting does not occur, even though the animal may continue to live for several months. Decapitation more than eight days after the blood meal does not interfere with molting; a headless adult is produced. If, however, the circulatory system of a bug decapitated eight days after a blood meal and that of a bug decapitated soon after the meal are joined, both bugs molt into adults. Clearly, some stimulus passes via the blood from the one insect to the other and induces molting. That stimulus must be a hormone, whose secretion by the head begins about eight days after a blood meal.

It was later shown that the **brain hormone**[2] (Fig. 9.21) functions by stimulating glands in the insect's prothorax (the part of the body immediately behind the head, to which the first pair of legs is attached). The prothoracic glands, in turn, secrete a hormone, often called **ecdysone** or molting hormone, that induces molting. It has been found to be involved in many other growth and development processes too. Apparently it acts on the genes of

[2] Brain hormone is also known as prothoracicotrophic hormone.

several types of cells, stimulating these cells to produce certain kinds of proteins, to grow, and to divide.

Wigglesworth became interested in the control factors that determine when a molt will result in an adult and when it will result in another immature stage. This is a particularly important question in insects like flies, beetles, and moths, which undergo a radical change from immature to adult characteristics—i.e. which undergo complete metamorphosis from grub to fly or from caterpillar to moth. Wigglesworth found that a third hormone is involved. This hormone, called **juvenile hormone,** is produced by a pair of glands (corpora allata) located just behind the brain and closely associated with it. When juvenile hormone is present in high concentration at the time of molting, another immature stage follows the molt. The pupal stage, which is the changeover stage between the last larval stage and the adult in insects like flies and moths, results from a low concentration of juvenile hormone. Juvenile hormone is absent in the pupa, and when it molts an adult results. Removal of the corpora allata from insects in the first or second immature stage results in pupation at the next molt, followed by a molt that results in a midget adult. Conversely, implantation of active corpora allata into insects about to undergo their final molt results in another immature stage instead of an adult; in this way several extra immature growth stages can be inserted into the insect's developmental sequence. These can be followed by pupation and a molt producing an unusually large adult when juvenile hormone is finally eliminated. The interactions, as now understood, between brain hormone, ecdysone, and juvenile hormone in molting are shown in Fig. 9.21.

The three hormones controlling molting have other functions as well. As we have already seen in our examination of plants, most hormones have a variety of actions. For example, though the molt from pupa to adult requires a near absence of juvenile hormone, secretion of it usually begins again after metamorphosis is complete; in many species a high concentration of juvenile hormone is necessary in females for deposition of yolk in the eggs and in males for formation of spermatophores.

Notice that the nervous system of insects is intimately involved in endocrine function. Of the three hormones we have discussed, one is secreted by the brain itself, one is secreted by glands closely associated with the brain, and secretion of the third is directly controlled by brain hormone. This intimate association between the nervous system and endocrine function, and the consequent lack of clear distinction between the two control systems, is characteristic for invertebrates.

HORMONES IN VERTEBRATE ANIMALS

It is not our purpose here to enumerate all known vertebrate hormones and their attributes. Instead, as in our discussion of plant hormones, we shall try to give you some idea of the way in which the diverse functions of a complex organism are regulated and indicate what is currently known of chemical control in vertebrates. Since far more is known about hormones in mammals than about those of any other group of animals, it is the mammalian hormonal system (especially the human system) that we shall emphasize here (Fig. 9.22; Table 9.1).

TABLE 9.1 *Important mammalian hormones*

Source	Hormone	Principal effects
Pyloric mucosa of stomach	Gastrin	Stimulates secretion of gastric juice; stimulates enzyme secretion by pancreas
Mucosa of duodenum	Secretin	Stimulates flow of pancreatic juice; inhibits gastrointestinal motility and gastric acid secretion
	Cholecystokinin	Stimulates release of bile by gallbladder
Pancreas	Insulin	Stimulates glycogen formation and storage; stimulates glucose oxidation; inhibits formation of new glucose; stimulates synthesis of protein and fat
	Glucagon	Stimulates conversion of glycogen into glucose
Adrenal medulla	Adrenalin	Stimulates elevation of blood-glucose concentration; stimulates "fight-or-flight" reactions
	Noradrenalin	Stimulates reactions similar to those produced by adrenalin, but causes more vasoconstriction and is less effective in conversion of glycogen into glucose
Adrenal cortex	Glucocorticoids (corticosterone, cortisol, cortisone, etc.)	Stimulate formation of carbohydrate from protein and fat, thus elevating glycogen stores and helping maintain normal blood-sugar levels
	Mineralocorticoids (aldosterone, deoxycorticosterone, etc.)	Stimulate kidney tubules to reabsorb more sodium and water and less potassium
	Cortical sex hormones	Stimulate secondary sexual characteristics, particularly those of the male
Thyroid	Thyroxin, triiodothyronine	Stimulate oxidative metabolism; help regulate growth and development
	Calcitonin	Prevents excessive rise in blood calcium
Parathyroids	Parathyroid hormone (PTH)	Regulates calcium-phosphate metabolism
Thymus	Thymosin	Stimulates immunologic competence in lymphoid tissues

TABLE 9.1   *Important mammalian hormones (Cont.)*

Source	Hormone	Principal effects
Testes	Testosterone	Stimulates development and maintenance of male accessory reproductive structures, secondary sexual characteristics, and behavior; stimulates spermatogenesis
Ovaries	Estrogen	Stimulates development and maintenance of female accessory reproductive structures, secondary sexual characteristics, and behavior; stimulates growth of the uterine lining
	Progesterone	Prepares uterus for embryo implantation and helps maintain pregnancy
Hypothalamus	Releasing hormones	Regulate hormone secretion by anterior pituitary
Posterior pituitary (storage organ for hormones produced by hypothalamus)	Oxytocin	Stimulates contraction of uterine muscles; stimulates release of milk by mammary glands
	Vasopressin	Stimulates increased water reabsorption by kidneys; stimulates constriction of blood vessels (and other smooth muscle)
Anterior pituitary	Growth hormone (STH)	Stimulates growth; stimulates protein synthesis, hydrolysis of fats, and increased blood-glucose concentrations
	Prolactin (PRL)	Stimulates milk secretion by mammary glands; participates in control of reproduction, osmoregulation, growth, and metabolism
	Melanophore–stimulating hormone (MSH)	Probably helps regulate salt and water balance; may influence certain types of behavior; (controls cutaneous pigmentation in poikilotherms)
	Thyrotrophic hormone (TSH)	Stimulates the thyroid
	Adrenocorticotrophic hormone (ACTH)	Stimulates the adrenal cortex
	Follicle-stimulating hormone (FSH)	Stimulates growth of ovarian follicles and of seminiferous tubules of the testes
	Luteinizing hormone (LH)	Stimulates conversion of follicles into corpora lutea; stimulates secretion of sex hormones by ovaries and testes
Pineal	Melatonin	Helps regulate production of gonadotrophins by anterior pituitary, perhaps by regulating hypothalamic releasing centers

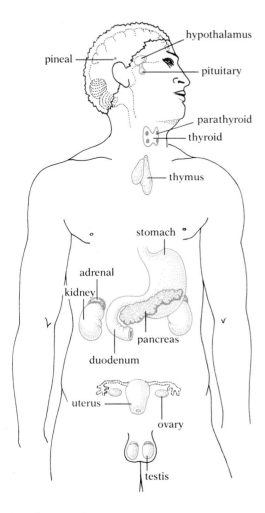

9.22 The major human endocrine organs

We have already said that hormones are specific chemical messengers that exert effects at points some distance removed from their sites of production. Hormones may, of course, diffuse from one place to another, but, as would be expected in animals with well-developed circulatory systems, most of their transport in vertebrates is by the blood. The majority of vertebrate hormones form weak bonds with plasma proteins and are transported in this bound form.

The tissues that produce and release hormones are termed **endocrine** tissues. The use of the word "endocrine"—i.e. secreting internally—is meant to convey that the hormones are secreted directly into the blood in the capillaries supplying the endocrine tissues and that no special ducts or tubes are involved. In fact, endocrine glands are often called the ductless glands.

HORMONAL CONTROL OF DIGESTION

The discovery of the chemical control system for secretion of gastric fluid by glands in the wall of the stomach provides an especially elegant example of biological research. We shall examine it in some detail.

The famous Russian physiologist Ivan P. Pavlov (1849–1936) showed in a series of now classic experiments that both the secretion of saliva and the first phase of gastric secretion, in which the secretion occurs before food actually reaches the stomach, are under nervous control. He also showed that, in the second phase of gastric secretion, food in the stomach leads somehow to release of gastric juice. The question he now sought to answer was whether the mere physical presence of food was sufficient to stimulate this release.

If a piece of fresh meat was inserted directly into the stomach of a dog without allowing the animal to sense the food, there was no secretion of gastric juice. Furthermore, stimulation of the stomach wall with a glass rod, with sand, or with other mechanical devices resulted in very little secretion. If, however, a piece of partly digested meat was inserted directly into the stomach, secretion promptly began. Apparently compounds released from the partly digested meat triggered the secretion. The conclusion to be drawn was that, under normal conditions, partial digestion during the first phase of gastric secretion, which is under nervous control, releases compounds that trigger the second phase of gastric secretion, which makes possible the continuation of gastric digestion after nervous stimulation has ceased.

How did the substances from partly digested meat stimulate release of gastric juice? An obvious possibility was that they might act by direct stimulation of the gastric glands. This was ruled out, however, when Pavlov divided the stomach surgically into two chambers and showed that partly digested meat in one chamber stimulated release of gastric juice in both chambers, even when the stomach had been completely isolated from nervous control. Direct contact between the meat substances and the glands was thus not necessary to induce secretion. Perhaps the substances from the partly digested meat were absorbed and carried to the gastric glands by the blood. This, too, was ruled out when it was shown that injection of the meat substances directly into the blood caused relatively little secretory response, and that, furthermore, little or no absorption occurred in the stomach.

At least one other promising possibility remained. Perhaps the meat substances triggered release of a hormone into the blood, and this hormone, in turn, stimulated the gastric glands to begin secreting. J. S. Edkins at St. Bartholomew's Hospital, London, showed in 1905 that, if a piece of mucosa stripped from the pyloric region of the stomach wall was ground up in the presence of meat substances and an extract was prepared from it, injection of the extract stimulated gastric secretion. Edkins concluded that the meat substances stimulated the mucosa of the pyloric region of the stomach to release a hormone, which he called *gastrin.* He believed that the gastrin was carried by the blood to the gastric glands and stimulated them to secrete gastric juice. Numerous later experiments brought confirmation for Edkins' conclusions. For example, when the circulatory systems of two dogs are interconnected, and partly digested food is placed in the stomach of one of the dogs, the gastric glands of both dogs begin secreting gastric juice; presumably gastrin from the one dog is carried by the blood to the other dog and stimulates its gastric glands.

W. M. Bayliss and E. H. Starling of University College, London, showed in 1902 that secretion of pancreatic juice is also under control of a hormone. This hormone, called *secretin,* is released by the mucosal cells of the duodenum when they are stimulated by the acidity of food coming from the stomach. Another hormone, *cholecystokinin,* released by the duodenum under stimulation by acids and fats, stimulates release of bile from the gallbladder.

THE PANCREAS AS AN ENDOCRINE ORGAN

Diabetes (or, more precisely, diabetes mellitus), a disease in which much sugar is excreted in the urine, has been known for centuries, but its causes did not begin to be understood until the latter half of the nineteenth century. In 1889 two German physicians, Johann von Mering and Oscar Minkowski, who were interested in the role of the pancreas as a producer of digestive enzymes, surgically removed the pancreas from a dog. A short time later it was noticed that the dog's urine was attracting an unusual number of ants. Analysis showed that the urine contained a high concentration of sugar. Furthermore, the dog soon developed other symptoms strikingly like those of human diabetes. Von Mering and Minkowski removed the pancreas from other dogs, and diabetes invariably developed. To eliminate the possibility that the extensive damage resulting from so severe an operation might be the causal factor, they performed operations in which all the damage usually associated with the operation was produced, but the pancreas was not actually removed. These dogs did not develop symptoms of diabetes. Clearly, the onset of diabetes was directly correlated with extirpation of the pancreas. But operations in which the pancreatic duct was destroyed without producing the disease made it clear also that diabetes was not correlated with absence of the pancreatic digestive enzymes. The unavoidable inference was that the pancreas participated in other body functions besides the digestive one.

Mounting evidence pointed to secretion by the pancreas of some substance that prevents diabetes in the normal animal, but all attempts at proof failed. Feeding bits of pancreas to diabetic dogs had no effect; if the pancreas contained a control chemical—a hormone—it was destroyed by digestive enzymes. Repeated efforts by numerous investigators to show that injection of an extract made from the pancreas would alter diabetic symptoms failed. The reason for these failures was soon realized: Grinding pancreatic tissue to produce the extracts mixed the hormone with the pancreatic digestive enzymes, which destroyed the hormone. But how could this be avoided?

It was known that the pancreas is a compound organ, i.e. that it contains several types of cells, which were thought to function independently. There are the cells involved in production and release of digestive enzymes, and there are other quite different cells, called *islet cells* or islets of Langerhans (Fig. 9.23). It seemed likely that the hormone so many people were searching for was produced by the islet cells (more specifically, by β islet cells). In a critical experiment that supported this hypothesis and opened the way for isolation of the hormone, it was shown that tying off the pancreatic duct results in atrophy of most of the pancreas but not in development of diabetes. Examination of the atrophied pancreas revealed that it was the enzyme-producing portion that had atrophied, while the islet cells had remained essentially intact. The hormone that prevented diabetes must have come from this portion of the pancreas.

The hormone, *insulin,* was finally isolated in 1922 by F. G. Banting and C. H. Best, working in the laboratory of J. J. R. MacLeod at the University of Toronto. They tied off the pancreatic ducts of a number of dogs, waited until the enzyme-producing tissue had atrophied, removed the degenerated pancreas and froze and macerated it in an isotonic medium (freezing prevents any remaining digestive enzymes from acting), filtered the solution, and quickly injected the filtered material into diabetic dogs. The dogs showed marked improvement. Banting and Best also obtained good results with extracts prepared from the pancreases of embryonic animals; since the islet cells develop in the embryo before the enzyme-producing cells, there are no

9.23   Drawing of an islet of Langerhans

The endocrine cells of the pancreas form an islet clearly distinct from the surrounding cells, whose function is secretion of digestive enzymes. The α islet cells secrete the hormone glucagon, the β islet cells insulin. $\times 220$.

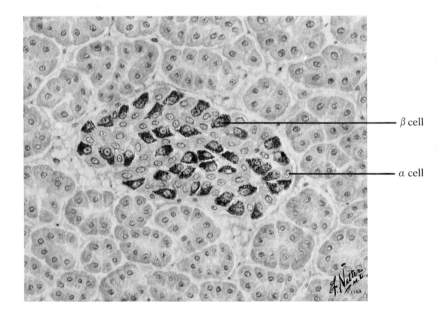

β cell

α cell

enzymes to destroy the insulin during the extraction procedure. Banting and MacLeod received the Nobel Prize in 1923 for this important work.

Banting and Best followed a procedure considered standard for demonstrating that a particular organ or tissue has an endocrine function. Let us outline the essential criteria of that procedure:

1. Removal or destruction of the organ in question should result in predictable symptoms presumed to be associated with absence of the hormone.
2. Administration of material prepared from the organ in question should relieve the symptoms.
3. It should be demonstrated that the hormone is present in both the organ and the blood, and the hormone should be extractable from each.

Fortunately administration of extracts of suspected organs has not always been as difficult as with the pancreas before Banting and Best solved the problem.

Insulin was crystallized by J. J. Abel of Johns Hopkins University in 1926. As we have already seen, it was the first protein for which the complete amino acid sequence was determined (by Frederick Sanger in 1954; see p. 56). More recently it has been discovered that the β islet cells of the pancreas first synthesize a much longer polypeptide chain called proinsulin, segments of which form the two disulfide-linked polypeptide chains of active insulin (Fig. 9.24). Someday, perhaps, methods will be devised for synthesizing insulin in commercial quantities. In the meantime, however, millions of people who would once have been doomed to invalidism and premature death will continue to lead relatively normal lives thanks to injections of insulin extracted from natural sources.[3]

High concentration of sugar in the urine is familiar as a major symptom of diabetes. How is insulin related to this symptom? Before attempting an answer, we must examine the symptom further. The presence of sugar in the urine of a diabetic means, not that the kidneys are functioning improperly, but that the blood-sugar concentration is higher than normal and that the kidneys are removing part of the excess. You will recall that the liver plays a critical role in regulating blood-sugar levels. When blood coming to the liver via the portal vein from the intestines contains a higher than normal concentration of sugar, the liver removes much of the excess and stores it as glycogen. Conversely, when blood coming to the liver is low in sugar, the liver converts some of its stored glycogen into glucose, which it adds to the blood. Other parts of the body, particularly the muscles and adipose (fatty) tissue, are also important elements in this regulatory system. When the blood-sugar concentration rises after a carbohydrate meal, part of the excess glucose is stored as glycogen in the muscles, and part is converted into fat and stored by adipose tissues; the rate of oxidation of glucose in the liver and muscles may also increase under these conditions.

This brief outline of the interplay between liver, muscles, and adipose

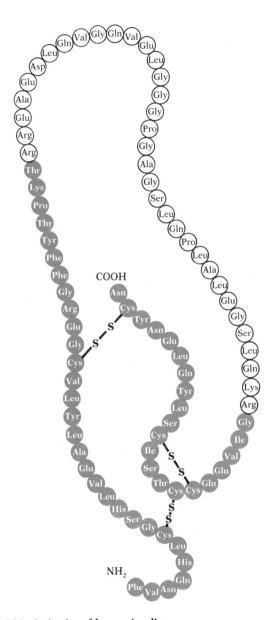

9.24 Activation of human insulin

The protein is synthesized initially as a single polypeptide chain with three internal disulfide bonds. Removal of a long section (white circles) by an activating enzyme leaves the active hormone as two separate short polypeptide chains (in color) held together by disulfide bonds.

---

[3] In recent years some sulfonylurea compounds (e.g. tolbutamide) and biguanide compounds (e.g. phenformin) have been found to have antidiabetic activity when administered orally. Unfortunately a variety of undesirable side effects of these drugs restrict their use to certain limited categories of patients.

tissues—all three target tissues for insulin—helps explain the following known actions of insulin, which have the net effect of reducing the concentration of glucose in the blood.

1. Insulin stimulates (probably by altering membrane permeabilities) absorption of more glucose from the blood by muscle cells and adipose cells, but it does not influence uptake of glucose by liver cells.
2. It promotes both oxidation of glucose and incorporation of glucose into glycogen in liver and muscle cells—actions whose effect is to reduce the supply of free glucose.
3. It inhibits metabolic breakdown of stored glycogen in liver and muscle cells.
4. It promotes synthesis of fats from glucose by adipose cells and also inhibits metabolic breakdown of fat.
5. It promotes uptake of amino acids by liver and muscle cells, and favors protein synthesis while inhibiting protein breakdown.

Notice that, although the first three actions are concerned with carbohydrate metabolism, the last two deal more directly with the metabolism of fats and proteins. But the promotion of fat and protein synthesis and the slowing of their catabolism force the cells to rely more heavily on glucose as a source of metabolic energy—the result being a reduction in the supply of free glucose. Thus we see once again that the metabolic pathways for all classes of nutrients form an interlocking system; alteration of one pathway unavoidably influences the others.

Too much insulin in the system, as from an overactive pancreas or from administration of too large a dose to a diabetic, can produce a severe reaction called insulin shock. The blood-sugar level falls so low that the brain, which has few stored food reserves of its own, becomes overirritable; convulsions may result, followed by unconsciousness and often death. A naturally occurring excess of insulin is, however, extremely rare. Far more common is a deficiency of insulin (or an insensitivity of the tissues to insulin), and it is this that is called diabetes. The liver and muscles do not convert enough glucose into glycogen, the liver produces too much new glucose, and utilization of carbohydrate in cellular respiration is impaired. The blood-sugar level rises above normal, and part of the excess glucose begins to appear in the urine. More water must be excreted as a vehicle for this glucose, and the diabetic thus tends to become dehydrated. The glycogen reserves become depleted as more and more glucose is poured into the blood and lost in the urine; yet the body still lacks sufficient energy, because of the impairment of carbohydrate metabolism. As the body begins to metabolize its reserves of proteins and fats—particularly the latter—the diabetic becomes emaciated, weak, and easily subject to infections. As if this were not enough, the excessive but incomplete metabolism of fats releases toxic substances that, seriously disturbing the delicately balanced pH of the body, often have a major part in the eventually fatal outcome of the untreated disease.

The pancreas secretes another polypeptide hormone besides insulin. This hormone, called *glucagon*, is produced in the α islet cells. Glucagon has effects opposite to those of insulin; it causes an increase in blood-glucose concentration. Several hormones produced by the adrenal glands also cause a rise in the blood-sugar level, and hormones from the pituitary may do so also. Again we see that the normal functioning of an organism depends on a delicate balance between opposing control systems; if one of these systems is disturbed and the proper balance destroyed, abnormalities result, and in severe cases these abnormalities may lead to disease or even death.

THE ADRENALS

The endocrine organs examined so far—stomach, duodenum, and pancreas—and the gonads, which will be discussed later, all have other major functions in addition to hormone secretion. They are multipurpose organs; each is an important component in both the endocrine system and some other system. By contrast, the adrenals (and also the thyroid, the parathyroids, and the pituitary) are endocrine glands whose only known function is hormone secretion.

The two adrenal glands, as their name implies (*ad-* means at and *renal* refers to the kidney), lie very near the kidneys (Fig. 9.22). Each adrenal in mammals is actually a double gland, composed of an inner corelike **medulla** and an outer barklike **cortex** (Fig. 9.25). The medulla and cortex arise in the embryo from different tissues, and their mature functions are unrelated. In fact, they remain as two separate pairs of glands in adult fish and amphibians. In reptiles, birds, and mammals the two sets of glands have evolved a close spatial relationship (in reptiles and birds the tissues of the two are actually intermingled, and do not form a distinct cortex and medulla, as in most mammals), but they remain functionally distinct. We shall therefore discuss the two components of the mammalian adrenals separately.

The adrenal medulla The adrenal medulla secretes two hormones, **adrenalin** (also known as epinephrine) and **noradrenalin** (norepinephrine), whose functions are similar but not identical. Both hormones have been isolated, identified, and synthesized in the laboratory (Fig. 9.26). They can be shown to produce a great variety of effects on the body. For example, adrenalin causes rise in blood pressure, acceleration of heartbeat, increased conversion of glycogen into glucose and release of glucose into the blood by the liver, increased oxygen consumption, release of reserve erythrocytes into the blood from the spleen, vasodilation and increased blood flow in skeletal and heart muscle, vasoconstriction and decreased blood flow in the skin and in the smooth muscle of the digestive tract, inhibition of intestinal peristalsis, erection of hairs, production of "gooseflesh," and dilation of the pupils.

At first glance, this list may seem to include a curious assortment of seemingly unrelated effects, but a more careful examination shows that these reactions occur together in response to intense physical exertion, pain, fear, anger, or other heightened emotional states; they have sometimes been called fight-or-flight reactions. It has been suggested, therefore, that adrenalin (and to a lesser extent noradrenalin) helps mobilize the resources of the body in emergencies (by stimulating reactions that combine to increase the supply of glucose and oxygen carried by the blood to the skeletal and heart muscles) and helps inhibit those functions not immediately important during the emergency (such as digestion, which might otherwise compete with the skeletal muscles for oxygen).

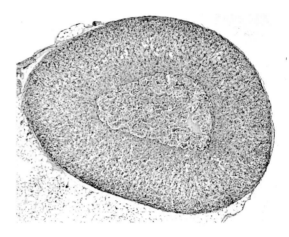

9.25 Cross section of the adrenal gland of a mouse
The medulla (central region) is entirely surrounded by cortex. [From C. D. Turner and J. T. Bagnara, *General endocrinology*, 6th ed., © 1976 by the W. B. Saunders Company, Philadelphia, Pa.]

9.26 Structural formulas of adrenalin, noradrenalin, and tyrosine
The two hormones are derived from the amino acid tyrosine.

9.27   Some steroids secreted by the adrenal cortex
Very slight differences in side chains can result in markedly different properties.

This would seem to be a reasonable hypothesis, but the importance of the adrenal medulla in producing the fight-or-flight reactions under normal circumstances can be seriously questioned on a number of grounds. For example, complete removal of the adrenal medullae causes little noticeable change in an animal; the animal still shows normal fight-or-flight reactions to appropriate stimuli. As we shall see in the next chapter, the portion of the nervous system called the sympathetic system stimulates the same fight-or-flight reactions, and most of the available evidence indicates that the sympathetic system is far more important in an animal's response to emergencies.

It is probably as an antagonist to insulin that adrenalin fulfills its most important normal function. It elevates blood sugar by stimulating the liver to produce glucose from its glycogen reserves, and it acts on muscles to transform their glycogen stores into lactic acid, which is converted into glucose after being transported to the liver by the blood.

The adrenal cortex   A person can live normally without the adrenal medullae, but not without the cortices. These are essential for life, and their removal is soon fatal. Death is preceded by a severe disruption of ionic balance in the body fluids, lowered blood pressure, impairment of kidney function, impairment of carbohydrate metabolism with a marked decrease in both blood-glucose concentration and stored glycogen, loss of weight, general muscular weakness, and a peculiar browning of the skin. These same symptoms are seen in varying degrees in individuals whose adrenal cortices are insufficiently active, a condition known as Addison's disease.

The numerous symptoms of adrenal cortical insufficiency listed here are not related to a single hormone. The adrenal cortex is, in fact, an amazing endocrine factory, producing so many different hormones that scientists still have no idea of the total number. All the cortical hormones are chemically extremely similar; all are steroids, often differing from each other by only one or two atoms of hydrogen or oxygen (Fig. 9.27). Yet these differences, minor as they may appear, give the various hormones strikingly different properties. The chemical specificity displayed in many life functions never ceases to amaze even experienced biologists.

Note that these hormones are chemically unlike the other hormones we have discussed. In mammals, only the hormones of the adrenal cortex and those of the gonads and other reproductive structures are steroids. The hormones of other endocrine organs, as far as is known, are amino acids (or, like adrenalin and noradrenalin, compounds derived from amino acids), short polypeptide chains, or proteins.

It is beyond the scope of this book to deal with all the known cortical steroids individually. We shall simply try here to give you some insight into the immense importance of these compounds, which may be grouped into three categories on the basis of their functions: (1) those that act primarily in regulating carbohydrate and protein metabolism, called glucocorticoids; (2) those that act primarily in regulating salt and water balance, called mineralocorticoids; and (3) those that function primarily as sex hormones.

Hormones in the first category, *glucocorticoids* (e.g. cortisol, corticosterone, cortisone), cause a rise in blood sugar and an increase in liver glycogen; both effects are probably due to an increased rate of conversion of protein into carbohydrate. The hormones also inhibit oxidation of glucose while promoting mobilization of fat reserves. In short, the glucocorticoids tend to elevate blood-sugar levels by stimulating the body to draw on its noncarbohydrate energy sources. When administered to a person with Addison's disease, they restore the blood-sugar level to normal. They thus act as antagonists to insulin.

Hormones in the second category, **mineralocorticoids** (especially aldosterone), stimulate the cells of the convoluted tubules of the kidneys to decrease reabsorption of potassium and increase reabsorption of sodium, actions that lead also to increased reabsorption of chloride and water. The reabsorption of these substances, in turn, causes a rise in blood volume and blood pressure. Animals deprived of mineralocorticoid soon begin excreting large quantities of urine containing high concentrations of sodium; their blood volume decreases and their blood pressure falls. If not given hormone-replacement therapy, the animals quickly die.

Hormones in the third category are very similar both chemically and functionally to the sex hormones produced by the gonads. They are probably not very important under normal circumstances, but tumors or other disturbances of the adrenal cortex may cause excessive secretion, especially of male hormone, with masculinizing effects on females or precocial sexual development of males.

The medical use of some of the cortical hormones has an interesting history. Soon after cortisone was isolated in 1935, it was found to have a remarkable ability to increase a test animal's resistance to exposure, cold, poisons, and other physiological stresses. The first test of the hormone as a therapeutic agent for human beings was performed at the Mayo Clinic on a young woman suffering from severe rheumatoid arthritis. All previous attempts to relieve her symptoms had failed, and she could now hardly move. But by the third day after injections of cortisone were begun she could move easily, and by the eighth day all her symptoms were essentially gone. If the injections were stopped, however, the symptoms promptly returned. Trials of cortisone on other arthritic patients yielded similar results.

Next, various investigators began trying cortisone on a host of diseases, many of them unrelated. And most of these investigators reported dramatic relief of symptoms, though seldom any actual cures. For example, all pneumonia symptoms of a boy with severe lobar pneumonia disappeared within 24 hours after administration of hormone (in this case the hormone used was not actually cortisone but a pituitary hormone, ACTH, that stimulates release of cortisone by the adrenal cortex). But the bacteria that cause pneumonia remained in large numbers in his body. Similarly, all disease symptoms of patients with tuberculosis disappeared several days after treatment with hormone was begun, but tuberculosis bacteria still swarmed in their bodies. In each case disease symptoms would immediately return if treatment was suspended.

You can imagine the excitement such results stirred among doctors (and in the newspapers). It was hoped that ways would be found to produce actual cures with cortisone and related hormones, but, even if that proved impossible, relief of symptoms for everything from rheumatism to cancer with a single drug seemed to signal the beginning of a new era in medicine. It was thought that stress opens the body to disease symptoms and that corti-

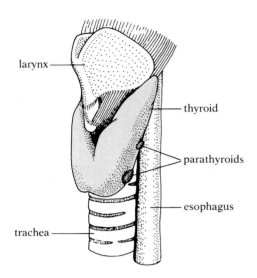

larynx

thyroid

parathyroids

esophagus

trachea

9.28   The thyroid and parathyroid glands in three-quarters view

sone helps the body withstand stress conditions. Thus administration of extra cortisone, by giving the body greater ability to withstand stress, would minimize its susceptibility to the damaging disease symptoms normally elicited by infections or by metabolic disturbances.

Cortisone and related hormones may to some extent actually function in this manner. They may indeed be involved in the body's reaction to stress, and stress may well play a fundamental role in disease. But, unfortunately, they have not proved to be the panacea first envisioned. When administered over a long period of time, they often cause side effects as bad as or worse than the condition being treated: high blood pressure, excessive growth of hair, mental aberrations, lowered resistance to certain infections such as poliomyelitis and tuberculosis, peptic ulcers, brittle bones that are easily fractured, etc. The hormones are now most frequently used to facilitate healing where administration need be repeated only a few times, or to give partial relief from the symptoms of arthritis and other diseases of connective tissues (they apparently cause changes in the collagenous fibers of such tissues). They are also sometimes used to treat severe allergic diseases, particularly asthma, and some types of lymphatic diseases. And they are used for temporary relief of severe symptoms in emergency situations. Because physicians must weigh the possible harmful side effects against the hoped-for symptomatic relief, they seldom prescribe enough to give full symptomatic alleviation of a chronic condition, preferring to have the patient endure mild symptoms in order to minimize the harmful side effects.

The dilemma presented by the cortical hormones serves as an example of a general problem faced by physicians every day. Most drugs—and other treatments, for that matter—have potential harmful side effects. Physicians must therefore always balance possible good against possible harm, and they must always remember that even the safest drugs are dangerous when used in excessive quantity or at the wrong time. The body is, after all, a finely tuned machine, with interactions between its parts so intricate that they still largely defy analysis. There is always a risk of damage to the machine when it is subjected to treatment with chemicals that almost always affect more different functions than can be predicted. Medicine has a long way to go before it can eliminate a very large element of guesswork from its practice, but it has come a long way, too, and the future looks encouraging as the increasing tempo of biological discovery provides the basis for new medical applications.

THE THYROID

Most vertebrates have two thyroid glands, located in the neck; in humans the two have fused to form a single gland (Fig. 9.28). There is convincing evidence that the vertebrate thyroids evolved from ventral pouches of the pharynx. These pouches probably functioned originally as channels in which food particles were strained from water currents flowing into the mouth and out through the gill slits. In the vertebrates, however, the developing pouches soon lost all connection to the pharynx and became both structurally and functionally independent of the digestive system. This is but one example of a common evolutionary occurrence—the development of a functionally new structure from an ancestral structure with an apparently unrelated function.

Years ago a condition known as goiter, in which the thyroid may become so enlarged that the whole neck looks swollen and deformed (Fig. 9.29), was very common in some areas of the world, such as the Swiss Alps and the Great Lakes region of the United States. Goiter is often associated with a group of other symptoms, including dry and puffy skin, loss of hair, obesity, a slower than normal heartbeat, physical lethargy, and mental dullness. No cause for this condition was known. Then in 1883 a Swiss surgeon, who believed that the thyroid had no important function, removed the gland from a number of his patients. Most of these patients developed all the symptoms usually associated with goiter, except the swelling of the neck. The results suggested that the normal thyroid must secrete some chemical that prevents these symptoms. The curious fact that patients with no thyroid and patients with the excessively large thyroid of a goiter showed the same complex of symptoms could be explained if the malformed gland of the goiter was, despite its large size, secreting too little hormone. By the 1890s patients with goiters or the other symptoms of hypothyroidism[4] were being successfully treated with injections of thyroid extract or with bits of sheep thyroid in their diets. But nothing more specific was known as yet about the hypothetical thyroid hormone itself. True, a German chemist, E. Baumann, discovered in 1896 that the thyroid contains iodine, an element previously unknown in the body. But little attention was paid to his discovery.

In 1905 David Marine of Western Reserve University noticed that many people in Cleveland had goiters. A high percentage of the dogs also had goiters. Even many of the trout in the streams had goiters. Marine suspected that the goiters might be caused by an insufficiency of iodine in the food and water. When he administered tiny traces of iodine in water to his experimental animals, their goiters and other symptoms disappeared. In 1916 Marine tried his treatment on approximately 2,500 schoolchildren in Akron, Ohio. He fed these children iodized salt. Another 2,500 children used as controls were fed uniodized salt. At the end of a specified period he found only two cases of goiter among the children who had eaten iodized salt, whereas there were 250 cases among the controls. Though it took years to convince a skeptical public, use of iodized salt finally became widespread, and hypothyroidism caused by insufficient iodine in the soil and water now seldom occurs in the United States or Europe.

The requirement for iodine became more understandable when a thyroid hormone, now known as **thyroxin** or T_4, was isolated in 1914 and synthesized in the laboratory in 1927. It proved to be an amino acid containing four atoms of iodine (Fig. 9.30). Later, another thyroid compound, identical to thyroxin except that it contained only three atoms of iodine, was found. This substance, called **triiodothyronine** or T_3, is three to five times more active than thyroxin, but is secreted in smaller amounts. Because T_4 and T_3 have virtually identical effects on target cells, they are usually considered together, under the designation "thyroid hormone" or TH.

The most characteristic effect of TH is stimulation of increased oxidative metabolism of most tissues of the body. The hormone apparently stimulates production of RNA in the nucleus, which soon leads to increased synthesis of certain enzymes, including mitochondrial respiratory enzymes, that fa-

[4] The prefix *hypo-* means less than normal, while the prefix *hyper-* means more than normal. Thus hypothyroidism means less than normal thyroid activity, and hyperthyroidism means more than normal thyroid activity.

9.29 A young Tanzanian woman with goiter
The scars on the woman's neck are the result of folk-medicine treatment.   [Courtesy M. C. Latham, Cornell University.]

9.30 Structural formula of thyroxin
Triiodothyronine has the same formula, except that the upper left iodine atom is replaced by hydrogen.

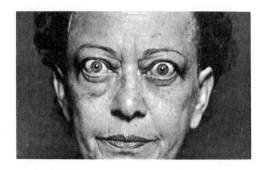

9.31   Exophthalmia
[From A. J. Carlson *et al., The machinery of the body,*
by permission of The University of Chicago Press,
copyright © 1961 by The University of Chicago.]

cilitate an elevated basal metabolic rate (BMR). Thus hyperthyroidism—excessive secretion of TH—produces many symptoms that you might predict: a higher than normal body temperature, profuse perspiration, high blood pressure, loss of weight, irritability, and muscular weakness. It also produces one very characteristic symptom that you might not predict, because it lacks an obvious connection to a higher than normal metabolic rate: exophthalmia, marked by a startling protrusion of the eyeballs (Fig. 9.31). Though hyperthyroidism can sometimes be controlled with antithyroid drugs, it is more often treated by surgical removal of part of the gland or by partial destruction with radioactive iodine.

When hypothyroidism—the opposite of hyperthyroidism—is caused by malfunction of the thyroid gland itself rather than by dietary iodine insufficiency, it is treated by administration of thyroid hormone. The untreated condition is particularly serious when it occurs in newborn children. Such children, grotesque in their development, are called cretins. They are dwarflike, and they never mature sexually. They have very low intelligence, seldom achieving a mental age of more than four or five years at the most. Prevention of cretinism by early administration of hormone to babies showing deficiency symptoms is surely one of the triumphs of modern medicine.

Although many of the symptoms of hypothyroidism—e.g. slow heartbeat, obesity, physical lethargy, and mental dullness—may well be consequences of a low BMR, other symptoms, especially some seen in cretinism, cannot easily be explained in this way. An example is the highly abnormal distribution of protein in the bodies of cretins, who have an excessive amount of glycoprotein in the skin—the cause of their puffy appearance—and an unusually high concentration of protein in the blood plasma, but whose kidneys and liver are severely deficient in protein and therefore markedly underdeveloped. Administration of TH relieves all these symptoms. It seems clear, then, that TH plays an important role in regulating protein synthesis and distribution.

The effect of TH on protein metabolism is but one manifestation of its more general role in the regulation of many aspects of development. Thus most vertebrates cannot develop normal adult form and function without the hormone. Not only is TH necessary for the protein synthesis required for proper growth; it is also essential for functional maturation of the testes and ovaries, and it acts synergistically with growth hormone from the pituitary gland in promoting skeletal development. In many lower vertebrates TH is necessary for metamorphosis and for molting.

In 1961 another thyroid hormone, called **calcitonin,** was discovered. It is functionally unrelated to TH, its chief effect being the prevention of an excessive rise of calcium concentration in the blood. It thus acts as an antagonist to the parathyroid hormone, discussed below. In lower vertebrates calcitonin is produced by separate glands (the ultimobranchial glands); the corresponding tissue in mammals becomes incorporated into the thyroid during embryonic development.

THE PARATHYROIDS

The parathyroid glands in humans are small pealike organs, usually four in number, located on the surface of the thyroid (Fig. 9.28). They were long thought to be part of the thyroid or to be functionally associated with it. Now, however, it is known that their close proximity to the thyroid is misleading; both developmentally and functionally, they are totally separate.

The parathyroid hormone, usually designated PTH, functions in regulating the calcium-phosphate balance between the blood and the other tissues. Consequently it is an important element in maintaining the relative constancy of the internal fluid environment of the body, a subject discussed at length in Chapter 8 (we have already seen that such hormones as insulin, glucagon, and calcitonin are also important in this regard). PTH increases the concentration of calcium, and decreases the concentration of phosphate in the blood by acting on at least three organs: the kidneys, the intestines, and the bones. It inhibits excretion of calcium by the kidneys and intestines, and it stimulates release of calcium into the blood from the bones (which contain more than 98 percent of the body's calcium and 66 percent of its phosphate). But calcium in bone is bonded with phosphate, and breakdown of bone releases phosphate as well as calcium. PTH compensates for this release of phosphate into the blood by stimulating excretion of this material by the kidneys. Actually, it overcompensates, causing more phosphate to be excreted than is added to the blood from bone; the result is that the concentration of phosphate in the blood drops as the secretion of PTH increases.

Naturally occurring hypoparathyroidism is very rare, but the parathyroids are sometimes accidentally removed during surgery on the thyroid. The result is a rise in the phosphate concentration, and a drop in the calcium concentration, in the blood (as more calcium is excreted by the kidneys and intestines and more is incorporated into bone). This change in the fluid environment of the cells produces serious disturbances, particularly of muscles and nerves. These tissues become very irritable, responding even to very minor stimuli with tremors, cramps, and convulsions. Complete absence of PTH is usually soon fatal unless very large quantities of calcium are included in the diet. Injections of PTH are effective in preventing the symptoms.

Hyperparathyroidism sometimes occurs naturally when the glands become enlarged or develop tumors. PTH is then produced in such quantity that the opposing action of calcitonin is no longer able to maintain a proper balance. The most obvious symptom of this condition is bones that are weak and easily bent or fractured, because of excessive withdrawal of calcium from the bones.

THE THYMUS

One of the functions of the thymus, a gland in the neck region particularly prominent in young animals, is production of a hormone important in stimulating immunologic competence in the plasma cells of the spleen, lymph nodes, and other lymphoid tissues. This hormone, called **thymosin,** has been the subject of intensive research in the last few years. There will be further reference to the thymus and its hormone in Chapter 16 (see p. 690).

THE PITUITARY AND THE HYPOTHALAMUS

The pituitary (also called the hypophysis) is a small gland lying just below the brain. Like the adrenals, the pituitary is a double gland (Fig. 9.32). It

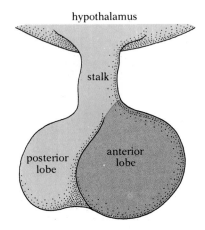

9.32   The pituitary gland

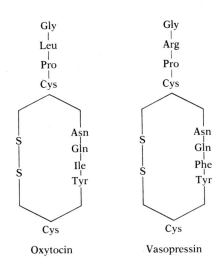

9.33   Structural formulas of oxytocin and vasopressin
The two hormones differ by only two amino acids, but this difference accounts for their distinctive activities.

consists of an anterior lobe, which develops in the embryo as an outgrowth from the roof of the mouth, and a posterior lobe, which develops as an outgrowth from the lower part of the brain. The two lobes eventually contact each other as they grow, and the anterior lobe partly wraps itself around the posterior lobe. In time, the anterior lobe loses its original connection with the mouth, but the posterior lobe retains its stalklike connection with a part of the brain called the **hypothalamus.** Despite their intimate spatial relationship, the two lobes remain fully distinct functionally, and we shall consider them separately here.

The posterior pituitary   Two hormones, *oxytocin* and *vasopressin,* are released by the posterior pituitary. As Fig. 9.33 shows, they are chemically very similar. Each contains eight amino acids (counting cystine as a single amino acid). Six occur in both compounds. Only two are different, yet they suffice to give the two hormones very different properties.

Oxytocin acts mainly on the muscles of the uterus, causing them to contract. The hormone is probably involved in childbirth, though it is not essential. Injections of oxytocin are sometimes used to induce labor when pregnancies have gone long past term.

Vasopressin causes constriction of the arterioles, and a consequent marked rise in blood pressure. It also stimulates the kidney tubules to reabsorb more water. This function—the source of its alternative name, "antidiuretic hormone"—is doubtless the more important under normal conditions. A human totally lacking vasopressin would have to excrete more than 20 liters of urine daily! The well-known diuretic effect of ethyl alcohol is due to its tendency to suppress vasopressin release.

We said earlier that the posterior pituitary originates as an outgrowth of the hypothalamus of the brain. Even in the adult it retains a stalklike connection with the hypothalamus (Fig. 9.32). There is now persuasive evidence that oxytocin and vasopressin do not originate in the posterior pituitary, but are produced in the hypothalamus and flow along nerves in the stalk to the posterior pituitary, where they are stored. The storage organ releases the hormones on stimulation by nervous impulses from the hypothalamus. If water is withheld from an experimental animal until intense thirst develops, it can be shown by staining that the downward flow of hormone along the nerves becomes very pronounced. The increased osmotic concentration of the blood, resulting from desiccation, has apparently stimulated the hypothalamus to increase its secretion of vasopressin, which can then stimulate the kidneys to conserve water by increasing reabsorption. Cutting the stalk between the hypothalamus and the posterior pituitary results in an accumulation of dark-staining material (presumably the hormone) at the cut end of the stalk, and stops release of the hormone by the posterior pituitary. It seems clear, then, that the hypothalamus, not the posterior pituitary, is the true endocrine organ in this case. Its action is an example of hormonal secretion by nervous tissue.

The anterior pituitary   The anterior pituitary (also called the adenohypophysis) is an immensely important organ that produces hormones of widely varying and far-reaching effect. At least seven hormones are secreted by the anterior pituitary in humans.

Prolactin (or lactogenic hormone) is the most versatile of all the pituitary hormones. It stimulates milk production by the female mammary glands shortly after birth of a baby; in its absence milk production soon ceases. It also plays a variety of roles in reproduction, osmoregulation, growth, and metabolism of carbohydrates and fats.

Another versatile pituitary hormone, **growth hormone** (also called somatotrophic hormone, STH), plays a critical role in promoting normal growth, especially in combination with thyroid hormone. It is a very powerful inducer of protein anabolism, favoring both cellular uptake of amino acids and their incorporation into protein, while inhibiting the protein-catabolizing influence of the glucocorticoids. In addition, it inhibits insulin, an action tending to elevate blood-sugar concentration, and it stimulates hydrolysis of fats in adipose tissues, with the effect of increasing fatty acid concentrations in the plasma.

If the supply of growth hormone is seriously deficient in a child, growth will be stunted and the child will be a midget. Oversupply of the hormone in a child results in a giant. (The tallest pituitary giant on record reached a height of 8 feet 11.1 inches.) Both pituitary midgets and pituitary giants have relatively normal body proportions, and their appearance is not unattractive. If, however, oversecretion of growth hormone begins during adult life, only certain bones, such as those of the face, fingers, and toes, can resume growth. The result is a condition known as acromegaly, characterized by disproportionately large hands and feet and distorted features—a greatly enlarged and protruding jaw, enlarged cheekbones and eyebrow ridges, and a thickened nose.

The anterior pituitary also secretes a number of very important hormones that exert controlling action on other endocrine organs. These hormones are **thyrotrophic hormone,** which stimulates the thyroid; **adrenocorticotrophic hormone** (abbreviated ACTH), which stimulates the adrenal cortex; and at least two **gonadotrophic hormones,** or gonadotrophins (follicle-stimulating hormone, FSH, and luteinizing hormone, LH), which act on the gonads. Proper growth and development of these endocrine glands depend on adequate secretion of the appropriate trophic (i.e. stimulatory) hormone from the pituitary; if the pituitary is removed or becomes inactive, these organs atrophy and function at very low levels. It is easy to understand why the pituitary is often called the master gland of the endocrine system.

The interaction between the anterior pituitary and the other endocrine glands over which it exerts control is an example of feedback, a type of interaction very common in living systems. Thyrotrophic hormone, released by the pituitary when the concentration of thyroxin in the blood is low, stimulates increased production of thyroxin by the thyroid, but the resulting rise in concentration of thyroxin in the blood inhibits secretion of more thyrotrophic hormone by the pituitary. In other words, the pituitary responds to a low thyroxin level in the blood by sending a chemical messenger that stimulates increased activity by the thyroid. But once the thyroid becomes more active, the increased amount of thyroxin produced tells the pituitary that release of thyrotrophic hormone can now be reduced. There is thus a feedback of information from the thyroid to the pituitary. The pituitary exerts control over the thyroid, and the thyroid, in turn, exerts some control over the pituitary. Each sends chemical messengers to the other. But note that the message from the pituitary to the thyroid is a stimulatory one, while the return message from the thyroid to the pituitary is an inhibitory one; the feedback is negative. The pituitary tends to speed up the system, and the thyroid tends to slow it down. The interaction between the two opposing forces produces a delicately balanced system.

The interaction of the pituitary with the adrenal cortex or the gonads is similar to its interaction with the thyroid. The pituitary responds to low levels of cortical hormones by secreting more ACTH and to low levels of sex hormones by secreting more gonadotrophic hormone. The resulting rise in concentration of cortical hormones or of sex hormones inhibits further secretion by the pituitary.

The anterior pituitary, important as it is as a regulator of other endocrine glands, does not, so far as is known, participate in the control of the pancreas, the adrenal medullae, or the parathyroids. Both the pancreas and the adrenal medullae are controlled in part by the autonomic nervous system (the pancreas is also influenced by adrenalin), and the parathyroids are thought to be regulated primarily by the concentration of calcium ions in the blood.

The control function of the hypothalamus   The delicately balanced feedback interaction between the anterior pituitary and other endocrine glands is not the only factor that regulates the regulator.

Suppose an animal detects the lengthening of days in spring, and its gonads secrete more sex hormone. How is its perception of the stimulus—increasing day length—able to affect its endocrine glands? Perception of stimuli involves the nervous system; however, there are no nervous connections either to the anterior pituitary or to endocrine glands such as the gonads. But nervous tissue, as we saw in the interaction of the hypothalamus and the posterior pituitary, is capable of hormonal secretion. There is now proof that the endocrine glands are stimulated by chemical messengers from the nervous system itself.

In such situations as an animal experiencing changes in day length, it turns out that the hypothalamus is one of the parts of the brain most involved. Now, the hypothalamus is located just above the pituitary. Although there is no direct physical connection between it and the anterior pituitary, there is an unusual connection between their blood supplies. Arteries to the hypothalamus break up into capillaries, and these capillaries eventually join to form several veins leading away from the hypothalamus. But unlike most veins, these do not run directly into a larger branch of the venous system; instead, they pass downward into the anterior pituitary and there break up into a second capillary bed. We have encountered two other places in the body where the circulation depends on two beds of capillaries arranged in sequence: the kidney nephrons, where one bed forms the glomerulus and the other envelops the tubules, and the hepatic portal system, where one bed is in the wall of the intestine and the other is in the liver. In both places the special type of circulation reflects an important functional arrangement. In the hepatic portal system, for example, many substances picked up by the blood in the first capillary bed are removed from the blood in the second capillary bed. The portal system linking the hypothalamus and the anterior pituitary seems to function in a similar fashion. Thus the relationship between the body's two principal control systems—nervous and endocrine—becomes much easier to understand.

The hypothalamus is now known to produce a variety of special peptide hormones called *releasing hormones,* which, carried directly to the anterior pituitary by the portal vessels, regulate the secretory activity of the anterior pituitary. Thus corticotrophic releasing hormone (CRH) from the hypothalamus stimulates release of ACTH from the pituitary, thyrotrophic releasing hormone (TRH) stimulates release of thyroid hormone, growth-hormone releasing hormone (GRH) stimulates release of growth hormone, etc. A few of the hypothalamic hormones are inhibitory rather than stimulatory; e.g. prolactin release–inhibiting hormone (PIH) inhibits release of prolactin by the anterior pituitary. We shall not attempt here to list all the known hypothalamic hormones, but we should emphasize the extreme importance of the discovery that the nervous system can directly influence the endocrine system, that nervous and hormonal control are parts of a single integrated coordinating system.

It now becomes possible to reconstruct the sequence of responses triggered in our hypothetical animal by the perception of changing day length. The animal detects the light with its sense organs; nervous impulses are transmitted from the sense organs to the hypothalamus of the brain; the hypothalamus secretes gonadotrophic releasing hormone (GnRH) into the blood; the releasing hormone stimulates the anterior pituitary to increase its secretion of gonadotrophic hormones; the gonadotrophic hormones stimulate the gonads to secrete more sex hormone; and the sex hormone helps prepare the animal physiologically to begin activities appropriate to the breeding season.

The hypothalamus is not only the point of entry for information transmitted to the endocrine system from the nervous system; it is also one of the major points for the transmission of feedback information from the endocrine system. Let us take the previously discussed negative feedback effect of a rise in thyroxin on the anterior pituitary as an example. To some extent probably, the thyroxin exerts negative feedback on the pituitary directly, by inhibiting its secretion of thyrotrophic hormone, but to a large extent it does so indirectly, by inhibiting the hypothalamus from secreting thyrotrophic releasing hormone (Fig. 9.34). Similarly, much of the negative feedback action of other hormones is via the hypothalamus.

THE PINEAL

The pineal, a lobe in the roof of the rear portion of the forebrain, has long intrigued investigators by its glandular appearance. Only recently, however, has it been demonstrated to have an endocrine function.

In some lower vertebrates the pineal is eyelike and responds to light both by generating nervous impulses and by secreting a hormone called *melatonin,* which lightens the skin by concentrating the pigment granules in melanophores (pigment-containing cells). It has been shown that in these animals the pineal is intimately involved (presumably through melatonin) in the control of circadian rhythms—cycles of activity repeated approximately every 24 hours.

The mammalian pineal, too, secretes melatonin, but it has no light-sensi-

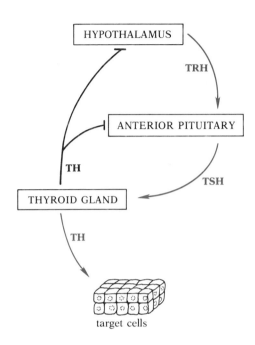

9.34 Feedback control of the thyroid gland
Thyrotrophic releasing hormone (TRH) from the hypothalamus stimulates the anterior pituitary to release thyrotrophic hormone (TSH), which stimulates the thyroid gland to secrete thyroxin (TH). The thyroxin stimulates target cells throughout the body, but it also inhibits both the hypothalamus and the anterior pituitary. (Color arrows indicate stimulatory influences, barred black lines inhibitory influences.)

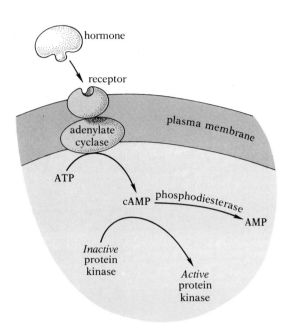

9.35 Model for control of hormonal activity via the adenylate cyclase system
The hormone molecule can bond loosely with a specific receptor protein built into the plasma membrane. Formation of the hormone-receptor complex activates the enzyme adenylate cyclase, which is also an integral part of the plasma membrane. The activated adenylate cyclase catalyzes synthesis of cAMP from ATP inside the cell. The cAMP, in turn, triggers activation of an enzyme called protein kinase; it is eventually broken down to AMP by the enzyme phosphodiesterase.

tive cells, and mammals have no melanophores. Though the pineal is a part of the brain, its principal (perhaps only) innervation originates outside the skull cavity in a cervical (neck) ganglion. Thus the mammalian pineal and its hormone have been a puzzle.

Evidence accumulated in the last few years suggests that the pineal may function as a neurosecretory transducer, converting neural information about light conditions into hormonal output. The conversion would take place as follows: Information about the light-dark cycle received by the eyes goes first (via the brain) to the cervical ganglia and is then conveyed by sympathetic nervous pathways to the pineal; the pineal responds by secreting melatonin in inverse proportion to the amount of light; the melatonin, in turn—probably operating through the hypothalamus—influences the secretion of gonadotrophic hormones by the anterior pituitary. The usual effect of melatonin is apparently an inhibitory one; in autumn, for example, as the days get shorter, increased production of melatonin tends to turn off gonadotrophic secretion, with the result that the gonads regress and the animal enters a nonreproductive phase.[5]

Melatonin may exert effects on other parts of the brain besides the hypothalamus, particularly those concerned with locomotor rhythms, feeding rhythms, and other biological rhythms. But until more research is done, all proposals concerning the functioning and effects of the pineal and its hormone remain highly speculative.

THE MECHANISM OF HORMONAL ACTION

The role of cyclic AMP The 1960s saw some exciting progress in the attempt to learn how hormones act on their target cells. A major contribution was that of E. W. Sutherland and T. W. Rall of Western Reserve University, who were investigating the mechanism by which glucagon and adrenalin stimulate liver cells to release more glucose into the blood. They discovered that these hormones stimulate an increase in the intracellular concentration of a compound called cyclic adenosine monophosphate—*cyclic AMP,* or cAMP for short. They found that this compound leads, in turn, to activation of an enzyme necessary for breakdown of glycogen to glucose. Subsequent research has demonstrated that, in addition to glucagon and adrenalin, a large number of other hormones act on their target cells either to increase or to decrease the concentration of cAMP. This work, far-reaching in its implications, earned Sutherland, now at Vanderbilt University, a Nobel Prize in 1971.

As you have doubtless deduced from its name, cAMP is a compound related to ATP (adenosine triphosphate) (see Fig. 4.9, p. 147). Very widely distributed in nature, it has been found in almost all animal tissues studied (vertebrate and invertebrate) and in bacteria. It is synthesized from ATP in living cells by a reaction catalyzed by an enzyme called *adenylate cyclase,* which appears to be built into the cell membrane.

[5] The sequence of events is probably different in animals that breed in autumn. Moreover, melatonin may sometimes stimulate the testes instead of inhibiting them; whether the hormone is gonadotrophic or antigonadotrophic probably depends on when in the circadian cycle it is released. According to some investigators, the agent influencing the pituitary is not melatonin, but a peptide produced by the pineal.

As evidence accumulated that many hormones, among them glucagon and adrenalin, do not actually enter their target cells, but rather form weak bonds with receptor sites on the cell membrane, a so-called **two-messenger model** of hormonal control was proposed. According to this model, an extracellular first messenger, which is the hormone itself, goes from an endocrine gland to the target cell and there stimulates production of an intracellular second messenger, which is often cAMP. More particularly, the binding of the hormone with a highly specific receptor site on the outer surface of the membrane of the target cell activates adenylate cyclase, which catalyzes production of more cAMP on the inner surface of the membrane (Fig. 9.35). The increased amounts of cAMP then interact with cytoplasmic enzyme systems and thus initiate the cell's characteristic responses to the hormonal stimulation. In other words, the initial extracellular signal (the hormone or first messenger) is converted into an intracellular signal (cAMP or second messenger) that the chemical machinery of the cell can more readily understand.

Let us return to the stimulation of glucose production in the liver by glucagon or adrenalin as an example of the mechanism by which the increased amount of cAMP influences the cell. In this case the cAMP activates an enzyme of a class called protein kinases. The activated protein kinase, in turn, activates a second enzyme, which activates a third enzyme, which catalyzes the first reaction in the breakdown of glycogen to glucose. We can diagram this chain of events, beginning with formation of the hormone-receptor complex, as follows:

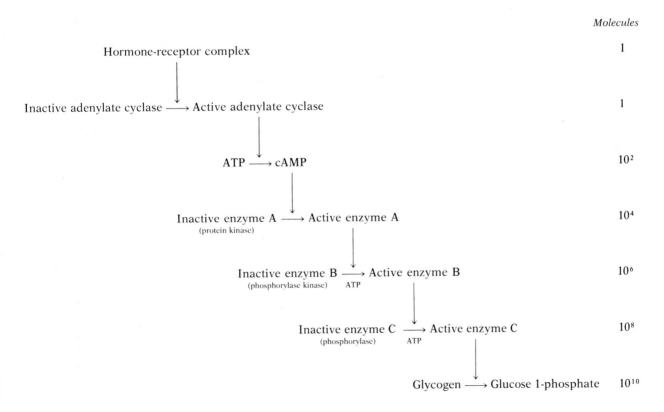

	Molecules
Hormone-receptor complex	1
Inactive adenylate cyclase ⟶ Active adenylate cyclase	1
ATP ⟶ cAMP	10^2
Inactive enzyme A ⟶ Active enzyme A (protein kinase)	10^4
Inactive enzyme B ⟶ Active enzyme B (phosphorylase kinase) ATP	10^6
Inactive enzyme C ⟶ Active enzyme C (phosphorylase) ATP	10^8
Glycogen ⟶ Glucose 1-phosphate	10^{10}

This cascade of enzyme-catalyzed reactions makes possible a very great *amplification* of the effects of the original hormone-binding event. Because an enzyme can be used over and over again, a single molecule of active adenylate cyclase may catalyze production of about 100 molecules of cAMP. Each molecule of cAMP may, in turn, catalyze production of roughly 100 molecules of active enzyme A, and so on; i.e. each successive step may result in a hundredfold amplification. The net result is that a single molecule of glucagon or adrenalin may lead to release of as many as 10^{10} (10 billion) molecules of glucose, and it may do so within only one or two minutes. We see, then, why only very small quantities of hormone are required for normal function.

The above sequence of reactions is not the only result of stimulation of a target cell by glucagon or adrenalin. You will recall from our earlier discussion of those hormones that release of glucose is just one of a variety of ways in which they bring about a rise in blood-sugar levels. Another of their important actions, for example, is to inhibit synthesis of glycogen. This action, like the one already discussed, depends on protein kinase activated by cAMP:

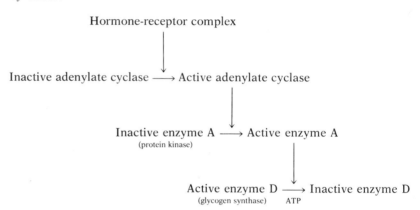

In other words, protein kinase (enzyme A), which we previously saw activating an enzyme necessary for glucose production, is here inactivating an enzyme and hence blocking the incorporation of glucose into glycogen. There is convincing evidence that protein kinase is an essential link in all hormonal actions mediated by the adenylate cyclase system.

We have indicated that either glucagon or adrenalin can initiate the processes we have examined. These two hormones are chemically very different, however. How could they have the same effect on adenylate cyclase? The answer appears to be that they bind to different receptor molecules on the membrane, but that both types of receptors are capable of activating adenylate cyclase. Thus the two hormones have additive effects. But this is true only for the liver; the two hormones do not have the same effect on muscle. Acting via adenylate cyclase, adrenalin can stimulate production of glucose from glycogen by muscle cells, but glucagon cannot do so, apparently because the membranes of muscle cells lack receptors for glucagon.

From the above example of glucose regulation, we can draw several important conclusions. First, it is the presence or absence of receptors on the cell membrane that determines whether or not a particular cell type is influenced by a given hormone, i.e. whether it is a target of that hormone. Second, a cell may respond in the same way to two or more different hormones if it has cyclase-activating receptors for each of those hormones; the hormones themselves are specific only for the receptors, not for the reactions their binding triggers via adenylate cyclase activation.

These rules of hormone action help explain how the many different hormones that depend on the adenylate cyclase system (including—in addition to glucagon and adrenalin—gastrin, secretin, parathyroid hormone, calcitonin, the trophic hormones of the anterior pituitary, and at least some of the hypothalamic releasing hormones) can have their own distinct sets of target cells, even though their immediate effect in every case is activation of adenylate cyclase. It need only be assumed that different types of cells have different hormone-specific receptor sites. Thus cells in the thyroid might have receptors on which only thyrotrophic hormone could react, while cells in the adrenal cortex might have receptors on which only ACTH could react. Such an arrangement would ensure that the thyroid would be stimulated by thyrotrophic hormone and not by ACTH, and that the adrenal cortex would be stimulated by ACTH and not by thyrotrophic hormone.

Much more difficult to answer is the question why different cells respond so differently to changes in their cAMP content. It is by no means clear why an increase in cAMP should cause increased thyroxin production by thyroid cells, increased hydrocortisone production by adrenal cortex cells, increased glucose production by liver cells, and increased protein synthesis by uterine cells. Two assumptions have to be made. The first is that a vast array of different chemical processes in cells can be regulated by cAMP; there is abundant evidence that this is a valid assumption. The second is that the response of a given cell to a change in cAMP concentration depends on its own chemical makeup, which in turn depends on its previous developmental history. Thus thyroid, adrenal, liver, and uterine cells, having followed different developmental pathways, must have different chemical profiles. Hence a hormonally induced rise in cAMP will take place against vastly different enzymatic backgrounds in each of these four cell types, and the effects of the rise may be correspondingly different.

A variety of other chemicals besides hormones influence cells by their effects on cAMP levels. Some of these (e.g. histamine) work through adenylate cyclase, but others exert their effect by acting on the enzyme **phosphodiesterase,** which breaks down cAMP. In a normal cell a certain amount of this enzyme, which keeps the concentration of cAMP in check, is always present. Thus, if adrenalin induces a rise in the cAMP concentration, phosphodiesterase causes a return to more normal levels as soon as the hormone is no longer present. It is clear that a chemical capable of either inhibiting this enzyme (e.g. caffeine) or activating it (e.g. nicotine) can markedly influence the cAMP levels in the cell.

The role of cAMP as a second messenger so impressed some researchers when it was first discovered that they envisioned it as a possible universal mediator of the action of hormones and other hormonelike compounds. This idea was soon shown to be incorrect. For example, recent evidence suggests that the principal effect of insulin (whose action is antagonistic to that of glucagon and adrenalin) is not an alteration of cAMP concentration, but a rise in the concentration of a related substance, cGMP (cyclic guanine monophosphate) that has effects opposite to those of cAMP. Several other con-

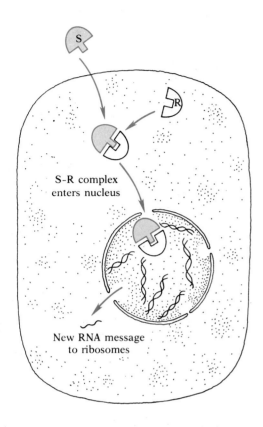

9.36 Model for the mode of action of steroid hormones

The hormone (S), being lipid-soluble, can penetrate the plasma membrane and enter the cytoplasm. There it binds with a receptor protein (R), and the complex then enters the nucleus, where it influences synthesis of RNA by the genes. A new message (in the form of new RNA) is thus sent to the ribosomes, which begin synthesizing the protein coded by the new RNA. The new protein, perhaps an enzyme, will then influence the chemical activity of the cell.

trol chemicals (especially acetylcholine in the nervous system) are thought to act via cGMP, but the details of the cGMP system, and its relationship to the adenylate cyclase system, are not yet well understood.

Thyroxin is another hormone whose action is not mediated by cAMP. Though the precise mechanism of its activity has not yet been established, it seems likely that this hormone, which can easily pass through membranes and enter cells, alters the activity of genes in the nucleus or influences mitochondrial enzyme activity, or both.

Steroid hormones, such as the ecdysone of insects or the hormones produced in vertebrates by the adrenal cortex and the gonads, have a mode of action that does not involve adenylate cyclase or cAMP. Instead of reacting with a receptor on the outer surface of the target-cell membrane, these hormones, which can easily move through membranes, penetrate into the cytoplasm of the target cell. There the steroid (S) binds to a specific cytoplasmic receptor molecule (R), which is probably a protein. The complex (S-R) then moves into the nucleus, where it helps regulate the activity of specific genes, in this way influencing what RNA message is transcribed from the DNA and exported from the nucleus to the cytoplasm (Fig. 9.36). In other words, by interacting with the genetic material of the target cells, steroid hormones help determine what instructions for protein synthesis (especially enzyme synthesis) are sent from the nucleus to the ribosomes.

The plant hormones constitute another group of control substances that probably do not act via the adenylate cyclase system. Indeed, cAMP does not seem to be present in significant quantities in higher plants. Some plant hormones (e.g. auxins) probably function, at least in part, by direct activation of important enzymes. Others (e.g. gibberellins) probably help regulate gene activity; i.e. they influence DNA transcription. And still others (e.g. cytokinins) appear to bind to ribosomes, and thus help regulate synthesis of proteins at that level.

Prostaglandins Since 1957 much attention has been focused on a group of substances called prostaglandins, which are secreted by most animal tissues and exhibit a bewildering array of actions. These include stimulation or relaxation of smooth muscle, dilation or constriction of blood vessels, stimulation of intestinal motility, modulation of synaptic transmission in the nervous system, stimulation of inflammation responses, and enhancement of the perception of pain. It has recently been discovered that the effectiveness of aspirin in combating inflammation and pain is due, at least in part, to its inhibition of prostaglandin synthesis.

Derived from long-chain fatty acids, prostaglandins are extremely potent; a few micrograms can produce a maximal response. They are released from tissues by a great variety of stimuli, including hormones, nervous stimuli, mechanical stimuli, oxygen deprivation, and inflammation. In some cases they may circulate in the blood like normal hormones and exert their effects at distant locations. In other cases they may also act at distant target sites, but without blood transport; an example is the prostaglandins in semen (secreted by the seminal vesicles), which cause contractions of the uterine muscles. In still other cases the prostaglandins may exert their principal effect within the cells where they are produced.

Studies of the effects of prostaglandins on a variety of target cells have revealed that they often mimic the effects of the appropriate stimulatory hormones, apparently by a mechanism correlated with an increase in intracellular cAMP. Since hormonal stimulation of these same cells results in an increase in their intracellular prostaglandins, it has been suggested that these substances may function to augment the stimulation of the adenylate cyclase system. In other words, the binding of a stimulatory hormone to the receptor site would produce two effects: activation of adenylate cyclase and stimulation of prostaglandin release. The prostaglandin would, in turn, facilitate the activity of the adenylate cyclase. Alternatively, the prostaglandin might act as a coupling factor between the hormone-receptor complex and the adenylate cyclase.

Prostaglandins are not always stimulatory to the adenylate cyclase system; in some cases they are inhibitory. Which effect they have depends partly on which of the more than 14 different prostaglandins is involved, partly on the distinctive characteristics of the target cell.

Research on prostaglandins is still so young that new data may well force extensive revision of our understanding of the mechanism of hormone–target cell interactions. The next few years will surely see much research both on their role in modulating hormone activity via the adenylate cyclase system and on their innumerable independent effects.

HORMONAL CONTROL OF VERTEBRATE REPRODUCTION

Reproduction is the central theme of life. All the other aspects of living discussed in this book—nutrient procurement, gas exchange, internal transport, waste excretion, osmoregulation, growth, hormonal and nervous control, and behavior—can be viewed, in a sense, as processes that enable organisms to survive to reproduce. It has been said that "the hen is the egg's way of producing another egg," and the idea is equally applicable to human beings; we are, in a way, elaborate devices for producing eggs and sperm, for bringing them together in the process of fertilization, and for giving birth to young. We shall not attempt here to discuss all aspects of animal reproduction; such fundamental topics as genetics, differentiation, and growth will be the subjects of later chapters. This section will be concerned only with the physiology of vertebrate (particularly mammalian) reproduction as an example of the complex interplay between a variety of different control mechanisms.

THE PROCESS OF SEXUAL REPRODUCTION

Sexual reproduction in higher animals always involves the union of two parental cells called *gametes,* one an egg cell (ovum) and the other a sperm cell (spermatozoon). Occasionally the same individual produces both gametes, which unite in a process known as self-fertilization; the process is most common among internal parasites such as tapeworms, whose chances of locating another individual for cross-fertilization are often poor. However, most animals use cross-fertilization (Fig. 9.37), even when, as in earthworms, each individual is hermaphroditic (i.e. possesses both male and female sexual organs). Sexual reproduction among vertebrates always involves cross-fertilization, and it will be exclusively with this type of reproduction that we shall be concerned here.

Sexual reproduction, as we have said, depends on the bringing together of

9.37 Two earthworms mating
The worms are hermaphroditic, each possessing both male and female reproductive organs. They are coupled at two mating points; at one the upper worm is acting as male and the lower worm as female, at the other they reverse these roles. Some white fluid can be seen at each mating point. [Oxford Scientific Films, Bruce Coleman Inc.]

9.38 Frogs spawning
The smaller male clasps the female in an embrace called amplexus and sprays semen over the eggs she releases. Many hundreds of eggs are produced. [Hans Pfletschinger, Peter Arnold, Inc.]

9.39 Internal fertilization by American coppers (*Lycaena phlaesa americana*)
The male butterfly inserts sperm into the reproductive tract of the female. Fertilization occurs within the body of the female, and the eggs are invested with a protective shell before being laid. [L. West, Bruce Coleman Inc.]

an egg cell and a sperm cell, which then unite in the process of fertilization to form the first cell of the new individual. There are two basic ways in which egg cells and sperm cells are brought together: external fertilization, where both types of gametes are shed into the surrounding medium and the sperm swim or are carried by water currents to the eggs; and internal fertilization, where the eggs are retained within the reproductive tract of the female until after they have been fertilized by sperm inserted into the female by the male.

External fertilization is limited essentially to animals living in aquatic environments, because the flagellated sperm must have fluid in which to swim and the eggs, in the absence of a protective coat or shell, which would prevent the sperm from penetrating and fertilizing them, would become desiccated on land. Almost all aquatic invertebrates, most fishes (but not sharks), and many amphibians use external fertilization. As you would expect, shedding eggs and sperm into the water of a lake or stream is an uncertain method of fertilization; many of the sperm never locate an egg and many eggs are never fertilized, even if both types of gamete are shed at the same time and in the same place, as is usually the case. Consequently animals using external fertilization generally release vast numbers of eggs and sperm at one time (Fig. 9.38). And they often go through elaborate behavioral sequences (in which hormonal control is very important) that ensure concurrence in both time and space in the release of gametes by the two sexes.

Most land animals, both invertebrate and vertebrate, use internal fertilization (Fig. 9.39). In effect, the sperm cells are provided with the sort of fluid environment that is no longer available to them outside the animals' bodies. Thus the sperm remain aquatic and swim through the film of fluid always present on the walls of the female reproductive tract. Once fertilized, the egg is either enclosed in a protective shell and released by the female or held within the female's body until the embryonic stages of development have been completed. Internal fertilization requires, of course, very close physiological and behavioral synchronization of the sexes, and this synchronization involves extensive hormonal control.

Let us summarize briefly the characteristic reproductive methods employed by the major classes of vertebrates. Fish, being aquatic, almost always use external fertilization and thus, of course, lay eggs with no shell. Although they usually go through elaborate behavioral rituals that help synchronize the release of gametes, huge numbers of eggs and sperm are released at each mating and the wastage of gametes is enormous. Amphibians (frogs, salamanders, etc.) evolved from fish, and they too generally use external fertilization; they must therefore return to the water or to a very moist place on land to lay their eggs. Some salamanders have evolved a behavioral sequence in which the male releases a membranous packet (spermatophore) containing sperm that the female picks up with her cloaca. These amphibians have thus evolved a primitive type of internal fertilization, and some of them can mate on land, but their eggs must still be laid in very moist places.

The reptiles, modern representatives of which are snakes, lizards, and turtles, evolved from ancestral amphibians. They were the first vertebrates to be fully emancipated from the ancestral dependence on the aquatic environment for reproduction. As would be expected, they use internal fertilization, and they lay eggs enclosed in tough membranes and shells. Since

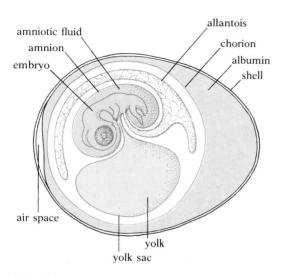

amniotic fluid
amnion
embryo
allantois
chorion
albumin
shell
air space
yolk
yolk sac

9.40   The embryonic membranes in a bird's egg
Although everything shown here is part of the "egg" in common parlance, the chorion is the outer boundary of structures derived from the true egg cell. The thick layer of albumin, a protein, is outside the cell.

internal fertilization entails much less wastage of egg cells than external fertilization, only a few egg cells are released during each reproductive season. Birds evolved from one group of ancient reptiles, and they too employ internal fertilization and lay eggs with shells.

The eggs of land vertebrates such as reptiles and birds have four different membranes in addition to the shell. These are the amnion, the allantois, the yolk sac, and the chorion (Fig. 9.40). The *amnion* encloses a fluid-filled chamber housing the embryo, which can thus develop in an aquatic medium even though the egg as a whole may be laid on dry land. The *allantois* functions as a receptacle for the urinary wastes of the developing embryo, and its blood vessels, which lie near the shell, function in gas exchange. The *yolk sac,* as its name indicates, encloses the yolk, which is food material used by the developing embryo. The *chorion* is an outer membrane surrounding the embryo and the other membranes.

Like reptiles and birds, mammals use internal fertilization, but (with a few rare exceptions) no shell is deposited around the fertilized egg and it is not laid. Instead, the early embryo with its membranes becomes implanted in a specialized chamber of the female genital tract, and there embryonic development is completed. The young animal is then born alive. The remainder of our discussion here will be concerned with mammalian, and in particular human, reproduction.

THE HUMAN AND OTHER MAMMALIAN REPRODUCTIVE SYSTEMS

The genital system of the human male   The male gonads, or sex organs, are the *testes*—oval glandular structures that form in the dorsal portion of the abdominal cavity from the same embryonic tissue that gives rise to the ovaries in females. In the human male the testes descend about the time of birth from their points of origin into the *scrotal sac* (or scrotum), a pouch whose cavity is initially continuous with the abdominal cavity via a passageway called the *inguinal canal.* After the testes have descended through the inguinal canal into the scrotum, the canal is slowly plugged by growth of connective tissue, so that the scrotal and abdominal cavities are no longer continuous.

Sometimes the inguinal canal fails to close properly; and even when it does, it remains a point of weakness and is easily broken open again when subjected to excessive strain, as when a man lifts a heavy object. The opening resulting from insufficient closure or from later rupture is known as an inguinal hernia; it is the most common type of hernia in human males. If the hernia is large, it must be repaired surgically to prevent a loop of the intestine from slipping through the opening into the scrotal sac, where the intestine may become caught so tightly that its blood supply is cut off and gangrene results. Inguinal hernia is largely a human hazard attributable to two-legged stance, which places much strain on the lower abdomen; such hernias are very infrequent in mammals that walk on four legs. In fact, in some mammals the inguinal canal remains partly open, and the testes move back into the abdominal cavity during the nonreproductive season.

Each testis has two functional components: the *seminiferous tubules,* in which the sperm cells are produced, and the *interstitial cells,* which secrete male sex hormone. The seminiferous tubules of the human are not functional at the temperatures characteristic of the abdominal cavity; if the testes fail to descend, the germinal epithelium of the tubules eventually degenerates. If, however, the testes descend normally into the scrotal sac, where the temperature is approximately 1.5°C cooler, the germinal epithelium of the seminiferous tubules becomes functional at the time of puberty. Mature sperm cells pass from the seminiferous tubules via many tiny ducts into a much-coiled tube, the *epididymis,* which lies on the surface of the testis (Figs. 9.41, 9.42). The sperm are stored in this organ until they are activated by secretions produced by it (sperm that have not passed through the epididymis are nonmotile); they are then released during copulation.

A long sperm duct, the *vas deferens,* runs from each epididymis through the inguinal canal and into the abdominal cavity, where it loops over the bladder and joins with the *urethra* just beyond the point where the urethra arises from the bladder. The urethra, in turn, passes through the *penis* and empties to the outside. Notice, then, that the urethra in the mammalian male is a common passageway used by both the excretory and reproductive systems; urine passes through it during excretion and semen passes through it during sexual activity. In more primitive vertebrates the relationship between the excretory and reproductive systems is even closer. In frogs, for example, sperm cells pass from the testes into the kidneys and down the excretory ducts to the cloaca; the reproductive system has no separate vasa deferentia. In the vertebrates there has been an evolutionary trend toward increasing liberation of the reproductive system from its ancestral dependence on the excretory system. There is far more separation in mammalian males than in fish or frogs, but the two systems still share the urethra and thus do not have separate openings to the outside. As we shall see, only in mammalian females has complete separation arisen.

As sperm pass through the vasa deferentia and urethra, seminal fluid is added to them to form *semen,* which is a mixture of seminal fluid and sperm cells. The seminal fluid is secreted by three sets of glands: the *seminal*

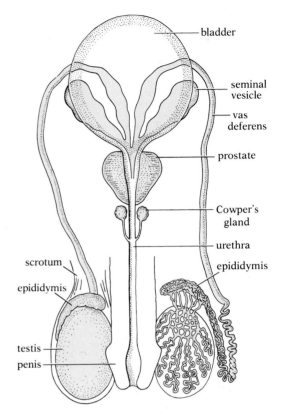

9.41 Reproductive tract of the human male: frontal view

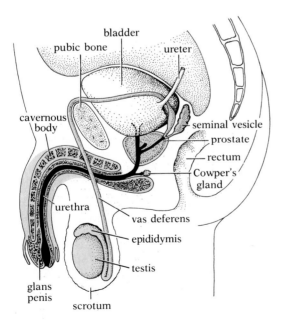

9.42 Reproductive tract of the human male: lateral view

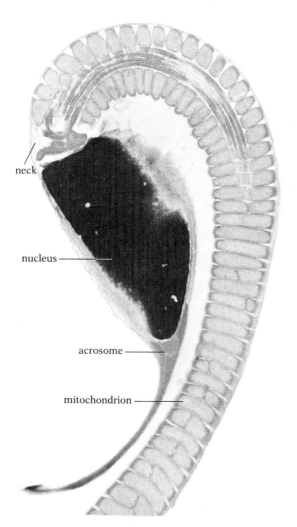

neck

nucleus

acrosome

mitochondrion

9.43 Electron micrograph of longitudinal section of part of a sperm cell from a kangaroo rat
The sperm head, which contains the acrosome (an enzyme-filled organelle) and the nucleus, is connected by a short neck to a portion of the flagellum tightly packed with spirally arranged mitochondria. There are no mitochondria in the long distal portion of the flagellum, which is not shown here. $\times 13,500$.
[Courtesy D. M. Phillips, Rockefeller University.]

vesicles, which empty into the vasa deferentia just before these join with the urethra; the *prostate,* which empties into the urethra near its junction with the vasa deferentia; and the *Cowper's glands,* which empty into the urethra at the base of the penis (Figs. 9.41, 9.42). Seminal fluid has a variety of functions: (1) It serves as a vehicle for transport of sperm; (2) it lubricates the passages through which the sperm must travel; (3) as an effectively buffered fluid, it helps protect the sperm from the harmful effects of the acids in the female genital tract; (4) it contains much sugar (mostly fructose), which the active sperm can use as a source of energy. The tiny sperm cells can store very little food themselves; they depend on an external source of nutrients for the respiratory production of the ATP necessary to keep their flagella active. The base of a sperm flagellum is an amazing power plant, packed with mitochondria, in which energy can be extracted from the sugar absorbed from the seminal fluid (Fig. 9.43).

During sexual excitement the arteries leading into the penis dilate, and the veins from the penis constrict, in response to stimulation by nerves of the autonomic system. Much blood is pumped under considerable pressure through the arteries into the spaces in the spongy erectile tissue of which the penis is largely composed (Fig. 9.42). The engorgement of the penis by blood under high arterial pressure causes it to increase greatly in size and to become hard and erect, thus preparing it for insertion into the female vagina during copulation (also called coitus). Note that erection of the penis does not involve activity of skeletal muscles, but is entirely a vasomotor phenomenon.

When the penis is sufficiently stimulated by friction during copulation, nervous reflexes involving pathways of the sympathetic system cause waves of contraction in the smooth muscles of the walls of the epididymides, vasa deferentia, seminal glands, and urethra. These contractions move sperm from the epididymides through the vasa deferentia, combine seminal fluid from the various glands with the sperm, and expel the semen from the urethra. An average of about 100 million sperm cells in about 3.5 ml of semen are released during one ejaculation by a human male. As a contraceptive measure, the sperm can be prevented from entering the female reproductive tract by use of a condom, a rubber sheath worn over the penis.

Hormonal control of sexual development in the male The testes begin secreting small amounts of the male sex hormone, *testosterone* (Fig. 9.44), during embryonic development. The presence of this hormone in the embryo is crucial to the differentiation of male structures. The level of production remains low until the time of puberty, and no sperm cells are produced.

The factors governing onset of puberty are not well understood. It has been proposed that the hypothalamus—which during childhood is sufficiently sensitive to testosterone to be inhibited by even low levels of the hormone—becomes less sensitive and begins sending more gonadotrophic releasing hormone (GnRH) to the anterior pituitary, stimulating it to increase its secretion of the two gonadotrophic hormones, **LH** and **FSH.** The LH induces the interstitial cells of the testes to produce more testosterone. The testosterone, together with FSH, induces maturation of the seminiferous tubules and stimulates them to begin sperm production (spermatogen-

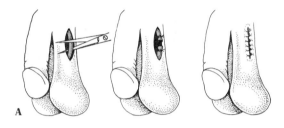

9.44   The structural formulas of male sex hormone (testosterone) and female sex hormone (progesterone) Differing in only one side group, these two steroids have amazingly different effects on the body.

esis). Maintenance of spermatogenesis over long periods requires the continued presence of testosterone (and the LH that induces it) and FSH.

Once testosterone appears in appreciable quantity in the system, it stimulates maturation of the accessory reproductive structures and also triggers the complex of changes in the secondary sexual characteristics normally associated with puberty: growth of the beard, growth of pubic hair, deepening of the voice, development of larger and stronger muscles, etc. If the testes are removed (castration) before puberty, these changes in the secondary sexual characteristics never occur. If castration is performed after puberty, there is some retrogression of the adult sexual characteristics, but they do not disappear entirely. Castration after puberty abolishes the sex urge in many animals, but not in man, where psychological factors are of much greater importance than in other animals. Unlike castration, cutting the vasa deferentia (vasectomy)—an operation that prevents movement of sperm into the urethra and is sometimes used as a birth-control measure (Fig. 9.45A)—causes no retrogression of sexual characteristics, because there is no alteration of hormone levels. However, recent evidence indicates that vasectomy sometimes triggers an auto-immune response that can cause damage to the circulatory system; hence the safety of this operation has been put in doubt.

The genital system of the human female The female gonads are the *ovaries,* which are located in the lower part of the abdominal cavity, where they are held in place by large ligaments. Like the testes, the ovaries have the two main functions of producing gametes (in this case egg cells) and secreting sex hormones. At the time of birth, a girl's ovaries already contain a huge number of primordial egg cells, or *oocytes;* estimates range from 100,000 to 1,000,000. Since, during the approximately 30 years of her reproductive life, a woman ovulates about 13 times per year, producing one mature ovum each time, it follows that usually fewer than 400 oocytes ever mature and leave the ovaries. The rest eventually degenerate, and ordinarily none can be found in the ovaries of women past the age of about 50.

Each oocyte is enclosed within a cellular jacket called a *follicle.* The oocyte fills most of the space in the small immature follicle. In the process of maturation, however, the follicle grows bigger relative to the oocyte and

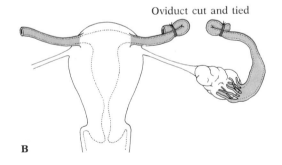

9.45   Surgical birth-control procedures (A) Vasectomy. A small incision is made in the wall of the scrotum (on each side); the vas deferens (color) is cut and tied, and the incision is sutured.   (B) Tubal ligation. Each of the oviducts is cut and tied, so that eggs cannot descend and no fertilization can occur.

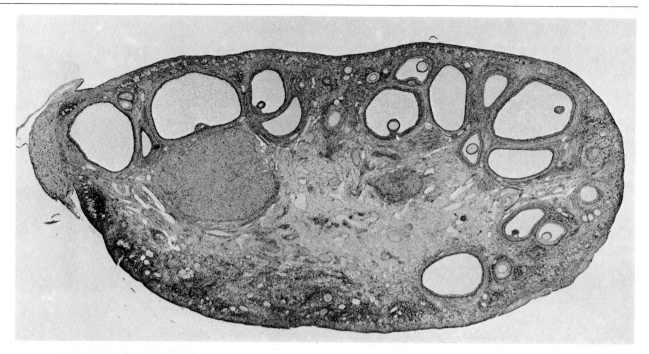

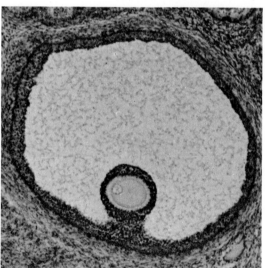

9.46   Photographs of sections of cat ovary
Top: Follicles of different sizes are shown in a section
of the entire ovary. The more mature follicles have a
large cavity, with the egg cell embedded in a pedestal
of epithelial cells that projects into the cavity.
Bottom: Enlarged view of a nearly mature follicle with
its egg cell.   [Courtesy Thomas Eisner, Cornell
University.]

develops a large fluid-filled cavity (Fig. 9.46); the oocyte, embedded in a
mass of follicular epithelial cells, protrudes into the cavity. A ripe follicle
bulges from the surface of the ovary; when ovulation occurs, its outer wall
ruptures and both the liquid and the detached oocyte are expelled. In the
human, only one oocyte is normally released at each ovulation. There is no
apparent regularity as to which ovary will ovulate at any given period.

Ovulation releases the oocyte from the ovary into the abdominal cavity.
From there, it is usually promptly drawn into the large funnel-shaped end of
one of the **oviducts** (or Fallopian tubes), which partly surround the ovaries
but are not continuous with them (Fig. 9.47). Cilia lining the funnel of the
oviduct produce currents that help move the oocyte into the oviduct. If
sperm are present to meet the oocyte while it is still in the upper third of the
oviduct, the penetration of a sperm through the membrane of the oocyte
stimulates it to complete its maturation into a true egg cell (ovum), and
almost immediately thereafter the nuclei of the sperm and ovum fuse in the
process of fertilization. A method of permanent sterilization of females,
sometimes used as a birth-control measure, is to cut and tie the oviducts
(tubal ligation), so that sperm cannot reach the oocyte and no ovum can

move down the oviduct into the uterus (Fig. 9.45B). Like vasectomy of the male, this operation causes no change in hormone production.

Each oviduct empties directly into the upper end of the **uterus** (or womb). This organ, which is about the size of a fist, lies in the lower portion of the abdominal cavity just behind the bladder (Fig. 9.48). It has very thick muscular walls and a mucous lining containing many blood vessels. If an egg is fertilized as it moves down the oviduct, it becomes implanted in the wall of the uterus, and there the embryo develops until the time of birth. One method of birth control involves insertion of a plastic ring or spiral into the uterus (Fig. 9.49A). Such intrauterine devices (IUD) seem to be very effective in preventing pregnancy, probably by preventing implantation in the uterus. However, the devices sometimes cause irritation and/or bleeding in the uterus; hence some women cannot tolerate them, and their safety for prolonged use is in doubt.

At its lower end the uterus connects with a muscular tube, the **vagina,** which leads to the outside. The vagina acts as the receptacle for the male penis during copulation. The great elasticity of its walls makes possible not only the reception of the penis, but the passage of the baby during childbirth.

The uterus and vagina do not lie in a straight line, as Fig. 9.47 might seem to indicate. Instead, the uterus projects forward nearly at a right angle to the vagina, as shown in Fig. 9.48. The **cervix,** a muscular ring of tissue at the mouth of the uterus, protrudes into the vagina. Devices that block the mouth of the uterus by covering the cervix are widely used in birth control. One such device, the diaphragm, is a shallow rubber cup with a spring around its rim. It is inserted into the vagina and positioned so that it covers the entire cervical region (Fig. 9.49B). It is very effective in preventing sperm from entering the uterus, particularly if it is used in conjunction with spermicidal jellies or creams.

The opening of the vagina in young human females is partly closed by a thin membrane called the **hymen.** Traditionally the hymen has been regarded as the symbol of virginity, to be destroyed the first time sexual intercourse takes place. Frequently, however, the membrane is ruptured during childhood, by disease or by a fall or as a result of strenuous physical exercise.

The external female genitalia are collectively termed the **vulva.** The vulvar region is bounded by two folds of skin, the labia minor and the labia major, that enclose the vestibule. The vagina opens into the rear portion of the vestibule, and the urethra opens into the midportion of the vestibule. Note, then, that in the adult mammalian female there is no interconnection between the excretory and reproductive systems, and that the urethra carries only excretory materials. During embryonic development the vagina does open into the urethra, but as development proceeds, the junction moves posteriorly until the vagina acquires its own opening to the exterior and the old interconnection disappears.

In the anterior portion of the vestibule, in front of the opening of the urethra, is a small erectile organ, the **clitoris,** which forms from the same embryonic tissue that gives rise to the penis in the male (see Fig. 17.49, p. 737). Like the penis, it becomes engorged with blood during sexual excitement and is a major site of stimulation during copulation.

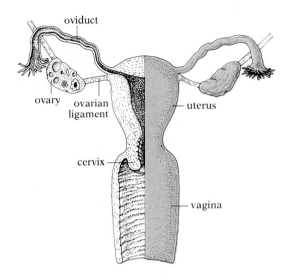

9.47 Reproductive tract of the human female: frontal view
The wall of one side has been dissected away to reveal the interior structure.

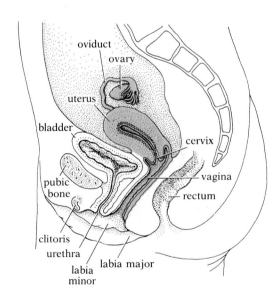

9.48 Reproductive tract of the human female: lateral view

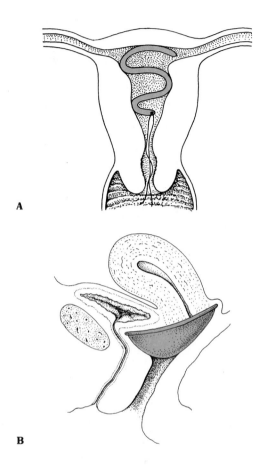

9.49   Two birth-control devices
(A) An IUD in place in the uterus. The strings that run through the cervix permit the woman to make sure the IUD has not been expelled.   (B) Diaphragm in position in the vagina. The device covers the mouth of the cervix, so that no sperm can enter the uterus.

Hormonal control of the female reproductive cycle   As in the male, puberty in the female is thought to begin when the hypothalamus loses its sensitivity to the low levels of sex hormone in childhood and starts secreting more GnRH, which stimulates the anterior pituitary to release increased amounts of FSH and LH. These gonadotrophic hormones cause maturation of the ovaries, which then begin secreting the female sex hormones, *estrogen* and *progesterone.* The estrogen stimulates maturation of the accessory reproductive structures (e.g. increase in the size of the uterus and vagina) and development of the female secondary sexual characteristics: growth of pubic hair, broadening of the pelvis, development of the breasts, change in the distribution of body fat, and some change in voice quality. The changing hormonal balance also triggers the onset of *menstrual cycles.* We shall be particularly concerned here with the menstrual cycles as an example of the complex interplay between several hormones and between the endocrine and nervous systems.

Rhythmic variations in the secretion of gonadotrophic hormones in the females of most species of mammals lead to what are known as *estrous cycles*—rhythmic variations in the condition of the reproductive tract and in the sex urge. The females of most species will accept the male in copulation only during those brief periods of the cycle near the time of ovulation when the uterine lining is thickest and the sex urge is at its height. During such periods the female is said to be "in heat" or in estrus. Many mammals have only one or a few estrous periods each year, but some, like rats, mice, and their relatives, may be in estrus as often as every five days. If fertilization does not occur, the thickened lining of the uterus is gradually reabsorbed by the female's body; ordinarily no bleeding is associated with this process.[6]

The reproductive cycle in humans and other higher primates differs in several important ways from that of other mammals. There is no distinct heat period, the female being to some degree receptive to the male throughout the cycle. And the thickened lining of the uterus is not completely reabsorbed if no fertilization occurs; instead, part of the lining is sloughed off during a period of bleeding known as menstruation. Human menstrual cycles average about 28 days; there are consequently about 13 of them each year. This is an extremely rough average, however, extensive variation occurring from person to person and from period to period in the same person.

Let us trace the sequence of events in the menstrual cycle, assuming a period of 28 days. It is customary in medical practice to consider the first day of menstruation as the first one of the cycle. From a biological point of view, however, it is more appropriate to regard the end of the period of bleeding as the beginning of the new cycle. At this point, the uterine lining is thin and there are no ripe follicles in the ovaries (Fig. 9.50). Under the influence of FSH from the anterior pituitary, several follicles in the ovaries begin growing (recall that the initials FSH stand for follicle-stimulating hormone). One of the follicles soon gains ascendancy, and the others cease growing. Influenced by the synergistic action of FSH and LH, the growing follicles begin secreting the first of the two female sex hormones, estrogen. The estrogen, in turn, stimulates the lining of the uterus to thicken. This follicular or growth

---

[6] The bleeding sometimes observed in dogs is not menstruation; it occurs at an entirely different time in the cycle, namely during the period of heat.

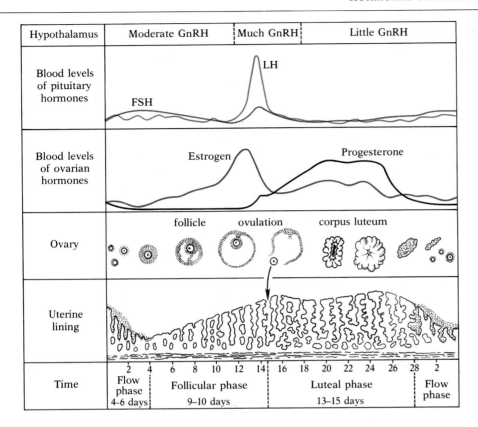

Hypothalamus	Moderate GnRH	Much GnRH	Little GnRH
Blood levels of pituitary hormones			
Blood levels of ovarian hormones			
Ovary			
Uterine lining			
Time			

9.50 The sequence of events in the human menstrual cycle

The hormone levels shown are only approximate. They are based on recent radioimmunoassay determinations, the results of which vary considerably.

phase of the cycle lasts, on the average, about nine to ten days after cessation of the previous menstrual flow.

As the follicles grow, they produce more and more estrogen (Fig. 9.50). The eventually high levels of estrogen (or, according to some investigators, the slowing of the rate of estrogen production as it reaches its peak) apparently stimulate, probably via GnRH from the hypothalamus, an abrupt surge of LH from the pituitary. This LH surge triggers ovulation by the ascendant follicle, which by this time is mature and bulging from the surface of the ovary.[7] The mechanism of ovulation is unknown. It is often stated that the pressure of the follicular fluid causes the wall to burst, but there is evidence that, on the contrary, this pressure may actually decline slightly just before ovulation. It does seem clear that hormonally induced thinning of the follicular wall is one important factor in ovulation. Whatever the mechanism, ovulation marks the end of the follicular or growth phase of the menstrual cycle.

Following ovulation, LH induces changes in the follicular cells that convert the old follicle into a yellowish mass of cells rich in blood vessels. The new structure formed from the ruptured follicle under the influence of LH—luteinizing hormone—is called the ***corpus luteum*** (Latin for yellow body). The corpus luteum continues secreting estrogen, though not as much

[7] Some investigators think that a slight rise in progesterone just prior to ovulation acts synergistically with LH to trigger the ovulatory event.

as was secreted by the follicle just prior to ovulation.[8] But the corpus luteum also secretes the second female sex hormone, progesterone (see Fig. 9.44).

Progesterone functions in preparing the uterus to receive the embryo. Acting on the uterine lining, which has already become much thicker under the stimulation of estrogen during the follicular phase, it causes maturation of the complex system of glands in the lining. The luteal phase of the menstrual cycle is, in fact, sometimes called the secretory phase, though this name is slightly misleading, since there is also some glandular activity in the uterine lining during the latter part of the follicular phase. Repeated experiments have shown that implantation of a fertilized ovum in the uterus cannot occur in the absence of the changes in the uterine lining produced by progesterone. Progesterone truly is the hormone of pregnancy.

In addition, high levels of progesterone inhibit initiation of the next cycle, though the precise mechanism of this inhibition is not entirely clear. The progesterone doubtless acts on the hypothalamus to suppress GnRH release, thereby limiting FSH and LH secretion by the pituitary and preventing the LH surge necessary for ovulation. But the progesterone may also act directly on the immature follicles in the ovary, inhibiting their growth and estrogen secretion. As long as progesterone is present in quantity in the system, then, there is little follicular growth. This inhibiting action of progesterone is, of course, an important element in regulating the duration of the menstrual cycle. It is also the basis for the action of birth-control pills. These pills contain synthetic compounds similar to progesterone and estrogen. Taken regularly, they inhibit secretion of FSH and LH (probably by suppressing GnRH release by the hypothalamus) and thus prevent follicular growth and ovulation—and consequently conception.[9]

If no fertilization occurs during a normal cycle, the corpus luteum begins to atrophy about eleven days after ovulation, and its secretion of progesterone falls. When this happens, the thickened lining of the uterus can no longer be maintained, and reabsorption of part of the lining begins. Unlike most mammals, humans and other higher primates cannot reabsorb all the extra tissue laid down during the follicular and luteal phases of the cycle; part of it must be sloughed off during menstruation, which lasts about four or five days.

The fall of progesterone levels, resulting from atrophy of the corpus luteum, frees the hypothalamus from inhibition and allows it to stimulate the pituitary to increase secretion of FSH. The immature follicles are also freed from inhibition and, under the influence of the FSH, begin growing, and a new cycle begins. The sequence of events in a normal menstrual cycle is depicted diagrammatically in Fig. 9.50.

From the above account it is apparent that the critical event in resetting the system is probably the fall in progesterone levels as a result of atrophy of the corpus luteum. But what causes this atrophy? No sure answer can yet be given, which means that a fundamental aspect of the timing of menstrual

[8] In rats maintenance of the corpus luteum requires prolactin in addition to LH, but the current evidence strongly suggests that in most mammals, including humans, LH alone suffices for both the formation and the secretory activity of the corpus luteum.

[9] Since the sex hormones, like most hormones, have multiple effects, women considering use of birth-control pills should become familiar with the possible harmful side effects, and should review their own medical histories for contraindications (e.g. vascular problems, migraine headaches) that may make use of the pills inadvisable.

cycles remains in doubt. Before sensitive methods for measuring hormone concentrations in the blood were available, it was thought that LH secretion declined gradually after ovulation, and that when levels fell low enough the corpus luteum could no longer be maintained; but it is now known, as shown by Fig. 9.50, that LH levels fall precipitously immediately after ovulation, long before the corpus luteum declines. Convincing recent evidence indicates that in cows certain prostaglandins, produced by the nonpregnant uterus and carried to the ovary by a special system of portal blood vessels, act to cut off progesterone secretion. But there is no such portal system in human beings; nor is there solid evidence that the human uterus sends any inhibiting substance to the ovaries. It has been suggested that the ovary itself produces prostaglandins that turn off progesterone secretion, but the evidence here is still scant.

During the flow phase, particularly in the first few days, secretion of progesterone and estrogen by the old corpus luteum is at a low level, and the follicles of the new cycle have not yet begun producing significant amounts of estrogen. Since a woman's body is accustomed from puberty to functioning in the presence of sex hormones, their withdrawal at the end of the luteal phase of each menstrual cycle is often accompanied by physiological and psychological disturbance, including irritability, depression, and sometimes nausea; abdominal cramps, caused by strong contractions of the uterus, are also common.

Emotional stress sometimes also accompanies the **menopause,** a period lasting a year or two at the end of a woman's reproductive life. The menopause usually comes sometime between the ages of 40 and 50. It is apparently attributable mainly to declining sensitivity of the ovaries to the stimulatory activity of gonadotrophins. The ovaries atrophy, the remaining follicles disappear, and secretion of estrogen and progesterone falls to low levels. Consequently there is no cyclic thickening of the uterine lining, and hence no menstruation. The changing hormonal balance during menopause may cause physiological and psychological disturbances until a new physiological balance has been established. Sometimes physiological difficulties persist indefinitely; a fairly common condition is postmenopausal osteoporosis, a thinning and weakening of the bones thought to be due to estrogen deficiency.

We have seen that ovulation in human beings occurs roughly midway in the menstrual cycle. This ovulation is spontaneous; it does not depend on a copulatory stimulus. In some mammals, such as rabbits and cats, however, ovulation is reflex-controlled; the nervous stimulus of copulation is necessary for the hypothalamus to trigger the release of LH by the pituitary that will lead to ovulation. In such reflex ovulators it is possible to predict with great precision just when ovulation will occur; in the rabbit, for example, it occurs about 10½ hours after copulation. Such precision is not possible with spontaneous ovulators like humans. Yet predictions of the time of ovulation are important in the practice of the rhythm method of birth control.

The rhythm method is based on the premise that fertilization can take place only during a very short period in each menstrual cycle. Copulation without risk of pregnancy should be possible during all other parts of the cycle. Present evidence indicates that human egg cells begin to deteriorate about 12 hours (maximum 24 hours) after ovulation and can no longer be fertilized after that time. In other words, fertilization can occur only if fertile sperm are in the upper third of the oviduct during the 12 to 24 hours immediately following ovulation; conception is not possible during the other 27 days of a 28-day cycle.

Immediately, the question of the fertile life of sperm in the female reproductive tract becomes pertinent. We have said that one ejaculation releases about 100 million sperm cells into the vagina. Conditions in the vagina are very inhospitable to sperm, and vast numbers are killed before they have a chance to pass the cervix. Millions of others die or become infertile in the uterus and oviducts, and millions more go up the wrong oviduct or never find their way into an oviduct at all. The journey to the upper portion of the oviducts is an extremely long and hazardous one for objects so tiny. There is evidence that movement of sperm from the vagina to the upper portion of the oviduct does not depend solely on the sperm's own swimming motions; contractions of the female genital tract, which may be accelerated by prostaglandins present in the seminal fluid and by oxytocin released by the posterior pituitary, may actually carry the sperm part of the way to their destination. Current evidence indicates that, in the female genital tract, human sperm cells remain fertile only about 48 hours or less after their release. The time may be much longer in some other animals. William A. Wimsatt of Cornell University has shown that some bats mate in the fall, but do not ovulate until after hibernation in the spring; the sperm must therefore remain fertile for several months. And in some invertebrates, such as bees and ants, fertile sperm may actually be stored in the body of the female for years, so that one mating provides a queen honeybee with enough sperm to last her throughout her reproductive life, during which time she produces thousands of offspring.

Since the fertile life of the human egg cell lasts at most one day and that of the human sperm at most two days, there is a period of about three days during which copulation can result in conception (the day when the egg is fertile and the two preceding days). But which three days? It is inability to answer this question precisely that makes the rhythm method of birth control unreliable. It is known that the three days come about midway through the menstrual cycle, but how long a given cycle will be cannot be predicted accurately and its midpoint therefore cannot be determined. Many women have very irregular cycles that vary by as much as 8–15 days or more. Even women whose cycles are very regular will have some that vary by as much as four or five days during the course of a year. Sickness or emotional upset frequently delays ovulation and prolongs the cycle by altering the hypothalamic control of gonadotrophin secretion. The least variable part of the cycle seems to be the luteal phase, the interval between ovulation and the onset of the next menstrual flow. This phase apparently averages between 14 and 15 days, and when a cycle is unusually short or long it is primarily the duration of the follicular phase that has changed. But the fact that the duration of the luteal phase is relatively stable is of little help in predicting the day of ovulation; since there is no way of knowing beforehand when the next menstrual flow will begin, it is not possible to count back 14 or 15 days to determine the ovulation date.

Hormonal control of pregnancy Our discussion so far has assumed that conception did not occur and that each cycle was terminated by a menstrual flow. Let us now assume that the egg cell is fertilized sometime during the 12 hours after ovulation. Only one of the millions of sperm cells released into the vagina actually penetrates the egg cell and fertilizes it. As soon as that one cell has fertilized the egg, the outer membrane of the egg changes in consistency and becomes impenetrable to the other sperm cells, which soon die. The fertilized egg, or *zygote,* moves down the oviduct, probably carried by fluid whose movement is caused by contractions of the circular muscles in the walls of the oviduct. The rate of movement is partly controlled by estrogen. During the days of transit, cell divisions begin and an embryo is formed.

The human embryo becomes implanted in the wall of the uterus 8–10 days after fertilization. During the interval between fertilization and implantation, the embryo is nourished by its limited supply of yolk and by materials secreted by the glands of the female genital tract. The delay before implantation varies from species to species; e.g. it is about 20–22 days in sheep, 35 days in cows, and as long as 56 days in horses. In a few species there is a much longer delay, during which development of the embryo proceeds very slowly or even ceases; among mammals with delayed implantation are brown bears (about five months), pine martens (six months), American badgers (two months), and armadillos (14 weeks).

After implantation in the uterine lining, the embryonic membranes form the *umbilical cord,* through which blood vessels contributed by the allantois run to a large structure, the *placenta,* formed from the embryonic membranes (primarily the chorion) and from the adjacent uterine tissue (Fig. 9.51). Within the placenta the blood vessels of the embryo and those of the mother lie very close together, but they are not joined and there is no mixing of maternal and fetal blood. Exchange of materials takes place in the placenta by diffusion between the blood of the mother and that of the embryo; nutritive substances and oxygen move from the mother to the embryo, and urinary wastes and carbon dioxide move from the embryo to the mother.

We saw earlier that progesterone is essential for maintenance of the uterine lining during implantation and pregnancy. But we saw also that in a normal menstrual cycle, when there is no fertilization, the corpus luteum soon begins to atrophy and cuts off the supply of progesterone, with the result that menstruation occurs. Clearly, this sequence of events cannot be allowed to take place after conception, or the uterine lining with the implanted embryo would be sloughed off and lost. It can be shown that when conception occurs, the corpus luteum does not atrophy, but lasts through most of the term of pregnancy. How is this possible? Apparently the chorionic portion of the placenta soon begins secreting a gonadotrophic hormone (human chorionic gonadotrophin, or HCG). This hormone preserves the corpus luteum, which continues to secrete progesterone and thus sustains the pregnancy.

So much chorionic gonadotrophin is produced in a pregnant woman that much of it is excreted in the urine. Many commonly used tests for pregnancy are based on this phenomenon. Urine from the subject is injected into a test animal such as a rat, rabbit, or frog. If HCG is in the urine, it induces development of corpora lutea and changes in the vagina of female rats 72 to 96 hours after injection. The rabbit test is faster; pregnancy urine induces corpora lutea formation within 24 hours after injection. And frog tests are

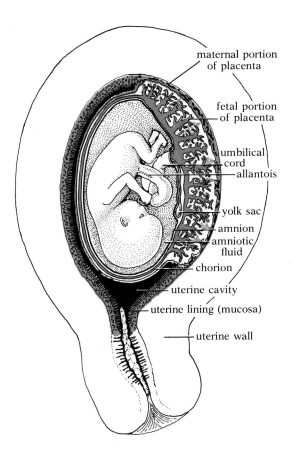

maternal portion of placenta

fetal portion of placenta

umbilical cord

allantois

yolk sac

amnion

amniotic fluid

chorion

uterine cavity

uterine lining (mucosa)

uterine wall

9.51 Pregnant uterus of a woman
The placenta consists of both maternal and fetal components, the latter derived from the chorion. Blood, spurting from vessels in the maternal placenta, bathes the villi of the fetal placenta. Nutrients absorbed by the villi are carried via the umbilical cord to the embryo, which lies in the amniotic sac, bathed by amniotic fluid.

even faster; pregnancy urine induces ovulation by females and release of sperm by males within 6–8 hours after injection. Recently developed immunological tests for HCG take less than an hour to run.

Although the corpus luteum is essential during early pregnancy, it can be shown that in humans it is no longer necessary after the first two months or so. Removal of the ovaries after this time does not terminate the pregnancy, evidently because both estrogen and progesterone, which can still be detected in the urine after excision of the ovaries, are produced by another organ. Apparently the placenta begins to secrete these hormones early in pregnancy, and once this secretion has reached a sufficiently high level the placenta itself can maintain the pregnancy in the absence of progesterone from the corpus luteum. This is not true of the rabbit, where removal of the ovaries just a few days before the end of pregnancy invariably results in abortion unless progesterone is artificially administered.

Hormonal control of parturition and lactation   Much research is still needed to clarify the complex interactions of hormones that control the birth process (parturition). The best evidence to date indicates that a rise in prostaglandin production by the placenta plays a central role in initiating parturition. The major stimulus for this rise in late pregnancy appears to be an increase in estrogen secretion (also by the placenta), which is itself a result, at least in part, of increased secretion of estrogen precursors by the fetal adrenal cortex. In some mammals, but not in primates, the terminal placenta begins converting progesterone, which severely inhibits muscular contraction, into estrogen; the resulting fall in plasma progesterone levels may make the uterine muscles more susceptible to stimulation. Since oxytocin, secreted by the posterior pituitary (and perhaps also by the placenta), is known to have a powerful stimulatory effect on uterine contractility, it seems likely that this hormone, too, contributes to the induction of labor.

Another hormone important in parturition is *relaxin,* which is secreted during pregnancy by the ovaries and the placenta. This hormone has the effect of loosening the connections between the bones of the pelvis and thereby enlarging the birth canal and facilitating parturition. Relaxin also aids in softening and dilating the cervix. The activity of relaxin is enhanced by estrogens.

Like the hormonal control of parturition, that of milk secretion is complicated and not completely understood. Growth and development of the mammary glands in humans seem to be controlled by a complex interaction between estrogen, progesterone, thyroxin, insulin, growth hormone, prolactin, glucocorticoids, and human placental lactogen (a potent prolactinlike hormone produced by the placenta). Initiation and maintenance of lactation by mature mammary glands following parturition seem to be controlled primarily by prolactin and glucocorticoids. These hormones apparently become effective in inducing lactation when the high levels of sex hormones, which inhibit lactation, disappear at the time of parturition.

The actual release of milk from the mammary glands involves both neural and hormonal mechanisms. The stimulus of suckling, or, in conditioned cows, of seeing the calf or hearing rattling milk pails, causes nervous stimulation of that part of the hypothalamus that stimulates release of oxytocin stored in the posterior pituitary. The oxytocin, in turn, induces constriction of the many tiny chambers in which the milk is stored in the mammary glands. The constriction forces the milk into ducts that lead to the nipple. Adrenalin inhibits this milk-ejection process.

SUGGESTED READING

GALSTON, A. W., and P. J. DAVIES, 1970. *Control mechanisms in plant development.* Prentice-Hall, Englewood Cliffs, N.J. *A brief elementary text, which remains timely, though the treatment of a few topics is slightly out of date.**

GARDNER, L. I., 1972. "Deprivation dwarfism," *Scientific American,* July. (Offprint 1253) *Inadequate secretion of pituitary hormones, especially growth hormone, as a probable cause of stunted growth in some emotionally deprived children.*

GILLIE, R. B., 1971. "Endemic goiter," *Scientific American,* June. *The intriguing history of this disorder; its current distribution as primarily a disease of the poor.*

GUILLEMIN, R., and R. BURGUS, 1972. "The hormones of the hypothalamus," *Scientific American,* November. (Offprint 1260) *The discovery of the hypothalamic releasing hormones and their role in regulating the anterior pituitary.*

LEVINE, S., 1971. "Stress and behavior," *Scientific American,* January. (Offprint 532) *The major role played in learning by the pituitary and adrenal hormones that regulate responses to stress.*

McEWEN, B. S., 1976. "Interactions between hormones and nerve tissue," *Scientific American,* July. (Offprint 1341) *The influence of the steroid hormones on the infant's development of brain circuits that control later behavior.*

O'MALLEY, B. W., and W. T. SCHRADER, 1976. "The receptors of steroid hormones," *Scientific American,* February. (Offprint 1334) *The mechanism by which steroid hormones are thought to act on their target cells.*

SCHNEIDERMAN, H. A., and L. I. GILBERT, 1964. "Control of growth and development in insects," *Science,* vol. 143, pp. 325–333.

van OVERBEEK, J., 1968. "The control of plant growth," *Scientific American,* July. (Offprint 1111)

WILLIAMS, C. M., 1950. "The metamorphosis of insects," *Scientific American,* April. (Offprint 49) *Fascinating piece by one of the pioneers in research on the hormonal control of insect development.*

WURTMAN, R. J., and J. AXELROD, 1965. "The pineal gland," *Scientific American,* July. (Offprint 1015) *The long search for the function of the pineal.*

* Available in paperback.

Chapter 10

NERVOUS CONTROL

As we saw in the last chapter, there is an intimate relationship between endocrine and nervous control systems in multicellular animals. Together these systems make possible the integration of function essential to animal behavior. Although plants, as we have seen, far from being passive, are complex, dynamic organisms that grow, change, react to external stimuli, and move—indeed, it is no exaggeration to say that plants behave—their behavior is fundamentally different from that of animals. The difference is largely one between two ways of life—sedentary on the one hand, mobile on the other. It is most obviously a difference in speed. Much of the behavior of plants depends on variations in growth rates or changes in the turgidity of cells, both inherently rather slow ways of bringing about movement. The behavior patterns of animals do not rely on such processes, for animals have evolved tissues specialized for production of rapid movement, notably the muscles. Correlated with the difference in speed—a result of the very different ways movement is produced—are basic differences in the control systems involved.

Hormonal control is a relatively slow process. Even when a hormone is transported in the phloem or in the bloodstream, there is an appreciable delay between the release of the hormone and its arrival at the target cells. Response to the stimulus that induced secretion of the hormone is therefore not immediate; there is a lag of seconds or often of minutes. But, given the inherent slowness of the mechanisms of plant movement, the delay involved in hormonal control is insignificant. Hormonal control is likewise adequate for animals when instantaneous response is not needed, as in control of digestion, salt and water balance, metabolism, and growth. But when rapid response is required, as in the movements produced by skeletal muscles, hormonal control is inadequate. It is here that nervous control is essential. A nerve impulse can move nearly 100 m/sec, thus reducing the interval be-tween stimulus and reponse to milliseconds. Evolution of nerve and muscle tissues, then, was basic to the evolution of active multicellular animals as we know them today.

EVOLUTION OF NERVOUS SYSTEMS

Irritability   Irritability—the capacity to respond to stimuli—is a universal characteristic of protoplasm. Any manifestation of irritability—any reaction to stimulus—ordinarily involves three principal components: (1) reception of the stimulus; (2) conduction of a signal; and (3) response by an effector.

A stimulus is usually an environmental change of some sort. Any such change is potentially a stimulus, but whether it actually becomes one de-pends, first, on the presence of protoplasm; second, on the capacity of the protoplasm to detect the change. Many environmental changes never func-tion directly as stimuli, so far as we know, because no protoplasm can detect them. The reason human beings can use radio and TV radiations in com-munication is that they have learned to convert undetectable energy changes into detectable ones—sound vibrations and light—which function as the actual stimuli.

Conduction of a signal, the second component in a reaction to stimulus, is a capacity inherent in protoplasm itself. If a stimulus can produce a change in protoplasm at some point (i.e. if it can be received), neighboring regions of protoplasm will almost surely be influenced to some extent; the initial change will spread. The spread may be limited and slow, or it may be extensive and rapid. It is this tendency for protoplasmic changes to spread from the point of origin—for changes at one point to induce changes at neighboring points—that was the raw material for evolution of nervous con-duction.

The response, the third component, is the action prompted by receipt of a stimulus. What different organisms do on receiving the same stimulus may be very different. And what a single organism does on receiving the same stimulus on two different occasions may also be different. Thus the response depends more on the characteristics of the organism than on the charac-teristics of the stimulus. The response is not produced directly by the stimu-lus—the way the breaking of a vase would be by hitting it with a hammer.

In unicellular organisms all three components of a reaction to stimulus—stimulus reception, signal conduction, and response—obviously occur within the protoplasm of a single cell. Any conduction is simply from one part of the cell to another, most often by means of the natural slow conductivity of the cytoplasm or of the plasma membrane. Similarly, stimulus reception is often general and nonlocalized.

There are some unicellular (or, better, acellular) organisms, however, that show more specialization. In ciliate Protozoa, for example, the bases of the cilia are interconnected by a system of fibrils that may function as a coordi-nation network. A specialized region of the cell (motorium) has been found to serve as a coordinating and relay center for impulses traveling along these fibrils; removal of this region or cutting of the fibrils disrupts the normally coordinated movement of the cilia. Similarly, many Protozoa possess or-ganelles specialized for stimulus reception, such as light-sensitive eyespots

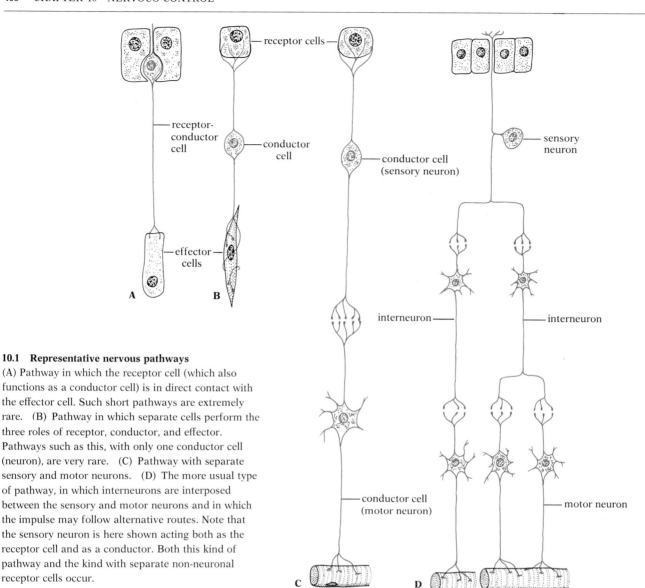

10.1 Representative nervous pathways
(A) Pathway in which the receptor cell (which also
functions as a conductor cell) is in direct contact with
the effector cell. Such short pathways are extremely
rare. (B) Pathway in which separate cells perform the
three roles of receptor, conductor, and effector.
Pathways such as this, with only one conductor cell
(neuron), are very rare. (C) Pathway with separate
sensory and motor neurons. (D) The more usual type
of pathway, in which interneurons are interposed
between the sensory and motor neurons and in which
the impulse may follow alternative routes. Note that
the sensory neuron is here shown acting both as the
receptor cell and as a conductor. Both this kind of
pathway and the kind with separate non-neuronal
receptor cells occur.

and sensory bristles. And many have specialized effector organelles such as
cilia, flagella, and contractile fibrils. Clearly, even at the acellular level of
complexity, specialized structures have evolved for sensory reception, conduction, and rapid response.

Simple nervous pathways All animal groups above the level of the sponges
have some form of nervous system, though in some groups it is very primitive. In the tentacles of some coelenterates, we see the simplest possible type
of true nervous pathway—one composed of only two specialized cells, a

receptor-conductor cell and an effector cell (Fig. 10.1A). Such a pathway allows little flexibility of behavior, because there are no alternative pathways for the impulse to travel, and because the lack of interconnection between this pathway and other parts of the nervous system precludes central control. Most nervous pathways, even in coelenterates, comprise at least three separate cells. In some cases these three cells represent the three components of reaction to stimulus; there is a receptor cell specialized for reception of a particular kind of stimulus, a conductor cell specialized for conducting impulses over long distances, and an effector cell (often a muscle cell) specialized for giving a response (Fig. 10.1B). More complex pathways may involve any number of additional conductor cells interposed between the receptor cell and the effector cell (Fig. 10.1C, D). Once the pathways include several conductor cells, response can be more flexible, because more than one route is usually open to the impulse coming from the receptor; any one of several alternative effectors, or all of them, may be activated. In general, the more conductor cells in the circuitry, the more flexible the response.

The typical nerve cell, or **neuron,** consists of an enlarged region, the **cell body,** which contains the nucleus, and one or more long processes, or **nerve fibers,** which extend from the cell body and may measure as much as 2 m (Fig. 3.59, p. 133). Neurons leading from receptor cells (and also combination receptor-conductor neurons) are called **sensory neurons;** those leading to effector cells are called **motor neurons;** and those lying between the sensory and motor neurons are called **interneurons.**

A junction between adjacent neurons is know as a **synapse.** The neurons usually do not contact each other at the synapse (Fig. 10.1); their fibers come very close to each other, but (except in a few special cases) a tiny gap remains between them.

Nerve nets and radial systems The simplest form of organized nervous system, seen in coelenterates of the hydra type, consists of separate receptor, conductor, and effector cells. The conductor cells do not, however, form definite pathways, but interlace to form a diffuse **nerve net** running throughout the body (Fig. 10.2). There is apparently no central control. Impulses can move in either direction along most of the fibers (a few fibers are unidirectional). They simply spread slowly from the region of initial stimulation to adjacent regions, becoming less intense as they spread. The stronger the initial stimulus, the farther the impulses will spread. Reactions are mostly limited to local contractions and to discharge of nematocysts. Such a system, lacking the potential for central coordination of complex reactions, can produce only a limited behavioral repertoire (though the behavior of hydra is more complex than might be predicted; see Fig. 11.1, p. 466).

Some degree of centralization is seen in other coelenterates, such as jellyfish, which have two nerve rings in the "bell" portion of the body. The other nerve cells tend to funnel into these rings, and conduction from one side of the animal to the other is thus much more rapid than would be possible with the randomly oriented pathways of a simple nerve net. This centralization is reflected in the swimming movements of jellyfish, which consist in contraction of the whole bell in a rhythmic coordinated fashion.

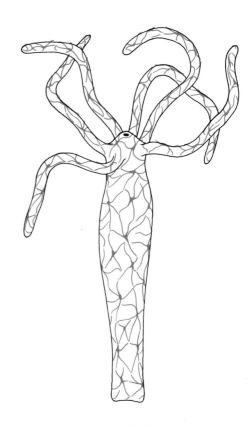

10.2 Nerve-net system of hydra
There is no central nervous system. (The gaps between the fibers of adjacent nerve cells are not as wide as shown here.)

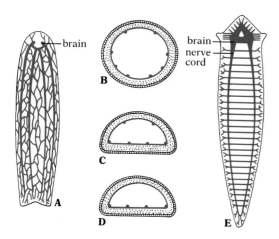

10.3 Nervous systems in flatworms

(A) Submuscular plexus of *Amphiscolops*. This is essentially a nerve net, but several longitudinal cords are present within it; the so-called brain is only a very tiny thickening at the anterior end of these cords. (B) Cross section of a somewhat more advanced flatworm with eight well-developed cords: a dorsal pair, a lateral pair, a ventrolateral pair, and a ventral pair. (C–D) There are fewer cords in (C), and still fewer in (D). (E) Ventral nerve plate of planaria, with two cords and a moderately developed brain.   [Modified from L. H. Hyman, *The invertebrates*, vol. 2, McGraw-Hill Book Co., copyright © 1951; and from R. Buchsbaum, *Animals without backbones*, University of Chicago Press, copyright © 1948 by The University of Chicago. Used by permission of the publishers.]

Apparently, radial symmetry, like that of coelenterates, severely restricts the evolutionary potential for extensive centralization of nervous systems. The behavior of even relatively advanced radial animals such as echinoderms—whose nervous systems are characterized by central nerve rings, unidirectional conduction, well-coordinated stimulus-response patterns, and some learning ability—is simple compared with that of many bilaterally symmetrical animals.

Evolutionary trends in bilateral nervous systems   The major trends in the evolution of nervous systems in bilaterally symmetrical animals can be detected even in the lowly flatworms. Let us summarize these trends briefly:

1. The nervous system became increasingly centralized by formation of major longitudinal nerve cords (the *central nervous system*) through which most pathways between receptors and effectors had to pass and in or near which most neuron cell bodies came to lie.

2. Conduction along nervous pathways became restricted to one direction only. A distinction thus developed between sensory fibers leading toward the central nervous system (afferent fibers) and motor fibers leading away from the central nervous system (efferent fibers).

3. Nervous pathways within the central nervous system became increasingly complex by interpolation of large numbers of interneuron—a development making for increased flexibility of response.

4. Cells performing different functions became increasingly segregated within the nervous system, and distinct functional areas and structures developed.

5. Increasing ascendancy of the front end of the longitudinal cords led to formation of a *brain,* which became more and more dominant—a process called *cephalization.*

6. The number and complexity of sense organs increased.

These trends are not yet very distinct in the most primitive flatworms (those thought to be most like the ancient ancestral forms); such flatworms have only a nerve net much like that of hydra. Some slightly more advanced flatworms (less like the ancestral forms) show the beginnings of a condensation of major longitudinal cords within their nerve nets (Fig. 10.3A); there are often as many as eight of these cords, located ventrally, dorsally, and laterally (Fig. 10.3B). Still more advanced flatworms show a reduction in the number of longitudinal cords (Fig. 10.3C, D), the most advanced representatives having only two, both located ventrally.

Those flatworms with the most primitive development of longitudinal cords show very little evidence of any special structure at the anterior end that could be called a brain (Fig. 10.3A); biologists have, however, charitably labeled "brain" the tiny swellings present there. Flatworms at more advanced stages show a much better developed brain (Fig. 10.3E), though even this brain exerts only limited dominance over the rest of the central nervous system.

Natural as it may seem, the near universal location of an animal's brain in its head (i.e. at the anterior end of its body) is not, in fact, inevitable; other arrangements are conceivable. The evolutionary and functional explanation for this location is almost certainly to be found in the animal's direction of movement. The anterior end is usually the part of a bilateral animal that first

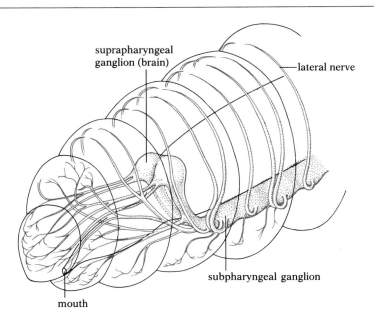

suprapharyngeal
ganglion (brain)

lateral nerve

subpharyngeal ganglion

mouth

10.4   Portions of earthworm and insect nervous systems
Left: The two parts of the double ventral nerve cord separate at the anterior end and encircle the pharynx. The ring of nervous tissue thus formed consists of the paired suprapharyngeal ganglia, the paired subpharyngeal ganglia, and the cords connecting them. The suprapharyngeal ganglia are customarily regarded as the brain, but they are only slightly larger than the segmental ganglia of the ventral nerve cord.
Below: Double ganglion and paired ventral nerve cord of a living fly larva, viewed with Nomarski optics. × 210.   [Below: Courtesy Thomas Eisner, Cornell University.]

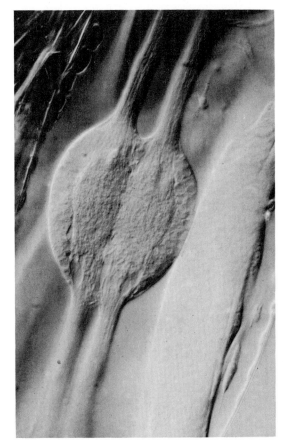

encounters new stimuli as the animal moves. Natural selection therefore favored development of a particularly high concentration of sense organs in this region, which, in turn, led to enlargement of the anterior ends of the longitudinal nerve cords.

The most primitive version of the brain was probably almost exclusively concerned with funneling impulses from the sense organs into the cords. Then, because of the adaptive advantage of shortening the pathway these impulses have to follow before reaching the main coordination areas of the central nervous system, natural selection must have favored grouping those areas toward the anterior ends of the cords. Thus the brain came to be more than a sensory funneling area; as coordination increasingly became concentrated in it, it became more and more dominant over the rest of the central nervous system. This dominance has its greatest development in mammals, especially human beings.

The evolutionary trends whose beginnings can be seen so clearly in flatworms, where intermediate stages between a nerve-net system and a centralized system can be studied in living animals, have their most extensive development in the vertebrates and, among invertebrates, in the annelids and arthropods. In all these animals there is a high degree of centralization, and the old nerve net (which is presumed to be the ancestral system) is represented only by vestiges in a few parts of the body where sluggish movements—e.g. the peristaltic contractions of the mammalian intestine—are controlled by slow diffuse conduction.

In annelids and arthropods the central nervous system is a pair of ventrally located longitudinal cords, in which the cell bodies of the neurons form masses called *ganglia* and the fibers, gathered into huge bundles, function as through-conducting systems (Fig. 10.4). Thus in primitive annelids and arthropods prominent ganglionic masses, a pair in each body segment, are connected by bundles of fibers running between the segments; almost all the cell bodies are located in ganglia. The brain is simply another ganglion located in the animal's head. It is little if any larger than the segmental ganglia; its dominance over the other ganglia is noticeable, but limited in comparison to that of the vertebrate brain.

More advanced arthropods, particularly some of the insects, show more concentration of coordination in the front end; abdominal ganglia have frequently moved forward and fused with the thoracic ganglia. The brain, however, has remained relatively small, and the thoracic ganglia continue to perform many vital coordinating functions. The persistence of thoracic ganglia in insects is probably correlated with at least two important functional arrangements of their bodies: the attachment of both their legs and wings on the thorax, which makes a concentration of motor coordinating centers in the thorax advantageous; and the location of many of their sense organs on the legs or thorax rather than on the head (e.g. taste receptors are on the feet of flies and some other insects, and the ears, when present, are often on the sides of the thorax).

The central nervous sytem (spinal cord plus brain) of vertebrates differs in several important ways from those of annelids and arthropods:

1. The vertebrate spinal cord is single; it is located dorsally; and it forms in the embryo as a tube with a hollow central canal, a remnant of which survives in the adult (see Fig. 10.10). The cords of annelids and arthropods, on the other hand, are double (two cords lying side by side and often partly fused); they are located ventrally; and they are always solid.

2. The vertebrate spinal cord is not so obviously organized into a series of alternating ganglia and connecting tracts.

3. Although many coordinating functions in vertebrates are still performed by the spinal cord, there has been extensive development of a brain, which exerts far more dominance over the entire nervous system than the brain of any annelid or arthropod. The vertebrate brain, in short, is the master control center for almost all bodily functions.

NERVOUS PATHWAYS IN VERTEBRATES

Even though some of the most exciting current neurobiological research is carried out on invertebrates, especially arthropods and molluscs, our models in this discussion of basic nervous pathways will be vertebrates—in great part because of the fascination inherent in the system that more than any other makes human beings human.

NEURONS OF VERTEBRATES

Vertebrate neurons may have one, two, or more fibers. When a neuron has more than one fiber, those fibers are customarily of two types: **dendrites,** which usually receive excitation from other cells and conduct toward the cell body; and **axons,** which conduct impulses away from the cell body (Fig. 10.5).

Dendrites are usually rather short, and any one neuron often has many of them. They frequently branch profusely and have a spiny appearance. When stained and examined under an ordinary light microscope, their cytoplasm

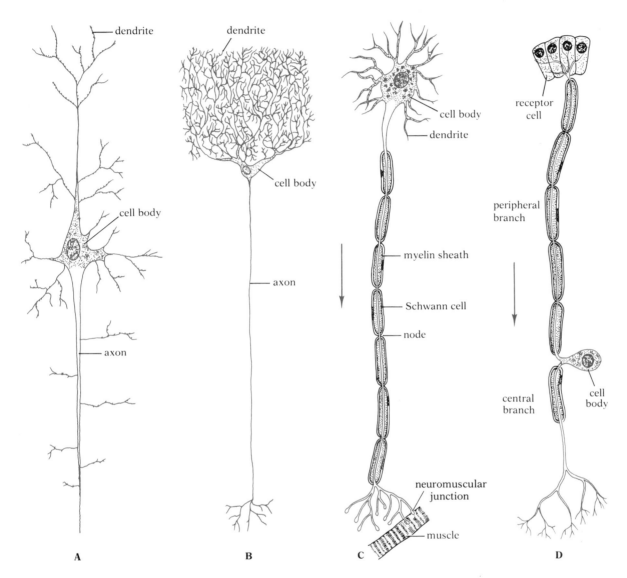

| A | B | C | D |

10.5 A variety of neuron types in human beings

(A) The dendrites, unlike the axon, often have a spiny look.   (B) The dendrites of certain brain cells branch profusely, giving the cell a treelike appearance.   (C) Motor neurons have long axons that run from the central nervous system to the effector (in this case muscle); these axons are frequently, but not always, myelinated. Note the presence of many granules in the cell body and dendrites and their absence from the axon.   (D) Many sensory neurons have only one fiber, which branches a short distance from the cell body, one branch (peripheral) running between the receptor site and the dorsal-root ganglion in which the cell body is located, and the other branch (central) running from the ganglion into the spinal cord or brain. Except for its terminal portions, the entire fiber is structurally and functionally of the axon type, even though the peripheral branch conducts impulses toward the cell body. A sensory neuron of this type thus has no true dendrites, although the peripheral branch is often called a dendrite because of the direction in which it conducts impulses.

appears to contain much dark granular material (Nissl bodies), which the electron microscope reveals as extensive arrays of flattened cisternae of the endoplasmic reticulum, with a great many associated ribosomes (Fig. 10.6).

There is usually only one axon per neuron, and it is often longer than the dendrites; it may branch extensively, but does not have a spiny appearance and does not contain Nissl bodies. Until recently it was thought that the most fundamental distinction between dendrites and axons—aside from the histological differences just mentioned—was that the dendrites receive excitation from other cells whereas axons do not, and that axons can stimulate other cells whereas dendrites cannot. But it is now known that this distinction does not always hold; dendrites sometimes form synapses on other dendrites, and axons sometimes receive synapses, especially near their bases or near their distal terminals.

Within the central nervous system the neurons are intimately associated with vast numbers of satellite cells called *neuroglia,* or simply glia for short. The glial cells usually envelop the individual neurons, except at synaptic junctions, and it has therefore been proposed that they act as insulation, preventing "cross talk" between adjacent neurons. Other suggestions about their possible functions have been made: that they provide nutrients to the neurons; that they reclaim substances secreted by the neurons and pass them back for reuse; that they play an important part in the development of the nervous system, providing a framework for the growth of the neurons. Though the neuroglia may exercise one or more of these functions, the

10.6   Electron micrograph of Nissl bodies
At a magnification of × 42,000 it can be seen that the granular material known as Nissl bodies, which is visible under the light microscope, actually consists of flattened cisternae of the endoplasmic reticulum plus numerous ribosomes; many of the ribosomes are not directly associated with the cisternal membranes. [Courtesy S. L. Palay, Harvard Medical School.]

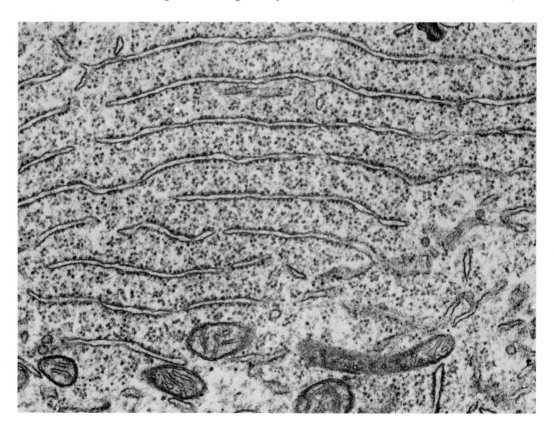

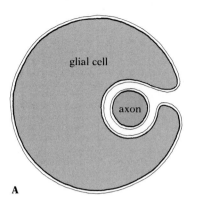

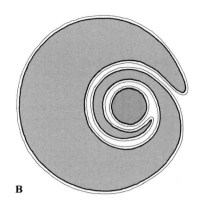

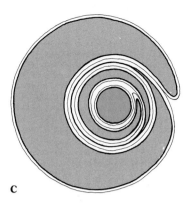

A   B   C

10.7 Development of the myelin sheath
(A) Initially the unmyelinated axon lies in an inpocketed area of the glial cell.   (B) The inpocketed area begins to coil.   (C) The glial-cell membrane is wound ever more tightly around the axon, forming what is known as a myelin sheath.

evidence remains equivocal. Much more research is needed before the role of these cells, the most abundant in the brain, is fully understood—research important not only in the study of brain function, but also in medicine, because the neuroglia are the principal source of tumors of the central nervous system.

Though the function of many neuroglial cells is in doubt, that of certain such cells is well known. These are specialized glial cells whose membranes wrap around and around the axons of many neurons in the central nervous system of vertebrates, forming a heavily lipid *myelin sheath* (Fig. 10.7). Outside the central nervous system many vertebrate axons are enveloped by satellite cells called *Schwann cells* (Fig. 10.8), which often give rise to a myelin sheath (Fig. 10.9). Myelin sheaths, which are interrupted at regular intervals called nodes (Fig. 10.5)—points where one Schwann cell ends and another begins—function in speeding up the conduction of impulses in the axons they envelop.

REFLEX ARCS

A reflex arc is a simple neural pathway linking a receptor and an effector. Such arcs are important functional units of the nervous system. The reflexes they produce, which are responses to specific stimuli, are usually rapid and relatively automatic.

The simplest reflex arcs in vertebrates involve only two neurons, sensory and motor. A classic example is the stretch reflex of the lower leg (known popularly as the knee-jerk reflex), commonly checked by physicians during physical examinations (Fig. 10.10). In this reflex, which normally functions in regulating leg movements and in maintaining posture, the receptors are stretch receptors, which are spindles formed by an intimate association between the terminal branches of a fiber of a sensory neuron and a specialized portion of a muscle. The sensory neuron is a very long one, running all the way from the stretch receptor in the knee to the spinal cord. Its cell body is located in a *dorsal-root ganglion* (or spinal ganglion), which lies just outside the spinal cord near its dorsal[1] surface. The axon enters the cord

[1] The terms "dorsal(ly)" and "ventral(ly)" as used here refer to the usual vertebrate standing on four legs. In humans the dorsal becomes the posterior side, and the ventral becomes the anterior side.

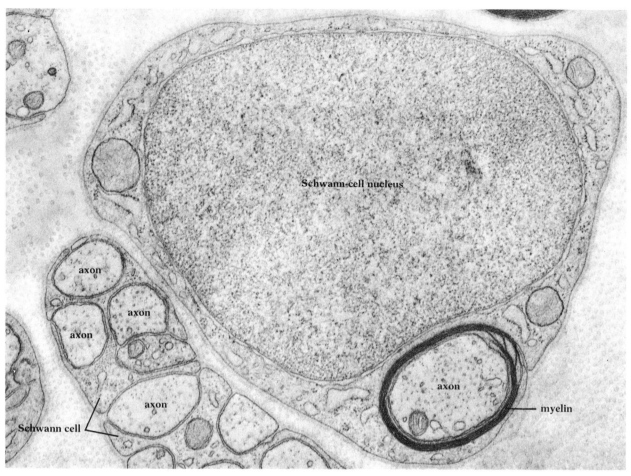

Schwann-cell nucleus

axon

axon

axon

axon

Schwann cell

axon

myelin

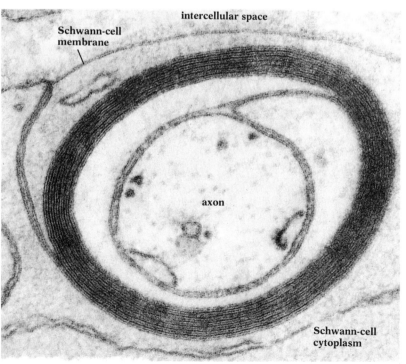

intercellular space

Schwann-cell membrane

axon

Schwann-cell cytoplasm

10.8 Electron micrograph of cross section of part of a nerve from a guinea pig

The axons of the neurons are enveloped by Schwann cells; at lower left a single Schwann cell envelops several unmyelinated axons. The axon at lower right has a myelin sheath, formed from an invaginated coiled portion of the Schwann-cell membrane; note the large nucleus of the Schwann cell. The numerous small circular structures in the spaces between the Schwann cells (as in extreme lower right corner) are cross sections of collagen fibrils. × 36,000.   [Courtesy H. deF. Webster, Harvard Medical School.]

10.9 Electron micrograph of myelin sheath

This micrograph shows conclusively that the myelin sheath is a coil of the Schwann-cell membrane. × 112,000.   [Courtesy J. David Robertson, *Ann. N.Y. Acad. Sci.*, vol. 94, 1961.]

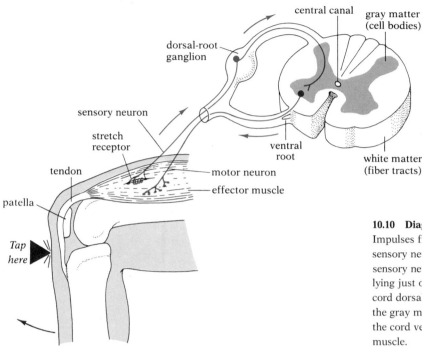

10.10 Diagram of the knee-jerk reflex arc
Impulses from the stretch receptor travel along the sensory neuron to the spinal cord. The cell body of the sensory neuron is located in a dorsal-root ganglion lying just outside the cord. The sensory axon enters the cord dorsally and synapses with the motor neuron in the gray matter of the cord. The motor neuron leaves the cord ventrally and carries impulses to the effector muscle.

dorsally, where it synapses with the dendrites or cell body of a motor neuron within the gray matter of the spinal cord. The axon of the motor neuron then exits ventrally from the spinal cord and runs all the way to the effector cells, which in this case are muscle fibers in the leg. Thus, when the physician taps the knee and a stretch receptor is stimulated, impulses travel up a sensory neuron to the spinal cord and back down a motor neuron to the leg, where they stimulate muscle fibers, which contract, causing the leg to jerk. A minimum of three cells is involved in this reflex arc: a receptor-sensory neuron, a motor neuron, and an effector cell.

Using this very simple reflex as a model, we can make several generalizations about all spinal reflex arcs:

1. There is never more than one sensory neuron in a single pathway (there may, of course, be many such pathways running side by side and subserving the same function), but that neuron may be exceedingly long; consider a sensory neuron in an elephant going all the way from its foot to the spinal cord.

2. The cell body of a sensory neuron is always outside the spinal cord in a dorsal-root ganglion.

3. The axons of sensory neurons always enter the spinal cord dorsally.

4. The axons of motor neurons always leave the spinal cord ventrally.

So far we have not made reference to nerves in our discussions of neural pathways. A ***nerve*** is a compound structure consisting of a number of neuron fibers bound together. Although there may be thousands of fibers in a single nerve, each is insulated from all the others and conducts impulses independently of the others. A nerve, therefore, is much like a telephone cable containing many functionally separate telephone wires; the many independent communications pathways have been packaged together for

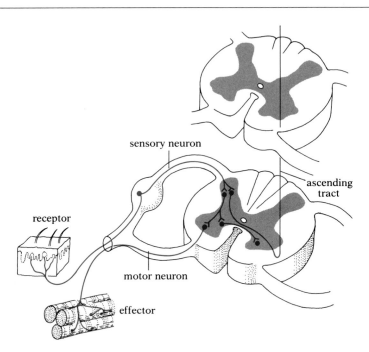

10.11   Diagram of a reflex arc with interneurons
The sensory neuron synapses with several interneurons in the gray matter of the cord. Some of these interneurons may synapse directly with motor neurons on the same side, but some cross to the other side of the cord and there synapse with other motor neurons and with additional interneurons that run in ascending tracts through the cord to the brain.

structural convenience. Thus the sensory and motor neurons of the knee-jerk reflex run through the same nerve, even though they carry impulses in opposite directions. A nerve containing both sensory and motor fibers is called a mixed nerve. All the nerves connected to the spinal cord are mixed nerves. There are 31 pairs in human beings; all of them branch repeatedly, giving rise to smaller nerves that innervate most parts of the body below the head. Some nerves connect directly to the brain rather than to the spinal cord; in humans there are 12 pairs of these cranial nerves, some purely motor, some purely sensory, and some mixed.

We have so far described reflexes in their simplest possible form; now we should re-examine them in their far more typical complexity. Very few reflex pathways involve only two neurons in series. At least one interneuron is usually interposed between the sensory neuron and the motor neuron (Fig. 10.11), and it is common for many interneurons to be involved even in relatively simple reflex arcs. It is important to remember that reflex arcs always interconnect with other neural pathways. For example, there are always interconnections with pathways leading to the brain, and the brain can send impulses that modify the reflexes. If you know the doctor is going to strike your knee, you can be ready for him, and consciously either partly inhibit the reflex or augment it.

Suppose you decide to try to inhibit the knee-jerk reflex. What do you do? You send many excitatory commands to the leg muscles that oppose, or antagonize, the extensor muscles. Thus, when your knee is struck and impulses are sent by way of the knee-jerk reflex arc to the extensor muscles, those muscles cannot bring about much jerk (extension) of the leg; there are too many other muscles pulling against them. But you can do more than this. You can send inhibitory commands to the motor neurons of the extensor muscles themselves. In other words, you can make it more difficult to stimulate the motor neurons to those muscles. This is an important point:

Neural commands are not all excitatory. In the nervous system as in the endocrine system, it is the integration of excitatory and inhibitory commands that makes possible precise control. Thus the normal knee-jerk reflex really has two components: stimulation of motor neurons leading to extensor muscles in the leg, and inhibition of motor neurons leading to flexor muscles in the leg, so that extensor muscles contract and flexor muscles at the same time partly relax.

Consider for a moment another spinal reflex. You are walking barefoot and you step on a thorn with your right foot. Impulses from pain receptors in your foot immediately ascend along sensory neurons to the spinal cord, pass along interneurons in the cord, and descend along motor neurons to the muscles of your leg. These descending impulses result in excitatory stimulation of the flexor muscles. At the same time, impulses moving along other interneurons in the cord exert inhibitory influence on the motor neurons to the extensor muscles of the leg. The result is that the flexor muscles contract, the extensor muscles partly relax, and your foot is quickly raised off the thorn. But if you raise your right foot, your left foot must bear the entire weight of your body. Some of the interneurons carrying impulses from the pain receptors in your right foot cross over to the left side of the spinal cord and synapse with motor neurons leading to your left leg. Excitatory impulses are sent to the extensor muscles of the left leg and the motor neurons to the flexor muscles are inhibited, with the result that the left leg remains firmly planted on the ground as the right leg is raised off the thorn. But balancing on one leg involves muscles in the sides of the leg in addition to those most concerned with extending and flexing it; the interneurons in the cord also synapse with neurons to these other muscles, eliciting commands that bring them into the proper functional relationship to the extensors and flexors as you throw your weight onto that leg.

Now, we have already said that interneurons from any reflex arc will synapse with other interneurons leading to the brain. Some of these interneurons carry impulses to parts of the brain concerned with conscious activity, and you become aware of what is happening. Other interneurons carry impulses to a part of the brain concerned with your general state of awareness, and you become more alert and "wide-awake." And still other interneurons carry impulses to the part of the brain called the cerebellum, telling it, in effect, what motor activities have been initiated. As the muscles of the legs start responding to the stimuli reaching them, stretch receptors in those muscles send impulses along sensory pathways to the spinal cord and up to the cerebellum, telling it how the muscles are actually responding. The cerebellum compares the information from these two sources, and if there is a discrepancy between what has been ordered and what is actually happening, it sends impulses to parts of the brain that function in modifying muscular activity.

We could continue adding layers of complexity almost indefinitely; we have mentioned here only a few of the many pathways stimulated by impulses coming to the cord from the pain receptors in your foot. Evidently a "simple reflex" is not really simple; even apparently automatic responses to stimuli may involve intricate and complex coordinating mechanisms. Our discussion may suggest how the linking together of sufficient numbers of neural pathways can result in the many complex behavior patterns of animals.

THE AUTONOMIC NERVOUS SYSTEM

The central nervous system serves as a coordinating system for two kinds of pathways: somatic and autonomic. **Somatic** pathways, exemplified by the reflex arcs discussed above, usually innervate skeletal muscle and include one sensory and one motor neuron, both lying largely outside the central nervous system. They involve, at least potentially, some conscious control of the reflex or an awareness that the reflex has occurred. **Autonomic** pathways, by contrast, are not ordinarily under the control of the will and usually function without our being aware of them. They innervate the heart, some glands, and the smooth muscle in the walls of the digestive tract, respiratory system, excretory system, reproductive system, and blood vessels. The pathways of the autonomic nervous system differ structurally from somatic pathways in having two motor neurons instead of one.

The autonomic nervous system (ANS) is separated into two parts, both structurally and functionally. These are called the sympathetic and the parasympathetic systems (Fig. 10.12).

In the **sympathetic system** the cell bodies of the first motor neurons lie in the thoracic and lumbar portions of the spinal cord. The axons of these neurons exit ventrally from the cord and run to ganglia lying near the cord, where they synapse with second motor neurons whose cell bodies lie in the ganglia. Thus the synapse between the first and second motor neurons occurs in a ganglion that is at a distance from the target organ, and the axon of the second motor neuron is quite long.

Two principal structural differences distinguish the **parasympathetic system** from the sympathetic system. First, the cell bodies of the first motor neurons of the parasympathetic system lie in the brain and in the sacral region of the spinal cord. Second, the synapses between first and second motor neurons of the parasympathetic system occur in the immediate vicinity of the target organs, or even inside those organs; the axon of the second motor neuron is thus relatively short.

Most internal organs are innervated by both sympathetic and parasympathetic fibers, with the two systems functioning in opposition to each other. Thus if the sympathetic system excites a particular organ, the parasympathetic system usually inhibits that organ, and vice versa. In general, the sympathetic system produces the same effects as the hormones of the adrenal medulla, i.e. the so-called fight-or-flight responses. Present evidence suggests that this nervous mechanism is far more important than the endocrine mechanism in preparing an animal for emergency situations. As for the effects produced by the parasympathetic system, you can easily figure them out for yourself by simply reversing the fight-or-flight responses. The condition of an organ innervated by the autonomic nervous system is determined at any given moment by the relative amounts of stimulation coming to it via each of the two parts of the system.

REFLEX CONTROL OF BREATHING AND HEARTBEAT

For two examples of reflex control at a somewhat more complex level than the reflexes already discussed, let us look briefly at the nervous control of

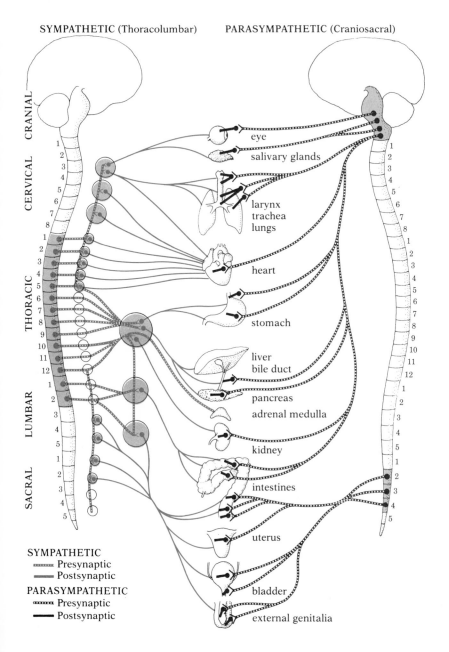

SYMPATHETIC (Thoracolumbar)   PARASYMPATHETIC (Craniosacral)

CRANIAL

CERVICAL

THORACIC

LUMBAR

SACRAL

eye
salivary glands
larynx
trachea
lungs
heart
stomach
liver
bile duct
pancreas
adrenal medulla
kidney
intestines
uterus
bladder
external genitalia

SYMPATHETIC
▭▭▭ Presynaptic
▬▬▬ Postsynaptic
PARASYMPATHETIC
▭▭▭ Presynaptic
▬▬▬ Postsynaptic

10.12   The autonomic nervous system

Of the 12 cranial and 31 spinal nerves, four cranial nerves (see gray area of brain at upper right, where they emerge) and about half the spinal nerves (see colored and gray segments, where they emerge from the cord) contribute neurons to the autonomic nervous system, which innervates internal organs.

The ANS is customarily divided into two parts: the sympathetic and the parasympathetic systems. The pathways of both usually have two motor (efferent) neurons; a first (presynaptic) neuron exits from the central nervous system and synapses with a second (postsynaptic) neuron that innervates the target organ.

The presynaptic neurons of the sympathetic system exit from the thoracic and upper lumbar regions of the spinal cord, and synapse with the postsynaptic neurons in a series of small ganglia (circles) lying near the cord or in larger ganglia in the abdominal cavity; the postsynaptic neurons then run from the ganglia to the target organs. The presynaptic neurons of the parasympathetic system exit from the medulla of the brain and from the sacral region of the spinal cord. These are very long neurons that run all the way to the target organ, where they synapse with short postsynaptic neurons.

Most but not all internal organs are innervated by both the sympathetic and the parasympathetic system.

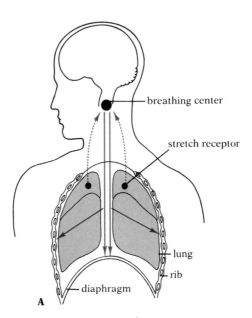
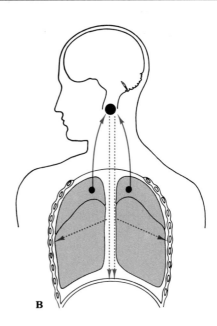

A

B

10.13 Nervous control of breathing

Solid arrows indicate pathways along which many impulses are moving; dashed arrows, pathways along which few impulses are moving.   (A) Impulses from the breathing center in the medulla of the brain go by motor neurons to the rib muscles and diaphragm. The muscles thus stimulated contract, moving the rib cage up and out and pulling the diaphragm downward. The resulting increase in volume of the thoracic cavity causes air to be drawn into the lungs in the process of inhalation.   (B) When the chest is expanded and the lungs are full of air, stretch receptors in the lungs are stimulated. These receptors send inhibitory commands to the breathing center and cause it to send fewer impulses to the rib muscles and diaphragm. When the muscles cease receiving excitatory commands from the breathing center, they relax, thereby reducing the volume of the thoracic cavity and expelling air from the lungs in the process of exhalation.   [Modified from G. G. Simpson, C. S. Pittendrigh, and L. H. Tiffany, *Life: An introduction to biology.* Copyright, © 1957, by Harcourt Brace Jovanovich, Inc., and reproduced with their permission.]

breathing, which involves somatic reflex arcs, and of heartbeat, which involves autonomic control.

Control of breathing   We saw in Chapter 6 that the effectors for the breathing movements are the muscles of the rib cage and diaphragm. These are skeletal muscles, and they can be controlled voluntarily to some extent. Ordinarily, however, the breathing movements occur automatically. The control center for these automatic movements is in a portion of the brain called the medulla oblongata, which is the portion of the brain to which the spinal cord attaches. When the breathing center in the medulla sends impulses to the muscles of the rib cage and diaphragm, those muscles contract, moving the rib cage up and out and pulling the diaphragm downward (see Fig. 6.18, p. 251). The resulting increase in volume lowers the pressure in the thorax, and air is drawn into the lungs. As the lungs expand, however, stretch receptors in their walls are stimulated, and they send impulses to the breathing center in the medulla (Fig. 10.13). These impulses result in inhibition of the breathing center, and it sends fewer excitatory commands to the muscles, with the result that the muscles relax and exhalation occurs. But when the muscles have relaxed, the stretch receptors are no longer stimulated and send fewer inhibitory commands to the breathing center. The breathing center, now free of inhibition, begins sending excitatory commands to the muscles of the rib cage and diaphragm, and inhalation occurs again. The interaction of the breathing center in the medulla with the stretch receptors in the lungs—another example of negative feedback—thus results in the alternation of inhalation and exhalation of the typical breathing cycle.

You ordinarily perform the rhythmic alternation of inhalation and exhalation without any conscious effort—in fact, without even being aware of it; the activity is automatic. It is clear, then, that somatic reflexes such as this one cannot be distinguished from autonomic reflexes simply on the basis of conscious or unconscious performance. The difference lies in the potential for conscious control of the effectors involved in the reflex. In this case the effectors are skeletal muscles, and you can exert some conscious control over them if you choose. For example, you can make other parts of your brain send inhibitory commands to the breathing center in the medulla, blocking the sending of excitatory commands to the muscles used in breathing. In other words, you can voluntarily hold your breath. But, as you know, you cannot hold your breath indefinitely; your conscious control of the breathing cycle is limited. This is because carbon dioxide (or, probably more correctly, the increased concentration of hydrogen ions that accompanies a rise in CO_2) acts as a powerful stimulant to the breathing center in the medulla. The longer you hold your breath, the higher the concentration of CO_2 (and H^+) in the cerebrospinal fluid of the brain rises, until eventually its excitatory effect on the breathing center surpasses the inhibitory effect of your conscious effort to cease breathing. When this happens, you must breathe again, despite your efforts not to do so.

It may seem strange that the index by which the breathing center determines when accelerated breathing is necessary should be the concentration of the waste product CO_2 rather than decreased concentration of oxygen.[2] But the effect is the same as though O_2 concentration were the determinant factor, for CO_2 production by the cells of the body is proportional to their consumption of O_2; thus increased concentration of CO_2 reliably reflects decreased concentration of O_2. This reliance of the breathing mechanism on CO_2 concentration does, however, create some curious problems in the artificial situations to which human beings are increasingly subjecting themselves. For example, if the ventilating system of a space ship failed to supply enough O_2 but removed CO_2 so fast that a concentration sufficient to stimulate the breathing center could not build up, the astronauts might actually black out from O_2 deprivation without ever having begun rapid breathing, because their breathing centers would not "know" that an increased breathing rate was necessary.

Control of rate of heartbeat   The nervous control of the rate of heartbeat provides a good example of autonomic control. Here the effector, which is heart muscle, is not potentially under conscious control (though you may consciously alter its rate of contraction indirectly, as by deliberately working yourself into an emotional or excited state).

As we saw in Chapter 7, nervous impulses are not necessary for the initiation of the heartbeat, such initiation being under the direct control of the S-A node in the wall of the heart itself. But the rhythm of the S-A node can be modified by impulses coming to it via two sets of nerves: a set of sympathetic nerves, which exert an excitatory effect on the S-A node, and a set of parasympathetic nerves, which exert an inhibitory effect. Although the sympathetic nerves to the heart issue from the central nervous system in the thoracic region of the spinal cord, the impulses they carry originate in a cardiac-accelerating center in the medulla of the brain. Similarly, the impulses carried by the parasympathetic nerves to the heart originate in a cardiac-decelerating center in the medulla.

[2] The concentration of O_2 does, in fact, exert some influence on respiratory rate in human beings, but the effect of CO_2 (and H^+) is so much greater that the O_2 concentration can be largely ignored under most circumstances.

If blood pressure is high, pressure receptors in the carotid artery of the neck and in the arch of the aorta (and to a lesser extent in other arteries) are stimulated. Impulses travel from them to the medulla, where the sympathetic pathways from the cardiac-accelerating center are inhibited and the parasympathetic pathways to the heart from the cardiac-decelerating center are activated. The result is that the heart rate is slowed. But as blood pressure falls, there is less stimulation of the pressure receptors, which therefore send fewer impulses to the medulla. The sympathetic pathways, freed of inhibition, begin carrying more impulses from the accelerating center to the S-A node, and the parasympathetic pathways begin carrying fewer impulses. This pattern of autonomic firing may be enhanced through stimulation of the accelerating center by impulses from chemoreceptors in the carotid artery and aorta, which respond to lowered O_2 and elevated CO_2 in the blood. It may also be enhanced through stimulation of the accelerating center by impulses from receptors in the wall of the right atrium, which respond to the filling of the atrium with blood. The result is acceleration of the heart rate.

The actual rate of heartbeat partly depends, then, on the relative activity of the accelerating and decelerating centers in the medulla, and the activity of these centers partly depends, in turn, on the amount of excitation they receive from the stretch receptors and the chemoreceptors in the arteries. We say "partly," because the two centers in the medulla are also significantly influenced by impulses from other parts of the brain. For example, when a person suddenly sees something frightening, his response, a quickening of the pulse, is prompted by impulses from his visual sense organs as relayed through higher centers of the brain to the medulla.

TRANSMISSION OF NERVOUS IMPULSES

The transmission of impulses along neural pathways may be as fast as 30–90 m/sec. In attempting to explain how nervous impulses are transmitted, we shall consider separately conduction along nerve cells and transmission across the synaptic gaps between cells.

CONDUCTION ALONG NEURONS

General features of the nerve impulse   Nerves will respond to a great variety of stimuli, such as a mild electric shock, a pinch, or an abrupt change of pH. Electrical stimuli are the most often used in laboratory experiments, for a number of reasons: The intensity and duration of such stimuli can be precisely measured; an electrical stimulus can be terminated almost instantaneously; and mild electrical stimuli do little if any damage to nerve cells.

Let us suppose that we are working with an isolated axon. We have touched two electrodes to the surface of the axon at points several centimeters apart. These electrodes are connected to recording equipment, which enables us to detect any electrical changes that may occur at the points on the axon with which the electrodes are in contact. Now we apply an extremely mild electrical stimulus. Nothing happens; our recording equipment shows no change. We increase the intensity of the stimulus and try again. This time our equipment tells us that an electrical change occurred at the point in contact with the first electrode and that a fraction of a second later a similar electrical change occurred at the point in contact with the second electrode. We have succeeded in stimulating the axon, and a wave of electrical change has moved down the axon from the point of stimulation, passing first one electrode and then the other. Let us suppose that our recording equipment has enabled us to measure the intensity of this electrical change. We next apply a still more intense stimulus. Again we record a wave of electrical change moving down the fiber, but the intensity and speed of this electrical change are the same as those recorded from the previous stimulation. Again we increase the intensity of the stimulus, but again the recorded change shows the same intensity and speed.

We have learned several important facts from this experiment: (1) A nervous impulse can be detected as a wave of electrical change moving along an axon. (2) A potential stimulus must be above a critical intensity (and duration) if it is actually to stimulate an axon; this critical intensity is known as the *threshold* value, and it differs for different neurons. (3) Increasing the intensity of the stimulus above the threshold value does not alter the intensity or speed of the nervous impulse produced; i.e. the axon fires maximally or not at all, a type of reaction commonly called an ***all-or-none response.***

Immediately an important question comes to mind. If an axon exhibits the all-or-none property with respect to intensity of impulse and speed of conduction, how do animals normally detect the intensity of a stimulus? They do this in several ways. First, an axon does not exhibit an all-or-none response with respect to frequency. The more intense the stimulus, the more frequent the impulses moving along the axon (up to a maximum value, of course). Second, because different neurons have different thresholds, a more intense natural stimulus ordinarily stimulates more neurons; individual neurons exhibit all-or-none properties, but nerves (which are composed of many neurons) do not. Apparently the brain interprets both a greater frequency of impulses coming to it via individual neurons and a greater number of stimulated neurons as indicating greater intensity of the stimulus.

The nature of the impulse   When it was discovered over a century ago that a nerve impulse involves electrical changes, scientists assumed that the impulse was a simple electric current flowing through a nerve, just as other currents flow through wires. It was soon shown, however, that the speed of a nerve impulse is far slower than the speed of electricity and, further, that the cytoplasmic core of nerve fibers offers so much resistance to simple electric currents that they die out after moving only a few millimeters through a nerve. The fact is that any resistance at all, however low, would cause a simple electric current to diminish in strength as it moved. Yet if we measure a nerve impulse at various points along the axon of a neuron, we find that it remains the same; its strength does not decrease with distance. Crushing or poisoning an axon may destroy its ability to conduct nerve impulses without appreciably altering its electrical conductivity. In short, evidence of many sorts has led to the conclusion that impulse conduction depends on activity by the living cell, activity that almost certainly involves chemical processes. The impulse, then, is not an electric current, but an electrochemical change propagated along the neuron.

The basic outlines of the modern theory of nerve action were proposed in 1902 by Julius Bernstein of the University of Halle in Germany. It was known

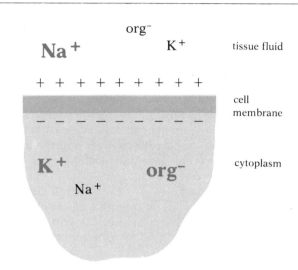

10.14   The polarization of the nerve-cell membrane
The concentration of sodium ions (Na⁺) is much
greater in the tissue fluid outside the cell, and the
concentration of potassium ions (K⁺) is much greater
inside the cell. There is also a much greater
concentration of negative organic ions (org⁻) inside the
cell. The net effect of these unequal distributions of
ions is that the inside of the cell is negative (about −70
millivolts) relative to the outside.

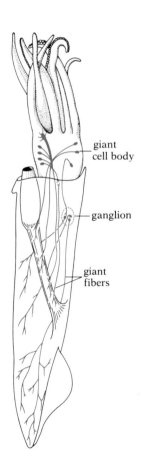

giant
cell body

ganglion

giant
fibers

10.15   Giant nerve fibers of a squid

by that time that the concentrations of certain ions inside nerve cells and in
the surrounding fluids are very different, the concentration of sodium (Na⁺)
being very low inside the cell and the concentration of potassium (K⁺) and
negative organic ions being very high (we saw in Chapter 8 that this is true to
some extent of most cells). It had also been shown that the unequal distribution of ions results in an electric potential difference across the cell
membrane in the resting state, the inside carrying a negative charge of about
−70 millivolts relative to the outside (Fig. 10.14). Bernstein suggested that
the permeability of the nerve-cell membrane varies for various ions, and
that it is the great selectivity of the membrane that maintains the separation
of ions and the resulting electric potential.

Bernstein further suggested that during the passage of an impulse the
selectivity of the membrane is momentarily destroyed. As a result, the ions
can move freely, and the electric potential difference across the membrane
falls to zero. In other words, the membrane is momentarily unable to maintain a separation of charge. To Bernstein this depolarization was basic to the
nerve impulse, which he saw as a wave of depolarization moving along the
nerve-cell membrane.

Although Bernstein's hypothesis was widely accepted by biologists, for
many years little experimental evidence either for or against it could be
obtained. Nerve fibers were simply too small for accurate measurements of
the changes taking place inside them during impulse conduction. At last, in
1933, J. Z. Young, then at Oxford University, discovered that squids possess
several giant nerve fibers, which may be as much as a millimeter in diameter. These fibers run in the body wall of the squid (Fig. 10.15) and innervate
muscles that enable the animal to propel itself backward at very high speed
by explosively expelling water from its mantle cavity through a funnel near
its head.

The great size of these squid nerve fibers makes them conduct impulses
very rapidly. In general, the greater the diameter of a nerve fiber, the faster it
conducts; conduction per unit length increases as the square of the diameter. Fibers larger than normal, though rarely as large as those of the squid,
are also found in many other invertebrates in parts of the body where very

rapid conduction is important; vertebrates have evolved myelinated fibers as an alternative adaptation for increasing conduction speed. Use of the giant nerve fibers of the squid, which finally made it possible for biologists to study in detail the events that occur during conduction, opened a whole new era of discovery in the important field of neurophysiology.

In 1939 H. J. Curtis and K. S. Cole in Woods Hole, Massachusetts, and A. L. Hodgkin and A. F. Huxley in Plymouth, England, developed a technique for inserting into a squid nerve fiber a very thin microelectrode consisting of a glass tube filled with salt solution or metal. The microelectrode was inserted into one end of the fiber and pushed down it a distance of 1–3 cm (Fig. 10.16). With the microelectrode in place, it could be shown that the electric potential across the nerve membrane did indeed change during the passage of an impulse, as Bernstein had predicted. But one thing was wrong: The potential changed too much; for an instant, the inside of the fiber became positive relative to the outside. The membrane did not simply become depolarized; it actually reversed its polarization momentarily. The conclusion had to be accepted, then, that stimulation does not destroy the selectivity of the membrane, as Bernstein thought, for had it done so the result would have been simple depolarization. Instead, it must radically alter this selectivity. In other words, the stimulated point on the membrane remains selective, but its selectivity is different. Hodgkin and Bernhard Katz proposed in 1947 that the membrane must initially allow far more positive Na^+ ions to enter the cell than it allows other ions to leave. In a long series of now classic experiments, Hodgkin and Huxley then went on to work out a detailed quantitative description of the ionic conductance changes (including those of both Na^+ and K^+) during impulse conduction in the squid. Their findings have been shown to be applicable to other animals as well.

Let us summarize the sequence of events during impulse conduction according to the Hodgkin-Huxley model:

The membrane of the resting neuron is polarized, with the inside negative relative to the outside. The concentration of Na^+ ions is much higher outside, and the concentration of K^+ ions is much higher inside. Stimulation causes the membrane to undergo an initial great increase in permeability to Na^+ ions. These rush across the membrane into the cell, both because of their natural tendency to diffuse from regions of their higher concentration to regions of their lower concentration and because they are attracted by the negative charge inside the cell. The inward flux of Na^+ is so great, however, that for a moment the inside actually becomes positively charged relative to the outside. A fraction of a second later, the membrane becomes highly permeable to K^+ ions, which rush out of the cell because their concentration is higher inside than outside and because they are repelled by the momentary high positive charge inside the cell. This exit of positively charged K^+ ions restores the electric charge inside the cell to its original negativity. In short, the inside surface of the membrane is initially negative; it becomes positive when Na^+ ions flood inward and then negative again when K^+ ions rush outward. It is this rapid cycle of changes, occurring at successive points in sequence, that constitutes the nerve impulse as it flows along a neuron.

The impulse is propagated along the neuron because the cycle of changes at each point alters the permeability of the membrane at the adjacent point and initiates a similar cycle there (Fig. 10.17). The effect is much like that of

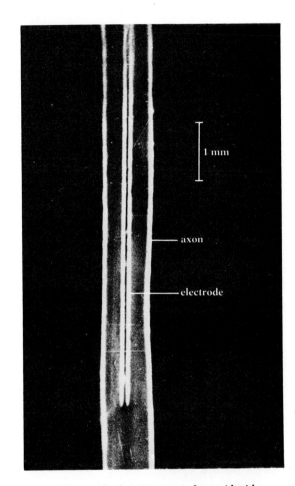

10.16 Photograph of a giant axon of a squid with a glass-tube microelectrode inside it
[Courtesy A. L. Hodgkin, *J. Physiol. (London)*, vol. 131, 1956.]

falling tenpins; when the bowling ball hits the first pin, that pin falls and strikes the next pin, which falls and hits the pin behind it, and so forth. If one is skillful or lucky, hitting one pin with the ball will cause all the other pins to fall in succession. In a nerve fiber stimulation at only one point produces reversal of polarization at that point and initiates a wave of polarization changes.

The Hodgkin-Huxley model of impulse conduction explains why the nerve impulse, unlike a simple electric current moving through a wire, does not decrease in strength as it moves along the fiber. The impulse is constantly being regenerated; the electrical change, or *action potential,* at each successive point is a new event, equal in magnitude to the similar events at the preceding points. An electric current is generated by a battery or dynamo and is simply carried passively by the wire, but a nerve impulse derives its energy from the path along which it travels, being generated anew at each successive point along the fiber. Myelinated fibers conduct impulses faster than nonmyelinated fibers because the action potential jumps from node to node along such a fiber instead of moving smoothly along its entire surface.

The sodium-potassium pump   If impulse conduction involves inward flow of Na$^+$ followed by outward flow of K$^+$, how does the neuron re-establish its original ionic balance? In other words, how does it get rid of the extra Na$^+$ and regain the lost K$^+$? If the initial ionic distribution were not restored, the

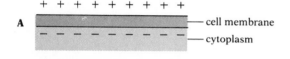

A

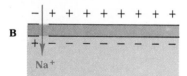

B

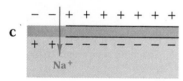

C

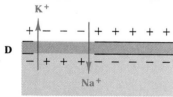

D

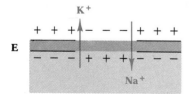

E

10.17 Model of propagation of a nerve impulse
(A) The interior of a resting nerve fiber is negative relative to the exterior because the ratio of negative to positive ions is higher inside the cell than outside. The interior has a high concentration of potassium ions (K$^+$), the exterior a high concentration of sodium ions (Na$^+$). (B) When a fiber is stimulated, the membrane, previously relatively impermeable to Na$^+$ ions, becomes highly permeable to them at the point of stimulation, and a large number rush into the cell (arrow). The result is a reversal of polarization at that point, the inside of the fiber becoming positive relative to the outside. (C) Meanwhile, because the change in membrane permeability at the point initially stimulated has altered the permeability at adjacent points, Na$^+$ ions begin rushing inward at those points. (D) An instant later, the membrane at the initial point of stimulation becomes highly permeable to K$^+$ ions; a large number of these rush out of the cell, and the inside of the fiber once again becomes negative relative to the outside. (E) The cycle of changes at each point alters the permeability of the membrane at adjacent points and initiates the same cycles of changes there; Na$^+$ ions rush into the cell, and K$^+$ ions rush out a moment later. The movement of this cycle of changes along the nerve fiber is what we call a nerve impulse.

neuron would eventually lose its ability to conduct impulses, but we know for a fact that a normal neuron can continue to conduct impulses indefinitely, with only a very brief refractory period (on the order of 0.5–2 milliseconds) after each impulse.

Since expelling the Na^+ means making it move against the concentration and electrostatic gradients, and since regaining the lost K^+ means making it move against its concentration gradient, some form of active transport across the membrane must be involved. It has been demonstrated that the membrane of the neuron incorporates an ATP-driven sodium-potassium exchange pump of the sort described in Chapter 8. Such a pump, as we saw in that chapter, renders many cells capable of active extrusion of Na^+ ions and active intake of K^+ ions. The neuron, then, is a type of cell that has evolved an extraordinarily effective version of the ionic pump so widely found in other cells. The development of such a pump, combined with the evolution of extreme susceptibility to induction of membrane-permeability changes by external stimuli, has been the basis for the high degree of specialization attained by neurons for the conduction of impulses.

A powerful sodium-potassium pump would explain how the cell can maintain its low concentration of Na^+ ions and its high concentration of K^+ ions. But if the pump simply exchanged positive Na^+ ions for positive K^+ ions, i.e. if it were electrically neutral, it would not give rise to a potential difference across the membrane. You will recall, however, that the sodium-potassium pump is not, in fact, electrically neutral; according to current models, it exchanges Na^+ and K^+ ions according to a 3:2 ratio; i.e. for each three Na^+ ions pumped in one direction, only two K^+ ions are pumped in the other direction (see p. 332). Now, if the pump in the membrane of a neuron pumps three Na^+ ions out of the cell for every two K^+ ions pumped in, the net effect is a flow of positive charge out of the cell. In other words, the pump is electrogenic; it will build up a separation of charge across the membrane.

A second factor contributes to maintenance of polarity across the nerve-cell membrane: The membrane of the resting cell is about 50 times more permeable to K^+ than to Na^+. Hence K^+ leaks back out of the cell faster than Na^+ leaks into the cell. This excess of outward-leaking positive K^+ ions over inward-leaking positive Na^+ ions results in a net outward movement of positively charged ions. There can be no correspondingly large outward movement of negatively charged ions, since the organic ions, which account for most of the negative charge inside the cell, cannot cross the membrane. Thus a separation of charge could be maintained even if the electrogenic sodium-potassium pump were inoperative for a short time. Note, then, that under normal conditions, polarization of the neuronal membrane depends on three interacting factors: the metabolically driven sodium-potassium pump, the greater leakage of K^+ ions than of Na^+ ions, and the inability of the negative organic ions to leave the cell.

TRANSMISSION ACROSS SYNAPSES

The nature of synaptic transmission   We have said that the axon of one neuron usually synapses with the dendrites or cell body of other neurons. Since the terminal portion of an axon ordinarily branches repeatedly, a single axon may synapse with many other neurons, and it usually synapses at numerous points with each of these neurons (Fig. 10.18). Each tiny branch

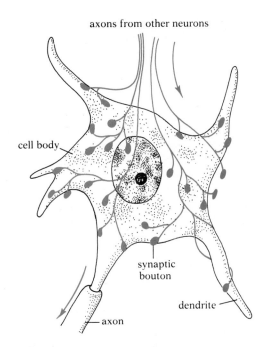

10.18   Synapses on a motor neuron
Many axons, each of which branches repeatedly, synapse on the dendrites and cell body of a single motor neuron. Each branch of an axon terminates in a swelling called a synaptic bouton.

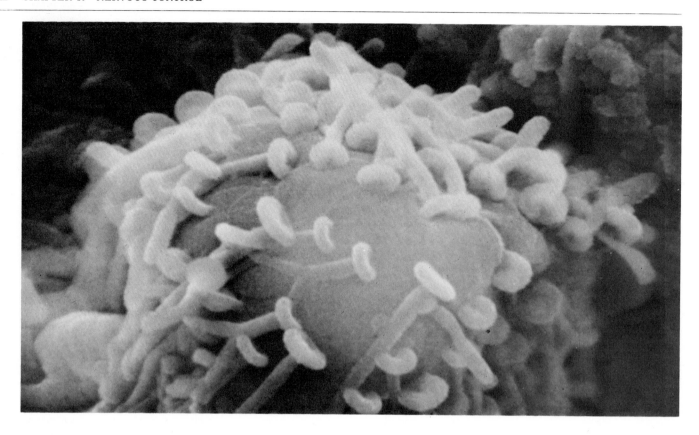

10.19   Scanning electron micrograph of synaptic boutons from a sea hare (*Aplysia californica*)
The synaptic boutons terminating the numerous axons are in contact with the cell body of a postsynaptic neuron. Note that it is the edge of the bouton, not its flattened end, that characteristically forms the synapse. × 14,450.   [Courtesy E. R. Lewis *et al.*, *Science*, vol. 165, 1969. Copyright 1969 by the American Association for the Advancement of Science.]

of an axon usually ends with a small swelling called a ***synaptic bouton*** (or sometimes a synaptic knob) (Fig. 10.19).

In a few cases there is a gap junction (see p. 227) between the membrane of the bouton and the membrane of the adjoining cell. Such a junction permits direct electrical coupling between the two neurons; i.e. an impulse traveling down the axon of the first neuron can pass unhindered to the next neuron. Since electrical synapses obviate delay in the transmission of impulses, they tend to occur in places in the nervous system where speed of conduction is of special importance. They also provide a high degree of certainty that an impulse in the first neuron will give rise to an impulse in the second neuron.

The vast majority of synapses, however, are not electrical, but chemical. A space about 20 nm wide, called the synaptic cleft, occurs between the bouton of the first (presynaptic) neuron and the membrane of the second (postsynaptic) neuron. Transmission across this cleft is by a diffusible ***transmitter*** chemical released from tiny ***synaptic vesicles*** in the bouton of the presynaptic neuron (Fig. 10.20). When an impulse traveling down the axon of the presynaptic neuron reaches the bouton, the bouton membrane becomes more permeable to calcium ions (Ca^{++}), which then diffuse into the bouton from the surrounding fluid. The Ca^{++} ions in some way stimulate synaptic vesicles in the bouton to move to the terminal membrane of the bouton, fuse with it, and then rupture, releasing transmitter chemical into the synaptic cleft. As we shall see in Chapter 11, Ca^{++} ions play an analogous role in triggering muscle contraction and ciliary action.

The transmitter molecules released into the cleft diffuse across it and bind by weak bonds to highly specific receptor sites on the ***postsynaptic membrane*** of the next neuron; the receptors are thought to be proteins forming an integral part of the membrane. The binding of transmitter to receptor results in alteration of the neuron's membrane potential, and a new impulse may be generated in that cell.

The fact that transmission depends on an influx of Ca^{++} into the bouton, followed by movement of the vesicles and then diffusion of transmitter across the cleft, means that synaptic transmission is much slower than impulse conduction along the neurons. For this reason the time it takes a message to traverse a neural pathway is always longer than would be calculated on the basis of the length of the pathway and the speed of conduction along the fibers. In general, the more synapses in a neural pathway, the slower the average speed of transmission per unit distance along the pathway.

It is the chemical synapses that make transmission of impulses along the neural pathways of higher animals one-way. An individual neuron can conduct impulses in both directions. If, for example, we stimulate an axon at a point between its base and its terminus, an impulse will move in both directions along the axon from the point of stimulation. But the impulse moving back toward the cell body and dendrites will die when it reaches the end of the cell; it cannot bridge the gap to the next cell.

For synapses between neurons outside the central nervous system, the transmitter chemical is ***acetylcholine.*** After this chemical has diffused across the synaptic cleft and exerted its effect at receptor sites on the postsynaptic membrane of a dendrite or the cell body of the next cell, it is promptly destroyed by an enzyme called ***cholinesterase.*** This destruction is of critical importance; if the acetylcholine were not destroyed, it would continue its stimulatory action indefinitely and all control would be lost. In fact, many insecticides, such as the organophosphates (also known as nerve gases), are cholinesterase inhibitors. They block destruction of acetylcholine, with the result that the animal's nervous system soon runs wild; the insect goes into uncontrollable tremors and spasms, and death ensues.

Besides serving as the synaptic transmitter in the peripheral nervous system, acetylcholine is one of a number of transmitter chemicals identified in the central nervous system. Others include ***noradrenalin*** (a substance also produced as a hormone by the adrenal medulla), ***serotonin, dopamine,*** and gamma-aminobutyric acid (***GABA***). A variety of other substances, too, are currently being investigated as likely transmitters.

The action of transmitter substances   Let us now examine more closely the effects of transmitter substances on postsynaptic membranes. When such a substance has diffused across the synaptic cleft, how does it affect the polarization of the postsynaptic membrane of one of the dendrites or the cell body of the next neuron? Let us suppose, first, that we are dealing with an excitatory transmitter substance. Apparently the binding of such a transmitter to receptor slightly increases the permeability of the postsynaptic membrane to Na^+ ions. The resulting increased inward flow of Na^+ ions

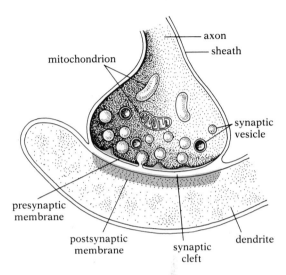

10.20   The synapse
Each synaptic bouton at the end of an axon encloses numerous synaptic vesicles containing transmitter substance. When vesicles release this substance into the synaptic cleft, the substance diffuses across the cleft and alters the polarization of the postsynaptic membrane of the dendrite or cell body of the next cell.

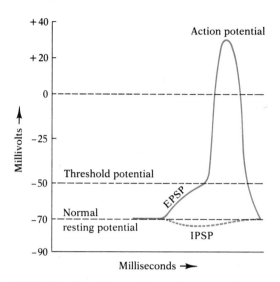

10.21   The effect of transmitter substance on the membrane potential of a neuron

The normal resting potential of a typical neuron is about −70 millivolts. An excitatory transmitter substance slightly reduces that polarization (i.e. makes the inner surface of the membrane less strongly negative), thereby creating an excitatory postsynaptic potential (EPSP). If the EPSP reaches the threshold level, which is usually about −50 mV, an impulse is triggered; a sudden inrush of Na^+ ions causes the inside of the cell to become positive (the action potential). A fraction of a second later, K^+ ions rush out of the cell, and the inside of the cell again becomes negative.

If the transmitter substance had been inhibitory, the membrane would have become hyperpolarized (to perhaps −75 mV), a condition called an inhibitory postsynaptic potential (IPSP) (dashed curve), and no action potential would have resulted; the neuron would slowly have returned to its resting potential after release of the transmitter had ceased.

slightly decreases the polarization of the neuron; i.e. the inside becomes less negative relative to the outside, a condition known as an excitatory postsynaptic potential (**EPSP**) (Fig. 10.21). If the EPSP is sufficiently great, it may spread to the base of the cell's axon (the axon hillock) and there trigger an impulse, which will move down the axon to the next synapse. What happens, in effect, is that excitatory transmitter substances produce a short circuit in the postsynaptic membrane potential, and this triggers the nerve impulse.

The transmitter substances released by the axons of some neurons have the opposite effect—an inhibitory one. They increase the polarization of the postsynaptic membrane and thus make the neuron harder to fire. Apparently these substances produce their inhibitory effects by making the postsynaptic membrane more permeable to chloride (or sometimes to K^+). If more chloride ions enter the cell (or if more K^+ ions leave), the membrane becomes more polarized; i.e. the inside of the cell becomes more negative relative to the outside, a condition known as an inhibitory postsynaptic potential (**IPSP**). More than the usual number of excitatory impulses would be needed to reduce the polarization of such an inhibited neuron to the threshold level for triggering an impulse.

How synaptic transmitter substances exert their effects on the permeability of the postsynaptic membrane is still a matter of speculation. It has been hypothesized that the binding of excitatory transmitter causes a conformational change in the receptor protein (or some other membrane protein associated with it) that opens a channel in the membrane through which Na^+ and K^+ ions can pass. Similarly, the binding of inhibitory transmitter may open a channel through which chloride ions (or K^+ ions) can pass. Note that it must be the nature of the receptor (or its relationship to other components of the membrane) that determines whether the binding of transmitter has an excitatory or an inhibitory effect, because a single transmitter (e.g. acetylcholine) may be excitatory at some synapses and inhibitory at others.

Both the excitatory and the inhibitory synapses so far discussed act on the postsynaptic cell in influencing whether or not it will fire an impulse. There is now abundant evidence that some synapses are also subject to **presynaptic inhibition.** In such cases axonic terminals of inhibitor interneurons impinge on the boutons of the presynaptic cell (Fig. 10.22) and act to reduce the number of vesicles of the presynaptic cell that release transmitter. Presynaptic inhibition has the advantage of reducing excitatory input to the postsynaptic cell from one particular source without altering the responsiveness of the cell to other inputs.

The integrative function of neurons   Synapses are points of resistance in nervous pathways. Thus an impulse may travel to the end of the axon of one neuron but die there, because not enough excitatory transmitter is released to initiate an impulse in the next neuron of the pathway. The transmitter molecules released as a consequence of the impulse do slightly decrease the polarization of the postsynaptic neuron (i.e. they do produce a slight EPSP), but not enough to cause the neuron to fire. Ordinarily, excitatory transmitter from many different synapses must impinge on the neuron within a short space of time if a sufficient EPSP is to be built up to trigger an impulse. These cooperating synapses may all be located at the terminals of branches of a single axon, or they may be at the terminals of the axons of several different neurons. The transmitter molecules from each individual synapse produce a slight EPSP; acting together, the transmitter molecules from all the synapses produce a large enough resultant EPSP to pass the threshold level and trigger an impulse. This additive phenomenon is called **summation.** Summation may be temporal as well as spatial; if new transmitter molecules arrive before the effects of previous transmitter molecules have disappeared, the effects may cumulate and trigger an impulse.

But summation is algebraic; if both excitatory and inhibitory transmitters impinge on a single cell at the same time, then their effects are added according to sign. Each excitatory molecule causes a slight decrease in polarization (plus), bringing the cell closer to the threshold potential, and each inhibitory molecule causes a slight increase (minus), removing it further from the threshold potential. The net result—an EPSP making the cell more likely to fire, or an IPSP making it less likely to fire—depends, then, on the cell's combining all the excitatory and inhibitory information it is receiving. The cell's response is not to any single unit of incoming information, but to the whole pattern of information (both spatial and temporal) impinging on it (Fig. 10.23).

Another name for this process is **integration.** The cell integrates all the signals that converge on it (which, in the case of an interneuron or a motor neuron, may be coming from thousands of different interneurons or sensory neurons), and either fires an impulse or remains silent (thereby blocking the signals). In short, all the incoming information is integrated to produce a simple yes-no decision by the cell. It is presumably by combining many such yes-no decisions at a host of different cells that the nervous system processes the information it receives.

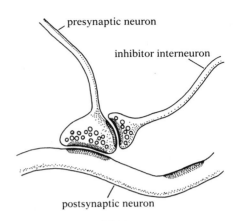

10.22 Presynaptic inhibition
An inhibitor interneuron forms an axo-axonic synapse on the bouton of the presynaptic neuron. When the inhibitor cell releases its transmitter substance (which is often GABA), fewer vesicles in the bouton of the presynaptic neuron will release their own transmitter substance, with the result that the postsynaptic neuron receives less stimulation from this particular pathway. Unlike postsynaptic inhibition, which reduces the responsiveness of the postsynaptic neuron to all excitatory inputs, presynaptic inhibition leaves unchanged the responsiveness of the postsynaptic neuron to inputs from other sources.

10.23 The integrative function of a neuron
The neuron receives both excitatory synapses (arrows) and inhibitory synapses (barred lines) from many different sources. The synapses vary in their distance from the base of the axon (called the axon hillock), where impulses are generated. Whether or not the neuron will fire an impulse is determined at any given moment by the algebraic sum of all the individual EPSPs and IPSPs arriving at the axon hillock from the various synapses. But unlike an axonic impulse, which shows no decrement with distance, depolarizations (EPSPs) or hyperpolarizations (IPSPs) decrease in magnitude as they spread along a dendrite or the cell body. Hence impinging interneurons such as **a** and **b,** which synapse near the axon hillock, can more easily influence the neuron's firing than interneurons such as **c** and **d,** which synapse on the cell at a distance from the hillock; only if these latter interneurons fire at a very high rate are they likely to have a major effect. In short, the geometry of the synapses on a neuron biases the integration process; inputs from some interneurons are given greater weight than inputs from other interneurons.

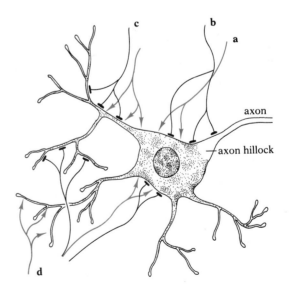

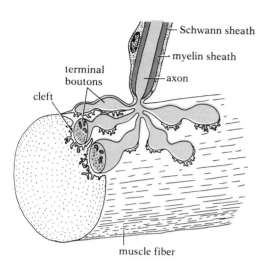

10.24 Neuromuscular junction
The end of the axon and the specialized adjacent portion of the muscle fiber together constitute the neuromuscular junction. As in a synapse between two neurons, there is a cleft between the two cells.

Synapses play an extremely important role in determining the routes impulses will follow through the nervous system. They are, in effect, the regulatory valves of the nervous system. It has been suggested that learning may involve the potentiation of certain synapses as a result of their use. With the synapses broken in, so to speak, later impulses would encounter less synaptic resistance if they followed the same pathways as earlier impulses.

Because synapses act as control valves in the nervous system, and because their proper function depends on a very delicate balance between transmitter substance, deactivating enzyme, and membrane sensitivity, it is not surprising that synaptic malfunctions have been implicated in several mental disorders—among them schizophrenia—or that many neurological drugs exert their effects at synapses. Used with caution and under medical supervision, some of these drugs can give relief from anxiety and tension or from neurological diseases involving biochemical lesions of synapses. But used improperly, the same agents can induce symptoms strikingly reminiscent of those seen in certain mental disorders, and in some cases the symptoms may be long-lasting or even permanent.

Neurological drugs can alter synaptic function in a variety of ways. On the one hand, they may turn off certain synapses by (1) interfering with synthesis of the appropriate transmitter substance; (2) blocking uptake of the transmitter into the synaptic vesicles; (3) preventing release of the transmitter from the vesicles; or (4) blocking the receptor sites on the postsynaptic membranes, so that the transmitter has no effect even if released. On the other hand, drugs may induce excessive and uncontrolled firing of postsynaptic cells by (1) stimulating massive release of transmitter substance from the vesicles; (2) mimicking the effect of the transmitter; or (3) inhibiting destruction of the transmitter once it has done its job, like the insecticides mentioned previously.

The physiological mode of action of a drug may help explain the behavioral symptoms it induces. Consider amphetamine (of which a well-known form is Dexidrin) and reserpine. Amphetamine, a stimulant, causes increased release of noradrenalin in the brain. Reserpine, a tranquilizer, blocks uptake of noradrenalin into synaptic vesicles and hence prevents its release. Thus the contrasting behavioral symptoms induced by these drugs are correlated with their opposite effects on some of the same synapses.

Recent research has provided partial explanations for the action of some other important neurological drugs. Nicotine acts as a stimulant because it mimics the effect of acetylcholine. Chlorpromazine, a commonly used tranquilizer, inhibits transmission of impulses at both acetylcholine- and noradrenalin-mediated synapses by combining with receptor sites on the postsynaptic membranes and thus blocking the transmitter substances. LSD (lysergic acid diethylamide) produces its characteristic derangement of certain mental functions by combining indiscriminately with receptor sites for serotonin. The mode of action of marihuana *(Cannabis)* remains unknown.

TRANSMISSION BETWEEN THE MOTOR AXON AND THE EFFECTOR

Just as there is a gap at the synapses between successive neurons in a neural pathway, there is also a gap between the terminus of an axon and the effector it innervates. When the effector is skeletal muscle, the gap is usually located within a specialized structure, the **neuromuscular junction** (or motor

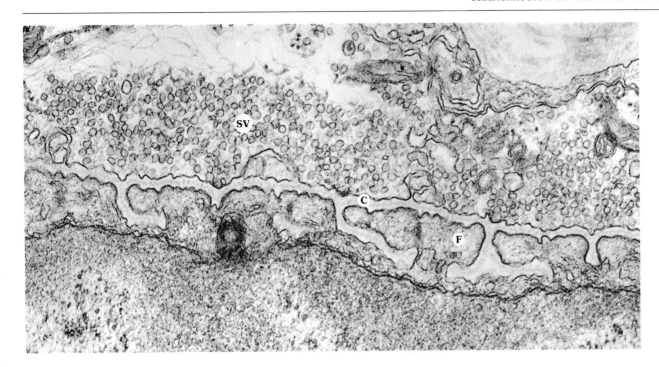

end plate), formed from the end of the axon and the adjacent portion of the muscle surface (Figs. 10.24, 10.25). Transmission across this gap is via transmitter chemicals, just as in synapses between neurons. Indeed, the process is so similar that the junction between a neuron and a muscle is often called a synapse; however, we shall reserve the term "synapse" for junctions between neurons.

The transmitter at neuromuscular junctions of somatic pathways is acetylcholine. Acetylcholine is also released by parasympathetic neurons at their junctions with effectors. But sympathetic motor fibers release noradrenalin (or, in a few cases, adrenalin). The fact that parasympathetic and sympathetic terminals employ different transmitter chemicals helps explain why these two components of the autonomic nervous system produce different responses by effectors.

We have said previously that the sympathetic nervous system and the hormones of the adrenal medulla have amazingly similar effects on the body, both eliciting the so-called fight-or-flight syndrome of responses. It now becomes clear why their effects are so similar: They release the same chemicals, and the reason they do apparently lies in a fascinating functional and evolutionary relationship between them. As we have seen, autonomic pathways typically include two motor neurons. There is one exception to this rule. The sympathetic pathway to the adrenal medulla has only one motor neuron (see Fig. 10.12). Apparently the adrenal medulla forms from presumptive nervous tissue (i.e. tissue destined to become nerve) in the embryo, and is actually itself the highly specialized second motor neuron of this sympathetic pathway. In other words, evolution has converted what once was a motor neuron of the sympathetic nervous system into an endocrine gland specialized for secretion in quantity of the same substances that all second motor neurons of the sympathetic system secrete.

10.25 Electron micrograph of portion of a neuromuscular junction of a frog

The upper half of the micrograph shows part of the terminus of an axon containing numerous synaptic vesicles (SV). The lower half shows part of a cell of the cutaneous pectoris muscle. There is a distinct cleft (C) between the two cells. Note the prominent junctional folds (F) on the surface of the muscle cell. × 43,000. [Courtesy A. W. Clark, *J. Cell Biol.*, vol. 69, 1976.]

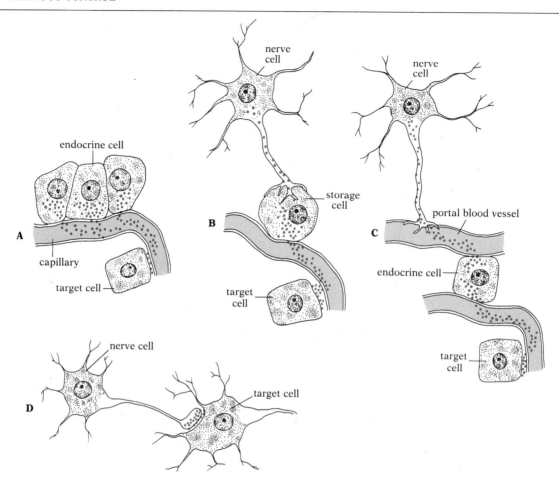

10.26   Four arrangements for chemical control compared

(A) A typical endocrine cell secretes its hormone directly into the blood, which carries the hormone to the target cells.   (B) Some hormones secreted by nerve cells in the hypothalamus are stored in cells of the posterior pituitary, and later released into the blood.   (C) Releasing hormones secreted by nerve cells in the hypothalamus are carried by the blood in a special portal system to endocrine cells of the anterior pituitary, which respond by secreting their own hormones into the general circulation.   (D) A more typical nerve cell secretes its transmitter substance directly onto the target cell; no transport by the blood is involved.

Here, then, is another example of the close interrelationship between the nervous system and the endocrine system, and of the resemblance between their mechanisms of action (Fig. 10.26). Neurosecretory activity like that of the hypothalamus in its interaction with the pituitary, and like that of the insect brain, which produces the hormone that regulates the prothoracic gland, is not an unusual or isolated phenomenon. Neurosecretion is fundamental to nerve action. Impulse transmission across all gaps in neural pathways depends on it. It is not surprising, therefore, that natural selection has sometimes given nervous tissue endocrine functions, and has given chemicals secreted by nervous tissue the properties of hormones.

SENSORY RECEPTION

Specialized receptor cells are the body's principal means of gaining information about the surrounding environment. They are the first elements in the reflex arcs whose structure and function we have been discussing.

Receptors may be either portions of nerve cells, as are many sensory endings in the skin; or they may be specialized cells in intimate contact with nerve cells, as are the taste cells of the tongue. In general, each type of receptor is responsive to a particular kind of stimulus: stretching, pressure, heat or cold, certain kinds of chemicals, vibrations, light. Most receptors will not respond to stimuli other than those for which they are specialized. Each type of receptor functions as a transducer, converting the energy that constitutes the particular stimulus to which it is attuned into the electrochemical energy of the nerve impulse.

MECHANISM OF RECEPTOR FUNCTION

By what process do the environmental phenomena that constitute stimuli cause receptors to initiate impulses? We have seen that excitatory transmitter substances at synapses induce impulses in the postsynaptic neuron by reducing the polarization of the membrane of that neuron to a critical level, i.e. by inducing an EPSP that reaches the threshold potential for impulse generation. A similar process is involved in the case of sensory receptors; the stimulus causes sufficient depolarization of the membrane of the receptor cell to cause it to initiate an impulse.

In 1950 Bernhard Katz of University College, London, demonstrated that the stretching of a muscle spindle produces a local depolarization of the receptor cell. When the depolarization, called the *generator potential,* reaches a threshold level, it triggers an action potential in the nerve fiber (Fig. 10.27). An increase in the intensity of the stimulus (in this case, increased stretching) causes a proportional rise in the generator potential. This rise, in turn, and the rate at which it occurs, determine the frequency of the triggered impulses. Thus the output from the receptor conveys a measure of the strength of the stimulus.

The basic question of sensory reception is, of course: How does the stimulus produce the generator potential? Since the generator potential is the result of partial depolarization of the cell membrane (i.e. a reduction of the negativity at the inner surface of the membrane), we would predict—reasoning from the analogy of synaptic stimulation—that the membrane becomes more permeable to Na^+ ions, which thus flow inward. The question then becomes: How does the stimulus increase the permeability of the membrane to Na^+ ions?

For the present, at least, only tentative answers can be given. Since different receptors are stimulated by different stimuli, the mechanism may well vary. In the case of vision, light energy is known to cause chemical changes in receptor pigments, and these changes are presumed to initiate chemical reactions that produce some sort of transmitter substance. The transmitter substance, in turn, acting in about the same way as synaptic excitatory transmitter substances, may depolarize the membrane of the next cell in the pathway. In the case of stretch receptors and pressure receptors, mechanical distortion of the membrane is thought to produce the permeability changes directly, either by opening channels through the membrane that allow Na^+ ions to leak inward, or by increasing the size of channels otherwise too small for the passage of Na^+ ions. The direct relationship between strength of stimulus and magnitude of generator potential could be explained by assuming that stronger stimuli distort a greater area of membrane, thereby opening more channels and allowing more ions to pass through the membrane.

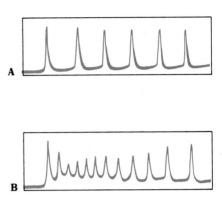

10.27   Relation of impulse pattern to intensity of stimulation of a stretch receptor

(A) A neuromuscular spindle is stretched slightly. The stimulus thus applied produces a generator potential (shown here as an upward shift in the base line) that triggers a series of impulses (action potentials).

(B) The same neuromuscular spindle is stretched more. The generator potential (base line) rises higher, and the frequency of the impulses increases.   Thus the frequency of impulses is seen to be a function of the intensity of the generator potential, which in turn is a function of the strength of the stimulus.   [Redrawn from B. Katz, *J. Physiol.* (*London*), vol. 111, 1950.]

SENSATIONS

The sensations you experience when touch receptors are stimulated are very different from those you experience when cold receptors are stimulated. Yet both types of receptors send the same sort of message to the central nervous system; in each case the message consists of action potentials moving along neurons, the intensity of the stimulus being signaled by the frequency of impulses along individual neurons and by the number of parallel neurons stimulated. How, then, do these messages, which qualitatively are essentially identical, give rise to the entirely different sensations of touch and cold? The difference lies neither in the receptors nor in the messages they send, but in the destinations of the messages in the brain. The sensation is not in any way an inherent property of the stimulus or of the neural message; it is a creation of the brain.

Each type of receptor sends impulses to a particular part of the brain; cold receptors, touch receptors, and pain receptors each send impulses to a different part of the brain. You experience the sensation of cold whenever excitatory impulses arrive at a part of the brain concerned with cold, and you experience the sensation of pain whenever excitatory impulses arrive at a part of the brain concerned with pain. It doesn't matter where the impulses originate or what stimulus initiates the impulses. It only matters what part of the brain is stimulated.

Normally, of course, impulses from a given receptor go to the appropriate part of the brain, because of the way the nerve circuitry is organized. But the circuitry can be changed experimentally. Thus if the fibers leading from one of your touch receptors and from one of your cold receptors were cut and rerouted, so that impulses from the touch receptor would go to the cold area in the brain and vice versa, every time you were touched and a touch receptor fired, you would experience the sensation of cold, and every time a cold object stimulated a cold receptor to fire, you would experience the sensation of touch. In other words, you would experience a sensation "inappropriate" to the stimulus, because the impulses would be going to the wrong part of the brain. But note that this inappropriateness would not change the quality of the sensation; the sensations would be the normal ones of touch and cold, even though they were in response to the wrong stimuli. The sensation is simply the brain's interpretation of incoming stimuli, and depends on the part of the brain stimulated. As a dramatic illustration of this point, consider that, if nerve fibers from your eyes could be crossed with nerve fibers from your ears, you would hear lightning and see thunder.

It is also the brain that is responsible for the localization of the sensation. Each part of the body has its own sensory area in the brain. Thus fibers from your big toe run to one area in your brain, fibers from your ankle run to another area, and fibers from the thumb run to still another. Again, it is the part of the brain to which the impulses go, not the stimulus or the receptor or the message itself, that determines localization of the sensation. If the fibers from the pain receptors in your big toe were crossed with those from your thumb, and if your big toe were then pricked with a needle, you would experience a sensation of pain in your thumb and would promptly examine your thumb for the cause of the trouble. In other words, the pain sensation is a creation of your brain, and it exists only in your brain, but it is referred by the brain to some other part of the body, where it then seems to you to exist.

Let us cite an actual experiment that makes this point with particular force. A group of scientists cut and crossed the fibers from pain receptors in the right and left rear feet of a white rat. They then made a small wound on the bottom of the right rear foot. Each time the rat put its weight on this foot the pain receptors there fired. But the impulses from these receptors were carried by the altered nerve circuitry to the part of the brain specialized as the sensory area for the left rear foot. The rat's brain, therefore, interpreted the signals as coming from the left rear foot, and the animal began to walk on three legs, keeping the left rear foot raised off the ground. This, of course, only intensified the stimulation of the pain receptors in the wounded right foot, because now that foot had to bear more weight than before. Presumably the rat thereupon experienced an even more intense sensation of pain in the left foot, and it raised that foot higher and higher, all to no avail. No matter how long an experiment such as this was continued, the rat never learned that raising its right foot would alleviate the pain. Its central nervous system continued to interpret the pain as coming from the left foot. A human being in such a situation would eventually learn to raise the right foot; it would still seem to him that the pain was in the left foot, but he could consciously interpret pain in the left foot as a signal to raise the right foot. He could, in other words, alter his response by an effort of will.

Another striking illustration of the fact that both the quality and the localization of sensation are determined exclusively by the brain is the phenomenon of so-called phantom-limb pains. People who have had an arm or a leg amputated sometimes complain of pains in the limb that isn't there. Apparently the stumps of the nerve fibers that formerly ran from the amputated limb to the brain become irritated in some manner and send impulses to the brain. Since those impulses go to the sensory area of the brain concerned with the amputated limb, the impulses are interpreted by the brain as coming from the lost limb. That the person knows his left arm, say, is missing doesn't alter the fact that the sensation he experiences is one of pain in the left arm. His sensation is presumably the same as might be experienced by a person whose left arm is actually hurt.

Notice that in our discussion of the rat whose nerve circuitry was altered we said that it "presumably" experienced a sensation of pain. The word "presumably" is very important here. We cannot be sure whether the rat experienced any sensation at all, much less what sort of sensation. Remember, sensation is a conscious experience. Nervous function that does not involve consciousness (autonomic pathways, for example) can produce automatic reactions to stimuli, but cannot produce sensation. No sensation is experienced when a receptor in an artery is stimulated by a high concentration of carbon dioxide in the blood, but the body responds to this stimulation. We have no way of knowing whether a rat has any such thing as consciousness; hence we cannot know whether it experiences sensation. We know that it responds to stimuli, and we know that many of its responses are of a kind we tend subjectively to identify with consciousness, but this is not the same thing as an objective proof of consciousness. Indeed, many scientists have insisted that, since we cannot prove the existence of consciousness in other animals, we should assume that it is absent. To many biologists,

A

Free nerve ending (pain)

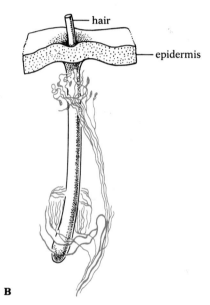

— hair

— epidermis

B

Nerve plexus around hair (touch)

C

Pacinian corpuscle (deep pressure)

10.28   Three receptors of the skin and the senses they mediate

however, particularly in recent years, it has seemed more consistent with the known facts about nervous systems and behavior to assume that at least some other animals besides humans are capable of some conscious activity. But no biologist would attribute to other animals the same degree or the same type of consciousness as that of human beings. Other animals have brains that differ markedly from the human brain, and these structural differences surely reflect functional differences. Thus, even if a rat has some degree of conscious awareness and can experience a true sensation of pain, that sensation is very likely qualitatively different from the human sensation of pain.

Sensations may well differ qualitatively even from one human being to another. That is why we said that the person experiencing phantom-limb pains in his left arm "presumably" has the same sensations as someone whose left arm is actually hurt. When two persons talk about a sharp jabbing pain in the left arm, we can be reasonably certain that their sensations are similar, because their nervous systems are similar, but we cannot be certain that the sensations are identical. The two persons have learned to apply the same words and descriptions to their individual sensations, but the possibility of qualitative differences remains, and can never be ruled out. We might show that the two stimuli are identical, and that the receptors and the neural messages are too, but these do not determine the sensation.

THE MAJOR SENSES

The traditional view is that human beings have five senses: touch, taste, smell, vision, and hearing. The expression "a sixth sense," referring to an uncanny faculty for obtaining information by other means than the average person's senses, is meaningless from a biological point of view; everyone has not only a sixth sense, but a seventh, eighth, ninth, tenth, etc. In short, the traditional five senses are only a few of many senses with which all human beings are endowed.

SENSORY RECEPTORS OF THE SKIN

There are numerous types of sensory receptors in the skin (see Fig. 3.59, p. 133, which shows some receptors in place). These receptors are concerned with at least five different senses, although the traditional classification recognizes only one—touch. These five senses are touch, pressure, heat, cold, and pain. Some of the skin receptors, particularly those concerned with pain, are simply the unmyelinated terminal branches of neurons (Fig. 10.28A). Others are nets of nerve fibers surrounding the bases of hairs; these fibers, which are particularly important in the sense of touch, are stimulated by the slightest displacement of the tiny hairs present on most parts of the body. Other skin receptors are more complex, consisting of nerve endings surrounded by a capsule of specialized connective-tissue cells (Fig. 10.28C).

The relative abundance of the various types of receptors differs greatly; e.g. pain receptors are nearly 27 times more abundant than cold receptors, and cold receptors are nearly 10 times more abundant than heat receptors. The receptors are not distributed evenly over the entire surface of the body; e.g. touch receptors are much more numerous in the fingertips than in the skin of the back, as might be expected in view of the normal functions of those two parts of the body. You can easily survey for yourself the distribution of different skin receptors on various parts of the body by using a very fine stiff bristle for touch, a thin needle to test for pain, and a very finely pointed sliver of ice to test for cold.

THE PROPRIOCEPTIVE AND VISCERAL SENSES

Unlike the receptors in the skin, which function in receiving information from the outside environment, some other receptors widely dispersed over the body function primarily in receiving information about the condition of the body itself. Though the senses these receptors mediate are not included in the traditional classification of five, they are of immense importance in the life of the individual. Among these receptors are the stretch receptors (proprioceptors) in the muscles and tendons (Fig. 10.29), which we mentioned earlier when discussing the knee-jerk reflex. They are sensitive to the changing tensions of muscles and tendons, and send impulses to the central nervous system, informing it of the position and movements of the various parts of the body.

Other dispersed receptors include those of the so-called visceral senses, located in the internal organs. Examples are the previously mentioned receptors in the carotid artery that are sensitive to carbon dioxide concentrations in the blood and to blood pressure. The firing of such visceral receptors seldom results in sensation (i.e. we are not aware of their action); the responses to their stimulation are usually mediated by the autonomic system. Sometimes, however, stimulation of visceral receptors produces conscious sensations, such as thirst, hunger, and nausea.

THE SENSES OF TASTE AND SMELL

The receptors of taste and smell are chemoreceptors; i.e. they are sensitive to solutions of certain types of chemicals, which can bind to them by weak bonds. The two senses are much alike, and when we speak of a taste sensation we are, as often as not, referring to a compound sensation produced by stimulation of both taste and smell receptors. One reason why hot foods often have more "taste" than cold foods is that they vaporize more, the vapors passing from the mouth upward into the nasal passages and there stimulating smell receptors. And one reason why we cannot "taste" foods well with a cold is that, with nasal passages inflamed and coated with mucus, the smell receptors are essentially nonfunctional. In other words, much of what we call taste is really smell. Conversely, some vapors entering our nostrils pass across the smell receptors and down into the mouth, where they stimulate taste receptors. In each case, taste and smell, chemicals must go into solution in the film of liquid coating the membranes of receptor cells before they can be detected. The major functional difference between the two kinds of receptor is that taste receptors are specialized for detection of chemicals present in quantity in the mouth itself, while smell receptors are more specialized for detecting vapors coming to the organism from distant sources; they are thus much more sensitive than taste receptors, as much as 3,000 times more in some instances.

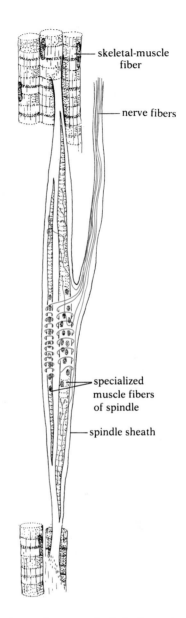

skeletal-muscle fiber

nerve fibers

specialized muscle fibers of spindle

spindle sheath

10.29 A stretch receptor in skeletal muscle
The terminal branches of sensory nerve fibers are intimately associated with several specialized muscle fibers that form an apparatus called a neuromuscular spindle.

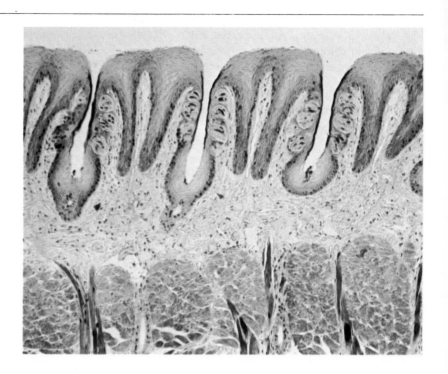

10.30   Photograph of section of rabbit tongue
The taste buds are located in the walls of the deep narrow pits.   [Courtesy Eastman Kodak Company, 1974.]

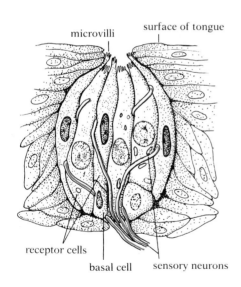

10.31   The structure of a taste bud
Each taste bud contains specialized receptor cells bearing sensory microvilli that are exposed in pits on the tongue surface. The ends of sensory neurons (color) are closely associated with the receptor cells. [After Murray, 1973.]

Taste   The receptor cells for taste are located in *taste buds* (Fig. 10.30) on the upper surface of the tongue and, to a lesser extent, on the surface of the pharynx and larynx. The receptor cells themselves are not neurons, but specialized cells with microvilli on their outer ends (Fig. 10.31). The ends of nerve fibers lie very close to these receptor cells, and when a receptor cell is stimulated, it generates impulses in the fibers.

There are apparently four basic taste senses: sweet, sour, salt, and bitter. The receptors for these four basic tastes have their areas of greatest concentration on different parts of the tongue—sweet and salt on the front, bitter on the back, and sour on the sides. A few substances stimulate only one of the four types of receptors, but most stimulate two, three, or four types in varying degrees. The sensations we experience are thus produced by a blending of the four basic sensations in different relative intensities.

The observation made earlier, that the sensations experienced by different people in response to the same stimuli may not be the same, can readily be confirmed for taste. The same substance can give rise to sensations of sweet in one person, if it stimulates primarily his sweet receptors, to sensations of bitter in a second person, if it stimulates primarily his bitter receptors, and to no sensation at all in a third, if it fails to stimulate any of his receptors. It is possible for someone to have particularly sensitive sweet receptors and unusually insensitive sour receptors; such a person might not like sugary food, because it would stimulate his sweet receptors excessively and arouse a sickeningly sweet sensation, but he might be very fond of lemon candy so sour that most people would reject it, because his sour receptors, being unusually insensitive, would be hardly stimulated by this candy. Although psychological factors play a large role in determining which foods people like and which they don't, it is also true that the different reactions of two persons to the same food may be, at least in part, the result of real biological differences between them. The sense of taste thus demonstrates with particular force that sensations are not an inherent property of the stimulus, but reside exclusively in the nervous system itself.

Smell   The receptor cells for the sense of smell (olfaction) in humans are located in two clefts in the upper part of the nasal passages (Fig. 10.32A). Unlike the receptor cells of taste, the olfactory receptors are true neurons. The cell bodies of most of these neurons lie embedded in the epithelial layer of the walls of the olfactory area of the nasal chamber (Fig. 10.32B). Dendrites run from the cell bodies to the surface of the epithelium, where they bear a cluster of modified cilia, which apparently function as the receptor sites.

Many attempts have been made to identify a group of primary odors from which all more complicated odors can be derived. But the olfactory sense has not proved as easily analyzable as taste. As far back as 1895 a system was devised in which all odors were explained in terms of nine basic odors. It soon became clear, however, that this system did not reduce odors to their fundamental components. For example, it is known that most sense receptors exposed to a constant and unchanging stimulus for a long period of time become adapted to the stimulus (i.e. they cease responding to it or respond only weakly); but when olfactory receptors adapted to one odor were then suddenly exposed to another odor presumed to be in the same one of the nine classes they often responded well. The inference was that different receptors must be involved and that the odors were not really made up of the same basic components. Many other classifications of fundamental odors were proposed during the next sixty years, but most proved just as unsatisfactory. In the 1950s, however, John E. Amoore proposed a system that seemed to go a long way toward giving us a consistent and meaningful theory of olfaction.

Chemical analysis of substances that smell much alike reveals that they are often not at all similar chemically. This fact baffled scientists for many years, because it seemed reasonable to expect substances capable of binding to the same receptors to have similar functional groups. Amoore, while still an undergraduate at Oxford University, worked out an explanation. Noting the great dissimilarities in chemical structure of substances that smelled alike, he decided that there must be some other basic similarity in the apparently dissimilar molecules. It had been suggested in 1949 by R. W. Moncrieff of Scotland that the spatial configuration of molecules, which is crucial in enzyme-substrate reactions and antigen-antibody reactions, might be more important than their chemical structure in determining which smell receptors they can stimulate. Amoore, after examining the size and shape of approximately 600 organic compounds whose odors had been well described, concluded that the evidence supported Moncrieff's hypothesis. In 1952 he published a stereochemical analysis of olfaction classifying all odors into seven basic primary odors: camphoraceous, musky, floral, pepperminty, ethereal (like ether), pungent, and putrid. He specified the size and shape a substance must have to stimulate the receptors for each of these primary odors. Thus, according to Amoore, camphoraceous-smelling mole-

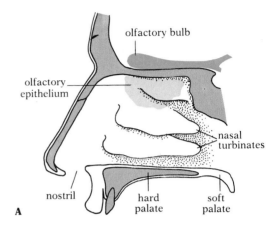

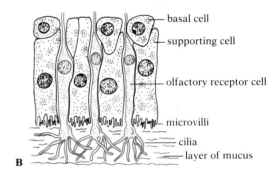

10.32   Human olfactory receptors
(A) The receptor cells are located in the olfactory epithelium, whose distribution on the wall of the upper portion of the nasal cavity is indicated in light color. The olfactory bulb of the brain receives information from the receptor cells.   (B) The cell bodies of most of the receptor cells are in the olfactory epithelium; their sensory cilia protrude into a layer of mucus on the surface of the epithelium.

cules are roughly spherical, with a diameter of about 0.7 nm; musky-smelling molecules are disc-shaped and about one nanometer in diameter; floral-smelling molecules are in the form of a disc with a flexible elongate side group attached; ethereal-smelling molecules are thin and rod-shaped; pepperminty-smelling molecules have a wedge shape; and so forth. The receptor sites on the sensory cells must have shapes into which such molecules can fit, just as a key fits into a lock. Amoore thought the receptor cells must be of seven types, each with receptor sites appropriate to one of the classes of molecules. A molecule could stimulate more than one kind of receptor cell if one part of the molecule could fit into one type of receptor site and another part into a different type of receptor site.

The results of research undertaken since 1952 on Amoore's hypothesis have been ambiguous. Amoore and his co-workers at Georgetown University have reported that they can synthesize molecules of certain shapes, predict what odor they will have, and then demonstrate (using panels of trained smellers) that the odor is indeed the one predicted. They have also reported that they have been able to design and synthesize molecules totally different in chemical structure, but with similar sizes and shapes, and to show that even trained odor testers cannot tell them apart. Other researchers, however, have not been able fully to duplicate Amoore's results. Moreover, a number of physical chemists have maintained that the shapes of some compounds taken to be similar by Amoore are actually not similar at all. For the present, then, the stereochemical hypothesis must be considered interesting but not entirely convincing, at least in its present form.

The olfactory sense, like all senses, may well differ in different species of animals. It is common knowledge that many animals have a far more acute sense of smell than humans. Most mammals, for example, depend much more on olfaction than humans do. Some moths have incredibly sensitive smell receptors, the males being capable of detecting the females at a distance of several miles. If the sense of smell depends on a limited number of primary odors, it seems entirely possible that these are not the same in all animals. Mammals and moths, for instance, are so different in their other characteristics that one would expect their smell receptors to function differently too.

THE SENSE OF VISION

A vast array of plants and animals have independently evolved structures specially sensitive to the wavelengths of radiation we know as light. These wavelengths are peculiarly suited to serve as carriers of information to the organism—frequently information from very distant sources. Longer wavelengths, such as those of the far infrared and radio portions of the electromagnetic spectrum, do not have enough energy to produce quickly the kinds of chemical reactions on which many biological functions, including vision, photosynthesis, and phytochrome conversions, depend. Shorter wavelengths, such as those of the X-ray and gamma-ray portions of the spectrum, have so much energy that they are destructive to many of the kinds of molecules of which organisms are composed. It is no accident, therefore, that so many unrelated organisms have convergently evolved specialized mechanisms for using portions of the same narrow band of wavelengths as a source of information about their environment.

10.33 Two animals with relatively simple lensed eyes
Left: The scallop has two rows of eyes along the
margin of its mantle, just inside the shell. The black
spot on each blue eye is the cornea.   Right: The
jumping spider has two large eyes directed forward,
and smaller ones on each side of its head.   The eyes of
the scallop and the spider, though structurally simpler
than those of most vertebrates, have lenses capable of
focusing incoming light on the receptor cells of the
retina.   [Oxford Scientific Films, Bruce Coleman Inc.;
and Oxford Scientific Films.]

Light receptors of animals   Almost all animals respond to light stimuli.
Even Protozoa react quickly to changes in light intensity, often moving away
from brightly lit areas. In fact, the single cells of many Protozoa have a
special region that serves as a sensitive detector of light. This region contains
a pigment that undergoes chemical changes when exposed to light energy,
which "tell" the protozoan that light is present. Most multicellular animals
have evolved specialized light receptor cells, but the basic mechanism of
detection is the same as in protozoans: Light energy produces changes in a
light-sensitive pigment. This pigment is usually a protein to which a portion
of a carotenoid molecule is attached. Carotenoids, you will recall, are yellow
light-sensitive pigments in plants. Animals obtain the carotenoid in their
diets as a vitamin (vitamin A), since it can be synthesized only by plants.

The light receptors of many invertebrates do not function as eyes, in the
usual sense of that word, They do not form images, but simply indicate to
the animal whether or not light is present and, frequently, whether the
intensity of the light is increasing or decreasing. Some of these receptors
give the animal no clue to the direction in which the light source lies, and the
animal responds by essentially random movements. However, many light
receptors are arranged in such fashion that direction becomes an additional
type of information accessible to the animal. The simple eyespots of planaria
(see Fig. 1.16, p. 20) are an example; the sensory cells in these organs are
stimulated primarily by light coming from above and slightly to the front.
Receptor organs capable of detecting in which direction a light source lies
are often equipped with numerous sensory cells oriented at different angles;
light coming from different directions will stimulate different cells. These
more complex eyes can likewise often detect movement, because light from
a moving object stimulates different cells in succession.

Still more complex eyes commonly include a lens capable of concentrating light on the receptor cells, thereby increasing the sensitivity of the eye to
light of weak intensity (Fig. 10.33). Lenses also greatly increase the ability to
detect direction and movement by focusing the light from each source onto

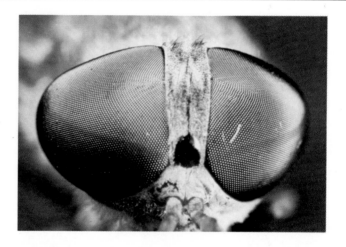

10.34   The compound eyes of a horsefly
Each eye is composed of a huge number of
ommatidia.   [J. A. L. Cooke, Oxford Scientific Films.]

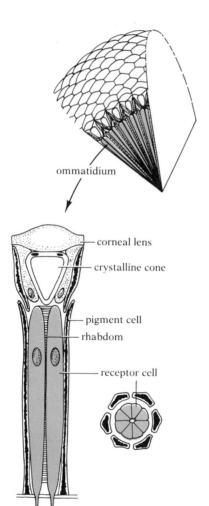

10.35   Ommatidium from the compound eye of a typical diurnal insect
Top: Section of compound eye.   Bottom: Longitudinal
and cross sections of one ommatidium.   The lens and
crystalline cone focus incoming light rays into the
rhabdom, which is a translucent cylinder formed from
highly specialized microvilli of the eight receptor cells;
a photosensitive pigment is located in the microvilli.
The pigment cells surrounding the ommatidium
contain a dark pigment that prevents passage of light
from one ommatidium to another.   [Modified from
R. E. Snodgrass, *Principles of insect morphology,*
McGraw-Hill Book Co., copyright © 1935. Used with
permission of McGraw-Hill Book Co.]

only a few receptor cells. Doubtless evolving initially as structures for concentrating the incident light, lenses made possible the later evolution of
image-forming eyes in some molluscs, most arthropods, and most vertebrates.

The independent evolution of image-forming eyes by different groups of
animals has produced two fundamentally different kinds of eyes: camera-
type eyes, such as those of some molluscs (e.g. octopus, squid) and vertebrates, and compound eyes, such as those of insects and crustaceans.

A *camera-type eye* uses a single-lens system to focus light on a surface
containing many receptor cells packed close together; the receptor surface,
called the *retina,* thus functions in a manner analogous to a piece of film.
The light pattern focused on the retina produces differential stimulation of
different receptor cells, just as a light pattern focused on a piece of film
produces different amounts of chemical reaction at different points on the
film.

A *compound eye,* on the other hand, uses many closely packed lenses, each
associated with only a few sensory cells (Fig. 10.34). Each lens with its
associated receptor cells forms a functional unit called an *ommatidium* (Fig.

10.35). Image formation depends on the light pattern falling on the surface of the compound eye; this light pattern determines which ommatidia will be stimulated and at what intensity. Thus there is no structure strictly analogous to the retina of a camera eye, the critical surface being the outer surface of the compound eye itself, composed of the closely packed individual lenses. Since each ommatidium points in a slightly different direction, each is stimulated by light coming from different points in the surrounding area. The insect's brain must integrate the messages coming to it from many ommatidia to produce (presumably) an image that represents the sum of many separate much smaller images.

It is commonly said that the mosaic nature of the stimuli coming to an insect's brain from its compound eyes must result in a rather crude image sensation that retains its mosaic characteristics. This is indeed the kind of image produced if the outer surface of a compound eye is detached and a photograph is taken through it. But we should be wary of extrapolating from the nature of the receptor organ to the nature of the animal's sensations (if, indeed, we can even apply the word "sensation" to an animal that may have no conscious awareness in the human sense). We have repeatedly emphasized the overriding importance of the brain as the source of sensation, and we do not know what the insect's brain does with the information coming to it from the compound eye. Surely it does not function simply like the camera that is used to take a picture through the outer portion of the insect's eye.

In fact, the evidence shows that the eye itself modifies incoming information before sending it to the brain. In other words, eyes do not function in so simple a manner as models based on cameras or mosaics imply. They do not just convert a pattern of light input into an exactly matching pattern of nerve impulses to the brain. Thus stimulation of one ommatidium influences the pattern of impulses sent to the brain by neighboring ommatidia, and stimulation of receptor cells in one part of the retina of a camera-type eye influences the pattern of impulses conveyed by nerve fibers from other receptor cells of the retina. In short, there is cross-coupling of sensory elements within the eye itself, and stimulation of any one element influences neighboring elements. Such information-modifying functions of the eye strengthen differences in neural activity originating from differently lighted parts of the eye, with the result that contrast is heightened and contours become sharper.

When the modified information pattern from the eye reaches the brain, it is further modified by the intricate interactions of the nerve cells there. The result, in the case of a compound eye, is almost certainly a clearer image (perhaps totally unlike a mosaic) than would be predicted on the basis of the structure of the receptor organ alone.

Structure of the human eye   The adult human eye is globe-shaped and has a diameter of approximately one inch (Fig. 10.36). It is encased in a tough but elastic coat of connective tissue, the ***sclera.*** The anterior portion of the sclera, called the ***cornea,*** is transparent and more strongly curved, and functions as the first element in the light-focusing system of the eye. Just inside the sclera is a layer of darkly pigmented tissue, the ***choroid,*** through which many blood vessels run; the choroid is important both as the structure providing a blood supply to the rest of the eye and as a light-absorbing

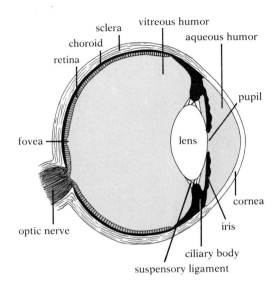

10.36 Diagrammatic section through the human eye

layer that (like the black inner surface of a camera) helps prevent internally reflected light from blurring the image. Just behind the junction between the main part of the sclera and the cornea, the choroid becomes thicker and has smooth muscles embedded in it; this portion of the choroid is called the *ciliary body.* At a point anterior to the ciliary body, the choroid leaves the surface of the eyeball and extends into the cavity of the eye as a ring of pigmented tissue, the *iris.* The iris contains smooth muscle fibers arranged in both circular and radial directions; when the circular muscle fibers contract, the opening in the center of the iris, called the *pupil,* is reduced; when the radial muscles contract, the pupil is dilated. The iris thus regulates the size of the opening admitting light (the pupil) in about the same way as the diaphragm of a camera regulates the lens aperture.

The *lens,* which functions as the second element in the light-focusing system, is suspended just behind the pupil by a *suspensory ligament* attached to the ciliary body. The lens and its suspensory ligament thus divide the cavity of the eyeball into two chambers, each filled with transparent fluid or semifluid material. The material in the chamber between the cornea and the lens is watery and is called the aqueous humor. The material in the chamber behind the lens is viscous and gelatinous; it is called the vitreous humor.

The *retina,* which contains the receptor cells for the sense of vision, is a thin tissue covering the inner surface of the choroid. It is composed of several layers of cells. The receptor cells are of two types, called rod cells and cone cells (Fig. 10.37). The *rod cells* are more abundant toward the periphery of the retina. They are exceedingly sensitive and enable us to see in light too dim to stimulate cone cells. They cannot detect colors, however, and the images to which they give rise are coarse and poorly defined. The *cone cells,* which are more abundant in the central portion of the retina, are used for vision in bright light. They give rise to detailed well-defined images, and they enable us to detect color.

The rods and cones synapse in the retina with short sensory neurons (bipolar cells), which themselves synapse in the retina with longer neurons (ganglion cells) whose axons, bundled together as the optic nerve, run to the visual centers of the brain (Fig. 10.38). The presence of several sets of syn-

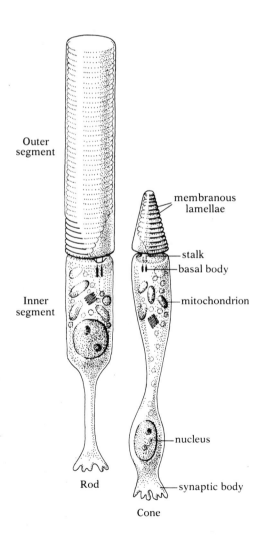

10.37   Rod and cone cells from a human retina
The outer segment and the stalk connecting it to the inner segment develop as a highly specialized cilium. Electron microscopy reveals that a basal body is located at the inner end of the stalk, and that the nine peripheral microtubules characteristic of all cilia run from it through the stalk into the outer segment; as is true of most cilia that have lost their motile properties, the two central microtubules of motile cilia are absent. The visual pigment is located in the numerous membranous lamellae of the outer segment. The pigments of rods and cones are different; those of cones are responsible for color vision.

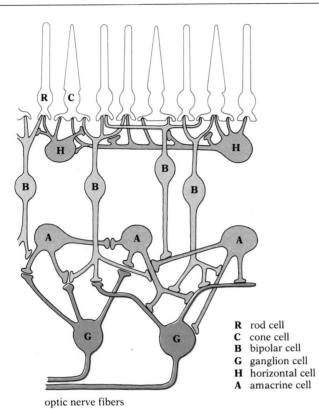

R rod cell
C cone cell
B bipolar cell
G ganglion cell
H horizontal cell
A amacrine cell

optic nerve fibers

10.38 The cells of the human retina

The receptor cells (rods and cones) synapse at their base with bipolar cells, which make up the middle layer of the retina. The bipolar cells, in turn, synapse with ganglion cells, which form the third layer; the axons of the ganglion cells form the optic nerve, which runs from the eye to the brain. Hence information that follows the most direct route to the brain moves from receptor cell to bipolar cell to ganglion cell to brain. Processing of information can occur within the retina because often several receptor cells synapse with a single bipolar cell and several bipolar cells synapse with a single ganglion cell. Besides convergence of information, there is lateral transfer of information from pathway to pathway via horizontal cells (each of which receives synapses from many receptor cells and synapses on many bipolar cells and on other horizontal cells) and via amacrine cells (which both receive synapses from and synapse on bipolar cells, and also synapse on many ganglion cells). Note that the three layers of the retina are arranged anatomically in reverse order from what might be expected; the receptor cells are in the back of the retina, and light must pass through the ganglion-cell and bipolar-cell layers to reach them.

10.39 The light-driven change of retinal from one isomer into another

With the absorption of light energy, the shape of the side chain is altered.

apses within the retina itself enables the eye to modify extensively the information transmitted from the receptor cells to the brain.

The light sensitivity of rods and cones Both rods and cones contain light-sensitive pigments. In rods the pigment, which is built into the membranes of the flattened vesicles in the outer segment (Fig. 10.37), is called *rhodopsin.* It consists of a protein (opsin) bonded with a light-absorbing prosthetic group called retinal, which is a derivative of vitamin A. When a molecule of rhodopsin is struck by a photon of light, the retinal is converted into a slightly different isomer (Fig. 10.39); this conversion is entirely light-driven,

Of all the special senses of vertebrates, the one most thoroughly studied with regard to information processing is the visual system. Yet even here much is still in question; pathways have been traced in detail only through seven of what must be a far larger number of levels.

The receptor cells—rods and cones—function like light meters, each cell responding to light falling directly on its outer segment. Collectively they determine for every point on the retina whether that point is illuminated and, if it is, how brightly. From the receptor cells the information follows pathways to bipolar cells and then to ganglion cells, still in the retina (Fig. 10.38). Pathways from several receptor and bipolar cells often converge in a single ganglion cell; in addition, neighboring pathways influence one another as information is passed between them laterally via horizontal and amacrine cells. Thus the information sent to the brain by the ganglion cells has already been processed and abstracted. It is very different from the primary information gathered by the receptor cells. Indeed, some of the primary information—particularly about absolute intensities of illumination—has been discarded, emphasis having shifted to differences in illumination between neighboring retinal areas.

The sensitivity of each ganglion cell can be described in terms of its *receptive field,* defined as a circular area of the retina which, when illuminated, influences the firing of that cell. Characteristically, the receptive fields are composed of two parts, a small central area and a ringlike peripheral area (figure A). Presumably the central area represents that portion of the retina from which the cell receives information by direct receptor cell–bipolar cell pathways, the periphery that portion of the retina from which the cell receives information that has moved laterally via horizontal or amacrine cells. Two types of ganglion cells have been recognized: *on*-center cells, in which illumination of the central area of the receptive field speeds up the firing of the cell while illumination of the periphery slows it down; and *off*-center cells, in which illumination of the central area slows down the firing while illumination of the periphery speeds it up.

The information conveyed to the brain from the ganglion cells via the optic nerve goes first to a group of cells known collectively as the lateral geniculate nucleus in a part of the brain called the thalamus (figure B). The lateral geniculate appears to function primarily as a relay station, since the receptive fields of its cells are essentially like those of ganglion cells; i.e. little additional processing seems to occur here.

From the lateral geniculate nucleus, the information moves to the visual area of the cerebral cortex (figure B), which consists of hundreds of millions of cells. Fortunately for our understanding, these cells are of relatively few types. One type has receptive fields for bars of light oriented horizontally, another for bars oriented vertically, a third for bars oriented at various intermediate angles (figure C). There is also a type that has angular or square receptive fields, and a type that responds best to moving edges.

In the cortex the information is processed through a succession of levels. At each level inputs from more and more of the lower-level cells converge to give rise to fields of progressively greater complexity of pattern. It is presumably by this procedure of hierarchical abstraction—of using at each successive level the inputs from more and more functionally related lower-level cells—that the brain arrives at its interpretation of the patterns of light the receptor cells of the retina originally detected.

A An *on*-center receptive field of a ganglion cell
The firing rate of the cell accelerates when a small circular area of the retina (color) is illuminated and slows down when a ring (gray) around that area is illuminated.

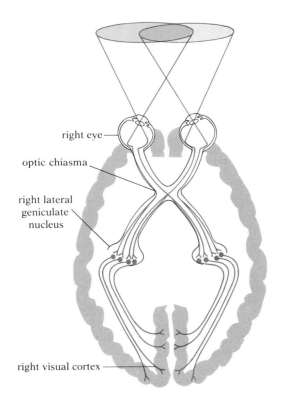

right eye

optic chiasma

right lateral
geniculate
nucleus

right visual cortex

B The visual pathway in primates, as seen from below

Because the neural pathways of the two eyes undergo partial crossing-over in the optic chiasma, the right side of each retina projects to the right lateral geniculate nucleus and then to the right visual cortex, while the left side of each retina projects to the left lateral geniculate nucleus and then to the left visual cortex. The result is that both halves of the visual cortex receive information about the region where the visual fields (colored and gray ovals) of the two eyes overlap.

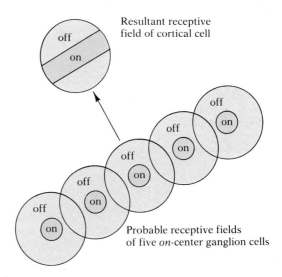

Resultant receptive field of cortical cell

off

on

off

off

off

off

on

on

on

on

Probable receptive fields of five *on*-center ganglion cells

A word of caution: Our knowledge of the processing of visual information in vertebrates comes chiefly from studies with animal subjects. The procedure is to direct points of light (or light shapes) onto specific areas of the retina in the eye of a living animal and to determine the responses of retinal cells or brain cells by means of intracellular electrodes. Most of the experiments on receptive fields in the visual cortex, as pioneered by D. H. Hubel and T. N. Wiesel of Harvard University, were performed on cats and monkeys. Although we may assume that most of the principles derived from this work on other animals apply to human vision as well, some differences of detail are only to be expected.

C A diagonal-bar receptive field of a cell in the visual cortex

A cell that responds to a bar of light oriented at an angle (in this case 30° from the vertical) probably receives input from a series of *on*-center ganglion cells whose receptive fields are arranged in a diagonal line across the retina.

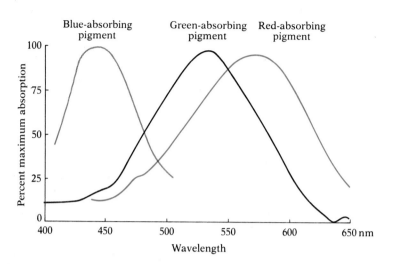

10.40   Absorption spectra of the three cone pigments of primates
The red-absorbing pigment actually has its maximum sensitivity in the yellow region of the spectrum.   [After A. F. MacNichol.]

and no enzyme is required. The isomerization of the retinal leads, in turn, to conformational changes in the protein and to hydrolysis of the bond between the retinal and the protein. The result is a bleaching of the rhodopsin and a subsequent change in the polarity of the plasma membrane of the rod cell (probably by means of release of calcium ions from the flattened vesicles), which triggers release of transmitter substance at synapses between the rod cell and bipolar cells (or horizontal cells). In the dark the retinal is reconverted into the original isomer and rebonded to opsin to regenerate functional rhodopsin.

The mechanism of cone vision is much more complex, and our understanding of color perception is still elementary. Experiments (performed at Johns Hopkins University, Harvard University, and the University of Pennsylvania) indicate that there are three functional types of cones, each containing a different pigment. All three pigments have retinal as their prosthetic group, but their protein components differ slightly. Each of the three pigments is sensitive to wavelengths of light covering a broad band of the visible spectrum, but each has its maximum absorption in a different portion of that spectrum (Fig. 10.40). Thus the three pigments can be designated as blue-absorbing, green-absorbing, and red-absorbing. This evidence for a three-color, three-receptor mechanism of cone reception agrees with a theory of color vision long supported by psychological experiments.

Human beings are so accustomed to their own color vision that they tend to assume other animals see colors in the same way. Humans and other primates are, however, rather unusual among mammals in possessing color vision; most mammals, having evolved as primarily nocturnal animals, probably see only in shades of gray. A bull's reaction is not to the color red, but to the movement of the matador's cape; the cape could be almost any color without significantly altering the results. Apparently many fishes and reptiles and most birds do have color vision; in this characteristic humans bear a greater resemblance to these animals, to which they are only distantly related, than to the other mammals, to which they are more closely related.

Insects, too, often have color vision—a very important attribute for the ones that feed on flowers. Many of them, however, do not have the same visible spectrum as humans. Their eyes usually cannot detect light of the longest wavelengths seen by humans; hence a room in which there is only pure red light will be in total darkness to many insects. But these insects can see light of wavelengths in the near end of the ultraviolet band, which humans cannot see (Fig. 10.41).

Refraction and accommodation   Since the pattern of illumination of retinal receptor cells is fundamental to image formation, high-resolution image vision depends on precise focusing of incoming light beams on the retina, just as clear high-resolution photography depends on precise focusing of incoming light on the film. If focusing is not good, the image is blurred. The object of focusing is to bring together at one point on the receptor surface all rays of light originating from a single point source. Suppose, for example, that you are looking at the face of another person. If you are to experience a clear image of his face, all light rays reflected from each point on the face must be brought together at a single point on your retina; thus all rays reflected from the point of the chin must be brought together at one point on your retina, all rays from the tip of the nose at another point on your retina, all rays from the center of the forehead at still another point on your retina, and so forth. In short, the projection of a true image of the observed object

10.41   Primrose willow *(Ludwigia peruviana)* as seen by humans (top) and with its ultraviolet pattern revealed (bottom)
When photographed with ultraviolet-sensitive film, flowers often reveal a pattern of markings that form a bull's-eye in the center, where the reproductive parts (and the nectar) are found. Since insects, unlike humans, see ultraviolet wavelengths, the reproductive parts are underscored for them, so to speak. The device fulfills a critical function for the plant, in that an insect drawn to the reproductive parts may serve the plant as a pollinating agent, carrying any pollen adhering to its body to the next flower it visits.

   A word of caution: All that the ultraviolet photograph can show is the pattern presumably visible to the insect; it cannot tell us how the insect experiences that pattern—if only because it does not show yellow, which insects also see, or how yellow and ultraviolet wavelengths might mix for an impression of color.   [Courtesy Thomas Eisner, Cornell University.]

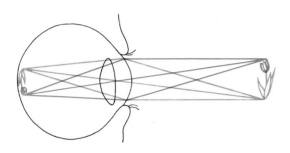

10.42   Image formation on the retina
Incoming light rays from each point on the object being viewed are bent by the cornea in such a manner that they come together at a single point on the retina. Thus an image of the object is projected on the retina. Note, however, that the image is inverted.

10.43.   Difference in degree of divergence of incoming light rays from near and far sources
Light rays from a point source travel outward in all directions. A cornea near the point source (Position 1) will thus be struck by many strongly divergent light rays (rays A through I). A cornea farther away (Position 2) will be missed by the most divergent rays (A–C and G–I) and will be struck by rays traveling at much smaller angles to each other (D–F). If the cornea is 6 m or more from the point source, the light rays reaching it will be traveling almost parallel to each other.

onto your retina requires a lens system capable of bending incoming rays of light and focusing them on the retina (Fig. 10.42), just as the lens system of a movie projector focuses onto the screen light that has come from the film.

The lens system of the human eye has two principal components, the cornea and the lens. It is actually the cornea, not the lens, that does most of the bending of incoming light rays. The great importance of the lens stems from the fact that it is the alterable part of the system; it makes possible adjustments in the focus depending on the distance of the object being viewed.

Let us consider the differences between the focusing of light coming from close objects and from distant objects. Light rays reflected from a given point travel away from that point in all directions (Fig. 10.43). If the cornea of an eye is located near the point source (Position 1), it will be struck by many light rays, some of which will be traveling at strongly divergent angles from each other. If, however, the cornea is located farther away from the point source (Position 2), many of the most divergent light rays, which would have struck it at Position 1, will miss it entirely, and the rays that do strike it will be traveling at only very slight angles to each other. If the cornea is 6 m (20 ft) or more from the point source, the rays of light that strike it can be considered for all practical purposes as traveling parallel to each other. From these facts it follows that if all the divergent rays of light from a near object are to be brought into focus on the retina, they must be strongly bent by the lens system; much less bending is necessary to bring into focus the essentially parallel rays coming from a distant object.

Most *refraction* (bending) of light from distant objects is performed by the cornea. But the cornea cannot refract the strongly divergent light from near objects sufficiently to bring the image into clear focus on the retina. It is here that the lens becomes important. The lens can produce enough refraction in addition to that produced by the cornea to bring the image into focus.

The lens is an elastic biconvex structure. It is attached, you will recall, to the ciliary body by the suspensory ligament. When the eye is viewing distant objects (i.e. objects more than 6 m away), a considerable tension on the suspensory ligament stretches the lens, which is thus flattened and becomes less convex. Now, the less convex a lens is, the less refractive power it has. Hence, when the lens is maximally stretched, it exerts very little influence on incoming light rays; under these circumstances refraction is primarily a function of the cornea. When, however, the eye is viewing a near object, the tension on the suspensory ligament is partly relaxed, and the front surface of

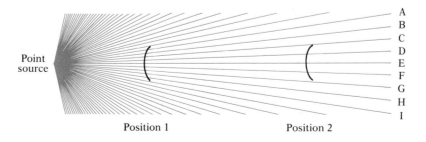

Farsighted eye

Nearsighted eye

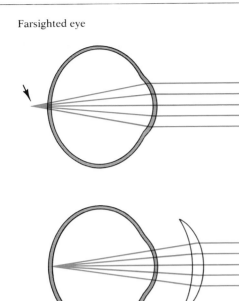

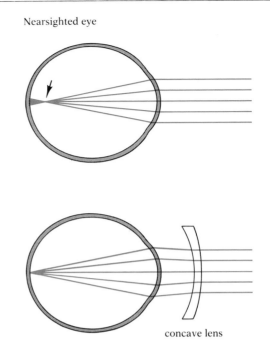

convex lens

concave lens

the lens bulges outward as a result of its natural elasticity. In other words, the reduced tension on the lens allows it to become more convex, which means that its refractive power increases. Tension on the lens is relaxed just enough for its refractive power to reach the point where it adequately supplements the refractive power of the cornea, and the image is brought into clear focus on the retina. The nearer the viewed object is to the eye, the more the tension on the lens is relaxed. This process of correcting the focus of images of near objects by changes in the shape of the lens is called ***accommodation.***

Accommodation depends on the action of the smooth muscles in the ciliary body. These muscles are arranged in such fashion that when they contract they pull the points of attachment of the suspensory ligament forward closer to the iris. The ligament consequently forms a circle with a smaller radius than before and, tension on it being thus reduced, allows the curvature of the lens to increase.

Since a camera lens, unlike a human lens, cannot change shape, focusing is accomplished by changing the distance between the lens and the film. The same arrangement is found in some animals with camera-type eyes. Fish, which lack ciliary muscles, have, instead, muscles that can move the whole lens forward and backward within the eye. And some molluscs change the length of the entire eyeball, thereby changing the distance between the cornea and the retina.

Structural defects in the shape of the eye are fairly common in human beings. Figure 10.44 shows how excessively short eyeballs result in farsightedness (hypermetropia), and excessively long eyes in nearsightedness (myopia). Another common defect, astigmatism, stems from unequal curvature of the cornea; it can be corrected with a lens ground unequally to compensate for the irregularities of the cornea.

10.44   Farsighted and nearsighted eyes
The farsighted eye is so short that the focal point would be behind the retina; i.e. the cornea and lens cannot bend the light rays enough to bring them together on the retina. Since rays from a distant source are almost parallel, they need less bending than the divergent rays from a near source; hence the person can see distant objects more clearly. Farsightedness is corrected by a convex lens (one that is thicker in the center than at the edges). Such a lens bends the incoming light rays before they reach the cornea, thus aiding the cornea in bringing them together on the retina.

The nearsighted eye is so long that the focal point is in front of the retina, and the rays have started to diverge again by the time they reach the retina, producing a blurred image. The strongly divergent rays from a near source require more bending; the focal point for them is therefore nearer the retina than the focal point for the almost parallel rays from a distant source. Hence the person can see near objects more clearly. Nearsightedness is corrected by a concave lens (one that is thinner in the center than at the edges). Such a lens spreads the incoming parallel rays from a distant source, thus making them strike the cornea at divergent angles just as the rays from a near source would do.

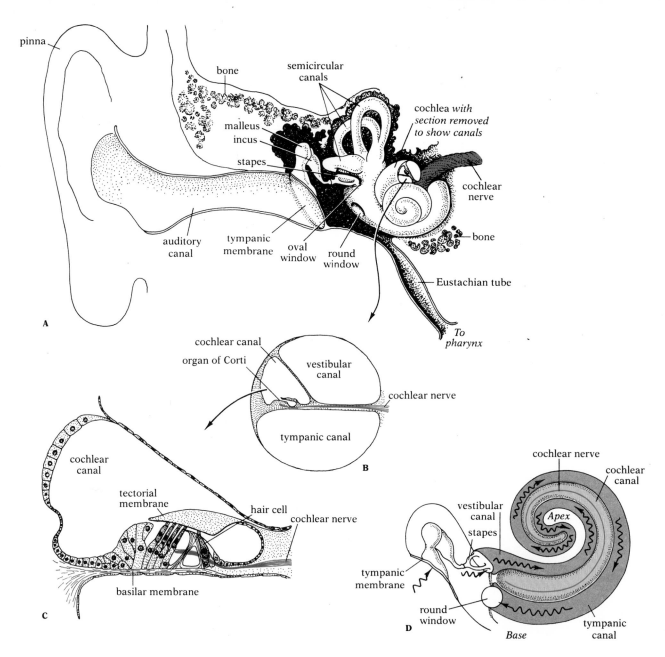

10.45   The human ear
(A) The major parts of the outer, middle, and inner ear (see text for description).
(B) Enlarged cross section through one unit of the coil of the cochlea, showing the relationship between the vestibular, cochlear, and tympanic canals and the location of the organ of Corti.   (C) Enlarged diagram of the organ of Corti, which rests on the basilar membrane separating the cochlear and

tympanic canals. When the basilar membrane vibrates and moves the sensory hair cells up and down, the hairs rub against the tectorial membrane overhanging them. The resulting deformation of the hairs produces a generator potential in the hair cells, which triggers impulses in sensory neurons running from the organ of Corti to the brain   (D) Diagram of the relationship between the middle ear and the cochlea, here

pictured partially uncoiled to show its canal system more clearly. When the stapes moves against the oval window, the fluid in the vestibular and tympanic canals oscillates (wavy arrows); the oscillation is made possible by the flexible round window, which permits relief of pressure. High-frequency sounds stimulate hair cells near the base of the cochlea, and low-frequency sounds stimulate hair cells near the apex.

THE SENSE OF HEARING

Receptors of the sense of hearing are specialized for detection of vibrations. The human ear is sensitive to vibrations of an amazing range of frequencies—from about 16 to 20,000 Hertz (cycles per second) in young people (some children can even hear frequencies as high as 40,000 Hertz, but ability to hear high frequencies declines steadily with age). Some other animals can hear much higher frequencies; dogs respond to whistles at 30,000 Hertz, which few humans can hear, and bats and some moths can hear frequencies of 100,000 Hertz or higher. Like smell and vision, hearing enables an animal to gain information from distant parts of its environment.

Structure of the human ear   The human ear (Fig. 10.45A) is divided into three parts: the *outer ear,* the *middle ear,* and the *inner ear.* The outer ear consists of the ear flap, or pinna, and the auditory canal. At the inner end of the auditory canal is the *tympanic membrane,* more commonly called the eardrum.

On the other side of the tympanic membrane is the chamber of the middle ear. This chamber is connected to the pharynx via the *Eustachian tube,* which functions as a duct making possible the equalization of air pressure between the outer and middle ear. When you ascend a high hill, the reduced pressure at the higher altitude results in a lower pressure in the auditory canal than in the middle-ear chamber, and the tympanic membrane is stretched outward. The pressure is equalized when air escapes from the middle ear through the Eustachian tube into the pharynx. When you descend quickly from a high altitude, the reverse process occurs. As you know, passage of air through the Eustachian tube is facilitated by swallowing, yawning, or coughing.

Three small bones[3] arranged in sequence extend across the chamber of the middle ear from the tympanic membrane to a membrane called the *oval window.* Another membrane, the *round window,* lies just below the oval window in the wall of the middle-ear chamber.

On the inner side of the oval and round windows is the inner ear, a complicated labyrinth of interconnected fluid-filled chambers and canals. The upper group of chambers and canals is concerned with the sense of equilibrium and will be discussed later. The lower portion of the inner ear consists of a long tube coiled like a snail shell. This is the *cochlea,* which is the organ of hearing. Inside the cochlea are three canals (Fig. 10.45B–D): the vestibular canal, which begins at the oval window; the tympanic canal, which connects with the vestibular canal and ends at the round window; and the cochlear canal, which lies between the other two. All three canals are filled with fluid.

The sensory portion of the cochlea, called the *organ of Corti* (Fig. 10.45C), projects into the cochlear canal from the *basilar membrane,* which forms the lower boundary of the cochlear canal. The organ of Corti consists of a

---

[3] These are the malleus, incus, and stapes, also commonly known respectively as the hammer, anvil, and stirrup. Only the stapes is present in the ear of amphibians, reptiles, and birds (where it is given a different name). The malleus and incus of mammals evolved from bones that are parts of the jaws in these other vertebrates.

layer of epithelium on which lie rows of specialized receptor cells bearing sensory hairs at their apexes. Dendrites of sensory neurons terminate on the surfaces of the hair cells. Overhanging the hair cells is a gelatinous structure, the *tectorial membrane,* into which the hairs project. When vibrations of the basilar membrane cause the sensory hairs to move up and down against the less mobile tectorial membrane, the deformations of the hairs thus produced apparently give rise to a generator potential in the hair cells, and these cells, in turn, stimulate the sensory neurons.

Reception of vibratory stimuli   Let us now trace briefly the steps involved in the reception of vibratory stimuli by the ear. Vibrations in the air pass down the auditory canal of the outer ear and strike the tympanic membrane, causing it to vibrate at the same frequency as the impinging air waves. These vibrations are transmitted across the cavity of the middle ear to the oval window by the three small middle-ear bones, which are so arranged as to constitute a lever system that diminishes the amplitude of the vibrations but increases their force. Furthermore, since the area of the tympanic membrane is nearly 30 times greater than that of the oval window, the pressure transmitted from the tympanic membrane is brought to bear on a much smaller area. Thus the result of transmission across the middle ear is the transformation of small pressures on the surface of the tympanic membrane into pressures as much as 22 times greater on the oval window. This increase in the force of incoming stimuli is very important in enabling us to detect very faint sounds—vibrations that, unmodified, would be insufficient to stimulate the receptor cells.

Vibrations of the tympanic membrane, then, cause movements of the middle-ear bones, which in turn produce movements of the membrane of the oval window. These movements in their turn produce movements of the fluid in the canals of the cochlea. Each time the membrane of the oval window is bent inward by the pistonlike action of the bone attached to it, fluid is pushed from the vestibular canal to the tympanic canal, with the result that the membrane of the round window bulges outward; when the membrane of the oval window then oscillates outward, fluid moves in the opposite direction and the membrane of the round window bulges inward. (Note that the immense resistance of fluids to compression would make movements of the oval window and fluid impossible, were it not for the movable round window, which acts as a pressure-release valve.)

The movements of the fluid in the cochlea are at the same frequencies as those of the air that entered the outer ear. The pressure waves in the fluid of the cochlea cause the basilar membrane to move up and down and rub the hairs of the hair cells against the tectorial membrane, thus deforming them. Stimulated through this deformation, the hair cells then stimulate the sensory neurons, which carry impulses to the auditory centers in the brain.

The characteristics of sounds   We can ordinarily distinguish three characteristics of the sounds we hear: pitch, volume (intensity), and tone quality. A satisfactory theory of hearing must be able to explain all three.

Pitch is a function of frequency; low-frequency vibrations stimulate a sensation of low pitch, and high-frequency vibrations stimulate a sensation of high pitch (remember that the sensation of pitch is not a property of the vibratory stimuli, but an interpretation of frequency produced by the brain). Low-frequency vibrations stimulate hair cells near the apex of the cochlea; high-frequency vibrations, hair cells near the base of the cochlea; and intermediate frequencies, hair cells of correspondingly intermediate regions of the cochlea. Thus the hair cells, like the keys of a piano graduated from low pitch to high pitch, are arranged in sequence, from those stimulated by low frequencies at the apex to those stimulated by high frequencies at the base. The neurons from each region along the length of the cochlea lead to slightly different areas in the brain. The pitch sensation we experience depends on which of these areas of the brain is stimulated.

Volume is a function of the amplitude of vibrations. Thus very intense vibrations cause the fluid of the cochlea to oscillate at greater amplitude, and the correspondingly greater amplitude of oscillation of the basilar membrane produces more intense stimulation of the hair cells. The result is the transmission of more impulses to the brain per unit time. The brain interprets this increased stimulation as loudness.

If a violin, a piano, and a clarinet all play a note at the same pitch and volume, each will sound different. This difference is called tone quality. It is apparently the result of stimulation of hair cells in other regions of the cochlea besides the main region stimulated. The secondary vibrations that produce such stimulation are know as harmonics or overtones. Different instruments (and different voices) produce different patterns of harmonics. Tone quality is thus the interpretation put by the brain on the pattern of hair cells stimulated.

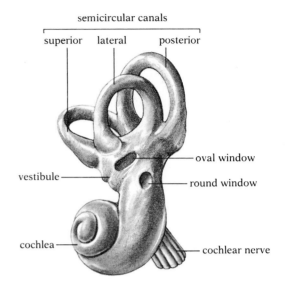

10.46   Labyrinth of the left human ear

THE SENSES OF EQUILIBRIUM AND ACCELERATION

The upper portion of the labyrinth of the inner ear is composed of three *semicircular canals* and a large vestibule that connects them to the cochlea (Fig. 10.46). Inside the vestibule are two chambers, the *utriculus* and the *sacculus.* Each contains a bed of sensory hair cells, upon which rest crystals of calcium carbonate, called otoliths. Changes in the position of the head cause the otoliths to exert more pull on some hair cells than on others, thereby stimulating them more. The relative strength of the pulls informs the cerebellum of the brain about the position of the head relative to gravity at any given moment.

Similar sensory devices are found in some other animals besides vertebrates. For example, crayfish and lobsters have equilibrium organs that consist of sand grains resting on beds of hair cells. When the animal molts, it loses the lining of the equilibrium organ, and with it the sand grains. Normally, the animal shovels new sand grains into the organs when molting has been completed. If a crayfish is kept in a tank containing iron filings instead of sand, it will replace the lost sand with filings. If a strong magnet is then held near the crayfish, it will orient to the magnet as though it were the pull of gravity; the crayfish can thus be made to swim upside down, because its organs of equilibrium tell the central nervous system that "down" is "up."

The semicircular canals are concerned with sensing acceleration (changes in the speed or direction of motion). Each of the three canals is oriented in a different plane of three-dimensional space (Fig. 10.46). At the base of each canal is a small chamber containing a tuft of sensory hair cells. When the head is moved or rotated in any direction, the fluid in the canals lags behind, because of its inertia, and thus exerts increased pressure on the hair cells. This pressure stimulates the hair cells to initiate impulses to the cerebellum of the brain. The brain, by integrating the different amounts of stimulation it receives from each of the three canals, can then determine very precisely the direction and rate of the acceleration.

LATERAL-LINE SYSTEMS

One of the most important sensory systems for fishes, the lateral-line system, is absent (except for some very specialized derivatives) in terrestrial vertebrates. It consists of clusters of sensory hair cells lying at intervals in grooves on the head and sides. Apparently the sensory cells are stimulated when pressure waves or currents in the water bend the hairs. This sensory system enables the fish to measure its progress through the water during locomotion, informs it of localized water disturbances, and allows "distant touch"—the detection of moving objects, such as predators or prey, or of stationary objects that reflect water movements produced by the fish itself. Fish also often stimulate the lateral-line system of other fish by fanning water against the receptors with their tails during courtship or combat.

There is reason to believe that the sensory-hair apparatus of hearing and equilibrium in terrestrial vertebrates evolved from elements of the ancestral lateral-line system. Like the lateral-line sense of fishes, both the sense of hearing and the sense of acceleration in terrestrial vertebrates are based on detection by hair cells of pressure currents in liquids. Hearing on land has involved the evolution of structures capable of converting weak pressure waves in air into stronger pressure waves in liquid. No such conversion is necessary in fishes, and, in fact, it is difficult to say if they hear in the strict sense. Their ear is entirely internal and is primarily an organ of equilibrium. But fishes certainly detect pressure waves easily, and some can even detect pitch (i.e. discriminate between different frequencies), but is this a hearing sense or a lateral-line sense? The question is probably meaningless; the animal detects the stimulus and can respond to it, and that, from a biological point of view, is what counts.

THE BRAIN

As we saw earlier, the brain of invertebrate animals is usually much smaller in relation to the size of the body than the brain of vertebrates, and its dominance over the rest of the central nervous system is usually less pronounced. The brain of many invertebrates, such as earthworms and insects, consists of a ring of nervous tissue encircling the anterior portion of the digestive tract (see Fig. 10.4). The two small ganglia lying on the upper side of the digestive tract are the only parts of this ring to which the name "brain" is given. Such a brain is usually not much larger than the other ganglia of the longitudinal ventral nerve cords.

Some interesting experiments with earthworms have shed light on the interrelationship between the brain and ventral cords of these animals. Earthworms were put in the long arm of a simple Y-maze. When the worms, moving down the passage, came to the fork, they received a mild electric shock if they turned left, but if they turned right they came to a moist

10.47 Brain of _Tritonia diomedea_, a marine mollusc
The brain of the mollusc is composed of a group of ganglia; within each ganglion the individual cells can be distinguished by their orange markings. So consistent are the positions of the cells from animal to animal that it has been possible to map them with precision and, with the help of microelectrode probes, to study the firing pattern of each cell in detail under different types of stimulation. In this way much can be learned about the circuitry of the central nervous system in these animals, and the interaction of nerve cells in producing their behavior. × 140.   [Courtesy A. O. D. Willows, University of Washington.]

chamber containing food. After many trials in the maze, the earthworms learned to turn right in the majority of cases. Then the heads of half the trained worms were tied off and amputated (with care taken to prevent infection or excessive bleeding). The other worms were kept as controls. After enough time had elapsed for the experimental worms to recover from the operation, they were again put in the maze. They ran the maze almost as well as when they had their heads! A memory of their earlier training had apparently been stored in the ventral nerve cord, and when the head was removed, the cords took complete command. Eventually some of the worms regenerated heads, and they then lost their ability to run the maze and had to be trained all over again. Apparently the new head, which had no memory of the earlier training, took dominance over the nervous system and suppressed the memory in the cords. Such experiments seem to indicate that the earthworm's brain, when present, has a limited dominance over the cords, but that the cords share enough of the brain's most basic functions for the animal to survive and behave almost normally without the brain.

The attempt to understand how brain cells interact to produce behavior patterns has been greatly aided in recent years by study of invertebrates, especially some of the large sluglike marine molluscs of the genera _Aplysia_ and _Tritonia_. Many of the brain cells of these molluscs are large enough, and sufficiently constant in their position, to be mapped visually (Fig. 10.47) and then probed with microelectrodes. In this way it has been possible to work out a "wiring diagram" for the feeding response, for example, that indicates just which neurons control the manipulation and swallowing of food, and how the neurons influence one another during the behavior.

The brains of the most primitive vertebrates are not much more dominant than those of invertebrates, but they do show the beginnings of the evolutionary trends that have made extensive brain development one of the most prominent characteristics of vertebrates. The immense importance of the vertebrate brain makes it worthy of more extensive treatment than we can give it here. The following is a brief introduction to some of the main aspects of the subject.

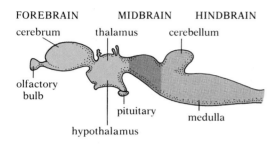

FOREBRAIN   MIDBRAIN   HINDBRAIN

10.48 Diagram of the principal divisions of the vertebrate brain
[Modified from A. S. Romer, *The vertebrate body,* Saunders, 1962.]

EVOLUTION OF THE VERTEBRATE BRAIN

The most primitive vertebrate brains and the partly developed brains of embryos consist of three irregular swellings at the anterior end of the longitudinal nerve cord. These three regions undergo much modification in the course of development of the more advanced vertebrates, showing specially thickened areas in their walls and distinctive outgrowths in other places. Despite these changes, however, the original three divisions of the brain can still be recognized even in the most advanced vertebrates, including humans. The three divisions are the **forebrain,** the **midbrain,** and the **hindbrain** (Fig. 10.48).

The brain contains a series of hollow compartments, or ventricles, which are continuous with the central canal of the spinal cord and which, like the canal, are filled with cerebrospinal fluid. The fluid is kept circulating by the beating of cilia on the epithelial cells that line the ventricles. Both the brain and the spinal cord are wrapped in three protective membranes knows as **meninges.**[4] The spaces between the three meninges are filled with cerebrospinal fluid, which cushions the nervous tissues against damage by the bones in which they are encased.

Very early in its evolution, the brain underwent modifications that set the stage for later evolutionary trends. Briefly, the modifications were these:

1. The ventral portion of the hindbrain, or **medulla oblongata,** became specialized as a control center for some autonomic and somatic pathways concerned with visceral functions (recall our discussions of breathing and heartbeat) and as a connecting tract between the spinal cord and the more anterior parts of the brain, while the anterior dorsal portion of the hindbrain became much enlarged as the **cerebellum,** a structure concerned with balance, equilibrium, and muscular coordination (see p. 411).

2. The dorsal part of the midbrain became specialized as the **optic lobes,** visual centers associated with the optic nerves.

3. The forebrain became divided into an anterior portion consisting of the **cerebrum,** with its prominent olfactory bulbs, and a posterior portion consisting of the **thalamus** and **hypothalamus.**

Later evolution has made few changes in the hindbrain, though the cerebellum has become larger and more complex in many animals. The truly major evolutionary change has been the steady increase in size and importance of the cerebrum, with a corresponding decrease in relative size and importance of the midbrain (Fig. 10.49).

The ancestral cerebrum was only a pair of small smooth swellings concerned chiefly with the sense of smell. As in the spinal cord, the gray matter (cell bodies and synapses) was mostly internal. The synapses functioned predominantly as relays between the olfactory bulbs and more posterior parts of the brain; little, if any, correlation occurred in the cerebrum. The cerebrums of modern fishes are still little more than relay stations, although the areas of gray matter are more massive. In amphibians, which evolved

---

[4] The three meninges are the pia mater, which lies on the surface of the brain and spinal cord; the dura mater, a tough membrane on the inner surface of the skull and vertebrae; and the arachnoid, a very fragile membrane lying between the other two.

from ancestral fish, there was an expansion of the gray matter and a multiplication of synapses between neurons. No longer was the cerebrum only a relay station; it now functioned as a correlation center for impulses coming to it from various sensory areas of the brain. Slowly, much of the gray matter moved outward from its initially internal position, until it came to lie on the surface of the cerebrum. This surface layer is known as the *cerebral cortex* ("cortex" means bark).

In certain advanced reptiles a new component of the cortex, called the *neocortex* (or neopallium), arose at a point on the anterior surface of the cerebrum. Mammals, which evolved from reptiles of this type, show the greatest development of the neocortex. Even in primitive mammals the neocortex has expanded to form a surface layer covering most of the forebrain. This does not mean that the old cortex of the ancestral brain has been reduced; it has simply been pushed to an internal position by the immense increase in relative size of the neocortex.

As the neocortex, a major coordinating center for sensory and motor functions involving all senses and all parts of the body, continued to expand, both by relative increase in its total size and by folding, which increased its surface area, it became more and more dominant over the other parts of the brain. The midbrain had been the chief control center in the earliest vertebrates. Then the thalamus portion of the forebrain became a major coordinating center, first sharing this function with the midbrain, then becoming dominant. Finally, with the rise of the neocortex and its preempting of many control functions from both the midbrain and the thalamus, the midbrain was left as a small connecting link between the hindbrain and the forebrain; it remains a control center for a few local reflex mechanisms and some of the simpler visual functions; it also plays a role in control of emotions.

We have repeatedly said that the neurons from different sensory receptors lead to different parts of the brain. These parts can be located and mapped by carefully following the paths of stained nerve fibers. The results can be checked by electrical procedures, in which sensory receptors are stimulated and recordings identify the area of the brain receiving the incoming impulses. Much can also be learned by determining what deficits ensue when carefully delimited parts of the brain are destroyed in experimental animals. Similar research approaches allow mapping of the motor areas.

It has been shown by mapping the brains of various mammals that the proportion of the total area of the cerebrum devoted to sensory and motor functions differs greatly from one species to another (Fig. 10.50). In general, the larger and the more convoluted the cerebral cortex, the smaller the proportion devoted exclusively to sensory and motor activities. Human beings represent the extreme example of this trend; in them the so-called associative areas constitute by far the largest proportion of the cerebral cortex. It is, of course, precisely this characteristic, with its behavioral consequences, that most clearly distinguishes humans from other animals.

Like the mammals, the birds are very advanced vertebrates, and like the mammals, they evolved from reptiles, but not the ones that gave rise to the mammals. Modern birds have relatively large cerebrums, which are clearly important coordination centers, but the arrangement of the various components of their cerebrums departs markedly from the mammalian arrangement, even though functionally analogous regions are present.

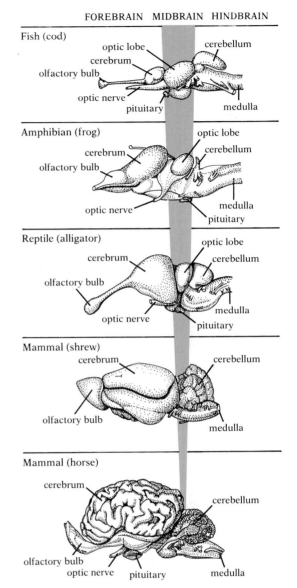

10.49 Evolutionary change in relative size of midbrain and forebrain in vertebrates
In this evolutionary sequence the relative size of the midbrain shows a marked decrease, and that of the forebrain a very considerable increase. [Modified from A. S. Romer, *The vertebrate body*, Saunders, 1962; and G. G. Simpson, C. S. Pittendrigh, and L. H. Tiffany, *Life: An introduction to biology*, copyright, © 1957, by Harcourt Brace Jovanovich, Inc., used by permission of the publishers.]

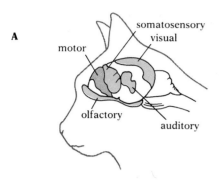

A

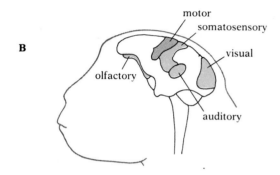

B

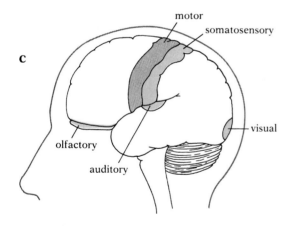

C

10.50 Proportion of cerebral cortex devoted to sensory and motor functions in three mammals
(A) In the cat sensory (color) and motor (gray) areas constitute a major portion of the cortex.   (B) In the monkey the proportion of cortex devoted to association areas (white) is much greater than in the cat.   (C) In humans the sensory and motor areas occupy a relatively small percentage of the cortex, most of the cortical area being devoted to association.

THE MAMMALIAN FOREBRAIN

Much of our information about the function of the various parts of the brain has reference particularly to such animals as rats, cats, monkeys, and chimpanzees, on which experiments can be performed. But about the human brain, too, a considerable body of information has accumulated, most of it derived from electrical stimulation during brain surgery and from observation of the effects of tumors and of accidental damage or destruction of parts of the brain. We cannot hope to summarize the whole fund of current knowledge here, but a few comments must be made about the forebrain, which plays so vital a part in all our lives.

The thalamus   The thalamus (Fig. 10.51A) is a major sensory-integration center in lower vertebrates, and although in humans many of the more complex aspects of sensory integration have been taken over by the cerebrum, the thalamus continues to be an important element in such integration.

The thalamus also contains part of an extremely important neural formation known as the *reticular system,* which is a complexly interconnected feltwork of neurons that runs through the brainstem of the medulla and midbrain as well as the thalamus. Every sensory pathway traveling to the higher centers of the brain sends side branches to the reticular system, and every descending motor pathway does so too. In short, the reticular system receives inputs from "wiretaps" on all incoming and outgoing communication channels of the brain. There are also pathways leading from the reticular system to a great variety of areas in the cortex, brainstem, and spinal cord.

Probes of the reticular system by means of microelectrodes have revealed that many of its neurons are "unspecific," in the sense that the same neuron may respond to stimulation of pain receptors in the foot, touch receptors on the hand, sound receptors in the ear, light receptors in the eye, etc. Apparently a major function of this curious part of the brain is activation of the rest of the brain upon receipt of stimuli. It acts as an arousal system, much as an alarm clock arouses you in the morning; in fact, the alarm clock wakes you only if it stimulates the reticular system sufficiently. Direct stimulation of the cortex does not awaken the brain; it must be called into action by the reticular system. The more stimulation the reticular system receives from the neural pathways it "wiretaps," the more arousal signals it sends to the rest of the brain. One reason why it is much easier to fall asleep when the bedroom is dark and quiet is that there are fewer incoming stimuli to trigger the arousal system. Barbiturates, which are commonly used in sleeping pills, block the reticular activating system and thus facilitate deep sleep. Destruction of the reticular system results in a permanent coma.

But relatively indiscriminate arousal is not the only important function of the reticular system. Apparently it also monitors all incoming stimuli, compares them with one another, and then magnifies some and suppresses others. Remember that at any given instant stimulation is coming into your central nervous system from receptors all over your body; touch receptors are being stimulated by your clothes, pressure receptors are being stimulated in your feet if you are standing (or somewhere else if you are sitting), hundreds of different vibrations are stimulating the hair cells of the cochlea, and light from literally thousands of objects is bombarding your rods and cones. You are aware of only a few of these stimulations at any given moment. If you were aware of all of them at once and tried to respond to them, you would become completely disorganized—literally go haywire. The reticular system, then, is an indispensable filter that lets only a few of the major inputs reach the higher centers of the brain and elicit reactions.

Not only does the reticular system filter out some stimuli at the level of the brain itself, but it also sends inhibitory impulses down the spinal cord to block some incoming impulses before they ever reach the brain. Thus if you are watching something intently, you may be completely unaware of a touch, or of whether you are warm or cold. And you may also be completely unaware of all objects in your visual field except the one on which you are concentrating. Or if you are extremely absorbed in what you are doing, you may be unaware that you have suffered an injury; soldiers in battle are often surprised to find that, hours before, they received wounds that would have caused them excruciating pain under normal circumstances—yet they never felt any pain at all.

The reticular system can do more than modify and even block incoming sensory impulses; it can similarly modify outgoing motor impulses both at the level of the brain and at the level of the spinal cord, thereby increasing the magnitude of some muscular responses and decreasing that of others.

The hypothalamus   The hypothalamus is the part of the brainstem just ventral to the thalamus. Among its major functions, as we saw in Chapter 9, are synthesis of the hormones stored in the posterior pituitary and secretion of the releasing hormones that help regulate the anterior pituitary. The hypothalamus is thus the crucial link between the neural and endocrine systems.

The hypothalamus is also the most important control center for the visceral functions of the body. Stimulation with microelectrodes has permitted localization of centers in the hypothalamus that control hunger, thirst, body temperature, water balance, blood pressure, reproductive behavior, pleasure, hostility, pain, etc. It is possible to fit experimental animals with electrodes inserted into various of these control centers and then, by turning the electricity on or off, make the animal feel hungry or sated, cold or hot, angry or benign. Cats wired in this way may be friendly one moment and in a rage, with fur erect, eyes wide, and claws out, the next moment, depending on whether the area controlling rage is being stimulated or not. Rats with electrodes in their pleasure centers will spend much of the time pressing levers that turn on the current; the sensation is apparently one they cannot resist. Cats that have just eaten a large meal will resume gulping food as soon as stimulation to their hunger centers is turned on. Often these various centers may be only a few millimeters apart; e.g. stimulation can sometimes be shifted from extreme pleasure to extreme pain or fright by moving the electrode only 0.02 inch.

Anger, hostility, pleasure—these and other terms used in the preceding paragraph pertain to the emotions. The hypothalamus, according to the prevailing evidence, functions as a major integrating region for emotional

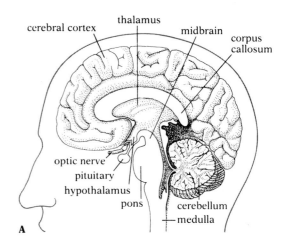

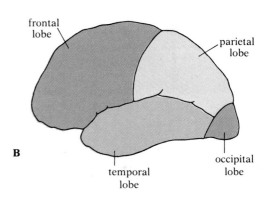

10.51 The human brain
(A) Sagittal section (longitudinal section through the midline) showing major parts of the brain.
(B) Diagram of left side of the cerebral cortex showing its four main lobes. These lobes are duplicated in the right hemisphere of the brain.

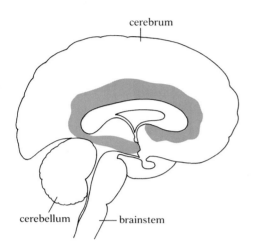

cerebrum

cerebellum   brainstem

10.52   The limbic system of the human brain
The limbic system (color) is not an anatomically
distinct structure, but rather a group of brain areas
that are related functionally in giving rise to feelings
and emotions.

responses. And far from being the only such region, it is apparently one of a
large group of brain structures that, though anatomically distinct, are functionally interrelated in mediating the emotions. These structures, known
collectively as the *limbic system,* ring the anterior end of the brainstem (Fig.
10.52); most are derived from the forebrain, but some originate from the
midbrain.

The cerebral cortex   Because the cerebral cortex (which includes the neocortex) has been identified with intellectual capacity, we tend to think of it as
synonymous with the brain. But we have seen that other parts of the brain
play a critical role in almost all our activities. The cortex has, in fact, been
viewed by some workers as an organ of elaboration and refinement of functions that, in its absence, could be performed to some extent by other parts
of the brain. This certainly seems to be true in lower vertebrates. A frog
whose entire cerebrum has been removed shows almost no behavioral
changes and can see as well as before. A decorticated rat shows no obvious
motor defects, and, though its ability to distinguish complex visual patterns
is impaired, it can tell light from dark and can respond to movement. A cat
without its cerebral cortex can move around sluggishly, swallow, react appropriately to pain stimuli, say miaow, and even purr, but it has the appearance of an automaton, seeming to be essentially unconscious. Monkeys and
human beings are almost completely disabled by loss of their cerebral cortices. Monkeys retain only a very crude ability to detect light and are severely paralyzed. Human beings become totally blind and suffer extensive
paralysis; though they can carry out such vegetative functions as breathing
and swallowing, they usually soon die. Apparently, then, no bodily processes
vital to life are controlled exclusively by the cortex, but as the cortex
evolved, older parts of the brainstem established new connections with it,
each part coming to be assisted in its functions by its own piece of the cortex.
Gradually, more and more of the function of some of these older parts was
transferred to the cortex until, as in human vision, the role of the cortex
became predominant or even essential.

We said earlier that the proportion of the cortex taken up by purely motor
and sensory areas is smaller in humans than in other animals (see Fig. 10.50).
Probing with electrodes shows, however, that within these limited areas
each part of the body is represented by its own control area. These areas are
not arranged in random fashion, but form a regular pattern. For example,
the point on the cortex that controls the thumb is adjacent to the point that
controls the first finger, and this point is adjacent to the point that controls
the second finger, and so forth; all points for finger control are near the
points controlling the palm and the wrist. Thus a map of the somatic sensory
area of the cortex (Fig. 10.53) gives, on the right side, a picture of the entire
left side of the body and, on the left side, a picture of the entire right side of
the body (we do not know why the fiber tracts running between the brain
and the spinal cord cross to the opposite side, so that the left brain controls
the right side of the body and the right brain the left side). Similar pictures
are obtained by mapping the motor area of the cortex.

The cortical pictures of the body are not accurate in their proportions,
however; they are distorted and grotesque, because the area of the cortex
devoted to each part of the body is proportional not to the size of the part

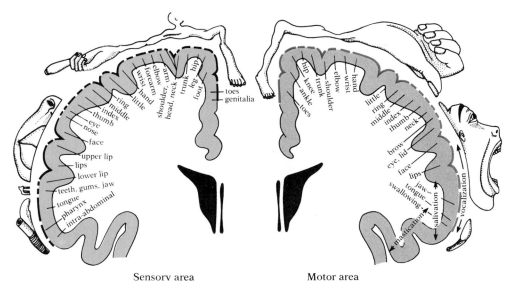

Sensory area   Motor area

10.53 Function maps of the surface of the somatic sensory area (left) and motor area (right) of the human cerebral cortex
Note that the area of cortex devoted to each body part is proportional to the importance of the sensory or motor activities of that part, not to its size; hence the face and hand are especially prominent. [Modified from W. Penfield and T. Rasmussen, *The cerebral cortex of man*, Macmillan, 1950.]

but to its sensory or motor capabilities. Thus, in humans, little motor area is devoted to muscles of the back, but an enormous area is devoted to the muscles of the hand and of the mouth, two of the most active portions of the human animal. The motor area for the hands is larger than their sensory area, whereas the sensory area for the lips is larger than their motor area. As you might expect, allocation of brain area is different in different animals and reflects the special characteristics of each; the snout of a pig is represented by a large area of the cortex, and the skin around the nostrils of a horse has nearly as much cortical area as all the rest of the body put together.

If human sensory and motor areas can be mapped, is it likewise possible to map the association areas so prominent in the human cortex? This is more difficult, but some progress has been made. For example, Wilder Penfield of the Montreal Neurological Institute located three speech areas on the cortex (Fig. 10.54). Curiously enough, these are almost always restricted to the left hemisphere in the normal brain; most functions are symmetrically represented in the two hemispheres, but speech seems to be an exception. When we say speech areas, we do not mean areas that control muscular motions involved in talking; such motions are controlled by motor areas. The speech areas are concerned with the thought processes underlying speech. Thus, when Penfield stimulated one point in the speech area of a certain patient, showed him a picture of a foot, and asked him to name it, the patient said he knew what it was—it "is what you put in your shoes"—but he simply couldn't think of the word. As soon as stimulation by the electrode ceased, he said "foot." Another patient under stimulation could not name a comb, but when asked what he does with it, he said, "I comb my hair." Even after saying this, he could not think of the noun "comb" when shown a picture of one and asked to name it.

Most of Penfield's mapping of cortical areas in the brain was done during operations on patients suffering from severe epilepsy. Epileptics periodic-

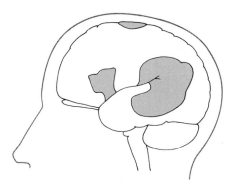

10.54 The three speech areas of the human cortex
In most cases the speech areas occur only in the left hemisphere. [Redrawn from W. Penfield and L. Roberts, *Speech and brain mechanisms*, © 1959 by Princeton University Press and used with their permission.]

ally suffer seizures varying in severity from a momentary tingling or sense of numbness in part of the body to general muscular convulsions and unconsciousness. Epilepsy usually results from injuries to the cortex. An epileptic attack occurs when spontaneous and uncontrolled discharges spread from the damaged region into the surrounding healthy tissue. If the discharges spread only a short distance, the attack is mild. If they spread over the entire brain, a catastrophically severe seizure results. Penfield and others developed techniques for cutting away a flap of bone from the roof of the patient's skull, stimulating the exposed brain surface with electrodes to locate the diseased region and also to determine the areas essential to proper speech, sensory, or motor function, which should not be removed, and then excising the diseased region. Such operations are performed while the patient is under only local anesthesia and is fully conscious (this is possible because, interestingly enough, there are no pain receptors in the brain itself). During the mapping phase of the operation, the surgeon stimulates various parts of the brain while the patient, who cannot see the surgeon and does not know when stimulation is on or off, talks with an observer who asks him questions and records his responses.

Brain surgery, particularly on patients suffering from severe epilepsy or other types of brain damage, has not only allowed mapping of motor, sensory, and speech areas, but has also made possible investigations of memory storage. When Penfield stimulated certain regions of the temporal lobes of the cerebral cortex (Fig. 10.51B), he at times caused patients to experience remarkably detailed recollections of past events, events frequently not remembered for many years. One patient relived an episode from her childhood. Another heard her small son playing in the yard, accompanied by sounds of automobiles, barking dogs, and all the other usual neighborhood noises. Another watched in her mind a scene from a play she had not seen in years. Another heard Christmas carols being sung in the church in her home town in Holland. Still another patient was certain the doctors were playing a record of a song in the operating room; she could not be convinced that there was really no phonograph—the memory was not a memory to her, it was reality.

This attitude, in fact, was a general one. Penfield says, "The patients have never looked upon an experiential response as a remembering. Instead of that it is a hearing-again and seeing-again—a living-through of past time." There is evidence that under stimulation the brain plays back the stored data at the same speed as the one at which the events actually happened. If the stimulation that gave the woman the experience of listening to the phonograph record was stopped and then almost instantly begun again, the record always went back and started at the beginning; it would not resume playing where it had left off when the stimulation ceased. Memories of similar "realness" and detail have sometimes been elicited by hypnosis. A classic case is that of an elderly bricklayer who, when hypnotized, described in detail the bumps on the individual bricks in a wall he had laid in his twenties and had not seen in years; the wall was checked and the bumps the bricklayer described were there!

Present evidence indicates that complex memories like those studied by Penfield, which involve the simultaneous presence of many different sensory qualities, are not stored in one place as a single memory. Instead, the visual component is stored in one place, the auditory component in another place, the smell component in a third place, and so forth; each sensory component is apparently stored in the same general part of the brain as that used in its recognition.[5] The memory traces (stores) for all these sensory components must be interconnected and their activation synchronized, so that the so-called memory areas in the temporal lobes can act as playback mechanisms, recombining the component sensory modalities to create the integrated memory sensation.

What is memory? We cannot say as yet. We assume it is a neuronal circuit potentiated by prior use, probably as a result of some sort of change in the synapses that interconnect the neurons participating in that memory trace. Presumably, the more impulses are sent along a given memory trace, the easier it becomes for later impulses to travel the same circuit. Potentiation induced by repeated use of a circuit would help explain why practice improves performance.

If our memory stores far more detail than we normally suspect (as the data from hypnosis and from experiments such as Penfield's suggest), if our permanent record of past events is far more complete than our normal conscious recollection of them, a fantastic amount of information must be stored in each human brain. How? We do not know, but we can make a few calculations to help us understand the problems involved in memory storage.

John Von Neumann has calculated that if the brain keeps a complete record of all electrochemical signals generated in all sensory receptors during a lifetime of 60 years, and if a record of the electrochemical activity of the approximately 10 billion neurons in the brain is also stored somehow, then the storage load must be about 2.8×10^{20} bits of information.[6] Reasoning from computer technology, we assume that storage of one bit requires one on-off switch. Von Neumann's calculations lead to the conclusion that every neuron must have a memory capacity equivalent to 30 billion (3×10^{10}) on-off switches! But if the individual neurons actually had such a capacity, they could not be the basic storage elements, because, by definition, a basic storage element may store only one bit of information.

Fortunately for our calculations, the anatomical and physiological evidence indicates that the whole neuron need not be the basic storage element. If memory resides in the strength of synapses, and if, as some workers maintain, each neuron in the brain receives an average of 1,000 synapses, then 1,000 times as many storage elements are available as there would be if the whole neurons were the basic elements. But if Von Neumann's calculations are correct, the individual synapses would still need a storage capacity of 30 million (3×10^7) bits—which is not permissible if the synapses are the basic storage elements.

Some workers have suggested that the many glial cells packed in the spaces between the neurons may play a role in memory. There is evidence

[5] Actually, each component is probably stored in more than one location, so that if one is damaged another may take over. In addition, each memory trace is stored in each half of the cortex; this duplication does not occur if all cross connections between the two hemispheres are severed.

[6] A bit is the smallest unit of information used in information theory. It corresponds to the result of a choice between two equally probable events.

that they interact electrochemically with the neurons. Perhaps changes in the electrochemical activity of these cells, which crowd to within a few hundredths of a millimeter of the neurons, participate in memory storage by altering the conductivity of the neuronal dendrites. If this is so, the storage capacity of the brain would be many times greater than calculations based only on the number of neurons and synapses show. But evidence for such a role of glial cells in memory is far from convincing.

Many scientists have questioned whether it is necessary to assume storage of every electrical event, as Von Neumann did. The fact that the experiential recall stimulated by a neurosurgeon is incredibly detailed does not mean that all sensory details are included. Indeed, it seems highly unlikely that they should be. Psychological tests indicate that a person can take in and deal consciously with only about 25 bits of information per second, even under the most favorable conditions. Most of the rest of the incoming stimulation is suppressed (we have already discussed the role of the reticular system in such suppression). Furthermore, the rate of 25 bits per second occurs only during a small fraction of the time, when the person is maximally alert; certainly such a rate is not maintained during sleep.

Even when incoming information is recognized and dealt with, it may not be stored indefinitely. There appears to be a distinction between short-term memory and long-term memory. We may, for example, look up a number in the telephone book and remember it for the time necessary to dial it, yet be unable to recall it an hour later; the number was held briefly in short-term memory, but never entered long-term memory. If the brain keeps no permanent record of items that go only into short-term memory, then the storage problem is much less severe than Von Neumann thought. But does our normal inability to recall items that were only in short-term memory really mean they are lost entirely from the brain? It seems unlikely that details about bumps on individual bricks were fixed in the long-term memory of the bricklayer; yet under hypnosis those details were found to have been stored. Is the difference between short-term and long-term memories, then, more one of accessibility than of storage? The plain fact is that we don't know the answers to such questions.

We would expect that stimulation, if it is to produce a memory, must cause chemical changes in the neurons. It has been found that neuronal activity does, in fact, increase the amount of RNA in the nerve cells. Furthermore, the RNA content of nerve cells is among the highest of any cells in the body. Since nucleic acids are complex polymers capable of encoding in their structure an enormous amount of information, the intriguing possibility arises that the potentiation of nerve circuits that establishes memory traces involves encoding information in the RNA of the nerve cells. More likely, since RNA plays a major role in protein synthesis, the increase in RNA may simply reflect an increase in the production of protein, and it may be the protein, not the RNA, that directly potentiates memory. Evidence for this possibility comes from experiments in which the antibiotic puromycin, which blocks protein synthesis, was injected into the brains of mice. The puromycin prevented new learning without affecting the things the mice had already learned. Similar experiments with fish indicate that puromycin blocks long-term, but not short-term memory, which suggests that memory formation is a two-step process, with only the second step—the fixation in long-term memory—requiring protein synthesis.

Our knowledge of how the brain functions is still in its infancy; there is a very long way to go. But progress is being made at an ever accelerating pace, and the day can be foreseen when human beings will understand their own understanding, when the "higher faculties" that set human beings apart can be explained in physicochemical terms. And with this knowledge will come, inevitably, increased power to control the functioning of the human mind (power that is already much further developed in its potential than most people realize). The way this knowledge is used may well determine, more than any other factor, what the future of our species will be. The time may come when the problems associated with the ability to control the human brain will outweigh the problems associated with the harnessing of nuclear energy.

SUGGESTED READING

AMOORE, J. E., J. W. JOHNSTON, and M. RUBIN, 1964. "The stereochemical theory of odor," *Scientific American*, February. (Offprint 297)   *The explanation of olfaction first developed by Amoore in 1952.*

"The brain," special issue of *Scientific American*, September 1979.   *Eleven articles by prominent investigators.*

DICARA, L. V., 1970. "Learning in the autonomic nervous system," *Scientific American*, January. (Offprint 525)   *The modifiability by learning of even such "involuntary" functions as heartbeat and intestinal contraction.*

FRENCH, J. D., 1957. "The reticular formation," *Scientific American*, May. (Offprint 66)   *The network of cells responsible for arousing and maintaining consciousness and for choosing between important and unimportant sensory messages.*

GAZZANIGA, M. S., 1967. "The split brain in man," *Scientific American*, August. (Offprint 508)   *The processing of information by each half of the cerebral cortex independently in individuals whose cortices no longer have cross connections.*

GESCHWIND, N., 1972. "Language and the brain," *Scientific American*, April. (Offprint 1246)   *The organization of the language areas in the human brain as revealed by speech disorders resulting from brain damage.*

GLICKSTEIN, M., and A. R. GIBSON, 1976. "Visual cells in the pons of the brain," *Scientific American*, November. (Offprint 573)   *Cells that help relay information from the visual cortex to the cerebellum, thus playing a role in the visual control of bodily movements.*

HUBEL, D. H., 1963. "The visual cortex of the brain," *Scientific American*, November. (Offprint 168)   *Report by one of the investigators who performed the now classic early experiments on visual processing in the cat.*

JULIEN, R. M., 1975. *A primer of drug action.* Freeman, San Francisco.   *Medically accurate and easy to understand. This discussion of psychoactive drugs, covering not only sedatives, tranquilizers, stimulants, opiates, psychedelics, and marijuana, but also alcohol, nicotine, and caffeine, is free of the mysticism and emotionalism often called forth by this subject.*

KATZ, B., 1966. *Nerve, muscle, and synapse.* McGraw-Hill, New York.   *An excellent little book about impulse conduction and synaptic transmission.*

KEYNES, R. D., 1958. "The nerve impulse and the squid," *Scientific American*, December. (Offprint 58)   *The role of the giant axon of the squid in the development of the modern understanding of impulse conduction.*

LAND, E. H., 1959. "Experiments in color vision," *Scientific American*, May. (Offprint 233)   *A model of color vision that differs from the standard one, by the inventor of the Polaroid camera.*

LESTER, H. A., 1977. "The response to acetylcholine," *Scientific American*, February. (Offprint 1352)   *How the receptors for acetylcholine are thought to function.*

MacNICHOL, E. F., 1964. "Three-pigment color vision," *Scientific American*, December. (Offprint 197)   *The experimental basis for the three-pigment model.*

MELZACK, R., 1961. "The perception of pain," *Scientific American*, February. (Offprint 457)   *Truly fascinating discussion.*

MICHAEL, C. R., 1969. "Retinal processing of visual images," *Scientific American*, May. (Offprint 1143)   *Retinal integration in frogs and ground squirrels—a process that rivals cortical integration in primates.*

MILLER, W. H., F. RATLIFF, and H. K. HARTLINE, 1961. "How cells receive stimuli," *Scientific American*, September. (Offprint 99)   *How sensory receptors work, with special attention to the eyes of the horseshoe crab.*

NATHANSON, J. A., and P. GREENGARD, 1977. " 'Second messengers' in the brain," *Scientific American*, August. (Offprint 1368)   *The possible role of cAMP and related compounds in mediating the function of neural transmitters.*

NICHOLLS, J. G., and D. VAN ESSEN, 1974. "The nervous system of the leech," *Scientific American*, January. (Offprint 1287)   *A type of neural system that has relatively few cells and is therefore a rewarding subject for experimentation.*

OLDS, J., 1956. "Pleasure centers in the brain," *Scientific American*, October. (Offprint 30)   *A fascinating description of some classic experiments by the man who conducted them.*

ROSENZWEIG, M. R., E. E. BENNETT, and M. C. DIAMOND, 1972. "Brain changes in response to experience," *Scientific American*, February. (Offprint 541)   *The contrasting effects of rich and dull environments on brain anatomy and chemistry in rats.*

SNYDER, S. H., 1977. "Opiate receptors and internal opiates," *Scientific American*, March. (Offprint 1354).   *How morphine and some morphinelike substances (enkephalins) normally synthesized by certain nerve cells exert their effect on the brain.*

WILLOWS, A. O. D., 1971. "Giant brain cells in mollusks," *Scientific American*, February. (Offprint 1212)   *On cells large enough to be identified and mapped visually. Electrode recording reveals how they interact to produce a behavior pattern.*

Chapter 11

EFFECTORS

Effectors are the parts of the organism that do things, carrying out the organism's response to stimuli. Their actions are as various as secretion by glands, production of light by fireflies, phototropic and geotropic responses in plants, cytoplasmic streaming in both plant and animal cells, and, most familiarly, movements of animals.

As we saw in a previous chapter, plants are capable of slow movements in response to light and gravity, movements produced by differential growth rates controlled by hormones. Many plants are also capable of some types of rapid movement. For example, some leaves droop or fold at night and expand again in the morning. The flowers of many plants open and close in a regular fashion at different times of day. The leaves of sensitive plants *(Mimosa)* fold and droop within a few seconds after being touched. The leaves of the Venus's-flytrap rapidly close around insects that have landed on them (see Fig. 5.11, p. 203). The seed pods of some plants snap open at maturity, vigorously expelling their seeds.

All these movements are far too rapid to depend on differential-growth changes. Another mechanism, turgor-pressure change, is involved. Leaves droop when certain of their cells lose so much water that they are no longer turgid enough to give rigidity to the leaf. Flowers fold when specially sensitive cells arranged in rows along the petals lose their turgidity, and they open again when these cells regain their turgidity. Rapid changes in turgidity in special effector cells located along the hinge of the leaf of the Venus's-flytrap are responsible for that plant's curious behavior. Similarly, rapid turgidity changes in specialized effector cells at the bases of the leaflets and petioles of the sensitive plant are responsible for that plant's response to a touch or other mechanical stimulus.

It is clear, then, that active movement is not exclusively a characteristic of animals. Nonetheless, it remains true that the most elaborate mechanisms for producing locomotion are found in the animal kingdom, and it is on these that we shall concentrate in this chapter (disregarding the many effector actions, such as glandular secretion, that occur without gross movement of the animal).

Effectors are often controlled by the nervous system; unlike sensory receptors and conductor cells, however, effector cells are not themselves part of the nervous system, even though they are the last components of reflex arcs. Among the numerous effector systems not under nervous control are the nematocysts of coelenterates, cilia and flagella, and some smooth muscles of vertebrates.

As was suggested in Chapter 3, the underlying mechanism of the various kinds of effector action—from cytoplasmic streaming to muscular movement—depends on either microfilaments or microtubules.

THE STRUCTURAL ARRANGEMENTS OF MUSCLES AND SKELETONS

The most prominent effectors in all multicellular animals except the sponges are the muscles—tissues composed of specialized contractile cells. All cells possess some ability to contract. But as multicellular animals became larger and more complex, and as division of labor among their cells and tissues increased, they evolved elongate cells specialized for contraction, which became the principal effectors of movement in higher animals.

ANIMALS WITH HYDROSTATIC SKELETONS

The first multicellular animals (disregarding the sponges) were doubtless small, perhaps on the order of one millimeter in length. They probably swam

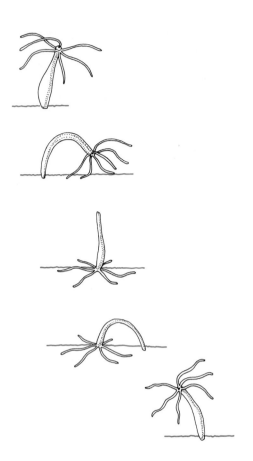

11.1   The somersaulting locomotion of hydra

11.2   The looping locomotion of a leech
The leech attaches its posterior end to the substrate by a sucker (A) and extends its body forward by contraction of the circular muscles (B). It attaches its front end by a second sucker (C) and then detaches the posterior end (D), drawing it forward by contraction of the longitudinal muscles (E). This type of locomotion, also used by the well-known "inchworm," is possible only because the force of the muscular contractions can be applied against the noncompressible body contents.

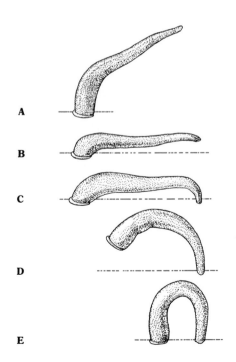

A

B

C

D

E

by means of cilia. Even today the smallest flatworms and the tiny larvae of many coelenterates depend primarily on cilia as their locomotory effectors. But cilia are practical only in very small organisms. As animals evolved larger size, they evolved contractile tissues that first supplemented and then supplanted the cilia as the chief effectors of locomotion. Though coelenterates have only very primitive contractile fibers, which do not constitute distinct tissues, rhythmic contractions of such fibers in the bell of a jellyfish enable it to swim weakly, and contractions of other fibers enable a jellyfish or a hydra to move its tentacles. The hydra is even able to move by turning somersaults (Fig. 11.1), a rather surprising type of movement in an animal with such primitive nerve and muscle cells.

Some animals, such as flatworms or snails, glide forward slowly as waves of contraction pass along their longitudinal muscles. With each wave, points on the lower surface of their bodies may advance a fraction of a millimeter; these points may then grip the substrate and, as the wave passes backward, act as anchors toward which more posterior parts of the lower surface are drawn. In this type of locomotion, little use is made of the circular muscles of the body or of the hydrostatic properties of the body contents.

In many animals the muscle fibers of the body wall are arranged in prominent longitudinal and circular layers. The fibers in these two layers are antagonistic to each other; i.e. they produce opposite actions. Contraction of the longitudinal muscles shortens the animal, contraction of the circular muscles lengthens it. Because the semifluid body contents resist compression, and thus function as a *hydrostatic skeleton*, the body volume remains constant and the shortening is accompanied by a compensating increase in diameter while the lengthening is accompanied by a compensating decrease in diameter. The looping locomotion characteristic of some worms, and particularly well exemplified by leeches (Fig. 11.2), is a type of movement that depends on the resistance of the hydrostatic skeleton.

The most complete exploitation of the potentialities of hydrostatic skeletons is seen in certain annelid worms, such as earthworms. Here the body cavity is partitioned into a series of separate fluid-filled chambers (see Fig. 8.11, p. 322). Correlated with this *segmentation* of the body cavity is a similar segmentation of the musculature; each segment of the body has its own circular and longitudinal muscles. It is thus possible for the animal to elongate one part of the body while simultaneously shortening another part. For a worm with an unsegmented body cavity, it would not be so easy to perform a variety of localized movements, because changes in the fluid pressure would be freely transmitted to all parts of the body. Segmentation carries with it, of course, a necessity for some degree of segmental organization of the nervous system and for serial repetition of other organs such as the nephridia.

Present evidence indicates that segmentation of the annelid type evolved as an adaptation for burrowing. The compartmented hydrostatic skeleton aids movement by peristaltic waves, which can develop considerable thrust against the substrate and push aside the soil particles. Though there are many unsegmented worms that burrow by peristaltic wave motions, none can develop as much thrust or burrow so continuously and effectively as segmented worms.

Many marine annelids possess paired lateral flaps on each segment called parapodia (Fig. 11.3), which may have evolved first as devices for producing

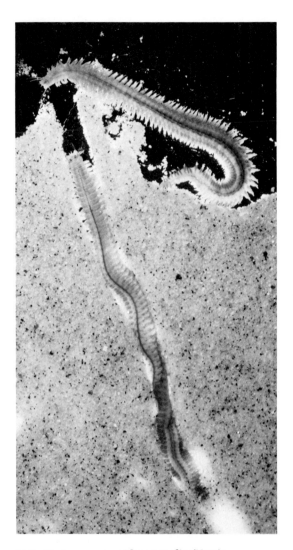

11.3 Marine worms with parapodia *(Nereis diversicolor)*
The worms get tiny particles of food and an adequate supply of oxygen as the waving parapodia produce water currents in the tubelike burrows the animals inhabit. Richly supplied with blood vessels, the parapodia also function as gills (see Fig. 6.10, p. 245). [Courtesy Hervé Chaumeton.]

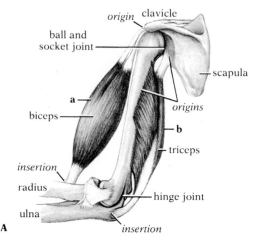

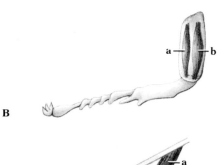

11.4 Mechanical arrangement of muscle and skeleton in a human arm and insect legs
(A) When the biceps (**a**) of the human arm contracts, the arm is flexed (bent) at the elbow. The triceps (**b**) has the opposite action; when it contracts, the lower arm is extended.   (B) the comparable muscles (**a** and **b**) in the insect leg have the same action, even though the muscles are inside the skeleton, because the fulcrum of the joint is distal to the insertions of the muscles.   (C) In another insect leg, muscles **a** and **b** have reverse actions: **a** is an extensor and **b** is a flexor. Even though both these muscles and the corresponding ones in the human arm span the joint, they have opposite actions; the reason is that the insect muscles are inside the skeleton while the human muscles are outside.

respiratory and feeding currents inside the tubes in which many rather sedentary annelids live; they often still function in this way and may also serve as primitive gills and as locomotory appendages. Those annelids whose parapodia are particularly well adapted for locomotion have, in addition to muscles arranged in longitudinal and circular layers, many large muscles running at odd angles. In part these muscles pull against other muscles that act against the hydrostatic skeleton, but in part they also pull against the body wall, which in these annelids has a cuticle much tougher and less pliable than that of earthworms. The cuticle is so tough, in fact, that these worms, because their girth is fixed and there can be little alternating swelling and constriction, generally do not move by peristaltic contractions. Many of them lack the high degree of internal segmentation of earthworms and their relatives, though they retain segmentation of the musculature of the body wall. It was probably from ancestral annelid worms of this general type that the arthropods evolved.

ANIMALS WITH HARD JOINTED SKELETONS

Exoskeletons vs. endoskeletons   The arthropods and the vertebrates are much the most mobile of the multicellular animals. Both groups possess paired locomotory appendages—legs and sometimes wings. Neither depends on a hydrostatic skeleton as the mechanical resistance against which their muscles act; each has evolved, instead, a hard jointed skeleton, with most of the skeletal muscles so arranged that one end is attached to one section of the skeleton and the other end to a different section (Fig. 11.4). Thus, when the muscle contracts, it causes the skeletal joint between its two points of attachment to bend. In many ways, then, the skeletal and muscular systems of arthropods and vertebrates show striking functional similarities. But these two great groups of animals evolved (if we read the evidence correctly) from entirely different ancestral stocks; they represent the highly successful products of two very different evolutionary lines. It is not surprising, therefore, that an examination of the bases of their striking similarities reveals equally significant differences. The two groups of animals have evolved many similar adaptations to the same functional problems, but they have arrived at those adaptations in entirely different ways.

The most obvious difference between the skeletal systems of arthropods and vertebrates is that the arthropods have an *exoskeleton*—a hard body covering with all muscles and organs located inside it—whereas vertebrates have an *endoskeleton*—a framework embedded within the organism, with the muscles outside. Besides functioning as structures against which muscles can pull, both types of skeleton are important in providing shape and structural support for large animals, particularly animals living on land, where the buoyancy of water is not available for support; in this respect they are analogous to the rigid xylem, which is a critical factor in enabling land plants to attain large size. Exoskeletons, which are composed of noncellular material secreted by the epidermis, function also as a protective armor for the softer body parts and as a waxy barrier preventing excessive water loss by terrestrial arthropods. The rib cage of the vertebrate endoskeleton protects the organs of the thorax, and the skull and vertebral column protect the brain and spinal cord.

Exoskeletons obviously impose difficulties in overall growth, and periodic molting of the exoskeleton and deposition of a new one are necessary to permit increase in size. Further, the mechanics of an exoskeletal system are such as to impose limitations on the possible size of the animal. The immense bulk that would be required in an exoskeleton strong enough to support an insect as large as a human being would pose insuperable mechanical problems. This is certainly one reason why arthropods, as varied and successful a group as they are, have never even approached the size of many vertebrates. In small animals, however, exoskeletons and endoskeletons are about equally effective, and in very small ones exoskeletons are probably superior.

The vertebrate skeletal and muscular systems   Vertebrate skeletons are composed primarily of bone and/or cartilage, two types of connective tissue mentioned earlier (see pp. 132–133).

Cartilage is firm, but not as hard or as brittle as bone. In all vertebrate embryos it is the primary component of the skeleton, and in some adult vertebrates—notably the sharks, skates, and rays—a cartilaginous skeleton persists throughout life. But in most vertebrates bone progressively replaces the cartilage as development proceeds; some cartilage is usually retained, however, where firmness combined with flexibility is needed, as at the ends of ribs, on the articulating surfaces in skeletal joints, in the walls of the larynx and trachea, in the external ear, and in the nose.

Some bones are partly "spongy," consisting of a network of hardened bars with the spaces between them filled with marrow. Other bones are more compact, their hard parts appearing as an almost continuous mass with only microscopic cavities in them. The shafts of typical long bones, like those of the upper arm and thigh, consist of compact bone surrounding a large central marrow cavity. In adults the marrow in the cavities of the shafts of long bones is primarily of the yellow fatty variety, while the marrow in the flat bones of the ribs and skull and in the ends of long bones is primarily of the red variety and is active in the production of blood cells. There is no sharp distinction beween the two types of marrow, however, and they may grade into each other. Even the most characteristic red marrow contains about 70 percent fat.

Compact bone is composed of structural units called *Haversian systems* (Fig. 11.5). Each such unit is irregularly cylindrical and is composed of concentrically arranged layers of hard inorganic matrix surrounding a microscopic central Haversian canal. Blood vessels and nerves pass through this canal. The scattered irregularly shaped bone cells lie in small cavities located along the interfaces between adjoining concentric layers of the hard matrix. Exchange of materials between the bone cells and the blood vessels in the Haversian canals is by way of radiating canalicules that penetrate and cross the layers of hard matrix.

Vertebrate skeletons are customarily divided into two components: (1) the axial skeleton, which is the main longitudinal portion, composed of the skull and the vertebral column with its associated rib cage; and (2) the appendicular skeleton, which includes the bones of the paired appendages (fins, legs, wings) and their associated pectoral and pelvic girdles (Fig. 11.6). Some bones are joined together by immovable joints or sutures, as in the case of

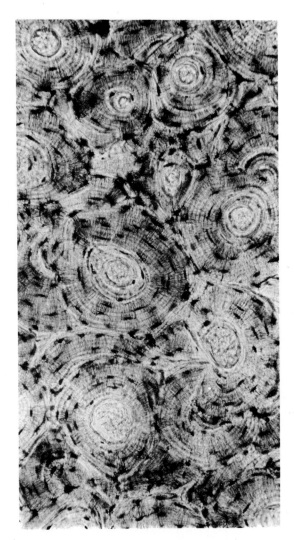

11.5   Photograph of cross section of bone, showing Haversian systems

Each Haversian system is seen as a nearly round area. The light circular core of each system is the Haversian canal, through which blood vessels pass. Around the Haversian canal is a series of concentrically arranged hard lamellae. The elongate dark areas between the lamellae are cavities, called lacunae, in which the bone cells are located. The numerous very thin dark lines running radially from the central canal across the lamellae to the lacunae are canalicules through which tissue fluid can diffuse. × 160.   [Courtesy Thomas Eisner, Cornell University.]

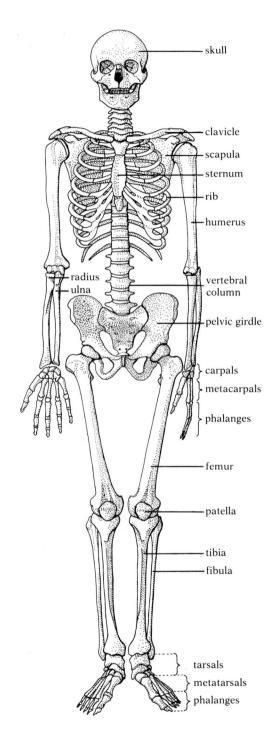

11.6   Human skeleton

the numerous small bones that together constitute the skull. But many others are held together at movable joints by *ligaments*. Skeletal muscles, attached to the bones by means of *tendons*, produce their effects by bending the skeleton at these movable joints. The force causing the bending is always exerted as a pull by contracting muscles; muscles cannot actively push. Reversal of the direction in which a joint is bent must be accomplished by contraction of a different set of muscles.

If a given muscle is attached to two bones with one or more joints between them, contraction of the muscle generally causes movement of only one of the two bones, the other being held relatively rigid by other muscles. The end of muscle attached to the essentially stationary bone—generally the proximal end in limb muscles—is called the *origin,* and the end of the muscle attached to the bone that moves—generally the distal end in limb muscles—is called the *insertion* (see Fig. 11.4A). The movable bones behave like a lever system with the fulcrum at the joint. A single muscle sometimes has multiple origins and/or insertions, which may be on the same or on different bones. The action resulting from contraction of any specific muscle depends primarily on the exact positions of its origins and insertions and on the type of joint between them.

Actually, under normal circumstances, muscles do not contract singly. The nervous sytem does not send impulses to one muscle without sending impulses to other nearby muscles. Thus the various muscles operate in antagonistic groups; if one group of muscles is strongly contracted, an antagonistic group is exerting an opposing pull, and these stretched muscles are ready to reverse the direction of the movement. In addition, other muscles (synergists) serve to guide and limit the principal movement. To understand fully the action of a muscle, therefore, it is necessary to know, in addition to its own origins and insertions, the positions, actions, and relations of its antagonists and synergists. What compounds the difficulty of obtaining a complete functional picture is a tendency to identify and study the muscles individually, although they usually act in complex coordinated sequences, producing motions that are the result of integration of numerous component motor patterns. Even the simplest action, e.g. taking a step, involves a complicated pattern of activity by a large number of muscles.

The various types of skeletal joints and the numerous muscle-joint arrangements exemplify the basic mechanical principles of pulleys, levers, braces of various types, etc., and make a fascinating subject of study for anyone interested in engineering and mechanics, but they are beyond the scope of this book.

The types of muscle   Three types of muscle tissue are recognized in vertebrates: skeletal muscle, smooth muscle, and heart or cardiac muscle.

Skeletal muscle (also called voluntary or striated muscle) produces the movements of the limbs, trunk, face, jaws, eyeballs, etc. It is by far the most abundant tissue in the vertebrate body. Most of what we commonly call meat is skeletal muscle. Each skeletal-muscle cell—or fiber, as it is usually called—is roughly cylindrical, contains many nuclei (i.e. it is coenocytic), and is crossed by alternating light and dark bands called *striations* (Fig. 11.7A). The fibers are usually bound together by connective tissue into bundles, which in turn are bound together by connective tissue to form muscles. A muscle, then, is a composite structure made up of many bundles of muscle fibers, just as a nerve is composed of many nerve fibers bound together. Skeletal muscle is innervated by the somatic nervous system.

Two types of skeletal muscle are often recognized: red muscle (or slow-twitch muscle) and white muscle (or fast-twitch muscle.) *Red muscle* has a rich blood supply, numerous mitochondria, and much *myoglobin,* a compound similar to hemoglobin that forms a loose combination with oxygen and stores it in the muscle. Red muscle oxidizes fatty acids as its primary source of energy. It contracts rather slowly, but is capable of long-term activity without appreciable fatigue. By contrast, *white muscle* has a more limited blood supply, few mitochondria, and a low myoglobin content. It depends almost entirely on anaerobic breakdown of glycogen for its energy supply. It is capable of very fast contractions and can develop great tension, but only for a short period, because it fatigues rapidly. Recent evidence indicates that red and white muscle are interconvertible; the factor that determines which set of properties a given muscle fiber will develop is the frequency of its nervous stimulation.

Smooth muscle (also called visceral muscle) forms the muscle layers in the walls of the digestive tract, bladder, various ducts, and other internal organs. It is also the muscle present in the walls of arteries and veins. The individual smooth-muscle cells are thin, elongate, and usually pointed at their ends (Fig. 11.7B). Each has a single nucleus. The fibers are not striated. They interlace to form sheets of muscle tissue rather than bundles. Smooth muscle is innervated by the autonomic nervous system. In some parts of the body (e.g. the intestinal wall and uterus), adjacent smooth-muscle cells are interconnected by gap junctions; the result is electrical coupling, which ensures contraction of the whole muscle as a single unit.

The functional differences between vertebrate skeletal muscle, which is primarily concerned with effecting adjustments to the organism's external environment, and vertebrate smooth muscle, which brings about movements in response to internal changes, are reflected in differences in their physiological characteristics. Cells of skeletal muscle are innervated by only one nerve fiber; they contract when stimulated by nerve impulses and relax when no such impulses are reaching them. Smooth-muscle cells, by contrast, are usually innervated by two nerve fibers, one from the sympathetic system and one from the parasympathetic system; they contract in response to impulses from one of the fibers and are inhibited from contracting by impulses from the other. Skeletal muscles cannot function normally in the absence of nervous connections and actually degenerate when deprived of their innervation, but smooth muscle (like cardiac muscle) can often contract without any nervous stimulation, as is commonly the case in peristaltic contractions of the intestine.

There are other contrasts: The action of skeletal muscle is more rapid, but smooth muscle can remain contracted longer. Skeletal muscle is more sensitive to electrical stimuli than smooth muscle, but the latter is more sensitive to chemical stimuli. Skeletal muscle has a definite resting length, whereas smooth muscle does not; yet smooth muscle contracts more readily in response to stretching.

Cardiac muscle, the tissue of which the heart is composed, shows some characteristics of skeletal muscle and some characteristics of smooth mus-

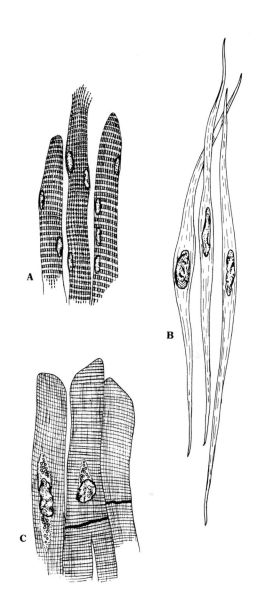

11.7 Fibers of vertebrate muscle
(A) Portions of three skeletal-muscle fibers. Each has many nuclei and is crossed by alternating light and dark bands, or striations.   (B) Smooth-muscle fibers. (C) Cardiac muscle. The thick black lines, called intercalated discs, are now known to be the places where one cell ends and the next begins. The cells of cardiac muscle often bifurcate, thus forming a complex three-dimensional network.

cle. Its fibers, like those of skeletal muscle, are striated. But like smooth muscle, it is innervated by the autonomic nervous system, and its activity is more like that of smooth muscle. Where two separate cardiac-muscle fibers (cells) meet, their adjacent membranes are so tightly appressed and so complexly interdigitated, and have so many desmosomes and other fibrous reinforcements, that for many years these areas were not recognized as cellular junctions. The sites of these junctions are visible under light microscopes as dark-colored discs (Fig. 11.7C).

The above descriptions of muscle types do not apply in all respects to the muscles of invertebrate animals. To give some examples: All the muscles of insects are striated, even those in the walls of their internal organs; many other invertebrates possess only smooth muscles. Some invertebrates have evolved arrangements enabling them to perform the same action in two different ways. Thus scallops, which swim by opening and closing their shells in a flapping motion, have two sets of shell-closing muscle fibers: a striated set, whose fast short-term action is used in swimming, and a smooth set, whose slower but longer-lasting action is used for holding the shells tightly closed when the scallop is at rest or is attacked by a predator.

THE PHYSIOLOGY OF MUSCLE ACTIVITY

Since more is known at present about the processes involved in contraction of skeletal muscle than of smooth muscle, most of the discussion below will be restricted to skeletal muscle.

THE GENERAL FEATURES OF SKELETAL-MUSCLE CONTRACTION

Individual muscle fibers resemble individual nerve cells in firing only if an impinging stimulus is of threshold intensity, duration, and rate. Like nerve cells too, vertebrate muscle fibers seem to exhibit the all-or-none property. If an excised vertebrate muscle fiber is administered a stimulus above the threshold value, the same degree of contraction is obtained whatever the value of the stimulus, provided that it is not so strong as to damage the cell.

You know, of course, that you can use the same muscles to perform tasks as different as lifting a pencil and lifting a 10 kg weight; and it is easy to demonstrate in the laboratory that an individual muscle is capable of giving graded responses, depending on the strength of the stimulation. Suppose a frog leg muscle is attached to a device (e.g. a kymograph) that will measure the extent of contraction upon stimulation. If a stimulus barely above the threshold intensity is administered, the muscle gives a very weak twitch (Fig. 11.8). If a slightly stronger stimulus is applied after a few seconds' delay, the muscle gives a slightly stronger twitch. Increasing the strength of the stimulus elicits an ever stronger contraction from the muscle, until the point is reached where further increases in the stimulus do not increase the strength of the response. The muscle has reached its maximal response.

How can these results be explained if muscle fibers give all-or-none responses? One possible explanation is that the threshold values of the different muscle fibers of which a muscle is composed are not the same. It must be kept in mind, too, that different muscle fibers may be innervated by different nerve fibers, and that these may not all fire at the same time.

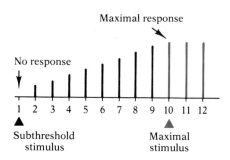

11.8   Response of a muscle to stimuli of various intensities
The numbers indicate the intensities at which stimuli are administered, and the height of the bars shows the strength of muscle response. Stimulus 1 is very weak and elicits no response; i.e. it is subthreshold. Stimulus 2 is somewhat stronger and proves to be above threshold, for the muscle contracts. Each stimulus from 3 to 10 is slightly stronger than the preceding, and each elicits a correspondingly stronger muscle contraction. Stimuli 11 and 12 are stronger than 10, but the muscle gives no greater response, indicating that 10 elicits a maximal response.

Consequently, although single fibers give an all-or-none response to stimuli, an increase in the strength of the stimulus above the threshold level may elicit a greater response from the whole muscle by stimulating more muscle fibers. Ultimately, however, all the fibers will be stimulated to respond, and the muscle will thus have reached a maximal response; increasing the intensity of the stimulus to supramaximal levels will not increase the response, there being no more fibers to stimulate. It must be stressed that the description given here applies only to vertebrate skeletal muscles. The striated-muscle fibers of invertebrates seldom exhibit the all-or-none property; the strength of their contraction is proportional to the frequency of stimulation.

The experiment described above requires an appreciable time delay between stimuli; each response is induced by a single brief stimulus, and the muscle must have sufficient time to relax fully before it is stimulated again. Such a muscular response is called a ***simple twitch.*** Let us examine its characteristics. Suppose that an isolated frog muscle has been attached to a kymograph apparatus (Fig. 11.9) in such a manner that every time the muscle is stimulated by an electric shock it moves a lever that writes on the revolving drum of the kymograph. The stronger the contraction of the muscle, the higher the lever will be pulled. If a single adequate stimulus is administered to this muscle, there is a brief interval during which no contraction occurs. This is the ***latent period*—**the interval between stimulation of the muscle and the commencement of the shortening process—which usually varies between 0.0025 and 0.004 second (Fig. 11.10). The latent period is followed by the ***contraction period,*** and this phase is followed immediately by a ***relaxation period.*** These three phases comprise a single simple twitch of the muscle.

Now let us suppose that a series of very frequent stimuli is applied to the muscle. In this case the muscle will not have completely relaxed after contracting in response to one stimulus when the next stimulus arrives. When this happens, a contraction is elicited that is greater than either stimulus alone would produce (Fig. 11.11). There has been a ***summation*** of contractions, the second adding to the first. If the initial stimulus was submaximal, the summation may be due in part to recruitment of additional muscle fibers by the second stimulus. But this is not the whole story. Summation can occur even if the individual stimuli are of maximal intensity, i.e. even if all fibers in the muscle are activated by each stimulus. Some change in the physiological condition of the muscle fibers during activity must account for

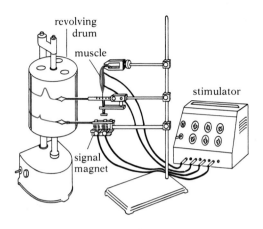

11.9 Kymograph apparatus for studying muscle contraction

The drum of the kymograph, which is covered with paper, revolves at a constant speed. The muscle is mounted in such a way that when it contracts it raises a stylus that writes on the revolving drum. Wires lead from a stimulator to the muscle and also to a signal magnet that can deflect a second stylus. At the moment when a stimulus is sent to the muscle, the signal stylus is deflected, producing a blip in the trace it is drawing on the revolving drum; one such blip is shown on the trace in the picture. This blip indicates exactly when the stimulus was administered. The stimulus causes the muscle to contract, raising the stylus to which it is attached and producing a corresponding rise in the trace the stylus is drawing on the revolving drum. As the muscle relaxes, the stylus is lowered and the trace it is drawing falls. Thus the trace drawn on the kymograph drum gives us a record of the contraction pattern of the muscle.

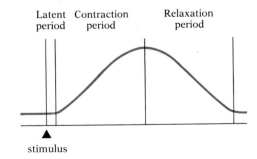

11.10 Kymograph record of a simple twitch

The duration of the latent period, which is too short to be measured even by the most sensitive kymograph, is much exaggerated here.

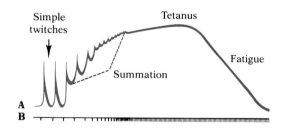

11.11 Kymograph record showing summation and tetanus

When the stimuli (line B) are widely spaced, the muscle has time to relax fully before the next stimulus arrives, and simple twitches result. (Because the drum is revolving much more slowly than in Fig. 11.10, each simple twitch is recorded as a sharp spike on the trace.) As the frequency of the stimuli increases, the muscle does not have time to relax fully from one contraction before the next stimulus arrives and causes it to contract again. The result is summation—contractions that are stronger (and hence produce taller spikes in the trace) than any single simple twitch. If the stimuli are very frequent, the muscle may not relax at all between successive stimulations; the resulting strong sustained contraction is called tetanus. If the very frequent stimulation continues, however, the muscle may fatigue and be unable to maintain the contraction.

the increased strength of contraction. It is not strictly true, then, that individual muscle fibers give an all-or-none response. Each fiber will respond maximally to a single isolated stimulus if that stimulus is above the threshold level, but the fiber may respond a little more strongly if it is given a rapid series of stimuli, because its initial contraction produces chemical changes that make it more irritable. These chemical changes may be in part due to the increase in temperature resulting from the work of contraction.

When stimuli are given with extreme rapidity, the muscle cannot relax at all between successive stimuli. Consequently the individual contractions become indistinguishable and fuse into a single sustained contraction known as *tetanus* (do not confuse this normal muscular response with the bacterial infection known also as lockjaw). For the same reasons that apply to summation, of which tetanus is a form, a tetanic contraction is greater than a maximal simple twitch of the same muscle. Normally, a high percentage of our actions involve tetanic contractions rather than simple twitches, because a volley of nerve impulses is sent to the muscle. If, however, a tetanic contraction is maintained too long, the muscle will begin to fatigue, and the strength of its contraction will fall, even though the stimuli continue at the same intensity. Fatigue is probably due in part to an accumulation of lactic acid, in part to depletion of stored energy reserves, and in part to other chemical changes.

Some muscles are never completely relaxed, but are kept in a state of partial contraction called muscle tone or *tonus.* Tonus is maintained by alternate contraction of different groups of muscle fibers, so that no single fiber has a chance to fatigue.

THE MOLECULAR BASIS OF CONTRACTION

As you would expect, the energy for muscle contraction comes from ATP. But so little ATP is actually stored in the muscles that a few muscular twitches might quickly exhaust the supply, were it not for another high-energy phosphate stored in the muscles. In vertebrates and some invertebrates, particularly echinoderms, this compound is *creatine phosphate,* formed by linkage of a phosphate group to creatine by a high-energy bond. Many invertebrates use a similar compound, arginine phosphate. These two compounds—creatine phosphate and arginine phosphate—are called *phosphagens.* The phosphagens cannot supply energy directly to the contraction mechanism of muscle, but they can pass their high-energy phosphate groups to ADP to form ATP, and the ATP can then act as the direct energy source for contraction. Enough high-energy phosphate is stored in the muscle to enable it to contract strongly during the several seconds' delay before the machinery of glycolysis and cellular respiration can be speeded up.

If the demands on the muscles are not great, much of the energy used to replenish the supply of phosphagens and ATP may come from the complete oxidation of nutrients to carbon dioxide and water. During the unavoidable delay before adjustments of the respiratory and circulatory systems increase the oxygen supply to the active muscles, some of the O_2 for oxidative phosphorylation in red muscles may come from oxymyoglobin. Myoglobin, you will recall, forms a loose combination with O_2 while the O_2 supply is plentiful and stores it until the demand for O_2 increases.

But during violent muscular activity, such as strenuous exercise or the lifting of a very heavy object, the energy demands of the muscles (especially white muscles) may be greater than can be met by complete respiration unaided, because sufficient oxygen cannot be gotten to the tissues fast enough. Under these circumstances lactic acid fermentation occurs; the muscles obtain the extra energy they need from anaerobic processes and thus incur what physiologists call an **oxygen debt.** Some of the lactic acid accumulates in the muscles, but much of it is promptly transported by the blood to the liver. When the violent activity is over, a period of hard breathing or panting helps supply the liver with the large quantities of O_2 it requires as it reconverts the lactic acid into pyruvic acid, some of which it oxidizes, using the energy thus obtained to resynthesize glycogen from the rest of the lactic acid. In this manner the oxygen debt is paid off.

If ATP is the immediate source of energy for muscle contraction, how is the ATP coupled to the contraction process and what is that process? Analysis of muscle shows that the major components of its contractile parts are two proteins, **actin** and **myosin.** V. A. Engelhardt and M. N. Ljubimova of the Academy of Sciences in Moscow demonstrated in 1939 that myosin can function as an enzyme that catalyzes removal of the terminal phosphate group from ATP. This is, of course, precisely the energy-liberating reaction we would expect to find coupled to the contraction process. Might myosin function in the intact muscle both as one of the contractile elements and as the enzyme that makes energy available for contraction? Proteins often exhibit dual functions, acting simultaneously as structural elements and as enzymes.

Evidence for the dual function of myosin in muscle contractions comes from experiments performed by Albert Szent-Györgyi in Woods Hole, Massachusetts. Szent-Györgyi showed that if actin and myosin are separately extracted from muscle, purified, and put into solution together, they will combine spontaneously to form a loose complex known as **actomyosin.** If the actomyosin complex is then precipitated and artificial fibers are prepared from it, the fibers will contract when exposed to ATP. Neither actin nor myosin alone contracts. Clearly, then, the actomyosin complex is a contractile material. And, clearly, the complex (actually its myosin component) can itself liberate from ATP the energy necessary for its own contraction, since in this experiment no other possible enzyme is present.

If actomyosin really is the contractile component of muscle, by what process does it contract? One way for a protein complex of this sort to shorten would be to fold more tightly. This hypothesis, once favored by many workers, has long since been abandoned, both on anatomical and on physiological grounds. Let us examine the anatomical evidence.

We have seen that a skeletal muscle is characterized by striations of light and dark bands and that it is composed of numerous muscle fibers (cells) bound together by connective tissue. Examination of these fibers under very high magnification reveals that they, in turn, are composed of numerous long thin myofibrils, each about 1–2 μm in diameter, with mitochondria in the cytoplasm between them. The myofibrils show the same pattern of cross striations as the fibers of which they are a part (Fig. 11.12). There is an alternation of fairly wide light and dark bands, which have been called **I bands** and **A bands** respectively. In the middle of each dark A band is a

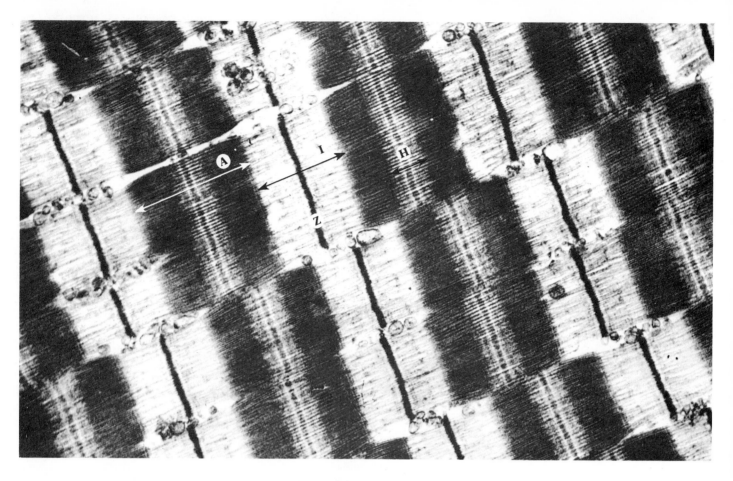

11.12 Electron micrograph of skeletal muscle from a rabbit
The myofibrils run diagonally across the micrograph from lower left to upper right; each looks like a ribbon crossed by alternating light and dark bands. The wide light bands are I bands; there is a narrow Z line in the middle of each. The wide dark bands are A bands, each of which has a lighter H zone across the middle. × 20,000. [Courtesy H. E. Huxley, Cambridge University.]

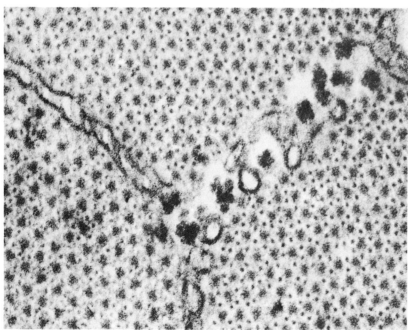

11.13 Electron micrograph of cross section of frog skeletal muscle
The thin actin filaments are arranged in a hexagonal pattern around the thick myosin filaments. Parts of three myofibrils are seen here, separated by sarcoplasmic reticulum. × 145,000. [Courtesy H. E. Huxley, Cambridge University.]

region lighter than the rest of the A band, but darker than the I bands—the **_H zone._** In the middle of the light I band is a very dark thin line called the **_Z line._** The entire region of a myofibril from one Z line to the next is called a **_sarcomere._** The sarcomeres are the functional units of muscular contraction.

Might the striations so characteristic of skeletal muscle be a structural reflection of the functional contractile units? According to a variety of evidence, yes. Chemical analysis shows that myosin is concentrated in the A bands and that actin is concentrated in the I bands. Furthermore, A. F. Huxley of Cambridge University showed in 1954 that the relative widths of the bands change as the fiber contracts; the I bands and H zones become narrower, but the A bands change very little, with the result that the A bands are moved closer together.

At about the time A. F. Huxley was performing his experiments under a high-power light microscope, H. E. Huxley (no kin to A. F.) of University College, London, began studying muscle with the electron microscope, and found that within each myofibril there are two types of filaments, thick ones and thin ones, arranged in a very precise pattern (Fig. 11.13). The two are interdigitated, with the thick ones located exclusively in the A bands and the thin ones primarily in the I bands, but extending some distance into the A bands. This distribution explains the different appearances of the A bands, I bands, and H zones. Each dark A band is precisely the length of one region of thick filaments; it is darkest near its borders, where the thick and thin filaments overlap, and lighter in its mid-region or H zone, where only the thick filaments are present (Fig. 11.14). Each light I band corresponds to a region where only the thin filaments are present. The Z line is a structure to which the thin filaments are anchored at their midpoints and against which they exert their pull during contraction; it also functions to hold the filaments in proper register. A major component of the Z line is a protein called **_α-actinin._**

The observations of the two Huxleys led each of them independently to propose a new theory of muscle contraction—that, instead of folding, the filaments telescope together by sliding past each other. If the filaments slide together, the zone of overlap between thick and thin filaments will increase until the thin filaments from the I bands on the two sides of an A band

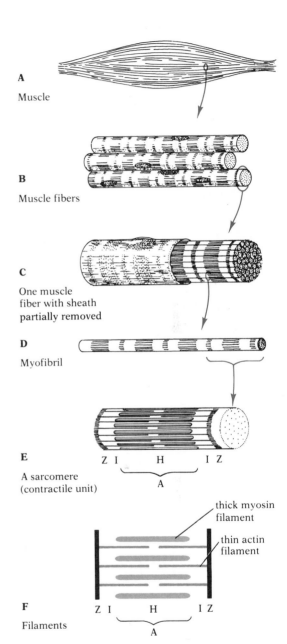

A
Muscle

B
Muscle fibers

C
One muscle
fiber with sheath
partially removed

D
Myofibril

E
Z I H I Z
A
A sarcomere
(contractile unit)

F
Z I H I Z
A
Filaments

thick myosin
filament

thin actin
filament

11.14 The component parts of skeletal muscle
The pattern of light and dark bands visible in myofibrils under high magnification (D; see also Fig. 11.12) is due to the interdigitation of actin and myosin filaments in each myofibril. As shown in (F), the A band corresponds to the length of the thick myosin filaments (color); the lighter H zone is the region where only the thick filaments occur, while the darker ends of the A band are regions where thick and thin filaments overlap. The I band corresponds to regions where only thin actin filaments occur. The Z line is a structure to which the thin filaments are fastened at their midpoints.

Relaxed

Moderately contracted

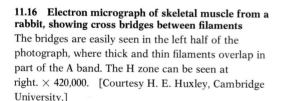

Strongly contracted

11.15 Arrangement of actin (gray) and myosin (color) filaments in a sarcomere in relaxed and contracted states

actually meet and overlap slightly; this sliding together will reduce the width of the H zone and even obliterate it entirely if the thin filaments meet (Fig. 11.15). The sliding together will also pull the Z lines closer together and greatly reduce the width of the I bands. But the width of the A bands will be minimally altered, since these correspond to the full length of the thick filaments, which remains the same (except, perhaps, for a slight crumpling due to contact with the Z lines under conditions of extreme contraction). Thus the sliding theory accounts for the changes observed in sarcomeres. Still to be answered was the question how the sliding is brought about.

Analysis shows that the thick filaments are composed of myosin and the thin filaments primarily of actin. We have already seen that myosin and actin must unite to form the actomyosin complex before acquiring contractile properties. Some sort of connection must exist, then, between the thick myosin filaments and the thin actin filaments. Indeed, electron micrographs show what appear to be small cross bridges between the filaments (Fig. 11.16). The evidence suggests that they are portions of the thick filaments (Fig. 11.17A). A single thick filament is a bundle of myosin molecules, each of which is composed of an elongated tail portion and a pair of globular heads (Fig. 11.17B, C). When a myosin molecule is fragmented by treatment with proteolytic enzymes, it is the head regions that exhibit both actin-binding activity and ATPase activity—a strong indication that they are the cross bridges.

According to the Huxley sliding-filament theory, the cross bridges act as hooks or levers that enable the myosin filaments to pull the actin filaments (Fig. 11.18). The cross bridges are regarded as movable (perhaps at the point where the globular head of the molecule joins the tail portion), and they are thought to bend toward the actin, hook onto it at specialized receptor sites, and then bend in the other direction, pulling the actin with them; they would then let go, bend in the original direction once more, hook onto the actin at a

11.16 Electron micrograph of skeletal muscle from a rabbit, showing cross bridges between filaments
The bridges are easily seen in the left half of the photograph, where thick and thin filaments overlap in part of the A band. The H zone can be seen at right. × 420,000. [Courtesy H. E. Huxley, Cambridge University.]

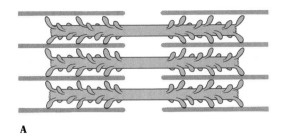

A

B

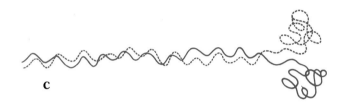

C

11.17 Molecular structure of myosin filaments
(A) Each myosin filament is linked to the adjacent thin filaments by numerous cross bridges. (B) The myosin filament is composed of a bundle of elongate molecules, each with a double club-shaped head, which acts as the cross bridge. (C) Each of the myosin molecules is thought to be composed of two intertwined polypeptide chains.

new active site, and again pull. In other words, the sliding together of the filaments would be effected by a ratchet mechanism. The necessary energy would come from hydrolysis of ATP by the myosin.

Although the remarkably precise arrangement of filaments found in striated-muscle cells has not been found in smooth-muscle cells, the evidence suggests that such cells contract by essentially the same mechanism. Actomyosin can be extracted from these muscles, and the electron microscope reveals filaments within them (Fig. 11.19).

THE CONTROL OF CONTRACTION

Like the membrane of a resting neuron, that of a resting muscle fiber is polarized, the outer surface being positively charged in relation to the inner one. Stimulatory transmitter substance released by a nerve axon at a neuromuscular junction (see Fig. 10.24, p. 426) causes a momentary reduction of this polarization. If the reduction reaches the threshold level, an impulse, or action potential, is triggered and propagated over the surface of the fiber. It has been known for a long time that the action potential somehow activates the contraction process, but only in recent years have scientists come to understand the mechanism by which the two are coupled.

In 1949 L. V. Heilbrunn and Floyd J. Wiercinski of the University of Pennsylvania injected a wide variety of substances into muscle fibers and found that the only one that caused contraction was a salt of calcium. It was reasonable to suppose, therefore, that calcium ions (Ca^{++})—of which some

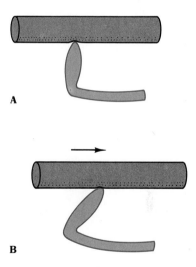

A

B

11.18 The action of a cross bridge in muscle contraction
(A) Each cross bridge, which is part of a myosin filament (color), hooks onto a thin filament (gray).
(B) It then undergoes a conformational change (here conceived as a bending to the right), which causes the thin filament to slide relative to the myosin. The cross bridge subsequently detaches from the thin filament, flips back to its original conformation, hooks on at a new binding site, and goes through the entire process again.

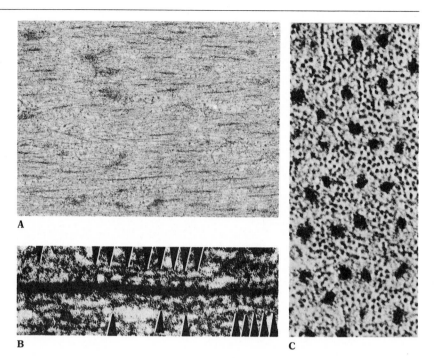

11.19   Electron micrographs of smooth muscle from a vein of a rabbit

(A) Longitudinal section (× 38,000). Filaments can be seen, but their arrangement is not as regular as in skeletal muscle, and they are not partitioned into sarcomeres.   (B) Under very high magnification (× 162,000), cross bridges can be made out (the locations of some are indicated by the dark wedges). (C) Cross section (× 186,000). The thick myosin filaments (ca. 15 nm) are surrounded by numerous thin actin filaments (ca. 5–8 nm). The ratio of about 15:1 of thin to thick filaments contrasts with the 10:1 ratio typical of striated muscle. The disposition of the thin filaments around the thick filaments does not appear as orderly as in striated muscle.   [Courtesy A. P. Somlyo, University of Pennsylvania.]

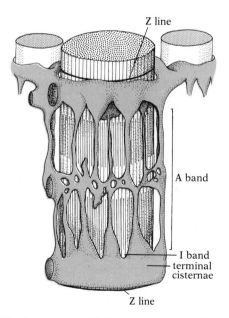

11.20   A sarcomere with its associated sarcoplasmic reticulum (in color)

normally flow into the fibers when they are stimulated—are the substance causing the cross bridges to become active. When this idea was first put forward, however, at least two major objections were raised against it. First, contraction of a vertebrate muscle fiber requires the essentially simultaneous shortening of all its many myofibrils, but the myofibrils in the center of a fiber are so far from the surface that Ca^{++} ions from outside could not possibly diffuse fast enough to reach them in the short interval between stimulation of the fiber and its contraction. Second, there is evidence that not enough Ca^{++} enters the cell to account for the sustained contraction resulting from rapid volleys of nerve impulses.

It was the rediscovery in 1955 of an extensive network of tubules in muscle fibers that opened the way to a solution of the problem of the coupling of excitation and contraction. We say "rediscovery," because the tubules had been well known to histologists before the First World War, but had been forgotten until Stanley Bennett and Keith R. Porter, working at the Rockefeller Institute, detected them again by the new technique of electron microscopy. The tubules were soon found to comprise two separate but functionally related systems: the *sarcoplasmic reticulum,* which does not open to the exterior, and the *T system* (for transverse system), which does open to the outside.

The sarcoplasmic reticulum is the muscle cell's highly specialized version of the ubiquitous endoplasmic reticulum. Its membranous canals form a cufflike network around each of the sarcomeres of the myofibrils (Fig. 11.20). At the ends of each sarcomere, in the I bands, the reticulum forms a series of sacs call *terminal cisternae.* The terminal cisternae at the distal end of one sarcomere and those at the proximal end of the next sarcomere beyond it are very close together, but lying between them, at the level of the Z line, is usually a tubule of the T system (Fig. 11.21). Though the terminal cisternae and the T tubules are in direct contact, there is no interconnection between their lumina (cavities) and hence no mixing of their contents.

The tubules of the T system can be shown to be deep invaginations of the cell membrane (Fig. 11.21). Therefore when an action potential is propagated across the cell surface, it also penetrates into the interior of the fiber via the membranes of the T tubules. The action potential moves much faster than diffusing ions, fast enough so that the stimulus for contraction can reach all the myofibrils at nearly the same instant and the myofibrils near the surface and those in the center of the fiber can contract together.

The intimate association discovered between the T tubules and the terminal cisternae of the sarcoplasmic reticulum suggested that an action potential moving along the membrane of a T tubule might alter the properties of the adjacent cisternal membranes. It was soon found that this is indeed what happens. The cisternae contain very large amounts of Ca^{++}, which, as we have already learned, causes myofibrils to contract. The action potential induces a sharp, very marked increase in the permeability of the cisternal membranes to Ca^{++} ions, allowing these to escape in large numbers (Fig. 11.22). It is this suddenly released intracellular Ca^{++} that is the direct stimulant for contraction.

The question then becomes: How does the calcium trigger contraction of the muscle fibers? To answer, we must look more closely at the structure of the thin filaments. As we have already seen, the main protein in the thin filaments is actin; in addition, these filaments contain two important regulatory proteins, tropomyosin and troponin.

The actin molecules are globular and form two helically intertwined rows along which run the long thin molecules of the first regulatory protein, ***tropomyosin*** (Fig. 11.23). In the resting muscle the tropomyosin prevents the actin from binding with cross bridges from the thick myosin filaments—probably by masking its binding sites for myosin. The molecules of the second regulatory protein, ***troponin,*** are also globular and occur in pairs near every seventh pair of actin units; each molecule has three binding sites: one for actin, one for tropomyosin, and one for Ca^{++} ions.

Whenever Ca^{++} ions are released from the sarcoplasmic cisternae, they are picked up by the calcium-binding sites of the troponin, and the troponin reacts to this binding with a conformational change. The conformational change effects a shift in the position of tropomyosin, as a result of which

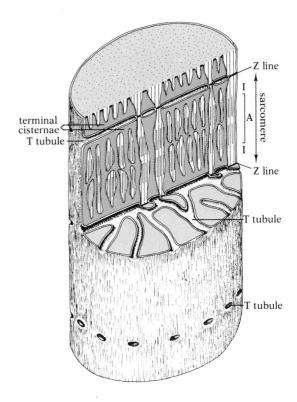

11.21 Portion of a muscle fiber, showing the relationship between the sarcoplasmic reticulum and the T-system

The longitudinal section reveals the intimate association between the terminal cisternae of the sarcoplasmic reticulum (color) and the T tubules. The cross section at the level of a Z line shows that the tubules of the T system are invaginations of the plasma membrane.

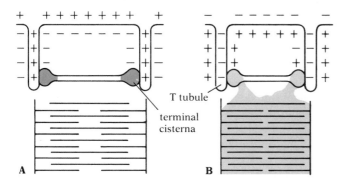

11.22 The role of calcium in stimulation of muscle contraction

(A) A sarcomere in the resting (relaxed) condition. Ca^{++} ions (color) are stored in high concentration in the terminal cisternae of the sarcoplasmic reticulum. (B) The polarization of the membranes of the T tubules is momentarily reversed during an action potential (impulse), and this reversal of polarization induces release of the Ca^{++} ions, which spread over the sarcomere and stimulate contraction.

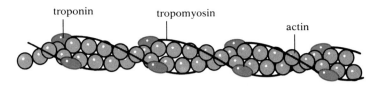

11.23   Molecular structure of a thin filament
Globular molecules of actin form two helically coiled rows. Molecules (likewise globular) of the regulatory protein troponin are evenly spaced along the rows of actin, and the long thin molecules of another regulatory protein, tropomyosin, run along the length of the rows.

tropomyosin ceases its inhibition of actin. The actin thus becomes free to bind with cross bridges from the myosin, and the contraction process is initiated.

We now have all the elements of the current model of cross-bridge action and its stimulation. Let us see how they fit together. In a resting muscle the cross bridges—i.e. the globular myosin heads of the thick filaments—have picked up ATP, but they cannot bind to the thin filaments, because tropomyosin is inhibiting the binding sites on the actin molecules (Fig. 11.24A). When stimulation from a motor nerve triggers an action potential and the action potential, transmitted along the T tubules, penetrates into the interior of the muscle fibers, the terminal cisternae of the sarcoplasmic reticulum release Ca^{++} ions. Some of these ions bind to troponin, which thereupon undergoes a conformational change displacing the tropomyosin and exposing the myosin-binding sites of the actin to the cross bridges (Fig. 11.24B). Formation of the actomyosin complex activates the ATPase site of the myosin, and ATP is hydrolyzed to ADP and P_i. This reaction brings about a conformational change in the myosin cross bridges, and the conformational change results in the power stroke enabling the filaments to slide along each other (Fig. 11.24C). When new ATP molecules bind to the myosin cross bridges, displacing the ADP and P_i formed by the previous reaction, the cross bridges break their attachment to actin and flip back to their original conformation (Fig. 11.24D).

As long as free Ca^{++} remains available, as when the nerve continues to stimulate the muscle, the cycle of cross-bridge binding, power stroke, and recovery flip can occur over and over again, as the muscle continues to contract. But if nervous stimulation ceases, the muscle relaxes, because a calcium pump in the cisternal membrane quickly moves the Ca^{++} back into the cisternae; the troponin-tropomyosin system can then resume its inhibition of the myosin-binding sites on the actin.

Note that in a relaxed muscle the myosin cross bridges are not attached to the thin filaments. The opposite occurs in vertebrates soon after death. The supply of ATP is quickly depleted, and since a cross bridge cannot separate from the actin until it picks up a new molecule of ATP, most of the cross bridges remain tightly bound to the actin of the thin filaments. The muscles are thus locked into a rigid position known as rigor mortis.

It is solely to vertebrate muscle (excluding some smooth muscle) that the above description applies in all its details. The muscle of some invertebrates (especially molluscs) lacks troponin, the binding of Ca^{++} being a function of the myosin. The muscle of still other invertebrates has both regulatory systems; i.e. both troponin and myosin can bind Ca^{++} and initiate the action of the cross bridges.

EFFECTORS OF NONMUSCULAR MOVEMENT

MOVEMENTS PRODUCED BY MICROFILAMENTS

We have seen that muscular contractions depend on microfilaments—the thick myosin filaments and the thin actin filaments. Do other types of cellular movement depend on microfilaments as well? In many cases the answer is an emphatic yes. Not only do many eucaryotic cells—nearly all, perhaps—contain microfilaments, but the most common of these are composed of actin, which often constitutes 2–4 percent (sometimes as much as 15 percent) of the total cellular protein. The technique of immunofluorescence has made it possible to visualize the arrangement of the microfilaments, as Fig. 3.35 (p. 113) shows for actin. The patterned arrays seen in the sarcomeres of striated muscle are not seen in nonmuscle cells, but the arrangements are scarcely less orderly than those in smooth muscle. Figure 11.25 shows how tropomyosin appears in a nonmuscle cell; it is probably associated with the actin filaments just as it is in muscle.

In addition to actin, nonmuscle cells usually contain myosin, although the quantity is about one tenth as great (relative to actin) as the quantity found in striated muscle; in this respect, again, there is a resemblance to smooth muscle. As in muscle, the complexing of actin and myosin activates ATPase activity of the myosin, and contractions are stimulated by an increase in free intracellular calcium. You will recall that Z lines, the structures that actin filaments pull against in striated muscle, are composed largely of α-actinin. This protein is regularly found in the plasma membrane of nonmuscle cells, where it apparently provides points of attachment for the actin microfilaments (Fig. 11.26). Evidently, the same fundamental mechanism that is re-

11.24 Model for the stimulation of muscle contraction
(A) In a resting muscle the myosin cross bridges, to which ATP is already bound, cannot bind to the actin in the thin filament, because the binding sites are masked by tropomyosin. (B) The binding of Ca++ ions to troponin causes a conformational change that slightly displaces the tropomyosin. The active sites of the actin are thus exposed, and the cross bridges bind to the actin. (C) The binding of each myosin cross bridge to actin activates the ATPase site of the myosin, and the bound ATP is hydrolyzed to ADP and inorganic phosphate (P_i). This hydrolysis of ATP provides energy for a conformational change in the cross bridge, whose bending, or power stroke, produces a sliding of the filaments one along the other. (D) If new ATP is available, it binds to the cross bridge, displacing the ADP and P_i, and causing the cross bridge to release the actin and flip back to its original conformation.

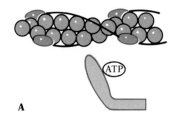

A

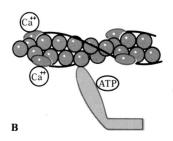

B

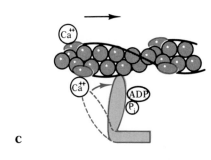

C

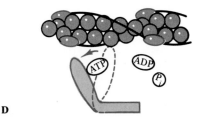

D

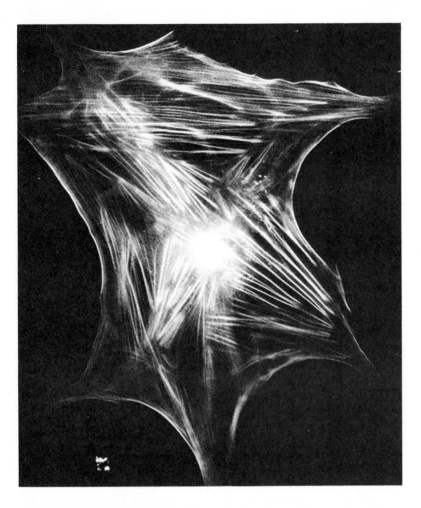

11.25 Immunofluorescent micrograph of tropomyosin in a nonmuscle cell of a rat

× 280. [Courtesy Elias Lazarides, California Institute of Technology.]

11.26 Immunofluorescent micrographs of α-actinin (left) and actin (right) in a fibroblast cell of a rat

Left: The α-actinin appears to be located at discrete points on the plasma membrane of the cell. Since in striated muscle α-actinin is the principal protein in the Z lines, to which the actin filaments are attached, it seems likely that the α-actinin in the plasma membrane provides the anchor points for actin in nonmuscle cells. Right: The fact that actin filaments run between distinct points on the membrane supports this idea. × 800. [Courtesy Elias Lazarides, California Institute of Technology.]

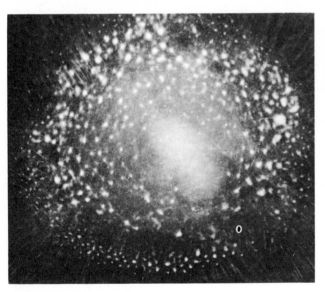

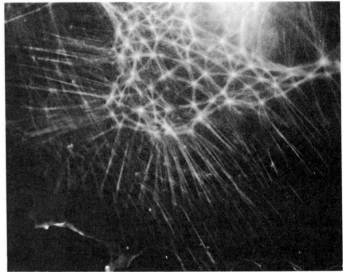

sponsible for muscular movement is at work in many other kinds of movement as well, a prime example of the unity in diversity so typical of the living world; only in procaryotic cells are actin and myosin absent.

A striking example of actomyosin-mediated nonmuscular movement in eucaryotes is found in the microvilli. As shown by Fig. 5.34 (p. 226), a bundle of microfilaments runs through the core of the microvilli of intestinal epithelial cells. Anchored by α-actinin at the tip of a microvillus, these microfilaments can cause the microvillus to move back and forth in a gentle waving motion. There is now evidence that transient microvilli, which probably function as sensory devices, form in many other types of cells too; they can suddenly pop out from the cell surface and be retracted with equal rapidity, probably through the agency of actomyosin microfilaments. Less dramatic, but of great importance in the lives of cells, are the numerous actomyosin-mediated changes of cell shape that help mold the tissues during embryological development.

A particularly important form of movement apparently mediated by an actomyosin system is cytoplasmic streaming. Whether this process occurs in plant cells (see Fig. 7.1, p. 258), in the movement of vertebrate white blood cells, or in the formation of pseudopodia by an amoeba (see Fig. 3.15, p. 93), it appears to depend on actin microfilaments. In each case treatment of the cell with cytochalasin B, a reagent that blocks the normal activity of actin microfilaments, results in a cessation of streaming. The exact mechanism of movement is still unclear, however.

Let us look in more detail at amoeboid movement. As in all cases of cytoplasmic streaming, movement is restricted to a central core of the cytoplasm, the sol-like endoplasm (Fig. 11.27). When the endoplasm reaches the advancing end of the pseudopodium, it spreads peripherally, forming a stiffer gel-like nonmoving layer, the ectoplasm. There is continuing conversion of ectoplasm into endoplasm at the rear of the advancing cell, and conversion of endoplasm into ectoplasm at the front. Endoplasm generated at the rear slides forward over the lower layer of stationary ectoplasm, which rests on the substratum, and then becomes converted into ectoplasm, while the rearmost ectoplasm is simultaneously being converted into endoplasm and becoming mobile once more. The resulting movement of the cell is much like that of a bulldozer on its caterpillar track.

As to the motive force for the endoplasm, it is not clear whether the endoplasm is squeezed forward by the contractions of microfilaments near the rear of the pseudopodium or whether, on the contrary, it is pulled by contractions near the advancing end of the pseudopodium. Possibly both mechanisms operate. But since immunofluorescent studies reveal that the actin microfilaments are much more abundant in the rear region than in the advancing portion of the pseudopodium, the push-from-behind model seems somewhat more likely.

MOVEMENTS PRODUCED BY MICROTUBULES

In addition to microfilament-mediated movements, there is a class of microtubule-mediated movements in eucaryotic cells. We saw in Chapter 3 that microtubules are composed of many molecules of a protein called tubulin

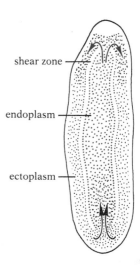

shear zone

endoplasm

ectoplasm

11.27 Amoeboid movement
The endoplasmic core slides forward between the peripheral layers of stationary ectoplasm. As the core moves forward, ectoplasm at the rear of the cell is converted into endoplasm, while endoplasm at the front of the cell is converted into ectoplasm.

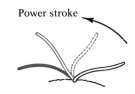

Power stroke

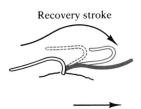

Recovery stroke

Organism moves to right

11.28   The stroke of a cilium
In the power stroke the stalk is extended fairly rigidly
and swept back by bending at its base. The recovery
stroke brings the cilium forward again, as a wave of
bending moves along the stalk from its base; at no
time during the recovery stroke is much surface
opposed to the water in the direction of movement.

11.29   Characteristic motion of a flagellum
Successive waves of bending that move along the
flagellum, from its base toward its tip, push against the
water and propel the animal forward (in this case down
the page).

(see Fig. 3.38, p. 115). Microtubules are primarily eucaryotic organelles; they occur only very rarely in procaryotic cells.

Let us turn to the beating of cilia and flagella as perhaps the most obvious examples of microtubule-mediated movement (Figs. 11.28, 11.29). As we have already seen, the stalks of all eucaryotic cilia and flagella are alike in exhibiting the 9 + 2 arrangement of internal microtubules, in which nine, usually double, microtubules near the periphery of the stalk surround two single ones located near the center (see Fig. 3.41, p. 116). Recent electron-microscope studies have revealed many more details about the relationship among the nine microtubules. The two single microtubules are surrounded by a sheath, to which each of the outer doublets is connected by a radial spoke; the outer doublets are linked, and each has two clawlike arms (Fig. 11.30).[1]

In a series of definitive experiments, Ian Gibbons of Harvard University showed that a protein called **dynein,** found in cilia, has ATPase activity and, further, that when all the dynein is extracted from cilia the doublets are left armless, and when the dynein is restored the arms reappear—a clear demonstration that the arms are the site of ATPase activity for the cilia. The resemblance of this arrangement to the localization of ATPase activity in the cross bridges of muscle is obvious. The dynein arms may well function as cross bridges between adjacent doublet microtubules. Current models of ciliary movement postulate that they do, that the arms provide the basis for a ratchetlike mechanism that enables ciliary microtubules to "walk along" or slide over one another. The models postulate, further, that because of the shear resistance within the cilium, due in large part to the radial spokes that bind all the doublets to the central sheath, the sliding of the doublets past one another brings about a bending of the ciliary stalk (Fig. 11.31).

The remarkable resemblance in mode of action between the tubulin-dynein system of cilia and flagella and the actin-myosin system of muscle might suggest that the two are evolutionarily related, that one is derived from the other. But there is no evidence to support this view. The amino acid sequences in tubulin and dynein show no similarities to those of actin and myosin. It must be concluded, therefore, that these two systems evolved their similar mechanism for producing ratchet-driven sliding motion independently—a truly impressive example of convergent evolution.

Microtubule-mediated movement is not limited to cilia and flagella. Thus in the process of nuclear division (mitosis), to be discussed in Chapter 13, a spindle of microtubules functions in moving chromosomes as new nuclei are organized (see Fig. 13.10, p. 566). And the transport of mitochondria and membranous vesicles through the axon of a neuron, from the cell body to the axonal tip (a process known as axoplasmic flow), can also be shown to depend on microtubules;[2] treatment of the axon with colchicine, a reagent that causes breakdown of microtubules and blocks assembly of new ones, inhibits the transport. According to current hypotheses, the organelles being

---

[1] The flagella seen in some bacteria lack microtubules and, consequently, the internal structure described here. The mechanism of their movement, fundamentally different from that of eucaryotic flagella, will be described in Chapter 21.

[2] Microtubule-mediated axoplasmic flow is relatively rapid (usually 100–400 mm per day). There is also a slow type of axoplasmic flow (usually 1–5 mm per day) that is not mediated by microtubules.

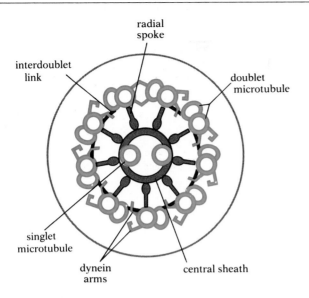

11.30 Internal structure of a cilium seen in cross section
See text for description.

transported are passed along the microtubules by arms composed of protein with ATPase activity, in a manner strongly reminiscent of the action of the dynein arms in cilia and flagella.

SUGGESTED READING

CARLSON, A. J., V. JOHNSON, and H. M. CAVERT, 1961. *The machinery of the body,* 5th ed. University of Chicago Press, Chicago. *Exceptionally clear elementary treatment of human physiology. Chapter 10 contains a good summary of muscle physiology at the nonmolecular level.*

COHEN, C., 1975. "The protein switch of muscle contraction," *Scientific American,* November. (Offprint 1329) *A more detailed account than given in this text of the way calcium, troponin, and tropomyosin interact to control muscle contraction.*

HUXLEY, H. E., 1958. "The contraction of muscle," *Scientific American,* November. (Offprint 19) *A description by Huxley of his sliding-filament model, written shortly after he developed it.*

HUXLEY, H. E., 1969. "The mechanism of muscular contraction," *Science,* vol. 164, pp. 1356–1366.

LAZARIDES, E., and J. P. REVEL, 1979. "The molecular basis of cell movement," *Scientific American,* May. (Offprint 1427)

MURRAY, J. M., and A. WEBER, 1974. "The cooperative action of muscle proteins," *Scientific American,* February. (Offprint 1290) *The structure and mode of action of the four major proteins that make up the microfilaments of muscle.*

SATIR, P., 1974. "How cilia move," *Scientific American,* October. (Offprint 1304)

SMITH, D. S., 1965. "The flight muscles of insects," *Scientific American,* June. (Offprint 1014) *The special arrangements of muscles enabling insect wings to beat hundreds of times per second.*

WESSELLS, N. K., 1971. "How living cells change shape," *Scientific American,* October. (Offprint 1233) *On the way many nonmuscle cells, especially embryonic ones, change shape or travel from one place to another by means of microfilaments and microtubules.*

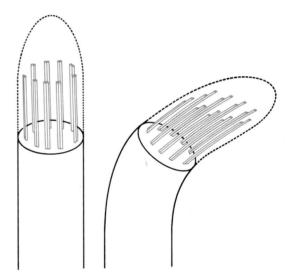

11.31 Bending of a cilium produced by sliding of microtubules past each other
The microtubules on the concave side of the bend (right) have slid tipward. Because the tubules are all interconnected, their changes in register can be accommodated only if the stalk of the cilium bends.

Chapter **12**

ANIMAL BEHAVIOR

The behavior of an animal—what it does and how it does it—constitutes one of its pre-eminent attributes: the product, as we saw in the preceding chapters, of the functions and interactions of its various control and effector mechanisms.

Although we broach the subject of animal behavior in the firm conviction that it is as valid a part of biology as anatomy or physiology, we must admit that the subject is such an enormous one that even a superficial treatment of all its many aspects, as they have been elucidated both by psychologists and by biologists, is far beyond the scope of this book. Accordingly, only a few aspects of behavior will be mentioned here—primarily those most directly related to the other topics discussed in this book and most often studied by scientists trained in biology.

A highly complex field, behavior demands the application of the most refined techniques of neurophysiology, endocrinology, effector physiology, anatomy, ecology, evolutionary biology, and mathematical analysis. It is, in short, a synthetic study, which depends for its own advances on advances in a great variety of other disciplines. It is thus not surprising that rigorous scientific investigation of behavior is a relatively recent development and that our knowledge of it is still in its infancy.

Even the mere detailed description of the behavior of an animal is a difficult task—not only because it requires incredibly careful observation over a long period of time, but also because our language lacks adequate words to convey accurately what has been seen. Almost all words have some sort of human connotation, imply some type of human motivation and purpose, which may well be irrelevant to the behavior of other animals; and we must constantly guard against unwarranted attribution of human characteristics to other species.

Some people feel so strongly about the danger of anthropomorphic or teleological[1] thinking in the study of the behavior of other animals that they try to avoid all words that betray a human bias. The result is often almost unreadable prose—and still such attempts usually fail, because less common words embodying a human point of view are simply substituted for the more familiar ones. An attempt has been made in this book to avoid the most obviously and blatantly anthropomorphic and teleological expressions. But the book has been written in English, and English (like all human languages) inevitably reflects the human activities and human interpretations around which it developed. The limitations of language being what they are, we have not tried to perform the hopeless task of scrubbing our text clean of all words or expressions that someone might take as implying human or supernatural purposiveness.

ANALYSIS OF BEHAVIOR

For centuries men have observed the behavior of animals and have explained it in terms of their own experience. Observing insects fly away at the approach of a man, they have commented that the insects see the man, are afraid of him, and fly away in order to avoid him. Observing earthworms squirm when pierced by fishhooks, they have explained that the hooks hurt the worms, causing them to writhe in pain. Observing adult birds feeding their young, they have declared that the birds love their babies and want to feed and protect them. But such descriptions are unacceptable; we have no evidence that being afraid, feeling pain, loving, wanting to do something, as those expressions apply to human beings, are meaningful when applied to insects, earthworms, or birds. Such descriptions are projections to animals of the sensations human beings would experience in similar circumstances, but as we emphasized in discussing the nervous system, sensations are products of the brain and depend on conscious awareness. Insects and earthworms have a brain so different from the human brain that extrapolations from one to the other demand the utmost caution. It is highly debatable whether insects and earthworms experience conscious awareness in any sense, and it is altogether unlikely that they experience it in the human sense. The same arguments apply, though with slightly less force, to comparisons between birds and human beings.

What, then, can we say about the insects that fly away and the earthworms that squirm on the fishhooks and the birds that feed their young? There is a centuries-old principle of logic called *Occam's razor* that says, if several explanations are all compatible with the evidence at hand, the simplest should be considered the most probable, or, to put it another way, explanations should be no more complicated than necessary. Application of this principle to our examples helps eliminate the excessively anthropomorphic interpretations first considered, in favor of simpler explanations. The insect receives visual stimuli from an approaching person, impulses travel along

[1] Anthropomorphism is attributing human characteristics to nonhuman beings and things. Teleology is the doctrine that the processes of nature are purposive and directed toward some goal. Also incompatible with scientific thinking about behavior is the so-called vitalist doctrine, which holds that life processes are not exclusively determined by the laws of the physical universe, but depend on some nonphysical "vital force."

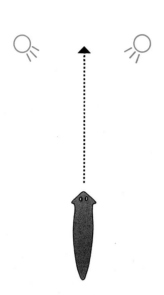

12.1 Phototaxis of a planarian
Two equally bright lights are located at equal distances from the worm. The animal moves toward a point midway between the two lights.

reflex circuits and stimulate the muscles of flight, and the insect flies away. This description is compatible with our observations, and it is undeniably simpler than the description of the insect as consciously experiencing fear and consciously deciding to fly away. "Fear" and "decide" are words that should not be applied to insects to describe their behavior. The same is true of the earthworm; "pain" may be a meaningless word when applied to such an animal. The hook certainly stimulates receptors that initiate impulses along nerve circuits, and these impulses result in squirming, but we have no basis for assuming conscious awareness, no basis for saying that the worm feels pain. Squirming, as a reflex response to stimulation of sensory receptors by fishhooks or other objects, has an obvious adaptive advantage. It might help the worm avoid further physical damage; no conscious awareness or volition need be associated with such an adaptive response. Even the more complex behavior of a mother bird feeding her young can be explained in terms of responses to stimuli within the context of the physiological condition of the bird at the time; no "love" or "desire" need be assumed.

Application of Occam's razor to the study of animal behavior was strongly urged by C. Lloyd Morgan of Bristol, England, who declared in 1893: "In no case may we interpret an action as the outcome of the exercise of a higher psychical faculty, if it can be interpreted as the outcome of the exercise of one which stands lower in the psychological scale." In other words, we should interpret the behavior of other animals in terms of the simplest neural mechanisms that can explain the observed actions. This principle, often called **Morgan's canon,** rules out conscious thought, deliberate decision, purposive determination, or foresight as explanations of animal behavior; it prohibits anecdotal descriptions and anthropomorphic interpretations. Yet, as we shall see, interpretation of all animal behavior in terms of the lowest possible psychical faculty has recently been questioned as sometimes not being the simplest interpretation, and hence as sometimes running counter to Occam's razor.

FUNDAMENTAL ELEMENTS OF BEHAVIOR

"Animal tropisms" and taxes   The revolt begun by Morgan against the old anthropomorphic approach to animal behavior found a strong supporter in the German physiologist Jacques Loeb, who from 1891 worked in the United States. Loeb was intrigued by the discoveries being made at the time in the study of plant tropisms, which reduced plant behavior to a very low psychical level indeed. The plant turns because unequal stimulation on opposite sides produces certain physiochemical changes that result in the turning response. Loeb set out to show that animal behavior rests on a similar basis, that "animal tropisms" are the elements of which more complex behavioral responses are composed.

Numerous experiments have shown that in some cases animals orient so that the critical stimulus registers equally on their left and right sides. Consider grayling butterflies, which escape from predators by flying toward the sun, thereby causing the predators to be partly blinded. A grayling orients to the sun by turning to that position in which both eyes are equally stimulated. If one eye is experimentally blinded, the butterfly flies in circles; it cannot orient toward the sun because it cannot achieve equal stimulation of the two eyes. Similarly, planarians will move toward a light by orienting in such a manner that both eyes are equally stimulated. If two equally bright lights a short distance apart are placed near a planarian, the animal will orient toward a point midway between them, thus attaining equal stimulation of the two eyes (Fig. 12.1). Admittedly, these oriented behaviors bear a superficial resemblance to plant tropisms, but the underlying mechanisms depend on nervous reflexes rather than hormonally controlled differential growth patterns, and Loeb's designation of them as "animal tropisms" is seldom adopted today; the term "tropism" is reserved for plants, and a simple continuously oriented movement in animals is called *taxis*. In the above-mentioned examples, the grayling butterfly and the planarian are exhibiting phototactic behavior.

Loeb, in pursuing his concept of "animal tropisms," put great stress on the factor crucial in plant tropisms, the relative amount of stimulation on different sides of the organism. But many taxes do not depend on a comparison within the animal's central nervous system of the incoming stimuli from two sides. We have seen that orientation of the grayling butterfly toward the sun depends on both eyes, but male graylings can orient toward a female and follow her in flight even if one eye has been blinded. Similarly, a dragonfly with only one eye can orient toward and pursue its prey. Clearly, these taxes involve mechanisms different from those in the taxes previously considered.

Kineses   Responses to simple stimuli need not involve any orientation relative to the stimulus. An example is the tendency of paramecians to congregate at a fairly precise distance from a bubble of carbon dioxide (Fig. 12.2). There is no evidence that the paramecians actually swim toward the bubble, but when in the course of their random swimming movements they chance to enter a region where the CO_2 has made the water mildly acidic, they swim more slowly. As a result, they gradually collect in the region of mild acidity. But they do not congregate in the more strongly acidic zone immediately surrounding the bubble of CO_2; whenever a paramecian moves too close to the bubble, it reacts to the stronger acidity (lower pH) by stopping, swimming backward, turning through an angle of roughly 30 degrees, and proceeding forward again. The direction of the turn is not related to the stimulation by the acid. If the paramecian again gets too close to the CO_2 bubble while swimming along its new course, it repeats the sequence of movements just described, turning through another 30-degree angle, and thus avoids the zone of excessive acidity. Avoidance of the zone of high acidity immediately adjacent to the bubble of CO_2 and the slower random swimming in the region of mild acidity together result in a congregation of paramecians where conditions are optimal. Presumably the adaptive value of this behavior is that the decay bacteria on which paramecians often feed generally lower the pH of the water surrounding them; it is therefore advantageous to the paramecians to be near regions of lowered pH, but not so near as to be damaged themselves. Such behavior, involving movements not oriented by the eliciting stimulus, is called a *kinesis.*

Another example of a kinesis is seen when sow bugs (small terrestrial crustaceans that normally live in moist places under stones or logs) are exposed to very dry conditions. Under such circumstances the animals become more active, moving about in an essentially random fashion. If, as they

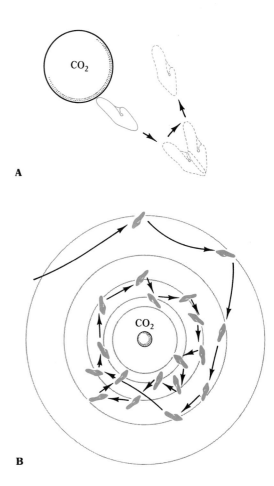

A

B

12.2 Response of a paramecian to a bubble of CO_2
(A) A paramecian encounters a bubble of CO_2, backs away, turns through an angle of about 30 degrees, and swims forward again.   (B) The paramecian tends to remain at a rather precise distance from the bubble through a combination of avoidance reactions when it is too close to the bubble and slower swimming when it is in mildly acidic regions. See text for fuller explanation.   [Redrawn from N. Tinbergen, *The study of instinct,* Oxford University Press, 1951, after Jennings.]

move, they encounter even drier conditions, they become still more active, but if they enter moister regions they move less actively; in very moist places they may completely cease moving. The result of this kinesis is that the animals tend to congregate in moist spots where they are in little danger of desiccation.

Reflexes and the more complex behavior patterns   Most taxes and kineses involve several reflexes, because the whole body is usually turned in response to the stimulus. This fact indicates that although taxes and reflexes are extremely interesting behavior patterns, of great importance in the lives of lower animals, they hardly qualify as the fundamental units on which all behavior is built. Should we then regard reflexes as the fundamental behavioral units? In a sense, yes. Reflexes are certainly more general and more fundamental than taxes or kineses. And it is true that there is no difference in kind between simple reflexes and more complex reactions; every conceivable intermediate stage exists between the simplest reflex pathway and the most complicated neural pathway. It is possible to view even the most complex behavior as the result of an intricate interaction among many enormously complex reflexes. But such a use of the term "reflex" is rather unproductive; we cannot study complex behavior in the same way as simpler reflexes. And applying the term "reflex" so broadly virtually makes it synonymous with "behavior," which doesn't help at all. It is therefore customary to restrict "reflex" to relatively simple and essentially automatic responses to stimuli.

The more complicated behavior patterns, which cannot profitably be interpreted simply as reflexes, kineses, or taxes, predominate in higher animals. When these behavior patterns are rather rigid and stereotyped, as is often the case in such invertebrates as worms and insects, and even in many vertebrates, they have often been known as **_instincts._** Though instincts were once thought to be strictly innate, the current view is that they often include learned elements and are sometimes alterable by experience, but only within rather narrow species-specific limits.

In the higher vertebrates, especially the mammals, behavior becomes increasingly modifiable by learning. Taxes and kineses, as they are usually understood, are almost nonexistent in the higher mammals, and simple reflexes, though still important, constitute only a very small portion of the total behavioral repertoire. Fairly rigid "instinctive" behavior is still present, to be sure, but highly variable behavior that few would care to call instinctive is also common. Explaining such behavior in terms of very simple elements, such as those we have considered, has proved unsatisfactory.

It is a mistake to try to apply Morgan's canon too rigidly to the complex behavior patterns of higher animals, particularly the mammals. Morgan's canon freed the study of animal behavior from the stranglehold of the uncritical anthropomorphic approach of earlier times, enabling it to develop as a valid branch of science. But, on the debit side, overzealous application led workers to underestimate the capabilities of the complex central nervous systems of higher animals. Horses and dogs, for example, have large well-developed brains with the same parts as human brains, and it is a misapplication of Occam's razor to insist on interpreting all the behavior of these animals in terms of simple stimulus-bound and automatic responses, as

Morgan's canon would have us do. It is actually simpler to assume that the brains of horses and dogs function more or less like human brains, and that these animals can think. The notion has even been entertained recently that invertebrates (e.g. bees, squids) might have some sort of awareness. Obviously, horses and dogs don't think exactly the way humans do, nor do invertebrates have the human type of awareness. The nervous systems of these animals are structurally different from those of humans, and this fact is reflected in their characteristically different behavior. But since the differences between their nervous systems and those of humans are mostly quantitative rather than qualitative, it may be too strict an application of Morgan's canon to deny them all capacity of even rudimentary thinking or awareness.

INHERITANCE AND LEARNING IN BEHAVIOR

During much of the first half of this century, a controversy raged over the relative importance of inheritance and learning in animal behavior. Some psychologists, particularly some members of the "white rat" school in America, went so far as to deny that inheritance plays any significant role in behavior. Such a view seemed preposterous to biologists, familiar as they were with the limitations as well as the capabilities of the nervous and effector systems of the various animal groups. Clearly, the developmental potential for particular nervous pathways and effectors is inherited, and an animal can exhibit only those behavior patterns for which it has the appropriate neural and effector mechanisms. Furthermore, learning itself depends on inherited neural pathways; if the necessary connections are lacking, no amount of experience can establish a given behavior pattern. But if some psychologists went to extremes in their emphasis on learning, many biologists studying animal behavior went to the opposite extreme. They greatly exaggerated the role of inheritance, claiming that it determines even the precise details of complex behavior patterns, and they unreasonably ignored the obvious importance of learning.

Fortunately, much of the furor of this old "nature vs. nurture" controversy has now subsided, and both psychologists and biologists increasingly recognize that inheritance and learning are both fundamental in determining the behavior of higher animals and that the contributions of these two elements are inextricably intertwined in most behavior patterns. As in so many instances in the history of science, insistence on an either-or approach proved sterile.

The interaction of inheritance and learning   One way of viewing the interaction of inheritance and learning in animal behavior is to regard inheritance as determining the limits within which a particular type of behavior can be modified and to regard learning as determining, within those limits, the precise character of the behavior. In some cases, as in the simplest reflexes, the limits imposed by inheritance leave little room for modification by learning; the available neural pathways and effectors rather rigidly determine the response to a given stimulus. In other cases the inherited limits may be so wide that learning plays the major role in determining the behavior elicited by a given stimulus.

A dramatic example of the interaction between inheritance and learning is the song of the European Chaffinch. W. H. Thorpe of Cambridge University raised Chaffinches in isolation and found that such birds were unable to sing a normal Chaffinch song. One might immediately conclude that the song is wholly learned. The matter is more complicated, however. Thorpe demonstrated that young Chaffinches raised in isolation and permitted to hear a recording of a Chaffinch song when about six months old would quickly learn to sing properly. But young Chaffinches permitted to hear recordings of songs of other species that sing similar songs did not ordinarily learn to sing those other songs. Apparently Chaffinches must learn to sing by hearing other Chaffinches, but they inherit an ability to recognize and respond to the songs of their own species. This conclusion was reinforced by experiments in which Thorpe played Chaffinch songs backward to his young birds; they responded to these, even though to our ears such a recording sounds completely unlike a typical Chaffinch song. Here, then, is an instance where the inherited limits ordinarily preclude an animal's learning something uncharacteristic of its species. The songs of individual Chaffinches differ slightly, and the differences are probably due, in part, to learning, but they are minor compared to those between the songs of different species.

Clearly, to ask whether Chaffinch song is inherited or learned is to ask a meaningless question. The either-or alternative is not applicable to this situation. The song is neither wholly inherited nor wholly learned; it is both inherited and learned. Chaffinches inherit the neural and muscular mechanisms responsible for Chaffinch song, and they inherit, apparently, the ability to recognize a Chaffinch song when they hear it; they also inherit severe limits on the type of song they can learn, but the experience of hearing another Chaffinch sing is necessary to trigger their inherited singing abilities into action, and in this sense their song is learned.

Human speech obviously involves a much greater learning component than Chaffinch songs. Human languages differ from one another much more, and the inherited limits for human language must be far less restrictive. But this does not mean that inheritance plays no role in human speech. The very fact that all efforts to train apes to speak have failed is evidence enough that human speech depends on human genes.[2] The relative roles of inheritance and learning are very different in human speech and in Chaffinch song, but for the one as for the other a simple either-or choice is unrealistic. It is meaningful and useful to ask questions concerning the respective roles of inheritance and learning in any specific behavior pattern (though precise answers to these questions are often hard to obtain), but it is not meaningful or useful to insist that either is sufficient without the other.

In general, the inherited limits within which behavior patterns can be modified by learning are much narrower in the invertebrates than in the vertebrates, and narrower in the so-called lower vertebrates (fish, amphibians, reptiles, and birds) than in the mammals. This is a valid generalization, but it does not mean that the limits of *all* insect or avian behavior are narrower than the limits of *all* mammalian behavior. In any animal the

---

[2] The failure to teach apes to speak more than a few simple words is probably due in part to their having very little tendency to imitate sounds, in part to anatomical limitations. Their larynx is not in the same position, relative to the pharynx, as the larynx of human adults. As we shall see later, the inability of apes to speak does not necessarily mean that they are incapable of using language.

limits for different behavior patterns are different; each animal, whether insect, bird, or mammal, has some behavioral traits that are rather rigidly determined by inheritance, with very little possibility of modification by learning, and other behavioral traits capable of much modification.

Such a difference can be observed in Herring Gulls. The adult birds nesting in colonies learn to recognize their own young about five days after they hatch; thereafter, if young of the same age from another nest are substituted for their own, the adults will not accept them and will neglect or even kill them. Yet these same gulls show amazingly little aptitude for learning to recognize their own eggs. They can be given substitute eggs quite different in color, pattern, shape, or size and will accept them without hesitation. These gulls, then, exhibit great aptitude for learning to recognize individual young so alike in appearance that human beings can tell them apart only with great difficulty, if at all, but they show very little aptitude (at least in the experimental context) for learning to recognize eggs so different that human beings can distinguish them at a glance. From an evolutionary point of view, this difference is readily understandable. There must have been far more selection pressure for evolution of recognition of young, which might stray from their own nest under normal circumstances, than for recognition of individual eggs, which in nature seldom wander from nest to nest.

Some difficulties in studying learning   It is probably impossible to separate completely what is inherited and what is learned in any behavior pattern. Behavior is not a simple combination of these two elements, but the outcome of a fusion between them. So far as the part played by learning can be distinguished, however, the learning process is one well worth studying. Before we examine some commonly recognized categories of learning, we should mention several factors that complicate its study.

1. It is often difficult to determine whether improvement in the performance of a behavior pattern is due to experience or simply to greater maturity or to a different physiological condition. For example, observations that young birds just leaving the nest cannot fly well, but improve rapidly over the next few days, have led to the widespread belief that the birds must learn to fly and that they improve with practice. But, as repeated experiments have demonstrated, when young birds are reared in narrow tubes or other devices that prevent them from flapping their wings and are released at an age when they normally would have already "learned" to fly, they are able to fly as well as control birds raised under normal conditions. In other words, it is not practice that causes the flight of a newly fledged bird to improve; it is greater maturity. Numerous other examples could be cited of improvement that appears to be a result of learning but is really a result of maturation.

Injections of hormones often cause behavior patterns to change in ways previously thought to be produced only by learning experiences. This is a further demonstration that one must be exceedingly careful to rule out physiological changes as a possible cause for a particular behavioral change before asserting that it is due to learning.

2. An animal may readily learn something in one context and be completely incapable of learning the same thing in another context. Thus one may erroneously conclude that the animal cannot learn something when, in fact, the negative results are simply a product of the experimental situation used.

As an example, let us consider the behavior of a female of the digger wasp *Ammophila*. This wasp digs two or three burrows, provisions each with a caterpillar she has paralyzed, and lays an egg in each. After the eggs have hatched and the larvae have begun to consume the caterpillars, she spends several days provisioning the burrows with additional caterpillars before she closes them and leaves them forever. G. P. Baerends, then of the University of Leiden in the Netherlands, investigated the factors that determine how much food will be brought to each burrow. He showed that the wasp inspects all her burrows early in the morning and then spends the rest of the day provisioning the one that contained the least food. If, after completion of her inspection visit, food is experimentally added to the least provisioned burrow, her behavior is unchanged; she continues trying to stuff more caterpillars into the already overflowing burrow. If an experimenter were unaware of the inspection visit, and performed an experiment of this type during the late morning, he might well conclude that the wasp instinctively spends a full day provisioning each burrow and that she is incapable of learning by observation of the burrow how much food is needed. Here, then, is a situation where the wasp can learn by inspection in the context of the first visit of the day, but cannot learn the same thing in the context of later visits that day.

3. An animal may be able to learn certain behavior patterns only during a rather limited *sensitive phase* in its life. If it does not encounter the necessary learning situation during the sensitive phase, it may never learn the behavior. Exposure to the learning situation before or after the sensitive phase may be ineffective in producing learning. For example, as Thorpe demonstrated in his studies of the development of Chaffinch song, unless young Chaffinches hear a Chaffinch song during a certain period in their development, they never learn to sing properly, despite frequent later exposure to singing Chaffinches. Sensitive phases are seldom so rigid in human beings, but there is abundant evidence that various types of learning ability are greatest at certain ages. For example, children between the ages of two and ten can learn languages far more easily than adults.

4. It is not always possible to tell immediately whether or not learning has taken place. There may be considerable delay between the ***latent learning*** that occurs on exposure to the learning situation and the performance of a behavior pattern that shows the effects of learning. For example, if young Chaffinches only a few weeks old are allowed to hear a tape recording of a singing adult for a few days and are then raised in isolation, they will sing a nearly normal Chaffinch song when they first begin to sing the following spring. Exposure to the song during their first summer, long before they themselves are old enough to sing, results in latent learning, but proof does not come until months later.

5. Comparisons between the learning capabilities of different species may be ill-founded and are often misleading. As might be deduced from the differences in their nervous systems, the superficially similar learning they exhibit may actually involve different underlying mechanisms and fulfill entirely different functions in the lives of the animals. Thus, as shown by T. C. Schnierla of Columbia University, rats and ants can learn to run the same maze, but they do so in very different ways; rats appear to learn a "map" of the maze as a whole, whereas ants appear to learn the maze as a series of

separate problems, one at each choice point. Mastery of the maze tends to improve the performance of rats when they are subsequently placed in new mazes, but it actually seems to hinder the performance of ants in new mazes. In other words, rats not only learn the particular maze but can also generalize to some extent from this experience and thus develop increased competence at maze running in general, whereas ants learn only the particular maze and this achievement makes their behavior in new mazes less flexible.

Despite the obvious dangers of the comparative approach, many papers have been published purporting to measure the relative intelligence of animals as different as fish, pigeons, rats, and monkeys by exposing them to the same problem-solving situation. Results from such experiments are highly suspect, because animals as different as these have entirely different modes of life and are likely to have evolved radically different levels of response to any given type of stimulus, depending on its importance in their lives under natural conditions. Such experiments are thus usually a measure not so much of differences in intelligence as simply of differences in response to the particular stimuli in the particular context of the experiment.

Because of the bias of the experimenters, the "intelligence tests" are commonly set up in a context that is biologically most meaningful to mammals, and hence the mammals are found to be the most "intelligent"; if the experiments were set up in a context biologically most meaningful to fish, quite different results might be obtained. This does not mean that one cannot establish a definition of intelligence consistent with its application to human beings and then try to test animals for "intelligence" as thus defined, but it does mean that the test should be adapted to the animal in question and that the biological implications of the results should be considered in their interpretation. It is certainly true that from the point of view of their total behavior mammals surpass other vertebrates in learning potential.

Types of learning   There are many classifications of learning. We shall restrict discussion here to some of the categories recognized by one of the most commonly used systems. All forms of learning involve, by definition, relatively enduring changes in behavior due to experience rather than maturation. More transient changes, such as those due to sensory adaptation, fatigue, fluctuations in physiological condition, and differences in motivation, are not considered to be learning.

1. One of the simplest types of learning is *habituation,* a gradual decline in response to "insignificant" stimuli on repeated exposure to them without any positive *reinforcement* (reward) or punishment. In effect, it is a learning to ignore stimuli that are unimportant in the life of the animal. Its relative durableness distinguishes it from sensory adaptation or fatigue.

2. *Conditioning* (sometimes called associative learning) is the association, as a result of reinforcement, of a response with a stimulus with which it was not previously associated. The simplest form of conditioning is seen in conditioned reflexes, first studied scientifically by the great Russian physiologist Ivan Pavlov. Pavlov rang a bell each time he fed meat to a group of dogs, and eventually the salivary reflex of the dogs became conditioned to the auditory stimulus of the ringing bell. Pavlov could then ring the bell, and the dogs would salivate even if they could not see, smell, or taste meat. A new reflex had been established, presumably by facilitation of neural pathways pre-

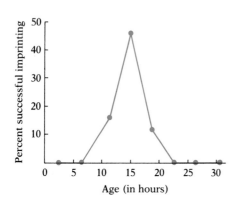

12.3   The sensitive phase for imprinting the following response of ducklings
Individual ducklings of various ages (in hours) were exposed to a moving decoy for one hour. They were then tested to see whether they had become imprinted on that object. The results showed that imprintability was at its peak when the birds were 15 hours old. Exposure to a suitable object too early or too late was ineffective.   [From E. H. Hess, *Science*, vol. 130, July 17, 1959.]

viously unused; a stimulus elicited a reflex response that it had never elicited before the training. The new stimulus (sound of ringing bell) had apparently been associated in the dogs' nervous systems with the original stimulus (sight, smell, or taste of meat), and the same response was now given to both.

Conditioning is not restricted to behavior patterns as simple as reflexes. Animals may be conditioned to perform such complex activities as running, pushing levers, opening doors, and performing complicated tricks. Much of the training of domesticated animals such as cats, dogs, and horses is based on conditioning; the animals learn to associate stimuli such as whistles or spoken commands with responses not normally elicited by such stimuli.

Conditioning is often used in testing the sensory capabilities of animals. To determine, for example, how sensitive a dog's hearing is to sounds of different frequencies, one can condition a dog to respond in a certain way to sounds of a particular frequency, say 1,200 Hertz (cycles per second), and to respond in a different way to sounds of all other frequencies. By then exposing the dog to sounds of frequencies differing only slightly from 1,200 Hertz, and noting the responses, the experimenter can determine the smallest differences in frequency to which the dog is sensitive. (It has been found that dogs can discriminate between sounds differing by as little as 2 Hertz.)

3. Very common is ***trial-and-error learning*** (sometimes called operant conditioning). An animal does something. If the result is rewarding, it may do the same thing again. If the result is not rewarding, or is disagreeable, it may, after several trials, learn not to do the same thing anymore. This is, of course, a type of learning with which we are all personally familiar. We learned as children not to touch hot stoves because we tried it and were burned. We learned to eat candy by trying it and liking the taste. We learned to play games like baseball or tennis by trying the various actions involved and profiting from our mistakes and successes. Much psychological research has been devoted to studying the sorts of influences that speed up or slow down trial-and-error learning, as in the running of mazes by experimental animals.

4. ***Imprinting,*** in which an animal learns to make a strong association with another organism or sometimes an object, is characterized by a short sensitive phase, which in all cases so far known occurs early in the animal's life (Fig. 12.3). The concept of imprinting was first formulated in 1935 by the great Austrian zoologist Konrad Lorenz, one of the fathers of modern ethology (the study of animal behavior from a biological point of view). He was studying birds such as geese, chickens, and partridges whose young, being precocial, are able to move around and feed themselves soon after hatching. He found that the young of such species will follow the first moving object they see, and form a strong and lasting attachment to it, particularly if the object emits a sound. In effect, they adopt this object as their parent. Ordinarily the first moving vocal object such a young bird sees is its mother, and imprinting on her has obvious survival value. But under experimental conditions the young bird may be imprinted on a toy train or a box pulled around by a string (especially if the box contains a loudly ticking clock or some other sound-producing device), or even on a dog, cat, or human being. Once the sensitive phase, usually only about 36 hours, has passed and the young birds have been imprinted on such surrogate mothers, they cannot be imprinted on any other object, including their true mother. Such filial im-

12.4 A male Zebra Finch (left) courting a female Bengalese Finch (right)
The male was raised by Bengalese Finch foster parents, and was imprinted on them.   [Courtesy Klaus Immelmann, University of Bielefeld.]

printing is important in establishing a bond between the young and their mother under natural conditions.

Imprinting may also play an important role in establishing proper species recognition and interaction, especially with regard to choice of a mate. As an illustration of the lasting effects of sexual imprinting, we can mention experiments by Klaus Immelmann, now of the University of Bielefeld in Germany, in which young Zebra Finch males were raised by foster parents of another species (Bengalese Finch). On reaching sexual maturity, the male Zebra Finches invariably courted female Bengalese Finches when given a choice between them and females of their own species (Fig. 12.4). If one of the male Zebra Finches was put in a cage with only a female Zebra Finch, he would eventually court her and even successfully rear young with her, but if then once again given a choice between female Bengalese Finches and female Zebra finches, the male would revert to his preference for Bengalese females. The experience of being paired with a female of his own kind (even for as long as seven years) had not extinguished the effects of the male's early imprinting. The sensitive phase for such sexual imprinting does not occur until well after the sensitive phase for filial imprinting has ended; it is therefore clear that these are two separate phenomena and not merely different manifestations of a single learning event.

Imprinting is an interesting example of the interaction of inheritance and learning. Inheritance determines the sensitive phase, the class of objects to which the response may be directed, the tendency to respond promptly and strongly to the first object in that class to which the animal is exposed, and the near irrevocability of the attachment once it is formed, while learning establishes the tie between the animal and the particular object on which the imprinting occurs.

5. *Insight learning* is most prevalent in the higher primates, particularly human beings; some workers prefer to call it reasoning, and to distinguish it from learning as such. Essentially, insight is the ability to respond correctly the first time to a situation different from any previously encountered. Through insight, an animal is able to apply its prior learning in other situations to a new situation and, in effect, solve the new problem mentally without the necessity of overt trial and error (Fig.12.5).

12.5 Lack of insight in a raccoon
Tied to stake A, the animal cannot quite reach the food dish as long as its leash is looped around stake B. In this situation a human or a chimpanzee would immediately turn, walk around stake B, and go to the food, without any previous experience of such a situation. But the raccoon will not perform the task correctly at first; it must find the solution by trial and error, though once it has done so it will learn very quickly.   [Modified from *High school biology*, Rand McNally, 1963. Used by permission of the Biological Sciences Curriculum Study.]

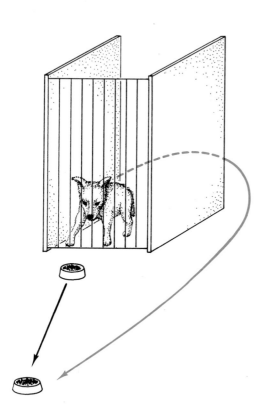

12.6 The effect of experimental details on insightful behavior by a dog
When the dog sees food immediately on the other side of a fence, it tries to get through the fence rather than detouring around the walls of the corridor. But if the food is moved farther away from the fence (black arrow), then the dog quickly turns and goes around the wall to the food (color arrow). The animal gives no indication of insight in the first situation, but does show insight in the second.   [After Köhler.]

As with so many other kinds of behavior, experiments on insight learning may yield very different results when seemingly small details of the experimental design are altered. Consider, for example, an experiment in which a hungry dog, in surroundings it knows well, finds itself in a corridor with a fence at one end, through which food can be seen immediately on the other side (Fig. 12.6). The dog will usually strive to get through the fence to the food, as though it did not remember the surroundings or the way out of the corridor to the site of the food. But if the food is moved away from the fence, as shown in the figure, the dog will promptly run around the wall and reach the food. In other words, in a situation where the dog and the food are very close, but separated by the fence, the dog appears to make no use of its previously learned knowledge of the area; it shows no insight. But when the distance between the dog and the food is greater than some critical value (which can be determined experimentally), then the dog does draw on its prior learning to solve this new problem; it does show insight. The moral, of course, is that one must always be wary of drawing general conclusions from results that may hold only for a very limited set of conditions.

It is important to distinguish insight from simple *generalization,* which is a characteristic feature of most learning. A dog conditioned to salivate on hearing a sound of 1,200 Hertz will also salivate to some extent if it hears sounds of 1,000 or 1,400 Hertz. Or a pigeon trained to peck a red button to receive food may peck an orange one if a red one is not available. These are examples of generalization, the ability of an animal conditioned to one stimulus to respond in the same way to other similar stimuli. In simple generalizations the new stimuli differ very little from previously experienced ones; in manifestations of insight they differ much more.

Insight is the ability to respond correctly to new stimuli that may be qualitatively as well as quantitatively different from previous ones. It often requires the mental putting together of several elements originally learned separately. For example, a hungry chimpanzee, released in a room with various boxes scattered around the floor and with a bunch of bananas hanging from the ceiling above his reach, will often survey the situation for a short time and then begin gathering the boxes and piling them on top of each other under the bunch of bananas. He can then climb on top of the boxes and reach the bananas (Fig. 12.7).

Many other more restricted types of learning are recognized by some scientists, including such complicated processes as concept formation, principle learning, and symbolism. These, as you would expect, are for the most part limited to higher mammals, particularly primates. It is beyond the scope of this book to discuss all such forms of learning, though we shall return to the subject of symbolism later in this chapter.

MOTIVATION

Motivation can be defined as the internal state of an animal that is the immediate cause of its behavior. All forms of learning are dependent on adequate motivation in the animal; so, too, is the performance of learned or inherited behavior patterns. But to say that behavior is dependent on motivation is to say no more than that behavior depends on its causes. It is when we try to say something more meaningful, more precise, about motivation that we encounter difficulty.

Students of behavior often speak of motivation in terms of drives or tendencies. They say that an animal experiences hunger drives, thirst drives, sex drives, attack drives, escape drives, etc.[3] But what determines these drives or tendencies? This is not an easy question, and we can only give a partial answer to it. Obviously, no one factor can be said to determine a drive; many factors must interact in intricate ways. In gross terms, however, we can say that all of the following play important roles in higher animals: the general health of the animal, its maturational state, hormones, integrative activity of the central nervous system, sensory stimuli, and previous experience that led to learning.

Clearly, the relative contributions of these elements vary with the behavior pattern. Thus hormones and learning are of little significance in simple behaviors like the knee-jerk reflex, where sensory stimuli are the predominant causal elements, but hormones are of enormous importance in the reproductive behavior of cows, which will not receive the bull in copulation when the level of their sex hormones is low. Similarly, the relative contributions of the motivational elements vary in different animal species, even when they are performing analogous behavior patterns. Thus the relative contribution of hormones to the sex drive is considerably smaller in primates, particularly humans, than in rats, dogs, or cows, while the relative contributions of sensory stimuli, activity of the cerebral cortex, and learning are greater.

All available evidence points to the hypothalamus as the part of the brain that acts as the principal controller of behavioral drives in higher vertebrates. Located here are the excitatory and inhibitory centers for the various drives. Arousal of a particular drive must result from activation of its hypothalamic excitatory centers through the action of hormones, incoming impulses from sensory receptors, and impulses from other parts of the brain. Similarly, reduction or satiation of a drive must result from activation of its hypothalamic inhibitory centers and from reduced stimulation of its excitatory centers.

The fact that positive reinforcement facilitates learning implies that behavior can be goal-directed—not generally in the sense that animals consciously decide on a goal and then strive to attain it, but rather that such behavior tends to lead to the fulfillment of biological needs of the animal, that it tends to be biologically functional and adaptive. This is true of both innate and learned behavior patterns. Let us examine feeding behavior as an example of motivated behavior directed toward meeting certain biological needs.

When the hypothalamic excitatory centers for hunger become activated, usually as a result of such influences as a low glucose concentration in the blood, the strength of the animal's hunger drive rises, or, to put it another

12.7 Insight learning by a chimpanzee
Unable to reach the bananas by jumping, the chimp solves the problem by making a stack of boxes from which he can reach the bananas.

---

[3] Much caution is necessary when one uses such terminology. To ascribe all the complex behavior involved, for example, in finding, capturing, and eating food to activation by a single "hunger drive" seems to imply that all the diverse components of this behavior depend on the same underlying physiological mechanisms. But they do not. The concept of a unitary drive such as hunger may be useful for descriptive purposes, but as an explanation of behavior it may easily lead to sterile theorizing and formulation of the wrong questions.

way, behavior of a sort that may result in feeding becomes more highly motivated. This behavior may include both learned and innate components, and it may involve many sorts of activities. Thus the animal will usually begin generalized ***appetitive behavior,*** manifested first by restlessness and a tendency to move around in what appears to be a rather random manner, and later by activities that become more obviously coordinated into some form of "searching" behavior. If the appetitive behavior results in the finding of food, it may be followed by a second type of behavior, called the ***consummatory act,*** which consists in eating the food and entails a variety of motor patterns such as tearing, chewing, and swallowing.

Once the food has been eaten, the animal's behavior may well undergo radical changes; searching activities may cease, and some completely different behavior, such as courtship, or nest building, or preening, or sleeping, may begin. We can say, then, that if the appetitive behavior caused by one drive results in attainment of its biological goal, i.e. the performance of the consummatory act, the strength of that drive is then diminished, and the animal's next behavior pattern will be determined by whatever other drives are now strongest. Attainment of a goal tends to satiate the corresponding drive and leave the animal free to respond to other drives. But attainment of the goal also acts as positive reinforcement to the successful appetitive behavior patterns. Thus, if the behavior is of a sort that can be modified by learning, the animal will be more likely to perform similar activities when the drive is again high than to perform activities that did not lead to goal attainment.

Sometimes the goal of motivated behavior is less obvious than in feeding. Feeding or drinking supplies materials necessary to the continued existence of the organism; hence such behavior can be said to help maintain homeostasis—an equilibrium among the animal's various physiological functions and between the animal and its environment. Similarly, escape from pain plainly contributes to homeostasis, because pain functions as a warning to the animal that the stimuli involved may well be harmful. But what about mating, or nest building, or care of the young? Such behavior patterns involve strong drives, and they obviously contribute to continuation of the organism's genetic lineage. But can they be said to contribute to maintenance of stability in the individual organism? We must apparently answer in the affirmative. Such behavior seems to fulfill biological needs built into the animal's nervous system; if those needs are not met, they apparently constitute a source of internal instability. Behavior, like the regulatory mechanisms of organisms already studied, seems to function, then, in maintaining homeostasis. Under natural conditions anything that contributes to homeostasis constitutes a positive reinforcement, and anything that contributes to instability constitutes a negative reinforcement.

THE CONCEPT OF SIGN STIMULI

Let us now turn to the stimuli that, detected by the sensory receptors, are capable of triggering behavior patterns. We shall restrict ourselves here to behavior patterns that are comparatively stereotyped, since these are generally easier to analyze and have been studied extensively by ethologists.

When we examined the provisioning behavior of a female *Ammophila* digger wasp, we saw that that behavior is not materially altered by the stimuli the female encounters on her provisioning visits to her burrow following the early-morning inspection visit. She surely sees and feels and smells the extra caterpillars put in the burrow by the experimenter, but her provisioning behavior remains unchanged. It is as though the behavior, once triggered by the proper stimulus during the inspection visit, must run its course regardless of later stimuli. Or consider the case of male *Aedes* mosquitoes attracted to a tuning fork producing a sound similar in pitch to that of a female mosquito in flight (Fig. 12.8). The males can surely detect by sight and smell that a tuning fork is different from a female mosquito, yet the sound continues to attract them. Similarly, males of some species of insects will attempt to copulate with bits of paper on which female scent has been placed, even though they can certainly see and feel the difference between the paper and a female insect.

Observations of this type lead to the conclusion that an animal responds at any given time only to a limited number of the thousands of stimuli its receptors are detecting. In a somewhat less rigid way, the same is true of human beings. A man driving an automobile along a highway sees hundreds, or perhaps thousands, of objects every minute—trees, telephone poles, houses, grass, clouds, etc.—but (one devoutly hopes) he does not respond to all of them. His driving behavior is determined largely by certain limited classes of stimuli, including the visual stimuli that indicate the width and curvature of the road, the visual stimuli from oncoming automobiles, and the auditory stimuli from automobile horns. Furthermore, a careful analysis would doubtless show that a driver responds to only a few of the clues that could provide him with information about road conditions; in other words, he ignores not only irrelevant stimuli but many potentially relevant ones as well.

From an evolutionary standpoint it is readily understandable that animals should be prompted to action by only a few of the many stimuli they encounter. An animal cannot possibly respond to all the stimuli impinging on its receptors. It must be selective. In instinctive behavior natural selection seems to have led to the evolution of special behavioral sensitivity to a few stimuli that under natural conditions would be reliable clues to the situation in which an animal found itself. Thus the condition of a wasp's burrow at the time of the inspection visit would normally be a reliable clue; under natural circumstances extra caterpillars seldom materialize in the burrow by another agency than the wasp's. Similarly, tuning forks and bits of paper specially impregnated with the scent of a female insect are not common in the insect cosmos. In short, the animal's seemingly blind, rigid response to certain stimuli to the exclusion of others is biologically functional and adaptive most of the time; it is only when an unpredictable factor like a human being intervenes that things go wrong.

Ethologists have called the stimuli that are particularly effective in triggering behavior *sign stimuli* (when the sign stimuli are emitted by a member of the same species, they are called social releasers, or simply releasers). The animal must possess neural mechanisms selectively sensitive to the sign stimuli; it is these *releasing mechanisms,* as they are called, that initiate the behavior when they are activated by the sign stimuli appropriate to them.

In summary, then, the ethologists' model holds that certain limited parts

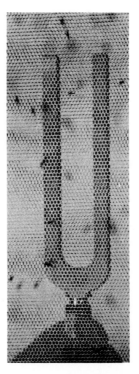

12.8 Photographs showing response of male mosquitoes to a tuning fork
Left: The fork is silent.   Right: The fork is vibrating and its sound is attracting males.   [Courtesy E. R. Willis, Illinois State University.]

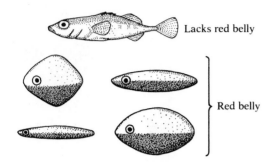

12.9 Models of male stickleback
The realistically shaped model lacking the red belly
was attacked by male sticklebacks much less often
than the oddly shaped models with red bellies.
[Modified from N. Tinbergen, *The study of instinct,*
Oxford University Press, 1951.]

of the animal's environment are especially potent in triggering behavioral
responses, i.e. in providing sign stimuli to which releasing mechanisms in
the animal are sensitive; if motivation is sufficiently high, the releasing
mechanisms then activate the neural pathways necessary for performance
of the appropriate behavior patterns. It should be emphasized that the term
"releasing mechanism," as currently used, is not meant to apply to any
particular part of the nervous system or to any particular type of neural
function; indeed, the releasing mechanisms for different behavior patterns
may well have entirely different neurological bases. The term is simply a
nonspecific designation for neural mechanisms selectively sensitive to certain stimuli—the sign stimuli—and functioning in the coupling of these stimuli to behavior patterns regularly associated with them.

Let us examine some of the actual behavior patterns analyzed by ethologists in formulating their view of behavior.

Niko Tinbergen and his associates at Oxford University studied fighting
between male stickleback fish in spring. In the spring the throat and belly of
the males become intensely red. It seemed probable, therefore, that the red
color was an important stimulus. The investigators presented their subjects
with a series of models, some quite like actual male sticklebacks except that
they lacked the red coloration, and some showing little resemblance to
actual sticklebacks except that they were red on the lower surface (Fig. 12.9).
The male fish attacked the red-bellied models, despite their un-fishlike appearance, much more vigorously than they did the fishlike ones that lacked
red. Surely the sticklebacks could see the other characteristics of the models,
but they reacted essentially only to the releasing stimuli from the red belly.

David Lack observed a similar reaction in European Robins. He showed
that a male Robin would threaten a bundle of red feathers more readily than
it would a whole mounted young Robin lacking a red breast (Fig. 12.10). The
red breast was apparently such a strong releaser that its effect outbalanced
that of the obviously un-birdlike features of the tuft of feathers.

G. H. Brückner of the University of Rostock, Germany, studied a situation
in which a sound acts as a releaser while associated visual stimuli have little
if any effect. He showed that a hen reacts to her chick's distress calls, but not

12.10   Models of European Robin
A mounted young Robin with a dull-brown breast
(A) was attacked much less than a tuft of red feathers
(B).   [Redrawn from N. Tinbergen, *The study of
instinct,* Oxford University Press, 1951, after Lack.]

A   B

to its distress actions (Fig. 12.11). A chick fastened to a peg under a sound-proof glass dome can struggle agitatedly in full view of the hen without eliciting any reaction from her, but if she hears the chick's call, she reacts vigorously, even if the chick is hidden from view.

It has been shown repeatedly that once the sign stimulus for a particular behavior pattern has been carefully analyzed, it is often possible to design *supernormal stimuli* even more effective than the natural one. For example, Tinbergen showed that the size of Oystercatcher eggs is important in determining the releasing properties of the eggs for the adult birds. An adult Oystercatcher provided with normal Oystercatcher eggs and with larger eggs of other species will usually react preferentially to the largest egg, even if she cannot possibly hatch it (Fig. 12.12); for her the large egg provides a supernormal stimulus.

Though the behaviors released by sign stimuli may be rigid and stereotyped, they nonetheless depend, as all behaviors do, not only on the stimulus but also on proper motivation, which in turn depends, as we have already seen, on such factors as health, maturation, hormones, previous learning experience, recent sensory stimulation, and so on. This is an important point. A potent sign stimulus (even a supernormal one) will not invariably elicit a response. A well-fed animal exhibiting appetitive behavior for nest building is most unlikely to respond to sign stimuli of food. And while a worker honeybee serving as a guard at the hive is easily stimulated to attack and sting an approaching person, that same bee, when it later serves as a forager, gathering nectar from flowers, will usually sting only if physically disturbed.

As many experiments have shown, the intensity of sign stimulus necessary to trigger a behavior pattern is inversely proportional to the animal's motivation to perform the behavior in question. For example, many predatory animals ordinarily respond only to prey of a certain size range; smaller or larger potential food items are ignored. But when the predators are extremely hungry, they become more responsive to smaller and larger prey, even sometimes attacking animals stronger than themselves.

The role of motivation in the response to sign stimuli is dramatically

12.11 Difference in response by a hen to her chick's visual and vocal distress signals
(A) The hen ignores the chick if she cannot hear its calls, even though its actions are clearly visible.
(B) Distress calls elicit vigorous reaction from the hen even when she cannot see the chick.   [Modified from N. Tinbergen, *The study of instinct*, Oxford University Press, 1951.]

12.12 Oystercatcher reacting to giant egg
She chooses it instead of her own egg (foreground) or a Herring Gull's egg (left).   Redrawn from N. Tinbergen, *The study of instinct*, Oxford University Press, 1951.]

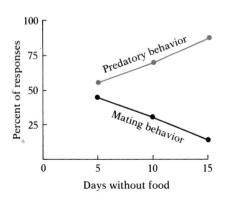

12.13   Change in response of a male jumping spider to a model as a function of the number of days he has gone without food
The model is one that can release either prey capture or mating, depending on the motivational state of the spider. The longer he has gone without food, the more likely he is to exhibit predatory behavior instead of mating behavior.   [After Drees.]

12.14   A young cuckoo throwing an egg of its foster parents out of the nest

illustrated by a motivational conflict that can be artificially induced in jumping spiders. Among the prey used by the male spiders are some insects (e.g. flies) that in size and form closely resemble female spiders. It is therefore possible to construct models capable of releasing either attempted prey capture or attempted mating, depending on the motivational state of the male. As Fig. 12.13 shows, the longer the male has gone without food, the more likely it is that the model will release prey-capture behavior rather than mating behavior.

Some responses to sign stimuli, just like some kinds of learning, will occur only during a sensitive phase of the animal's life. A good example is found in the behavior of hatchling cuckoos. The European species of cuckoos are nest parasites, laying their eggs in the nests of other birds and letting those other birds incubate them and feed the young when they hatch. As soon as a young cuckoo hatches, it uses its back and stubby wings to lift the eggs (or even the newly hatched young) of its foster parents to the rim of the nest and throw them out (Fig. 12.14). When O. Heinroth studied this behavior, using artificial eggs, he found that the young cuckoos responded to the sign stimulus of the eggs only during the first day after hatching; eggs put in the nest the second day released no response whatever.

Having examined some of the elements that help determine behavioral responses, let us now turn to a brief examination of some of the major classes of animal behavior. We shall focus especially on communication, orientation, feeding, and social behavior—a selection not intended to imply that other types of behavior (e.g. defense against predators or bodily-maintenance behavior, which includes preening, bathing, etc.) are not equally important in the life of the animal.

ANIMAL COMMUNICATION

The ability to communicate with one another is not restricted to members of species that live in societies, such as bees, ants, termites, and human beings; animals that live in less complex social groupings also communicate, and even the least social of animals must communicate with other individuals at certain critical times, such as the time of mating. The highly varied methods of communication animals have evolved utilize particularly the senses of hearing, sight, and smell. A brief examination of some of these methods will not only aid in understanding a subject that is in itself fascinating, but will also illuminate some of the most fundamental principles of animal behavior.

COMMUNICATION BY SOUND

Being vocal animals ourselves, we are very familiar with the use of sound as a medium of communication. No other species has a sound language that even approaches the complexity and refinement of human spoken languages. But many other species can communicate an amazing amount of information via sound, information on which the life of the individual may depend.

Sound communication in insects   We have already mentioned two examples of sound communication: Male *Aedes* mosquitoes are attracted by the

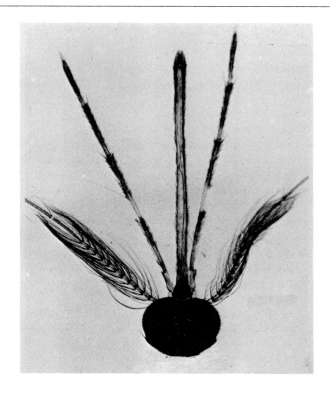

12.15 Photographs of head of male mosquito
Left: A very young male with the antennal hairs lying against the shaft. Such an animal is nearly deaf.
Right: Mature male with the antennal hairs erect. The antennae are now very sensitive to sounds of the proper frequencies.   [Courtesy Thomas Eisner.]

buzzing sound produced by the female's wings during flight (or by devices such as tuning forks that emit sounds of a similar pitch), and hen chickens respond in a characteristic fashion to the distress calls of their chicks. Let us return to the first of these examples.

The head of a male mosquito bears two antennae, each covered with long hairs. When sound waves of certain frequencies strike the antennae, these are caused to vibrate in unison. The vibrations stimulate sensory cells packed tightly into a small segment at the base of each antenna. The male responds to such stimulation by homing in on the source of the sound, thus locating the female and copulating with her. A striking demonstration of the adaptiveness of this communication system is that during the first 24 hours of the adult life of the male, when he is not yet sexually competent, the antennal hairs lie close to the shaft (Fig. 12.15, left) and thus make him nearly deaf. Only after he becomes fully developed sexually do the hairs stand erect (Fig. 12.15, right), enabling the antennae to receive sound stimuli from the female. Thus the male does not waste energy responding to females before he is sexually competent, but once he becomes competent he has a built-in system for locating a mate without random searching. Furthermore, his built-in receptor system is species-specific; it is stimulated by sounds of the frequency characteristic of females of his own species, not by the frequencies characteristic of other species of mosquitoes. Hence the sound produced by the female's wings functions both as mating call and species recognition signal.

Many other insects use sound in a similar way. For example, male crickets use calls produced by rasping together specialized parts of their wings. These calls function in species recognition, in attracting females and stimulating their reproductive behavior, and in warning away other males. So species-specific are the calls of crickets (Fig. 12.16) that in several cases closely related species can best be told apart by human beings on the basis of the calls; the species may be almost indistinguishable on an anatomical basis, but have distinctly different calls.

Sound communication in frogs   The calls of frogs serve functions similar to those of cricket calls, and like these they are very species-specific. The male frogs attract females to their territories by calling. In one experiment C. M. Bogert of the American Museum of Natural History recorded the call of male toads. He then captured 24 female toads and released them in the dark in the vicinity of a loudspeaker over which he was playing the recorded male call. Thirty minutes later, he turned on the lights and determined the position of each female. Nineteen of the females had moved nearer the loudspeaker, four had moved farther away, and one had escaped. Eighteen of the 19 females that had moved toward the speaker were physiologically ready to lay eggs; of the four females that moved away from the speaker, three had already laid their eggs and the other was not yet of reproductive age. In a control experiment Bogert showed that 24 females released in the dark near a silent speaker had scattered randomly in all directions by the time the lights were turned on 30 minutes later. This demonstration that females ready for mating are strongly attracted by the male's vocalizations while other females are not serves to emphasize that the effectiveness of a sign stimulus depends, among other things, on the condition of the recipient of the stimulus; the call of the male frog is an effective stimulus for movement by the female toward the male only if the female's reproductive drive is high, largely as a result of a high level of sex hormones.

Robert R. Capranica, of Cornell University, and his colleagues have shown that female cricket frogs will respond to the mating calls of males from their own or nearby populations, but not to those of males of the same species from more distant populations. This curious parochialism has a physiological basis. The ears of the female frogs are sensitive to only a very narrow band of frequencies, and the frequency to which they are maximally sensitive is different for frogs from different localities. The calls of the males also vary geographically in the frequency of their maximal intensity; i.e. the males have different local dialects (Fig. 12.17). A female's lack of response to the call of a male from a distant locality may be due to a mismatch between her ears and his call; she may be deaf to his calling frequency. Because mating depends on response of the females to the males, it is clear that the females' auditory apparatus and the males' vocal apparatus must always evolve in tandem. Thus, in any given locality, the males call at the frequency to which the females in that locality are most sensitive. This is a particularly dramatic example of a general phenomenon—the coupled evolution of social releasers and of the sensory and releasing mechanisms responsive to them.

Bird songs   Of all the familiar animal sounds—the buzzing of mosquitoes, the calling of crickets and frogs, the barking, roaring, purring, grunting of various mammals—perhaps none, with the exception of human speech, has received so much attention as the singing of birds. It has been celebrated in poetry, copied in musical compositions, mimicked by whistlers, adored by

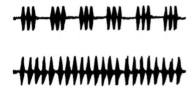

12.16   Song patterns of two species of crickets
The pattern of trills of the song of *Gryllus armatus* (top trace) is clearly different from that of *G. rubens* (bottom trace).   [Data from R. R. Hoy, Cornell University.]

lovers, and enthusiastically welcomed by those longing for spring. The popular "explanation" for bird song is simple: The bird is happy and sings with joy, welcoming the morning and the spring and expressing love for his mate. Biologists need to cast a skeptical eye on these anthropomorphic fancies.

Objective investigation has demonstrated that bird song functions primarily as a species recognition signal; as an individual recognition signal (there are usually small individual differences in the song, which other individuals can learn to recognize); as a display that attracts females to the male and contributes to the synchronization of their reproductive drives (increasing sexual motivation and decreasing attack and escape motivations); and as a display important in defense of territory. In its defensive function a bird's singing is certainly no indication of happiness or joy; if such human-oriented concepts could properly be applied to birds, the singing would more accurately be taken as an indication of combativeness.

The role of singing in the establishment and defense of territories is an especially interesting one. A territory may be defined as an area defended by one member of a species against intrusion by other members of the same species (and occasionally against members of other species). A male bird chooses an unoccupied area and begins to sing vigorously within it, thus warning away other males. The boundaries between the territories of two males are regularly patrolled, and the two may sing loudly at each other across the border. Although during early spring there is often much shifting of boundaries as more and more males arrive and begin competing for territories, later in the season the boundaries usually become fairly well stabilized and each male knows where they are. During the period when the boundaries are being established, it is often the males that can sing loudest and most vigorously that successfully retain large territories or even expand their territories at the expense of other males that sing less loudly and vigorously.

Experiments performed on thrushes by William C. Dilger of Cornell University illustrate especially well the role of singing in territorial defense. Dilger set up a loudspeaker in the territory of a male Wood Thrush, and placed a stuffed thrush near it. Wires led from the loudspeaker to a tape player in a blind in which Dilger could sit. When Dilger played a recording of a singing male Wood Thrush, the male bird in whose territory the loudspeaker was located responded as though another male had entered its territory. If the volume at which the recording was played was very low, the defending male attacked the stuffed bird. If the volume was high, the defending bird retreated. By alternately turning the volume up and down, Dilger could make the defending male move alternately toward or away from the stuffed bird and loudspeaker, almost as one might work a yo-yo. So precisely was he able to control the defending bird's movements by this method that he could make it teeter on one leg, its conflicting attack and escape drives almost exactly balancing each other.

Notice that agonistic[4] encounters between individuals of the same species may often be resolved by vocal and/or visual displays without any physical combat. It is, in fact, rather rare that individuals of the same species engage

[4] Agonistic behavior embraces all aspects of behavior exhibited during hostile encounters, including threat, attack, appeasement, and fleeing.

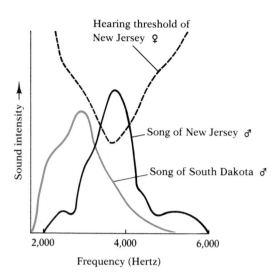

12.17   The auditory sensitivity of a female cricket frog to the call of males from the same population and from a different population

The New Jersey female's hearing (dashed curve) is maximally sensitive (i.e. has the lowest threshold) at just those frequencies where the sound intensity, or loudness, of the New Jersey male's call (black curve) is greatest; in other words, the female's hearing and the male's calling are almost perfectly matched when the two come from the same population. No proper match exists when the male comes from another population (South Dakota), which employs a different dialect (color curve); because there is no overlap between the female's auditory sensitivity and the male's call, the female cannot hear the male and shows no response to him.   [Data from R. R. Capranica, Cornell University.]

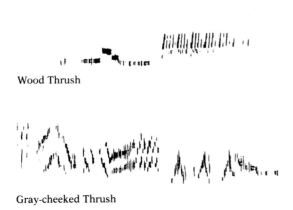

Wood Thrush

Gray-cheeked Thrush

Swainson's Thrush

12.18 Sound spectrographs of the songs of three species of thrushes used in Dilger's experiments
Each spectrograph shows, along the vertical axis, the frequencies of sound contained in the song and, along the horizontal axis, the time sequence of the sounds. Although all three kinds of thrushes look much alike, having brown backs and spotted breasts, their songs are markedly different, as these spectrographs show. [Data from the Library of Natural Sounds, Laboratory of Ornithology, Cornell University.]

in combat serious enough to cause significant damage. The adaptive importance of a nonviolent settlement of differences is obvious. Agonistic encounters occur frequently, and if they often led to serious physical damage, even many of the winners might, in the long run, be losers. It is not surprising, therefore, that most animals have evolved other methods of resolving conflicts, methods usually involving displays by which the combatants convey to each other the intensity of their attack motivation. The individual showing the higher attack motivation is ordinarily the winner. The animals can, in effect, tell without fighting which would be the probable winner if they were to fight; hence an actual fight is unnecessary. In our example, the vigor and loudness with which male Wood Thrushes sing convey to other Wood Thrushes very precise information concerning the motivational states of the singers. A singing duel can, therefore, resolve a boundary dispute without fighting.

Dilger's experiments also illustrate the importance of song in species recognition. Several other species of thrushes breed in the area where the experiments on Wood Thrushes were conducted. Some of these species look much like Wood Thrushes; they have brown backs and spotted breasts. Dilger showed that a male Wood Thrush would attack models of any of these species set up in its territory, provided the models were silent. It would even attack models of other species unrelated to thrushes if these happened to be brown with spotted breasts. If, however, the models were set up on a loudspeaker and Dilger played a recording of the song appropriate to the model, the Wood Thrush would pay the models of other species little attention. Apparently it could not distinguish visually (in this context) between the various brown thrushlike birds with spotted breasts, and responded to all, if they remained silent, as though they were invading members of its own species. But it could distinguish between the songs of the other species and that of its own species, and would not attack a model "singing" the song of some other species, even if it was brown and had a spotted breast. Auditory stimuli were thus shown to be far more important than visual stimuli in species recognition by thrushes (Fig. 12.18). This is frequently the case among birds living in dense woods and thickets, where vision is obstructed; visual displays are often more important for birds living in more open habitats.

Like the frogs studied by Capranica, many bird species exhibit regional dialects in their songs. But the underlying basis for these dialects is not the same as in frogs, and the females are certainly not deaf to the songs of males from other areas. In most cases the dialects are not genetically inherited, as they appear to be in frogs, but are learned. Young birds growing up in a given area hear the local version of the species song and mimic it when they themselves begin singing. Such song dialects are therefore rather like the regional dialects of human languages.

COMMUNICATION BY CHEMICALS

Since most communication between human beings involves auditory or visual signals, we tend to think of these as inherently the most appropriate methods of communication in animals. But we should not let the limitations of our own senses prevent us from recognizing that the olfactory sense is immensely important in the lives of many animals and constitutes a basis for effective communication.

Many animals secrete substances that influence the behavior of other members of the same species. Such substances are called **pheromones.** Most pheromones can be classified in one or the other of two groups: those that act as releasers, triggering a more or less immediate and reversible behavioral change in the recipient, and those that act as primers, initiating more profound physiological changes in the recipient, but not necessarily triggering any immediate behavioral reaction.

Releaser pheromones   We have seen how a male mosquito is attracted to the female by the sound produced by her beating wings. Many insects, and also other types of animals, use chemical sex attractants instead of mating calls. For example, female silkworm moths release a sex attractant so powerful that males are attracted to a single female from distances of two miles or more, even though each female releases less than 0.00000001 gram (0.01 microgram) of the attractant chemical. The males must be able to detect and respond to incredibly minute amounts of the attractant. The chemical acts as a releaser to which the male moth responds by flying upwind, thus moving toward the female. Only in the immediate vicinity of the female is the concentration of pheromone sufficient to establish a gradient; when the male comes into this region, he stops flying upwind and follows the gradient instead, thus locating the female. It can be shown that in this behavior the male responds only to the chemical releaser, not to visual stimuli: He will be attracted to a female in a gauze cage even if he cannot see her, but he will not be attracted to a female clearly visible in a tightly sealed glass cage from which none of the pheromone can escape.

Sex-attractant pheromones are now known from many other insects, and the chemistry of some has been worked out. Research into pheromones is currently a very active field, because it is possible, in some cases, to use sex attractants in nontoxic species-specific insect control, in place of toxic and relatively nonspecific insecticides. For example, the sex attractant of the gypsy moth, an insect that, when present in large numbers, is capable of causing severe damage to forest trees, can be used to lure the moths into traps. In other cases spraying an area with an insect sex attractant so jams the communication channels of the insects that effective mating is greatly reduced and their numbers are kept in check.

The trail substances of ants constitute another class of releaser pheromones (Fig. 12.19). A foraging ant returning to the nest from a food source intermittently touches the tip of her abdomen to the ground and secretes a tiny amount of trail substance. Other worker ants can follow the trail to the food source; these ants, too, will lay trail substance as they return with food to the nest. The better the source of food, the more ants are attracted to the trail, which thus grows stronger. Workers that do not find food do not lay trail; hence when the food has been consumed no more trail is laid and, since the trail pheromone is very volatile, the trail disappears within a few minutes. It has been demonstrated that trail substances are species-specific; no two species have been found to secrete the same substance. This specificity is, of course, biologically adaptive, because it ensures that workers will not mistakenly follow trails of other ant species that may cross their own.

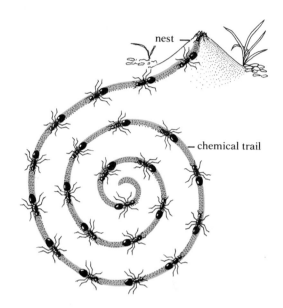

12.19   Ants following a spiral path of trail substance laid down by an experimenter

12.20 A pronghorn buck scent-marking a plant with his facial glands

Other releaser pheromones of ants include alarm substances and death substances. The latter provide a particularly good example of the extent to which much insect behavior is stimulus-bound and rigidly stereotyped. Certain long-chain fatty acids and their esters released from a dead and decomposing ant act as pheromones that stimulate worker ants to pick up the carcass and carry it to a refuse pile outside the nest. When living ants are experimentally painted with these substances, workers pick them up and dump them on the refuse pile. The hapless victims of the experimenter promptly return to the nest, only to be thrown back on the refuse pile, again and again. The workers can surely see and feel that the object they are carrying is struggling in a most undead manner, but they disregard the evidence from all their other senses, respond only to the death pheromone, and continue to treat the painted ant as a carcass to be taken to the refuse pile.

Our examples have been drawn from insects, but releaser pheromones are by no means restricted to them. They occur in many animal groups, notably mammals, most of which, as you know, rely on olfaction much more than humans do. Male mammals can often tell by smell when the female is in heat, because she secretes certain pheromones at that time. Besides playing a part in sexual recognition and reproduction, mammalian pheromones often serve in marking territories and home ranges. You are familiar with the way dogs and many other mammals use their urine as a marking substance. Many species (e.g. hamsters, deer) use pheromones secreted by special scent glands for the same purpose (Fig. 12.20).

Releaser pheromones, which must have arisen long before auditory or visual communication, probably represent one of the most primitive forms of biological communication. Thus the free-swimming gametes of many protistans locate each other by chemical cues; and the flagellated sperm cells of many nonseed plants and of animals that use external fertilization are attracted to the egg cells by pheromones secreted by the eggs.

Primer pheromones Primer pheromones produce relatively long-term alterations in the physiological condition of the recipient and thus change the effects that later stimuli will have on the recipient's behavior; they do not necessarily produce any immediate behavioral change.

As various experiments have shown, pheromones affect the estrous cycles and pregnancy of mice. In one series of experiments W. K. Whitten of the Australian National University demonstrated that the estrous cycles of female mice in a laboratory colony can be initiated and synchronized by the

12.21 Courtship displays of a male Mallard
(A) Normal swimming posture.   (B) The "Head-Up-Tail-Up" display.   (C) Part of the "Down-Up" display.   (D) The "Grunt-Whistle" display.   [Adapted from K. Z. Lorenz, "The evolution of behavior," *Sci. Am.*, December 1958. Copyright © 1958 by Scientific American, Inc. All rights reserved.]

odor of a male mouse, even if the male cannot be seen or heard. In another, Helen Bruce of the National Institute for Medical Research in London showed that the pregnancy of a newly impregnated female mouse will be blocked by the odor of a strange male mouse. The blockage will not occur if the olfactory bulbs of the female mouse's brain are removed. The pheromone from the strange male probably exerts its effect by inhibiting in some manner the secretion of the pituitary gonadotrophins necessary for development of the corpus luteum. In a third series of experiments, conducted in 1955, S. van der Lee and L. M. Boot in the Netherlands showed that crowding of female mice results in disturbance and even blockage of estrous cycles. The fact that removal of the olfactory bulbs restores the cycles to normal indicates that pheromones are involved. It seems possible, therefore, that pheromones help regulate population density in some species.

Another type of primer pheromone is seen in social insects such as ants, bees, and termites. These pheromones, which are ingested rather than simply smelled, play an important role in caste determination. An example is queen substance, a pheromone secreted by honeybee queens that prevents the workers from developing reproductive capabilities and also from rearing new queens. If a queen dies, the concentration of pheromone circulating in the hive declines, and workers develop reproductive capabilities and also begin building large queen cells in which they will rear new queens. The number of soldiers in a termite colony is similarly regulated by pheromones secreted by the fully developed soldiers.

12.22   Courtship display of a male American Goldeneye duck
The bird throws his head back while simultaneously kicking up a spray of water.   [Courtesy D. G. Allen.]

COMMUNICATION BY VISUAL DISPLAYS

Displays in reproductive behavior   A display may be defined as a behavior that has evolved specifically as a signal. According to this definition, a song or a call is a display. Many animals have also evolved a variety of often complex actions that function as signals when seen by other individuals; such displays frequently include vocal elements. For example, at mating time in the spring many male Prairie Chickens assemble on a courting stage, usually a small knoll or open grassy area. Each male then begins to shuffle his feet and run back and forth with wings drooped, neck erect, and bill pointed straight down, while at the same time making deep pumping noises. The females congregate on the periphery of the courting stage and watch the displaying males for a while. Finally each female chooses one of the males for her mate (among most birds the female does the choosing).

Other examples may often be observed—male pigeons strutting, with tail spread and dragging on the ground, neck fluffed, and wings lowered, or courting songbirds in spring going through odd and seemingly senseless antics (more than one kindhearted person has felt sorry for the poor "demented" birds). If you have the chance, watch a flock of ducks on a pond in early spring. You may see a male give a loud whistle and raise both head and tail as high in the air as he can while also raising his wings (Fig. 12.21B), or he may raise his stern in the air, dip his head in the water, and then abruptly raise it and whistle (Fig. 12.21C), or he may put his bill in the water and then quickly flick his head to the side, toss an arc of droplets into the air while arching his body upward, and follow this acrobatic feat with a whistle and a grunt (Fig. 12.21D). Similar displays are performed by oceanic ducks (Fig. 12.22).

12.23 Courtship display of a male Ruffed Grouse
The bird mounts his drumming log and beats his wings rapidly and forcefully, producing a loud sound that attracts females. He also fans out his strikingly patterned tail and spreads his neck ruff—other devices that appear to make him more attractive to females. [Courtesy D. G. Allen.]

Such carryings-on may seem senseless to a casual observer, but biologically they are far from senseless. They function in synchronizing the sexual physiology of the male and female and in making the female more receptive to the male. They also assure that the female will choose a male of her own species as her mate; the males of each species—there may be more than one in the same area—give somewhat different displays. Special structures or bright patches of color, which clearly evolved in association with the displays as part of the signal system, are often elaborately exposed (Fig. 12.23).

The flashing lights of fireflies are an example of another, rather different, type of visual signal system. The pattern of flashes given by the flying male differs from species to species and thus serves as a species-recognition signal (Fig. 12.24). The female responds to a male of her own species by giving a brief flash of her own, to which the male responds by coming to rest and initiating mating.

Complex behavior patterns are often composed of several separate activities occurring in sequence, with each activity serving as the initiator of the next. The mating behavior of the three-spined stickleback provides a good example. In spring the male fish has bright red underparts, and the female's abdomen is swollen by the large number of eggs it contains. Tinbergen and his associates have shown that the female is attracted by the male's red belly and that the male is stimulated by the sight of the female's swollen abdomen. When a female swims into the territory of a male in reproductive condition, she often adopts an unusual head-up posture (Fig. 12.25). The combination of her swollen abdomen and courting posture acts as a releaser for the male to swim toward her in a curious zigzag fashion. The combination of the male's red belly and zigzag dance acts as a releaser for the female to swim toward the male in the head-up posture. Her approach acts as a releaser for the male to turn and swim rapidly toward the nest that he has already constructed. His swimming away, in turn, stimulates the female to follow him, which stimulates the male to make a series of rapid thrusts with his snout into the nest entrance and then to turn on his side and raise his dorsal spines. This "showing of the nest entrance" behavior of the male acts as a releaser for the female to enter the nest. Her occupation of the nest, in turn, stimulates the male to thrust his snout against her rump in a series of quick rhythmic trembling movements, which induce the female to spawn. It can be shown that, without the stimulus of the male's tremble-thrusts, the female is incapable of spawning; the stimulus can be effectively duplicated, however, by prodding her with a glass rod or other hard object. Once the female has spawned, the fresh eggs stimulate the male to fertilize them.

Most of the links in the chain of reactions in stickleback courtship depend on displays that provide visual releasing stimuli. Though the sequence described here is the usual one, it is not absolutely rigid and variations are possible. If, however, the sequence or the way in which the displays are performed is altered too much, the consummatory acts—spawning and fertilization—will not occur. Since the mating behavior of each species of stickleback differs in certain critical elements, it is unlikely that a male of one species and a female of another will get through enough of the ritual for spawning and fertilization to occur. This elaborate behavior functions, therefore, both in bringing together and synchronizing the two sexes in the mating act and in avoiding mating errors.

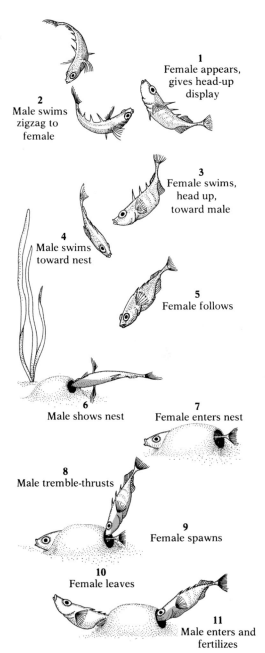

1 Female appears, gives head-up display

2 Male swims zigzag to female

3 Female swims, head up, toward male

4 Male swims toward nest

5 Female follows

6 Male shows nest

7 Female enters nest

8 Male tremble-thrusts

9 Female spawns

10 Female leaves

11 Male enters and fertilizes

12.24 The flashing patterns of the males of three species of fireflies
The females recognize the flashing pattern of males of their own species and respond to them.

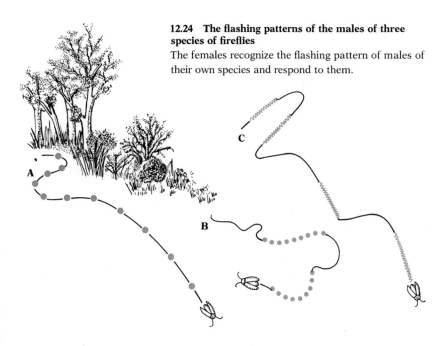

12.25 Courtship behavior in the three-spined stickleback
See text for full description. [Modified from N. Tinbergen, *The study of instinct*, Oxford University Press, 1951.]

Displays in agonistic behavior   The examples of visual displays given so far have been drawn from reproductive behavior, but visual displays are also of great importance in many other aspects of an animal's life. For instance, a dog wags its tail as a *greeting display*. Two dogs or cats may give *threat displays*, with hackles raised, teeth bared, ears laid back, body raised as high off the ground as possible, and movements stiff-legged and exaggerated. The loser of such an encounter responds with *appeasement displays*—fur sleeked, tail tucked under, head down and often turned away from the antagonist, legs bent. Such visual displays function like vocalizations in similar situations: They communicate the current balance between the individual's attack and escape motivations (Fig. 12.26).

Analogous agonistic displays can be observed in many animals. A high attack motivation is often conveyed, as it is in dogs, by directing the face straight at the antagonist and spreading and raising the body, making it look as large as possible (Figs. 12.27B, 12.28, 12.29, 12.30). Appeasement displays

12.26   Agonistic behavior of male bison
Top: Three males, all with their tail "flags" raised—an indication of high aggressive motivation—charge one another.   Bottom: A few moments later the male at right has lowered his tail and tucked it between his legs, a signal of diminished aggressive motivation; he is the likely loser of this encounter.   [Courtesy D. G. Allen.]

Long Call

Facing Away

B

Upright

C

Choking

12.27 Agonistic displays of Black-headed Gull
(A) The first response given by a male on his territory
when another male approaches is the "Long Call," in
which the body is tilted downward, the wing butts are
lifted, the head is thrust forward, and a characteristic
call is given.   (B) If the other male continues to
approach, the defending bird may move to meet the
intruder at the boundary of his territory, where he
gives the "Upright" display, lifting his still-folded
wings, stretching his neck upward, and pointing his bill
downward.   (C) If the intruder performs counterthreat
displays, the defender may then adopt the "Choking"
posture, tilting his body head down in an almost
vertical position and moving his head in a series of
quick up-and-down jerks.   (D) "Facing Away" is an
appeasement display in which a gull turns his head so
that the other bird cannot see the beak and eyes or the
black facial mask.   [Adapted from N. Tinbergen, "The
evolution of behavior in gulls," *Sci. Am.*, December
1960. Copyright © 1960 by Scientific American, Inc. All
rights reserved.]

12.28 Highly motivated threat display of a wolf
The animal makes himself look bigger (by standing
high on his legs and raising his hair and ears) and
stares directly at his opponent.

12.29 A toad directing a threat display at a snake
Even this small animal encountering a deadly predator
makes himself look as big as possible as he confronts
his opponent.   [Redrawn from Eibl-Eibesfeldt.]

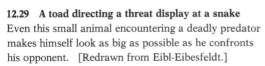

12.30 Threat displays of an Australian crayfish (left) and a baboon spider (right)
Note the similarities of these displays with those of the gull (Fig.12.27B) and the wolf (Fig.12.28). Again, the animals adopt a pose that makes them look as large as possible, and they face the object of their threat, exposing their weapons.   [Courtesy Michael Morcombe (left); S. C. Bisserot, Bruce Coleman Inc. (right).]

usually involve making the body appear as small as possible and turning the face away from the antagonist or exposing to the antagonist the appeaser's most vulnerable spot (Fig. 12.31); such appeasement displays tend to inhibit further attack by the antagonist.

COMMUNICATION IN HONEYBEES

A last example of communication must be mentioned. This is the amazing ability of scout honeybees to inform the workers in the hive of the quality of a food source and its direction and distance from the hive. This communication depends on displays that include auditory, visual, chemical, and tactile elements.

The work of Karl von Frisch of the University of Munich, Germany, on the language of bees is a biological classic.[5] Von Frisch had long been interested in the ability of bees to distinguish between different colors and scents. In the course of his experiments, he would set up in the vicinity of a hive a table with sheets of paper on which he had smeared honey. He would then have to wait—sometimes for several hours—for the bees to find the honey. He noticed that when one bee finally discovered the feeding place, many others appeared at the table within a short time. It seemed likely that the first bee had somehow informed the others of the existence of the new feeding place.

In order to see what happens in the hive when a scout bee returns from a new food supply, von Frisch set up an observation hive with glass sides.

[5] In 1973 Karl von Frisch, as well as Konrad Lorenz and Niko Tinbergen, whose investigations have also been mentioned in this chapter, were jointly awarded the Nobel Prize, in recognition of their pioneering studies of animal behavior.

When a bee landed at the new feeding place and began to feed, von Frisch daubed a spot of paint on her thorax so that he could recognize her when she returned to the hive. He discovered that the returning bee first feeds several other bees and then performs a dance on the surface of the honeycomb. The dance consists of circling first to the left, then to the right, and repeating this pattern over and over with great vigor (Fig. 12.32A). Von Frisch named this the round dance. The dance excites other bees in the vicinity of the dancer, and they begin to follow her, with their antennae held close to her. Suddenly, however, they turn away one by one and leave the hive; a short time later they appear at the feeding place. Apparently the round dance is a display that informs the other bees of the existence of the food supply.

Von Frisch wanted to know exactly what sorts of information the round dance conveys. He fed several bees at a dish containing sugar water scented with honey that he had located 10 m west of the hive. He also put out dishes of sugar water to the north, south, and east of the hive. Other bees began appearing in approximately equal numbers at all four dishes a few minutes after the bees that had been fed at the west dish began performing a round dance in the hive. There was no evidence that the round dance indicated direction, and other similar experiments brought no evidence that it indicated distance. It seemed simply to say, "Fly out and seek in the neighborhood of the hive." Von Frisch did find, however, that if each of the dishes of sugar water was scented with a different flower, the other bees came in significantly greater numbers to the dish that the dancer had visited. He showed that these bees determined what scent to search for in two ways: They smelled the body of the dancer by holding their antennae near her, and they detected the odor in the droplets of material she fed to them.

Even though von Frisch had satisfied himself that the round dance indicates neither direction nor distance, he began to suspect that at times bees are able to communicate this type of information, presumably in some other way. In 1944 he performed the following experiment. He set up two dishes of sugar water, one at 10 m from the hive and the other at 300 m. Each was scented with lavender oil. He then fed a few bees at the dish 10 m from the hive; shortly thereafter numerous bees appeared at this dish, but only a few appeared at the distant dish. When he reversed the procedure and fed forager bees at the dish 300 m from the hive, other bees appeared in large numbers at this dish, but only a few appeared at the nearer dish. Distance was clearly being communicated in some manner.

12.31 A dog making a full appeasement display to another dog
The loser in an agonistic encounter exposes his vulnerable underparts to the victor, who is thus inhibited from further attack.

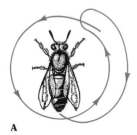

A

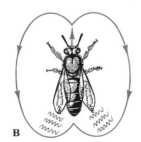

B

12.32 Round dance and waggle dance of scout honeybee
(A) In the round dance the bee circles first one way and then the other, over and over again. The dance tells other bees that there is a source of food near the hive. (B) In the waggle dance the scout runs forward in a straight line while waggling her abdomen, circles, runs forward again, circles in the other direction, and runs forward again. The orientation of the run indicates the direction from the hive of the food source, and the number of turns per unit time indicates the distance. [Redrawn from K. von Frisch, *Bees: Their vision, chemical senses, and language.* Copyright 1950, Cornell University. Used by permission of Cornell University Press.]

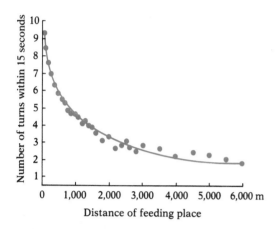

12.33   Graph showing relationship between distance of food source and number of turns per unit time in the waggle dance
The number of turns decreases with distance. [Modified from K. von Frisch, *Bees: Their vision, chemical senses, and language.* Copyright 1950, Cornell University. Used by permission of Cornell University Press.]

Von Frisch observed that while the foragers from the dish 10 m from the hive danced the famliar round dance, the foragers from the dish 300 m away danced a different dance, one that he described as the waggle dance (Fig. 12.32B). The bee runs a short distance in a straight line while waggling her abdomen from side to side very rapidly, then circles, runs forward again, circles in the other direction, and runs forward again. She repeats this dance many times. Von Frisch found an inverse relationship between the distance of the food source and the number of turns per unit time in this waggle dance (Fig. 12.33). He concluded that the number of turns tells the other bees the distance to the food. Experiments by other investigators suggest that sounds made by the bee during the dance may also be important in communicating distance.

Von Frisch found that bees can also communicate the direction of a food source. The location of the food relative to the position of the sun is indicated by the direction of the straight portion of the waggle dance. In the dark hive a run straight up the vertical comb means that the food lies in the direction of the sun; a run straight down the vertical comb means that the food lies in the opposite direction from the sun; a run at an angle indicates that the food is to be found at that angle to the sun (e.g. a run 30 degrees to the right of vertical indicates that the feeding place is 30 degrees to the right of the sun). In other words, bees use the force of gravity as a symbol for the sun when inside the dark hive (Fig. 12.34). When, however, they perform the waggle dance on a horizontal surface outside, they abandon the symbolism and orient the dance relative to the sun itself.

It has been shown that bees can detect the sun's position and orient themselves by it even if they cannot see the sun itself but only a small patch of blue sky, because they are able to analyze the polarization of the light reflected from the particles in the atmosphere. Since the plane of polarization of light from any point in the sky is related in a fixed manner to the position of the sun and to the position of the observer, the bees can tell, by this means, where the sun is, and can thus continue foraging on cloudy days as long as some blue sky remains visible.

SYMBOLISM IN COMMUNICATION

As more and more of the kinds of behavior once thought unique to human beings are found to occur in other animals—insight learning is an example—some writers have argued that at least symbolic communication remains a uniquely human capacity. Yet if, as we have just seen, honeybees are able to communicate the direction of a food source by using gravity as a symbol for the sun, can we still view symbolic communication as exclusive with our own species? Although some biologists would answer no, others who hold out for the qualitative as well as the quantitative distinctiveness of human behavior have sometimes insisted that the apparent rigidity of meaning of the honeybee dance makes it fundamentally different from the plastic, often individual and subjective use of words as symbols by human beings.

Some recent work with chimpanzees challenges a human monopoly on symbolic communication in even this more limited sense. Whereas earlier efforts to teach chimps to speak all failed (they lack the proper anatomy to form human words; see p. 494*n*), approaches employing sign language have met with some success. B. T. Gardner and R. A. Gardner of the University of Nevada have taught several chimps (among whom Washoe, especially, has made a name for herself) to use the sign language of the deaf, and some of their animals appear to have acquired vocabularies of nearly 200 words, which they can use in simple sentences.

D. Premack and A. J. Premack, now of the University of Pennsylvania, have taught chimpanzees to use pieces of colored plastic as symbols for words and phrases, and their animals, too, appear able to construct simple sentences. Duane Rumbaugh and his colleagues at Emory University report that they have used a computer to teach chimps to read and write. Other investigators as well, using a variety of approaches, report some success in teaching language to chimpanzees and to gorillas.

Do the animals really have thoughts, ideas, mental images, and the other types of mental experiences reflected in human language? Or have they merely learned to use signs or objects in a way that mimics human use of language, in order to earn rewards from the experimenters? Not all specialists in the field are convinced that the chimpanzees of the Gardners, the Premacks, Rumbaugh, and others are using language in the fullest sense. Yet there is some evidence that the animals can at times put together symbols in new meaningful ways, producing combinations they have not previously encountered. Such observations suggest that they may form concepts, which they then symbolize for communication; demonstrating such a process, however, is another matter. At any rate, enough evidence has already accumulated to jolt us in our smug assumption that symbolic communication is the exclusive province of human beings.

BIOLOGICAL CLOCKS

That honeybees dancing the waggle dance in a dark hive can communicate directions in terms of the sun's position means that they must be able to compensate for the apparent motion of the sun across the sky during the course of a day. If they continued to orient their dance to convey the location of a food source relative to the position of the sun as they last saw it, they would soon be giving the other bees erroneous information and these would depart in a wrong direction. This does not happen. The dancer slowly shifts the orientation of her dance relative to gravity, so that she is always indicating the direction of the food in terms of the position of the sun at that moment. To make such adjustments, she must have an accurate internal sense of time, a "biological clock." She must also somehow be programed to shift her bearing at roughly 15 degrees per hour, which is the average rate of change in the sun's azimuth (direction from the observer) during the day.

Bees also quickly learn what time of day each species of nectar-yielding flower opens and visit it punctually every day; they are not telling time from external cues, but depend on their own internal clocks, for they can do just as well when kept under conditions of constant light and temperature.

Numerous other examples of timekeeping by living organisms could be mentioned. The leaves of many plants show regular movements in a cycle of approximately 24 hours, even if kept under constant conditions. Many flowers open and close in a similar 24-hour cycle. Adult insects of many species emerge from the pupa at a particular time of day, whatever the age of the

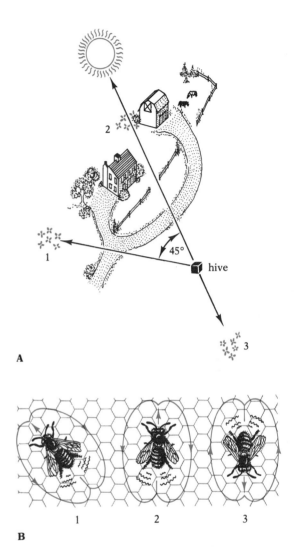

12.34 An example of bee language
(A) Three different sources of food were located (1) 45 degrees to the left of the sun as seen from the hive; (2) straight toward the sun; and (3) straight away from the sun.   (B) The dances performed on the vertical comb in the darkened hive by scout bees from these food sources were oriented 45 degrees to the left of vertical (1), straight up (2), and straight down (3). In short, the vertical direction on the comb symbolizes the sun.

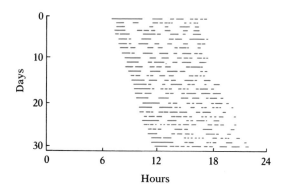

12.35 Activity rhythm of a hamster in constant light
The horizontal bars (color) for each day indicate the periods when the hamster was active. The animal's circadian rhythm is slightly longer than 24 hours. Hence the period of its activity in constant light, when it has no external time cue, starts at a slightly later hour each day; i.e. the activity period drifts rightward, increasingly out of phase with real time.

pupa or the conditions it has experienced. Certain crabs become darker in the morning and lighter in the afternoon, even under constant conditions, when no light cues are available. Many animals show activity rhythms that vary with a period of approximately 24 hours, even if the animals are kept under constant conditions and have no known external indication of the actual daily environmental cycle. Karl C. Hamner of the University of California at Los Angeles took hamsters, fruit flies, cockroaches, cockleburs, soybean plants, and bread molds to the South Pole and placed them on a turntable set to rotate at exactly the same speed as the earth but in the opposite direction. In other words, he exposed these organisms to conditions where no earthly indications of daily time exist. Yet the regular 24-hour cyclic activities of these organisms continued. Rhythms of this sort, with a period of approximately 24 hours, are called *circadian rhythms* (from the Latin *circa*, about, and *dies*, day). They are an important element in determining the motivational state and behavioral responsiveness of animals.

Manifestations of biological clocks are not restricted to the sorts of gross reactions of whole organisms so far mentioned. There is ample evidence that clock phenomena are characteristic of all living things, whether individual cells or whole multicellular plants or animals. Aspects of cellular function that vary in approximately 24-hour cycles include enzyme activity, osmotic pressure, respiration rate, growth rate, membrane permeability, bioluminescence, sensitivity to light and temperature, and reactions to various drugs. Physicians are becoming increasingly aware that the proper dosage of a drug may be very different at different times of day; in some cases, what constitutes a beneficial dose at one time may actually be lethal at another.

Although many manifestations of the internal clock have been studied in detail, very little is known so far about the mechanism that actuates it. It is known that the clock cannot depend on the usual sort of enzyme-catalyzed reactions, because it shows amazingly little temperature dependence—a point already discussed in connection with the role of the clock in the control of flowering (see p. 354). It is also widely thought that the basic period of the clock is innate. Thus Erwin Bünning of the University of Tübingen, Germany, has shown that exposure of the fruit fly *Drosophila* to constant conditions for 15 consecutive generations fails to eliminate the essentially 24-hour rhythms of this insect. Some investigators, however, think the basic period is determined by subtle geophysical variables that the standard experiments under so-called constant conditions have failed to eliminate; among the possibilities suggested are rhythmic variations in geomagnetism, gravitation, electric fields, and cosmic radiation.

Even if the clock is innate, it does not follow that it is unresponsive to environmental conditions; in point of fact, it is strongly influenced by them. Under normal conditions the clock is constantly being reset (re-entrained) by the environmental cycle. If an organism is kept under constant conditions for a long time, its clock will slowly get more and more out of phase with the environmental diurnal cycle (Fig. 12.35). Thus, if an animal whose activity rhythm has an innate period of 24 hours and 15 minutes (the innate period often deviates slightly from 24 hours) is kept for 20 days under constant conditions, at the end of that time its activity cycle will be five hours out of phase with the actual diurnal cycle. In plants and many lower animals, most cycles fail to operate if the organism has been exposed to constant light or constant darkness throughout its life; an environmental stimulus is necessary to trigger the beginning of such cycles, but often a single stimulus is sufficient—it need not be repeated for the cycles to continue.

The fact that environmental conditions can reset the biological clock means that it can be experimentally manipulated. For example, if an animal with a pronounced circadian activity rhythm is put in a lightproof room for five days and exposed to an artificial day that begins and ends six hours later than the natural day, and if the animal is then exposed to constant light, it can be observed that, despite the absence of external time cues, an activity rhythm of roughly 24 hours continues. But this activity rhythm is now in phase with the artificial day, not with the true day—an indication that the animal's internal clock has been successfully reset. A similar phase displacement is experienced by a person who flies from Chicago to Paris. At first he is disconcertingly out of synchrony with the people around him; he feels like sleeping when they are wakeful, hungry when their minds are on higher things. After three or four days, however, his internal clock has shifted to Paris time, and his problems are over.

It has been shown that when an organism is exposed to new conditions that shift the setting of its biological clock, the clocks of its various cells or organs do not all necessarily shift together. Some may adjust to the new conditions in only two or three days, while others may take a week or more. Thus it is possible for the different organs of an individual to be thrown out of phase with each other. For example, an endocrine gland might be in the phase of its maximal secretion of hormone and the target organ, at that time, in a phase of relative unresponsiveness to the hormone. Or an enzyme system might be potentially most active at a time when its substrate was not available. Such uncoordinated phase shifts of the various biological clocks in an individual may well lead to serious physiological disturbances and perhaps to disease.

Air and space travel, varying light regimes as a result of artificial lighting, less orderly activities, and other conditions characteristic of modern life may be altering the balance between the phases of our various clocks with profound effects of which we are as yet only dimly aware. Biological clocks will be an important field of research in the years ahead, as human beings increasingly subject themselves to conditions tending to disrupt the balance between the cycles of their individual cells and organs.

ORIENTATION BEHAVIOR

Intriguing to scientist and nonscientist alike is the rather widespread ability of animals to find their way from place to place by means clearly not random. Some types of orientation are now fairly well understood, while others, many of them long familiar, remain unexplained and present a continuing challenge to biologists.

ORIENTATION BY VISUAL LANDMARKS

Many animals orient by visual landmarks. An example is the female digger wasp *Philanthus triangulum,* who can locate her nest with amazing accuracy

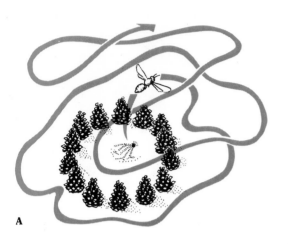

12.36   Use of visual landmarks by female digger wasp in locating nest entrance
(A) A ring of pine cones was placed around the nest entrance while the wasp was inside. When she emerged, she made an orientation flight around the nest and then flew off.   (B) While the wasp was gone, the ring of pine cones was moved a short distance. The returning wasp went to the center of the ring of cones rather than to the true nest entrance.   [Redrawn from N. Tinbergen, *The study of instinct,* Oxford University Press, 1951.]

when she returns from hunting. Tinbergen and Kruyt, who studied this phenomenon, put a ring of pine cones around the nest of a wasp while she was inside digging it (Fig. 12.36A). When the wasp emerged, she flew around the nest for about six seconds, presumably making a study of the locality. She then departed and was out of sight of the nest for 90 minutes. While she was gone, the experimenters moved the ring of pine cones to a spot about 30 cm away (Fig. 12.36B). When the wasp returned with captured prey, she went to the ring of pine cones, not to the actual site of the nest. This experiment was repeated many times, always with the same results. The wasp could never locate her nest until the original situation was restored.

But what about wasps that nest in open flat expanses of sand where there are no obvious landmarks close to the nest? Experiments have repeatedly shown that in these instances distant landmarks, such as a tree on the horizon, play an important role; the wasp is remarkably accurate at judging her position relative to such distant landmarks.

Orientation by landmarks is also important in the ability of mammals such as dogs, cats, or horses to return home from several miles away. Such animals often have a very good map sense, and only a few familiar landmarks or even rather subtle features such as the pitch of the land may give them enough cues to orient themselves. There has, however, been much exaggeration of this ability. Most reports of pets returning home from distant places completely unknown to them suffer from imaginative embellishment, and the feats performed must be attributed to random wandering. The popular press focuses on the very few dogs and cats that manage to return from long distances and ignores the thousands that don't return from even modest distances when abandoned by their owners in the countryside.

ORIENTATION BY ECHOLOCATION

Echolocation by bats is probably the most widely known example of sonar orientation—the determination by an animal of its position relative to objects around it through detection of the echoes produced when sounds it emits strike the objects. In 1793 the great Italian naturalist Lazzaro Spallanzani observed that captive bats released in a completely dark room could fly about without hitting the walls or the furniture. Thinking that the bats might have extremely acute vision and be able to see by light not detectable by himself, Spallanzani put black hoods over their heads and found that they could no longer avoid obstacles. This seemed to indicate that vision was indeed important in obstacle avoidance. In later experiments, however, when Spallanzani removed the eyes of several bats, he found that they could orient well, but that if he plugged their ears they could not. It seemed likely, therefore, that bats use sound cues. But what sound? Flying bats seem remarkably silent.

It was not until 1938 that Donald R. Griffin, then of Harvard University, solved the mystery. He showed that flying bats produce a great variety of vocalizations, some at frequencies as high as 100,000 Hertz. The problem had been that these sounds are too high-pitched for the human ear to detect. When picked up by a sensitive microphone and transformed to lower frequencies, these sounds can be heard by us as a series of short clicks. Their extremely high frequencies and correspondingly very short wavelengths suit them particularly well for echolocation, because they can be reflected from smaller objects and because they spread less widely and are less diffuse than sounds of longer wavelengths. They thus permit more sensitive detection and more precise localization of objects; some bats can avoid wires only 0.2 mm in diameter. The bats therefore have no difficulty finding their way through the pitch-dark caves in which many of them live; their own sonar system tells them the precise location of all obstacles (Fig. 12.37).

Bats have a remarkable ability to detect the echoes of their own clicks, even in the midst of an enormous amount of noise. Attempts to confuse them by exposing them to intense vibrations while they are flying regularly fail. It is biologically important, of course, that they should be able to analyze sounds and eliminate noise. They often fly in flocks of thousands, and it is essential that each bat be able to distinguish the echoes of its own sounds from those of all the other bats.

The sounds used by bats in echolocation have come to have some interesting interspecies communication properties. Many bats use echolocation to detect insects, which they capture in flight. Among these insects are a number of species of moths, of which some have evolved simple ears capable of detecting the high-pitched clicks of the bats. Many such moths, when they hear an approaching bat, begin evasive tactics and often escape. Some other species of moths that are distasteful to bats respond to bat calls by producing high-pitched sounds of their own, which apparently say to the bat, "Yes, I'm here, but you won't like me." Bats almost invariably respond by turning away.

By no means restricted to bats, echolocation is also used, for example, by Central American Oilbirds *(Steatornis caripensis)*, which live in deep caves and come out to feed at night. By emitting a steady stream of clicking sounds (which, unlike the clicks of bats, can be heard by humans) and detecting the echoes from obstacles, these birds can fly rapidly through the near total darkness of the caves without bumping into walls or columns. If their ears are plugged, they cannot fly through a dark cave, but can do so if some light is provided, an indication that they are capable of using both auditory and visual cues.

Many marine organisms probably use echolocation to some extent, Dolphins are an example. When they swim through a tank of turbid water, they emit frequent high-pitched clicking sounds and, while they do, can readily avoid all submerged obstacles. Their reliance on echolocation rather than vision in this behavior is confirmed by the ease with which they avoid submerged sheets of clear plastic and negotiate obstacle courses when blindfolded. Echolocation is also involved in their speedy and unerring location of food thrown into murky water and their ability to chase and capture fish while blindfolded.

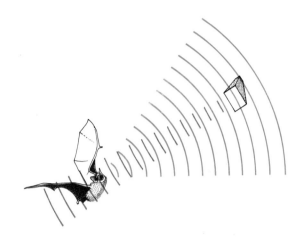

12.37 Echolocation by a bat
The ultrasonic chirps (color waves) of the bat are reflected by objects in its path, and the returning echoes (gray waves) give the bat information about the size and position of the objects.

ELECTRIC ORIENTATION

The capacity of the electric eel to generate powerful electric impulses, which it can use to stun its prey, has been well publicized. Less widely known is the capacity of six groups of fishes with much smaller and less powerful electric organs to produce weak electric fields that help them in orientation and in location of prey. By means of receptors sensitive to changes of as little as

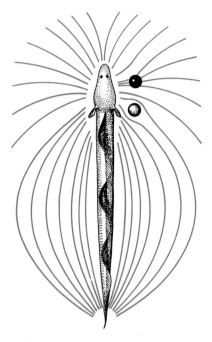

12.38 The electric fish *Gymnarchus* surrounded by the electric field it has produced
Nearby objects distort the field. Good conductors (black ball) concentrate lines of electric force on a small area of the fish's body, whereas bad conductors (light ball) spread the lines.   [Redrawn from J. Alcock, *Animal behavior*, Sinauer, 1975.]

0.03 microvolt per centimeter, these fishes can detect any distortions produced by nearby objects in the electric fields they generate around themselves (Fig. 12.38). Not only can they tell the difference between conducting and nonconducting objects, but they can also learn to discriminate between very similar structures. H. W. Lissmann of the University of Cambridge was able to train one electric fish *(Gymnarchus niloticus)* to distinguish between two ceramic pots that differed only in that one contained a glass rod 0.8 mm wide and the other a glass rod of the same length but 2.0 mm wide. These amazing animals, then, have what might be called a system of distant touch, which permits them both to find their way about and to locate prey during the night in often muddy water. As might be expected of such a system, it is used in communication too.

Also of great interest to biologists is the recent discovery that elasmobranch fishes (sharks, skates), though they do not generate electric fields themselves, have extremely sensitive receptor organs (derived from their lateral-line systems) that permit them to locate prey by detecting the exceedingly weak electric fields associated with the prey's body. Ad. J. Kalmijn of the Woods Hole Oceanographic Institution has found that these fish can locate electrodes buried beneath the sand when the field produced by the electrodes is as low as 0.01 microvolt per square centimeter.

ORIENTATION BY SMELL

A cue often used by mammals, particularly dogs, is scent. The ability of some dogs to trail a person, a rabbit, or a deer, and to backtrail when they find themselves in unfamiliar territory, is well known. Our own poor olfactory sense makes it hard for us to imagine the sensory world most mammals experience—a world in which olfactory information is received from hundreds of sources every minute, in which nearly every object is most clearly identified by its own distinctive smell. Hence we often fail to appreciate the sensory basis for much of the behavior of such animals. By way of compensation, we tend to ascribe to other mammals special senses that we do not possess; usually (and perhaps always) the sensory differences are quantitative rather than qualitative.

Some examples of orientation by olfactory cues have already been mentioned: the location of a female by the male silkworm moth, the pursuit of scent trails by foraging ants. One of the most dramatic examples of olfactory (or perhaps we should simply say chemical) orientation is the spawning runs of salmon. A young salmon hatches in a freshwater stream and shortly thereafter swims downstream to the ocean, where it ranges widely for several years while growing and maturing sexually. But no matter how far it has ranged, when the time for it to spawn approaches, it returns to the same freshwater stream in which it was hatched. The validity of this statement was demonstrated by an extensive study in which several Canadian investigators marked 469,326 young sockeye salmon in a tributary of the Fraser River. Nearly 11,000 of these were recovered in the same stream several years later when the fish had reached spawning age. But not a single marked fish was ever found to have entered any other stream.

Arthur D. Hasler and his associates of the University of Wisconsin have shown that salmon have chemoreceptors so acute that the fish can be conditioned to differentiate between even very dilute rinses prepared from different water plants. or between water from different streams. These workers have shown, moreover, that salmon whose olfactory tissues have been destroyed or whose nasal sacs have been plugged with cotton cannot distinguish between odors or locate the right spawning stream. Hasler has therefore hypothesized that salmon use their sense of smell in migrating back to their parental stream, and that the adult salmon must remember the odor of the stream in which they hatched and respond positively to it when their reproductive drive becomes high.

Recent field experiments by Hasler and his group have lent strong support to this hypothesis. The basic design was to imprint young salmon to a chemical not normally encountered in nature, and then to try to use this chemical to lure the salmon, when adult, into a stream where they had never previously been. A total of 26,000 salmon were held for 30 days during their smolting[6] period (when they are most easily imprinted to odors) in a tank supplied with water from a source different from the intended test stream. To this water was added a very tiny amount of a chemical called morpholine, a stable soluble compound that had been shown to be neither naturally attractive nor repellent to the fish. The salmon were then released into Lake Michigan. Eighteen months later, when the fish were ready to spawn, the researchers artificially created what they hoped would be a "home stream" by adding tiny quantities of morpholine to the water in a small tributary stream of Lake Michigan. Sure enough, many of the imprinted salmon did enter the stream containing morpholine, whereas very few of the 28,000 control fish (held in tanks without the chemical) did so.

To make doubly certain of their results, Hasler's group then exposed 5,000 more smolts to morpholine and released them into Lake Michigan, but this time, when the salmon were adult and ready to spawn, no morpholine was added to the tributary stream. Now only a few fish entered the test stream— evidence that its attractiveness to salmon in the earlier tests must have been due to the morpholine and not to any natural property of its own. Thus Hasler's hypothesis that salmon return to the stream where they were raised by detecting its odors, to which they were imprinted as smolts, seems firmly established (Fig. 12.39). There is even some evidence that only two days of exposure to a chemical during the sensitive phase for imprinting is sufficient to establish the system.

AVIAN ORIENTATION AND NAVIGATION

Migration and homing in birds   A subject of wonder for centuries has been the way many birds can travel thousands of kilometers from the place of their birth in the north to the wintering grounds of their species in the south, and back again the following spring. These migrations are not the result of random wandering. The members of each species usually follow a precise route characteristic of that species, often flying hundreds of kilometers each day or each night, and a few species make their entire migratory journey of several thousand kilometers nonstop. For example, the Pacific Golden Plover, a bird that cannot land on water, flies each fall from Alaska to its winter home in the Hawaiian Islands (Fig. 12.40). Even some small songbirds, such

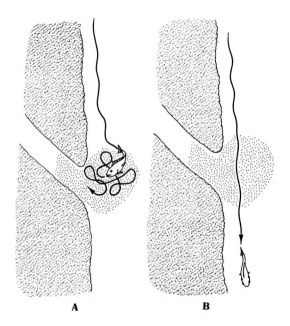

A   **B**

12.39 Tracks of salmon when they encounter morpholine
In one series of experiments (not discussed in the text), Hasler and his associates used ultrasonic tracking to plot the paths followed by adult salmon as they swam southward along the coast of Lake Michigan in search of a spawning stream. Morpholine was released in the area indicated in color.   (A) Track typical of fish imprinted on morpholine. On entering the morpholine-scented area, the fish stopped their southward migration and began to circle.   (B) Track of control fish, not imprinted on morpholine. They typically swam through the morpholine-scented area without pausing.

---

[6] Smolting is a stage in the development of young salmon in which they turn silver-colored and begin swimming in a migratory fashion.

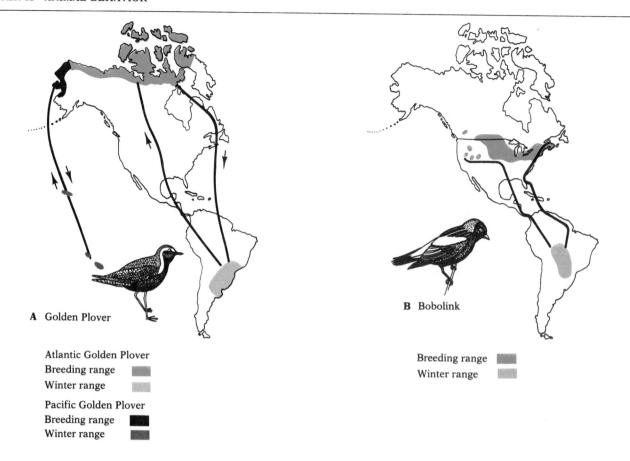

A Golden Plover

Atlantic Golden Plover
Breeding range
Winter range

Pacific Golden Plover
Breeding range
Winter range

B Bobolink

Breeding range
Winter range

12.40 Distribution and migration of the Golden Plover and Bobolink

(A) The Pacific Golden Plover flies over open ocean from its breeding grounds in Alaska to its winter range in Hawaii, the Marquesas Islands, and the Low Archipelago. The Atlantic Golden Plover follows one route south in autumn to its winter range in South America and another route north in spring.
(B) Several separate colonies of Bobolinks have become established west of the main breeding ranges in the northern United States. The birds from these colonies do not fly directly to the winter range in South America; instead, they first migrate east to the ancestral flyway and then turn south.   [Modified from *Migration of birds*, U.S. Fish and Wildlife Service, 1950.]

as warblers, depart in autumn from Nova Scotia or Cape Cod and fly non-stop over the Atlantic Ocean to South America.

It is plain that migration depends on something besides orientation by landmarks. The Golden Plover just mentioned and many other species fly over the open ocean, where there are few landmarks. Furthermore, the young birds have never before made the journey and could not be familiar with any landmarks. In some species they do not leave the northern breeding grounds until several weeks after the older birds have departed; yet they set out unerringly along the same route that the members of their species have followed for centuries.

The navigational abilities of birds are not restricted to the migratory seasons. Homing pigeons are renowned for their ability to return to their home lofts when released at distant points; they are frequently raced over distances from 150 to 1,000 km, often averaging speeds of 80 or 90 km (50–60 miles) an hour and sometimes making a 1,000 km flight from an unfamiliar release point to their home lofts in a single day. Other birds can perform similar feats; e.g. a Manx Shearwater captured on the west coast of England and flown by plane to Boston, where it was released, was back in its nest in England 12 days later, having flown 5,000 km across open ocean.

The sun and stars as compasses In 1950, the same year as von Frisch showed that bees can tell directions from the sun, Gustav Kramer of the Max

Planck Institute in Wilhelmshaven, Germany, suggested that celestial cues might be used by orienting birds. He and his colleagues were soon able to confirm this surmise by observations of starlings and several other migratory species in cages. The birds became unusually active at migration time, and if they could see the sun they began to fly at the side of the cage that lay in the direction of their normal migratory route. If the apparent position of the sun was altered with mirrors, the direction of the birds' attempted flights was altered accordingly. If the sun was obscured by heavy clouds, the birds became disoriented.

Further proof that starlings and other birds can tell compass directions from the sun came from experiments showing that a bird in a circular cage with food chambers arranged around its periphery can be trained to go to a chamber lying in a particular direction (Fig. 12.41). The cage can be moved to different localities or the bird can be put in the cage at different times of day without in any way interfering with its ability to locate the proper chamber. If it has been trained to eat at the northwest chamber, it will go to the northwest chamber in each experiment. In other words, it makes no difference whether the bird sees a sun low in the east in the morning, a higher sun nearer noon, or a sun in the west in the afternoon; each of these can tell the bird where northwest is. This means, of course, that the bird must use its internal clock to combine the time of day with the sun's position in order to calculate directions.

The role of the internal clock was confirmed when birds trained to orient toward the northwest in an experimental cage, using the sun as their cue, were put in a lightproof room for 4–6 days and exposed to an artificial day that began and ended six hours later than the natural day. When they were then again placed in the experimental cage and exposed to the sun, they oriented toward the northeast instead of the northwest. Because their clock had been shifted six hours out of phase with the natural day, they interpreted the sun's position erroneously; they misread the sun compass.

Experiments of this sort were first extended to free-flying pigeons by K. Schmidt-Koenig, now of the University of Tübingen, Germany. The procedure is to divide a flock of pigeons of similar previous experience into two groups, A and B. The internal clocks of group A are left set to true sun time, but the clocks of group B are reset six hours earlier or later. Then both groups are taken to the same point far from home, and the birds are released singly. The direction in which each bird vanishes from sight is recorded. If the sun is visible, the usual result is that the birds of group A vanish, on the average, in a direction lying roughly homeward, whereas those of group B vanish 90 degrees to the right or left of that direction (depending on whether their clocks have been reset six hours slow or fast) (Fig. 12.42). It is clear, then, that pigeons normally compensate for the sun's apparent motion, and that when their internal clocks are adjusted to an incorrect setting they misread the sun compass.

The sun is not the only celestial body that can provide directional information for orienting birds. Some birds that migrate at night, such as European warblers, can tell directions from the stars. When these birds are in closed cages indoors, their flutterings at migration time ("migratory restlessness") are disoriented. But when they are put in an outdoor cage with a glass top, which allows them to see the night sky but nothing else, their activity is oriented in the direction in which their species would normally

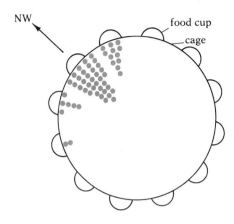

12.41 Choice of food cups by starlings trained to go to the northwest cup in a circular cage
Each dot indicates the choice of one starling. Most birds go to the correct cup or to one of the neighboring cups as long as the sun is visible, even if all other visual cues are altered.   [After K. Hoffmann.]

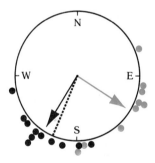

12.42 Effect of a six-hour-fast clock shift on the initial bearings chosen by homing pigeons on a sunny day
Each dot indicates the bearings chosen by one bird; black dots represent control birds, color dots experimental birds. The dashed line marks the proper homeward direction. The mean bearing of the clock-shifted birds (color arrow) is about 90 degrees to the left of the mean bearing of the control birds (black arrow). The experimental birds have been clock-shifted a quarter of a day, and they have made an error of a quarter of a circle in reading the sun compass.   [After W. T. Keeton.]

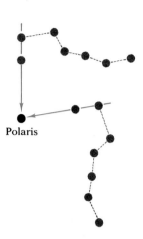

12.43 An example of how north can be located by star patterns

If one draws an arrow through the two end stars in the cup of the Big Dipper (Ursa Major), it will point toward Polaris, the North Star. Although the position of the constellation changes during the night, the same stars always determine an arrow pointing toward Polaris; hence directions can be determined without need of time compensation. Many different star patterns could be used for finding direction in this way. [After S. T. Emlen.]

migrate. It seems that the night sky provides the warblers with enough cues for orientation.

In a long series of experiments with Indigo Buntings in a planetarium, Stephen T. Emlen of Cornell University has shown that at the time of fall migration the birds will orient in a direction that is south according to the stars represented, whether it is true south or not—a conclusive demonstration that they can orient by the stars. But, in reading the star compass, the buntings do not rely on their internal clocks to compensate for the movement of the stars during the night; instead, they use pattern information to locate directions, a process that does not require time compensation (Fig. 12.43). In other words, the two celestial compasses—sun and stars—are read in entirely different ways by the birds.

Emlen has found that there is a short sensitive phase in which the young buntings must learn to read the star compass. Buntings that have not seen the starry sky during the few weeks before the start of their first migratory season in September never learn to use the star compass, no matter how often they see the stars thereafter.

Emlen has also elegantly demonstrated the importance of a bird's physiological state in determining its orientational response to directional cues. By manipulating the photoperiods of the buntings, he was able to bring one group into autumnal migratory condition and a second group into spring migratory condition at the same time. He then tested both simultaneously under a spring planetarium sky, and found that the birds in autumnal condition oriented southward and those in spring condition northward. In other words, their different hormonal states resulted in opposite responses to the same external stimuli. Just how such a switch could be produced in the nervous system is a question that research has yet to answer.

Magnetic cues in orientation The discovery that birds and many other animals can tell compass directions by observing the sun or the stars, while a major step forward in our attempt to understand orientation behavior, leaves still unexplained homing behavior of the sort seen in pigeons or the Manx Shearwater mentioned above. Telling compass directions, such as south in fall and north in spring, is not the same thing as telling the homeward direction after being transferred far from home. If you were to blindfold a person, take him 500 km from home, remove the blindfold, give him a compass, and tell him to start walking straight home, he would not be able to do it unless he also had what is called map information, i.e. information that told him what direction he must select in order to go home.

What sort of environmental cues might a bird use to tell homeward direction? According to one hypothesis popular during the 1950s, a pigeon could compute its homeward direction by determining both its latitudinal and its longitudinal displacement from the sun. However, evidence from many kinds of experiments indicates that the sun is used only as a simple compass, and the idea that it provides birds with information for bicoordinate navigation has been abandoned.

More recently, even the idea that a sun or star compass is essential to avian navigation has been discarded. Numerous researchers using radar to observe the migratory movements of wild birds have reported frequent instances of well-oriented migration under heavy cloud cover, when celestial cues were not available. Furthermore, William T. Keeton of Cornell University has shown that, when pigeons whose internal clocks have been set six hours fast are released under heavy overcast, both they and the control birds (on normal time) orient homeward; there is no indication of the 90-degree difference in initial bearings regularly seen in releases under sunny skies (Fig. 12.44). This means that, in the absence of the sun compass, pigeons can orient by some process that does not require time compensation.

Since clock-shift experiments under clear skies demonstrate that pigeons use information from the sun when it is available, and Keeton's experiments show that they can orient without it when necessary, it must be concluded that the pigeon navigational system employs a variety of cues, some of them redundant. This redundancy has probably been responsible for some of the difficulty in unraveling the mystery of avian orientation. For example, it undoubtedly complicated tests designed to determine whether birds can use magnetic cues in orientation. Experiments by Keeton have shown that magnets glued to the backs of experienced pigeons often result in disorientation when the birds are released under total overcast. Although the possibility that birds might use the earth's magnetic field as one source of directional information had been suggested off and on for more than a century, previous attempts to determine whether magnets interfere with homing yielded mostly negative results. The reason is that the experiments were conducted under clear skies, when the sun compass was available to the birds; the effects of the magnets are most pronounced when the sun is absent. Perhaps sun cues and magnetic cues can be used interchangeably by the birds, and it is only when neither are available that experienced birds become disoriented.

The suggestion that magnetic cues may be used by pigeons in orientation is supported by field experiments conducted by Charles Walcott of the State University of New York, Stony Brook, and by laboratory training experiments performed by M. Bookman at the Massachusetts Institute of Technology. Tests with European Robins and warblers by W. Wiltschko of the University of Frankfurt, Germany, point to use of magnetic cues by these species. It may be that a magnetic sense will prove to be widespread in birds. It must be emphasized, however, that no one has yet demonstrated a sensory basis for detection of magnetic stimuli by birds, though the presence in the heads of pigeons of a small deposit of magnetite, a magnetic mineral, was reported in 1979 by Walcott, J. Gould, and J. Kirschvink (the latter two of Princeton University).

Orientational responses to magnetic cues have also been reported in a variety of insect species. Of such responses, the most conclusively demonstrated is a systematic perturbation of the angle of the waggle dance of honeybees by the earth's magnetic field, discovered by Martin Lindauer, now of the University of Würzburg, Germany. Its mechanism, like the mechanism of magnetic detection by birds, has yet to be elucidated. A small deposit of ferromagnetic material in the honeybee thorax was found by Gould and his co-workers in 1978.

Birds and insects are not alone in possessing a magnetic sense. Magnetic responses (usually orientational) have been reported for a variety of invertebrates and for several species of fishes and amphibians. Especially interesting is the demonstration by Kalmijn that elasmobranch fishes can use

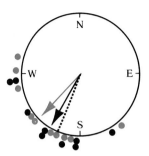

12.44 Effect of a six-hour-fast clock shift on the initial bearings chosen by homing pigeons on a totally overcast day
When the sun is not visible, the experimental birds choose bearings (color dots) not significantly different from those of control birds (black dots); the 90-degree deflection of their bearings seen on sunny days (compare Fig. 12.42) is not evident. In the absence of the sun compass, the birds appear to orient by some other system, which does not require time compensation. [After W. T. Keeton.]

12.45   A leopard pursuing a baboon
Like many other carnivorous animals, the leopard
must invest much energy in capturing its prey.   [John
Dominis, *Life* Magazine, © 1965 Time Inc.]

12.46   A spider in its web
The spider devotes much of its prey-capturing activity
to spinning a web rather than to chasing prey.   [S. J.
Krasemann, National Audubon Society, Photo
Researchers, Inc.]

their electric sense to detect earth-strength magnetic fields by detecting the
electric field induced when they swim through the magnetic field. Also of
great interest is the discovery by R. P. Blakemore in 1975 of a group of
magnetotactic bacteria (in the Northern hemisphere they always swim
northward), and the more recent demonstration by Kalmijn and Blakemore
that their detection mechanism depends on minute ferromagnetic dipole
particles formed by the bacteria in their cells. Since two very different
mechanisms of magnetic detection have now been found—one in the elasmobranch fishes and another in the magnetotactic bacteria—the old argument that no living creature could possibly detect a magnetic field as weak
as that of the earth has finally been laid to rest, and we can hope that the
mechanism of detection in birds, insects, and other animals may soon be
discovered.

The problems that remain   Although the studies of animal orientation have
led to the exciting discovery of a whole new sensory capacity of animals—the
magnetic sense—the problem of explaining goal-directed navigation, such as
homing by birds, remains. The bulk of current evidence can be interpreted
as indicating that magnetic cues provide the birds with yet another compass,
but we have already seen that a compass is not enough.

In the last few years a host of other cues have been found to play some
role in avian orientation. Thus Floriano Papi and his colleagues at the University of Pisa in Italy have presented evidence that olfactory cues may
sometimes be important. Several investigators have called attention to meteorological parameters (atmospheric pressure patterns, wind direction,
wind turbulence, etc.) that might provide directional information. Melvin
Kreithen and Keeton at Cornell have found evidence that pigeons have
several previously unsuspected sensory capabilities (barometric-pressure

12.47 A python in a tree, waiting for prey to come near
[Hans Reinhard, Bruce Coleman Inc.]

detection, polarized-light detection, ultraviolet-light detection, infrasound detection) that might help them find their way. But despite all the progress in this field of research during the last two decades, a full solution to the mystery continues to elude scientists.

FEEDING BEHAVIOR

The behavior patterns associated with feeding vary greatly depending both on the animal's mode of life and on the sort of food it uses. Sessile animals, such as sea anemones, corals, and oysters, which rely on their food to come to them, do not have to search for food, as many mobile animals do, but it is essential that they position themselves in a favorable location. Herbivorous animals may have to spend time searching for suitable food plants, but once the plants are found no elaborate stalking and capturing behavior is necessary. For some carnivorous animals, on the other hand, the stalk and the capture may require as much energy as the search (Fig. 12.45). Rather than spend time in stalking, other carnivorous animals may invest energy in luring prey or in constructing prey-capturing devices, such as the web of a spider (Fig. 12.46), or they may merely seek out a place of concealment near a spot where prey are likely to pass and then lie in wait (Figs. 12.47, 12.48). Clearly, the enormous diversity of feeding methods makes impractical any attempt to enumerate them here. Instead, we shall turn to some behavioral strategies associated with feeding.

12.48 A leaflike praying mantis awaiting its prey
After remaining quite still until some unsuspecting insect lands nearby, the mantis moves with startling speed to capture it. The mantis' cryptic appearance not only helps it in catching prey, but also in avoiding notice by its own predators.   [Norman Myers, Bruce Coleman Inc.]

12.49   A larva of the arctid moth *Utethesia ornatrix* inside a seed pod of its poisonous host plant
The rattle-box plant (*Crotalaria spectabilis*) contains a bitter-tasting, very toxic compound (a pyrrolizidine alkaloid) that protects it from most herbivores, including both insects and vertebrates. But the *Utethesia* moth is not only able to eat the plant's seeds (the part of the plant where the poison is most concentrated), but it can also sequester the poison for its own protection against predators. Many other plants produce compounds (the so-called secondary plant substances) that protect them against being eaten, and many herbivorous insects are food plant–specific, because they have evolved specialized enzymes for detoxifying the protective compounds of their particular food plants. In the rattle-box plant the protective compound does not become toxic until acted on by enzymes (the mixed-function oxidases) of a herbivore; *Utethesia* larvae somehow avoid activating the toxin.   [Information and photograph, courtesy Thomas Eisner, Cornell University.]

Although no hard-and-fast boundary lines can be drawn, most animals can be characterized as either ***specialist*** or ***generalist*** feeders. As the terms suggest, specialists feed on a very limited number of foods, while generalists have more catholic tastes. The specialist is usually highly adapted for efficiently locating, capturing, and metabolizing its specific food (Fig. 12.49). So long as the food item remains available, the specialist may flourish, but if that item should disappear, then it may pay a stiff penalty for having, in its evolution, "put all its eggs in one basket." The generalist cannot be so exquisitely adapted for utilization of any one food item, but because its behavioral repertoire includes the ability to search for and use many different kinds of foods, its existence is less tied to the availability of any one.

The problem of a mobile carnivorous animal that will take at least a modest number of different food items is to maximize the net energy return from its feeding endeavors. Let us designate the energy gained from the food eaten as E_g, the energy invested in finding, capturing, and eating the food as E_i, and the time invested as t. The animal must try to make the quantity

$$\frac{E_g - E_i}{t}$$

as large as possible; its magnitude will depend on the animal's decisions about where to search, how long to search in the selected area, what prey to search for, and whether to pursue a prey individual once found.

Searching in places where appropriate food is most likely to be found is clearly an important aspect of efficient feeding behavior. In other words, foraging should be a decidedly nonrandom process. In some instances specificity in foraging might be largely innate, as in some insects that, immediately upon reaching maturity, fly into a certain type of environment where they will stay for the rest of their short lives. The environment toward which they are so strongly attracted is sometimes one they themselves have never previously experienced; they merely respond in a stereotyped fashion to sign stimuli to which they appear to be sensitively attuned. Or the specificity of foraging may be due largely to learning, and may change as the availability of the various kinds of food items changes. As you might expect, automatic specificity of foraging tends to be most common among specialists, whereas specificity based on learning is more typical of generalists.

The rapidity with which an animal can learn where food is to be found and restrict its searching accordingly has been verified in wild crows by Harvey Croze of Oxford University. Croze put bits of meat under red mussel shells on a beach, and found that once a pair of crows had discovered the meat they returned to the same area of the beach daily and even defended it against intrusion by other crows. Similar observations have often been made by people with feeders in their gardens; once a feeder has been located by, let us say, chickadees and cardinals, those birds become regular return visitors, even though they may previously have seldom been seen in the garden.

If an area has consistently been a rich source of food, animals that have regularly visited it may continue to do so, even if the food supply starts dwindling. Presumably the assurance of finding at least some food, and the fact that no time is lost in searching for the area, make its continued exploitation a winning strategy still. But as the area declines further, a point is eventually reached when the yield is so low that the animal must move on, if net energy gain is to be maximized. Just when this point will be reached, i.e. when it will pay to leave a known but deteriorating food supply to search for an unknown better one, will depend on many factors, including (1) the probability that the food supply in other areas will be richer than in the one to be left; (2) the patchiness of the food (i.e. whether it is likely to be scattered over much of the countryside or tightly clumped in only a few places); (3) its predictability (e.g. whether it will almost always be present during the proper season wherever a certain recognizable habitat occurs, or whether it will instead be randomly distributed in space or in time or in both); and (4) the intensity of competition the animal is likely to encounter if it leaves its own area and tries to invade a new one.

What prey to search for involves much the same considerations as where to search. Some degree of specificity usually increases the efficiency of the search, even for generalists. By concentrating the search on one or a few kinds of prey, the animal can become maximally responsive to the sign stimuli from those prey, while avoiding distraction from other less important stimuli. Such an animal is said to have formed a *search image.* For example, the crows studied by Croze appear to have formed such a specific search image for the red mussel shells under which the bits of meat were hidden that they ignored red razor clam shells; even red grooved mussel shells (a surface-texture variant) received less attention than standard red mussel shells.

Formation of search images is not restricted to carnivorous animals; it has repeatedly been found in seed eaters as well. So long as one kind of seed is nutritious and reasonably abundant, a seed-eating animal increases its efficiency by focusing its search on that kind of seed; its loss in not taking nutritious seeds of other kinds is less than its gain from being able to find

12.50 Flocking defense against a falcon by flying starlings

Starlings usually fly in a loose formation, and continue to do so even in the presence of a falcon if they are above the falcon. But if they are below, they form a tight flock. The falcon cannot stoop at a bird in such a flock without risking serious damage to itself by crashing into other birds. The falcon will stoop only if an individual starling becomes separated from the flock.

more seeds of the sought-for kind per unit time. Thus, as was observed by Marian Dawkins of Oxford University, chicks fed rice dyed the same color as the background improved rapidly in their ability to detect the rice, but if they were fed conspicuous grains their ability to detect the cryptic rice fell sharply; finding the rice appeared to depend on the chicks' maintaining a sharp search image for it.

Both the formation of search images and the tendency to restrict the search to particular places are thought to be reasons why edible prey are almost never taken in direct proportion to their relative abundance. When a prey species is rare, predators generally feed on it less than would be expected if they sampled randomly, but when a prey species becomes moderately abundant the predators take it more often than would be expected. If, for example, a chickadee is searching for small brown insects on the buds of a certain kind of tree, it will not likely often encounter a small black insect that occurs occasionally on the tree trunks; hence the black insect will be eaten less often than a random-search model would predict. But if the black insect becomes abundant, the chickadee may change its foraging strategy, form a new search image (i.e. become especially responsive to the sign stimulus of a small black spot), and concentrate on the tree trunks, with the result that it now eats more of the black insects than predicted by the random-search model.

Finally, how should a predator decide whether a prey individual, once located, should be pursued? For a predator whose prey is nonmobile or very easily caught, the answer is that each individual should ordinarily be taken, because the time invested in capturing the prey is minimal compared to the time invested in searching. But for a predator whose prey is very hard to catch, owing to superior speed or agility, many prey individuals should be ignored, because the energy required to catch them would be greater than the energy gained from eating them. Only when circumstances are favorable, as when the prey can be taken by surprise, or when the prey is young, sick, or enfeebled by age, should the predator initiate pursuit. That such selectivity of pursuit is indeed found in many predators has repeatedly been verified (Fig. 12.50).

From this brief discussion of a few of the factors that govern the feeding behavior of animals, we see that the behavior can often be analyzed and understood in terms of energetic considerations. In short, the behavior is adaptive for the animal; it makes biological sense.

SOCIAL BEHAVIOR (SOCIOBIOLOGY)

In some animal species the individuals pass much of their lives without any cooperative activity, their intraspecific interactions (aside from mating) being largely restricted to antagonistic ones. But in many species some degree of intraspecific cooperation is evident. The cooperative interaction may be relatively simple and of limited duration, as in the winter flocks formed by chickadees (which are not found in flocks during the breeding season); such flocking probably aids in locating patchily distributed food supplies and in spotting and eluding predators. Defense against predators is especially apparent in schools of swimming fish or flocks of flying birds (Fig. 12.50). An example of transient cooperation with a different function is the huddling together of coveys of quail in winter, which enables the birds to conserve body heat and thus withstand low temperatures that would kill isolated birds. At the other end of the cooperativity scale are such highly evolved and long-lasting societies as those of honeybees and human beings, in which almost every aspect of each individual's life depends on the activities of others.

A prerequisite to any sort of cooperative interaction is the ability of the individuals to communicate with one another. We have already examined some of the many types of communication in animals, and it is no surprise to find that, whatever the mode of communication—auditory, visual, olfactory, or some other—the richness and complexity of that communication tend to be correlated with the degree to which an animal is social.

SPATIAL FACTORS THAT INFLUENCE SOCIALITY

The social behavior of any given species of animal is strongly influenced by the way the individuals of that species are organized in space, which is in turn strongly dependent on the distribution of the food supply and on the animal's method of exploiting it. We shall here mention three aspects of spatial organization: individual distance, territory, and home range.

In many species each individual may be said to carry around itself a small volume of space that it tends to treat as uniquely its own. The animal shows signs of agitation, or even overt aggression, when another individual comes too close and breaches this *individual distance.* One consequence of the effort of animals to maintain their individual distances from all others is the remarkably regular spacing often seen where many individuals are gathered together, as in a flock of birds lined up on a telephone wire or a school of minnows swimming in formation. Maintenance of relatively inviolate individual distance is often seen in animals that are only moderately social; it tends to be compromised (though still present) in highly social animals such as baboons, chimpanzees, and many other primates, where mutual grooming and other forms of bodily contact are common; and it is apparently nonexistent in social insects, where intimate bodily contact within the nest is virtually continual.

A much larger unit of space is the *territory,* defined as an area actively defended from conspecifics by an animal or a group of animals; we have already discussed the role of displays in territorial defense. Several types of territories may be recognized. One type is an area within which mating, nesting, and feeding during the breeding season all occur. Another is an area for mating and nesting, feeding occurring elsewhere. Still others are areas used for mating only, areas used for nesting only, areas used for roosting or feeding only, and so forth. Whatever the type, territoriality seems to function in spacing individuals in such a manner that the most severe aspects of competition and individual antagonism are minimized and social stability is improved.

Among songbirds the most common type of territory is one that contains both the nest and the feeding area. It must be relatively large if it is to embrace resources sufficient to support the adults and their voracious young, which often consume thousands of food items each day. Probably because birds with this type of territory are rather widely spaced, they tend

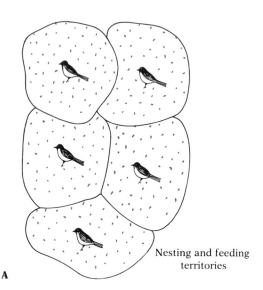

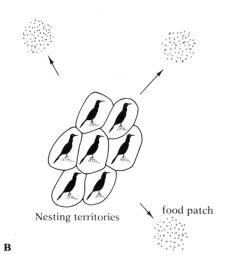

12.51 Territory used for both feeding and nesting compared with territory used only for nesting
(A) When food is distributed more or less evenly throughout the area, it may be energetically efficient for each individual to occupy and defend its own territory for both feeding and breeding. (B) If food occurs in unpredictable patches, often far from good nesting sites, it may be more efficient for the individuals to occupy only small nesting territories in a colony and to forage as a group. This type of organization leads to more intimate and structured social systems. [Redrawn by permission from E. O. Wilson, *Sociobiology,* Harvard University Press, 1975.]

not to have elaborate cooperative social systems. By contrast, birds that must range far from their nesting area in search of an unpredictable food supply usually defend only very small territories immediately surrounding their nests, which may be located in dense colonies, as in the case of many sea birds such as gulls and terns (Fig. 12.51). By living in colonies, they can most fully exploit the best nesting areas and can benefit from concerted defense against predators.

A still larger unit of space is the *home range,* the total area in which an animal (or group of animals) travels in the course of its normal activities. In some cases the home range of an animal may be identical to its territory, but often it embraces additional areas that are not defended and may be shared with other territory holders (Fig. 12.52). Many large mammals, especially herbivorous ones that must forage widely to find enough food, have no territory, but only a home range. As might be expected, the size of the home range tends to vary with the size of the animal. Thus the home range of meadow mice averages about 270 m², that of cottontail rabbits about 5.6 hectares (14 acres), which is about two hundred times as much, and that of whitetail deer about 1–5 km² (0.5–2 square miles).

Both individual distance and territoriality, when they occur, make it difficult for one individual to approach another. But obviously such approach is necessary if mating is to occur; a major function of courtship displays is diminution of the agonistic motivations aroused when the partners approach each other. In parental tending of young, too, there must be close physical contact. It is not surprising, then, that most higher forms of sociality—characterized by considerable division of labor and often by *altruism* (behavior, possibly self-destructive, that is performed for the benefit of others)—probably arose out of family groups. In these groups, the barriers to intimate contact have already been breached, and, of course, the individuals have a genetic investment in one another; they share many genes, and natural selection would especially favor cooperation and altruism among them.[7]

INSECT SOCIETIES

In the social insects the unit of social organization is invariably the family, which typically consists of a single reproductive female, the queen, and her daughters (also sometimes her sons). In bees and wasps the daughters serve as workers; in ants some also serve as soldiers. In termites both daughters and sons usually function as workers or soldiers. The workers and soldiers are nonreproductive individuals; the society so influences their development that they are physically unlike the queen and constitute separate sterile castes.

Subsocial and primitively eusocial bees   To understand how the elaborate social systems of honeybees or ants may have arisen, let us consider first some members of the order Hymenoptera (bees, wasps, and ants) in which sociality is less highly developed.

---

[7] Note that altruistic behavior among carriers of the same gene favors the survival of that gene, if not the survival of the individual carriers. The concept of altruism will be more thoroughly treated in Chapter 18 (p. 786).

A female bee of the subsocial genus *Halictus* constructs an underground comb of up to 20 cells, lays an egg in each, provisions the cells with food, and then closes the nest, after which she may remain on guard until her offspring emerge. In some species the young bees leave and build their own nests elsewhere, but in some they remain in the parental nest, enlarging it and laying their eggs there. Though there are no castes, and each female is capable of reproduction, there is a breakdown of spatial barriers; and the communal sharing of a nesting site can easily be imagined as the evolutionary forerunner of more complex social systems in other bees.

At a more advanced stage of sociality are the bumblebees. Here again the nest is founded in spring by a single female. But unlike the offspring of the *Halictus* bee, the young that hatch from the founding bumblebee's eggs do not become reproductives in their own right; they serve as workers, enlarging the nest, gathering nectar and pollen, and caring for the young that hatch from later eggs laid by the founder. The founder, who remains in the nest as the queen, now devotes almost all her energy to egg laying. Eventually there may be several hundred, or even a thousand, bees in the colony. As the season nears its end, some unfertilized eggs are laid that give rise to males, and some of the fertilized eggs give rise to queens when the young that hatch from them are treated in a special fashion. These new reproductive individuals fly out and mate. The bees in the old hive die with the coming of winter, but the fertilized young queens hibernate and found their own nests the next spring. Because bumblebees have division of labor correlated with sterile castes, they are said to be eusocial (i.e. truly social), but their colonies, which must be founded anew each year because they cannot survive the winter, are by no means as complex as those of honeybees.

Honeybee societies   Containing as many as 80,000 individuals, a colony of honeybees is much larger than one of bumblebees, and it is much longer-lasting, because of its capacity to overwinter. A single queen may live for seven years or more. She continues to lay eggs fertilized by sperm she has stored since she mated during her nuptial flight as a young queen. During the course of her life she lays hundreds of thousands of fertilized eggs, most of which hatch into larvae that develop into the worker bees that carry out the tasks of the colony.

During the course of their development worker bees may perform a variety of tasks. They usually serve as nurse bees for roughly the first two weeks after metamorphosis, first incubating the brood and preparing brood cells, and later feeding the larvae. Then they may become house bees for a week or two, acting as storekeepers, housecleaners, wax secreters, or guards. Finally, they may become field bees for four or five weeks, foraging for nectar and pollen. The individual has no choice of roles in this system; her role is determined by a combination of caste, stage of development, and the conditions of the hive.

As would be expected in a social organization so large and intricate, communication within a beehive is extensive and complex. We have already mentioned the well-known dances by which scout bees inform the foragers of the location of food sources, but communication by dances is also used in other contexts—notably in the establishment of new colonies. When a colony has become large, the queen may leave, taking with her many of the workers

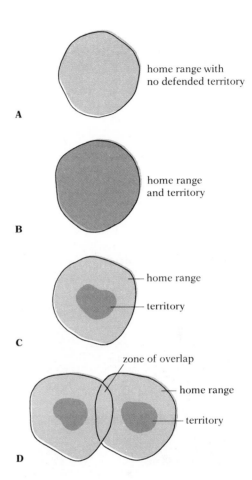

12.52   Relationships between home range and territory
(A) Some animals have a home range but no defended territory.   (B) In some cases an animal's home range may be the same as its territory.   (C) More often the home range is larger than the territory.   (D) Home ranges often overlap, but territories seldom do.

12.53 A bee swarm
The bees form a huge ball in the fork of a tree. Scout bees fly out from the swarm in search of possible new nesting sites.   [Courtesy E. S. Ross.]

12.54 Ants tending aphids on a thistle
The ants guard the aphids (small purplish-black insects), move them to new feeding spots when necessary, and periodically milk them by stroking them with their antennae. The stroking stimulates the aphids to release honeydew, a sweet sticky fluid used by the ants as food.   [Jane Burton, Bruce Coleman Inc.]

in a dense swarm. The swarm usually lands on some object, such as a branch of a tree and remains there for some days while scout bees fly out in search of potential new nesting sites (Fig. 12.53); maintenance of the swarm depends on secretion of a swarming pheromone by the queen. When a scout finds a possible site, she returns to the swarm and performs a dance telling where the site is located. The dancer also informs the other bees, by the vigor of her dancing, how good the site is.

Now, it will often happen that several different scouts will have located different potential nest sites, and each will dance about her own site. As long as there is no agreement among the scouts, the swarm makes no decision. The scouts must fly back out and reassess their own sites, inspect one another's sites, and perhaps find new possibilities. Moreover, workers stimulated by the dances go out to inspect the sites, and, as would be expected, they go predominately to the sites whose scouts danced longest and most vigorously. If these recruits agree with the high assessment of the quality of a site, they too return to the swarm and dance, recruiting still more bees to make inspection flights. Slowly, then, a consensus emerges, as more and more bees dance for one site and the scouts that had previously danced for other locations cease to display. Finally, when the consensus is clear, the swarm makes its decision and flies to the chosen spot to begin a new colony. Such apparent democracy of decision making is almost unique in the animal world.

What, you may ask, happens to the part of the colony left behind in the old hive after the queen and her swarm of attendants depart? Usually, before the queen leaves, she cuts back on her secretion of a pheromone—queen substance—the suppression of which frees the workers from inhibition against rearing new queens. They build some bigger cells, and the larvae in these cells develop as queens. When a new queen emerges, the other developing queens are killed, and then the young queen goes on her nuptial flight, attracting males (drones, which hatched from unfertilized eggs) by means of the queen substance. After mating, the queen returns to the hive and begins the steady egg production that will be her principal duty for the rest of her life.

Note that the previous paragraphs mentioned two pheromones—queen substance and swarming pheromone. These are but two representatives of an array of chemicals that play a fundamental role in the communication necessary for coordination of activities within the hive. Among the many other pheromones are a trail substance, alarm substances, various kinds of attractants, and a hive-identification substance. In addition, constant regurgitative feeding among the workers provides all members of the colony with continuous information about the nutritional condition of the whole colony.

Ant societies An ant nest, whose organization is similar to that of a honeybee hive in many respects, likewise contains a queen and numerous sterile workers. In addition to the ordinary workers, there is often a soldier caste (sometimes more than one); like the workers, the soldiers are sterile.

Worth special mention are the farming activities of some ants. There are, for example, dairying species that keep herds of aphids, small insects that secrete a sweet liquid called honeydew (Fig. 12.54). The ants carry the aphids to new feeding spots and guard them against predator insects; at regular intervals they stimulate the aphids to secrete honeydew by stroking them with their antennae. Other species of farming ants grow fungi. These ants collect leaves, dung, and other organic material, chew it to a proper consistency, and place it in moist chambers in their nest. On this material they grow fungi, which they use as food. The particular species of fungi grown have become so dependent on the ants for their propagation that they can live nowhere else but in the nests of their hosts.

Certain species of ants have evolved a curious behavior called slave making. They raid the nests of certain other species of ants, and carry captive pupae back to their own nest. These pupae develop into workers that become fully functioning parts of the slave maker's colony. Note that use of the term "slavery" for the behavior of these ants is somewhat misleading, because it is not comparable to slavery in human beings. The ants always enslave *other species* of ants, whereas human slavery is of members of the same species. Slavery in ants is perhaps more comparable to use of chimpanzees by humans.

Figure 12.55 shows one example of the amazing cooperation exhibited by ants in performing tasks that it would be impossible for any single individual to perform.

12.55 Driver ants transporting a grub in a "living cradle"
The ants form the cradle, and also the chain by which it is suspended, with their own bodies. [M. P. L. Fogden, Bruce Coleman Inc.]

VERTEBRATE SOCIETIES

We have seen that an insect society is a special kind of family group—a queen plus her offspring. Most vertebrate societies, too, are based either on families or on extended kinship groups, but their organization is very different from that of insect societies. Typically, a vertebrate social group consists of a leader (usually a male), together with his mates and offspring (and perhaps their mates and offspring). As we have already mentioned, such a grouping benefits both from reduced aggression over spacing and from the greater tendency for cooperativity and altruism to evolve among individuals with shared genes.

Vertebrate societies differ from insect societies in several fundamental ways. Division of labor is not based on biologically determined castes. Most adult members of the society take part in reproduction (or at least try to do so); there is no setting aside of one or two individuals to serve as reproductive machines for the entire group. The organization depends on the ability of individuals to recognize one another; no such requirement appears to exist in insect societies. And the behavior of social vertebrates, though it may contain many instinctive elements, is nonetheless far less rigid and less stereotyped than the behavior of social insects.

Social hierarchies   Among vertebrates particularly, an important aspect of social organization is the hierarchy. For example, a newly formed flock of hens will within a few days establish a series of dominance-subordination relationships that will effectively order the individuals within the flock. These relationships are based on the first few hostile encounters between each pair of birds. Bird A may be particularly large and aggressive, and may win her encounters with all other individuals of the flock. She thus becomes the dominant bird, and in future encounters the other hens will often move away from her without attempting to resist. Similarly, bird B may become established as dominant over all other birds in the flock except A. In other words, B can peck C or D without their pecking back, but B submits to pecking from A without pecking back. In this way, a flock of eight hens might come to have a *peck order* that could be diagrammed as follows:

$$A \longrightarrow B \longrightarrow C \longrightarrow D \longrightarrow E \longrightarrow F \longrightarrow G \longrightarrow H$$

where A is dominant over all the other birds, and has the "right" to peck them without being pecked back, and H is dominant over none. Sometimes the peck orders are less simple. For example, D might have been particularly aggressive when she happened to have her initial encounters with A, and A may have been tired or ill at the time, with the result that, although D is subordinate to B and C, and B and C are subordinate to A, D may be dominant to A:

$$A \longrightarrow B \longrightarrow C \longrightarrow D \longrightarrow E \longrightarrow F \longrightarrow G \longrightarrow H$$

Among the numerous factors that help determine an animal's position in a social hierarchy are age, size, strength, sex, health, hormonal condition, seniority in the group, and location of the first encounter (in territorial animals an individual fighting on its own established territory is almost always the victor, apparently because it is so much more highly motivated to attack than the invader). Once established, however, a *peck right* system like that in chickens tends to become more or less fixed, and an old or ill individual may continue to dominate more vigorous associates long after it has lost the capacity to defeat them in physical combat. The reason is probably that, once the dominance-subordination relationship for each combination of individuals has been established, the subordinate individual makes little effort to retaliate when pecked by the dominant one and thus fails to discover that conditions have changed. A peck-right system thus tends to inhibit social mobility.

In some other animals the hierarchical system is a more fluid one. In pigeons, for example, a subordinate bird will often give at least token resistance to a dominant one. In such a *peck dominance* system, changes in the physical condition of individuals are more quickly discovered and exploited, and there is thus more social mobility than in a peck-right system.

A social hierarchy, once established, tends to give order and stability to the relationships within the group. There is less tension, and there is less fighting and disturbance. When, for example, one flock of chickens is allowed to establish a stable peck order and another is kept disrupted by frequent additions of new members and removal of old ones, the birds in the unstable flock eat less, gain less weight, lay fewer eggs, fight more, and suffer more wounds.

Within a social hierarchy the dominant individuals usually enjoy many perquisites that are denied the subordinate individuals. When food or water is scarce, they usually have first access to what there is. They do most of the mating and hence leave more offspring; in baboons, for example, subordinate males may be allowed to copulate with females when they are not in full estrus and the likelihood of fertilization is low, but ordinarily only the highest-ranking male may mate with a female in full estrus.

One might well ask why a subordinate individual remains in the group, to suffer such abuse. The answer seems to be that, in species that are fully social, an individual must usually be part of a group to find food, avoid predators, and have any chance at all of reproducing. Its chances may not be good within the group, but since it has no chance whatever outside, belonging remains its best strategy. Moreover, if it survives long enough, it has a chance of succeeding to a dominant position as the dominant individuals, which are usually older, die off.

Mating systems   Several kinds of mating systems are found among the various groups of vertebrates. *Monogamy,* in which one male mates with one female, is common in birds, but relatively rare in mammals. Of the two kinds of *polygamy,* in which one male mates with many females *(polygyny)* or one female mates with many males *(polyandry),* polygyny is by far the more common. Many species form no pair bonds at all, and their mating is often called *promiscuous,* though this term may be misleading if one takes it to mean random and disorderly mating; the choice of mates is usually far from a random process (Fig. 12.56).

To understand the evolution of different mating systems in different species, we must first recognize that one of the principal determining factors is what has been called *parental investment*—the cost to the parent, in terms of its ability to produce more offspring, of behavior enhancing the likelihood that a particular offspring will survive and reproduce. Thus a fish that releases thousands of eggs into the surrounding water and then swims off has much less parental investment in any one of the eggs than a bird that constructs an elaborate nest, lays only a few eggs in it, and then remains to

12.56 Sage grouse on their mating grounds
Top: In this species many males (white breasts) display simultaneously on a compact mating area called a lek; each male, however, actively defends his own spot within that mating area. The females (smaller, browner birds) congregate in the lek and, after observing the males' displays, choose the ones from whom they will solicit copulation.   Bottom: A male in full display. [Top: L. L. Rue III, Photo Researchers, Inc. Bottom: Bob and Clara Calhoun, Bruce Coleman Inc.]

incubate the eggs and tend the young when they hatch. By studying the parental investment characteristic of different vertebrate species, together with the way of life they pursue (the places where they occur, the food they use, the predators that feed on them, etc.), we can better understand the patterns of occurrence of the different types of vertebrate mating systems.

For example, why should monogamy be more common in birds than in mammals? It must be recognized, first, that male birds are physiologically as well endowed for the time-consuming and energy-draining tasks of incubating the eggs and feeding the young as female birds. It must be realized, too, that many kinds of young birds require a thousand or more items of food per day. Meeting their needs often demands the cooperation of both male and female parent; without cooperation, neither may have surviving offspring. It is therefore evolutionarily to each parent's advantage to form a monogamous pair bond and make a large investment in the young.

In mammals, by contrast, where the young are nourished on milk, the male's role is reduced, especially among large herbivorous species, where there is often little or no carrying of food to the young. If the males play any role at all after fertilizing the eggs, it is usually to guard the female and the young or to defend a suitable territory. But a dominant male may be able to accomplish this task for several females at once; hence the chance of his leaving descendants is enhanced by polygyny.

Notice that in mammals the male and female have very unequal parental investments in the young. The female must carry the fetus from conception until birth, and must suckle the young for an extended period following birth. She is therefore more limited than the male in how many young she can produce, and it may be evolutionarily to her advantage to remain with one male who can provide good protection and a territory with a rich food supply for her young. Her chances for leaving offspring hinge to a large extent on her choice of a mate; effective reproductive strategy calls for her being very selective.

Following this same type of reasoning, we can understand why, among birds, species whose young are born helpless and in need of much parental care (e.g. most small songbirds) are more often monogamous than species whose young are precocial and able to help care for themselves almost immediately (e.g. chickens and turkeys). In the latter, the males need make little parental investment and can maximize their chances of leaving progeny by mating with many females (Fig. 12.56).

Sometimes exceptions to the usual patterns are especially revealing of important underlying principles. Such is the case with the well-known Redwing, a species in the blackbird family. The young Redwings are helpless when they hatch and must be fed; yet Redwings are polygynous. To understand why, we must look at the marshy habitat where they live. Marshes can produce huge quantities of insect food for Redwings, but they tend to do so unevenly, some patches being extremely productive, others almost nonproductive. A male whose territory is in a highly productive patch can support several females and their young, whereas a male with a nonproductive (usually peripheral) territory may be unable to support even a single female. In view of this situation, it is advantageous to a female to select for a mate a male with a rich territory, even if that male already has other mates. That such a reproductive strategy really does maximize the female's reproductive success is shown by the data in Fig. 12.57.

This analysis of vertebrate mating systems underscores an important point—that in mating behavior, as in all the other sorts of behavior we have examined, an evolutionary approach helps reveal order in what might otherwise be a bewildering diversity.

THE EVOLUTION OF BEHAVIOR

Implicit throughout this chapter has been the assumption that behavior is a biological attribute, one to be investigated like anatomy or physiology and subject to the same kinds of evolutionary processes. The corollary assumptions—that behavior is adaptive and that natural selection brings about an increase in well-adapted and a decrease in poorly adapted behavior patterns in the population—are basic to the elucidation of many types of behavior. The following example will serve to show that behavior patterns, like anatomical or physiological characteristics, can be understood in terms of their adaptive roles.

Black-headed Gulls always remove broken eggshells from their nests after the young have hatched. Tinbergen and his associates at Oxford were interested in discovering the biological significance of this behavior pattern. They found that it is not limited to the time of hatching; it can be elicited by placing shells or other conspicuous objects in the nest at any time during the breeding season. It seemed that the more conspicuous the object was, and the more like real eggshell, the greater was its stimulus potential. These observations made the investigators suspect defense against visual predators. They constructed artificial nests in which they placed eggs and other conspicuous objects; in the control nests they put only eggs. On later examination they found that many more of the nests containing conspicuous objects had been robbed of their eggs, primarily by Carrion Crows and Herring Gulls. They also found that nests containing washed whole eggs were robbed less often than nests containing unwashed shell fragments. In this case olfactory predators, primarily hedgehogs and foxes, were the culprits. These experiments served to show that, on the average, gulls that

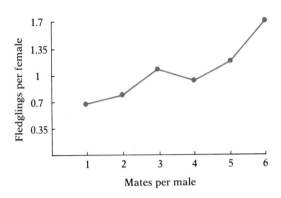

12.57 Reproductive success of female Redwings as a function of the number of females mated to the same male

On the average, females mated to males that have more than one mate fledge more young; indeed, those whose males have five or six mates fledge the most young. It appears to be evolutionarily advantageous for female Redwings to choose the males with the best territories, even if those males already have mates. [Redrawn from J. Alcock, *Animal behavior*, Sinauer, 1975.]

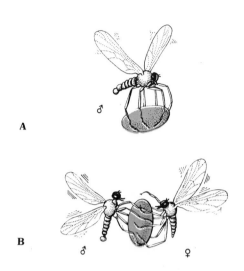

12.58   Male balloon fly carrying a "present" (A) and presenting it to a female (B)
[Redrawn from J. Alcock, *Animal behavior*, Sinauer, 1975.]

remove broken eggshells and other conspicuous objects from their nests will successfully raise more young than gulls that do not. The value of this behavior pattern in raising offspring must have constituted an effective selection pressure favoring its evolution.

If behavior patterns evolve and can be studied in terms of the selection pressures that produce them, it follows that we should be able to make reasonable conjectures concerning the ancestral behavior patterns from which newer behavior patterns have arisen, just as we can make inferences about the ancestral structures from which our hands or other structures have evolved. And we should be able to study the genetics of behavior just as we study the genetics of other characters. Both of these approaches to behavior have been much used in recent years.

THE DERIVATION OF BEHAVIOR PATTERNS

Evidence from comparison of species   One fruitful way of studying the evolutionary derivation of behavior is to compare the behavior patterns of a number of related species. Because these patterns often represent different stages of development of the same basic behavior, they may give a clue to the ancestral condition. Let us look at several examples.

There are certain species of flies (family Empididae) in which the males always present the females with a silken balloon before mating (Fig. 12.58). This is a curious bit of behavior, and we could hardly guess what selection pressures brought it about, and from what ancestral beginnings, were it not for the fact that there are other species of empidid flies still living today that exhibit various stages of development of this courtship pattern. In many species of empidid flies where the male does not give anything to the female when courting her, she sometimes captures and devours him. In other slightly more advanced species, the male captures prey, and presents this to the female, and then mates with her while she is occupied with eating the prey. The selection pressure for this behavior seems easy to explain: Males that divert the females' attention by giving them prey succeed in mating and escaping more frequently than those that do not. A second, less obvious adaptive advantage may be that the male, by providing the female with a protein-rich meal, helps ensure that her eggs (which he will presumably inseminate) will mature rapidly.

The male of still more advanced species of empidid flies captures prey, wraps it in a ball of silk, and presents this to the female. Presumably the fact that the female is occupied longer in opening the balloon and eating the prey gives the male more time to accomplish copulation. In still more advanced species, the male encloses only a tiny prey or fragments of prey in the balloon, and the female does not actually eat the prey. In other words, at this stage in the evolution of the courtship behavior, the balloon has replaced the prey as the important element, becoming part of a display that functions in making the female more receptive to the male. We can understand, therefore, how in still more advanced species an empty balloon or even some other bright object such as a rose petal can suffice. By examining a whole group of species in this way, we can get a reasonably good idea of the derivation of behavior that would otherwise seem odd and enigmatic, and we can identify the probable selection pressures involved.

Another good example is the evolution of nest building in the parrot genus *Agapornis*, which has been studied by William C. Dilger, whose experiments with Wood Thrushes we mentioned earlier. These small African parrots are unusual in building nests; most parrots simply lay their eggs on the bare floor of a cavity in a tree. The various species of *Agapornis* exhibit several stages in the evolution of nest-building behavior. Females of the most primitive living species use their very sharp bills to cut small irregular bits of bark, leaf, wood fiber, or paper. They thrust these small pieces of material amid their feathers at any point on the body. When a number of pieces are lodged among their feathers, they fly to the nest cavity and unload them. The nest they make from this material is only a simple pad. Females of a somewhat more advanced species cut long regular strips of material and tuck the ends of these into the feathers of their rumps (Fig. 12.59). Flying to the nest cavity with three or four such strips dangling in the air behind them, they construct more elaborate nests with a deep cup for reception of the eggs. Females of the most advanced species no longer tuck nesting material amid their feathers, but carry it to the cavity one piece at a time in their bills. This means that they can carry sticks and other stronger material and hence can construct very elaborate roofed nests with two chambers and a passageway.

We have, then, what appear to be a number of evolutionary trends: (1) from carrying material tucked all over the body, to carrying it only in the rump feathers, and finally to carrying it in the bill; (2) from cutting small irregular pieces, to cutting long regular ones, and finally to using twigs in addition to the strips; (3) toward increasing complexity of the nests from a simple pad, to a well-formed cup, and finally to an elaborate roofed structure. If the more advanced species were the only ones still living today, there would be little hope of understanding the evolution of nest-building behavior in these parrots. But the survival of more primitive species and the comparative studies they make possible not only help clarify the evolution of this particular behavior, but also give new insight into the evolution of animal behavior in general.

Evidence from behavioral analysis   Through detailed analysis of the motor patterns involved in a particular behavior and of the precise context in which each component of the behavior occurs, and through comparison of the results with those from similar analyses of other behaviors, a biologist can often convincingly identify the actions from which the behavior in question is derived.

Consider, for example, the evolution of courtship or agonistic displays. Such displays usually occur in situations where the animal is in a state of conflict between two or more drives. A courting animal is motivated to mate, but the close approach required by mating violates its individual distance and, in a territorial animal, the exclusivity of its defended territory as well; both circumstances give rise to attack motivation. Since the other individual feels threatened in a similar way and may evince aggressive tendencies, the courting animal also experiences some motivation to flee. Analogously, an animal in an agonistic encounter experiences conflict between attack and escape motivations. The balance between an animal's contending motivations is very important in determining its behavior in such situations.

The ambivalence of an animal experiencing conflict between motivations

A

12.59 A male lovebird cutting, tucking, and carrying nesting material
(A) The captive male *Agapornis roseicollis* cuts strips of nesting paper by punching a line of holes with his supersharp beak. (B) He then tucks the strips amid specialized rump feathers. (C) The strips dangle behind as he flies to his nest. In the wild, the birds cut leaves and bark, and use these as nesting material. [Courtesy D. G. Allen.]

C

B

often thwarts the complete performance of the behavior most appropriate to any one of the separate motivations. As a consequence, the animal frequently performs acts that seem out of context in the given situation, or performs an appropriate act but directs it at the wrong object, or starts to do something but doesn't finish doing it. Because such apparently irrelevant or redirected or incomplete actions occur so often in conflict situations, they have commonly been the raw material from which natural selection has molded new displays that communicate the motivations of the animals and thus help resolve the conflict situation. Detailed analysis can help reveal what actions were the raw material for any particular display.

Let us look more closely at the types of actions mentioned above as raw materials for displays. The first type, irrelevant action—often called ***displacement activity***—is an action performed in what appears to be the wrong context. A male Wood Thrush responding to Dilger's loudspeaker may be in such a state of conflict between attack and escape that he suddenly starts to scratch his head or to preen his wings or to eat, or he may even go to sleep. He has to do something, it seems, and if he cannot do something appropriate he does something inappropriate. In a similar way, when you are nervous over making some decision, you may drum your fingers on the desk or crack your knuckles or doodle. Such actions performed by the male thrush or by you may indicate to a perceptive observer something about its or your current motivational state. Natural selection tends to exaggerate and ritualize such actions, increasing their information content and thus giving them a true display function. In other words, actions that were irrelevant in their original context come to have great relevance, and become appropriate and integral parts of other kinds of situations. Displacement activities are most often actions that ordinarily function in the daily maintenance of the individual, such as preening, scratching, yawning, stretching, and eating (Fig. 12.60).

A second type of action that may serve as raw material for evolution of displays is one performed in the appropriate context but directed at the wrong objects; such an action is often called ***redirected activity.*** A bird thwarted in attacking an intruder on its territory may peck vigorously at the branch on which it stands instead. A dog prevented from attacking an antagonist may bite viciously at a nearby object. A man angry at his boss may come home and yell at his children. Like displacement activities, redirected activities may indicate to a perceptive observer the motivational state of the performer, and natural selection can ritualize them and cause their incorporation into displays.

A third source of motor patterns for displays includes actions called ***intention movements,*** which are the incomplete initial stages of other actions. A bird crouches slightly before it flies. A man clenches his fist before striking an opponent. Even if the bird does not actually fly or if the man does not actually strike his opponent, the crouch or the clenched fist may serve as a signal to another individual, and may therefore come to have a ritualized display function quite apart from its initial significance as simply a movement preparatory to further action.

In sum, displacement activities, redirected activities, and intention movements all have some inherent information content; i.e. they reveal something about the performer's motivational state even before being incor-

12.60 Displacement scratching compared with normal scratching of the head
In *Agapornis*, displacement scratching is often directed at the hard bill (A), whereas normal scratching for bodily maintenance is directed at the fleshy parts of the head and neck (B). In both cases the foot is brought forward over the wing; some other species of birds scratch from under the wing.

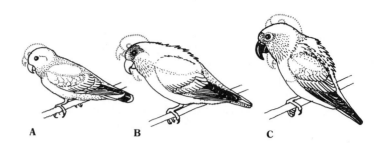

A   B   C

12.61   Ritualization of head bobbing in courtship feeding in *Agapornis*
In order to regurgitate food to feed to the female, the male must bob his head up and down.   (A) In a primitive species of *Agapornis, A. cana,* this action is performed in a simple way, the amplitude of the movement being small and the speed slow.   (B) In *A. roseicollis,* a somewhat more advanced species, the movement is faster and has greater amplitude, functioning as a display.   (C) In *A. personata fischeri,* a still more advanced species, the movements are not only of great amplitude and performed very rapidly, but the ritualization has progressed to the point where the male never actually feeds the female after performing the head bobbing; the action now only has a display function.   [Redrawn from W. C. Dilger, *Z. Tierpsychol.,* vol. 17, 1960.]

porated into displays. Their incorporation into more formal displays involves their *ritualization,* a process by which they generally become more exaggerated and stereotyped (Fig. 12.61), often being performed in a way that would not be at all appropriate in their original context (Fig. 12.60). The highly stereotyped manner in which the component parts of a display are performed is probably important in ensuring unambiguous communication; if there were much random variation (from individual to individual, or in a single individual from time to time), then the exact meaning of the message might not be clear and reliable for the observer.

THE GENETICS OF BEHAVIOR

Another way of trying to learn more about the evolution of behavior is to use some of the traditional procedures of geneticists, especially hybridization experiments.

Dilger, who studied the carrying behavior of *Agapornis* in this way, succeeded in hybridizing the species that carries strips tucked in the rump feathers with the species that carries material in the bill. The hybrids clearly showed effects produced by the genes of both parents. They cut strips and tried time and again to tuck them, but without success. Sometimes they failed to let go of the strip; after a lengthy bout of tucking, when their heads came forward again, the strip was still held tight in their bills. Sometimes they let go of the strip, only to see it fall from the rump to the floor. It was as though the genes of one parent made them try to tuck, but the genes of the other parent prevented them from doing it correctly. Eventually the hybrids learned that they could not tuck, and began carrying nesting material in their bills, but even the oldest and most experienced still gave at least a perfunctory flick of her head over her shoulder before flying to the nest with the material in her bill.

David Bentley of the University of California, Berkeley, and Ronald R. Hoy, now of Cornell University, hybridized two species of crickets whose songs had distinctly different patterns. The songs sung by the hybrids had patterns intermediate between those of the parental species, but differed from one another according to which species was paternal and which maternal (Fig. 12.62). Perhaps the most interesting result of Bentley and Hoy's work was the discovery that hybrid females were attracted preferentially to the songs of hybrid males rather than to the songs of either parental species. Thus the intriguing possibility arises that song production by males and

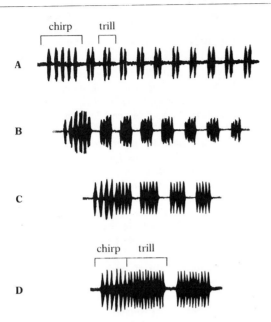

12.62 Song patterns of two species of crickets and of hybrids between them

The songs of *Teleogryllus oceanicus* (A) and *T. commodus* (D) differ markedly from each other. The songs of the hybrids (B and C) are intermediate, but differ among themselves depending on which parent species was paternal and which maternal. Hybrid females were found to prefer the songs of hybrid males. [Data from R. R. Hoy, Cornell University.]

song receptivity by females are linked genetically and transmitted together, thus ensuring that females will always respond to males of their own kind.

Dilger's experiments on *Agapornis* parrots and Bentley and Hoy's on crickets involved hybridization of two different species. But there have also been many experiments using crosses within a single species. For example, W. C. Rothenbuhler worked with two strains of honeybees—one called a hygienic strain, because the workers uncap cells in which larvae have died and remove the corpses, and one called unhygienic, because the workers do not clean out cells containing dead larvae. When Rothenbuhler crossed the two strains, all their progeny were unhygienic, but when he back-crossed these progeny to the hygienic parental strain, bees with four different behavior patterns were produced in roughly equal frequencies: (1) bees that uncapped cells containing dead larvae and removed the larvae; (2) bees that uncapped the cells but did not remove the larvae; (3) bees that did not uncap cells but would remove dead larvae if the cells were uncapped by the investigator; and (4) bees that neither uncapped cells nor removed dead larvae, even if the cells were opened. Rothenbuhler was therefore able to conclude that uncapping behavior and larvae-removing behavior are controlled by separate genes and inherited independently.[8]

Similar genetic studies have been carried out on other behavior patterns in honeybees and also on behavior patterns in other species, especially in the fruit fly *Drosophila*, a favorite experimental subject of geneticists. The conclusion from such studies has been that, despite the complicating modifications of behavior due to learning, the underlying genetic basis of many behavior patterns is amenable to the same sort of analysis as the genetic basis of anatomical or physiological traits.

[8] Students acquainted with Mendelian genetics will recognize that the results of Rothenbuhler's back cross are those that would be expected in a dihybrid cross involving two alleles of each gene, with dominance.

SUGGESTED READING

ALCOCK, J., 1975. *Animal behavior: An evolutionary approach.* Sinauer, Sunderland, Mass. *Good introductory textbook at an intermediate level of difficulty.*

BARASH, D. P., 1977. *Sociobiology and behavior.* Elsevier, New York. *Readable elementary introduction to a currently popular field.**

BENTLEY, D., and R. R. HOY, 1974. "The neurobiology of cricket song," *Scientific American,* August. (Offprint 1302) *An excellent illustration of how a variety of approaches—genetic, developmental, neurophysiological, and ethological—can be combined to interpret an animal's behavior.*

BERTRAM, B. C. R., 1975. "The social system of lions," *Scientific American,* May.

BUCK, J., and E. BUCK, 1976. "Synchronous fireflies," *Scientific American,* May. *A remarkable behavior pattern in certain species of fireflies in the tropical regions of Asia and the Pacific, in which thousands of male fireflies flash in unison.*

DILGER, W. C., 1962. "The behavior of lovebirds," *Scientific American,* January. (Offprint 1049) *A summary of Dilger's classic study of courtship and nest-building behavior in Agapornis parrots.*

EATON, G. G., 1976. "The social order of Japanese macaques," *Scientific American,* October. (Offprint 1345)

EMLEN, S. T., 1975. "The stellar-orientation system of a migratory bird," *Scientific American,* August. (Offprint 1327) *Summary of Emlen's elegant planetarium experiments on the star-compass orientation of Indigo Buntings.*

FRISCH, K. VON, 1971. *Bees: Their vision, chemical senses, and language,* rev. ed. Cornell University Press, Ithaca, N.Y. *A beautifully written little book that summarizes many of von Frisch's most important experiments.**

GRIFFIN, D. R., 1976. *The question of animal awareness.* Rockefeller University Press, New York. *A forthright challenge to the traditional belief that only humans can think and experience awareness.*

GUHL, A. M., 1956. "The social order of chickens," *Scientific American,* February. (Offprint 471) *Clear treatment of the peck-right system of social dominance.*

HÖLLDOBLER, B., 1971. "Communication between ants and their guests," *Scientific American,* March. (Offprint 1218) *The fascinating behavioral interactions between ants and a variety of other insects, especially beetles, that live in ant nests.*

KEETON, W. T., 1974. "The mystery of pigeon homing," *Scientific American.* December. (Offprint 1311)

KUMMER, H., 1971. *Primate societies.* Aldine-Atherton, Chicago. *Short and easy to read.**

LEHRMAN, D. S., 1964. "The reproductive behavior of ring doves," *Scientific American,* November. (Offprint 488) *Especially good account of the role hormones play, together with external stimuli, in determining behavior.*

LORENZ, K., 1952. *King Solomon's ring.* Crowell, New York. *Delightful accounts of animal behavior by one of the fathers of modern ethology.**

LORENZ, K., 1966. *On aggression.* Harcourt, Brace & World, New York. *A stimulating book that has given rise to much controversy concerning the possible biological basis of aggressive behavior in human beings.**

MANNING, A., 1979. *An introduction to animal behavior,* 3rd ed. Addison-Wesley, New York. *An excellent introductory textbook.**

MORSE, R. A., 1972. "Environmental control in the beehive," *Scientific American,* April. (Offprint 1247) *How honeybees regulate the temperature and humidity of the hive, eliminate polluted air, remove wastes, and control for parasites and pathogens.*

PENGELLEY, E. T., and S. J. ASMUNDSON, 1971. "Annual biological clocks," *Scientific American,* April. (Offprint 1219)

* Available in paperback.

PREMACK, A. J., and D. PREMACK, 1972. "Teaching language to an ape," *Scientific American*, October. (Offprint 549) *The Premacks' efforts to teach language to a chimpanzee by use of plastic symbols, and the evidence that the animal learned not only words, but also concepts of class and sentence structure.*

SCHNEIDER, D., 1974. "The sex-attractant receptor of moths," *Scientific American*, July. (Offprint 1299) *Evidence that the male silk moth can detect a single molecule of the female's sex-attractant pheromone.*

WATTS, C. R., and A. W. STOKES, 1971. "The social order of turkeys," *Scientific American*, June. *On the peculiar social organization of wild turkeys in Texas. Brothers remain together for life, and the status of each individual is determined both by his rank within the group of brothers and by the rank of his group among other such groups in the flock.*

WILEY, H., 1978. "The lek mating system of the sage grouse," *Scientific American*, May. (Offprint 1390)

WILSON, E. O., 1975. "Slavery in ants," *Scientific American*, June. (Offprint 1323)

PART III

THE PERPETUATION OF LIFE

Chapter 13

CELLULAR REPRODUCTION

All living things are composed of cells, and all cells arise from previously existing ones. This formulation of the cell theory, so simple and familiar as to seem almost commonplace, acquires another dimension when one considers the far from simple process of cell division—the intricate sequence of events that must take place before one cell can become two.

THE CENTRAL IMPORTANCE OF THE NUCLEUS

As you know, the nucleus is the control center of the cell. It contains the chromosomes, which bear the genes—the units of information that, passed from generation to generation, determine the characteristics of each new organism and direct its myriad activities.

Experimental removal of the nucleus can provide a clear demonstration of the role it fulfills in the ongoing life of the cell. When the nucleus of an amoeba, for example, is carefully removed—by surgical procedure or with a micropipette—so as not to cause any significant damage to the cytoplasm, the cell can continue functioning for days or even weeks. Apparently the enzymes and organelles already formed can maintain the cell for that length of time. Eventually, however, the cell begins to run down: Pseudopod formation diminishes and movement becomes sluggish, there is little if any feeding, the cell may become spherical, and finally death results. If a new nucleus is inserted into the cell within the first two or three days after the initial operation, the amoeba may resume activity, but if the new nucleus is inserted after six days, there is no response; irreversible degeneration of the cell has apparently already taken place. Similar results have been obtained when nuclei have been removed from a variety of other types of cells, both plant and animal. Clearly, then, the nucleus is essential for the continued life of the cell; normal function depends on it.

13.1   Photograph of *Acetabularia mediterranea*
[Courtesy Turtox/Cambosco, Macmillan Science Co., Inc.]

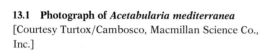

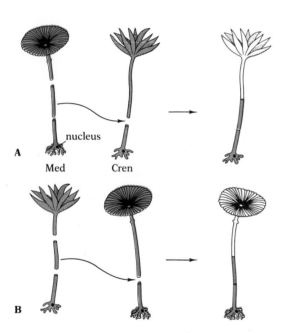

13.2   Hämmerling's experiments on *Acetabularia mediterranea* (med) and *A. crenulata* (cren)
(A) A piece of med stalk (color) grafted onto a cren base (gray) regenerates a cren cap (after first regenerating a cap of intermediate type, not shown here).   (B) A piece of cren stalk (gray) grafted onto a med base (color) regenerates a med cap.   In each case the characteristics of the regenerated cap (white) are determined by the part containing the nucleus.

The importance of the nucleus in the transmission of hereditary information has been shown in a variety of ways. Joachim Hämmerling of the Max Planck Institute for Marine Biology in Berlin carried out his demonstrations with the unicellular green alga *Acetabularia*. This curious plant, though it is composed of only one cell, consists of a branching rootlike base, a stalk, and a cap. The nucleus is in the base. Hämmerling worked with two closely related species, *Acetabularia mediterranea* and *Acetabularia crenulata*—med and cren for short. Med has a disc-shaped cap with scalloped edges (Fig. 13.1), cren one deeply dissected into nearly separate lobes.

Hämmerling showed that if he amputated the stalk and cap of a cren individual and grafted a med stalk without a cap onto the cren base, a new cap was eventually regenerated, intermediate in appearance between med and cren. If this cap was amputated, the second cap regenerated was a typical cren one (Fig. 13.2A). In other words, this second cap did not take its characteristics from the med stalk, on which it grew, but from the cren base, which contained the nucleus. The reciprocal experiment yielded similar results: A cell consisting of a med base and cren stalk regenerated a med cap (Fig. 13.2B). The nucleus had obviously been the controlling factor. The intermediate appearance of the first cap regenerated must be explained by the retention in the stalk of some influence from the original nucleus; by the time the second cap was formed, this influence had given way to that of the nucleus currently in the cell.

If the nucleus carries the information or blueprint for the development of the new individual, it follows that when the cell divides the nuclear information must be transmitted in orderly fashion to both of the new cells. The division cannot be a simple splitting of the information into two halves, because this would give neither of the new cells a satisfactory blueprint. Just as you cannot have two buildings erected by cutting one blueprint in two and giving half to each of two contractors, so two new cells cannot develop if each receives only half the necessary information from the parental cell. If you wanted two contractors to erect identical buildings simultaneously, you would logically duplicate the necessary blueprint and give a complete copy to each. The same applies to the cell. If a parental cell is to divide and produce two viable new cells, it must first make a complete copy of the genetic information in its nucleus and then, as it divides, give one complete copy to each daughter cell. In other words, division of the nucleus is not simply a process of halving; it is a process of duplicating genetic information and distributing the duplicates.

In the procaryotic cells of bacteria and blue-green algae, cell division appears much less complex than in the eucaryotic cells of true plants and animals. First, the single circular chromosome replicates—the new chromosome, like the old, attaching to the plasma membrane (Fig. 13.3). Next, new plasma membrane and wall material form near the midpoint of the parental cell and slowly grow inward, cutting through both the cytoplasm and the nucleoid (Fig. 13.4). Thus each new daughter cell receives a complete chromosome, a full complement of genetic information. This process is known as transverse fission or **binary fission.**

Some primitive eucaryotic cells (e.g. dinoflagellates), even though they have numerous chromosomes packaged in a membrane-bounded nucleus, divide by a process remarkably like the binary fission of procaryotes. The chromosomes, which are attached to the inner surface of the nuclear membrane, are first replicated and then separated into two groups by growth of the membrane between the points of attachment of the original chromosomes and their replicates. In this way two new nuclei are organized. Although microtubules are intimately associated with the nuclear membrane, they play no direct role in moving the chromosomes; however, they seem to help stiffen the nuclear membrane and thus facilitate its partitioning of the chromosomes.

In most higher eucaryotes the mechanism of organizing the chromosomes into new daughter nuclei differs markedly from the one just described. An elaborate series of nuclear changes, observable under the ordinary light microscope, occurs. The remainder of this chapter is largely an account of these changes as they are currently understood.

MITOTIC CELL DIVISION

Cell division in eucaryotic cells involves two fairly distinct processes that often, but not always, occur together: division of the nucleus and division of the cytoplasm. The process by which the nucleus divides to produce two new nuclei, each with the same number of chromosomes as the parental nucleus, is called **mitosis** (or karyokinesis). The process of division of the cytoplasm is called **cytokinesis.** We shall discuss mitosis and cytokinesis separately below.

MITOSIS

Many lines of evidence converge to show conclusively that the units of genetic information, the genes, are located on the chromosomes. The

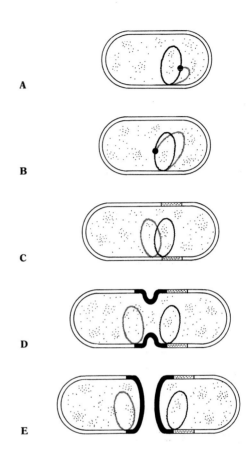

A

B

C

D

E

13.3   Binary fission of a procaryotic cell
(A) The circular chromosome of a young cell is attached to the plasma membrane near one end; it has already begun replication (the partially formed new chromosome is shown in color). (B) Replication is about 80 percent complete. (C) Chromosomal replication is complete, and the new chromosome now has an independent point of attachment to the membrane. During replication additional membrane and wall (stippled) was formed. (D) More new membrane and wall (black) has formed between the points of attachment of the two chromosomes. Part of this growth forms invaginations that will give rise to a septum cutting the cell in two. (E) Fission is complete, and two daughter cells have formed.

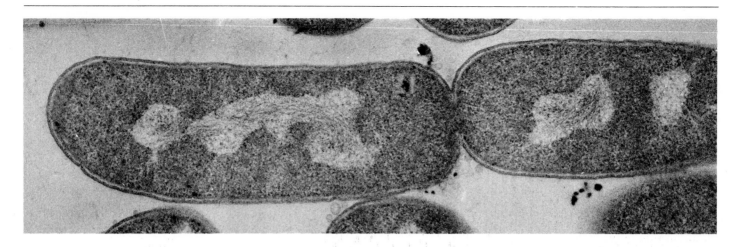

13.4 Electron micrograph of a dividing bacterial cell New wall is growing inward, cutting this cell of *Escherichia coli* in two. The nucleoid (white areas inside cells) has already divided. × 48,750. [Courtesy R. G. E. Murray, University of Western Ontario.]

number of chromosomes is usually constant for normal somatic[1] cells of all individuals of the same species. For example, the nuclei of human somatic cells contain 46 chromosomes (Fig. 13.5), those of the fruit fly *Drosophila melanogaster* contain 8, and those of onion 16. Thus, if each daughter nucleus is to get a complete set of genetic instructions, each must receive a full set of chromosomes exactly like the set initially present in the parental nucleus; only in this way can a basic constancy of chromosome number and gene content in the somatic cells be maintained. Thus nuclear division entails, first, precise duplication of the genetic material and, second, distribution of a complete set of the material to each daughter cell.

For convenience, it is customary to separate each mitotic cycle, from one cell division to the next, into a series of stages, each designated by a special name. Although each stage will be discussed separately here, it should be kept in mind that the entire process is a continuum, not a series of discrete occurrences.

Interphase The nondividing cell is said to be in the interphase state. The nucleus is clearly visible as a distinct membrane-bounded organelle, and one or more nucleoli are usually prominent. But chromosomes, as ordinarily pictured, are not visible in the nucleus; there are no distinct rodlike bodies such as are seen so easily in a dividing cell (see Fig. 3.23, p. 99). The interphase chromosomes are so thin and tangled that they cannot be recognized as separate entities. They appear only as an irregular granular-looking mass of chromatin material.

In interphase animal cells, but not in most plant cells, there is a special region of cytoplasm just outside the nucleus that contains two small cylindrical bodies oriented at right angles to each other. These are the ***centrioles*** (see Fig. 3.39, p. 115, and Fig. 3.43), which will move apart and become associated with the poles of the mitotic apparatus of the dividing cell. In many animal cells this separation of the centrioles occurs just before the onset of mitosis, but in some cells it occurs during interphase, long before mitosis begins. No centrioles have yet been detected in the cells of most seed

[1] "Somatic," as used here, refers to all cells in the body except reproductive cells—eggs and sperm.

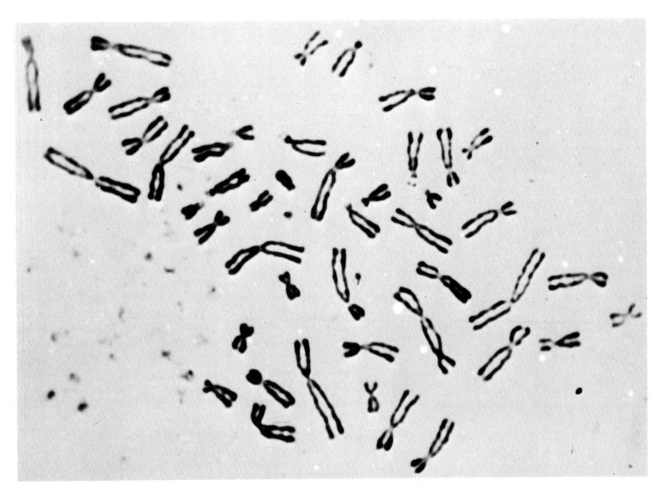

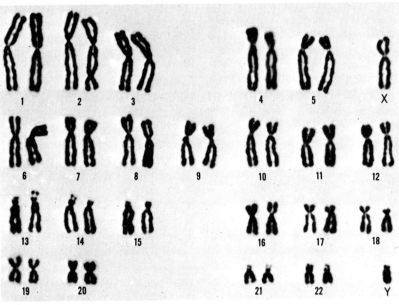

13.5 Photograph of chromosomes of a human male
At bottom, the chromosomes have been arranged in
pairs and numbered according to the accepted
convention. A human somatic cell contains 23 pairs of
chromosomes, including a pair of sex chromosomes
(for a male, X and Y). [Courtesy M. W. Shaw,
University of Michigan.]

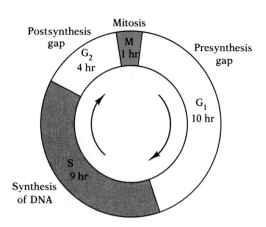

13.6   The cell cycle
This particular cycle assumes a period of 24 hours, but some cells complete the cycle in less than an hour and others take many days. Similarly, the ratios of the four phases of the cycle vary, with G_1 exhibiting the most variation.

plants, but they do occur in some algae, fungi, bryophytes (mosses, liverworts, etc.), and ferns in association with the production of motile sperm.

In the past, interphase cells were commonly called resting cells, but that terminology has been rejected as grossly inappropriate. An interphase cell is definitely not resting; it is carrying out all the innumerable activities of a living, functioning cell—respiration, protein synthesis, growth, differentiation, and so forth. Furthermore, it is during interphase that the genetic material is replicated in preparation for the next division sequence.

The replication of the genetic material does not, however, begin immediately after completion of the last division sequence. There is a gap in time, designated the G_1 stage, before genetic replication—i.e. the synthesis of new DNA—takes place during what is known as the S stage. Another time gap, designated G_2, separates the end of replication from the onset of mitosis proper. These three subdivisions of interphase, together with mitosis (the M stage), constitute what is called the *cell cycle* (Fig. 13.6). Cells in tissue culture show striking morphological changes as they pass through the cell cycle, but the significance of these changes, especially in the intact organism, is not well understood.

The duration of the complete cell cycle can vary greatly. Though usually lasting 10–30 hours in plants and 18–24 hours in animals, it may be as short as 20 minutes or as long as several days or even weeks. Although all the stages can vary in duration to some degree, by far the greatest variation occurs in the G_1 stage. At one extreme very rapidly dividing embryonic cells may pass so quickly through the G_1 stage that it can hardly be said to exist at all, whereas at the other extreme some cell types become arrested in the G_1 stage and may never divide again; differentiated skeletal-muscle cells and nerve cells, for example, are arrested in the G_1 stage and normally do not divide. There are, however, a few cases in which nondividing cells are arrested in the G_2 stage; i.e. they have replicated their DNA, but do not divide. The heart-muscle cells of human adults are an example of cells in G_2 arrest.

Both G_1 and G_2 arrest appear to be due to failure to produce some essential control chemical. Thus, if the nucleus from a cell in G_1 arrest is transplanted into a cell that is just entering the S stage, the transplanted nucleus will promptly be activated and itself enter the S stage, apparently because it has been stimulated by a control substance present in the cytoplasm of the host cell. Similarly, when a cell in G_2 arrest is fused with a mitotic cell, its chromosomes soon begin to condense and the cell enters mitosis. From such experiments it is deduced that the missing control substance in G_1 arrest is necessary for initiation of DNA synthesis, and that the missing control substance in G_2 arrest acts as a chromosomal condensing factor.

In nearly all cultured animal cells, production of these control substances depends on stimulation by **growth factors** (also called mitogenic factors or mitogens) from blood serum. These factors, most of which are peptides or small proteins, often have very specific target tissues. Thus there is a nerve growth factor (NGF) essential for mitosis of embryonic sympathetic nerve cells, an epidermal growth factor (EGF) that causes proliferation of epithelial cells, a fibroblast growth factor (FGF) that stimulates division of fibroblasts and of the cells lining small blood vessels, and so forth.

Countering the action of the growth factors is a class of mitosis-inhibiting substances called **chalones.** Like the growth factors, the various chalones, most of which are peptides or glycoproteins, are very specific in their target tissues; they affect only the type of tissue in which they themselves were produced. Thus the first chalone discovered (in 1960, by W. S. Bullough and E. B. Laurence of Birbeck College, London) was one produced by skin cells that inhibits mitosis by neighboring skin cells. When skin cells are damaged, they synthesize less chalone, and cells in the vicinity of the wound are freed of inhibition and begin dividing, producing the new tissue necessary for wound healing. Once an adequate number of healthy cells has been produced, they secrete enough chalone to re-establish mitotic inhibition and turn off the wound-healing process. This feedback mechanism, which limits cell proliferation in normal tissues, does not function properly in malignant tissues; cancerous cells appear to have lost most of their responsiveness to chalones and other mitotic inhibitors, with the result that they do not go into G_1 arrest, but continue dividing in an uncontrolled fashion.

Let us now assume that the balance between growth factors and chalones favors cell division and that the normal cell we are examining has passed through the G_1, S, and G_2 stages of interphase and is entering mitosis proper—itself a process so complex that it is customarily divided into four stages.

Prophase Prophase is, in a sense, a preparatory stage that readies the nucleus for the crucial separation of two complete sets of chromosomes into two daughter nuclei. As the two centrioles of an animal cell move toward opposite sides of the nucleus, the initially indistinct chromosomes begin to condense into visible threads, which become progressively shorter and thicker and more easily stainable with dyes. When the chromosomes first become visible during early prophase, they appear as long, thin, intertwined filaments, but by late prophase the individual chromosomes can be clearly discerned as much shorter rodlike structures. As the chromosomes become more distinct, the nucleoli become less distinct, often disappearing altogether by the end of prophase (Fig. 13.7). It seems probable that the condensing of the chromosomes is largely a matter of their becoming coiled into a tight helix, which is then coiled again.

1. INTERPHASE

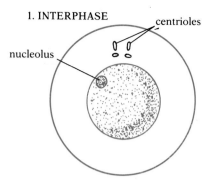

Chromosomes not seen as distinct
  structures.
Nucleolus visible.

2. EARLY PROPHASE

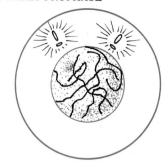

Centrioles moving apart.
Chromosomes appear as long thin
  threads.
Nucleolus becoming less distinct.

3. MIDDLE PROPHASE

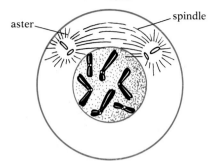

Centrioles farther apart,
  spindle beginning to form.
Each chromosome to be seen as
  composed of 2 chromatids held
  together by their centromeres.

4. LATE PROPHASE

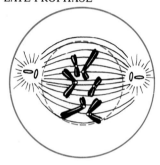

Centrioles nearly at opposite sides
  of nucleus.
Spindle nearly complete.
Nuclear membrane disappearing.
Chromosomes move toward equator.
Nucleolus no longer visible.

5. METAPHASE

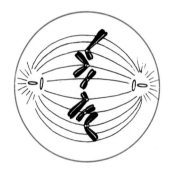

Nuclear membrane has disappeared.
Centromeres of each double-stranded
  chromosome attached to spindle
  microtubules at spindle equator.

6. EARLY ANAPHASE

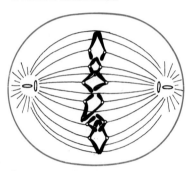

Centromeres have uncoupled and
  begun moving toward opposite
  poles of spindle.

7. LATE ANAPHASE

The 2 sets of new single-stranded
  chromosomes nearing respective
  poles.
Cytokinesis beginning.

8. TELOPHASE

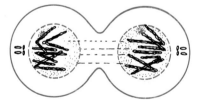

New nuclear membranes forming.
Chromosomes become longer, thinner,
  and less distinct.
Nucleolus reappearing.
Centrioles replicated.
Cytokinesis nearly complete.

9. INTERPHASE

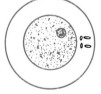

Nuclear membranes complete.
Chromosomes no longer visible.
Cytokinesis complete.

13.7 Mitosis and cytokinesis in an animal cell

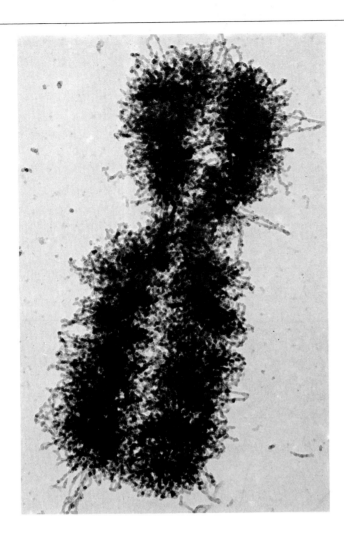

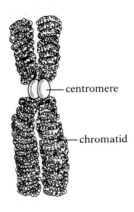

13.8   A late-prophase chromosome
Above: A diagrammatic drawing. The chromosome
consists of two identical chromatids united at their
centromeres. The chromatin (DNA plus protein) of
each chromatid is supercoiled.   Left: Electron
micrograph of a human chromosome. The massed coils
of chromatin are easy to see, but the centromeres
cannot be distinguished here, though their location can
be fixed as the point where the two chromatids are
joined. $\times$ 21,000.   [Left: Courtesy G. F. Bahr,
Federation Proceedings, vol. 34, 1975; micrograph by W.
Engler.]

The shortening of the chromosomes during cell division has one obvious
advantage. In their shorter form, they can be moved about freely without
becoming hopelessly tangled. You may well ask, then, why the chromosomes do not simply remain in their shorter coiled form even during interphase. One probable reason is that the genetic material functions as a
surface on which other molecules are synthesized, and only in the long, thin,
uncoiled form is the surface sufficiently exposed to function in this way.

Examined under very high magnification, an individual chromosome
from a late-prophase nucleus can be seen to consist of two separate strands
called ***chromatids*** (Fig. 13.8). The replication that occurred during interphase produced half the DNA of each chromatid by using the other half as a
model, by a process to be discussed at length in a later chapter. Thus the two
chromatids of a prophase chromosome are genetically identical. Somewhere along its length each chromatid has a special region called a ***centromere.*** The centromeres of the two chromatids of a double-stranded chromosome are held tightly together, by forces so far unidentified.

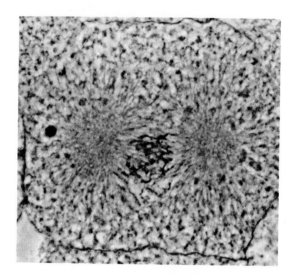

13.9 Prophase in a cell from a whitefish blastula
The two large asters are prominent, and a clump of
chromosomes can be seen between them. × 960.
[Courtesy E. V. Gravé.]

As the two centrioles of an animal cell move apart, a system of microtubules, which radiate in all directions (Fig. 13.7: 2), appears near each
centriole. The microtubules[2] are divided into two groups: the **spindle microtubules,** which form a basketlike structure between the two centrioles; and
the blindly ending astral microtubules, collectively called an **aster,** which
radiate in other directions from each centriole (Fig. 13.9). Even though most
higher-plant cells lack centrioles, they develop a spindle much like that in
animal cells, but no asters are formed.

During late prophase the double-stranded chromosomes, which at first
were distributed essentially at random within the nucleus, begin to move
toward the middle, or equator, of the biconical spindle formed by the spindle microtubules. Prophase ends when the nuclear membrane has disappeared and each chromosome has moved to the mid-plane of the spindle,
attached by its centromeres to spindle microtubules.

Metaphase During the brief stage termed metaphase, the chromosomes
are arranged on the equatorial plane of the spindle, and in side view appear
to form a line across the middle of the spindle (Fig. 13.10). Each double-
stranded chromosome is attached by its centromeres to microtubules of the
spindle, and it is actually the centromeres that are lined up precisely along
the equatorial plane, while the arms of the chromosomes may extend in any
direction (Fig. 13.7: 5). Metaphase ends when the centromeres of the chromatids forming each double-stranded chromosome uncouple, and each of
the former chromatids thus becomes an independent single-stranded chromosome—in other words, when the total number of independent chromosomes in the nucleus is doubled. (Note that, in determining the number of
independent chromosomes present in a nucleus, one counts the number of
independent centromeres, not the number of chromatids.)

[2]Until very recently these microtubules were thought to be fibers. Hence the references to
"spindle fibers" and "astral fibers" in the literature.

13.10 Metaphase in a cell from a whitefish blastula
The chromosomes are lined up along the equator of the
spindle, midway between the two asters. × 960.
[Courtesy E. V. Gravé.]

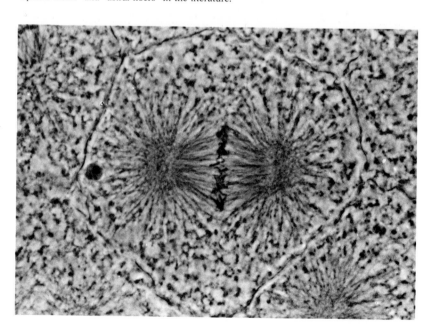

Anaphase   Now begins the separation of two complete sets of chromosomes, the critical event for which the previous stages were the preparation. At the beginning of anaphase the two new single-stranded chromosomes derived from each original double-stranded chromosome begin to move away from each other, one going toward one pole (centriole) of the spindle and the other going toward the opposite pole. In higher organisms a chromosome cannot move during anaphase unless its centromere is attached to a spindle microtubule; it can be demonstrated that both the centromere and the microtubule are necessary for the movement. As the chromosomes move apart, it becomes apparent that the microtubules to which they are attached are chromosome-to-pole microtubules. They thus differ from unoccupied microtubules of the spindle, which are longer pole-to-pole microtubules.

A cell in early anaphase can be recognized by the fact that the chromosomes are in two equal groups a short distance apart. A cell in late anaphase contains two groups of chromosomes that are more widely separated, the two clusters having almost reached their respective poles of the spindle (Fig. 13.11; see also Fig. 13.7:7). Cytokinesis often begins during late anaphase.

Telophase   Telophase is essentially a reversal of prophase. The two sets of chromosomes, having reached their respective poles, become enclosed in new nuclear membranes as the spindle disappears (Fig. 13.12). These new nuclear membranes are assembled from membranous vesicles derived from the endoplasmic reticulum. Then the chromosomes begin to uncoil and to resume their interphase form, while the nucleoli slowly reappear (Fig. 13.7: 8). The centriole of each nucleus is usually replicated during late telophase. Cytokinesis is often completed during telophase. Telophase ends when the new nuclei have fully assumed the characteristics of interphase, thus bringing to a close the complete mitotic process.

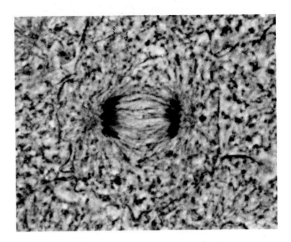

13.11   Late anaphase in a cell from a whitefish blastula
Two groups of chromosomes have moved apart from each other and are approaching their respective poles of the spindle. × 960. [Courtesy E. V. Gravé.]

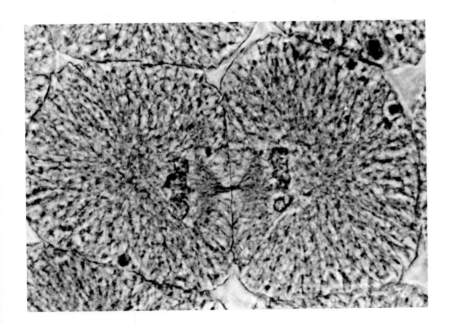

13.12   Telophase cells from a whitefish blastula
The two groups of chromosomes are being organized into new nuclei. Cytokinesis is already complete, new membranes having partitioned the original cell into two. × 960. [Courtesy E. V. Gravé.]

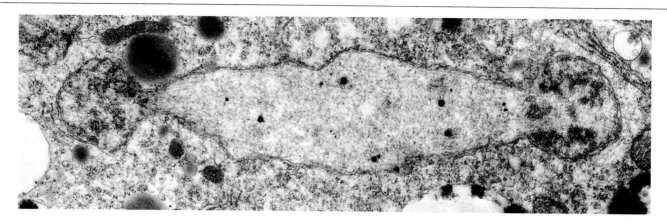

13.13 Persistent nuclear membrane during mitosis in the fungus *Catenaria*
In this fungus the spindle forms inside the nucleus. Here two new telophase nuclei are being pinched off by constrictions of the nuclear envelope. The middle part of the envelope, between the new nuclei, will eventually disintegrate. × 24,000. [Courtesy M. S. Fuller, *Mycologia*, vol. 60, 1968.]

Note that when the individual chromosomes fade from view at the end of telophase they are single-stranded, but that when they reappear during prophase in the next cell cycle they are double-stranded—the replication, which cannot be observed with a light microscope, having occurred during the S stage of interphase. Thus, if prophase begins with a nucleus containing, say, six double-stranded chromosomes, telophase ends with two new nuclei each containing six single-stranded chromosomes. Mitosis separates the strands of the initially double-stranded chromosomes. Thus each new daughter cell has the same number and kinds of chromosomes as the parental cell and therefore has the complete set of genetic information necessary to determine its characteristics and to guide its activities.

A word of caution: The description of mitosis given here holds for most higher plants and animals (with the differences noted concerning presence or absence of centrioles and asters). But it does not hold in all details for all eucaryotic cells. For example, in many protozoans, fungi, and algae, the spindle forms inside the nucleus and the nuclear membrane never disappears; when the chromosomes have been separated into two clusters at telophase, the new nuclei are simply pinched off (Fig. 13.13).

The movement of the chromosomes As we saw in Chapter 11, microtubules play a central role in the movement of cilia and flagella; in the mitotic spindle we again encounter microtubules in association with movement—this time the movement of the chromosomes toward opposite poles of the cell in anaphase. Yet a fully satisfactory explanation for the separation of the chromosomes has so far proved elusive; several proposed mechanisms are currently under investigation.

One proposal, put forward by J. R. McIntosh of the University of Colorado, is that the chromosome-to-pole microtubules may slide along the pole-to-pole microtubules by means of dynein-like cross bridges similar to those in cilia and flagella. There is evidence that an ATPase is present in the spindle; perhaps it is located in the hypothesized cross bridges.

The recent discovery that actin is also present in the spindle has suggested to some researchers the possibility of an actin-myosin system of sliding microfilaments, as in muscle. If such a system exists, the spindle microtubules might function merely as guides for the movement.

A third possibility, espoused by S. Inoué of the University of Pennsylvania, is that the movement is produced by disassembly of the chromosome-to-pole microtubules near the poles. You will recall that each microtubule is composed of many helically stacked molecules of the globular protein tubulin (see Fig. 3.38, p. 115). Suppose that one molecule of tubulin is removed from a microtubule at its pole end. The effect might be to pull all the remaining tubulin molecules toward the pole, shortening the tubule by a distance equivalent to the lost molecule. Removal of a second molecule might then again pull the remaining ones poleward, further shortening the tubule. Thus, with more and more tubulin molecules removed from the pole end of the tubule, the tubule would be progressively shortened, and the end attached to the chromosome would be drawn closer and closer to the pole. This model is consistent with the observation that chromsome-to-pole microtubules do indeed become steadily shorter (but not thicker) during anaphase.

Whatever the mechanism that produces the movement, it is clear that the chromosomes do not merely move along the microtubules, but that the microtubules themselves move. Thus, if a beam of ultraviolet light is used to produce a visible defect in a spindle microtubule, the defect will be seen to move poleward at the same rate as the chromosomes in that cell.

CYTOKINESIS

We said earlier that division of the cytoplasm often accompanies division of the nucleus and that it often begins in late anaphase, reaching completion during telophase. But this is not always the case. Mitosis without cytokinesis is common in some algae and fungi, producing coenocytic plant bodies (bodies with many nuclei, but with no, or few, cellular partitions). It regularly occurs during certain phases of reproduction in seed plants and certain other vascular plants. It is also common in a few lower invertebrate animals with coenocytic bodies. During the early development of insect eggs, mitosis without cytokinesis produces hundreds of nuclei in a limited amount of cytoplasm; later, cytokinesis cuts up this cytoplasm to produce many new cells in a very short time.

Cytokinesis in animal cells   Division of an animal cell normally begins with the formation of a *cleavage furrow* running around the cell (Fig. 13.14). When cytokinesis occurs during mitosis, the location of the furrow is ordinarily determined by the orientation of the spindle, in whose equatorial region the furrow forms (Fig. 13.7: 7). The furrow becomes progressively deeper, until it cuts completely through the cell (and its spindle), producing two new cells.

Very little is known about the mechanism of formation of the cleavage furrow. Since the location of the furrow is usually related to the position of the centrioles and spindle, an early hypothesis was that some of the astral microtubules of the mitotic apparatus are attached to the cell surface and pull the surface inward to form the cleavage furrow. However, cytokinesis is known to occur in many instances long after mitosis is complete and the spindle has disappeared; it is also significant that removal of the entire mitotic apparatus from a sea-urchin egg by micromanipulation does not

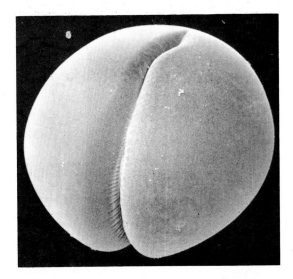

13.14 Scanning electron micrograph of a dividing frog egg
The cleavage furrow is not yet complete (see top of cell). Note the puckered stress lines in the furrow. × 40.   [Courtesy H. W. Beams and R. G. Kessel, *Am. Scientist*, vol. 64, 1976. Reprinted by permission of *American Scientist*, journal of Sigma Xi, The Scientific Research Society.]

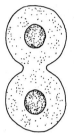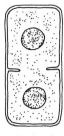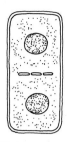

Animal cell   Algal cell   Higher-plant cell

13.15 Three mechanisms of cytokinesis
(A) Cytokinesis in animal cells typically occurs by a pinching-in of the plasma membrane. (B) In many algal cells cytokinesis occurs by an inward growth of new wall and membrane. (C) In higher plants cytokinesis typically begins in the middle and proceeds toward the periphery, as membranous vesicles fuse to form the cell plate. [Adapted from C. J. Avers, *Cell biology*, © 1976 by Litton Educational Publishing, Inc. Reprinted by permission of D. Van Nostrand Company.]

13.16 Cell division in plants
(A) Prophase. The chromosomes are in a single clump. (B) Metaphase. (C) Anaphase. Two groups of chromosomes are moving to opposite poles of the spindle; note the absence of asters. (D) Telophase. A cell plate has begun to form across the equator of the spindle. [Courtesy Carolina Biological Supply Co.]

inhibit furrow formation. A more recent finding that there is a dense ring of actin microfilaments (and probably also myosin microfilaments) at the site of the cleavage furrow makes it seem likely that microfilament activity of the sort known to be involved in many other types of cell movement is involved here as well—especially as drugs such as cytochalasin, which destroy microfilaments, cause cytokinesis to cease.

Cytokinesis in plant cells Since plant cells possess relatively rigid cell walls, which cannot develop cleavage furrows, it is not surprising that cytokinesis should differ in plant and in animal cells. In many fungi and algae new plasma membrane and wall grow inward around the wall midline, until the growing edges meet and completely separate the daughter cells (Fig. 13.15). In higher plants a special membrane, called the *cell plate,* forms halfway between the two nuclei (at the equator of the spindle if cytokinesis accompanies mitosis; Fig. 13.16D). The cell plate begins to form in the center of the cytoplasm and slowly becomes larger until its edges reach the outer surface of the cell and the cell's contents are cut in two. We see, then, that cytokinesis of higher-plant cells progresses from the middle to the periphery, whereas cytokinesis in animal cells progresses from the periphery to the middle.

The cell plate forms from membranous vesicles that first line up and then unite (Fig. 13.17). The vesicles are thought to be derived in part from the

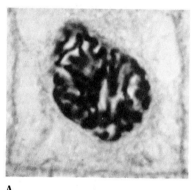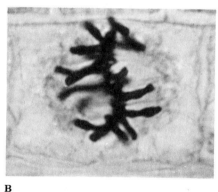

A   B

C   D

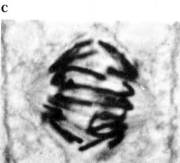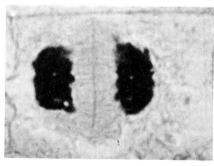

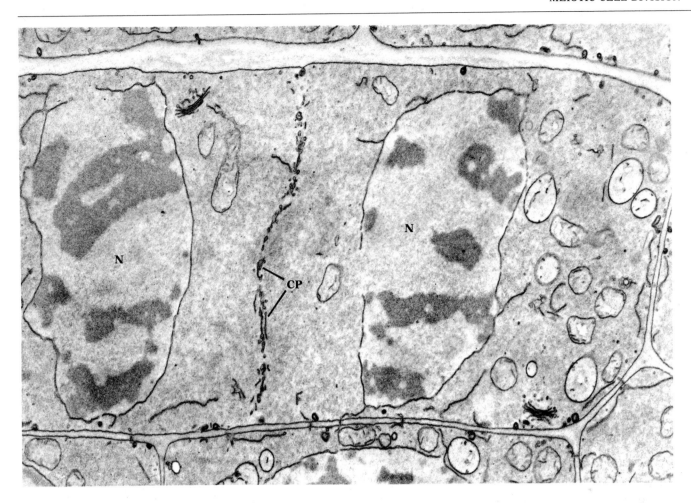

Golgi and in part from the endoplasmic reticulum. When the membranes of the vesicles have fused with one another and, peripherally, with the old plasma membrane, they constitute the partitioning membranes of the two newly formed daughter cells. The contents of the vesicles, trapped between the daughter cells, give rise to the middle lamella and to the beginnings of primary cell walls.

MEIOTIC CELL DIVISION

Notice that the nuclei of the cells shown in Fig. 13.7, belonging to a hypothetical organism, contain six chromosomes. But notice also that those six chromosomes are of only three types, there being two chromosomes of each type. Cells of this sort, with two of each type of chromosome, are said to be *diploid.* The somatic cells of the human body, as well as those of most higher plants and animals, are diploid. We have said that each somatic cell of a human being has 46 chromosomes and that each somatic cell of a fruit fly has eight; since these cells are diploid, it follows that the chromosomes in a human somatic cell are of 23 types and those in a fruit-fly somatic cell are of

13.17 Electron micrograph of a late-telophase cell in corn root, showing formation of cell plate
Mitosis has been completed, and the two new nuclei (N) are being formed; the chromosomes (dark areas in the nuclei) are no longer visible as distinct structures, but the nuclear membranes are not yet complete. A cell plate (CP) is being assembled from numerous small vesicular structures. At the lower end of the nucleus on the right, a length of endoplasmic reticulum can be seen that appears to run from the nuclear membrane to the cell wall, through the wall, and into the adjacent cell. × 17,600. [Courtesy W. G. Whaley *et al., Am. J. Botany,* vol. 47, 1960.]

1. EARLY PROPHASE I

Chromosomes become visible as long, well-separated filaments; they do not appear double-stranded, though other evidence indicates that replication has already occurred.

2. MIDDLE PROPHASE I

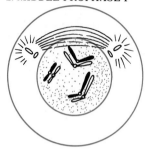

Homologous chromosomes synapse and become shorter and thicker.

3. LATE PROPHASE I

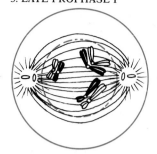

Chromosomes become clearly double-stranded.
Nuclear membrane begins to disappear.

4. METAPHASE I

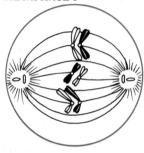

Each synaptic pair moves to the equator of the spindle as a unit.

5. ANAPHASE I

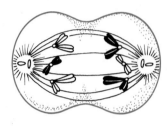

Centromeres do not uncouple. Double-stranded chromosomes move apart to opposite poles.

6. TELOPHASE I

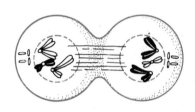

New haploid nuclei form. Chromosomes are double-stranded when they fade from view.

7. INTERKINESIS

No replication of genetic material occurs.

8. PROPHASE II

9. METAPHASE II

10. ANAPHASE II

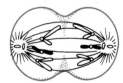

11. TELOPHASE II

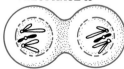 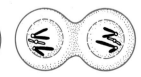

12. INTERPHASE

13.18   Meiosis in an animal cell

four types. Mitosis, as we have seen, produces new cells with exactly the same chromosomal endowment as the parental cell. Hence, whenever a human somatic cell divides mitotically, the new cells produced have 46 chromosomes of 23 types, there being two of each type.

Mitosis, then, serves to maintain a constant number of chromosomes in somatic cells. What would be its effect in reproductive cells? As you know, in sexual reproduction two cells (the egg and the sperm) unite to form the first cell (zygote) of the new individual. If those two cells (gametes) were produced by normal mitosis in human beings or fruit flies or our hypothetical organism, the zygote produced by their union would have double the normal number of chromosomes, and at each successive generation the number would again double, until the total chromosome number per cell approached infinity. This does not happen: The chromosome number normally remains constant within a species. At some point, therefore, a different kind of cell division must take place, a division that reduces the number of chromosomes by half, so that when the egg and sperm unite in fertilization the normal diploid number is restored. This special process of reduction division is called **meiosis.**   In all multicellular animals meiosis occurs at the time of gamete production. Consequently each gamete possesses only half the species-typical number of chromosomes. It is important to note that in the reduction division of meiosis the chromosomes of the parental cell are not simply separated into two random halves; the diploid nucleus contains two of each type of chromosome, and meiosis partitions these chromosome pairs so that each gamete contains one of each type of chromosome. Such a cell, with only one of each type of chromosome, is said to be **haploid.** When two haploid gametes unite in fertilization, the resulting zygote is diploid, having received one of each chromosome type from the sperm of the male parent and one of each type from the egg of the female parent.

THE PROCESS OF MEIOSIS

Complete meiosis involves two successive division sequences, which result in four new haploid cells. The first division sequence accomplishes the reduction in the number of chromosomes; the second separates chromatids. The same four stages as in mitosis—namely prophase, metaphase, anaphase, and telophase—are recognized in each division sequence.

First prophase   Many of the events in prophase I of meiosis resemble those in the prophase of mitosis. The individual chromosomes come slowly into view as they coil and become shorter, thicker, and more easily stainable. The nucleoli slowly fade from view, and, finally, the nuclear membrane disappears and the spindle is organized. Radioactive-tracer studies show that the replication of the genetic material occurs during the interphase that precedes prophase I, as in mitosis (but during early prophase of meiosis the chromosomes are not recognizably double-stranded).

The chief difference between prophase of meiosis and mitosis is that in meiosis the members of each pair of chromosomes (homologous chromosomes) move together and come to lie side by side in an intimate association (Fig. 13.18: 2). They do not fuse, but often intertwine. This pairing process is known as **synapsis.** Each chromosome being visibly double-stranded by this time, a synaptic pair can be seen to consist of two identical double-stranded

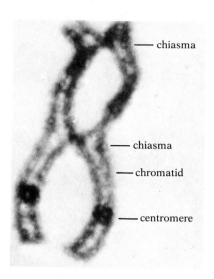

13.19 Photomicrograph showing two chiasmata between homologous chromosomes
Each chromosome is clearly recognizable as double-stranded. The centromeres, too, are clearly visible. Crossing-over is taking place at two points—the two chiasmata. Note that the crossing-over involves only one chromatid from each chromosome. These chromosomes, of a plethodontid salamander from Costa Rica, are from a spermatocyte in prometaphase of meiosis I. [Courtesy James Kezer, University of Oregon.]

chromosomes lying next to each other; they are sometimes referred to as a tetrad, because they make up a total of four chromatids. Toward the end of prophase the synaptic pair moves as a unit to the equator of the spindle.

First metaphase In mitosis the two chromosomes of each different type are completely independent of each other in their movements. They do not synapse, and each individual chromosome moves on its own to a separate set of chromosome-to-pole microtubules. Hence, at metaphase of mitosis in our hypothetical organism of Fig. 13.7, each of the six chromosomes occupies a different set of microtubules; similarly, each of the 46 chromosomes in a mitotically dividing human cell occupies a different set of microtubules. But in meiosis the two chromosomes of each type synapse, as we have seen, and move onto the spindle as a single unit. Hence, at metaphase I of meiosis in our hypothetical organism (Fig. 13.18: 4), only three sets of spindle microtubules are occupied, not six. An entire synaptic pair (two chromosomes, four chromatids) is attached to each of these three sets of microtubules.

First anaphase In mitosis metaphase ends and anaphase begins when the tightly coupled centromeres of each double-stranded chromosome uncouple and the two independent single-stranded chromosomes thus formed move away from each other toward opposite poles of the spindle. But in meiosis I two independent chromosomes with separate centromeres are attached to each occupied set of microtubules from the beginning of metaphase. Hence there is no uncoupling of centromeres. The two double-stranded chromosomes of each synaptic pair move away from each other toward opposite poles during anaphase. This means that, in our hypothetical organism, only three chromosomes move to each pole (Fig. 13.18: 5), in contrast to the six that move to each pole in mitosis. Notice, however, that because the synaptic pairing was not random, but involved the two homol-

TABLE 13.1 *Comparison of mitosis and first division sequence of meiosis*

Phase	Mitosis	Meiosis I
Prophase	No synapsis; double-stranded chromosomes move to spindle individually	Synapsis; double-stranded chromosomes move to spindle in pairs (tetrads)
Metaphase	Each chromosome attached to separate microtubules; centromeres in chromatids of double-stranded chromosomes uncouple	The two chromosomes of each homologous pair attached to the same microtubules; centromeres in chromatids of double-stranded chromosomes do not uncouple.
Anaphase	Separation of new single-stranded chromosomes derived from one original double-stranded chromosome	Separation of old double-stranded chromosomes of each synaptic pair
Telophase	Formation of two new nuclei, each with same number of chromosomes as parental nucleus; chromosomes single-stranded when they fade from view	Formation of two new nuclei, each with half the chromosomes present in parental nucleus; chromosomes double-stranded when they fade from view
Interphase, Interkinesis	Replication of genetic material, with formation of new chromatids	No replication of genetic material and hence no new chromatids

ogous chromosomes of each type, the two daughter nuclei get not just any three chromosomes, but rather one of each of the three types.

First telophase   Telophase of mitosis and meiosis are essentially the same, except that each of the two new nuclei formed in mitosis has the same number of chromosomes as the parental nucleus, whereas each of the new nuclei formed in meiosis has half the chromosomes present in the parental nucleus (Fig. 13.18: 6). At the end of telophase of mitosis, the chromosomes are single-stranded when they fade from view; at the end of telophase I of meiosis, the chromosomes are double-stranded when they fade from view.

Interkinesis   Following telophase I of meiosis, there is a short period called interkinesis, which is similar to an interphase between two mitotic division sequences except that no replication of the genetic material occurs and hence no new chromatids are formed (replication is unnecessary, since the chromosomes are already double-stranded when interkinesis begins).

Table 13.1 summarizes the differences between mitosis and meiosis I.

Second division sequence of meiosis   The second division sequence of meiosis, which follows interkinesis, is essentially a mitotic one from the standpoint of mechanics, though the functional result is different (Fig. 13.18: 8–11). The chromosomes do not synapse; they cannot, since the nucleus is haploid and there are no homologous chromosomes. Each double-stranded chromosome moves onto the spindle independently, and its centromeres attach to a set of chromosome-to-pole microtubules. At the end of metaphase II the centromeres uncouple, and the new single-stranded chromosomes thus formed move away from each other toward opposite poles of the spindle during anaphase II. The new nuclei formed during telophase II are haploid like the nuclei formed during telophase I, but their chromosomes are single-stranded instead of double-stranded.

In summary, then, the first meiotic division produces two haploid cells containing double-stranded chromosomes. Each of these cells divides in the second meiotic division; thus a total of four new haploid cells containing single-stranded chromosomes are produced.

Since meiosis I achieves the reduction of chromosome number from diploidy to haploidy, why, it may be asked, must a second division sequence occur? For an answer, we must take a closer look at synapsis during prophase I.

In prophase I the two homologous chromosomes of each pair move together, attracted to each other by a mechanism so far unidentified. While they lie side by side, it sometimes happens that a chromatid from one double-stranded chromosome and a chromatid from the other double-stranded chromosome break at corresponding points and exchange parts. This process is called **crossing-over.** Later in prophase the chromosomes can be seen to be joined at a discrete point called a **chiasma** (plural: chiasmata), where the crossing-over took place (Fig. 13.19).

Now, suppose that the chromatids of the first chromosome bear a gene A near one end and a gene B near the other end, and that the chromatids of the second chromosome bear slight variants of these genes, designated a and b (Fig. 13.20). If crossing-over occurs—i.e. if chromatids from these chromo-

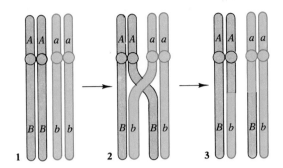

13.20   Schematic model of crossing-over
(1) The two homologous double-stranded chromosomes, one bearing alleles A and B and the other alleles a and b, lie side by side in synapsis. (2) Portions of an AB chromatid and an ab chromatid exchange positions. (3) After breakage of the crossed chromatids, the fragments fuse in the exchanged configuration, with the result that one chromatid of the first chromosome bears alleles A and b and one chromatid of the second chromosome bears alleles a and B.

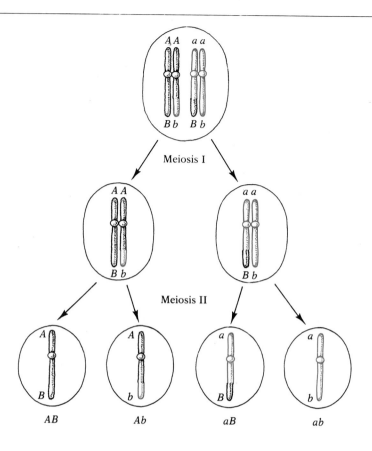

13.21 The products of meiosis when crossing-over has occurred

Top: A cell in which crossing-over has taken place during prophase I. Middle: The chromosomes in the cells produced by meiosis I have unlike chromatids; in the one cell are *AB* and *Ab* chromatids and in the other *aB* and *ab* chromatids. Bottom: Meiosis II separates the chromatids, with the result that four genetically different cells are produced.

somes exchange parts as a result of breakage-fusion at a point between the two genes in question—the first chromosome will end up with one original chromatid bearing *A* and *B* and one crossed-over chromatid bearing *A* and *b*. Similarly, the second chromosome will have one chromatid bearing *a* and *b* and one bearing *a* and *B*. In short, each of the four chromatids will be different from the others.

We have said that meiosis II is mechanically like mitosis; i.e. the individual events in these two processes are the same. But mitosis functions to separate identical chromatids, producing new cells with the same genetic endowment. By contrast, meiosis II separates chromatids that often differ, because crossing-over has taken place, in which case it produces four genetically different cells (Fig. 13.21). It is plain, then, that meiosis II is necessary to complete the segregation of recombinant genetic entities. Crossing-over, with meiosis II, increases the number of different genetic combinations that can be produced.

THE TIMING OF MEIOSIS IN THE LIFE CYCLE

Meiosis in the life cycle of animals With rare exceptions, higher animals exist as diploid multicellular organisms through most of their life cycle. At the time of reproduction, meiosis produces haploid gametes, which, when their nuclei unite in fertilization, give rise to the diploid zygote. The zygote then divides mitotically to produce the new diploid multicellular individual. The gametes—sperm and egg cells—are thus the only haploid stage in the animal life cycle (see Fig. 13.25C).

In male animals sperm cells (spermatozoa) are produced by the germinal epithelium lining the seminiferous tubules of the testes (Fig. 13.22). When one of the epithelial cells undergoes meiosis, the four haploid cells that result are all quite small, but approximately equal in size (Fig. 13.23). All four soon differentiate into sperm cells with long flagella, but with very little cytoplasm in the head, which consists primarily of the nucleus. This process of sperm production is called *spermatogenesis.*   In female animals the egg cells are produced within the follicles of the ovaries by a process called *oogenesis.* When a cell in the ovary undergoes meiosis, the haploid cells that result are very unequal in size. The first meiotic division produces one relatively large cell and a tiny one called a first *polar body* (Fig. 13.24). The second meiotic division of the larger of these two cells (secondary oocyte) produces a tiny second polar body and a large cell that soon differentiates into the egg cell (or ovum). The first polar body may or may not go through the second meiotic division. Thus, when a diploid cell in the ovary undergoes complete meiosis, only one mature ovum is produced (Fig. 13.23); the polar bodies are essentially nonfunctional. By contrast, a diploid cell undergoing complete meiosis in the testis gives rise to four functional sperm cells.

The advantage of the unequal cytokinesis of oogenesis is obvious. By this mechanism an unusually large supply of cytoplasm and stored food is allotted to the nonmotile ovum for use by the embryo that will develop from it. In fact, the ovum provides almost all the cytoplasm and initial food supply for

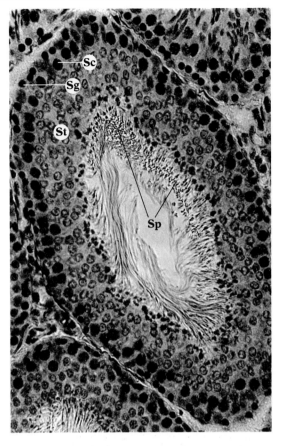

13.22   Photograph of cross section of rat seminiferous tubule, showing spermatogenesis
The dark-stained outermost cells in the wall of the tubule are spermatogonia (Sg), which divide mitotically, producing new cells that move inward. These cells enlarge and differentiate into primary spermatocytes (Sc), which divide meiotically to produce secondary spermatocytes and then spermatids (St). The spermatids differentiate into mature sperm cells, or spermatozoa (Sp), whose long flagella can be seen in the lumen of the tubule in this photograph. [Courtesy Thomas Eisner, Cornell University.]

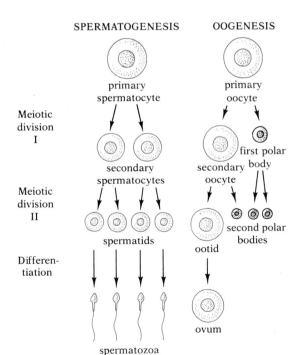

13.23   Schematic illustration of spermatogenesis and oogenesis in an animal
The first polar body does not divide in all animals.

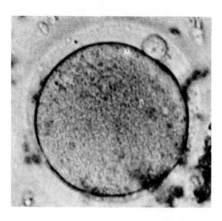

13.24 Human egg cell with polar body
The polar body is the small circular structure on top of the egg cell. At lower right is a concentration of follicle cells. × 250. [Courtesy J. F. Kennedy, San Diego, Calif.]

the embryo. The tiny, highly motile sperm cell contributes, essentially, only its genetic material.

Meiosis in the life cycles of plants   That meiosis produces gametes in animals does not mean that it must do so in all organisms. There is no inherent reason why the cells resulting from meiosis must be specialized for sexual reproduction. And, indeed, they are not specialized in this way in most plants. Meiosis in plants usually produces haploid reproductive cells called *spores,* which often divide mitotically to develop into haploid multicellular plant bodies.

Let us briefly examine the life cycle of a hypothetical plant in which meiosis produces spores rather than gametes (Fig. 13.25B). While the plant is in the diploid portion of its life cycle, certain cells in its reproductive organs divide by meiosis to produce haploid spores (stage 1). These spores divide mitotically and develop into haploid multicellular plants (stage 2). The haploid multicellular plant eventually produces cells specialized as gametes (stage 3). Notice that the gametes are produced by mitosis, not meiosis, because the cells that divide to produce the gametes are already haploid. Two of these gametes unite in fertilization to form the diploid zygote (stage 4), which divides mitotically and develops into a diploid multicellular plant (stage 5). In time, this plant produces spores, and the cycle starts over again.

As we shall see in some detail in a later chapter, the various groups of plants vary greatly in the relative importance of the diploid and haploid phases in their life cycles. In some, the two phases are approximately equal, as shown in Fig. 13.25B (stages 2 and 5). A very few plants have cycles almost like the animal life cycle shown in Fig. 13.25C; i.e. stages 1 and 2 are absent, and the haploid phase of the cycle is represented only by the gametes. In the flowering plants stages 1 and 2 have not been abandoned altogether, but stage 2 has been reduced to a tiny three-to-eight-cell entity that is not free-living, and the plant spends most of its life cycle as a multicellular diploid organism (stage 5).

At the opposite extreme are some primitive plants whose life cycles are almost completely the reverse of those of animals (Fig. 13.25A). There is no stage 5. The diploid zygote (stage 4) formed by the union of gametes promptly undergoes meiosis to produce four haploid spore cells (stage 1), each of which divides mitotically and develops into a haploid multicellular stage (stage 2) in which the organism passes most of its life. This multicellular plant eventually produces, by mitosis, cells specialized as gametes (stage 3), which unite to form the diploid zygote and start the cycle over again. In such an organism, then, the haploid phase of the cycle (particularly stage 2) is dominant, and the only diploid stage—the zygote—is very transitory.

Examination and comparison of the three life cycles illustrated in Fig. 13.25 allow us to make several important generalizations:

1. Meiosis produces either gametes, specialized for sexual reproduction, or spores, specialized for asexual reproduction.
2. Only diploid cells can divide by meiosis, but both haploid and diploid cells can divide by mitosis.
3. If mitosis produces a multicellular organism after fertilization but before meiosis, that organism is diploid.

4. If mitosis produces a multicellular organism after meiosis but before fertilization, that organism is haploid.

5. The haploid phase of the life cycle of multicellular animals is represented only by the gametes (stages 1 and 2 being absent).

6. Most multicellular plants include all five stages in their life cycles, but the relative importance of these varies greatly.

7. With some major exceptions, the haploid stages are dominant in the more primitive plants, the diploid stages in the more advanced plants.

THE ADAPTIVE SIGNIFICANCE OF SEXUAL REPRODUCTION

Characteristic of the vast majority of both plants and animals, sexual reproduction is so widespread that we tend to regard it as a universal attribute of life; but it is not. Many plants and animals reproduce asexually, i.e. give rise to new individuals by mitotic cell division. This type of reproduction is undeniably simpler than sexual reproduction, in which the complicated processes of meiosis and fertilization must alternate. We can legitimately ask, therefore, why asexual reproduction is not adequate for most organisms, why natural selection has so often favored sexual reproduction, with all the complexities it entails.

We have seen that mitosis produces new cells with a genetic endowment identical to that of the parental cell; each new cell gets a set of chromosomes copied from the parental set. Hence the characteristics of offspring produced by mitosis will be essentially the same as those of the parent. In other words, asexual reproduction does not give rise to genetic variation; instead, reproduction without meiosis and fertilization gives rise to individuals genetically identical to the parent and to one another. Such a group of genetically identical individuals is called a **clone**.

Let us consider a specific example. Suppose that our hypothetical species of Fig. 13.7 were to reproduce asexually (i.e. by mitosis). Suppose, further,

A PRIMITIVE PLANT

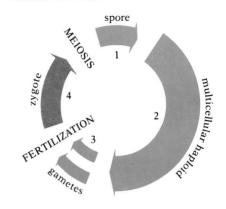

B INTERMEDIATE PLANT

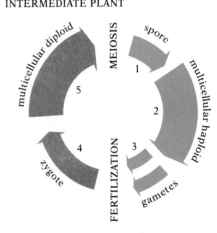

C ANIMAL

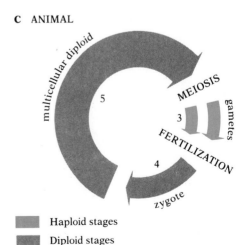

Haploid stages

Diploid stages

13.25   Three types of life cycles
(A) In some very primitive plants the diploid phase is represented only by the zygote, which quickly divides by meiosis to produce haploid spores, which may divide mitotically to produce a multicellular haploid plant.   (B) In most multicellular plants there are two multicellular stages, one haploid and one diploid (stages 2 and 5); the relative importance of these two stages varies greatly from one plant group to another (the cycle shown here is an intermediate one in which stages 2 and 5 are nearly equal). In flowering plants the multicellular diploid stage (stage 5) is the major one, and the multicellular haploid stage (stage 2) is much reduced, being represented by a tiny organism with very few cells.   (C) Animals and a very few plants have a life cycle in which meiosis produces gametes directly—the spore stage (stage 1) and the multicellular haploid stage (stage 2) being absent.

that a certain individual of this species has gray fur and a long tail, and that the gene determining the gray coat color is located on its V-shaped chromosomes while the gene determining the long tail is located on its J-shaped chromosomes. When this individual reproduces asexually, all its offspring will get identical chromosomes. They will be members of a clone, and will consequently have gray fur and long tails.

Now, suppose that another individual of this same species has a different form of the gene for coat color—one determining black rather than gray fur—on its V-shaped chromosomes and a different form of the gene for tail length—one determining a short rather than a long tail—on its J-shaped chromosomes. If this individual reproduces asexually, all its offspring will have black fur and short tails like their parent.

Each new generation, then, will include some individuals with gray fur and long tails and some with black fur and short tails. But asexual reproduction provides no way to produce new individuals with the combination of gray fur and short tails or the combination of black fur and long tails. In short, none of the offspring can have genetic combinations different from those of the parent; mitosis provides no way in which the genes of one individual can be recombined with the genes of another individual.

But now let us look at the chromosomes of a diploid individual produced by sexual reproduction. The first cell (zygote) of this individual was formed by the union of two haploid gametes, an egg cell from one parent and a sperm cell from a second parent. Each of these gametes contributed one chromosome of each type characteristic of the species. Thus, if our hypothetical organism of Fig. 13.7 was produced by sexual reproduction, it received one V chromosome, one J chromosome, and one I chromosome from its father and one V chromosome, one J chromosome, and one I chromosome from its mother. Hence, if one parent has gray fur and a long tail and the other has black fur and a short tail, some of the offspring may well have gray fur and short tails or black fur and long tails. Some of the genes of one parent have recombined with genes from the other parent to produce progeny with characteristics different from those of either parent. Unlike mitotic asexual reproduction, sexual reproduction augments variation in the population by recombining chromosomes and the genes they bear.

Suppose our hypothetical diploid organism produces gametes. Since it received one of each type of chromosome from its mother and one of each type from its father, each of its cells contains the following chromosomes: $V_m V_p J_m J_p I_m I_p$ (where m designates maternal chromosomes and p paternal chromosomes). When a given cell in its reproductive tissue undergoes meiosis, one of the resulting haploid cells does not necessarily get all the maternal chromosomes and the other all the paternal chromosomes. Indeed, it is far more likely that each should get some combination of maternal and paternal chromosomes, for the disposition of a particular pair of homologous chromosomes on the meiotic spindle is purely a matter of chance (e.g. in Fig. 13:18: 4, the white V chromosome might just as easily have moved to the right of the equator and the black one to the left). All the following combinations in the haploid nucleus are equally probable:

$$V_p J_p I_p \qquad\qquad V_p J_m I_m$$
$$V_p J_p I_m \qquad\qquad V_m J_p I_m$$
$$V_p J_m I_p \qquad\qquad V_m J_m I_p$$
$$V_m J_p I_p \qquad\qquad V_m J_m I_m$$

That is, cells with any of the eight chromosomal combinations listed here can be produced by meiosis from a parental cell with the chromosomal makeup $V_m V_p J_m J_p I_m I_p$. A parental cell that contained four kinds of chromosomes instead of three could produce even more combinations. The number of possible chromosomal combinations obtainable by meiosis from a human cell with 23 kinds of chromosomes is huge.

The number of possible chromosomal combinations for the cells produced by meiosis in any given organism is equal to (2^n), where n is the number of types of chromosomes, i.e. the haploid number (3 in our hypothetical organism and 23 in a human). But this expression neglects crossing-over; it assumes intact original chromosomes. Hence calculations based on it greatly underestimate the actual number of combinations that can be produced. If, in our hypothetical organism, only one of the three synaptic pairs underwent a single crossing-over (e.g. if the V chromosomes crossed over, with a resulting production of some haploid cells with V_p' and some with V_m' chromosomes, in addition to cells with V_p and V_m chromosomes), then the number of possible chromosomal combinations would increase from 8 to 16. Since a vast number of different types of crossing-over can occur, a single individual of our hypothetical organism could, over time, produce hundreds or even thousands of genetically different haploid cells by meiosis.

If crossing-over is disregarded, the number of possible chromosomal combinations in the diploid zygote formed by union of gametes from two parents is equal to $(2^n)^2$, where n is the haploid number. Thus a mating between two of our hypothetical organisms could potentially produce diploid offspring with 64 possible chromosomal combinations—a number likely to be increased to many thousands by crossing-over. (We leave it to you to calculate the corresponding overwhelming numbers for humans.)

Sexual reproduction, as we have seen, makes possible genetic recombination and thus increases the variation among individuals. What makes such variation advantageous? Genetic variability is the raw material that natural selection acts upon in promoting the survival and reproduction of well-adapted individuals and the extinction of ill-adapted ones. If a certain type of organism produces genetically varied descendants, the chances that some of these will be well adapted to new environmental conditions, and can survive to reproduce, will be much greater than if they are genetically uniform. The advantages of sexual reproduction, then—and of the genetic variability it favors—are linked to the requirements of changing environmental conditions.

Asexual reproduction, by contrast, may be advantageous when environmental conditions are constant. Under stable conditions an asexual individual whose genetic endowment suits it particularly well for the exploitation of its environment may rapidly produce a clone of progeny equally well suited to exploiting that environment.

On balance, given the continual fluctuations that characterize most environments both over the short term and over the long term, sexual reproduction, with its potential for genetic change and hence for evolutionary adaptation, has evidently been the more advantageous: It has come to be the basic mode of reproduction in the majority of organisms.

It is interesting to note that some kinds of organisms exploit the benefits of both sexual and asexual reproduction. For example, female pea aphids, small insects abundant on alfalfa and many other plants, reproduce parthenogenetically[3] in early spring and rapidly build up a dense population while weather conditions are favorable (cool and moist) and predators are few, but later in the season, as the weather becomes less favorable (hot and dry) and as predators become more numerous, they mate and lay fertilized eggs.

The artificial propagation by asexual means of valuable decorative or crop plants has become a standard practice of modern horticulture. By means of cuttings taken from a prize individual, a clone of plants all possessing the same genetic endowment as the original plant (and presumably the potential for developing the same prized characteristics) is developed. For example, when a new variety of rose is put on the market, all the plants offered for sale are members of a clone; they have all been produced asexually from a single rose bush selected for its unusually fine characteristics. Perpetuation of the new variety requires continued asexual propagation; if the plants were allowed to reproduce sexually, the original favorable combination of genes would be lost. Cloning of superior domestic animals, which could be immensely valuable to agriculture, would be far more difficult than cloning plants, and it is a question whether the procedure will ever be economically feasible, especially on a large scale.

[3] Parthenogenesis is the production of offspring without fertilization.

SUGGESTED READING

GIBOR, A., 1966. "*Acetabularia:* A useful giant cell," *Scientific American,* November. (Offprint 1057)   *On some of the classical experiments in which this unicellular alga was used to study the role of the nucleus in controlling differentiation.*

MAZIA, D., 1974. "The cell cycle," *Scientific American,* January. (Offprint 1288)   *The stages of interphase and mitosis proper; experiments conducted to ascertain the characteristics of these stages and the controls governing them.*

SWANSON, C. P., and P. L. WEBSTER, 1977. *The cell,* 4th ed. Prentice-Hall, Englewood Cliffs, N.J.   *Small book covering many aspects of cell biology at an intermediate level. Chapters 7 and 8 contain clear, well-illustrated descriptions of mitosis and meiosis.*

Chapter 14

PATTERNS OF INHERITANCE

It is time now that we offer evidence for our oft-repeated, but so far unsupported, assertion that there are units called genes located on the chromosomes in the nucleus, and that these units, passed on from generation to generation, exert control over the characteristics of organisms. The experiments cited in the last chapter, demonstrating the importance of the nucleus in heredity and in normal cell function, could be explained adequately without reference to such structures as chromosomes or genes. It is evidence from breeding experiments that first led to a formulation of the modern concepts of chromosomal inheritance, and it is this evidence that we shall examine here; the more recent biochemical evidence will be treated in Chapter 15.

MONOHYBRID INHERITANCE

EXPERIMENTS BY MENDEL

Let us begin our examination of the chromosomal theory of inheritance by considering a series of experiments performed on ordinary garden peas during the years 1856 to 1868 by a modest Austrian monk named Gregor Mendel. Mendel published the results of these experiments in 1866.

Mendel's results Of the numerous characteristics of garden peas that Mendel studied, seven were particularly interesting to him. He noticed that each of these seven characteristics occurred in two contrasting forms (Table 14.1). Thus the seeds were either round or wrinkled, the flowers were either red or white, the pods were either green or yellow, etc. When Mendel cross-

P generation
Red flower   White flower

F₁ generation   100% red

F₂ generation   75% red, 25% white

14.1   Results of Mendel's cross of red-flowered and white-flowered peas

pollinated plants with contrasting forms of one of these characteristics, all the offspring were alike and resembled one of the two parents. When these offspring were crossed among themselves, however, some of their offspring showed one of the original contrasting traits and some showed the other. In other words, a trait that had been present in the grandparental generation but not in the parental generation reappeared.

Let us examine more closely some of Mendel's crosses. When plants with red flowers were crossed with plants with white flowers, all the offspring—the F_1 generation[1]—had red flowers (Fig. 14.1). Similarly, when plants characterized by round seeds were crossed with plants characterized by wrinkled seeds, all the offspring had round seeds. Apparently one form of each characteristic had taken precedence over the other; i.e. red color had taken precedence over white in the flowers, and round form had taken precedence over wrinkled in the seeds. Mendel termed the traits that appear in the F_1 offspring of such crosses (in these examples, red flowers or round seeds) *dominant characters,* and the traits that become latent in such crosses (in these examples, white flowers or wrinkled seeds) *recessive characters.*

When Mendel allowed the F_1 peas from the cross involving flower color, all of which were red, to breed freely among themselves, their offspring (the F_2 generation) were of two types; there were 705 plants with red flowers and 224 with white flowers. The recessive character had reappeared in approximately one fourth of the F_2 plants. Similarly, when F_1 peas from the cross involving seed form, all of which had round seeds, were allowed to breed freely among themselves, their F_2 offspring were of two types; 5,474 had round seeds and 1,850 had wrinkled seeds. Again, the recessive character had reappeared in approximately one fourth of the F_2 plants. The same thing was true of crosses involving the other five characters that Mendel studied (Table 14.1); in each case the recessive character disappeared in the F_1 generation, but reappeared in approximately one fourth of the plants in the

[1] The parental generation in such a cross is customarily designated the P generation. The offspring are designated the F_1 generation (meaning first filial generation), and the offspring of the F_1 generation are designated the F_2 generation (meaning second filial generation). Mendel did not use these terms, but we shall throughout our discussion of his experiments.

TABLE 14.1   *Mendel's results from crosses involving single character differences*

P characters	F_1	F_2	F_2 ratio
1. Round × wrinkled seeds	All round	5,474 round : 1,850 wrinkled	2.96 : 1
2. Yellow × green seeds	All yellow	6,022 yellow : 2,001 green	3.01 : 1
3. Red × white flowers	All red	705 red : 224 white	3.15 : 1
4. Inflated × constricted pods	All inflated	882 inflated : 299 constricted	2.95 : 1
5. Green × yellow pods	All green	428 green : 152 yellow	2.82 : 1
6. Axial × terminal flowers	All axial	651 axial : 207 terminal	3.14 : 1
7. Long × short stems	All long	787 long : 277 short	2.84 : 1

F_2 generation. We can summarize the results of the experiment involving flower color as follows:

$$P \qquad \text{red} \quad \times \quad \text{white}$$
$$\downarrow$$
$$F_1 \qquad \qquad \text{all red}$$
$$\downarrow$$
$$F_2 \qquad \tfrac{3}{4}\ \text{red} \qquad \tfrac{1}{4}\ \text{white}$$

Mendel's conclusions   From experiments of this type, Mendel drew the important conclusion that each pea plant possesses two hereditary factors for each character, and that when gametes are formed the two factors segregate and pass into separate gametes, so that each gamete possesses only one factor for each character. Each new plant thus receives one factor for each character from its male parent and one for each character from its female parent. The fact that two contrasting parental traits, such as red and white flowers, can both appear in normal form in the F_2 offspring indicates that the hereditary factors must exist as separate particulate entities in the cell; they do not blend or fuse with each other. Thus the cells of an F_1 pea plant from the cross involving flower color contain, according to Mendel, one factor for red color and one factor for white color, the factor for red being dominant and the factor for white being recessive. But their existence together in the same nucleus does not change the factors; the red and white factors do not alter each other. They remain distinct, and segregate unchanged when germ cells are formed.

You will have noticed that Mendel's conclusions are consistent with what we know about the chromosomes and their behavior in meiosis. The diploid nucleus contains two of each type of chromosome. Presumably each of the two chromosomes of any given pair bears genes for the same characters; hence the diploid cell contains two doses, which may or may not be identical, of each type of gene; these are the two hereditary factors for each character that Mendel described.[2] Since the members of each pair of chromosomes segregate during meiosis, gametes contain only one chromosome of each type and hence only one dose of each gene, just as Mendel deduced. In short, Mendel's theories seem rather obvious in the light of the events of cell division. But it should be remembered that Mendel did his work before the details of cell division had been learned, before, in fact, the significance of chromosomes for heredity had been discovered. He arrived at his conclusions purely by reasoning from the patterns of inheritance he detected in his experiments, without any reference to the structural components of the cell or its nucleus. The chromosomal theory of inheritance, therefore, rests today on two independent lines of evidence, one from breeding experiments, the other from microscopic examination of the nucleus. That these two lines of evidence should agree amazingly well is testimony to the strength of the theory.

The particulate theory of heredity is very different from the view, prevalent in Mendel's day, that each character of the offspring is a blend of the characters of the parents, a view that many people still hold today and

[2] It will be necessary to consider other definitions of the gene in the next chapter. For the moment, however, we can define a gene (Mendel's "factor") as a hereditary unit, located at a specific place or locus on a chromosome, that determines a particular character of the organism.

express when, for example, they speak of the mixing of blood. We now know, of course, that the blood is not the hereditary material and that expressions like "related by blood" have little biological meaning. The blood of an individual is not simply a half-and-half mixture of his parents' blood or a mixture of four equal parts from his grandparents. His blood is distinctly his own, and its characteristics are determined by the particular hereditary factors for blood that he happens to have received from his ancestors— which ancestors is a question of chance.

We should not give you the impression that the publication of Mendel's classic paper in 1866 immediately changed the prevalent ideas about inheritance, giving birth to a flourishing new branch of biology. Far from it. The scientific community of his day was unprepared for so radical a view of heredity, and it paid little heed to Mendel's results or theories. His paper was soon forgotten. It was not until 1900, after the details of cell division had been worked out and the scientific community was in a more receptive mood, that Mendel's paper was rediscovered almost simultaneously by three men, each of whom had independently performed experiments that led to the same conclusions. These men were Hugo De Vries in Holland, Carl Correns in Germany, and Erich von Tschermak-Seysenegg in Austria.

A modern interpretation of Mendel's experiments   Let us now re-examine Mendel's cross of garden peas of different colors, interpreting his results in modern terminology. Mendel was working with two forms of the gene for flower color, one that produced red flowers and another that produced white flowers. When a gene exists in more than one form, in this way, the different forms are called *alleles.* In the gene for flower color in peas, the allele for red is dominant while the allele for white is recessive. It is customary to designate genes by letters, using capital letters for dominant alleles and small letters for recessive alleles. We may thus designate the allele for red flowers in peas as C and the allele for white flowers as c. Now, a diploid cell contains two doses of each gene, one on each of two homologous chromosomes. Such a cell may thus have two doses of the same allele or one dose of one allele and one dose of another allele. Thus cells of a pea plant may contain two doses of the allele for red flowers (C/C), or two doses of the allele for white flowers (c/c), or one dose of the allele for red and one dose of the allele for white (C/c). Cells with two doses of the same allele (C/C or c/c) are said to be *homozygous.* Those with one each of two different alleles (C/c) are said to be *heterozygous.* (The slash between letters is meant to indicate that the two doses of a gene are on separate chromosomes.)

Note that one cannot tell by visual inspection whether a given pea plant is homozygous dominant (C/C) or heterozygous (C/c), because the two types of plants will look alike; both will have red flowers. In other words, where one allele is dominant over another, the dominant allele takes full precedence over the recessive allele, and a heterozygous organism exhibits the trait determined by that dominant allele; one dose of the dominant allele is as effective as two doses in determining the character trait. This means that there is often no one-to-one correspondence between the different possible genic combinations *(genotypes)* and the possible appearances *(phenotypes)* of the organisms. Thus, in the example of flower color in peas discussed here, there are three possible genotypes, *C/C*, *C/c*, and *c/c*, but only two possible phenotypes, red and white.

We can now apply this understanding of genes to Mendel's pea cross and rewrite the summary of p. 585 as follows:

P		*C/C*	×	*c/c*	
		red		white	
			↓		
F₁		*C/c*	×	*C/c*	
		red		red	
			↓		
F₂	*C/C*	*C/c*		*c/C*	*c/c*
	red	red		red	white

Here we have shown both the genotypes and the phenotypes of the plants in the three generations. Mendel began with a cross in the parental generation between a plant with a homozygous dominant genotype (red phenotype) and a plant with a homozygous recessive genotype (white phenotype). All of the F₁ progeny had red phenotypes, because all of them were heterozygous, having received a dominant allele for red (*C*) from the homozygous dominant parent and a recessive allele for white (*c*) from the homozygous recessive parent. But when the F₁ individuals were allowed to cross freely among themselves, the F₂ progeny they produced were of three genotypes and two phenotypes; one fourth were homozygous dominant and showed red phenotypes, two fourths were heterozygous and showed red phenotypes, and one fourth were homozygous recessive and showed white phenotypes. Thus the ratio of genotypes in the F₂ was 1 : 2 : 1, and the ratio of the phenotypes was 3 : 1.

How do we figure out the possible genotypic combinations in the F₂? This is an easy matter in a monohybrid cross (a cross involving only one character) such as this. All individuals in the F₁ generation are heterozygous (*C/c*); i.e. they have one of each of the two types of alleles. Each of these two alleles is located on a different one of the two chromosomes of a homologous pair. In meiosis these two chromosomes synapse, move onto the spindle as a unit, and then separate, moving to opposite poles, so that the chromosome bearing the *C* allele is incorporated into one new haploid nucleus and the chromosome bearing the *c* allele is incorporated into the other new haploid nucleus. This means that half the gametes produced by such a heterozygous individual will contain the *C* allele and half the *c* allele. When two such individuals are crossed (Fig. 14.2), there are four possible combinations of their gametes:

   C from male parent, *C* from female parent
   C from male parent, *c* from female parent
   c from male parent, *C* from female parent
   c from male parent, *c* from female parent

The first of these four possible combinations produces homozygous dominant offspring (red); the second and third produce heterozygous offspring (also red); and the fourth produces homozygous recessive offspring (white). Since each of these four combinations is equally probable, we would expect,

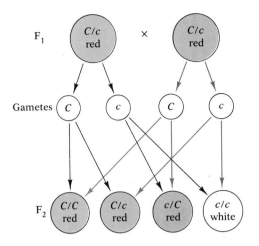

14.2   Gametes formed by F₁ individuals in Mendel's cross for flower color, and their possible combinations in the F₂

if a large number of F_2 progeny are produced, a genotypic ratio close to 1 : 2 : 1 and a phenotypic ratio close to 3 : 1, just as Mendel found.

An easy way to figure out the possible genotypes produced in the F_2 is to construct a so-called Punnett square. Along a horizontal line, write all the possible kinds of gametes the male parent can produce; in a vertical column to the left, write all the possible kinds of gametes the female parent can produce; then draw squares for each possible combination of these, as follows:

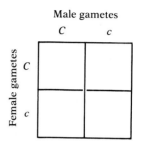

Next, write in each box, first, the symbol for the female gamete and, second, the symbol for the male gamete. Each box will then contain the symbols for the genotype of one possible zygote combination from the cross in question. A glance at the completed Punnett square in Fig. 14.3 shows that the cross yields the expected 1 : 2 : 1 genotypic ratio and, since dominance is present, the expected 3 : 1 phenotypic ratio.

Extensive investigation of a vast array of plant and animal species by thousands of scientists has demonstrated conclusively that the results Mendel obtained from his monohybrid crosses, and the interpretations he placed on them, are not limited to garden peas but are of general validity. Whenever a monohybrid cross is made between two contrasting homozygous individuals, regardless of the character involved, the expected genotypic ratio in the F_2 is 1 : 2 : 1. And whenever there is dominance, the expected phenotypic ratio is 3 : 1. If these ratios are not obtained in large samples, it can be assumed that some complicating condition is present, which must be identified.

THE TEST CROSS

We have seen that in a monohybrid cross involving dominance the homozygous dominant progeny and the heterozygous progeny have the same phenotype and cannot be distinguished by inspection. But it is often of practical importance to distinguish between individuals with these two genotypes—in breeding, for example. Homozygous individuals breed true; i.e. matings between two such individuals produce offspring all of a single genotype and phenotype like their parents. But heterozygous individuals do not breed true; as we have seen, a cross between two such individuals may produce offspring of three genotypes and two phenotypes. Consequently the identification of homozygous individuals is of the utmost importance to an animal or plant breeder who is trying to establish true-breeding strains of animals or plants. This is no problem when the breeder is interested in a recessive character, because homozygous recessive organisms can readily be recog-

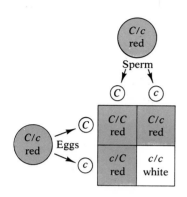

14.3 Punnett-square representation of the same information as shown in Fig. 14.2

nized by their phenotypes. But if the breeder is interested in establishing a strain that is true-breeding for a dominant character, he needs a test that will enable him to tell whether a given individual is homozygous dominant or heterozygous.

The test used for this purpose is a cross between the individual of unknown genotype and a homozygous recessive individual, which can be recognized by its phenotype. Suppose, for example, we want to know whether a particular red-flowered pea plant has a genotype of C/C or C/c. The most we can say from simple inspection is that it is red and hence must have at least one C allele; in other words, half of its genotype for flower color is known, and half is unknown and can be designated by a dash: $C/-$. We now cross this plant with a white-flowering plant, which can only have the genotype c/c. The results should give us the information we seek. If the plant in question has $C/\underline{C}$ as its genotype, then the cross will turn out as follows:

$$C/\underline{C} \quad \times \quad c/c$$
$$\text{red} \qquad \text{white}$$
$$\downarrow$$
$$C/c \quad C/c \quad C/c \quad C/c$$
$$\text{red} \quad \text{red} \quad \text{red} \quad \text{red}$$

If, however, the plant has the other possible genotype, C/c, then the cross will turn out as follows:

$$C/\underline{c} \quad \times \quad c/c$$
$$\text{red} \qquad \text{white}$$
$$\downarrow$$
$$C/c \quad C/c \quad c/c \quad c/c$$
$$\text{red} \quad \text{red} \quad \text{white} \quad \text{white}$$

In other words, the unknown half of the test plant's genotype may be either a C allele or a c allele. If it is a C allele, then the full genotype of the plant is C/C and such a plant, when test-crossed with a homozygous recessive (c/c) plant, will produce progeny all of which are heterozygous (C/c) and show a red phenotype (Fig. 14.4A). If, on the other hand, the unknown is a c allele, then the full genotype of the plant is C/c and such a plant, when test-crossed with a homozygous recessive plant, will produce progeny of which half are heterozygous and show a red phenotype and half are homozygous recessive and show a white phenotype (Fig. 14.4B). In practice, we make the cross $C/- \times c/c$ and observe the results. If we obtain a large number of progeny from this cross and all of them show the dominant phenotype (in this case red flowers), the chances are great that the test plant's genotype is C/C. If, however, we obtain progeny of which some show the dominant phenotype (red) and some the recessive phenotype (white), we know that the test plant's genotype is C/c. We expect a 1 : 1 ratio in this case, but if our results happened to depart considerably from this expected ratio, we should still feel certain that the test plant's genotype is C/c. The reasons are obvious: If any of the progeny show the recessive phenotype, their genotype must be c/c, which means that they received a c allele from each parent; hence the test plant must have a c allele in its genotype. Since we already know that it has at least one C allele, we combine these two pieces of information to write its genotype as C/c.

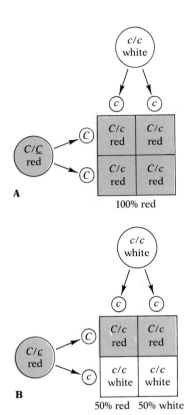

14.4 A test cross
Homozygous dominant *(C/C)* and heterozygous *(C/c)* red-flowered individuals are to be distinguished by crossing them with a homozygous recessive *(c/c)* white-flowered individual.   (A) If the red-flowered plant is homozygous, all the progeny of the test cross will be heterozygous *(C/c)* and will have red flowers.   (B) If the red-flowered plant is heterozygous, 50 percent of the progeny will be heterozygous *(C/c)* and will have red flowers, and 50 percent will be homozygous recessive *(c/c)* and will have white flowers.

INTERMEDIATE INHERITANCE

The seven characters of peas that Mendel used in the experiments reported in his paper were all of the kind in which one allele shows complete dominance over the other. Many characteristics in a variety of organisms show this mode of inheritance. But many others do not. In fact, there is evidence that Mendel himself studied some characters that do not exhibit dominance, even though he did not take them into account in establishing his theory. Inheritance in which heterozygous individuals clearly show effects of both alleles is termed intermediate inheritance.

In many cases of intermediate inheritance, heterozygous individuals have a phenotype that is actually intermediate between the phenotype of individuals homozygous for one allele and the phenotype of individuals homozygous for the other allele. For example, crosses between homozygous red snapdragons and homozygous white snapdragons yield pink snapdragons. When these pink plants are crossed among themselves, they yield red, pink, and white offspring in a ratio of 1 : 2 : 1, as follows:

$$
\begin{array}{ccccccc}
P & & R/R & \times & R'/R' & & \\
& & \text{red} & & \text{white} & & \\
& & & \downarrow & & & \\
F_1 & & R/R' & \times & R/R' & & \\
& & \text{pink} & & \text{pink} & & \\
& & & \downarrow & & & \\
F_2 & R/R & R/R' & & R'/R & R'/R' \\
& \text{red} & \text{pink} & & \text{pink} & \text{white}
\end{array}
$$

Notice that when there is no dominance both alleles are designated by a capital letter, the one being distinguished from the other by a prime, as here, or by a superscript (e.g. C^r for red, C^w for white).

In other cases of intermediate inheritance, the heterozygous phenotype is not so obviously intermediate between the two homozygous phenotypes. For example, in a certain strain of chickens a mating between a black chicken and a splashed white chicken produces offspring all of which have a distinctive appearance called blue Andalusian. A cross between two blue Andalusians produces black, blue Andalusian, and splashed white offspring in a ratio of 1 : 2 : 1, as follows:

$$
\begin{array}{ccccc}
P & & \text{black} & \times & \text{white} \\
& & & \downarrow & \\
F_1 & & \text{blue} & \times & \text{blue} \\
& & & \downarrow & \\
F_2 & \text{black} & \text{blue} & \text{blue} & \text{white}
\end{array}
$$

Here two alleles, one for black and one for white, are interacting to produce a phenotype somewhat different from the gray that might be expected as the intermediate between black and white.

To summarize, the inheritance pattern in incomplete dominance differs in the following ways from that in complete dominance: (1) The F_1 offspring of a monohybrid cross between parents each homozygous for a different allele have a phenotype different from both parents, and (2) the F_2 phenotypic ratio is 1 : 2 : 1 (just like the genotypic ratio) rather than 3 : 1.

DIHYBRID AND TRIHYBRID INHERITANCE

We have limited our discussion so far to crosses involving a single phenotypic character; or rather we have so far chosen to ignore all but one character. The latter is a more accurate statement, because all crosses involve many more than one character. Organisms contain thousands of genes, and it would be impossible to set up a cross in which only one character was allowed to vary. Let us now turn to crosses involving two characters (dihybrid crosses), three characters (trihybrid crosses), or more.

THE BASIC DIHYBRID RATIO

Mendel's experiments on garden peas were not limited to single characters, but sometimes involved two or more of the characters listed in Table 14.1. For example, he crossed plants characterized by round yellow seeds with plants characterized by wrinkled green seeds. The F_1 plants all had round yellow seeds. When these plants were crossed among themselves, the resulting F_2 progeny showed four different phenotypes:

315 had round yellow seeds
101 had wrinkled yellow seeds
108 had round green seeds
32 had wrinkled green seeds

These numbers represent a ratio of about 9 : 3 : 3 : 1 for the four phenotypes.

This experiment demonstrated that a dihybrid cross can produce some new plants phenotypically unlike either of the original parental plants; here the new phenotypes were wrinkled yellow and round green. It demonstrated, in other words, that the genes for seed color and the genes for seed form do not necessarily stay together in the combinations in which they occurred in the parental generation. The obvious inference was that the genes for seed color are on the chromosomes of one homologous pair and the genes for seed form on the chromosomes of a different pair and that, consequently, the genes for the two characters segregate independently during meiosis.

The 9 : 3 : 3 : 1 phenotypic ratio is characteristic of the F_2 generation of a dihybrid cross (with dominance) in which the genes for the two characters are *independent* (i.e. located on nonhomologous chromosomes). Each independent gene behaves in a dihybrid cross exactly as in a monohybrid cross. If we view Mendel's F_2 results as the product of a monohybrid cross for seed color (ignoring seed form), we find that there were 416 yellow seeds (315 + 101) and 140 green seeds (108 + 32), which closely approximates the 3 : 1 F_2 ratio expected in a monohybrid cross. Similarly, if we treat the experiment as a monohybrid cross for seed form and ignore seed color, the F_2 results also show a phenotypic ratio of approximately 3 : 1. The dihybrid F_2 ratio of 9 : 3 : 3 : 1 is thus simply the product of two separate and independent 3 : 1 ratios.

Let us examine in somewhat more detail a cross of this type, using the symbols *R* for the allele for round seed and *r* for the allele for wrinkled seed, and the symbols *G* for the allele for yellow seed and *g* for the allele for green seed. In the summary of Mendel's cross below, the dash means that it does

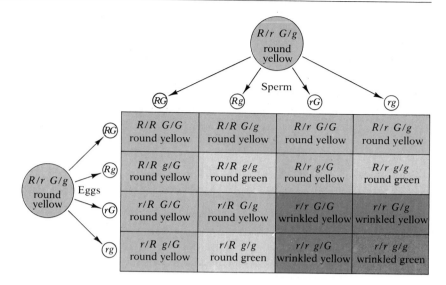

14.5 **Punnett-square representation of a mating of two F₁ individuals in one of Mendel's dihybrid crosses in peas**

not matter phenotypically whether the dominant or the recessive allele occurs in the spot indicated.

| P | | $R/R\ G/G$ | × | $r/r\ g/g$ | |
| | | round yellow | | wrinkled green | |

↓

| F₁ | | $R/r\ G/g$ | × | $R/r\ G/g$ | |
| | | round yellow | | round yellow | |

↓

| F₂ | $9\ R/-\ G/-$ | $3\ r/r\ G/-$ | $3\ R/-\ g/g$ | $1\ r/r\ g/g$ |
| | round yellow | wrinkled yellow | round green | wrinkled green |

The round yellow parent could produce gametes of only one genotype, *RG*. The wrinkled green parent could produce only *rg* gametes. When *RG* gametes from the one parent united with *rg* gametes from the other parent in the process of fertilization, all the resulting F₁ offspring were heterozygous for both characters ($R/r\ G/g$) and showed the phenotype of the dominant parent (round yellow). Each of these F₁ individuals could produce four different types of gametes, *RG*, *Rg*, *rG*, and *rg* (Fig. 14.5). When two such individuals were crossed, there were 16 possible combinations of gametes (4 × 4), as shown in the figure. These 16 combinations included nine genotypes ($R/R\ G/G$, $R/R\ G/g$, $R/R\ g/g$, $R/r\ G/G$, $R/r\ G/g$, $R/r\ g/g$, $r/r\ G/G$, $r/r\ G/g$, $r/r\ g/g$), which determined four phenotypes in the ratio of 9 : 3 : 3 : 1. What would the phenotypic ratio have been if both characters had exhibited intermediate inheritance rather than dominance-recessiveness? What would have been the ratio if one character had exhibited dominance-recessiveness and the other intermediate inheritance?

One way to calculate the genotypic and phenotypic ratios in a dihybrid cross is to construct a Punnett square and then count the number of boxes representing each genotype and phenotype (Fig. 14.5). This method, satisfactory in a monohybrid cross and only moderately laborious in a dihybrid cross, becomes prohibitively tedious in a trihybrid cross or a cross involving even more than three characters. There is an alternative procedure that is much easier. It is based on the principle that *the chance that a number of independent events will occur together is equal to the product of the chances that each event will occur separately.*

Suppose we want to know how many of the 16 combinations in Mendel's cross will produce the wrinkled yellow phenotype. We know that wrinkled is recessive; hence it is expected in $\frac{1}{4}$ of the F_2 individuals in a monohybrid cross. We know that yellow is dominant; hence it is expected in $\frac{3}{4}$ of the F_2 individuals. Multiplying these two separate values ($\frac{1}{4} \times \frac{3}{4}$) gives us $\frac{3}{16}$; three of the 16 possible combinations will produce a wrinkled yellow phenotype. Similarly, if we want to know how many of the combinations will produce a round yellow phenotype, we multiply the separate expectancies for two dominant characters ($\frac{3}{4} \times \frac{3}{4}$) to get $\frac{9}{16}$.

To use a more complex example, suppose we want to find out, for a trihybrid cross involving the two seed characters and flower color, what fraction of the F_2 individuals (produced by allowing $C/c\ R/r\ G/g$ individuals to cross among themselves) will exhibit a phenotype combining red flowers, wrinkled seeds, and yellow seeds. The separate probability for red flowers in a monohybrid cross is $\frac{3}{4}$, that for wrinkled seeds is $\frac{1}{4}$, and that for yellow seeds is $\frac{3}{4}$. Multiplying these three values ($\frac{3}{4} \times \frac{1}{4} \times \frac{3}{4}$) gives us $\frac{9}{64}$. This tells us that of the 64 possible combinations in a trihybrid cross nine will produce the phenotype here specified. What proportion would show a phenotype combining white flowers, round seeds, and green seeds?

GENE INTERACTIONS

Some dihybrid crosses involve two genes that are inherited independently, but exert their phenotypic effect on the same character. When two or more genes interact in determining a single character, the ratios obtained from crosses in which they are involved are sometimes different from the basic ratio. Let us look at a few types of gene interaction.

Complementary genes Complementary genes are genes that are mutually dependent; neither can exert its phenotypic effect unless the other does also. For example, purple color in the flowers of sweet peas occurs only when both the dominant allele for a gene C and the dominant allele for another gene P are present together. In the absence of either dominant allele, the flowers are white. A dihybrid cross involving these characters can be summarized as follows:

P		$C/C\ P/P$	$\times$	$c/c\ p/p$	
		purple		white	
			$\downarrow$		
F_1		$C/c\ P/p$	$\times$	$C/c\ P/p$	
		purple		purple	
			$\downarrow$		
F_2	9 $C/\!-\ P/\!-$	3 $C/\!-\ p/p$		3 $c/c\ P/\!-$	1 $c/c\ p/p$
	purple	white		white	white

Pea

Rose

Walnut

Single

14.6 Comb types in chickens
For a discussion of the genetics of this character, see text. [Redrawn from E. J. Gardner, *Principles of genetics*, Wiley, 1960.]

Note that in a normal dihybrid cross the four groups of genotypes shown in the F_2 would produce four phenotypes in a ratio of 9 : 3 : 3 : 1. But here they have produced only two phenotypes in a ratio of 9 : 7, which is quite different from any ratio we have previously encountered. The last three phenotypic classes of the normal 9 : 3 : 3 : 1 ratio have been combined.

In sweet peas experiments have established the mechanism for this curious gene interaction, where neither dominant allele can exert its effect unless the other dominant is also present. *C* regulates the production of a necessary raw material for anthocyanin, a purple pigment, and *P* regulates the conversion of this raw material into anthocyanin. It is not enough for the plant to have a *C* allele and make the raw material if it does not also have *P* and hence the capacity to convert the raw material into purple pigment. Conversely, it is not enough for the plant to have a *P* allele if it does not also have *C* and hence the capacity to produce the raw material. In either case the flower is white because no anthocyanin can be synthesized.

Epistasis When one gene has the effect of masking the phenotypic expression of another gene, the first gene is said to be epistatic to the second. For example, in guinea pigs the gene for production of melanin is epistatic to that for deposition of melanin. The first gene has two alleles: *C*, which causes the pigment to be produced, and *c*, whose effect is nonproduction of pigment; hence a homozygous recessive individual, *c/c*, is an albino. The second gene has an allele *B* that causes deposition of much melanin, which gives the guinea pig a black coat, and an allele *b* that causes deposition of only a moderate amount of melanin, which gives the guinea pig a brown coat. Neither *B* nor *b* can cause deposition of melanin if *C* is not present to make the melanin. We can summarize a cross involving these two genes as follows:

P		C/C B/B	×	c/c b/b	
		black		albino	
			↓		
F_1		C/c B/b	×	C/c B/b	
		black		black	
			↓		
F_2	9 $C/- B/-$	3 $C/- b/b$		3 c/c $B/-$	1 c/c b/b
	black	brown		albino	albino

Instead of an F_2 phenotypic ratio of 9 : 3 : 3 : 1, this cross has yielded a ratio of 9 : 3 : 4. The last two phenotypic classes of the normal 9 : 3 : 3 : 1 ratio have been combined.

It is important to distinguish epistasis from dominance, which it superficially resembles. Dominance is the phenotypic expression of one member of a pair of alleles at the expense of the other. Epistasis is the masking by one gene of the phenotypic effect of another entirely different gene. Dominance refers to interaction between alleles, epistasis to interaction between nonallelic genes.

Collaboration Sometimes two genes influencing the same character interact to produce single-character phenotypes that neither gene alone could produce. Such collaborative interaction is seen in the control of the form of the comb in chickens (Fig. 14.6). One gene, *R*, produces rose comb, while its recessive allele, *r*, produces single comb. Another gene, *P*, produces pea comb, while its recessive allele, *p*, also produces single comb. When *R* and *P* occur together, they collaborate to produce walnut comb, a type of comb that neither could produce alone. Rose comb is characteristic of Wyandotte chickens, and pea comb is characteristic of Brahma chickens. A cross between a Wyandotte and a Brahma could be summarized as follows:

have same recessive trait

P 　　　　　　　　　 *R/R p/p*　×　*r/r P/P*
　　　　　　　　　　　 rose　　　　　 pea
　　　　　　　　　　　　　　　 ↓
F₁ 　　　　　　　　　 *R/r P/p*　×　*R/r P/p*
　　　　　　　　　　　 walnut　　　　 walnut
　　　　　　　　　　　　　　　 ↓
F₂ 　　　9 *R/– P/–*　　 3 *R/– p/p*　　 3 *r/r P/–*　　 1 *r/r p/p*
　　　　　 walnut　　　　 rose　　　　　 pea　　　　　 single

Modifier genes Probably no inherited characteristic is controlled exclusively by one gene pair. Even when only one principal gene is involved, its expression is influenced to some extent by countless other genes with individual effects often so slight that they are very difficult to locate and analyze. An example is eye color in human beings.

Human eye color can be regarded as controlled by one gene with two alleles—a dominant allele, *B*, for brown eyes, and a recessive allele, *b*, for blue eyes. Brown-eyed people (*B/B* or *B/b*) have branching pigment cells containing melanin in the front layer of the iris. Blue-eyed people (*b/b*) lack melanin in the front layer; the blue is an effect of the black pigment on the back of the iris as faintly seen through the semi-opaque front layer.

This description of the inheritance of eye color on the basis of a single-gene system assumes only two phenotypes, brown and blue. And it is, in fact, possible to assign most people to one or the other of these two phenotypic classes. But it is common knowledge that eyes exhibit endless variations in hue; everyday terminology recognizes eyes as gray (genetically a form of blue) or black (genetically a form of brown), to name only the most familiar variations. It is obvious, then, that an explanation of eye color in terms of a single-gene system is an oversimplification. Many modifier genes are also involved, some affecting the amount of pigment in the iris, some the tone of the pigment (which may be light yellow, dark brown, etc.), some its distribution (even over the whole iris, or in scattered spots, or in a ring around the outer edge of the iris, etc.). Two blue-eyed people may occasionally have a brown-eyed child, because one of them, in whom the lack of pigmentation is due to the action of modifier genes, actually has the genotype *B/b* (instead of *b/b*).

Another example of the action of modifier genes is seen in the size of the spots of Beagle dogs (Fig. 14.7). A very large number of modifier genes is involved; by itself none produces very marked effects, but in combination they can radically alter the dogs' appearance.

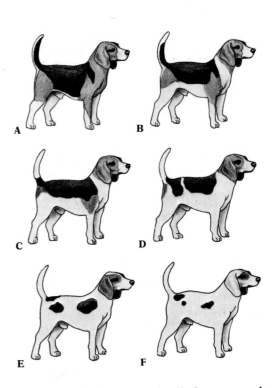

14.7 Variation in spotting of Beagle dogs as a result of the action of many modifier genes

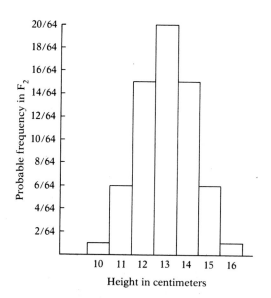

14.8 Bar graph showing probable frequency of phenotypes in F$_2$ generation of a hypothetical cross involving multiple-gene inheritance

The phenotypes have the so-called normal distribution. For discussion of this cross, see text.

Multiple-gene inheritance The phenotypic characteristics we have discussed so far vary in a discontinuous fashion; i.e. a limited number of relatively distinct phenotypes can be recognized. Pea flowers are either red or white, pea seeds either yellow or green, chicken combs either walnut or rose or pea or single. True, modifiers may blur the boundaries of the classes, as in human eye color or spotting in dogs, but a fairly limited number of phenotypes—blue vs. brown or spotted vs. unspotted—can still be meaningfully discussed. Many characters, however, vary in a more continuous fashion; human height, human skin pigmentation, and corn ear length are just three of many possible examples. What can be said about the genetic basis of characteristics such as these?

One explanation commonly given is that two or more separate genes affect the same character in the same way in an additive fashion. Suppose that the height of a hypothetical plant species is controlled by three separate genes, each occurring in two allelic forms, of which one contributes one centimeter in height while the other has no effect on height. We may designate the alleles that contribute to height as A^1, B^1, and C^1, and the alleles that do not contribute as A^0, B^0, and C^0. Suppose that the minimum height against which these genes work is 10 cm. Then a plant with the genotype $A^0A^0B^0B^0C^0C^0$ would be 10 cm tall, because it has no genes for increased height. At the other extreme would be a plant with the genotype $A^1A^1B^1B^1C^1C^1$, which would be 16 cm tall, because it has six superscript-one genes each of which contributes an additional centimeter to the 10 cm base. Between these two extremes would be plants that have from one to five superscript-one genes, with heights from 11 to 15 cm.

If we calculate the probable frequency of each of the F$_2$ height classes in a cross between an $A^0A^0B^0B^0C^0C^0$ individual and an $A^1A^1B^1B^1C^1C^1$ individual, we shall find that these frequencies, when graphed (Fig. 14.8), approximate a normal curve; the extremes of 10 and 16 cm are very rare, the intermediate heights being much more common. This distribution is very similar to the distribution of phenotypic frequencies actually found in examples of continuously varying characteristics. The more genes are added to such a model, the smoother the curve of phenotypic frequencies becomes. Since modifier genes and environmental influences also affect the phenotype, plants with the genotype $A^1A^0B^1B^1C^1C^0$ will not all be precisely 14 cm tall. Some will be slightly less than 14 cm, and some will be slightly more. The other height classes will likewise show internal variation. Hence there will be individuals of almost every conceivable height between the two extremes of 10 and 16 cm; they will not occur in seven clearly distinct classes differing from each other in precise multiples of one centimeter.

The additive effect produced by the interaction of two or more genes has been proposed as an explanation for many continuously varying characters, including those mentioned earlier (human height and skin color and corn ear length). In many ways the explanation is not completely satisfactory. For example, it is difficult to establish how many genes are involved in any given case; the assumption that all the genes contribute equally is an obvious oversimplification, and the often profound effects of environment complicate the analysis of the data. Nevertheless, the multiple-gene hypothesis is frequently a useful one, and until some better model is proposed geneticists will doubtless continue employing it.

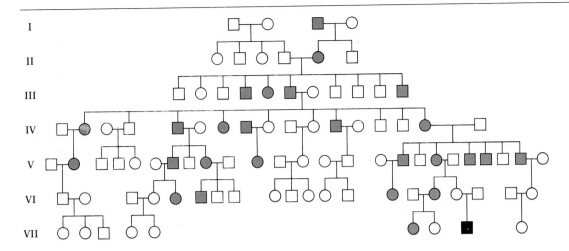

Because genes interact, it is possible for an individual to carry a dominant gene that is not expressed phenotypically. Complementary or epistatic genes, for example, or a large number of modifier genes, may suppress the expression of a dominant gene. And when a gene is expressed, there are many possible degrees of intensity of expression. We can speak, therefore, of incomplete penetrance and variable expressivity. *Penetrance* is the percentage of individuals that, carrying a given gene in proper combination for its expression, actually express that gene's phenotype. *Expressivity* denotes the manner in which the phenotype is expressed.

Incomplete penetrance and variable expressivity are illustrated by a gene in human beings that causes blue sclera, a condition in which the whites of the eyes appear bluish. This gene usually behaves as a simple dominant, and anyone who possesses the gene, whether in heterozygous or homozygous form, might be expected to show the blue-sclera phenotype. But it has been found that only about nine out of ten people who have the gene actually show the phenotype. We can say, therefore, that the penetrance of this gene is about 90 percent (or 0.9). Among those showing the phenotype, the expressivity is variable, the intensity of the bluish coloration ranging from very pale whitish blue to very dark blackish blue.

Figure 14.9 shows a portion of the pedigree of the Hancock family of Virginia, which has for many generations had many members with syndactyly of the ring and little fingers (the two fingers are joined by a web of muscle and skin; Fig. 14.10). This character, like blue sclera, exhibits variable expressivity: A few individuals have three fingers webbed; most have two fingers fully webbed; some have two fingers partially webbed; and a few have a crooked finger with no webbing.

Like blue sclera, too, syndactyly exhibits incomplete penetrance. Of the more than 50 persons in this family known to have had syndactyly, all but one had a parent who also had syndactyly. Of the more than 75 who had parents with normal fingers, only one had syndactyly. This pattern of inheritance is very close to the one expected for a dominant character; as a rule, only individuals with a parent showing the character will show the character

14.9   Pedigree of syndactyly in seven generations of one family
Squares represent males, and circles represent females. Color indicates syndactyly and white indicates normal condition. The character appears to be inherited as a simple autosomal dominant; i.e. there is no correlation with sex, and individuals with syndactyly have a parent who also shows this trait. There is one exception: an individual (black square) who has syndactyly even though neither of his parents shows the trait. Presumably the gene was present in his mother without expression.

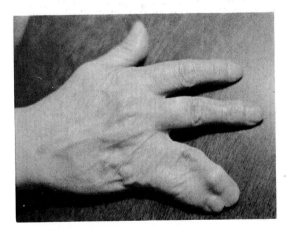

14.10   Photograph showing syndactyly
[Courtesy S. B. Moore.]

14.11   Effect of temperature on expression of a gene for coat color in the Himalayan rabbit
(A) Normally, only the feet, tail, ears, and nose are black.   (B) Fur is plucked from a patch on the back, and an ice pack is applied to the area.   (C) The new fur grown under the artificially low temperatures is black. Himalayan rabbits are normally homozygous for the gene that controls synthesis of the black pigment, but the gene is active only at low temperatures (below about 33°C).   [Modified from A. M. Winchester, *Genetics*, Houghton Mifflin, 1972.]

themselves, because anyone carrying the dominant gene ordinarily shows its phenotype. (By contrast, a recessive character often appears in persons neither of whose parents showed the character, because they were heterozygous.) The pedigree in Fig. 14.9 makes it appear, then, that syndactyly in this family is inherited as a simple dominant, and that the abnormality in the one man both of whose parents had normal fingers was probably due to incomplete penetrance of the gene for syndactyly; the man's mother must have been carrying the gene, even though she did not exhibit its phenotype.

Both the examples cited make it clear that penetrance and expressivity are aspects of the same phenomenon. Lack of penetrance of the gene for blue sclera produces white color, which is simply one extreme of the expressivity gradient from white through very pale blue to dark blue. Lack of penetrance of the gene for syndactyly is simply one extreme of the expressivity gradient whose other extreme is full webbing of three fingers.

As we have indicated, incomplete penetrance and variable expressivity may be due to peculiarities of the genetic background against which the gene in question must act. This explanation points up a general truth: *The action of any gene can be fully understood only in terms of the overall genetic makeup of the individual organism in which it occurs.*

Penetrance and expressivity may also be affected by environmental influences. For example, *Drosophila* homozygous for the gene for vestigial wings have wings that are only tiny stumps if they are reared at normal room temperatures (about 20°C), but if they are reared at temperatures as high as 31°C, their wings grow almost as long as normal wings. Himalayan rabbits are normally white with black ears, nose, feet, and tail (Fig. 14.11), but if the fur on a patch on the back is plucked and an ice pack is kept on the patch, the new fur that grows there will be black; the gene for black color can express itself only if the temperature is low, which it normally is only at the body extremities. A human being with genes for great height will not grow tall if raised on a starvation diet. Numerous other examples could be cited.

We see, then, that the expression of a gene depends both on the other genes present (the genetic environment) and on the physical environment (temperature, sunlight, humidity, diet, etc.). We don't inherit characters. We inherit only genes, only potentialities; other factors govern whether or not the potentialities are realized. All organisms are products of both their inheritance and their environment.

THE EVALUATION OF EXPERIMENTAL RESULTS

The meaning of predicted ratios   We have mentioned ratios such as 1 : 2 : 1 or 3 : 1 or 1 : 1 or 9 : 3 : 3 : 1, expected in the results of various types of crosses. We should now examine more closely what a prediction of this kind means. Does it mean that in a cross such as $C/c \times c/c$ the expected ratio is exactly 1 : 1, that exactly equal numbers of red-flowering and white-flowering pea plants are expected to be produced? No, it simply means that the chances that any given plant produced by this cross will have red flowers as opposed to white flowers are fifty-fifty, and that if large numbers of offspring are produced by this cross the ratio of red-flowering to white-flowering plants is expected to approximate 1 : 1. The same goes for the sex of human children. The chances that any particular new baby will be a boy instead of a girl are roughly fifty-fifty, and we expect that in any large number of young children, such as all those born in a fair-sized city this year, the ratio of boys to girls will closely approximate 1 : 1. But we are not particularly surprised if any given family has four sons and one daughter or six daughters and one son. In short, in any small sample large departures from the predicted results are not unusual or surprising.

Let us examine the basis for the prediction of a 1 : 1 ratio in the progeny of the $C/c \times c/c$ cross. All gametes produced by the homozygous recessive parent in this cross will be alike with respect to the gene for flower color; all will bear the c allele. It is therefore the heterozygous parent in this cross that determines which of the two possible genotypes (C/c or c/c) the offspring will have.

The diploid cells of the heterozygous parent have one C allele and one c allele in their nuclei. When these cells undergo meiosis, half the new haploid cells receive the chromosome bearing the C allele and half receive its homologous chromosome bearing the c allele. Therefore half the gametes produced by this plant carry the C allele, and half carry the c allele. Which of these two types of gametes will be involved in any given fertilization is, apparently, purely a matter of chance. Consequently the probability that any given gamete from the homozygous recessive parent in this cross will be fertilized by a gamete carrying the C allele and produce a red-flowering offspring is 0.5 (which may also be expressed as a percentage or as a fraction: 50 percent or $\frac{1}{2}$). The probability for fertilization by a gamete carrying the c allele, which would result in a white-flowering offspring, is likewise 0.5. If 100 fertilizations occur, we would expect about 50 red-flowering offspring (0.5×100) and about 50 white-flowering offspring. This is the basis for the predicted 1 : 1 ratio.

The whole matter hinges on the segregation in meiosis of the C and c alleles of the heterozygous parent and the incorporation of these alleles into gametes that have an equal probability of fertilizing the gametes of the other parent. But since the type of gamete involved in each individual fertilization is purely a matter of chance, it would actually be surprising to get an exact fifty-fifty split in a sample of 100, even though that is the prediction.

The need for statistical analysis   Now suppose we actually made a cross in which the predicted phenotypic results are 1 : 1. Suppose, further, that our actual results are 45 of one phenotype and 55 of the other. A ratio of 45 : 55 is fairly close to 1 : 1, and we might well conclude that our actual results are close enough to the predicted results to justify our assuming that the deviation is due to chance alone. But suppose in another similar cross we get actual results of 5 and 15. This is noticeably further from the predicted 1 : 1 ratio. Should we conclude that here, too, the deviation can reasonably be attributed to chance, or should we conclude that there is some genetic explanation for the deviation, that our original prediction should have been different?

Scientists in all fields of research constantly encounter the same fundamental question—whether the deviations they observe in their experimental results are significant or not. They cannot rely simply on a guess. They cannot just say, "That looks pretty close to what I predicted," or "That looks rather far off, perhaps I overlooked something." To help them arrive at a decision, they can refer to generally agreed upon standards based on the mathematical probability that the observed deviations in their sample could occur by chance alone. This type of mathematical treatment of data is known as statistical analysis.

One point is immediately apparent: In determining statistical significance, the size of the sample is of critical importance. A large deviation from the predicted results is not unusual in small samples, but the same percentage of deviation in a large sample may be very surprising. We said earlier that a family with four sons and one daughter would not be considered unusual, even though the predicted sex ratio is approximately 1 : 1. But if out of 500 children born in a city in a particular year, 400 were boys and only 100 were girls, we would strongly suspect something amiss. When the predicted ratio is 1 : 1, an observed ratio of 4 : 1 is not alarming in a sample of only 5, but in a sample of 500 it is cause for re-examination. Or, to take another example, if we toss a coin 10 times, we can logically predict 5 heads and 5 tails, but we are not greatly surprised if in fact we get 7 heads and 3 tails (a ratio of 7 : 3 instead of 1 : 1). But if we toss the coin 100 times and get 70 heads and only 30 tails (again a 7 : 3 ratio), we suspect that something is wrong either with the coin or with the way it is being tossed. In other words, a ratio of 7 : 3 seems reasonable enough when the sample is small, but unreasonable when the sample is large. Clearly, then, tests of significance must take into account both the amount of deviation and the sample size.

Statisticians have devised many mathematical tests for evaluating experimental or observational data. Though these tests differ in their form and in the sorts of data to which they can validly be applied, all are simply ways of calculating the probability that the deviations of the observed values are due to chance alone. According to the convention usually followed, deviations for which the probability that they are due to chance is higher than 0.05 (one in 20) are regarded as not statistically significant; they are presumed to be chance deviations that can be disregarded, and the observed results are held to fit the predictions reasonably well. If the probability that the deviations are due only to chance is equal to or less than one in 20, then the deviations are considered statistically significant; i.e. it is concluded that the results do not fit the predictions and that some modification of the predictions may be necessary.

MULTIPLE ALLELES

From our discussion so far, you may have gotten the impression that genes can have only two allelic forms, which may exhibit either a dominant-recessive or an intermediate relationship to each other. This is not so. Genes may exist in any number of allelic forms. Of course, under normal circumstances, the maximum number of alleles for each gene that any individual can possess is two, because he has only two doses of each gene. But many other alleles may be present in the population to which he belongs.

Eye color in *Drosophila* The first example of multiple alleles was discovered in the little fruit fly, *Drosophila melanogaster*—from the genetic point of view, one of the most extensively studied species in existence. These flies normally have red eyes. But individuals with white, eosin, wine-colored, apricot, ivory, or cherry eyes also occur. It has been shown that each of these eye colors is controlled by a different allele of the eye-color gene. About two dozen alleles for eye color have been discovered, and others may well exist. The allele for the wild-type eye (red) is dominant over all the rest, but when any two of the others occur together in a heterozygous fly they produce an intermediate phenotype.

Human A-B-O blood types   A well-known example of multiple alleles in human beings—and a relatively simple one, since only a few alleles are involved—is that of the A-B-O blood series, in which four blood types are generally recognized: A, B, AB, and O. The erythrocytes in type A blood bear antigen A; in type B, antigen B; in type O, neither A nor B; in type AB, both A and B.

These two kinds of cellular antigen, A and B, react with certain antibodies, anti-A and anti-B, that may be present in the blood plasma. If red cells bearing a particular antigen and plasma containing the corresponding antibody come to be mixed in a person's blood, the cells agglutinate (clump) and he may die. Normally, a person with antigen A on his red cells does *not* have anti-A in his plasma, but he does have anti-B; a person with antigen B on his cells has only anti-A in his plasma; a person lacking both antigens A and B (type O blood) has both anti-A and anti-B; and a person with both antigens A and B (type AB blood) lacks both types of antibodies (see Table 14.2). Normally, in short, a person's plasma contains antibodies corresponding to any antigens his own cells do not bear. This is one of the few cases where the body normally synthesizes antibody against an antigen to which it has not actually been exposed.

The presence of these antigens and antibodies in the blood has important implications for blood transfusions. Because the antibodies present in the plasma of blood of one type will tend to react with the antigens on the erythrocytes of other blood types and cause clumping, it is always best, when transfusions are to be given, to obtain a donor who has the same blood type as the patient. It has been found, however, that when such a donor is not available blood of another type may be used, provided that the *plasma of the patient* and the *erythrocytes of the donor* are compatible; in other words, one can usually ignore the erythrocytes of the patient and the plasma of the donor. The reason is that, unless the transfusion is to be a massive one or is to be made very rapidly, the donor's plasma is sufficiently diluted during transfusion so that little or no agglutination occurs. This means that type O blood can be given to anyone because its erythrocytes have no antigens and thus are obviously compatible with the plasma of any patient; type O blood is sometimes called the universal donor. But type O patients can receive transfusions only from type O donors, because their plasma contains both anti-A and anti-B and thus is obviously not compatible with the erythrocytes of any other class of donor. Conversely, people with type AB blood, whose plasma contains no antibodies, are universal recipients, but cannot act as donors for any except type-AB patients. Table 14.3 summarizes these transfusion relationships.

At first glance, you might suppose that two independent genes are involved in the A-B-O system, one determining whether the A antigen is present and another whether the B antigen is present. But this is not the case. It

TABLE 14.2   *Antigen and antibody content of the blood types of the A-B-O series*

Blood type	Blood contains	
	Cellular antigens	Plasma antibodies
O	None	anti-A and anti-B
A	A	anti-B
B	B	anti-A
AB	A and B	None

TABLE 14.3   *Transfusion relationships of the A-B-O blood groups*

Blood group	Can act as donor to	Can receive blood from
O	O, A, B, AB	O
A	A, AB	O, A
B	B, AB	O, B
AB	AB	O, A, B, AB

TABLE 14.4 *Genotypes of the A-B-O blood types*

Blood type	Genotype
O	i/i
A	I^A/I^A or I^A/i
B	I^B/I^B or I^B/i
AB	I^A/I^B

TABLE 14.5 *Frequencies of A-B-O blood groups in selected populations*

Population	O	A	B	AB
United States whites	45%	41%	10%	4%
United States blacks	47	28	20	5
African Pygmies	31	30	29	10
African Bushmen	56	34	8	2
Australian aborigines	34	66	0	0
Pure Peruvian Indians	100	0	0	0
Tuamotuans of Polynesia	48	52	0	0

has been shown that the inheritance of the A-B-O groups is best described by a theory of three alleles,[3] here designated I^A, I^B, and i. Both I^A and I^B are dominant over i, but neither I^A nor I^B is dominant over the other. Accordingly the four blood-type phenotypes correspond to the genotypes indicated in Table 14.4.

Blood typing is often used as a source of evidence in paternity cases in court. For example, a man with type O blood could not possibly be the father of a child with type A blood whose mother is type B. The child's true father must be either type A or type AB, because the child must have received its I^A allele from its father; an O man has no such allele. Similarly, a man with type AB blood could not possibly be the father of a type O child, because the child must have received an i allele from its father, but an AB man has no such allele. Note that this test of possible paternity is strictly a negative one. It is sometimes possible to say definitely that a particular man is not the father of a particular child, but it is never possible on this basis to say that a particular man is the father of a particular child. If a man with type A blood is accused of being the father of a type A child whose mother is type B, blood typing does not help settle the question. The child's father is either type A or type AB, but since there are millions of type A and type AB men in the world the man's type A blood does not constitute evidence that he is the father.

As indicated in Table 14.5, the frequencies of the various A-B-O blood types vary in populations of different ancestral extraction. Anthropologists have found data on the frequencies of blood types useful in tracing the prehistoric movements and derivations of the various subgroups of the human species. Notice that in many human populations the most frequent phenotype is type O, which corresponds to the homozygous recessive genotype, and that in these populations, accordingly, the i allele is more common than the I^A or I^B alleles. Here we have a good illustration of an important fact: Whether an allele is dominant or recessive does not determine whether it will be common or rare in the population. Many beginning students have the mistaken impression that dominant alleles are the common ones and recessive alleles the rarer ones. "Dominant" and "recessive" describe the way the alleles interact when they occur together in a heterozygous individual; they do not indicate which allele determines the more advantageous phenotype. Natural selection tends to increase the frequency of the allele that determines the more adaptive phenotype, whether that allele is dominant or recessive, and it tends to decrease the frequency of the allele that determines the less adaptive phenotype; it is the relative adaptiveness of the phenotype that determines which allele is the more common.

Human Rh blood factors Another series of blood antigens is designated Rh (which stands for rhesus monkey, the animal in which the antigens were first discovered). The Rh series includes at least nine antigens, and some authorities believe that the inheritance of these antigens is best explained in terms of multiple alleles; if this view is correct, the Rh alleles would constitute one of the largest series of multiple alleles known in human beings (much larger series are known in some other species). Other authorities prefer to explain

---

[3] Actually, four alleles are now known. What we here designate as the I^A allele is really two different but very similar alleles, I^{A1} and I^{A2}. This means that there are actually six blood types (O, A_1, A_2, B, A_1B, A_2B), rather than four. For our purposes, however, this complication can be ignored.

Rh inheritance in terms of three or more separate genes that lie close together on the same chromosome and have very similar functions. Which, if either, of these two views is correct bears on the question of the nature of the gene and may, in some instances, also have clinical implications, but for a basic understanding of the Rh blood groups we can use a far simpler system.

Two phenotypic classes may be recognized: Rh-positive and Rh-negative. Rh-positive individuals have the Rh antigen on their erythrocytes, while Rh-negative individuals do not (more precisely, their antigen is so weak that it can be disregarded). About 85 percent of the Caucasian population in the United States is Rh-positive, and 15 percent is Rh-negative. Rh-negative is much rarer in people of Mongoloid or Negroid extraction.

For convenience let us assume that Rh-positive is the phenotype produced by a dominant gene *Rh*, and that Rh-negative is the phenotype produced by its recessive allele *rh*.[4] This assumption allows us to treat the genetics of Rh blood types like a simple monohybrid cross, even though we know that the Rh-positive phenotype can be produced by any one of a number of slightly different alleles that produce slightly different antigens. On this assumption, the possible genotypes and their phenotypes are:

Phenotypes	*Genotypes*
Rh+	*Rh/Rh* or *Rh/rh*
Rh−	*rh/rh*

Whereas a person's blood plasma may contain anti-A or anti-B antibody without stimulation by the corresponding antigen, it will not contain any antibody against the Rh antigen unless the body is sensitized to the antigen by exposure to it. Thus the blood of a normal unsensitized Rh-negative person contains neither Rh antigen on its cells nor Rh antibody in its plasma. But if an Rh negative person is mistakenly given a transfusion of Rh-positive blood, this exposure to the antigen will sensitize him and stimulate his plasma cells to begin synthesizing antibodies. If the same patient is later given a second transfusion of Rh-positive blood, the antibodies now present in his plasma will react with the antigens in the donated blood, and the patient may die. It is important, therefore, that patients and donors be carefully typed for both the A-B-O blood group and the Rh factor.

The Rh factor also has medical importance in certain cases of pregnancy. If an Rh-negative woman (genotype *rh/rh*) marries an Rh-positive man (genotype *Rh/Rh* or *Rh/rh*), some of their children may be heterozygous and have the Rh-positive phenotype. Often this causes no trouble; since there is no direct connection between the circulatory systems of the mother and the fetus in the normal placenta, there can be no mixing of blood. But in the late stages of some pregnancies and during the birth process, some seepage of blood between the two circulatory systems may occur. If the blood of an Rh-positive fetus seeps into the circulatory system of an Rh-negative mother, the effect is the same as though she had been given a transfusion of Rh-positive blood. Her plasma cells are stimulated to begin synthesizing Rh antibody. Ordinarily there is no immediate harm, because the baby is born before the mother's blood contains appreciable quantities

[4] Notice that we are here using a two-letter symbol for a single gene. Thus *Rh* is the symbol for a single dominant allele, *rh* the symbol for a single recessive allele. Do not make the mistake of thinking that the *r* and the *h* stand for different things.

The inheritance of Rh blood groups provides an especially good example of the complexity often found when a system is studied in detail. We discuss it here lest you assume that the inheritance of all characteristics is as simple as in the examples cited in this chapter.

Soon after Karl Landsteiner and Alexander Wiener discovered the Rh factor in 1940, it became clear to them that a simple grouping of people as Rh-positive and Rh-negative was not fully satisfactory. More than one Rh antigen was plainly involved: While some persons classified as Rh-positive had only one Rh-positive antigen, others had two. As more and more antigens were discovered over the years, the genetics of the Rh factor grew increasingly complicated. By way of explanation some authorities hypothesized multiple alleles of a single gene, others at least three different genes located very close together on the same chromosome (linked genes).

The table, which compares the two approaches as usually formulated, is now itself known to be an oversimplification (though a useful one), because many additional minor variants have been found (e.g. C^w and D^u alleles are encountered fairly often, the latter especially among African blacks). Thus a 1968 revision of the single-gene system (the Wiener system) lists 37 different alleles as necessary to account for the phenotypes then known. And one proposed system describes the CDe combination as

$$R^{1,2,-3,-4,5,-6,7,-8,-9,-10,-11,12,13,14,15,16,17,18,19,-20,-21}$$

It is easy to see why this system has not come into general use!

Most instances of Rh disease seem to arise when an Rh-negative woman—cde/cde according to the simplified three-gene system, r/r according to the single-gene system—bears an Rh-positive child, whose genotype may be designated CDe/cde, cDE/cde, cDe/cde, or CDE/cde (or, alternatively, R^1/r, R^2/r, $R^°/r$, or R^z/r). According to the three-gene system, D is decisive as the most antigenic factor; factors C and E are much less antigenic, though on some occasions they too may induce formation of maternal antibodies.

It must be emphasized that the gene frequencies shown in the table refer only to Caucasians. They differ markedly in persons of other racial backgrounds. For example, in Orientals the most common alleles are R^1 (nearly 0.60) and R^2 (nearly 0.40); r, the second most common allele in Caucasians, has a frequency of only 0.01. By contrast, allele $R^°$ is most frequent (about 0.50) in American blacks, followed by r (about 0.25), R^2 (about 0.10), and R^1 (about 0.05).

Designations of the genes controlling Rh factor

Original system	Single-gene system (after Wiener)	Three-gene system (Fisher-Race)	Frequency in U.S. Caucasians
Rh	R^1	CDe	0.43
	R^2	cDE	0.15
	$R^°$	cDe	0.027
	R^z	CDE	Rare
rh	r	cde	0.38
	r'	Cde	0.006
	r''	cdE	0.005
	r^y	CdE	Rare

of antibody. But if this sensitized mother later bears a second or third Rh-positive fetus, and if seepage again develops, antibodies from the maternal blood may enter the fetal circulation and react with the fetal cells, causing a disease known as erythroblastosis fetalis (or simply Rh disease), which, unless treated promptly, is often fatal to the baby.

It must be emphasized that erythroblastosis fetalis does not occur in all second or later pregnancies of Rh-negative women married to Rh-positive men. First, the husband may be heterozygous, in which case the chances are fifty-fifty that any given child will have Rh-negative blood like the mother and cause no trouble. Second, in the absence of seepage there will be no trouble; there are cases on record of Rh-negative women who have borne more than ten healthy Rh-positive children. Third, only one of the numerous forms of the Rh antigen (determined by an allele designated as $Rh°$) is usually involved in erythroblastosis fetalis, and other forms will cause no trouble.

In recent years a method of immunizing Rh-negative mothers so that they will not form Rh antibodies has been developed. If an Rh-negative woman bears an Rh-positive child, she is injected immediately after delivery with anti-Rh, which kills any fetal cells that may have gotten into her circulation before those cells can stimulate production of antibodies. The injected anti-Rh soon disappears, and the woman is left in the unsensitized condition, able to bear a second Rh-positive child without danger. Even if a fetus should develop erythroblastosis fetalis, physicians have learned that if the newborn infant is promptly given a massive transfusion that replaces all its blood (contaminated with antibodies from the mother) with blood free of antibodies, the chances are good that it will live and develop normally.

MUTATIONS AND DELETERIOUS GENES

A variety of influences can cause slight changes in the chemical structure of a gene. Such changes are called mutations. The rate at which any particular gene undergoes mutation is ordinarily extremely low. But every individual organism has a very large number of different genes, and the total number of genes in all the individuals of a species is vast indeed. Hence mutations are constantly occurring within a species, pure chance determining in which individual any given mutation will occur. Now, every living organism is the product of billions of years of evolution and is a finely tuned, smoothly running, astoundingly intricate mechanism, in which the function of every part in some way influences the function of every other part. By comparison, the best Swiss watch is simple indeed. If you were to take such a watch, remove its back, and make some random change in its parts, the chances are very great that you would make it run worse rather than better. A random change in any delicate and intricate mechanism is far more likely to damage it than to improve it. Mutational changes in genes being random, it is easy to understand why the vast majority of new mutations are deleterious. Only very rarely is a new mutation beneficial.

Heterozygous vs. homozygous effects   When a new deleterious allele arises by mutation, natural selection can act against it only if it causes some change in the organism's characteristics. Selection acts directly on pheno-

14.12   A creeper hen
Her legs are very short and she cannot walk normally.
Such a hen is heterozygous for a gene that is lethal
when homozygous.   [From a photograph by C. D.
Mueller in A. M. Winchester, *Genetics*, Houghton
Mifflin, 1972.]

types and only indirectly on genotypes. Because dominant deleterious mutations will be expressed phenotypically, they can be eliminated from the population rapidly by natural selection. But many new mutations are recessive to the normal alleles. And since the probability that the same mutation will occur twice in the same individual is vanishingly slight, most new alleles occur in combination with the normal allele in diploid cells; i.e. the diploid cell is heterozygous, containing one normal allele and one new allele produced by mutation. If a new allele is recessive, and if it occurs in heterozygous condition, it can have little immediate phenotypic effect on the organisms that possess it; its deleterious effects cannot be fully expressed. Natural selection therefore cannot eliminate it from the population very rapidly. Deleterious alleles that are not dominant may be retained in the population in heterozygous condition for a long time.

When two individuals both carrying the same deleterious recessive allele in heterozygous condition mate, about one fourth of their progeny will be homozygous for the deleterious allele, and these homozygous offspring will have the harmful phenotype. The phenotype may even kill the organism. An allele whose phenotype, when expressed, results in the death of the organism is called a ***lethal***. The occurrence of lethals can modify the phenotypic ratios obtained in the progeny of some crosses, as the following example shows:

In chickens one allele of a certain gene, when it occurs in heterozygous condition with the normal allele, causes the chicken to be a "creeper," with short crooked legs (Fig. 14.12). When two creeper chickens are crossed, their offspring are of two phenotypes, normal and creeper, in a ratio of approximately 1 : 2. Now, this is a ratio different from any we have previously encountered. It occurs because about one fourth of the incubated eggs fail to hatch, the embryos dying early in their development. If these dead embryos are regarded as a third phenotypic class, then the cross can be said to have produced a phenotypic ratio of 1 : 2 : 1, the typical one for the F_2 generation of a monohybrid cross where dominance is lacking. The ratio of 1 : 2 seen in the live chicks is the result of the lethality of the creeper allele when it occurs in homozygous condition.

In numerous instances alleles harmful or even lethal when homozygous are actually beneficial when heterozygous. For example, in England there is a breed of cattle called Dexter, a good beef producer, for which it is impossible to establish a pure-breeding herd because some of its most desirable characteristics are caused by the heterozygous expression of a gene that is lethal when homozygous.

An example in human beings is the gene for ***sickle-cell anemia*** in Africa. When homozygous, this gene results in a serious abnormality of the red blood cells, which are curved like a sickle and bear long filamentous processes (Fig. 14.13). These abnormal cells tend to form clumps and to clog the smaller blood vessels. The resulting impairment of the circulation leads to severe pains in the abdomen, back, head, and extremities, and to enlargement of the heart and atrophy of brain cells. In addition, the tendency of the deformed red blood cells to rupture easily brings about severe anemia. As might be expected, victims of sickle-cell anemia usually suffer an early death. Individuals heterozygous for the sickle-cell gene sometimes show mild symptoms of the disease, but the condition is not serious.

It might be supposed that natural selection would operate against the propagation of any gene so obviously harmful and that such a gene would be held at very low frequency in the population. This seems to be true among blacks in the United States. But the gene is surprisingly common in many parts of Africa, being carried by as much as 20 percent of the black population. What is the explanation? A. C. Allison of Oxford, England, has found that individuals heterozygous for this gene have a much higher than normal resistance to malaria. Since malaria is very common in many parts of Africa, the gene must be regarded as beneficial when heterozygous. Thus, in Africa, there is selection for the gene because of its heterozygous effect on malarial resistance and selection against it because of its homozygous production of sickle-cell anemia. The balance between these two opposing selection pressures determines the frequency of the gene in the population.

The gene for sickle-cell anemia is a dramatic example of a gene that has more than one effect. Such a gene is said to be **_pleiotropic._** Pleiotropy is, in fact, the rule rather than the exception. All genes probably have many effects on the organism. Even when a gene produces only one perceptible phenotypic effect, it doubtless has numerous physiological effects more difficult to detect.

The effect of inbreeding   The conditions under which genes cause deleterious phenotypes explain the danger of matings between closely related

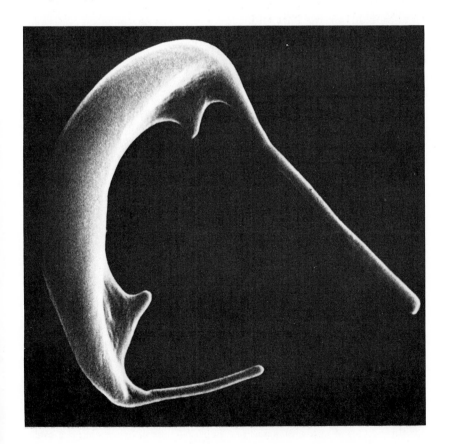

14.13   Scanning electron micrograph of a sickled erythrocyte
The sickled red blood cell looks dramatically different from the normal disc-shaped ones (see Fig. 7.35, p. 300).   [Courtesy M. C. Bessis, Institute of Cellular Pathology, Kremlin Bicêtre, France.]

human beings. Everyone probably carries in heterozygous combination many genes that would cause harmful effects if present in homozygous combination, including some lethals. But because most of these deleterious genes originated as rare mutations, and are limited to a tiny percentage of the population, the chances are slight that two unrelated persons will be carrying the same deleterious recessive genes and produce homozygous offspring that show the harmful phenotype. The chances are much greater that two closely related persons will be carrying the same harmful recessives, having received them from common ancestors, and that, if they mate, they will have children homozygous for the deleterious traits. In short, close inbreeding increases the percentage of homozygosity, as Fig. 14.14 shows. You can see from the graph that brother-sister matings and matings between double first cousins cause rapid increases in homozygosity, and that matings between first cousins cause moderate increases.

Diploidy vs. haploidy Before leaving this discussion of mutation, harmful genes, and the implications of homozygosity vs. heterozygosity, we might speculate for a moment on the adaptive significance of the fact that the diploid stages of the life cycle have such marked dominance over the haploid stages in most higher plants and animals. Why should natural selection have favored diploidy rather than haploidy? One possible reason is that new harmful mutations immediately exert their effect in a haploid organism; they cannot be masked by a dominant allele. But in diploid species an organism can survive a harmful mutation if the new allele is recessive. Moreover, the mutant gene can be carried for generations in the population, perhaps exerting a beneficial effect when heterozygous. Such a gene may also serve as a latent source of variation; the day may come when the environment or the genetic makeup of the organism has changed so much that the gene is no longer deleterious even when homozygous.

SEX AND INHERITANCE

SEX DETERMINATION

The sex chromosomes We have said repeatedly that a diploid individual has two of each type of chromosome, identical in size and shape, and hence two doses of each gene. But we must now qualify that statement somewhat. In most higher organisms where the sexes are separate (i.e. where males and females are separate individuals), the chromosomal endowments of males and females are different, and one or the other of the two sexes has one chromosomal pair consisting of two chromosomes that differ markedly from each other in size and shape. These are the *sex chromosomes,* which play a fundamental role in determining the sex of the individual. All other chromosomes are called *autosomes.*

Let us look first at the chromosomes of *Drosophila* and human beings. In each case the sex chromosomes are of two sorts: one bearing many genes, conventionally designated the *X chromosome,* and one of a different shape and bearing only a few genes, designated the *Y chromosome.* Females characteristically have two X chromosomes, and males have one X and one Y. The diploid number in *Drosophila* is eight (four pairs); a female therefore has three pairs of autosomes and one pair of X chromosomes, and a male

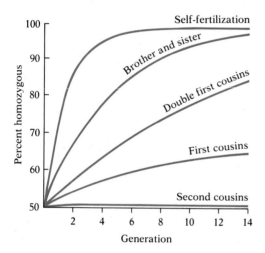

14.14 Graph showing percentage of homozygotes in successive generations under different inbreeding systems

It is assumed in this graph that the initial condition is two alleles of equal frequency (hence the 50 percent homozygosity at the start). If a similar graph were drawn for very rare recessive alleles, the rises in homozygosity would be much steeper, particularly in the three curves for cousin-to-cousin matings. Thus matings between first or second cousins, though of negligible effect in the situation graphed here, may greatly increase the percentage of homozygosity of rare, perhaps deleterious recessive alleles. [Modified from Sewall Wright, *Genetics,* vol. 6, 1921.]

has three pairs of autosomes and a pair of sex chromosomes consisting of one X and one Y (Fig. 14.15). The diploid number in human beings is 46 (23 pairs); a female therefore has 22 pairs of autosomes and one pair of X chromosomes, and a male has 22 pairs of autosomes plus one X and one Y (Fig. 13.5, p. 561).

When a female produces egg cells by meiosis, all the eggs receive one of each type of autosome plus one X chromosome. When a male produces sperm cells by meiosis, half the sperm cells receive one of each type of autosome plus one X chromosome and half receive one of each autosome plus one Y chromosome. In short, all the egg cells are alike in chromosomal content, but the sperm cells are of two types occurring in equal numbers (Fig. 14.15). When fertilization takes place, the chances are approximately equal that the egg will be fertilized by a sperm carrying an X chromosome or by a sperm carrying a Y chromosome. If fertilization is by an X-bearing sperm, the resulting zygote will be XX and will develop into a female. If fertilization is by a Y-bearing sperm, the resulting zygote will be XY and will develop into a male.[5] We see, therefore, that the sex of an individual is normally determined at the moment of fertilization and depends on which of the two types of sperm fertilizes the egg.

Since meiosis in the male must produce exactly equal numbers of X sperm and Y sperm, we would predict that in a large population equal numbers of boys and girls would be conceived. But many studies have shown that in the United States about 114 boys are conceived for every 100 girls.[6] Why this departure from the predicted results? Apparently X sperm and Y sperm differ in their physical attributes; e.g. Y sperm, because they are smaller than X sperm, may be lighter and may swim faster. Since the sex ratio has been shown to vary from country to country and even from family to family, many factors, including genetic background and the age and health of the parents, doubtless affect the characteristics of the two types of sperm. These factors must produce differences in their ability to survive the conditions in the female genital tract, which alter with age and changing health, or in their rate of movement from the vagina to the point of conception in the oviduct, or in their capacity to penetrate the outer surface of the egg. Recently techniques have been devised that may enable animal breeders to make a partial separation of the two types of sperm in bull semen. The day will almost certainly come when the two types of sperm in human semen can be separated in the laboratory, thus making it possible, through artificial insemination, to select the sex of a child.

The role of the Y chromosome in sex determination   Occasionally the members of a homologous pair of chromosomes fail to separate properly in meiosis and both move to the same pole. The effect of such nondisjunction,

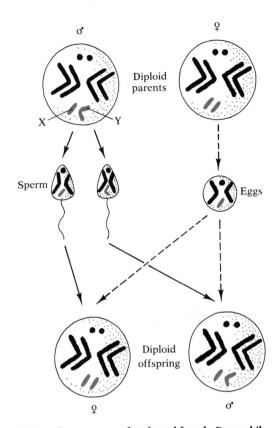

14.15 Chromosomes of male and female *Drosophila melanogaster*
There are three pairs of autosomes and one pair of sex chromosomes. Males (symbolized by ♂) have one X chromosome and one Y chromosome; since these separate at meiosis, half the sperm carry an X and half carry a Y. Since females (♀) have two X chromosomes, all eggs have one X. The sex of the offspring depends on which type of sperm fertilizes the egg.

[5] The XY system, where XX is female and XY is male, is characteristic of many animals, including all mammals. It is also found in many plants with separate sexes. Birds, butterflies and moths, and a few other animals have just the opposite system, where XX is male and XY is female (to distinguish this system from the usual XY system, the symbols Z and W are often substituted—ZZ being male and ZW being female). A completely different mechanism of sex determination exists in the Hymenoptera (bees, wasps, ants, etc.), where the males hatch from unfertilized eggs and are haploid while the females hatch from fertilized eggs and are diploid.

[6] Since more males than females die during embryonic development, the ratio at birth is about 106 : 100, and a higher rate of mortality of males after birth brings the numbers of the two sexes into balance by about the age of 10; thereafter females outnumber males in any age group.

as it is called, is that one of the daughter cells receives one too many chromosomes and the other daughter cell receives one too few. If a human gamete carrying an extra chromosome is involved in fertilization, the zygote produced has 47 chromosomes (the normal two of most types plus three of one type) instead of the normal 46. Sometimes it is the sex chromosomes that fail to separate in meiosis, and XXY individuals may result. Since XXY individuals in *Drosophila* are essentially normal females, it was suggested that only the X chromosomes function in sex determination—one X producing a male and two a female, regardless of the presence or absence of a Y chromosome.

This idea seemed supported by the finding that some animals, such as grasshoppers, lack the Y chromosome entirely; the females have two X chromosomes and the males a single unpaired X, the female thus having one more chromosome than the male (in some species of grasshoppers the total numbers are 22 and 21 respectively). Such a system of sex determination is known as the XO system, where O designates the absence of a chromosome.

For many years it was thought that the mechanism of sex determination in humans is the same as in *Drosophila*, but we now know that this is not the case. Apparently the human Y chromosome bears genes with strong male-determining properties, and it is the presence of the Y that determines maleness and its absence that determines femaleness. Thus, whereas an XXY *Drosophila* is a female, an XXY human is a male.

SEX-LINKED CHARACTERS

Many genes occur on the X chromosome and not on the Y chromosome. Such genes are said to be sex-linked. The inheritance patterns for the characteristics controlled by sex-linked genes are completely different from those for characteristics controlled by autosomal genes, for obvious reasons. The females have two doses of each sex-linked gene, one from each parent, but the males have only one dose of each sex-linked gene, and that one dose always comes from the mother, since the father contributes a Y chromosome instead of an X to his sons. Hence, in the male, all sex-linked characteristics are inherited from the mother only. And since the male has only one dose of each sex-linked gene, recessive genes cannot be masked, and recessive sex-linked phenotypes occur much more often in males than in females.

Sex linkage was discovered in 1910 by the great American geneticist Thomas Hunt Morgan of Columbia University. It was Morgan who began the systematic use of *Drosophila* in genetic studies. This little fruit fly made it possible to perform in a few months experiments that it would have taken Mendel years to perform on peas. *Drosophila* can be easily and economically cultured in large numbers in the laboratory. Subject to a remarkable number of easily detectable genetic variations, they can produce a new generation every 10 or 12 days. Much of the modern knowledge of genetics derives from work on this tiny insect.

Let us examine the first sex-linked trait discovered by Morgan: white eye color in *Drosophila*. This trait is controlled by a recessive allele *r*. The normal red eye color is controlled by a dominant allele *R*. If a homozygous red-eyed female is crossed with a white-eyed male, all the F₁ offspring, regardless of sex, have red eyes, since they receive from their mother an X chromosome bearing an allele for red. In addition, the F_1 females receive from their father an X chromosome bearing an allele for white eyes, but the allele for red, being dominant, masks its presence. The F_1 males, like the females, receive from their mother an X chromosome bearing an allele for red eyes. But unlike the females, they receive no gene for eye color from their father, who contributes a Y chromosome instead of an X (in writing the genotype of a male for a sex-linked character, the Y is customarily shown in order to indicate clearly that no second X chromosome is present and hence that there is no second dose of the sex-linked gene). We can summarize this cross as follows (♀ denotes females, ♂ males):

P R/R × r/Y
                    red-eyed ♀   white-eyed ♂
                                    ↓
F_1 R/r × R/Y
                    red-eyed ♀   red-eyed ♂
                                    ↓
F_2 R/R r/R R/Y r/Y
           red-   red-   red-   white-
           eyed ♀   eyed ♀   eyed ♂   eyed ♂

Notice that when the F_1 flies of this cross are allowed to mate among themselves, the F_2 flies thus produced show the customary 3 : 1 phenotypic ratio of a monohybrid cross where dominance is present. But notice also that this 3 : 1 ratio is rather different from the 3 : 1 ratio obtained in a cross involving autosomal genes. In an autosomal cross there is no correlation of phenotype with sex, but in this cross all F_2 individuals showing the recessive phenotype are males. In other words, an autosomal cross gives a 3 : 1 F_2 ratio for both females and males, but this cross yielded females of a single phenotype and males with a 1 : 1 phenotypic ratio.

Now let us examine the reciprocal cross, where the parental generation consists of homozygous white-eyed females and red-eyed males. We can summarize this cross as follows:

P r/r × R/Y
                    white-eyed ♀   red-eyed ♂
                                    ↓
F_1 r/R × r/Y
                    red-eyed ♀   white-eyed ♂
                                    ↓
F_2 r/r R/r r/Y R/Y
           white-   red-   white-   red-
           eyed ♀   eyed ♀   eyed ♂   eyed ♂

Notice that the phenotypic makeup of both the F_1 and the F_2 generations differs both from a normal autosomal cross and from the reciprocal cross for this same sex-linked trait. In the F_1 all females show the dominant phenotype, all males the recessive phenotype. In the F_2, instead of a 3 : 1 ratio, there is a 1 : 1 ratio in each sex. Comparison of the two reciprocal crosses makes it clear that when a sex-linked trait is involved in a cross the results depend on which parent shows the trait (or carries the gene for the trait). By contrast, in crosses involving autosomal genes it does not matter

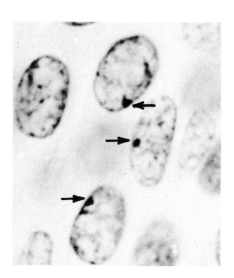

14.16   Nuclei from epidermal cells of a human female
The arrows indicate the Barr bodies. × 2,600.
[Courtesy M. L. Barr, *Canadian Cancer Conference*, vol. 2, 1957 (Academic Press, New York).]

which parent possesses the gene in question; the results of autosomal reciprocal crosses are identical.

Two well-known examples of recessive sex-linked traits in human beings are red-green color blindness and hemophilia ("bleeder's disease"). Color blindness occurs in about 8 percent of white men in the United States and in about 4 percent of black men. It occurs in only about one percent of white women and about 0.8 percent of black women. It is expected, of course, that more men than women will show such a trait, because a man needs only one dose of the gene to show the phenotype, and he can inherit this one dose from a heterozygous mother who is not herself color-blind. But for a woman to be color-blind, she must have two doses of the gene (i.e. be homozygous); not only must her father be color-blind, but her mother must be either color-blind or a heterozygous carrier of the gene. Since the gene is not very common in the population, it is not likely that two such people will marry; hence the low number of color-blind women.

The statement that females have two doses of each sex-linked gene whereas males have only one, though technically correct, requires qualification. In most interphase somatic cells of females, one of the X chromosomes condenses into a tiny dark object called a ***Barr body*** (Fig. 14.16). The genes on this condensed X chromosome are inactive. Thus, in a normally functioning female cell, there is only one active dose of each sex-linked gene. (Even in cells with abnormal numbers of sex chromosomes—e.g. XXX or XXXX—only one X chromosome is functional, the others all being condensed into Barr bodies.) Why, then, are sex-linked recessive traits expressed in females only when homozygous? And why, if there is only one active dose of a sex-linked gene in the cells of both males and females, are the patterns of inheritance in the two sexes markedly different?

The explanation js that it is not always the same X chromosome that condenses into a Barr body in the different somatic cells of a given individual. Let us call the two X chromosomes X_1 and X_2. Characteristically, about half the cells in any given female show active X_1 chromosomes, the other half active X_2 chromosomes, with no discernible pattern as to which chromosome, X_1 or X_2, is active in which cell. For example, consider the distribution of alleles of the sex-linked gene for production of the enzyme glucose 6-phosphate dehydrogenase; one of these alleles determines production, the other nonproduction. When the erythrocytes of women heterozygous for these two alleles were examined, roughly half the erythrocytes revealed normal enzyme activity, the other half a complete lack of it. Apparently female cells differ in their effective genetic makeup as far as sex-linked traits are concerned; women are, in a sense, genetic mosaics for sex-linked traits. But, it seems, as long as half the cells are normal in a woman heterozygous for traits such as red-green color blindness or hemophilia, she will not be color-blind or hemophilic.

It has long been known that, because of hormonal imbalances during embryonic development, some human beings exhibit an anatomical and physiological sex different from their genetic sex. For example, an XY individual might appear to be a female. With the discovery of Barr bodies, a simple technique for determining the genetic sex of individuals became available, one of great aid in studying such developmental anomalies. If Barr bodies are visible in the nuclei of cells examined microscopically, the individual from whom the cells came can be identified as genetically female;

if there are no Barr bodies, the individual can be identified as genetically male. This technique has recently been used to determine the sex of fetuses before birth, where such information would be medically helpful; the cells examined are cells sloughed off by the fetus into the amniotic fluid. The technique has other applications. You may have read of Barr body examinations being administered to entrants in athletic events for women.

GENES ON THE Y CHROMOSOME

Genes on the Y chromosome, but not on the X, are termed ***holandric*** (Fig. 14.17). The phenotypic traits they control appear, of course, only in males. There are apparently very few genes on the human Y chromosome; we have mentioned the maleness determiners, which, probably absent in *Drosophila*, are thought to be on the Y chromosome in humans.

In some species a few genes occur on both the X and the Y chromosome; their inheritance patterns are the same as for autosomal genes. No such genes have been conclusively demonstrated in humans.

SEX-INFLUENCED CHARACTERS

Not all genes for characters commonly associated with sex are sex-linked. As we saw above, sex-linked genes may control characters not customarily regarded as "sexual." And many genes that do control "sexual" characters are located on the autosomes. For example, a number of genes that control growth and development of the sexual organs, such as the penis, the vagina, the uterus, or the oviducts, or that control distribution of body hair, size of breasts, pitch of voice, or other secondary sexual characteristics, are autosomal and are present in individuals of both sexes. That their phenotypic expression is different in the two sexes indicates that they are sex-limited, not that they are sex-linked. Apparently the sex hormones influence the activity of the genes, either inhibiting or stimulating them.

LINKAGE

The patterns of inheritance we have discussed so far amply support Mendel's so-called first law, or Law of Segregation—that in each individual the genes occur in pairs,[7] and that in the formation of gametes the members of each pair separate and pass into different gametes, so that each gamete has only one of each type of gene. And all dihybrid and trihybrid crosses we have discussed agree with Mendel's so-called second law, or Law of Independent Segregation, which states that when two or more pairs of genes are involved in a cross the members of one pair segregate independently of the members of all other pairs. Many crosses, however, fail to obey this second law, as numerous experiments have demonstrated. What is the explanation?

The chromosomal basis of linkage   Mendel knew nothing about chromosomes; he derived his principles exclusively from data from crosses, and either he was lucky enough to have worked only with dihybrid crosses that yielded 9 : 3 : 3 : 1 phenotypic ratios in the F_2 generation or, more likely, he

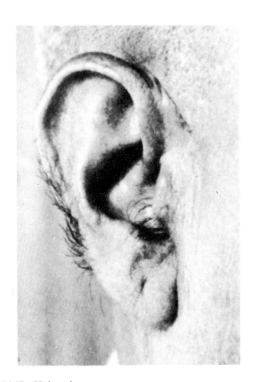

14.17   Hairy pinna
This trait is thought to be determined by a gene on the Y chromosome.   [Courtesy K. R. Dronamraju, in E. J. Gardner, *Principles of genetics*, copyright © 1975 by John Wiley & Sons, Inc. Reprinted with their permission.]

---

[7] To be more precise, we should say that in each *diploid* individual the genes occur in pairs, since the haploid individuals of lower plants obviously have only one dose of each gene.

chose to ignore crosses that did not fit this ratio (or obvious modifications of it). But shortly after the rediscovery of Mendel's paper in 1900, W. S. Sutton of Columbia University pointed out the striking accord between Mendel's conclusion that hereditary factors (genes) occur in pairs in somatic cells, and separate in gametogenesis, and the recent cytological evidence that somatic cells contain two of each kind of chromosome and that these chromosomes segregate in meiosis. Sutton interpreted this agreement as powerful evidence that the chromosomes are the bearers of the genes.

But, as a moment's thought will convince you, the supposition that the genes are located on the chromosomes is incompatible with the unmodified form of Mendel's second law. *Drosophila* have only four pairs of chromosomes, garden peas only seven, humans only 23, and other species similarly limited numbers of pairs. Since each species has thousands of different genes, there must be many different genes on each chromosome. And since it is whole chromosomes that segregate independently in meiosis, only genes located on different chromosomes can segregate independently. Genes located on the same chromosome cannot separate and hence must move together during meiosis; such genes are said to be ***linked.*** Application of Mendel's second law, therefore, must be restricted to pairs of genes located on different chromosomes, i.e. to genes that are not linked.[8]

One of the first examples of linkage was reported in 1906 by William Bateson and R. C. Punnett of Cambridge University. They crossed sweet peas that had purple flowers and long pollen with ones that had red flowers and round pollen. All the F_1 plants had purple flowers and long pollen, as expected (it was already known that purple was dominant over red and that long was dominant over round). The F_2 plants from this cross did not show the expected 9 : 3 : 3 : 1 ratio, however, but a highly anomalous one. Next, Bateson and Punnett tried a test cross, crossing the F_1 plants back to homozygous recessive plants (with red flowers and round pollen). Their results were equally anomalous. Using the symbols *B* for purple, *b* for red, *L* for long, and *l* for round, we can summarize them as follows:

<table>
<tr><td align="center">*BbLl*</td><td align="center">×</td><td align="center">*bbll*</td><td></td></tr>
<tr><td align="center">purple-long</td><td></td><td align="center">red-round</td><td></td></tr>
<tr><td></td><td align="center">↓</td><td></td><td></td></tr>
<tr><td align="center">7 *BbLl*</td><td align="center">1 *Bbll*</td><td align="center">1 *bbLl*</td><td align="center">7 *bbll*</td></tr>
<tr><td align="center">purple-
long</td><td align="center">purple-
round</td><td align="center">red-
long</td><td align="center">red-
round</td></tr>
</table>

You will recall that, in a test cross, the phenotypic ratio of the offspring depends on the genotype of the parent showing the dominant phenotype, since the recessive parent produces only one kind of gamete. In this example the homozygous recessive red-round parent could produce only *bl* gametes.

---

[8] Mendel reported on crosses involving seven characters (see Table 14.1), the genes for each of which segregated independently of all the others. If Mendel had studied any eighth character he would have found (since garden peas have only seven pairs of chromosomes) that its genes segregated *with* those of one of the other seven characters and hence did not bear out his conclusions. It is hard to believe that Mendel had the incredible luck of choosing precisely seven characters all of which were independent. One suspects that he must have studied other characters, but that he ignored them when he wrote his paper.

Hence it was the gametes of the heterozygous purple-long parent that must have determined the phenotype of the offspring. According to the Law of Independent Segregation, this parent should have produced four kinds of gametes *(BL, Bl, bL, and bl)* in equal numbers. When united with the *bl* gametes from the homozygous recessive parents, *BL* gametes should have given rise to purple-long offspring, *Bl* gametes to purple-round, *bL* to red-long, and *bl* to red-round, and these four phenotypes should have occurred in equal numbers, i.e. in a 1 : 1 : 1 : 1 ratio. But the result Bateson and Punnett actually obtained—a ratio of 7 : 1 : 1 : 7—makes it appear that the heterozygous parent produced far more *BL* and *bl* gametes than *Bl* and *bL* gametes.

It was not until 1910 that T. H. Morgan, who had obtained similar results from *Drosophila* crosses, provided the explanation accepted today. He postulated that the anomalous ratios were caused by linkage. Thus we should write the genotypes of the parents in Bateson and Punnett's test cross *BL/bl* and *bl/bl*, to show, by the positions of the slashes, that *B* and *L* are on one chromosome and *b* and *l* on another (we would have written these genotypes *B/b L/l* and *b/b l/l* if the genes were not linked).

Now, if in Bateson and Punnett's cross the genes for purple and long and the genes for red and round were linked, we might expect the *BL/bl* parent in the test cross to have produced only two kinds of gametes, *BL* and *bl*, and the test cross to have yielded offspring of only two phenotypes, purple-long and red-round, in equal numbers. Yet the cross also yielded some purple-round and red-long offspring. How could the *BL/bl* parent have produced *Bl* and *bL* gametes? Morgan suggested that some mechanism occasionally breaks the original linkages between purple and long and between red and round and establishes in a few individuals new linkages between purple and round and between red and long; i.e. it makes possible the production of *Bl* and *bL* gametes. The mechanism of this recombination is crossing-over, a process we discussed in Chapter 13 (see Fig. 13.20, p. 575). Crossing-over increases the number of genetic combinations any given cross can produce.

Chromosomal mapping   If we assume, as Morgan did, that the probability of breakage is approximately equal at any point along the length of a chromosome, then the greater the distance between two linked genes, the greater the frequency with which they will cross over, because there are more points between them at which a break may occur. Or, to be more precise, the frequency of crossing-over between any two linked genes will be proportional to the distance between them. The percentage of crossing-over can therefore serve as a tool for mapping the locations of genes on chromosomes.

This percentage gives us no information about the absolute distances between genes—these distances cannot be stated in nanometers—but it does give us relative distances. By convention, one unit of map distance on a chromosome is the distance within which crossing-over occurs one percent of the time. In Bateson and Punnett's test cross two out of 16 of the offspring were recombinant products of crossing-over. Two is 12.5 percent of 16; hence the genes controlling flower color and pollen shape in the sweet peas of this cross are located 12.5 map units apart.

Suppose we know that linked genes *B* and *L* are 12.5 map units apart. And suppose we find another gene, *A*, linked with these, that crosses over with gene *L* 5 percent of the time. How do we determine the order of the genes? The order could be *B–A–L:*

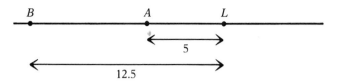

or it could be *B–L–A:*

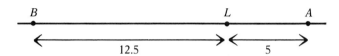

Obviously, the way to decide between these two alternatives is to determine the frequency of crossing-over between *A* and *B*. If this frequency is 7.5 percent (12.5 − 5), then we know that the first alternative is correct; if it is 17.5 percent (12.5 + 5), then we know that the second alternative is correct. In this way, by determining the frequency of crossing-over between each gene and at least two other known genes, it is possible to build up a map showing the arrangement of many different genes on a chromosome (Fig. 14.18).

One important implication of the fact that the frequencies of crossing-over permit the mapping of genes should be noted. All known frequencies

14.18 Chromosome map of *Drosophila melanogaster* Only a few of the many known genes are shown. The figures indicate their position in crossover map units from the zero end of the chromosome. Neither the Y chromosome nor the tiny fourth chromosome is shown.

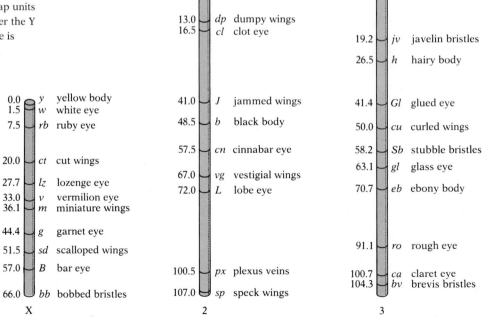

agree with a model of the chromosome in which the genes are sequentially arranged along the chromosome. Therefore one possible definition of the gene is that it is a region (or locus) on a chromosome that controls one or more characteristics of the organism. If two characters are always linked, and recombination by crossing-over never occurs, then we assume that they are controlled by the same point on the chromosome, i.e. by the same gene. If, on the other hand, crossing-over does occur between them, even if extremely seldom, then we can say that they must be controlled by different points on the chromosome, i.e. by different genes. Crossing-over has thus been the classical test of whether two characters are controlled by one chromosomal locus (gene) or by two separate chromosomal loci (genes). According to this view, the gene is the smallest unit of recombination. We shall examine other possible definitions of the gene later.

Giant chromosomes   Morgan's propositions—that the genes are bound together in sequence in a limited number of paired linkage groups, that genes belonging to homologous linkage groups can undergo orderly recombination by crossing-over, and that the frequency of crossing-over reveals both the linear order and the relative distances apart of the genes in each linkage group—were derived from breeding experiments and did not depend on knowledge of the chromosomes. Morgan and most other geneticists of his day were convinced of the validity of Sutton's theory that the genes are located on the chromosomes, but they recognized that the evidence for Sutton's theory was entirely circumstantial. They knew that genes occur in pairs, and that chromosomes do too. They knew that the members of each pair of genes separate at meiosis, as do the chromosomes. And they knew that the number of linkage groups in each species examined corresponds to the number of pairs of chromosomes (there are four linkage groups and four pairs of chromosomes in *Drosophila*, and there are ten linkage groups and ten pairs of chromosomes in corn). But no one had ever actually demonstrated the presence of a gene on a chromosome.

Eventually, stronger evidence could be adduced for the chromosomal theory of genes. Some of it came from study of the giant chromosomes in the salivary glands of the larvae of many flies, including *Drosophila*. These chromosomes are more than 200 times larger than normal chromosomes, and they can easily be studied in detail through an ordinary microscope. It has been suggested that the giant chromosomes are the product of repeated replication of the chromosomal material during interphase without accompanying separation of the strands by mitosis, and that each chromosome is therefore composed of a very large number of identical strands lying side by side. These chromosomes were first described in 1881, but it was not until 1930 that geneticists first paid serious attention to them and realized their potentialities for study of the arrangement of genes.

When stained appropriately, these giant chromosomes have a banded appearance (Fig. 14.19). The bands, which differ in width and the spacings between them, enable the worker who is thoroughly familiar with them to recognize with great precision the various regions of the chromosome. It has been possible by detailed comparative studies of chromosome abnormalities to determine the location of individual genes in relation to the bands. Cytological studies of giant chromosomes have provided a second way of

14.19   Photograph of giant chromosomes from salivary gland of *Drosophila melanogaster*

Note the pattern of banding by which different parts of the chromosomes can be identified.   [Courtesy Science Software Systems, Inc., West Los Angeles.]

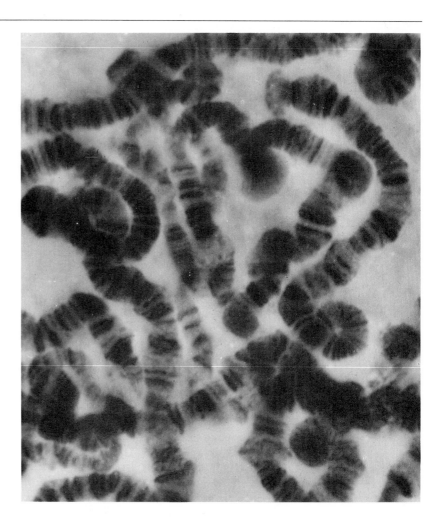

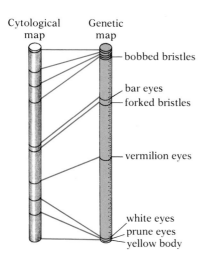

14.20   Comparison of cytological and genetic (crossover) maps of a portion of the X chromosome in *Drosophila melanogaster*

The two methods of mapping yield the same sequence for the genes, but the spacing is very different. [Modified from *Biological science*, Houghton Mifflin, 1963. Used by permission of the Biological Sciences Curriculum Study.]

mapping the arrangement of the genes on the chromosomes. Such mapping has fully corroborated that based on crossing-over frequencies with regard to the sequence of the genes; but it has not done so with regard to the spacing between them (Fig. 14.20). Apparently breaks do not occur with equal facility at all points along the chromosomes; some parts of the chromosome are more susceptible to breakage than other parts. However, even though maps based on cytological examination give a more accurate picture of the relative distances between the genes, maps based on crossing-over frequencies are still constructed, because they are more useful than cytological maps in predicting the results of crosses involving linked genes.

CHROMOSOMAL ALTERATIONS

Structural alterations   Besides crossing-over, in which corresponding segments are interchanged between homologous chromosomes, there are other kinds of chromosomal rearrangements that occur much less frequently. Some involve interchange of segments between nonhomologous chromosomes; others involve alterations within a single chromosome.

In one form of alteration, called *translocation,* portions of two nonhomologous chromosomes are interchanged, with a consequent modification in linkage groups. Suppose, for example, that the bar chromosomes in a certain species bear the genes *ABCDEFG* and that the J-shaped chromosomes bear the genes *LMNOPQRST.* If the *EFG* end of the bar chromosome and the *ST* end of the J chromosome were interchanged, the result would be a shorter bar chromosome bearing only the genes *ABCDST* and a longer J chromosome bearing the genes *LMNOPQREFG.* By changing the linkage relationships of genes, translocations can have important effects on phenotypes.

Note that in the example just cited the fragments broken off the original chromosomes were not lost to the cell, because they fused onto other chromosomes. But sometimes deleted portions do not fuse onto any chromosome and are lost entirely. This type of chromosomal alteration is called a *deletion.* Any chromosomal fragment that does not have a centromere will fail to move along the spindle during cell division and hence will not be incorporated into either daughter nucleus. One of the harmful effects of intense radioactivity on somatic cells is the large amount of chromosomal breakage it produces, with the consequent loss of important genes on deleted fragments. The occurrence of a deletion can often be detected cytologically in cells undergoing meiosis, because when the aberrant chromosome synapses with its normal homologue the normal chromosome buckles at the point where material is missing from the aberrant chromosome (Fig. 14.21).

Sometimes a piece breaks off one chromosome and fuses onto the end of the homologous chromosome. Such an alteration is called a *duplication.* An example would be loss of the *ABC* portion from a chromosome bearing the genes *ABCDEFGH* and fusion of this portion onto the homologous chromosome. The chromosome in which the deletion occurred would thus bear only the genes *DEFGH,* while the chromosome undergoing the duplication would bear the genes *ABCABCDEFGH.*

Another common form of chromosomal alteration, involving only one chromosome, is called an *inversion.* A portion of a chromosome breaks out, turns around, and fuses back in its original position, but with its ends reversed. Suppose a chromosome bears the genes *RSTUVWXYZ,* in that order. If the *UVWX* segment were to break out and become reattached in reverse order, the result would be a chromosome with the gene sequence *RSTXWVUYZ* (Fig. 14.22). Some such inversions have little phenotypic effect, since the same genes are still all present on the same chromosome. But sometimes there is a phenotypic change, apparently the result of what is known as the *position effect,* whereby the expression of a given gene is influenced by the genes close to it on the chromosome. In the original sequence in the example, gene *U* was located between genes *T* and *V* and was at a distance from gene *Y;* but in the inverted sequence, *U* is between *V* and *Y* and is at a distance from *T.* Any effect *T* may have had on the expression of *U* may be reduced as a result of the inversion, while an effect of *Y* on *U* may be increased. Another effect of this inversion would be to increase the frequency of crossing-over between *U* and *T* and to decrease it between *U* and *Y.* There is evidence that natural selection tends to favor inversions that reduce crossing-over between genes that act together to produce a beneficial phenotype, and to operate against inversions that disrupt such advantageous groupings. Inversions may thus play an important evolutionary role in conserving favorable gene groupings.

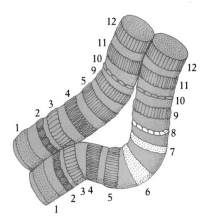

14.21 Synapsis of a deletion chromosome with its normal homologue
The upper chromosome has lost the region containing bands 6, 7, and 8. Consequently, when it synapses with a normal chromosome, that chromosome buckles outward opposite the deletion region.

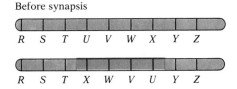

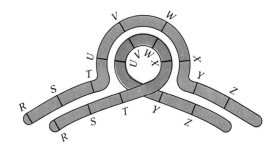

14.22 A hypothetical inversion and its effect on synapsis
Top: The upper chromosome has the normal gene sequence; the lower chromosome has undergone an inversion—the *UVWX* part has broken out, turned around, and fused back in. Bottom: When the normal chromosome and the chromosome with the inversion synapse during meiosis, they form complicated loops that bring corresponding genes into adjacent positions.

In humans most trisomies are lethal. Trisomy-18 (Edwards' syndrome) and trisomy-13 (Patau's syndrome), for example, produce physical malformations, and mental and developmental retardation, so severe that most afflicted infants die within a few weeks after birth. Because trisomies of most other autosomes result in spontaneous abortion, they are not found in live births. Two kinds of trisomy—trisomy-21 (Down's syndrome) and trisomies of the sex chromosomes—are exceptional in that their victims may survive.

Down's syndrome (formerly often called mongolian idiocy), in which three chromosomes of type 21 occur in the individual's cells, was the first clinical condition ever referred to a chromosomal abnormality. It is associated with a variety of characteristic physical features (e.g. broad head, rounded face, perceptible epicanthic folds of the eyes, a flattened bridge of the nose, protruding tongue, small irregular teeth, short stature) and also mental retardation (the modal I.Q. is about 42). The incidence of Down's syndrome is related, in part, to the age of the mother. It occurs in less than one out of 1,000 births to women under 20; it is more than seven times more common in births to women 35–39 years old, more than twenty times more common when the mother is 40–44, and more than fifty times more common when she is 45 or older. A similar association with the age of the mother is seen in Edwards' and Patau's syndromes.

Trisomy of the sex chromosomes can take several forms. In one, called Klinefelter's syndrome, the chromosomal makeup is XXY, and the individuals are males. Though the symptoms of the condition are variable, and some of those affected are nearly normal, most show a variety of physical abnormalities and mental retardation; moreover, they often suffer from thyroid dysfunction, chronic pulmonary distress, and diabetes.

Males with a second type of sex-chromosome trisomy, the XYY syndrome, generally show fewer and less severe abnormalities, though they often have poorly developed genitalia and subnormal intelligence. Because the incidence of XYY individuals in penal institutions is often significantly higher than in the general population, some investigators have suggested that men with the XYY condition are predisposed to violence, but the evidence for this conclusion is very weak.

Women with triple-X syndrome (XXX) usually have underdeveloped sexual characteristics and often subnormal intelligence, but since their abnormalities are not obvious most live a relatively normal life.

These trisomic conditions, as well as many other kinds of genetic or chromosomal diseases, can be detected during embryonic development by the process of *amniocentesis,* in which amniotic fluid containing sloughed embryonic cells is withdrawn from the uterus with a long needle inserted through the woman's abdominal wall. The embryonic cells are then examined for abnormalities. If severe ones are found, the parents have the option of having the pregnancy aborted.

Changes in chromosome number   As we saw earlier, the separation of chromosomes in cell division does not always proceed normally, and chromosomes that should have moved away from each other to opposite poles of the spindle move instead to the same pole and become incorporated into the same daughter nucleus. The result of such *nondisjunction* may be production of an organism with an extra chromosome (or sometimes two, three, or more extra chromosomes). The presence of three chromosomes of one type in an otherwise diploid individual is called *trisomy.*

Occasionally cell division may be so aberrant that all the chromosomes move to the same pole, giving rise to a daughter cell with twice the normal number of chromosomes. If this happens during meiosis, the gamete produced is diploid instead of haploid. If such a gamete unites at fertilization with a normal haploid gamete, a triploid zygote results; if it unites with another diploid gamete, also produced by aberrant meiosis, a tetraploid zygote results. Cells or organisms that have more than two complete sets of chromosomes (i.e. that are triploid, tetraploid, hexaploid, etc.) are said to be *polyploid.*

Polyploidy has apparently occurred rather often in plants, in which it has sometimes given rise to new species adaptively superior to the original diploid species under certain environmental conditions. Polyploidy can be stimulated in the laboratory by treating plants with certain chemicals that cause nondisjunction during cell division. This procedure has been used in the production of many of the new strains of cultivated plants developed in the last few decades. Polyploidy is very rare in animals and has probably not been an important factor in the origin of new animal species.

GENETICS PROBLEMS

The best way to gain an understanding of genetics is to work with it. The fundamental principles discussed above will become clearer to you, and you will grasp them more surely, if you carefully think through the following problems, which illustrate the various patterns of inheritance treated in this chapter. Additional problems will be found in the Study Guide accompanying this book and in the genetics textbooks listed at the end of the chapter.

1. In squash a gene for white color (W) is dominant over its allele for yellow color (w). Give the genotypic and phenotypic ratios for the results of each of the following crosses:

$$W/W \quad \times \quad w/w$$
$$W/w \quad \times \quad w/w$$
$$W/w \quad \times \quad W/w$$

2. If pollen from the anthers of a heterozygous white-fruited squash plant is placed on the pistil of a yellow-fruited plant, show, using ratios, the genotypes and phenotypes you would expect the seeds from this cross to produce.

3. In human beings, brown eyes are usually dominant over blue eyes. Suppose a blue-eyed man marries a brown-eyed woman whose father was blue-eyed. What proportion of their children would you predict will have blue eyes?

4. If a brown-eyed man marries a blue-eyed woman and they have ten children, all brown-eyed, can you be certain that the man is homozygous? If the eleventh child has brown eyes, what will that show about the father's genotype?

5. A brown-eyed man whose father was brown-eyed and whose mother was blue-eyed married a blue-eyed woman whose father and mother were both brown-eyed. The couple has a blue-eyed son. For which of the individuals mentioned can you be sure of the genotypes? What are their genotypes? What genotypes are possible for the others?

6. If the litter resulting from the mating of two short-tailed cats contains three kittens without tails, two with long tails, and six with short tails, what would be the simplest way of explaining the inheritance of tail length in these cats? Show genotypes.

7. When Mexican Hairless dogs are crossed with normally-haired dogs, about half the pups are hairless and half have hair. When, however, two Mexican Hairless dogs are mated, about a third of the pups produced have hair, about two thirds are hairless, and some deformed puppies are born dead. Explain these results.

8. In peas a gene for tall plants (*T*) is dominant over its allele for short plants (*t*). The gene for smooth peas (*S*) is dominant over its allele for wrinkled peas (*s*). Calculate both phenotypic ratios and genotypic ratios for the results of each of the following crosses:

$$
\begin{array}{ccc}
T/t\ S/s & \times & T/t\ S/s \\
T/t\ s/s & \times & t/t\ s/s \\
t/t\ S/s & \times & T/t\ s/s \\
T/T\ s/s & \times & t/t\ S/S
\end{array}
$$

9. In hogs a gene that produces a white belt around the animal's body is dominant over its allele for a uniformly colored body. Another independent gene produces fusion of the two hoofs on each foot (an instance of syndactyly); it is dominant over its allele, which produces normal hoofs. Suppose a uniformly colored hog homozygous for syndactyly is mated with a normal-footed hog homozygous for the belted character. What would be the phenotype of the F_1? If the F_1 individuals are allowed to breed freely among themselves, what genotypic and phenotypic ratios would you predict for the F_2?

10. In watermelons the genes for green color and for short shape are dominant over their alleles for striped color and for long shape. Suppose a plant with long striped fruit is crossed with a plant heterozygous for both these characters. What phenotypes would this cross produce and in what ratios?

11. In the fruit fly *Drosophila melanogaster*, vestigial wings and hairy body are produced by two recessive genes located on different chromosomes. The normal alleles, long wings and hairless body, are dominant. Suppose a vestigial-winged hairy male is crossed with a homozygous normal female. What types of progeny would be expected? If the F_1 from this cross are permitted to mate randomly among themselves, what progeny would be expected in the F_2? Show complete genotypes, phenotypes, and ratios for each generation.

12. Suppose a hairy female heterozygous for vestigial wing is crossed with a vestigial-winged male heterozygous for the hairy character. What will be the characteristics of the F_1?

13. In some breeds of dogs a dominant gene controls the characteristic of barking while trailing. In these dogs another independent gene produces erect ears; it is dominant over its allele for drooping ears. Suppose a dog breeder wants to produce a pure-breeding strain of droop-eared barkers, but he knows that the genes for silent trailing and erect ears are present in his kennels. How should he proceed?

14. A dominant gene, *A*, causes yellow color in rats. The dominant allele of another independent gene, *R*, produces black coat color. When the two dominants occur together (*A/– R/–*), they interact to produce gray. Rats of the genotype *a/a r/r* are cream-colored. If a gray male and a yellow female, when mated, produce offspring approximately $\frac{3}{8}$ of which are yellow, $\frac{3}{8}$ gray, $\frac{1}{8}$ cream, and $\frac{1}{8}$ black, what are the genotypes of the two parents?

15. What are the genotypes of a yellow male rat and a black female that, when mated, produce 46 gray and 53 yellow offspring?

16. In Leghorn chickens colored feathers are due to a dominant gene, *C*; white feathers are due to its recessive allele, *c*. Another dominant gene, *I*, inhibits expression of color in birds with genotypes *C/C* or *C/c*. Consequently both *C/– I/–* and *c/c –/–* are white. A colored cock is mated with a white hen and produces many offspring, all colored. Give the genotypes of both parents and offspring.

17. If the dominant gene *K* is necessary for hearing, and the dominant gene *M* results in deafness no matter what other genes are present, what percentage of the offspring produced by the cross *k/k M/m* × *K/k m/m* will be deaf?

18. What fraction of the offspring of parents each with the genotype *K/k L/l M/m* will be *k/k l/l m/m*?

19. Suppose two *D/d E/e F/f G/g H/h* individuals are mated. What would be the predicted frequency of *d/d E/E F/f g/g H/h* offspring from such a mating?

20. If a man with blood type B, one of whose parents had blood type O, marries a woman with blood type AB, what will be the theoretical percentage of their children with blood type B?

21. Both Mrs. Smith and Mrs. Jones had babies the same day in the same hospital. Mrs. Smith took home a baby girl, whom she named Shirley. Mrs. Jones took home a baby girl, whom she named Jane. Mrs. Jones began to suspect, however, that her child had been accidentally switched with the Smith baby in the nursery. Blood tests were made: Mr. Smith was type A, Mrs. Smith type B, Mr. Jones type A, Mrs. Jones type A, Shirley type O, and Jane type B. Had a mixup occurred?

22. Suppose that gene *b* is sex-linked, recessive, and lethal. A man marries a woman who is heterozygous for this gene. If this couple had many normal children, what would be the predicted sex ratio of these children?

23. Red-green color blindness is inherited as a sex-linked recessive. If a color-blind woman marries a man who has normal vision, what would be the expected phenotypes of their children with reference to this character?

24. A man and his wife both have normal color vision, but a daughter has red-green color blindness, a sex-linked recessive trait. The man sues his wife for divorce on grounds of infidelity. Can genetics provide evidence supporting his case?

25. Suppose a pigeon breeder finds that about one fourth of the eggs produced by one of his prize pairs do not hatch. Of the young birds produced by this pair, two thirds are males. Give a possible explanation for these results. (Remember the mechanism of sex determination in birds; see p. 609, footnote 5.)

26. It is exceedingly difficult to determine the sex of very young chickens, but it is easy to tell, by visual observation, whether or not they are barred. The barred pattern is inherited as a sex-linked dominant. Set up a cross allowing the sex of all chicks to be determined when they hatch. (Remember that chickens are birds.)

27. In cats short hair is dominant over long hair; the gene involved is autosomal. Another gene, B^1, which is sex-linked, produces yellow coat color; its allele B^2 produces black coat color; and the heterozygous combination B^1/B^2 produces tortoise-shell coat color. If a long-haired black male is mated with a tortoise-shell female homozygous for short hair, what kind of kittens will be produced in the F_1? If the F_1 cats are allowed to interbreed freely, what are the chances of obtaining a long-haired yellow male?

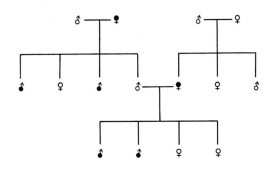

28. The diagram shows three generations of the pedigree of deafness in a family. Black circles indicate deaf persons. An arrow on a circle indicates a male, a cross below a circle a female. State whether the condition of deafness in this family is inherited as

    a. a dominant autosomal characteristic
    b. a recessive autosomal characteristic
    c. a sex-linked dominant characteristic
    d. a sex-linked recessive characteristic
    e. a holandric characteristic

29. In *Drosophila melanogaster* there is a dominant gene for gray body color and another dominant gene for normal wings. The recessive alleles of these two genes result in black body color and vestigial wings respectively. Flies homozygous for gray body and normal wings were crossed with flies that had black bodies and vestigial wings. The F_1 progeny were then test-crossed, with the following results:

Gray body, normal wings	236
Black body, vestigial wings	253
Gray body, vestigial wings	50
Black body, normal wings	61

Would you say that these two genes are linked? If so, how many units apart are they on the chromosome?

30. In rabbits a dominant gene produces spotted body color, and its recessive allele solid body color. Another dominant gene produces short hair, and its recessive allele long hair. Rabbits heterozygous for both characteristics were mated with homozygous recessive rabbits. The results of this cross were as follows:

Spotted, short hair	96
Solid, short hair	14
Spotted, long hair	10
Solid, long hair	80

What evidence for linkage is shown in this cross? Give the percentage of crossing-over and the map distance between the genes.

31. In *Drosophila melanogaster* the genes for normal bristles and normal eye color are known to be about 20 units apart on the same chromosome. Individuals homozygous dominant for these genes were mated with homozygous recessive individuals. The F_1 progeny were then test-crossed. If there were 1,000 offspring from the test cross, how many of the offspring would you predict would show the crossover phenotypes?

32. The crossover frequency between linked genes *A* and *B* is 40%; between *B* and *C*, 20%; between *C* and *D*, 10%; between *C* and *A*, 20%; between *D* and *B*, 10%. What is the sequence of the genes on the chromosome?

33. Suppose that nondisjunction resulted in the production of new individuals with the following chromosomal abnormalities: XO, XXX, XYY, XXXX, XXXY, XXXXY. Indicate the expected phenotypic sex corresponding to each of these chromosomal combinations if it occurred (*a*) in a human; (*b*) in a *Drosophila*. How many Barr bodies would there be in human cells showing each of these combinations?

SUGGESTED READING

CLARKE, C. A., 1968. "The prevention of 'rhesus' babies," *Scientific American,* November. (Offprint 1126) *Immunological techniques that make it possible to prevent Rh disease.*

FRIEDMANN, T., 1971. "Prenatal diagnosis of genetic disease," *Scientific American,* November. *The technique of amniocentesis in diagnosing some debilitating genetic diseases at an early stage in embryonic development.*

GARDNER, E. J., 1975. *Principles of genetics,* 5th ed. Wiley, New York. *Clearly written elementary text with good practice problems.*

McKUSICK, V. A., 1971. "The mapping of human chromosomes," *Scientific American,* April.

MENDEL, G., 1948. *Experiments in plant hybridisation.* Harvard University Press, Cambridge, Mass. *Translation of Gregor Mendel's original paper, which was first published in 1866.*

MITTWOCH, U., 1963. "Sex differences in cells," *Scientific American,* July. (Offprint 161) *The discovery of Barr bodies and other differences between male and female cells, which permit easy distinction between them.*

WINCHESTER, A. M., 1972. *Genetics: A survey of the principles of heredity,* 4th ed. Houghton Mifflin, Boston. *An elementary text with simple explanations and good practice problems.*

Chapter 15

THE NATURE OF THE GENE
AND ITS ACTION

In the last chapter we postulated the existence of physical units of inheritance, the genes, which interact with environmental influences to determine the phenotypic characteristics of the organism; and we cited evidence for their arrangement in linear sequence along the chromosomes. Presenting, as we did, a model of the gene based almost entirely on deductions from inheritance patterns, we obviously left some fundamental genetic problems untouched. We made no reference to the chemical nature of the genes; we did not say how the genetic material—DNA, as we saw earlier—is replicated, how genes influence the characteristics of the organism, or how they regulate the myriad activities of a living cell. In effect, we treated the genes as though they were a string of beads of unknown composition somehow exerting almost magical control over the visible attributes of living things.

Not many years ago this was all any textbook could do. Nothing was known of the chemical makeup of genes; nor was it known how they are replicated or how they influence the phenotype. But that is all changed. The enormous advances of the last thirty years in our understanding of the control mechanisms of living cells constitute nothing short of a revolution, a revolution with such far-reaching implications that many scientists have claimed it will eventually prove more important to the future of the human species than the birth of the atomic age in physics.

THE CHEMICAL NATURE OF THE GENETIC MATERIAL

THE DISCOVERY OF DNA AND ITS FUNCTION

The composition of chromosomes If the genes were located on the chromosomes, an obvious first step in ascertaining the chemical nature of genes was to determine the types of compounds present in the chromosomes. In

1869 Friedrich Miescher, a Swiss biochemist, showed that when cells are treated with pepsin to render the proteins soluble, the nuclei shrink but remain essentially intact. He showed, further, that the same nuclear material that can withstand peptic digestion also behaves totally unlike protein when treated with a variety of other reagents, and that it contains phosphorus in addition to the carbon, oxygen, hydrogen, and nitrogen that would be expected if it were a protein. This material has since been named nucleic acid. Further research has shown that cells contain several sorts of nucleic acids, some not restricted to the nucleus; the type studied by Miescher was DNA (deoxyribonucleic acid).

In 1914 Robert Feulgen, a German chemist, devised a method of selectively staining DNA a brilliant crimson. When, ten years later, Feulgen applied his technique to whole cells, he found that the nuclear DNA is restricted to the chromosomes. Other workers have since used Feulgen staining to measure the DNA content of nuclei from many types of cells. They have shown conclusively that all the somatic (body) cells of a given organism ordinarily contain the same amount of DNA—despite the fact that cells from such different tissues as liver, kidney, heart, nerve, and muscle differ drastically in the amounts of other substances they contain—and, further, that egg and sperm cells contain only half as much DNA as the somatic cells. Since biologists had already concluded that mitosis distributes a complete set of genes to every somatic cell, regardless of its eventual role, and that meiosis distributes to every gamete cell exactly half the amount of genetic material found in the somatic cells, the discovery that the amount of DNA is usually constant in all somatic cells within a species, but is halved in the gametes, suggested that DNA might be the essential material of the genes.

But many workers refused to take this possibility seriously. It was known that the chromosomes of most organisms contain both protein and DNA, and most biologists assumed that the protein must be the genetic material, because, in their view, only protein had the chemical complexity necessary to encode so much information. Discovery that the amount of structural protein in the chromosomes, unlike the amount of DNA, is not constant, but varies with the overall activity of the cell, did not, by and large, cause this long-held conviction to be relinquished.

Today, however, biologists are convinced that DNA is the genetic material. To understand how this shift has come about, we must leave our old friends *Drosophila* and peas, and turn to microorganisms such as bacteria, viruses, and molds.

DNA vs. protein In 1928 Fred Griffith, an English medical bacteriologist, published a paper describing some experiments, now considered classic, on pneumococci, the bacteria that cause pneumonia. Griffith studied the effects on mice of a virulent strain of bacteria (S) and a nonvirulent strain (R). He showed that mice injected with live strain-R bacteria survived, mice injected with live strain-S bacteria soon died, but mice injected with heat-killed strain-S bacteria survived. These results (Table 15.1) were readily understandable. But the results of another of his experiments were thoroughly perplexing: Mice injected with a mixture of live strain-R and heat-killed strain-S bacteria died. How could a mixture of nonvirulent and dead bacteria have killed the mice? Griffith examined the bodies of the dead mice and

TABLE 15.1 *Griffith's results*

Bacteria injected	Reaction of mice
Live strain R	Survived
Live strain S	Died
Dead strain S	Survived
Live strain R plus dead strain S	Died

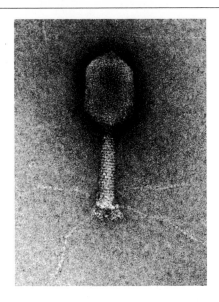

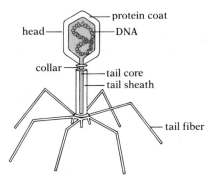

15.1 Electron micrograph and interpretive drawing of a bacteriophage
Micrograph: × 200,000. [Courtesy M. Wurtz, University of Basel.]

found that they were full of live strain-S bacteria! Where had they come from? After many careful experiments, he became convinced that somehow the live strain-R bacteria had been transformed into live strain-S bacteria by material from the dead strain-S cells. The transformed bacteria, when cultured, reproduced new strain-S bacteria. Presumably, hereditary material from the dead bacteria had entered the live strain-R cells and changed them into strain-S cells.

By 1931 other workers had shown that the rodent host was not essential for bacterial *transformation;* it could occur just as well in test-tube cultures. Two years later James L. Alloway of the Rockefeller Institute showed that not even whole strain-S cells were necessary; live strain-R cells in a test tube could be transformed into strain-S bacteria by a cell-free extract of S bacteria. Apparently the hereditary characteristics of the virulent strain were transmitted to the cells of the nonvirulent strain by some substance that could withstand both the killing and extracting of the cells in which it had originally been contained.

The work of Griffith, Alloway, and others in the late 1920s and early 1930s was interesting to other biologists, but its significance was not appreciated at that time. It was not until 1944 that O. T. Avery, Colin MacLeod, and Maclyn McCarty of the Rockefeller Institute demonstrated through purification techniques that the transforming principle was DNA; nothing else was necessary. From our present perspective this seems like strong evidence that DNA rather than protein or a nucleoprotein complex was the essential genetic material, but at that time many scientists remained unconvinced.

During the next ten years, however, the evidence for DNA steadily became stronger. At least 30 different examples of bacterial transformation by purified DNA were described. And strong evidence came from another source, the studies by Alfred D. Hershey and Martha Chase, of the Carnegie Laboratory of Genetics, of a special type of virus that attacks the bacterium *Escherichia coli,* which is abundant in the human digestive tract. This type of bacteria-destroying virus is called *bacteriophage*—phage for short.

It has been established by a variety of methods that free virus particles, or *virions,* are made up of two chief components, a protein coat and a nucleic acid core. The electron microscope has revealed that phage viruses are structurally more complex than many other types (Fig. 15.1). Their protein coat is divided into a head region and an elongate tail region made up of a hollow core, a surrounding sheath, and six distal fibers. All the DNA is in the head. Electron micrographs show that when a phage attacks a bacterial cell it becomes attached by the tip of its tail to the wall of the bacterial cell (Fig. 15.2); apparently the tip of the tail contains a protein that reacts specifically with receptor sites of the bacterial wall. There is no evidence that the protein coat of the phage ever actually enters the bacterial cell; yet within a few minutes after the phage becomes attached to the wall, new phage appear within the bacterium and the bacterial cell soon ruptures, releasing hundreds of new phage into the surrounding medium. It can be shown that these new phage are genetically identical with those that initiated the infection. Hereditary material must have been injected into the bacterial cell by the phage attached to its wall, and this hereditary material must have usurped the metabolic machinery of the bacterium and put it to work manufacturing new phage.

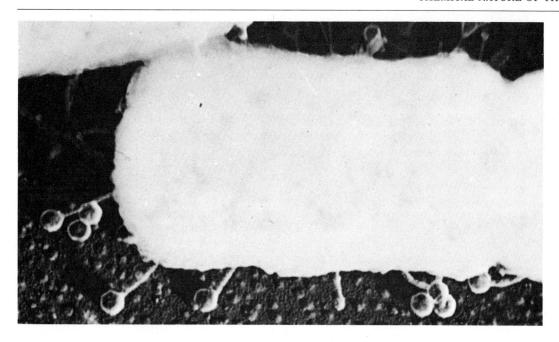

Hershey and Chase designed an experiment to determine whether the infecting phage injects into the bacterium only DNA or only protein or some of both. They made use of the fact that DNA contains phosphorus but no sulfur, whereas protein contains sulfur but no phosphorus. They cultured phage on bacteria grown on a medium containing a radioactive isotope of phosphorus (P^{32}) and a radioactive isotope of sulfur (S^{35}). The phage incorporated the S^{35} into their protein, and they incorporated the P^{32} into their DNA. Hershey and Chase then infected nonradioactive bacteria with the radioactive phage. They allowed sufficient time for the phage to become attached to the walls of the bacteria and inject hereditary material. Then they agitated the bacteria in a blendor in order to detach what remained of the phage from their surfaces. Analysis of these remains showed that they contained S^{35} but no P^{32}, an indication that only the empty protein coat had been left outside the bacterial cell. Analysis of the bacteria showed that they contained P^{32} but no S^{35}, an indication that only DNA had been injected into them by the phage. Since new phage were produced in these bacteria, DNA alone was sufficient to transmit to the bacteria all the genetic information necessary for their production. This experiment, reported in 1952, supported the earlier conclusions based on transformation experiments that nucleic acids, not proteins, constitute the genetic material.

THE MOLECULAR STRUCTURE OF DNA

We have already seen that a molecule of DNA is composed of building blocks called nucleotides, each of which is itself composed of a five-carbon sugar bonded to a phosphate group and a nitrogenous base (Fig. 15.3). There are four kinds of nucleotides in DNA, which differ from one another in their nitrogenous bases. Two of the bases, **adenine** and **guanine,** are purines,

15.2   Electron micrograph of part of a bacterial cell with numerous phage particles attached
$\times$ 108,750.   [Courtesy T. F. Anderson, E. L. Wollman, and F. Jacob, *Ann. Inst. Pasteur,* vol. 93, 1957.]

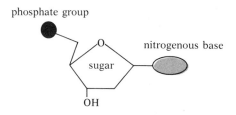

15.3   Diagram of a nucleotide from DNA
A phosphate group and a nitrogenous base are attached to deoxyribose, a five-carbon sugar.

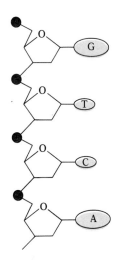

15.4   Portion of a single chain of DNA

Nucleotides are hooked together by bonds between their sugar and phosphate groups. The nitrogenous bases (G, guanine; T, thymine; C, cytosine; A, adenine) are side groups.

which are double-ring structures; the other two, **cytosine** and **thymine,** are pyrimidines, which are single-ring structures (see Fig. 2.45, p. 64). In a DNA molecule the nucleotides are arranged in sequence, held together by covalent bonds between the sugar of one nucleotide and the phosphate group of the next nucleotide beyond it; the nitrogenous bases are arranged as side groups off the chains (Fig. 15.4). DNA molcules ordinarily exist as double-chain structures, with the two chains held together by hydrogen bonds between their nitrogenous bases; such bonding can occur only between cytosine and guanine or between thymine and adenine (Fig. 15.5). Finally, the ladderlike double-chain molcule is coiled into a double helix (Fig. 15.6).

Determining the structure of so complicated—and important—a molecule as DNA had become an irresistible challenge to many scientists. In 1950 almost nothing was known about the spatial arrangement of the atoms within the DNA molecule; nor was it known how this molecule could contain within it the necesary information for replicating itself and for controlling cellular function. About this time, several workers began applying the techniques of X-ray diffraction analysis to DNA. Outstanding among them was Maurice H. F. Wilkins of King's College, London, who succeeded in obtaining much sharper X-ray diffraction patterns than had previously been obtainable. These diffraction patterns revealed three major periodicities in the crystalline DNA: one of 0.34 nm, one of 2.0 nm, and one of 3.4 nm.

Now began a collaboration whose outcome would rank as one of the major milestones in the history of biology. James D. Watson and Francis H. C. Crick, working at Cambridge University, decided to try to develop a model of the structure of the DNA molecule by combining what was known about the chemical content of DNA with the information gained from Wilkins' X-ray diffraction studies, as well as with data on the exact distances between bonded atoms in molecules, the angles between bonds, and the sizes of atoms. Watson and Crick built scale models of the component parts of DNA

15.5   Portion of a DNA molecule uncoiled

The molecule has a ladderlike structure, with the two uprights composed of alternating sugar and phosphate groups and the cross rungs composed of paired nitrogenous bases. Each cross rung has one purine base (large oval) and one pyrimidine base (small oval). When the purine is guanine (G), then the pyrimidine with which it is paired is always cytosine (C); when the purine is adenine (A), then the pyrimidine is thymine (T). Adenine and thymine are linked by two hydrogen bonds (striped bands), guanine and cytosine by three. Note that the two chains run in opposite directions, i.e. that the free phosphate is at the upper end of the left chain and at the lower end of the right chain.

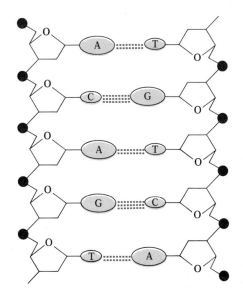

and then attempted to fit them together in a way that would agree with the information from all these separate sources.

They were certain that the 0.34 nm periodicity discovered by Wilkins corresponded to the distance between successive nucleotides in the DNA chain and the 2.0 nm periodicity to the width of the chain. But what about the 3.4 nm periodicity? To explain it, they postulated that the chain of nucleotides was coiled in a helix. The 3.4 nm periodicity would thus correspond to the distance between successive turns of the helix and indicate how tight the chain was wound. Since 3.4 is exactly ten times the distance between successive nucleotides, each turn of the helix would have to be ten nucleotides long.

Having made these essential assumptions about the meaning of the X-ray diffraction data, Watson and Crick then tried to correlate them with the information from other sources. They immediately ran into a discrepancy. They calculated that a single chain of nucleotides coiled in a helix 2.0 nm wide with turns 3.4 nm long would have a density only half as great as the known density of DNA. An obvious inference was that the DNA molecule is composed of two nucleotide chains rather than one. Now they had to determine the relationship between the two chains within the double helix. They tried several arrangements of their scale model and found that the one that best fitted all the data was one in which the two nucleotide chains were wound in opposite directions around a hypothetical cylinder of appropriate diameter, with the purine and pyrimidine bases oriented toward the interior of the cylinder (Fig. 15.6). With the bases oriented in this manner, hydrogen bonds between the bases of opposite chains could supply the force to hold the two chains together and to maintain the helical configuration. In other words, the DNA molecule, when unwound, would have a ladderlike structure, with the uprights of the ladder formed by the two long chains of alternating sugar and phosphate groups, and with each of the cross rungs formed by two nitrogenous bases loosely bonded to each other by hydrogen bonds (Fig. 15.5).

Watson and Crick soon realized that each cross rung must be composed of one purine base and one pyrimidine base. Their scale model showed that the available space between the sugar-phosphate uprights was just sufficient to accommodate three ring structures. Hence two purines opposite each other occupied too much space, because each had two rings for a total of four, and two pyrimidines opposite each other did not come close enough to bond properly, because each had only one ring. This left four possible pairings: adenine–thymine, adenine–cytosine, guanine–thymine, and guanine–cytosine. Further examination revealed that, although adenine and cytosine were of the proper size to fit together into the available space, they could not be arranged in a way that would permit hydrogen bonding between them; the same was true of guanine and thymine. Therefore neither adenine–cytosine nor guanine–thymine cross rungs could occur in the DNA molecule. This left only adenine–thymine and guanine–cytosine. Both of these base pairs seemed to fulfill all requirements. It did not seem to matter in which order the bases occurred: Thymine–adenine was as satisfactory as adenine–thymine, and cytosine–guanine was as satisfactory as guanine–cytosine. The essential requirement seemed to be that adenine and thymine always be paired with each other and that guanine and cytosine always be paired. This

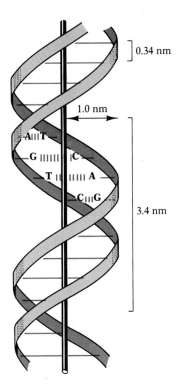

15.6 The Watson-Crick model of DNA
The molecule is composed of two polynucleotide chains held together by hydrogen bonds between their adjacent bases. The double-chained structure (shown here wound around a hypothetical rod) is coiled in a helix. The width of the molecule is 2.0 nm; the distance between adjacent nucleotides is 0.34 nm; and the length of one complete coil is 3.4 nm.

15.7   Replication of DNA

As the two polynucleotide chains of the old DNA (black) uncoil, new polynucleotide chains (color) are synthesized on their surfaces. The process produces two complete double-chained molecules each of which is identical in base sequence to the original double-chained molecule.

pairing explained an earlier finding by E. Chargaff and his colleagues at Columbia University that all samples of DNA, regardless of other differences in their composition, are alike in always containing exactly equal amounts of adenine and thymine nucleotides, as well as exactly equal amounts of guanine and cytosine nucleotides. The amounts of adenine and thymine are always equal because these two bases are always paired, and, similarly, the amounts of guanine and cytosine are always equal because these are always paired.

In summary, then, the Watson-Crick model of the DNA molecule shows a double helix in which the two chains, composed of alternating sugar and phosphate groups, are bonded together by hydrogen bonds between adenine and thymine from opposite chains and between guanine and cytosine from opposite chains. This model, in essentially the same form in which it was first proposed in 1953, has been consistently supported by later research, and it has received general acceptance. Watson, Crick, and Wilkins have been awarded the Nobel Prize for their critically important work.

THE REPLICATION OF DNA

The theory of Watson and Crick   DNA, if it is the genetic substance, must have built into it the information necessary to replicate itself and to control the cell's attributes and functions. One of the most satisfying things about the Watson-Crick model of DNA is that it immediately suggests a way in which the first of these two requirements may be met.

Since the DNA of all organisms is alike in being a polymer composed of only four different nucleotides, the essential distinction between the DNA of one gene and the DNA of another gene must be—aside from the total number of nucleotides—the sequence in which the four possible types of cross rungs (adenine–thymine, thymine—adenine, guanine—cytosine, and cytosine—guanine) occur. The basic question of genetic replication is, then: Assuming that an adequate supply of the four nucleotides is already synthesized in the cells, what tells the cell's biochemical machinery how to put these nucleotide building blocks together in exactly the sequences and quantities characteristic of the DNA already present in the cell?

Watson and Crick pointed out that if the two chains of a DNA molecule are separated by rupturing the hydrogen bonds between the base pairs, each chain provides all the information necessary for synthesizing a new partner identical to its previous partner. Since an adenine nucleotide must always pair with a thymine nucleotide, and since a guanine nucleotide must always pair with a cytosine nucleotide, the sequence of nucleotides in one chain tells precisely what the sequence of nucleotides in its complementary chain must be. In other words, if the cell could separate the two chains in its DNA molecules—much as one might unzip a zipper—it could line up nucleotides for a new chain next to each of the old chains, putting each type of nucleotide opposite its proper partner. As the nucleotides were arranged in the proper sequence, they could be bonded together to form the new chain. Thus, separating the two chains of a DNA molecule and using each chain as a template or mold against which to synthesize a new partner for it would result in two complete double-chained molecules identical to the original molecule (Fig. 15.7).

Experimental support for the theory   Satisfying as it was, this explanation of DNA replication was pure speculation, unsupported by any experimental evidence, when it was first put forward by Watson and Crick in 1953. Since then, convincing evidence has come from the work of a number of investigators. We have space to mention only two supporting experiments here.

In 1957 Arthur Kornberg and his associates at Washington University in St. Louis reported that they had established a system in which DNA synthesis would take place in a test tube. They extracted DNA polymerase, the enzyme necessary for DNA synthesis, from cells of *Escherichia coli* and combined it with a plentiful supply of the four nucleotides as raw material; these nucleotides had been activated by ATP and also contained a radioactive isotope of carbon (C^{14}). After they had added some primer DNA, as a potential template, to this mixture and incubated the preparation for a suitable period, they found, on analysis, that some new DNA containing C^{14} was present. The labeled nucleotides had been built into DNA chains. This synthesis would not occur unless the primer DNA was present. Kornberg was able to show that the ratio of adenine and thymine to guanine and cytosine in the new DNA was precisely the same as that in the primer DNA. In other words, all the evidence indicated that the newly synthesized DNA was identical with the primer DNA—that the primer DNA had functioned as a template. Kornberg received a Nobel Prize for this work.

Kornberg's experiment seemed to demonstrate that DNA synthesis cannot take place in the absence of primer DNA, which presumably provides the information necessary to guide the synthesis. It also strongly suggested that new DNA is always a copy of the primer DNA. But it did not actually provide any evidence that the copy mechanism works in the way proposed by Watson and Crick. More direct support for their model came from an experiment reported in 1958 by Matthew S. Meselson and Frank W. Stahl of the California Institute of Technology.

These workers grew *Escherichia coli* for many generations on a medium in which the nitrogen source contained only the heavy isotope N^{15}. Eventually all the DNA in these bacteria contained the heavy isotope instead of the normal isotope N^{14}. Then the nitrogen source was abruptly changed from N^{15} to N^{14}. Cell samples were removed at regular intervals thereafter, and the DNA was extracted from them and subjected to a complicated procedure designed to separate DNA of different densities.

The experiment showed that when cells containing only heavy DNA (i.e. DNA in which both chains had only N^{15} in their purine and pyrimidine bases) were allowed to undergo one division on the N^{14} medium, the DNA of the new cells was intermediate in density between heavy DNA and light DNA. In other words, the nitrogen in the DNA of the new cells was half N^{15} and half N^{14}. This is precisely what would be expected if the two chains of the heavy parental DNA separated and acted as templates for the synthesis of new partners from nucleotides containing only N^{14} (Fig. 15.8). Each new DNA molecule should be composed of one heavy chain from the parent and one light chain newly synthesized, the molecule thus having intermediate density.

To prove that all the N^{15} really was in one chain of the intermediate-density DNA and all the N^{14} in the other chain, Meselson and Stahl subjected the DNA to a treatment that breaks the hydrogen bonds between the bases and

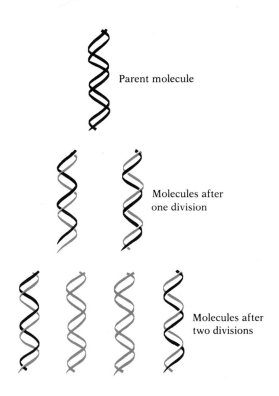

15.8   Diagram of results of the Meselson-Stahl experiment
The parent DNA molecule (black chains) contained only heavy nitrogen. After one division, the DNA had an intermediate density, an indication that half the nitrogen in each molecule was heavy and half was light; the two heavy parental chains had separated, and each had acted as the template for synthesis of a complementary light chain (color). Even after several additional duplications, the two original heavy parental chains remained intact.   [Redrawn from M. S. Meselson and F. W. Stahl, *Proc. Nat. Acad. Sci.*, vol. 44, 1958.]

separates the chains. Sure enough, this procedure produced heavy chains and light chains; the two isotopes had not been distributed randomly throughout the DNA molecule, but had been localized each in one of the chains, just as would be predicted from the Watson-Crick theory.

Recently electron microscopy has made it possible to visualize DNA in the process of replication. Curiously enough, the replication does not begin at the end of the molecule, as had been expected (and as shown in Fig. 15.7). Rather, it starts at an internal point, forming an eye-shaped region

as it moves in both directions (Fig. 15.9A). Often replication is proceeding in several places at once on the template DNA (Fig. 15.10).

THE ONE GENE–ONE ENZYME HYPOTHESIS

We have seen that the Watson-Crick model of DNA provides a plausible explanation for genetic replication, and that there is convincing experimental evidence in support of that explanation. Now we must ask what the chemistry of heredity can tell us about the way genes control cellular function. How can the DNA gene be related to phenotypically expressed inherited traits? How can nucleic acid in the nucleus control what goes on in the cytoplasm?

Bases for the hypothesis   Alkaptonuria is a hereditary disease in human beings in which the urine is very dark-colored. More than a hundred years ago it was shown that the darkening of the urine is caused by the presence in it of a chemical called alkapton. Normal individuals possess an enzyme that catalyzes the oxidation of alkapton to carbon dioxide and water, but those who have alkaptonuria lack this enzyme and must excrete alkapton in its undegraded form. In 1909 Archibald Garrod, an English physician, showed that alkaptonuria is inherited as a simple Mendelian recessive. Apparently the normal gene is necessary for production of the enzyme that oxidizes alkapton; persons homozygous for the mutant form of the gene do not produce the enzyme.

Phenylketonuria is another rare disease related to alkaptonuria. Its victims have extremely high concentrations of the amino acid phenylalanine in the blood. They excrete phenylpyruvic acid in the urine, and they are invariably severely feeble-minded. Normal individuals convert excess phenylalanine into tyrosine, which is then further metabolized, but phenylketonurics lack the enzyme that catalyzes this reaction. Like alkaptonuria, this disease is inherited as a simple Mendelian recessive. Again, it is clear that the normal gene is necessary for production of an enzyme that persons homozygous for the mutant allele cannot produce.[1]

Alkaptonuria, phenylketonuria, and other similar examples that could be cited should have suggested to geneticists that there is a close relationship between genes and enzymes. But for many years most geneticists failed to

[1] When phenylketonuria (PKU) is detected within the first four months of life and a diet instituted that contains only a bare minimum of phenylalanine (some is essential for protein synthesis), the neurological effects that lead to mental deficiency can usually be avoided. Some states now require testing for PKU in newborn infants.

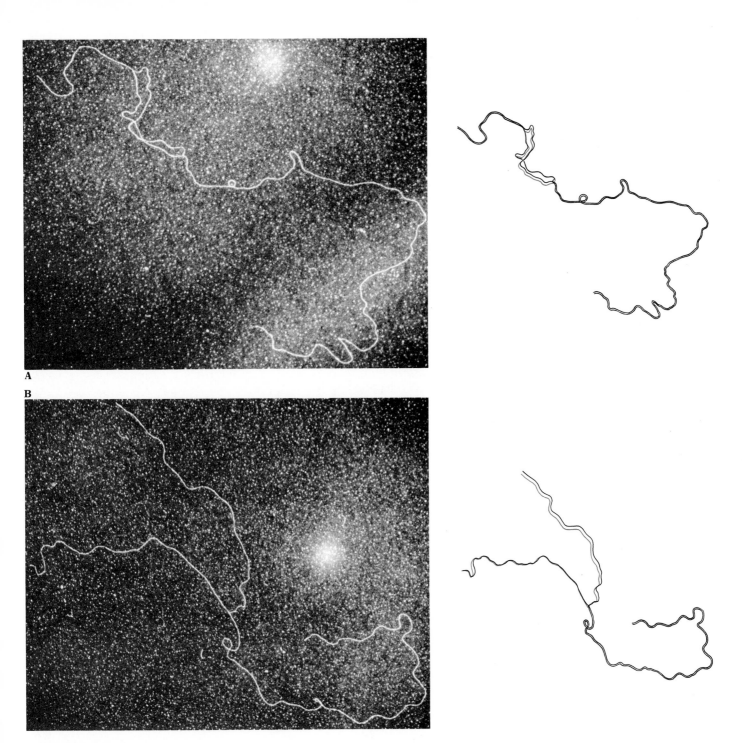

15.9 Electron micrographs of a DNA molecule in the process of replication

(A) A region where the two strands of the DNA have uncoupled can be seen at upper left. As the interpretive sketch shows, new DNA strands (color) are being synthesized in this so-called eye region. The replication proceeds in both directions; hence the eye grows at both its ends. × 19,850.

(B) Replication has proceeded so far that it has reached one end of the DNA template, and the eye has broken open. The DNA shown here, from the bacteriophage T7, is about 40,000 nucleotides long. × 16,900. [Courtesy J. Wolfson and D. Dressler, *Proc. Nat. Acad. Sci.*, vol. 69, 1972.]

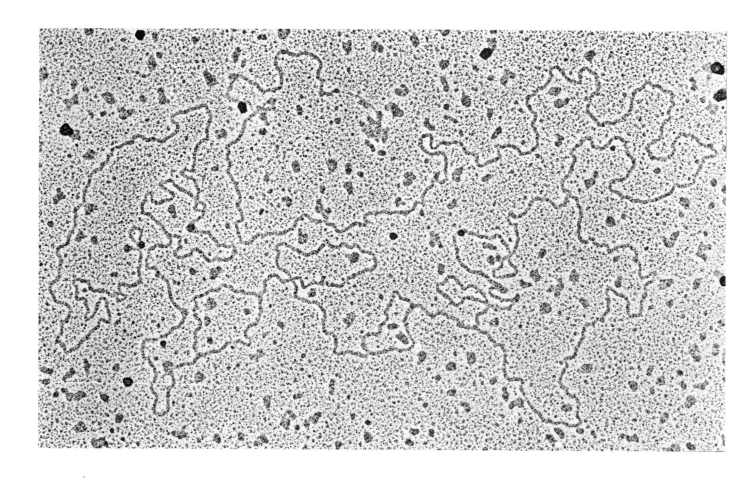

15.10 Electron micrograph of a DNA molecule on which replication has been initiated at four sites
As the sketch shows, new strands of DNA (color) are growing in each of the four eye regions. × 62,595. [Photo: Courtesy D. R. Wolstenholme, *Chromosoma*, vol. 43, 1973; sketch adapted from Wolstenholme, in D. Mazia, "The cell cycle," *Sci. Am.*, January 1974.]

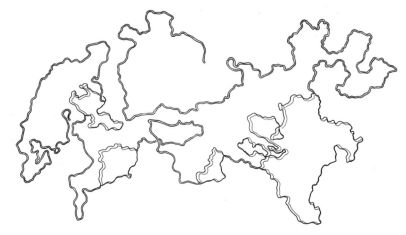

make the connection. Then in the 1930s several workers studied the genetics of eye-pigment formation in a series of *Drosophila* eye-color mutants. They were able to show that synthesis of normal eye pigment involves a series of reactions that produce identifiable intermediate compounds and that each of these reactions is under genetic control. Mutations result in alterations of these chemical reactions and thus change the pigment produced by them. A relationship between genes and specific enzymes was strongly suggested.

But concepts of gene action remained rather nebulous until the early 1940s, when George W. Beadle and Edward L. Tatum, then at Stanford University, formulated their "one gene–one enzyme" hypothesis: that the production of each enzyme in the cell is controlled by a single gene. Genes, they proposed, exert their control over cellular function by directing the synthesis of the enzymes that control the chemical reactions of the cell; these reactions, in turn, determine the phenotypic characteristics.

Evidence from *Neurospora*   Beadle and Tatum formulated their one gene–one enzyme hypothesis on the basis of the results they obtained from their pioneering experiments on red bread mold, *Neurospora crassa*, which in the last four decades had taken its place alongside such organisms as *Drosophila* and corn among the genetically most extensively studied living things. It has many advantages as an experimental organism: (1) It is easily cultured in the laboratory. (2) It has a life cycle of only ten days. (3) It sometimes reproduces sexually, but it also readily multiplies asexually; hence new strains can be propagated without the genetic change that would result from sexual recombination. (4) It is normally haploid, which means that recessive genes cannot be hidden by dominant alleles. And (5) when it produces spores by meiosis, the products of each meiosis are packaged together in a spore case in a definite order (Fig. 15.11), which makes it possible to analyze directly the genotypic frequencies produced by a cross, without the necessity of relying solely on statistical analysis (Fig. 15.12).

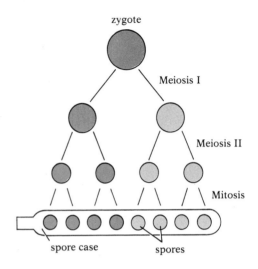

15.11 Formation of ascospores in *Neurospora*
One of the important advantages of using *Neurospora* in genetics experiments is that the arrangement of the spores in the spore case (ascus) shows the order in which the divisions of the nuclei actually took place. For example, if alleles for contrasting traits (dark and light color) are segregated by meiosis, the spores carrying them will be grouped in the spore case and can be removed and germinated separately. Thus the experimenter can be certain of the precise ratio of all the products of a given meiosis without resorting to statistical analysis. [Modified from J. A. Ramsay, *The experimental basis of modern biology*, Cambridge University Press, 1965.]

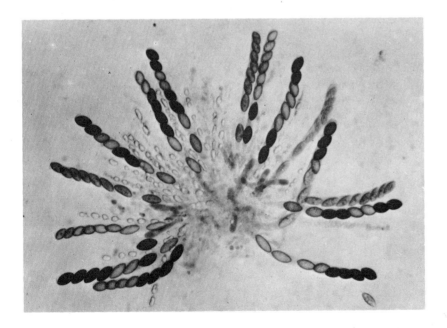

15.12 Photograph of a squashed *Neurospora* perithecium
By visually assaying the sequence of light- and dark-colored spores in each spore case (ascus), the results of each cross can be determined. Note the different ratios of light and dark spores in the various spore cases; each case contains the products of a different cross. [Courtesy R. L. Phillips, University of Minnesota.]

Wild-type *Neurospora* can be grown on a minimal nutrient medium of known composition. Beadle and Tatum used X rays to induce mutations affecting the nutrient requirements of the mold. Their procedure was as follows (Fig. 15.13):

Asexual spores were irradiated to induce mutations, and then the spores were germinated on a complete medium (one composed of the normal minimal medium plus all known vitamins and amino acids and several alternative carbon sources). This procedure ensured the survival and growth of most nutritional mutants, because the medium supplied compounds they might no longer be able to synthesize for themselves.

Once sufficiently large colonies had been established on the complete medium, a small portion of each colony was transferred onto minimal medium. This step revealed that some of the colonies were unable to grow on the minimal medium, hence that they must be mutant strains requiring some nutrient not supplied by this medium. The mutant strains having thus been located, the next step was to try to determine which of the supplements present in the complete medium, but not in the minimal medium, was needed by each strain.

First, small portions of each mutant colony were transferred from the complete medium to each of three new media: minimal plus all known vitamins, minimal plus all amino acids, and minimal plus alternative carbon sources (usually glucose as a substitute for sucrose). By observing on which of these three media the mutant would grow, it was possible to learn whether the unknown nutrient needed by the mutant was a vitamin, an amino acid, or a simpler carbon source. Suppose, for example, that a particular mutant failed to grow on minimal plus amino acids or minimal plus alternative carbon, but did grow on minimal plus vitamins. It could be concluded that the nutrient needed was a vitamin. Now samples of this mutant strain could be transferred to a series of media each of which consisted of minimal medium plus one vitamin. If the mold grew only on a medium containing vitamin B_6, the nutrient it needed was clearly vitamin B_6. The mutation must have destroyed the mold's capacity to synthesize this compound for itself, probably by failing to provide an enzyme necessary for the synthesis.

Beadle and Tatum isolated a large number of different mutant strains by this method. In every case where the mutant had lost the ability to synthesize a particular nutrient, they found that the character was inherited in a pattern that indicated the action of only one gene. It was this finding that led them to formulate the one gene–one enzyme hypothesis. Since enzymes are proteins, this was, in effect, a one gene–one protein hypothesis, a proposal that each gene controls the synthesis of a protein. Beadle and Tatum were awarded the Nobel Prize for their important work.

In the years since the one gene–one enzyme hypothesis was formulated, numerous examples have been found in which mutations of any one of two or more genes result in the same nutrient requirement, or in which mutation of a single gene results in two new nutrient requirements. Although these examples might appear, at first, to contradict the hypothesis, further study has shown that in many such cases the one-to-one relationship between genes and enzymes still holds. Mutation of two different genes can cause the same new nutrient requirement if, for example, the product of the reaction

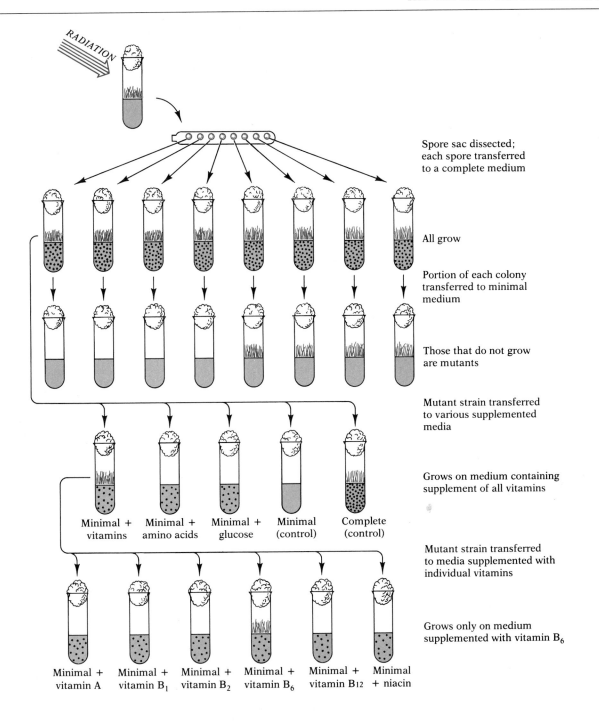

15.13 The experimental procedure of Beadle and Tatum

For discussion, see text.

catalyzed by the enzyme associated with the first gene is the necessary substrate for the reaction catalyzed by the enzyme associated with the second gene, as follows:

$$A \xrightarrow{\text{enzyme 1}} B \xrightarrow{\text{enzyme 2}} C$$

Absence of either enzyme might result in an inability to synthesize compound C, which might then have to be supplied preformed in the nutrient medium. Similarly, mutation of a single gene may result in two new nutritional requirements if the reaction catalyzed by the enzyme associated with the gene produces two different essential products, as follows:

$$D \xrightarrow{\text{enzyme}} E + F$$

In this case both E and F may have to be supplied preformed in the nutrient medium if the enzyme is absent.

A modification of the hypothesis   Studies of the genetics of sickle-cell anemia, a hereditary blood defect, have shown that it results from mutation of a single gene. In 1956 Vernon M. Ingram of Cambridge University demonstrated that the abnormal behavior of the erythrocytes in this disease is due to the substitution of one amino acid (valine) for another (glutamic acid) in the hemoglobin molecule. The molecule is made up of two identical half molecules; each half numbers about 300 amino acids and consists of two different polypeptide chains. Sickle-cell hemoblobin differs from normal hemoglobin in only one of the 300 amino acids in each half molecule. Yet this slight difference is sufficient to produce a serious, often fatal, disease.

This point is in itself remarkable, but also remarkable is the fact that mutation of a gene in the nucleus could cause such a precise change in a protein of the cytoplasm. Ingram's discovery served to strengthen the idea that genes encode the information that directs the synthesis of proteins, and it thus seemed to support the concept of a one gene–one protein relationship. But soon thereafter it was found that another hereditary blood disease, which involves alteration of a different polypeptide chain in the hemoglobin molecule, segregates independently from sickle-cell anemia and must, therefore, be controlled by a different gene. Here was an instance, then, where two different genes were responsible for the same protein.

This and numerous other examples indicate that when a protein is composed of two or more chemically different and independent polypeptide chains, each chain is determined by its own gene. Thus the hypothesis takes the new form: one gene–one polypeptide. But note that when a protein composed of two polypeptide chains is synthesized initially as a single chain, as is true for insulin (see Fig. 9.24, p. 363), then it is determined by a single gene; in such cases the chains are not considered independent, because they are held together by disulfide bonds (a part of their primary structure), not by weak bonds alone.

THE SYNTHESIS OF PROTEINS

If it is assumed that genes control the cell's activities by controlling the synthesis of protein enzymes, the next question becomes: By what mechanism is this gene-controlled synthesis of proteins accomplished?

RNA, TRANSMITTER OF INFORMATION FROM DNA TO RIBOSOMES

The problem of sequence We saw in an earlier chapter that polysaccharides such as starch and cellulose are long polymers. But those polymers are built up by serial repetition of identical units—molecules of glucose—and the order in which the building blocks are linked presents no problem. Proteins, too, are long polymers. But they are composed of some 20 different units—the amino acids—and the sequence in which those units are linked is of critical importance. It is the need to specify this sequence that makes protein synthesis so much more complex than polysaccharide synthesis.

If genes specify amino acid sequences for proteins, and if DNA is the stuff of which genes are made, then the sequence of bases in the DNA must in some way indicate the sequence in which amino acids are to be linked in protein synthesis. The problem then becomes one of translating nucleotide base sequence into amino acid sequence. One of the first points that had to be established was whether the translation is normally direct or indirect, i.e. whether protein is synthesized on the DNA template itself or whether some intermediate agent is involved. There had long been evidence that, even though some protein synthesis occurs on ribosomes in the nucleus, most of it takes place on cytoplasmic ribosomes. Since, with a few notable exceptions, DNA is largely restricted to the nucleus, it seemed likely that the protein is not synthesized along a DNA template as new DNA is. The information, presumably coded in nucleotide sequences in the DNA, must be transmitted from the nucleus to the sites of protein synthesis in the cytoplasm. How?

The transcription of information from DNA to RNA Several workers showed in the early 1940s that cells in tissues where protein synthesis is particularly active, such as the vertebrate pancreas, contain large amounts of ribonucleic acid (RNA). Since this nucleic acid is present in only limited quantities in cells that do not produce protein secretions, such as those in muscle and kidney, there seemed to be a definite correlation between protein synthesis and RNA. It had long been known that RNA, unlike DNA, occurs in the cytoplasm as well as in the nucleus; radioactive-tracer experiments revealed that RNA is synthesized in the nucleus and moves from the nucleus into the cytoplasm. All these lines of evidence pointed to the possibility that RNA might be the chemical messenger between the DNA of the nucleus and the protein-synthesizing cytoplasmic ribosomes.

Although RNA and DNA are very similar compounds, they differ in three important ways, as we saw in Chapter 2: (1) The sugar in RNA is ribose, whereas that in DNA is deoxyribose (Fig. 15.14). (2) RNA has **uracil** where DNA has thymine. (3) RNA is ordinarily single-stranded, whereas DNA is usually double-stranded.

Despite these differences, it was obvious that DNA could easily act as a template for the synthesis of RNA. The synthesis would proceed in essentially the same way as that of new DNA. The two strands of a DNA molecule would uncouple and RNA would be synthesized along one of the DNA strands (Fig. 15.15). For every adenine in the DNA template a uracil ribonucleotide would be added to the growing RNA strand; for every thymine in DNA an adenine ribonucleotide would be added, for every guanine a cyto-

15.14 Ribose and deoxyribose
The two five-carbon sugars differ only at the site shown in color, where deoxyribose lacks an oxygen atom that is present in ribose.

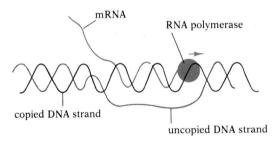

15.15 Transcription of a gene
As the polymerase molecule moves along the DNA, it catalyzes transcription of only one of the two DNA strands. Since the other strand is never transcribed, it is clear that only one strand carries the genes. What the function of the second strand might be—aside from acting as a template for production of a complementary strand when the DNA replicates—is obscure; by binding loosely to the gene-bearing strand, it may help mask its functional groups from mutagenic agents.

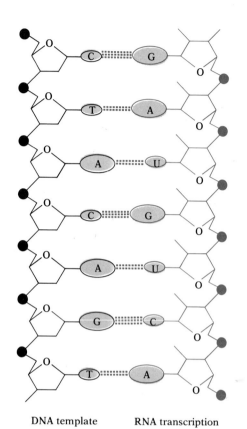

DNA template   RNA transcription

15.16   Transcription of RNA on a DNA template
The sugar in RNA (ribose) is slightly different from
that in DNA (deoxyribose), and uracil (U) takes the
place of the thymine in DNA.

sine, and for every cytosine a guanine (Fig. 15.16). In other words, the sequence of deoxyribonucleotides in the single DNA strand would determine
the sequence of ribonucleotides for the synthesis of an RNA molecule. This
process is called **transcription;** the information encoded in the nucleic acid
language of the DNA gene is transcribed into the RNA dialect. Once synthesized, the new RNA molecule could travel from its place of synthesis on the
chromosome of the nucleus to its functional location in the cytoplasm.

DNA, then, would act as a template or mold for the synthesis not only of
new DNA just before a cell undergoes division, but also of RNA when the cell
is metabolically active but is not dividing. In corroboration of this model it
has been possible to purify an enzyme (RNA polymerase) that can synthesize RNA in a test tube if single-stranded DNA (prepared artificially in the
laboratory) is added as a template. The RNA thus synthesized has a base
content complementary to that of the DNA, as the model predicts.

At least three major types of RNA are synthesized in the nucleus and
transported to the cytoplasm. One type, called **messenger RNA** (mRNA), is
responsible for carrying instructions for protein synthesis from the DNA in
the nucleus to the ribosomes in the cytoplasm. The indications are that each
molecule of mRNA is synthesized along a chromosomal gene (or genes) and
then moves through a "pore" in the nuclear membrane to the cytoplasm,
where it complexes with one or more ribosomes and acts as the template for
synthesis of protein molecules.

The role of the ribosomes in protein synthesis   The ribosomes, which are
essential for protein synthesis in living cells, are themselves partly composed
of RNA of a second type, called **ribosomal RNA** (rRNA). For a while it
seemed reasonable to suppose that the rRNA, which is complexed with a
large number of proteins in an extraordinarily complicated fashion to form
the ribosome, might provide important instructions for determining what
proteins are to be produced. But it soon became clear that this is not the
case. The rRNA, which is synthesized on the DNA of the nucleolus (see Fig.
16.7, p. 678), probably serves chiefly to bind other forms of RNA to the
ribosome temporarily. Each ribosome appears to function as an unspecific
device capable of synthesizing any protein for which it is supplied the necessary information by mRNA. It is thus analogous to a modern manufacturing machine that receives its instructions from a tape fed into it.

Each molecule of mRNA ordinarily becomes associated with four, five, or
more ribosomes, which move along it in sequence. As each ribosome moves
along the mRNA, it "reads" the information coded in the mRNA and builds a
polypeptide chain according to that information (Fig. 15.17). In other words,
as each ribosome moves along the nucleotide chain of the mRNA, it manufactures a polypeptide chain by translating the nucleotide sequence into an
amino acid sequence at a rate of about 35 amino acids per second. The
complex of ribosomes associated with a single strand of mRNA has been
called a **polyribosome** (or sometimes simply a polysome) (Fig. 15.18).

Each ribosome, when it is attached to mRNA, is made up of two subunits,
one about twice as large as the other (Fig. 15.17). The smaller subunit is
apparently responsible for binding the mRNA. The larger subunit contains
the enzyme that catalyzes the peptide bonding of amino acids. The two
subunits exist in the cytoplasm as separate entities when not carrying out
protein synthesis, i.e. when not complexed with mRNA.

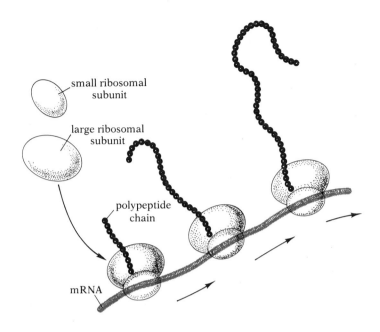

small ribosomal
subunit

large ribosomal
subunit

polypeptide
chain

mRNA

15.17   Synthesis of polypeptide chains by the ribosomes

Free ribosomes exist as two separate subunits. They form a complex upon initiation of protein synthesis on mRNA. As the ribosomes move along the mRNA, they "read" the coded information and synthesize a polypeptide chain according to that information.

THE GENETIC CODE

The messenger RNA that moves from the nucleus to the ribosomes carries information for protein synthesis. But if a protein is actually to be synthesized according to this information, the nucleic acid message must be translated into an amino acid sequence. Notice that this *translation* step represents a far more fundamental change than the transcription step discussed above; if transcription involved switching from one dialect of the nucleic acid language to another, translation involves switching from the nucleic acid language (written in nucleotide sequences) to the protein language (written in amino acid sequences).

To understand the process of translation, it is necessary to understand first how information about amino acids is encoded by the nucleic acid sequence in mRNA. There are only four different nucleotides in mRNA (corresponding to four nucleotides in DNA), which must be capable of coding for at least 20 different amino acids. Hence the coding unit (or "word") —the *codon*, as it is called—cannot be only one nucleotide long, because such a system could code for only four amino acids. Nor can the codon be two nucleotides long, because only 16 combinations (4^2) would be possible. If the codon is three nucleotides long, there are 64 possible combinations (4^3), which is more than enough to code for 20 amino acids. Therefore the codon must be at least three nucleotides long (on the assumption, for which there is ample experimental evidence, that all codons are the same length). Since nucleic acids must code an immense amount of information within a small space, natural selection may be expected to have favored the shortest codon system that would do the job—namely one no more than three nucleotides long. To be sure, what is expected and what actually is are not always the same. In this instance, however, convincing evidence, to be described later, supports the concept of a triplet codon.

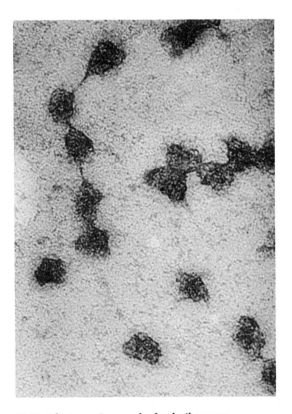

15.18   Electron micrograph of polyribosomes

Each polyribosome consists of a cluster of ribosomes in sequence along a strand of mRNA (seen here as a dark thread connecting the ribosomes). × 400,000. [Courtesy Alexander Rich, Massachusetts Institute of Technology.]

If there are 64 possible triplet codons, but only 20 amino acids to be specified, the question arises whether the extra 44 triplets ever occur and if they occur, whether they are meaningful or nonsensical. It is now known that all but three of the possible 64 triplets specify some amino acid. Thus there are synonyms in the genetic code. In some cases more than one triplet has the same meaning; e.g. the codons ACU, ACC, ACA, and ACG all specify the amino acid threonine. Such a code, in which some words are synonymous, is said to be a degenerate code.

The finding that the genetic code uses three-letter words to designate the various amino acids spurred an intensive effort to decipher the code, to construct a dictionary that would give the meaning of each triplet codon. Eventually this effort, which followed many different experimental routes, culminated in the dictionary reproduced in Table 15.2.

TRANSFER RNA AND ITS ROLE IN THE TRANSLATION OF MESSENGER RNA

Messenger RNA, synthesized on DNA in the nucleus, moves to the ribosomes and functions as a template for protein synthesis; the sequence of its nucleotide bases provides the information that determines the sequence of amino acids in the protein. Now, do the amino acids interact directly with the codons of the mRNA, or might some intermediate agent or adapter molecule be involved? It took years of investigation before this question could be answered with assurance.

In 1957 Mahlon Hoagland and his associates at Harvard University demonstrated that amino acids become attached to RNA *before* they arrive at the ribosomes. This RNA is neither mRNA nor ribosomal RNA, but a third type consisting of relatively small molecules that are not contained in organelles. It has been shown that each molecule of this RNA binds a single molecule of

TABLE 15.2 *The genetic code (messenger RNA)*

First base in the codon	Second base in the codon				Third base in the codon
	U	C	A	G	
U	Phenylalanine	Serine	Tyrosine	Cysteine	U
	Phenylalanine	Serine	Tyrosine	Cysteine	C
	Leucine	Serine	*Termination*	*Termination*	A
	Leucine	Serine	*Termination*	Tryptophan	G
C	Leucine	Proline	Histidine	Arginine	U
	Leucine	Proline	Histidine	Arginine	C
	Leucine	Proline	Glutamine	Arginine	A
	Leucine	Proline	Glutamine	Arginine	G
A	Isoleucine	Threonine	Asparagine	Serine	U
	Isoleucine	Threonine	Asparagine	Serine	C
	Isoleucine	Threonine	Lysine	Arginine	A
	Methionine	Threonine	Lysine	Arginine	G
G	Valine	Alanine	Aspartic acid	Glycine	U
	Valine	Alanine	Aspartic acid	Glycine	C
	Valine	Alanine	Glutamic acid	Glycine	A
	Valine	Alanine	Glutamic acid	Glycine	G

For several years the most profitable approach for assigning meaning to codons was to produce synthetic messenger RNA in a test tube, using a special enzyme that does not require a DNA template, but simply links nucleotides together in random order. If the only nucleotide made available to the enzyme was uracil, then a long polyuracil chain was synthesized:

UUUUUUUUUUUU

Similarly, if only adenine was made available, a polyadenine chain was formed. When poly-U was used in place of normal mRNA in a cell-free system for protein synthesis, a polypeptide chain composed of only phenylalanine resulted—a clear indication that UUU codes for phenylalanine. Similarly, AAA was shown to code for lysine, and CCC for proline.

Understandably enough, it was more difficult to assign triplets composed of two or three different nucleotides. If, for example, uracil nucleotide and guanine nucleotide were made available to the enzyme in a 2:1 ratio, mRNA in which the codons GUU, UGU, and UUG predominated was synthesized. It could be shown that when this mRNA was used as a template in polypeptide synthesis, the amino acids cysteine, valine, and leucine were the principal ones incorporated into the polypeptide, but it was impossible to determine by this technique which of the three triplets coded for which of the amino acids.

In 1964 Philip Leder and Marshall W. Nirenberg of the National Institutes of Health developed a technique for forming a complex between ribosomes and RNA trinucleotides (three nucleotides bonded together in sequence) of known composition. They showed that these trinucleotides would act as though they were short pieces of mRNA, and that transfer RNAs would couple with them. For example, if they used a trinucleotide with the composition UUU, phenylalanine tRNA would couple with it. Since it is relatively easy to synthesize trinucleotides with a particular base sequence, each of the 64 possible triplets could be synthesized, complexed with ribosomes, and exposed to a mixture of tRNAs. By establishing which tRNA coupled with which trinucleotide, it became possible to determine the meaning of many codons. A few trinucleotides proved to be not entirely specific in their binding; thus some ambiguity in codon assignments remained.

Shortly after the trinucleotide-binding technique became available, procedures for producing synthetic mRNA polymers with known repeating sequences (e.g. AAGAAGAAG . . .) were developed, and these helped resolve the remaining ambiguities in the code.

The information in Table 15.2 represents a synthesis of what was learned from the various experimental approaches.

A glance at the table reveals that all but two amino acids (methionine and tryptophan) are coded by more than one codon. It also shows that synonymous codons generally differ from one another only in the third of their three nucleotides (the exceptions are the codons for arginine, serine, and leucine, of which there are six each). Evidently the first two nucleotides in a codon are much more important than the third in determining specificity. Furthermore, U and C are always equivalent when they are in the third position, and A and G are equivalent in the third position in all but two cases (UG– and AU–).

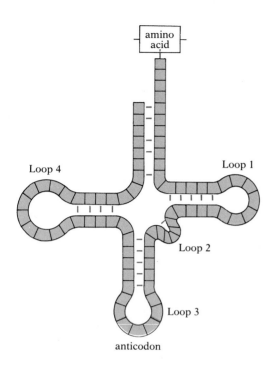

15.19   Cloverleaf model of transfer RNA

The single polynucleotide chain folds back on itself, forming five regions of complementary base pairing, four loops with unpaired bases, and an unpaired terminal portion to which the amino acid can attach. The unpaired triplet that acts as the anticodon (color) is on the third loop. Loop 1 is thought to function in binding the tRNA to the ribosome (probably by binding to rRNA), and loop 4 probably binds an activating enzyme. (The molecule is here shown flattened, with all coiling eliminated.)

amino acid and transports it to the ribosome; hence the name *transfer RNA* (tRNA).

At least one form of tRNA is specific for each of the 20 amino acids; the amino acid arginine, for example, will combine only with tRNA specialized to transport arginine, the amino acid leucine will combine only with tRNA specialized to transport leucine, and so on. All the different forms of tRNA share certain characteristics: a length of no more than 73 to 93 nucleotides; a single chain folded into a cloverleaf shape, with base pairing in some regions, but not in others (Fig. 15.19); termination with the nucleotide sequence CCA. When an amino acid bonds with its specific tRNA, it does so at the CCA end.

According to the model now generally accepted, specific enzymes are responsible for matching each amino acid with the appropriate tRNA. These enzymes, which recognize the structure of a particular amino acid and its tRNA, attach the amino acid to the tRNA, using ATP as an energy source. The tRNA carries the amino acid to a ribosome that is moving along a strand of mRNA (Fig. 15.20). There the tRNA becomes attached to the mRNA by base pairing of the Watson-Crick type. We have seen that the codon for each amino acid is a sequence of three bases on the single-stranded mRNA. Each tRNA has, in an exposed position, an unpaired triplet of bases called an *anticodon,* which is complementary to an mRNA codon for its particular amino acid.

If the codon CCG (i.e. cytosine, cytosine, guanine) codes for the amino acid proline, then, according to the model, one type of tRNA for proline will have the anticodon sequence GGC, which is complementary to CCG; similarly, if GUA (guanine, uracil, adenine) codes for valine, one type of tRNA for valine will have the anticodon CAU, which is complementary to GUA.[2] Whenever a molecule of this type of proline tRNA with an attached molecule of proline approaches a molecule of mRNA, its exposed GGC triplet can bond to the mRNA at only those points where the mRNA has a CCG triplet; similarly, the exposed CAU triplet of valine tRNA can bond to the mRNA only at points where a GUA triplet occurs.

Thus the sequence of three-base codons on the mRNA determines the precise order in which the different tRNAs with their attached amino acids will bond to the messenger molecule. This sequential binding of tRNAs to the mRNA automatically brings the amino acids to the ribosomes in the proper sequence. As each amino acid is brought to the ribosome by its tRNA, it is linked by a peptide bond to the length of polypeptide that has already been synthesized. Once a tRNA molecule has donated its amino acid to the growing polypeptide chain, it leaves the mRNA and moves away to pick up another load of amino acid and repeat the process. We see, then, that during protein synthesis a ribosome progresses step by step along the mRNA, picking up a correctly charged (i.e. loaded) tRNA and forming a peptide bond at each step. When the peptide chain has been completed, it is released from the ribosome to assume its functional role in the cell.

[2] In biochemistry the convention is to write both codons and anticodons beginning with the nucleotide nearest the free phosphate end of the molecule. This means that anticodons, as written, are complementary to codons read backwards. We have here deliberately violated this convention for the sake of ease of understanding.

THE READING OF THE CODE BY THE RIBOSOMES

Recognizing the proper codons   We have said that as a ribosome moves along a molecule of messenger RNA it translates the trinucleotide codons into amino acids that are added to the growing polypeptide chain. But how does the ribosome read the codons? Does it read them in an overlapping or nonoverlapping manner? Suppose that a particular piece of mRNA has the sequence of nucleotides

AUGCAGGUA

If the code were a completely overlapping one, this piece of mRNA would code for seven amino acids; AUG would indicate the first amino acid, UGC

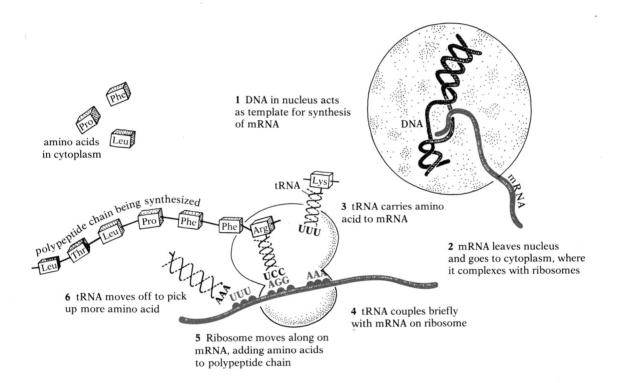

1 DNA in nucleus acts as template for synthesis of mRNA

amino acids in cytoplasm

polypeptide chain being synthesized

3 tRNA carries amino acid to mRNA

2 mRNA leaves nucleus and goes to cytoplasm, where it complexes with ribosomes

6 tRNA moves off to pick up more amino acid

4 tRNA couples briefly with mRNA on ribosome

5 Ribosome moves along on mRNA, adding amino acids to polypeptide chain

15.20   Model for control of protein synthesis by the genes
Messenger RNA is synthesized, one chain of the DNA gene serving as the template. This mRNA then goes into the cytoplasm and becomes associated with ribosomes. The various types of tRNA in the cytoplasm pick up the amino acids for which they are specific and bring them to a ribosome as it moves along the mRNA. Each tRNA bonds to the mRNA at a point where a triplet of bases complementary to an exposed triplet on the tRNA occurs. This ordering of the tRNA molecules automatically orders the amino acids, which are then linked by peptide bonds. Synthesis of the polypeptide chain thus proceeds one amino acid at a time in an orderly sequence as the ribosomes move along the mRNA. As each tRNA donates its amino acid to the growing polypeptide chain, it uncouples from the mRNA and moves away into the cytoplasm, where it can be used again. Note: The various molecules and organelles shown here are not drawn to scale with respect to one another or to the cell.

the second, GCA the third, CAG the fourth, etc., as follows:

AUG

UGC

GCA

CAG

AGG

GGU

GUA

If the code were a nonoverlapping one, this piece of mRNA would code for only three amino acids: AUG, CAG, and GUA—thus:

AUGCAGGUA

What sort of evidence would permit a choice between these alternatives? Notice that in a completely overlapping code each nucleotide (except the first two and the last two) would be included in the codons for three amino acids; thus the first G in our hypothetical piece of mRNA would be in the following three codons: AUG, UGC, GCA. By contrast, each nucleotide in a nonoverlapping code would be included in only one codon; e.g. the first G would appear only in the AUG codon. It is clear that if the code is completely overlapping, a mutation substituting one nucleotide for another in a nucleic acid strand should result in a change of three adjacent amino acids in the protein determined by this nucleic acid; but if the code is nonoverlapping, such a substitution should change only one amino acid in the protein. Now, Ingram's discovery that the hemoglobin of sickle-cell anemia differs from normal hemoglobin in only one amino acid in each half molecule would seem to indicate that a single mutation can result in an alteration of a single amino acid; it is apparently not necessary for three adjacent amino acids to be changed together.

A further consideration is that an overlapping code would limit the order in which amino acids could be arranged in a protein. Thus the amino acid determined by AUG could be followed only by an amino acid whose codon began with UG; there would be a maximum of only four such amino acids (determined by UGA, UGG, UGU, and UGC).[3] The amino acid coded by AUG could never be followed by one coded by CCG or by one coded by UAU, etc. But all attempts to demonstrate such limitations on amino acid sequence in proteins have failed. Scientists are now agreed that the genetic code is nonoverlapping.

The assumption that the code is nonoverlapping raises another kind of problem. If the codons in our hypothetical piece of mRNA are AUG, CAG, and GUA, how does the ribosome "know" that it should read these particular triplets and not such triplets as UGC and AGG? To put it another way, how does the ribosome recognize the proper triplet "words" if all these are strung together in one long RNA molecule? One possibility would be some sort of punctuation between the codons, as follows:

AUG,CAG,GUA

But no chemical basis for such punctuation has been found.

---

[3] In fact, one of these four codons—UGA—does not code for an amino acid; it is a so-called termination codon (see p. 651).

Another explanation has been proposed by Crick. He suggests that the sequence of triplets is "read" from a fixed point, i.e. that the synthetic machinery of the ribosome begins at one end of the chain of nucleotides and reads along the chain, taking each successive nonoverlapping triplet in turn. Thus, if AUG in our hypothetical example is at the front end of the chain (the end with a terminal phosphate group), the ribosome must read it first and must read CAG next and GUA next:

$$\underbrace{AUG} \quad \underbrace{CAG} \quad \underbrace{GUA}$$

Crick and his associates at Cambridge University performed a series of ingenious experiments (reported in 1961) that strongly supported both his suggestion and the idea that the codon is three nucleotides long. They treated bacteriophage T4 wih compounds called acridines, which tend to delete or insert nucleotides in DNA.

Let us suppose that the mRNA transcribed from a particular gene[4] begins with the following sequence of bases:

AUGCGUACCUUAUGGCUGCGUUUUGGC

This sequence should be read as follows:

$$\underbrace{AUG} \quad \underbrace{CGU} \quad \underbrace{ACC} \quad \underbrace{UUA} \quad \underbrace{UGG} \quad \underbrace{CUG} \quad \underbrace{CGU} \quad \underbrace{UUU} \quad \underbrace{GGC}$$

If Crick's hypothesis of reading from one end is correct, we would expect that deleting a single nucleotide near the front end would cause complete disruption of function in the nucleic acid, because the ribosome would read incorrect triplets from that point on to the other end; i.e. there would have been a so-called *frame shift* of the reading process. For example, if a mutation deleted the fourth nucleotide from the left (C) in our hypothetical sequence, then the ribosome—proceeding, let us assume, from left to right—would read the sequence as follows:

$$\underbrace{AUG} \quad \underbrace{\cancel{C}GU \quad ACC \quad UUA \quad UGG \quad CUG \quad CGU \quad UUU \quad GGC}$$

In other words, a completely different chain of amino acids would be synthesized; only the first amino acid would be the correct one.[5] Similarly, if a mutation added a nucleotide near the beginning of the chain, we would expect the nucleic acid to lose its function. For example, if an A were added in our hypothetical chain immediately after the first triplet, the ribosome would read the sequence as follows:

$$\underbrace{AUG} \quad \underbrace{A \quad CGU \quad ACC \quad UUA \quad UGG \quad CUG \quad CGU \quad UUU \quad GGC}$$

Again, the altered series of nucleotides would code for a polypeptide chain completely different from that coded by the original one. Crick found that when his acridine treatment produced single deletion or addition mutations

---

[4] We are here using a modified Beadle and Tatum definition of a gene as the length of DNA that codes for a single polypeptide chain.

[5] It is possible for a frame shift in the reading process to change the meaning, rather than destroy it. In the virus φχ174, in which two genes may overlap, the one gene is read as a frame shift of the other. Thus the mRNA sequence GAAGGAG may be read GAA GGA G, as part of one gene, and G AAG GAG, as part of another gene. How widespread overlapping genes are, or indeed whether they even occur in cellular organisms, is not yet known.

near the front end of the nucleic acid of a particular gene, that gene's normal function was destroyed just as predicted.

Crick also predicted that if a deletion or addition mutation occurred near the terminal end of the nucleotide chain (near the right end in our example), the gene's function would often be only slightly altered, because the ribosome would read correct triplets along most of the length of the gene. For example, if the fourth nucleotide from the right were deleted, the cell would read the sequence as follows:

$$\underset{\smile}{\text{AUG}} \quad \underset{\smile}{\text{CGU}} \quad \underset{\smile}{\text{ACC}} \quad \underset{\smile}{\text{UUA}} \quad \underset{\smile}{\text{UGG}} \quad \underset{\smile}{\text{CUG}} \quad \underset{\smile}{\text{CGU}} \quad \underset{\smile}{\text{UU}\cancel{\text{X}}} \quad \text{GGC}$$

The first seven triplet codons would be normal and would code for the normal amino acids; the eighth codon would have been changed from UUU to UUG and would code for an incorrect amino acid; and the last codon, normally GGC, would not be read at all because only two nucleotides (G and C) would remain. Thus the chain of nine amino acids coded by the normal gene would have been altered to a chain of eight amino acids, of which the eighth was incorrect. If the active site of this polypeptide were near the altered end, such a mutation would probably cause loss of all function. But if the active site were nearer the other end or the middle of the polypeptide chain, such a mutation might alter, but not entirely destroy, the enzyme's activity. In Crick's experiments, when the bacteriophage gene was altered by a single deletion or addition near the terminal end, some activity remained. (The alteration may look large in our example where the gene codes for a polypeptide chain only eight or nine amino acids long, but an actual gene would generally code for a protein that is several hundred amino acids long.)

We have seen that a single deletion mutation near the front end of the nucleotide chain destroys function. Crick reasoned that if this result is due to the fact that the ribosome reads incorrect codons beyond that point, and if the codons really are triplets, three deletion mutations close together should partly restore activity, because the ribosome would read correct codons beyond the point of the third deletion. For example, if three nucleotides were deleted, then the cell would read the sequence as

$$\underset{\smile}{\text{AUG}} \quad \cancel{\text{XGX}} \quad \underset{\smile}{\cancel{\text{X}}\text{CC}} \quad \underset{\smile}{\text{UUA}} \quad \underset{\smile}{\text{UGG}} \quad \underset{\smile}{\text{CUG}} \quad \underset{\smile}{\text{CGU}} \quad \underset{\smile}{\text{UUU}} \quad \underset{\smile}{\text{GGC}}$$

The first triplet and the last six would be the same in the altered version and in the normal one, and would code for the correct amino acids. The second and third triplet in the normal sequence (CGU and ACC) would have been replaced by a single GCC triplet. Thus the mutation would have resulted in a nucleic acid that codes for a polypeptide chain or enzyme similar to the normal one.

Such an altered enzyme might be expected to show at least a little of the activity characteristic of the normal enzyme, and Crick found that this is indeed the case. One or two deletions near the front end of the nucleotide chain destroyed activity, but a third deletion mutation in the same vicinity partly restored it. Similarly, three addition mutations located close together resulted in partial activity. And the combination of one deletion and one addition, if these were close together, gave partial activity. Thus Crick's experiments not only gave strong support to his hypothesis that the codons are read from a given point, but they also strengthened the widely accepted proposition that the codon is three nucleotides long.

Spacing Let us now turn to another important question about the genetic code: What designates the ending of one message and the beginning of another?

The chromosomes of viruses and bacteria, which are composed exclusively of nucleic acid, give every indication of being single long DNA molecules, with no break where one gene ends and another begins. There is a strong likelihood that all the DNA of a eucaryotic chromosome, too, is a single immensely long folded and coiled molecule. And there is convincing evidence in procaryotes that several adjacent genes are often transcribed together; thus a single mRNA molecule often encodes instructions for synthesis of several different proteins.

The end of the mRNA nucleotide sequence coding for a single polypeptide chain is now thought to be indicated by one of three so-called termination codons—codons whose sole function is to signal the end of the portion of the mRNA molecule that contains the instructions for synthesis of a single protein. Thus, in the sequence

$$\ldots \text{ GAA AAC AAA UAU } \text{UAA} \text{ GUG CGG UGC}$$

the termination codon UAA would signal the end of the stretch of mRNA specifying a protein whose last amino acid is the one coded by UAU. In other words, a termination codon functions as a period in the genetic language.

A SUMMARY OF THE MODEL

At the risk of being repetitious, let us now summarize the model of protein synthesis outlined here. When the double-stranded DNA that constitutes a particular gene is activated, its two nucleotide chains partially separate, and one of the chains acts as the template for synthesis of a molecule of single-stranded messenger RNA. This mRNA leaves the nucleus and moves into the cytoplasm, where it becomes associated with a cluster of ribosomes.

The mRNA, carrying information in the form of a nonoverlapping degenerate code, acts as the template for synthesis of polypeptide chains. As the ribosomes move along the mRNA, they read it three nucleotides at a time, starting at the end with the terminal phosphate group. Amino acids to be incorporated into the polypeptide chains are activated by ATP and picked up by molecules of transfer RNA, of which one type or more is specific for each of the 20 amino acids.

Each molecule of tRNA has an exposed anticodon (a sequence of three unpaired bases) complementary to the mRNA codon (also a sequence of three unpaired bases) that codes for its particular amino acid. After picking up an amino acid from the cytoplasmic pool, the tRNA moves to a ribosome and attaches to the mRNA at a point where the appropriate codon occurs. This ordering of the tRNAs along the mRNA molecule also orders the amino acids attached to them. As the amino acids are moved into the proper sequence in this way, peptide linkages are formed between them. Their amino acids once unloaded, the tRNAs uncouple from the mRNA and move

15.21 Isomerization of thymine
On rare occasions a hydrogen atom (color) changes position from the nitrogen to the oxygen. In its normal form thymine pairs with adenine, but in its enol form it pairs with guanine instead.

away to pick up another load. When a ribosome reaches a termination codon, it releases the completed polypeptide chain.

According to this model, then, the flow of information proceeds as follows: The DNA of the gene determines the mRNA, which determines protein enzymes, which control chemical reactions, which produce the characteristics of the organism.

MUTATIONS

Types of mutation Alterations in the DNA that change its information content and thus produce new alleles are called mutations. As a nucleotide sequence coding for one polypeptide chain, a gene is subject to a number of types of mutation.

We mentioned two while discussing Crick's experiments on how the genetic code is read: deletion of nucleotides from the sequence and addition of extra nucleotides to the sequence. As we have seen, additions or deletions (the latter are the more common) often result in production of inactive enzymes, because of frame shifts in the translation process. Of course, inactivation would also be the likely result of a large deletion even if there were no frame shift (i.e. even if the number of nucleotides deleted were a multiple of three), because so many amino acids would be missing from the enzyme synthesized.

A third type of mutation is ***base substitution*** (also called point mutation), the exchange of one nucleotide for a different one. A codon that normally has the base composition CGG, say, might be changed to CAG; this new triplet would code for a different amino acid. As a rule, base substitutions that produce termination triplets seriously reduce or destroy the activity of the enzyme coded by the gene, because only a portion of the normal polypeptide is manufactured. Base substitutions that result in replacement of one amino acid by another usually have less severe effects on the activity of the enzyme; its activity may be somewhat reduced or its sensitivity to environmental conditions may be altered, but rarely is there complete inactivation.

Base-substitution mutations are often due to the action of mutagenic agents, but sometimes they occur spontaneously as a result of mispairing of bases during DNA replication. As we have seen, C must normally pair with G, and A with T. But occasionally one of the bases may exist transiently in a slightly different isomeric form, one in which a hydrogen atom has changed position (Fig. 15.21). Since the hydrogen-bonding capacity of such an isomer is different from that of the usual form of the base, the base will pair with a different partner during replication. For example, when thymine has undergone this sort of isomerization, it will pair with guanine instead of adenine. If the isomerization has occurred in a strand of DNA that is functioning as a template for synthesis of a new strand, then that new strand will have a guanine in a position where there would normally be adenine.

Another type of mutation has been found in microorganisms, and probably occurs in higher organisms as well. This is ***intragenic recombination***, the recombination of parts of genes. For example, suppose that a chromosome bears allele *M*, a nucleotide chain perhaps 1,000 nucleotides long, and that the codon sequence of part of this chain is

TAC CTG AAA CGG AAA ATC GCA TTT

Now, suppose that a homologous chromosome bears a slightly different allele, *m*, and that the part of its base sequence corresponding to that shown above for *M* is

TAC CCG AAA CGG AAA ATC GGA TTT

Note that the two alleles differ in the second and seventh codons. If recombination of the first four bases in *M* with the last four bases in *m* were to occur, the new allele thus produced would have the base sequence

TAC CTG AAA CGG AAA ATC GGA TTT

It differs from allele *M* in its seventh codon and from allele *m* in its second codon.

Many scientists at first ruled out a process of breakage-fusion at points within a gene (in our hypothetical case at the point between the fourth and fifth triplets) as the mechanism of intragenic recombination, even though such a process was the accepted explanation for crossing-over (i.e. interchange of whole genes) between homologous chromosomes. They assumed that in higher organisms the break must occur at points on the chromosomes between DNA molecules, not within the molecules. But later, when it became apparent that the DNA of a given chromosome is probably a single very long molecule, even in higher organisms, it began to look more likely that intragenic-recombination mutations do indeed involve breakage and repair of DNA, i.e. that crossing-over can occur within the length of DNA that codes for a single protein. Most investigators now think that crossing-over can occur between any two nucleotides in the DNA molecule.

Mutagenic agents High-energy radiations, including both ultraviolet light and ionizing radiation such as X rays, cosmic rays, and emissions from radioactive materials, can cause genetic mutations. In addition, a variety of chemicals have been found to be mutagenic. A normal spontaneous mutation rate for a single gene would be one mutation in every 10^6–10^8 replications, but the rate can be greatly increased by unusual exposure to mutagenic agents.

Ionizing radiations sometimes induce simple point mutations, but they also frequently produce large deletions of genetic material, presumably because their high-energy particles collide with the DNA molecules and cause breaks to occur.

Ultraviolet light most often exerts its mutagenic effect by causing abnormal bonding of thymine bases. When two adjacent thymine units absorb ultraviolet light and are thus energized, they often bond to each other, forming a thymine dimer (Fig. 15.22). Because both thymines are in the same DNA strand, formation of a dimer results in inactivation of the DNA; not only can no mRNA be transcribed from it, but—more important for dividing cells—DNA replication cannot take place.

Some mutagenic chemicals produce their effects by directly converting one base into another. For example, nitrous acid (HNO_2) is a very powerful

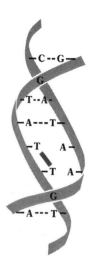

15.22 A thymine dimer within a DNA molecule
Two adjacent thymine nucleotides are bonded with each other, rather than with the adenine nucleotides of the complementary strand. Such a mutation, often induced by ultraviolet light, inactivates the DNA.

The discovery, about 1974, of a group of enzymes of microbial origin that break DNA chains at very specific points proved invaluable in tackling what had seemed, until then, an almost insuperable task—determining the nucleotide sequence of large DNA molecules. Called *restriction endonucleases,* these enzymes will cleave a strand of DNA only in places where certain specific nucleotide sequences occur. For example, the enzyme Eco RI recognizes the DNA sequence GAATTC, and the enzyme Hind III cuts DNA at the nucleotide sequence AAGCTT. Cleavage of DNA with a variety of restriction endonucleases produces fragments short enough for direct determination of nucleotide sequences.

In analyzing the sequence of nucleotides in the DNA fragments (which must be less than 100 nucleotides long), one end of each fragment in a homogeneous preparation is first labeled with a radioactive isotope of phosphorus. The fragments are then treated with a reagent that can break the chains wherever a particular base—let's say adenine—occurs; the treatment is kept mild enough for the chains to be broken at only a few of the many possible positions. The resulting smaller fragments are next sorted according to their lengths by gel electrophoresis. If, among those pieces of DNA that retain their radioactive label, fragments are found that are 6, 18, 37, and 43 nucleotides long, adenines must be present at positions 6, 18, 37, and 43. Similar treatment with enzymes that break the chains next to thymine, cytosine, or guanine permits determination of the positions where those bases occur. Finally, comparison of the results with those obtained when the process is begun with other restriction endonucleases—special attention being paid to overlapping sequences—makes it possible to work out the entire nucleotide sequence of a piece of DNA several genes long.

The sequencing of the nucleotides of genes has already yielded some surprising facts. Thus, while the existence of nongenic nucleotide sequences between adjacent genes had been known for years, the discovery in 1977 of nongenic sequences within genes was completely unexpected. Confirmation came from many sources. In 1978 two research teams, working independently, prepared a map of the human genomic (i.e. coding) region containing the genes for both the δ and the β polypeptide chains of hemoglobin.* They established that each gene is interrupted by a sequence less than 1,000 nucleotides

* The δ chains of hemoglobin occur in a minority of the hemoglobin molecules as a substitute for the β chains.

mutagen that deaminates cytosine, changing it into uracil.[6] Other chemical mutagens, of a type called base analogues, are themselves incorporated into nucleic acids in place of one of the normal bases. An example is 5-bromouracil, an analogue of thymine. When a strand of DNA contains a unit of 5-bromouracil instead of thymine, it is prone to errors of replication, because the 5-bromouracil will sometimes pair with guanine rather than with the correct adenine.

It seems certain that the changes just described occur far more frequently than was once imagined, but the vast majority of them are promptly rectified. It appears that proofreading, or editing, goes on simultaneously with DNA replication and that, when an error is detected, repair enzymes excise the incorrect region of the growing DNA molecule and splice in the correct version.

There is now a wealth of evidence for a close correlation between the mutagenicity of a chemical and its carcinogenicity, i.e. between the potential of a chemical for producing genetic mutations and its cancer-inducing ac-

[6] Nitrous acid also converts adenine into hypoxanthine and guanine into xanthine.

long and that the entire genomic region consists of the following parts: (1) the nucleotide sequence coding for the first portion of the δ chain; (2) an intervening sequence; (3) the sequence for the last portion of the δ chain; (4) an intergenic region about 7,000 nucleotides long; (5) the sequence coding for the first portion of the β chain; (6) an intervening region; and (7) the sequence for the last portion of the β chain. Nongenic sequences within genes are now known from a variety of organisms and are probably present in many eucaryotic cells, but their function is still unclear.[†]

Powerful tools in basic research on the structure of the gene, restriction endonucleases may find another important application in the diagnosis of certain human genetic diseases. They can be used, for example, in prenatal tests for δβ-thalassemia, a rare hereditary form of anemia. In accordance with a technique developed by Stuart

H. Orkin and his co-workers at Children's Hospital in Boston, δβ-thalassemia can be diagnosed early in a pregnancy by taking a small sample of amniotic fluid containing fetal cells, extracting the DNA from the cells, and treating the DNA with a restriction endonuclease such as Eco RI. Normally this enzyme yields four fragments containing δ- or β-globin gene sequences, but when a fetus is homozygous for δβ-thalassemia three of these fragments are missing.

Eco RI is likewise used in prenatal tests when O_{Arabia}, a mutant form of hemoglobin, is suspected. The gene for this mutant form has the DNA sequence AAATTC instead of the normal GAATTC; Eco RI, which cuts the normal gene wherever GAATTC occurs, cannot cut the mutant one that produces O_{Arabia}, and the fragments expected when the gene is normal fail to appear.

There is reason to think that sickle-cell anemia, in which the glutamic acid at position 6 in the β polypeptide chain is changed to valine ($β^{6\ Glu→Val}$), will also soon be detectable prenatally by use of restriction endonucleases. The mutation that causes substitution of valine for glutamic acid alters the DNA from GAGGAG to GTGGAG, and a recently discovered enzyme (Mnl I) that cuts DNA wherever the sequence GAGG occurs should be useful in detecting this mutation.

[†] Recent evidence suggests that, when a gene with an intervening region is transcribed, the intervening region is transcribed too. The product is a long RNA molecule, called mRNA precursor, that contains nonmessage portions. These portions are then cut out and the remaining functional regions spliced together to form the final mRNA product. In other words, the DNA is first transcribed, and then the product is edited before it is sent to the ribosomes to direct protein synthesis.

tivity. This fact strongly suggests that many cancers are caused, at least in part, by mutations in somatic cells. Thus we see that alterations of the DNA are important not only in germ cells, where they may affect future offspring, but also in somatic cells, whose metabolism or growth they may disrupt, causing disease or degeneration.

The connection between mutagenicity and carcinogenicity is the basis for the widely used Ames test,[7] in which such chemicals as environmental pollutants, reagents used in industrial processes, proposed new drugs, and food additives are screened for potential carcinogenicity. The compound to be tested is added to a culture of about one billion bacteria; a special mutant strain of *Salmonella* is used, which requires histidine as a nutrient. When the mixture of bacteria and chemical is incubated on a medium deficient in histidine, some cells undergo a mutation that is the reverse of their original mutation, and they thus regain the ability to synthesize histidine and to grow on the deficient medium. After several days a count is made of the number of so-called reverse-mutant colonies derived from these cells. The increase in

[7] Named for Bruce N. Ames of the University of California, Berkeley.

the mutation rate of the bacteria over the normal spontaneous level is then used to predict the likely cancer-inducing potency of the chemical.

In early screening tests some substances known to be potent carcinogens gave negative results in the Ames test. It was soon realized that many chemicals not themselves mutagens or carcinogens are transformed in the mammalian body, especially in the liver, into derivatives that are mutagenic and carcinogenic. For this reason newer versions of the Ames test add liver homogenate to the bacterial culture, so that the bacteria will be exposed both to the original chemical and to its metabolic derivatives.

NOW WHAT IS A GENE?

We have used the word "gene" in a variety of ways in this and the last chapter. Mendel did not use this term, but spoke of characters and their corresponding "factors" or "elements" in the germ cells. From 1902 onward it was generally agreed that the factors of inheritance, first called genes by Wilhelm Johannsen, a Danish botanist, in 1911, are associated with the chromosomes.

The work of T. H. Morgan with *Drosophila* led to a conception of the gene as the smallest unit of recombination. As indicated in the last chapter, two characters were regarded as determined by different genes if they could be recombined, and as determined by the same gene if they could not be recombined. Physically, the genes were thought of as tiny particles arranged in linear sequence on the chromosomes. Recombination between linked genes was explained by a crossing-over mechanism when the chromosomes broke at points between the particles.

Though defined essentially as the unit of recombination, the gene in classical genetics was also regarded as the unit of mutation and as the unit of function, i.e. as the smallest unit whose alteration by mutation would change the phenotype and as the smallest unit of control over the phenotype. This concept of the gene was satisfyingly unified. But from what is now known of the structure of the genetic material and its function it is clear that the units of recombination, mutation, and function are not identical.

All available evidence indicates that recombination by crossing-over can take place between any two nucleotides—which would make the unit of recombination only one nucleotide long. As we saw earlier, some mutations change only one nucleotide in a nucleotide chain; hence the unit of mutation is also only one nucleotide long, and is the same as the unit of recombination. The codon is larger, being three nucleotides long. And longest of all is the unit of function—sometimes called the cistron—which is usually the nucleotide chain determining one polypeptide chain. Since an average polypeptide chain contains about 300–500 amino acids, the average cistron must be 900–1,500 nucleotides long, and sometimes even longer.

According to a strict application of Morgan's definition, the gene, as the unit of recombination, would be only one nucleotide long. Such a gene would also be the unit of mutation. But it would be a very small entity, difficult to analyze and lacking in the functional attributes traditionally assigned to the gene. A second alternative would be to regard the gene and the codon as synonymous. The gene would then be a slightly larger entity, but it would lack almost all the attributes of the traditional gene. At the other extreme, the definition of the gene implicit in the one gene–one enzyme hypothesis of Beadle and Tatum would make the gene synonymous with the cistron; it would then be the unit of function, but not of recombination or mutation. Equating the gene with the cistron has the advantage of emphasizing physiological activity and of postulating a less complex relationship between genes and phenotypic characteristics. Today biochemical geneticists follow the one gene–one polypeptide principle and regard the gene as equivalent to the cistron. It is this definition, according to which genes are fairly large and complex entities, that we used in discussing the mechanism of gene action and mutations.

Although our discussion of the biochemical gene has focused on the function of determining a polypeptide chain, it must not be overlooked that some functional units of DNA, instead of determining polypeptide chains, act as templates for transfer RNA or ribosomal RNA. The functional definition of the gene is usually taken to include such units; thus it would perhaps be more accurate to say that the gene is the length of DNA that codes for one functional product. In addition, some parts of the DNA molecule serve exclusively in the control of genetic transcription. These are usually called "regions" rather than genes.

Besides the DNA in genes and control regions, there is much DNA of unknown function. Some of this nongenic DNA occurs between adjacent genes, but there are also other large regions of apparently nongenic DNA in chromosomes, especially near the centromeres. In some cases, as much as one fifth of all the chromosomal DNA may be nongenic.

It must be emphasized that, while most geneticists accept the biochemical definition of the gene, they often find it inapplicable in work on higher plants or animals. When they speak of the genes for vestigial wings or forked bristles in *Drosophila*, or of the genes for attached ear lobes or curly hair in human beings, or of the genes for tall plants or wrinkled seeds in garden peas, they have no proof that the entities in question would agree with the biochemical definition of the gene. For only a very small number of "genes" is anything known about the DNA of which they are composed or about the enzymes associated with their phenotypic expression. For practical purposes, most genes of higher organisms are—and will continue to be for many years—Morgan-type genes, determined by recombination of phenotypic traits. It may seem strange that geneticists continue to employ different definitions of the gene for different purposes. The fact is that, at the present stage of knowledge, one definition is more useful in one context and another in another context; a rigid terminology would only hamper the formulation of current ideas and research aims.

EXTRACHROMOSOMAL INHERITANCE

THE GENETIC ROLE OF CYTOPLASMIC ORGANELLES

Our attention in this and the preceding chapter has been focused on the chromosomal genes as the hereditary units that determine the characteristics of organisms. But are the chromosomal genes the only hereditary constituents of cells? It was recognized as early as 1909 that a few inherited traits do not obey Mendelian laws—notably certain traits apparently inherited exclusively from the mother, in which the sperm cell apparently plays no role whatever. Since almost all the cytoplasm of a zygote is contributed by the egg cell, it was suggested that such traits might be controlled by cytoplasmic rather than nuclear factors. It has only been in recent years, however, that much interest has centered on the possibility of cytoplasmic inheritance.

It is now known that such cytoplasmic structures as plastids and mitochondria are self-replicating, and that their characteristics are at least partly determined by hereditary control factors located within themselves. For example, if all plastids (and proplastids) are removed from a single-celled organism like *Euglena* (which can survive heterotrophically), neither the altered cell nor its asexually produced descendants can regain chloroplasts; nuclear genes cannot restore these important structures to the cell. But such a cell can be surgically reinfected with chloroplasts, and the new chloroplasts will divide and perpetuate themselves. If all the plastids in a strain of *Euglena* with normal chloroplasts are destroyed (by treatment with such chemical agents as streptomycin or antihistamine) and the cell is then reinfected with certain types of abnormal plastids, the abnormal plastids will multiply and will be passed on to the cell's descendants indefinitely. Since the nuclear genes are not altered by this experiment, the hereditary perpetuation of the plastid abnormality must be due to factors inherent in the plastids themselves.

Numerous experiments with a great variety of species of flowering plants have shown that many characteristics of the chloroplasts are inherited exclusively from the maternal plant, as would be expected if their controlling factors are cytoplasmic. It must be emphasized, however, that some characteristics of chloroplasts are inherited in a Mendelian manner and are apparently dependent on chromosomal genes. In other words, a cytoplasmic role in the inheritance of structures like plastids and mitochondria does not exclude influence by the nuclear genes on the phenotype of these organelles. The total phenotype is probably a result of the combined effects of nuclear and cytoplasmic determiners.

The indications are that, like nuclear genes, the units of inheritance in plastids and mitochondria are composed of DNA. But the number of cytoplasmic genes must be small compared to the number of nuclear genes, because the great bulk of the cell's DNA is in the nucleus.

PLASMIDS

Bacterial cells often contain, in addition to their principal chromosome, other small circular pieces of DNA that are free in the cytoplasm and replicate independently of the principal chromosome. These structures, called plasmids, have greatly contributed to our understanding of bacterial genetics and are now playing a central role in some of the most exciting new techniques of genetic analysis. Let us look briefly at several kinds of plasmids.

The sex factor in *Escherichia coli*   As we saw earlier, it was demonstrated in 1944, by Avery, MacLeod, and McCarty, that when live bacteria are transformed by materials extracted from dead bacteria of another strain, the transforming principle is DNA; genetic material from the dead bacteria becomes incorporated into the control apparatus of the live bacteria. This discovery of genetic exchange, even though the exchange might be regarded as accidental, reopened an old question—whether bacteria have any sort of normal genetic exchange and recombination comparable to those associated with sexuality in higher organisms. For years it had been assumed that bacteria lacked any form of sexuality, but now that assumption needed to be re-examined.

In 1946 Joshua Lederberg and Edward L. Tatum, then at Yale University, published the results of an elegant experiment they had performed to determine whether genetic recombination occurs naturally in *Escherichia coli*. These scientists used techniques similar to those developed earlier by Beadle and Tatum when working on *Neurospora*. They isolated two mutant strains of *E. coli*, each of which lacked the ability to synthesize a particular pair of nutrients; the first mutant was unable to synthesize nutrients A and B, and the second mutant was unable to synthesize nutrients C and D.[8] The mutants could be grown only on a medium that supplied the nutrients they could not synthesize. But when the two mutant strains were mixed on a minimal medium (one lacking all four critical nutrients), some healthy colonies were formed and these could subsist indefinitely on the minimal medium. Apparently the individuals in these colonies could synthesize all four nutrients. It seemed that they had somehow inherited both the first strain's ability to synthesize C and D and the second strain's ability to synthesize A and B. In short, they appeared to be the result of recombination of traits from the two original mutant strains. Lederberg and Tatum demonstrated that direct contact between the cells of the two strains was necessary for this recombination to occur. Since such contact is not necessary for transformation, they concluded that this recombination was not the result of transformation.

A few years later it was shown by several researchers that the recombination demonstrated by Lederberg and Tatum is brought about through a process called **conjugation**, which is analogous to sexual mating in higher organisms. Two bacterial cells come to lie very close to each other, and a narrow cytoplasmic bridge, or pilus, forms between them. This bridge is visible under the electron microscope (Fig. 15.23). Genetic material can pass through the bridge from one cell to the other.

It has been found that conjugation can occur only between cells of different mating types. A type F^- cell cannot conjugate with another F^- cell. Similarly, F^+ cannot conjugate with F^+. But F^- and F^+ can conjugate with each other. Apparently F^+ cells differ from F^- cells in containing in their cytoplasm a so-called **sex factor,** which is composed of DNA. This factor can easily be transferred from F^+, which acts as donor and is thus analogous to a male, to F^-, which acts as recipient and is thus analogous to a female. The products of $F^- \times F^+$ crosses are always F^+. If this were the only type of cross possible, one would expect that all cells would eventually be F^+ and that no further conjugation could take place. But conjugation in other types of crosses does not always convert the recipient cell from female into male, as we shall see.

---

[8] In the experiments by Lederberg and Tatum, nutrients A and B were the amino acid methionine and the vitamin biotin; C and D were the amino acids threonine and leucine. Similar experiments have since been performed with other pairs of nutrients.

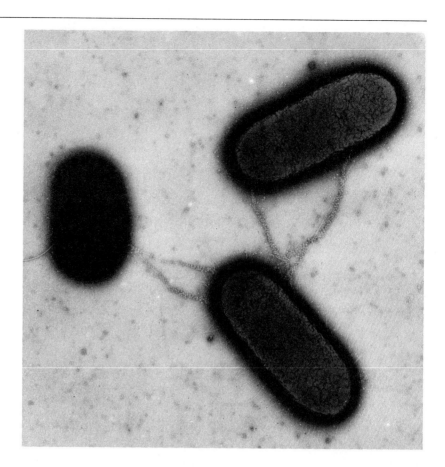

15.23 Conjugating bacteria
Long cytoplasmic bridges, or pili, connect the lower cell with two others with which it is conjugating simultaneously. × 19,500. [Courtesy L. G. Caro, University of Geneva, and R. Curtiss, University of Alabama.]

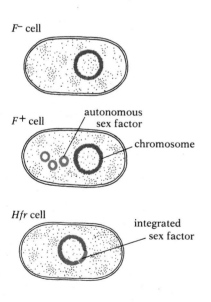

15.24 *F⁻*, *F⁺*, and *Hfr* cells of *Escherichia coli* compared
F⁻ cells lack the sex factor. *F⁺* cells have sex factor free in the cytoplasm. *Hfr* cells have sex factor integrated with the chromosome.

F^+ strains usually include a very few cells that, when they conjugate, do not ordinarily transfer sex factor to the F^- cells, but do transfer chromosomal DNA. These are called *Hfr* cells (for high-frequency recombination). Pure strains of *Hfr* bacteria have been isolated, and it is these that are generally used in experiments on conjugation. Because sex factor is not usually transferred, crosses of the $F^- \times Hfr$ type do not usually convert the F^- cells into F^+ or *Hfr*; the female remains a female.

Bacterial chromosomes, you will recall, differ from those of eucaryotic organisms in being composed almost exclusively of DNA and in being circular. But both electron microscopy and studies of the genetic results of conjugation indicate that, when chromosomal material is transferred from an *Hfr* cell to an F^- cell during conjugation, that material is linear, not circular. These observations have led to the following hypothesis:

The sex factor, which is a type of plasmid, is absent from F^- cells, but present in the cytoplasm of F^+ cells, where, presumably, several copies of it usually occur in each cell (Fig. 15.24). In this free cytoplasmic form the sex factor can easily be transmitted from cell to cell; for this reason mixing of F^- and F^+ cells usually results in conversion of many of the F^- cells into F^+ cells. At times, however, according to the hypothesis, the sex factor becomes integrated into the chromosome and the result is an *Hfr* cell; the integrated sex factor suppresses all autonomous sex factors. When an *Hfr* cell and an *F*⁻ cell begin conjugation, synthesis of a linear chromosome on the template of the circular *Hfr* chromosome is initiated at the point where the sex factor is located. This newly synthesized linear chromosome moves (by an unknown mechanism) through the conjugation bridge into the *F*⁻ cell (Fig. 15.25). Since the linear chromosome always moves during conjugation with the sex factor at the rear end, and since transfer of the chromosome is seldom complete, an *F*⁻ cell crossed with an *Hfr* cell does not ordinarily receive the sex factor and is therefore generally not converted into an *F*⁺ or an *Hfr* cell.

When two conjugating cells break apart before transfer of the linear chromosome is complete, as they usually do, the *F*⁻ cell retains whatever portion of the chromosome has already entered it. Thus the normally haploid cell becomes temporarily partially diploid; it has its own original chromosome plus part of a chromosome donated to it by the *Hfr* cell. Crossing-over soon occurs between the *F*⁻ cell's own chromosome and the recently acquired chromosomal fragment, so that some genes from the *Hfr* cell become part of the *F*⁻ chromosome. The remaining chromosomal fragment is then destroyed, and the *F*⁻ cell is restored to the haploid condition.

The genes transferred during conjugation always enter the *F*⁻ cell in a regular sequence characteristic of the particular *Hfr* strain being studied. Some genes are always transferred, even in conjugations that are very brief, while others are transferred only when conjugations last longer. This indicates that, for a given strain, the same end of the chromosome always enters the *F*⁻ cell first. Thus, if the genes on an *Hfr* chromosome were arranged in the sequence

ABCDEFGHIJKLMN

and if the *A* end were the leading end, then a short conjugation might transfer only genes *ABC*, a moderately long conjugation might transfer genes *ABCDEFGH*, and a longer conjugation might transfer genes *ABCDEFGHIJKL*. E. L. Wollman and François Jacob of the Pasteur Institute in Paris developed a technique for separating conjugating bacteria by agitating them in a blendor. They showed that the number of genes transferred is proportional to the length of time the conjugating pair is left undisturbed, which means that the chromosome moves at a constant speed. Hence the sequence and distance apart of the genes on a bacterial chromosome can be mapped by disrupting conjugation at carefully measured intervals after the mixing of *F*⁻ and *Hfr* strains and then determining which genes have entered the *F*⁻ cells (Fig. 15.25).

Comparison of the maps obtained for different strains of *Hfr* has revealed that the genes are always arranged in the same order, but that each different *Hfr* strain has a different gene at the anterior end of its chromosome. It seems that the sex factor is inserted into the circular bacterial chromosome at a different place in each strain. Since chromosome replication begins at the point where the sex factor is inserted, the linear chromosomes synthesized by the different strains during conjugation will have different genes at the front end (Fig. 15.26).

As this brief summary of sexuality in bacteria makes plain, the sex factor differs in its behavior from every other component of the cell so far dis-

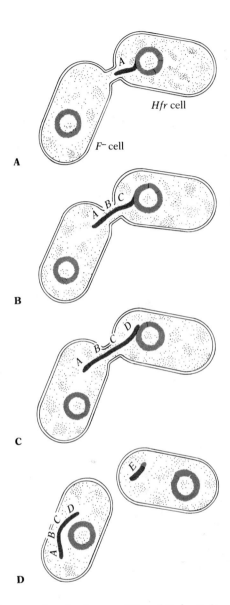

15.25 Conjugation between *Hfr* and *F*⁻ bacteria
As a linear chromosome is synthesized on the circular chromosome of the *Hfr* cell, it moves into the *F*⁻ cell. Conjugation usually stops before the entire chromosome has moved across; hence the sex factor (color) ordinarily remains in the *Hfr* cell. Since the chromosome moves at a fairly steady rate, disrupting conjugation at measured times after its inception permits mapping of the genes on the bacterial chromosome.

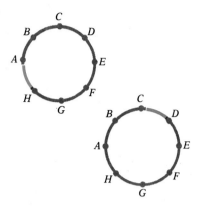

15.26   Model of integration of sex factor with the bacterial chromosome
The sex factor (color) is inserted into the circular chromosome at different points in different strains. Since the DNA replication that results in formation of a linear chromosome in conjugation begins at the sex factor, it follows that, although the order of the genes is the same from strain to strain, the genes at the anterior end of the linear chromosome are different. Thus the linear chromosome synthesized from the chromosome at left will move during conjugation with the gene sequence *ABC*..., while the one synthesized from the chromosome at right will move with the sequence *DEF*....

cussed. At times it is apparently free in the cytoplasm as a very tiny piece of circular DNA. At other times it is attached to the chromosome and behaves like other chromosomal genes. Jacob and Wollman have given the name *episomes* to plasmids that exhibit this dual behavior. All episomes may exist in the cell in either of two states: the autonomous or detached state, in which their replication is independent of chromosomal replication, and the integrated state, in which they are attached to the chromosome and replicate synchronously with it. Episomes are nonessential to the individual cell, as is illustrated by F⁻ strains of *E. coli*, which lack the sex factor; but when they are present, their effects may be very important.

Proviruses   Earlier in this chapter we mentioned experiments by Hershey and Chase showing that bacteriophage viruses inject their DNA contents into the bacterial cells they attack, but leave their protein coats outside. When virulent phage (i.e. phage that kill the bacteria they invade) do this, the viral DNA promptly takes control of the bacterial cell's metabolic machinery and puts it to work manufacturing new viral DNA and new viral protein. These two components are then assembled into new infective virions. About 20–25 minutes after the initial injection of the viral DNA, the bacterial cell lyses, or bursts, releasing as many as one hundred or more new virions, which may then attack other bacterial cells and start the *lytic cycle* over again (Fig. 15.27).

In the early 1920s it was discovered that when certain strains of bacteria are mixed with certain other strains the latter undergo lysis. The strains that cause other strains to lyse are called lysogenic strains. The explanation for lysogeny remained a mystery for many years. Then in 1953 André Lwoff and his colleagues at the Pasteur Institute in Paris found that, if they exposed lysogenic bacteria to ultraviolet light or X rays or various chemicals, the bacteria would lyse within an hour, releasing large numbers of infectious virions. Apparently the lysogenic bacteria had been carrying viruses within their cells, but these viruses had not become active and had not usurped the cell's metabolic machinery until exposed to the inducing action of the ultraviolet light, X rays, or chemicals. Presumably, however, the viruses within an occasional lysogenic bacterial cell become active normally, and the cell lyses. This release of infective viruses by a tiny fraction of the lysogenic cells would explain why mixing a lysogenic strain with a susceptible strain causes lysis of the cells of the susceptible strain.

The discovery by Lwoff that viruses can be present in an inactive state inside their host cells showed that some viruses must be temperate rather than virulent. Virulent phages invariably kill their hosts. Temperate phages may or may not kill their hosts, depending on a variety of conditions. When they do not kill their hosts, their injected DNA becomes associated with the bacterial chromosome at a definite location, and the *lysogenic cycle* begins. While integrated into the bacterial chromosome, the viral DNA behaves as an additional part of that chromosome: It is replicated with the rest of the chromosome; it can be transferred from one cell to another during conjugation; its genes can undergo genetic recombination with bacterial genes; and its genes can even produce phenotypic effects in the host bacterium, such as modifications of colony morphology, changes in the properties of the cell wall, and changes in the production of antigens and enzymes. It has

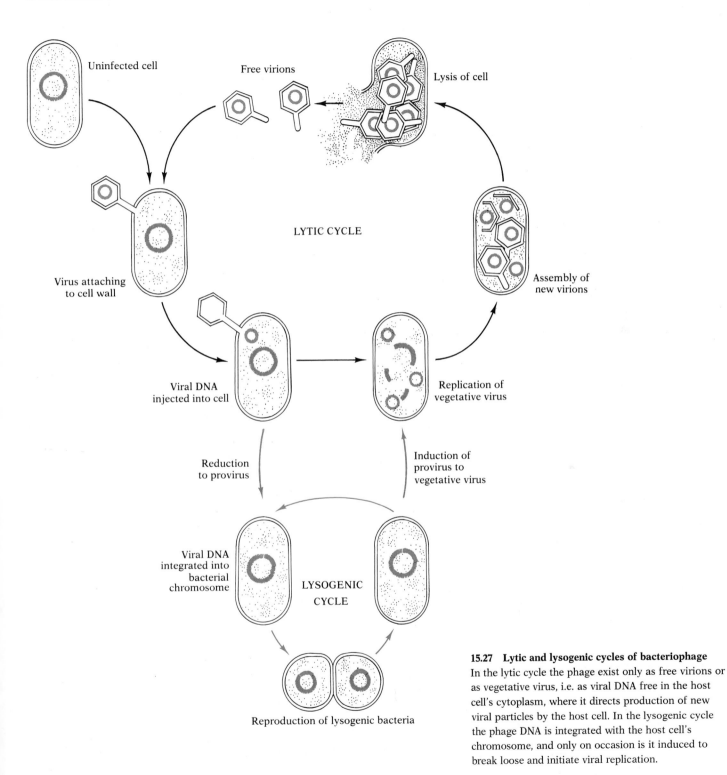

15.27 Lytic and lysogenic cycles of bacteriophage
In the lytic cycle the phage exist only as free virions or as vegetative virus, i.e. as viral DNA free in the host cell's cytoplasm, where it directs production of new viral particles by the host cell. In the lysogenic cycle the phage DNA is integrated with the host cell's chromosome, and only on occasion is it induced to break loose and initiate viral replication.

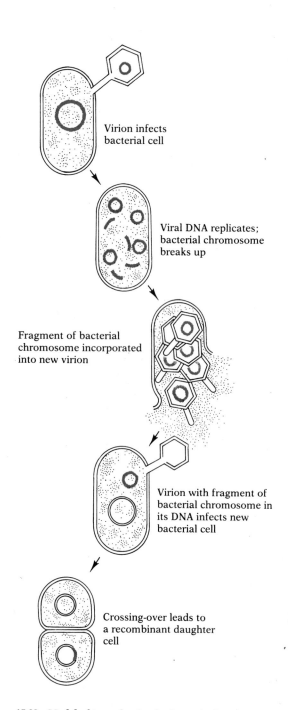

Virion infects
bacterial cell

Viral DNA replicates;
bacterial chromosome
breaks up

Fragment of bacterial
chromosome incorporated
into new virion

Virion with fragment of
bacterial chromosome in
its DNA infects new
bacterial cell

Crossing-over leads to
a recombinant daughter
cell

15.28   Model of transduction by bacteriophage

been shown, for example, that diphtheria bacteria can produce the toxin that causes the disease only if they are carrying a specific type of viral gene. And viral genes confer on the bacterium in which they reside immunity from further infection by the same type of virus. We see, then, that the DNA of temperate viruses has the properties of an episome for the bacterium. It can exist in an autonomous or vegetative state, replicating independently and eventually destroying the cell, or it can exist in the integrated state, as a **provirus,** functioning and replicating as a portion of the bacterial chromosome (Fig. 15.27).

Not only do viral genes sometimes act as bacterial genes, but the reverse is also true. When temperate viruses are in the vegetative state and have put the bacterial cell to work making more viruses, small fragments of the bacterial chromosome may become enclosed in the new viral coats. If a temperate virus carrying bacterial DNA in this manner infects a new host, it will inject both viral and bacterial DNA into this new host. Sometimes the injected bacterial genes undergo recombination with the new host's genes (presumably by a process like crossing-over). The virus has thus acted as a vehicle for transferring genes from one bacterial cell to another (Fig. 15.28). This process, called **transduction,** was first described in 1952 by Norton D. Zinder and Joshua Lederberg at the University of Wisconsin. Since the likelihood that two bacterial genes will be cotransduced depends on how close together they are, assessing the frequency of cotransduction has proved a very sensitive way to map bacterial chromosomes.

As you will have gathered from this discussion, it is not easy to distinguish between bacterial and viral genetics. A piece of DNA functioning as a bacterial gene at one time may function as a viral gene a little later. Thus it is not always meaningful to distinguish between bacterial and viral genes. If viral DNA can act as episomes in the cells of higher organisms (as it almost certainly can) and if viruses can transduce genes in higher organisms (as they probably can, though this has not yet been proved), then the genetics of viruses and the genetics of higher organisms also become intertwined. Some of the genes that produce important phenotypic effects in the human body may be proviral. And the genes of some viruses may have been derived originally from human chromosomes. If these possibilities make genetics far more complex and confusing than it seemed a few decades ago when our knowledge did not go much beyond the Mendelian laws, they also make this subject one of the most exciting and promising of modern biology.

Nonepisomal plasmids   Most bacterial cells contain, in addition to episomes, plasmids that do not have the ability to integrate into the main chromosome. In other words, unlike episomes, these plasmids exist only in autonomous form. Some such plasmids carry only one or two genes, while others may be as much as one fifth the size of the main chromosome and carry many genes.

Especially important are plasmids with genes for resistance to various antibiotics, including streptomycin, tetracycline, and ampicillin (see Fig. 3.46, p. 121). When bacteria with resistance plasmids are exposed to one of these antibiotics, the plasmids immediately begin replicating, and a cell that originally had only two or three may soon have a thousand or more.

Plasmids have recently come into the scientific limelight, because of their role in a new technique for inserting genes from one kind of organism into cells of another kind of organism. This ***recombinant DNA*** technique has become a subject of public debate, because of concern over accidentally producing new pathogens or developing "genetic monsters." Let us take a look at how the technique works.

Plasmids, like DNA from bacterial chromosomes, can transform bacterial cells. In other words, if bacterial cells are placed in a medium containing free plasmids, some of the cells will pick up plasmids, just as some of the nonvirulent pneumococci studied by Griffith picked up DNA that had been released into the medium by dead virulent pneumococci (see p. 627). Thus nonresistant bacteria can be transformed into resistant ones by exposure to a medium in which bacteria with plasmids for antibiotic resistance have been killed. Recombinant DNA technology makes use of this transforming potential of plasmids: Purified plasmids are modified by the addition of foreign genetic material; when bacterial cells then pick up the modified plasmids, they acquire the foreign genes.

To follow the procedure in more detail: Bacterial cells containing plasmids are broken up and their DNA is extracted. The DNA is then subjected to differential centrifugation to separate the plasmids from the bacterial chromosomes. The purified plasmids are next exposed to a restriction endonuclease, an enzyme that cleaves the DNA circle next to very specific nucleotide sequences (Fig. 15.29). The plasmid DNA, which is now linear with "sticky" ends where it was cleaved, is then mixed with fragments of foreign DNA prepared with the same restriction endonuclease and therefore equipped with "sticky" ends complementary to those of the plasmid DNA. In such a mixture the plasmid DNA and the foreign DNA will spontaneously anneal by complementary base pairing, re-forming a circle in the process. The backbones (i.e. chains of alternating phosphate and sugar groups) of the DNA circle can then be sealed with a special enzyme (DNA ligase). The end result is plasmids that contain a graft of foreign genetic material. All that remains to be done is to mix these plasmids with bacterial cells (usually treated with a calcium salt to make them more permeable). The bacterial cells will pick up the modified plasmids, which include both the original plasmid genes and the foreign genes.

Recombinant DNA technology makes it possible not only to study the activity of genes from eucaryotic cells in the simpler environment of a procaryotic host cell, but also, in some cases, to use the host cells as chemical factories to produce substances of medical importance. For example, if a gene for insulin were inserted in plasmids with antibiotic resistance, and then these plasmids were used to transform nonresistant bacterial cells,[9] the cells would replicate the modified plasmids hundreds, perhaps thousands, of times upon exposure to the antibiotic. Possessing so many copies of the insulin gene, the bacterial cells could then produce enormous quantities of insulin, which could be harvested for use in treating diabetics.

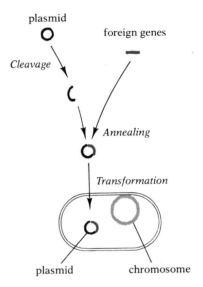

15.29 The recombinant DNA technique
Plasmids, removed from a donor cell, are cleaved by a restriction endonuclease. They are then mixed with fragments of DNA (color) from another cell, produced with the same restriction endonuclease. The plasmid DNA and the foreign DNA anneal (join) at their sticky ends, forming a circle. The resulting new plasmids, bearing foreign genes, are added to medium containing live bacterial cells. The cells pick up the modified plasmids, and are thus transformed into cells possessing the foreign genes.

---

[9] Transformation of nonresistant cells into antibiotic-resistant ones is, according to the rules governing this research, done only in types of bacteria in which resistance commonly occurs normally, never in bacteria that would be unlikely to acquire resistance under normal conditions.

The same sort of procedure could potentially be used to produce large quantities of other hormones that are difficult to synthesize, or to produce antibodies against a variety of diseases. Looking further into the future, some researchers have even suggested using recombinant DNA technology to transfer genes for C_4 photosynthesis to some crop plants that normally depend exclusively on C_3 photosynthesis, for the sake of greater productivity under less than optimal growing conditions. Others have proposed transferring genes for the fixation of atmospheric nitrogen to crop plants, thus eliminating the need for application of nitrogenous fertilizers.

Recognition of the risk of undesirable side effects of recombinant DNA technology has led to strict regulation of how the technology is to be used. Laboratories must be equipped with facilities for sterile handling and for physical containment similar to those long used in medical microbiology laboratories where dangerous pathogens are handled. In addition, there is a consensus that the microorganisms used in recombinant DNA research should be from strains carrying crippling mutations that make them incapable of surviving outside the laboratory.

SUGGESTED READING

BROWN, D., 1973. "The isolation of genes," *Scientific American*, August. (Offprint 1278) *How a molecule of RNA can be used to locate and purify the stretch of DNA that coded for it.*

CAMPBELL, A. M., 1976. "How viruses insert their DNA into the DNA of the host cell," *Scientific American*, December. (Offprint 1347)

CLOWES, R. C., 1973. "The molecule of infectious drug resistance," *Scientific American*, April. (Offprint 1269) *Experiments demonstrating that the bacterial genes for drug resistance are carried on plasmids and can be transmitted from one bacterium to another.*

COHEN, S. N., 1975. "The manipulation of genes," *Scientific American*, July. (Offprint 1324) *The recombinant DNA technique explained in considerable detail.*

CRICK, F. H. C., 1954. "The structure of the hereditary material," *Scientific American*, October. (Offprint 5) *On the discovery of the structure of DNA—an account written by Crick himself just two years after first publication of the monumental Watson-Crick model.*

CRICK, F. H. C., 1962. "The genetic code," *Scientific American*, October. (Offprint 123) *The experiments with which Crick helped confirm that the genetic code is nonoverlapping and that the codon is three nucleotides long.*

DEVORET, R., 1979. "Bacterial tests for potential carcinogens," *Scientific American*, August. (Offprint 1433) *The relationship between mutagenicity and carcinogenicity. The tests described include the Ames test.*

FIDDES, J. C., 1977. "The nucleotide sequence of a viral DNA," *Scientific American*, December. (Offprint 1374) *How the base sequence of the entire chromosome of the bacterial virus φχ174 was worked out.*

KAPLAN, M. M., and R. G. WEBSTER, 1977. "The epidemiology of influenza," *Scientific American*, December. (Offprint 1375) *On the possibility that genetic recombination between human and animal strains of the influenza virus may be responsible for the appearance of new subtypes of the virus.*

RICH, A., and S. H. KIM, 1978. "The three-dimensional structure of transfer RNA," *Scientific American*, January. (Offprint 1377) *The way the three-dimensional structure of tRNA was worked out, and how that structure helps in understanding the mechanism of function of tRNA.*

WATSON, J. D., 1968. *The double helix.* Atheneum, New York.   *A fascinating account of the discovery of the structure of DNA by one of the two discoverers. Watson spares neither himself nor others in giving a rare behind-the-scenes look at the dynamics of research in a very competitive field.*

WATSON, J. D., 1976. *Molecular biology of the gene,* 3rd ed. Benjamin, Menlo Park, Calif. *A particularly well-written text on molecular genetics. Makes even difficult topics easy to understand.*

Chapter 16

DEVELOPMENT: ASPECTS OF CELLULAR CONTROL

By the process of mitosis every cell in the body of a multicellular organism apparently inherits, in the form of the nucleotide sequences of its DNA, the full set of genetic information that is the evolutionary endowment of its lineage; but though every cell receives the same set of instructions as every other cell in the organism's body, it neither looks nor functions like every other cell. A mesophyll cell in the leaf of a plant and an epidermal cell in its root are structurally different and have different functional capabilities; though their DNA is identical, they synthesize different sets of proteins. It is plain that they have followed different developmental paths.

Just as cells from different parts of an organism differ in structure and function, so a single cell differs in its activities in accordance with its stage of development and in response to changed conditions. An intestinal cell must perform very different biochemical functions depending on whether an animal has just eaten or whether it has gone for a long time without food; it must synthesize different enzymes in each case.

What are the control mechanisms that determine for each cell when and how it will act on the genetic instructions it has inherited? That is the question that must be answered if we are to understand how the genes of an organism direct its development and regulate its continuing activity.

THE CONTROL OF GENE TRANSCRIPTION

One obvious way in which two cells with the same genetic endowment might come to perform different tasks would be for each cell to act on a different part of the genetic message, just as one carpenter, following the building plans for a house, might erect the walls of the living room, and another, using the same set of plans, might lay flooring in a bedroom. In other words, cells may be selective as to which of their many inherited instructions they actually use. They may make the proteins specified by some of their genes and not those specified by others of their genes. One way to exert such selectivity would be to transcribe some genes, but not others—to synthesize messenger RNA along the DNA of some genes, but not of others.

CONTROL OF GENE TRANSCRIPTION IN BACTERIA

Bacterial cells, which have but one chromosome that has little protein complexed with it, are genetically far simpler than eucaryotic cells. For this reason, and because bacteria can be grown rapidly in huge numbers under controlled conditions in the laboratory, much of our current knowledge of the control of gene transcription comes from studies of these microorganisms, especially *Escherichia coli.*

The Jacob-Monod model of gene induction   Studying enzyme synthesis in *Escherichia coli,* the French biochemists François Jacob and Jacques Monod conceived a model, as awesome as it was suggestive, of gene regulation in bacterial cells. They worked mainly with the enzyme *β*-galactosidase, which, along with other enzymes, catalyzes the breakdown of lactose.

Normally, they found, production of the enzyme depends on the presence of an inducer—lactose or, more likely, a derivative of lactose; in other words, the normal enzyme was an ***inducible enzyme.*** But they found a mutant strain in which the same enzyme was a ***constitutive enzyme;*** i.e. it was an enzyme whose production was continuous, apparently uninfluenced by control substances such as inducers.

By recombination experiments, Jacob and Monod were able to demonstrate the participation of two genes in the production of *β*-galactosidase: a ***structural gene,*** which specifies the amino acid sequence of the enzyme, and a ***regulator gene,*** which controls the activity of the structural gene. They proposed that the regulator gene, which is located at some distance from the structural gene, normally directs the synthesis of a ***repressor*** protein that inhibits transcription of the structural gene. The mutant allele of the regulator gene, they concluded, lacks the ability to direct synthesis of repressor; hence it cannot prevent transcription of the structural gene, which is thus left free to direct continuous protein synthesis.

Later, Jacob and Monod discovered that a special region of DNA contiguous to the structural gene for *β*-galactosidase determines whether transcription of the structural gene will be initiated; they called this special region the ***operator*** (Fig. 16.1).[1] The operator, they established, also determines whether two other structural genes located in sequence with that for *β*-galactosidase will be transcribed; these genes code for enzymes that, like *β*-galactosidase, catalyze reactions involved in the breakdown of lactose. Jacob and Monod called the combination of the operator and its three associated structural genes an ***operon.*** They suggested that the repressor protein coded by the regulator gene exerts its effect on the operon by combining with the operator and inhibiting it.

Still later, yet another region of control DNA, the ***promoter,*** which is

[1] Because the operator, a region of control DNA, apparently does not code for any product, it does not qualify as a gene, according to the now generally accepted definition.

located next to the operator and may actually overlap it, was found to be essential for transcription of the structural genes in the operon. The promoter is the site on the DNA where RNA polymerase, the enzyme that catalyzes synthesis of mRNA on the DNA template, binds by weak bonds to the DNA in order to initiate transcription.

The current version of the Jacob-Monod model of control of the so-called *lac* operon—the operon responsible for the synthesis of enzymes that catalyze the breakdown of lactose—can be summarized as follows:

Before transcription of the three structural genes can begin, RNA polymerase must bind to the promoter region of the operon. This binding is blocked if the repressor protein (coded by the regulator gene) has already bound to the operator region (Fig. 16.1A). As long as active repressor is present in the nucleus, no binding of polymerase—and hence no transcription—can occur.

A OPERON REPRESSED

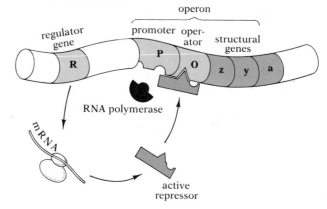

B OPERON DEREPRESSED

16.1 The *lac* operon, example of an inducible operon
(A) The operon consists of an operator region, a promoter region, and three structural genes (z, y, and a). The regulator gene, located some distance away, codes for mRNA, which is translated on the ribosomes and determines synthesis of repressor protein. When the repressor protein binds to the operator, it blocks the promoter's binding site for RNA polymerase and thus prevents transcription of the structural genes.   (B) Binding of inducer (in this case lactose) to the repressor may inactivate the repressor. Under these conditions the RNA polymerase becomes free to bind to the promoter region and initiate transcription of the structural genes. These, transcribed as a unit, determine production of polycistronic mRNA. The mRNA complexes with ribosomes in the cytoplasm and is translated into three enzymes (enzyme I is β-galactosidase), each of which catalyzes one of the reactions of the biochemical pathway of metabolic breakdown of lactose.

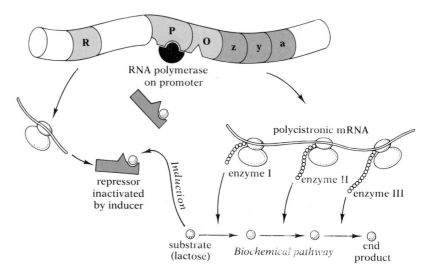

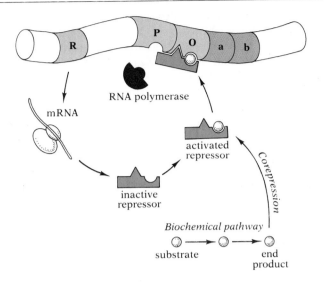

16.2 A repressible operon
The repressor protein coded by the regulator gene is initially inactive. Only if it binds with a corepressor molecule (often the end product of a biochemical pathway) can it then bind to the operator and block transcription of the structural genes. (In the absence of corepressor, the RNA polymerase could bind to the promoter, and transcription could proceed unhindered.)

If inducer is present, however, it will bind to the repressor, causing a conformational change in the repressor that prevents its binding to the operator; in short, the inducer inactivates the repressor.[2] Now free to bind to the promoter, RNA polymerase can initiate transcription of the structural genes and production of mRNA (Fig. 16.1B). The mRNA, carrying the instructions of all three structural genes (i.e. it is polycistronic), leaves the nuclear region and complexes with ribosomes in the cytoplasm, where its information is translated as the three enzymes necessary for lactose metabolism are synthesized.

According to the Jacob-Monod model, then, the condition of the operator region is the key to whether or not the *lac* operon will be activated. If repressor protein is bound to the operator, there will be no transcription. If no repressor is bound to the operator (because the repressor has been inactivated by inducer), transcription can proceed freely.

Notice that the three jointly controlled structural genes of the *lac* operon specify enzymes with closely related functions. It is characteristic for the structural genes of an operon to determine the enzymes of a single biochemical pathway; thus the whole pathway can be regulated as a unit. The adaptive advantage of such coordinated control is obvious.

Gene corepression In the years since the Jacob-Monod model was first proposed, it has become apparent that not all operons are regulated in the same way as the *lac* operon, which is an inducible operon, i.e. one that must be turned on by an inducer substance. Many operons are, instead, continuously on unless turned off by *corepressor* substances. In such cases the repressor protein coded by the regulator gene is inactive when first produced. Only if a corepressor substance binds to it, can it bind to the operator and block RNA polymerase binding (Fig. 16.2). Enzymes coded by structural

[2] It is important to remember that cytoplasmic substances are called inducers or corepressors according to whether their effect is to induce or repress gene activity, not according to whether they activate or inactivate repressor substance.

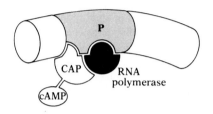

16.3 Induction of gene transcription by the cAMP-CAP complex
Binding of cAMP-CAP to the promoter region of the *lac* operon facilitates binding of RNA polymerase. The cAMP necessary to initiate this process is most abundant in the cell when glucose is in short supply and metabolism of lactose as an alternative source of energy becomes important.

genes in operons controlled in this way are called ***repressible enzymes.*** Unlike inducible enzymes, which are synthesized only if their operon is turned on by an inducer, repressible enzymes are automatically synthesized unless their operon is turned off by a corepressor.

An inducer is usually either the first substrate in the biochemical pathway being regulated or some substance closely related to that substrate. Thus the inducer for the *lac* operon is lactose or a substance derived from it. It is not surprising, therefore, to find that a corepressor is usually the end product of the biochemical pathway being regulated, or a closely related substance. For example, the enzymes for synthesis of amino acids are repressible enzymes, and it is the amino acids themselves that act as corepressors. Thus *E. coli* grown on histidine-deficient medium will steadily produce the enzymes necessary for synthesizing histidine, but if histidine is added to the medium, production of the enzymes soon stops. In both substrate induction and end-product corepression, gene transcription is regulated by the cellular substances most affected by the transcription—a truly elegant functional arrangement.

Positive control of gene transcription Yet another complication has emerged since the Jacob-Monod model was originally proposed. Some operons, it seems, are regulated in more than one way. The *lac* operon is a case in point.

This operon tends to be induced (i.e. turned on) when lactose is abundant in the cell. But if the cell also contains an abundance of glucose, the operon remains off; the cell preferentially metabolizes glucose. Only when the concentration of glucose falls will it turn to lactose as an alternative source of energy, and a second mechanism for turning on the structural genes concerned with lactose metabolism comes into play. As the concentration of glucose falls, the concentration of cyclic AMP rises. The cAMP binds with an activator protein known as CAP (catabolic gene activator protein), and the cAMP-CAP complex then binds to the promoter region of the operon. The binding of this complex facilitates the binding of RNA polymerase to the promoter and thereby makes possible the transcription of the structural genes (Fig. 16.3).

Notice that, whereas induction turns on gene transcription in an essentially negative way, by preventing repressor from blocking RNA polymerase binding, control by the cAMP-CAP complex is positive. The binding of the complex to the promoter region of the DNA directly activates the promoter and stimulates the initiation of transcription via polymerase binding.

We have here outlined three different systems for control of gene transcription in bacterial cells—substrate induction, end-product corepression, and cAMP-CAP activation. These are three of the best understood systems, but there are others, some of them considerably more complex and involving the interaction of as many as three or more control substances and, sometimes, other DNA control regions in addition to the regulator genes, operators, and promoters discussed here. In short, there appears to be an amazing variety of control systems even in so apparently simple a cell as that of *Escherichia coli.*

CONTROL OF GENE TRANSCRIPTION IN EUCARYOTIC CELLS

Although the various models of transcription control in bacteria have provided suggestions and insights for the study of such control in eucaryotic cells, they have not, as was originally hoped, been found directly applicable to eucaryotic cells. This is not really surprising when one remembers that the DNA of bacterial chromosomes is only very lightly complexed with protein whereas the DNA of eucaryotic chromosomes combines with a variety of proteins to form a nucleoprotein material called *chromatin.* Surely the protein of the chromatin plays a major role in gene regulation.

The role of chromosomal proteins and chromosomal RNA   Early attention focused on a class of basic proteins called *histones,* which contribute roughly the same amount of material, by weight, to the chromosomes as DNA. The histones appear to be important structural elements of the chromosomes, and there is evidence that they play a key role in the coiling and uncoiling of chromosomes during the various stages of the cell cycle. But it has become apparent that the histones cannot be responsible for selective regulation of genes, because their total quantity as well as their amino acid sequences are remarkably constant, both in all developmental stages of a given cell and in all cells of a given organism; histones are even very similar in cells as far apart evolutionarily as those from a mammal and those from a pea plant. This is not to say that the histones play no role in gene regulation; indeed, they are now thought to provide a general masking of the genes that insulates them from transcription. But some other agent must provide the selectivity that determines which genes will be unmasked at any given time.

More recently, evidence has mounted that a second class of chromatin proteins, the so-called *nonhistone acidic proteins,* may be selective agents in gene regulation. In marked contrast to histones, these acidic proteins exhibit a rich diversity, not only from organism to organism, but also from tissue to tissue within a given organism, and even within a single cell at various times, depending on its developmental stage and on its current functional condition. Moreover, they seem to contribute very little to chromatin structure, which suggests that their role is enzymatic and regulatory.

It is important to remember that some RNA is also associated with eucaryotic chromosomes. Some investigators think it may function in controlling gene transcription, either directly or through interaction with the acidic chromatin proteins.

In summary, it is impossible at present to say with certainty just what substances are the selective agents of gene control in eucaryotic cells. Whatever they are, their activity must be modulated by chemicals entering the nucleus from the cytoplasm—chemicals that lead to induction (derepression) or repression of certain genes, though they probably do not function in precisely the same way as bacterial inducers and corepressors.

The chromosomal organization of eucaryotic genes   Soon after the giant salivary chromosomes of *Drosophila* were discovered, it was noticed that some chromosomal regions stain only faintly when treated with basic dyes whereas other regions stain intensely. The nonstaining regions were called *euchromatin* and the staining ones *heterochromatin.* Differentiation of chromosomes into euchromatic and heterochromatic regions was later found to be a general characteristic of eucaryotic cells. As better genetic maps of chromosomes were constructed, it became clear that the euchromatic regions contain active genes whereas the heterochromatic regions are inactive.

At first, some investigators thought that the heterochromatic regions are simply devoid of genes, functioning merely as structural elements of the chromosome. This is probably true of a large region of heterochromatin located in the vicinity of the centromere, where curious repetitive base sequences in the DNA may play a role in the attachment of the chromosome to the spindle during cell division. But most heterochromatic regions, it has turned out, are not regions where genes are absent, but regions where genes (often long series of genes) are inactive. Thus, for example, when an X chromosome of a human female is converted into a Barr body, the entire chromosome becomes heterochromatic. Many regions completely heterochromatic in adults were euchromatic at earlier stages in the development of the organism.

The existence of extensive regions of heterochromatin indicates a tendency in eucaryotic organisms for functionally related genes (e.g. genes concerned with a particular stage of embryonic development) to be grouped together in the same region of the chromosome, where they can be jointly repressed (heterochromatinized). Despite this general grouping of genes, however, the clumping of small sets of related genes into contiguous operons that is so characteristic of bacteria has not been found in eucaryotes. No really convincing evidence has been adduced, moreover, for the existence in eucaryotes of operator regions of the bacterial type. Most eucaryotic mRNA carries information from only one gene (i.e. it is monocistronic); eucaryotic genes are not usually transcribed in groups, because they are not organized as operons.

Sometimes, however, the eucaryotic mRNA transcribed from a single gene is very long, consisting of as many as 6,000 nucleotides, and the long protein synthesized when such mRNA is translated on the ribosomes may promptly be cut by cellular proteases into several shorter proteins. The fact that one gene may code for several (as many as seven) enzymes in eucaryotes contradicts the original one gene–one enzyme hypothesis. However, the newer formulation—one gene–one polypeptide chain—can be stretched to cover this situation, because the one gene coded for only one very long original polypeptide. At any rate, whereas in bacterial cells synthesis of a group of functionally related enzymes can be controlled as a unit because their genes are organized in an operon, in eucaryotic cells synthesis of a group of enzymes can sometimes be controlled as a unit because all the enzymes are coded by a single gene.

Other instances of apparent simultaneous control of synthesis of functionally related enzymes in eucaryotes cannot be explained in this way, however. In these cases the enzymes are coded by different genes, which may be at a distance from one another on the chromosome or may even be on different chromosomes. Indeed, the genes that specify two polypeptide

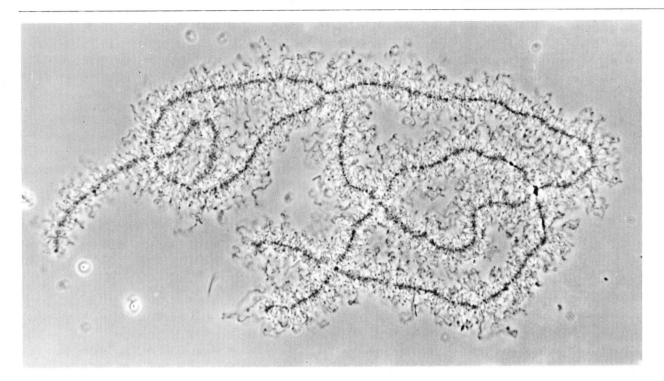

chains of a single protein (e.g. the α and β chains of hemoglobin) may be on different chromosomes. The "rules" for coordinated induction and repression in such cases are still unknown.

Lampbrush chromosomes and chromosomal puffs as visible evidence of gene activity It is known that just before the first meiotic division the oocytes of many vertebrates synthesize very large amounts of mRNA for later use; the mRNA is then ready during the early stages of development, after fertilization, when the chromosomes are so busy with DNA replication in support of rapid cell division that they are largely unavailable for transcription into mRNA. This is but one instance of changing chromosomal activity in a cell at different stages of its development. In vertebrate oocytes this changing activity has certain visible manifestations, which can be observed with a phase-contrast microscope. The parts of the chromosome bearing genes that are being repeatedly transcribed are looped out laterally from the main chromosomal axis, while the parts bearing repressed genes are tightly compacted; chromosomes in this condition are called *lampbrush chromosomes* (Fig. 16.4).

Some investigators think that each loop is a single gene, others that it is a series of copies, arranged in sequence, of a single gene. Whichever is true, it is certain that the loops represent those parts of the genome that code for the particular proteins the early embryo will have to produce in quantity. Since some of these proteins will not be needed later in development, some of the chromosomal regions so evident as lampbrush loops in the oocyte will presumably become heterochromatin for most of the life of the organism.

16.4 Lampbrush chromosome from an oocyte of the spotted newt *Triturus viridescens*
The many feathery projections from the chromosome are regions bearing genes that are being repeatedly transcribed. $\times$ 476. [Courtesy J. G. Gall, *Methods in cell physiology*, ed. by D. M. Prescott, Academic Press, 1966, vol. II.]

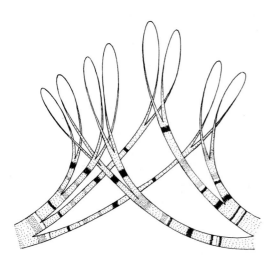

16.5 Drawing of part of a chromosome puff
Spread apart in the puff, the loops of the many parallel chromosomal fibrils expose maximum surface for synthesis of mRNA.

The giant chromosomes of the salivary glands of larval flies (Diptera) have been interpreted as lateral arrays of partially replicated chromosomes (polytene chromosomes) that have remained stuck together. If the cells have indeed gained multiple copies of their genes in this way, they would be capable of more rapid synthesis of large amounts of RNA.

When certain regions of a giant chromosome are especially active, all the parallel DNA molecules form lampbrush-type loops in those regions (Fig. 16.5). The resulting clusters of lateral loops, called *chromosomal puffs,* are clearly visible under the microscope (Fig. 16.6). The locations of puffs are different on chromosomes in different tissues, and they are different in the same tissue at different stages of development, although at any given time all the cells of any one type in any given tissue show the same pattern of puffing.

If the puffs indicate the location of active genes—as the correlation of puffing pattern with developmental stages strongly suggests—we would expect them to be sites of active synthesis of mRNA. It can easily be demonstrated that unpuffed bands of the chromosome contain essentially only DNA and protein and that puffs contain, in addition, much RNA. That this RNA is being synthesized at the puff and has not simply accumulated there after synthesis elsewhere can be shown by injecting radioactive uridine into fly larvae. Uridine is used by the cells as a precursor of uracil, a nitrogenous base that is incorporated into RNA but not into DNA. When this experiment is performed, radioactivity soon appears in the puffs (and in the nucleoli, which synthesize large amounts of ribosomal RNA), but it does not appear in other parts of the cell until much later—an indication that RNA is indeed being synthesized at the site of the puffs. Furthermore, it can be shown that the RNA made in one puff differs chemically from the RNA made in a puff at a different position on the chromosome, as would be expected if each gene codes for a different mRNA.

Puffs thus provide a way of determining visually whether or not changes in the extranuclear environment can alter the pattern of gene activity. As expected, they can. For example, if ecdysone, the hormone that causes molting in insects, is injected into a fly larva, the chromosomes rapidly undergo a shift in their puffing pattern, taking on that characteristically found at the time of molting in normal untreated individuals. If treatment is stopped, the puffs characteristic of molting disappear. If treatment is begun again, they reappear. Or, to give another example, if chromosomes from one type of cell are transplanted into a different type of cell or into the same type of cell at a different developmental stage, they quickly lose the puffs characteristic of the donor cells and develop puffs characteristic of the type and stage of the recipient cells. Puffing can be prevented entirely by treatment with actinomycin, which is known to be an inhibitor of nucleic acid synthesis.

Quantitative vs. qualitative regulation   If, as we have seen, regulatory factors can act on the genes, the question arises whether the regulatory changes are entirely quantitative, or whether they might be qualitative as well. In short, do environmental influences regulate only how much mRNA a gene makes, or do they also influence what message the mRNA conveys?

A classic experiment performed in 1932 by Hans Spemann[3] and Oscar E. Schotté at the University of Freiburg suggests the answer to this question. In salamanders the ectodermal tissue of the mouth is induced to form teeth by the endodermal tissue with which it is in contact. In frogs the corresponding ectoderm forms horny jaws instead of true teeth. Spemann and Schotté transplanted a piece of ectoderm from the flank of a frog embryo to the mouth region of a salamander. The transplanted tissue developed into horny jaws. It had been induced by the salamander endoderm, but instead of forming salamander teeth it formed the corresponding structures of a frog, namely horny jaws. Apparently all the salamander inducers could do was to turn on the appropriate genes; they could not alter what those genes would make.

Numerous other experiments have yielded similar results, and it is now generally accepted that environmental influences act only by turning on or off the synthetic activity of the various genes; there is no evidence that they in any way determine what those genes will make. Each gene can make one and only one kind of mRNA, as coded by the sequence of its nucleotides; regulators of the genes determine if and when each gene will synthesize its particular mRNA.

GENE AMPLIFICATION AS A CONTROL MECHANISM

Ordinarily a single copy of each gene per haploid chromosome set is sufficient to meet the needs of a cell for the substance coded by the gene. The rate of transcription of a single gene is such that in four days it may produce 100,000 messenger RNA molecules, which can lead to the synthesis of as many as 10 billion (10^{10}) protein molecules. But in some cases the demand for a product is so great that a single gene cannot meet it. For example, so much ribosomal RNA is required by a eucaryotic cell for construction of ribosomes (which may account for as much as one fourth of the total cellular mass) that such cells manufacture between 100 and 1,500 serially arranged copies of the genes for rRNA (Fig. 16.7). These extra copies of the rRNA genes (with their complexed protein) usually loop out from the main axis of one of the chromosomes as a mass of material that we call the nucleolus. The point where the nucleolus is attached to the chromosome,

[3] Spemann, whom we shall meet again, was awarded the Nobel Prize in 1935 for his fundamental work in embryological development.

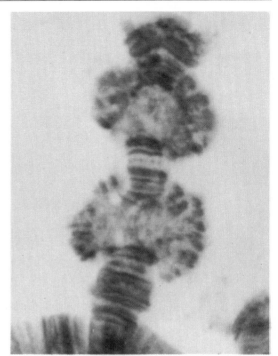
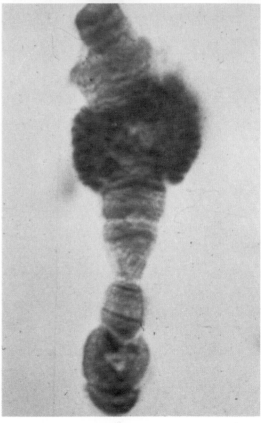

16.6 Puffing patterns in giant chromosomes from the salivary gland of a midge
Top: There are two prominent puffs. The two major structural elements of the chromosome can easily be distinguished here, because the DNA has been stained reddish brown and the protein greenish brown.
Bottom: Another chromosome shows a different pattern of puffing. Here the DNA has been stained blue and the RNA reddish violet. Note the concentration of RNA at the puff, where active transcription is taking place.   [Top: Ulrich Clever. Bottom: Claus Pelling, Max Planck Institute for Biology, Tübingen.]

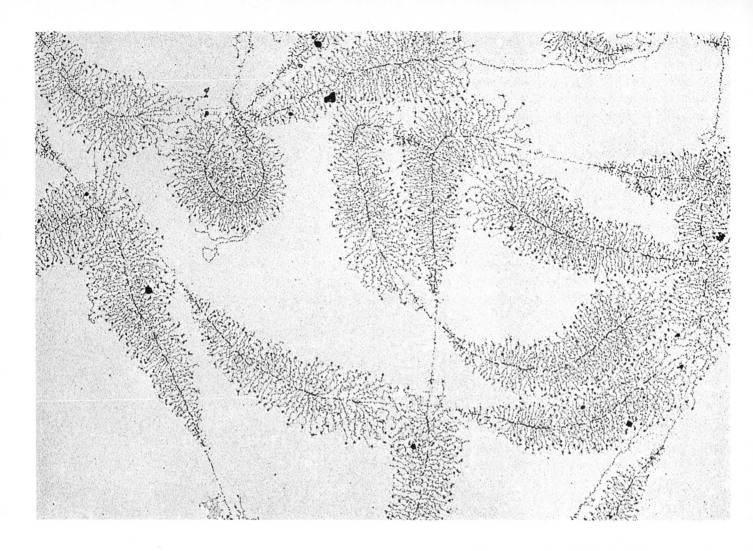

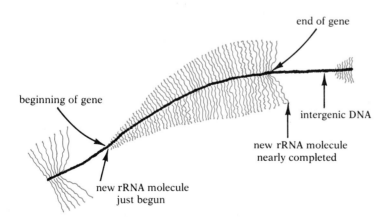

beginning of gene

end of gene

intergenic DNA

new rRNA molecule
nearly completed

new rRNA molecule
just begun

16.7 Transcription of rRNA on DNA in the nucleolus of an oocyte of the spotted newt *(Triturus viridescens)*
Top: Electron micrograph of portion of the nucleolus. The continuous strand of DNA bears multiple copies of the genes for rRNA. Strands of rRNA at progressive stages of synthesis feather out from each gene. Because the successive genes are separated by regions of intergenic DNA, where no rRNA is attached, one can tell more or less exactly where on the DNA molecule each gene begins and ends. $\times$ 28,000.   Bottom: Diagram of rRNA synthesis on one gene. Many molecules of rRNA (color) are being synthesized as rRNA polymerase molecules move along the DNA of the gene. Strands of rRNA attached to the DNA near the beginning of the gene are still short, their synthesis having just begun; strands of rRNA attached at successive points along the gene are progressively longer.   [Top: Courtesy O. L. Miller, Jr., and B. R. Beatty, *J. Cell. Physiol.*, vol. 74, 1969.]

called the nucleolar organizer region, is the location of the original rRNA genes from which the copies were made. There are also multiple copies of the genes for transfer RNA and for histones, though these do not form recognizable structures such as the nucleolus.

Possession of multiple copies of the genes for rRNA, tRNA, and histones is the normal condition for eucaryotic cells and represents no special response to changing developmental or functional conditions. However, in the oocytes of vertebrates, which must produce all the ribosomes that will supply the embryo until it has completed most of its early cell-cleavage stages (up to gastrulation), such huge quantities of rRNA are needed that even the many copies of the rRNA genes in the nucleolus cannot meet the demand. In the oocytes, therefore, additional copies of the rRNA genes are made—sometimes as many as two million additional copies—and these are organized into large numbers of extrachromosomal nucleoli that can easily be seen even under an ordinary light microscope. The production of these extra copies of the rRNA genes is called *gene amplification.*

When amplification of the rRNA genes was first discovered about ten years ago, the possibility was immediately suggested that a similar amplification might occur for other sets of genes at various stages of embryonic development. Selective amplification would be one way of influencing the amount of mRNA of particular types that would be produced in the nucleus. But, contrary to expectation, additional instances of gene amplification have not been found; it must therefore provisionally be concluded that cells use the amplification process only for genes that code for rRNA and not for those that code for mRNA.

POST-TRANSCRIPTIONAL CONTROL

Our discussion of cellular control mechanisms has so far focused on regulation of the amount of messenger RNA synthesis, whether by control of DNA transcription or by gene amplification. But there are numerous other points in the flow of information within the cell where control could be exerted.

Control of conversion of mRNA in the nucleus   Though we have often spoken of the synthesis of mRNA along the DNA template, the fact is that the immediate product of DNA transcription in eucaryotic cells is apparently not functional mRNA but an mRNA precursor. The individual molecules of this pre-mRNA are far larger than the molecules of functional mRNA. Enzymes in the nucleus must cut away large portions of the pre-mRNA molecule before it leaves the nucleus; indeed, probably less than 10 percent of the molecule survives this pruning. It was formerly thought that the portions removed were exclusively on one end of the pre-mRNA molecule, but recent evidence indicates that a middle portion, which separates parts of the molecule that must be incorporated into the functional mRNA, is also sometimes removed. Hence conversion may involve not only the cutting away of one end of the molecule, but sometimes also the cutting out of a central portion, followed by the splicing together of the remaining portions (see p. 655*n*); also, for reasons not yet understood, a long chain of adenine nucleotides (as many as 200) is usually added to the end of each mRNA before it leaves the nucleus (Fig. 16.8).

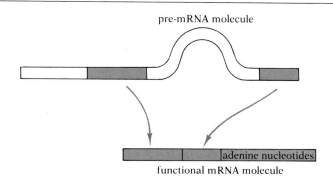

pre-mRNA molecule

adenine nucleotides

functional mRNA molecule

16.8   Conversion of pre-mRNA into mRNA
The immediate product of transcription of a gene is a pre-mRNA molecule. One end of the molecule (white) is cut off; often excised as well is a segment of the molecule separating two portions of the eventual mRNA. Finally, a long chain of adenine nucleotides is added.

The enzymatic conversion of pre-mRNA into functional mRNA would seem to be a likely point at which control might be exerted. For example, the pre-mRNA transcribed from some genes might immediately be converted into mRNA and exported to the cytoplasm, whereas the pre-mRNA transcribed from other genes might be held in the nucleus for later conversion. It is even possible that some pre-mRNA is never converted and is eventually degraded.

Having outlined these promising possibilities, we are forced to an anticlimactic admission. The subject is still so new that little evidence has been gathered as yet on whether conversion of pre-mRNA into mRNA in the nucleus really is an important point of cellular control.

Control of translation and degradation of mRNA   For synthesis of protein along a molecule of mRNA to begin, the mRNA must complex with ribosomes. But much of the mRNA exported to the cytoplasm by the nucleus of an oocyte does not complex until after the egg has been fertilized. Presumably fertilization stimulates formation of polyribosomal complexes, but how it does so is still a mystery. One suggestion is that the nonfunctioning mRNA is loosely bound to protein, which falls away under the altered chemical conditions in the cell following fertilization. Whatever the mechanism, it seems reasonable to suppose that a similar process operates in other kinds of cells to control translation of mRNA. A cell that might have to respond to sudden environmental changes might very well benefit from having a supply of the appropriate kinds of mRNA available in the cytoplasm in dormant form, ready to be mobilized very quickly when needed. At present, however, there is very little evidence to indicate how widespread this mechanism of control may be.

Control might also be exerted by regulation of the rate at which mRNA is broken down in the cell. It is a fact that mRNA has a very variable life-span. Sometimes it is destroyed in a matter of hours. At other times it is highly durable; if *Acetabularia* can generate an entire new cap in the absence of any new nuclear commands (Fig. 13.2, p. 558), then the mRNA present in the cell at the beginning of the experiment must survive throughout the regeneration period, directing synthesis of the needed proteins.

Control by alteration or destruction of proteins   Although some proteins are functional as soon as they are synthesized, other proteins are synthesized in a nonfunctional form and must be altered before they become functional. The digestive enzymes pepsin, trypsin, and chymotrypsin, for example, are initially secreted in inactive form (see pp. 231, 232), as is the hormone insulin (p. 363); activation involves the enzymatic deletion of part of the polypeptide chain. Regulation of the rate at which such proteins are converted into their functional form provides an opportunity for control of some cellular processes.

Control can also be exerted by altering the rates at which various enzymes are destroyed in the cell. Some enzymes are used only briefly before being destroyed, whereas others may never be destroyed. Moreover, a given enzyme may have very different survival times in cells from different tissues; e.g. an enzyme studied in rats was found to last about 1.6 days in heart muscle, about 31 days in skeletal muscle, and about 16 days in liver.

Control of protein activity   The activity of proteins is responsive to many kinds of environmental influences, ranging from the physical condition of the medium to chemical control of enzymatic activity. We have encountered numerous instances in previous chapters: the influence of temperature and pH on the activity of enzymes (p. 71); the influence of light on the conversion of phytochrome from the P_r to the P_{fr} state and vice versa (p. 353); the effect of inhibitors and allosteric modulators on the catalytic properties of enzymes (p. 75); the inactivation of enzymes early in a biochemical pathway through feedback inhibition by products of the pathway, as in glycolysis, where the allosteric enzyme catalyzing production of fructose diphosphate is negatively modulated by the ATP and citric acid formed later in the pathway (pp. 164, 170); the activation or deactivation of enzymes by other enzymes that phosphorylate or dephosphorylate them, a mechanism seen in the enzymes whose regulation is mediated by cyclic AMP and protein kinases (p. 376). It is clear that, like transcription and translation, enzymatic catalysis provides ready points of cellular control.

Some proteins—hemoglobin is one and collagen is another—have a capacity for self-assembly, which offers an additional possibility for control by environmental influences. The α and β chains of hemoglobin are coded by different genes and their mRNAs are separately translated, but after the two kinds of chains have been synthesized, and have assumed their secondary and tertiary conformations, they spontaneously come together in the proper orientation to form the four-chained functional hemoglobin molecule (see Fig. 2.42, p. 61). The process of aggregation is sensitive to many factors, including pH and the concentrations of certain ions; even very slight changes in intracellular conditions can profoundly alter the rate at which functional hemoglobin is assembled.

Self-assembly by extracellular collagen, which is so abundant in most animal tissues, produces many different arrangements. The individual collagen molecules synthesized in the cells are secreted into the intercellular spaces, where they assemble into orderly arrays of fibrils. The arrays differ from tissue to tissue, depending on the reinforcement needed by the tissue to fulfill its functions. Apparently the different chemical environments of the tissues influence the process of self-assembly and thus account for the differences in the arrays.

Summary of cellular control mechanisms   The crucial question to which we have attempted to suggest some answers in the preceding sections is how a single cell may develop into the dynamic complex of anatomic and biochemical relationships that constitutes any living organism.

As we have seen, a host of environmental influences interact with inherited information—information derived from hundreds of millions of years of evolution—in guiding both the development and the moment-to-moment functional capabilities of the organism. Control can be applied at many different points in the path of information flow from the genes to their phenotypic expression. It can be applied in a variety of ways at the stages of gene transcription, pre-mRNA conversion, mRNA translation, protein modification, enzyme catalysis, and protein self-assembly (Fig. 16.9).

Undeniably, we have learned much in recent years about the nature of genes and the mechanisms of their action. But when we consider the immense complexity of a living being—the interlocking relationships among countless units at multiple levels of organization, the precise regulation of thousands of ongoing responses to ever changing conditions—our newly acquired knowledge seems almost paltry, unequal to negotiating the huge developmental span from genes to functional organisms.

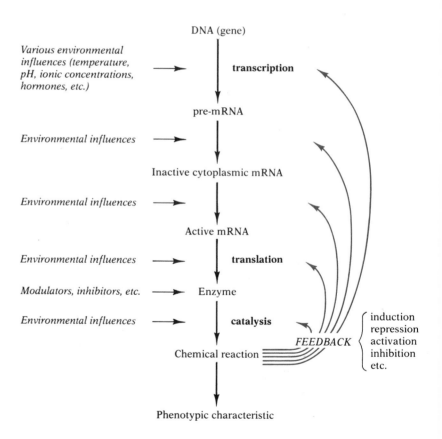

16.9   Summary of cellular control mechanisms

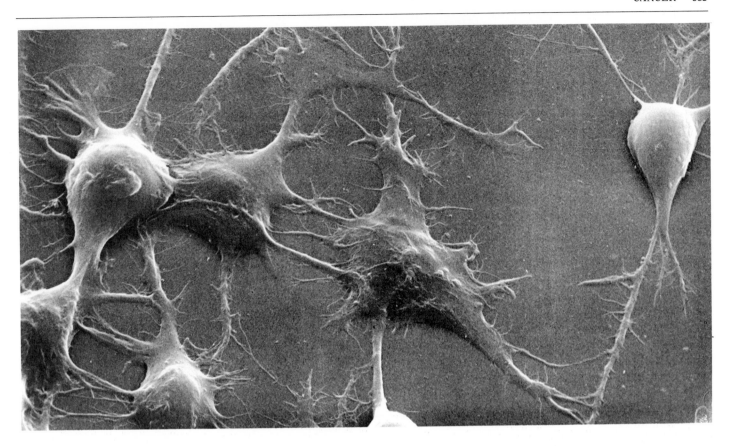

CANCER: A FAILURE OF NORMAL CELLULAR CONTROLS

The most distinctive feature of cancer cells is their unrestrained proliferation, which results in formation of malignant tumors and often in the spread of the cells from the original site of growth to many other parts of the body. Clearly, these are cells in which the normal controls are no longer working properly. It is the hope of biologists that they may learn more about how normal cellular controls work by studying how they fail—by investigating how the normal controls can become so deranged as to permit cancerous growth. There is the underlying hope, too, that such study will eventually lead to better ways of preventing or treating cancer.

Study of cells in culture   One of the most profitable ways of studying the properties of specific kinds of cells is to remove them from the body of the organism and grow them in tissue culture in the laboratory. The usual procedure is to seed a large number of cells onto (or into) a sterile culture medium to which a variety of nutrients and growth-stimulating factors have been added. While such a culture, no matter how elaborately set up and controlled, is far from the normal environment of the cells, some of the cells may nevertheless survive and engage in many of their usual developmental and functional activities, including cell division (Fig. 16.10). Most cultured lineages of animal cells die out after a number of cellular generations (the

16.10   Phase-contrast photomicrograph of cultured neuroblastoma cells from a mouse
These cells, from a malignant tumor of the nervous tissue (neuroblastoma), are growing on a solid surface; they have many elongate processes resembling axons and dendrites. × 1,600.   [Courtesy Gunter Albrecht-Buehler, Cold Spring Harbor Laboratory, and Frank Solomon, Massachusetts Institute of Technology.]

number tends to be specific for the species and tissue of origin), but a few acquire the ability to continue dividing indefinitely. Prominent among the kinds of cell lineages potentially immortal in culture are cancerous ones; hence study of such cells in culture has been vigorously pursued in recent years.

A recently developed technique called **somatic cell hybridization,** in which two cultured cells, often of different origins, are fused together, has made possible exciting new approaches to the study of cellular control systems. The fusing of cells became relatively easy after it was discovered that a virus called Sendai modifies the surfaces of cells in a way that makes them stick to each other and then coalesce. Fortunately Sendai viruses inactivated by exposure to ultraviolet light continue to promote cell fusion; hence the cells can be studied without the complications of a virus infection.

When two cells are fused their nuclei usually remain separate, but occasionally when cell division begins only a single spindle forms. The result is that each daughter cell receives a single nucleus containing two sets of chromosomes, one from each of the two parental lines, which may be different species or different tissues from the same species. In this way hybrid cells containing human chromosomes in combination with chromosomes from a variety of other animal species have been produced.

The cell-fusion technique has been especially valuable in cancer research, because study of fused cancerous and noncancerous cells, and of fused cancerous cells of two different types, helps reveal how the altered control system of one cell influences the control system of the other cell.

The characteristics of cancer cells   One approach to trying to understand what causes cells to become cancerous is to look for consistent differences between cancerous and normal cells. We shall briefly mention a few of the results of this search.

One characteristic of cancer cells is that they almost always have an abnormally large number of chromosomes. For example, the cells of the human cancer line called HeLa,[4] by far the most widely studied line of cultured human cells, typically have 70–80 chromosomes instead of the normal 46. Interestingly enough, noncancerous cell lines that pass the "crisis stage" in culture and become potentially immortal also have extra chromosomes; thus it seems likely that possession of extra chromosomes somehow frees cells from some of the normal constraints on proliferation.

When normal cells are nutritionally limited and stop dividing, they ordinarily stabilize in the G_1 stage of the cell cycle (see Fig. 13.6, p. 562). Cancer cells, however, appear to lack the control mechanism for going into G_1 quiescence, and the various cells stop more or less randomly in all phases of the cell cycle. But cells that stop in the "wrong" phases of the cycle are much more susceptible to various poisons or other stress-causing factors, and tend to die off more rapidly than normal cells. This difference between cancer cells and normal cells is the basis for some cancer treatments that attempt to kill the cancer cells without killing many normal cells at the same time. That

---

[4]The HeLa cell line is derived from a carcinoma of the cervix of a young black woman named Henrietta Lacks, who died of her cancer in 1951. This was the first stable, vigorously growing line of cultured human cells used in cancer research. Today HeLa cells are found growing in medical and research laboratories the world over.

the cell cycle should be upset in cancer cells makes it understandable that the vast majority of cancers occur in tissues where much cell division takes place (e.g. epithelial tissues).

Besides nuclear differences between cancer cells and normal cells, there are a host of differences pertaining to the cell surface. Since it is at the surface of the cell that extracellular environmental influences must first act, it seems that cancer cells suffer not only from genetic defects, but also from insufficient environmental modulation of their activities. Which usually comes first—the genetic disturbance or the lack of responsiveness to the environment—is by no means clear.

Cultured cancer cells tend to have a rather spherical shape, one seen in normal cells only during a short period in early G_1 immediately following mitosis. This peculiarity of shape is probably due to the abnormally small number of structure-stabilizing microfibrils in the cells, which also makes the cells more mobile than normal cells.

Excessively mobile, cancer cells also fail to show normal contact inhibition of movement when they encounter other cells. Contact inhibition, as we saw in Chapter 3, is mediated by the glycolipids and glycoproteins of the cell coat. It is not surprising that cancer cells typically have fewer glycolipids and glycoproteins in their cell coats than normal cells, and that those they have are qualitatively different.

The anomalous glycolipids and glycoproteins are probably also correlated with the absence in cancer cells of the normal contact inhibition that slows or stops cell division in contiguous cells (see p. 98), and with the apparent inability of cancer cells to recognize other cells of their own tissue type. Whereas normal cells from two different tissues (e.g. liver and kidney) mixed in culture tend to sort themselves out and reaggregate according to tissue type, cancerous cells do not do so. This absence of normal cellular affinities is probably the reason why malignant cells can spread from their tissue of origin into many other parts of the body—a process known as metastasis.

That the surfaces of cancer cells characteristically bear antigens not found on normal cells constitutes another abnormality of their cell coats. In most human beings the body's normal immunological mechanisms probably respond to the antigens, destroying new cancer cells as fast as they arise. Thus cancerous tumors may be traceable not only to a breakdown of normal cellular controls, which causes some cells to become malignant, but also to a failure of the normal immunologic response to such cells.

Many other characteristics distinctive of cancer cells might be cited. For example, cancer cells consume much more glucose than they can metabolize efficiently, and as a result excrete much lactic acid; and they secrete abnormally large quantities of proteolytic enzymes, which may alter both their own surfaces and the surfaces of normal cells near them. Whichever characteristics are singled out, it is important to remember the difficulty of establishing whether they are primary disturbances that create the cancerous condition or merely secondary responses to the primary metabolic changes.

Viruses as a cause of cancer   Evidence has mounted over the years that some cancers are caused by viruses. Viruses were suspected as causal agents long before any proof was available, but now a number of cancer-inducing viruses are known and have been extensively studied. The difficulty was that viruses could seldom be detected in cancer cells. We now know the reason: The viruses had become integrated as provirus with the host cells' chromosomes and had, in effect, disappeared from view. We have seen earlier that somatic mutations may be one cause of cancer (see p. 654), and we now see that the introduction of new genetic material into somatic cells by viruses can play a similar role. In short, there must be a number of ways in which the genetic control system of a cell can be cast into the cancerous state. What is remarkable is that the end results are so similar, no matter how the cancers are induced.

When a cancer virus invades a cell of a species that is its normal host, it usually causes the cell promptly to make new virus particles (i.e. it triggers the lytic cycle; see p. 663); such a cell, which can replicate the virus, is said to be **permissive** for that virus. But when the virus enters a cell of another species incapable of making that particular kind of virus (i.e. a cell that is **nonpermissive**), it may then become integrated as provirus and **transform** the cell into the cancerous condition. For example, SV40, a virus that normally multiplies in monkeys, can transform cells of the much-studied 3T3 line of cultured mouse cells, but cannot multiply in them, whereas polyoma virus, which is a mouse virus, multiplies in 3T3 cells, but does not transform

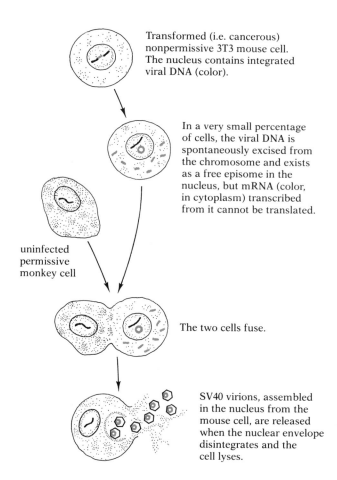

16.11   Demonstration of the presence of virus in cancerous cells

Fusion of transformed 3T3 mouse cells with uninfected monkey cells demonstrates the presence of SV40 virus in the transformed cells. Virions could not be produced by the 3T3 cells because they lack some factor necessary for translation of the mRNA transcribed from the viral DNA. But when 3T3 cells and monkey cells are fused, the monkey cells provide the missing translation factor; thus new virions are assembled in the 3T3 nucleus and then released when the cell lyses.

Transformed (i.e. cancerous) nonpermissive 3T3 mouse cell. The nucleus contains integrated viral DNA (color).

In a very small percentage of cells, the viral DNA is spontaneously excised from the chromosome and exists as a free episome in the nucleus, but mRNA (color, in cytoplasm) transcribed from it cannot be translated.

uninfected permissive monkey cell

The two cells fuse.

SV40 virions, assembled in the nucleus from the mouse cell, are released when the nuclear envelope disintegrates and the cell lyses.

them. It seems, then, that virus-induced cancers result from a mismatch between a virus and the host cell it has invaded.

The technique of cell fusion has made it possible in some cases to confirm the presence of provirus in cancerous cells. For example, when a 3T3 cell transformed into the cancerous state by SV40 virus is fused with a normal uninfected monkey cell permissive for SV40, production of new SV40 virions begins within 24 hours in many of the fused cells (Fig. 16.11). Though neither the 3T3 mouse cells nor the monkey cells contained visible virus before the fusion, some property of the monkey cell enabled the SV40 DNA from the nuclei of the transformed mouse cells to "come out of hiding" and trigger viral multiplication.

THE IMMUNE RESPONSE

The cellular control system for the immune response of vertebrates is an especially intriguing one. As we saw in Chapter 7, an important defense against disease in vertebrate animals is their ability to manufacture antibodies that can inactivate or destroy invading antigens, which are large molecules (usually either proteins or polysaccharides) ordinarily foreign to the organism's own body; the antigens may be free in solution, as is true for toxins secreted by some microorganisms, or built into the outer surface of viruses or cells (e.g. bacteria). The antigens stimulate certain cells to produce highly specific antibodies against them.

The vast potential number of different antigens, each of which can stimulate synthesis of a different proteinaceous antibody, imposes special requirements on the cellular control system. That system must somehow enable an antigen to turn on synthesis of one out of billions of possible antibodies—i.e. it must promote remarkable selectivity—and it must also provide some way of distinguishing between molecules that are part of the organism's own body and very similar foreign molecules.

A demonstration of immunologic reaction   Immunologic reactions have been the subject of much study since the English physician Edward Jenner discovered in 1796 that people develop immunity to smallpox if they are artificially injected with material that induces a very mild form of the disease (actually cowpox). Further dramatic demonstrations of the immune reaction were made by Louis Pasteur in France during the latter half of the nineteenth century.

The object of one of Pasteur's many investigations was a disease of cattle and sheep called anthrax, which was ravaging the herds of Europe at that time. Persuaded that a certain type of bacterium caused the disease, he exposed bacteria of this type to temperatures that were high enough to weaken but not kill them. When he injected these weakened bacteria into healthy sheep, the sheep became slightly ill, but thereafter they exhibited immunity to further infection by this disease. To convince the skeptics of his day, Pasteur arranged a demonstration attended by his most influential contemporaries. With these as witnesses, he injected weakened bacteria into 25 sheep, leaving 25 others uninjected as controls. Several weeks later, with the witnesses again assembled, he gave all 50 sheep a massive injection of fully active bacteria—more than enough to kill any normal healthy sheep. A

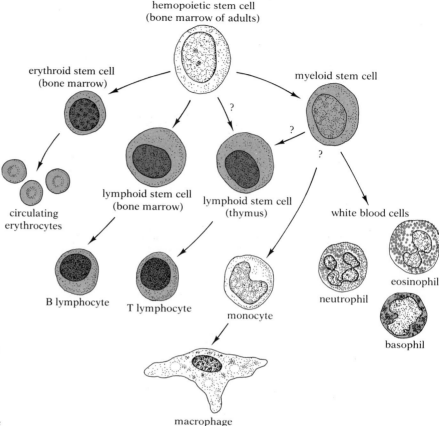

16.12   Hemopoiesis: the origin of blood cells
Hemopoietic stem cells give rise to at least four types
of more specialized stem cells: those from which red
blood cells (erythrocytes) are derived; those from
which B lymphocytes are derived; those that migrate
to the thymus and give rise to T lymphocytes; and
those that give rise to the various other kinds of white
blood cells.

few days later, all 25 control sheep were dead, while all 25 of the treated
sheep were alive and healthy.

In modern terminology, we would say that Pasteur's sheep reacted to the
antigens of the weakened anthrax bacteria by producing antibodies against
them. But what does this mean? By what process does exposure to an
antigen stimulate an organism to make antibodies?

The cells of the immune response   The cells that respond to the presence of
a foreign antigen are certain white blood cells called *lymphocytes,* which
occur not only in the blood but also in the lymph and in lymphoid tissues of
the lymph nodes, spleen, liver, tonsils, thymus, and bone marrow, and in
more limited lymphoid areas associated with the lungs and the intestinal
tract. The lymphocytes are derived from lymphoid stem cells located in the
thymus and bone marrow (or, in birds, in the bursa of Fabricius, a lymphoid
organ near the cloaca) (Fig. 16.12). The lymphoid stem cells, in turn, are
derived from hemopoietic stem cells (cells that give rise also to the other
types of white blood cells and to erythrocytes) located in the red bone
marrow (or in the yolk sac of the early embryo or the liver and spleen later in
embryonic development).

Before it is stimulated by an antigen, a circulating lymphocyte is a small, metabolically quiescent cell that can move freely between the blood and the lymphatic tissues by squeezing between the endothelial cells of the walls of blood vessels. Upon stimulation by an antigen, the small lymphocyte grows larger and then begins dividing (Fig. 16.13). Let us assume, for the moment, that the lymphocyte we are following is a *B lymphocyte,* i.e. one derived from a stem cell in the bone marrow. Cell division by a stimulated B lymphocyte gives rise over a period of several days to numerous *plasma cells,* and it is primarily these cells that secrete the antibody molecules that attack the stimulating antigen. A stimulated B lymphocyte also gives rise to other lymphocytes like itself, which serve as *memory cells* and make possible (in a manner to be described later) more rapid response if the same antigen should be encountered again.

The antibodies form weak bonds with the antigen molecules. Because each antibody molecule can bond with two antigen molecules, the antibodies tend to agglutinate the antigens (Fig. 16.14), which may then be engulfed by the large phagocytic *macrophages* present in the lymphoid tissues. If the antigens are on the surfaces of cells, the antigen-antibody combination may attract other protein components of the blood serum, called *complement,* which can bind to the bound antibody. Complement not only provides binding sites of its own for macrophages, thus facilitating phagocytosis, but may also make holes in the plasma membrane of the foreign cell and thus bring about its death.

It was formerly thought that the only role of the immune system was to produce circulating antibodies, i.e. provide what is called *humoral immunity,* but more recently it was discovered that many immunological responses are due to antibodies that remain bound to the surfaces of the

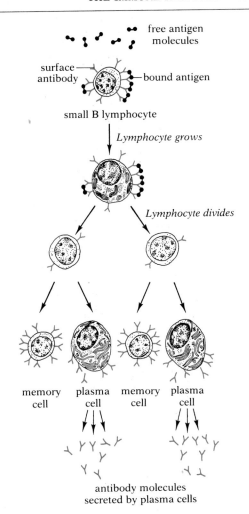

16.13 Stimulation of B lymphocyte by antigen
When antigen molecules bind to surface receptors (thought to be bound antibodies) of a small B lymphocyte, the lymphocyte first grows larger and then begins a series of cell divisions (only two are shown here). Some of the cells produced by this proliferation are memory cells that resemble the original lymphocyte; others become specialized as plasma cells, which secrete antibodies.

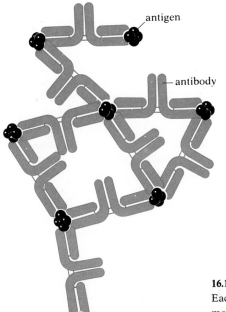

16.14 Agglutination of antigen molecules by antibody
Each antibody molecule can bind to two antigen molecules; hence the antigens can be held together in large clumps.

lymphocytes that produce them. In such cases it is the whole antibody-bearing cell that attacks the antigen. The cells responsible for this *cell-mediated immunity* have been found to be the lymphocytes derived from stem cells in the thymus. These *T lymphocytes* appear morphologically identical to the B lymphocytes from bone marrow, and like the B lymphocytes they first enlarge and then divide when stimulated by antigen. A stimulated T lymphocyte gives rise to both immunologically active T cells and memory cells.

Free molecular antigens and viruses tend predominantly to stimulate humoral immunity, while cell-bound antigens tend to stimulate cell-mediated immunity, but there are frequent exceptions, and in many cases both types of immunity are triggered. In fact, there is a growing body of evidence that the functions of B cells and T cells are not as separate as was once thought. Many T cells appear to play a regulatory role in the responses of B cells to antigens; thus T cells may sometimes act as helpers in B-cell responses or as suppressors of B-cell responses. The thymus appears to play a key role in conferring these regulatory capabilities on the lymphocytes it produces, and it probably maintains, by means of the hormone it secretes, a continuing influence on the T lymphocytes even after they have moved out into the general circulation. The mechanisms of T cell–B cell interactions are still poorly understood.

The mechanism of stimulation by antigen   Antigens are always large molecules, but not all of the molecule functions directly in stimulating lymphocytes. Apparently only certain regions of the molecule, probably about ten amino acids long, serve as *antigenic determinants,* i.e. as sites of interaction with the receptors on the lymphocytes. A single antigen molecule may thus bear many antigenic determinant regions, which may all be alike and stimulate production of only one kind of antibody, or may be of several different types and stimulate production of a corresponding number of different antibodies.

Because each kind of antigen stimulates synthesis of antibodies specific for it—and because it seemed unreasonable that a cell should possess genes determining proteinaceous antibodies against all the billions of possible antigens in the world, most of which it would never encounter in its lifetime—it was long assumed that the antigen itself must be responsible for providing the lymphocyte with the necessary instructions for synthesizing antibody against it. The lymphocyte might, for example, manufacture some sort of precursor protein whose three-dimensional conformation would then be determined by the antigen, acting as a mold or template. But as we have seen, the conformation of a protein is a natural consequence of its primary structure, i.e. of its amino acid sequence, and it would seem unlikely that some other conformation, imposed upon the protein by the antigen, would be stable. It has been shown, in fact, that if the conformation of an antibody is temporarily destroyed by denaturation, and if the three-dimensional pattern is then allowed to re-form in the absence of antigen, the antibody regains full specificity; clearly, its conformation does not depend on antigen. Moreover, if the antigen acted as a conformational template for synthesis of antibody by a lymphocyte, it would most probably enter the cell, but ex-

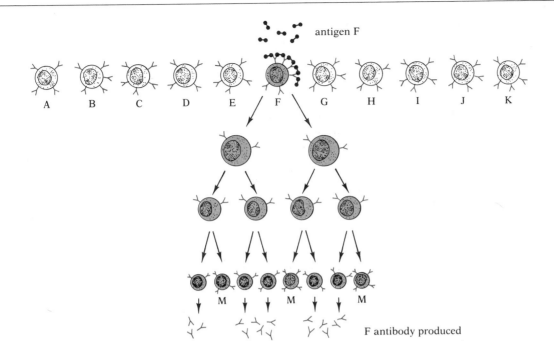

antigen F

A   B   C   D   E   F   G   H   I   J   K

M   M   M

F antibody produced

periments with radioactive antigen show that very little enters the lympho-cytes; the antigen appears to exert its effect at the cell surface. For these and numerous other reasons, then, the so-called instructive hypothesis of an-tibody induction has been abandoned.

The hypothesis now most widely accepted postulates the existence in the organism of an enormous number of slightly different lymphocytes, each specialized for potential production of a different antibody. Estimates of how many different kinds of lymphocytes would be required to endow a cell with the capability of synthesizing antibody against virtually any antigen run as high as 10 billion (10^{10}) or more. The presumption is that the antigen reacts only with those very few lymphocytes capable of synthesizing an-tibody specific for it, and that, rather than provide instructions for the lymphocytes, the antigen merely stimulates the lymphocytes to begin proli-ferating. In other words, the antigen functions to induce amplification of the appropriate lymphocyte types (Fig. 16.15). Each stimulated lymphocyte gives rise to a clone of cells (a group of genetically identical cells descended from a common ancestral cell), according to this interpretation, which is known as the hypothesis of **clonal selection.** When an antigen stimulates a given B lymphocyte, it is calculated, each of the plasma cells in the clone to which the lymphocyte gives rise may transcribe as many as 20,000 mRNA molecules from its gene for antibody, enabling each of the plasma cells to secrete 2,000 identical antibody molecules per second.

Supposing the clonal-selection hypothesis is correct, how does the selec-tion operate? How does an antigen recognize the correct lymphocyte? Re-cent evidence provides a ready answer: Each lymphocyte, whether of the B

16.15 The clonal-selection hypothesis of antigenic stimulation
Antigens are thought able to react with only a few very specific lymphocytes from among the billions of kinds of lymphocytes in the organism's body. In this example antigen F can bind only to the lymphocyte of type F; it does not affect the other lymphocytes (top row). The F lymphocyte, stimulated by the binding of antigen, proliferates to form a clone of genetically identical cells. Some cells of the clone are memory cells (M); others are cells that actively secrete F antibody.

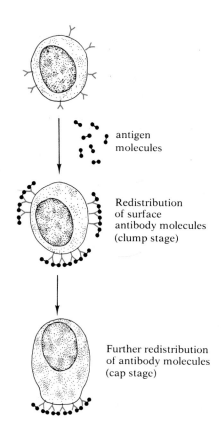

16.16   Mechanism of stimulation of a lymphocyte by antigen

The binding of each antigen molecule to more than one surface antibody may cause a change in the distribution of the antibodies that triggers activation of the lymphocyte. The redistribution of the surface antibody molecules, which are presumed to be membrane proteins, depends on the fluidity of the plasma membrane, envisioned by the fluid-mosaic model of the membrane.

or the T type, has exposed antibodies on its surface, presumably as membrane proteins (the bound antibody of B lymphocytes is thought to be identical with the free antibody its descendant plasma cells will later secrete). An antigen can bind only to those lymphocytes whose surface antibodies have binding sites specific for its antigenic determinant regions. Just how this binding of antigen induces the lymphocyte to begin dividing is not known, but it is thought that the cross-linking of the surface antibodies by the antigen molecules changes the distribution of the antibodies on the cell, and that this change of distribution may provide the signal that activates the cell cycle (Fig. 16.16)

When the antigen selectively stimulates certain lymphocytes to proliferate, the clones produced include not only antibody-producing plasma cells, if the cells stimulated were B lymphocytes, or immunologically active T cells, if the cells stimulated were T lymphocytes, but also many memory cells—lymphocytes genetically like the ones originally stimulated. The memory cells may continue to be present throughout the remainder of the organism's life; it is they that confer on the organism a lasting immunity to the stimulating antigen. Several days are required after the first exposure to antigen for a clonal expansion sufficient to provide significant defense. But on later exposure to the same antigen each of the many memory cells can produce a clone; thus production of antibodies or active T cells (or both) can quickly accelerate, providing almost immediate immunological defense.

The antibody molecule   Antibodies, whether circulating or membrane-bound, are globulin proteins—predominantly gamma globulins. Each antibody molecule consists of four polypeptide chains: two identical "heavy" chains and two identical shorter "light" chains; the chains are linked by disulfide bonds (Fig. 16.17).

About three fourths of each heavy chain and half of each light chain show great constancy of amino acid sequence; the variability lies mostly in the remaining portions of each chain—at the free amino ends. The binding sites for antigens (two on each antibody molecule) are at the ends of the variable portions. Each binding site is a pocket or cleft bounded partly by the heavy chain and partly by the light chain (Fig. 16.18).

The genetic basis of antibody diversity   Let us assume that the clonal-selection hypothesis is correct, and that an organism really does have billions of different kinds of lymphocytes, each bearing on its surface a different antibody. How can so many different kinds of antibodies be determined? There are two opposing schools of thought on this question. According to the *germline* hypothesis, every lymphocyte (and indeed every cell in the organism's body) contains in its nucleus separate genes for all potential antibodies, and the reason why each lymphocyte (and its clonal descendants) makes only one antibody is that all but one of its genes for antibody are repressed. According to the *somatic mutation* hypothesis, each cell contains only a few antibody genes, but because there are regions on these genes that are extremely susceptible to mutation, virtually every lymphocyte carries its own unique mutant form of the genes.

At first, the idea that every cell carries genes for all possible antibodies may seem preposterous. There are so many different proteins and polysac-

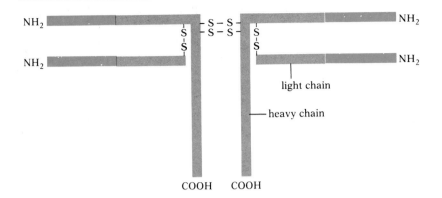

16.17 The arrangement of the four polypeptide chains in a molecule of gamma-globulin antibody
Variation between antibodies is due almost exclusively to differences in amino acid sequences within the regions shown in color, which are at the amino ends of the chains.

charides that the genes for antibodies against every one of them would take up all the DNA in a cell's nucleus. But does the cell really need to make a separate antibody for every possible antigen? Consider a shoe manufacturer who produces a very wide variety of different shapes and sizes of shoes; he will most likely be able to supply shoes that fit the feet of any customer. Similarly, a cell that makes, let us say, merely 10 billion different antibodies is likely to have at least one with a binding site on which a given antigen will fit. Now, if 10 billion (10^{10}) different antibodies are needed, and if the binding site of each is a function of both its light chains and its heavy chains, then only 100,000 (10^5) genes for light chains and 100,000 genes for heavy chains will suffice (assuming that any light chain can be combined with any heavy chain). In other words, 200,000 genes could code for 10 billion different antibodies ($10^5 \times 10^5 = 10^{10}$). It has been calculated that 200,000 genes would occupy less than 10 percent of the total DNA content of a typical vertebrate cell; thus the germline hypothesis cannot be dismissed out of hand.

Some kinds of experiments, however, cast doubt on the germline hypothesis. For example, when mice of an inbred line are stimulated by a single antigen, they all produce different sets of antibodies, despite their nearly identical genotypes.

It may turn out that neither the germline nor the somatic-mutation hypothesis is entirely correct. Both multiple germ lines and somatic mutations may well be involved. Thus it has been found that there are five major classes of heavy antibody chains, of which some, moreover, have an array of subclasses; it seems likely that each subclass is coded by its own gene. Similarly, there are two major classes of light chains, each of which contains several subclasses; again, each subclass is probably coded by its own gene. Finally, because identical variable regions (V) are sometimes found associated with different classes of constant regions (C), it is generally thought that the V and C regions are coded by separate genes. In other words, though there may not be 200,000 different antibody genes, there is surely a considerable number of them. Concerning the likelihood of accelerated somatic mutation, many investigators believe that certain hypervariable subregions within the V regions of both the light and heavy chains may be coded by unstable so-called hot spots in the DNA, where point mutations or intragenic-recombination mutations are extremely frequent. Some investigators

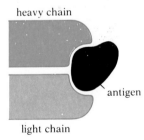

16.18 The antigen-binding site of an antibody molecule
The binding site is a pocket formed by the interaction of a heavy chain and a light chain.

have even suggested that the DNA coding for the hypervariable subregions may be small pieces of DNA that are spliced into larger "framework" genes by special enzymes. At any rate, recent DNA hybridization experiments indicate that the chromosomes in a given nucleus bear only a few (probably one to three) V-region genes; this small number—if it should be further substantiated—would mean that some mechanism for an extreme amount of somatic mutation must be operating to produce the extraordinary diversity of lymphocytes characteristic of vertebrates.

Note one implication of the idea of separate genes for the V and C regions of the globulin chains: In the special case of genetic determination of antibodies, the one gene–one polypeptide rule would give way to a two genes–one polypeptide formula.

Recognition of "self"   Since rejections of grafts or transplanted organs seem to hinge on a distinction between "self" and "not self," i.e. on the body's ability to tell which substances are part of itself and which are not, an obvious possibility for successful transplants is to use identical twins. Such individuals develop from the same fertilized egg cell and thus have identical genes, which presumably control the synthesis of identical substances in the two individuals. The body of one twin could therefore not distinguish between its own parts and parts taken from the other twin; both would be identified as "self." Numerous transplants between identical twins have been tried, and the results have borne out the predictions. The transplants have been accepted by the recipient's body. Most widely publicized have been cases in which a kidney from a healthy twin was transplanted to his twin dying of kidney failure (if the kidneys are normal and healthy, one is sufficient; hence the donor can give one away without impairing his own health).

But few people have identical twins who can come to their aid when they need a transplant. Clearly, the real hope for the future is transplantation from healthy unrelated donors or, for such organs as the heart, from the bodies of recently deceased persons or from other animals such as baboons or apes. Such transplants have, in fact, been tried, with X-ray or chemical suppression of the tissues involved in the immune reactions. Many such attempts have failed, and the others have been only partly successful. It is very difficult, with the techniques now available, to suppress the immune reaction sufficiently to prevent rejection of the transplant without at the same time damaging the patient in other ways. In any such attempt, of course, the patient must be given constant massive doses of antibiotic drugs, because his own defenses against infection have been suppressed.

Basic to a real understanding of immune reactions is more knowledge of the way immunologic capabilities develop and of the way the body distinguishes between "self" and "not self." Many advances have been made in this currently very active field, but information is still scanty.

In 1945 R. D. Owen of the California Institute of Technology discovered that in some cases nonidentical twin calves each had the antigens normal to their own erythrocytes and also the antigens characteristic of their twin's erythrocytes. The calves were healthy, but they were not destroying, or in any way rejecting, the antigens from their nonidentical twin. Their bodies should have identified their twin's antigens as "not self" and made antibodies against them. Why were they not doing so? Owen showed that the circulatory systems of the calves had been interconnected early in their embryonic development and that they must have exchanged blood-forming elements during this period. Apparently, at this early embryonic stage the calves exhibited immunologic tolerance to an exchange that, after birth, would have resulted in antibody formation. Such observations led to the hypothesis that the early embryo lacks immunologic mechanisms, that these develop later, and that when they develop all substances present in the body at that time are identified as "self." In other words, if foreign substances could be introduced into the embryo soon enough, they would be considered "self" when the immunologic mechanisms developed.

Numerous experiments have demonstrated that this prediction is correct. Mice injected as early embryos with cells from another mouse will accept grafts from that mouse after they are born. Such experiments have shown, further, that immunologic capabilities do not arise all at once, but develop gradually during and after embryonic development. Immunologic capabilities against some antigens may arise early in embryonic life, while tolerance to other antigens may last until several weeks after birth. That the defensive apparatus against different types of antigens should arise at different times is consistent with the idea that there are many different lineages of lymphocytes, each responsible for immunologic capabilities against different antigens. Presumably the lymphocytes of each type mature normally if their antigen is absent from the system. If, however, this antigen is present in quantity at some critical stage of their development, they fail to mature.

According to this view of the development of immunologic capabilities, then, substances can be identified as "self" or "not self" by the mature organism because the "self" substances were present at the critical time during development and caused suppression of the immunologic elements that started to develop against them, while "not self" substances were not present at the critical developmental moment and immunologic elements against them were consequently free to mature.

There are, however, some serious objections to this formulation. Some proteins produced only later in life—long after the supposed critical period in early development—are tolerated as "self" substances. It seems, then, that the capacity for developing self-tolerance continues to some degree throughout life. Moreover, tolerance is not necessarily permanent. Thus, when E. L. Triplett, then of the University of California, Santa Barbara, removed the pituitary glands of a group of tree frogs and then reimplanted them in the same individuals several weeks later, the glands were rejected; the frogs had lost their tolerance for a normal part of their own bodies. This experiment strongly suggests that tolerance does not involve destruction of immunologic elements against "self" substances, but merely their suppression—a suppression that may disappear when exposure to a "self" substance is interrupted.

Immunologic disease   Though we normally think of the immunologic mechanisms of the body as a defense against disease, there is increasing evidence that these very mechanisms are at times responsible for disease symptoms. In some cases the invading microbes or other foreign materials do not themselves cause much, if any, damage to the host. It is the immunologic response of the host that actually does the damage; allergies are examples of this type of disease. Several viral diseases are known in which the virus itself does no harm, but the body's immunologic reaction produces such severe symptoms that death sometimes results.

Occasionally the immunologic mechanisms get out of control and become sensitized to some part of the animal's own body, causing interference with or even destruction of that part. In such cases the ability to distinguish between "self" and "not self" is impaired, and the body begins to destroy itself. Among the most intensively studied auto-immune diseases of human beings are rheumatoid arthritis, certain hemolytic anemias, rheumatic fever, ulcerative colitis, and glomerulonephritis. Auto-immune reactions have also been linked to some aspects of aging.

SUGGESTED READING

CAPRA, J. D., and A. B EDMUNDSON, 1977. "The antibody combining site," *Scientific American*, January. (Offprint 1350) *On the possible evolution of antibodies and the mechanism of their reaction with antigens: recent insights gained through new techniques of analysis.*

COOPER, M. D., and A. R. LAWTON, 1974. "The development of the immune system," *Scientific American*, November. (Offprint 1306)

CUNNINGHAM, A. J., 1978. *Understanding immunology*. Academic Press, New York. *Brief up-to-date survey of a rapidly changing field.**

CUNNINGHAM, B. A., 1977. "The structure and function of histocompatibility antigens," *Scientific American*, October. (Offprint 1369) *Antigens on the surfaces of cells— their role in rejection of transplanted organs, on the one hand, and in the body's defense against infection and cancer, on the other.*

HOOD, L. E., I. L. WEISSMAN, and W. B WOOD, 1978. *Immunology.* Benjamin/Cummings, Menlo Park, Calif. *Up-to-date treatment.**

MANIATIS, T., and M. PTASHNE, 1976. "A DNA operator-repressor system," *Scientific American*, January. (Offprint 1333) *Fine account of how operators, promoters, repressors, and other gene-control elements work.*

MAYER, M. M., 1973. "The complement system," *Scientific American*, November. (Offprint 1283) *On the way an intricate set of enzymes works with antibodies to make holes in the membranes of foreign cells, thereby destroying them.*

NICOLSON, G. L., 1979. "Cancer metastasis," *Scientific American*, March. (Offprint 1422)

NOTKINS, A. L., and H. KOPROWSKI, 1973. "How the immune response to a virus can cause disease," *Scientific American*, January. (Offprint 1263)

OLD, L. J., 1977. "Cancer immunology," *Scientific American*, May. (Offprint 1358) *The distinctive antigens on the surfaces of cancer cells—and the problem of mobilizing the immune system to combat cancer.*

RUDDLE, F. H., and R. S. KUCHERLAPATI, 1974. "Hybrid cells and human genes," *Scientific American*, July. (Offprint 1300) *The new technique of fusing human cells with cells from other mammals in mapping human genes and studying their regulation.*

STEIN, G. S., J. S. STEIN, and L. J. KLEINSMITH, 1975. "Chromosomal proteins and gene regulation," *Scientific American*, February. (Offprint 1315) *The possible roles of histones and nonhistone acidic proteins in determining which genes will be turned on.*

WATSON, J. D., 1976. *Molecular biology of the gene*, 3rd ed. Benjamin, Menlo Park, Calif. *See Chapter 14 for an excellent summary of genetic control systems, and Chapters 19 and 20 for similarly sound treatments of the fundamental principles of immunology and of the viral origins of cancer.*

* In paperback.

Chapter 17

DEVELOPMENT: FROM EGG TO ORGANISM

To biologists and nonbiologists alike, probably no aspect of biology is more amazing than the development of a complete new organism from one cell, a development so precisely controlled that the entire intricate organization of cells, tissues, organs, and organ systems characterizing the functioning adult comes into being with rarely a flaw. We have previously examined the genetic information that controls development and programs a mouse zygote to develop into a mouse, an oak zygote into an oak, and an earthworm zygote into an earthworm. We have also discussed possible cellular control mechanisms in development—how individual genes may be turned on or off in its course. Let us now first briefly examine a few representative patterns in animal and plant development and then try to relate these patterns to the control mechanisms we considered earlier.

DEVELOPMENT OF A MULTICELLULAR ANIMAL

As a first representative pattern of development, we shall take the principal events in the development of a multicellular animal, making no attempt to discuss this development in great detail or even to mention all the important events. Our purpose is simply to survey the kinds of events that any model of developmental control must seek to explain.

FERTILIZATION

We saw in Chapter 9 that certain cells are early set aside as egg primordia in the ovary of a female animal. These cells grow to an unusually large size, and when they then undergo meiosis, the divisions are unequal and almost all the cytoplasm is retained in the ripe ovum, the other haploid cells being the

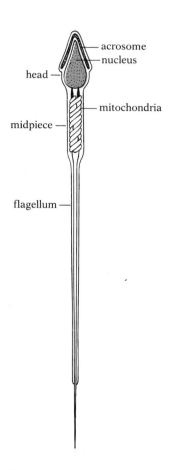

17.1 A human sperm cell

Three sections are easily distinguishable: the head, containing the nucleus; the midpiece, in which many mitochondria are tightly packed; and the long flagellum. The chromosomes in the nucleus are complexed with exceedingly alkaline (arginine-rich) histones and are thus supercondensed into an almost crystalline, genetically inactive form. In the apex of the head, immediately in front of the nucleus, is the acrosome, a membrane-bounded vesicle derived from the Golgi apparatus. There is only a tiny amount of cytosol in the cell.

tiny polar bodies that soon deteriorate (see Fig. 13.24, p. 578). The sperm, on the other hand, is an unusually small cell with very little cytoplasm. The ovum thus furnishes most of the initial cytoplasm for the embryo (hence the much greater importance of the mother than of the father in the transmission of cytoplasmically inherited traits).

The penetration of the sperm into the egg cell stimulates the egg to begin development into an embryo. Note we said that *penetration* is the trigger, not the fusion of the sperm nucleus with the egg nucleus, even though this fusion is the actual event of fertilization. Apparently, true fertilization is not necessary to induce embryonic development in many animals, even animals that do not normally reproduce parthenogenetically (i.e. without fertilization of the egg). It is easy to induce unfertilized frog eggs, for example, to begin development in the laboratory by pricking them with a fine needle dipped in blood. A few such eggs will develop into viable normal-appearing tadpoles.[1] Adult rabbits have been produced from unfertilized eggs by similar procedures. (What would be the sex of all such parthenogenetically produced rabbits?) Unfertilized eggs can even be stimulated to begin developing by giving them a mild electric shock, or by changing the salt concentration in the surrounding fluid, or by just shaking them, but the development aborts after only a few divisions, apparently because some cellular factor (provided by blood cells in the method mentioned above) is necessary for mitosis to proceed normally.

Let us look more closely at the process of fertilization of an egg cell by a sperm cell. In many mammalian species the egg is initially enclosed in a thin protective layer of follicle cells. This layer, a barrier to the sperm, is loosened by hyaluronidase, an enzyme secreted by the sperm; similar enzymes are used by bacteria in penetrating host tissues. But even after the follicular cell barrier has been loosened, the sperm cell (attracted to the egg by chemicals released by it[2]) still encounters a membranous barrier to its penetration of the egg.

At this point the acrosome, a membrane-bounded vesicle located in the apex of the head of the sperm cell (Fig. 17.1; see also Fig. 9.43, p. 386) comes into play. Its membrane fuses with the plasma membrane of the sperm cell and forms a tube that penetrates through the jellylike coat on the egg-cell membrane and then fuses with a microvillus of the egg cell (Fig. 17.2); the fusion is probably facilitated by enzymes released from the acrosome. Once this fusion of sperm and egg membranes has occurred, the sperm nucleus can move into the cytoplasm of the egg.

As soon as the sperm and egg membranes have fused and the sperm nucleus has moved into the egg, a host of changes begin in the egg. Almost immediately, vesicles in the outer region of the egg cytoplasm discharge their strongly acidic glycoprotein contents into the region around the cell. These glycoproteins promptly undergo polymerization to form a nearly crystalline **fertilization membrane** surrounding the egg. The fertilization membrane, and a mucous hyaline layer between it and the plasma mem-

---

[1] Most embryos developed from unfertilized eggs are haploids, which invariably die after reaching, at most, the tadpole stage. Such embryos as survive to reach adulthood are the few that undergo spontaneous chromosomal doubling, becoming diploid (though the two sets of chromosomes are identical), and the even fewer that become $4n$, $6n$, or $8n$.

[2] Attractant chemicals released by gamete cells are called gamones. The best-known gamone is fertilizin, a chemical released by the egg cells.

brane, act as a barrier against entry of additional sperm cells. Another process initiated immediately after penetration is completion of oogenic meiosis if that process is not already complete (it is completed in some species prior to sperm penetration, but not in others, where it is arrested at metaphase of meiosis I or II, depending on the species). Among the many other changes regularly seen in the egg after penetration are striking alterations in the permeability of the plasma membrane (especially a much enhanced permeability to inorganic phosphate) and an increased rate of oxygen consumption.

All the changes we have enumerated, from formation of the fertilization membrane to increase in the metabolic rate, are triggered, not by the presence of the sperm nucleus in the egg cell, but rather by the interaction of the sperm with the plasma membrane of the egg. Thus, if a sperm nucleus is injected into an unfertilized egg with a micropipette, none of the changes take place and both the egg and the sperm nuclei remain inert; they show no tendency to move toward each other or to fuse. Some sort of signal from a receptor in the plasma membrane appears to be essential.

Important as are all the events discussed so far, they do not constitute fertilization in the genetic sense; true fertilization is the union of the two gamete nuclei. This union depends on some attraction of the sperm nucleus by the egg nucleus, the nature of which is still unknown. If sperm are induced to penetrate immature sea-urchin eggs in which the chromosomes are still condensed, no attraction of the sperm nucleus occurs; nor is there any tendency for the supercondensed chromosomes of the sperm nucleus to undergo decondensation. In short, the mature egg nucleus must be responsible both for attracting the sperm nucleus and for inducing the deconden-

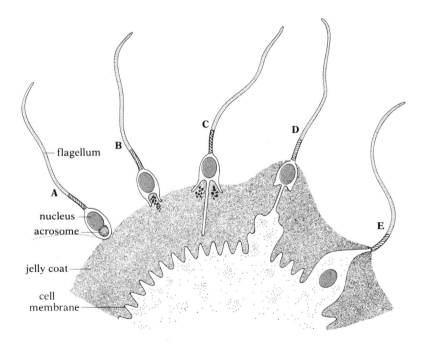

17.2   The fertilization process
(A) A sperm cell comes into contact with the jelly coat surrounding an egg cell.   (B) The membrane of the acrosome fuses with the plasma membrane of the sperm and ruptures, releasing the acrosomal contents, which include enzymes that act on the jelly coat and on the membrane of the egg cell.   (C) A membranous tube forms, which pushes through the jelly coat.
(D) The tube fuses with an enlarged microvillus of the egg cell; there is now no membranous barrier between the contents of the sperm cell and of the egg cell.
(E) The sperm nucleus moves into the egg cell.

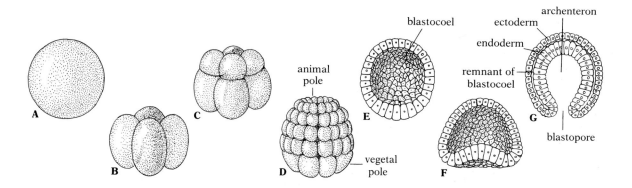

17.3 Early embryology of amphioxus

(A) Zygote. (B–D) Early cleavage stages forming a morula (C) and then a blastula (D). (E) Longitudinal section through a blastula, showing the blastocoel. (F–G) Longitudinal sections through an early and a late gastrula. Notice that the invagination is at the vegetal pole of the embryo, where the cells are largest.

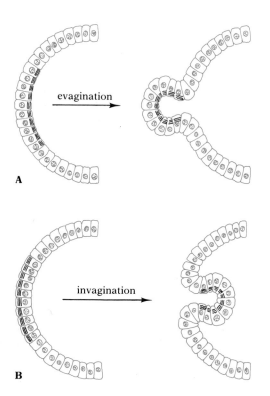

sation of its chromosomes. Decondensation of both sperm and egg chromosomes is necessary if the zygote nucleus produced by fusion of the two gamete nuclei is to carry out DNA replication preparatory to the first mitotic division of embryonic development.

EMBRYONIC DEVELOPMENT

Early cleavage and morphogenetic stages   In normal development the zygote begins a rapid series of mitotic divisions immediately after fertilization has taken place. These early cleavages are not accompanied by protoplasmic growth. They produce a grapelike cluster of cells called a *morula,* which is little if any larger than the single egg cell from which it is derived (Fig. 17.3C). The cytoplasm of the one large cell is simply partitioned into many new cells that are much smaller.

During this early cleavage stage of development, the nuclei cycle very rapidly between chromosomal replication (the S period of the cell cycle) and mitosis (the M period); the G_1 and G_2 periods are practically absent. Such rapid cycling, in which G_1 and G_2 are skipped, is made possible by the fact that the ovum already contains both huge quantities of the DNA polymerase necessary for catalyzing repeated chromosomal replication and most of the mRNA required for synthesis of proteins during early cleavage (at least 50 percent of the proteins synthesized are histones for the new chromosomes; the proteins of growth are produced in only negligible amounts, as would be expected). The brief interval taken up by transcription—because so little new

17.4 The mechanism of some morphogenetic movements in cells

Contraction of microfilaments (black), asymmetrically positioned in the cells, may change the shapes of the cells and produce evaginations (A), invaginations (B), or other alterations of the arrangement of cells in a developing organ. The contractile microfilaments are of the same type as those involved in amoeboid movement and cytokinesis.

mRNA is needed—allows the rapid cycling between the S and M periods. But note that, since control of this cleavage stage of embryonic development depends largely on mRNA synthesized in the oocyte prior to fertilization, the paternal genes have little input until later in development; the course of the early cleavages is determined almost exclusively by the maternal genes.

We have already said that ova are remarkably large cells. They are so large, in fact, that the ratio of nuclear to cytoplasmic material would be too low for proper control of ordinary cellular activities. The early cleavages of embryonic development, with their minimal cell growth, thus help restore a more normal ratio of nuclear to cytoplasmic material.

As cleavage continues, the newly formed cells (blastomeres) of many species begin to secrete a fluid into the center of the mass of cells. As a result, the blastomeres come to be arranged in a sphere surrounding a fluid-filled cavity called a *blastocoel* (Fig. 17.3E). An embryo at this stage is termed a *blastula.*

Next begins a series of complex movements important in establishing the definitive shape and pattern of the developing embryo. The establishment of shape and pattern in all organisms is called *morphogenesis* (meaning the genesis of form). Morphogenetic movements of cells in large masses always occur during the early developmental stages of animals.

The mechanism of these movements is still very poorly understood. There are often changes in the shapes of the cells, probably effected by contractile microfilaments or by some microtubular apparatus. The changes in shape may be relatively small (Fig. 17.4), or they may be extensive (Fig. 17.5). Possibly important in some of the movements are changes in the adhesive affinities of the cells for neighboring cells or, in the case of epithelial cells, for the basement membrane on which these cells sit. It may be relatively easy for a group of cells that adhere tightly to each other, but have very little affinity for a layer of cells lying under them, to slide, as a group, over the surface of that underlying layer, just as it is easy for individual cancer cells, which lack cellular affinities, to move across the surfaces of healthy cells.

Since the pattern of cleavages and cell movement is greatly influenced by the amount of yolk (stored food) in the egg, we shall examine, first, the pattern in an animal whose eggs have little yolk, then that in animals whose eggs have more yolk.

In amphioxus (see Fig. 25.79, p. 1047), a tiny marine chordate whose egg has very little yolk, the movements that occur after formation of the blastula (when it is composed of about 500 cells) convert it into a two-layered structure called a *gastrula.* The process of gastrulation begins when a small depression, or invagination, starts to form at a point on the surface of the blastula where the cells are somewhat larger than those on the opposite side (Fig. 17.3F). The differences in cell size are not very great in amphioxus embryos; they are more pronounced in many other animals. The smaller cells make up the *animal hemisphere* of the embryo. The larger cells make up the *vegetal hemisphere.* It is at the pole of the vegetal hemisphere that the invagination of gastrulation typically occurs. As gastrulation proceeds, and more and more cells move to the point of invagination and then fold inward, the invagination becomes larger and larger. Eventually the invaginated cell layer comes to lie almost against the outer layer, thus nearly obliterating the old blastocoel (Fig. 17.3G). The resulting gastrula is a two-layered cup, with a

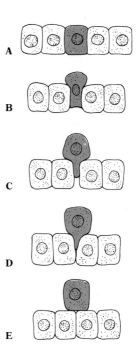

17.5   Change in the shape of a cell as a mechanism for altering its position

(A) The colored cell, like its neighbors, is part of an epithelial layer.   (B) The cell begins to constrict horizontally, but simultaneously elongates, forming a globular protuberance that projects outside the epithelial layer.   (C–D) The horizontal constriction, accompanied by elongation, continues as the lower portion of the cell begins to shorten.   (E) The eventual result of the cell's changes of shape is a shift from its original place in the epithelial layer to a new position.

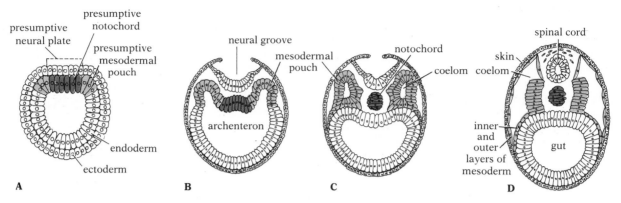

17.6 Neurulation in amphioxus

The cross sections through amphioxus embryos show the progressive formation of mesoderm and the neural tube. (A) When gastrulation has been completed, the dorsal part of the endoderm has already differentiated as presumptive mesoderm, i.e. as presumptive notochord (dark color) and mesodermal pouches (light color); similarly, part of the ectoderm has differentiated as presumptive neural plate. (B–C) The notochord and mesodermal pouches form as evaginations from the endoderm, and the neural plate invaginates from the ectoderm as it begins to form the spinal cord. (D) In this later embryo both the spinal cord and the mesoderm are taking their definitive form. Notice that there is a cavity (the coelom) in the mesoderm.

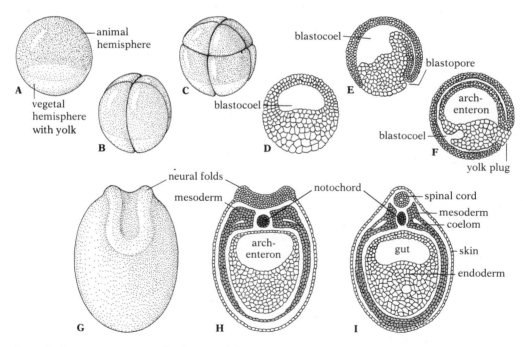

17.7 Early embryology of a frog

The large amount of yolk in the frog egg causes its pattern of gastrulation to differ from that in amphioxus. (A) Zygote. (B–C) Early cleavage stages. Note that the first horizontal cleavage is nearer the animal pole, much larger cells thus being produced at the vegetal pole. (D) Longitudinal section of a blastula. (E–F) Longitudinal sections of two gastrula stages. (G) An early neurula, showing the neural folds and neural groove. (H) Cross section of a neurula after formation of mesoderm. (I) Cross section of a later embryo, showing definitive spinal cord.

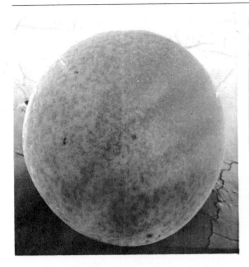
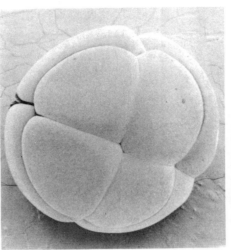
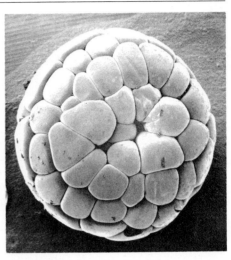

new cavity that opens to the outside via the **blastopore,** which is at the point where invagination first began. The new cavity, called the **archenteron,** will become the cavity of the digestive tract, and the blastopore will become the anus.

Gastrulation, as it occurs in amphioxus, first produces an embryo with two primary cell layers, an outer **ectoderm** and an inner **endoderm.** A third primary layer, the **mesoderm,** soon begins to form between the ectoderm and the endoderm. In amphioxus the mesoderm originates as pouches pinched off the endoderm (Fig. 17.6). In many other animals it arises from inwandering cells derived primarily from the area around the blastopore where the ectoderm and endoderm meet.

In the amphioxus egg, where the distinction between animal and vegetal hemispheres is only slight owing to the small amount of yolk in the vegetal hemisphere, the early cleavages are nearly equal (i.e. the new cells are of nearly the same size) and gastrulation can occur in a direct and uncomplicated manner. Many eggs have far more yolk in their vegetal hemisphere, and this deposit of stored food imposes complications and limitations on such processes as cleavage and gastrulation. Generally, the more yolk an egg contains, the more cleavage tends to be restricted to the animal hemisphere and the more gastrulation departs from the pattern in amphioxus.

Frog eggs, which contain far more yolk than those of amphioxus but much less than those of most birds, may serve as examples of eggs with an intermediate yolk mass. The first two cleavages, which are perpendicular to each other, cut through both the animal and vegetal poles, producing cells of roughly the same size (Fig. 17.7B). But the next cleavage is horizontal and located decidedly nearer the animal pole (Fig. 17.7C); thus the four cells produced at the animal end of the egg are considerably smaller than the four at the vegetal end. From this stage onward, more cleavages occur in the animal hemisphere of the embryo than in the vegetal hemisphere as the blastula develops. As in amphioxus, there is very little increase in total size during these early cleavage stages (Fig. 17.8).

Early in its second day of development, the frog embryo begins gastrulation. Simple invagination at the vegetal pole is not mechanically feasible,

17.8   Scanning electron micrographs of frog egg and some early cleavage stages
Left: Unfertilized egg.   Middle: 8-cell stage.   Right: 32 to 64-cell stage.   All three micrographs are at the same magnification: × 46½. Note that there has been no overall growth in size during these cleavage stages; the 32 to 64-cell embryo is no larger than the egg cell.   [Courtesy R. G. Kessel and C. Y. Shih, *Scanning electron microscopy in biology,* Springer-Verlag, 1974.]

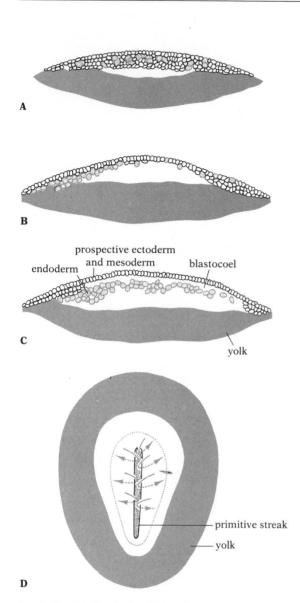

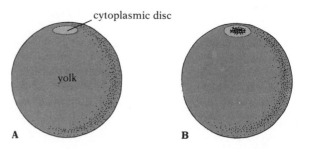

17.9 Egg and early-cleavage embryo of a chick
(A) The zygote. A small cytoplasmic disc lies on the surface of a massive yolk. (B) Early cleavage. There is no cleavage of the yolk.

17.10 Gastrulation in the chick embryo
(A) Longitudinal section through a blastula. Larger yolk-laden cells (color) are intermixed with smaller cells. (B) The larger cells begin to accumulate on the lower surface of the cell mass. (C) The layer of larger cells separates from the layer of smaller cells to become the endoderm; the cavity between the two layers is the blastocoel. (D) Surface view of a gastrula. Involution of cells along the midline of the embryo during gastrulation (color arrows) produces a clearly visible primitive streak, which is essentially a very elongate blastopore.

because of the large mass of inert yolk. Instead, portions of the cell layer of the animal hemisphere move down around the yolk mass and then fold in at the edge of the yolk. This involution begins at what will be the dorsal side of the yolk mass, forming initially a crescent-shaped blastopore at the edge of the yolk (Fig. 17.7E). This infolding slowly spreads to all sides of the yolk, so that the crescent blastopore is converted into a circle. Movement of the other cells around the yolk eventually encloses this material almost completely within the cavity of the archenteron.

Birds' eggs contain so much yolk that the small disc of cytoplasm on its surface is dwarfed by comparison. No cleavage of the massive yolk is possible, and all cell division is restricted to the small cytoplasmic disc (Fig. 17.9). (Note that the yolk and the small lighter-colored cytoplasmic disc on its surface constitute the true egg cell. The white albumin of the egg is outside the cell.) The gastrulation process is of necessity greatly modified in such eggs. Neither invagination of the vegetal pole as in amphioxus nor involution around the edges of the vegetal pole as in a frog can occur. Instead, the endoderm splits away from the underside of the ectoderm (Fig. 17.10A–C). Then cells from the upper layer (primordial ectoderm and mesoderm) involute along the longitudinal midline of the embryo to form the mesoderm. This involution gives rise to a clearly visible line or streak on the surface of the cytoplasmic disc (Fig. 17.10D); the streak is, in effect, a very elongate blastopore.

The fates of cells in different parts of the three primary layers of vertebrates have been determined by staining them with dyes of different colors and then following their movements. As you might expect, the ectoderm eventually gives rise to the outermost layer of the body—the epidermal portion of the skin—and to structures derived from the epidermis, such as hair, nails, the eye lens, many glands, and the epithelium of the nasal cavity, mouth, and anal canal. As you might also expect, the endoderm gives rise to the innermost layer of the body—the epithelial lining of the digestive tract and of other structures derived from the digestive tract, such as the respiratory passages and the lungs, the liver, the pancreas, the thyroid, and the bladder. The mesoderm gives rise to most of the tissues in between, such as muscle, connective tissue (including blood and bone), and the notochord (a dorsally located supportive rod found in all chordates, at least in the embryological stages).

One major tissue located topographically between the skin and the gut does not develop from the mesoderm. This is the nervous tissue, which, curiously enough, is derived from the ectoderm. Soon after gastrulation, the ectoderm becomes divided into two components, the epidermis and the neural tube. A sheet of ectodermal cells lying along the midline of the embryo above the newly formed digestive tract and developing notochord bends inward in a process called **neurulation,** and forms a long groove extending most of the length of the embryo (Figs. 17.6, 17.7G–H). The dorsal folds that border this groove then move toward each other and fuse together, converting the groove into a long tube lying beneath the surface of the back. This neural tube becomes detached from the epidermis above it, and in time differentiates into the spinal cord and brain (Figs. 17.6D, 17.7 I).

We see, then, that the morphogenetic movements of gastrulation and neurulation give shape and form to the embryo, and bring masses of cells into the proper position for their later differentiation into the principal tissues of the adult body. In effect, the movements mold the embryonic mass into the structural configuration on which differentiation will superimpose the finer detail of the finished organism.

Later embryonic development   Much must happen to convert a gastrula into a fully developed young animal ready for birth: The individual tissues and organs must be formed; an efficient circulatory system must quickly come into function (Fig. 17.11); in a vertebrate the four limbs must develop; the elaborate system of nervous control must be established; and so forth. The complexity and precision characterizing these developmental changes are staggering to contemplate. To give but one example: Approximately 43 muscles, 29 bones, and many hundreds of nervous pathways must form in

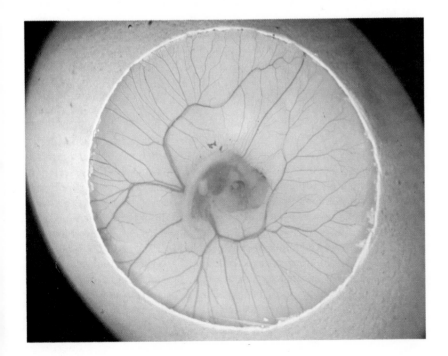

17.11 Chick embryo after four days of incubation
The tiny embryo lies on the surface of the yolk. It has a functional circulatory system, including a beating heart, even at this early stage of its development. Note the long branching blood vessels that run out of the embryo into the yolk; they transport nutrients to the embryo.   [Oxford Scientific Films.]

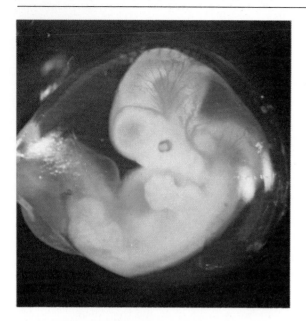

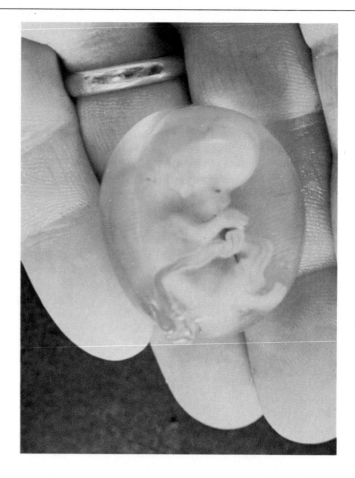

17.12 Human embryos after 5½–6 weeks (left) and 8 weeks (right) of development
Both embryos are seen inside the fluid-filled amnion. The 5½–6-week embryo ($\times$ 2) shows the beginnings of eyespots, but no distinctive face; note its mittenlike hands and feet, with no separation between the digits. By contrast, the 8-week embryo ($\times$ 1.4) has a distinct face, and there is separation between its fingers. [Photos © Donald Yeager, RBP, 1977.]

each human arm and hand. To function properly, all these components must be precisely correlated. Each muscle must have exactly the right origins and insertions; each bone must be jointed to the next bone beyond it in a certain way; each nerve fiber must have all the proper synaptic connections with the central nervous system and must terminate on the right effector cells. Incredibly sensitive mechanisms of developmental control must operate if such an intricate structure can arise from a mass of initially undifferentiated cells. Yet the developmental processes that produce all these later embryonic changes are the same ones we have seen at work in the early embryo—cell division, cell growth, cell differentiation, and morphogenetic movements. Bursts of mitotic activity in some areas and cessation of cleavage in other areas alter the balance between the parts. Special patterns of cell growth produce important changes in size and shape. Through differentiation cells may lose particular capacities, but may become more efficient at performing other functions. Foldings and pouchings establish the primordia of lungs and glands, of eyes and bladder. Even cell death plays an important role in the normal development of the living animal; e.g. fingers and toes become separated by the death of the cells between them (Fig. 17.12), and initially solid cords of mesodermal tissue become hollow blood vessels by death of their innermost cells.

In human beings the egg is fertilized soon after ovulation, while it is still in the upper portion of the oviduct. It immediately begins cleavage and has reached the blastula stage by the time it becomes implanted in the uterine wall 8–10 days after fertilization. During the implantation process the amnion, which will enclose the embryo in a fluid-filled chamber throughout the rest of its development, begins to form.

By the twenty-third day two other embryonic membranes—the chorion and the allantois—have collaborated with maternal tissues of the uterus to give rise to a functional placenta; the chorion contributes the fingerlike villi of the fetal portion of the placenta, and the allantois contributes the blood vessels (see Fig. 9.51, p. 395). By this time, the neural groove is complete; the mesoderm, too, is well developed, and individual segments (somites) can easily be distinguished. The tubular embryonic heart has begun to pulsate weakly.

By the end of the first month of development the embryo, which now exhibits arm and leg buds, is roughly 5 mm (⅕ inch) long. The hands and feet are still mittenlike, with no separations between the individual fingers and toes.

During the second month the embryo grows to a length of approximately 30 mm (a bit more than one inch) and to a weight of about one gram (0.03 ounce). By the end of the month (i.e. after eight weeks of development), the embryo, now called a *fetus,* is recognizable as a human. It has a flat face with widely separated eyes, and the digits (fingers and toes) are well separated (Fig. 17.12, right). The cerebral cortex has begun to show cellular differentiation, and organization of the sense organs is well advanced. The muscles are sufficiently differentiated for some movement to be possible. Ossification of the skeleton has begun. The liver, which is serving as the chief producer of blood cells, is proportionately much larger than it will be later.

It is during the first two months of development that the embryo is especially sensitive to a wide variety of factors that can cause serious malformations. For example, if during that period the mother should contract "German measles" (more properly called rubella), this normally mild viral disease can result in a malformed heart, cataracts of the eyes, or deafness in the fetus. The disastrous effects on the fetus of the tranquilizer thalidomide when taken by the mother early in pregnancy have been widely publicized.

By the end of 14 weeks the fetus is about 107 mm long, and the body proportions approach those expected at birth. The sex of the fetus can be determined externally. The cerebral hemispheres have begun to overlap the rest of the brain, and the sense organs are almost complete. The heart, too, has taken on nearly its final form. Spontaneous movements are frequent.

At this point in development the basic pattern of all the physiological systems has been established. The remainder of fetal development is largely a combination of further refinement and elaboration and of growth in overall size (over 90 percent of fetal weight gain takes place in the last four months). In time, the bone marrow takes over from the liver primary responsibility for hematopoiesis; the spinal cord and then the brain become myelinated; the eyes become light-sensitive, and later the ears become audiosensitive; many new brain cells are produced; and numerous new neural circuits are established and become functional up to and past the time of birth.

By the end of 24 weeks the fetus has some chance of survival outside the uterus if it is given respiratory assistance and kept in an incubator. However, it is still so small (about 0.66 kg— 1½ pounds) and so poorly developed that it is subject to many special medical problems, which threaten its life for months. If, on the other hand, birth occurs at full term (on the average, about 280 days after the beginning of the mother's last regular menstrual period), the chances of survival are high.

Several very important developmental changes in the circulatory system occur at the time of birth: The placental circulation is cut off. The ductus arteriosus (shunt between the pulmonary artery and the aorta) is closed. The lungs are inflated for the first time. Blood is forced into the pulmonary system. And production of fetal hemoglobin soon gives way to production of the adult type.

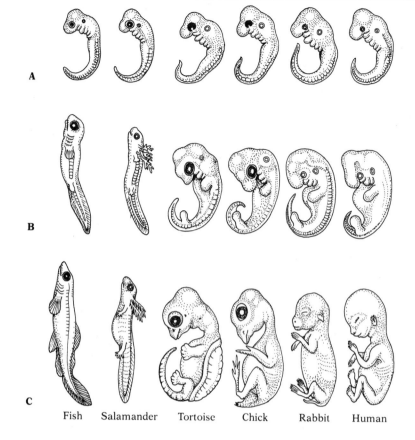

17.13 Vertebrate embryos compared at three stages of development

At stage A, all the embryos—whether fish, amphibian, reptile, bird, or mammal—strongly resemble one another. Later, at stage B, the fish and the salamander are noticeably different, but the other embryos are still very similar; note the gill pouches in the pharyngeal region and the prominent tail. By stage C, each embryo has taken on some of the features distinctive of its own species.   [Redrawn from G. J. Romanes, *Darwin and after Darwin*, Open Court Publishing Co., 1901.]

It is beyond the scope of this book to discuss in detail the many events that occur during later embryonic development. Yet it is these events that mold morphologically similar gastrulas into a fish in one instance, a rabbit in another, and a human being in still another, depending on the genetic endowment of the gastrula in question; the developmental events are programed differently for each species. An understanding of how such different programs arise and how they are carried out is one of the important goals of developmental biologists.

One interesting aspect of the differences in the developmental programs of different species should be mentioned here—namely that the early embryos of most vertebrates closely resemble one another. For example, the early human embryo, with its well-developed tail and a series of gill pouches in the pharyngeal region, looks very much like an early fish embryo (Fig. 17.13); and it looks even more like an early rabbit embryo. It is only as development proceeds that the distinctive traits of each kind of vertebrate become apparent.

POSTEMBRYONIC DEVELOPMENT

The extent to which an animal has developed by the time of birth varies greatly among different species. Some young animals are entirely self-suffi-

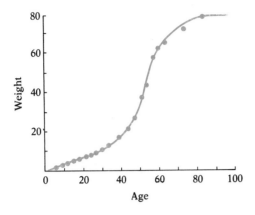

17.14   A typical S-shaped growth curve
[Redrawn from D'A. W. Thompson, *On growth and form*, Cambridge University Press, 1942 (from G. Backman, after Stefanowska).]

cient from the time they are born and neither need nor receive parental care. Others can run about and feed themselves as soon as they are born, but still benefit from a limited amount of parental care; baby chicks and ducks are examples. Still other animals are born while still at an early stage of development, and are nearly helpless and totally dependent on parental care; baby human beings and baby robins are examples (just-hatched baby robins are blind, almost devoid of feathers, and unable to stand).

The extent of development at birth is often (though not always) a reflection of the length of the embryonic period, which is usually correlated in animals that lay eggs with the amount of yolk the eggs contain. Among birds particularly, species that have a short incubation period for the eggs characteristically have altricial—poorly developed—young (see Fig. 25.97, p. 1058), while species that have a longer incubation period characteristically have precocial (well-developed) young; e.g. robins incubate their eggs only 13 days, while chickens have a 21-day incubation period. Regardless of their state of development at birth, however, all animals continue to undergo major developmental changes during their postembryonic life.

Growth   Although postembryonic development seldom involves any major morphogenetic movements, there is some cell multiplication and cell differentiation. But the preponderant factor by far in many animals is growth in size. Usually growth begins slowly, becomes more rapid for a time, and then slows down again or stops. This pattern yields the characteristic S-shaped growth curve shown in Fig. 17.14. It must be emphasized that, although the general shape of this curve holds for most organisms, its details vary in important ways from species to species. The slope of the curve is different for different species, depending on whether they grow very rapidly for a shorter time or more slowly for a longer time (compare the rate of increase in weight of, say, a calf and a child). The shape of the curve is seldom as smooth as it appears in a generalized growth curve, because so many factors can affect the rate of growth. For example, in most mammals growth slows down for a while immediately after weaning, and it often varies greatly during puberty; such irregularities are reflected in bumps and dips in the curve (Fig. 17.15). An especially marked departure from the smooth generalized curve is seen in the growth of arthropods. These animals

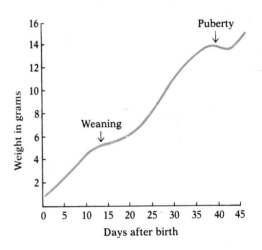

17.15   Growth in weight of a mouse
The rate of growth is slower at weaning and at puberty.   [Modified from D'A. W. Thompson, *On growth and form*, Cambridge University Press, 1942 (after W. Ostwald).]

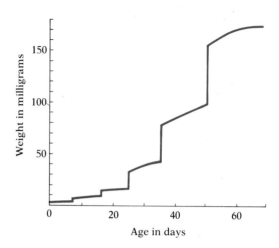

17.16   Growth in weight of an insect
The growth spurts of the water boatman occur at the time of molt, when the old exoskeleton has been shed and the new one has not yet fully hardened. [Modified from V. B. Wigglesworth, *The principles of insect physiology,* Methuen, 1947.]

can undergo only limited growth between molts, because the hard exoskeleton that encases their bodies can be stretched only slightly. However, at each molt there is a sharp burst of growth during the short period after the old exoskeleton has been shed and before the new one has hardened. The resulting growth curve shows a steplike pattern (Fig. 17.16).

Growth does not occur at the same rate and at the same time in all parts of the body. It is obvious to anyone that the differences between a baby chick and an adult hen or rooster, or between a newborn baby and an adult human being, are differences not only in overall size but also in body proportions. The head of a young child is far larger in relation to the rest of its body than that of the adult. And the child's legs are much shorter in relation to its trunk than those of the adult. If the child's body were simply to grow as large as an adult's while maintaining the same proportions, the result would be a most unadultlike individual (Fig. 17.17). Normal adult proportions arise because the various parts of the body grow at quite different rates or stop growing at different times (Fig. 17.18).

Two closely related species that differ in size are frequently also quite different in body proportions, not because of any basic difference in the growth patterns of the two species but simply because a slight increase in

17.17   Changes in body proportions during human fetal and postnatal growth
The head grows proportionately much more slowly than the limbs.   [Modified from *Morris' human anatomy,* ed. by C. M. Jackson, Blakiston, 1925.]

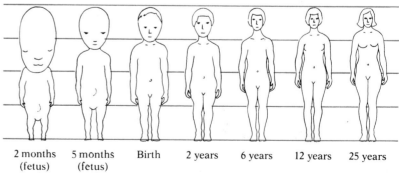

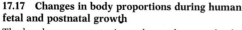

2 months (fetus)   5 months (fetus)   Birth   2 years   6 years   12 years   25 years

overall body size automatically results in disproportionate increases in some parts of the body. In elk, for example, the size of the antlers increases much faster than the overall body size. Consequently, if species A grows slightly larger than species B, then A will have proportionately much larger antlers.

Larval development and metamorphosis   Growth in size is not always the principal mechanism of postembryonic development. Many aquatic animals, particularly those leading sessile lives as adults, go through *larval* stages that bear little resemblance to the adult (Fig. 17.19). The series of sometimes drastic developmental changes that convert an immature animal into the adult form is called *metamorphosis.* It often involves extensive cell division and differentiation, and sometimes even morphogenetic movement; growth alone could not accomplish the transformation of a larva into an adult.

In many aquatic animals dispersal of the species depends on the larval stage; the tiny larvae either swim or are passively carried by currents to new locations, where they settle down and undergo metamorphosis into sedentary adults. In other species, such as frogs, where the adult is not sedentary, the adaptive significance of the larval stage (the tadpole in the case of frogs) seems less a matter of dispersal than of exploiting alternative food sources during the developmental stages (tadpoles feed primarily on microscopic plant material, while adult frogs are carnivorous and take fairly large prey).

Although a larval stage occurs in the life history of many aquatic animals, probably the most familiar larvae are those of certain groups of terrestrial insects, including flies, beetles, wasps, butterflies, and moths. The young fly, wasp, or beetle is a grub that bears no resemblance to the adult. The young butterfly or moth is a caterpillar. In the course of their larval lives, these insects molt several times and grow much larger, but this growth does not bring them any closer to adult appearance; they simply become larger larvae (Fig. 17.20). Finally, after they have completed their larval development, they enter an inactive stage called the *pupa,* during which they are usually enclosed in a case or cocoon. During the pupal stage most of the old larval tissues are destroyed, and new tissues and organs develop from small discs of cells that were present in the larva but never underwent much development. The adult that emerges from the pupa is thus radically different from the larva; it is almost a new organism built from the raw materials of the larval body.

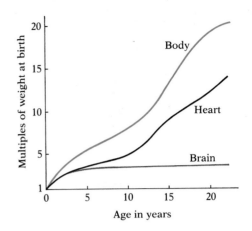

17.18   Graph showing differences in relative growth of body, heart, and brain of a man
[Modified from D'A. W. Thompson, *On growth and form,* Cambridge University Press, 1942 (from Quetelet's data).]

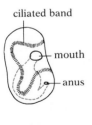

ciliated band

mouth

anus

Ciliated larva   Bipinnaria larva   Brachiolaria larva

17.19   Three larval stages of the sea star *Asterias vulgaris*
The gastrula develops into the ciliated larva, which changes into the bipinnaria larva, which changes into the brachiolaria larva, which metamorphoses into the characteristically shaped sea star.

A Monarch butterfly egg

D Full-grown larva going into pupation

B First-stage larva eating its own egg shell

C Later-stage larva eating molt skin

E Early chrysalis

17.20 Developmental stages of a Monarch butterfly, an insect with complete metamorphosis

From the egg (A) hatches a small larva (B), which goes through a series of molts, growing larger at each (C). Eventually the full-grown larva attaches to a plant (D) and goes into pupation, forming a case around itself (E); in moths and butterflies the encased pupa is called a chrysalis. During pupation most of the larval tissues are broken down and new adult tissues are formed; in the Monarch some of the adult structures can be seen through the case of the late chrysalis (F). The newly formed adult emerges from the pupal case and rests while its wings become inflated (G). It then flies off to feed at flowers—here milkweed (H)—before mating and laying eggs (always on milkweed) to start the cycle over again.   [D. Overcash (A, B, C, H), L. West (D), and E. R. Degginger (E, F, G), Bruce Coleman Inc.]

F Late-stage chrysalis

G Inflation of wings by newly emerged butterfly

H Adult butterfly at flower

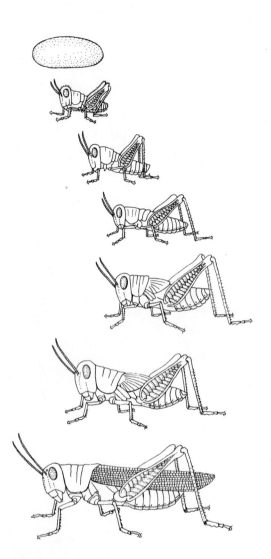

17.21 Gradual metamorphosis of a grasshopper
The insect that emerges from the egg (top) goes
through several nymphal stages that bring it gradually
closer to the adult form (bottom).

Insects with a pupal stage and the type of development just described are
said to undergo **complete metamorphosis.** The sharp distinction between the
larval and adult stages in such insects has meant evolution in two markedly
different directions; in general, the larva is more specialized for feeding and
growth, while the adult is more specialized for active dispersal and reproduction.

Complete metamorphosis is not characteristic of all insects. Many, such as
grasshoppers, cockroaches, bugs, and lice, undergo **gradual metamorphosis**
(Fig. 17.21). The young of such insects resemble the adults, except that their
body proportions are different (the wings and reproductive organs, especially, are poorly developed). They go through a series of molts during which
their form gradually changes and becomes more and more like that of the
adult, largely as a result of differential growth of the various body parts.
They have no pupal stage and experience no wholesale destruction of the
immature tissues.

Aging and death   Discussions of development often stop with the completely matured adult. But development in its full biological sense does not
cease then. The adult organism is not a static entity; it continues to change,
and hence to develop, until death brings the developmental process to an
end.

The term "aging" is applied to the complex of developmental changes that
lead, with the passage of time, to the deterioration of the mature organism
(Table 17.1) and ultimately to its death. For many years, little research was
devoted to aging, but now it has become a major field of investigation.
Modern scientific progress and improved medical techniques have greatly
increased our ability to protect ourselves against disease, starvation, and the
destructive forces of the physical environment. More and more people are
living to an advanced age. And as the life expectancy increases and the
proportion of the population in the upper age brackets rises, the changes
associated with aging become more obvious and more important to all of us.

Little is known about the factors involved in aging. The process seems to
be correlated with specialization of cells for one or a few highly specific
functions. Cells that remain relatively unspecialized in this sense (i.e. that
remain more versatile) and continue to divide do not age as rapidly (if at all)
as cells that have lost the capacity to divide. Thus bacteria and some other
unicellular organisms cannot be said to age. They may, in fact, be potentially
immortal, for any cell that is not destroyed eventually divides to produce
two young cells; division is thus a process of rejuvenation. Within the body
of a multicellular animal, tissues like muscle and nerve, in which cell division has normally ceased, slowly deteriorate, whereas tissues such as those
of the liver and pancreas, in which active cell division continues, age much
more slowly. Furthermore, animals that grow as long as they live seem to
show fewer symptoms of aging than, for example, mammals and birds,
which cease growing soon after they reach maturity. Within a single species
(e.g. rats), individuals whose period of growth and development is slowed
and extended by providing a very limited diet are usually older than normal
before they begin showing signs of aging.

The same pattern is seen in plants. Many annual plants practically cease
growing soon after the flowering period, when their fruit begins to form, and

they age and die soon afterward (in many cases, if the plants are experimentally prevented from flowering, they will live on far beyond their normal time of death). By contrast, some woody perennials, of which the giant sequoia and the redwoods are extreme examples, may live for hundreds or thousands of years, always continuing to grow. In effect, they circumvent the problem of aging by forming vast numbers of new cells each year while equally vast numbers of their older cells die. The specialized cells age and die, but the continued youthful activity of the meristems keeps the living part of the plant young and vigorous. As an organism, a sequoia may be over 4,000 years old, but it contains no living cells that are more than a few years old.

Clearly, then, the aging and death of individual cells and the aging and death of the multicellular organism as a whole are two rather different things. Paradoxical as it may seem at first, a plant may retain youthful attributes in part because many of its cells age and die. The death of these cells is functionally necessary for the continued life of the plant (recollect that it is only after the cells of the xylem die that they can function in internal transport). Similarly, the death of individual cells, as noted previously, plays an essential role in the development of an animal embryo and in the complete metamorphosis of some insects. And early death of individual red blood cells and epidermal cells is entirely normal even in a young healthy mammal. Aging of the whole organism, therefore, is a matter not simply of the death of its cells, but of the deterioration and death of those cells and tissues that cannot be replaced.

What makes irreplaceable tissues age? We don't know. However, we do know some of the factors that contribute to aging:

1. The replacement of damaged tissue by connective tissue places an increased burden on the remaining cells in that tissue. When cells in a tissue like muscle die as a result of disease or injury, no new muscle cells are formed to replace the ones lost. Wound healing in such cases involves growth of connective (scar) tissue, which serves as a patching material but cannot, of course, function like the original muscle cells. As more and more irreplaceable cells die, the increased burden placed on the remaining cells of that type may contribute to their aging.

2. Changing hormonal balance, as by a drop in the level of sex hormones, may disturb the function of a variety of tissues and perhaps cause them to function less well.

3. As cells become older, they tend to accumulate some metabolic wastes that they apparently cannot expel, and these wastes may contribute to the eventual deterioration of the cells.

4. The number of collagen fibers (and their thickness) increases in the intercellular ground substance of connective tissue, and the elastic fibers become thicker and less elastic, perhaps as a result of increased binding of calcium ions. Such changes, which are fundamentally a continuation of the course of development that began with the initial formation of connective tissue in the embryo, contribute to loss of elasticity in the skin, hardening of the arteries, and stiffening of the joints. Once set in motion, it seems, the development of connective tissue cannot be halted when it has reached an optimal level, but must continue inexorably, contributing to the destruction of the organism it helped build.

TABLE 17.1 *Average decline in a human male from ages 30 to 75*

Characteristic	Percent decline
Weight of brain	44
Number of axons in spinal nerve	37
Velocity of nerve impulse	10
Number of taste buds	64
Blood supply to brain	20
Output of heart at rest	30
Speed of return to normal pH of blood after displacement	83
Number of glomeruli in kidney	44
Glomerular filtration rate	31
Vital capacity of lungs	44
Maximum O_2 uptake during exercise	60

Although all these factors are doubtless involved in aging, they are not really explanations but symptoms. The real question is why these changes occur, and to this question scientists cannot as yet give a satisfying answer. Some investigators have suggested that somatic cells slowly cease to function and eventually die as a result of damage by radiation (particularly X rays and cosmic rays). However, all laboratory experiments indicate that radiation damage is greatest to actively dividing cells—the cells that age most slowly. Furthermore, the amount of radiation damage would be proportional to the chronological age of the cells, but we know that aging is a function of physiological age, not of chronological age; e.g. a five-year-old rat is physiologically very old indeed, and its tissues show pronounced symptoms of aging, whereas five-year-old tissues in a human being are not yet even mature.

Other workers have suggested that the ceaseless stresses of life wear down the body's ability to maintain homeostasis, eventually upsetting its physiological balance to such a degree that death results. But this hypothesis remains rather nebulous and difficult to translate into concrete terms.

Still other investigators put major emphasis on intrinsic rather than extrinsic factors. They believe that the changes characteristic of aging are programed in the genes just like the earlier developmental changes, and that, although extrinsic environmental factors doubtless influence aging, they do so only by speeding up or slowing down processes that would occur anyway. These processes may involve a decline in the production of important enzymes or an altered chemical balance or physical structure, with an ensuing loss of ability to perform certain functions; or they may involve development of auto-immune reactions (allergies against parts of the organism's own body) that result in destruction of essential tissues; or they may involve increased rupture of lysosomes and release of destructive hydrolytic enzymes within the cells.

A particularly interesting proposal is that the aging of cells is due to somatic-gene mutations that result in their making defective enzymes. According to this hypothesis, the different rates of aging in different species reflect different inherited capacities for DNA repair. Thus species that age slowly would be genetically well endowed with enzymes that repair DNA damaged by mutagenic agents; hence many of their mutations would be only temporary. Species that age rapidly would have only limited ability to repair mutated DNA.

Whatever the processes of aging, it seems clear that we shall not understand them fully until we know much more about how developmental processes in general are regulated by the interaction of inherited and environmental influences. The basic problems in understanding aging, then, are essentially the same as those of embryonic development or maturation. Aging is simply another aspect of the general phenomenon of development.

DEVELOPMENT OF AN ANGIOSPERM PLANT

The following examination of some of the major events in the development of angiosperm plants is not an attempt to describe these events fully or even to mention all the important ones. Like the preceding discussion of animal development, it is simply meant to give you some familiarity with the kinds of events that occur.

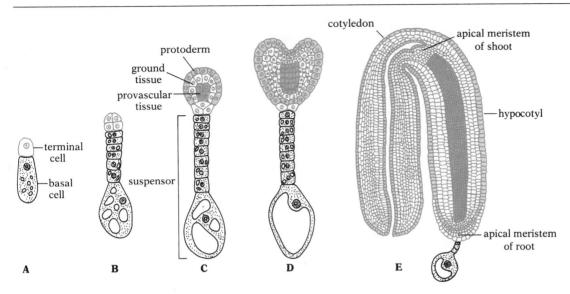

17.22 Embryonic development of shepherd's purse (Capsella)
(A) The first division of the zygote produces a smaller terminal cell (color) and a larger basal cell (black).
(B–C) Divisions of the terminal cell give rise to a globular embryo, whereas divisions of the basal cell give rise to a stalklike suspensor. Cells of the globular embryo soon differentiate to form three tissue types: protoderm on the surface, provascular tissue in the center, and ground tissue between the other two.
(D) The formerly globular embryo becomes heart-shaped as two mounds that will develop into cotyledons begin to form.   (E) Elongation of the cotyledons and the hypocotyl within the confines of the seed causes the embryo to fold back on itself.   [E: Redrawn from M. Schaffner, *The Ohio naturalist*, 1906.]

THE SEED AND ITS GERMINATION

The egg cell of an angiosperm plant is retained within the ovary of the maternal plant and is fertilized there by a sperm nucleus from a pollen grain. After fertilization, the zygote undergoes a series of mitotic divisions and develops into a tiny *embryo*. This embryo, together with a food-storage tissue called the *endosperm*, becomes enclosed in a tough protective *seed coat*. The resulting composite structure, made up of embryo, endosperm, and seed coat, is called a *seed.* The embryo in some species, such as peas and beans, absorbs all the endosperm before the seed is released from the parent plant. In other species, such as corn, the embryo does not absorb significant quantities of the endosperm until the seed begins to germinate. The embryonic stages of development usually do not last long, and by the time the ripe seed is released from the parent plant it is quite dry and its embryo has usually become dormant. The seed may last for months or years in this dormant state.

Development of the embryo   Soon after an egg cell has been fertilized, it begins to undergo a series of changes. Its wall, previously very thin, thickens; its endoplasmic reticulum and Golgi become more extensive; and new ribosomes are synthesized. Unlike the fertilized egg cell of an animal, however, it does not form a fertilization membrane.

The first cell division may occur a day or so after fertilization. It invariably gives rise to two cells of unequal size: a smaller nonvacuolated terminal cell and a larger vacuolated basal cell (Fig. 17.22A). The developmental fates of these two cells are very different. It is the terminal cell that will give rise to the embryo proper, while the basal cell, by means of about three transverse-division cycles, will form an elongate *suspensor* structure (Fig. 17.22B), which functions only while the embryo is in the seed—probably in moving nutrients into the embryo. Let us follow the embryonic development of *Capsella*, a much-studied dicot commonly called shepherd's purse.

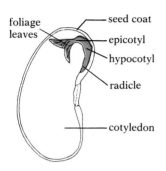

17.23 Diagram of a bean seed
One cotyledon (half bean) has been removed so that
the embryo (color) can be seen.

The terminal cell gives rise to a globular structure, in which three types of tissues begin to differentiate: a surface layer of **protoderm**, which will form epidermal tissue; an inner core of **provascular tissue**, which will form cambium and the vascular tissues; and a middle layer of **ground tissue**, which will form the cortex (Fig. 17.22C). Shortly thereafter, the radial symmetry of the globular-stage embryo begins to disappear as two mounds arise on the portion of the embryo opposite the suspensor (Fig. 17.22D). These mounds, which give the embryo a heart-shaped appearance, will become the **cotyledons**, or embryo leaves. The part of the embryo proper below the point of attachment of the cotyledons is called the **hypocotyl**; it will form the first part of the stem of the young plant.

As the cotyledons and hypocotyl of *Capsella* continue to elongate within the very limited space available to them in the seed,[3] the embryo begins to curve back on itself (Fig. 17.22E). Such curvature, though common in angiosperms, does not occur in all species.

As most of the cells of the embryo take on more and more of the characteristics of the tissues to which they will give rise, small clumps of cells at each end of the embryonic axis remain relatively undifferentiated. One clump, located just beyond the point of attachment of the cotyledons, will become the **apical meristem of the shoot**. The other clump, at the pole of the embryonic axis near the suspensor, will become the **apical meristem of the root tip** (Fig. 17.22E). In some species (e.g. bean), cell divisions in these meristems during embryonic development give rise to an **epicotyl**—a region of shoot above the point of attachment of the cotyledons, often bearing the first foliage leaves (the plumule)—and, at the other end of the embryonic axis, to a **radicle**, which will develop into the primary root (Fig. 17.23).

We have now followed the embryo to a point where its development normally slows down and enters a dormant stage, which will not be broken until the seed germinates. Just what causes this dormancy is not fully understood. Indeed, a number of factors probably interact to bring it on—among them, (1) decline in endosperm, which in some plants is entirely used up as its nutrients are transferred to the cotyledons; (2) lowered concentrations of gibberellin in the tissues surrounding the embryo; (3) changes in the cells of the seed coat, including progressive dehydration, deposition of pigments, and thickening and hardening of the cell walls—changes that tend to reduce light levels inside the seed; and (4) secretion of growth-inhibitor substances by the extraembryonic tissues of the seed.

This brief summary points up several major differences between the embryonic development of an angiosperm plant and that characteristic of most animals:

1. Unlike animals, plants have no early developmental stage in which cleavage is unaccompanied by cell growth; the two processes—cell division and growth—typically occur together.[4]

2. Plant cells do not migrate during morphogenesis. The cell walls constrain their shape, and their middle lamellae tend to tie each cell to its neighbors (the plasmodesmata between adjacent cells would be ruptured if the cells could slide along each other). In the plant embryo form and shape

---

[3] More accurately, it is the ovule within which the embryo develops and by which it is constrained.

[4] It is true, however, that the zygote is larger than any individual cell in the fully developed embryo, because the rate of cell division in the embryo is faster than the rate of cell growth.

are not established by movement of cells, as in animals, but by the patterns of cell division and growth.

3. In plants a few cells are early set aside to remain forever embryonic (meristematic).

4. The fully developed plant embryo does not possess, even in rudimentary form, all the organs of the adult plant; organogenesis (the formation of new organs) will continue throughout the life of the plant, as new roots, branches, leaves, and reproductive structures are formed during each growing season.

Seed germination Germination of a seed begins with the imbibition of much water, which greatly increases the volume of the seed (sometimes as much as 200 percent). The resulting hydration of the protoplasm increases enzymatic activity, and the metabolic rate of the embryo shows a marked rise. This higher metabolic rate makes possible resumption of active cell division, synthesis of new protoplasm, and increase in cell size by uptake of water. The growing embryo soon bursts out of the seed coat and rapidly assumes a characteristic plant form, with distinguishable shoot and root.

The hypocotyl (with attached radicle) is the first part of the embryo to emerge from the seed. It promptly turns downward no matter what the orientation of the seed may be. By the time the epicotyl begins its rapid development, the radicle, at the lower end of the hypocotyl, has already formed a young root system capable of anchoring the plant to the substrate and of absorbing water and minerals. In some dicots (which have, as the term "dicot" indicates, two cotyledons), the upper portion of the hypocotyl elongates and forms an arch, which pushes upward through the soil and emerges into the air (Fig. 17.24). Once the hypocotyl arch is exposed to light, it straightens, the cotyledons and the epicotyl thus being pulled out of the soil. The epicotyl then begins to elongate. In these dicots, of which the garden bean is an example, the shoot of the mature plant is mostly of epicotyl origin, but a short region (usually little more than one centimeter) at the base of the stem is derived from the hypocotyl.

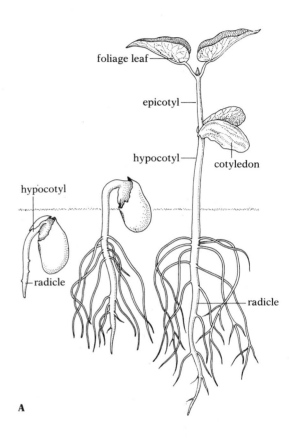

A

B

17.24 Germination and early development of a bean (a dicot)
(A) The hypocotyl and radicle emerge from the seed first (left). As the upper portion of the hypocotyl elongates, it forms an arch that pushes out of the soil into the air; the radicle gives rise to the first root system (middle). The hypocotyl straightens, pulling the cotyledons out of the ground as the epicotyl begins its development (right). (B) The photograph shows an actual germinating bean. [Photo: Courtesy G. R. Roberts.]

17.25 Germination and early development of a pea (a dicot)

The development of a pea differs from that of a bean in that no hypocotyl arch is formed and the cotyledons remain beneath the soil.

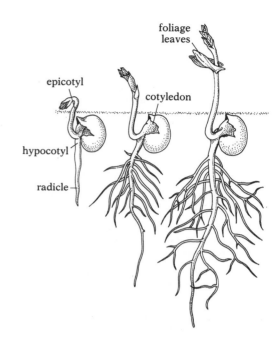

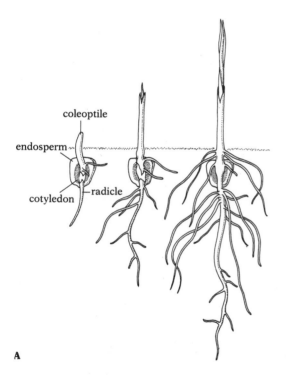

A

B

17.26 Germination and early development of corn (a monocot)

(A) The young shoot is enclosed within a coleoptile, a tubular protective sheath. (B) The photograph shows an actual germinating corn grain, with coleoptile and radicle. [Photo: Courtesy Carolina Biological Supply Co.]

Other dicots, of which the garden pea is an example, show a slightly different pattern of germination (Fig. 17.25). In these plants no hypocotyl arch forms and the cotyledons are never raised above ground. Instead, the epicotyl begins to elongate soon after the young root system has begun to form; it always grows upward and soon emerges from the soil. In such plants the entire shoot is of epicotyl origin. A similar pattern is seen in corn, a monocot (which has only one cotyledon but a large endosperm) (Fig. 17.26).

GROWTH AND DIFFERENTIATION OF THE PLANT BODY

The seedling plant grows in length rather slowly at first, then enters a longer period of much more rapid growth, and finally slows down again or even stops as it approaches maturity. If the height (or weight) of an annual plant is plotted against age, the resulting S-shaped growth curve is similar to the one characterizing the growth of animals (Fig. 17.14). The growth curves of perennial plants differ from those of annuals, because such organisms continue to grow to some extent throughout their lives while annuals cease growing after they reach maturity.

Growth of the shoot or root of a young plant involves both cell multiplication and cell elongation. In plants with only primary growth, both processes are ordinarily restricted to a relatively limited region near the apex of the shoot or root. Let us look first at the root.

Growth of the root The extreme tip of a root is covered by a conical root cap consisting of a mass of nondividing parenchyma cells (Fig. 17.27). These cells secrete a gelatinous substance that lubricates the surface of the root cap and facilitates the pushing of the root tip through the soil as the root elongates. As the tip moves through the soil, some of the cells on the surface of the root cap are abraded. They are replaced by new cells added to the cap by the apical meristem. Located just behind the cap, the apical meristem is usually restricted to an area a millimeter long or less and is composed of relatively small, actively dividing cells. Most of the new cells it produces are laid down on the side away from the root cap. These cells are left behind as the meristem lays down additional new cells in front of them and the tip continues to move through the soil. It is these new cells, derived from the apical meristem, that will form the primary tissues of the root.

Cell division and cell enlargement both occur in the meristem, but because the rate of division is high, the rate of enlargement is only sufficient to maintain an average cell size. However, as the new cells become further removed from the meristem by the deposition of additional intervening cells, mitotic activity in most of them slows down and cell enlargement becomes the dominant process. Most of this enlargement is elongation rather than increase in width. As we saw in Chapter 9, cell elongation is under the control of hormones, particularly auxins. Let us look at the process in more detail.

Since plant cells, unlike animal cells, are enclosed in a boxlike cell wall, cell growth is possible only if the box can be made bigger, i.e. the walls extended. It is in this process—the enlarging of the walls—that auxin is so important. The walls, you will recall, are composed primarily of polysaccharides, of which cellulose is the one we have mentioned most often. In primary walls (i.e. those of growing cells) the cellulose is present in the form

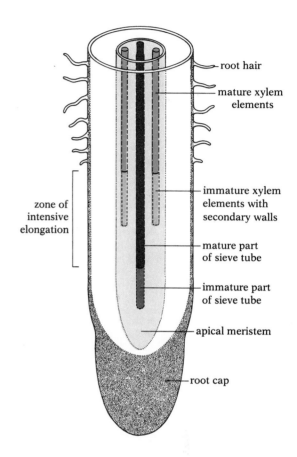

17.27 Diagram of longitudinal section of a root tip of a tobacco plant

New cells are produced by mitotic divisions in the meristem region just behind the root cap. There is a zone of especially intensive cell elongation not far behind the meristem. Both phloem and xylem are well differentiated in the region of the root hairs, back of the zone of intensive elongation.   [Modified from K. Esau, *Plant anatomy*, Wiley, 1965.]

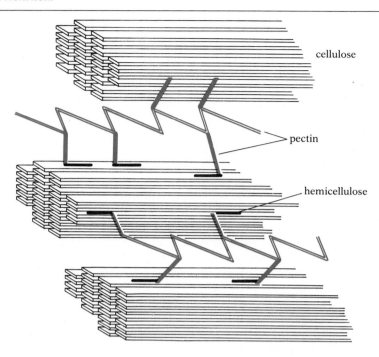

17.28   Molecular arrangement of polysaccharides in a primary cell wall
Each fibril is a bundle of parallel cellulose chains. The fibrils, which are cross-linked via pectin and hemicellulose, form a relatively rigid network.   [Modified from Peter Albersheim, "The walls of growing plant cells," *Sci. Am.*, April 1975.]

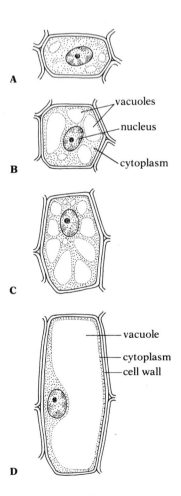

of long fibrils, each of which is thought to be a bundle of about 40 cellulose chains aligned parallel to one another and extensively cross-linked by hydrogen bonds (Fig. 17.28). The cellulose fibrils, in turn, are linked to other polysaccharides that make up the matrix of the wall. These matrix polysaccharides, especially those traditionally called pectin and hemicellulose,[5] hold the fibrils in a more or less rigid pattern. For cell growth to occur, two things must happen: Some of the cross linkages must be temporarily broken, so that the wall will become more plastic, and new wall material must be inserted.

A current model holds that auxin plays a fundamental role in inducing both greater wall plasticity and insertion of new wall material. Its effect on plasticity, one manifested very quickly (in less than ten minutes), is attributed to an auxin-activated membrane-bound ATPase that acts as a proton

---

[5] The classical pectin fraction of the wall is now thought to be composed largely of two materials, arabinan-galactan and rhamnogalacturonan. The classical hemicellulose fraction is xyloglucan.

17.29.   Elongation growth of a plant cell
As more and more water moves into the cell vacuoles, the walls are stretched, but in only one dimension. Almost no synthesis of new cytoplasm occurs during this kind of growth. The increased volume of the cell is taken up by the expanding vacuoles, which eventually fuse into a single large vacuole (D); in the mature cell the thin band of peripheral cytoplasm may constitute less than 10 percent of the cell's volume.

pump, transporting hydrogen ions from the cytoplasm into the wall. With an increased concentration of hydrogen ions in the wall, some of the hydrogen bonds between the polysaccharides break, the fibrils loosen, and the walls become more pliable.

Because the cell contents are hypertonic relative to the extracellular fluids, and the sole constraint on further movement of water into the cell has been resistance by the walls, more water begins to move into the cell as soon as the walls become more plastic and less resistant to stretching. As more and more water floods into the cell, especially into the vacuole, the hydrostatic pressure inside the cell causes more and more stretching of the walls, until the volume of the cell has been increased as much as a hundredfold. But note that this enormous increase in cell size has been achieved with little or no synthesis of new cytoplasm; the cytoplasm has been restricted to a thin layer next to the walls, the greater part of the cellular volume being filled by the expanded cell vacuole (Fig. 17.29). Cell growth in plants thus differs greatly from cell growth in animals, where most increase in size is due to formation of new cytoplasm rather than to vacuolation.

The effect of auxin on the second aspect of wall enlargement—insertion of new wall material—is much slower. It is thought to depend on synthesis of new mRNA coding for the enzymes that catalyze addition of further units to the complex polysaccharide structure of the wall.

We have emphasized that growth in length of the root or stem is due largely to cell elongation (enlargement of cells in other dimensions is, of course, important in the morphogenesis of other parts of the plant). There is presumably some mechanism whereby the effects of auxin can be targeted against the side walls of the cell and not against the end walls, so that the stretching is in only one dimension. The unidimensional stretching is reflected in the way the alignment of the cellulose fibrils changes during cell elongation (Fig. 17.30), but no convincing explanation for the asymmetry has yet emerged. Perhaps the ATPase pumps are missing from the membranes at the ends of the cells, or perhaps the walls themselves differ in some hitherto undiscovered fashion.

The zone where cell elongation predominates in the root is just behind the meristem. It usually extends only a few millimeters along the root (Fig. 17.27). For example, in a corn seedling the fastest elongation occurs about 4 mm from the root tip, and cells more than 10 mm behind the tip have completed their elongation. The situation is similar in bean seedlings (Fig. 17.31). The elongation in this zone has the effect of pushing the root tip through the soil faster than if it were driven only by the production of new cells in the zone of multiplication.

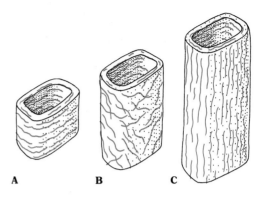

17.30   Change in the orientation of fibrils of the cell wall as a plant cell elongates
(A) Before the start of elongation the cellulose fibrils are arranged horizontally.   (B) As elongation begins, and the wall is stretched, the fibrils are displaced and tipped toward a more vertical alignment.   (C) In the fully elongated cell the fibrils in the outer, older wall layers are oriented vertically. The recently deposited fibrils in the inner newer layers are horizontal.

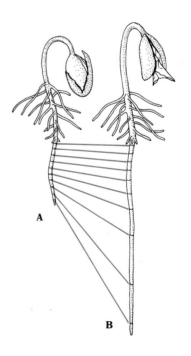

17.31   Growth of a bean root
Note that the parts of the root immediately behind the tip in the earlier stage (A) are those that have undergone the most elongation by the later stage (B).   [Modified from V. A. Greulach and J. E. Adams, *Plants: An introduction to modern botany*, Wiley, 1962.]

vascular cell

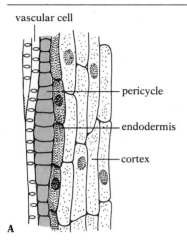

pericycle

endodermis

cortex

A

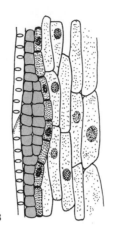

B

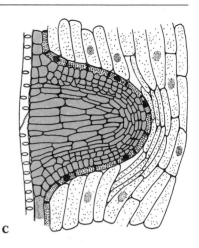

C

17.32   Origin of a new lateral root from the pericycle of an older root

(A) A new lateral root starts to form when a group of cells in the pericycle layer (color) begin to enlarge in a radial direction.   (B) These cells then divide tangentially.   (C) Continued divisions and elongations in this new direction produce a growing mass that pushes through the outer tissue layers of the old root.   Note that a new root forms by a change in the orientation of cell elongation and cell division, not by the morphogenetic movements of cells expected in a comparable developmental event in an animal. [Redrawn from *Biologie: Ein Lehrbuch,* ed. by G. Czihak *et al.,* 2nd ed., Springer-Verlag, 1978.]

Though cell division has slowed down in the regions behind the apical meristem, it has not ceased entirely. Indeed, the pattern of persistent cell division contributes importantly to the development of the characteristic organization of the root. For example, it is persistent meristematic activity in the pericycle that gives rise to new lateral roots, which, as they grow, push out through the overlying endodermal, cortical, and epidermal tissues to enter the soil (Figs. 17.32, 17.33).

Growth of the stem   Growth of the main stem axis is basically similar to growth of the root. New cells are produced by an apical meristem (which is not, however, covered by any structure analogous to the root cap), and these cells then elongate, pushing the apex upward. An obvious difference between growth of the stem and of the root is the lateral production of leaves by the growing tip of the stem. At regular intervals an increase in the rate of cell division under a localized region of the sloping surface of the apical meristem of a stem gives rise to a series of swellings that function as leaf primordia (Fig. 17.34). The point at which each leaf primordium arises from the stem is called a **node,** and the length of stem between two successive nodes is called an **internode** (see Fig. 3.52, p. 127). Most increase in length of the stem results from elongation of the cells in the young internodes.

At the tip of the stem is a series of internodes that have not yet undergone much elongation. The tiny leaf primordia that separate these internodes curve up and over the meristem, with the older, larger ones enveloping the younger, smaller ones (Fig. 17.35). The resulting compound structure, consisting of the apical meristem and a series of unelongated internodes enclosed within the leaf primordia, is called a **bud.** In shoots that have periodic (episodic) growth, the bud is protected on its outer surface by overlapping scales, which are modified leaves that grow from the base of the bud.

When a dormant bud "opens" in the spring, the scales curve away from the bud and then fall off, and the internodes that were contained within the bud begin to elongate rapidly. As the nodes become further and further separated, mitotic activity (mostly in one plane) in the leaf primordia gives rise to young leaves; the pattern of cell divisions, characteristic for each species, determines whether the leaves will be entire or lobed, simple or compound.

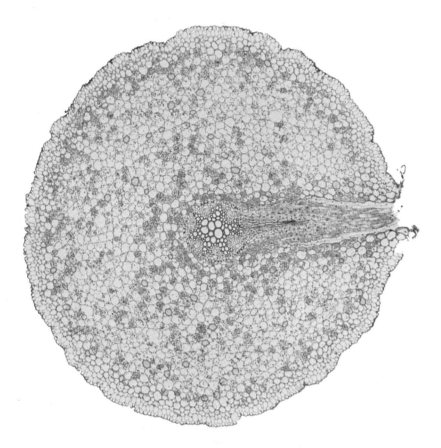

17.33 Origin of a lateral root in *Ranunculus*
The new lateral root has pushed through the cortical and epidermal layers of the old root and is ready to enter the soil.   [Courtesy Carolina Biological Supply Co.]

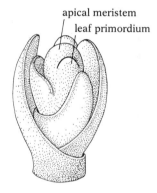

apical meristem
leaf primordium

17.34 A bud
[Used by permission from J. D. Dodd, *Course book in general botany,* © 1977 by The Iowa State University Press.]

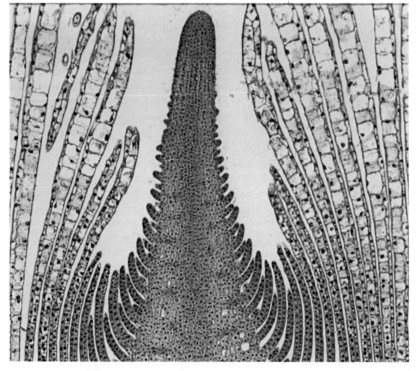

17.35 Photograph of sectioned *Elodea* bud
The stem tip forms a bud as upcurving leaf primordia enclose the apical meristem.   [Courtesy Carolina Biological Supply Co.]

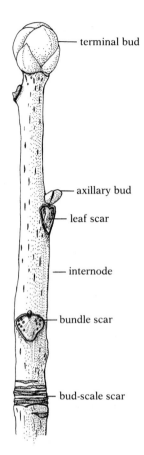

terminal bud

axillary bud

leaf scar

internode

bundle scar

bud-scale scar

17.36   Portion of a stem
The bud-scale scar shows where the dormant terminal bud of the previous winter was; the length of stem between the bud-scale scar and the terminal bud is one year's growth. Leaf scars show where petioles were attached to the stem. The bundle scars within each leaf scar show where the vascular bundles passed into the petiole. The axillary bud will give rise to a branch stem during the next growing season.

Before the leaf is fully formed, a small mound of meristematic tissue usually arises in the angle between the base of each leaf and the internode above it. Each of these new meristematic regions gives rise to a lateral or **axillary bud** with the same essential features as the terminal buds already described (Fig. 17.36). Elongation of the internodes of the lateral buds produces branch stems. Here, then, is another important way in which growth of the stem differs from growth of the root; in the root there are no surface meristems analogous to the lateral buds, and, as we have seen, the lateral roots arise from cell division in deeper tissue layers.

Differentiation of tissues   Cell division and cell elongation cannot alone produce all the essential features of the fully developed plant, of course. All the new cells produced by the apical meristems are fundamentally alike. Yet some of these cells will become collenchyma, some will become xylem vessels, some will become sieve cells, and so forth. The process whereby a cell changes from its immature form to some one mature form is called *differentiation.*

In the growing root or stem, cells begin to differentiate into the various tissues of the plant while they are still within the meristematic region. As the predominant activities of cell division and cell enlargement run their course, the cells are left to mature into their final form. Thus, although the pattern of histogenesis (formation of tissues) is laid out very close to the tip of the elongating axis, it remains useful to recognize successive regions along the axis that are related mainly to cell division, cell elongation, and cell maturation (Fig. 17.27).

Three concentric areas can be distinguished even in the region just behind the meristem of a root when it is viewed in cross section (Fig. 17.37A). These are (1) an outer protoderm; (2) a wide area of parenchymatous ground tissue located beneath the protoderm; and (3) an inner core of provascular tissue composed of particularly elongate cells. Just as in the embryo, the protoderm rapidly matures into the epidermis, the ground tissue into the cortex and endodermis, and the provascular core into the primary tissues of the stele: primary xylem, primary phloem, pericycle, and vascular cambium (Fig. 17.37B–C). Differentiation in the growing stem follows a similar pattern, except that there are usually two areas of ground tissue—one between the protoderm and the provascular cylinder, which gives rise to the cortex and endodermis, and a second inside the provascular cylinder, which becomes the pith.

As we saw in Chapter 7, increase in circumference of the root or stem depends on formation of secondary tissues composed of cells derived from lateral meristems, particularly the vascular cambium. As cells of the vascular cambium undergo mitosis, many new cells are produced on the inner face of the cambium and these differentiate into secondary xylem, while other new cells are produced on the outer face of the cambium and differentiate into secondary phloem (Fig. 17.37D–E). As more and more secondary vascular tissue is formed and the circumference increases steadily, the old epidermis and cortex are broken and sloughed off. These are replaced by a secondary protective tissue, the cork, composed of cells derived from a new lateral meristem, the **cork cambium,** which forms from a layer of the old cortex or from the pericycle or even from the older phloem, depending on the species.

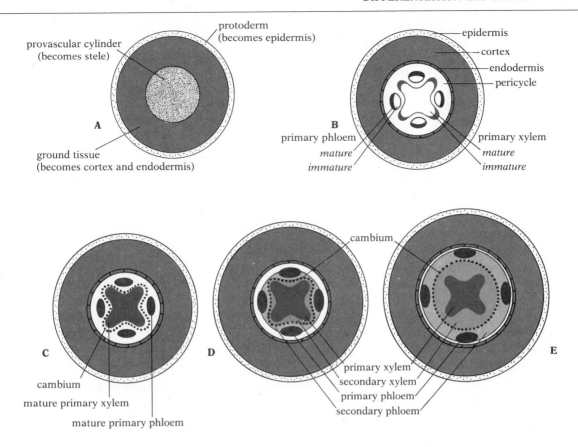

17.37 Differentiation in a young root

(A) Cross section just back of the meristem. Three distinct concentric areas can already be detected.
(B) At a slightly later stage of development the protoderm has differentiated into epidermis; the ground tissue has differentiated into cortex and endodermis; and the provascular cylinder has begun to differentiate into primary xylem and primary phloem.
(C) Differentiation in the provascular cylinder is complete, and the cambium is about to become active.
(D) Divisions in the cambium have given rise to secondary xylem and secondary phloem, which are located between the primary xylem and primary phloem.   (E) The areas of secondary tissue continue to thicken as more and more new cells are produced by the cambium.

THE PROBLEM OF DIFFERENTIATION AND PATTERN FORMATION

We have already seen that, because mitosis gives each new daughter cell a complete set of chromosomes exactly like that in the parental cell and because nuclear genes are located in precise sequences on the chromosomes, all the somatic cells in a multicellular organism ordinarily have identical genotypes. There is no evidence that any genes are normally lost or gained in the course of a cell's development;[6] there is, indeed, every reason

---

[6] There are some apparent exceptions to the finding that differentiation does not normally involve gain or loss of genes. We shall mention three:

1. In a very few species, of which gall midges are an example, the somatic cells have fewer chromosomes than the germ cells. The zygote of a gall midge has about 40 chromosomes. At the time of the fifth cleavage in the embryo, only eight of the chromosomes move properly on the mitotic spindles in most cells; the other 32 break up, and their material becomes dispersed throughout the cytoplasm and is lost. Thus somatic cells of the gall midge have only eight chromosomes, and only the few cells early set aside to form the germ cells retain the full 40. However, this reduction of the number of chromosomes occurs in all somatic cells and hence is not correlated with their differing developmental fates.

2. In some species the somatic cells become polyploid while the germ cells that will give rise to gametes remain diploid. But since all the somatic cells are alike, the extra chromosomes they possess cannot be the cause of their following different paths of development.

3. Mammalian liver cells are often polyploid, but it can be shown that a cell can differentiate into a fully functional liver cell without being polyploid.

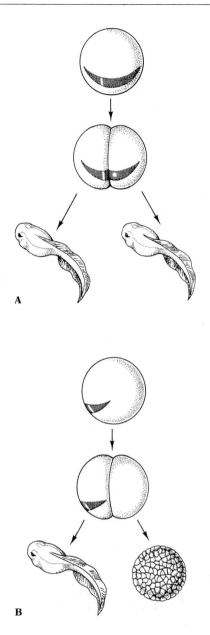

17.38 Importance of the gray crescent in the early development of a frog embryo

(A) If the two cells produced by a normal first cleavage, which passes through the gray crescent (color), are separated, each develops into a normal tadpole.   (B) If the first cleavage is experimentally oriented so that it does not pass through the gray crescent, and the two daughter cells are separated, the cell with the crescent develops normally, but the other cell develops into an unorganized cellular mass.

to think that the genetic endowment of a nerve or muscle cell is exactly the same as that of a liver or bone cell in the same individual. Hence the factors that—through gene transcription, mRNA translation, or enzyme activity—influence which of a given cell's many genetic potentialities will be realized must play a fundamental role in cellular differentiation.

A satisfactory model of development must explain how the two types of control factors—genetic and nongenetic—interact to produce differentiated cells, tissues, and organs. It must give us some basis for understanding how the epicotyl of a germinating seed becomes negatively geotropic while the genetically identical hypocotyl becomes positively geotropic. It must give us some framework for explaining why cells produced to the outside by vascular cambium develop into phloem while genetically identical cells produced to the inside develop into xylem. It must help us understand why some ectodermal cells in an animal embryo fold inward and form nervous tissue while other genetically identical ectodermal cells remain on the surface and form skin. And it must give us insight into how genetically identical cells can form a grub at one time and an adult beetle at another. This is an immense challenge to developmental biologists, who have a long way to go before they can fit all the pieces of the puzzle together for a clear and complete picture of development. But much has already been learned; many of the pieces have been discovered, and some have been fitted together. Let us look at a few of the results.

THE POLARITY OF CELLS AND THE PLANE OF CLEAVAGE

If embryonic cells cannot be distinguished on the basis of genetic content, it is logical to look for differences in their cytoplasm. We have already seen that the cytoplasm of the unfertilized egg is often not homogeneous. Most animal eggs contain stored food material, or yolk, which, since it is usually concentrated in one part of the cell, establishes a distinction between animal and vegetal hemispheres; i.e. it brings about a polarization of the cell. Another distinction between the two hemispheres is that the animal hemisphere often has much more pigment in the cytoplasm than the vegetal hemisphere. There is abundant evidence for other materials similarly restricted to certain regions of the cytoplasm—not only in the egg cells of animals but also in those of plants. When an egg cell divides, therefore, the cytoplasmic materials it contains may be distributed unevenly between the two daughter cells, depending on the orientation of the plane of cleavage.

Let us examine an actual example. As soon as the egg cell of a leopard frog is penetrated by a sperm cell, some of the contents of the egg shifts position and a crescent-shaped grayish area appears on the egg opposite the point where the sperm entered. The material in this so-called *gray crescent* will play a very prominent role throughout embryonic development. The first cleavage of the frog zygote normally passes through the gray crescent, so that each daughter cell receives half (Fig. 17.38A). If these two cells are separated, each will develop into a normal tadpole. But if the plane of the first cleavage is experimentally made to pass to the side of the gray crescent, the result of separating the daughter cells will be very different; the cell that contains the gray crescent will develop into a normal tadpole, but the other cell will form only an unorganized mass of cells (Fig. 17.38B). In other words, the way in which the material of the gray crescent is distributed is of utmost importance to the developmental potential of the cells.

As the above experiment confirms, the normal first cleavage of the frog egg is **indeterminate;** the new cells have the full developmental potential of the original zygote, and all pathways of differentiation remain open to them. Indeterminate cleavage is characteristic of most echinoderms and vertebrates, including humans. That early cleavage should be indeterminate in humans was to be expected, of course, because only animals with indeterminate cleavage can give birth to identical twins.[7]

In some other groups of animals, such an annelids and molluscs, the normal first cleavage partitions critical cytoplasmic constituents asymmetrically, and is therefore a **determinate** cleavage; when the daughter cells are separated, they do not have equivalent developmental potentialities. For example, in some molluscs, such as mussels and sea snails, a protuberance called the polar lobe develops on the fertilized egg cell just before the first cleavage occurs. The plane of cleavage is oriented in such a way that one of the two daughter cells receives the entire polar lobe (Fig. 17.39). If the two daughter cells are separated, the one with the lobe material (which was drawn back into the main body of the cell soon after division was accomplished) will form a complete normal embryo, but the one with no lobe material forms a highly aberrant embryo lacking all mesodermal structures. Something in the polar-lobe material must be essential for formation of mesoderm, and its asymmetric distribution makes the first cleavage of the egg a determinate one.

Eggs such as those of frogs, which give rise to daughter cells each capable of developing into a complete embryo, have sometimes been called **regulative eggs,** because the early embryos they produce can regulate their development to compensate for missing portions of cytoplasmic materials. Eggs such as those of molluscs, in which the regional cytoplasmic differences are so marked that the embryos they produce cannot compensate for missing portions of cytoplasmic materials, have been called **mosaic eggs.** But this terminology is a little misleading. Both kinds of eggs are mosaics in the sense that the various developmentally important materials are concentrated in different parts of the egg cytoplasm. While this segregation of materials may be more pronounced in some eggs than in others, the variable that causes an indeterminate first cleavage in one case and a determinate one in the other is the orientation of the cleavage plane relative to the segregated materials, as Fig. 17.38 makes clear.

Even in the species where the first few cleavages are indeterminate, a stage is eventually reached where asymmetric partitioning of cytoplasmic substances gives rise to cells with different developmental potentialities. For example, in sea urchins (which are echinoderms), the first two cleavages are vertical and indeterminate; they run parallel with the axial gradient of materials between the animal and vegetal poles and thus partition the materials symmetrically. But the third cleavage is horizontal, giving rise to an

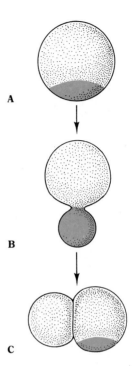

17.39 Cleavage of the egg of the sea snail *Ilyanassa* (A) At one pole of the zygote is a region of clear cytoplasm (color)   (B) Just before the first cleavage, this polar cytoplasm moves into a large protuberance, the polar lobe.   (C) The first cleavage partitions the zygote in such a way that the entire polar lobe goes to one of the daughter cells. The lobe recedes during interphase (as shown here), but it will form again prior to the next division sequence. Only the cell that receives the polar-lobe material can give rise to mesoderm.

---

[7] Identical twins develop from isolated cells derived from the same zygote. Nonidentical (fraternal) twins develop from separate zygotes when two egg cells are released from the ovaries at the same time and are fertilized by different sperm cells. Consequently, identical twins are genetically identical, whereas fraternal twins are no more alike genetically than any two siblings. Fraternal twins are much more common than identical twins.

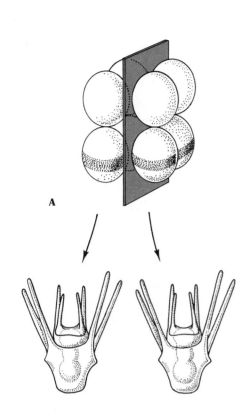

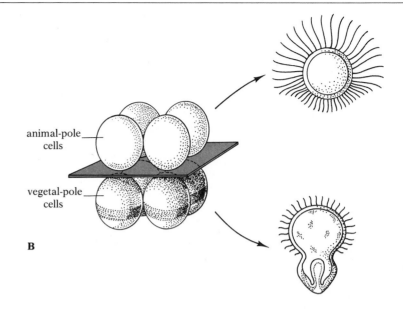

17.40   Experimental separation of cells after the third cleavage in the embryo of a sea urchin
(A) If the embryo is partitioned meridionally, so that each half receives both animal-pole and vegetal-pole cells, each half develops into a normal larva.   (B) If the embryo is partitioned horizontally, so that one half receives only animal-pole cells and the other half only vegetal-pole cells, the animal half develops into an abnormal blastulalike larva with overdeveloped cilia, and the vegetal half develops into an abnormal larva with an overdeveloped digestive cavity.

eight-cell embryo with four cells containing animal-pole substances and four containing vegetal-pole substances. This third cleavage is a determinate one; if the animal and vegetal halves of the embryo are separated, the animal half will give rise to an abnormal blastulalike larva with overdeveloped cilia, and the vegetal half will give rise to a different type of abnormal larva with an overdeveloped digestive cavity (Fig. 17.40B). By contrast, if the eight-cell embryo is divided meridionally (i.e. along its animal-vegetal axis), each half will develop into a small but normal larva (Fig. 17.40A).

It is clear, then, that the cytoplasmic substances to which a nucleus is exposed must play a prominent role early during embryological development in activating some genes and repressing others. Because the egg cell itself contains different substances in different regions of its cytoplasm—whether these substances are arranged in a pole-to-pole gradient (as they apparently are in sea urchins) or in a less symmetrical pattern (as they are in frog and sea-snail eggs, with their gray crescents and polar lobes)—cells with different cytoplasmic environments for their nuclei are produced early in embryological development (at the first cleavage in some eggs, at the third cleavage or later in others). By producing different regulatory effects on the genes in the nuclei, the cytoplasmic substances restrict the future course of development of the cells and their progeny. Nonuniform distribution of cytoplasmic substances in dividing cells, together with the orientation of the planes of cleavage, therefore becomes the key to the initial pattern of development.

But what factors determine the original distribution of substances in the cytoplasm of the egg cell? The detailed mechanisms responsible for patterning the egg cytoplasm are still largely unknown, but some of the stimuli that play a role in guiding the patterning have been studied. We have already mentioned that the point of penetration of a frog egg by a sperm cell determines where on the egg the gray crescent will form. An example in which a physical parameter of the environment plays a crucial role is seen in the spore of the primitive vascular plant *Equisetum* (see Fig. 23.33, p. 977).

An *Equisetum* spore exhibits little polarity initially (Fig. 17.41A), but when it is exposed to light its chloroplasts (and some other less easily observed cytoplasmic materials) migrate toward the incident light, while the nucleus (and other substances) moves in the opposite direction (Fig. 17.41B). This response to the direction of the incident light establishes an axis of polarity for the spore, one that determines the direction of the first cleavage. The first cleavage produces a large chloroplast-rich cell at the lighted end of the spore axis and a small chloroplast-poor cell at the other end (Fig. 17.41C). The large cell will later give rise to the structure (prothallium) that will form the plant shoot, and the small cell will give rise to the rootlike rhizoids.

The influence of a variety of physical factors on the establishment of embryological polarity has been demonstrated with particular clarity in the early development of the brown alga *Fucus* (Fig. 23.18. p. 967), which grows on intertidal rocks along Northern coasts. The zygote of *Fucus* is spherical at first and has no visible surface features that would indicate separate cytoplasmic regions. However, shortly after fertilization, the zygote becomes asymmetrical, as a small protuberance forms on one side (Fig. 17.42). This protuberance permanently determines the polarization of the embryo that will develop from the zygote. The first cleavage is always oriented in such a way that the side of the zygote bearing the protuberance becomes one daughter cell and the side without the protuberance becomes the other daughter cell. The cell with the protuberance always proceeds to form the holdfast, a rootlike structure that anchors the plant to the rocks. The other cell always forms the more erect part of the plant.

D. M. Whitaker of Stanford University studied the factors that determine where on the *Fucus* zygote the protuberance will form. He found that a number of asymmetric environmental conditions play a role, the most important being illumination. The protuberance forms on the side away from the light. When lighting is equal on all sides, but a temperature gradient is maintained across the zygote, the protuberance forms on the warmer side. Similarly, in a pH gradient, the protuberance forms on the side with the lower pH. If illumination, temperature, and pH are all equal on all sides, but other zygotes are present in the culture, the protuberance forms on the side nearest the other zygotes. Finally, if the zygote is spun in a centrifuge, the protuberance forms on the side farthest from the center of the centrifuge. The adaptive significance of some of these reactions is understandable; e.g. a holdfast growing away from light and toward the pull of gravity would normally be going in the right direction to encounter the rocks at the bottom of the water.

Before leaving the subject of polarity, we should point out that polarity is by no means restricted to the egg cell and early embryo; it is, in fact, an extremely common aspect of organismic organization. We have encountered a number of earlier examples: The epithelial cells of animals manifest a strong polarity in the differing permeability characteristics of their outer and inner membranes (p. 128). The ion-pumping activity of the specialized epithelial cells of osmoregulatory structures depends on the polarity of the cells (p. 330). In plants, transport of auxin in a stem is strongly polar (Fig. 9.11, p. 342).

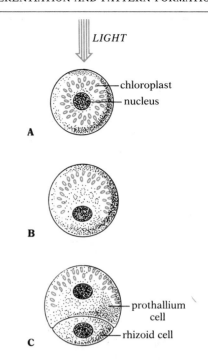

17.41 Light induction of polarity in the spore of *Equisetum*
(A) The unpolarized spore is exposed to a directed source of light (here pointing downward).
(B) Chloroplasts move toward the lighted side, and the nucleus moves toward the opposite side.   (C) The first cleavage is unequal and is oriented so as to produce a large chloroplast-rich prothallium cell and a smaller chloroplast-poor rhizoid cell.   [Redrawn from *Biologie: Ein Lehrbuch*, ed. by G. Czihak *et al.*, 2nd ed., Springer-Verlag, 1978.]

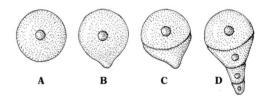

17.42 Early development of *Fucus* egg
(A) The zygote.   (B) Formation of a protuberance on one side of the zygote.   (C–D) The orientation of the first few cleavages is determined by the location of the protuberance.

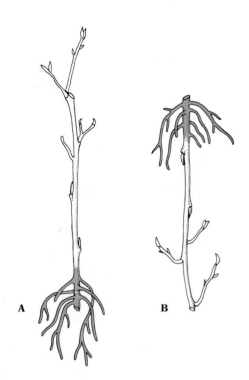

17.43 Polarity of a stem cutting
A cutting of a willow stem will produce new roots at its
physiological basal end and new buds and shoots at its
physiological upper end, whether it is oriented
normally with respect to gravity (A) or upside down (B).

That the structures of some developed organisms continue to show polarity in their developmental potential can be seen especially well in the regeneration of plant stems. Thus cuttings of willow stems will form roots at their physiological basal ends and buds at their apical ends, regardless of their orientation in space (Fig. 17.43). They will do so no matter how short the piece of stem may be, apparently as a result of the polarity of the individual cells within the stem.

TISSUE INTERACTIONS IN DEVELOPMENT

The immediate environment of the nucleus of a developing cell is the cytoplasm of that cell, which, as we have seen, can exert a profound influence on differentiation. But developing cells are also subject to influences external to their own cytoplasm. Depending on their location in the embryo, they are likely to be exposed to different combinations of environmental factors, which may help determine their developmental direction.

As countless experiments have shown, developing cells are indeed influenced by neighboring cells. For example, they interact via diffusible chemicals, and both their motility and their mitotic activity are inhibited when they come into contact with one another. Developing cells are likewise influenced by hormones and by many parameters of their physical environment.

The role of organizers and inducers   We have already seen that the gray crescent is important in determining the potentialities of the cells produced during early cleavage in the frog embryo. Let us now return to the gray crescent and examine its role in the course of later development.

The gray crescent is located on the frog egg near the boundary between the animal and vegetal hemispheres. It is along this boundary that involution of cells occurs during gastrulation. In the early gastrula the cells derived from the gray-crescent portion of the egg thus become the ***dorsal lip of the blastopore.*** These cells soon move inward and form the ***chordamesoderm,*** which is at first located in the roof of the newly forming archenteron (Fig. 17.6B), but soon detaches from the archenteron to form the notochord and other mesodermal structures (Fig. 17.6C). The chordamesoderm also seems to exert a very important influence on the ectodermal cells lying over it; those ectodermal cells invaginate during neurulation, forming the neural tube, which differentiates into the brain and spinal cord, but neurulation does not occur if the chordamesoderm is missing.

In a series of classic experiments performed in 1924, Hans Spemann of the University of Freiburg, Germany, who had previously demonstrated the importance of the gray crescent in early cleavage, and his colleague Hilde Mangold turned their attention to the dorsal lip of the blastopore in the early gastrula stage of a salamander embryo. They transplanted the dorsal lip from its normal position on a light-colored embryo to the belly region of another darker-colored embryo. After the operation, gastrulation occurred in two places on the recipient embryo—at the site of its own blastoporal lip and at the site of the implanted lip. Eventually two nervous systems were formed, and sometimes even two complete embryos developed, joined together ventrally (Fig. 17.44). Most of the tissue in both embryos was dark-colored, an indication that the transplanted blastoporal lip had altered the course of development of cells derived from the host. Similar transplants of tissues from other regions of the embryos failed to produce comparable results. The dorsal lip of the blastopore must play a crucial role in determining the form of the early embryo, probably by inducing the formation of the neural groove, which in turn is important in establishing the longitudinal axis of the embryo and in inducing formation of other structures.

Spemann and Mangold called the dorsal lip of the blastopore the *organizer.* They envisioned the entire developmental process as one in which a succession of principal organizer regions, each taking over where the previous one left off, control the differentiation of the major tissues and organs. We now know that induction is not limited to a small number of organizer regions, but is a general phenomenon; inductive tissue interactions are the rule rather than the exception.

Some of the first definitive studies on embryonic induction in an animal were performed in 1905 by Warren H. Lewis of Johns Hopkins University. Lewis worked on the development of the eye lens in frogs. In normal development the eyes form as lateral outpockets from the brain. When one of these outpockets, or optic vesicles as they are called, comes into contact with the epidermis on the side of the head, the contacted epidermal cells promptly undergo a series of changes and form a thick plate of cells that sinks inward, becomes detached from the epidermis, and eventually differentiates into the eye lens (Figs. 17.45, 17.46). Lewis cut the connection between one of the optic vesicles and the brain before the vesicle came into contact with the epidermis. He then moved the vesicle posteriorly into the trunk region of the embryo. Despite its lack of connection to the brain, the vesicle continued to develop, and when it came into contact with the epidermis of the trunk, that epidermis differentiated into a lens. The epidermis on the head that would normally have formed a lens failed to do so. Clearly,

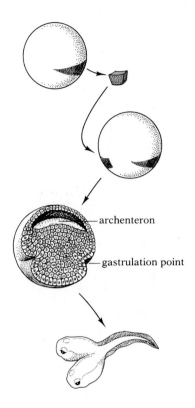

— archenteron

— gastrulation point

17.44 Spemann's experimental transplantation of the gray crescent

When Spemann transplanted gray-crescent material (color) from one salamander zygote to another, the zygote with two gray crescents proceeded to form an embryo with two gastrulation points—one at the dorsal lip of the blastopore, derived from its own gray crescent, and one at the second dorsal lip, derived from the transplanted gray crescent. A double larva was the result.

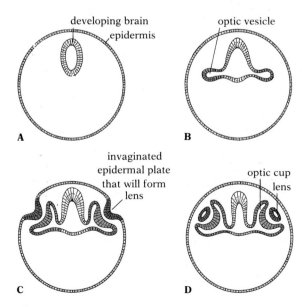

developing brain
epidermis
optic vesicle
invaginated epidermal plate that will form lens
optic cup
lens

A   B   C   D

17.45 Development of optic vesicles and their induction of lenses in a frog

See text for discussion.

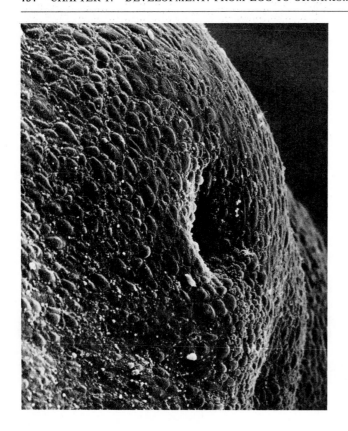
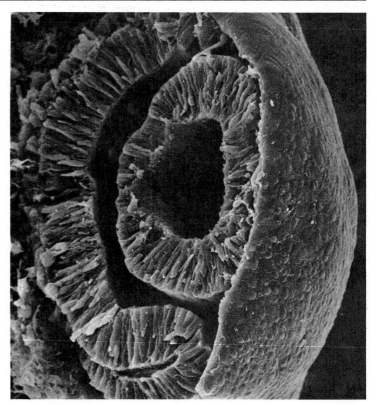

17.46   Scanning electron micrographs of development of the eye lens of a chick
Left: The pit in the surface ectoderm marks the point where the presumptive lens tissue has sunk inward.   Right: A section through the eye region at a later stage shows the lens as a hollow vesicle (it will eventually become solid). There is no longer any pit in the surface ectoderm, which will become the cornea. × 750.   [Courtesy K. W. Tosney, as printed in N. K. Wessells, *Tissue interactions and development,* copyright © 1977 by Benjamin/Cummings, Menlo Park, Calif.]

the differentiation of epidermal tissue into lens tissue depends on some inductive stimulus from the underlying optic vesicle. Later experiments have shown that if a barrier is inserted between the vesicle and the epidermis no lens develops.

Other experiments have shown that the regulation is not all one-way; the lens, once it begins to form, also influences the further development of the vesicle. If epidermis from a species with normally small eyes is transplanted to the sides of the head of a species with large eyes, the eyes that are formed do not have a large cup (formed from the vesicle) and a small lens, as might be expected. Instead, both the cup and the lens are intermediate in size and correctly proportioned to each other. Obviously, each influences the other as they develop together. Feedback is just as important in the control of embryonic development as it is in nervous or hormonal control of the fully-formed organism.

Some cell-to-cell interactions may be an effect of cell contact, as in contact inhibition, but most developmental interactions are probably chemically mediated; i.e. one cell is influenced by a substance secreted by another. Such chemical mediation can be demonstrated experimentally. For example, if the presumptive epithelial tissue and the presumptive connective tissue of an embryonic mouse pancreas are separated, the epithelial tissue will not differentiate properly in the absence of the connective tissue. But if the two groups of cells are grown in tissue culture and separated by a filter that will allow macromolecules, but not whole cells or organelles, to pass, then the epithelium will differentiate into normal pancreas tissue. Obviously, the necessary induction has occurred, and it must be attributed to a diffusible chemical that could pass through the filter.

In this case the chemical turned out to be a soluble precursor of collagen. Indeed, collagen or a precursor has proved to play a key role in many other inductive processes that have been studied. As chemical analyses have demonstrated, however, the potent inducer from the chordamesoderm of the frog embryo is not collagen, but some still unidentified protein.

Of the many substances that can act as intercellular inducers in embryonic development, some play a so-called instructive role, others play a more permissive role. The instructive inducers restrict the developmental potential of the target cell and thus help determine the course of differentiation. The permissive inducers facilitate the expression of potentialities already determined. For example, a rudimentary organ may form fully committed cells during embryology, but will not complete its development and become functional until acted on by a permissive inducer, which in some instances may be a hormone.

We must stress that the so-called instructive inducers do not give their target cells instructions about the design of the tissues or organs they are to form; they instruct the cells only in the sense that, through repression of some genes and derepression of others, they tell the cells what part of their genetic endowment they are to use. A dramatic example of this principle is the experiment in which Spemann and Schotté transplanted frog ectoderm to the mouth region of a salamander and found that, though induced by salamander endoderm, the transplanted tissue formed the typical horny jaws of a frog (see p. 677).

Hormones in development   Plant hormones—alternatively known as growth substances, you will recall—play a key role in nearly all phases of development. They may fulfill both instructive and permissive functions, as can be shown experimentally when varying concentrations of auxin and cytokinin are added to the basic growth medium of a culture of pith from the stem of a tobacco plant. At some concentrations the hormones merely enable cell division to proceed, a mass of undifferentiated cells called a callus being formed (Fig. 17.47, left). Here the hormones are acting permissively; they are in no way restricting the developmental potential of the cells. At other relative concentrations the hormones induce formation of new roots or new stems; because they stimulate the cells to differentiate in a particular way, they are acting instructively.

The inductive action of the hormones varies as a result of differences in other factors influencing the cells. Thus, if pith cells are cultured in a thin layer on a medium containing two parts per million of auxin, they do not divide, but grow to an unusually large size. When the pith tissue is molded into a cylinder, however, and the same concentration of auxin is inserted into the apical end of the cylinder, the cells undergo many divisions and some of them differentiate into xylem. The difference in response to the same inducing stimulus must be attributed to the different locations—superficial or internal—of the cells.

In vertebrates hormones play a predominantly permissive role in development, but are not, on that account, any less important than instructive inducers. What good would it do an organism if cells were set aside as a

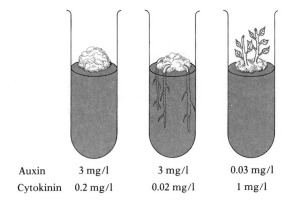

| Auxin | 3 mg/l | 3 mg/l | 0.03 mg/l |
| Cytokinin | 0.2 mg/l | 0.02 mg/l | 1 mg/l |

17.47   The effect of auxin and cytokinin on the development of cultured pith cells from a tobacco root
At certain concentrations of the hormones, only an amorphous callus is formed. At other concentrations roots or shoots develop from the callus.   [Redrawn from *Biologie: Ein Lehrbuch,* ed. by G. Czihak *et al.,* 2nd ed., Springer-Verlag, 1978.]

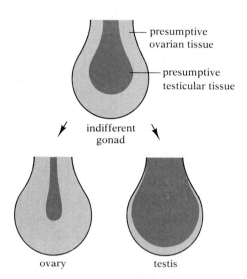

presumptive
ovarian tissue

presumptive
testicular tissue

indifferent
gonad

ovary   testis

17.48   Development of ovary or testis from the indifferent gonad of an amphibian
Depending on the hormonal condition of the embryo, one or the other of the two types of tissue in the indifferent gonad gains developmental ascendancy and the other type is repressed.

presumptive tissue or organ—i.e. underwent determination in response to instructive inducers—and their developmental potential could never be expressed for lack of a necessary permissive hormone? The essential interplay between instructive inducers and permissive hormones can be studied in the gonads of amphibians. Two distinct kinds of cells are early set aside to become gonads in amphibians: peripherally located cortical cells and centrally located medullary cells (Fig. 17.48). The cortical cells have the potential for forming ovarian tissue, and the medullary cells for forming testicular tissue. Only when sex hormone acts on the sexually uncommitted embryonic gonad, does one or the other of its potentialities gain ascendancy.

Similarly, the same embryonic primordia give rise to the accessory sexual organs of both sexes in humans (Fig. 17.49). Whether these primordia form male or female structures depends on whether or not male sex hormone is present in the embryo at the critical stages of embryonic development. In the absence of male hormone, the individual develops as a female, no matter what its genetic sex (XX or XY). In birds, where the female is the heterogametic sex, the situation is reversed; the sexual organs will follow a masculine developmental course unless female hormone is present at the critical embryonic stages. Normally, the hormonal condition of the embryo will be the one appropriate to its genetic sex, but in rare instances this is not the case, and an individual will develop organs that, though appropriate to its genetic sex, are poorly formed, or it may develop organs that, though nearly normal, belong to the other sex.

Morphogenetic fields   Not all chemicals influencing the developmental directions of cells are stimulatory. Many, in fact, are inhibitory, or else they are stimulatory to some cells and inhibitory to others. Auxin is an example; in normal development, it stimulates differentiation of xylem but inhibits development of lateral buds.

There is abundant evidence, from both plants and animals, that as a particular tissue or organ develops it releases substances that inhibit formation of the same kind of tissue or organ in the immediate area. For example, frog embryos cultured with pieces of tissue from an adult frog heart do not develop a normal heart, and embryos cultured with pieces of adult brain do not develop a normal brain. If a leg-forming region from the embryo of a salamander is transplanted (before it has actually taken the form of a leg) to a spot immediately adjacent to a leg-forming region on another embryo, only one leg develops; the leg field has sufficient capacity for self-regulation to ensure that a single normal-sized leg develops, despite the presence of twice the normal number of leg-forming cells. Similar principles seem to operate in determining the position of a new leaf on a growing stem. Each new leaf emerges at a point separated by some critical distance from the nearest older leaves and from the shoot apex. If one of the nearby older leaves is removed, the new leaf develops at a point closer to the position of the removed leaf than it would normally occupy. It seems that each leaf is surrounded by a morphogenetic field that somehow prevents the formation of a new leaf within it, but that, within itself, is capable of regulation. The presence of morphogenetic fields, first proposed by C. M. Child of the University of Chicago, seems to be a general phenomenon.

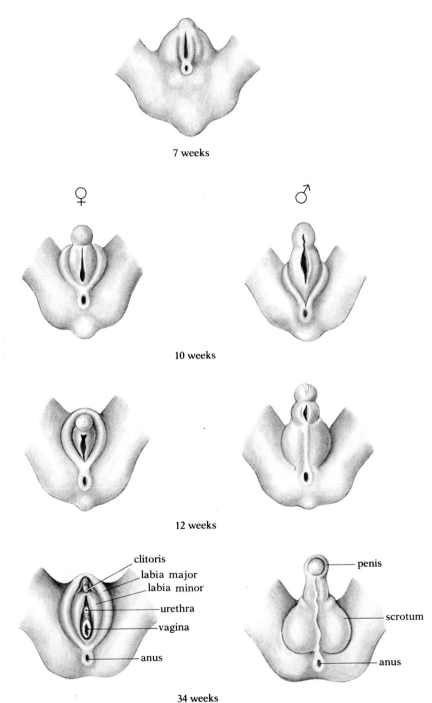

17.49 Development of the external genitalia of human beings

At 7 weeks the genitalia of male and female fetuses are virtually identical. At 10 weeks the penis of the male is slightly larger than the clitoris and labia minor, which form from the same primordium in the female. At 12 weeks these differences are more pronounced, and the male scrotum has formed from the tissue that becomes the labia major in the female. At 34 weeks the distinctive features of the genitalia of the two sexes are fully apparent. It is largely the concentration of male sex hormone that determines which of these developmental pathways will be followed.

Pattern formation No less important than cell differentiation and tissue formation in the later development of animals is pattern formation. It is crucial, for example, in the development of a vertebrate forelimb, that the bones be arranged in proper order, from humerus at the base to phalanges at the distal end.

Curt Stern of the University of California, Berkeley, has shown in some elegant experiments with *Drosophila* that cells must have a way of telling where they are, i.e. that there must be some sort of positional information available to them. Stern worked with a special mutant strain of *Drosophila* that forms legs on the head where antennae should be. He took the embryonic cells of one of these curiously located legs and grafted them onto the tip of the antenna of an embryo of a normal *Drosophila* at an early stage of the antenna's development. The cells proceeded to differentiate into the *end* structures of a leg. They could tell somehow that they were at the end of an appendage, even though that appendage was an antenna instead of a leg, and they responded by forming the only sort of end structure their genes permitted, namely the end of a leg.

Lewis Wolpert of Middlesex Hospital Medical School, London, in a model based on his investigations of the development of the wing in chicks, has

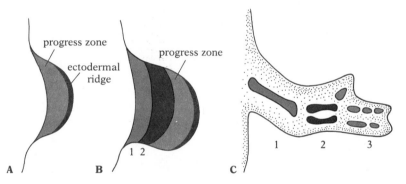

17.50 Pattern formation in the development of the chick wing

Left: Scanning electron micrograph of the ectodermal ridge of a chick wing bud. × 675.   Right: Diagram of three stages of development of the wing bud, according to Wolpert's model.   (A) Just behind the ectodermal ridge is a progress zone, where new cells are produced.   (B–C) The first band of cells derived from the progress zone will become the humerus section of the wing; the second band will become the section containing the radius and ulna; and a third band, derived from the progress zone late in the development of the wing bud, will become the distal part of the wing.   [Left: Courtesy K. W. Tosney, as printed in N. K. Wessells, *Tissue interactions and development*, copyright © 1977 by Benjamin/Cummings, Menlo Park, Calif.]

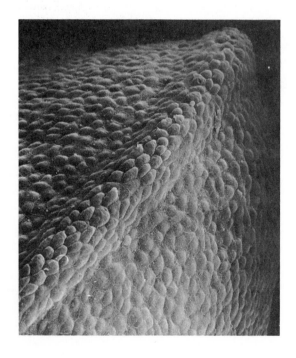

suggested how cells might acquire the positional information they need to ensure proper pattern formation. His model predicates a bicoordinate system: a proximodistal axis (from the body to the end of the extremity) and an anterioposterior (front to rear) axis.

The proximodistal coordinate Wolpert proposes is linked with the progress zone for development of the wing, an area associated with an ectodermal ridge running across the tip of the limb bud. New cells are produced in the progress zone and left behind as the area is pushed farther and farther away from the body—much as the apical meristem of a plant shoot is gradually pushed upward. A cell could derive proximodistal positional values by measuring the time it spent in this progress zone (Fig. 17.50). Cells left behind by the progress zone very early in development of the wing would have a low positional value, which might tell them they should develop into the basal part of the wing (the humerus portion); cells left behind a bit later would have an intermediate positional value, appropriate to formation of the middle part of the wing (the radius and ulna portion); and cells left behind late in the development (i.e. that had spent a long time in the progress zone) would have a high positional value, appropriate to formation of the distal part of the wing. As to the second coordinate, Wolpert proposes that it is a front-to-rear diffusion gradient of a control substance (a morphogen) secreted by a small group of cells at the rear margin of the wing bud. With these two coordinates, the cells could "read" their position with sufficient accuracy to ensure their differentiation into an appropriate structure.

To test this model, Wolpert grafted the progress zone (i.e. the ectodermal ridge and the associated area of actively dividing cells) from an early wing bud onto the end of an older bud whose own progress zone had been removed. The result was development of a wing with two humerus sections and two radius-ulna sections (Fig. 17.51). As Wolpert interprets this result, the cells of the graft had no way of telling they were so far out on the wing that they should form only its distal parts; because they had spent so little time in the progress zone, some of these cells read their position as being near the wing base, and developed into structures appropriate to such a position.

Whether Wolpert's model is accurate, or whether one of the several other models that have been proposed (models usually based on two or three differently oriented chemical gradients) is more nearly correct, the fact remains that cells do apparently derive positional information from some source. Only when that source has been further elucidated shall we have any clear understanding of pattern formation.

Effects of the physical environment   Cell-to-cell and tissue-to-tissue interactions are not the only important elements in guiding embryonic development. A number of physical factors may likewise affect the activity of a developing cell. Among these are temperature, light, humidity, gravity, and pressure. We have previously observed the influence of temperature on the development of pigmentation in Himalayan rabbits and on the vestigial-wing character in *Drosophila* (see p. 598), and we saw in our discussion of plant hormones that light, through its effects on auxin and phytochrome, plays a critical indirect role in such developmental phenomena as increase

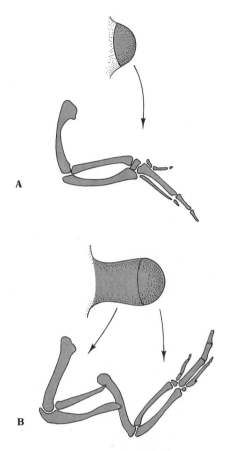

17.51   Results of Wolpert's experiment with a grafted progress zone of a chick wing bud

(A) A normal wing developed from an intact wing bud.   (B) The wing that developed when an early wing bud (color) was grafted onto the original bud (gray) after the basal part of the wing had already begun to develop. The wing has extra humerus and radius-ulna sections (color).

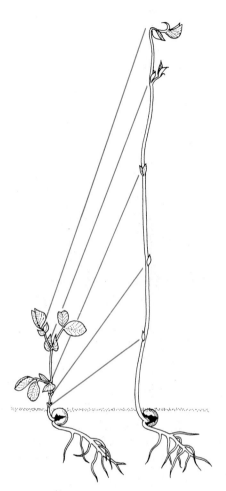

17.52   Development of a pea seedling in the dark
The seedling at left grew normally in the light, but the one at right grew in total darkness. It is abnormally tall and spindly (the color lines link its nodes with corresponding ones of the normal seedling), and it has no chlorophyll.   [Redrawn from *Biologie: Ein Lehrbuch*, ed. by G. Czihak *et al.*, 2nd ed., Springer-Verlag, 1978.]

in shoot length and initiation of flowering. Numerous other examples could be cited. If a seedling plant is grown exclusively in the dark, no mature chloroplasts form; the entire shoot remains colorless, though it usually grows abnormally tall (Fig. 17.52), and the plant eventually dies. Light is an essential factor in inducing the development of chloroplasts and the synthesis of chlorophyll.

A particularly striking example of light-mediated determination of development is seen in ferns. When a spore germinates, it first gives rise to a filamentous structure called a protonema, but normally, as soon as the protonema reaches the light, the pattern of growth changes and a platelike structure called a prothallium develops. It can be shown that the transition from protonema to prothallium depends on light of short wavelengths (blue light). If the germinating spore is kept in the dark or in light of longer wavelengths (e.g. red light), only a protonema will develop (Fig. 17.53A), but if the spore is kept in blue light (or in white light, which contains sufficient blue-light energy), a prothallium will develop (Fig. 17.53B).

The question arises whether, after the cells have been stimulated by blue light to differentiate into prothallial cells, they are capable of reverting to protonema-type development. To find out, it is only necessary to move the prothallium from the blue light into red light. Very soon a cell at the apex of the prothallium gives rise to a filament that in all respects resembles the protonema normally produced only by a germinating spore (Fig. 17.53C). Here, then, is a case where an external factor—the wavelength of light—determines which of two markedly different patterns of development will occur.

THE GRADUALNESS OF DIFFERENTIATION

Let us now see what sort of picture emerges of the course of development from the facts and ideas discussed so far.

If, as we must assume, all the cells of a single organism usually have the same genetic potential, it follows that other factors determine which potentialities are expressed. The first restrictions on the potential of an embryonic cell are often the result of qualitative (and sometimes also quantitative) differences in the early cleavages, which give the nuclei of the different cells different cytoplasmic environments and thus bring about the activation of different genes. As development proceeds, the extracellular environment of the cells becomes less uniform. For example, some cells are located more internally and hence are exposed to less illumination, more pressure, and probably a different pH. Moreover, the various cells differ with regard to their immediate cellular neighbors and hence with regard to the influences—both contactual and chemical—exerted on them by those neighbors. Such differences in the environments of the various cells further intensify the differences in their developmental directions.

As the cells and tissues become more and more differentiated, and perhaps (in animals) undergo morphogenetic movements, they exert an increasing influence on all other cells in their vicinity via chemicals they secrete. Some of the chemicals block pathways the neighboring cells might otherwise have followed; other chemicals tend to induce the neighboring cells to follow alternative pathways. Still other chemicals may provide, through gradients in their concentration, positional information that helps impose pattern on the development.

As cells and tissues respond developmentally to the host of influences impinging on them, they in their turn alter the environment of the cells and tissues in their vicinity. And so the snowballing effect continues. Each step in the development of one cell alters the influence that cell will have on all other cells. The environment of each cell is constantly changing as development proceeds, and the changes in the environment profoundly affect the activity of the genes.

C. H. Waddington of the University of Edinburgh has suggested that differentiating cells can be compared with balls rolling down a slope cut by many valleys (Fig. 17.54). Each ball rolls into one of the valleys. This valley soon branches into two or more separate valleys, and each of these eventually branches in its turn. At each point of branching, the ball enters one of the alternative valleys. As it passes each intersection, the number of alternative pathways still open to it diminishes. Since the ball cannot roll uphill, it cannot normally retrace its course and take a different route. Finally it reaches the bottom of the slope at a point determined by the particular alternative pathways it took at each point of branching.

In a similar way, an early embryonic cell may follow any of a large number of different developmental pathways. Once it has differentiated as ectoderm, however, it cannot ordinarily go back and form a mesodermal or endodermal structure. It has passed the first branching point in the developmental landscape. Now it can form any ectodermal structure, which still leaves it many alternatives. But soon it passes a second important branching point; it either sinks inward as the neural groove forms and differentiates into nervous tissue, or it remains on the surface and differentiates as epidermis. Suppose it follows the former course. Many alternatives are still open to it. It may form part of the brain or part of the spinal cord; it may become part of the somatic or part of the autonomic nervous system; it may differentiate as a multipolar, a bipolar, or a unipolar neuron. But as each

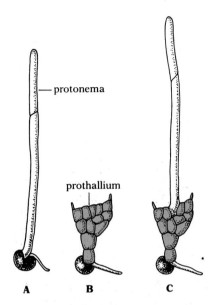

17.53 Effect of different wavelengths of light on the early development of a fern
(A) A germinating fern spore kept in darkness or in red light will give rise to a filamentous protonema.   (B) In blue (or white) light a spore will give rise to a platelike prothallium.   (C) When a developing prothallium is moved into red light, one of the cells at the summit of the prothallium changes its course of development and produces a protonema.   [Redrawn from *Biologie: Ein Lehrbuch*, ed. by G. Czihak *et al.*, 2nd ed., Springer-Verlag, 1978.]

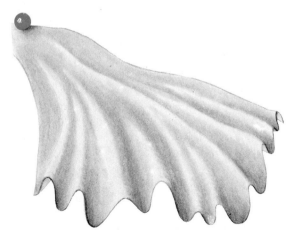

17.54 Developmental landscape model
As the ball rolls down the hill past each branch in the valleys, its potential becomes progressively more restricted.

branching point is passed, the total number of alternatives still ahead diminishes, until finally the cell has become some one kind of fully differentiated cell. Differentiation is thus a matter of progressive determination, a gradual restricting of development to one of the many initially possible pathways.

Evidence from tissue transplants and from reaggregation of cells   The gradualness of differentiation, as depicted by Waddington's model, can be demonstrated by transplantation experiments.

If in a very early amphibian gastrula a piece of tissue from the part of the presumptive ectoderm that will later participate in the formation of the spinal cord is exchanged with a piece of tissue from the part of the ectoderm that will form epidermis, each transplanted tissue develops into whatever is appropriate for its new location; the tissue that would have formed nerve now forms epidermis, and vice versa. Location is the important factor at this stage.

Entirely different results are obtained if the same kind of exchange is made a short time later. The piece that would have formed nerve in its original location also forms nerve in its new location, and the piece that would have formed epidermis forms epidermis in its new location. Something has happened in the late gastrula that has caused the cells from the two different parts of the ectoderm to set out on different developmental pathways. In the late gastrula, you will recall, chordamesoderm derived from the dorsal lip of the blastopore lies under the part of the ectoderm that will form the nerve cord and brain, but there is no chordamesoderm under the part of the ectoderm that will form epidermis. It is induction from the chordamesoderm that has started differentiation of some of the ectoderm into nervous tissue, while the rest of the ectoderm, responding to other influences, has begun differentiation into epidermis. Each tissue has lost some of its former potential; thus the presumptive epidermis is no longer competent to respond to the nerve-inducing secretions of the chordamesoderm.

Now suppose that presumptive spinal cord, instead of being moved to an epidermal location, is moved to the location where the optic vesicles will form. We know from the previous experiment that the tissue has been determined as nerve, but will it form only spinal cord or can it still form other parts of the nervous system? If this experiment is performed very soon after neurulation has begun, the transplanted tissue will form optic vesicle; its location in the developing nervous system is the important factor. But if the experiment is performed after the nervous system has begun to assume its definitive form, the transplanted tissue will form spinal cord, regardless of its new location. It has already passed another critical decision point in its differentiation, and altering its location does not reverse its development. We see, then, that the tissue first differentiates as nervous tissue and later as a particular type of nervous tissue. Differentiation is gradual; it does not occur all at once.

The gradualness of differentiation can also be demonstrated by the reaggregation of separated cells. In 1907 H. V. Wilson of the University of North Carolina discovered that if he dissociated the cells of the body of a sponge by pressing it through a fine sieve, the dispersed cells migrated over the surface of the culture dish until they encountered one another and clumped together to form multicellular aggregates. These aggregates, if suitably cultured, grew into complete new sponges with the architecture characteristic of the sponge before disruption by the sieve. Later workers showed that if the cells of two sponges of different species (colored differently) were dispersed and then mixed together, they would sort themselves and aggregate by species. These results led other workers to wonder if the capacity of cells for recognizing and aggregating with their own kind was a general phenomenon not limited to sponges. It was soon found that the ability of cells to recognize other cells of similar tissue type is a general phenomenon, but that the recognition of species, demonstrated in sponges, is not so general. Thus retina cells from a chick and retina cells from a mouse will clump together to form a new composite retina.

When the cells of the mesodermal and ectodermal layers of a gastrula are dispersed (by mild digestion with trypsin) and then mixed with one another, they move together to form a clump with the ectodermal cells on the outside and the mesodermal cells on the inside. Clearly, ectodermal cells can recognize other ectodermal cells and mesodermal cells can recognize other mesodermal cells; and mesoderm and ectoderm can recognize each other and establish the proper inside-outside relationship. Evidently the two types of cells, mesodermal and ectodermal, are already partly differentiated, even at this early stage of development.

If the same experiment is performed with cells from a later embryo, mesodermal cells aggregate not just with any other mesodermal cells, but with mesodermal cells from the same part of the embryo as they. Thus presumptive liver cells clump together and form liver lobules; presumptive heart cells clump together and form lumps of beating heart muscle; presumptive kidney cells clump together and constitute characteristic kidney capsules and tubules; and so on. In other words, in this later embryo, the cells have differentiated further and are no longer simply mesodermal as against ectodermal; they are now determined as future liver or heart or kidney cells. They have moved farther down the valleys of Waddington's developmental landscape.

IS DIFFERENTIATION IRREVERSIBLE?

We have seen that, as development of a multicellular organism proceeds, the individual cells become more and more committed to one particular course of differentiation. Since this differentiation reflects changes in the activity of the cell's genetic material, the question arises whether some of those changes might be irreversible. We know that not all the changes can be permanent, because the cell in its day-to-day, indeed moment-to-moment, activities must be able to alter its pattern of genetic transcription and mRNA translation in response to the constantly fluctuating requirements of its metabolic machinery for enzymes, but other more profound changes that occurred at critical decision points in the cell's development might conceivably be permanent.

Reversibility of differentiation in plant cells   Some years ago F. C. Steward and his colleagues at Cornell University succeeded in growing whole new carrot plants from single mature cells removed from a carrot root. As Fig. 17.47 suggests, the same thing can be done with tobacco. There is, in fact, a wealth of evidence that even fully differentiated plant cells (except, of course, ones like sieve cells that have lost their nuclei) can be made to revert to the embryonic state and resume totipotency (unlimited developmental potential). For plants, then, the question whether differentiation is irreversible can be answered in the negative: The changes undergone by the nuclei of plant cells during differentiation appear to be entirely reversible.

Nuclear-transplant experiments with animal cells   We saw earlier that the presumptive epidermis and the presumptive neural tissue of an amphibian embryo reach a stage soon after gastrulation when, if their positions are interchanged, one can no longer assume the role of the other. Does this mean that the nuclei have been so altered that they cannot back up and follow a different developmental course? Or might it mean that the whole cells are no longer receptive to the inducing factors encountered in the new locations? One way to find out is to remove the nucleus from a more differentiated cell and insert it into a less differentiated cell, where it will presumably be exposed to cytoplasmic factors more conducive to dedifferentiation.

Some of the earliest nuclear-transplant experiments were performed on frog embryos in 1953 by Robert W. Briggs and Thomas J. King of the Institute for Cancer Research in Philadelphia. Their method was to prick a frog egg with a glass needle to stimulate it to begin developing, remove its nucleus,[8] and then, with a micropipette, insert another nucleus obtained from a partly differentiated cell of a developing embryo. They found that when the transplanted nucleus was one from the animal pole of a blastula the egg usually developed normally—an indication that nuclei in blastula-stage cells have not undergone any permanent alteration and retain the capacity to direct all aspects of development. But if the transplanted nucleus was from a late gastrula, the eggs usually did not develop normally; the embryos produced (except for a small percentage) were severely abnormal, or their development was arrested at some early stage. It seemed that the nuclei of gastrula cells had undergone some stable change, which persisted even if the nuclei were subjected to serial transplants through four or five generations. Did this mean that the nuclei of differentiated frog cells, unlike those of plant cells, permanently lose some of their former developmental potential?

This has been a difficult question to answer because, even though most nuclei from late-gastrula cells in Briggs and King's experiments seemed unable to direct development of a normal embryo, a very few such nuclei could do so. Moreover, in later experiments with the African clawed toad *(Xenopus)*, J. B. Gurdon of Oxford University found that a high percentage of nuclei taken from intestinal cells of swimming tadpoles, or even from skin cells of adults, could direct completely normal development when injected into enucleated egg cells. The fate of these nuclei from fully differentiated cells should have been far more rigidly determined than that of the gastrula nuclei of Briggs and King; yet they seemed able to revert completely to the egg-stage condition and begin development all over again.[9]

[8] In most recent experiments of this type the nucleus of the egg cell is usually destroyed with X rays.

[9] The possibility of artificial cloning arises for animals in which transplanted nuclei can resume totipotency. Many nuclei, all taken from cells of a single individual, could be transplanted into egg cells. The new individuals produced from these eggs would all be genetically identical copies of the individual from which the nuclei were taken.

A somatic cell-fusion experiment   An especially dramatic demonstration of reactivation of a nucleus comes from the new technique of cell fusion. You will recall that fully differentiated mammalian erythrocytes have lost their nuclei. The erythrocytes of birds retain theirs, but these nuclei are completely repressed; the chromosomes are tightly condensed as heterochromatin, and no mRNA is synthesized. But if an erythrocyte from a chicken is fused with a human HeLa cell (see p. 684), the erythrocyte nucleus swells, parts of the chromosomes are converted into euchromatin, and the nucleus begins producing mRNA. Indeed, the formerly totally inactive nucleus even resumes active mitosis. It seems, then, that the drastic repression experienced by the nuclear material in the course of differentiation of a chicken erythrocyte is reversible.

Regeneration of lost body parts   Almost all embryonic animals have extensive regenerative capacities. Some animals retain these after they reach maturity, while others lose them. An adult sea star or hydra can be chopped into many pieces and each piece can regenerate all necessary parts to become a whole individual. An earthworm can regenerate a new head or tail. Half a planarian can regenerate the other half. Salamanders and lizards can regenerate new tails. But adult birds and mammals cannot regenerate whole new organs; regeneration in these animals is mostly limited to the healing of wounds, though some internal organs (e.g. the liver) have impressive regenerative capabilities as long as some of the organ is left to initiate the process.

It was shown in the early 1950s that regeneration of structures such as limbs normally depends, at least in vertebrates, on the nerve supply in the regenerative region. For example, if a leg of a salamander is amputated, one of the first things that happen during the healing of the wound is penetration of the wound tissue by nerve fibers growing outward from the spinal cord. Meanwhile the scar tissue that first formed begins to disappear, and a mound of dedifferentiated cells, called a blastema, forms at the end of the limb stump. Gradually, as mitotic activity and cellular redifferentiation take place within the blastema, the mound comes to look more and more like the normal limb bud of an embryonic salamander (Fig. 17.55). It slowly elongates, and after several weeks a distinct elbow and digits appear.

This rudimentary limb continues to grow, and its cells continue to differentiate into muscle, tendon, bone, connective tissue, etc., until finally it has become a fully functional new leg. But if the nerves leading to the stump of an amputated leg are removed, no regeneration occurs, and the stump itself shrivels and disappears. Regeneration of the leg will take place only if an abundant supply of nerve fibers is present. Marcus Singer, now at Case Western Reserve University, showed that it makes no difference whether the fibers are motor or sensory, or whether they are somatic or autonomic; it is only the total number per unit volume of tissue that is important. Adequate innervation is also necessary for normal maintenance; if the nerve supply to a limb is partly destroyed (as often happens in human spinal injuries), the muscles of that limb usually begin to atrophy.

Discoveries such as these naturally raised the question whether loss of regenerative capacities in the course of development is a result of inadequate nerve supply. To test this possibility, Singer performed a revealing experiment on frogs, which he reported in 1954.

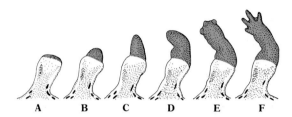

17.55 Regeneration of a salamander arm
[Adapted from M. Singer, "The regeneration of body parts," *Sci. Am.*, October 1958. Copyright © 1958 by Scientific American, Inc. All rights reserved.]

By contrast with adult frogs, young tadpoles have considerable regenerative capabilities. A young tadpole whose leg is amputated quickly grows a new one, but an older tadpole can only partly regenerate a lost leg, and an adult frog cannot regenerate a leg at all. It had been shown that the number of nerve fibers in a tadpole's leg does not increase as rapidly as the volume of other tissues in the leg, and that this relative decline in the number of nerve fibers exactly parallels the gradual loss of regenerative capacity. Singer amputated a front leg of a frog. Then he dissected the large sciatic nerve from the corresponding hind leg without cutting its connections to the spinal cord. Next, he pulled this nerve forward under the skin of the thigh, flank, and abdomen to the stump of the amputated front leg and implanted the ends of the nerve in the stump. With this increased nerve supply, the stump proceeded to regenerate a new leg. Singer concluded that every part of the body probably has the latent power to regenerate and needs only sufficient inductive stimulation to do so.

It has generally been held that the regeneration-inducing activity of nerves is due to their secretion of some trophic chemical, and that adult frogs have too few limb nerves to provide enough of the chemical unless the supply is augmented, as in Singer's experiment. Many studies, particularly of regeneration in invertebrates, have substantiated that nerve cells in regions of regeneration contain a larger than normal number of secretory granules.

An alternative explanation of regeneration induction is favored by some investigators. Since limb regeneration can be induced in adult frogs simply by repeated mechanical trauma to the amputation stump, and since a major effect of such trauma is production of local electric currents, these investigators suggest that the primary initiating stimulus for regeneration may be electric currents of a particular sort resulting from the injury.

It has now been demonstrated experimentally that formation of a blastema, with dedifferentiation and increased mitotic activity, can be stimulated by very weak, nonuniform electric currents. Responsiveness of the cells to the electric currents is greatly enhanced by the hormone prolactin. Since the magnitude of the current generated at an amputation site is directly proportional to the amount of innervation at the site, the proponents of the electric-induction hypothesis suggest that the regeneration brought about by Singer's nerve-transplant experiments may have been triggered by increased electrical stimulation rather than increased release of an inducing chemical, though such a chemical may well be involved at a later stage of the process.

Which of these hypotheses is closer to the facts cannot yet be decided; indeed, both chemical and electrical factors may play a role in regeneration. Be that as it may, limb regeneration, with its dedifferentiation and redifferentiation of cells, clearly points to the likelihood that the nuclei of differentiated animal cells have not undergone any irreversible change. Resumption of full embryonic potential seems to be more difficult for them than for the nuclei of differentiated plant cells, but it probably always remains a possibility.

SUGGESTED READING

Brachet, J., 1974. *Introduction to molecular embryology.* Springer, Heidelberg. *A clearly written small book.**

Bryant, P. J., S. V. Bryant, and V. French, 1977. "Biological regeneration and pattern formation," *Scientific American,* July. (Offprint 1363)   *On basic principles of the organization and growth of complex structures in animals.*

Ebert, J. D., and I. Sussex, 1970. *Interacting systems in development,* 2nd ed. Holt, Rinehart & Winston, New York.   *An excellent short book that gives equal time to plants and animals.**

Epel, D., 1977. "The program of fertilization," *Scientific American,* November. (Offprint 1372)   *The numerous changes that occur in an egg cell as soon as a sperm cell reaches it.*

Gierer, A., 1974. "Hydra as a model for the development of biological form," *Scientific American,* December. (Offprint 1305)   *On the physicochemical basis of pattern development.*

Gurdon, J. B., 1968. "Transplanted nuclei and cell differentiation," *Scientific American,* December. (Offprint 1128)

Wessells, N. K., 1977. *Tissue interactions and development.* Benjamin, Menlo Park, Calif.   *A superb little book, which deals only with animals.**

Wolpert, L., 1978. "Pattern formation in biological development," *Scientific American,* October. (Offprint 1409)   *Wolpert's model for pattern development, based on studies of the chick wing.*

* Available in paperback.

PART IV

THE BIOLOGY OF POPULATIONS
AND COMMUNITIES

Chapter **18**

EVOLUTION

The subject of evolution was briefly introduced in Chapter 1 as a unifying principle for the study of all the topics that were to follow. That introduction was necessarily far from complete. But the succeeding chapters have provided a background that now enables us to take a closer look at the mechanisms of evolution.

EVOLUTION AS CHANGE IN THE GENETIC MAKEUP OF POPULATIONS

Fundamental to the modern theory of evolution are two concepts—that the characteristics of living things change with time and that the change is directed by natural selection. The change we are here discussing is not change in an individual during its lifetime, although such change is a universal and important attribute of life; it is change in the characteristics of populations over the course of many generations. An individual cannot evolve, but a population can. The genetic makeup of an individual is set from the moment of conception; most of the changes during its lifetime are simply changes in the expression of the developmental potential inherent in its genes. But in populations both the genetic makeup and the expression of the developmental potential can change. The former—change in the genetic makeup of a population in successive generations—is evolution.

GENETIC VARIATION AS THE RAW MATERIAL FOR EVOLUTION

A population is composed of many individuals. With rare exceptions, no two of these are exactly alike. In human beings we are well aware of the uniqueness of the individual, for we are accustomed to recognizing different persons at sight, and we know from experience that each has distinctive ana-tomical and physiological characteristics, as well as distinctive abilities and behavior traits. We are also fairly well aware of individual variation in such common domesticated animals as dogs, cats, and horses. But we tend to overlook the similar individual variation in less familiar species such as robins, squirrels, earthworms, sea stars, dandelions, and corn plants. Yet even though this variation may be less obvious to our unpracticed eye, it exists in all such species.

The members of a population, then, share some important features, but differ from one another in numerous ways, some rather obvious, some very subtle. It follows that if there is selection against certain variants within a population and selection for other variants within it, the overall makeup of that population may change with time.

Sources of genetic variation We discussed the various sources of genetic variation in Chapters 14 and 15. Ultimately, of course, all new genes arise by mutation. But once a variety of alleles is in existence, recombination be-comes a mechanism that provides almost endless genotypic variation in the population. The variation resulting from meiosis and recombination at fer-tilization in biparental organisms has already been cited as of immense importance. Crossing-over, translocation, and other chromosomal aberra-tions are further mechanisms of recombination. And in microorganisms, and perhaps in higher organisms, the processes of transformation and transduction constitute still other mechanisms of recombination. Together, mutation and recombination provide the genetic variability on which natu-ral selection can act to produce evolutionary change.

It must be kept in mind, however, that natural selection can act on genetic variation only when it is expressed as phenotypic variation. A completely recessive allele never occurring in the homozygous condition would be to-tally masked from the action of natural selection. But few if any alleles are, in fact, completely recessive against all genetic backgrounds; most alleles that we ordinarily call recessive probably have some very slight phenotypic effect.

Any phenotypic variation within a population may give rise to reproduc-tive differentials between individuals, whether or not the variation reflects corresponding genetic differences. Thus variations produced by exposure to different environmental conditions during development, or produced by disease or accidents, are subject to natural selection. But even though the action of natural selection on all types of variations alters the immediate makeup of a population, it is only its action on variations reflecting genetic differences that has any long-term effect on the population. Variation that is exclusively phenotypic is not raw material for evolutionary change.

Our discussion of genetics will have made it clear that development of athletic prowess by extensive practice, or development of intellectual powers by education, or maintenance of health by correct diet and prompt medical treatment of all ailments cannot alter the genes in the germ cells. The gametes will carry the same genetic information they would have car-ried in the absence of athletic or mental training or proper health care. In short, selection that acts on variations produced exclusively by practice, education, diet, or medical treatment cannot bring about biological evolution (though it might lead to cultural evolution).

There is also some genetic variation that is not raw material for evolutionary change. This is variation produced by somatic mutations. It would be possible, for example, for an important mutation to occur in an ectodermal cell of an early animal embryo. All the cells descended from the mutant cell would be of the mutant type. The result might be a major change in the animal's nervous system, but the change could not be passed on to the animal's offspring. The ectodermal cells are not the ones that give rise to gametes. Mutations in somatic cells cannot alter the genes in the germ cells, which are the cells that will produce the gametes. Hence selection that acts on variations produced by somatic mutations cannot result in evolutionary change in sexually reproducing organisms.

Lacking genetic data, many prominent biologists of the last century and of the early part of this century assumed that exclusively phenotypic variation could serve as evolutionary raw material; that it cannot is far from obvious to many nonbiologists even today. The theory of *evolution by natural selection* proposed by Darwin and Wallace had an influential rival during the nineteenth century in the concept of *evolution by the inheritance of acquired characteristics*—an old and widely held idea often identified with Jean Baptiste de Lamarck (1744–1829), who was one of its more prominent supporters in the early 1800s.

The Lamarckian hypothesis was that somatic characteristics acquired by an individual during its lifetime could be transmitted to its offspring. Thus the characteristics of each generation would be determined, in part at least, by all that happened to the members of the preceding generations—by all the modifications that occurred in them, including those caused by experience, use and disuse of body parts, and accidents. Evolutionary change would be the gradual accumulation of such acquired modifications over many generations. The classic example (though by now rather hackneyed) is the evolution of the long necks of giraffes.

According to the Lamarckian view, ancestral giraffes with short necks tended to stretch their necks as much as they could to reach the tree foliage that served as a major part of their food. This frequent neck stretching caused their offspring to have slightly longer necks. Since these also stretched their necks, the next generation had still longer necks. And so, as a result of neck stretching to reach higher and higher foliage, each generation had slightly longer necks than the preceding generation.

The modern theory of natural selection, on the other hand, proposes that ancestral giraffes probably had short necks, but that the precise length of the neck varied from individual to individual because of their different genotypes. If the supply of food was somewhat limited, then individuals with longer necks had a better chance of surviving and leaving progeny than those with shorter necks. This means, not that all the individuals with shorter necks perished or that all with longer necks survived to reproduce, but simply that a slightly higher proportion of those with longer necks survived and left offspring. As a result, the proportion of individuals with genes for longer necks increased slightly with each succeeding generation.

Although the hypothesis of evolution by inheritance of acquired characteristics is rejected by most modern biologists, it was a logical and reasonable one when first proposed. It has simply not stood the test of further scientific research. In Lamarck's day (as in Darwin's), nothing was known about the mechanism of inheritance. Mendel had not yet performed his experiments on garden peas. It was not illogical, therefore, to assume that a change in any part of the body could be inherited.

As long ago as the days of ancient Greece, it had been suggested that particles from all parts of the body, pangenes, come together to form the eggs and semen. This Greek idea of *pangenesis,* if it could have been proved, would have provided a genetic basis for the Lamarckian hypothesis. If a long-distance runner built up his leg muscles, and if the pangenes in them were thereby altered, then when the runner formed semen the pangenes for leg muscles in his semen would be the altered type and would confer larger leg muscles on his children. One of the most telling points against Lamarckianism is the refutation of the idea of pangenesis on which it depends. We now know that somatic cells do not affect the genotype of the germ cells; immense alterations of the somatic cells can be brought about without in any way influencing the hereditary information in the gametes. The hypothesis of inheritance of acquired characteristics is therefore no longer tenable.

THE GENE POOL AND FACTORS THAT AFFECT ITS EQUILIBRIUM

The gene pool   To understand evolution as change in the genetic makeup of populations in successive generations, it is necessary to know something about population genetics. Our study of the genetics of individuals in Chapter 14 was based on the concept of the genotype, which is the genetic constitution of an individual. Our study of the genetics of populations will be based in a similar manner on the concept of the gene pool, which is the genetic constitution of a population. The gene pool is the sum total of all the genes possessed by all the individuals in the population.

You will recall that the genotype of a diploid individual can contain a maximum of only two alleles of any given gene. But there is no such restriction on the gene pool of a population. It can contain any number of different allelic forms of a gene. The gene pool is characterized with regard to any given gene by the frequencies, or ratios, of the alleles of that gene in the population. Suppose, to use a simple example, that gene A occurs in only two allelic forms, A and a, in a particular sexually reproducing population. And suppose that allele A constitutes 90 percent of the total of both alleles while allele a constitutes 10 percent of the total. The frequencies of A and a in the gene pool of this population are therefore 0.9 and 0.1. If those frequencies were to change with time, the change would be evolution. Evolution, as change in the genetic makeup of populations, is, more precisely then, a change in the allelic frequencies (or genotypic ratios) within gene pools. Hence it is possible to determine what factors cause evolution by determining what factors can produce a shift in allelic frequencies.

Let us examine more carefully our hypothetical population in which allele A has a frequency of 0.9 and allele a has a frequency of 0.1. How can we calculate the genotypic ratios that will be present in this population? If we assume that all possible genotypes have an equal chance of surviving, this calculation is not hard to make. If the ratio of A to a in the entire population is 9 : 1, then the ratio of sperm cells carrying allele A to sperm cells carrying allele a is 9 : 1. And the ratio of egg cells with allele A to egg cells with allele a is also 9 : 1. Using this information, we can set up a Punnett square much like those we used for crosses between two individuals:

	Sperm	
	0.9 A	0.1 a
0.9 A	0.81 A/A	0.09 A/a
0.1 a	0.09 a/A	0.01 a/a

Eggs

Notice that the only difference between this and a Punnett square for a cross between individuals is that here the sperm and eggs are not those produced by a single male and a single female, but those produced by all the males and females in the population, with the frequency of each type of sperm and egg shown on the horizontal and vertical axes respectively. Filling in the square (by combining the indicated alleles and multiplying their frequencies) tells us that the frequency of the homozygous dominant genotype (A/A) in the population will be 0.81, the frequency of the heterozygous genotype (A/a or a/A) 0.18, and that of the homozygous recessive genotype (a/a) 0.01.

We have thus fully characterized the gene pool of the present generation of our hypothetical population with regard to alleles A and a and the genotypes they form. Now we want to know whether the frequencies we have found will change in successive generations—in short, whether the population will evolve.

The Hardy-Weinberg Law and conditions for genetic equilibrium   It is easily assumed that the more frequent allele (in our hypothetical case, A) will automatically increase in frequency while the less frequent allele (a) will automatically decrease in frequency and eventually be lost from the population. But this assumption is incorrect. The rarity of a particular allele in a population does not doom it to automatic disappearance, as we can verify by using the known frequencies for one generation to compute the frequencies for the next generation.

We have said that the allelic frequencies in the gene pool of the present generation of our hypothetical population are 0.9 and 0.1 and that the genotypic ratios are 0.81, 0.18, and 0.01. If we use these figures to set up a Punnett square and compute the corresponding values for the next generation, we find that they are the same ones we started with. The allelic frequencies of the next generation are also 0.9 and 0.1, and the genotypic ratios are 0.81, 0.18, and 0.01. We could perform the same calculation for generation after generation, always with the same results; neither the allelic frequencies nor the genotypic ratios would change. We must conclude, therefore, that evolutionary change is not usually automatic, that it occurs only when something disturbs the genetic equilibrium. This was first recognized in 1908 by G. H. Hardy of Cambridge University and W. Weinberg, a German physician, working independently. According to the Hardy-Weinberg Law, *under certain conditions of stability both allelic frequencies and genotypic ratios remain constant from generation to generation in sexually reproducing populations.*[1]

Let us examine the "certain conditions" that the Hardy-Weinberg Law says must be met if the gene pool of a population is to be in genetic equilibrium. They are as follows:

1. The population must be large enough to make it highly unlikely that chance alone could significantly alter allelic frequencies.
2. Mutations must not occur, or else there must be mutational equilibrium.
3. There must be no immigration or emigration.
4. Reproduction must be totally random.

With regard to the first condition, a population would have to be infinitely large for chance to be completely ruled out as a causal factor in the changing of allelic frequencies. In reality, of course, no population is infinitely large, but many natural populations are large enough so that chance alone would not be likely to cause any appreciable alteration in the allelic frequencies in their gene pools. Any breeding population with more than 10,000 members of breeding age is probably not significantly affected by random change. But allelic frequencies in small isolated populations of, say, less than 100 breeding-age members are highly susceptible to random fluctuations, which can easily lead to loss of an allele from the gene pool even when that allele is an adaptively superior one. In such populations, in fact, there are relatively few alleles with intermediate frequencies; apparently the tendency is for most alleles either to be soon lost or to become fixed as the only allele present. In other words, small populations tend to have a high degree of homozygosity, while large populations tend to be more variable. Thus chance may cause evolutionary change in small populations, but since this change, called *genetic drift,* is not much influenced by the relative adaptiveness of the different alleles, it is essentially an indeterminate evolution, as likely to take one direction as another. (A special type of genetic drift, called the founder effect, will be discussed on p. 795.)

The second condition for genetic equilibrium—either no mutation or mutational equilibrium—is never met in any population. Mutations are always occurring. There is no known way of stopping them. Most genes probably undergo mutation once every hundred thousand to hundred million duplications; the rate of mutation for different genes varies greatly. As for mutational equilibrium, very rarely, if ever, are the mutations of alleles for the same character in exact equilibrium; i.e. the number of forward mutations per unit time is rarely exactly the same as the number of back mutations.[2] The result of this difference is a *mutation pressure* tending to cause a slow shift in the allelic frequencies in the population. The more stable allele will tend to increase in frequency, and the more mutable allele will tend to decrease in frequency, unless some other factor offsets the mutation pressure. But even though mutation pressure is almost always present, it is probably seldom a major factor in producing changes in allelic frequencies

---

[1] For this statement to hold true for genotypes, the initial genotypic frequencies must be in equilibrium. Thus, if genotypes A/A and a/a were present in a population, but A/a were missing (let's suppose, because of human intervention), then all three genotypes would appear in the next generation, and thereafter the genotypic frequencies would remain constant.

[2] By convention, the mutation from the more common allele to the less common one is called the forward mutation, and the reverse is called the back mutation.

in a population. Acting alone, mutation would take an enormous amount of time to produce much change (except in the origin of polyploidy, which we shall discuss later). And, furthermore, mutation is random; its trend is often in a direction different from that in which other factors are causing the population actually to evolve. Mutations increase variability and are thus the ultimate raw material of evolution, but they seldom determine the direction or nature of evolutionary change.

According to the third condition for genetic equilibrium, a gene pool cannot accept immigrants from other populations, for these might introduce new alleles, and it cannot suffer loss of alleles by emigration. A high percentage of natural populations, however, probably experience at least a small amount of gene migration, generally called *gene flow,* and this factor, which enhances variation, tends to upset Hardy-Weinberg equilibria. But there are doubtless populations that experience no gene flow, and in many instances where flow does occur it is probably sufficiently slight to be essentially negligible as a factor causing shifts in allelic frequencies. We can conclude, therefore, that this third condition for genetic equilibrium is sometimes met in nature.

The fourth condition for genetic equilibrium in a population is that reproduction be totally random. Reproduction in this context does not mean simply the mating process, but the vast number of factors that contribute to the reproductive continuity of the population: selection of a mate, physical efficiency and frequency of the mating process, fertility, total number of zygotes produced at each mating, percentage of zygotes that lead to successful embryonic development and birth, survival of the young until they are of reproductive age, fertility of the young, and even, in some cases, survival of postreproductive adults when their survival affects either the chances of survival of the young or their reproductive efficiency. If reproduction is to be totally random, all these factors must be random; i.e. they must be independent of genotype. This condition is probably never met in any real population. The factors mentioned here are always correlated in part with genotype; i.e. an organism's genotype influences its selection of a mate, the physical efficiency and frequency of its mating, its fertility when mated with organisms of other genotypes, the total number of zygotes it produces at each mating, the percentage of successful births of its embryos, the survival and fertility of its young, and its postreproductive survival. In short, there is probably no aspect of reproduction that is totally uncorrelated with genotype. Nonrandom reproduction is the universal rule. And nonrandom reproduction, in the broad sense in which the term has been defined here, is synonymous with natural selection. Natural selection, then, is always operative in all populations; there is always *selection pressure* acting to disturb the Hardy-Weinberg equilibrium.

In summary, of the four conditions necessary to achieve the genetic equilibrium described by the Hardy-Weinberg Law, the first (large population size, i.e. no genetic drift) is met reasonably often, the second (no mutation pressure) is never met, the third (no migration, i.e. no gene flow into or out of the population) is met sometimes, and the fourth (random reproduction, i.e. no selection pressure) is never met. It follows that complete equilibrium in a gene pool is not expected. Evolutionary change is a fundamental characteristic of the life of all populations, including human populations.

Why go to so much trouble to explain the Hardy-Weinberg Law only to

On p. 755 we used a Punnett square to calculate the ratios of the genotypes produced by alleles A and a, whose respective frequencies in a hypothetical population were given as 0.9 and 0.1. The same results can be obtained more rapidly by using an algebraic formula.

Expansion of the binomial expression $(p + q)^2$, where p is the frequency of one allele (in our case, A) and q the frequency of the other allele (a), yields the formula for the Hardy-Weinberg equilibrium:

$$p^2 + 2pq + q^2 = 1$$

Substituting the allelic frequencies 0.9 and 0.1 for p and q respectively, we obtain

$$
\begin{array}{ccccccc}
p^2 & + & 2pq & + & q^2 & = & 1 \\
(0.9)(0.9) & + & 2(0.9)(0.1) & + & (0.1)(0.1) & = & 1 \\
0.81 & + & 0.18 & + & 0.01 & = & 1
\end{array}
$$

The three terms of the Hardy-Weinberg formula indicate the frequencies of the three genotypes:

$$
\begin{array}{rcl}
p^2 & = \text{frequency of } A/A & = 0.81 \\
2pq & = \text{frequency of } A/a & = 0.18 \\
q^2 & = \text{frequency of } a/a & = 0.01
\end{array}
$$

These are, of course, the same results we obtained using the Punnett square.

In this example we have assumed that we know the allelic frequencies and want to compute the corresponding genotypic frequencies. But the Hardy-Weinberg formula permits many other sorts of calculations as well. Suppose, for example, that we know a certain disease caused by a recessive allele d occurs in 4 percent of a certain population and that we want to find out what percent are heterozygous carriers of the disease. Since the disease occurs only in homozygous recessive individuals, it follows that the frequency of the d/d genotype is 0.04. Letting q^2 stand for the frequency of d/d in the formula, we can write

$$q^2 = 0.04$$

The frequency of allele d, then, is the square root of 0.04:

$$q = \sqrt{0.04} = 0.2$$

If the frequency of allele d is 0.2, the frequency of allele D must be 0.8, because the two frequencies must always add up to 1 (that is, $p + q = 1$).

Substituting the frequencies of both alleles in the Hardy-Weinberg formula, we can compute the frequencies of the genotypes:

$$
\begin{array}{ccccccc}
p^2 & + & 2pq & + & q^2 & = & 1 \\
(0.8)(0.8) & + & 2(0.8)(0.2) & + & (0.2)(0.2) & = & 1 \\
0.64 & + & 0.32 & + & 0.04 & = & 1
\end{array}
$$

Since the term $2pq$ stands for the frequency of the heterozygous genotype, which is what we wanted to know originally, our answer is that 0.32 or 32 percent of the population are heterozygous carriers of the allele d that causes the disease we are studying.

Let us now see how this type of reasoning can be applied to calculate changes in allelic frequencies when only phenotypic frequencies can be measured directly. Suppose we find, one spring, in a large population of freely interbreeding plants that 59 percent have yellow blossoms (known to be a dominant phenotype) and 41 percent white blossoms (a recessive phenotype). We return to the same place in the spring of the following year, after a very severe winter, and find 64 percent with yellow blossoms and 36 percent with white blossoms. Clearly, plants with the dominant allele survived better, but we want to know exactly how much the allelic frequencies changed.

Since white blossoms indicate a recessive phenotype, we know that the frequency of the genotype y/y was 0.41 initially. Setting $q^2 = 0.41$, we calculate that $q = 0.64$ (approximately), which means that the frequency of allele y was 0.64. The frequency of the dominant allele Y was therefore 0.36 ($1 - 0.64 = 0.36$). The next spring the frequency of white blossoms had fallen to 36 percent, which gives $q^2 = 0.36$ and $q = 0.60$. The frequency of allele y is thus 0.60, and the frequency of Y must be 0.40. In summary, the frequency of allele y has changed from 0.64 to 0.60 and the frequency of Y from 0.36 to 0.40 in one year.

The above examples all involved only two alleles. Similar procedures, even if considerably more complicated mathematically, can be used for situations involving multiple alleles. Thus the Hardy-Weinberg formula for a triallelic situation requires expansion of the trinomial $(p + q + r)^2$, where r is the frequency of the third allele. Similarly, a quadriallelic situation requires expansion of $(p + q + r + s)^2$.

show that it describes a situation that never occurs in nature? For a compelling reason: The Hardy-Weinberg Law sets up a null hypothesis—no evolution—which, because it can easily be disproved, provides an indirect demonstration that all populations must constantly be evolving.

THE ROLE OF NATURAL SELECTION

Changes in individual allelic frequencies caused by natural selection   Let us now return for a moment to our hypothetical population in which the initial frequencies of the alleles A and a are 0.9 and 0.1 and the genotypic frequencies 0.81, 0.18, and 0.01. We have seen that although these frequencies will not change automatically with the passage of time—changing only when something disturbs the genetic equilibrium—mutation pressure (to a slight extent) and selection pressure (to a greater extent) are always disturbing this equilibrium; other factors at work are often gene flow and, if the population is small, genetic drift. We shall here restrict discussion to what is by far the most important of these factors in guiding evolution: selection pressure.

Suppose that selection acts against the dominant phenotype in our example, and that this negative selection pressure is strong enough to reduce the frequency of A in the present generation from 0.9 to 0.8 before reproduction occurs. (Of course there will be a corresponding increase in the frequency of a from 0.1 to 0.2, since the two frequencies must total 1.0.) Now let us set up a Punnett square and calculate the genotypic ratios that will be present in the zygotes of the second generation:

	Sperm	
	0.8 A	0.2 a
0.8 A	0.64 A/A	0.16 A/a
0.2 a	0.16 a/A	0.04 a/a

Eggs (left label, rows 0.8 A and 0.2 a)

We find that the genotypic ratios of the zygotes in the second generation are different from those in the parental generation; instead of 0.81, 0.18, and 0.01, the ratios are 0.64, 0.32, and 0.04 (Fig. 18.1). If selection now acts against the dominant phenotype in this generation, and thereby again reduces the frequency of A, the genotypic ratios in the third generation will be different from those of both preceding generations; the frequency of A/A will be lower and that of a/a higher. If this same selection pressure were to continue for many generations, the frequency of A/A would fall very low and the frequency of a/a would rise very high. Thus natural selection would have caused a change from a population in which 99 percent of the individuals showed the dominant phenotype and only one percent the recessive phenotype to a population in which very few showed the dominant phenotype and most showed the recessive phenotype. This evolutionary change of the phenotype most characteristic of the population would have occurred without the necessity of any new mutation, simply as a result of natural selection.

Rather than deal with a hypothetical example, let us cite an actual situation in which selection has produced a radical shift in allelic frequencies. Soon after the discovery of the antibiotic activity of penicillin, it was found that *Staphylococcus aureus*, a bacterial species that can cause numerous infections, including boils and abscesses, quickly developed resistance to the drug. Higher and higher doses of penicillin were necessary to kill the bacteria, and the resistant bacteria became a serious problem in hospitals. Clearly, under the influence of the strong selection exerted by the penicillin, the bacterial population had evolved. Many studies have shown that the drug does not induce mutations for resistance; it simply selects against nonresistant bacteria (Fig. 18.2). Some genes determining metabolic pathways that confer resistance to penicillin are already present in low frequency in most populations, having arisen earlier as a result of random mutations. Individuals possessing these genes are thus **preadapted** to survive the antibiotic treatment, and, since it is they that reproduce and perpetuate the population (the nonresistant individuals having been killed), the

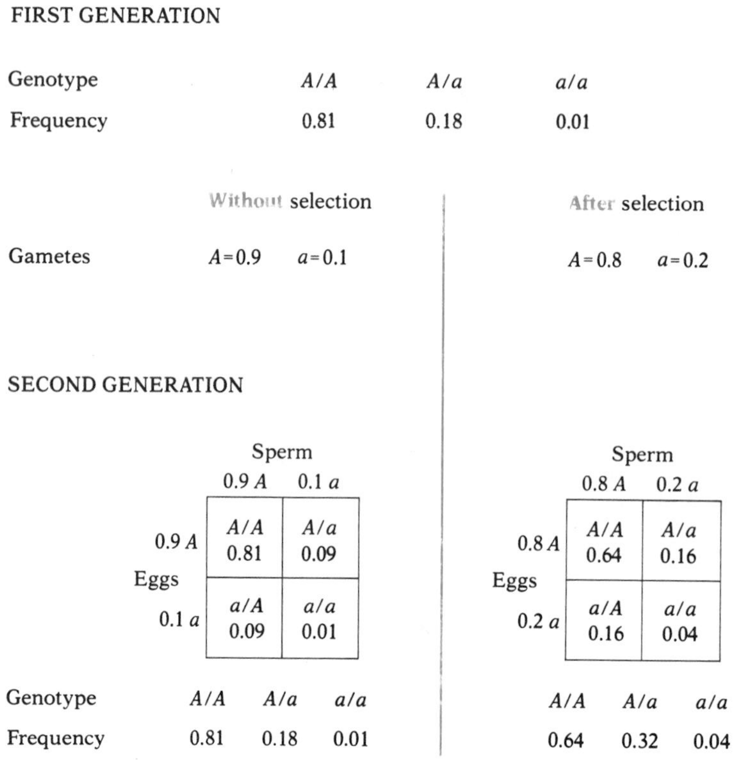

FIRST GENERATION

Genotype	A/A	A/a	a/a
Frequency	0.81	0.18	0.01

	Without selection	After selection
Gametes	A=0.9 a=0.1	A=0.8 a=0.2

SECOND GENERATION

Sperm
0.9 A   0.1 a

Eggs
	0.9 A	0.1 a
0.9 A	A/A 0.81	A/a 0.09
0.1 a	a/A 0.09	a/a 0.01

Sperm
0.8 A   0.2 a

Eggs
	0.8 A	0.2 a
0.8 A	A/A 0.64	A/a 0.16
0.2 a	a/A 0.16	a/a 0.04

Genotype	A/A	A/a	a/a	A/A	A/a	a/a
Frequency	0.81	0.18	0.01	0.64	0.32	0.04

18.1   Comparison of genotypic frequencies under selection and no selection

Left: In the absence of selection, the genotypic frequencies in a Hardy-Weinberg population are the same in the second and all later generations as in the first generation.

Right: When there is selection, the genotypic frequencies change from one generation to the next. In this hypothetical case, discussed in more detail in the text, the frequencies of A and a in the gene pool of the first generation were initially 0.9 and 0.1, but

natural selection altered these to 0.8 and 0.2 before reproduction occurred. Consequently the initial genotypic frequencies in the second generation are different from the initial ones in the first generation.

next generation shows a marked resistance to penicillin. If such genes were not already present in a population exposed to penicilin, no cells would survive and the population would be wiped out.

The primacy of selection pressure does not mean that new mutations cannot improve the resistance; in fact, continued selection with penicillin usually leads to gradually increased resistance, which is almost certainly due in part to new mutations that enhance resistance and to selective increase in the frequencies of these new mutant genes. But the mutations are not produced by the same conditions that select for the mutant genes once they arise. That mutations beneficial in an environment containing penicillin should arise when this drug is administered is purely a matter of chance; the same mutations would probably arise in the absence of penicillin.

Evolution of drug resistance in bacteria is not entirely comparable to evolution in biparental organisms, because intense selection can change gene frequencies much more rapidly in haploid asexual organisms than in biparental ones. The recombination that occurs at every generation in a biparental species often re-establishes genotypes eliminated in the previous generation; this does not happen in asexual organisms. Nevertheless, even very small selection pressures can produce major shifts in gene frequencies in biparental populations when the time scale is one in which 50,000 years is a rather brief period. J. B. S. Haldane has shown that if the individuals carrying a given dominant allele benefit by 0.001 in their capacity to survive (e.g. if 1,000 A/A or A/a individuals survive to reproduce for every 999 a/a individuals that survive to reproduce), then the frequency of the dominant allele could increase from 0.00001 to 0.99 in only 23,400 generations. In other words, a positive selection pressure of only 0.001 could cause a very rare allele to become very common in only 23,400 generations. "Only" in conjunction with 23,400 generations may sound incongruous, but remember that many plants and animals have at least one generation a year, and that in very few species is the generation time more than 10 years (humans are among the few exceptions). Hence 23,400 generations often means less than 23,400 years and rarely more than 234,000 years. Both of these are relatively short time-spans when measured on the geologic time scale. Even a selection pressure as low as 0.0001 (one part in 10,000) would be a major factor in a breeding population of 5,000 or more individuals on such a time scale. Recent evidence suggests, however, that many selection pressures in nature are much larger than 0.001; hence major changes in allelic frequencies sometimes probably take less than a century, perhaps even less than 50 years.

Directional selection of polygenic characters   So far, we have discussed situations in which we have posited only two clearly distinct phenotypes determined by two alleles of a single gene. But in reality the vast majority of characters on which natural selection acts are influenced by many different genes, most of which have multiple alleles in the population; the expression of many characters, moreover, is influenced considerably by environmental conditions. Consequently such characters usually show variation with a whole range of values, which often tend to have a frequency distribution that, when graphed, approximates the so-called normal or bell-shaped curve (Fig. 18.3).

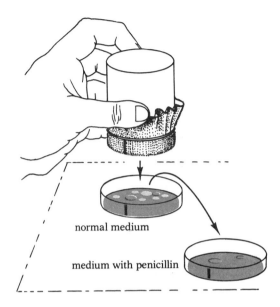

18.2   An experiment showing that mutation for resistance to penicillin is spontaneous
Bacterial cells were cultured on a normal agar medium; many colonies developed (upper culture dish). Then a block wrapped with velveteen was pressed against the surface of the culture to pick up cells from each of the colonies. The block was next pressed against the surface of a second culture dish containing sterile medium to which penicillin had been added; care was taken to align the transfer block and the culture dishes according to markers on the block and dishes (black lines). The cells from most of the colonies on the original dish failed to grow on the penicillin medium, but those of a few colonies (two are shown here) did grow. Had the cells of those two colonies spontaneously become penicillin-resistant before being transferred to the penicillin medium, or did exposure to the penicillin induce a mutation for resistance?

Because the transfer block had been aligned with each dish in the same way, according to the markers, it was possible to tell precisely which original colonies had given rise to the two colonies on the penicillin medium. Cells could therefore be taken from those original colonies, which had never been exposed to penicillin, and tested for resistance. They were found to be resistant. Hence the mutation for resistance must have arisen spontaneously; it was not induced by exposure to the drug.

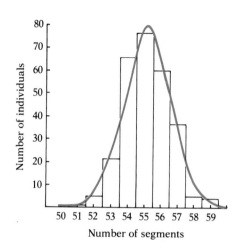

18.3 Frequency distribution of number of segments in a sample of millipeds of the species *Narceus annularis*

The pattern of variation in number of segments (shown by the vertical bars) approximates, but does not exactly fit, the bell-shaped normal curve of probability.

If the environmental conditions should change, with a consequent shift in the selection pressure, we would expect that as a result of changing allelic frequencies the curve of phenotypic variation would shift also. A hypothetical case is illustrated in Fig. 18.4. Graph 18.4A shows the annual rainfall at which the plants in a particular population would grow best (let us assume that the conditions under which a plant grows best are genetically determined). The actual annual rainfall in the area where this population occurs averages 40 cm, as indicated by arrow 1. The population contains a very few plants (S) that would grow best if the annual rainfall were about 32 cm and a very few (W) that would grow best if the annual rainfall were about 48 cm. Plants that would grow best if the annual rainfall were about 36 cm (T) or 44 cm (V) are fairly common in the population. But by far the most numerous are plants (U) that grow best when the annual rainfall is about 40 cm.

Now let us suppose that the average annual rainfall in the area in question slowly increases over a period of years until it is 44 cm (arrow 2). Under these new environmental conditions, the V plants (which grow best when the annual rainfall is about 44 cm) will do better than before; a higher percentage of them can be expected to survive and reproduce, and their frequency should increase. Similarly, the W plants (which grow best when the annual rainfall is about 48 cm) will now grow better than formerly, and they too should increase in frequency. Conversely, the T plants and the U plants will not grow as well as formerly, and they should decrease in frequency. And the S plants, only a few of which managed to survive when the annual rainfall was 40 cm, would now be so poorly adapted to the prevalent conditions that none could survive. These changing frequencies, produced by the shift in the selection operating on the population, would give rise to the new curve shown in Fig. 18.4B.

If the average annual rainfall continues to increase over a period of years until it reaches 48 cm (arrow 3), the W plants and X plants should increase in frequency, the U plants and V plants should decrease in frequency, and the T plants should disappear; these shifts would give rise to the curve shown in Fig. 18.4C. If the average annual rainfall then slowly increases to 52 cm (arrow 4), it should cause further shifts in frequencies, producing the curve shown in Fig. 18.4D.

The changing environmental conditions, then, have given rise to what is called *directional selection,* which has caused the population to evolve along a particular functional line. If the population had not been sufficiently variable genetically to have the potential to change when the environment changed, it would have been much reduced, or it might even have become extinct.

The creative role of natural selection   Notice that in our hypothetical plant population the directional selection did not just skew the curve to the right, as might happen when the variation in the gene pool is small. Instead, it caused the extremes as well as the mean (average) of the population to shift to the right. The shift was eventually so great, in fact, that a class of plants (X) not even present in the original population became the largest class. But, you might ask, if X plants, Y plants, and Z plants were not present initially, how did they arise in the descendent populations? One possibility is that, purely by chance, new mutant genes arose that made their possessors grow better in wetter habitats; such mutant genes would have been strongly selected for and would rapidly have spread through the population. But if moisture preference is influenced by many different genes, as is highly likely, new classes such as X, Y, and Z could arise without the necessity of any new mutations, simply through the separate increase in frequency of many different genes, which would then be more likely to occur together and produce a new phenotype.

Haldane has calculated how long it would take for a new phenotype to be created in this way. He has shown that if each of 15 independent genes is present in one percent of the individuals of a population, then all 15 genes will occur together in only one of 10^{30} individuals.[3] But there has never been a population of higher organisms containing anywhere near 10^{30} individuals (in fact, individual higher plants have never totaled 10^{30} at any time during the history of life on earth). Hence the chances that all 15 genes would occur together in even one individual in a real population are exceedingly small— zero for all practical purposes. In other words, it can be assumed that the phenotype produced by the combined action of all 15 genes does not exist in the population. But, according to Haldane, if there is moderate natural

[3]Very large numbers are customarily written as powers of 10. Thus 10^{30} is equivalent to 1 followed by 30 zeros. Similarly, 10^6 would be 1 followed by 6 zeros, or one million.

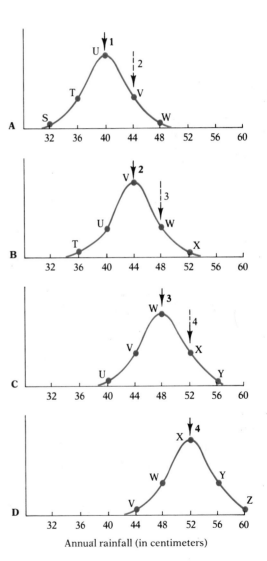

18.4 Evolutionary change of a hypothetical plant population in response to directional selection by changing rainfall
The various phenotypes (S, T, U, etc.) in a hypothetical plant population reflect different genetically determined growth responses to annual rainfall. The four curves show the frequency distributions of the phenotypes at different times.(A, B, C, D), under conditions of average annual rainfall indicated by the solid black arrows. (A) shows the frequency distribution of phenotypes over a long period of annual rainfall averaging about 40 cm. When the annual rainfall slowly increases to 44 cm (broken arrow 2), the frequency distribution of phenotypes slowly shifts to that shown in (B), where it tends to remain as long as the average annual rainfall continues at 44 cm (solid arrow 2). But if rainfall again slowly increases, this time to 48 cm (broken arrow 3), the frequency distribution shifts to position (C). Another increase in rainfall to 52 cm (arrow 4) results in still another shift in the population to condition (D). The changing rainfall has exerted directional selection on the plant population, shifting the curve of relative phenotypic frequencies more and more to the right.

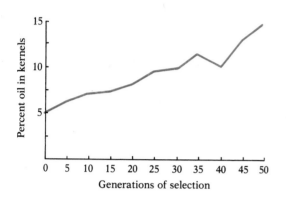

18.5 Results of 50 generations of selection for high oil content in corn kernels
[From C. M. Woodworth *et al.*, *Agron. J.*, vol. 44, 1952.]

selection for each of the 15 genes, it would take only about 10,000 years for each gene to increase from a frequency of one percent to a frequency of 99 percent. Once each gene is present in 99 percent of the population, 86 percent of the individuals in it will have all 15 genes and hence will show the phenotype that was previously nonexistent in the population. Thus selection, even in the absence of new mutation, can produce new phenotypes by combining old genes in new ways.

An actual illustration of the sort of change outlined in the hypothetical example above is provided by a long-term selection experiment performed on corn by agronomists at the University of Illinois. These agronomists selected for high oil content of the corn kernels; the directional selection was continued for 50 generations. There was a steady increase in oil content throughout most of this period (Fig. 18.5). The kernels of the original stock of corn plants averaged about 5 percent oil; those of the plants in the fiftieth generation after selection averaged about 15 percent (higher than any individuals in the first generation), and there was no indication that a maximum had been reached.

That this steady change over the course of 50 generations must have been due to the formation of new gene combinations through selection rather than to the occurrence of a series of new mutations can be seen from a few simple calculations. The agronomists raised between 200 and 300 corn plants in each generation. In 50 generations, then, they raised between 10,000 and 15,000. But the usual rate of mutation per gene in corn is never greater than one in 50,000 plants, and it is usually lower. Hence it is unlikely that even one mutation contributing to an increase in oil content occurred during the experiment, and it is certain that there was no series of such mutations. Therefore the gradual increase in oil content during the 50 generations of directional selection must have been due to the formation of new gene combinations, not to mutations.

The new gene combinations produced by selection sometimes have the result of changing formerly recessive alleles into dominant ones. It is important to remember that an allele is not automatically either dominant or recessive; it is the genetic background against which it must function (i.e. the other genes present in the same individual) that determines its activity. When that background changes through shifting gene frequencies, the enzyme-making activity of the individual allele may also change, because every gene influences the activity of every other gene to some extent. We saw earlier that most new mutant alleles are recessive, and that they may be carried in low frequency in the population indefinitely without being expressed phenotypically. When, generations later, selection alters the genetic background in such a manner that the allele becomes dominant (i.e. becomes more active), the phenotype it produces (which is new to the population, even though the gene is not) provides new variation as raw material upon which selection may act. Thus the evolution of dominance as a result of selection is in a very real sense a creative process.

To summarize: In biparental populations selection (whether natural or artificial) determines the direction of change largely by altering the frequencies of genes that arose through random mutation many generations before, thus establishing new gene combinations and gene activities that produce new phenotypes. Mutation is not usually a major directing force in

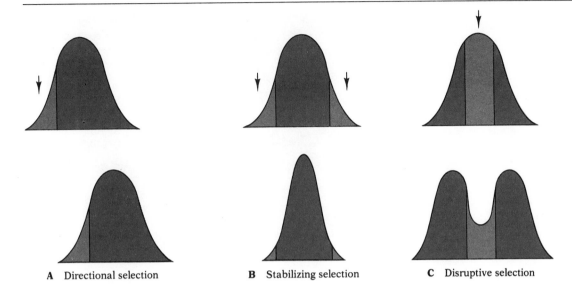

A Directional selection   **B** Stabilizing selection   **C** Disruptive selection

evolution; the principal evolutionary role of new mutations consists in replenishing the store of variability in the gene pool and thereby providing the potential upon which future selection can act.

Disruptive selection   It can sometimes happen that a polygenic character of a population is subject to two (or more) opposing directional selection pressures. Suppose, for example, that a certain population of birds shows much variation in bill length. Suppose, further, that as conditions change there are increasingly good feeding opportunities for the birds with the shortest bills and also for those with the longest bills, but decreasing opportunities for birds with bills of intermediate length. The effect of such selection will be to divide the population into two distinct types, one with short bills and one with long bills. The combined action of the opposing directional pressures is thus to disrupt the population (Fig. 18.6C).

The conservative role of natural selection   Besides a creative role, natural selection also plays an extremely important conservative role. Each species, in the course of its evolution, comes to have a constellation of genes that interact in very precise ways in governing the developmental, physiological, and biochemical processes on which the continued existence of the species depends. Anything that disrupts the harmonious interaction of its genes is usually deleterious to the species. But in a sexually reproducing population, favorable groupings of genes tend to be dispersed and new groupings formed by the recombination that occurs when each generation reproduces. Most of these new groupings will be less adaptive than the original grouping (although a few may be more adaptive). And the vast majority of new mutations tend to disrupt rather than enhance the established harmonious relationships among the genes. If unchecked, recombination and random mutation would therefore tend to destroy the favorable gene groupings on which the success of members of the population rests. Selection, by con-

18.6 Directional, stabilizing, and disruptive selection compared
Each graph indicates the relative abundance of individuals of various heights in a population. In each instance the original condition of the population is shown above, the later condition, after the specified selection, below.   (A) Directional selection acts (arrow) against individuals exhibiting one extreme of a character (here the shortest individuals—represented by the gray area under the upper curve). The eventual result (bottom curve) is that the distribution of heights in the population has shifted to the right; i.e. the population has evolved in the direction of greater height.   (B) Stabilizing selection acts against both extremes; i.e. it culls individuals that deviate too far from the mean condition, thus decreasing diversity and preventing evolution away from the standard condition.   (C) By contrast, disruptive selection acts against individuals in the mid-part of a distribution, thereby favoring both extremes; in our example both the shortest and the tallest individuals would be favored, but individuals of medium height would be selected against. In effect, there are two opposing directional selection pressures: against individuals that are only moderately short and against individuals that are only moderately tall. The result is a tendency for the population to split into two contrasting subpopulations.

18.7   A pair of wood ducks
The showy coloration of the male contrasts with the
drab brown of the female.   [Courtesy D. G. Allen.]

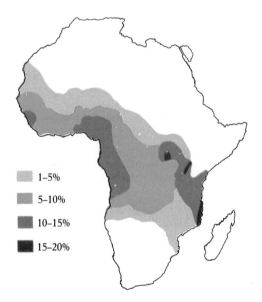

1–5%

5–10%

10–15%

15–20%

18.8   The distribution of sickle-cell anemia in Africa
The various shades of color indicate the percentage of
the population in each area that suffers from the
disease.

stantly acting to eliminate all but the most favorable gene combinations,
counteracts the disrupting, disintegrating tendency of recombination and
mutation and is thus the chief factor maintaining stability where otherwise
there would be chaos; this sort of selection is appropriately called ***stabilizing
selection*** (Fig. 18.6B).

**Effective selection pressure as the algebraic sum of numerous separate
selection pressures**   Probably many characteristics benefit the organisms
that possess them in some ways and harm them in others. The evolutionary
fate of such characteristics depends on whether or not the various positive
selection pressures produced by their advantageous effects outweigh the
negative selection pressures produced by their harmful effects. If the algebraic sum (an addition taking into account plus and minus signs) of all the
separate selection pressures is positive, the trait will increase in frequency,
but if the algebraic sum is negative, the trait will decrease in frequency.

As an example of the determination of a complex character having both
beneficial and deleterious effects, let us consider the selection pressures on
plumage in ducks. Many closely related species of dabbling ducks (Mallard,
Pintail, Gadwall, etc.) occur together in most of North America. Hybridization takes place among them because the females sometimes err in their
selection of mates.[4] Since the hybrids are apparently less viable than the
parental type, there has been strong selection for showy male plumage,
distinctive for each species, that helps reduce the number of mating errors.
No such selection pressure has operated on the females. These brownish
nondescript birds closely resemble one another and are probably much less
easily seen by predators than the males, whose bright plumage doubtless
makes them more subject to predation—a liability that must cause strong
selection against such plumage. But the positive selection resulting from
reduced mating errors is apparently greater than the negative selection

[4]In most species of birds, including the ducks, it is the female that chooses the mate. The male
displays until some female chooses him. Hence it is more important that the female be able to
recognize the males of her own species than that the male be able to recognize the females.

resulting from predation, and the showy plumage of the males has been maintained and even enriched (Fig. 18.7). The situation is different, however, on some isolated islands where only one species of dabbling duck exists and where, therefore, no mating errors can occur. Here the negative selection predominates, and the males have lost their showy plumage and resemble the more protectively colored females.

Just as polygenic traits often have both advantageous and disadvantageous effects, the alleles of a single gene also may have multiple effects (pleiotropy), and it is most unlikely that all of them will be advantageous. Whether an allele increases or decreases in frequency is determined, as in the case of polygenic traits, by whether the sum of the various positive selection presures acting on it is greater or smaller than the sum of the negative selection pressures.

Many instances are known in which the effects of a given allele are more advantageous in the heterozygous than in the homozygous condition. In some parts of Africa, for example, the allele for sickle-cell anemia occurs in human beings much more often than we might expect in view of its highly deleterious effect when homozygous (Fig. 18.8). This is because the allele, when heterozygous, confers on the possessor a partial resistance to malaria. The equilibrium frequency of the sickle-cell allele is thus determined by at least four separate selection pressures: (1) the strongly negative selection pressure on the recessive homozygotes, who suffer the full debilitating effects of sickle-cell anemia; (2) the weaker negative selection pressure on the heterozygotes as a result of their mild anemia; (3) the negative selection pressure on the dominant homozygotes, who are more susceptible to malaria; and (4) the fairly strong positive selection pressure on the heterozygotes as a result of their resistance to malaria.

Balanced polymorphism   Polymorphism is the occurrence in a population of two or more distinct forms, or morphs, of a genetically determined character. For example, Mendel's peas were polymorphic with regard to the color of their flowers; some plants had red flowers, others white. Human populations are polymorphic with regard to blood groups; the same population usually includes type A, type B, type AB, and type O individuals. Polymorphism is common in wild populations too. Thus several species of snails occur in banded and unbanded forms. The red fox has both red and silver-colored morphs. The fish *Xiphophorus maculatus* shows a variety of patterns of spotting, of which three are illustrated in Fig. 18.9. In some instances the genetic basis of the polymorphism is known, as in Mendel's peas and human blood groups. But in many other instances, especially when the traits are polygenic, it is not.

When the relative frequencies of the different morphs in a population are stably balanced over time, we speak of ***balanced polymorphism***. Sometimes the balance is maintained because the polymorphism itself is advantageous. Thus, if a polymorphic species lives in an environment subdivided into many local areas where different conditions prevail, one of the morphs may do better in one area, others may do better in other areas. If the various morphs are all present among an individual's descendants, these can exploit more completely the subdivisions of the variable environment. Sometimes one morph is adaptively superior at one time of year, and another is superior

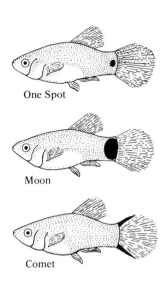

One Spot

Moon

Comet

18.9 Polymorphism in the fish *Xiphophorus maculatus*
Notice the differences in the black spots at the base of the tail.   [Redrawn from M. Gordon, *Advances in Genet.*, vol. 1, 1947.]

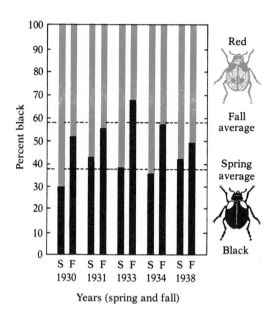

18.10 Polymorphism in the ladybird beetle *Adalia bipunctata*

The frequency of black morphs (black bars) is much higher in the fall than in the spring; conversely, the red morphs (color bars) are much more frequent in the spring than in the fall. Apparently the black morphs are better adapted for life in the summer, and the red morphs are better adapted for life in the winter.   [Modified from B. Wallace and A. M. Srb, *Adaptation*, Prentice-Hall, 1964, after Timoféeff-Ressovsky.]

at another time of year (Fig. 18.10); thus an individual's descendants may have a better chance of survival if they are polymorphic than if all belong to a single form. Polymorphism of this type, accordingly, may be the result of a kind of disruptive selection; each of several contrasting forms is being selected for.

But there are many instances of balanced polymorphism in which some of the contrasting morphs, instead of being themselves selected for, are the unavoidable by-products of selection for a heterozygous phenotype. As we saw with sickle-cell anemia in Africa, the heterozygotes are sometimes better adapted than the homozygotes—a condition known as ***heterozygote superiority*** (or sometimes heterosis or overdominance).

Heterozygote superiority favors balanced polymorphism, because it makes for retention in the population of both alleles of a given gene at frequencies higher than would be predicted from the selection acting on the homozygous phenotypes. Thus, if A/a individuals are adaptively superior to both A/A and a/a individuals, both allele A and allele a will be retained in fairly high frequency in the population; neither allele will be eliminated, as might tend to happen if one of the homozygotes were far superior. Therefore all three possible genotypes—A/A, A/a, and a/a—will occur frequently in each generation, and if each of these produces a noticeably different phenotype the population will be polymorphic. The relative frequencies of A and a, and of the three morphs they produce, will be determined, as in sickle-cell anemia, by the balance between the several selection pressures acting on the system.

ADAPTATION

Every organism is, in a sense, a complex bundle of immense numbers of adaptations. We have already examined a host of adaptations in earlier chapters, adaptations concerned with nutrient procurement, gas exchange, internal transport, regulation of body fluids, hormonal and nervous control, effector activity, reproduction, development, etc. We should pause here to say more explicitly what is meant by adaptation.

An adaptation is any genetically controlled characteristic that increases an organism's fitness. ***Fitness,*** as the term is used in evolutionary biology, is an individual's (or allele's or genotype's) probable genetic contribution to succeeding generations. An adaptation, then, is a characteristic that enhances an organism's chances of perpetuating its genes, which usually means the leaving of descendants. Notice that we did not say adaptations increase the organism's chances of surviving, as is sometimes erroneously stated. While an adaptation, if it is to enhance the leaving of descendants, will ordinarily also enhance prereproductive survival, it will not necessarily enhance postreproductive survival. In many species it is, in fact, adaptive for the adults to die soon after they have completed reproduction.

Adaptations may be structural, physiological, or behavioral. They may be genetically simple or complex. They may involve individual cells or subcellular components, or whole organs or organ systems. They may be highly specific, of benefit only under very limited circumstances, or they may be general, of benefit under many and varied circumstances.

A population may become adapted to changed environmental conditions with extreme rapidity. A good example is provided by a study published in 1937 by W. B. Kemp of the Maryland Agricultural Experiment Station. The owner of a pasture in southern Maryland had seeded the pasture with a mixture of grasses and legumes. Then he divided the pasture into two parts, allowing one to be heavily grazed by cattle, while protecting the other from the livestock and leaving it to produce hay. Three years after this division, Kemp obtained specimens of blue grass, orchard grass, and white clover from each part of the pasture and planted them in an experimental garden where all the plants were exposed to the same environmental conditions. He found that the specimens of all three species from the heavily grazed half of the pasture exhibited dwarf, rambling growth, while specimens of the same three species from the ungrazed half exhibited vigorous, upright growth. In only three years the two populations of each species, known to have been identical initially because one batch of seed was used for the entire pasture, had become markedly different in their genetically determined growth pattern. Apparently the grazing cattle in the one half of the pasture had devoured most of the upright plants, and only plants low enough to be missed had survived and set seed. There had been, in short, intense selection against upright growth in this half of the pasture and correspondingly intense selection for the adaptively superior dwarf, rambling growth. By contrast, in the other half of the pasture, where there was no grazing, upright growth was adaptively superior, and dwarf plants were unable to compete effectively.

Let us look now at some other particularly striking examples of adaptation, which will help clarify the processes by which adaptations come into being.

ADAPTATIONS FOR POLLINATION IN FLOWERING PLANTS

The flowering plants depend on external agents to carry pollen from the male parts in the flowers of one plant to the female parts in the flowers of another plant (Fig. 18.11). The flowers of each species are adapted in shape, structure, color, and odor to the particular pollinating agents on which they depend, and they provide an especially clear illustration of the adaptiveness of evolution. Evolving together, the plants and their pollinators became more finely tuned to each other's peculiarities—a process often termed *coevolution.* We have encountered previous examples of coevolution: the dairying ants and the aphids they tend, and the ruminant mammals and their symbiotic microorganisms.

There is indeed a striking correlation between the pollinators and the species they pollinate. Bees are attracted by bright colors and by sweet, aromatic, or minty odors; they are active only during the day, and they usually alight on a petal before moving into the part of the flower containing the nectar and pollen. Bee flowers have showy, brightly colored petals that are usually blue or yellow but seldom red (bees can see blue or yellow light well, but they cannot see red at all); they usually have a sweet, aromatic, or minty fragrance; they are generally open only during the day; and they often have a special protruding lip on which the bees can land.

18.11 A variety of plant pollinators
(A) A sphinx moth inserting its long
proboscis into a flower to obtain nectar. The
proboscis is likely to be dusted with pollen,
which the moth may carry to the next flower
it visits.   (B) A bird, called the Honeyeater,
probing its bill into a flower as pollen adheres
to its face.   (C) A honey possum feeding on
nectar from flowers.   (D) A bat
(*Leptonycteris*) at a flower, its face liberally
dusted with pollen.   [Courtesy D. Cavagnaro
(A); M. Morcombe (B, C); D. J. Howell,
Purdue University (D).]

A   B   C   D

Hummingbirds, on the other hand, can see red well but blue only poorly; they have a weak sense of smell; and they ordinarily do not land on flowers, but hover in front of them while sucking the nectar. Flowers pollinated primarily by hummingbirds are usually red or yellow, are nearly odorless, and lack any protruding landing platform.

In contrast to both bees and hummingbirds, moths are most active at dusk and during the night, and the flowers they pollinate are mostly white and are open only during the late afternoon and night. These flowers often have a very heavy fragrance that helps guide the moths to them.

The bases of the petals of flowers pollinated by bees, birds, or moths are often fused to form a tube whose length corresponds closely to the length of the tongue or bill of the particular species most important as the pollinator of that plant (Fig. 18.12).

Unlike bees and moths, the short-tongued flies (which feed primarily on carrion, dung, humus, sap, and blood) are attracted by rank rather than sweet odors, and they rely very little on vision in locating food. The flowers of plants that depend on these flies for pollination are usually dull-colored and very ill-smelling. These flowers are sometimes shaped in such a way that they temporarily entrap the flies that enter them; thus they ensure that the flies become covered with pollen before they escape and fly to another flower. Trapping mechanisms are also common in flowers pollinated by beetles.

A particularly dramatic example of adaptation for pollination is seen in some species of orchids, where the flowers resemble in both shape and color The females of certain species of wasps or bees (Fig. 18.13). The male insect is stimulated to attempt to copulate with the flower and becomes covered with pollen in the process. When he later attempts to copulate with another flower, some of the pollen from the first flower is deposited on the second. So

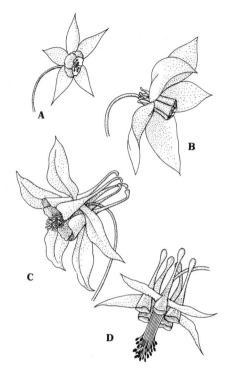

18.12   Characters of columbine flowers correlated with their pollinators

(A) *Aquilegia ecalcarata*, pollinated by bees.   (B) *A. nivalis*, pollinated by long-tongued bees.   (C) *A. vulgaris*, pollinated by long-tongued bumblebees. (D) *A. formosa*, pollinated by hummingbirds.   The length and curvature of the nectar tubes of the flowers are correlated with the length and curvature of the bees' tongues and the hummingbirds' bills.   [Adapted from V. Grant, "The fertilization of flowers," *Sci. Am.*, June 1951. Copyright © 1951 by Scientific American, Inc. All rights reserved.]

18.13   An orchid flower that resembles a fly
The flowers of this species *(Trichroceros antennifera)* look enough like female flies to attract some male flies to land on them. The males thus become dusted with pollen, which they may carry to other flowers.   [J. H. Carmichael, Jr., Bruce Coleman Inc.]

18.14   A bombardier beetle *(Brachinus)* spraying its defensive secretion
Note how accurately the beetle can aim its spray at the offending object, in this case forceps tweaking a rear leg (left) or a front leg (right).   [Courtesy Daniel Aneshansley and Thomas Eisner, Cornell University.]

complete is the deception that sperm have actually been found inside the orchid flowers after a visit by the male insect.

Flowers pollinated by wind rather than animals characteristically lack bright colors, special odors, and nectar. In fact, most of them have no petals, and their sexual parts are freely exposed to the air currents. The pollen grains produced by these flowers are particularly small and light, and it is not unusual for them to be blown hundreds of miles.

We see, then, that the characteristics of flowers are not simply pleasing curiosities of nature that serve no practical function. They are important adaptations evolved in response to fundamental selection pressures. Our aesthetic enjoyment should not make us oblivious to the essential role they play in the survival of the plant species.

DEFENSIVE ADAPTATIONS OF ANIMALS

Defensive secretions of arthropods   Many arthropods possess glands whose secretions act as repellents against predators. In some species the secretions are merely released as a liquid ooze when the animal is disturbed, while in others the secretion is forcibly expelled as a spray that may be aimed very precisely toward the source of the disturbance. The secretions are usually odorous and irritating, particularly if they hit a sensitive part of the predator, such as the mouth, nose, or eyes. For example, the bombardier beetle *Brachinus* sprays a noxious compound from the tip of its abdomen, which it can aim precisely enough to hit a single ant attacking one of its legs (Fig. 18.14). Ants hit by the spray are instantly repelled. The spray is also effective against some vertebrate predators.

In some cases defensive chemicals are not actually synthesized by the arthropod that uses them, but are obtained from the plants on which the arthropods feed. A particularly striking example is shown in Fig. 18.15. The larvae of the sawfly *Pseudoperga guerini* feed on *Eucalyptus* leaves, from which they obtain repellent chemicals that they sequester in their crops.

It must be emphasized that arthropods with defensive secretions are sometimes injured or even killed before a predator is repelled. Such events, however, do not indicate that the defensive secretion is an adaptive failure. A defensive mechanism need not be one hundred percent effective to give the individuals that possess it a significantly greater chance to survive to reproduce. In this case the insects with defensive secretions have a far better chance of being dropped by predators before being seriously injured than if they had no secretion. Moreover, some predators may learn from one or two unpleasant experiences to avoid the prey in question on sight alone.

Cryptic appearance   Many animals blend into their surroundings so well as to be nearly undetectable. Frequently their color matches the background almost perfectly (Fig. 18.16). In some cases the animals even have the ability to alter the condition of their own pigment cells and thus change their appearance to harmonize with their background (Figs. 18.17, 18.18). Often, rather than match the color of the general background, the animals may resemble inanimate objects commonly found in their habitat, such as leaves (Fig. 18.19), twigs or sticks (Fig. 18.20), shells (Fig. 18.21), or even bird droppings (Fig. 18.22).

Careful studies have confirmed that, as had been assumed, cryptic appearance is an adaptive characteristic that helps animals escape predation. One such study was conducted by F. B. Sumner of the Scripps Institution of Oceanography in California. Sumner investigated predation by Galápagos penguins on mosquito fish (*Gambusia partuelis*), which can contract or expand their pigment cells to become lighter or darker, depending on their background. He established that the penguins caught 70 percent of the fish that contrasted with their background but only 34 percent of the fish that resembled their background. Sumner also exposed mosquito fish to large sunfish and found that the sunfish captured 53 percent of the contrastingly colored prey, but only 25 percent of the cryptically colored prey. In a similar

18.15   Defensive behavior of the sawfly *Pseudoperga guerini* .
Left: The mother sawfly is protecting her very young larvae, which have not yet developed their own chemical defenses. She would not fly away, no matter how much the photographer disturbed her.   Right: After her death, the larvae can protect themselves by regurgitating a repellent substance (yellow droplets), which they have obtained from the *Eucalyptus* leaves on which they feed. In the circle they form, only the individuals on the periphery, all of which face outward, release the repellent when disturbed.   [Courtesy Thomas Eisner, Cornell University.]

18.16 Cryptic coloration of animals living in dense growths of the brown alga *Sargassum* off Bermuda
Left: The sargassum crab is the same color as the alga, and its rounded body resembles the floats of the alga.   Right: The sargassum angler fish looks so much like the alga that even when one knows it's there one has difficulty locating it.   [Oxford Scientific Films, Bruce Coleman Inc.; and Oxford Scientific Films.]

18.17 Color change by the frog *Hyla versicolor*
The brownish individual has been on the tree trunk for some time and matches it well. The green one has just been moved there from a pond, where it was among green duckweed; it has not had time to change color.   [Courtesy V. N. Rockcastle, Cornell University.]

18.18 Flounders on two different backgrounds
The fish can change color to match the background, whether it is light-colored (top) or dark (bottom).   [Allan Power (top) and Jane Burton (bottom), Bruce Coleman Inc.]

18.19 A leaflike mantis
The mantis (at top in picture) looks strikingly like the green leaves below it. Even the relationship of its thorax and abdomen reflects the way the leaves often occur in pairs. Compare Fig. 12.48, p. 533.   [Peter Ward, Bruce Coleman Inc.]

18.20 A moth that looks like a broken twig
The perfection of the cryptic appearance of this moth *(Phalera bucephala)* is a triumph of evolutionary adaptation.   [Jane Burton, Bruce Coleman Inc.]

18.21 A nudibranch mollusc that strikingly resembles a shell
The nudibranch, which does not itself have a shell, is at lower right; the shell is at upper left.   [Rod Borland, Bruce Coleman Inc.]

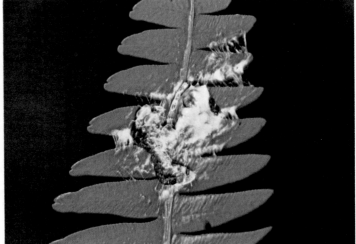

18.22 A bird-dropping spider from Borneo
The spider and the silk it spreads around itself strongly resemble the droppings of a bird.   [M. P. L. Fogden, Bruce Coleman Inc.]

experiment F. B. Isely of Trinity University in San Antonio, Texas, studied predation by chickens, turkeys, and native birds on grasshoppers of various colors on differently colored backgrounds. He found that 88 percent of the nonprotected grasshoppers were eaten whereas only 40 percent of the cryptically colored ones were eaten.

One of the most extensively studied cases of cryptic coloration is the so-called industrial melanism of moths. Since the mid-1880s many species of moths have become decidedly darker in industrial regions. This is actually a case of polymorphism where the less frequent of two forms has become the more frequent, and vice versa; the originally predominant light form in these species of moths has given way, in industrial areas, to the dark (melanic) form. In the Manchester area, in England, the first black specimens of the species *Biston betularia* were caught in 1848; by 1895 melanics constituted about 98 percent of the total population in the area. It has been calculated that for such a remarkable shift in frequency to have occurred in so short a time the melanic form must have had at least a 30 percent advantage over the light form.

In 1937 E. B. Ford of Oxford University proposed the following explanation for this striking evolutionary change. The various species of moths exhibiting the rapid shift to melanism, though unrelated to one another, all habitually rest during the day in an exposed position on tree trunks or rocks, being protected from predation only by their close resemblance to their background. In former years the tree trunks and rocks were rather light-colored and often covered with light-colored lichens. Against this background the light forms of the moths were astonishingly difficult to see, whereas the melanic forms were quite conspicuous (Fig. 18.23). Under these conditions it seems likely that predators such as birds captured melanics far more easily than the cryptically colored light moths. The light forms would thus have been strongly favored, and they would have occurred in much higher frequency than melanics. But with the advent of extensive industrialization, tree trunks and rocks were blackened by soot, and the lichens, which are particularly sensitive to such pollution, disappeared. In this altered environment the melanic moths resembled the background more closely than the light moths. Thus selection would have been reversed and would now have favored the melanics, which would consequently have increased in frequency.

You will note that Ford's hypothesis stresses the role of differential predation by birds on the two color forms of the moths. But this idea ran into opposition from both entomologists and ornithologists. Both insect and bird

18.23 Cryptic coloration of peppered moths
Top: Light and dark morphs of *Biston betularia* at rest on a lichen-covered tree trunk in unpolluted countryside. The light moth is very difficult to see (it is slightly below and to the right of the dark moth). Bottom: Light and dark morphs on a soot-covered tree trunk near Birmingham, England. Here the light form is the easier to see. [From the experiments of Dr. H. B. D. Kettlewell, Oxford University.]

specialists agreed that birds capture very few resting moths and are un-selective as to the color of the few they do capture. But though they agreed that Ford's hypothesis could not be correct, they could present no reliable data to support their views. They were arguing on the basis of casual un-controlled observations, made, it's true, by many persons over a period of many years.

Finally, in the mid-1950s, H. B. D. Kettlewell of Oxford University decided to perform a series of carefully planned field experiments designed to settle the matter one way or the other. He released approximately equal numbers of the light and melanic forms of *Biston betularia* onto trees in a rural area in the county of Dorset, England, where the tree trunks were light-colored, lichens were abundant, leaf washings revealed very little pollution, and the wild population of the moth was about 94.6 percent light-colored. A direct watch on the resting moths was maintained, with the help of binoculars, from blinds where the observer could not be seen by potential bird preda-tors. It was found that several species of insectivorous birds do prey by sight on the moths and that, of 190 moths observed to be captured by birds, 164 were melanics and only 26 were light forms. Furthermore, of approximately 500 marked individuals of each color form released in another experiment, roughly twice as many light moths as melanic moths were recaptured in traps set up in the Dorset woods, an indication that more of the light moths had survived. It was plain that predation was an important selective factor among the moths, and that melanics were more subject to predation under the conditions prevailing in the Dorset woods than light-colored moths.

Taken alone, however, these experiments did not prove conclusively that the factor favoring the light moths over the melanics was their resemblance to their background. The results could be explained by assuming, for exam-ple, that the birds preferred the melanics because of some difference in flavor. Therefore Kettlewell duplicated the experiments under the reverse environmental conditions—in woods near Birmingham, England, where the tree trunks were blackened with soot, lichens were absent, leaf washings revealed heavy pollution, and the wild population of the moth was about 85 percent melanic. The results of these experiments were the reverse of those in the Dorset experiments. Now birds were observed to capture nearly three times as many light moths as melanics, and roughly twice as many melanics were recaptured in traps. Here also predation was an important selective factor, but here it was the melanics that had the adaptive advantage. These experiments by Kettlewell prove that birds do hunt by sight, and that those moths that most closely resemble the background on which they rest have much the best chance of escaping predation.

Warning coloration   Whereas some animals have evolved cryptic color-ation, others have evolved colors and patterns that contrast boldly with their background and thus render them clearly visible to potential predators (Fig. 18.24). Many of these animals are in some way disagreeable to predators; they may taste bad, or smell bad, or sting, or secrete poisonous substances. In other words, they are animals that a predator will usually reject after one or two unpleasant encounters with them. Such animals benefit by being gaudily colored and conspicuous because predators that have experienced their unpleasant features learn to recognize and avoid them more easily in

18.24 Aposematic coloration
The bright color of the poison-arrow frog (from which South American Indians obtain poison for their arrow tips) makes it easily recognizable by predators, which carefully avoid it.   [Courtesy Michael Hopiak, Cornell Laboratory of Ornithology.]

the future. Their flashy appearance is protective because it warns potential predators that they should stay away; such a warning appearance is said to be *aposematic.*

The warning is sometimes so effective that, after unpleasant experiences with one or two warningly colored insects, some vertebrate predators simply avoid all flashily colored insects, whether or not they resemble the ones they encountered earlier. G. D. H. Carpenter demonstrated this by offering over 200 different species of insects to an insectivorous monkey. The monkey accepted 83 percent of the cryptically colored insects but only 16 percent of the warningly colored ones, even though many of the insects belonged to species the monkey had probably not previously encountered. It is possible that such avoidance of aposematic insects does not depend entirely on learning. Predators that have a genetic predisposition to avoid brightly colored prey would, one supposes, have an adaptive advantage over predators that waste time and energy pursuing inedible prey; hence there may be a tendency for predators to evolve an avoidance response to warningly colored prey.

Mimicry Species not naturally protected by some unpleasant character of their own may closely resemble (mimic) in appearance and behavior some warningly colored unpalatable species. Such a resemblance is adaptive; the mimic species suffers little predation because predators cannot distinguish it from its unpleasant models. This phenomenon is called **Batesian mimicry.**

Convincing evidence for the effectiveness of this type of mimicry in protecting the mimic species comes from the elegant experiments of Jane van Z. Brower, then at Oxford University. In one experiment she fed specimens of a butterfly species called the Viceroy (*Limenitis archippus*) to a group of caged jays, which accepted the Viceroys readily. Then she offered the jays specimens of a distasteful butterfly called the Monarch (*Danaus plexippus*), which the Viceroy mimics (Fig. 18.25). After a few trials, the jays refused to eat the Monarchs. When they were again offered Viceroys, the jays now also refused to eat these, even though they had earlier eaten them readily. A few unpleasant encounters with the Monarchs (Fig. 18.26) had caused the jays to reject their mimics, the Viceroys.

In another series of experiments Brower produced an artificial model-mimic system with starlings as the predators and mealworms, which starlings ordinarily eat voraciously, as prey. The first step was to paint a green band on some of the mealworms, an orange one on the others. Some of the green-banded worms were then dipped into a solution of a chemical very distasteful to starlings; these worms were used as the "models." The rest of the green-banded worms were dipped only into distilled water and therefore remained palatable; these worms were the "mimics." None of the orange-banded worms were made distasteful. Green was chosen as the "warning color" for the distasteful worms because in nature this color, usually associated with cryptically colored rather than with warningly colored species, was not likely to have been associated by the birds with an unpleasant experience. It was demonstrated that the paint itself was not distasteful to the birds.

Each starling was given 10 trials per day for about 16 days, each trial consisting of two mealworms—one orange-banded and one green-banded.

18.25 An example of Batesian mimicry
The Monarch butterfly (top) is a distasteful species. The Viceroy (bottom) mimics the Monarch; species in the group to which the Viceroy belongs ordinarily have a quite different appearance.   [D. Overcash (top) and J. Shaw (bottom), Bruce Coleman Inc.]

Of the green-banded worms, different groups of birds received a different proportion of "mimics" (palatable) to "models" (unpalatable because of having been dipped into the distasteful chemical); the proportion of mimics for the different groups was, respectively, 10, 30, 60, and 90 percent for the 16 days of the experiment.

After a few unpleasant encounters with the models, the birds learned to recognize and avoid green-banded worms as distasteful, with the result that the mimics among them also escaped predation, particularly when the percentage of mimics presented to the starlings was 60 percent or less. Even when the percentage of mimics was as high as 90 percent and the percentage of models only 10 percent, 17 percent of the mimics escaped predation. With one exception, all the starlings readily ate the orange-banded worms throughout the experiments—a demonstration that the birds had not simply learned to avoid all mealworms and that their avoidance of the palatable green-banded ones (mimics) was due to the resemblance of these worms to the unpalatable green-banded ones (the models).

Brower's original experimental proof of the adaptiveness of mimicry, published in 1960, stimulated a vigorous new quantitative approach to this intriguing subject. And her experiments made it necessary to revise some long-held ideas about mimicry. For example, it was generally believed that for Batesian mimicry to be effective the model species must be more common than the mimic species, i.e. that in any given trial a predator must be more likely to get an unpalatable than a palatable mouthful. Otherwise, it was thought, the predator would not readily learn to associate the appearance of the prey with an unpleasant rather than a pleasant experience.

Brower's experiments showed, however, that, if the model is distasteful enough, mimicry is still very effective when the ratio of mimics to models is as high as 60 to 40 and that mimics benefit to some degree even when the

18.26 Effect on a Blue Jay of eating a Monarch butterfly
A jay with no previous experience of the distasteful chemical (a powerful cardiac glycoside) in Monarchs tears a Monarch apart, swallows it, begins looking sick, and finally vomits. After a few such experiences jays will avoid Monarchs, and will also avoid Viceroy butterflies, which, though not themselves distasteful, mimic the appearance of Monarchs.   [Courtesy L. P. Brower, Amherst College.]

18.27 A fish with a fake eyespot near the tail
This fish *(Chaetodon plebeius)* from New Caledonia has a prominent fake eyespot; its real eye is much less conspicuous, because of the stripe that runs through it. [Courtesy Douglas Faulkner.]

ratio is 90 to 10. In other words, the proportion of mimics to models does not have a set maximum value beyond which the effectiveness of the mimicry starts breaking down; the value of mimicry in any given case depends on how distasteful the model is, how tasty the mimic is, and how many alternative sources of food are available to the predators. That just a few unpleasant specimens among many outwardly similar pleasant ones should be able to discredit the whole lot is understandable enough. If in a particular bag of candy, all pieces of which looked alike, one out of ten pieces had an extremely unpleasant taste, we should probably quickly learn to avoid taking any candy from that bag on the principle that the risk of so disagreeable an experience is not worth taking.

So far we have discussed only Batesian mimicry—mimicry of a distasteful species by one not distasteful, i.e. mimicry based on deception. A second kind of mimicry, called **_Müllerian mimicry,_** involves the evolution of a similar appearance by two or more distasteful species. In this type of mimicry each species is both model and mimic. Each species has some defensive mechanism, but if each had its own characteristic appearance, the predators would have to learn to avoid each species separately; the learning process would thus be a more demanding one, and some individuals of each prey species would have to be sacrificed to the learning process. If, however, several protected species evolve more and more toward one appearance type, they come to constitute a single prey group from the standpoint of the predators, which accordingly learn avoidance more easily. If avoidance involves any genetic predisposition, then resemblances among the prey would facilitate more rapid selection for improved prey-recognition mechanisms in the predators.

Ambiguity of body orientation Cryptic appearance and mimicry function in deceiving potential predators; so do a variety of appearances that make it difficult for the predator to tell which is the front end of the prey animal. For example, the real eye is often obscured by the pattern of coloration on the animal's head, but a prominent fake eyespot is located near the animal's tail (Fig. 18.27). Or the tail looks more like a head than the head itself (Fig. 18.28). The result is that the predator often aims its attack incorrectly, expecting the prey to move in one direction while it is in fact escaping in the opposite direction.

SYMBIOTIC ADAPTATIONS

The term "symbiosis" is used in a variety of ways in the biological literature. Some authors apply it only to cases where two species live together to their mutual benefit. We shall use the word in a broader sense.

Etymologically, symbiosis simply means "living together," without any implied value judgments. This is the meaning it was given when it was first introduced into biology, and this is the meaning it will have in this book. We shall, however, recognize three categories of symbiosis. The first is **commensalism,** a relationship in which <u>one</u> species <u>benefits</u> while the other receives <u>neither benefit nor harm</u>. The second is **mutualism,** where <u>both</u> species benefit. The third is **parasitism,** where one species benefits and individuals of the other species are harmed. We can summarize the distinctions as follows (a plus sign means benefit, a minus sign harm, and a zero no significant benefit or harm):

Relationship	Species A	Species B
Commensalism	+	0
Mutualism	+	+
Parasitism	+	−

Commensalism   The advantage derived by the commensal species from its association with the host often involves shelter, support, transport, or food, or several of these. For example, in tropical forests numerous small plants, called epiphytes, usually grow on the branches of the larger trees or in forks of their trunks (see Fig. 19.42, p. 871). These commensals, among which species of orchids and bromeliads are prominent, are not parasites. They use the host trees only as a base of attachment and do not obtain nourishment from them. They apparently do no harm to the host except when so many of them are on one tree that they stunt its growth or cause limbs to break. A similar type of commensalism is use of trees as nesting places by birds.

Sometimes it is difficult to tell what benefit is involved in a commensal relationship. For example, certain species of barnacles occur nowhere except attached to the backs of whales, and other species of barnacles occur nowhere except attached to the barnacles that are attached to whales. Just what advantages either of these groups of barnacles enjoys is not clear. They do, of course, get a relatively unoccupied base for attachment, and they get transport that increases the dispersal of their offspring. But it is hard to see how these benefits alone would have sufficed for the evolution of such specificity.

In some cases of commensalism, however, the benefit is dramatically obvious. For example, certain species of fish regularly live in association with sea anemones, deriving protection and shelter from them and sometimes stealing some of their food (Fig. 18.29). These fish swim freely among

18.28 Ambiguity of body orientation in the leaf-tailed gecko lizard
Because the expanded tail looks much like a head, a predator finds it hard to tell in which direction the gecko will run.   [Courtesy Australian News and Information Bureau.]

18.29   Two anemone fish living among the tentacles of a sea anemone from the Palau Islands
[Courtesy Douglas Faulkner.]

the tentacles of the anemones, even though those tentacles quickly paralyze other fishes that touch them. The anemones regularly feed on fish; yet the particular species that live as commensals with them sometimes actually enter the gastrovascular cavity of their host, emerging later with no apparent ill effects. The physiological and behavioral adaptations that make such a commensal relationship possible must be quite extensive. Another striking example is a small tropical fish that lives in the respiratory tree of a particular species of sea cucumber. The fish emerges at night to feed and then returns to its curious abode by first poking its host's rectal opening with its snout and then quickly turning so that it is drawn tail first through the rectal chamber into the respiratory tree. Still another example is a tiny crab that lives in the mantle cavity of oysters. The crab enters the cavity as a larva and eventually grows too big to escape through the narrow opening between the two valves of the oyster's shell. It is thus a prisoner of its host, but a well-sheltered prisoner. It steals a few particles of food from the oyster, but apparently does it no significant harm.

Mutualism   Symbiotic relationships beneficial to both species are common. Figure 18.30 illustrates two instances of the widespread phenomenon called cleaning symbiosis, which is patently mutualistic. We mentioned several other examples of mutualism in earlier chapters: the relationship between a termite or a cow and the cellulose-digesting microorganisms in its digestive tract, or between a human being and the bacteria in his intestine that synthesize vitamin B_{12}. The plants we call lichens are actually formed of an alga and a fungus united in such close mutualistic symbiosis that they give the appearance of being one plant (see Fig. 24.8, p. 997). Apparently the fungus benefits from the photosynthetic activity of the alga, and the alga benefits from the water-retaining properties of the fungal walls.

As is apparent from this discussion of commensalism and mutualism (particularly the former), the division of symbiosis into three subcategories is in many ways arbitrary. Commensalism, mutualism, and parasitism are all parts of a continuous spectrum of possible interactions. We called the relationship between epiphytes and their hosts commensalism, but said that sometimes the host is harmed. In such cases, should we call the relationship parasitism? We spoke of commensalistic interactions between certain fish and sea anemones, and pointed out that both the fish and the anemones had probably evolved special physiological and behavioral adaptations making their relationship possible. Would the anemones have evolved such adaptations unless the relationship was beneficial to them—a mutualistic rather than a commensalistic relationship? We saw that oyster crabs steal food from their host. Can we be certain that this is not significantly detrimental to the host? By what measure do we decide how much harm is significant and how much is not? Should we call this relationship parasitism? We labeled the association of algae and fungi to form lichens mutualism. But is it certain that the algae derive significant benefit from the association? After all, we know that the algal species in lichens are perfectly capable of free living whereas the fungal species usually are not. Might we have been too ready to say that the algae are shielded from desiccation by the fungi? Should we classify lichens as commensalistic associations, or even as parasitic ones?

It really isn't very important which category we apply to most of these cases. The categories are only devices to help us organize what we know about nature. What is important is to keep in mind how commensalism, mutualism, and parasitism grade into each other, and to recognize that each case of symbiosis is different from all others and must be studied and analyzed on its own merits.

Parasitism   Just as there are no sharp boundaries between parasitism and commensalism, or even between parasitism and mutualism, there is no strict delimitation between parasitism and predation. Mosquitoes and lice both suck the blood of mammals; yet we usually call only the latter parasites. Foxes and tapeworms may both attack rabbits, but foxes are called predators and tapeworms are called parasites. The usual distinction is that a predator eats its prey quickly and then goes on its way, while a parasite passes much of its life on or in the body of a living host, from which it derives food in a manner harmful to the host (Fig. 18.31). Obviously this is not always a clear distinction. How long must one organism live on the body of another to be classed as a parasite? But though there will always be intermediate cases, it is profitable to distinguish between predation and parasitism, because each of these is a mode of existence followed by many kinds of organisms and each involves its own characteristic sorts of adaptations.

Parasites are customarily divided into two types: external parasites and internal parasites. The former live on the outer surface of their host, usually either feeding on the hair, feathers, scales, or skin of the host or sucking its blood. Internal parasites may live in the various tubes and ducts of the host's body, e.g. the digestive tract or respiratory passages or urinary ducts; or they may bore into and live embedded in tissues such as muscle or liver; or, in the case of viruses and some bacteria and protozoans, they may actually live inside the individual cells of their host.

18.30 Cleaning symbiosis
Top: A wrasse picks food particles from the teeth of a squirrelfish near a reef off New Caledonia.   Bottom: Two oxpeckers search for parasitic insects on an impala, in Africa.   In both cases the symbiosis is mutualistic: The cleaner obtains food, and the cleanee gets rid of matter that could endanger its health.   [Top: Douglas Faulkner. Bottom: L. L. Rue III, Bruce Coleman Inc.]

18.31 A caterpillar (tomato worm) with numerous pupal cases of a parasitic wasp attached to its body When the wasp larvae emerge they will feed on the host caterpillar.   [R. P. Carr, Bruce Coleman Inc.]

Internal parasitism is usually marked by much more extreme specializations than external parasitism. The habitats available inside the body of another living organism are completely unlike those outside, and the unusual problems they pose have resulted in evolutionary adaptations entirely different from those seen in free-living forms. For example, internal parasites have often lost organs or whole organ systems that would be essential in a free-living species. Tapeworms, for instance, have no digestive system. They live in their host's intestine, where they are bathed by the products of the host's digestion, which they can absorb directly across their body wall without having to carry out any digestion themselves.

Because of their frequent evolutionary loss of structures, internal parasites are often said to be degenerate. "Degenerate," of course, implies no value judgment, but simply refers to the lack, common in parasites, of many structures present in their free-living ancestors. From an evolutionary point of view, loss of structures useless in a new environment is an instance of positive adaptation. Such loss is just as much an evolutionary advance, a specialization, as the development of increased complexity in some other environment.

Specialization does not necessarily mean increased structural complexity; it only means the evolution of characteristics particularly suited to some special situation or way of life. In internal parasites—or cave animals, which frequently lack eyes—the development and maintenance of structures that no longer serve a useful function would require energy that the organism might use to more advantage in some other way. And some useless structures, such as eyes in both internal parasites and cave animals, might well be a handicap in these special environments, because they would be a likely point of infection. It is readily understandable, therefore, that natural selection might favor those individuals in which such useless organs are either relatively small or lacking entirely. The concept of evolutionary loss of unused structures as a result of differential survival and reproduction (natural selection) of different genotypes should not be confused with the Lamarckian idea that use and disuse can directly influence the size of the structures in an organism's progeny.

Structural degeneracy is far from being the only sort of special adaptation commonly seen in internal parasites. They often have body walls highly resistant to the destructive enzymes and antibodies of the host. Tapeworms, for example, are constantly bathed by the potent digestive juices of their host; yet their enzyme-resistant cuticle protects them from being digested. And tapeworms have very specialized heads, with hooks and suckers, that enable them to anchor themselves and avoid being expelled by the often vigorous peristaltic contractions that move the other contents of the host's intestine (see Fig. 25.26, p. 1014).

Perhaps the most striking of all adaptations of internal parasites concern their life histories and reproduction. Individual hosts don't live forever. If the parasitic species is to be perpetuated, therefore, a mechanism is needed for changing hosts. At some point in their life cycle, then, all internal parasites move from one host individual to another. But this is seldom simple. Rarely can a parasite move directly from one host to another of the same species. For example, consider the life cycle of a beef tapeworm.

The eggs of a beef tapeworm living in a man's intestine are shed in the host's feces. A cow eats plants contaminated with human feces, and the tapeworm eggs hatch in the cow's intestine. The young larvae bore through the wall of the cow's intestine, enter a blood vessel, and are carried by the blood to a muscle, where they encyst (become surrounded by a bladderlike case and lie inactive). If a man then eats the raw or insufficiently cooked beef, the tapeworm larvae become activated in his intestine; the head of the larva attaches to the intestinal wall, and a mature worm develops. The beef tapeworm thus passes through two hosts during its life cycle: an intermediate host (cow), in which it undergoes some of its early development, and a final host (man), in which it matures.

As life cycles of internal parasites go, the one just described is rather simple. It is not unusual for a life cycle to include two or three intermediate hosts and/or a free-living larval stage. But such a complex development makes the chances that any one larval parasite will encounter the right hosts in the right sequence exceedingly poor. The vast majority die without completing their life cycle. It is therefore understandable that internal parasites should characteristically produce huge numbers of eggs. Although they may be structurally degenerate in other ways, they usually have extremely well-developed reproductive structures. In fact, some internal parasites seem to be little more than a sac of highly efficient reproductive organs.

In the course of their evolution, parasites usually develop special features of behavior and physiology that make them better adjusted to the particular characteristics of their host and more efficient at competing with other parasites. Thus they often tend to become more and more host-specific. Where an ancestral organism may have parasitized all species in a particular family, each of its various descendants may parasitize only one species of host at each stage in its development. This evolutionary trend toward specificity is manifest in another way too. Parasites often tend to become more specific also with regard to the part of the host's body they can inhabit. Some external parasites may live only on the head of their host, others only on the back, and still others only on the legs. Internal parasites inhabiting the digestive tract of their host are often restricted to one small section of that tract. G. A. School of McGill University examined the distribution of eight species of nematode worms in the intestines of the European tortoise. He found that the different species of worms differed in their linear distribution within the intestine (some being most common near the anterior end and others in the middle third of the intestine), and that they also differed in their radial distribution (some being more abundant near the mucosa and others being distributed more generally throughout the lumen).

It must be remembered that as the parasites evolve the host species is also evolving, and that there is strong selection pressure for its evolution of more effective defenses against the ravages of its parasites. There is thus a constant interplay between host and parasite. As the one evolves better defenses, the other evolves ways of counteracting them—these counteractions leading to pressure on the host to evolve still better defenses, against which the parasite may then evolve new means of surviving, and so forth. Although this sort of coevolution will continue as long as the host-parasite relationship exists, a dynamic balance is usually reached eventually, the host surviving without being seriously damaged and the parasite prospering moderately well too. It is, in fact, decidedly disadvantageous to the parasite to kill its host. We have said that many parasites are host-specific. If they should cause the extinction of their host, then they themselves would also become extinct. Therefore the balance eventually reached is not due solely to the evolution of better defenses by the host. It is also due to the parasite's evolving in such a way that it becomes better adjusted to its host and causes less serious disturbance.

Probably most long-established host-parasite relationships are balanced ones. Relationships that result in serious disease in the host are usually relatively new, or they are relationships in which a new and more virulent form of the parasite has recently arisen, or in which the host showing the serious disease symptoms is not the main host of the parasite. American Indians, for example, suffered severely when exposed to pathogens first brought to North America by European colonists, even though some of those same pathogens caused only mild symptoms in the Europeans, who had been exposed to the pathogens for many centuries and in whom the host-parasite relationship had nearly reached a balance. Many examples are known where humans are only occasional hosts for a particular parasite and suffer severe disease symptoms, although the wild animal that is the major host shows few ill effects from its relationship with the same parasite.

ALTRUISTIC BEHAVIOR

In our discussion of social behavior in Chapter 12 we mentioned that many highly social animals regularly exhibit what has been called altruistic behavior—behavior that may reduce the personal reproductive success of the individual exhibiting it while increasing the reproductive success of others. But how would such behavior evolve? It would seem, on its face, to be maladaptive; i.e. rather than enhance the altruist's chances of perpetuating its genes it would seem to reduce those chances. In other words, individuals whose genetic makeup predisposes them to altruistic behavior should have less reproductive success, it would seem, than individuals with genes for more selfish behavior, and the genes for altruism, therefore, should eventually disappear from the population. Explaining the evolution of altruistic behavior has been one of the major challenges of sociobiology.

The altruistic behavior of a parent toward its offspring (e.g. in defending them against predators) is a special case. When a parent endangers its own life and its chances of producing more offspring in the future, it may not, in fact, be reducing the likelihood of its personal reproductive success at all. When an animal already has a large parental investment in its young, its best reproductive strategy may be to risk defending those young rather than abandon them and gamble on being able to replace them in the future.[5] In other words, such apparently altruistic behavior, rather than being maladaptive, may actually increase the altruist's individual fitness.

The matter becomes more complicated when the beneficiaries of the altruistic actions are not the altruist's own progeny. Yet many examples of such actions among social animals have been documented. Among monkeys and apes, childless adults often care for the infants of others for short periods; they may hold the young, groom them, or play with them. African wild dogs sometimes bring fresh meat to the den and give it to other adults that remained with the cubs. In several species of birds (e.g. the Florida Scrub Jay), breeding pairs are regularly assisted by helpers, which are other birds (usually rather young) that are not mated and have no territories of their own. How can the evolution of this kind of altruistic behavior be explained?

In recent years much attention has focused on what is called **kin selection** as an explanation for the evolution of many kinds of altruistic behavior. This explanation turns on an evaluation of the fitness of an individual in terms of its success in perpetuating its genes. Now, in sexually reproducing populations an individual shares genes not only with its offspring but also with other members of its kinship group, i.e. with its siblings, its nieces and nephews, its cousins, etc. The individual therefore has some degree of evolutionary investment in all its genetic relatives. It follows that altruistic behavior that benefits the altruist's kin may enhance the altruist's own **inclusive fitness,** which is the sum of the individual's personal reproductive success and the success of its relatives devalued in proportion to their genetic distance. Inclusive fitness is thus a measure of the individual's total genetic representation in the gene pools of succeeding generations.

Let us consider for a moment the genetic distance of an individual from a few of its closer relatives. In a sexually reproducing population an individual shares 50 percent of its genes with each of its own offspring, an average of 25 percent of its genes with its nieces or nephews, and an average of 12.5 percent of its genes with its cousins. In other words, if an individual with no offspring is represented in the next generation only by three nieces and nephews, and another individual is represented by one offspring but no other relatives, then the individual with no offspring actually has the greater genetic representation. Thus, there may be positive selection of genes for altruism toward relatives if the probable cost to the altruist is low, the benefit to the relative high, and the coefficient of relationship between the individuals (i.e. the percentage of shared genes) high. It is not surprising, therefore, that the majority of documented instances of altruism occur

---

[5] It need hardly be stated that such terms as "strategy," "risk taking," "gambling," etc., habitually used in evolutionary contexts, do not imply any conscious weighing of alternatives by the animals. The animals are not concerned with ways of maximizing their genetic contribution to future generations! The terms merely reflect the probabilistic aspects of the selection pressures leading to the evolution of certain traits.

within social groupings where all the members are likely to be rather closely related. The extreme example, of course, is the altruism of the nonreproductive workers in an insect society such as a honeybee hive. All the workers are progeny of the hive's queen and hence share 50 percent of their genes with her. What's more, not only are all the workers siblings but, because the male bees (the drones) that fathered them were haploid and hence produced genetically identical sperm, the workers may actually share 75 percent of their genes with one another.[6]

Although kinship selection can account for the evolution of many cases of altruism, it cannot account for all. It has been suggested that in some instances reciprocity is the explanation, i.e. that an altruistic act (which may not be especially dangerous) by one individual may increase the chances that the other individual will repay by a similar altruistic act in the future. Such a mechanism, of course, could operate only in societies where there is individual recognition. In still other instances, what at first appears to be completely altruistic behavior proves, on closer analysis, to serve selfish ends. Thus helpers in the breeding territory of a mated pair of birds may actually be helping themselves. Some are offspring of the pair, in which case kin selection may be operating. But others are no relation, and they, by gaining access to a desirable territory—one they may take over if anything happens to the breeders—are probably improving their own personal fitness.

THE POSSIBILITY OF NONADAPTIVE CHARACTERS

The assertion is often made that certain characters present in high frequencies in populations are nonadaptive, i.e. that they lack any survival value. What is the basis for such assertions?

Let us consider one example. A critic of the idea that cryptic coloration is protective counted the cryptically colored and noncryptically colored insects in the crops and stomachs of insectivorous birds. Finding that large numbers of both types of insects had been eaten, he concluded that the appearance of cryptically colored insects does not protect them and must be nonadaptive, i.e. that natural selection does not favor such an appearance. This worker, however, had failed to obtain data on the relative abundance of the two types of insects before they were caught. It is only by comparison with such data that stomach counts can be reliably evaluated. For example, if there were five cryptically colored insects for every two noncryptically colored ones in the area of the study, and if the birds' stomachs contained three cryptically colored ones for every two noncryptically colored ones, then the cryptically colored insects must indeed have been protected, because in proportion to their original numbers fewer of them had been caught. As we have repeatedly pointed out, an adaptation need not be, and in reality never is, one hundred percent effective; adaptations increase the probability, not the certainty, of genetic contribution to the next generation.

It has been argued that, even though well-developed characters such as precise mimicry or functional wings or warm-bloodedness or the mammalian middle-ear structure may be adaptive, incomplete stages leading to them

---

[6] This unusually close genetic relationship between workers holds only in the Hymenoptera, where males are haploid. In termites (order Isoptera), where males are diploid, workers share only 50 percent of their genes.

cannot be adaptive. Such characters, it has been maintained, must arise all at once as major evolutionary jumps or, if they evolve gradually by the accumulation of many small changes, then the process of accumulation must be directed by some force other than natural selection.

The fossil record, laboratory experiments, and careful field studies, however, all reveal that major evolutionary changes seldom arise all at once, that the usual pattern is one of gradual accumulation of many small changes as gene frequencies and combinations shift over the generations. Mutations with large phenotypic effects, which would be necessary to produce major changes in one big jump, are practically always lethal or at least detrimental to survival, because they constitute such a great disturbance to the delicately balanced systems in which they arise. The question therefore becomes: Are the numerous tiny changes that produce the early stages of a major new character adaptive, or is their increase in frequency in the population caused by some directing force other than natural selection?

The answer seems to be that even the incipient stages in the evolution of a new character are adaptive. For example, even a very slight resemblance to a distasteful species may improve the chances for survival of an unprotected species enough for that slight resemblance, and any further changes that enhance it, to increase in frequency. Or even the faintest resemblance of an orchid flower to a female wasp may increase the chances of pollination enough for that resemblance to increase in frequency. The point to remember is that an adaptive value so slight as to be undetectable by us may yet be great enough to result in evolutionary change, given the time scale on which evolution occurs.

You will recall Haldane's calculations that a positive selection pressure of only one part in 1,000 could increase the frequency of a dominant allele from 0.00001 to 0.99 in only 23,400 generations, and that even a selection pressure as low as one part in 10,000 could be a major factor toward producing evolutionary change in a reasonably large population. But if a character that increased fitness by one part in 1,000 affected viability, we would need an experimental population of at least 16 million to detect it, even using our best statistical tests. For all practical purposes, such a large experimental population, particularly of higher organisms, is an impossible requirement. But the mere fact that we cannot verify the adaptiveness of a character does not entitle us to state categorically that it has none. Statements that a particular minor character of some organism is "obviously nonadaptive" still appear much too frequently in the biological literature. No character is "obviously" nonadaptive. This does not mean that there is no such thing as a nonadaptive character; it simply means that we can never be sure of the nonadaptiveness of any specific character.

But if we admit the theoretical possibility of nonadaptive characters (i.e. characters that are selectively nearly neutral), we must ask how they could increase in frequency and become characteristic of a population.[7] (We may exclude from this discussion once-adaptive vestigial organs that have not yet been completely eliminated by negative selection.) One possibility is, of

---

[7] Actually, it is highly improbable that any character can be entirely neutral against all the genetic backgrounds in a population and under all the varied environmental conditions to which any population is exposed. Hence all that is meant by nonadaptive is "very slightly advantageous" or "very slightly disadvantageous."

course, that nonadaptive characters may increase in frequency purely by chance (genetic drift); however, we saw earlier that major evolutionary changes would be unlikely to occur by chance alone in any except very small populations. A second possibility, and a far more likely one, is that the nonadaptive character is determined by the same gene (or group of genes) that determines some other character that is definitely adaptive. In other words, the nonadaptive character is an incidental effect of a pleiotropic gene. If the gene increases in frequency because one of its phenotypic effects is strongly favored by natural selection, then the other phenotypic characters determined by it will, of necessity, also increase in frequency, and some of these other characters may be nonadaptive. In fact, it is even possible for decidedly deleterious characters to increase in frequency as a result of pleiotropy; as we saw earlier in our discussion of balanced polymorphism, sickle-cell anemia has increased in parts of Africa because the allele that determines it also determines an advantageous resistance to malaria in the heterozygote. However, as pleiotropic genes increase in frequency, there is often an accompanying change in the genetic background, caused by natural selection, that tends to make the beneficial effects of the gene dominant and its deleterious effects recessive.

Examples of human-induced evolution in which characters not merely nonadaptive but decidedly maladaptive have been produced as side effects of intense directional selection for other traits are well known to all of us. For instance, intense selective breeding of dogs or cats for show qualities frequently produces animals so nervous and high-strung that they make unsatisfactory pets. Or intense selection of tomatoes or strawberries for resistance to disease, high yield, and ability to withstand the rigors of long-distance shipping results in fruit that is nearly tasteless. In such cases human intervention has radically altered the balance between the conflicting selection pressures operating in the wild; certain desirable characteristics sought in domestic plants and animals have been bought at the price of undesirable ones.

SPECIES AND SPECIATION

We have so far discussed only one major aspect of evolution, the gradual change of a given population through time. Now we must turn to another of its major aspects, the processes by which a single population may split, giving rise to two or more different descendent populations. But before we can discuss this topic meaningfully, we must examine more carefully the populations that we have so far casually taken for granted. With reference to sexually reproducing organisms, we can define a **population** as a group of individuals that share a common gene pool (i.e. that interbreed to a larger or smaller extent).

UNITS OF POPULATION

Demes A deme is a small local population, such as all the deer mice or all the red oaks in a certain woodland or all the perch in a given pond. Although no two individuals in a deme are exactly alike, the members of a deme do usually resemble one another more closely than they resemble the members of other demes, for at least two reasons: (1) They are more closely related genetically, because pairings occur more frequently between members of the same deme than between members of different demes; and (2) they are exposed to more similar environmental influences and hence to more nearly the same selection pressures.

It must be emphasized that demes are not clear-cut permanent units of population. Although the deer mice in one woodlot are more likely to mate among themselves than with deer mice in the next woodlot down the road, there will almost certainly be occasional matings between mice from different woodlots. Similarly, although the female parts of a particular red oak tree are more likely to receive pollen from another red oak tree in the same woodlot, there is an appreciable chance that they will sometimes receive pollen from a tree in another nearby woodlot. And the woodlots themselves are not permanent ecological features. They have only a transient existence as separate and distinct ecological units; neighboring woodlots may fuse after a few years, or a single large woodlot may become divided into two or more separate smaller ones. Such changes in ecological features will produce corresponding changes in the demes of deer mice and red oak trees. Demes, then, are usually temporary units of population that intergrade with other similar units.

Species   Notice that intergradation is between "similar" demes. We expect some interbreeding between deer mice from adjacent demes, but we do not expect interbreeding between deer mice and house mice or between deer mice and black rats or between deer mice and gray squirrels. Nor do we expect to find crosses between red oaks and sugar maples or even between red oaks and pin oaks, even if they occur together in the same woodlot. In short, we recognize the existence of units of population larger than demes and both more distinct from each other and longer-lasting than demes. One such unit of population is that containing all the demes of deer mice. Another is that containing all the demes of red oaks. These larger units are known as species.

For centuries it has been recognized that the variation between living organisms does not form a continuum—that there are, instead, many discontinuities in the variation, that plants and animals seem to be divided naturally into many separate and distinct "kinds," or species. This does not mean that all the individuals of any one species are precisely alike—far from it. Any two individuals are probably distinguishable from each other in a variety of ways. But it does mean that all the members of a single species share certain biologically important attributes and that, as a group, they are genetically separated from other such groups. That such groups exist in nature has been recognized even by primitive peoples. Ernst Mayr of Harvard University cites a tribe in New Guinea that had 136 different names for what biologists later showed to be 137 species of local birds; the natives had confused only two species.

But although the existence of discrete clusters of living things that can be called species has long been recognized, the concept of what a species is has changed many times in the course of history. One idea widely held by nonbiologists and once popular among biologists as well is that each species is a static, immutable entity typified by some ideal form, of which all the real

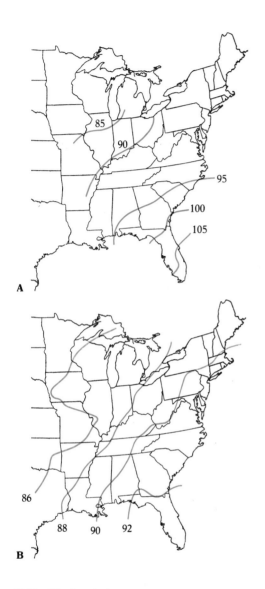

18.32 Clinal variation
(A) Map showing by means of isophene lines (lines connecting equal values) the geographic variation in the mean number of subcaudal scales of the snake *Coluber constrictor* (the racer). (B) Isophene map showing geographic variation in the apical taper of leaves of the milkweed *Asclepias tuberosa*. The numbers represent degree of apical taper. [Modified from W. Auffenberg, *Tulane Stud. Zool.*, vol. 2, 1955 (A), and from R. E. Woodson, *Ann. Missouri Botan. Garden*, vol. 34, 1947 (B).]

individuals belonging to that species are rough approximations; individual variation is supposed to result from the imperfection with which the individuals reflect the ideal characteristics. This static, typological concept contradicts all that we have learned about evolution. The modern concept of species rejects the notion that there is some immutable ideal type for every species. A species, in the modern view, is a genetically distinctive group of natural populations (demes) that share a common gene pool and that are reproductively isolated from all other such groups. Or, to word it another way, a species is the largest unit of population within which effective gene flow (exchange of genetic material) occurs or can occur.

Notice that the modern concept of species says nothing about how different from each other two populations must be to qualify as separate species. Admittedly, most species can be separated on the basis of fairly obvious anatomical, physiological, or behavioral characters, and biologists often rely on these in determining species. But the final criterion is always reproduction—whether or not there is actual or potential gene flow. If there is complete intrinsic reproductive isolation between two outwardly almost identical populations (i.e. if there can be no gene flow between them), then those populations belong to different species despite their great similarity. On the other hand, if two populations show striking differences, but there is effective gene flow between them, then those populations belong to the same species. Anatomical, physiological, or behavioral characters simply serve as clues toward the identification of reproductively isolated populations; they are not in themselves regarded as determining whether a population constitutes a species.

Intraspecific variation We have already discussed the sorts of variation that may occur between individuals of a single deme as a result of mutation and recombination, particularly the latter. And we have seen that this variation is very important biologically, whether it involves almost imperceptible and intergrading differences or striking polymorphic discontinuities. But there is another sort of intraspecific variation that we have not yet discussed. This is variation between the demes of a single species, i.e. variation correlated with geographical distribution.

There is usually so much gene flow between adjacent demes of the same species that differences between them are slight. Thus the ratio of alleles *A* and *a* may be 0.90 to 0.10 in one deme and 0.89 to 0.11 in the adjacent deme. But the farther apart geographically two demes are, the smaller the chance of direct gene flow between them, and hence the greater the probability that the differences between them will be more marked. If, for example, we collect a sample of 500 deer mice each from Plymouth County, Massachusetts, Crawford County, Pennsylvania, and Roanoke County, Virginia, we shall find numerous differences that will enable us to distinguish between the three populations quite readily—much more readily than we could between the populations in three adjacent counties in Massachusetts or in three adjacent counties in Pennsylvania. Some of this geographic variation may reflect chance events such as genetic drift or the occurrence in one deme of a mutation that would be favorable in all demes, but that has not yet spread to them. But much of the geographic variation probably reflects differences in the selection pressures operating on the populations as a result of the differences between the environmental conditions in their re-

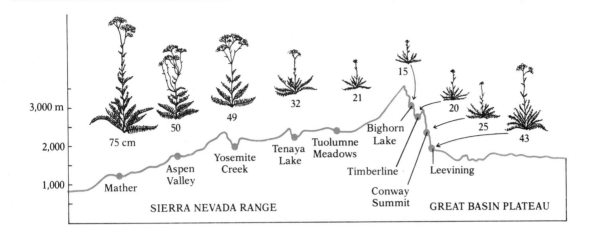

3,000 m

2,000

1,000

75 cm

50

49

32

21

15

20

25

43

Mather

Aspen
Valley

Yosemite
Creek

Tenaya
Lake

Tuolumne
Meadows

Bighorn
Lake

Timberline

Conway
Summit

Leevining

SIERRA NEVADA RANGE

GREAT BASIN PLATEAU

spective ranges. In other words, much geographic variation is adaptive. Each local population or deme tends to evolve adaptations to the specific environmental conditions in its own small portion of the species range. Such geographic variation is found in the vast majority of animal and plant species.

Environmental conditions often vary geographically in a more or less regular manner. There are changes in temperature with latitude or with altitude on mountain slopes, or changes in rainfall with longitude, as in many parts of the western United States, or changes in topography with longitude, e.g. as one moves from the Atlantic coast to the Central States. Such environmental gradients are usually accompanied by genetic-variation gradients in the species of animals and plants that inhabit the areas involved. Most species show north–south gradients in many characters, and east–west gradients are not uncommon (Fig. 18.32). Altitudinal gradients in various characters are also often found (Fig. 18.33).

When a character of a species shows a gradual variation correlated with geography, we speak of that variation as forming a *cline*. For example, many mammals and birds exhibit north–south clines in average body size, being larger in the colder climates farther north and smaller in the warmer climates farther south. Similarly, many mammalian species show north–south clines in the size of such extremities as the tails and ears, these parts being smaller in the demes farther north.[8] A single widespread species often has many characters that vary clinally, but the various clines frequently do not coincide in direction, location, or intensity; one character may show clinal variation from north to south, another from east to west, and still another from northwest to southeast.

Sometimes geographically correlated genetic variation is not as gradual as in the clines discussed above. There may be a rather abrupt shift in some character in a particular part of the species range. Suppose, for example, that the average height of a certain species of plant decreases very gradually as one moves northward from Florida through Georgia, South Carolina, and

18.33   Altitudinal cline in height of the milfoil *Achillea lanulosa*

The higher the altitude, the shorter the plants. This variation was shown to be genetic (i.e. not merely a phenotypic reaction) by collecting seeds from the locations indicated and planting them in a test garden at Stanford where all were exposed to the same environmental conditions. The differences in height were still evident in the plants grown from these seeds.   [Modified from J. Clausen *et al.*, *Carnegie Inst. Wash. Publ.*, no. 581, 1948.]

---

[8] Increase in average body size with increasing cold is so common in homeothermic animals that this tendency has been generalized as Bergmann's rule; the tendency toward decrease in the size of the extremities with increasing cold has been generalized as Allen's rule. The adaptive significance of these clines is obvious when one considers the role of surface-to-volume ratios in heat exchange.

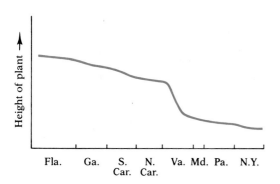

18.34   Graph showing stepped clinal variation in hypothetical plant
The horizontal spans correspond roughly to the north–south distances through the states. See text for discussion.

North Carolina, then decreases very rapidly in the counties of southern Virginia, and then decreases only slightly as one continues to move northward through northern Virginia, Maryland, Pennsylvania, and New York. Such variation forms a stepped cline; i.e. the abrupt shift in height in southern Virginia breaks the otherwise gradual geographic variation and constitutes a step in the north–south cline (Fig. 18.34). When such an abrupt shift in a genetically determined character occurs in a geographically variable species, some biologists designate the populations on the two sides of the step as *subspecies* or *races.* These terms are also sometimes applied to more isolated populations—such as those on different islands or in separate mountain ranges or, as in fish, in separate rivers—when the populations are recognizably different genetically but are believed potentially capable of interbreeding freely. Subspecies or races (the two terms, as used here, are equivalent) may be defined, then, as groups of natural populations within a species that differ genetically and that are partly isolated from each other reproductively because they have different ranges.

Note that two subspecies of the same species cannot, by definition, long occur together geographically, because it is only the limitation on interbreeding imposed by distance that keeps them genetically distinctive. If they occurred together, they would interbreed and any distinction between them would quickly disappear. Many biologists have argued against the formal recognition of subspecies or races. One reason is that the distinctions between them are often made arbitrarily on the basis of only one character—the fact that other characters may form entirely different patterns of variation being ignored (Fig. 18.35). Another reason is that most units so recognized probably have only a transitory existence as separate entities and do not, as was once thought, go on to become fully separate species.

As a result of intraspecific geographic variation (whether irregular, clinal, or racial), two populations belonging to the same species but occurring in two widely separated localities often show no more resemblance to each other than to populations belonging to other species. Such intraspecific dissimilarity serves to emphasize the point made earlier, that it is not the degree of morphological resemblance that determines whether or not two populations belong to the same species; it is whether they are reproductively isolated from each other. There are even instances where two widely separated populations are regarded as belonging to the same species even though the respective individuals, when brought together, are incapable of producing viable offspring. The reason they are considered members of the same species is that the populations are connected by an unbroken chain of intermediate populations that permit gene flow between them.

SPECIATION

In considering how species originate—the phenomenon of speciation—we shall concentrate particularly on divergent speciation, the process by which one ancestral species gives rise to two or more descendent species, which grow increasingly unlike (diverge) as they evolve.

The role of geographic isolation   Since species are defined in terms of reproductive isolation, the fundamental question of divergent speciation must be: How do two sets of populations that initially share a common gene pool come to have completely separate gene pools? That is, how does the possibility of effective gene flow between the two sets of populations disappear? How do barriers to the exchange of genes arise?

Most biologists agree that in the vast majority of cases (excluding speciation by polyploidy) the initiating factor in speciation is geographic separation. As long as all the populations of a species are in direct or indirect contact, gene flow will continue throughout the system and no splitting can occur, although various populations within the system may diverge in numerous characters and thus give rise to much intraspecific variation of the sorts discussed above. But if the initially continuous system of populations is divided by some geographic feature that constitutes a barrier to the dispersal of the species, then the separated population systems will no longer be able to exchange genes and their further evolution will therefore be independent. Given sufficient time, the two separate population systems will become more and more unlike each other as each evolves in its own way. At first, the only reproductive isolation between them will be geographic—isolation by physical separation—and they will still be potentially capable of interbreeding; according to the modern concept of species, they will still belong to the same species. Eventually, however, they may become genetically so different that there would be no effective gene flow between them even if they should again come into contact. When this point in their gradual divergence has been reached, the two population systems constitute two separate species.

There are at least three factors that will make geographically separated population systems diverge in time.

1. Chances are that the two systems will have somewhat different initial gene frequencies. Because most species exhibit geographic variation, it is highly unlikely that a geographic barrier would divide a variable species into portions exactly alike genetically; it would be much more likely to separate populations already genetically different, such as the terminal portions of a cline. Separation can occur in other ways than through the splitting of a once continuous distribution by a new geographic barrier. When, as often happens, a small number of individuals manage to cross an already existing barrier and found a new geographically isolated colony, these founders will, of course, carry with them in their own genotypes only a small percentage of the total genetic variation present in the gene pool of the parental population, and the new colony will thus have allelic frequencies very different from those of the parental population; this is a special form of genetic drift called the **founder effect.** Obviously, if from the moment of their separation two populations have different genetic potentials, their future evolution is likely to follow different paths.

2. Separated population systems will probably experience different mutations. Mutations are random (though some are more probable than others), and the chances are good that some mutations will occur in one of the populations and not in the other, and vice versa. Since there is no gene flow between the populations, a new mutant gene arising in one of them cannot spread to the other.

3. Isolated populations will almost certainly be exposed to different environmental selection pressures, since they occupy different ranges. The chances that two separate ranges will be identical in every significant environmental factor are essentially nil.

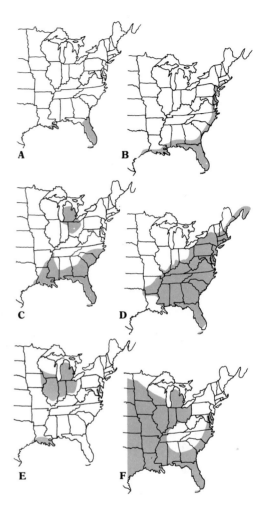

18.35 Discordant geographic variation in six characters of the snake *Coluber constrictor* in the eastern United States

Since no two of the characters vary together, selection of any one of them as a criterion for recognition of subspecies or races is largely arbitrary.   (A) Areas where red eyes are found in juveniles.   (B) Areas where red ventral spots are found on juveniles. (C) Areas where the loreal scale is in contact with the first supralabial scale in at least 10 percent of the specimens   (D) Areas where black adults are found.   (E) Areas where dark postocular stripes are found.   (F) Areas where full-grown adults have white chins. [Redrawn from W. Auffenberg, *Tulane Stud. Zool.*, vol. 2, 1955.]

In addition to these three causes of divergence, a fourth—genetic drift of the ordinary sort—would be important in small populations.

The barriers that can cause the initial spatial separation leading to speciation are of many different types. A barrier is any physical or ecological feature that prevents the movement across it of the species in question. What is a barrier for one species may not be a barrier for another. Thus a prairie is a barrier for forest species but not, obviously, for prairie species. A mountain range is a barrier to species that can live only in lowlands, a desert is a barrier to species that require a moist environment, and a valley is a barrier to montane species. On a grander scale, oceans and glaciers have played a role in the speciation of many plants and animals. Let us look at a few actual examples of geographic isolation leading to speciation.

One of the most frequently cited examples is that of the Kaibab squirrel, which occurs on the north side of the Grand Canyon, and of the Abert squirrel, which occurs on the south side. The two are clearly very closely related and doubtless evolved from the same ancestor, but they almost never interbreed at present, because they do not cross the Grand Canyon. Biologists are not agreed whether these two squirrels have reached the level of full species or whether they should be considered well-marked geographic variants of a single species, but the fact remains that the Grand Canyon has acted as a barrier separating the two sets of populations, and that those populations have, as a result, evolved divergently until they have at least approached the level of fully distinct species. The Grand Canyon also separates the range of the gray-tailed antelope squirrel from that of the closely related white-tailed antelope squirrel, and it separates the range of the rock pocket mouse from that of the long-tailed pocket mouse.

On islands of the Pacific, in many instances, two closely related species of snails, clearly descended from the same ancestral population, live in valley woodlands separated by treeless ridges that the snails apparently cannot cross. Blind cave beetles (genus *Pseudanophthalmus)* living in different

TABLE 18.1   *Intrinsic isolating mechanisms*

Mechanisms that prevent mating	1. Ecogeographic isolation 2. Habitat isolation 3. Seasonal isolation 4. Behavioral isolation 5. Mechanical isolation	Mechanisms operative in the parents, preventing fertilization
Mechanisms that prevent production of hybrid young after mating	6. Gametic isolation 7. Developmental isolation	
Mechanisms that prevent perpetuation of hybrids	8. Hybrid inviability 9. Hybrid sterility 10. Selective hybrid elimination	Mechanisms operative in the hybrids, preventing their success

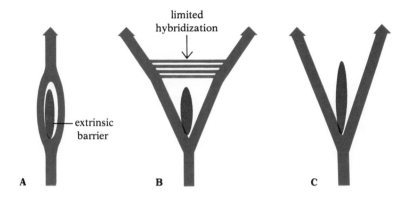

limited
hybridization

extrinsic
barrier

A   B   C

18.36 Model of geographic speciation
(A) an extrinsic barrier (geographic) divides a population, but the barrier breaks down before the two subpopulations have been isolated long enough to have evolved intrinsic reproductive isolating mechanisms; hence the populations fuse back together.   (B) Two populations are isolated by an extrinsic barrier long enough to have evolved incomplete intrinsic reproductive isolating mechanisms. When the extrinsic barrier breaks down, some hybridization occurs. But the hybrids are not as well adapted as the parental forms; hence there is a strong selection pressure favoring forms of intrinsic isolation that prevent mating, and the two populations diverge more rapidly until mating between them is no longer possible. This rapid divergence is called character displacement. (C) Two populations are isolated by a geographic barrier so long that by the time the barrier breaks down they are too different to interbreed.

caves in the eastern United States have often diverged to the level of full species. Two river systems only a few miles apart, but with no interconnections, often have their own species of minnows.

Intrinsic reproductive isolation   According to the model of divergent speciation outlined above, the initial factor preventing gene flow between two closely related population systems is ordinarily an extrinsic one—geography. Then, the model says, as the two populations diverge, they accumulate differences that will lead, given enough time, to the development of intrinsic isolating mechanisms—biological characteristics that prevent the two populations from occurring together or from interbreeding effectively when (or if) they again occur together. In other words, speciation is initiated when through external barriers the two population systems become entirely **allopatric** (come to have different ranges), but is not completed until the populations have evolved intrinsic mechanisms that will keep them allopatric or that will keep their gene pools separate even when they are **sympatric** (have the same range) (Fig. 18.36). Let us now examine the various kinds of intrinsic isolating mechanisms that may arise. One way of classifying them is shown in Table 18.1.

1. *Ecogeographic isolation*   Two population systems, initially separated by some extrinsic barrier, may in time become so specialized for different environmental conditions that even if the original extrinsic barrier is removed they may never become sympatric, because neither can survive under the conditions where the other occurs. In other words, they may evolve genetic differences that will maintain their geographic separation. An example is seen in two well-known tree species of the genus *Platanus: P. occidentalis* (the sycamore or buttonwood tree), which occurs in the eastern United States, and *P. orientalis* (the Oriental plane tree), which occurs in the eastern part of the Mediterranean region. They can be artificially crossed and the hybrids are vigorous and fertile. But each species is adapted to the climate in its own native range, and the climates in the two ranges are so different that neither species will long survive under natural conditions in the range of the other. Thus there are genetic differences that under natural conditions would prevent gene flow between the two species. Their separation is not merely geographic; it is both geographic and genetic.

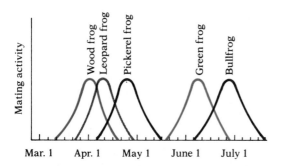

18.37   Mating seasons at Ithaca, New York, for five species of frogs of the genus *Rana*
The period of most active mating is different for each species. Where the mating seasons for two or more species overlap, different breeding sites are used.   [Modified from B. Wallace and A. M. Srb, *Adaptation*, © 1964. By permission of Prentice-Hall, Inc., Englewood Cliffs, N.J.]

18.38   A male fiddler crab *(Uca)*
The animal waves its large cheliped in the air as a courtship display. Details of the display differ among the various species of fiddler crabs.   [J. Shaw, Bruce Coleman Inc.]

2. ***Habitat isolation***   When two sympatric populations occupy different habitats within their common range, the individuals of each population will be more likely to encounter and mate with members of their own population than with members of the other population. Their genetically determined preference for different habitats thus helps keep the two gene pools separate. There are numerous examples of such habitat isolation. *Bufo woodhousei* and *B. americanus* are two closely related toads that can cross and produce viable offspring. But in those areas where the ranges of the two toads overlap, *B. woodhousei* normally breeds in the quieter water of streams, while *B. americanus* breeds in shallow rainpools. The dragonfly *Progomphus obscurus* lives in northern Florida, while its close relative *P. alachuensis* lives in southern Florida. The ranges of the two species overlap in north-central Florida, but there the two species occupy different habitats, *P. obscurus* being restricted to rivers and streams and *P. alachuensis* to lakes. In California the ranges of *Ceanothus thyrsiflorus* and *C. dentatus*, two species of wild lilacs, overlap broadly, but *C. thyrsiflorus* grows on moist hillsides with good soil, while *C. dentatus* grows on drier, more exposed sites with poor or shallow soil.

3. ***Seasonal isolation***   If two closely related species are sympatric, but breed during different seasons of the year, interbreeding between them will be effectively prevented. For example, *Pinus radiata* and *P. muricata*, two species of pine, are sympatric in some parts of California. They are capable of crossing, but do so rather seldom under natural conditions, because *P. radiata* sheds its pollen early in February while *P. muricata* does not shed its pollen until April. *Reticulitermes hageni* and *R. virginicus*, two closely related species of termites, are sympatric in southern Florida, but the mating flights of the former occur from March through May while those of the latter occur in the fall and winter months. Five species of frogs belonging to the genus *Rana* are sympatric in much of eastern North America, but the period of most active mating is different for each species (Fig. 18.37).

4. ***Behavioral isolation***   In Chapter 12 we discussed the immense importance of behavior in courtship and mating, particularly with respect to species recognition. We saw, for example, that ducks often have elaborate courtship displays, usually combined with striking color patterns in the males, and that these function in minimizing the chances that a female will select a male of the wrong species as a mate. We also noted the complex interplay of behavioral patterns during the courtship of stickleback fish. Each species of stickleback has its own courtship pattern, and where two species are sympatric, crosses rarely occur, because a courtship between members of different species involves so many wrong responses that the courtship is unlikely to proceed all the way to the spawning stage.

A particularly interesting example of the functioning of visual displays in species recognition has been reported by Joselyn Crane of the New York Zoological Society. She found twelve species of fiddler crabs of the genus *Uca* actively courting on the same small beach (only about 56 m²) in Panama. Each species had its own characteristic display, which included waving the large claw (cheliped), elevating the body, and moving around the burrow (Fig. 18.38). Crane found that the displays were so distinctive that she could recognize each species from a considerable distance merely by the form of its display.

Auditory stimuli are important in species recognition among many animals, particularly birds and insects, and help prevent mating between related species. In several instances specialists have noticed that two or three very different songs were sung by what had been considered members of a single species of cricket. On investigation, they found that each song was, in fact, sung by a different species, but the species were so similar morphologically that no one had previously distinguished among them. Despite the morphological similarity of these closely related crickets, they do not hybridize in nature even when they are sympatric, because females do not respond to the stridulation of a male of a different species. *Sibling species*— species so closely related that humans can hardly distinguish them, at least until some diagnostic character, such as the song of the crickets, is found— are a fairly common phenomenon.

5. *Mechanical isolation*   If structural differences between two closely related species make it physically impossible for matings between males of one species and females of the other to occur, the two populations will obviously not exchange genes. If, for example, one species of animal is much larger than the other, matings between them may be very difficult, if not impossible. Or if the genital organs of the males of one species and the females of the other do not fit, mating will be prevented. The observation that the copulatory organs vary greatly from species to species in many animal groups led long ago to the so-called "lock-and-key" hypothesis, which holds that the male and female genitalia are so precisely fitted to each other that even slight changes in the structure of either would make copulation impossible. This supposition has proved incorrect in many groups. For example, even though the male copulatory organs (gonopods) of millipeds are usually so different in different species (Fig. 18.39) that they have served as the basis for much of the classification of these animals, the female genitalia of related species are often almost indistinguishable—an indication that the major isolating mechanism between the species must be something other than a lock-and-key lack of fit. Nonetheless, a lock-and-key mechanism probably does operate in a few animal groups, among them several subfamilies of snails in which interspecific matings are physically very difficult.

Mechanical isolation is probably much more important in plants than in animals, particularly in plants that depend on insect pollinators. Consider milkweeds, for example. In these plants the pollen is contained in small sacs that stick to the legs of insects. The female part of each flower (stigma) has slits in it into which the sac must be inserted if pollination is to take place. This insertion is not easy, and few of the sacs carried from one flower to another on the legs of an insect are ever actually introduced into a stigma. Insertion of a pollen sac from one species of milkweed into the stigma of a different species is essentially impossible, because both the sacs and the slits in the stigma vary in shape from species to species. Thus even though several closely related species of milkweeds are sympatric in many parts of the world, hybridization between them is almost nonexistent. Consider also *Salvia apiana* and *S. mellifera*, two closely related species of sages that are sympatric in California. They are reproductively isolated by differences in habitat and breeding seasons and by the behavior of their pollinators. In addition, as has recently been demonstrated, mechanical features play a

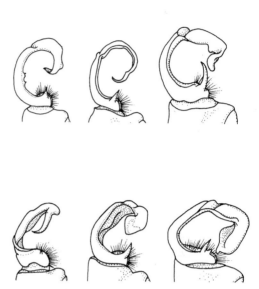

18.39   Male gonopods of six species of millipeds of the genus *Brachoria*
The shape of the distal portion of the male gonopods is distinctive for each species; yet the genitalia of the females of these species are nearly indistinguishable from one another. The animals being eyeless, the male gonopods do not function in visual species recognition; no functional or evolutionary explanation for the differences in these structures has been found thus far.

role. Whereas *S. mellifera* is pollinated by relatively small bees, the flowers of *S. apiana* can be entered only by very large bees whose weight is sufficient for the landing platform (lower lip of corolla) to unfold and permit free entrance into the flower.

6. *Gametic isolation* Even if individuals of two animal species mate or the pollen from one plant species gets onto the stigma of another, actual fertilization may not take place. For example, if cross-insemination occurs between *Drosophila virilis* and *D. americana*, the sperm are rapidly immobilized by the unsuitable environment in the reproductive tract of the female and they never reach the egg cells. In other species of *Drosophila* interspecific matings cause an antigenic reaction in the genital tract of the female; the walls of the vagina swell enormously and kill the sperm before they reach the eggs. In tobacco 68 different interspecific combinations are known in which no cross-fertilization will take place, even when the pollen is placed on the stigma, because the sperm nucleus from the pollen is unable to reach the egg nucleus in the ovule.

7. *Developmental isolation* Even when cross-fertilization occurs, the development of the embryo is often irregular and may cease before birth. The eggs of fish can often be fertilized by sperm from a great variety of other species, but development is usually arrested in the early stages. Crosses between sheep and goats produce embryos that die long before birth.

8. *Hybrid inviability* Hybrids are often weak and malformed and frequently die before they reproduce; hence there is no actual gene flow through them from the gene pool of the one parental species to the gene pool of the other parental species. An example of hybrid inviability is seen in certain tobacco hybrids, which form tumors in their vegetative parts and die before they flower.

9. *Hybrid sterility* Some interspecific crosses produce vigorous but sterile hybrids. The best-known example is, of course, the cross between a female horse and a male donkey, which produces the mule. Mules have many characteristics superior to those of both parental species, but they are sterile. No matter how many mules are produced, the gene pools of horses and donkeys remain distinct, because there is no gene flow between them. The same is true of horses and zebras, which can hybridize to produce sterile zebroids.

10. *Selective hybrid elimination* The members of two closely related populations may be able to cross and produce fertile offspring. If those offspring and their progeny are as vigorous and well adapted as the parental forms, then the two original populations will not remain distinct for long if they are sympatric, and it will no longer be possible to regard them as full species. But if the fertile offspring and their progeny are less well adapted than the parental forms, then they will soon be eliminated. There will be some gene flow between the two parental gene pools through the hybrids, but not much. The parental populations are consequently regarded as separate species.

Usually, if the parental populations are sympatric, they will rapidly evolve more effective isolating mechanisms. The reason is clear. Since the hybrids have lower fitness and tend to die out within a generation or two, individuals of the parent species that tend to mate with members of their own species will, in the long run, have more descendants than individuals that tend to mate with members of the wrong species. In other words, there will be selection for correct mating and selection against wrong mating. Gene combinations that lead to correct mate selection will increase in frequency, and combinations that lead to incorrect selection will decrease, until eventually all hybridization ceases. The tendency of closely related sympatric species to diverge rapidly in characteristics that reduce the chances of hybridization and/or minimize competition between them is called **character displacement** (see Fig. 18.36).

Situations in which only one of the ten isolating mechanisms described above is operative are extremely rare. Ordinarily two, three, four, or more all contribute to keeping two species apart. For example, closely related sympatric plant species often exhibit habitat and seasonal isolation in addition to some form of hybrid incapacity. In general, sympatric species, whether plant or animal, tend rapidly to evolve one or more of the forms of isolation that prevent mating (habitat, seasonal, behavioral, mechanical) rather than depend only on those forms that prevent the birth or perpetuation of hybrids. The reasons are similar to those mentioned in the preceding paragraph: Individuals that tend to mate with members of the wrong species will leave fewer descendants than those that mate with members of their own species. Wrong matings produce gamete wastage, whether fertilization takes place or not, or whether the hybrids are viable or not.

Speciation by polyploidy The model of speciation discussed above involves the gradual divergence of geographically separated populations. There is another way in which new species may arise—an almost instantaneous process that makes it entirely possible for a parent to belong to one species and its offspring to another. This process is speciation by polyploidy. It has apparently been common in plants but rare in animals.

One type of polyploid speciation, called autopolyploidy, involves a sudden multiplication of the number of chromosomes in an otherwise normal organism, usually as a result of the nondisjunction of chromosomes during meiosis. An example of this type of polyploidy was discovered by Hugo De Vries, one of the early geneticists, while he was making extensive studies of the evening primrose, *Oenothera lamarckiana*. This diploid species has 14 chromosomes. During De Vries' studies, a new form suddenly arose. This new form, to which he gave the species name of *Oenothera gigas*, had 28 chromosomes (i.e. it was tetraploid). It was reproductively isolated from the parental species because hybrids between *O. lamarckiana* and *O. gigas* were triploid (they received one of each type of chromosome from their *O. lamarckiana* parent and two of each type from their *O. gigas* parent), and triploid individuals, because of the highly irregular distribution of their chromosomes at meiosis, are sterile. It is characteristic of autopolyploidy that the polyploids are fertile and can breed with each other, but cannot cross with the diploid species from which they arose. Hence polyploid populations fulfill all the requirements of the modern definition of species—they are genetically distinctive, and they are reproductively isolated—although botanists do not always choose to give each such polyploid population a formal species name.

A second type of polyploid speciation, called allopolyploidy, involves a multiplication (usually a doubling) of the number of chromosomes in a hybrid between two species. This type of polyploidy has probably been far

18.40 Distribution of 28 species of millipeds of the genus *Brachoria*
The color dots indicate all known localities for these millipeds, many of which are sympatric. All the speciation in this genus must have occurred within this very limited area in the eastern United States.

more important in speciation than autopolyploidy. Suppose one species has a chromosome type *A*; if we assume that the species is diploid, its genotype relative to this chromosome will be *A/A*. Suppose another related species has no *A* chromosome, but in its place a type *B*; its diploid chromosomal genotype will thus be *B/B*. Hybrids between these two species will have the genotype *A/B*. But if the *A* and *B* chromosomes are sufficiently different, they will not synapse in meiosis. Hence the hybrids will not be able to produce normal gametes and will be sterile. But if the number of chromosomes in the hybrids is doubled in some way, then their genotype will be *A/A B/B*. In this case the two *A* chromosomes will synapse in meiosis, as will the two *B* chromosomes. Hence the *A/A B/B* individuals will be able to produce normal gametes and will be fertile. They will, in effect, be diploid, since they will have only two of each type of chromosome. But they will have a complete diploid set from each of the parental species. The *A/A B/B* individuals will be able to breed freely among themselves, but they will be unable to cross with either of the parental species. Consequently the allopolyploid population must be regarded as a distinct species.

Allopolyploid plants are frequently larger and more vigorous than the parental diploid plants. They are often more aggressive invaders of new habitats. When conditions become unfavorable and the environment is rapidly changing, the diploids are usually the first to become extinct. Hence allopolyploid speciation has probably played an important role in the perpetuation of some plant groups during periods of widespread environmental change.

Allopolyploidy has also proved of great importance in the production of valuable new crop plants. As soon as it was realized that many of our most useful plants, such as oats, wheat, cotton, tobacco, potato, banana, coffee, and sugar cane, are polyploids, plant breeders began trying to stimulate polyploidy, and obtained many new varieties. It was found that the chemical colchicine would readily induce polyploidy. One of the first artificially produced allopolyploids came from a cross made in 1924 between the radish and the cabbage. Unfortunately it had the root of the cabbage and the shoot of the radish. Other crosses have yielded more desirable results.

Adaptive radiation   One of the most striking aspects of life is its extreme diversity. A bewildering array of species now occupies this globe. And the fossil record shows that of the species that have existed at one time or another those now living represent only a tiny fraction (probably less than one tenth of one percent, all the other species being now extinct). Clearly, then, divergent evolution—the evolutionary splitting of species into many separate descendent species—has been an exceedingly frequent occurrence. Is it possible to account for such a degree of evolutionary radiation by the models outlined above? In particular, could opportunities for geographic isolation have been sufficient to lead to all the speciation not caused by polyploidy? After all, it is not unusual for a complex of four, five, or more closely related species to occur within a rather limited area. For instance, 28 species of the milliped genus *Brachoria*, many of them sympatric, are confined to a small portion of the deciduous forests of the eastern United States (Fig. 18.40). How could so much speciation have occurred in so small an area if geographic isolation was a necessary factor? In an attempt to answer such questions, let us turn to a particularly instructive and historically important example—the finches on the Galápagos Islands, which played a major role in leading Charles Darwin to formulate his theory of evolution by natural selection.

The Galápagos Islands lie astride the equator in the Pacific Ocean roughly 950 km west of the coast of Ecuador, to which country they now belong (Fig. 18.41). These islands have never been connected to South America or to any other land mass; nor have they been connected to each other. They apparently arose from the ocean floor as volcanoes somewhat more than a million years ago. At first, of course, they were completely devoid of life, and were thus an environment open to exploitation by whatever species from the South American mainland might chance to reach them. Relatively few species ever did so. The only land vertebrates present on the islands before human beings got there were six species of reptiles (one snake, a huge tortoise, and four lizards, including two very large iguanas), two species of mammals (a rice rat and a bat), and a limited number of birds (including two species of owls, one hawk, one dove, one cuckoo, one warbler, two flycatchers, one swallow, four mockingbirds, and the famous Darwin's finches).

The 14 species of Darwin's finches constitute a separate subfamily found nowhere else in the world. They are believed to have evolved on the Galápagos Islands from some unknown finch ancestor that colonized the islands from the South American mainland. It is readily understandable that the descendants of the geographically isolated colonizers should have undergone so much evolutionary change as to become, in time, very unlike their mainland ancestors. More perplexing at first glance is the manner in which the descendants of the original immigrants split into the separate populations that gave rise to today's 14 species.

The point to remember is that we are dealing not with a single island but with a cluster of more than 15 separate islands. The finches will not readily fly across wide stretches of water, and they show a strong tendency to remain near their home area. Hence a population on any one of the islands is effectively isolated from the populations on the other islands. We can suppose that the initial colony was established on some one of the islands where the colonizers, perhaps blown by high winds, chanced to land. Later, stragglers from this colony wandered or were blown to other islands and founded new colonies. The allelic frequencies in the new colonies differed from those in the original colonies from the moment they started, because of the founder effect. In time, the colonies on the different islands diverged even more, for the reasons already outlined in the model of geographic speciation (different mutations, different selection pressures, and, in such small populations as some of these must have been, genetic drift). What we might expect, therefore, is a different species, or at least a different race, on each of the islands. But this is not what has actually been found; most of the islands have more than one species of finch, and the larger islands have ten (Fig. 18.41, bottom). What is the explanation?

Let us suppose that form A evolved originally on Indefatigable Island and that the closely related form B evolved on Charles Island. If, later, form A had spread to Charles Island before the two forms had been isolated long enough to evolve any but minor differences, the two forms might have interbred freely and merged with each other. But if A and B had been separated long enough to have evolved major differences before A invaded Charles Island, then A and B might have been intrinsically isolated from

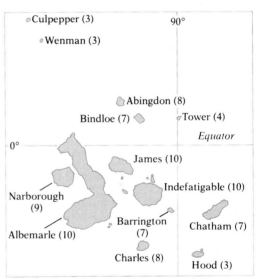

18.41 The Galápagos Islands

Top: The islands are located about 950 km off the coast of Ecuador. Cocos Island is about 700 km northeast of the Galápagos. Bottom: The islands shown in greater detail. The number in parentheses after each name indicates the number of species of Darwin's finches that occur on the island. The island names given here are the English ones, which Darwin used. The Ecuadorian government has renamed the majority of the islands, but most of the biological literature continues to use the older English names. [Modified from D. Lack, *Darwin's finches*, Cambridge University Press, 1947.]

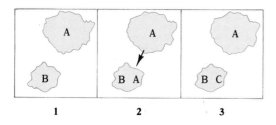

18.42 Model of speciation on the Galápagos Islands
(1) An ancestral form colonized two islands, and the two populations, being isolated from each other, eventually evolve into separate species A and B.
(2) Some individuals of A colonize B's island. The two species coexist, but intense competition between them leads to rapid divergent evolution.   (3) This rapid evolution of the population of A on B's island causes it to become more and more different from the original species A, until eventually it is sufficiently distinct to be considered a full species, C, in its own right.

each other (i.e. been full species), and they might have been able to coexist on the same island without interbreeding (Fig. 18.42). If they formed occasional hybrids, those hybrids might well have been less viable than the parental forms. Accordingly, natural selection would have favored individuals that mated only with their own kind, and this selection pressure would have led rapidly to more effective intrinsic isolating mechanisms preventing the gamete wastage involved in cross-matings. It has been shown, in fact, that Darwin's finches readily recognize members of their own species and show little interest in members of a different species.

We have now arrived at a point in our hypothetical example where Indefatigable is occupied by species A and Charles by both A and B. It would be highly unlikely that A and B could coexist indefinitely if they used the same food supply or the same nesting sites; the ensuing competition would be very severe, and the less well adapted species would tend to be eliminated by the other unless it evolved differences that minimized the competition. In other words, wherever two or more closely related species occur together, natural selection would favor character displacement, specifically the evolution of different feeding and nesting habits. This is precisely what we find in Darwin's finches. The 14 species form four groups (genera) (Fig. 18.43). One group includes six species that live primarily on the ground; of these, some feed primarily on seeds and others mostly on cactus flowers. Of the species that feed on seeds, some feed on large seeds, some on medium-sized seeds, and some on small seeds. The second group contains six species that live primarily in trees. Of these, one is a vegetarian and the others eat insects, but the insect eaters differ from one another in the size of their prey and in the way they catch them (Fig. 18.44). A third group contains only one species, which has become very unfinchlike and strongly resembles the warblers of the mainland. The fourth group contains only one species, restricted to

18.43 Darwin's finches
Birds 1–6 are the ground finches *(Geospiza)*, 7–12 the tree finches *(Camarhynchus)*. Bird 13 is the Warbler Finch *(Certhidea olivacea)*. The Cocos Finch *(Pinaroloxias inornata)* is not shown.
1   Large Cactus Ground Finch *(G. conirostris)*
2   Large Ground Finch *(G. magnirostris)*
3   Medium Ground Finch *(G. fortis)*
4   Cactus Ground Finch *(G. scandens)*
5   Sharp-beaked Ground Finch *(G. difficilis)*
6   Small Ground Finch *(G. fuliginosa)*
7   Tool-using Finch *(Camarhynchus pallidus)*
8   Vegetarian Tree Finch *(C. crassirostris)*
9   Large Insectivorous Tree Finch of Charles Island *(C. pauper)*
10 Large Insectivorous Tree Finch *(C. psittacula)*
11 Small Insectivorous Tree Finch *(C. parvulus)*
12 Mangrove Finch *(C. heliobates)*
[*Biological science: Molecules to man*, Houghton Mifflin, 1963. Reprinted by permission of the Biological Sciences Curriculum Study.]

18.44   The Tool-using Finch
One of the insectivorous tree finches of the Galápagos
has evolved most unusual feeding habits. Sometimes
called the Woodpecker Finch, it chisels into wood after
insects, but it lacks the long tongue that a woodpecker
uses to probe insects out of a crack. Instead, it pokes
into the crack with a cactus spine or twig that it holds
in its beak. This is one of the few known cases of use
of a tool by a bird.

Cocos Island, which is about 700 km northeast of the Galápagos Islands and
about 500 km from Panama. Correlated with the differences in diet among
the species are major differences in the size and shape of their beaks (Fig.
18.45). These characteristics of the beak are apparently the principal means
by which the birds recognize other members of their own species.

Now, if on Charles Island selection favored character displacement between species A and B, the population of species A on Charles Island would
become less and less like the population of species A on Indefatigable Island.
Eventually these differences might become so great that the two populations
would be intrinsically isolated from each other and would thus be separate
species. We might now designate as species C the Charles population derived
from species A. The geographic separation of the two islands would thus
have led to the evolution of three species (A, B, and C) from a single original
species. The process of island hopping followed by divergence could continue indefinitely and produce many additional species. It was doubtless
such a process, involving initial divergence on separate islands followed by
intensification of differences when sympatry later developed, that led to the
formation of the 14 species of Darwin's finches.

Now let us apply the principles learned from Darwin's finches to the case
of the 28 species of *Brachoria* millipeds confined to a small area in the
eastern United States (see Fig. 18.40). These animals live in the humus layer
on the floor of deciduous forests. They are rather sluggish and seldom move
very far. It would have been easy for populations to become isolated in local
forested areas separated by less hospitable regions. Such allopatric populations could have become sufficiently different so that, when conditions
changed and they became sympatric, they would behave as full species.
Clearly, the same sorts of processes as seen on islands can account for
radiation in a continental area. And on a somewhat larger geographic scale,
the same processes can account for the observed adaptive radiation in insects, fish, reptiles, birds, mammals, and many plant groups. In short, adaptive radiation on islands, like that of Darwin's finches, is dramatic and lends
itself particularly well to analysis, but it does not differ in principle from
adaptive radiation under other circumstances. Thus it helps show that the
model of speciation we have outlined can account for the great amount of
divergence necessary to produce the immense diversity among living things.
No other mechanism, except polyploidy in some plant groups, seems to be
needed.

It should be pointed out that the rate of evolutionary divergence is not
always constant. Almost surely, when the first colonizing finches reached the
Galápagos Islands, they encountered environmental conditions quite unlike
those they had left behind in South America. Hence the selection pressures
to which they were subjected were probably intense, and they must have
diverged very rapidly from the ancestral South American population. Later,
as they became increasingly well adapted to conditions on the Galápagos,
the rate of evolutionary change probably slowed down.

In general, when conditions change radically and organisms have new
evolutionary opportunities for which they are at least modestly preadapted,
they may undergo an evolutionary burst—a period of rapid adaptational
change—which may then be followed by a more stable period during which
any further evolutionary changes are merely fine-tuning of their already well-
adapted characteristics. Such bursts of rapid evolutionary divergence prob-

ably characterized the tremendous radiation of amphibians when they moved onto land for the first time, and the explosive radiation of the mammals when the dinosaurs died off, leaving many adaptive slots unoccupied.

THE SPECIES PROBLEM

It would be wrong to leave you with the impression that the modern definition of species can be applied without difficulty in all cases, or, indeed, that it is valid in all cases. The details of the definition itself are controversial and can provoke heated arguments among biologists. Although most of them accept the major ideas on which the definition rests, and although most of them, if they were to study the same set of natural populations, would probably agree in the great majority of instances which populations represent full species and which do not, there would be a small percentage of populations on which they could not agree. These are the cases where the modern definition of species is hard to apply or invalid. Let us examine a few such cases.

Asexual organisms   Since the modern definition of species assumes interbreeding, it obviously does not apply to asexual organisms. Can asexual organisms be said to form species in any sense? And would such species be comparable to sexual ones?

Asexual organisms do seem to form recognizable groups or kinds, even though the members of a group cannot exchange genes. Gaps, or discontinuities in the variation, occur between the various kinds just as they do between sexual species. One possible explanation for the groupings is that not all variants would be equally well adapted, and hence that only those individuals whose genotypes produce well-adapted phenotypes would survive in significant numbers. Since there would be a limited number of superiorly adapted "types," all individuals falling within the bounds of one such type would constitute a natural group that could be called a species and all individuals falling within the bounds of another adaptive type would constitute a second species. The asexual species thus determined would resemble sexual species in that the latter, too, represent adaptive peaks. The two types of species would, therefore, play comparable ecological roles, but they would nonetheless differ fundamentally from a genetic and evolutionary point of view.

There are relatively few groups, primarily among the microorganisms, in which gene exchange is totally absent. Among higher organisms the few instances of completely asexual reproduction usually represent evolutionary blind alleys of little importance in the overall history of life.

Fossil species   The modern definition of species can be applied strictly only to contemporaneous forms, i.e. forms occurring in only one time transect. It cannot be applied strictly to fossil forms, since it is obviously impossible to use the criterion of interbreeding when comparing a form with its ancestors of a million years earlier. All that paleontologists can do when comparing forms from different time transects is use morphological criteria and classify two forms as separate species when they differ to about the same degree as related forms known to be species on reproductive grounds. For practical purposes, paleontologists usually regard gaps in the fossil record as breaks

18.45   Beak differences in Darwin's finches on the central islands
The differences may appear slight at first glance, but they have important functional implications for the birds' diets and for species recognition in mating.   Top row (diagonally downward from left to right): *Geospiza magnirostris, G. fortis, G. fuliginosa.*   Second row: *G. difficilis debilirostris, G. scandens, Camarhynchus crassirostris.*   Third row: *C. psittacula, C. parvulus, C. pallidus.*   Fourth row: *C. heliobates, Certhidea olivacea, Pinaroloxias inornata.*   [After Swarth.]

between species, even though they are fully aware that no gaps actually occurred in the lineages of the organisms.

Populations at an intermediate stage of divergence   Our model of allopatric speciation assumes that geographically isolated populations will slowly diverge by essentially imperceptible stages until they have reached the level of full species. The intrinsic reproductive isolation that makes them full species itself evolves gradually. Hence there is no precise point at which the diverging populations suddenly reach the level of full species. It would never be possible to state objectively that two populations in one generation have not quite reached the level of fully separate species, but that the two populations of the very next generation have done so. There will be time transects in the history of any two diverging lineages when the populations are in a hazy intermediate state between obviously belonging to the same species and obviously belonging to two separate species. But our definition of species makes no provision for such intermediate stages. It assumes that two populations either exhibit intrinsic reproductive isolation or that they do not, i.e. that they either belong to different species or that they belong to the same species. It does not explicitly recognize the condition "nearly species" or "barely species." Consequently such intermediate stages, when they are encountered, must always pose a problem to biologists intent on rigid categorization of what in nature is a fluid system. But the existence of intermediate stages does not invalidate the concept of speciation, because that very concept, in its modern form, predicts them.

Allopatric species   One of the most obvious and frequently encountered problems in applying the modern definition of species arises when two populations are closely related and completely allopatric. Since they are allopatric, they are obviously not exchanging genes. But the definition is not based on extrinsic isolation; it is based on intrinsic isolation. There must be neither actual nor potential effective gene flow if the two populations are to be regarded as separate species. How can potential gene flow be determined? One way that immediately comes to mind is to release a large sample of individuals from one population in the range of the other and then see whether free interbreeding takes place and, if it does, whether the hybrids are as viable as the parents. But there are obvious reasons why wholesale introduction of foreign plants and animals is seldom desirable; in fact, in many cases it is illegal. An alternative would be to bring individuals from the two allopatric populations together in the laboratory and see if they will interbreed. Sometimes this procedure is useful. If one finds that the individuals will breed freely with other members of their own population but not with members of the other population, then it is reasonable to conclude that the two populations are intrinsically isolated and should be considered separate species.

But what if interbreeding occurs freely between members of different populations in the laboratory? Are the two populations then to be regarded as belonging to the same species? No, the interbreeding simply demonstrates that certain types of intrinsic isolation do not exist between the populations. It says nothing about other types of intrinsic isolation. For example, under natural conditions ecogeographic or habitat isolation may exist, but these might very well be inoperative under laboratory conditions.

Or behavioral isolation may be operative in nature but not in the laboratory; it is known that many species of animals that will have nothing to do with each other in the wild, because of important differences in their behavior patterns, will mate in the laboratory, where their normal behavior patterns break down. Therefore, when members of two different allopatric populations cross in the laboratory and produce viable offspring, the question whether they belong to the same or to different species remains unanswered. The question likewise remains unanswered, of course, for the many species that will not breed at all under laboratory conditions.

In many cases, then, there is no good test for determining whether two allopatric populations belong to the same or to different species. The usual practice in such cases is to determine the extent to which the two populations differ and then to compare this degree of difference with that seen in related sympatric species. If the differences between the allopatric populations are of the same order of magnitude as (or greater than) those that distinguish sympatric species, the allopatric populations are considered fully separate species; if the differences are less than those that usually distinguish sympatric species, the two allopatric populations are regarded as belonging to the same species.

But degree of difference leaves much to be desired as an index to the probable presence or absence of intrinsic reproductive isolating mechanisms between allopatric populations. The assumption that allopatric populations are intrinsically isolated only if they differ from each other at least as much as related sympatric species do is often not valid. Owing to the frequency of character displacement in sympatric species, the differences between them are apt to be greater than between allopatric species. Therefore, taking the differences between sympatric species as a yardstick against which to measure differences between allopatric populations probably often results in classifying allopatric populations as members of the same species when in reality they are intrinsically isolated and should be regarded as separate species.

THE CONCEPT OF PHYLOGENY

Evolution implies that many unlike species have a common ancestor and that all forms of life probably stem from the same remote beginnings. Thus one of the tasks it sets biologists is to discover the relationships among the species alive today and to trace the ancestors from which they descended. But because so many different kinds of evidence must be weighed, reconstructing the evolutionary history—the phylogeny—of any group of organisms always entails a strong element of speculation (Fig. 18.46).

DETERMINING PHYLOGENETIC RELATIONSHIPS

Sources of data   When systematists,[9] also known as taxonomists, set out to reconstruct the phylogeny of a group of species that they think are related, they usually have before them only the species living today. They cannot observe their phylogenetic history. To reconstruct it as closely as possible, they must make inferences based on observational and experimental data

[9] Systematics, or taxonomy, in the words of G. G. Simpson, is "the scientific study of the kinds and diversity of organisms and of any and all relationships among them."

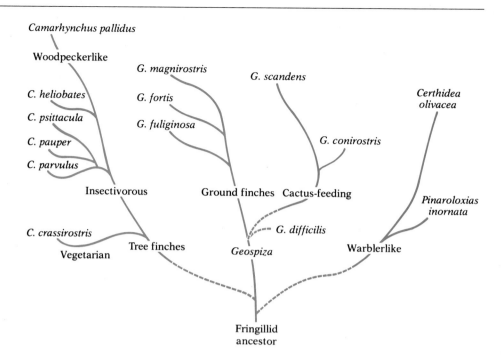

18.46   A phylogenetic tree for Darwin's finches
Biologists often resort to phylogenetic trees in attempting to show the hypothetical connections between related organisms. No phylogenetic tree should be taken too seriously, however, because the evidence is usually sufficient only to indicate the broad outlines of the tree, and much guesswork (educated guesswork, preferably) goes into the adding of details.   [Redrawn from D. Lack, *Darwin's finches,* Cambridge University Press, 1947.]

that appear relevant, if only remotely so. Those data can usually be interpreted in several ways, and only experience and good judgment can help in choosing among them.

The usual procedure in reconstructing phylogenies is to examine as many independent characters of the species in question as possible and to determine in which characters they differ and in which they are alike. The assumption is that the differences and resemblances will reflect, at least in part, their true phylogenetic relationships. Ordinarily, as many different types of characters as possible are used in the hope that misleading data from any single character will be detected by a lack of agreement with the data from other characters.

The most easily studied and widely used characters pertain to morphology—including external morphology, internal anatomy and histology, and the morphology of the chromosomes in cell nuclei. It is particularly helpful, of course, when morphological characters of living species can be compared with those of fossil forms. The fossil record is the most direct source of evidence about the stages through which past forms of life passed, but unfortunately that record is usually very incomplete and is subject to the same sorts of errors of interpretation as the characters of living species. For many groups of organisms there is no suitable fossil record for working out the relationships between species; at best, the fossils may suggest the broad outlines of the evolution of major groups. Even in those groups where fossils are abundant, only the hard parts of the organisms' bodies have usually been preserved. Nevertheless, in some groups, notably the horses, the fossil record has provided much phylogenetic information that could have been obtained from no other source (Fig. 18.47).

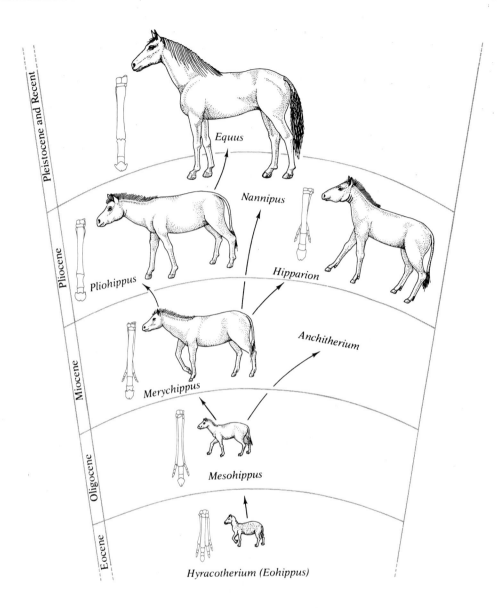

18.47 Evolution of horses
The fairly complete fossil record of horses has enabled paleontologists to work out a reasonable picture of the evolutionary history of the group. Emphasis here is on the direct ancestors of the modern horse; of the many branches that were evolutionary dead ends—i.e. left no modern descendants—only three are indicated. *Hyracotherium* lived in the Eocene epoch about 55 million years ago. It was a small animal, only about the size of a fox terrier. It had four toes on each front foot (shown here) and three on each rear foot. It was a browser, feeding on trees and bushes. *Mesohippus*, which lived during the Oligocene epoch about 35 million years ago, was a bit larger, and its front feet, like the rear feet, had only three toes. *Merychippus*, a grazer, lived during the Miocene about 25 million years ago. It had three toes on each foot, the middle one much larger than the others, which were short and thin and did not reach the ground. *Pliohippus*, of the Pliocene, often had only one toe on each foot, though in some individuals tiny remnants of other toes persisted. *Equus*, the modern horse, is much larger than the ancestors shown here. It has only one toe on each foot.   [Modified from *Horses* by G. G. Simpson. Copyright © 1951 by Oxford University Press, Inc. Renewed 1979 by G. G. Simpson. Used by permission.]

Another frequently used source of information is embryology. Morphological characters are often easier to interpret if the manner in which they develop is known. For example, if it can be shown that a particular structure in organism A and a structure of quite different appearance in organism B both develop from the same embryonic primordium, then the resemblances and differences between those structures in A and B take on a phylogenetic significance that they would not have if they developed from entirely different embryonic primordia. Embryological evidence often allows biologists to trace the probable evolutionary changes that have occurred in important structures and helps them reconstruct the probable chain of evolutionary events that led to the modern forms of life. For example, the fact that pharyngeal gill pouches appear briefly during the early embryology of mammals, including humans, is thought to indicate that the distant ancestors of land vertebrates were aquatic.

Life histories have also played an important role in phylogenetic studies. The stages through which plants pass during their life cycles are particularly important sources of information, as we shall see when we examine the algae and the vascular plants, for example.

The morphology of the adult and embryo, combined when possible with information from the fossil record and from life histories, has traditionally been the basic source of data on which phylogenetic hypotheses have been founded. In addition, comparative physiology, comparative behavior, and comparative ecology have supplied valuable information and will doubtless increase in importance as the effort to interrelate form and function grows. More recent has been the development of techniques for comparing the proteins of different species (e.g. those in the blood) by means of chromatography or electrical separation (electrophoresis). Where the complete amino acid sequence of a protein has been worked out in several species—cytochrome *c* is an example—very precise comparisons are possible.

A technique has even been developed for determining the degree of similarity between the DNA of two species. DNA is extracted from two organisms belonging to different species and treated in such fashion as to make its two strands uncouple. Next, single-stranded DNA from the two organisms is mixed under conditions that favor recoupling. The amount of coupling between strands from the different organisms—expressed as a percentage of the amount of coupling that occurs between strands from the same organism under the same conditions—gives a measure of the degree of genetic similarity between the two species.

Whenever new techniques for studying organisms are devised, the data they produce provide another source of information for the systematists. In modern biology it has become their task to gather together and correlate all that is known about the organisms under investigation, and to try to reconstruct, in the light of modern evolutionary theory, some sort of intelligible picture of the organisms and their relationships with one another. Systematists are thus just what the term implies: They are the ones who try to fit together into an orderly system all of the information gathered about the organisms by anatomists, paleontologists, cytologists, physiologists, geneticists, embryologists, ethologists, ecologists, biochemists, and still other specialists.

The problem of convergence Whether investigators are using traditional morphological data or making DNA comparisons, whether they are obtaining information from physiology or from behavior or from life histories, they are still faced with the problem of interpreting the similarities and differences they find. They must always ask themselves whether close similarities in a particular character really indicate close phylogenetic relationship or whether they simply reflect similar adaptation to the same environmental situation. The latter phenomenon is common in nature and is a frequent source of confusion in phylogenetic studies.

When organisms that are not closely related become more similar in one or more characters because of independent adaptation to similar environmental situations, they are said to have undergone convergent evolution and the phenomenon is called *convergence* (Fig. 18.48). Whales, which are mammals descended from terrestrial ancestors, have evolved flippers from the legs of their ancestors; those flippers superficially resemble the fins of fish, but the resemblances are due to convergence and they do not indicate a close relationship between whales and fish. Both arthropods and terrestrial vertebrates have evolved jointed legs and hinged jaws, but these similarities do not indicate that arthropods and vertebrates have evolved from a common ancestor that also had jointed legs and hinged jaws; there is good reason to think that these two groups of animals evolved their legs and jaws independently and that their legless ancestors were not closely related. The "moles" of Australia are not true moles but marsupials (mammals whose young are born at an early stage of embryonic development and complete their development in a pouch on the mother's abdomen); they occupy the same habitat in Australia as do the true moles in other parts of the world and have, as a result, convergently evolved many startling similarities to the true moles. The marsupial mole is but one of a vast array of Australian marsupials that are strikingly convergent with placental mammals of other continents (Fig. 18.49).

The preceding discussion makes it evident that when systematists find similarities between two species, they must try to determine whether the similarities are probably *homologous* (inherited from a common ancestor) or merely *analogous* (similar in function and often in superficial structure, but of different evolutionary origins). Thus the wings of robins and those of bluebirds are considered homologous; i.e. the evidence indicates that both were derived from the wings of a common avian ancestor. But the wings of robins and butterflies are only analogous, because, though they are functionally similar structures, they were not inherited from a common ancestor; they evolved independently and from different ancestral structures.

It is always important to indicate in what sense two structures are considered homologous or analogous. Thus the wings of birds and bats are not homologous as wings, for they evolved independently, but they contain homologous bones, both types of wings having evolved from the forelimbs of ancient vertebrates that were ancestors to both birds and mammals. In short, the wings of birds and bats are analogous as wings and homologous as forelimbs. Similarly, the flippers of whales and seals evolved independently of each other, but both evolved from the front legs of land-mammal ancestors. Thus the flippers are homologous in the sense that both are forelimbs,

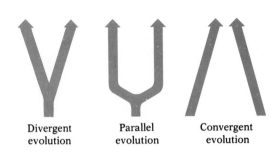

Divergent evolution   Parallel evolution   Convergent evolution

18.48 Patterns of evolution
In divergent evolution one stock splits into two, which become less and less like each other as time passes. In parallel evolution two related species evolve in much the same way for a long period of time, probably in response to similar environmental selection pressures. Convergent evolution occurs when two groups that are not closely related come to resemble each other more and more as time passes; this is usually the result of occupation of similar habitats and adoption of similar environmental roles.

A

B

with the same basic bone structure as that of other vertebrate forelimbs, but the modifications that make them flippers are analogous, not homologous.

Primitive vs. advanced Once all the data concerning a group of organisms have been collected and analyzed as to similarities and differences, and once similarities have been evaluated as to probable homology and analogy, reasonable guesses can be made regarding the degree of relationship between the various organisms. But in any attempt to reconstruct the evolutionary history of the organisms, it is necessary to consider development in time, or the direction in which the characteristics of the organisms have evolved.

Suppose, for example, that the species in one group of beetles have three segments in their tarsi (feet), while the species in another group have four, and the species in still another group five. Before speculation concerning the phylogeny of these beetles can begin, it must be determined whether the common ancestors of all three groups had three, four, five, or some other number of segments in their tarsi. Has there been an evolutionary gain or an evolutionary loss of segments, or both? In the language of the systematists: What is the primitive condition? "Primitive" means older, more like the ancestral condition. The contrasting term is "advanced," which means newer, less like the ancestral condition. Note that the terms "primitive" and "advanced" do not imply that one condition is superior to the other, nor that one is more complex than the other; they refer only to relative sequence in time.[10]

[10] For the beetles, all available evidence indicates that the larger number of tarsal segments is primitive and that the smaller number is advanced, i.e. that there has been an evolutionary reduction in the number of segments.

C

D

18.49 Australian marsupials that are convergent with placental mammals of other continents
(A) A marsupial mouse.   (B) A marsupial glider, convergent with placental flying squirrels.   (C) The tiger cat *(Dasyurus)*, a marsupial carnivore.   (D) A cuscus, a marsupial monkey of New Guinea with a prehensile tail.   [Chicago Zoological Park (Tom McHugh, Photo Researchers) (A); M. Morcombe (B, C); Jack Fields, Photo Researchers, Inc. (D).]

Two other terms, "specialized" and "generalized," are sometimes confused with "primitive" and "advanced." These terms refer to the relationships of organisms, or particular characteristics of organisms, to their environment. "Specialized" means adapted to a special, usually rather narrow, way of life. "Generalized" means broadly adapted to a greater variety of different types of environments and ways of life. Generalized characters are likely to be more primitive, and specialized ones more advanced, but this is far from being an absolute rule.

In general, within any particular group of organisms, the more primitive characters will be the ones that have evolved more slowly. Hence systematists have made a great effort to learn which characters tend to evolve slowly (i.e. to be conservative) and which tend to evolve rapidly (i.e. to be labile). The most reliable way of finding out is, of course, to examine the fossil record. However, many groups have fossil records so poor that the results of such an examination are unsatisfactory. In such cases it is necessary to deduce which characters are conservative and which are labile by study of contemporary organisms. One may assume, for example, that if a particular structure occurs in basically the same form in a great variety of only distantly related organisms, the characters of that structure are conservative and therefore provide more information about the primitive condition than characters of some structure that varies greatly from species to species. Unfortunately, it is seldom possible to generalize from one group of organisms to another as to which sorts of characters are conservative and which are labile. Labile and conservative characters differ from group to group and must be determined separately for each.

Sequence of amino acids in cytochrome c for 35 organisms

Group	Organism	1	2	3	4	5	6	7	8	9	10	11	12	13	14	15	16	17	18	19	20	21	22	23	24	25	26	27	28	29	30	31	32	33	34	35	36	37	38	39	40
Mammals	Human, chimpanzee	G	D	V	E	K	G	K	K	I	F	I	M	K	C	S	Q	C	H	T	V	E	K	G	G	K	H	K	T	G	P	N	L	H	G	L	F	G	R	K	T
	Rhesus monkey	G	D	V	E	K	G	K	K	I	F	I	M	K	C	S	Q	C	H	T	V	E	K	G	G	K	H	K	T	G	P	N	L	H	G	L	F	G	R	K	T
	Horse	G	D	V	E	K	G	K	K	I	F	V	Q	K	C	A	Q	C	H	T	V	E	K	G	G	K	H	K	T	G	P	N	L	H	G	L	F	G	R	K	T
	Donkey	G	D	V	E	K	G	K	K	I	F	V	Q	K	C	A	Q	C	H	T	V	E	K	G	G	K	H	K	T	G	P	N	L	H	G	L	F	G	R	K	T
	Cow, pig, sheep	G	D	V	E	K	G	K	K	I	F	V	Q	K	C	A	Q	C	H	T	V	E	K	G	G	K	H	K	T	G	P	N	L	H	G	L	F	G	R	K	T
	Dog	G	D	V	E	K	G	K	K	I	F	V	Q	K	C	A	Q	C	H	T	V	E	K	G	G	K	H	K	T	G	P	N	L	H	G	L	F	G	R	K	T
	Rabbit	G	D	V	E	K	G	K	K	I	F	V	Q	K	C	A	Q	C	H	T	V	E	K	G	G	K	H	K	T	G	P	N	L	H	G	L	F	G	R	K	T
	California gray whale	G	D	V	E	K	G	K	K	I	F	V	Q	K	C	A	Q	C	H	T	V	E	K	G	G	K	H	K	T	G	P	N	L	H	G	L	F	G	R	K	T
	Great gray kangaroo	G	D	V	E	K	G	K	K	I	F	V	Q	K	C	A	Q	C	H	T	V	E	K	G	G	K	H	K	T	G	P	N	L	N	G	I	F	G	R	K	T
Other vertebrates	Chicken, turkey	G	D	I	E	K	G	K	K	I	F	V	Q	K	C	S	Q	C	H	T	V	E	K	G	G	K	H	K	T	G	P	N	L	H	G	L	F	G	R	K	T
	Pigeon	G	D	I	E	K	G	K	K	I	F	V	Q	K	C	S	Q	C	H	T	V	E	K	G	G	K	H	K	T	G	P	N	L	H	G	L	F	G	R	K	T
	Pekin duck	G	D	V	E	K	G	K	K	I	F	V	Q	K	C	S	Q	C	H	T	V	E	K	G	G	K	H	K	T	G	P	N	L	H	G	L	F	G	R	K	T
	Snapping turtle	G	D	V	E	K	G	K	K	I	F	V	Q	K	C	A	Q	C	H	T	V	E	K	G	G	K	H	K	T	G	P	N	L	N	G	L	I	G	R	K	T
	Rattlesnake	G	D	V	E	K	G	K	K	I	F	T	M	K	C	S	Q	C	H	T	V	E	K	G	G	K	H	K	T	G	P	N	L	H	G	L	F	G	R	K	T
	Bullfrog	G	D	V	E	K	G	K	K	I	F	V	Q	K	C	A	Q	C	H	T	C	E	K	G	G	K	H	K	V	G	P	N	L	Y	G	L	I	G	R	K	T
	Tuna	G	D	V	A	K	G	K	K	T	F	V	Q	K	C	A	Q	C	H	T	V	E	N	G	G	K	H	K	V	G	P	N	L	W	G	L	F	G	R	K	T
	Dogfish shark	G	D	V	E	K	G	K	K	V	F	V	Q	K	C	A	Q	C	H	T	V	E	N	G	G	K	H	K	T	G	P	N	L	S	G	L	F	G	R	K	T
Insects*	Ailanthus silkmoth	G	N	A	E	N	G	K	K	I	F	V	Q	R	C	A	Q	C	H	T	V	E	A	G	G	K	H	K	V	G	P	N	L	H	G	F	Y	G	R	K	T
	Tobacco hornworm moth	G	N	A	D	N	G	K	K	I	F	V	Q	R	C	A	Q	C	H	T	V	E	A	G	G	K	H	K	V	G	P	N	L	H	G	F	F	G	R	K	T
	Screwworm fly	G	D	V	E	K	G	K	K	I	F	V	Q	R	C	A	Q	C	H	T	V	E	A	G	G	K	H	K	V	G	P	N	L	H	G	L	F	G	R	K	T
	Fruit fly (Drosophila)	G	D	V	E	K	G	K	K	L	F	V	Q	R	C	A	Q	C	H	T	V	E	A	G	G	K	H	K	V	G	P	N	L	H	G	L	I	G	R	K	T
Fungi*	Baker's yeast	G	S	A	K	K	G	A	T	L	F	K	T	R	C	E	L	C	H	T	V	E	K	G	G	P	H	K	V	G	P	N	L	H	G	I	F	G	R	H	S
	Candida krusei (a yeast)	G	S	A	K	K	G	A	T	L	F	K	T	R	C	A	E	C	H	T	I	E	A	G	G	P	H	K	V	G	P	N	L	H	G	I	F	S	R	H	S
	Red bread mold	G	D	S	K	K	G	A	N	L	F	K	T	R	C	A	E	C	H	G	E	G	G	N	L	T	Q	K	I	G	P	A	L	H	G	L	F	G	R	K	T
Plants*	Wheat	G	N	P	D	A	G	A	K	I	F	K	T	K	C	A	Q	C	H	T	V	D	A	G	A	G	H	K	Q	G	P	N	L	H	G	L	F	G	R	Q	S
	Sunflower	G	D	P	T	T	G	A	K	I	F	K	T	K	C	A	Q	C	H	T	V	E	K	G	A	G	H	K	Q	G	P	N	L	N	G	L	F	G	R	Q	S
	Mung bean	G	D	S	K	S	G	E	K	I	F	K	T	K	C	A	Q	C	H	T	V	D	K	G	A	G	H	K	Q	G	P	N	L	N	G	L	F	G	R	Q	S
	Cauliflower	G	D	S	K	A	G	E	K	I	F	K	T	K	C	A	Q	C	H	T	V	D	K	G	A	G	H	K	Q	G	P	N	L	N	G	L	F	G	R	Q	S
	Pumpkin	G	D	S	K	A	G	E	K	I	F	K	T	K	C	A	Q	C	H	T	V	D	K	G	A	G	H	K	Q	G	P	N	L	N	G	L	F	G	R	Q	S
	Castor bean	G	D	V	K	A	G	E	K	I	F	K	T	K	C	A	Q	C	H	T	V	E	K	G	A	G	H	K	Q	G	P	N	L	N	G	L	F	G	R	Q	S
	Sesame	G	D	V	K	S	G	E	K	I	F	K	T	K	C	A	Q	C	H	T	V	D	K	G	A	G	H	K	Q	G	P	N	L	N	G	L	F	G	R	Q	S
	Number of different amino acids	1	3	5	5	5	1	3	3	4	1	4	3	2	1	3	3	1	1	2	4	3	4	2	3	4	2	1	4	1	1	2	1	5	1	3	3	2	1	3	2

* In cytochrome *c* from insects, fungi, and plants, a few (4–8) amino acids are usually ahead of what is here labeled Position 1; these are omitted from the table.

Symbol	Amino acid	Symbol	Amino acid
☐	NONPOLAR	▨	POLAR
G	Glycine	S	Serine
A	Alanine	T	Threonine
V	Valine	C	Cysteine
L	Leucine	Y	Tyrosine
I	Isoleucine	N	Asparagine
M	Methionine	Q	Glutamine
F	Phenylalanine		
W	Tryptophan	▨	ACIDIC
P	Proline	D	Aspartic acid
		E	Glutamic acid
		▨	BASIC
		K	Lysine
		R	Arginine
		H	Histidine

G is printed in color, because glycine, despite its technically nonpolar R group, behaves like a polar amino acid.

CYTOCHROME *c* AS A TAXONOMIC CHARACTER

One protein much examined in recent years as an indicator of phylogenetic relationships is cytochrome *c*, an essential component of the respiratory chain in mitochondria. The complete amino acid sequence of this enzyme has been worked out for a variety of organisms, both plant and animal. The table shows the sequence for some of the species so far examined, with the various functional groupings of amino acids (as determined by their R groups) indicated by a color code.

Perhaps the first thing one notices from this table is that cytochrome *c* is remarkably similar in all the species, even though some of them have probably not had a common ancestor for more than a billion years. For example, all the cytochromes have the same amino acid sequence from positions 70 through 80. It has been found, in fact, that cytochrome *c* is an evolutionarily

			45			50			55			60			65			70			75			80			85			90			95			100	104

```
G Q A P G Y S Y T A A N K N K G I I W G E D T L M E Y L E N P K K Y I P G T K M I F V G I K K K E E R A D L I A Y L K K A T N E
G Q A P G Y S Y T A A N K N K G I T W G E D T L M E Y L E N P K K Y I P G T K M I F V G I K K K E E R A D L I A Y L K K A T N E
G Q A P G F T Y T D A N K N K G I T W K E E T L M E Y L E N P K K Y I P G T K M I F A G I K K K T E R E D L I A Y L K K A T N E
G Q A P G F T Y T D A N K N K G I T W K E E T L M E Y L E N P K K Y I P G T K M I F A G I K K K T E R E D L I A Y L K K A T N E
G Q A P G F S Y T D A N K N K G I T W G E E T L M E Y L E N P K K Y I P G T K M I F A G I K K K G E R E D L I A Y L K K A T N E
G Q A P G F S Y T D A N K N K G I T W G E E T L M E Y L E N P K K Y I P G T K M I F A G I K K T G E R A D L I A Y L K K A T K E
G Q A V G F S Y T D A N K N K G I T W G E D T L M E Y L E N P K K Y I P G T K M I F A G I K K K D E R A D L I A Y L K K A T N E
G Q A V G F S Y T D A N K N K G I T W G E E T L M E Y L E N P K K Y I P G T K M I F A G I K K K G E R A D L I A Y L K K A T N E
G Q A P G F T Y T D A N K N K G I I W G E D T L M E Y L E N P K K Y I P G T K M I F A G I K K K G E R A D L I A Y L K K A T N E

G Q A E G F S Y T D A N K N K G I T W G E D T L M E Y L E N P K K Y I P G T K M I F A G I K K K S E R V D L I A Y L K D A T S K
G Q A E G F S Y T D A N K N K G I T W G E D T L M E Y L E N P K K Y I P G T K M I F A G I K K K A E R A D L I A Y L K Q A T A K
G Q A E G F S Y T D A N K N K G I T W G E D T L M E Y L E N P K K Y I P G T K M I F A G I K K K S E R A D L I A Y L K D A T A K
G Q A V G Y S Y T E A N K N K G I T W G E E T L M E Y L E N P K K Y I P G T K M I F A G I K K K A E R A D L I A Y L K D A T S K
G Q A V G Y S Y T E A N K N K G I I W G D D T L M E Y L E N P K K Y I P G T K M I F T G L S K K K E R T N L I A Y L K E K T A A
G Q A A G F S Y T D A N K N K G I T W G E D T L M E Y L E N P K K Y I P G T K M I F A G I K K K G E R Q D L I A Y L K S A C S K
G Q A E G Y S Y T D A N K S K G I V W N N D T L M E Y L E N P K K Y I P G T K M I F A G I K K K G E R Q D L V A Y L K S A T S -
G Q A Q G F S Y T D A N K S K G I T W Q Q E T L R I Y L E N P K K Y I P G T K M I F A G L K K K S E R Q D L I A Y L K K T A A S

G Q A P G F S Y S N A N K A K G I T W G D D T L F E Y L E N P K K Y I P G T K M V F A G L K K A N E R A D L I A Y L K E S T K -
G Q A P G F S Y S N A N K A K G I T W Q D D T L F E Y L E N P K K Y I P G T K M V F A G L K K A N E R A D L I A Y L K Q A T K -
G Q A A G F A Y T N A N K A K G I T W Q D D T L F E Y L E N P K K Y I P G T K M I F A G L K K P N E R G D L I A Y L K S A T K -
G Q A A G F A Y T N A N K A K G I T W Q D D T L F E Y L E N P K K Y I P G T K M I F A G L K K P N E R G D L I A Y L K S A T K -

G Q A Q G Y S Y T D A N I K K N V L W D E N N M S E Y L T N P K K Y I P G T K M A F G G L K K E K D R N D L I T Y L K K A C E -
G Q A Q G Y S Y T D A N K R A G V E W A E P T M S D Y L E N P K K Y I P G T K M A F G G L K K A K D R N D L V T Y M L E A S K -
G S V D G Y A Y T D A N K Q K G I T W D E N T L F E Y L E N P K K Y I P G T K M A F G G L K K D K D R N D I I T F M K E A T A -

G T T A G Y S Y S A A N K N K A V E W E E N T L Y D Y L L N P K K Y I P G T K M V F P G L K K P Q D R A D L I A Y L K K A T S S
G T T A G Y S Y S A A N K N M A V I W E E N T L Y D Y L L N P K K Y I P G T K M V F P G L K K P Q E R A D L I A Y L K T S T A -
G T T A G Y S Y S T A N K N M A V E W E E K T L Y D Y L L N P K K Y I P G T K M V F P G L K K P Q D R A D L I A Y L K E S T A -
G T T P G Y S Y S A A N K N R A V I W E E K T L Y D Y L L N P K K Y I P G T K M V F P G L K K P Q D R A D L I A Y L K E A T A -
G T T A G Y S Y S A A N K N M A V Q W G E N T L Y D Y L L N P K K Y I P G T K M V F P G L K K P Q D R A D L I A Y L K E A T A -
G T T P G Y S Y S A A N K N M A V I W G E N T L Y D Y L L N P K K Y I P G T K M V F P G L K K P Q D R A D L I A Y L K E A T A -
```

1 3 3 6 1 2 3 1 2 5 1 1 2 6 4 3 2 6 1 7 4 5 2 2 5 3 1 1 3 1 1 1 1 1 1 1 1 1 1 3 1 5 1 2 2 1 6 9 2 1 7 1 2 2 2 2 2 6 4 4 5 4

conservative protein; its amino acid sequence has changed at a considerably slower rate (about 20 million years for one percent change) than, for example, the amino acid sequence of hemoglobin (5.8 million years) or that of fibrin (only one million years). The minimal change in cytochrome *c* suggests that only minor alterations can be tolerated if the enzyme is to continue functioning properly. Notice that even at points along the chain where there is variability, the amino acids are often functionally similar ones (e.g. they are all nonpolar or all polar).

If we compare various species in the table with one another, we find that the number of differences in amino acids usually agrees reasonably well with the presumed evolutionary distances among the species. Thus the mammals differ less among themselves than any of them differ from the fish. Human beings and chimpanzees do not differ at all; both differ by one amino acid from

the rhesus monkey, by an average of 10.4 amino acids from the other mammals, by an average of 14.5 from the reptiles, by 18 from the amphibian, and by an average of 22.5 from the fishes. This is an accurate reflection of the generally accepted evolutionary sequence of fish ⟶ amphibian ⟶ reptile ⟶ mammal.

Because cytochrome *c* is evolutionarily so conservative, its value as a taxonomic character is limited to studies of the relationships among evolutionarily distant organisms. It cannot be used in comparing families or genera. But other more rapidly changing proteins, such as fibrin, may prove useful in those cases.

The table is adapted from M. O. Dayhoff, ed., *Atlas of protein sequence and structure* (Washington, D.C.: National Biomedical Research Foundation, 1972), vol. 5, and R. E. Dickerson, "The structure and history of an ancient protein," *Scientific American*, April 1972.

PHYLOGENY AND CLASSIFICATION

Over a million species of animals and over 325,000 species of plants are known. To deal with this vast array of organic diversity, we obviously need some sort of system by which species can be classified in a logical and meaningful manner. Many different kinds of classification are possible. We could, for example, classify flowering plants according to their color: all white-flowered species in one group, all red-flowered species in a second group, all yellow-flowered species in a third group, etc. Or we could classify these same plants according to their average height: all species less than 3 cm tall in one group, all species more than 3 cm and less than 30 cm tall in a second group, all species more than 30 cm but less than one meter tall in a third group, etc. Or we could classify the same plants according to the environment in which they grow: all species that grow primarily in fields in one group, all species that grow primarily in forests in a second group, all species that grow primarily in lakes in a third group, etc. Each of these three systems of classification, and many others that could be devised, would impose a measure of order, but how meaningful, how informative, would these classifications be? Obviously the information they would convey—about flower color, about average height, about habitat—is of an incidental kind that would fail to set apart fundamentally different organisms.

The classification system used in biology today, by contrast, is an attempt to encode the evolutionary history of the organisms; it is thus often a means of conveying information about many of their characteristics to those familiar with the various taxonomic groups (taxons) to which the organisms are assigned.

The classification hierarchy   Suppose you had to classify all the people on earth on the basis of where they live. You would probably begin by dividing the entire world population into groups based on country. This subdivision separates inhabitants of the United States from the inhabitants of France or Argentina, but it still leaves very large groups that must be further subdivided. Next, you would probably subdivide the population of the United States by states, then by counties, then by city or village or township, then by street, and finally by house number. You could do the same thing for Mexico, England, Australia, and all the other countries (using whatever political subdivisions in those countries correspond to states, counties, etc. in the United States). This procedure would enable you to place every individual in an orderly system of hierarchically arranged categories, as follows:

Country
State
County
City
Street
Number

Note that each level in this hierarchy is contained within and is partly determined by all levels above it. Thus, once the country has been determined as the United States, a Mexican state or a Canadian province is

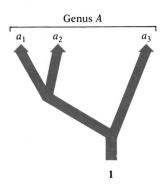
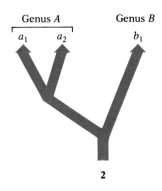
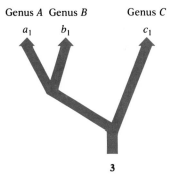

excluded. Similarly, once the state has been determined as Pennsylvania, a county in New York or one in California is excluded.

The same principles apply to the classification of living things on the basis of phylogenetic relationships. Again a hierarchy of categories is used, as follows:

> Kingdom
> Phylum or Division
> Class
> Order
> Family
> Genus
> Species

Each category (taxon) in this hierarchy is a collective unit containing one or more groups from the next-lower level in the hierarchy. Thus a genus is a group of closely related species (Fig. 18.50); a family is a group of related genera; an order is a group of related families; a class is a group of related orders; etc. The species in any one genus are believed to be more closely related to each other than to species in any other genus; the genera in any one family are believed to be more closely related to each other than to genera in any other family; and the families in any one order are believed to be more closely related to each other than to families in any other order; etc.

Table 18.2 gives the classification of six species. Notice that the table shows us immediately that the six species are not closely related, but that a human being and a wolf are more closely related to each other, both being Mammalia, than either is to a bird such as the Herring Gull. And it shows us that the mammals and the bird are more closely related to each other than to the housefly, which is in a different phylum, or to the moss or the red oak, which are in a different kingdom. These relationships are correlated with similarities and differences in the morphology, physiology, and ecology of the six species.

The six species of Table 18.2 are all well known, and the relationships between them are probably intuitively clear to you. But many species are not so well known, and the relationships between them not so clear. Much research may be necessary before they can be fitted into the classification system with any degree of certainty, and their assignment to genus or family (or even order) may have to be changed as more is learned about them.

18.50   Alternative generic grouping for three related species

Biologists try to group species in a way that will indicate their phylogenetic relationships. Thus a genus is a group of related species. But how closely related? There is no absolute answer to this question. Some biologists (the "lumpers") like large genera containing many subdivisions (subgenera or "species groups"); others (the "splitters") prefer small compact genera, containing only species that are very closely related. In the drawing, three alternative ways of grouping three related species are shown. The first recognizes only one genus, the second recognizes two, and the third recognizes three.

Hierarchical classification systems similar to the one in current use have been employed by naturalists for many centuries. The current system dates from the work of the great Swedish naturalist Carolus Linnaeus (1707–1778), who wrote extensively on the classification of both plants and animals. His system used kingdoms, classes, orders, genera, and species; the phylum and family categories were added later. Now, the rationale on which Linnaeus based his system was necessarily very different from the phylogenetic one employed today. He worked a century before Darwin, and he had no conception of evolution, doubtless conceiving of each species as an immutable entity, the product of a divine creation. He was simply grouping organisms according to similarities, primarily morphological. That his results were so similar to those obtained today is a reflection of the fact that morphological characters, being products of evolution, tell us much about evolutionary relationships. The Linnaean system, however, produces results quite different from those of the modern phylogenetic system whenever it has to deal with convergence or with gross morphological similarities that may be misleading as to actual phylogenetic relationships.

An outline of a modern classification of living things is given on pp. 1075–80.

Nomenclature The modern system of naming species also dates from Linnaeus. Before him, there had been little uniformity in the designation of species. Some species had a one-word name, others had two-word names, and still others had names consisting of long descriptive phrases. For example, the pre-Linnaean name for the common carnation was *dianthus floribus solitariis, squamis calycinis subovatis brevissimis, corollis crenatis*, and the name for the honeybee was *Apis pubescens, thorace subgriseo, abdomine fusco, pedibus posticis glabris utrinque margine ciliatis*. Linnaeus simplified things by giving each species a name consisting of two words: first, the name of the genus to which the species belongs and, second, a designation for that particular species. Thus the above-mentioned species of carnation became *Dianthus caryophyllus*, and the honeybee became *Apis mellifera*. Other species in the genus *Dianthus* have the same first word in their names, but each has its own specific designation (e.g. *Dianthus prolifer, Dianthus barbatus,*

TABLE 18.2 *Classification of six species*

Category	Haircap moss	Red oak	Housefly	Herring Gull	Wolf	Human
Kingdom	Plantae	Plantae	Animalia	Animalia	Animalia	Animalia
Phylum or Division	Bryophyta	Tracheophyta	Arthropoda	Chordata	Chordata	Chordata
Class	Musci	Angiospermae	Insecta	Aves	Mammalia	Mammalia
Order	Bryales	Fagales	Diptera	Charadriiformes	Carnivora	Primates
Family	Polytrichaceae	Fagaceae	Muscidae	Laridae	Canidae	Hominidae
Genus	*Polytrichum*	*Quercus*	*Musca*	*Larus*	*Canis*	*Homo*
Species	*commune*	*rubra*	*domestica*	*argentatus*	*lupus*	*sapiens*

Dianthus deltoides). No two species can have the same name.[11] Notice that the names are always Latin (or Latinized) and that the genus name is capitalized while the specific name is not.[12] Both names are customarily printed in italics (underlined if handwritten or typed). The correct name for any species, according to the present rules, is usually the oldest validly proposed name.[13]

The same Latin scientific names are used throughout the world. This uniformity of usage ensures that each scientist will know exactly which species another scientist is discussing. There would be no such assurance if common names were used, for not only does a given species have a different common name in each language, but it often has two or three names in a single language. For example, the plant *Bidens frondosa* is known by all the following English names: beggar-ticks, sticktight, bur marigold, devil's bootjack, pitchfork weed, and rayless marigold. To confuse matters further, a single common name is frequently applied to several species. For example, "gopher" is the name of a turtle in Florida and the name of a rodent in Kansas, and "raspberry" is the common name for more than a hundred species of plants.

[11] More precisely, no two species of plants can have the same name, and no two species of animals can have the same name. Since the International Rules of Botanical Nomenclature and the International Rules of Zoological Nomenclature are completely separate, it is possible for a plant and an animal to have the same name. There is also a separate International Bacteriological Code of Nomenclature.

[12] This rule always holds for zoological names, but specific botanical names are sometimes capitalized when they are derived from the name of a person or from other proper nouns.

[13] For purposes of priority, botanical naming dates from the publication of Linnaeus' *Species plantarum* in 1753, and zoological naming dates from the publication of the 10th edition of his *Systema naturae* in 1758.

SUGGESTED READING

BISHOP, J. A., and L. M. COOK, 1975. "Moths, melanism and clean air," *Scientific American*, January. (Offprint 1314)   *The lessening of air pollution in Britain and the diminishing frequency of melanics in some moth populations.*

CLARKE, B., 1975. "The causes of biological diversity," *Scientific American*, August. (Offprint 1326)

DAWKINS, R., 1976. *The selfish gene.* Oxford University Press, New York.   *On the idea that the gene, not the individual organism, is the unit of selection, the organism being merely the container, the robot vehicle, of its selfish genes.*

GRANT, V., 1951. "The fertilization of flowers," *Scientific American*, June. (Offprint 12)   *The special adaptations of flowers that help ensure their pollination.*

GRANT, V., 1977. *Organic evolution.* Freeman, San Francisco.   *Excellent up-to-date treatment, with special emphasis on speciation.*

KETTLEWELL, H. B. D., 1959. "Darwin's missing evidence," *Scientific American*, March. (Offprint 842)   *The story of the industrial melanism of the pepper moth in England.*

LACK, D., 1947. *Darwin's finches.* Cambridge University Press, New York.   *Clear account of the classic example of island speciation.*

MAYR, E., 1978. "Evolution," *Scientific American*, September. (Offprint 1400)

STEBBINS, G. L., 1977. *Processes of organic evolution,* 3rd ed. Prentice-Hall, Englewood Cliffs, N.J.   *Clearly written, well-balanced account.**

*Available in paperback.

Chapter 19

ECOLOGY

Ecology is usually defined as the study of the interactions between organisms and their environment. "Environment," given a very broad meaning here, embraces all those things extrinsic to the organism that in any way impinge on it. It includes not only light, temperature, rainfall, humidity, and topography, but also parasites, predators, mates, and competitors. Anything affecting a particular organism and not an integral part of it is part of that organism's environment.

Life is characterized by many different levels of organization. Much of the first half of this book dealt with life on the molecular, cellular, tissue, organ, organ-system, or individual level. Although ecology is often concerned with phenomena on these levels (e.g. the osmotic interactions between an organism and its environmental medium), the present chapter will deal primarily with three higher levels of organization: *populations,* which are groups of individuals belonging to the same species; *communities,* which are units composed of all the populations living in a given area; and *ecosystems,* which are communities and their physical environments considered together. Each of these designations may be applied to a small local entity or to a large widespread one. Thus the sycamore trees in a given woodlot may be regarded as a population, and so may all the sycamore trees in the eastern United States. Similarly, a small pond and its inhabitants or the forest in which the pond is located may be treated as an ecosystem.

The various ecosystems are linked to one another by biological, chemical, and physical processes. Inputs and outputs of energy, gases, inorganic chemicals, and organic compounds can cross ecosystem boundaries through meteorological factors such as wind and precipitation, geological ones such as running water and gravity, and biological ones such as the movement of animals. Thus the entire earth is itself a true ecosystem, in that no part is fully isolated from the rest. The global ecosystem is ordinarily called the ***biosphere.***

The biosphere contains all living organisms and their environment. It forms a relatively thin shell around the earth, extending only a few kilometers above and below sea level. Except for energy, it is self-sufficient; all other requirements for life, such as water, oxygen, and nutrients, are supplied by utilization and recycling of materials already contained within the system.

POPULATIONS AS UNITS OF STRUCTURE AND FUNCTION

We shall begin our study of ecology by focusing on the same level of organization that claimed our attention in the last chapter, namely populations—groups of individuals belonging to the same species or to the same local subdivision of a species. We shall be especially interested in the dynamics of populations and in the environmental factors that help regulate them.

POPULATION SIZE AND DISTRIBUTION

Ecologists sometimes need to know the number of individuals in a population. When they consider endangered species, for example, the size of the surviving population is of crucial importance in the design of proper management procedures. Thus conservation authorities must have an accurate count (or at least a good estimate) of just how many Whooping Cranes or blue whales still exist on the earth.

But more often ecologists are concerned, not with the total number of living individuals of a species, but with the density of the population in a given region, i.e. with the number of individuals per unit area or volume (e.g. 50 pine trees per acre, or 5,000 diatoms per liter of water). In some situations, especially when the size of the individuals in a population is extremely variable, ecologists find that biomass (i.e. the total mass of all the individuals), or its energy equivalent in calories, per unit area or volume is a more useful index to the population's importance in the ecosystem.

Estimating population density   However population density is expressed, measuring it is not always a simple procedure, and different methods may yield widely varying results. Making total counts in the area under study often works well when the organisms being counted are large or conspicuous, and not excessively abundant. An alternative procedure is to count only a limited number of small sampling blocks (quadrats) and then estimate the density for the whole study area from these (Fig. 19.1). It goes without saying that the sampling blocks must be as representative of the entire study area as possible, if large errors in estimating are to be avoided.

Another method sometimes used in estimating population densities of animals is the mark-and-recapture technique. Here a limited number of individuals (let's say 20) are captured, marked with a tag or dye or some other device, and then released back into the same population. At some later time a second group of animals is captured from the population, and the percentage of marked individuals is determined; if 10 percent of the animals

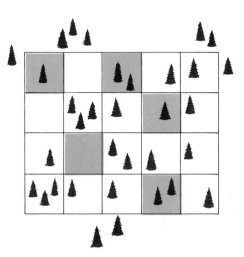

19.1 The quadrat sampling method
A grid is set up in the study area, and counts are made in a sampling of the squares thus delimited. The results permit an estimate of the population of the entire study area. The five squares sampled here (color) contain a total of six trees; hence it is estimated that the 20 squares of the entire grid will contain approximately 24 trees. In this example the selection of squares was random, but in some situations, as when obvious gradients run through the study area, other sampling designs may be more suitable.

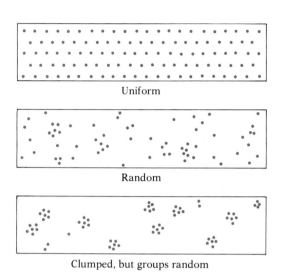

Uniform

Random

Clumped, but groups random

19.2 Uniform, random, and clumped distributions

in this group are marked, the investigator may conclude that the original 20 marked individuals represented only 10 percent of the population in the study area, and hence that the total population is about 200.

Spacing Within any given area the individuals of a population could be distributed uniformly, randomly, or in clumps (Fig. 19.2). Uniform distributions are not common; they occur only where environmental conditions are fairly uniform throughout the area and where, in addition, there is intense competition or antagonism between individuals. For example, creosote bushes are often spaced almost uniformly over a desert area because the roots of each bush give off toxic substances that prevent germination of seedlings in a circular zone around the base of the bush. It has been shown that other aromatic shrubs give off volatile growth inhibitors that prevent root growth of seedlings in the vicinity, and this emanation sometimes produces quite regular spacing of the shrubs.

Random distributions are also relatively rare. They occur only where the environmental conditions are uniform, where there is no intense competition or antagonism between individuals, and where, moreover, there is no tendency for the individuals to aggregate. Since the chances that all three of these conditions will be met simultaneously are obviously poor, ecologists must usually work with nonrandom distributions, which complicates the sampling procedures and statistical tests they may use.

Clumping is by far the most common distribution pattern for both plants and animals in nature, for several reasons:

1. The environmental conditions are seldom uniform throughout even a relatively small area. Variations in soil, in topography, in the distributions of other species, and in such microclimatic factors as moisture, temperature, and light may produce important habitat differences within the area. The organisms will obviously tend to occur in those spots where such conditions are most favorable.

2. Reproductive patterns often favor clumping. This is particularly true in plants that reproduce vegetatively and in animals where the young remain with their parent.

3. Animals often exhibit behavior patterns that lead to active congregation in loose groups or in more organized colonies, schools, flocks, or herds. And even sexual attraction produces a departure from the theoretical conditions necessary for a completely random distribution.

A clumped distribution may increase competition for nutrients, food, space, or light, but this deleterious effect is often offset by some beneficial ones. For example, trees growing together in a hedgerow on the Great Plains may compete more intensely for nutrients and light than if they were widely separated, but they may be better able to withstand strong winds. And the clump, which has less surface area in proportion to mass than an isolated tree, may be better able to conserve moisture and to modify its own microclimate and microhabitat in other ways. Aggregations of animals often reduce the rate of temperature change in their midst—an effect that is particularly important in cold weather. A group of animals may also have an advantage in locating food and in withstanding attacks by predators. Light-sensitive animals often survive exposure to illumination longer if they are in a group. Thus we see that the optimum density for population growth and survival is often an intermediate one; undercrowding may be as deleterious as overcrowding.

POPULATION GROWTH

The exponential growth curve   One way to understand the dynamics of real populations is to find out what to expect of a population under ideal conditions, and then try to determine how actual conditions modify this expected pattern. Let us assume we can study a population that is stationary, has a stable age distribution, faces no predation, parasitism, or competition, and exists in an environment with unlimited resources. For any species the population growth rate under such ideal conditions is enormous. You are doubtless familiar with the dramatic projections of what would happen if a pair of organisms produced a full complement of offspring and if all those offspring survived to produce a full complement of offspring in their turn, and so on for a number of generations. For example, one pair of houseflies starting to breed in April could have 191,010,000,000,000,000,000 descendants by August if all their eggs hatched and if all the resulting young survived to reproduce. One biologist[1] has made the following calculations concerning 100 sea stars (starfish) found living along a small sector of the Pacific coast just north of San Francisco:

> Assuming that half of these were females, and that each produced one million eggs (a modest estimate), the population in the next year would be about 50,000,000. These would include about 25,000,000 females, all of which would again produce about a million eggs each. It is obvious that, if the ordinary rate of reproduction were to continue for even a few generations with 100 percent survival of all offspring, soon the starfish would fill the seas and be pushed out across the lands by the sheer pressure of reproduction. Indeed, at the rate of reproduction here described, it would take only fifteen generations for the number of starfish to exceed the estimated number of electrons in the visible universe (10^{79})!

We mentioned in Chapter 1 that such calculations as these played an important role in leading Charles Darwin to formulate his theory of natural selection. Darwin himself made the following estimates on the reproductive potential of elephants:

> The elephant is reckoned the slowest breeder of all known animals, and I have taken some pains to estimate its probable minimum rate of natural increase; it will be safest to assume that it begins breeding when 30 years old, and goes on breeding till 90 years old, bringing forth six young in the interval, and surviving till 100 years old; if this be so, after a period of from 740 to 750 years there would be nearly 19 million elephants alive descended from the first pair.[2]

Someone has extended Darwin's calculations to show that in 100,000 years one pair of elephants would have so many living descendants that they would fill the visible universe.

---

[1] E. O. Dodson, *Evolution: Process and product*, Reinhold Publishing Corp., 1960, p. 4.

[2] Darwin attributed a somewhat greater life-span to elephants than they probably have; most do not live past 50, and the highest authenticated age is 77. Since they begin breeding at much under 30, however, Darwin's calculations remain essentially correct.

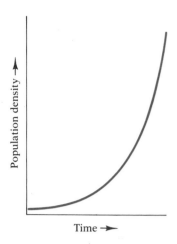

19.3 The exponential growth curve
The rate of increase steadily accelerates until, in theory, the population density would eventually be increasing at an infinitely high rate. Clearly, no real population can for long continue increasing exponentially.

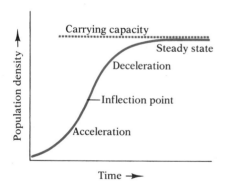

19.4 The logistic growth curve
Often called an S-shaped growth curve, the logistic curve shows an accelerating rate of growth at low densities, but eventually reaches an inflection point where the rate-change shifts from acceleration to deceleration. The deceleration continues as the population density approaches the carrying capacity of the environment. When the carrying capacity is reached, there is no further increase in density, and the population continues in steady state.

We have here mentioned only houseflies, sea stars, and elephants, but the same considerations apply to all organisms, whether plant or animal, unicellular or multicellular. All have the potential of explosive growth; i.e. under unlimited environmental conditions their growth curve would be exponential, as shown in Fig. 19.3. Let us examine such a curve more carefully. If I is the rate of increase in the number of individuals in the population, b the average birth rate, d the average death rate, and N the number of individuals in the population at a given moment, then we can write the equation for a population growth curve such as that of Fig. 19.3 as follows:

$$I = (b - d)N$$

From this formulation it is immediately apparent that a population will grow only if the average birth rate exceeds the average death rate, so that the term $(b - d)$ is greater than zero. Conversely, the population will decline if the average birth rate is less than the average death rate, so that the term $(b - d)$ is less than zero. If the birth and death rates are equal, i.e. if $b = d$, then $(b - d) = 0$, and therefore $I = 0$; such a population would be neither growing nor decreasing. In summary, then, it is the value of $(b - d)$, the difference between the birth and death rates, that determines whether a population will grow, be stable, or decline. This value, customarily called the **intrinsic rate of increase** of the population, is designated r. In other words, $b - d = r$. We can therefore rewrite the equation for the growth curve of Fig. 19.3 as follows:

$$I = rN$$

In the hypothetical sea star or elephant populations living under unlimited environmental conditions and undergoing a population explosion, births exceed deaths and r is therefore greater than zero. But the rate of population growth is a function not just of r but also of N, the population size. Since in each successive generation N is larger, it follows that the rate of increase, I, will also be larger with each generation. It is this accelerating rate of increase that accounts for the fact that the slope of the curve in Fig. 19.3 becomes steeper and steeper, until eventually the population would be expanding at infinite speed.

The logistic growth curve No real population is expanding at infinite speed, and we are not buried in sea stars or elephants. The reason must be that the exponential growth of real populations is checked. In many instances the result is a growth curve like that in Fig. 19.4, which shows initial rapid expansion of the population when it is at low densities, then decelerating growth at higher densities, and an eventual leveling off as the density approaches what is called the **carrying capacity** of the environment. This is the maximum density of population that the environment can support over a sustained period without lasting damage to the environment itself.

A population growth curve such as that in Fig. 19.4, often called an S-shaped or logistic curve, reflects a changing ratio between births and deaths. During the acceleration stage births greatly exceed deaths; during the deceleration stage the value of r is steadily falling (because the birth rate is declining, or the death rate is rising, or both), though it still is greater than one; and after the curve has leveled off (i.e. after population growth has ceased), births and deaths are in balance. But for such changes in the ratio of births and deaths to occur—for the birth rate to decline (or the death rate to increase) as the population density rises—something must be limiting population growth in such fashion that the effectiveness of the limitation is proportional to the population density; i.e. the limitation must become more severe as the density approaches the carrying capacity of the environment. If we denote the carrying capacity by K, we can express this **density-dependent limitation** as

$$\frac{K - N}{K}$$

Inserting this limiting term into the equation for exponential growth, we obtain

$$I = r\left(\frac{K - N}{K}\right)N$$

This insertion has converted the equation into one for logistic growth, in which the growth is no longer a function solely of the intrinsic rate of increase, r, and the population size, N, but also of the ratio between N and the carrying capacity, K.

Let us now see how insertion of the limiting term $(K - N)/K$ results in the S-shaped growth curve of Fig. 19.4. At low population densities, where N is much smaller than the carrying capacity, the value of the limiting term is essentially K divided by K, or approximately one, which means that growth is primarily a function of $r \cdot N$ and hence increases exponentially. But as growth continues, and N becomes larger and larger, the value of the limiting term steadily declines from one and acts as a brake on the rate of further growth. Finally, when the carrying capacity is reached and N equals K, the value of the limiting term becomes zero and no further growth is possible:

$$I = r\left(\frac{K - N}{K}\right)N = r \cdot 0 \cdot N = 0$$

A population that has reached this steady-state level, where births and deaths are in balance, is said to have **zero population growth.**

The growth curves for some real populations approach the idealized logistic curve almost exactly (Fig. 19.5), but more often they are only rough approximations and exhibit considerable fluctuation around the carrying capacity, sometimes overshooting it temporarily and sometimes falling well below it (Fig. 19.6). Note that whenever the population density fluctuates above the carrying capacity, so that N is greater than K, the limiting term $(K - N)/K$ takes on a negative value, with the result that I also becomes negative. This means that the population density will decrease instead of increasing, until it returns to the carrying capacity or below. In short, the limiting term provides feedback control, usually holding the population density near the steady-state level.

As can be seen from Fig. 19.4, the rate of increase of the population is greatest (i.e. the growth curve is steepest) at the inflection point of the curve (the point of changeover from an accelerating rate to a decelerating rate,

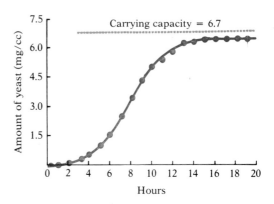

19.5 Growth curve of a laboratory population of yeast cells

This curve closely approximates the hypothetical logistic curve.   [Modified from T. Carlson, *Biochem. Z.,* vol. 57, 1913.]

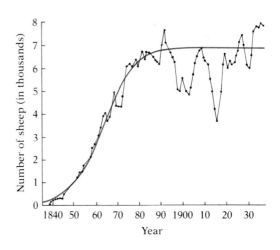

19.6 Growth curve of the sheep population of South Australia

The smooth curve (color) is the hypothetical logistic curve about which the real curve seems to fluctuate.   [Modified from J. Davidson, *Trans. Roy. Soc. S. Australia,* vol. 62, 1938.]

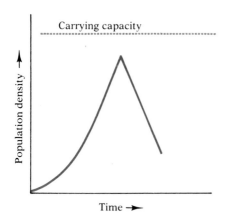

19.7   A growth curve in which an exponentially growing population suddenly declines before reaching the carrying capacity of its environment
This type of growth curve, sometimes called a boom-and-bust curve, is most characteristic of populations of small organisms limited by density-independent factors.

where the density is $K/2$), not when the population density has reached its higher steady-state level. It follows that the density at which the inflection point (sometimes called the point of ***maximum sustained yield***) occurs is of very great importance in the informed management of such organisms as game animals or commercially valuable fishes. If harvesting of the organisms reduces their population density only to the point of maximum sustained yield, one would expect no lasting damage to the resource population. If, however, the resource population is overexploited, so that its density is reduced too far below the point of maximum sustained yield, then the recovery of the population may be endangered. In short, cropping resource populations to their point of maximum sustained yield would seem to be an optimum strategy in terms of both human benefit and perpetuation of the resource.

Unfortunately things are rarely that simple. Conclusions based only on the logistic growth curve fail to take into account such important variables as age and size distributions in the population. For example, when a laboratory population of flour beetles was cropped to a level one tenth as dense as K, far below the point of expected maximum sustained yield at $K/2$, the sustained yield actually increased because, though the number of young beetles had been reduced, the number of reproductives had not been significantly altered. Moreover, in deciding on optimal cropping rates for resource populations, economic yield must also be considered. Thus, though there may be a maximum yield in terms of numbers of animals caught if a fish population is cropped to the $K/2$ level, economic yield may be greater if cropping is less severe, so that the fish will grow to larger size; fewer larger fish may bring more money to the fisherman than more smaller ones. In short, the point of optimal yield that managers of fish and game must strive to find is not necessarily the same as the point of maximum sustained yield.

The exponential growth curve with sudden crash   Not all growth curves of real populations assume the logistic form of Figs. 19.5 and 19.6. The population densities of many small short-lived organisms, or of organisms that live in disturbed or transient habitats, never reach the carrying capacity of the environment before they crash, often abruptly (Fig. 19.7). For example, a population of pea aphids (small insects that suck phloem sap) in a field of alfalfa in spring may grow at an exponential rate if the weather is cool and moist, but if the weather then becomes hot and dry most of the aphids will die, and hence the population density will fall precipitously to a very low level. This crash will occur when the weather changes even if the density of aphids is far below the carrying capacity of the alfalfa field. Note, then, that the weather is here exerting a ***density-independent limitation*** on the aphid population; its operation does not in any way depend on the density of the aphids. In addition to the weather, other environmental factors that may exert density-independent limitation on populations of organisms are sudden floods, fire, or physical disruption of the habitat.

Mortality and survivorship   Important determinants of the makeup of a population in addition to the birth rate and the death rate are the potential life-span, the average life expectancy, and the average age of reproduction. Thus, for example, if the potential life-span is long and the age of reproduction low, more generations will be living concurrently than if the potential life-span is short and the age of reproduction high. If the potential life-span is long but the average life expectancy is low, the age distribution in the population will be very different from what it would be if the potential life-span were long and the average life expectancy also long.

In attempting to understand the dynamics of a population, it is often useful to determine the mortality rates for the various age groups in a population. Such data show what stages in the life cycle are most susceptible to environmental control, and make it possible to compute the percentage of individuals that will still be alive at the end of each age interval. The results may be graphed as a survivorship curve (Fig. 19.8).

The curves in Fig. 19.8 illustrate several survivorship patterns. Curve I approaches the pattern that would be expected if all the individuals in the population realized the physiologically possible longevity. There would be full survival through all the early age intervals (as shown by the horizontal portions of the curve), and then all the individuals would die more or less at once and the curve would fall suddenly and precipitously. Curve III approaches the other extreme, where the mortality is exceedingly high among the very young but where any individual surviving the earliest life stages has a good chance of surviving for a long time thereafter. Between these two extremes is the condition represented by curve II, where the mortality rate at all ages is constant.

The survivorship curves for most wild-animal populations are probably intermediate between types II and III, and the curves for most plant populations are probably near the extreme of III. In other words, high mortality among the young is the general rule in nature.

Changes in environmental conditions may radically alter the shape of the survivorship curve for any given population, and the altered mortality rates, in turn, may have profound effects on the dynamics of the population and on its future size. For example, the chief cause for the enormous increase in the population of human beings has been a great reduction in mortality during the early life stages as a result of improvements in sanitation, nutrition, and medical care. These improvements have caused a shift in the human survivorship curve from one intermediate between types II and III in primitive societies to one approaching type I in the most advanced societies. There is no evidence that the shift is due to a rise in the human birth rate per reproductive-age female (in fact, the birth rate has almost certainly fallen) or, despite all the advances of modern medicine, to an increase in the potential life-span.

As Fig. 19.9 shows, the human population of the world increased very slowly for many thousands of years, even though the birth rate was probably high (primitive societies put great value on large families, which were essential in view of the very high mortality rates, particularly in infants). It has been estimated that there were approximately 5 million people on the entire earth ten thousand years ago, and that the number had risen to only about 250 million by 1 A.D. and to about 500 million by the year 1650. From these figures, one can see that until about three hundred years ago the human population doubled approximately every 1,600 years. Now, by contrast, the population of the world is doubling every 35 years; the present total is over 4 billion.

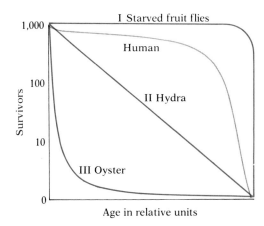

19.8   Four types of survivorship curves
For an initial population of 1,000 individuals, the curves show the number of survivors at different ages from birth to the maximum possible age for that species. Curves I, II, and III represent three basic types of survivorship; for explanation, see text.   [Modified from E. P. Odum, *Fundamentals of ecology*, Saunders, 1959, after Deevey.]

How much longer the human species can continue with such a rate of increase, with such an imbalance between births and deaths, is one of the most pressing questions of our time. Some have argued, in fact, that it is *the* most important question; all other aspects of the so-called ecological crisis—hunger, poverty, crowding, pollution, accumulation of wastes, destruction of the environment on which all life depends—are the inexorable consequences of a continuing rise in the number of human beings.

POPULATION REGULATION

As we have seen, regulation of population density can occur in different ways in different populations. For organisms whose population growth curves are of the S-shaped type (Fig. 19.4), density-dependent limitation is primary, and the maximum density at which such a population can remain stable is determined largely by the carrying capacity *(K)* of the environment. By contrast, for organisms with population growth curves of the type shown in Fig. 19.7 (sometimes called a boom-and-bust curve), density-independent limitation is more important, and the maximum density achieved before decline begins is primarily a function of the rapidity of the organism's reproduction.

Organisms of the first type—with S-shaped growth curves—are often called **K** *strategists.* They usually live in fairly stable and predictable environments. Because their fitness depends less on rapid reproduction than on their ability to exploit the limited environmental resources efficiently, they generally have longer life-spans and reproduce at a later life stage, usually produce fewer young but (among animals) often exhibit more parental care, tend to have a survivorship curve of types I or II, usually have a larger body size, and have greater competitive ability.

Organisms of the second type, whose population limitation is primarily density-independent, are called **r** *strategists.* They often live in variable, unpredictable, disturbed, or temporary environments. Because their mortality is often sudden and catastrophic, their fitness is heavily dependent on producing as many young as rapidly as possible, under conditions where competition is minimal. They have characteristically evolved high intrinsic rates of increase (i.e. high *r*) at the expense of lesser ability to compete under crowded conditions when resources are limited. They are generally small organisms with brief life-spans, early reproduction, and survivorship curves of type III.

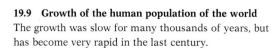

19.9 Growth of the human population of the world
The growth was slow for many thousands of years, but has become very rapid in the last century.

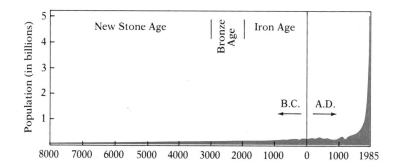

It is necessary to point out that the differences between *K* strategists and *r* strategists are not absolute but relative. All gradations between the two extreme types we have described are possible, depending on the relative mix of density-dependent and density-independent limitation to which a population is subjected.

We have already seen that density-independent limitation is ordinarily due to accidents of weather or physical disturbance. Let us now look more closely at the various environmental factors that can help limit populations in a density-dependent fashion.

Predation and parasitism as density-dependent limiting factors   Both predation and parasitism usually influence the prey (or host) species in a density-dependent manner. As the population of the host or prey increases, a higher percentage of the population is usually victimized, because the individuals—having perhaps been forced into less favorable situations or being weaker on account of the greater drain on available resources—become easier to find and attack. Furthermore, there is evidence that as the relative densities of prey shift, predators that take a variety of prey species tend to change their search images and alter their hunting patterns so as to concentrate on the most common species.

As the density of a prey species increases, the density of the predators feeding on it often increases also. This increase in predators, together with their increased concentration on the particular prey species, may be one factor that causes the density of the prey to fall again. But as the density of the prey falls, there is usually a corresponding, but slightly later, fall in the density of the predator. The result may be a series of density fluctuations like those shown in Fig. 19.10, where the fluctuations of the predator (the lynx in this case) closely follow those of the prey (the hare), but with a characteristic time lag. Such linked fluctuations of predator and prey seem to indicate that the major limiting factor for the predator is the availability of its food and that predation is probably one significant limiting factor for the prey. The length of the time lag between a change in one population and the response in the other influences the extent of the density fluctuations; in general, the longer the time lag, the greater the period and amplitude of expected oscillation in the system.

In stable predator-prey systems, the density-limiting function of the predation is often decidedly beneficial to the prey population, even though it is destructive to individuals. This cardinal fact of ecology is frequently overlooked by those to whom parasites are repugnant and predators evil. When people set out to protect the prey from their "enemies" (sometimes only to preserve them for their human ones) by killing the predators, the results are often very different from the ones expected. Released from the density-regulating influence of the predators, the prey species may experience such a population explosion that it damages the environmental resources on which its own continued existence depends, with consequent wholesale extermination of individuals of the "protected" species through starvation or disease.

Similar difficulties have sometimes arisen when predator-prey stability has inadvertently been destroyed by pesticides. For example, application of certain insecticides to strawberries in an attempt to destroy cyclamen mites

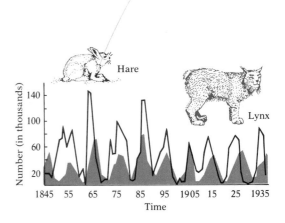

19.10   Fluctuations in abundance of lynx and snowshoe hare over a period of 90 years
The oscillations of the lynx population seem to follow those of the hare population. The index of abundance used here (vertical axis) is the number of pelts received by the Hudson's Bay Company.   [Redrawn from D. A. MacLulich, *Univ. Toronto Studies*, Biol. Ser., no. 43, 1937.]

that were damaging the berries killed both the cyclamen mites and the carnivorous mites that preyed on them. But the cyclamen mites quickly reinvaded the strawberry fields, while the predatory mites did not. The result was that the cyclamen mites, now free of their natural predators, rapidly increased in density and did more damage to the strawberries than if the insecticides had never been applied.

The history of the moose population on Isle Royale provides a good example of the role of predation in regulating prey populations. Isle Royale is an island in Lake Superior, 544 km² in area, which is maintained in its natural state by the U.S. National Park Service. The first moose colonized the island in 1908. In the absence of large predators on the island, and with no hunting allowed, the moose population grew rapidly, until it was so large by 1930 that it exceeded the carrying capacity of the environment and was overexploiting the island's vegetation. The result was a population crash. When the vegetation recovered, the moose population once more became dense, only to crash again in the 1940s. Then in 1948 timber wolves reached the island via the winter ice and began to prey on the moose, reducing their population to between 600 and 1,000 individuals.

At this point, the predator-prey system became stable, and a healthy population of about 1,000 moose has coexisted with a population of about 24 wolves (winter census figures) on Isle Royale each year thereafter. The wolves capture an average of one moose every three days, which keeps the moose population in check. But because the wolves, whose diet is about 85 percent moose, can usually catch only old, sick, or very young animals, they appear unable to increase their rate of hunting success. Hence the wolf population, too, is held in check. In this system, then, the prey species is predation-limited and the predator species is food-limited.

As the example of Isle Royale suggests, predator-prey relationships in a stable ecosystem, like long-established host-parasite relationships, tend to evolve toward a dynamic balance in which the predation is an important regulatory influence in the life of the prey species, but not a real threat to its survival. This sort of balance is, of course, beneficial to the predator as well as to the prey, because extinction or severe depression of the prey population would diminish the future resources of the predator and might even lead to its own extinction. Factors affecting the predator-prey balance include the relative numbers and sizes of the two species, the vulnerability of the prey to the predator, the extent to which the predator can (or does) use other food sources, and the amount of energy obtained by the predator from consuming one prey individual relative to the energy expended in its capture.

Intraspecific competition as a density-dependent limiting factor   Competition is one of the chief density-dependent limiting factors. The continued healthy existence of most organisms depends on utilization of some environmental resources that are in limited supply, such as food, water, space, or light. As the population density increases, the competition among members of the population for these limited resources becomes more intense and the deleterious results of the competition become more and more effective in limiting the population.

To take a familiar example, if flowers are planted too close together in a flower bed the plants will be weak and spindly and will produce few if any blossoms. Only if they are thinned, either artificially or by the natural death of the weakest individuals, will they grow well. The same sort of competition for space, light, water, and nutrients operates in a forest and keeps down the density of the trees. The decline of the lynx population following each decline in the hare population, illustrated in Fig. 19.10, and the moderate density of the wolf population on Isle Royale, are two examples where intense intraspecific competition for food may be regarded as the immediate limiting factor.

As another example, consider competition among individuals for breeding territories. Although territories tend to be larger when population pressure is low and smaller when it is high, there is usually a minimum requisite territory below which successful reproduction will not occur. The size of the minimum territory varies greatly from species to species, being rather large in animals that are only mildly social and sometimes quite small in highly social animals such as nesting gulls. But whatever the minimum size may be, suitable territories are clearly a limited resource, and competition for them, which will become increasingly intense as population density rises, will act as a density-dependent brake on population growth.

Interspecific competition as a density-dependent limiting factor   Interspecific competition is an interaction in which both parties are harmed. If we were to fit the concept into the list on p. 781, it would be symbolized by minus-minus. Ordinarily, interspecific competition occurs when two or more species use the same limited resource, such as food, water, sunlight, shelter, space, or nesting sites. Clearly, it will become more intense as the number of individuals depending on the limited resource rises; hence its effect on the competing species is density-dependent.

The more similar the requirements of the species involved, the more intense the competition. Or, as ecologists often put it, the more the niches of the species overlap, the more intense the interspecific competition. *Niche,* an important concept in ecology, is defined in a variety of ways. Here it will be used to denote the functional role and position of an organism in the ecosystem.

Niche should not be confused with *habitat,* which is the physical place where the organism lives. The characteristics of the habitat help define the niche, but they alone are far from sufficient. Also included is what the organism eats; how and where it finds and captures its food; what extremes of heat and cold, dry and wet, sun and shade, and other climatic factors it can withstand, and what values of these factors are optimal for it; at what time of year and what time of day it is most active; what its parasites and predators are; where, how, and when it reproduces; and so forth. In short, every aspect of an organism's existence helps define that organism's niche. It must be emphasized that niche is an abstraction and as such not subject to direct measurement. What can be measured are certain of the more important variables of an organism's niche.

The following graphic method of representing the niche of a species has been proposed: Consider one environmental variable, X_1, such as temperature; determine the high and low extremes the species in question can tolerate; plot these on a coordinate, and connect them by a line. Do the same

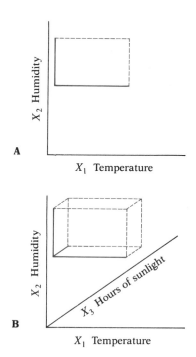

19.11 Graphical representations of the ecological niche of a species

(A) The ranges of toleration by a species for two environmental variables are plotted on a coordinate system, and a rectangular area is determined. (B) If a third variable is plotted at right angles to the other two, a volume is determined. Though the procedure is difficult to carry further graphically, it can be continued mathematically to define a species' niche in multidimensions—or as a so-called hypervolume. The more the niche hypervolumes of two species overlap, the more intense the competition between the species.

for a second environmental variable, X_2, such as humidity, and draw its line at right angles to the first one. These two lines together determine a rectangular surface (Fig. 19.11A).

Now determine the values for a third environmental variable, X_3 (e.g. hours of sunlight), and connect them by a line oriented at right angles to the other two. The three lines together now determine a volume within which every point corresponds to some combination of values for the three variables that would permit the species to survive (Fig. 19.11B).[3]

If you were to continue this procedure, adding more and more variables each determining a different dimension, you would obtain a multidimensional hypervolume. If you included a dimension for every variable relevant to the species in question, the resulting multidimensional hypervolume would represent the niche of the species.

We have said that the more the niches of two species overlap, the more intense the competition between them will be. This idea was enunciated with especial clarity by G. F. Gause of the University of Moscow in 1934. According to Gause's principle, also known as the *competitive exclusion principle,* two species cannot for long simultaneously occupy the same niche in the same place. But it is implicit in our definition of niche that no two species could ever occupy the same niche. To do so, they would have to be identical in every respect; hence they would be one species, not two. What, then, is the value of this principle? Its value is much like that of the Hardy-Weinberg Law, which also describes a situation that would never really occur. Both the Hardy-Weinberg Law and the competitive exclusion principle establish null conditions—points of departure, so to speak—and thus facilitate our understanding of what will happen when the null conditions do not hold. Thus, for example, the competitive exclusion principle helps us see why the different species living together in a stable community ordinarily occupy quite distinct niches, and it helps us understand the sorts of interactions that will result when two species with very similar (though not identical) niches occur together.

The more similar the two niches are, the more likely it is that both species will use in common and in the same way at least one limited resource (food, shelter, nesting sites, etc.), for which they will compete. According to the principle of limiting similarity—a modification of the competitive exclusion principle—there is a limit on the amount of niche overlap compatible with coexistence. Competition for the one most limited resource therefore usually leads to one (or two) of three possible outcomes:

1. The competitive superiority of one of the rival species may be such that the other is driven to extinction.

2. One species may be competitively superior in some regions, and the other may be superior in other regions with different environmental conditions, with the result that one is eliminated in some places and the other is eliminated in other places; i.e. sympatry disappears, but both species survive in allopatric ranges.

3. The two species may rapidly evolve in divergent directions under the strong selection pressure resulting from their intense competition. Natural selection would favor those individuals with characteristics differing from

---

[3] There is an important complication that partly invalidates this procedure. The environmental variables are seldom completely independent. Thus, for example, the temperature extremes a species can tolerate may be different at different humidities.

those of the other species, because such characteristics would tend to minimize competition. In other words, the two species would rapidly evolve greater differences in their niches. This is the phenomenon of character displacement, discussed in the preceding chapter.

Whether extinction, range restriction, character displacement, or a combination of the last two, will be the outcome in any given instance of intense interspecific competition is determined by a host of factors too complex to analyze here.

Let us examine a few well-studied examples of competition and its consequences. Gause himself worked with two closely related species of protozoans, *Paramecium caudatum* and *Paramecium aurelia*, in laboratory cultures. When the two species were cultured separately, the population curve for each species had a typical S shape (Fig. 19.12). But when they were cultured together, the population growth rate of *P. aurelia* was slower than normal and *P. caudatum* failed to survive.

Similar results were obtained by Thomas Park and his colleagues at the University of Chicago. They worked with flour beetles of the genus *Tribolium*. When *T. confusum* and *T. castaneum* were kept together in the same container of flour, one or the other species always became extinct. The conditions of temperature and humidity under which the competition took place greatly influenced which species would win. Thus *T. castaneum* usually won under hot-wet conditions, while *T. confusum* usually won under cool-dry conditions. The competition between the flour beetles was not for food, which was plentifully available for both. Apparently it was more a matter of competition for space. Crowding and the resultant conditioning of the medium had slightly different effects on the natality, mortality, and rate of development of the two species, and thus determined which would survive and which would become extinct.

J. H. Connell's study of barnacles on the Scottish coast provides a good example of competition between two species whose niches overlap but are sufficiently distinct for both species to survive by occupying slightly different habitats. A species of the genus *Chthamalus* occupies the upper part of the intertidal zone, and a species of *Balanus* occupies the lower part of the intertidal zone (Fig. 19.13). The boundary between the two distributions is roughly the level of the mean high neap tide.[4] Casual examination of the distribution of these two species of barnacles on the intertidal rocks would not reveal whether they were kept separate by competition or by different responses to physical factors such as the percentage of time out of water.

To find the answer, Connell artificially kept one study area clear of *Balanus* and another clear of *Chthamalus*. He found that the larvae of *Chthamalus* would settle and grow in the upper portion of the zone normally occupied by *Balanus* as long as *Balanus* was not there. In the reciprocal experiment, *Balanus* larvae would settle in the *Chthamalus* zone but could not survive there even if no *Chthamalus* were present. Apparently each species could occupy a portion of the intertidal zone (roughly between the levels of the mean high spring tide and the mean high neap tide for *Chthamalus* and below the mean tide for *Balanus*) from which the other was barred by physical factors. But in the intermediate zone where physical factors permitted each species to survive (roughly between the levels of

[4] Neap tides are the lowest tides of a lunar month.

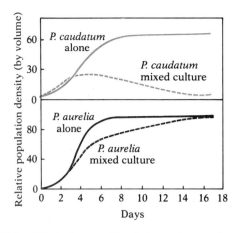

19.12 Effect of competition between two species of *Paramecium*
The solid curves show the growth of population volume of each species alone in a controlled environment with a fixed food supply. The dotted curves show the change in population volume of the same species when in competition with each other under the same conditions.   [Modified from G. F. Gause, *Science*, vol. 79, 1934.]

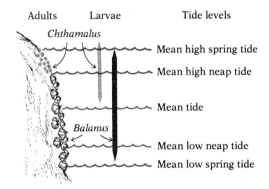

19.13 Effect of competition between two species of barnacles
Although there is a broad area in which the larvae of both species settle, competition eliminates most of the overlap by the time the adult stage is reached, *Chthamalus* being largely restricted to the zone above the level of the mean high neap tide and *Balanus* to the zone below this level. See text for more complete description.   [Modified from J. H. Connell, *Ecology*, vol. 142, 1961.]

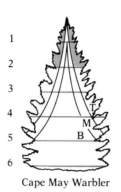
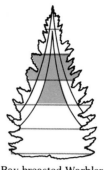

Cape May Warbler   Bay-breasted Warbler   Myrtle Warbler

19.14   Differences in the feeding niches of three species of warblers living in the same community
Color areas: parts of tree where the birds spend half their total feeding time. T: terminal parts of branches; M: middle; B: base. The height zones (1–6) are measured in 3 m units. The Cape May Warbler feeds on the tips of the highest branches of the spruce trees; the Bay-breasted Warbler feeds in the middle part of the tree; and the Myrtle Warbler feeds predominantly among the lower branches.   [Modified from R. H. MacArthur, *Ecology*, vol. 39, 1958.]

mean high neap tide and mean tide), it was competition that kept them separate, *Balanus* usually eliminating *Chthamalus*.

It is not always easy to detect the differences between the niches of two or more closely related sympatric species. At first glance, the species may appear to be occupying the same niche in a stable way and thus to discredit the exclusion principle. But closer study usually reveals differences of fundamental importance. When Robert MacArthur, then at Yale University, studied a community where several closely related species of warblers (small insect-eating birds) occurred together, he found that their feeding habits were significantly different. Myrtle Warblers fed predominantly among the lower branches of spruce trees; Bay-breasted Warblers fed in the middle portions of the trees; and Cape May Warblers fed toward the tops of the same trees and on the outer tips of the branches (Fig. 19.14). We noted similar differences in feeding habits between sympatric Galápagos finches.

Cormorants and shags are closely related sympatric species of birds whose habits and ecological requirements appear very similar. But David Lack, who studied them, found out that, although both nest on cliffs and feed on fish, cormorants nest on broad ledges and feed chiefly in shallow estuaries and harbors, whereas shags nest on narrow ledges and feed mainly at sea. Their niches, then, are very different, and there is little competition between them.

Emigration as a density-dependent limiting factor   In some animals crowding induces physiological and behavioral changes that result in increased emigration from the crowded region. Such changes can be observed in many species of aphids (small insects that suck the juices of plants). During seasons of the year when conditions are favorable, they are represented largely by wingless females reproducing parthenogenetically. But when conditions deteriorate and competition becomes intense, winged females develop and these may move out of the area in which they were born. Similarly, in the fungus fly *Oligarces paradoxus*, individuals reproduce parthenogenetically without ever fully maturing to the winged adult stage as long as there is an abundant supply of fresh food. But when the food supply deteriorates, in either quantity or quality, and competition becomes intense, the insects go through a complete developmental sequence leading to winged adults, which fly to other more hospitable areas.

One of the best-known examples of physiological and behavioral changes induced by crowding is that of the solitary and migratory phases seen in several species of locusts, particularly *Locusta migratoria* in Eurasia. Individuals of the migratory phase have longer wings, a higher fat content, a lower water content, and a darker color than solitary-phase individuals, and they are much more gregarious and more readily stimulated to marching and flying by the presence of other individuals. The solitary phase is characteristic of low-density populations, and the migratory phase of high-density ones. As the density of a given population rises, the proportion of individuals developing into the migratory phase also rises; the sight and smell of other locusts seem to play an important role in triggering this line of development. When the proportion of migratory-phase individuals has risen sufficiently high, enormous swarms emigrate from the crowded area, consuming nearly all the vegetation in their path and often completely devastating agricultural crops. A locust plague of this sort is described in the Old Testament (Exodus 10:14–15):

> And the locusts went up over all the land of Egypt, and rested in all the coasts of Egypt: very grievous were they . . .
> For they covered the face of the whole earth, so that the land was darkened; and they did eat every herb of the land, and all the fruit of the trees . . . and there remained not any green thing in the trees, or in the herbs of the field, through all the land of Egypt.

How seriously are we to take the suggestion that emigration to other planets might be the solution to the human population problem? Discounting for the moment the inhospitable conditions on planets close enough to be reasonably considered, simple calculations show that this plan is not in fact a practicable one. Bruce Wallace of Cornell University has pointed out that just to maintain the world's human population at its present level would require that rockets with a payload of 50 million tons depart each day. Building the number of rockets necessary to stabilize the population at this level for a full year might well require all the matter the earth is made of, which would leave those of us who stay at home with nothing to stand on!

Physiological mechanisms as density-dependent limiting factors   Some ecologists have suggested that physiological phenomena may be important in limiting populations. Thus it has long been known that very dense populations often experience severe disease epidemics. There is now evidence that this proneness to epidemics is not due solely to the greater ease with which the pathogens spread; a density-induced change in host resistance is apparently also involved. Numerous experiments have shown that increased population density is accompanied by a marked depression of both inflammatory responses and antibody formation, with a resultant increase in susceptibility to infection or parasitism. There is also an increased susceptibility to the harmful effects of various toxic substances.

The effects of increasing population density have been observed in numerous laboratory experiments with mice. There is hypertrophy of the adrenal cortex and degeneration of the thymus. Somatic growth is suppressed, sexual maturation delayed (or even totally inhibited at very high densities), and reproduction by mature mice diminished. The effects on reproduction include delayed spermatogenesis in males and, in females, prolonged estrous cycles, reduced rate of uterine implantation, and inadequate lactation. There is also evidence of increased intra-uterine mortality of the embryos. Furthermore, crowding of female mice before pregnancy can result in permanent behavioral disturbances in the young they later produce. There would seem, then, to be an endocrine feedback mechanism that can help regulate and limit population size by altering the reproductive rate. Presumably, as the density rises and aggressive behavior increases, endocrine disturbance rises and the reproductive rate falls; conversely, as the density and aggressive behavior decrease, the reproductive rate rises.

The idea that endocrine changes act as major density-dependent limiting factors in nature has been severely criticized on the ground that it is largely based on laboratory work with densities much higher than those that would actually occur in nature. To meet this criticism, attempts have been made to duplicate the intermittent crowding more likely to occur in nature. These have shown that mice exposed to a few short periods of crowding every day in the laboratory actually show greater hypertrophy of the adrenal cortex than mice exposed to continuous crowding. Nevertheless, endocrine changes have seldom been found in wild animals undergoing population stress, and the importance of such changes in population regulation remains unclear.

Some writers have suggested that endocrinological changes resulting from increased crowding might help stabilize human population densities. That seems improbable, because even if the stress of crowding does induce hormonal changes in humans these would be unlikely to become severe or widespread enough to have significant effects on population dynamics before the densities reached extremes that would prove fatal for other reasons. We must remember that a highly social species like ours must have evolved a far greater tolerance to crowding than most other animal species.

THE ECONOMY OF ECOSYSTEMS

A species does not exist as an isolated entity, but is always interacting in a variety of ways with other species in the community to which it belongs. Among common interactions, as we have seen, are the feeding of one kind of organism on another and competition for environmental resources. These and other interactions involve the flow of energy and materials within the community. But the flow of energy and materials cannot be understood apart from the physical environment, which together with the community makes up the ecosystem. We shall here examine some of the ways in which the movement of energy and materials binds together the community and the physical environment as a functioning system.

THE FLOW OF ENERGY

Life depends, ultimately, on radiation from the sun. The familiarity of this statement should not blunt one to its significance. With the exception of the relatively unimportant chemosynthetic organisms, all forms of life obtain their high-energy organic nutrients, either directly or indirectly, from photosynthesis. Photosynthesis, however, employs less than one tenth of one percent of the solar energy reaching the surface of the earth.

The total amount of energy that photosynthesis binds into organic matter

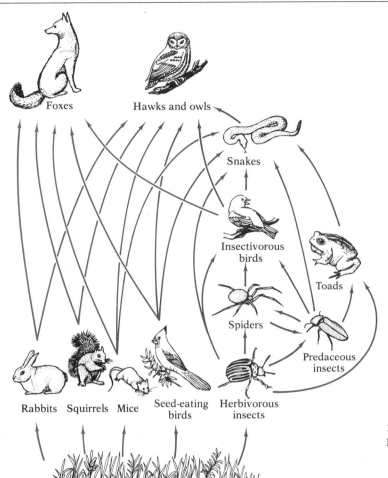

19.15 Diagram of a hypothetical food web
No real food web would be as simple as this one.

is called *gross primary productivity.* It is estimated that plants use from 15 to 20 percent of their gross productivity in their own respiration. What is left over is known as *net primary productivity.* The total net primary productivity of the biosphere, estimated at about 6×10^{20} gram calories of energy per year,[5] constitutes the energy base for heterotrophic life on earth. Heterotrophic organisms—bacteria, fungi, and animals—obtain the energy they need by consuming green plants, feeding on other heterotrophic organisms that fed on green plants, or consuming the dead bodies or waste products of other organisms.

The sequence of organisms through which energy may move in a community is customarily called a *food chain.* In most real communities there are so many possible food chains, so complexly intertwined, that together they form a community *food web* (Fig. 19.15). No matter how long a food

[5] The energy content of organic materials is determined by burning them in pure oxygen and measuring the heat liberated. Organic matter has a relatively uniform energy content of about 4.25 kcal per dry gram of plant tissue and 5.0 kcal per dry gram of animal tissue. Some ecologists, therefore, give primary productivity values in dry-weight units. In these terms, the total net primary productivity for the biosphere is about 164 billion dry tons of organic matter per year.

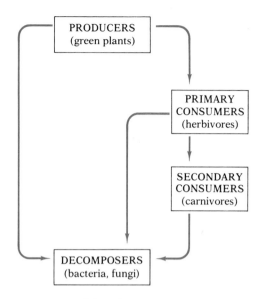

19.16   Diagram of the relationships between the principal trophic levels in an ecosystem
The green plants are the producers, which are eaten by the herbivores, the primary consumers. The primary consumers may in turn be eaten by parasites or carnivores, the secondary consumers. Producers or consumers may die and become food for decomposer organisms.

chain or how complex a food web may be, however, certain basic characteristics are always present. Every food chain or web begins with the autotrophic organisms (green plants in the vast majority of cases) that are the *producers* for the community. And every food chain or web ends with *decomposers,* the organisms of decay, which are usually bacteria and fungi that release simple substances reusable by the producers. The links between the producers and the decomposers are more variable. The producers may die and be acted upon directly by the decomposers, in which event there are no intermediate links. Or the producers may be eaten by *primary consumers,* the herbivores. These, in turn, may be either acted upon directly by decomposers or fed upon by *secondary consumers* such as carnivores or parasites or scavengers (Fig. 19.16).

Ecologists speak of the successive levels of nourishment in the food chains of a community as *trophic levels.* Thus all the producers together constitute the first trophic level, the primary consumers (herbivores) the second trophic level, the herbivore-eating carnivores the third trophic level, and so on. The species that comprise each trophic level differ from one community to another. Moreover, the trophic levels themselves are not hard-and-fast categories, since many species that eat a varied diet—especially omnivores—may function at two or more trophic levels within a single food web. For example, a chickadee, which eats seeds, herbivorous insects, and carnivorous insects, functions at the second, third, and fourth trophic levels. Despite these complications, the concept of trophic levels remains a useful one in community analysis.

At each successive trophic level there is loss of energy from the system. The loss is due partly to the consumer population's inability to harvest any more than a fraction of the food population; partly to failure of assimilation (most animals cannot utilize the cellulose walls of plant cells, for example); and partly to respiration and the consequent dissipation of energy as heat (the Second Law of Thermodynamics, you will recall, states that every energy transformation involves loss of some usable energy; p. 138). As a result, only about 10 percent of the energy at one trophic level can usually be passed on to the next level. In other words, the productivity (energy bound into new organic matter per unit area per unit time) at each trophic level is only about 10 percent of that at the preceding level. There is less productivity from the herbivores of a community than from the plants of that community, and there is less productivity from the carnivores than from the herbivores, and so on. Thus the distribution of productivity within a community can be represented by a pyramid, with the first trophic level (producers) at the base and the last consumer trophic level at the apex (Fig. 19.17). Because of the rapid fall in productivity from one trophic level to the next, there are seldom more than four or five levels in a food chain; the fifth level has no more than about 0.0001 the productivity of the first, and the productivity possible for any subsequent levels becomes too low to be effective.

The *pyramid of productivity* (also called the pyramid of energy flow) just described is a characteristic of all ecosystems. Several other attributes of ecosystems may fit a pyramidal model because they are related to the flow of energy through the system, but they are not themselves inherent properties of ecosystems and therefore often deviate from the pyramidal model. One example is the *pyramid of biomass* (Fig. 19.18A). In general, the decrease of energy at each successive trophic level means that less biomass can be supported at each level. Thus the total mass of carnivores in a given community is almost always less than the total mass of herbivores. The size, growth rate, and longevity of the species at the various trophic levels of a community are important in determining whether or not the pyramidal model will hold for the biomass of that community. Thus, in some aquatic communities where the producers are small algae with high metabolic and reproductive rates, there may be a greater biomass of consumers than of producers at any given moment, but the total mass of all the algae that live during the course of a year will be greater than the total mass of consumers that live during that year.

The interrelationships between the organisms at different trophic levels exert some influence on the size of the organisms. Thus carnivores are often larger and stronger than their herbivorous prey. And secondary carnivores are often larger than the primary carnivores on which they feed. Now, since total biomass tends to decline at successive trophic levels, it follows that, if the size of individuals increases at successive levels, the number of individuals must decline at each level (except at the decomposer level). Consequently some communities show a ***pyramid of numbers,*** there being fewer individual herbivores than plants, and fewer individual carnivores than herbivores (Fig. 19.18B). Indeed, it is entirely understandable why ***top predators*** (predators at the top of their food chains), such as lions or wolves or killer whales, are not themselves preyed on: There are too few of them, they are too widely scattered, and they contain too little energy to make the effort worthwhile.

Many communities, however, have no pyramid of numbers. For example, there may be many more individual insect consumers than plants, even though their biomass may be less, because plant-eating insects are often far smaller than their food plants; an extreme case would be a single large tree on which thousands of insects are feeding. Food chains involving parasites also tend to have reversed size relationships, because the parasites are smaller and usually more numerous than the host.

CYCLES OF MATERIALS

We have seen that energy is steadily drained from the ecosystem as it is passed along the links of a food chain. The system cannot continue functioning without a constant input of energy from the outside. In other words, there is no such thing as an energy cycle. But this is not the case with materials. The same materials can and must be used over and over again, and hence can be passed round and round through the ecosystem indefinitely. We can, therefore, speak of cycles of materials. Let us examine several such cycles.

The water cycle   When rainwater falls on the land, some of it quickly evaporates again into the atmosphere. Of the water that does not immediately evaporate, some is absorbed by plants or is drunk by animals, some runs off the surface of the land into streams and lakes, and some percolates down through the soil into the water table below (Fig. 19.19). The water in the streams and lakes and in the subsurface water table eventually finds

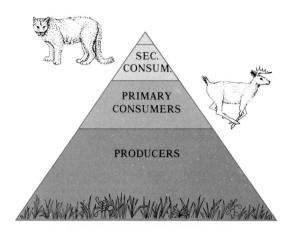

19.17   The pyramid of productivity
There is much more productivity at the producer level in an ecosystem than at the consumer levels, and there is more at the primary consumer level than at the secondary consumer level.

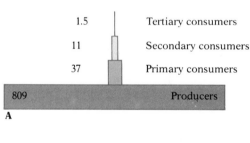

1.5	Tertiary consumers
11	Secondary consumers
37	Primary consumers
809	Producers

A

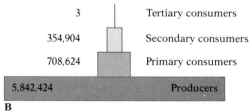

3	Tertiary consumers
354,904	Secondary consumers
708,624	Primary consumers
5,842,424	Producers

B

19.18   Examples of the pyramids of biomass and numbers
(A) Pyramid of biomass in the aquatic ecosystem of Silver Springs, Florida. Figures represent grams of dry biomass per square meter.   (B) Pyramid of numbers in a bluegrass field.   [Modified from E. P. Odum, *Fundamentals of ecology,* Saunders, 1959.]

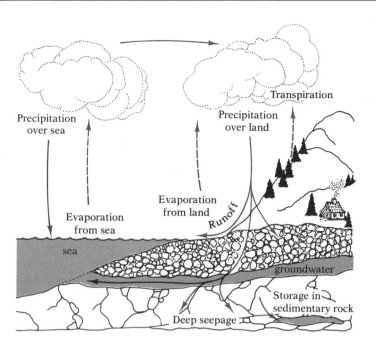

19.19 The water cycle
This diagram shows most of the major pathways of water movement through the ecosystem—but not the more recent ones created by human beings.

its way to the ocean. There is constant evaporation from streams, lakes, and oceans, and also from the bodies of plants and animals. The energy for most of this evaporation comes either directly or indirectly from solar radiation.

The endless cycling of water to earth as rain, back to the atmosphere through evaporation, and back again to earth as rain, maintains the various freshwater environments and supplies the vast quantities of water necessary for life on land. The water cycle is likewise a major factor in modifying temperatures, and it provides for the transport through ecosystems of many chemical nutrients. The tremendous importance of rainfall for terrestrial and freshwater life requires no elaboration here; one need only visualize a desert on the one hand and a lush tropical forest on the other.

The carbon cycle The carbon dioxide contained in the atmosphere and dissolved in water constitutes the reservoir of inorganic carbon from which almost all organic carbon is derived. It is photosynthesis, largely by green plants, that extracts the carbon from this inorganic reservoir and incorporates it into the complex organic molecules characteristic of living substance (Fig. 19.20). Some of these organic molecules are soon broken down again, and their carbon is released as CO_2 by the plants in the course of their own respiration. But much of it remains in the plant bodies until they die or are eaten by animals. The carbon obtained from plants by animals may be released as CO_2 during respiration, or it may be eliminated in more complex compounds in the body wastes, or it may remain in the animals until they die. Usually the wastes from animals and the dead bodies of both plants and animals are broken down (respired) by the decomposers, and the carbon is released as CO_2.

Notice that whether the carbon follows a short pathway involving only one or two trophic levels or a longer pathway involving three, four, or more, most of it eventually returns as CO_2 to the air or water whence it started. This is, then, a true cycle (or rather a complex of interlocking cycles); carbon is constantly moving from the inorganic reservoir to the living system and back again.

The pathways just outlined are all pathways through which carbon moves rather rapidly. Complete passage through the system may take only minutes or hours, or at most a few years. There are alternative pathways, however, that take much longer. The dead bodies of organisms occasionally fail to be decomposed promptly and are converted instead into coal, oil, gas, rock (particularly limestone), or diamond. Carbon in these forms may be removed from circulation for very long periods, perhaps permanently; but some of it may eventually return to the inorganic reservoir if the coal, oil, and gas (the fossil fuels) are burned or if the rocks are sufficiently weathered. Human beings have of course greatly accelerated the return of such carbon to the active cycle.

Of the CO_2 released by the burning of fossil fuels, about half remains in the atmosphere and the rest is absorbed by the ocean waters. The CO_2 reservoir in the atmosphere is also being increased through the oxidation of organic materials once incorporated in the plants that grew in areas subsequently cleared for roads or buildings or being used for agriculture. (Agricultural crops usually fix less CO_2 than the natural vegetation they displace, because they are highly productive only for a relatively short time. Meanwhile the reserve of organic litter built up in the soil by the native vegetation

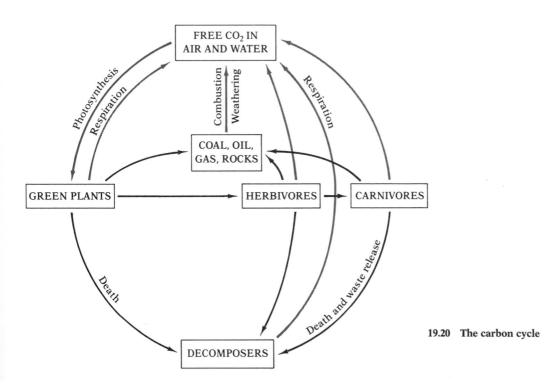

19.20 **The carbon cycle**

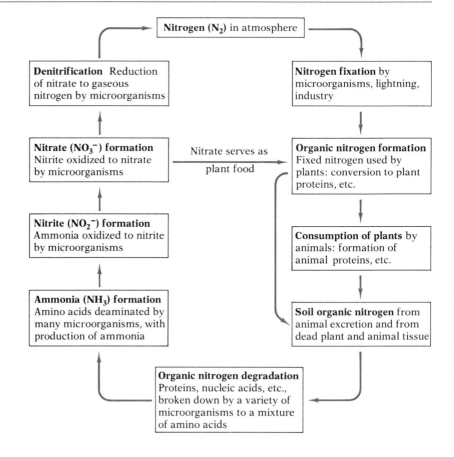

19.21 The nitrogen cycle
The many steps at which microorganisms play a major role show to what extent higher forms of life depend on bacteria, blue-green algae, and fungi for their own continued existence.   [Modified from M. J. Pelczar *et al.*, *Microbiology*, 4th ed., McGraw-Hill Book Co., copyright © 1977. Used by permission.]

slowly decomposes, releasing its carbon into the atmosphere as CO_2.) These human activities have increased atmospheric CO_2 levels by 15 percent in the last hundred years; it is entirely possible that they will double the CO_2 concentration in the next hundred years.

Now, CO_2 plays an integral role in the regulation of temperatures on the surface of the earth. Heat radiated from the earth is absorbed by CO_2 in the atmosphere and, radiated back to the surface, tends to warm the earth in a so-called greenhouse effect. Thus a rise in the CO_2 levels in the atmosphere should cause the temperature of the earth to increase, were it not that cloud patterns and the amount of water vapor and atmospheric particulate matter—which are also changing as a result of human activities—likewise influence the temperature balance of the earth. What the final outcome will be is a matter of considerable debate; predictions range from tropical heat to the return of another ice age.

The nitrogen cycle Another critical element in community metabolism is nitrogen, a constituent of the amino acids of which proteins are composed and of the nucleotides of which nucleic acids are composed. The reservoir of inorganic nitrogen is the gaseous N_2, which constitutes roughly 78 percent of the atmosphere. But N_2 has very little biological activity. It enters the bodies of all organisms, but comes back, in most cases, without having played any significant role in their life processes. Some microorganisms, however—a few bacteria and many blue-green algae—can use N_2 in the synthesis of substances usable by other organisms. This process is known as *nitrogen fixation.* Although some nitrogen fixation may also occur as a result of electrical discharges, such as lightning, and although more and more nitrogen is now supplied in commercial fertilizers manufactured by industrial processes, it is biological nitrogen fixation by microorganisms that provides most of the usable nitrogen for the earth's ecosystems (Fig. 19.21).

Some of the nitrogen-fixing bacteria (genus *Rhizobium*) live in a close symbiotic relationship with the roots of higher plants, where they occur in prominent *nodules* (Fig. 19.22). The legumes (plants belonging to the pea family—bean, clover, alfalfa, lupine, etc.) are particularly well known for their numerous root nodules, but plants of some other families have them also.[6] Other nitrogen-fixing microorganisms live free in soil or water. All of these nitrogen-fixing microorganisms can reduce N_2 to ammonia (NH_3), which is often in the form of ammonium ions (NH_4^+). They then either use the ammonium in the synthesis of organic nitrogen-containing compounds or excrete it into the soil or water in which they live.

The symbiotic bacteria in root nodules promptly release much of the fixed nitrogen they produce into the host plant's cytoplasm, primarily in the form of amino acids. It has been estimated that as much as 90 percent of the fixed nitrogen can be liberated into the host's cytoplasm in this way, there being little or no storage of fixed nitrogen within the bacteria or the nodules. Consequently legumes can grow well in soils very poor in available nitrogen. The bacteria in their nodules not only supply the plants with all the fixed nitrogen they need, but actually produce a surplus, some of which is excreted from the roots of the legumes into the soil. Thus legumes (and the other plants that have similar nodules) tend to increase the fertility of the soil in which they grow. Farmers often build up the nitrogen content of their fields by periodically planting them to legumes. As much as 400 pounds of nitrogen may be fixed in a single season by an acre of alfalfa.

Blue-green algae are the most important nitrogen-fixing microorganisms that live free in the soil or water, although some bacteria (e.g. *Azotobacter* and *Clostridium)* also play a role. These organisms, which may fix between 20 and 50 pounds of nitrogen per acre annually, release ammonia into the surrounding medium, and when they die the fixed nitrogen in their cells is broken down to ammonia by decomposer organisms. The decomposers act in the same way upon the organic nitrogen compounds in the bodies of green plants or animals or other microorganisms when they die, and upon the nitrogen compounds in the urine and feces of animals. Some of this free ammonia is picked up as ammonium ions by the roots of higher plants, particularly certain grasses and forest trees, and incorporated into more complex compounds. But most flowering plants use nitrate in preference to ammonia. Nitrate seems to be the main source of nitrogen for higher plants. It is produced from ammonia in the soil by nitrifying bacteria.

The process of *nitrification* is usually accomplished by two different groups of bacteria, working in sequence. The first group converts ammonium ions into nitrite (NO_2^-), and the second group converts this nitrite into

19.22 Photograph of roots of a legume (bird's-foot trefoil), showing nodules
[Courtesy Nitragin Co., Milwaukee, Wis.]

---

[6] Among the nonleguminous plants with symbiotic nitrogen-fixing microorganisms in nodules are species of alder *(Alnus),* oleaster *(Elaeagnus),* wax myrtle *(Myrica),* and New Jersey tea *(Ceanothus).*

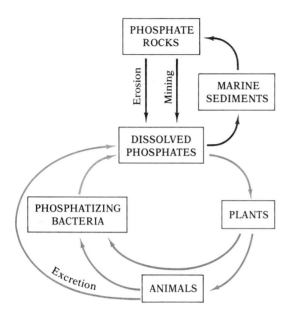

19.23 A simplified version of the phosphorus cycle
Phosphate from rock dissolves very slowly (unless the process is speeded up by human intervention). The dissolved phosphate can be used by plants, which may pass it to animals. Some of the phosphate is excreted by animals and goes immediately into the dissolved pool. When plants or animals die, phosphate is released from the organic compounds (e.g. nucleic acids) by phosphatizing bacteria.

Each year huge quantities of dissolved phosphate are carried into the sea in runoff water. Although the formation of new rocks from marine sediments, where the phosphorus eventually comes to rest, is a very slow process, it is unlikely that we shall soon run out of phosphate rock, because the known reserves are large.

nitrate (NO_3^-) and, releases it into the soil, where it can be picked up by the roots of plants.[7] Most of the nitrate in the roots is quickly incorporated into organic nitrogen compounds and then either stored, primarily in cell vacuoles, or transported to other parts of the plant body through the vascular tissue.

The nitrogen compounds in the plant body may eventually again be broken down to ammonia by decomposers when the plant dies or when an animal that ate the plant dies or excretes it. Notice that nitrogen can cycle repeatedly from plants to decomposers to nitrifying bacteria to plants without having to return to the gaseous N_2 state in the atmosphere. In this respect, the nitrogen cycle differs from the carbon cycle, where every turn of the cycle includes a return of CO_2 to the atmosphere.

Although nitrogen need not return to the atmosphere at every turn of the cycle, there is a steady drain of some of it away from the soil or water and back to the atmosphere. This is because some bacteria carry out a process of **denitrification,** converting ammonia or nitrite or nitrate into N_2 and releasing it. In short, the denitrifying bacteria remove nitrogen from the soil–organism part of the nitrogen cycle and return it to the atmosphere, while the nitrogen-fixing microorganisms do the reverse: They take nitrogen from the atmosphere and add it to the soil-organism part of the cycle.

The phosphorus cycle Another mineral essential to life is phosphorus. Like nitrogen, it is one of the chief ingredients in commercial fertilizers. Unlike carbon and nitrogen, for which the reservoir is the atmosphere, phosphorus has its reservoir in rocks (Fig. 19.23).

Under natural conditions much less phosphorus is available to organisms than nitrogen; in natural waters, for example, the ratio of phosphorus to nitrogen is about 1:23. However, the mining of roughly three million tons each year has greatly accelerated the movement of phosphorus from the rocks to the water–organism part of the cycle. This mineral, the normal limiting resource for algae in many freshwater lakes, is now being poured into the aquatic environment in enormous quantities in sewage and detergents and in runoff from inorganic fertilizers used in farming. One consequence is extensive algal blooms that cover the surface of the water with scum and foul the shores with stinking masses of rotting organic matter (Fig. 19.24).

Now, the increased photosynthetic productivity associated with the algal blooms might be expected to make more food available for higher links in food webs and thus be of benefit to the biotic community. But excessive growth of algae actually causes destruction of many of the higher links in the food webs. At the end of the growing season, many of the algae die and sink to the bottom, where they stimulate massive growth of bacteria the following year. There is so much bacteria-produced decomposition that the oxygen of the deeper colder layers of the lake becomes depleted, with the result that cold-water fish such as trout, cisco, whitefish, pike, and sturgeon die and are replaced by less valuable species such as carp and catfish. Deoxygenation of the water also causes chemical changes in the bottom mud that produce increased quantities of odorous, sometimes toxic, gases.

[7] Among the most important genera of nitrite bacteria are *Nitrosomonas, Nitrosocystis,* and *Nitrosospira.* Among the most important genera of nitrate bacteria are *Nitrobacter, Nitrocystis,* and *Bactoderma.*

These changes also further accelerate the ***eutrophication,*** or aging, process of the lake.[8]

Extensive use of tertiary treatment of sewage, which few municipalities now carry out, combined with a reduction of the phosphate content of detergents, would slow the death of our lakes, but wouldn't prevent it, because at least 30 percent of the polluting phosphorus comes from agricultural sources, where this mineral is essential to the production of the large crops needed to feed a burgeoning population. Furthermore, the tertiary sewage treatments devised so far, though quite effective in removing phosphates, are much less so in removing nitrates, which are probably the major natural limiting nutrient in estuaries. Thus a change from phosphate detergents to nitrogen detergents, which has become popular in the name of "better ecology," may simply shift the pollution problem from one place to another. Unless new detergents free of nutrients that are difficult to remove can be developed, or more effective sewage treatment devised, a return to soaps may become imperative.

---

[8] The term "eutrophication" was originally applied to the accumulation of nutrients and increase in organic matter that are a natural part of the aging of lakes. Recently it has been applied not so much to the natural process as to the greatly accelerated one resulting from human interference.

19.24   A field experiment demonstrating the importance of phosphorus in the eutrophication of a lake

The two basins of a lake were separated by a plastic curtain. The far basin was fertilized with phosphorus, carbon, and nitrogen. The near basin, used as a control, received only carbon and nitrogen. Within two months the far basin was covered by a heavy bloom of algae, whereas the control basin showed no change in organic production.   [Courtesy D. W. Schindler, *Science*, vol. 184, 1974. Copyright 1974 by the American Association for the Advancement of Science.]

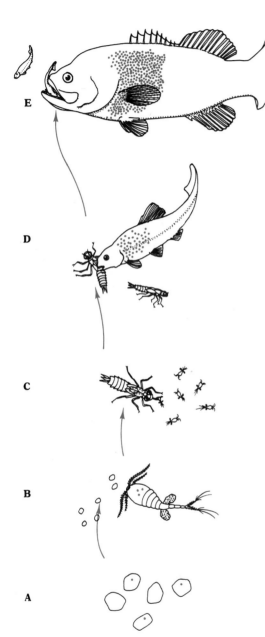

Commercial chemicals Modern industry and agriculture have been releasing vast quantities of new or previously rare chemicals into the environment. The U.S. Food and Drug Administration has estimated their number in the environment at about half a million, with 400–500 new ones created each year. The pathways through ecosystems are known for only a few. Some will probably be incorporated into the natural biogeochemical cycles and degraded to harmless simpler substances. But many are so different from any naturally occurring substances that we have no idea as yet what their eventual fate or their effects on the biosphere may be. Many of the by-products of industrial processes have so far not even been fully characterized chemically. Will some of these prove harmful to life? The answer is surely yes. But which will be harmful, and how harmful, and to what organisms? It is with these questions that much ecological research in the future must deal.

The matter is complicated by the fact that a substance may not be harmful in the form in which it is released, but may be changed by microorganisms, or by natural physical processes, into some other substance with vastly different properties. A case in point is the mercury pollution of bodies of water near some plastics factories. The mercury was originally released in an insoluble and nontoxic form that was thought to be stable. When it settled in the bottom mud, however, microorganisms converted it into methyl mercury, a water-soluble compound that accumulates in organisms. The mercury poisoning that resulted was most severe in human beings and other top predators.

The harmful effect on top predators of the mercury from plastics factories (and of the mercury, principally from fungicides, that has recently caused the fish in many streams and lakes to be declared unsafe for human consumption) is an example of a common phenomenon called **_biological magnification._** If a persistent chemical,[9] when ingested, is retained in the body rather than excreted, then that chemical will tend to become more and more

[9] A persistent chemical is one comparatively stable under natural biological and environmental conditions.

19.25 An example of biological magnification
(A) Some individuals of a single-celled plant species at the bottom of a food chain have picked up a small amount of a stable nonexcretable chemical (color). (B) *Cyclops*, a small crustacean, incorporates into its own tissues the chemical from all the infected plants it eats. (C) A dragonfly nymph stores all the chemical acquired from the numerous *Cyclops* it eats. (D) Further magnification occurs when a minnow eats many of the infected dragonfly nymphs. (E) When a bass, the top predator in this food chain, eats many infected minnows, the result is a very high concentration in its tissues of a chemical that was much less concentrated in the organisms lower in the chain.

concentrated as it is passed up the food chain. Figure 19.25 shows why this is so.

DDT (a persistent insecticide that has become so pervasive in the biosphere that it can now be found in the fatty tissues of nearly every living organism) has had more severe effects on predatory birds such as the Bald Eagle and Peregrine Falcon than on seed-eating birds because of biological magnification. Some investigators have reported that the reproductive rate of eagles and falcons has been calamitously reduced because DDT—and its metabolites, DDD and DDE—interfere with deposition of calcium in the eggshells, with the result that the thin-shelled eggs are easily broken and few birds hatch.

This brief summary of some of the biogeochemical cycles will have suggested the complexity of the movement of materials through an ecosystem, the interdependence of the different species within it, and, in particular, the fundamental and essential role played by microorganisms. Because the microorganisms are not observed as easily as the larger plants and animals, we tend to forget about them, or to think only of the harmful ones, especially those that cause diseases; and thus we overlook the others, many of which are indispensable to our continued existence.

THE ROLE OF SOILS

Soil is an essential link in the water cycle, on which terrestrial plants depend. The plants require the water itself, of course, but they also depend on the water cycle because the minerals they need—nitrogen and phosphorus, as well as calcium, sulfur, potassium, and other ions whose cycles we have not discussed—do not become available to them until dissolved in soil water. It follows that the properties of soils, including particle size, amount of organic material, and pH, play a very important role; these factors help determine the availability of water and minerals to the plants, and the rapidity with which these materials move through the soil.

Most soils are a complex of mineral particles, organic material, water, soluble chemical compounds, and air. In this complex the dominant components by far are the mineral particles, which are composed largely of compounds of silicon and aluminum. They vary in size from tiny clay particles (less than 0.002 mm in diameter), through silt particles (0.002–0.05 mm), to coarse sand grains (0.05–2.0 mm). The proportions of clay, silt, and sand particles in any given soil determine many of its other characteristics. For example, very sandy soils, which contain less than 20 percent of silt and clay particles, have many air-filled spaces, but they are so porous and their particles have so little affinity for water that it rapidly drains through them and they are unsuitable for growth of many kinds of plants. As the percentage of clay particles increases, the water retention of the soil also increases until, in excessively clayey soils, the drainage is so poor and the water is held so tightly to the particles that the air spaces become filled with water; few plants can grow in such waterlogged soil. Although different species of plants are adapted to different soil types, most do best in soils of the type known as *loam,* which contain fairly high percentages of each size of particle (e.g. 24 percent clay, 29 percent silt, 30 percent fine sand, and 17 percent coarse sand). In such a soil, drainage is good but not excessive and there is good aeration; the soil particles are surrounded by (or contain) a shell of water, but there are numerous air-filled spaces between them.

Loams usually also contain considerable amounts of organic material (roughly 3–10 percent), mostly of plant origin. As this material decomposes, inorganic substances required for good plant growth are released into the soil. The organic material thus contributes to soil fertility. But it also plays another very important role. Since it usually has a rather porous spongy texture, it helps loosen clayey soils and increase the proportion of pore spaces, thus promoting drainage and aeration. This is particularly true when the organic material is in the form of **humus,** which consists mostly of decomposition products from cellulose and lignin. It is interesting that humus has the opposite effect on sandy soils, where it tends to reduce pore size by binding the sand grains together, thereby increasing the amount of water held in the soil. Thus we see that organisms are not only influenced by their physical environment but that they, in turn, modify that environment.

The proportion of clay particles affects not only the physical structure of the soil and its aeration and water-holding capacity, but also the amounts and the availability to plants of certain minerals—in part because of the influence of the clay particles on water movement. If, for example, water percolates downward through the soil very rapidly and in large quantities, it will tend to leach many important ions from the soil, carrying them deep into the underlying rock layers, where roots cannot reach them. Nitrate ions are especially susceptible to leaching, and sulfate, calcium, and potassium ions may also be rapidly removed from the soil.

Excessive removal of calcium is particularly serious because it tends to make the soil become more acid. Although many plants grow best in slightly acid soils, most do not do well in strongly acid ones (however, some species, such as rhododendrons and cranberries, prefer very acid conditions). The acidity of the soil influences the solubility, and hence the availability, of iron, manganese, phosphate, and some other ions (see Fig. 5.9, p. 201), as well as the activity of soil organisms, many of which are inhibited by high acidity.

We have repeatedly spoken of the "availability" of ions to plants. Chemical analyses that give the total amount of the various ions present in soils can be misleading: A certain proportion of these ions is not free, and hence not available to the plants. A complex equilibrium generally exists between ions free in the soil water and ions adsorbed on the surface of colloidal clay and organic particles. Many factors, of which acidity is a prime example, can shift this equilibrium, either increasing the proportion of ions bound to the particles, and thus reducing availability, or increasing the proportion of free ions available in the soil solution.

Even air conditions may influence the ionic makeup of soil. In an extensive study of an experimental forest at Hubbard Brook, New Hampshire, it was found that the rain often contained appreciable quantities of sulfuric acid, probably because of industrial release of sulfur dioxide into the air. Hydrogen ions from the acid tended to displace nutrient cations, such as Ca^{++}, from negatively charged sites on the soil particles, with the result that these nutrients were leached more rapidly from the soil into streams and lakes. There is increasing evidence that such loss of soil fertility as a consequence of acid rain is distressingly widespread and represents a hitherto unrecognized cost of air pollution.

The various characteristics of soils have a bearing not only on how many and what kinds of plants are likely to grow in any given region, but also on the occurrence of soil animals. Earthworms, nematode worms, and millipeds, for example, are all sensitive to the structure, drainage, acidity, and chemical composition of soils. And animals that do not live in the soil are, of course, indirectly influenced in their distributions by soil types, because of their dependence on plants as a source of high-energy organic nutrients.

Conversely, the soils themselves can be fully understood only by considering the effects on them of the plants and animals living in and on them. Plant roots break up the soil in which they grow, and they remove substances from the soil and add other substances to it. The plant shoots shield the soil beneath them, thereby altering the patterns of rainfall, humidity, light, and wind to which the soil is subject. And when the plants die, their substance adds organic material to the soil, changing both its physical and chemical makeup. Microorganisms in the soil alter its composition profoundly. Soil animals, such as earthworms and millipeds, also have a marked effect; they constantly work the soil, breaking down its organic components and moving materials between soil layers.

Some of the effects of vegetation on the soil were dramatically demonstrated in the Hubbard Brook study. The investigators, after first obtaining accurate measurements of the nutrient input and output of a particular watershed over a period of several years, cleared the watershed of all its vegetation, and again monitored input and output. They found that the volume of runoff water promptly rose to levels many times greater than before (during one period it was actually 418 percent greater)—a predictable result of disrupting the water cycle by removal of all transpiring surfaces. Less predictable was an extraordinary loss of soil fertility. The runoff output of nitrate rose steeply (as much as 45 times higher than in undisturbed watersheds) (Fig. 19.26), and there was a drastic increase in net losses of such nutrients as potassium (21 times greater) and calcium (10 times greater). Apparently removal of the vegetation had so altered the chemistry of the soil that nutrients were bound less tightly to soil particles and hence were rapidly leached away.

It can be seen, then, that wholesale human destruction of vegetation results not only in increased erosion by wind and water, a long-familiar consequence, but also in severe loss of fertility in the soil that remains. The stability of the physical part of an ecosystem clearly depends on the production and decomposition of organic matter, and on an orderly flow of nutrients between the living and the nonliving components of the system.

The cutting down of forests is not the only way of ruining the soil; overgrazing and other poor farming practices have caused permanent damage in areas where forests never stood. The valleys of the Tigris and the Euphrates once supported the Sumerian civilization; later they were the granary of the great Babylonian Empire. But poor farming practices led to such extensive erosion and to such a buildup of salt in the soil that the amount of cultivable land today is under 20 percent; the ancient irrigation works are filled with silt, and so much soil has been washed into the Persian Gulf that the ancient seaport of Ur is now 240 km from the coast, its buildings buried under 10 m of silt. Similar conditions prevail in Syria, where human activity has reduced more than a million acres to desert. Closer to home, overgrazing and the plowing-under of native grasses led to the dust-bowl conditions of the 1930s

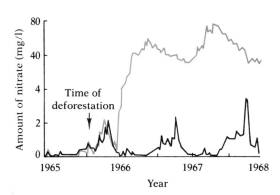

19.26   Change in the runoff of nitrate as a result of deforestation in the Hubbard Brook experimental forest
The color curve indicates the output of nitrate in stream water in the deforested area, and the black curve the output in an undisturbed area. Shortly after deforestation the output in the experimental area rose to about 40 times its previous level, whereas the output in the undisturbed control area remained at about the same level.   [Redrawn from G. E. Likens *et al., Ecological Monographs*, vol. 40, 1970, copyright 1970 by the Ecological Society of America.]

in parts of the United States, when clouds of topsoil were blown hundreds of kilometers and vast areas were left barren and useless. Such examples from the past show us that the ecological crisis is not entirely the result of modern technology, but it is disconcerting that, despite increased knowledge of soil dynamics, wholesale destruction of the earth's soils should continue.

Irrigation has been viewed as a way of greatly increasing the productivity of dry areas. However, it is all too often a short-term remedy, likely to be extremely destructive in the long run. In many cases it leads simply to accelerated erosion; in the United States, for example, an estimated 2,000 irrigation dams are now useless impoundments of silt, sand, and gravel. In other cases irrigation leads to rapid deposition of salt in the soil—salinization—until eventually there is so much salt that plants cannot grow. The salinization may occur because adding water to land overlying a salty water table causes the ground water to rise, the salt thus being carried into the topsoil, or because salts, originally present in low concentrations in the irrigation water, accumulate in the soil as the water evaporates. The Indus Valley of Pakistan, the largest irrigated region in the world—over 23 million acres watered by canals—fell victim to salinization. The irrigation system seemed very promising; the soil was good, and the addition of adequate water was expected to make it produce abundant crops. But as one observer has said, "The result was tragically spectacular. In flying over large tracts of this area one would imagine that it was an Arctic landscape because the white crust of salt glistens like snow." There are indications that the irrigation system made possible by the Aswan high dam in Egypt may similarly produce salinization of the soil; one ecologist has commented, "The Aswan high dam is designed to bring another million acres of land under irrigation, and it may well prove to be the ultimate disaster for Egypt."

One promising development is the use of "drip irrigation," a method of reducing salinization by dripping water onto the soil from overhead pipes at a rate calculated to ensure that virtually all of it penetrates into the soil instead of being lost by evaporation. Of particular value in drip irrigation is the use, where possible, of water from treated sewage; mineral nutrients are thus recycled to the land, where they are needed, rather than released into streams and lakes, where they are ecologically damaging.

This brief discussion of soils and what can happen to them illustrates how complex, and at the same time how fragile, the earth's ecosystems are. In the past, people have usually simply trusted to luck in their attempts to manipulate them. Someone has said that human beings refuse to try to understand any system they did not design themselves. Surely the time for indulging such attitudes is past. Gaining knowledge of how ecosystems function, and using such knowledge, have virtually become ethical imperatives if human beings are to continue to inhabit the earth.

THE BIOTIC COMMUNITY

In nature it is rare to find only one kind of organism living in a region. Usually a diversity of plant and animal species occupy the same general area, sharing some resources and competing for others. The biotic community they form can be considered a unit of life, with its own characteristic structure and functional interrelationships. We shall here examine some features of community dynamics.

DOMINANCE, DIVERSITY, AND COMMUNITY STABILITY

As we have seen, the species comprising a community influence one another in countless ways, for both good and ill. Thus a predator may, if overabundant, represent a threat to the continuation of a prey population, but under more balanced conditions the activity of the same predator may, by regulating the density of the prey population, help keep it vigorous and healthy. As we have seen, too, there are numerous interactions of the populations in the community with the physical environment. Thus plants extract mineral nutrients from the soil, but, when they die, their substance may contribute to the organic content of the soil.

Species interactions that are far from obvious may yet be of crucial importance to a community. Consider an experiment with an intertidal community of 15 species of marine invertebrates conducted by R. T. Paine of the University of Washington at Mukkaw Bay. The top predator in this community was a sea star, *Pisaster ochraceus*. When Paine excluded the sea star from one area but allowed it to remain in another, the result was a radical change in the species diversity of the first area. Of the original 15 species, only eight remained. One of two competing species of barnacles was eliminated, because the sites for its attachment were not cleared of other organisms by the sea star, and the other barnacle was a better competitor for what little space there was. In the same food web a sponge and its predator disappeared, although neither was preyed upon by *Pisaster;* apparently the sea star had some sort of indirect influence on them, probably through the clearing of space for the sponge. Dramatic as these effects of the removal of *Pisaster* were, they would have been virtually impossible to predict beforehand.

Efforts to eliminate undesirable species from a community often reveal hidden linkages to other organisms, and may dramatically demonstrate the complex interactions on which community stability rests. An unintended cautionary example was provided by the World Health Organization (WHO) in a campaign to eradicate malaria-carrying mosquitoes in the Borneo states of Malaysia, where as many as 90 percent of the population of some areas suffered from the disease. Mosquito control was achieved by spraying the insides of the village huts with DDT and Dieldrin, two powerful contact insecticides, and malaria was indeed eradicated. But soon the villagers began to notice that the thatch roofs of their huts were rotting and beginning to collapse. Investigation showed that the deterioration, which occurred only in huts sprayed with DDT, was due to the larvae of a moth that normally lives in small numbers in the thatch roofs. Whereas the thatch-eating moth larvae avoided food sprayed with DDT, the moth's natural enemy, a parasitic wasp, was very sensitive to it. The net result was a substantial increase in the population of the larvae eating the thatch.

That would have been an interesting story in itself, but there was yet another side effect potentially more serious. Cockroaches and a small house lizard, the gecko, are two normal inhabitants of the village huts. DDT-contaminated cockroaches were eaten by the geckos, which were in turn eaten by house cats (as were some cockroaches). The cats, poisoned by the accumulation of the insecticide, died. What ensued was a population explosion of rats, which are potential carriers of such diseases as leptospirosis, typhus,

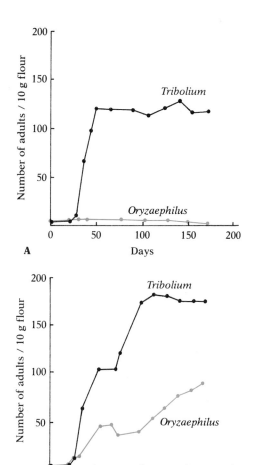

19.27 An experiment showing the effect of increased diversity of the physical environment on community stability

A. C. Crombie of the University of Cambridge performed an experiment with two species of beetles in a culture with a flour medium. *Tribolium* always overcame *Oryzaephilus*, which became extinct, in cultures with medium only (A). But if Crombie added pieces of fine glass tubing to the culture, both species survived and increased in numbers (B).   [Redrawn from R. H. MacArthur, *Geographical ecology*, Harper & Row, 1972.]

and plague. In an attempt to restore the cat population, WHO and the Royal Air Force undertook a remarkable venture, "Operation Cat Drop," in which they parachuted cats into the villages. With the cat population restored, the rats and the consequent threat of serious disease subsided.

Fortunately, not every pest-control effort entails such complications. Insecticides rapidly broken down in the environment and more selective in their toxicity have been developed, and these are less disruptive of biological communities. Much to be preferred, however, is biological control, e.g. the use of the natural predators, parasites, or pathogens of the pest.

Recognizing the crucial role played in a community by one or two species, such as the sea star at Mukkaw Bay or the parasitic wasp on Borneo, ecologists would like to be able to assign some sort of importance value to the species in a community. Such a procedure might help them predict what environmental impact a proposed human disturbance would have, for example. But what is the meaning of "importance" in this context? Many answers are possible, and each, unfortunately, leads to a different conclusion.

If the emphasis is placed on density, the highest importance value will be assigned to the densest species. But in a forest the densest plant species may be a small shrub growing below the large trees; yet the large trees, not the shrubs, are clearly the dominant species, in the sense of determining the nature of the whole community. An alternative would be to emphasize productivity, a much better measure of the species' position in the flow of energy and materials through the community, but this approach fails to give adequate recognition to the regulatory role played by predators such as the sea star or parasitic wasp mentioned above. Since these organisms belong to high trophic levels, their productivity is small; yet they were clearly key members of their communities, and their removal resulted in sweeping community changes. Productivity, or some combination of density and size, may work relatively well as a measure of importance in the channeling of energy through the community, but they do not necessarily lead to reliable predictions about the effect on the community of the removal of a relatively nonproductive species.

In each of our two examples the removal of a single species caused major community upheavals. Many ecologists have held that in more complex communities, where there are more species and therefore more alternative interactions, there will be greater stability—a superior ability to withstand perturbations such as the removal of a species. In such a community, for example, the predators often have many prey species available and, because they can switch from one to another, are less sensitive to variations in the abundance of any one. Reciprocally, the prey may be subject to several different predators, its density regulation thus being less dependent on any single predator. And primary production in a complex community is less likely to be strongly dominated by only one or two plant species.

It is a question, however, whether greater species diversity and complexity of interactions really make for increased community stability. The response of a simple community to a disturbance may be more violent and immediate, but such a community may recover more quickly and settle more rapidly into a new functional mode. The complex community, by contrast, may respond less dramatically to the disturbance, but, because of the multiplicity of interactions within it, the effects may continue to ripple through the system for a long time, causing numerous small, but nonetheless important, dislocations and distortions. Which community, then, is the more stable—the simpler one, which responds violently but recovers quickly, or the more complex one, which responds less violently but continues to show effects over a longer period? The answer, of course, lies in how one defines "stability." The point to remember is that species diversity and community complexity do indeed influence the way a community will respond to disturbance.

Another kind of complexity likewise needs mention here—complexity of the physical environment. In simple laboratory cultures of several competing species or of predator-prey systems, extinction of at least one of the species virtually always ensues (see Fig. 19.12). But if the physical environment in the culture is made more diverse, as by adding tunnels, blind alleys, and obstacles to travel, the wild fluctuations and oscillations typical of simple laboratory systems can often be reduced, and extinction postponed (Fig. 19.27). Thus structural diversity of the environment seems to favor community stability, probably by making possible a finer division and definition of niches.

Human intervention in biological communities nearly always has the effect of simplifying them, in terms both of reduced species diversity (Fig. 19.28) and of diminished structural complexity. The result, as would be predicted from all we have learned about community dynamics, is to make the communities far more prone to extreme fluctuations in response to changing conditions. Prime examples are the highly artificial communities created by modern agriculture, which has tended more and more to emphasize monoculture—the planting of a single crop species in enormous fields from which all other plant species are systematically excluded. These communities are very unstable and can be maintained only at the price of constant vigilance and the investment of much energy in curbing insect infestations and outbreaks of disease, in maintaining soil fertility, and in cleaning out or killing invading weeds. As experiments have demonstrated, multiple-species crops have greater resistance to pests and disease, but the difficulty and expense of adapting conventional farm machinery to the planting of such crops has made it impractical for large-scale farming so far.

Pollution, though less extreme in its simplification of communities than monoculture farming, also reduces species diversity and thereby shortens food chains and increases community instability. Top predators are frequently the first species to disappear. With the predators gone, one or more prey species may multiply unchecked and further degrade the habitat, often making it unfit for other species and sometimes actually rendering it inadequate for the support of even the exploding species itself.

ECOLOGICAL SUCCESSION

The nature and causes of succession   Succession is a more or less orderly process of community change. It involves replacement, in the course of time, of the dominant species within a given area by other species. A farmer's field is allowed to lie fallow. A crop of annual weeds grows in it during the first year. Many perennial herbs appear in the second year and become even more common in the third year. Soon, however, these are superseded as the dominant vegetation by woody shrubs, and they may in turn be replaced

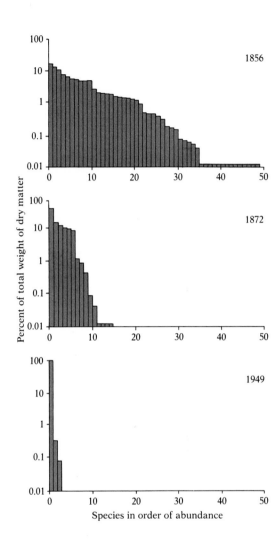

19.28   An experiment showing the effect of human intervention on species diversity in an ecosystem Nitrogen fertilizer has been applied yearly since 1856 to an experimental grass plot at the Rothamsted Experimental Station in England. When the experiment began, 49 species were growing in the plot (top). By 1872 the number had fallen to 15 (middle). Only 3 species remained in 1949 (bottom). This sequence resembles an early succession run backward. The community probably shrank because one ecological factor was emphasized at the expense of all the others.   [Redrawn from R. M. May, "The evolution of ecological systems," *Sci. Am.*, September 1978. Copyright © 1978 by Scientific American, Inc. All rights reserved.]

eventually by trees. Or a lake dries up, and its sandy bottom becomes covered with grass, which later gives way to trees.

But why the change? Why does succession occur? The cause cannot be climate, because succession will occur even if the climate remains the same. Climate may be a major factor in determining what sorts of species will follow one another, but the succession itself must result from other causes. The traditional view has been that the most important cause is the modification of the physical environment produced by the community itself. Most successional communities tend to alter the area in which they occur in such a way as to make it less favorable for themselves and more favorable for other communities. In effect, each community in the succession sows the seeds of its own destruction.

Consider the alterations initiated by pioneer communities on land. Usually these communities will produce a layer of litter on the surface of the soil. The accumulation of litter affects the runoff of rainwater, the soil temperature, and the formation of humus. The humus, in turn, contributes to soil development and thus alters the water relations, the pH and aeration of the soil, the availability of nutrients, and the sorts of soil organisms that will occur. But the organisms characteristic of the pioneer communities that produced these changes may not prosper under the new conditions, and they may be replaced by invading competitors that do better in an area with the new type of soil.

Though much of the modification of the habitat that produces succession is due to the action of living organisms, other forces may also contribute. Among these are physiographic changes. For example, greater depth in the channel of a stream may result in better drainage of a swamp; or an overflowing stream may deposit rich silt on nearby bottomland. Such changes will soon be followed by changes in vegetation.

There has been growing recognition that modification of the habitat—whether by action of organisms or by physiographic changes—is not the entire explanation for ecological succession. To some degree, changes in vegetation merely reflect the fact that some species are more easily dispersed and grow more rapidly than others. Thus it is only to be expected that annual herbs will be far more important members of a pioneer community on recently abandoned farmland than the seedlings of slow-growing trees, even though the trees may eventually become the dominant plants.

Successions in different places and at different times are not identical; the species involved are often completely different, and the climatic and substrate conditions vary widely. Moreover, the sequence of changes in ***primary successions,*** in which communities are established in newly formed habitats (e.g. on sand dunes or bare rock), is often longer and slower than in ***secondary successions,*** in which communities are re-established in areas where they were destroyed (e.g. in fields where the original forests were cleared for farming).

Some examples of succession   One of the first examples of primary succession studied in detail (by H. C. Cowles of the University of Chicago) was that on the sand dunes at the southern end of Lake Michigan. Lake Michigan once extended much farther south than it does today. As the lakeshore gradually receded northward, it left exposed a series of successively younger

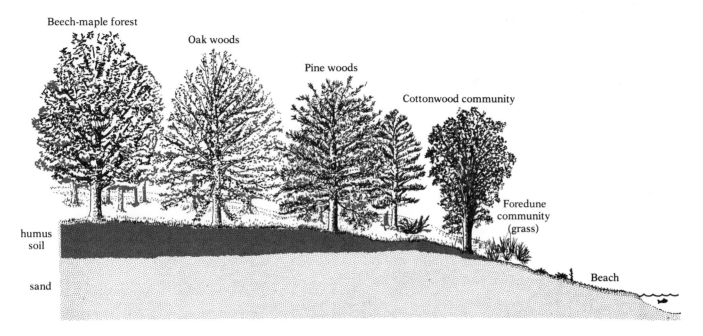

Beech-maple forest

Oak woods

Pine woods

Cottonwood community

Foredune community (grass)

humus soil

sand

Beach

19.29   Successional stages at the southern end of Lake Michigan

beaches and sand dunes. Thus someone who starts at the water's edge and walks south for several miles will pass through a series of communities (Fig. 19.29) that represent various successional stages beginning with bare beach and culminating in an old well-established forest dominated by beech and sugar-maple trees. Let us examine these stages in more detail.

1. The lower beach near the water's edge has no land life, because it is frequently exposed to the destructive action of waves.

2. The middle beach is ordinarily dry in summer, but is occasionally washed by the waves produced by severe winter storms. Conditions of life are very severe and exclude biennial and perennial plants, but a few succulent annuals similar to those that inhabit deserts grow there in summer.

3. Conditions on the upper beach are much less severe than on the lower and middle beaches and the flora is richer, but vegetation is still very sparse. Decay of driftwood has begun adding some organic matter to the substratum.

4. Behind the upper beach is the foredune community, a pioneer community dominated by sand-binding beach grasses. Tiger beetles, grasshoppers, and burrowing spiders are characteristic animals of this community.

5. The first tree-bearing stage in this succession is a cottonwood community, a loosely organized pioneer community on sands that have usually been partly bound by the roots of grasses, but are still subject to considerable shifting by the wind.

6. The cottonwood community grades into a community dominated by jack pine, juniper, and bearberry, where the soil has a considerable humus component.

7. The pine community grades, in turn, into a forest dominated by oaks.

8. The oak community grades into a forest dominated by sugar-maple and beech trees growing in a deep humus-rich moist soil.

A A newly formed pond near the beach has sandy borders bare of vegetation.

B After two years such a pond is ringed by low vegetation, including cottonwood saplings.

C A 50-year-old pond is bordered by mature cottonwood trees. So much sediment is produced by organisms growing in the pond that only a small area of water, choked with weeds, remains.

D After 150–250 years the area that was once a pond has become a meadow.

19.30 Succession in ponds on Presque Isle, Pennsylvania
[Courtesy E. J. Kormondy, *Smithsonian Mag.*, vol. 1, 1970.]

During the walk from the lakeshore to the beech-maple forest, an observant person would notice many changes in the physical environment—a progressive decrease in total light intensity at ground level, a decrease in wind velocity and in the rate of evaporation, an increase in soil moisture and in relative humidity, and an increase in the amount of humus in the soil and the amount of leaf mold on its surface. Presumably the series of communities and environmental changes seen contemporaneously as one moves laterally away from the lake duplicates approximately the series of successional stages through which the area now covered by the beech-maple forest must have passed since the time when it was a wave-washed beach.

Another much-studied type of succession is that in ponds (Fig. 19.30). Sediments washed from the surrounding land begin to fill the pond, and the dead bodies of planktonic organisms add organic material. Soon pioneer submerged vascular plants appear in the shallower water near the margins of the pond. Their roots hold the silt, and the lake bottom is built up faster where they are growing. In addition, as these plants die, their bodies accumulate faster than decomposers can break them down. Soon the water is shallow enough for broad-leaved floating pondweeds to displace the submerged species, which now become established in a zone farther out in the pond, where conditions are more favorable for them. But as the bottom continues to build up, the floating pondweeds are in their turn displaced by emergent species (plants that have their roots in the mud of the bottom, but their shoots extending into the air above the water), such as cattails, bulrushes, and reeds. These plants grow very close together and hold the sediment tightly, and their great bulk results in rapid accumulation of organic material. Soon conditions are dry enough for a few terrestrial plants to gain a foothold. Now an area that was formerly part of the pond is newly formed dry land. This entire sequence can sometimes be seen as a nearly continuous series of zones girdling a pond or lake. With the passage of the years, the pond becomes smaller and smaller as the zones move nearer and nearer its center. Eventually nothing of the pond remains.

Successions need not begin with land reclaimed from lakes or ponds as in the preceding examples. Consider a bare rock surface (Fig. 19.31). The first pioneer plants may be lichens, which grow during the brief periods when the rock surface is wet and lie dormant during periods when the surface is dry. The lichens release acids and other substances that corrode the rock. Dust particles and bits of dead lichen may collect in the tiny crevices thus formed, and pioneer mosses may gain anchorage there. The mosses grow in tufts or clumps that trap more dust and debris and gradually form a thickening mat. A few fern spores or seeds of grasses and annual herbs may land in the mat of soil and moss and germinate. These may be followed by perennial herbs. As more and more such plants survive and grow, they catch and hold still more mineral and organic material, and the new soil layer thus becomes thicker. Later, shrubs and even trees may start to grow in the soil that now covers what once was a bare rock surface.

Succession may also be observed in miniature on a log lying in a forest (Fig. 19.32).

Secondary succession on abandoned croplands, unused railway rights-of-way (Fig. 19.33), plowed grasslands, or cutover forests often proceeds relatively quickly in its initial stages, because the effects of the previous com-

19.31 An early successional stage on a bare rock surface
First to gain a foothold on rock are often lichens, shown here bearing fruiting bodies. Chemicals produced by the lichens corrode the rock surface and help prepare the way for later successional stages.   [Courtesy Carolina Biological Supply Co.]

19.32 A stage in the succession of organisms living on a fallen log
This log, in a forest in interior Alaska, is covered by a community of mosses and lichens. [B. Ruth, Bruce Coleman Inc.]

19.33 Secondary succession on an abandoned railway right-of-way
The ties are rotting, and vegetation is taking over in the formerly cleared area. [M. P. Gadomski, Bruce Coleman Inc.]

munities have not been wholly erased and the physical conditions are not as bleak as on a beach or a bare rock surface. Let us take an abandoned cornfield in Georgia as an example. The very first year it will be covered with annual weeds, such as ragweed, horseweed, and crabgrass. In the second year ragweed, goldenrod, and asters will probably be common, and there will be much grass. The grass will usually be dominant for several years, and then more and more shrubs and tree seedlings will appear. The first tree seedlings to grow well in the unshaded field will be pines, and eventually a pine forest will replace the grass and shrubs. But pine seedlings do not grow

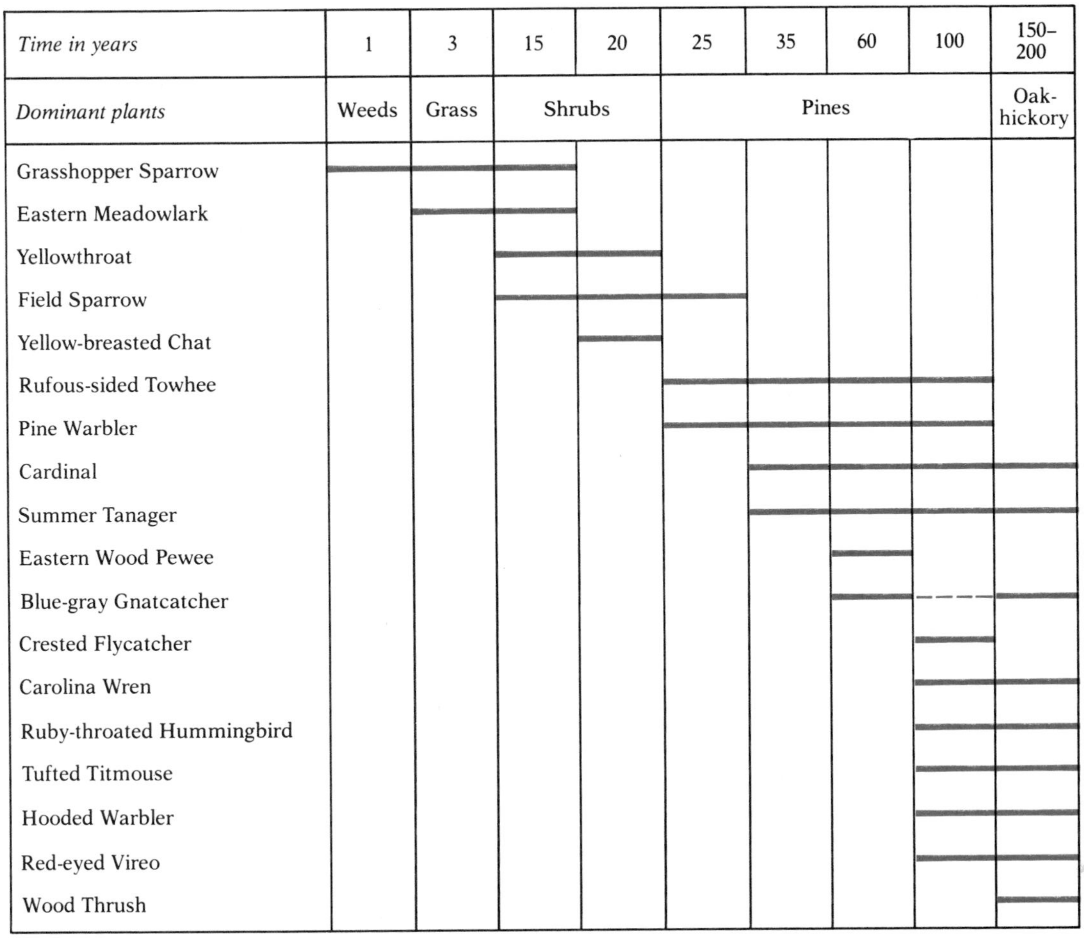

Time in years	1	3	15	20	25	35	60	100	150–200
Dominant plants	Weeds	Grass	Shrubs		Pines				Oak-hickory

Grasshopper Sparrow

Eastern Meadowlark

Yellowthroat

Field Sparrow

Yellow-breasted Chat

Rufous-sided Towhee

Pine Warbler

Cardinal

Summer Tanager

Eastern Wood Pewee

Blue-gray Gnatcatcher

Crested Flycatcher

Carolina Wren

Ruby-throated Hummingbird

Tufted Titmouse

Hooded Warbler

Red-eyed Vireo

Wood Thrush

well in the shade of older pines. Seedlings of oaks, hickories, and other deciduous trees are more shade-tolerant, and these trees will gradually develop in the lower strata of the forest beneath the old pines, eventually replacing them. The deciduous forest thus formed is more stable and will ordinarily maintain itself for a very long time.

Forests often become stratified into more or less distinct layers, each of which has its own populations and interactions. The forest floor, or herb layer, the shrub level, the short-tree level, and the canopy level are common strata of deciduous and tropical forests. The canopy species capture most of the sunlight, but much of the energy they assimilate must be used to build and maintain woody supporting tissues. The herb layer, on the other hand, receives as little as one percent of the available sunlight, but the plants have no wood, and can use all the energy from photosynthesis for maintenance and reproduction.

As can be seen from Fig. 19.34, changes corresponding to the succession of dominant plants also take place in the animal portion of the abandoned cropland community.

19.34 Bird succession on abandoned upland farmland in Georgia
The bars indicate when each of the bird species was present in a density of at least one pair per 10 acres. In the early (weed and grass) stages Grasshopper Sparrows and Eastern Meadowlarks were the dominant bird species. During the shrub stage Yellowthroats and Field Sparrows became dominant. Pine Warblers and Rufous-sided Towhees dominated the young pine forests, and Red-eyed Vireos, Wood Thrushes, and Cardinals dominated the oak-hickory forests. [Based on data in E. P. Odum, *Fundamentals of ecology*, Saunders, 1959.]

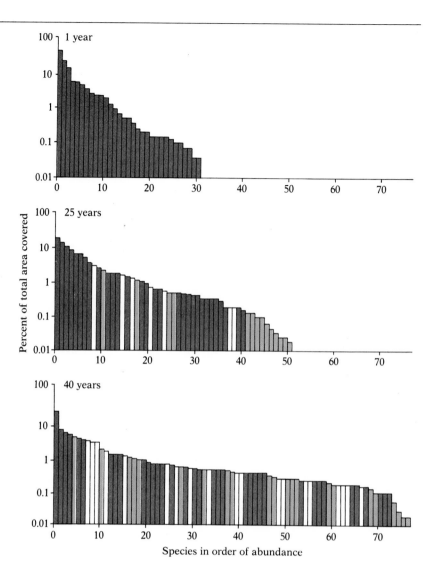

19.35 The changing pattern of species diversity and abundance in the early successional communities on an abandoned agricultural field in Illinois
After one year there were 31 species of plants, all herbs (gray bars), growing in the field. After 25 years there were 51 species, including some shrubs (white bars) and trees (colored bars). After 40 years there were 77 species, and shrubs and trees had greatly increased in relative abundance. [Redrawn from R. M. May, "The evolution of ecological systems," *Sci. Am.*, September 1978. Copyright © 1978 by Scientific American, Inc. All rights reserved.]

A summary of trends in many successions Despite numerous differences between the various successions, especially between primary and secondary successions, some generalizations tend to hold true in most cases where both autotrophs and heterotrophs are involved:

1. The species composition changes continuously during the succession (Fig. 19.35), but the change is usually more rapid in the earlier stages than in the later ones.

2. The total number of species represented increases initially, then sometimes declines slightly, and finally becomes more or less stabilized in the older stages. This trend applies particularly to the heterotrophs, whose variety is usually much greater in the later stages of the succession.

3. Net primary productivity increases until it reaches a stable high level (Fig. 19.36).

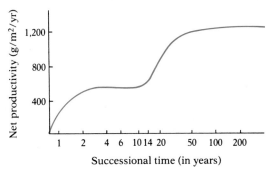

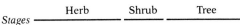

19.36   Change in net primary productivity during succession on an area cleared of an oak-pine forest in Brookhaven, New York
The first rise represents the invasion of the area by herbs. The later rise (after about 14 years) reflects the entrance of larger woody plants into the community.   [Redrawn from R. H. Whittaker, *Communities and ecosystems*, 2nd ed., Macmillan, 1975, after B. Holt and G. M. Woodwell.]

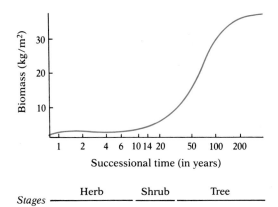

19.37   Change in biomass during succession in the Brookhaven study area
The total biomass remained low during the early years, when herbs were the dominant plants in the community, but increased later when shrubs and trees became more prominent.   [Redrawn from R. H. Whittaker, *Communities and ecosystems*, 2nd ed., Macmillan, 1975, after B. Holt and G. M. Woodwell.]

4. The store of inorganic nutrients held in the organisms and soil of the ecosystem increases, and an increasing proportion of this store is held in the tissues of plants.

5. Both the total biomass in the ecosystem (Fig. 19.37) and the amount of nonliving organic matter increase during the succession until a more stable stage is reached.

6. The height and massiveness of the plants in the community increase and lead to greater differentiation of vertical strata.

7. With more extensive aboveground plant cover, the microclimate within the community becomes increasingly determined by the community itself.

8. The food webs become more complex, and the relations between species in them become better defined or more specialized. As a result, the efficiency of resource utilization at the various trophic levels rises.

In summary, if these generalizations are correct, the trend of most successions is toward a more complex and longer-lasting ecosystem, in which less energy is wasted and hence a greater biomass can be supported without further increase in the supply of energy.

THE CONCEPT OF CLIMAX

If some disruptive factor does not interfere, most successions—as the previous discussion will have suggested—eventually reach a stage far more stable than those that preceded; the important species populations attain a steady state (i.e. a balance between births and deaths), and both energy flow and biomass likewise attain equilibrium, with gross primary productivity equaled by total respiration. The community of this stage is called the *climax community*. It has much less tendency than earlier successional communities to alter its environment in a manner injurious to itself. In fact, its more complex organization, larger organic structure, and more balanced metabolism enable it to buffer its own physical environment to such an extent that it can be self-perpetuating. Consequently it may persist for centuries, not being replaced by another stage so long as climate, physiography, and other major environmental factors remain essentially the same.

It must be emphasized, however, that a climax community is not static; it does slowly change, and will change rapidly if there are major shifts in the environment, either physical or biotic. For example, sixty years ago chestnut trees were among the dominant plants in the climax forests of much of eastern North America, but they have been almost completely eliminated by a fungal blight, and the present-day climax forests of the region are dominated by other species. Thus there can be no absolute distinction between climax and the other stages of succession; the difference between them is relative.

A view that held sway among some American ecologists during much of this century was that all succession in a given large climatic region will converge to the same climax type—that there is only one type of climax community for the region and that any sites dominated by other communities have not yet reached climax, no matter how stable and long-lasting they may seem. Thus a beech-maple forest was considered the climax for much of the northeastern United States, and a white spruce–balsam fir forest was considered the climax for much of Canada. The proponents of this so-called monoclimax hypothesis regarded plant communities as distinct entities comparable to species—describable in analogous ways and amenable to categorization and classification.

Most modern ecologists, however, reject this view. Pointing to evidence that each species is distributed according to its own particular biological potentialities, they contend that the aggregation of species characterizing any given community is the fortuitous product of local environmental conditions and of whatever plant and animal species happen to be available in the area. Since environmental conditions such as temperature, humidity, soil characteristics, topographic features, wind patterns, and so on, vary continuously in both space and time, it follows that vegetation likewise varies continuously in both space and time. Boundaries between communities are seldom distinct, because the distributions of the various species composing communities are not well correlated with one another. If a traveler along the Mississippi River Basin carefully noted which tree species occur where, he would see that the different species drop out or make their appearance at different places, with little apparent correlation with one another. There would be no place where he would notice any abrupt change in the community; yet the small changes from mile to mile would be cumulative, so that after traveling many miles he would find himself in a community with a species composition almost completely different from that of the community in which his journey began.

This view of communities as parts of a gradually changing continuum, whose characteristics at any specific place are uniquely determined by a combination of local physical conditions, local biotic factors, local species distributions, and a considerable element of chance, leads to the conclusion that similarities between climax communities in different places are only approximate. Hence there is no absolute climax for any region. Climax, according to this view, has meaning only in relation to the individual site and its environmental conditions. The climax for a given spot should be determined, not by referring to some theoretical regional climax, but by actually observing what populations replace others and then maintain themselves in a stable condition. Rather than try to fit all the climax communities of a large region into a single regional climax, it is more productive in most cases to emphasize study of the gradients between the individual local climaxes, in an attempt to learn how changes in the component parts of communities are correlated with changing environmental conditions.

BIOMES

Even though the present trend is away from definition of regional climaxes in terms of aggregations of particular dominant and subdominant species, most biologists find it convenient to recognize a limited number of major climax formations called biomes. Thus, instead of saying that nearly all local communities in the northeastern United States tend to converge toward a regional climax of beech-maple forest, they say simply that in a large portion of the northeastern United States the majority of local climaxes are deciduous forests of some type—meaning that the dominant form of plant life (plant formation) is usually deciduous trees. They do not insist that every site within the deciduous-forest biome be converging toward a deciduous forest; local conditions may sometimes determine climax communities in which grass or pine trees are dominant. Nor do they insist that all those sites within the biome that do have a deciduous-forest climax must be dominated by beech and maple; they expect the importance of individual species to vary from one place to another. In short, they think of biomes as no more than generalizations about the most common climax of the region.

Let us briefly survey some of the world's major biomes (Fig. 19.38).

Tundra In the far-northern parts of North America, Europe, and Asia is the tundra. It is the most continuous of the earth's biomes, forming a circumpolar band interrupted only narrowly by the North Atlantic and the Bering Sea. It corresponds roughly to the region where the subsoil is permanently frozen. The land has the appearance of a gently rolling plain, with many lakes, ponds, and bogs in the depressions (Fig. 19.39).

Even though the Siberian word "tundra" means north of the timberline, there are, in fact, a few trees on the tundra, but they are small, widely scattered, and clearly not the dominant vegetation except locally. Much of

Tundra, ice

Taiga

Temperate deciduous forest

Grassland

Scrub forest

Desert

Tropical rain forest

Temperate rain forest

Tropical seasonal forest

19.38 The major biomes of the world

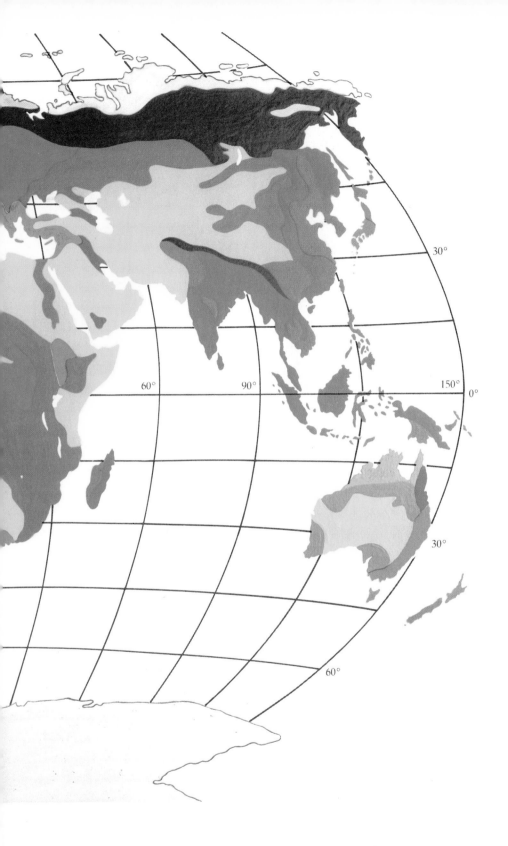

19.39   The tundra
Top: A tundra in Alaska, seen in autumn. Mount
McKinley is in the background.   Bottom: A caribou
bull, distinctive animal of the tundra.   [B. Ruth (top)
and L. L. Rue III (bottom), Bruce Coleman Inc.]

the ground is covered by mosses (particularly sphagnum), lichens (particularly so-called reindeer moss), and grasses. There are numerous small perennial herbs, which are able to withstand frequent freezing and which grow
rapidly during the brief cool summers, often carpeting the tundra with
brightly colored flowers.

Reindeer, caribou, arctic wolves, arctic foxes, arctic hares, and lemmings
are among the principal mammals; polar bears are common on parts of the
tundra near the coast. Vast numbers of birds, particularly shorebirds (sand-
pipers, plovers, etc.) and waterfowl (ducks, geese, etc.), nest on the tundra in
summer, but they are not permanent residents and migrate south for the

19.40 The taiga
Coniferous forests, such as this one in Norway, cover extensive areas in the northern part of North America, Europe, and Asia. [Paolo Koch, Photo Researchers, Inc.]

winter. Insects, particularly flies (including mosquitoes), are incredibly abundant. In short, far from being a barren lifeless land as many people imagine, the tundra teems with life. It is true, however, that though the number of individual organisms on the tundra is often very large, the number of species is quite limited.

Taiga South of the tundra, in both North America and Eurasia, is a wide zone dominated by coniferous forests. This is the taiga (Fig. 19.40), also called the boreal forest. Like the tundra, it is dotted by lakes, ponds, and bogs. And like the tundra, it has very cold winters. But it has longer and somewhat warmer summers, during which the subsoil thaws and vegetation grows abundantly.

The number of species living in the taiga is larger than on the tundra, but considerably smaller than in biomes farther south. Though conifers (including spruce, fir, and tamarack) are the most characteristic of the larger plants in the taiga, some deciduous trees (e.g. paper birch) are also common. Moose, black bear, wolves, lynx, wolverines, martens, porcupines, and many smaller rodents are important mammals in the taiga communities. Birds are abundant in summer.

19.41 A temperate deciduous forest in early autumn
[A. Avis, Bruce Coleman Inc.]

Deciduous forests The biomes south of the taiga do not form such definite circumglobal belts as the tundra and the taiga. There is more variation in the amount of rainfall at this latitude, and consequently more longitudinal variation in the types of climax communities that predominate.

In those parts of the temperate zone where rainfall is abundant and the summers are relatively long and warm, as in most of the eastern United States, most of central Europe, and part of eastern Asia, the climax communities are frequently dominated by broad-leaved trees. Such areas, in which the foliage changes color in autumn and drops, constitute the deciduous-forest biomes (Fig. 19.41). They characteristically include many more plant species than the taiga to the north, and show more vertical stratification. Among the common mammals in this biome are squirrels, deer, foxes, and bear.

Tropical rain forests Tropical areas with abundant rainfall are (or, more correctly, were) usually covered by rain forests, which include some of the most complex communities on earth. The diversity of species is enormous; a temperate forest is composed of two or three, or occasionally as many as ten,

dominant tree species, but a tropical rain forest may be composed of a hundred or more. One may actually have difficulty finding any two trees of the same species within an area of many acres.

The dominant trees in rain forests are usually very tall, and their interlacing tops form dense canopies that intercept much of the sunlight, leaving the forest floor only dimly lit even at midday (Fig. 19.42). The canopy likewise breaks direct fall of rain, but water drips from it to the forest floor much of the time, even when no rain is actually falling. It also shields the lower levels from wind and hence greatly reduces the rate of evaporation. The lower levels of the forest are consequently very humid. Temperatures near the forest floor are nearly constant. The pronounced differences in the microenvironmental conditions at different levels within such a forest result in a striking degree of vertical stratification; many species of animals and epiphytic plants (plants growing on the large trees) occur only in the canopy, others only in the middle strata, and still others only on the forest floor. Some vertical stratification is found in any community, particularly any forest community, but nowhere is it so extensively developed as in a tropical rain forest.

19.42 Tropical rain forest
Left: A forest in Costa Rica in February. The trees have remained green, even though it is "winter." Note the very dense vegetation.   Right: Epiphytic bromeliads (distinctive plants of the tropics) growing on the trunk and branches of a tree in western Mexico.   [John Shaw (left) and M. P. L. Fogden (right), Bruce Coleman Inc.]

19.43 A grassland in Arizona, with pronghorns, characteristic animals of this biome
[J. Couffer, Bruce Coleman Inc.]

Grasslands Huge areas in both the temperate and tropic regions of the world are covered by grassland biomes (Fig. 19.43). These are typically areas where either relatively low total annual rainfall (25–30 cm) or uneven seasonal occurrence of rainfall makes conditions inhospitable for forests but suitable for often luxuriant growth of grasses. The grasslands of temperate regions characteristically undergo an annual warm-cold cycle, whereas the grasslands of the tropics (often called savannas) undergo a wet-dry cycle instead.

Temperate and tropical grasslands are remarkably similar in appearance, although the particular species they contain may be very different. In both cases there are usually vast numbers of large and conspicuous herbivores, often including ungulates (e.g. bison and pronghorn antelope in the United States). Burrowing rodents or rodentlike animals are often common (e.g. prairie dogs in the western United States).

Deserts In places where rainfall is often less than 25 cm (10 inches) per year, not even grasses can survive as the dominant vegetation, and desert biomes occur. Deserts are subject to the most extreme temperature fluctuations of any biome type; during the day they are exposed to intense sunlight, and the temperature of both air and soil may rise very high (to 40°C or

higher for air temperature and to 70°C or higher for surface temperature), but in the absence of the moderating influence of abundant vegetation, heat is rapidly lost at night, and a short while after sunset searing heat has usually given way to bitter cold.

Some deserts, such as parts of the Sahara, are nearly barren of vegetation, but more commonly there are scattered drought-resistant shrubs (e.g. sagebrush, greasewood, creosote bush, and mesquite) and succulent plants that can store much water in their tissues (e.g. cacti in New World deserts and euphorbias in Old World deserts) (Fig. 19.44). In addition, there are often many small rapid-growing annual herbs with seeds that will germinate only when there is a hard rain; once they germinate, the young plants shoot up, flower, set seed, and die, all within a few days.

Most desert animals are active primarily at night or during the brief periods in early morning and late afternoon when the heat is not so intense. During the day they remain in cool underground burrows or in cavities in plants or, in the case of some spiders and insects, in the shade of the plants. Among the animals often found in deserts are rodents (e.g. the kangaroo rat), snakes, lizards, a few birds, arachnids, and insects. Most show numerous remarkable physiological and behavioral adaptations for life in their hostile environment.

19.44 Deserts
Left: Sand and limestone in the Algerian Sahara; vegetation is extremely sparse.   Right: Cacti and other thorny, drought-resistant plants, abundant in many deserts, growing in Picacho Park, Arizona.   [Left: Giorgio Gualco, Bruce Coleman Inc. Right: E. S. Ross.]

Altitudinal biomes We have seen that, moving north or south on the earth's surface, one may pass through a series of different biomes. The same thing is true if one moves vertically on the slopes of tall mountains. Climatic conditions change with altitude, and the biotic communities change correspondingly (Figs. 19.45, 19.46). Thus arms, or isolated pockets, of the taiga extend far south in the United States on the slopes of the Appalachian Mountains in the east and of the Rockies and Coast Ranges in the west. There are even tundralike spots on the highest peaks.

Aquatic ecosystems So far, we have concentrated on terrestrial ecosystems, but we should not forget that many of the earth's biotic communities are found in aquatic environments, and that these too vary in type with varying physical conditions. Thus the communities in lakes differ from those in the flowing waters of rivers and streams, and even those in a single stream differ from one another, depending on whether they are in rapids, where fast-flowing water is made turbulent by rocks and sudden falls, or in water flowing slowly and calmly over a smooth bottom.

Let us consider the largest bodies of water—the oceans. There are several

19.45 The correspondence between latitudinal and altitudinal life zones.

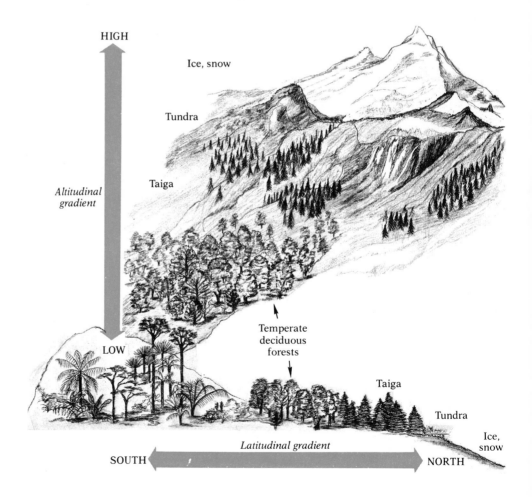

ways of classifying the ecosystems found in them (Fig. 19.47). It is often useful, for example, to distinguish between a *benthic division* comprising the ocean bottom together with all bottom-dwelling organisms and a *pelagic division* consisting of the water above the bottom and all the organisms in it—both those that are free-swimming (the nekton) and those that float (the plankton) and are carried passively with the water currents. If the criterion is the distribution of light in the water, it is possible to distinguish between an upper lighted *photic zone* and a deeper lightless *aphotic zone* (usually below about 200 m). Still another possible distinction is between the *neritic province* above the continental shelves and the *oceanic province* of the main ocean basin.

The various subdivisions of the oceans are not fully analogous to the terrestrial biomes, because they are biologically interdependent. Thus benthic communities in the aphotic zone, being made up entirely of heterotrophs (including animals, fungi, heterotrophic protists, and bacteria), must depend on the photosynthetic organisms of the photic portion of the pelagic

19.46 An alpine meadow in the Sierra Nevada Mountains of California, with paintbrush in bloom
Trees can be seen in the lower, more protected areas. Most of the meadow is above the timberline. The upper limit of tree growth in mountainous regions (usually between 3,000 and 3,500 m) is determined largely by temperature, but is also influenced by soil conditions and rainfall. More barren areas can be seen on the higher peaks in the distance.   [Bob and Clara Calhoun, Bruce Coleman Inc.]

division above them for the primary productivity that makes their existence possible; they receive their nutrients as a rain of dead organisms from the water above. When the flow of energy and materials is considered, then, the communities in the dark aphotic zone cannot be understood apart from those in the photic zone, and, more broadly, the benthic division cannot be understood apart from the pelagic division.

The most complex oceanic communities occur in the shallow waters of the neritic province, especially that portion called the *littoral zone,* which extends from the beach to a point where the water is deep enough no longer to be completely stirred by the action of waves or tides (Fig. 19.47). Here high primary productivity by both free-floating and bottom-anchored algae makes possible much niche diversification among herbivores. Because the littoral zone is subject to far more variation in temperature, water turbulence, salinity, and lighting than any other portion of the ocean, the littoral communities in different locations often vary greatly.

BIOGEOGRAPHY

In considering climax communities, we saw that the ranges of different species are never fully correlated, that the complex of environmental factors determining range boundaries is unique for each species. There is, however, one generalization that can be made regarding population dispersal, whatever the factors tending to limit it: Since the intrinsic rate of increase of populations always exceeds the carrying capacity of the environment, a population is always under pressure to expand its niche or to spread into new territories. There is a selective premium on genotypes that improve the capacity of individuals to occupy new habitats within the same community or to invade habitats in other communities. It is with the latter alternative—extension of the range into new areas, so that the species becomes part of new communities—that we shall be concerned here.

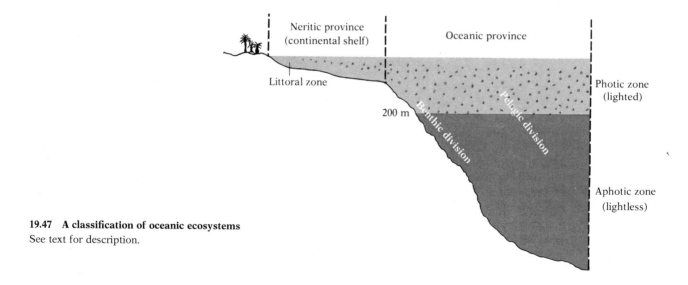

19.47   A classification of oceanic ecosystems
See text for description.

THE DISPERSAL OF SPECIES

Before a species can successfully spread into a new area, it must meet at least three major conditions: (1) It must possess the *physiological potential* to survive and reproduce in the new area. (2) It must have the *ecological opportunity* to become established in the new area. And (3) it must have *physical access* to the new area. Let us examine each of these conditions in more detail.

Physiological potential for survival in a new area   Any new area into which colonizing members of a species move will be different at least to some degree from the area the colonizers left. Hence the colonizing population will immediately be subject to selection pressure for evolution of better adaptations for the new environment. But since such evolutionary improvement comes after the fact of colonization, the colonization itself is possible only if the colonizers are already at least minimally preadapted for survival under the new environmental conditions. "Preadapted," of course, does not imply any foresight, any intentional preparation for the move into strange territory; it simply means that characteristics evolved in the previous habitat are at least minimally suited to the new habitat. For example, the colonizers must be able to use some source of food in the new area, and they must be able to withstand the rigors of the new climate.

Let us see how climate acts in limiting the distributions of organisms. We are all familiar with broad regional differences in temperature, humidity, rainfall, and other meteorological factors. Temperatures range from those of hot tropical regions to those of permanently frozen polar areas; rain may fall every day on a rain forest or only a few times a decade on some deserts. The average pattern of such meteorological factors over an extended period is called the climate of a region.

Although the climate of a desert is dry, at any given time rain may be falling. Thus we distinguish between climate, defined above, and weather, which is the immediate pattern of meteorological factors. Weather has obvious day-to-day effects on living organisms. Thus a late frost after a warm spell can destroy the potential productivity of a plant seedling; the reproductive success of many animals will be very different in unusually dry and unusually wet years. In other words, though the climate of an area is an important factor in determining distributions of species, the weather is what is important to the organism at any particular moment.

Indeed, it is often the extremes of the weather—e.g. the highest or lowest temperatures of the year, or the longest period without rainfall—that are most important in limiting the distribution of organisms. Thus a plant that cannot tolerate temperatures below 0°C will be unable to survive in a region renowned for its "warm" climate (with an average annual temperature of, say, 25°C) if the temperature falls below 0°C even one day a year. All other conditions in the region may be optimal for the plant, but the one condition it cannot tolerate excludes it from growing there.

Since the environmental conditions affecting an organism and limiting its distribution are those impinging directly on it, it is actually the microen-

19.48 A diagrammatic illustration of the Law of Tolerance

The species in question is most abundant in areas where the environmental variable is within the optimum range for that species. The species is rare in areas where it experiences physiological stress, whether because the environmental variable has too high a value or too low a value. And the species does not occur at all in areas beyond its upper and lower limits of tolerance.

vironmental conditions that ecologists must analyze for a full understanding of the physical factors with which the organisms they study must cope. Variations in the weather, for example, can be very local. Someone living only a few miles from the weather station may hear over the radio that the temperature is 10°C. Checking his own outdoor thermometer, he may find that it registers 7°C, a difference of 3° from the temperature at the station. This difference, which could easily be greater, may reflect a difference in microclimate. Within a single lot, variations of several degrees may occur between sheltered and exposed places, between points near the ground and points at various elevations above the ground. Since the same sorts of highly local variations may be found in humidity, wind velocity, amount of sunlight, soil type, etc., one should not expect to find exactly the same kinds of organisms living at all points within even a very small area. Plants and animals don't live under generalized regional conditions; they live under microenvironmental conditions that may vary radically over a given region. Thus when ecologists say that plant species A ranges from South Carolina in the south to central New York in the north, and from the Atlantic coast to central Ohio in the west, they do not mean that individuals of species A are to be found everywhere within that large area, but only that there are places within the overall range where the species occurs and that there are no places outside that range where it occurs.

Our focus has been on climate and weather, but the same sorts of considerations apply to other aspects of the environment. The climate of a region might be ideal for a particular species of plant, and the soil rich in nitrogen and potassium, but if there is less phosphorus in a part of the region than the plant requires, it cannot grow there. The importance of single environmental factors was recognized as long ago as 1840 by Justus von Liebig of the University of Giessen, Germany. Liebig's so-called **Law of the Minimum** states that growth of a plant will be limited by whichever requisite factor is most deficient in the local environment. Later, V. E. Shelford of the University of Illinois expanded Liebig's Law, applying it also to animals and taking into account that too much may be as bad as too little. Shelford's **Law of Tolerance** states that the distribution of a species will be limited by that environmental factor for which the organism has the narrowest range of adaptability (Fig. 19.48).

Although the principle behind both Liebig's and Shelford's laws is an important one in ecology, the assumption that a single factor is always limiting should be viewed with caution. In real-life situations (as contrasted with laboratory models), the various environmental factors interact in so many complex ways that it is often impossible to describe any one factor as *the* limiting one. For example, when one condition is not optimal (though tolerable) for a species, the limits of tolerance for some other factor may be reduced. Moreover, unless the Law of Tolerance is extended to such biotic limiting factors as predation and competition, in addition to the more physical ones included in its original formulation, it is of only restricted applicability. As an illustration, consider the case of Klamath weed (*Hypericum perforatum*).

This plant, which is poisonous to cattle, was brought from Europe to the western United States and quickly spread to millions of acres of valuable range land. To control the Klamath weed, a species of flea beetle that feeds on it was introduced. The beetle soon eliminated the weed from the range lands, though some small populations persist in shaded places in forests, where the beetle does not feed on it as much. It is clearly the beetle, not lack of tolerance to some physical factor of the environment, that today limits the range of the Klamath weed to forests; yet an observer who did not know the history of the weed-beetle interaction would hardly be able to determine what keeps the weed from spreading into the range lands and might erroneously suspect some climatic or nutritional factor.

Ecological opportunity to become established in a new area   This requirement often means that the colonizing species must encounter little competition at first. There must be an essentially unoccupied niche it can fill. The reason is simple: Even if the colonizer has the physiological potential for surviving in the new habitat, chances are that it will be less well adapted to the new conditions than species that have been in the area for a longer time. If one of the established species occupies a niche very similar to the one the colonizer could potentially fill, the established species may well have the competitive advantage and be able to prevent the colonizer from taking hold. This is not always the case, however; if conditions in the new range are very similar to those in the old, a colonizing species may sometimes be competitively superior to an established species and be able to supplant it.

The importance of available niches has been especially well illustrated in studies of island biogeography. One might reasonably expect that a small island would have less diversity of niches than a large one, and hence would support fewer species of any given group of organisms. This has, in fact, been found to be true; the number of species on islands has repeatedly been shown to be a function of the area of the islands (Fig. 19.49). Although the slopes of area-species curves differ, depending on ecological conditions and on the taxonomic group of organisms studied, a rough generalization is that for each tenfold increase in area the number of species approximately doubles; thus an island with an area of 100 km² will support roughly twice as many species as an island with an area of 10 km².

When an island has reached saturation, i.e. when it has as many species of a particular group of organisms as it can support, is all further immigration barred? The answer turns out to be no. When an island has reached equilibrium in terms of species diversity, immigration continues, but is balanced by extinction.

As Fig. 19.50 shows, the rate of new immigration falls as the number of species on the island rises; since more and more of the potential immigrant species from nearby islands or continents have already reached the island, a rising percentage of immigrant individuals represent species that are already there. Conversely, the rate of extinction rises as the number of species on the island rises. This is due partly to chance: If there are more species, then there are more possibilities for extinction. But another reason (the one that makes the extinction curve in Fig. 19.50 bend upward) is increasing interspecific competition, which is a reflection of the limited number of available niches on the island. The point where the immigration and extinction curves cross represents a state of equilibrium for species diversity; both immigration and extinction continue, but the number of species does not change significantly.

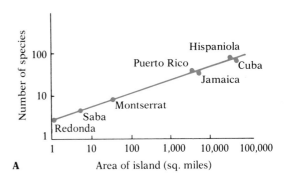

A

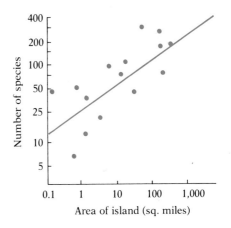

B

19.49   Two examples of area–species curves
(A) Number of species of amphibians and reptiles on islands in the West Indies.   (B) Number of plant species on the Galápagos Islands.   In both cases the actual counts of species increase roughly as a linear function of the area of the islands.   [(A) Redrawn from R. H. MacArthur and E. O. Wilson, *The theory of island biogeography*, copyright © 1967 by Princeton University Press. Used by permission. (B) Redrawn from C. J. Krebs, *Ecology*, 2nd ed., Harper & Row, 1978, after Preston.]

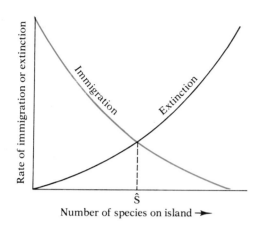

19.50　The equilibrium model of species diversity on an island

As the number of species on the island increases, the rate of new immigrations falls and the rate of extinctions rises. When the point is reached where immigration and extinction are equal (point where the two curves cross), the system has reached equilibrium; the number of species on the island is at saturation ($\hat{S}$); i.e. the island cannot for long support more species.

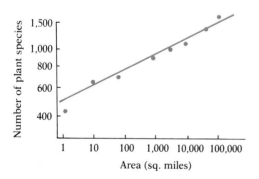

19.51　An area–species curve for flowering plants in England

Areas of different sizes were sampled, from a plot of only one square mile in Surrey to the whole of England.　[Redrawn from C. J. Krebs, *Ecology*, 2nd ed., Harper & Row, 1978, after Williams.]

Ecologists have so zealously studied island biogeography because it provides a clear model of the factors governing species dispersal in general. Thus freshwater lakes can be considered biogeographic islands; so can forests separated from one another by grasslands, or deserts separated by wetter areas. These habitat "islands," like real islands, obey the area-species rule: The greater the area, the greater the species diversity (Fig. 19.51).

Physical access to a new area　It is useless for a species to have the physiological potential and ecological opportunity for surviving in a new range if it has no way of getting there. Doubtless many common North American mammals could survive and prosper in Australia, but unless they have some way of reaching that continent this potential range extension will be unrealized.

There are many ways organisms may be dispersed from one place to another. Most obvious for many animals is active locomotion—walking, crawling, swimming, or flying. Even many sedentary marine animals have a free-swimming larval stage. Active locomotion may carry the members of a single generation only a short distance from their point of origin, but over many generations the cumulative effect may spread the species over hundreds or thousands of miles. This progressive dispersal in the course of a long series of generations may at first glance seem too slow to be of much significance, but it is not. Probably all organisms can disperse fast enough so that, if favorable habitats were continuous, they could spread over the entire earth in a relatively short time, as measured on the geological and evolutionary time scale.

Large and heavy animals as we are, we tend to think automatically of active locomotion as the principal means of dispersal of organisms. But for plants and for many very small animals, passive transport is the chief means of dispersal. For example, the seeds or spores of many plants may be blown very long distances, and insects, spiders, small molluscs, and other invertebrates have been known to be blown hundreds of miles by the strong winds of storms. Pilots sometimes encounter large numbers of insects being swept along by fast-moving air currents at high altitudes. Aquatic organisms may similarly be swept along by strong water currents. Even some fairly large terrestrial plants and animals may be carried across many miles of water on floating logs or rafts of matted vegetation. Many such logs and rafts are swept out to sea by large rivers like the Amazon and the Congo, particularly during floods. A raft about 10 m², composed of soil and decaying organic matter laced together by roots, was sighted in the Atlantic Ocean off the coast of North America in 1892. Many shrubs and several trees 10 m tall were growing on it. This raft, which looked for all the world like a floating island, is known to have drifted at least 1,600 km.

Some plants and small animals are dispersed by birds and mammals. For example, the seeds of many plants pass through the digestive tracts of higher animals without being harmed, and may germinate and grow if the animals' feces are deposited in a favorable place. Birds sometimes transport seeds long distances in this way. There is also some evidence that plant seeds and the eggs and larvae of some small aquatic animals may be transported on the feathers or feet of swimming or wading birds. Of course human beings are by far the most important agents of dispersal at the present time;

but they are relative newcomers on the biological scene, and older distribution patterns cannot be explained in terms of their activities.

The recent history of Krakatoa, a small island between Java and Sumatra in the southeast Pacific, constitutes a natural experiment in colonization of new territory by both plants and animals, and well illustrates the types of dispersal enumerated here. On August 27, 1883, a violent volcanic explosion destroyed much of the island and left the rest completely covered by a layer of hot ashes and pumice 6 to 60 m deep. There is no evidence that any life remained on the island. The island nearest to Krakatoa was 19 km away, but most of the life on it was destroyed by toxic gases and a thick layer of ashes produced by the explosion. The next-closest island was about 40 km away. In short, Krakatoa suddenly became an island devoid of life, separated by 40 km of open ocean from the nearest significant source of potential colonizers.

Nine months after the eruption a biologist visited the island and, though he searched diligently, the sole living thing he could find was a lone spider busily spinning a web; the spider had almost certainly been blown to the island. Only three years later numerous plants were found growing along the beaches, and several species of ferns and grasses were growing farther inland. The beach plants were of the kind found on the beaches of almost all tropical Pacific islands—plants whose seeds are highly resistant to seawater and are regularly carried long distances by ocean currents. The ferns found growing so soon on Krakatoa reproduce by means of very light spores that can easily be carried by even gentle air currents.

By 1896, thirteen years after the explosion, the island was fairly well covered with vegetation, but plants still were more abundant near the shores than in the interior. Most of the plants were ones distributed by sea currents and wind, but about 9 percent must have arrived by other intermediaries, probably birds. By 1906 the island was densely covered with plants, and there were 263 species of animals living there; most were insects, but there were four species of land snails, two species of reptiles, and 16 species of birds. Many of the insects either flew or were blown to the island, but some of them (and perhaps the reptiles also) probably arrived on floating logs or rafts. The number of bird species on Krakatoa reached saturation by 1921, but the number of plant species continued to rise at least until the late 1930s.

Organisms spread by wind and ocean currents, and actively flying animals such as birds, bats, and insects, continue to dominate Krakatoa today, but the percentage of organisms that must have arrived by other means has risen. If a person with no knowledge of the island's history were to visit Krakatoa, he would hardly guess that approximately one hundred years ago it was a lifeless mass of ash and steaming lava.

Had Krakatoa been farther away from areas that could act as sources of colonizers, the species diversity would have increased more slowly. In addition, equilibrium would have been reached at a lower species number, in accordance with a general rule of island biogeography: The more distant an island is from a major source of new colonists, the lower its species diversity will be at equilibrium (Fig. 19.52). In other words, distance alone eliminates many potential colonizers from the race.

Whatever the distance to be crossed, some species have readier physical access than others to a region that they might potentially colonize; thus it was easier for small animals that could be blown by wind or carried on rafts

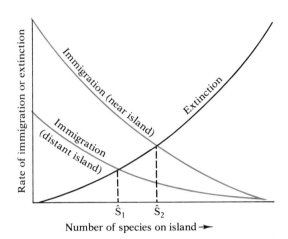

19.52 The effect of the distance of an island from potential sources of colonizers on the island's area–species curve

The rate of immigration is lower for a distant island than for a near island. Hence equilibrium (i.e. species saturation) is reached at a lower number of species ($\hat{S}_1$) for the distant island than for the near island ($\hat{S}_2$).

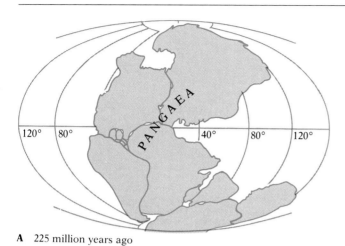

A   225 million years ago

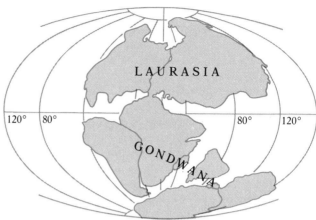

B   135 million years ago

19.53   The origin of the modern distribution of continents through continental drift
(A) Early in the Mesozoic era, about 225 million years ago, all the earth's major land masses were united in a single massive supercontinent called Pangaea.
(B) About 135 million years ago, at the start of the Cretaceous period, Pangaea had broken into a northern supercontinent, Laurasia, and a southern supercontinent, Gondwana; Gondwana itself had also begun to break up.   (C) By 65 million years ago the breakup of Gondwana was complete, and the future South America, Africa, Madagascar, India, Antarctica, and Australia were drifting apart.   (D) The present continental arrangement.

to get to Krakatoa than it was for large mammals to get there. It is obvious that the geological or ecological zones that intervene between any two regions will be much more effective as barriers to dispersal for some species than for others. A wide expanse of ocean may almost completely prohibit movement of horses or elephants, but coconut palms may cross it in fair numbers, because their large water-resistant seeds can float in seawater for many weeks without harm. A grassland separating two forested areas may be an almost insuperable barrier for some forest animals and prevent them from moving from one forest to the other. Other forest animals may have difficulty crossing the grassland, but may do so occasionally, and still others may cross the grassland freely. In short, what is a barrier to dispersal for one species may be a possible but difficult route for another or an easily negotiated path for a third. A knowledge of the sorts of routes and barriers that are effective for different species is a prerequisite to understanding the distribution patterns of organisms on the earth's surface.

BIOGEOGRAPHIC REGIONS OF THE WORLD

Even a thorough understanding of the ecology of living species—including their physiological potential for survival in other habitats, the ecological opportunities they would find in those habitats, the ways in which they can be physically dispersed, and the routes and barriers they would be likely to encounter—and a detailed knowledge of the present geography of the earth do not suffice in fully explaining all the distributions of plants and animals now found on the earth. The reason is simple: There is a historical element that must be considered. The earth itself and the organisms on it are constantly changing, and present distributions are in very large part the result of past conditions—conditions often very different from those now prevailing. For example, a knowledge of present conditions alone would be insufficient to explain why certain animals occur in South America, Africa, and southern Asia but nowhere else. Only by combining knowledge of present conditions with evidence from the fossil record and with geological evidence

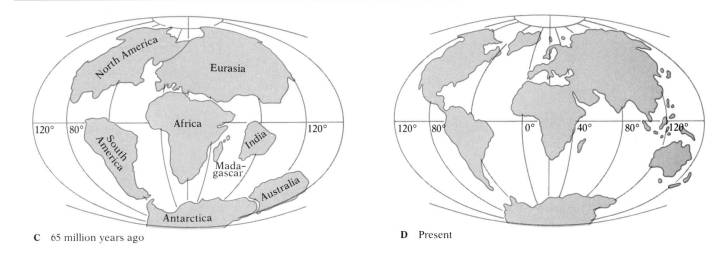

C   65 million years ago

D   Present

of the past shapes, connections, and climates of the earth's land masses and oceans can ecologists hope to gain insight into the present geography of life.

Continental drift   Not many years ago both geologists and biologists assumed that the distribution of the earth's major land masses had been more or less the same throughout the history of life. That assumption has recently been summarily dismissed by most geologists, who are now convinced that some 225 million years ago, early in the Mesozoic era (see Table 20.1, p. 906), there was a single massive supercontinent, called *Pangaea* (Fig. 19.53A).

Pangaea is thought to have broken up in the course of the Mesozoic, as the present-day continents (and some major islands, e.g. Madagascar, Greenland) drifted slowly apart. The first major break was an east-west one, separating a northern supercontinent called *Laurasia* (composed of the land masses that would later become North America, Greenland, and Eurasia minus India) from a southern supercontinent called *Gondwana* (composed of the future South America, Africa, Madagascar, India, Antarctica, and Australia). Soon thereafter Gondwana began to break up, as India drifted off to the north and an African–South American mass separated from an Antarctic-Australian mass. By the start of the Cretaceous period, roughly 135 million years ago, the distribution of the continents was probably that shown in Fig. 19.53B.

Some 70 million years later, about 65 million years ago, South America had split from Africa and was drifting westward; India had moved farther northward, but had not yet collided with the rest of Asia; and Australia had split from Antarctica and begun drifting northeastward (Fig. 19.53C). The Laurasian supercontinent was still intact, however; no split had occurred yet between North America and Eurasia. This split was one of the last to take place as the distribution of continents we know today slowly emerged (Fig. 19.53D).

The conclusion that the earth's major land masses (or plates, as geologists now call them) can drift from one place to another (at roughly 2.5 cm per year) is supported by evidence of many different kinds. Students of the new

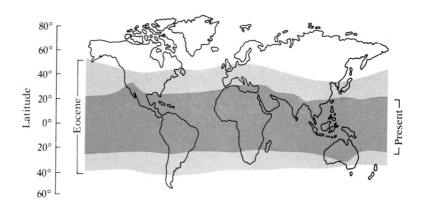

19.54   Distribution of tropical and subtropical forests during the Eocene (about 50 million years ago) and now

Warm conditions extended much farther north and south during the Eocene; e.g. they prevailed in most of what is now the United States and almost all of South America.   [Redrawn from John Napier, *The roots of mankind*, © 1970, Smithsonian Institution Press, Washington, D.C. Used by permission.]

19.55   Distribution of glaciation during the most recent Pleistocene ice age, 20,000–15,000 years ago

geological subdiscipline of plate tectonics have found that along some rifts in the earth's crust molten rock is slowly welling to the surface, forcing the plates on the two sides of the rift apart. At other places crust is slowly folding down and sinking into the earth's interior. It is this process of upwelling of new crust in some places and sinking of old crust in others that provides the force for continental drift. Evidence that such drift must have occurred comes from studies of the geological features of regions now separated but thought to have once been in contact; striking similarities are frequently found in such regions. For example, the rocks along part of the east coast of Brazil match almost exactly those in Ghana, on the west coast of Africa. The complementarity of the shapes of the east coast of South America and the west coast of Africa had, of course, suggested long ago that the two continents might once have been joined, but until plate tectonics supplied a plausible mechanism for continental drift the possibility had been discounted.

If continents move in the course of millions of years, so that their distances from the earth's poles and from the equator change, then their climates must have undergone major shifts. India, for example, has moved from a position next to Antarctica all the way across the equator to its present location in the tropics of the northern hemisphere. Australia, too, has moved steadily northward. But climatic changes due to causes other than continental drift have also occurred during the history of life on earth. Thus Antarctica, though probably always near the South Pole, has not always been the bleak ice-covered land it is today, for fossils of amphibians and reptiles have been found there. During at least part of the Mesozoic Antarctica must have been reasonably warm, and it was probably warm again about 50 million years ago, when tropical and subtropical climates were far more widespread on the earth than they are today (Fig. 19.54). By contrast, the earth was much colder only a few thousand years ago, during the periods of extensive glaciation (Fig. 19.55). Thus both the past configurations of the land masses and their past climates are important in explaining the distributions of organisms on the earth.

The island continents   The biota (flora and fauna) of the *Australian region* (Australia, New Zealand, and adjacent islands) is by all odds the most unusual found on any of the earth's major land masses. Many species common in Australia occur nowhere else. Conversely, many species widespread in the rest of the world are absent from Australia. As we have seen, Australia has had no land connection to Eurasia or Africa since Pangaea began to break up more than 135 million years ago. The ancestors of some of the more ancient groups of organisms now living in the Australian region were probably there before the breakup of Pangaea, but the ancestors of more recent groups must have crossed a water barrier to get there.

That water barrier, however, was dotted by islands (the East Indies, presently the nation of Indonesia), and organisms could have spread from one island to the next over a period of millions of years; it is not necessary to assume that a great expanse of ocean was crossed at any one time. Furthermore, most of the western islands of the East Indies were probably interconnected as an extension of the Asian land mass several times in the not too distant past; hence the distance from Asia to Australia has not always been as great as it is today.

Perhaps the best-known aspect of Australia's curious biota is its mammalian fauna, which is completely unlike that of any other continent. Other than wild dogs (dingoes), which were probably human imports of prehistoric times, the only placental mammals present in the Australian region before European explorers landed there were a number of species of rodents belonging to a single family and a variety of bats. Bats can, of course, fly across water barriers and would be expected to reach most oceanic islands. The rodents of Australia are apparently relatively recent arrivals, having come from Asia by island-hopping through the East Indies. Most of the ecological niches that on other continents would be filled by placental mammals are in Australia filled by marsupials.[10]

Apparently the marsupials reached Australia very early and, encountering no competition from placental mammals, underwent extensive evolutionary radiation. Since they were filling niches similar to those filled in the rest of the world by placentals and were thus subject to similar selection pressures, they evolved striking convergent similarities to the placentals. Certain of the marsupials resemble placental shrews, others placental jumping mice, weasels, wolverines, wolves, anteaters, moles, rats, flying squirrels, groundhogs, bears, etc. (see Fig. 18.49, p. 815). The uninitiated visitor to an Australian zoo finds it hard to believe when first looking at an assemblage of these animals that he is not seeing close relatives of the mammals familiar to him from other parts of the world. Not all marsupials look like their ecologically equivalent placentals, however; the kangaroos are markedly different (Fig. 19.56), though some of them play an ecological role very similar to that of horses and other large placental grazers. The Australian marsupials are a fascinating biological development, but most of them are probably doomed to extinction, at least in the wild. The changes brought by civilization have not been favorable for many of them. And most marsupials seem unable to compete successfully with the placental mammals introduced by human beings.

19.56 Kangaroos in flight at Lake Eucumbene, New South Wales
Though neither their appearance nor their behavior suggests it, kangaroos are Australia's ecological equivalents of ungulate grazers on other continents.   [Courtesy Australian News and Information Bureau.]

---

[10] Whereas the young of placental mammals undergo their entire embryonic development in the mother's uterus, the young of marsupial mammals develop only a short while in the uterus; after birth, they move to a pouch on the mother's abdomen, attach to a nipple, and there complete their development.

The South American continent, known to biologists as the **Neotropical region** (meaning "new tropics"), has had a history similar to that of Australia. It, too, has been an island continent unconnected to the other major land masses of the world through much of its history. And it, too, had an early mammalian fauna that included a great variety of marsupials, as the fossil record shows. The prevalence of marsupials in both Australia and South America probably means that these organisms were present throughout the region comprising Australia, Antarctica, and South America at a time when there were only small water breaks (if any) in this land mass (Fig. 19.53C) and when the climate was warmer than it is now.

Later, South America also had a variety of placentals that probably reached it during a short period of connection to North America via a Central American land bridge during the early part of the Age of Mammals (about 60 million years ago). After this land bridge disappeared, both the marsupials and the placentals evolved, in isolation, many characteristics convergent to those evolved by other placentals on the main World Continent (Fig. 19.57). There were times during this period of isolation lasting some 60 million years when the water barrier between South America and what is now northern Central America was not so wide. A few additional placental mammals chanced to get across into South America at such times; among these were the ancestors of the modern New World monkeys and a number of rodents.

Although, by five million years ago, there were 23 families of mammals in South America, not one of these was represented in North America. But then a land connection to North America (the Isthmus of Panama) was re-established, and many additional immigrants arrived in South America. Some species also moved in the opposite direction, from South America to North

19.57 Convergence of South American rodents and African ungulates
There are remarkable similarities between the two groups of animals shown here, even though rodents and ungulates comprise separate mammalian orders that are not closely related.

South American rodents

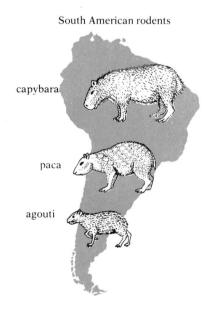

capybara

paca

agouti

African ungulates

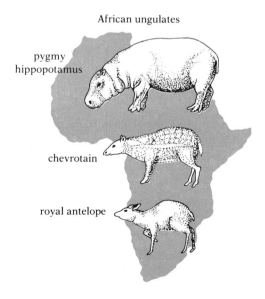

pygmy
hippopotamus

chevrotain

royal antelope

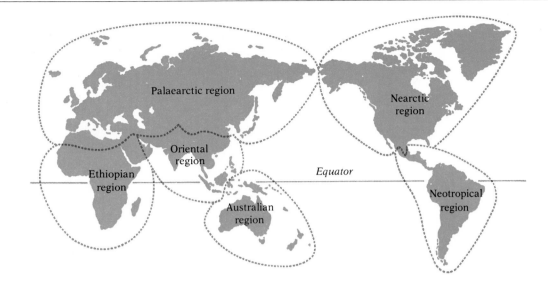

19.58   **Biogeographic regions of the world**

America; the opossum, the porcupine, and the armadillo are examples. But the movement was predominantly southward, apparently because the groups that had originated in the north were competitively superior and could displace the older South American fauna, much of which suffered extinction.

Central America has never been more than a narrow bridge. Because its climate and rugged terrain have not been hospitable to all northern species, many groups of organisms have never been able to move between the North American and South American continents. The Central American land bridge has been a selective or filter route, along which some species but not others could pass. For this reason, South America, like Australia, has been an island continent through much of its history, but its nearness to North America (Mexico) and its recent direct connection to North America via the Central American land bridge have given it a more diverse biota with more similarities to that of the World Continent.

The World Continent   Europe, Asia, Africa, and North America have formed a relatively continuous land mass, the so-called World Continent, throughout most of geological time. Consequently their biotas are more alike in many aspects than they are like those of the two island continents. Nevertheless, biologists customarily divide the World Continent into four biogeographical regions: the **Nearctic** ("new northern"), which is North America; the **Palaearctic** ("old northern"), which is Europe and northern Asia; the **Oriental,** which is southern Asia; and the **Ethiopian,** which is Africa south of the Sahara (Fig. 19.58).

The geological feature that looks like the most obvious present-day barrier between the Palaearctic and Ethiopian regions—the Mediterranean Sea—is not really the important barrier. Many species can move between Europe and North Africa by circling around the eastern end of the Mediterranean. The real barrier to species dispersal is the Sahara Desert. Africa north of the Sahara is part of the Palaearctic region.

The Oriental region—tropical Asia—is separated from Palaearctic northern Asia in most places by east–west mountain ranges, of which the Himalayas between India and China are part. These east–west mountains constitute important breaks between climatic regions, and they therefore act as both topographic and climatic barriers between cold-adapted and warm-adapted species. By contrast, north–south mountains such as those in North America tend to facilitate mixing of cold-adapted and warm-adapted species.

That the Palaearctic, Oriental, and Ethiopian regions constitute an essentially continuous land mass is obvious, but you may wonder why the Nearctic is considered part of the same land mass. The answer is that North America and Eurasia have been connected through much of their geological history (Fig. 19.53B, C). Even after they broke apart as the northern part of the Atlantic Ocean formed, a new connection, the Siberian land bridge between what are now Alaska and Siberia, provided a link; part of this bridge is beneath water at the present time. Because the climate of Alaska and Siberia is so forbidding, you might not expect many organisms to use a bridge in that region. But again present conditions are misleading. Fossils of many temperate and even subtropical species of plants and animals are abundant in Alaska. Indeed, all the evidence points to much movement, on many occasions, between Asia and North America via the Siberian land bridge. In fact, the Nearctic and Palaearctic regions remain biologically so similar that many biologists regard them as a single region, which they call the Holarctic.

The history of the major land masses and of their changing climates helps explain present distribution patterns. Consider the disjunct distribution referred to earlier, in which a species occurs in South America, Africa, and southern Asia. If the species belongs to a group of animals of very ancient origin (e.g. cockroaches), this distribution may indicate that the species occurred throughout the old Gondwana supercontinent and continued to survive in South America, Africa, and India after these drifted apart. But if the species belongs to a more recently evolved group (e.g. the majority of modern mammalian and bird families), which arose after Gondwana had broken up, it may be assumed that the species moved between the New World and the Old World via either the North Atlantic or the Siberian land bridge and between North and South America via Central America, and that it then became extinct in the north, either because of climatic changes or because of intense competition. The fossil record shows that this pattern of dispersal between southern regions by way of the northern continents has occurred again and again. For example, members of the camel family occur today in South America (llamas, alpacas, vicuñas, etc.), northern Africa, and central Asia, but fossils indicate that the family originated in North America, spread to South America via Central America and to the Old World via Siberia, and later became extinct in North America; hence the disjunct distribution we see today.

SUGGESTED READING

BELL, R. H. V., 1971. "A grazing ecosystem in the Serengeti," *Scientific American*, July. (Offprint 1228)   *The synchronization of the migrations of ungulates across the plains of Tanzania with the growth of certain grasses—a striking example of the precision with which organisms mesh within an ecosystem.*

BORMANN, F. H., and G. E. LIKENS, 1970. "The nutrient cycles of an ecosystem," *Scientific American*, October. (Offprint 1202)   *On the studies of the Hubbard Brook Forest, discussed in the text.*

BRILL, W. J., 1977. "Biological nitrogen fixation," *Scientific American*, March. (Offprint 922)   *How certain bacteria and blue-green algae act as the major suppliers of fixed nitrogen for the rest of the living world.*

COOPER, C. F., 1961. "The ecology of fire," *Scientific American*, April. (Offprint 1099)   *How some biotic communities depend on periodic burning for their continued existence, and why efforts to eliminate fires may be threatening some of our finest forests.*

DIETZ, R. S., and J. C. HOLDEN, 1970. "The breakup of Pangaea," *Scientific American*, October. (Offprint 892)

GATES, D. M., 1971. "The flow of energy in the biosphere," *Scientific American*, September. (Offprint 664)   *On the radiant energy the earth receives from the sun—and the relatively small percentage trapped by green plants and made available to biotic communities.*

GOSZ, J. R., R. T. HOLMES, G. E. LIKENS, and F. H. BORMANN, 1978. "The flow of energy in a forest ecosystem," *Scientific American*, March.

HALLAM, A., 1972. "Continental drift and the fossil record," *Scientific American*, November. (Offprint 903)

HAZEN, W. E., ed., 1975. *Readings in population and community ecology*, 3rd ed. Saunders, Philadelphia.   *A collection of 23 important articles by prominent ecologists.**

HORN, H. S., 1975. "Forest succession," *Scientific American*, May. (Offprint 1321)

KORMONDY, E. J., 1976. *Concepts of ecology*, 2nd ed. Prentice-Hall, Englewood Cliffs, N.J.   *Short, simple, clearly written summary of ecology.**

MYERS, J. H., and C. J. KREBS, 1974. "Population cycles in rodents," *Scientific American*, June. (Offprint 1296)   *On a possible link between fluctuations in the population density of some small rodents in the field and changes in the genetic makeup of the populations.*

RICHARDS, P. W., 1973. "The tropical rain forest," *Scientific American*, December. (Offprint 1286)   *On the human threat to its survival.*

RICKLEFS, R. E., 1976. *The economy of nature.* Chiron Press, Portland, Ore.   *Excellent ecology text, clear treatment at an intermediate level.*

SMITH, R. L., 1977. *Elements of ecology and field biology.* Harper & Row, New York.   *An elementary ecology text that emphasizes the major types of ecosystems and their characteristics.*

WAGNER, R. H., 1978. *Environment and man*, 3rd ed. Norton, New York.   *An excellent treatment of the pressing ecological problems facing civilization today.*

WENT, F. W., 1949. "The plants of Krakatoa," *Scientific American*, September.   *The reinvasion of the island of Krakatoa by plants after all life on it had been destroyed by a volcanic eruption.*

WHITTAKER, R. H., 1975. *Communities and ecosystems*, 2nd ed. Macmillan, New York.   *Excellent short text by one of the leading community ecologists of the day.**

*In paperback.

PART V

THE GENESIS AND
DIVERSITY OF ORGANISMS

Chapter 20

THE ORIGIN AND EARLY EVOLUTION OF LIFE

Few problems have exercised the human imagination like the problem of the origin of life. Religion, mythology, philosophy have proposed a great variety of answers to it, but different though these answers have been, most of them share the assumption that the phenomenon must be attributed to an agency outside nature, a creator. In the same way, the diversity of species was conceived as resulting from separate, deliberate acts of creation. It was not until the latter part of the nineteenth century that the theory of evolution was able to account for the origin of species without invoking a supernatural agency. Can twentieth-century science do the same for the origin of life itself?

THE ORIGIN OF LIFE

We discussed the principle of biogenesis—that life can arise only from life—in Chapter 3. And we cited Pasteur's classic experiment in support of this principle. Pasteur's work effectively put to rest, as far as most biologists were concerned, the long-held idea of spontaneous generation. No longer could men of science seriously entertain the notion that the maggots in decaying meat arise *de novo* from the meat, or that earthworms arise from the soil during heavy rains, or that mice arise from sweaty shirts placed in a dark corner and sprinkled with wheat, as Jean-Baptiste van Helmont had suggested, or even that microorganisms appear spontaneously in spoiling broth. It may seem strange, then, that in the last quarter of the twentieth century spontaneous generation should be a topic of major interest in biology. But there is a radical difference between the modern ideas of spontaneous generation and those of Pasteur's day. The modern theorists do not suggest that life can arise spontaneously under present conditions on earth; indeed, most of them are convinced that this cannot happen. What they do suggest is that life could and did arise spontaneously from nonliving matter under the conditions prevailing on the early earth, and that it is from such beginnings that all present earthly life has descended.

It is one of the purposes of this chapter to outline for you a theory of the origin of life now widely held by scientists. The basis for this theory was first enunciated clearly and forcefully by the Russian biochemist A. I. Oparin in 1936.[1] You should be aware, however, that although the broad outlines of the theory have wide support many of the details are disputed. And you should realize that we have no direct evidence concerning the origin of life. We cannot be sure how life did arise; we can only gather indirect evidence to show how it could have arisen and how it probably arose.

Formation of the earth and its atmosphere   We are not certain how the solar system formed; we have only hypotheses. But as astronomers probe ever deeper into the secrets of the universe and gather more evidence, these hypotheses become increasingly convincing. The one most widely held today is that the universe is about 20 billion years old, and that the sun and its planets formed between four and a half and five billion years ago from a cloud of cosmic dust and gas. Most of this material condensed into a single compact mass, producing enormous heat and pressure, which initiated thermonuclear reactions and converted the condensed mass into the sun. Within the remainder of the dust and gas cloud, which now formed a disc held in the gravitational field of the newborn sun, lesser centers of condensation began to form. These became the planets, of which the earth is one.

As the earth condensed, a stratification of its components took place, the heavier materials, such as iron and nickel, moving toward the center and the lighter substances becoming more concentrated nearer the surface. Among these lighter materials must have been hydrogen, helium, and the noble gases, which formed the first atmosphere. But unlike larger planets such as Jupiter and Saturn, the earth was too small, and its gravitational field too weak, to retain this first atmosphere; eventually all the gases escaped into space, leaving a bare rocky globe with neither oceans nor atmosphere.

As time passed, however, gravitational compression of the earth, together with radioactive decay, generated an enormous amount of heat, and the interior of the earth became molten. The effect was further stratification into a core of iron and nickel, a mantle some 4,700 km thick composed of dense silicates of iron and magnesium, and an outer crust 8–65 km thick composed mainly of lighter silicates. But the intense heat in the interior of the earth also tended to drive out various gases, which escaped primarily by volcanic action. These gases formed a second atmosphere for the earth.

We must know something about the probable early composition of this secondary atmosphere to understand the conditions under which life arose. The present atmosphere contains about 78 percent molecular nitrogen (N_2), 21 percent molecular oxygen (O_2), and 0.033 percent carbon dioxide (CO_2), as well as traces of rarer gases such as helium and neon. But available evidence indicates that when the atmosphere first formed it contained virtually no

[1] Oparin had published a brief explanation of his theory earlier (1924), but this book was never translated from the Russian. His ideas did not have a major impact on scientific thought until *The origin of life on earth* appeared in 1936 (first English edition, 1938).

free oxygen and was therefore not an oxidizing atmosphere, as the present one is. Two principal models for the composition of the early atmosphere have been proposed in recent years; both are compatible with the modern hypothesis of the origin of life.

According to the first model, the early atmosphere was a reducing one; i.e. it contained much hydrogen (H_2). Consequently the atmospheric nitrogen was probably in the form of ammonia (NH_3), the oxygen in the form of water vapor (H_2O), and the carbon primarily in the form of methane (CH_4). Note that methane is a simple hydrocarbon—an organic compound; hence, if this model is correct, the atmosphere of the early earth contained organic molecules long before there were any organisms.

The second current model of the earth's early atmosphere assumes that, instead of containing much ammonia or methane, it was made up primarily of the gases known to occur in the present-day outgassing of volcanoes. These gases are H_2O, CO, CO_2, N_2, and H_2; in such a mixture, hydrogen cyanide (HCN) is easily formed and would probably also have been present in the atmosphere. Note that this model, like the first, envisions an atmosphere in which there is no free oxygen.

Initially, most of the earth's water was probably present as vapor in the atmosphere, a condition leading to torrential rains. These would have filled the low places on the crust with water and given rise to the first oceans. As rivers rushed down the slopes, they must have dissolved away and carried with them salts and minerals of various sorts, which slowly accumulated in the seas. If methane and ammonia were present in the atmosphere, some of the methane and much of the ammonia (as ammonium ions, NH_4^+) probably also dissolved in the waters of the newly formed oceans.

Formation of small organic molecules   If the early earth had a reducing atmosphere, as Oparin suggested and as many astronomers and geochemists also believe, and the primitive seas contained a mixture of salts, ammonium, and methane, how were the more complex organic molecules formed? Methane may be organic, but it is far from being a sufficient base on which to build living things. A mixture of ammonia, methane, water, and hydrogen is thermodynamically stable. There is no tendency for these materials to react with each other to form other compounds. Yet for life to have arisen it would seem at the very least that the critical building-block materials, particularly amino acids and the purine and pyrimidine bases, would have been necessary. How might these compounds have been formed on the abiotic primitive earth?

If more complex organic compounds were produced by reactions in the stable mixture of ammonia, methane, water, and hydrogen, clearly some external source of energy must have been acting on the mixture. One possible source might have been solar radiation, including visible light, ultraviolet light, and X rays; of these, ultraviolet light would probably have been the most important. A second important possibility is energy from electrical discharges, such as lightning. A third is heat. There are other possible sources of energy, such as cosmic rays, radioactive disintegration of elements in the earth's interior, and volcanic explosions, but any role played by these in organic synthesis is likely to have been minor.

How can it be demonstrated that ultraviolet radiation or electrical discharges or heat or a combination of these are capable of causing reactions that produce complex organic compounds from a mixture of ammonia, methane, water, and hydrogen? An answer was provided in 1953 by Stanley L. Miller, who was then a graduate student working under Harold C. Urey at the University of Chicago. Miller set up an airtight apparatus in which the four gases could be circulated past electrical discharges from tungsten electrodes. He kept the gases circulating continuously in this way for one week, and then analyzed the contents of his apparatus. He found that an amazing number and variety of organic compounds had been synthesized. Among these were some of the biologically most important amino acids and also such substances as urea, hydrogen cyanide, acetic acid, and lactic acid. Lest it be objected that microorganisms had contaminated his gas mixture and synthesized the compounds, Miller in another experiment circulated the gases in the same way, but without any electrical discharges; no significant yield of complex organic compounds resulted. In still another experiment, he prepared the apparatus with the gas mixture inside and then sterilized it at 130°C for 18 hours before starting the sparking. The yields of complex compounds were the same as in his first experiments; a great variety of organic compounds were formed. Clearly, the synthesis was not brought about by microorganisms, but was abiotic—a synthesis in the absence of any living organisms, a synthesis under conditions presumably similar to those on the primordial earth. This experiment by Miller, which gave the first conclusive evidence that some of the steps hypothesized by Oparin could really occur, marked a turning point in the scientific approach to the problem of how life began.

In the years since 1953 many investigators have synthesized a great variety of organic compounds from reducing mixtures of gases in which the only initial carbon source was methane. Thus Miller's results have been confirmed and extended. Investigators who have used mixtures of gases characteristic of volcanic emissions, plus hydrogen cyanide—on the assumption that the early atmosphere contained these or similar gases—have also achieved positive results. Those who have turned to other energy sources, such as ultraviolet light or heat or both, have also obtained large yields; these results are important, because on the early earth ultraviolet light was probably more available as a source of energy than lightning discharges. Significantly, the amino acids most easily synthesized in all these experiments performed under abiotic conditions were the very ones that are most abundant in proteins today.

The wide variety of conditions under which abiotic synthesis of the organic compounds essential to life has now been demonstrated makes it safe to conclude that, even if conditions on the primitive earth were only roughly similar to the ones postulated in either of the models outlined earlier, these compounds actually appeared and became dissolved in the waters of the seas. In other words, synthesis of such organic compounds on the early earth was not only possible but highly probable.

It may be argued, however, that even if organic compounds were synthesized abiotically on the primordial earth, they would have been destroyed too fast to accumulate in quantities sufficient for the later origin of living things. After all, most of these organic compounds are known to be highly perishable. But why are they perishable? One reason is that they tend to react slowly with molecular oxygen and become oxidized. Another is that they are broken down by organisms of decay, primarily microorganisms. But the prebiotic atmosphere contained virtually no free oxygen, and there were no organisms of any kind. Therefore neither oxidation nor decay would have destroyed the organic molecules, and they could have accumulated in the seas over hundreds of millions of years. No such accumulation would be possible today.

Formation of polymers Let us suppose, then, that a variety of hydrocarbons, fatty acids, amino acids, purine and pyrimidine bases, simple sugars, and other relatively small organic compounds slowly accumulated in the ancient seas. That is still not a sufficient basis for the beginning of life. Macromolecules are needed, particularly polypeptides and nucleic acids. How might these polymers have formed from the building-block substances present in the "soup" of the ancient oceans? This question is not easy to answer, and several hypotheses are currently supported by different investigators.

Some think that the concentration of organic material in the seas was high enough for chance bondings between simpler molecules to give rise, over a period of hundreds of millions of years, to considerable quantities of macromolecules. They point out that, even though each such polymerization reaction is rather unlikely in the absence of protein enzymes, nevertheless on the time scale here involved enough rare and unlikely events may occur to produce, collectively, a major change. As George Wald of Harvard University has said, "Given so much time, the 'impossible' becomes possible, the possible probable, and the probable virtually certain."

Other investigators, however, have been unwilling to agree that organic material in the early oceans was sufficiently abundant for the occurrence of chance polymerizations. They have suggested concentration mechanisms that would have speeded up chemical reactions. One such mechanism might have been adsorption of the building-block compounds on surfaces such as those of clay minerals. Another might have involved the collecting of small amounts of dilute solution of building-block compounds in puddles on the beaches of lagoons and ponds. The heat of the sun might have evaporated much of the water, thus concentrating the organic chemicals, and provided energy for polymerization reactions. The resulting polymers might then have been washed back into the pond. Such a process could slowly have built up a supply of macromolecules in the pond. The hypothesis seems a reasonable one, for Sidney W. Fox of the University of Miami has shown that, if a nearly dry mixture of amino acids is heated, polypeptide molecules are indeed synthesized (particularly if phosphates are present). Alternatively, after condensation by evaporation, the energy for polymerization reactions in the puddles might have come from ultraviolet radiation rather than heat.

Although various concentrating mechanisms may well have played a role in prebiotic polymerization reactions, too much may have been made of the difficulty of incorporating building-block compounds into polymers, at least in the case of amino acids. In experiments of the Miller type, various re-

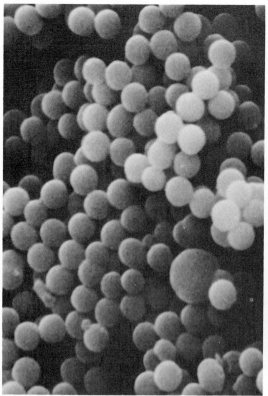 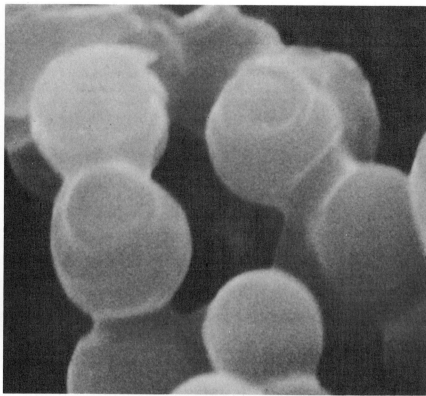

20.1 Scanning electron micrographs of proteinoid microspheres
Left: The spheres are remarkably uniform in size, though a few are as much as four times larger than the others. × 6,700.   Right: At higher magnification (× 20,800), connecting bridges between the spheres can be seen. The scars on the surfaces of some of the spheres indicate locations where bridges have been broken.   [Courtesy Sidney W. Fox, University of Miami, and Steven Brooke Studios, Coral Gables, Fla.]

searchers have observed that spherules yielding amino acids upon hydrolysis are formed after only 48 hours, whereas free amino acids do not appear in appreciable quantities until much later. Similarly, in some of their experiments on thermal synthesis, Fox and his associates obtained polymers first, and amino acids only later, *after hydrolysis* of the polymers. In short, abiotic synthesis may differ fundamentally from the more familiar biochemical syntheses in that polymers may be formed first, rather than the expected monomers.

Formation of molecular aggregates and primitive cells We have now reached a point in our model for the origin of life where the "soup" of the ancient seas, or at least that in some estuaries and lagoons, contains a mixture of salts and organic molecules, including polymers such as polypeptides and perhaps nucleic acids. We must next ask how the orderliness that characterizes living things could emerge from this mixture.

Oparin pointed out that, under appropriate conditions of temperature, ionic composition, and pH, colloids of macromolecules tend to give rise to complex units called *coacervate droplets*. Each such droplet is a cluster of macromolecules surrounded by a shell of water in which the individual water molecules are rigidly oriented relative to the colloidal particles. There is thus a definite demarcation or interface between the coacervate droplet and the liquid in which it floats. In a sense, the shell of oriented water molecules forms a membrane around the droplet.

Now, coacervate droplets have a marked tendency to adsorb and incorporate various substances from the surrounding solution; sometimes this tendency, which is a selective one, is so pronounced that the droplets may almost completely remove some materials from the medium. In this way the droplets may grow at the expense of the surrounding liquid. And coacervate droplets have a strong tendency toward formation of definite internal structure; i.e. the molecules within the droplet tend to become arranged in an orderly manner instead of being randomly scattered. As more and more different materials are incorporated into the droplet, a membrane consisting of surface-active substances may form just inside the shell of oriented water molecules, the permeability of the boundary of the droplet thus becoming even more selective than before. Thus, although coacervate droplets are not alive in the usual sense of the word, they do exhibit many properties ordinarily associated with living organisms. In fact, they look so much like organisms when viewed under a light microscope or even in an electron micrograph that experienced biologists have on occasion mistaken them for bacteria and attempted to assign them to species!

Fox, like Oparin, has envisioned the prebiological systems (prebionts) that led to development of the first cells as microscopic multimolecular droplets, but he suggests *proteinoid microspheres* rather than coacervate droplets. Fox's microspheres are droplets that form spontaneously when hot aqueous solutions of polypeptides are cooled. The microspheres may exhibit many properties characteristic of cells, including swelling in a hypotonic medium and shrinking in a hypertonic one, formation of a double-layered outer boundary, an internal movement reminiscent of cytoplasmic streaming, nonrandom movement of the whole unit if the microsphere is asymmetric and contains ATP, growth in size and increase in complexity, budding in a manner superficially similar to that seen in yeasts, and a tendency to aggregate in clusters of various types resembling those seen in many bacteria (Fig. 20.1).

Since either type of droplet, complex coacervate or proteinoid microsphere, is structurally organized and sharply separated from the external medium, the chemical reactions that take place within the droplet depend not only on the conditions of the medium, but also on the physicochemical organization of the droplet itself. Because various substances may be more concentrated in the droplet, the probability of their taking part in chemical reactions is increased; and because of the organization within the droplet, each reaction that takes place will influence other reactions in ways that are most unlikely when the substances are free in the external medium. Furthermore, catalytic activity of both inorganic substances such as metallic compounds and organic ones such as proteins is enhanced by the regular spatial arrangement of molecules within the droplet. In short, the special conditions within the droplet will exert selective and regulative influence over the chemical reactions taking place within the droplet.

Vast numbers of different prebiological systems of this kind may have arisen in the seas of the early earth. Most would probably have been too unstable to last long. But a few may have contained particularly favorable combinations of materials, especially complexes with catalytic activity, and may thus have developed unusually harmonious interactions between the reactions occurring within them. They might thus have survived longer and undergone relatively coordinated growth. As such droplets increased in size, they would have been more susceptible to physical fragmentation, which would have produced new smaller droplets with composition and properties essentially similar to those of the original droplet. These, in turn, would have grown and fragmented again. This primitive reproduction might not initially have been under the control of nucleic acids, even though these compounds could have been synthesized under abiotic conditions and may well have been incorporated into some of the prebionts. The nucleic acids would, of course, have been able to reproduce themselves exactly if a sufficient pool of nucleotides was present and if there were appropriate catalysts (even weak ones).

Clearly, if the new droplets formed upon fragmentation of the most stable prebionts could have incorporated more exact replicates of the favorable features of the parent droplet, their chances of survival would have been greater than without a genetic control system. In other words, chances of survival would have been increased if the sequence of nucleotides in nucleic acids could have come to code not only for more nucleic acid but for other components of the droplets, particularly proteins. As far as is known, nucleic acids have no catalytic properties and proteins have no duplicative properties; hence the establishment of a functional relation between the two was crucial to the origin of the type of life we know today.

Just how a correlation between nucleotide sequences in nucleic acids and amino acid sequences in proteins would have arisen remains an unsolved problem. There are, however, some suggestive leads to a possible solution. The first is that, so far as is known, the genetic code is universal; the codons have the same amino acid translations in all organisms, from microorganisms to human beings. That no variation should have arisen in the whole course of cellular existence argues against mere happenstance dictating which codon would go with which amino acid. Moreover, a careful look at the dictionary of the genetic code (p. 644) will show that all codons in which U is the second letter code for hydrophobic amino acids, and that all five electrically charged amino acids (see Fig. 2.34, p. 54) have codons in which the middle letter is a purine (A or G). In short, some chemical system seems to underlie the pairing of codons and amino acids. Although the chemical explanation for that system is unknown, the presumption is that, as it gradually came into being, it made possible both more accurate duplication during the reproductive process and more precise control over the chemical reactions taking place within the droplets. It is proposed, further, that a small percentage of prebionts with particularly favorable characteristics slowly developed into the first primitive cells. Notice that there was almost certainly no abrupt transition from "nonliving" prebionts to "living" cells. The attributes we normally associate with life were acquired gradually. At this stage, the boundary between living and nonliving is an arbitrary one.

Not all biologists accept the sequence outlined above. Some think it more likely that the first "living" things were self-replicating macromolecules such as nucleic acids—"naked genes," if you will. The first cells would then have arisen as these macromolecules slowly accumulated a shell of other substances (a primitive cytoplasm) around themselves. In other words, while the first model for the genesis of cells suggests that prebionts capable of a very primitive and imprecise form of reproduction arose first and then slowly developed a genetic control system, the suggestion here is that the control system arose first and that cytoplasm and a membrane then developed around it.

The closest things to naked genes found in nature today are viruses, which are nucleic acids with a thin shell of protein around them. Thus viruses might be modern representatives of one of the earliest steps in the evolution of cells from naked genes. This possibility has been disputed on the ground that all modern viruses are obligatorily parasitic—that they can reproduce only inside living cells—and consequently that they could not have existed before there were any cells. According to this view, viruses are of much more recent origin. But it has been pointed out that, although today the raw materials necessary for viral reproduction can be found only inside living cells (which is why viruses are obligate parasites), many of these substances would have been available in the "soup" hypothesized for the prebiological earth, and that viruses or something like them could, therefore, have been free-living as long as the "soup" existed. What this argument fails to take into account is that viruses depend on their host cells not only for raw materials, but also for metabolic machinery—enzymes, ribosomes, tRNA, etc; such machinery would not have been free in the prebiological medium.

Evolution of more complex biochemical pathways by primitive organisms   According to the view now most widely accepted, the earliest organisms were heterotrophs. They used as nutrients the carbohydrates, amino acids, and other organic compounds free in the environment in which they lived. In other words, they depended on previous abiotic synthesis of organic compounds. But as the organisms became more abundant and more efficient at removing nutrients from the medium, they must have begun to deplete the supply of nutrients. The rate of spontaneous formation of organic matter from inorganic raw materials was most likely never very high, and it must have taken many millions of years for a moderate supply of nutrients to accumulate. Now, that supply was probably being used up at an ever increasing rate and, with the drain on available resources becoming more and more severe, the competition between organisms must have increased too. Forms inefficient at obtaining nutrients doubtless perished; those more efficient survived in greater numbers. Natural selection would have favored any new mutation that enhanced the ability of its possessor to obtain or process food.

At first, the primitive organisms probably carried out relatively few complex biochemical transformations. They could obtain most of the materials they needed ready-made. But it would have been these very materials—materials that could be used directly with little alteration—that would have dwindled most rapidly. Hence there would have been strong selection for any organisms that could use alternative nutrients. Suppose, for example, that compound A, which was necessary for the life of cells, was initially available in the medium, but that its supply was being rapidly exhausted. If some cells possessed a mutant gene that coded for an enzyme a that catalyzed synthesis of A from another compound, B, in greater supply in the medium, then those cells would have had an adaptive advantage over cells that lacked the mutant gene. They could survive even when *A* was no longer available in the medium by carrying out the reaction

$$B \xrightarrow{\ a\ } A$$

But then there would have been increasing demand for free *B*, and the rate of its utilization would soon have exceeded the rate of its abiotic synthesis. Thus the supply of *B* would have dwindled, and there would have been strong selection for any cells possessing a second mutant gene that coded for an enzyme *b*, catalyzing synthesis of *B* from *C*. These cells would not have been dependent on a free supply of either *A* or *B*, because they could make both *A* and *B* for themselves as long as they could obtain enough *C*:

$$C \xrightarrow{\ b\ } B \xrightarrow{\ a\ } A$$

This process of evolution of synthetic ability, first proposed by N. H. Horowitz, might have continued until eventually most cells made all the *A* they required by carrying out a long chain of chemical reactions:

$$G \xrightarrow{\ f\ } F \xrightarrow{\ e\ } E \xrightarrow{\ d\ } D \xrightarrow{\ c\ } C \xrightarrow{\ b\ } B \xrightarrow{\ a\ } A$$

In this way the primitive cells would slowly have evolved more elaborate biochemical capabilities.

It seems most unlikely that much synthetic ability could have arisen unless the cells possessed some mechanism for handling chemical energy. The fact that all living things, from bacteria to human beings, employ ATP as their principal energy currency strongly suggests that use of ATP was an early evolutionary development. But what is the evidence that ATP was available to early organisms? The two organic precursors of ATP are adenine and ribose, a five-carbon sugar (see Fig. 4.9, p. 147). Both these compounds also occur in nucleic acids, and experiments have demonstrated that both can be synthesized abiotically under presumed prebiological conditions. Indeed, of the five nitrogenous bases that occur in DNA and RNA, adenine is the one that forms most easily under simulated prebiological conditions. It is probably no accident, therefore, that adenine is the base found in a host of biologically critical compounds—not only ATP (and ADP, AMP, and cAMP), but also the electron carriers NAD, NADP, and FAD. ATP would probably have been available for coupling exergonic and endergonic reactions as soon as the primitive cells could perform the reactions.

The earliest form of metabolism using ATP was almost surely fermentation, a process universal in living organisms today (of the three main energy-yielding systems—fermentation, respiration, and photosynthesis—only fermentation is found in all bacteria and all eucaryotes).

Evolution of autotrophy   Even though the primitive heterotrophs probably evolved more and more elaborate biochemical pathways that enabled them to use a greater variety of the organic compounds free in their environment, and even though some of them probably evolved other methods of feeding, such as saprophytism, parasitism, and predation, nevertheless life would eventually have ceased if all nutrition had remained heterotrophic. The reason is not only that nutrients must have been used up much faster than they were being synthesized, but also that the organisms themselves must have been altering the environment in ways that decreased the rate of abiotic synthesis of organic compounds. For example, their metabolism, which would have been fermentative in the absence of molecular oxygen, would have released carbon dioxide into the atmosphere. Abiotic synthesis of complex organic compounds from CO_2 is much less likely than from methane or hydrogen cyanide.

That life did not become extinct as the supply of free organic compounds dwindled is attributable to the evolution of photosynthetic pathways by some of the primitive organisms. The first such pathway may well have been that of cyclic photophosphorylation, in which light of visible wavelengths is used as an energy source in the direct synthesis of ATP by cells. Later, the much more complex pathways of noncyclic photophosphorylation and CO_2 fixation, in which energy from sunlight is used in synthesis of carbohydrate from CO_2 and water (or some other electron source) would have arisen, probably appearing first in organisms now interpreted as primitive blue-green algae. From this time onward, the continuation of life on earth depended on the activity of the photosynthetic autotrophs.

The evolution of photosynthesis based on water as the electron donor probably administered the coup de grâce to significant abiotic synthesis of complex organic compounds. An important by-product of such photosynthesis is molecular oxygen. The O_2 released by photosynthesis must have helped convert the atmosphere from a reducing to an oxidizing one, particularly since much of the free hydrogen present initially must by this time have escaped from the planet into space. Once the layer of ozone (O_3) now present high in the atmosphere had been formed by some of the O_2, this layer effectively screened out most of the ultraviolet radiation from the sun and allowed very little high-energy radiation to reach the earth's surface. In other words, living organisms, once they arose, changed their environment in a way that destroyed the conditions that had made possible the origin of life; they caused what has sometimes been called an oxygen revolution.

Once molecular oxygen became a major component of the atmosphere, both heterotrophic and autotrophic organisms could evolve the biochemical pathways of aerobic respiration, by which far more energy can be extracted from nutrient molecules than by fermentation alone. As we shall see shortly, the progressive increase in atmospheric O_2 was probably a major factor in determining the later course of evolution.

The possibility of life on other planets   If life could arise spontaneously from nonliving matter on the primordial earth, might it also have arisen elsewhere in the universe? Few scientists concerned with the origin of organisms think that life is unique to the planet earth; most are convinced that life has probably arisen many times in many places. They point out that no unduplicable event was necessary to the origin of life on earth. On the contrary, all the events now hypothesized and all the known characteristics of life seem to fall well within the general laws of the universe; i.e. they are natural phenomena susceptible of duplication. Indeed, some have argued recently that biochemical evolution (i.e. life) is an inevitable part of the overall evolution of matter in the universe. Given the immense size of the universe, they argue, it would actually be unreasonable to think that life is restricted to one small planet in one minor solar system.

One interesting series of calculations on the probability of life elsewhere in the universe was made some years ago by Harlow Shapley of Harvard University. Shapley said that at least 10^{20} stars are visible to us with present-day telescopes (to say nothing about the vast numbers beyond the reach of our telescopes). Many of these stars probably lack planets, but Shapley thought it reasonable to assume that at least one star in every thousand has a planetary system, which gives a total of 10^{17} stars with planets. Now, if it is assumed that life wherever it occurs is at least roughly similar to earthly life in its basic chemistry, then it must be concluded that only planets with moderate temperatures could support life. The planetary systems of many stars may not include such a planet. Shapley suggested that, by a modest estimate, at least one in a thousand of the stars with planets has an appropriate planet; this gives a total of 10^{14} stars with at least one planet of the right temperature. Now, not only must a planet have moderate temperatures if it is to support life, but it must also be within a certain size range to hold a suitable atmosphere. If one out of every thousand of the planets of the right temperature is also of the right size, this gives a total of 10^{11}. But even if a planet has an appropriate temperature and atmosphere, life still might not have arisen, for any number of reasons. Again Shapley used an estimate of one in a thousand, which gives 10^8 (one hundred million) planets on which life may well have arisen. Today biologists working in this field consider Shapley's estimate too conservative; they suggest that the figure should be 10^{16} or more.

If one admits the possibility that life has arisen on huge numbers of other planets, the question immediately arises whether civilizations as advanced as, or more advanced than, our own might have developed elsewhere. Here there is little on which to base an estimate. Nevertheless, many scientists believe that there are numerous civilizations in the universe more advanced than our own. Shapley's estimate was that there are at least 100,000 such civilizations; recent estimates have ranged as high as 5×10^{14} (five hundred trillion).

So confident are some scientists that these other civilizations exist that they have used huge radio receivers to listen for signals from outer space, and they have included in some recent space-probe capsules information designed to be interpretable by beings that, though unfamiliar with earthly languages, can decode messages written in mathematical terms. They think it possible that someday human beings may be able to communicate with intelligent organisms living on planets in other parts of the universe. It should be realized, however, that even messages traveling at the speed of light would take so many thousands or millions of years to reach their destination that contemporary organisms on planets in different solar systems could never communicate; a radio message received on earth today would have been sent by organisms that died long ago.

It is perhaps well to point out here that, although the universe may contain vast numbers of other intelligent creatures with technologically advanced civilizations, human beings are most unlikely to have evolved on any planet but earth. Totally separate evolutionary developments can never duplicate each other. Indeed, if human beings should become extinct on earth, and intelligent civilization-building life arise a second time, that life would surely not be human. Loren Eiseley of the University of Pennsylvania has put the matter eloquently: "There may be wisdom; there may be power; somewhere across space great instruments, handled by strong, manipulative organs, may stare vainly at our floating cloud wrack, their owners yearning as we yearn. Nevertheless, in the nature of life and in the principles of evolution we have had our answer. Of men, elsewhere and beyond, there will be none forever."

PRECAMBRIAN EVOLUTION

THE FOSSIL RECORD

The oldest known fossils (as of this writing) are from the Fig-Tree chert of eastern Transvaal, South Africa, and are dated at 3.1 billion years. They appear to be bacteria—procaryotic cells, as would be expected of the oldest fossils, since these cells are assumed to be more primitive than eucaryotic ones. Most authorities think it probable that the first cellular organisms, living at a time when all nutrition was heterotrophic, were of the bacterial type, and that it was the evolution of blue-green algae from bacteria roughly 2.3 billion years ago that initiated the oxygen revolution. These algae, like true plants but unlike photosynthetic bacteria, use water as the electron source in noncyclic photophosphorylation and hence release molecular oxygen as a by-product.

More and more fossils of procaryotic organisms from the first two and a half billion years or so of life are being found, and some fossils of eucaryotic cells are thought to date back at least 1.5 billion years. But the oldest geologic period from which fossils of higher forms of life are fairly abundant is the **Cambrian,** which began nearly 600 million years ago (Table 20.1). Many of the Cambrian fossils are of relatively complex organisms—how complex is suggested by the fact that most of the animal phyla extant today are represented. The few Precambrian fossils of eucaryotic organisms are mostly of simple algae and a few invertebrates, whose relationships are poorly understood. We therefore have very little evidence concerning a most fascinating evolutionary development—the early radiation of life and the origin of the major animal phyla and plant divisions.

It is not clear why there are so few Precambrian fossils. Among the many explanations that have been offered, one is that most Precambrian organisms were soft-bodied and hence did not readily form fossils; it is the hard parts of plants and animals that are most often fossilized, because these are the most likely to resist decay and become buried. This explanation, though doubtless partly correct, is insufficient. It seems most unlikely that the many hard-bodied animals present at the start of the Cambrian period arose suddenly at that time; they almost certainly descended from Precambrian ancestors that gradually developed shells or exoskeletons or other hard parts. And, indeed, a few recently discovered beds of Precambrian fossils, notably in Australia, have confirmed the existence of hard-bodied invertebrate animals before the Cambrian (Fig. 20.2). It is possible that the ranges of these

20.2 Fossil of a Precambrian animal (*Dickinsonia costata*) from South Australia
This animal, which appears to have been a segmented worm of some sort, lived about 600 million years ago. Reproduced natural size. [Courtesy M. F. Glaessner, University of Adelaide.]

TABLE 20.1 *The geologic time scale*

Era	Period	Epoch	Millions of years (approx.) from start of period to present	Plant life	Animal life
CENOZOIC	Quaternary	Recent	0.01	Increase in number of herbs	Rise of civilizations
		Pleisto-cene	2.5		First *Homo*
	Tertiary	Pliocene	7	Dominance of land by angiosperms	First humans
		Miocene	26		
		Oligocene	38		Dominance of land by mammals, birds, and insects
		Eocene	54		
		Paleocene	65		

———————— BUILDING OF ANCESTRAL ROCKY MOUNTAINS ————————

Era	Period	Millions of years	Plant life	Animal life
MESOZOIC	Cretaceous	136	Angiosperms arise and expand as gymnosperms decline	Last of the dinosaurs; second great radiation of insects
	Jurassic	190	Gymnosperms (esp. cycads and conifers) still dominant; last of the seed ferns	Dinosaurs abundant; first birds
	Triassic	225	Dominance of land by gymnosperms; further decline of lycopsids and sphenopsids	First mammals; First dinosaurs

———————— BUILDING OF ANCESTRAL APPALACHIAN MOUNTAINS ————————

Era	Period	Millions of years	Plant life	Animal life
PALEOZOIC	Permian	280	Precipitous decline of lycopsids, sphenopsids, and seed ferns	Great expansion of reptiles; decline of amphibians; last of the trilobites
	Carboniferous*	345	Great coal forests, dominated at first by lycopsids and sphenopsids, and later also by ferns and gymnosperms	Age of Amphibians; first reptiles; first great radiation of insects
	Devonian	395	Expansion of primitive tracheophytes; origin of first seed plants toward end of period; first liverworts	Age of Fishes; first amphibians and insects
	Silurian	430	Invasion of land by the first tracheophytes toward end of period	Invasion of land by a few arthropods
	Ordovician	500	Marine algae abundant	First vertebrates (Agnatha)
	Cambrian	570	Primitive marine algae (esp. Cyanophyta and probably Chlorophyta)	Marine invertebrates abundant (including representatives of most phyla)

———————— INTERVAL OF GREAT EROSION ————————

PRECAMBRIAN			Primitive marine life

* In North America, the Lower Carboniferous is often called the Mississippian period, and the Upper Carboniferous is called the Pennsylvanian period.

organisms were restricted to areas with rather special environmental conditions that made fossilization unlikely. Furthermore, we must remember that fossils may be destroyed by normal geological processes; the greater the lapse of time, the more likely the destruction. A high percentage of any fossils formed as long ago as the Precambrian is likely to have been destroyed long before there were human beings to study them.

THE BERKNER-MARSHALL HYPOTHESIS

A provocative hypothesis explaining the apparent great increase in living things at the beginning of the Cambrian has recently been put forward by L. V. Berkner and L. C. Marshall of the Southwest Center for Advanced Studies, Dallas, Texas. They propose that throughout the 1.7 billion years or so that photosynthetic organisms existed during the Precambrian, the concentration of molecular oxygen (O_2) in the atmosphere was still so low that living organisms could survive only at some depths in lakes or seas, because the intense ultraviolet radiation at the surface or on land would have been lethal. In addition, the low concentration of O_2 would have imposed severe limits on aerobic respiration. In short, the oxygen revolution was still far from complete.

At about the beginning of the Cambrian, 600 million years ago, according to Berkner and Marshall, the atmospheric concentration of O_2 passed the "first critical level"; i.e. it reached about one percent of present atmospheric levels—enough to cut the intensity of ultraviolet radiation so that life could exist anywhere in the earth's bodies of water, and also enough greatly to enhance the evolutionary potential of aerobic respiration. The result was the explosive evolutionary surge that marked the beginning of the Cambrian period.

The Berkner-Marshall hypothesis goes on to propose that the great increase in living things during the Cambrian meant a far greater abundance of green plants, and hence much faster photosynthetic production of O_2. The rapid rise in atmospheric concentration of O_2 that must have followed presumably reached a "second critical level" of about 10 percent of present levels some 420 million years ago (during the Silurian period). At this point, the shielding from ultraviolet radiation would have been sufficient to permit the spread of life onto land, and thus a second great evolutionary explosion would have begun.

THE ORIGIN OF EUCARYOTIC CELLS

The scarcity of fossils from the Precambrian period leaves us with virtually no direct evidence concerning the evolution of the first eucaryotic cells from procaryotic progenitors. The oldest fossil eucaryotes are already relatively complex, and hence shed little light on the problem of their derivation. Nonetheless, a growing understanding of the special properties of such characteristic eucaryotic organelles as chloroplasts and mitochondria has led to an intriguing hypothesis regarding the origin of the complicated kind of cell that is the structural unit of higher plants and animals. Although it is still too early to evaluate satisfactorily all aspects of the proposed model,

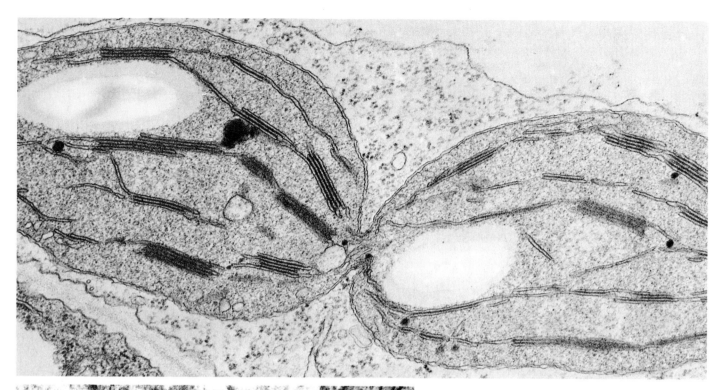

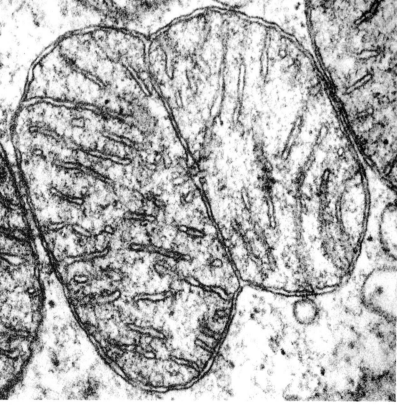

20.3   Electron micrograph of a dividing chloroplast in a tobacco leaf
The large white areas in each of the daughter chloroplasts are starch granules. × 36,000.   [Courtesy D. A. Stetler and W. M. Laetsch, *Am. J. Botany,* vol. 56, 1969.]

20.4   Electron micrograph of a dividing mitochondrion
The partition between the two daughter mitochondria is nearly complete. × 80,750.   [Courtesy W.J. Larsen, *J. Cell Biol.,* vol. 47, 1970.]

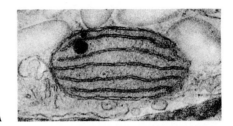

A

20.5   Chloroplast of a red alga (A) compared with a whole blue-green alga (B)
The photosynthetic lamellae in the chloroplasts of red and brown algae are not arranged in the stacks, called grana, that characterize the chloroplasts of most tracheophytes; they are more like the stroma lamellae of tracheophytes. In this respect a red algal chloroplast (A) resembles an entire blue-green algal cell; in such a cell (B), the lamellae, which show no granum arrangement either, are free in the cytoplasm, for there is no plastid membrane.   Both photos: × 30,000. [Courtesy R. E. Lee, University of the Witwatersrand (A), and M. Jost, Michigan State University (B).]

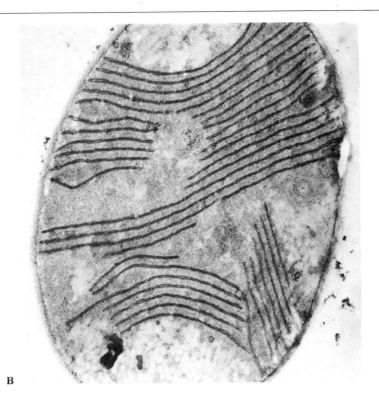

B

the weight of current evidence seems to indicate that at least its broad outlines are correct, and we shall discuss these briefly here.

As we saw in Chapter 15, plastids and mitochondria are self-replicating bodies (Figs. 20.3, 20.4). They contain genetic material, and they can carry out protein synthesis on their own ribosomes. In short, they have many features in common with free-living procaryotic organisms. Might these organelles be modern descendants of ancient procaryotic cells that became obligate endosymbionts of other cells and have evolved in concert with their hosts ever since? Several investigators, most prominently Lynn Margulis of Boston University, think the answer is yes.

It was noted as long ago as the nineteenth century, when chloroplasts were first studied under the microscope, that they resembled certain free-living blue-green algae, and the suggestion was made that they might have originated as such algae (Fig. 20.5). A bacterial origin for mitochondria was first proposed in the 1920s. More recent research findings have tended to support the suggestion that chloroplasts are derived from procaryotic algae and that mitochondria are derived from bacteria: (1) The enzymes for synthesis of DNA, RNA, and protein in chloroplasts and mitochondria are qualitatively like those of procaryotic cells and correspondingly unlike those in the rest of the plant cell. (2) Unlike the nuclear DNA of eucaryotic cells, which is coated with protein, the DNA of chloroplasts and mitochondria is nearly naked, like that of procaryotic cells. (3) The minichromosome of

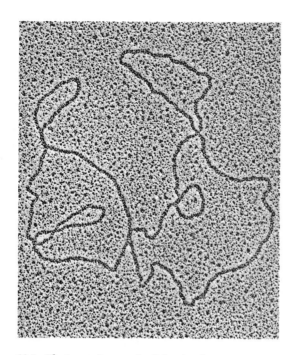

20.6   Electron micrograph of the circular chromosome of a mitochondrion from an oocyte of *Xenopus*

× 84,000.   [I. B. Dawid, National Institutes of Health, Bethesda, Md., and D. R. Wolstenholme, University of Utah.]

chloroplasts and mitochondria is usually circular, like the bacterial chromosome (Fig. 20.6). (4) The ribosomes of chloroplasts and mitochondria, like those of procaryotes, tend to be 15 percent smaller than the cytoplasmic ribosomes of eucaryotes. (5) Protein synthesis by the cytoplasmic ribosomes of eucaryotic cells is inhibited by certain drugs (e.g. cycloheximide) and not by others (e.g. chloramphenicol, erythromycin); procaryotic protein synthesis shows exactly opposite specificity, and chloroplasts and mitochondria resemble procaryotes in this regard. In summary, a growing list of characteristics attests to a possible derivation of these two organelles from procaryotic ancestors.

The capacity of unicellular organisms to live as endosymbionts inside the cells of other organisms—as the ancestors of chloroplasts and mitochondria are presumed to have done—can be amply demonstrated by present-day examples. Thus a variety of protozoans contain endosymbiotic single-celled algae, and the gastrodermal cells of *Chlorohydra* and of several species of sea anemones also contain algal cells (see Fig. 1.15, p. 20). Even some higher animals, especially several species of molluscs, regularly have intracellular algal symbionts. And intracellular symbiotic bacteria occur widely in both the plant and animal kingdoms.

Although both photosynthetic bacteria and blue-green algae probably arose long before the first eucaryotic cells, the blue-green algae seem the more likely progenitors of chloroplasts,[2] since they have the same basic type of chlorophyll *a* and use the same noncyclic photophosphorylation pathways. Mitochondria were probably derived from aerobic bacteria; the inner mitochondrial membrane may be viewed as the homologue of the bacterial plasma membrane, since the two are biochemically similar, particularly in their capacity for electron transport. The outer mitochondrial membrane resembles the endoplasmic reticulum, and hence was probably derived from the host cell during the period of common evolution. If chloroplasts and mitochondria did indeed arise as symbionts, the symbiosis has clearly been a mutualistic one.

Though the organelles discussed here contain DNA, and divide, grow, and differentiate partly on their own, they are not fully autonomous entities. Many of their proteins are specified by nuclear genes. Presumably the symbionts gave up much of their genetic control to the host during the hundreds of millions of years since they last lived as free cells. This surrender of control to the nuclear genes is probably the reason why the minichromosomes of chloroplasts and mitochondria are invariably less than one tenth the size of typical procaryotic chromosomes.

We began this section by saying that we would outline the most recent hypothesis concerning the origin of eucaryotic cells—literally, true nucleate cells—and then proceeded to discuss chloroplasts and mitochondria instead of the nucleus. The reason is that these organelles are just as fundamental to eucaryotic cells as the presence of a nuclear membrane, and it is primarily on them that recent conjecture has centered. There is little evidence bearing on the origin of the nuclear membrane and the associated endoplasmic reticulum. Since there is membrane flow in fully developed eucaryotic cells from the nuclear membrane to the ER to the Golgi apparatus (see Fig. 3.29, p. 107), the nuclear membrane may have been the first part of the intra-

[2] But see discussion of the Prochlorophyta on p. 936.

cellular membrane system to evolve, with the other parts arising from it later (Fig. 20.7), but how the nuclear membrane itself arose is still a mystery.

Despite its current prestige, the endosymbiont model of the origin of eucaryotic cells is far from proven. Indeed, a variety of other models have been proposed in recent years; two are described by Bogorad and by Uzzel and Spolsky in articles listed at the end of the chapter.

THE KINGDOMS OF LIFE

Because the fossil record is nearly blank for the long span of time when the basic pattern of organismic diversity was coming into being, ideas about the evolutionary relationships between the major phyla and divisions of organisms are rather vague. There is little evidence concerning the relationships between the major groups of algae. It is not known whether fungi evolved from photosynthetic green algae, or directly from heterotrophic organisms such as bacteria, or from some other stock. It is uncertain how Protozoa are related to multicellular plants or to multicellular animals.

Ignorance of the relationships between major groups of organisms apparently never hindered the age-old attempt to assign all living things to one or another of a few large categories called kingdoms. One of the oldest and most widely used classifications recognizes only two kingdoms—one for plants and the other for animals. This dichotomy works well as long as the organisms to be classified are the generally familiar ones. Dandelions, grasses, daffodils, roses, and oak trees can easily be recognized as plants; no fancy definitions or elaborate diagnostic keys are necessary. Similarly, cats, horses, chickens, earthworms, and houseflies can easily be recognized as animals. Things become a bit more difficult when the organisms in question are bread molds or sponges. These don't fit quite so neatly within the common intuitive concept of "plant" and "animal." Nevertheless, most biologists managed to convince themselves that bread molds, when their characteristics were carefully examined, seemed definitely more "plantlike" than "animal-like," despite their lack of chlorophyll, and that sponges are "animals," despite their sedentary way of life.

With the advent of electron microscopy and more detailed study of cellular structure, however, it soon became clear that the bacteria and the blue-green algae differ from all other forms of life in a very fundamental way: They are procaryotic, whereas all other organisms are eucaryotic. Though both bacteria and blue-green algae had traditionally been assigned to the plant kingdom, this procedure was plainly no longer acceptable, and the practice of setting them apart as a separate kingdom, the **Monera,** rapidly gained ground; it is a practice we shall follow.

More recently, there has been a growing realization that the fungi—traditionally regarded as plants—differ from true plants in a variety of basic characteristics. First, they are not photosynthetic; indeed, as heterotrophs, they are more animal-like than plantlike in their nutrition. Second, their cell walls are chemically different from those of most plants in that the primary component of the walls is usually not cellulose. Third, they are not multicellular in the sense that both plants and animals are; the partitions between adjacent fungal cells, if present at all, tend to be incomplete, the cytoplasm thus being continuous. In short, the fungi seem to represent a separate evolutionary development, coordinate with that of plants and that of an-

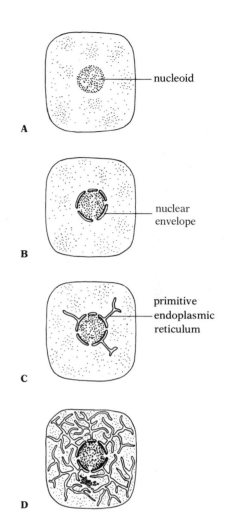

20.7 Model for the evolution of the nuclear envelope and the cytoplasmic membrane system

(A) The ancestral cell is presumed to have been procaryotic, i.e. to have had a nucleoid with no boundary membrane.   (B) Membranous vesicles may have formed as new structures surrounding the nuclear area, giving rise to the double nuclear envelope. (C) Evaginations from the outer membrane of the nuclear envelope may have given rise to the endoplasmic reticulum, which itself became elaborated (D) into an extensive cytoplasmic membrane system.

imals. We shall, therefore, follow the growing practice of recognizing a kingdom *Fungi.*

As we saw in Chapter 5, putting the fungi in a kingdom of their own makes it possible to understand the tripartite evolution of higher organisms in terms of exploitation of three different modes of nutrition—photosynthetic autotrophism by plants, absorptive heterotrophism by fungi, and ingestive heterotrophism by animals.

Recognizing the kingdoms Monera and Fungi resolves two serious anomalies in the older classifications. But another major problem remains: the classification of eucaryotic unicellular organisms. The ones zoologists have traditionally called Protozoa have been particularly troublesome to those who insist on a neat separation between plants and animals. This is true especially of the group of protozoans known as flagellates. These creatures have long flagella that enable them to swim actively in a manner intuitively felt to be "animal-like." Yet some of them possess chlorophyll and carry out photosynthesis, a characteristic ordinarily considered decidedly "plantlike." How can organisms such as these be classified? One possibility would be to rule arbitrarily that all those with chlorophyll are to be considered plants, and all lacking chlorophyll animals. But this seemingly simple procedure has serious drawbacks. Some green flagellates are clearly very closely related to colorless species; yet the proposed rule would put the green ones in a different kingdom from their colorless close relatives. Furthermore, there are some species of flagellates in which both green and colorless races sometimes occur; the suggested rule is unable to deal effectively with such species. Suppose, since most Protozoa lack chlorophyll and are rather animal-like, that one adopts a different rule and regards them all as animals, even if they are green. Again there are serious drawbacks. In many cases unicellular green flagellates are obviously very closely related to species of simple multicellular green algae. Evolutionary patterns within the green algae are best interpreted, in fact, if one includes unicellular green flagellates; transferring them from the algae to the animal kingdom is unsatisfactory, because it clearly separates closely related organisms.

Whatever the criteria chosen, it is impossible to make a clean separation between plants and animals at the unicellular level. The reason is obvious. Unicellular organisms (and some multicellular ones) are at an evolutionary stage where it is essentially meaningless to talk about "plants" and "animals"—artificial human categories not dictated by the rules of nature. At the lowest evolutionary levels, about the only distinction between plants and animals that stands up is this: Plants are living things studied by people who say they are studying plants (botanists), and animals are living things studied by people who say they are studying animals (zoologists). Facetious as this distinction sounds, it is basically accurate. The unicellular green flagellates may reasonably be classified as both plants and animals because they are studied by both botanists and zoologists, the first calling them algae and the second calling them Protozoa.

An alternative that has won much support is to assign the eucaryotic unicellular (and very primitive multicellular) organisms to a separate kingdom, designated *Protista.* As Table 20.2 shows, the limits of this kingdom vary greatly among the many different kingdom classifications that may be encountered in textbooks today. In some classifications the Protista take in only unicellular organisms (with or without the bacteria and blue-green algae); in others they include all multicellular algae (and at times fungi) as well. In other words, the category Protista is sometimes restricted to organisms that are unicellular during much or all of their lives, and sometimes it is broadened—the name then used may be Protoctista—to include all plantlike organisms whose bodies show relatively little distinction between tissues.

Both usages of Protista have the advantage of allowing a clear separation between the plant and the animal kingdom, all the troublesome forms being lumped together under Protista. But the first usage, which restricts Protista to unicellular organisms, separates green flagellates from their close multicellular relatives in the algae. And the second usage, which includes all algae in the Protista, leaves only bryophytes and vascular plants in the plant kingdom—an arrangement accepted by few botanists; moreover, it combines groups that are probably not at all closely related—notably the ciliate Protozoa and the brown algae—thus making the Protista a phylogenetically meaningless assemblage. In brief, both usages of Protista get around some of the difficulties inherent in the two-kingdom system, but both create other problems.

Nonetheless, there is a growing tendency to recognize a kingdom Protista, generally limited to groups entirely or primarily unicellular. The Protista so conceived figure in the five-kingdom system proposed by R. H. Whittaker of Cornell University (System 7 of Table 20.2), which likewise includes, in addition to Plantae and Animalia, the kingdoms Monera and Fungi. The in-

TABLE 20.2 *Various kingdom classifications*

System 1	System 2	System 3	System 4	System 5	System 6	System 7
PLANTAE	MONERA	PROTISTA	PROTISTA*	MONERA	MONERA	MONERA
Bacteria	Bacteria	Bacteria	Bacteria	Bacteria	Bacteria	Bacteria
Blue-green algae	Blue-green algae	Blue-green algae	Blue-green algae	Blue-green algae	Blue-green algae	Blue-green algae
Chrysophytes		Protozoa	Protozoa			
Green algae	PLANTAE	Slime molds	Chrysophytes	PROTISTA*	PLANTAE	PROTISTA
Brown algae	Chrysophytes		Green algae	Protozoa	Chrysophytes	Protozoa
Red algae	Green algae	PLANTAE	Brown algae	Chrysophytes	Green algae	Chrysophytes
Slime molds	Brown algae	Chrysophytes	Red algae	Green algae	Brown algae	Slime molds
True fungi	Red algae	Green algae	Slime molds	Brown algae	Red algae	
Bryophytes	Slime molds	Brown algae	True fungi	Red algae	Bryophytes	PLANTAE
Tracheophytes	True fungi	Red algae		Slime molds	Tracheophytes	Green algae
	Bryophytes	True fungi	PLANTAE	True fungi		Brown algae
ANIMALIA	Tracheophytes	Bryophytes	Bryophytes		FUNGI	Red algae
Protozoa		Tracheophytes	Tracheophytes	PLANTAE	Slime molds	Bryophytes
Multicellular	ANIMALIA			Bryophytes	True fungi	Tracheophytes
animals	Protozoa	ANIMALIA	ANIMALIA	Tracheophytes		
	Multicellular	Multicellular	Multicellular		ANIMALIA	FUNGI
	animals	animals	animals	ANIMALIA	Protozoa	True fungi
				Multicellular	Multicellular	
				animals	animals	ANIMALIA
						Multicellular
						animals

* When multicellular groups are included, the Protista are sometimes rechristened Protoctista.

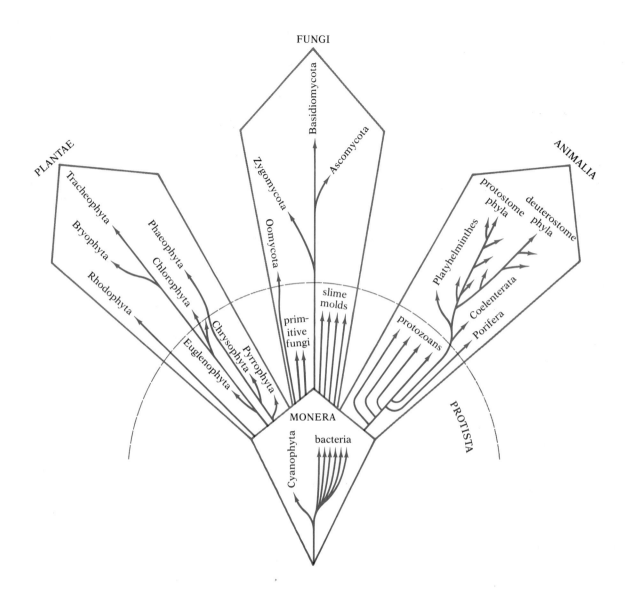

20.8 The five-kingdom classification system used in this text

It is believed that the procaryotic organisms (kingdom Monera) gave rise to many eucaryotic lineages. Of these, the ones terminating at a unicellular, colonial, or very primitive multicellular level constitute the kingdom Protista. Three major lineages that reached the higher multicellular level, each exploiting a different mode of nutrition, form the kingdoms Plantae, Fungi, and Animalia.

A possible four-kingdom system, very similar to the five-kingdom system diagramed here, does not recognize the Protista as a separate kingdom, but assigns the various groups of eucaryotic organisms to that one of the three major kingdoms to which they seem most closely allied. The difference between the two systems is mainly one of emphasis—whether to take the dashed line in the figure as a kingdom boundary or merely as an indicator of the transition from the protistan (i.e. unicellular) level of organization to the multicellular level within three eucaryotic kingdoms.

creasing acceptance won by that system prompted its adoption in this book. There is an alternative system, however, which has much to recommend it—a system that recognizes only four kingdoms and interprets the unicellular representatives of the plant, fungal, and animal kingdoms as being "on the protistan level."[3] As Fig. 20.8 shows, the difference between this system and Whittaker's lies essentially in the status conferred on the eucaryotic unicellular organisms.

The viruses have not been mentioned in this discussion of kingdom classification, because they differ in such fundamental ways from cellular organisms that biologists have been hesitant to regard them as living things. Yet they undeniably possess many properties in common with living organisms, and, living or not, their study has come to be a major aspect of modern biology. Perhaps the definition of life needs to be expanded to include noncellular entities that possess nucleic acid genes, in which case a separate kingdom could be established for the viruses.

[3] The four-kingdom system, with recognition of a protistan level within each of the three kingdoms of higher organisms (System 6 of Table 20.2), was proposed in 1974 by Gordon F. Leedale of the University of Leeds. The temptation to use it in this book was strong, but it seemed unwise to adopt in an introductory text a system that had not yet established itself among professional biologists.

SUGGESTED READING

BOGORAD, L., 1975. "Evolution of organelles and eukaryotic genomes," *Science*, vol. 188, pp. 891–898.   *An alternative to the endosymbiont hypothesis of the origin of eucaryotic cells.*

DICKERSON, R. E., 1978. "Chemical evolution and the origin of life," *Scientific American*, September. (Offprint 1401)

GLAESSNER, M. F., 1961. "Pre-Cambrian animals," *Scientific American*, March. (Offprint 837)   *Some of the fascinating Precambrian invertebrate fossils found in Australia.*

MARGULIS, L., 1971. "Symbiosis and evolution," *Scientific American*, August. (Offprint 1230)   *Clear exposition of the endosymbiont hypothesis of the origin of eucaryotic cells by one of its principal proponents.*

MILLER, S. L., and L. E. ORGEL, 1974. *The origins of life on earth.* Prentice-Hall, Englewood Cliffs, N.J.   *Easy-to-read introduction to the subject. Miller's experiments of 1953 initiated the modern era of research on the beginnings of life.*

OPARIN, A. I., 1969. *Genesis and evolutionary development of life.* Academic Press, New York.   *Excellent summary by one of the founding fathers of this field of biology.*

SCHOPF, J. W., 1978. "The evolution of the earliest cells," *Scientific American*, September. (Offprint 1402)   *The fossil evidence for the appearance of cellular life on earth; the special metabolic characteristics of bacteria and blue-green algae that enabled them to prosper under conditions that shut out most higher forms of life.*

UZZEL, T., and C. SPOLSKY, 1974. "Mitochondria and plastids as endosymbionts: A revival of special creation?" *American Scientist*, vol. 62, pp. 334–343.   *A critique of the endosymbiont hypothesis, and a proposal for an alternative model of the origin of eucaryotic cells.*

WHITTAKER, R. H., 1969. "New concepts of kingdoms of organisms," *Science*, vol. 163, pp. 150–160.   *The five-kingdom system proposed and explained.*

WHITTAKER, R. H., and L. MARGULIS, 1978. "Protist classification and the kingdoms of organisms," *Biosystems*, vol. 10, pp. 3–18.   *Alternative ways of applying the five-kingdom system.*

Chapter 21

VIRUSES AND MONERA

Viruses are not ordinarily given a place in formal classifications of living organisms. Yet biologists and laymen alike tend to think of them as living, using such expressions as "live virus vaccine" and "killed virus vaccine"—curious phrases to apply to nonliving things. The ambivalence is excusable. On the one hand, viruses lack all metabolic machinery and cannot reproduce in the absence of a host. On the other, they possess nucleic acid genes that encode sufficient information for the production of new viruses with the same characteristics as those that supply the templates; and reproduction with gene-controlled heredity is one of the most basic attributes of life.

Whether viruses are classed as living, as nonliving, or as something in between, it must be granted that investigation of viruses is intimately bound to the study of procaryotic organisms, especially bacteria. Not only do viruses and procaryotes have similar genetic material, but their life cycles are intertwined. We shall therefore consider both these groups in the present chapter.

VIRUSES

The discovery of viruses By the latter part of the nineteenth century the idea had become firmly established that many diseases are caused by microorganisms. Pioneer bacteriologists such as Louis Pasteur and Robert Koch had isolated the pathogens for a number of diseases that afflict human beings and their domestic animals. But for some diseases, notably smallpox, biologists, try as they might, could find no causal microorganism. As early as 1796 it had been known that smallpox could be induced in a healthy person by something in the pus from a smallpox victim, and Edward Jenner had demonstrated that a person vaccinated with material from cowpox lesions developed an immunity to smallpox. Yet no bacterial agent could be found.

A crucial experiment was performed in 1892 by a Russian biologist, Dmitri Iwanowsky, who was studying a disease of tobacco plants called tobacco mosaic. The leaves of plants with this disease become mottled and wrinkled. If juice is extracted from an infected plant and rubbed on the leaves of a healthy one, the latter soon develops tobacco mosaic disease. If, however, the juice is heated nearly to boiling before it is rubbed on the healthy leaves, no disease develops. Concluding that the disease must be caused by bacteria in the plant juice, Iwanowsky passed juice from an infected tobacco plant through a very fine porcelain filter in order to remove the bacteria; he then rubbed the filtered juice on the leaves of healthy plants. Contrary to his expectation, the plants developed mosaic disease. What could the explanation be?

Iwanowsky suggested two possibilities. Either bacteria in the infected plants secrete toxins, and it is these rather than the bacterial cells themselves that are present in infectious juice. Or the bacteria that cause this disease are much smaller than other known bacteria and can pass unharmed through a fine porcelain filter. When it was later demonstrated that the infectious material in filtered juice could reproduce in a new host, Iwanowsky abandoned his first explanation in favor of the second—that some type of extremely small bacterium was the causal agent of the disease. During the next several decades many other diseases of both plants and animals were found to be caused by infectious agents so small that they could pass through porcelain filters and could not be seen with even the best light microscopes. These microbial agents of disease came to be called filterable viruses, or simply viruses. They were still assumed to be very small bacteria.

There were, however, a few hints that viruses might be something quite different from bacteria. First, all attempts to culture them on media customarily used for bacteria failed. Second, the virus material, unlike bacteria, could be precipitated from an alcoholic suspension without losing its infectious power. But not until 1935 was it conclusively demonstrated that viruses and bacteria are two very different things. In that year W. M. Stanley of the Rockefeller Institute isolated and crystallized tobacco mosaic virus. If the crystals were injected into tobacco plants, they again became active, multiplied, and caused disease symptoms in the plants. That viruses could be crystallized showed that they were not cells but must be much simpler chemical entities.

The structure of viruses We saw in Chapter 15 that a free virus particle (virion) consists of a protein coat and a nucleic acid core. The protein coat, or *capsid,* may be complicated, as in the bacteriophage of Fig. 15.1 (p. 628), with its tail and long leglike fibers, or it may be a simple polyhedron or rod (actually a helix of protein molecules) (Fig. 21.1). Many animal viruses, some plant viruses, and a very few bacteriophages have a membranous *envelope* surrounding the capsid; sometimes this envelope is derived from the plasma membrane of the host cell in which the virion was produced, and sometimes it is synthesized in the host cell's cytoplasm, but in either case it usually

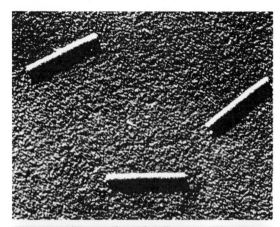

21.1 Electron micrographs of two types of viruses
Top: Tobacco mosaic virus, a rod-shaped virus. Bottom: *Tipula* iridescent virus, a polyhedral virus that attacks the larvae of a species of crane fly. [Courtesy Virus Laboratory, University of California, Berkeley.]

contains some virus-specific proteins. In addition to the proteins of the capsid and envelope, some viruses possess a limited number of enzymes (though never any multienzyme systems).

The nucleic acid is usually a single molecule[1] consisting of as few as 3,500 nucleotides[2] or as many as 600,000, depending on the kind of virus. If 1,000 nucleotides is taken as a reasonable estimate of the average length of a gene, then the total number of genes in a virus particle ranges from fewer than five to several hundred.

Viruses show a great diversity in the form of their nucleic acid. Thus many have double-stranded DNA, but some have single-stranded DNA. In either case the DNA may be linear or circular. Many other viruses differ fundamentally from all cellular organisms in having RNA genes. In most instances the RNA is single-stranded, but there are some RNA viruses in which it is double-stranded (this is the first time we have encountered double-stranded RNA).

The reproduction of viruses Because viruses lack multienzyme systems and cannot generate ATP, and because they lack raw materials for synthesis, they cannot reproduce themselves in the same sense as true living organisms. It is the host cell, not the virus, that manufactures new virus particles when the old virus provides the instructions. Hence viruses cannot be cultured on artificial media; they require living host cells. Pharmaceutical companies and research laboratories often grow them in fertilized chicken eggs or in cells in tissue cultures.

Before examining the diverse reproductive adaptations of viruses, let us briefly review our earlier discussion of the reproduction of bacteriophage viruses (pp. 628, 662). You will recall that a free phage particle becomes attached by the tip of its tail to the wall of a bacterial cell and that the phage nucleic acid is injected into the host while the protein coat remains outside. The energy for this injection comes from hydrolysis of about 140 ATP molecules bound (together with 140 Ca^{++} ions) in the tail of the phage. Once inside the bacterial cell, the phage DNA provides genetic information for synthesis of new viral DNA and protein. Among the viral proteins synthesized are not only structural proteins for new viral capsids, but also enzymes that aid in the synthesis and processing of viral components.

Eventually, after the new viral nucleic acid and proteins have been manufactured, they are assembled into virions and released by lysis of the bacterial cell through the agency of a phage-induced enzyme that attacks the wall of the bacterial cell. This series of events is known as the lytic cycle (see Fig. 15.27, p. 663).

In Chapter 15 we saw that at times viral DNA does not immediately take control of the host cell's metabolic machinery and put it to work making new virus particles, but is instead integrated into the bacterial chromosome and reproduced with the chromosome for an indefinite number of generations; it is then known as provirus. This so-called lysogenic cycle (see Fig. 15.27) occurs when a viral gene induces production by the host of a repressor substance that blocks expression of the viral genes governing reproduction.

---

[1] A few viruses (e.g. influenza virus) have segmented genomes; i.e. their nucleic acid is in several pieces (six to eight for influenza virus).

[2] Some so-called defective viruses have even fewer nucleotides.

As long as repression continues, the viral nucleic acid may remain integrated in the host's chromosome. The repressor substance also confers on the host cell immunity to virulent infection by other viruses of the same or similar types. Indeed, a standard test for finding out whether cells are carrying provirus of a given type is to attempt to infect them with the virus in question.

As this discussion makes clear, viruses can exist in three different states: as free infectious virions; as vegetative virus, i.e. as viral nucleic acid directing synthesis of new viral components in a host cell; and as provirus, i.e. as viral nucleic acid integrated with host DNA and replicated in synchrony with it.

Not all viruses inject their nucleic acid into their host cells in the manner of bacteriophages. Often, in fact, the entire virion enters the cell. Animal viruses of this type first become attached to the cell membrane at special glycoprotein adsorption sites for which they have a high affinity; much of the host specificity of such viruses comes from their differing affinities for the various adsorption sites. Most plant viruses depend on insect vectors to inject them through the thick cell walls into the host-cell cytoplasm. Once a virion has entered the cytoplasm of the host cell, its nucleic acid is promptly freed and the capsid is broken down by cellular enzymes.

If the nucleic acid of a plant or animal virus is DNA, it enters the nucleus of the host cell and serves as a template for the synthesis of both mRNA and new viral DNA.[3] But what if the viral nucleic acid is RNA? How can it provide the cell with the necessary instructions for making mRNA or more viral RNA? In general, there are two possibilities. Either the virus brings with it into the host cell special enzymes called **RNA transcriptases,** capable of using an RNA molecule as a template for synthesis of new RNA; or else part of the viral RNA first functions as mRNA on the host cell's ribosomes, providing instructions for synthesis of RNA transcriptase by the cell. In either case, then, the secret of replication by RNA viruses in the cytoplasm of their host cells (unlike DNA viruses, they do not enter the nucleus) is RNA transcriptase, which enables sequences of RNA nucleotides to play the role of genes. This is the first time we have encountered RNA $\rightarrow$ RNA synthesis; cells not infected with viruses never possess RNA transcriptase, as far as is known.

Some RNA viruses have another mechanism of replication. A group of RNA viruses called retroviruses do not carry out RNA $\rightarrow$ RNA transcription. Instead, their RNA is transcribed into DNA in a reaction catalyzed by an enzyme called **reverse transcriptase,** which the viruses bring with them into the host cells. The newly formed DNA then becomes integrated as provirus into the host's chromosomes, where it may remain for an extended period, even being passed to the host's descendants, generation after generation. (Since the retrovirus group includes some important cancer-causing viruses, transmission of the DNA with the host's genes may be one of the reasons why some kinds of cancers seem to run in families.) Sometimes the provirus remains replicatively silent (i.e. no new virions are produced), but in other instances the proviral DNA is transcribed to produce both new viral RNA

[3] The pox viruses, which have double-stranded DNA as their genetic material, are an exception. Their DNA does not enter the nucleus; instead, it functions entirely in the cytoplasm, and the viruses must bring their own DNA and RNA polymerases into the cells.

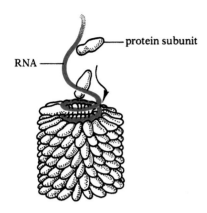

21.2   Part of a TMV virion, showing its structure
At the center of the cylindrical virion is a long RNA molecule. Protein subunits are packed in helical array around the RNA; they are added in sequence as a virion is assembled.

and mRNA coding for viral proteins; when new virions are assembled, they can be released from the cell by a budding process, which, unlike lysis of host cells by bacteriophage, does not kill the cell. As far as is known, RNA → DNA transcription is unique to the life cycle of the retroviruses; cells uninfected by retrovirus do not contain reverse transcriptase.

Let us now turn to the assembly of new virions in infected host cells. The traditional view has been that, once new viral components have been synthesized by the host cell, a new virion is produced by self-assembly of these components. This process is exemplified by tobacco mosaic virus (TMV), where 2,130 identical protein subunits have traditionally been envisioned as spontaneously forming a helix, with a long RNA molecule inside (Fig. 21.2).[4] Even the most complex bacteriophages, with capsids composed of more than 50 kinds of proteins, have been thought to form by self-assembly. But a growing body of evidence now suggests that the capsids of only a few of the simplest viruses (including TMV, but not the DNA phages) are formed entirely through self-assembly. Though some self-assembly seems to be involved, in most cases there is also some enzymatic processing of the proteins (e.g. exposure of new binding sites by cleavage of large polypeptides into smaller ones after they have been brought into position in the growing capsid structure).

The new virions, once assembled, are not always released by lysis of the host cell. Like the retroviruses already mentioned, many viruses are released by extrusion, a process something like budding whereby the virus becomes enveloped in a small piece of cell membrane (Fig. 21.3).

The origin of viruses   Viruses are not cells; even the most complex viruses, which carry with them an array of enzymes, lack the metabolic machinery for the basic cellular function of energy generation, and they never have the ribosomes required for protein synthesis. Therefore they fail to meet an essential requisite that, according to the definition in Chapter 3, must be met by living things. If we knew more about the origins of viruses it would be easier to decide whether a virus should be regarded as a type of organism or as a particularly interesting kind of nonliving material object. Of the various hypotheses of viral origin that have been proposed, the three outlined below have received serious consideration.

The first hypothesis suggests that viruses are organisms that have reached the extreme of evolutionary specialization for parasitism. Its proponents point out that loss of structures is commonly seen in internal parasites, and that intracellular parasites might conceivably lose everything but their nucleus. A virus particle resembles a cellular nucleus; hence they conclude that viruses arose from cellular ancestors by a process of gradual loss of all other cellular components. If this hypothesis is correct, then viruses should perhaps be regarded as degenerate cells and classified as living organisms.

The second hypothesis suggests that the ancestors of modern viruses were free-living noncellular predecessors of cellular organisms. When the organic nutrients of the primordial seas disappeared, those of their descendants that remained noncellular survived by becoming parasites on the cellular organ-

[4]Recent evidence suggests that TMV capsids may not, in fact, normally be assembled in the helical fashion traditionally depicted. In general, they are probably built up as stacks of two-layered discs, which later undergo a conformational change that generates the characteristic helix of the completed capsid.

isms that had arisen by that time. According to this view, modern viruses are representatives of an early "nearly living" stage in the origin of life.

The third hypothesis suggests that viruses are neither primitive nor specialized organisms, but fragments of genetic material derived from cellular organisms. They might originally have existed as DNA-containing cellular organelles, but that seems unlikely because no organelle bounded by anything resembling a capsid (which is unlike a membrane) is known. Alternatively, they might have begun as bare nucleic acid similar to episomes (which can exist in both free-cytoplasmic and integrated-nuclear states remarkably similar to the corresponding states of viruses), but later evolved the capacity of causing their host cells to synthesize a protein shell in which replicates of the nucleic acid could be enclosed when moving from cell to cell. If either variant of this hypothesis is correct, some viruses may have arisen as fragments of bacterial DNA, others as fragments of plant nucleic acid, and still others as fragments of the genetic material of higher animals. The host specificity of viruses might be a reflection of their origin; a given type of virus might be able to parasitize only species fairly closely related to the one from which the virus was originally derived.

A recent textbook on virology[5] characterizes viruses as follows: "Ultimately, the definition of a certain element—gene, plasmid, or virus—will depend on the duration of its joint evolutionary history with other components of the genome. . . . A virus is essentially a part of a cell. We observe and recognize as viruses those parts independent enough to pass from cell to cell."

Viral disease Among the many human diseases caused by viruses are chicken pox, mumps, measles, smallpox, yellow fever, rabies, influenza, viral pneumonia, the common cold, poliomyelitis (infantile paralysis), fever blisters, several types of encephalitis, and infectious hepatitis. We have already discussed the association of viruses with some kinds of cancers (pp. 685–687); in some cases cells may contain latent cancer viruses, acquired early in life, without actually becoming malignant until induced by an irritant, such as certain carcinogenic chemicals, tobacco smoke, frequent abrasion, or radioactivity.

Unfortunately most viral infections do not respond to treatment with the sulfur drugs or antibiotics that have been so effective against bacterial diseases. But some of the most pernicious virus diseases, notably smallpox and polio, are preventable by means of vaccines (see p. 931).

Immune responses, though important in preventing reinfection, ordinarily appear too late to account for recovery from viral diseases. The evidence suggests that at least one important factor in recovery is a protein called **interferon,** which is produced by host cells in response to invading viruses. The interferon cannot save those cells, but when it is released into the medium and encounters uninfected cells, it interacts with a receptor site on their membranes and confers on them a resistance to viral infection, probably by inducing production of another protein that blocks translation of the mRNA transcribed from the viral nucleic acid. In other words, the interferon acts as a messenger from infected cells to uninfected cells, telling them to mobilize their defenses against viral infection.

[5] S. E. Luria, J. E. Darnell, D. Baltimore, and A. Campbell, *General virology* (3rd ed.; New York: Wiley, 1978), p. 490.

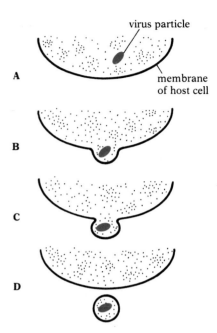

21.3 Extrusion of a virus from the host cell
In some cases a newly assembled virus particle becomes closely associated with the membrane of the host cell (A). The membrane then forms an evagination into which the particle moves (B, C). Finally the evaginated membrane is pinched off as an envelope around the free virus particle (D).

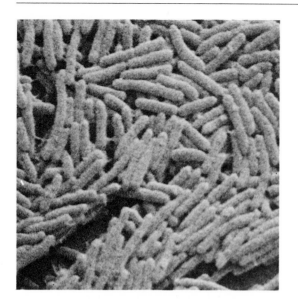

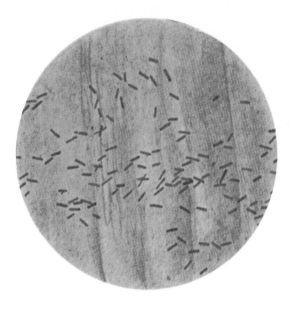

21.4 *Escherichia coli,* a rod-shaped bacterium
Top: Electron micrograph of a colony
(× 20,350).   Bottom: Stained cells as they appear
under a high-power light microscope.   [Courtesy David
Scharf, Peter Arnold, Inc. (top); Center for Disease
Control, Atlanta, Ga. (bottom).]

Shortly after its discovery in 1957, it was suggested that interferon might provide effective chemotherapy for viral diseases. This hope has not yet been realized.

Even though interferon induced by one type of virus blocks infection by virtually all other types in the host that produces it, it was long thought to be ineffective in other host species—the corollary being that interferons produced by other animals would be of no value in treating human beings. That assessment is now known to be wrong; there is, in fact, some interspecific reactivity. But the protein is produced in such minute quantities by living cells that in 1978 the cost of obtaining even as little as a millionth of an ounce was $1,500. Nevertheless a research campaign is now under way to try to learn more about the potential value of interferon as an antiviral (and perhaps anticancer) drug. Because the quantities required for combating viral diseases are much greater than can be obtained from natural sources, direct administration of interferon—assuming it proves useful in treatment—is not practicable until industrial synthesis of the protein becomes feasible.

VIROIDS

Several plant diseases once thought to be viral are now known to be caused by a newly discovered infectious agent—the viroids. Viroids are probably also responsible for at least one animal disease (scrapie in sheep). As they become better known, they are likely to be implicated in other diseases too, possibly including human ones.

The viroids consist entirely of RNA; they lack a protein coat of any sort. The RNA itself is much smaller than the nucleic acid genome of any known virus—only between 300 and 400 nucleotides long, which is barely enough to code for a single protein. This very limited length makes it likely that the viroid RNA exerts some kind of regulatory action—that, instead of functioning as mRNA, it probably stimulates production of more RNA like itself along one of the host cell's genes. Evidence tending to support this hypothesis is that production of new viroids takes place in the host's nucleus rather than the cytoplasm, where most RNA viruses are replicated.

With some misgivings, we have here treated the viroids as a separate group of semiliving entities coequal with the viruses. So little is known about them that their place in the biological scheme of things is still largely an open question; all that seems certain is that they are not viruses, at least not as the term is now understood. A reasonable classification must await the results of further research.

MONERA

The kingdom Monera includes three divisions: the Schizomycetes (bacteria), the Cyanophyta (blue-green algae), and the recently discovered Prochlorophyta. The cells of all three groups are procaryotic; i.e. they lack a nuclear membrane, mitochondria, an endoplasmic reticulum, Golgi apparatus, and lysosomes.

SCHIZOMYCETES

The discoverer of bacteria was Antoni van Leeuwenhoek (1632–1723), a Dutchman who was both an extraordinarily skilled maker of microscopes

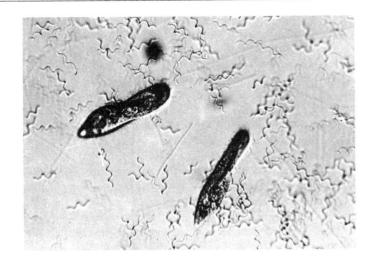

21.5 *Spirillum volutans,* a helically coiled bacterium
The two paramecians (which are Protozoa) are
included for size comparison.   [Courtesy
Turtox/Cambosco, Macmillan Science Co., Inc.]

and a careful and inquisitive observer. Though he was a merchant with little
formal education, his curiosity about the natural world led him to study an
enormous variety of microscopic objects—among them protozoans, algae,
yeasts, and bacteria.

Early studies of bacteria were largely of species belonging to a group now
called the eubacteria ("true" bacteria). All of these are rather similar in their
basic characteristics. But later other groups of organisms with quite different properties also came to be regarded as bacteria. Thus the modern
Schizomycetes constitute an assemblage of diverse forms probably not very
closely related to one another. The bacteria could easily be separated into
seven or more divisions that would be as distinct from one another as each is
from the blue-green algae.[6]

Bacterial cells   Most bacteria are very tiny, far smaller than the individual
cells in the body of a multicellular plant or animal. In fact, some bacteria
(e.g. the Rickettsiae and the Chlamydiae) are as small as some of the largest
viruses. However, even the smallest bacteria are fundamentally different
from viruses in that they are cellular. They always contain both RNA and
DNA, whereas viruses contain one or the other but never both; they have
ribosomes, whereas viruses do not; they always possess integrated multienzyme systems, whereas viruses carry, at most, only an assortment of individual enzymes; they can generate ATP and use it in the synthesis of many
other organic compounds, whereas viruses cannot; they provide both the
raw material and all the metabolic machinery for their own reproduction,
whereas viruses do not. No entity truly transitional between viruses and
cellular organisms is known to exist.

The cells of most bacteria have one of three fundamental shapes: spherical or ovoid, cylindrical or rod-shaped (Fig. 21.4), or helically coiled (Fig.
21.5). Spherical bacteria are called ***cocci*** (singular, coccus); rod-shaped ones
are called ***bacilli;*** and helically coiled ones are called ***spirilla.*** There are,
however, bacteria with still other shapes; the spirochetes, the group to which
the causative agent of syphilis belongs, are an example (Fig. 21.6).

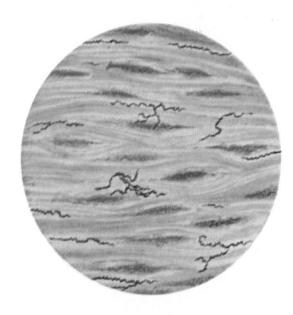

21.6 *Treponema pallidum* in human tissues
This organism causes syphilis in humans.   [Courtesy
Center for Disease Control, Atlanta, Ga.]

[6] Bergey's *Manual of determinative bacteriology* (8th ed., 1974) lists 19 groups of bacteria.

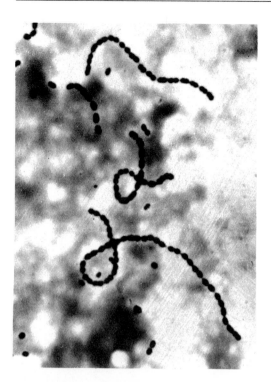

21.7 *Streptococcus lactis,* a bacterium common in milk

All streptococci are spherical (coccal) bacteria normally grouped in chainlike clusters.   [Courtesy Turtox/Cambosco, Macmillan Science Co., Inc.]

When cell division takes place, the daughter cells of some species remain attached and form characteristic aggregates. Thus cells of the bacterium that causes pneumonia are often found in pairs (diplococci). The cells of some spherical species form long chainlike aggregates (streptococci) (Fig. 21.7), while others form grapelike clusters (staphylococci) (Fig. 21.8). Each of the cells in a diplococcal, streptococcal, or staphylococcal aggregate is an independent organism, but some bacteria form either coenocytic or multicellular filaments. A basic difference between a chain of independent cells and a multicellular filament is that adjacent cells of a filament have a common cell wall. Many Actinomycetes, among other filamentous forms, strikingly resemble molds (which are fungi), but this is a case of evolutionary convergence and not an indication of close relationship, because the cells of Actinomycetes, like those of other bacteria, are procaryotic whereas those of fungi are eucaryotic (Fig. 21.9).

Like plant cells, most bacterial cells are enclosed in a cell wall, which protects the cell both from physical damage and from osmotic disruption. But the walls of bacteria and blue-green algae differ from those of plants in important aspects of their composition. The walls of eucaryotic cells derive their tensile strength largely from cellulose and related compounds (or from chitin in fungi), whereas those of procaryotic cells derive theirs from murein, a huge polymer composed of polysaccharide chains covalently cross-linked by short chains of amino acids; muramic acid, one of the principal polysaccharide constituents of murein, never occurs in the walls of eucaryotic cells. This important difference between the chemical composition of procaryotic and eucaryotic cells is the basis for the selective activity of some drugs, such as penicillin. Nontoxic to plants and animals (or to resting bacterial cells), penicillin is toxic to growing bacteria, because it inhibits formation of murein, and thus interferes with bacterial multiplication.

Most bacteria are hypertonic relative to the fluid medium in which they live; hence they would swell and burst if they did not have well-developed walls. However, many species can be cultured in the laboratory on a medium with an elevated osmotic pressure. When treated with penicillin under such conditions, the bacteria survive, growing as so-called L forms, which have very poorly developed walls. Their shape is usually very different from normal, as would be expected.

The L forms are human artifacts, but there are some naturally occurring bacteria that lack walls and can live only as parasites of plants and animals, i.e. in environments where their osmotic relations are such that the absence of a mechanically strong cell wall is not lethal. These are the mycoplasmas,[7]

[7] The mycoplasmas were formerly known as pleuropneumonialike organisms (PPLO).

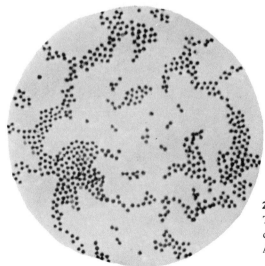

21.8 *Staphylococcus*
The round cells occur in irregularly shaped clusters.   [Courtesy Center for Disease Control, Atlanta, Ga.]

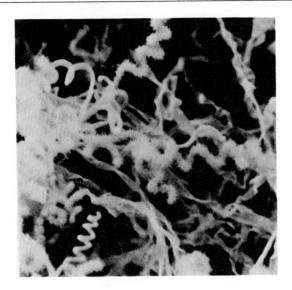

which are the smallest known cellular organisms. Not only are the cells at
the lower limit of resolution of the light microscope, but they also have less
than half as much DNA as most other bacterial cells. The informational
content of the DNA is probably near the lower limit of what is necessary to
code for the essential metabolic machinery of a cell. Some of the mycoplasmas cause severe human diseases.

Differences in the relative amounts of certain components in the walls of
different types of bacteria make the cells show characteristic reactions to a
variety of stains. Since there are few visible morphological characters that
can be used in identifying bacteria, diagnostic staining is an important
laboratory tool.

Many bacterial cells secrete polysaccharide mucoid materials that accumulate on the outer surface of the cell wall and form a *capsule.* The capsule
makes the cell more resistant to the defenses of host organisms; hence
encapsulated strains of a given bacterial species are more likely to cause
disease than unencapsulated strains.

Some of the eubacteria (mostly rod-shaped ones) can form special resting
cells called *endospores,* which enable them to withstand conditions that
would quickly kill the normal active cell. Each small endospore develops
inside a vegetative cell and contains DNA plus a limited amount of other
essential materials from that cell (Fig. 21.10). It is enclosed in an almost
indestructible spore coat. Once the endospore has fully developed, the remainder of the vegetative cell in which it formed may disintegrate. Because
of their very low water content and refractile coats, spores of many species
can survive an hour or more of boiling or an hour in a hot oven. They can be
frozen for decades or perhaps for centuries without harm. They can survive
long periods of drying. And they can even withstand treatment with strong
disinfectant solutions. When conditions again become favorable, the spores
may germinate, giving rise to normal vegetative cells that resume growing
and dividing. Fortunately few disease-causing bacteria can form endospores.

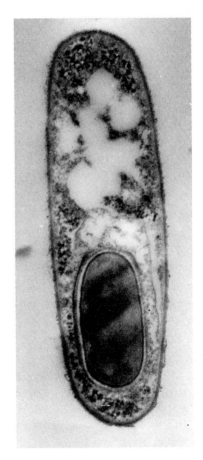

21.10   Electron micrograph of a sporulating bacillus
The spore is the dark oval at the lower end of the cell.
The developing spore coat is clearly visible. The white
areas in the upper end of the cell are not vacuoles, but
areas filled with fatty material. × 33,540.   [Courtesy G.
B. Chapman, *J. Bacteriol.*, vol. 71, 1956.]

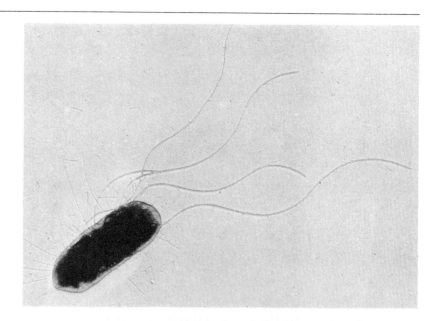

21.11   Electron micrograph of the bacterium _E. coli_, showing its long flagella
Though bacterial flagella are organelles of locomotion, they differ from the flagella of eucaryotic cells in lacking internal microtubules and, therefore, in their mechanism of movement. × 18,900.   [Courtesy Ginny Fonte, University of Colorado.]

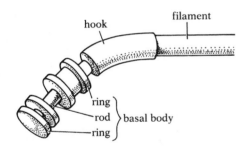

21.12   The basal portion of a bacterial flagellum
The electron micrograph at right shows the base of a flagellum from _E. coli_ (× 508,000). The interpretive drawing elucidates the complex system of rings, rod, hook, and filament. Unlike the flagella of eucaryotes, which beat back and forth, bacterial flagella rotate. Their complicated structure provides the necessary anchor, joints, and couplings for such rotational motion.   [Micrograph: Courtesy M. L. DePamphilis and Julius Adler, _J. Bacteriol._, vol. 105, 1971.]

Many bacteria are motile; i.e. they can move about actively. In some cases the motion is produced by the beating action of flagella (Fig. 21.11). Bacterial flagella are structurally very different from the flagella of eucaryotic cells. They are not enclosed within the cytoplasmic membrane; they do not contain the nine peripheral and two central microtubules found in all flagella of eucaryotic cells; and they do not contain the protein tubulin. They arise from basal granules, but these granules are much smaller than the basal bodies of eucaryotic cells and are probably not homologous with them. A bacterial flagellum has approximately the same diameter as one of the tubules from a eucaryotic flagellum. Its motion is not the whiplike beating characteristic of eucaryotic flagella, but rather a rotary motion made possible by an amazingly complex attachment, which is unlike anything seen elsewhere in the living world (Fig. 21.12).

Some bacteria (the myxobacteria) that lack flagella exhibit a peculiar gliding movement that does not involve any visible locomotor organelles; the mechanism of this movement has not yet been discovered.

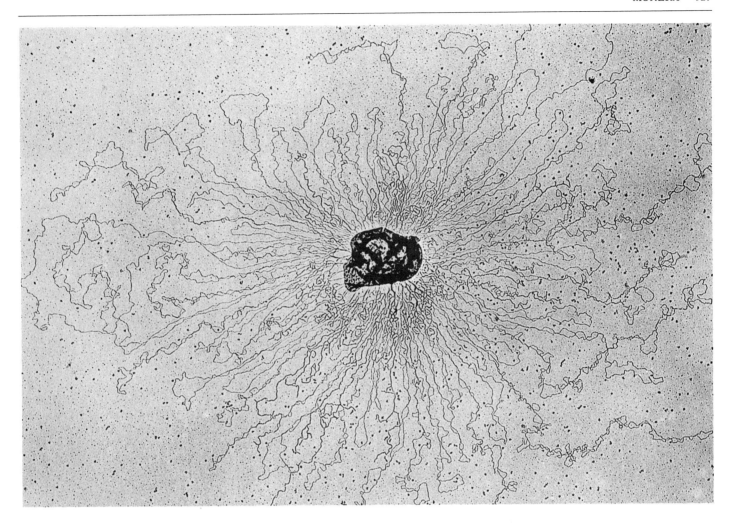

21.13   Electron micrograph of the DNA of a bacterial cell *(E. coli)* spread onto a surface
All the many coils of DNA seen here are parts of a single chromosome. It is hard to imagine how so very much DNA could have been packed into so small a cell. × 11,500. [From R. Kavenoff and O. A. Ryder, *Chromosoma*, vol. 55, 1976, by courtesy of the authors and L. P. Velten.]

Bacterial reproduction   As we have already seen, bacteria lack membrane-bounded nuclei, but electron microscopy reveals the presence of a nuclear area, called a nucleoid, in which genes are arranged in sequence along a single circular chromosome composed of DNA with only a thin coating of loosely bonded protein (see p. 559). The single chromosome may be surprisingly long, considering that it is packed into so tiny a cell (Fig. 21.13).

Most bacteria reproduce by *binary fission,* a type of cell division in which two equal daughter cells with characteristics essentially like those of the parent cell are produced without mitosis (see Fig. 13.4, p. 560). It has been established that when a bacterial cell undergoes fission, each daughter cell receives a full set of genes, i.e. a complete chromosome. The details of nuclear division in procaryotic cells have been discovered only recently (see Fig. 13.3). The DNA, which is attached to the cell membrane, is replicated, and the two chromosomes move apart into separate nuclear areas long before division of the cytoplasm occurs. When the plasma membrane grows inward, it simply partitions the binucleate parent cell into two daughter cells each of which already has its own nucleoid. In many bacteria a much-convoluted inward extension of the plasma membrane, called a mesosome, is seen at or near the site of division. Its functional role in the division process is not yet well understood.

Bacteria commonly have an enormous reproductive potential. Many species may divide as often as once every 20 minutes under favorable conditions. If all the descendants of a cell of this type survived and divided every 20 minutes, the single initial cell would have about 500,000 descendants at the end of 6 hours, and by the end of 24 hours the total weight of its descendants would be nearly 2,000,000 kg. Although increases of this magnitude do not actually occur, the real increases are frequently huge, which helps explain the rapidity with which food sometimes spoils or a disease develops.

Although the reproductive process itself is asexual, genetic recombination does occur occasionally, at least in some bacteria. As we saw in Chapter 15, three mechanisms of recombination are known: conjugation, in which part of a chromosome is transferred from a donor cell to a recipient; transformation, in which a living cell picks up fragments of DNA released into the medium from dead cells; and transduction, in which fragments of DNA are carried from one cell to another by viruses. When a normally haploid bacterial cell receives extra DNA by one of these procedures, it becomes partly diploid (usually only partly, because it is very rare for a cell to receive an entire extra chromosome). The diploidy is only temporary, however; haploidy is soon re-established by elimination of all genes that do not become incorporated into the new recombinant chromosome.

Bacterial nutrition Most bacteria are heterotrophic, being either saprophytes or parasites. Like animals, the majority of these bacteria are aerobic; i.e. they cannot live without molecular oxygen, which they use in the respiratory breakdown of carbohydrates and other food materials to carbon dioxide and water. Aerobic respiration is carried out with the help of an electron-transport system, built (since bacteria lack mitochondria) into the inner surface of the cell membrane or its invaginations.

To some bacteria that obtain all their energy by fermentation, however, oxygen is lethal. Such bacteria are called *obligate anaerobes. Clostridium botulinum,* the causal agent of the most dangerous kind of food poisoning, botulism, is an example of an obligate anaerobe; it grows well in tightly closed food containers that were not properly sterilized before being filled.

Other bacteria, called *facultative anaerobes,* can live either in the presence or in the absence of molecular oxygen. Some of the facultative anaerobes are simply indifferent to O_2, obtaining all their energy from fermentation whether O_2 is present or not; others obtain their energy by fermentation when O_2 is not available, but carry out respiration (via the Krebs cycle) when O_2 is present, and may grow faster under these conditions. Respiration is, of course, a much more efficient energy-yielding process than fermentation. Certain types of facultative anaerobes do not rely on fermentation when O_2 is absent; they continue to carry out complete respiration by using substitute inorganic substances (such as nitrates, sulfates, or carbonates) as ultimate electron acceptors in place of O_2.

Besides lactic and alcoholic fermentations—the most common fermentative processes in living organisms—at least ten other types of fermentation occur in different groups of bacteria. The products include acetic acid, butylene glycol, butyric acid, and propionic acid. All fermentations are alike,

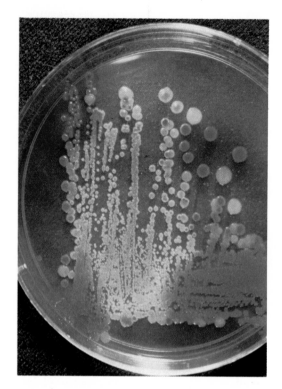

21.14 Colonies of two kinds of bacteria growing on the same culture medium, with different effects on the medium

Colonies of *E. coli* on Endo agar carry out fermentation, making the medium red, whereas colonies of *Salmonella* do not carry out fermentation and the medium is colorless.   [Courtesy Elliot Scientific Corp.]

however, in that they are energy-yielding biological oxidation-reduction sequences in which organic molecules serve as the final electron acceptors. ATP is generated in fermentations by phosphorylations on the substrate level.

Bacteria differ considerably in the sorts of molecules they can use as energy sources and in the specific amino acids and vitamins they require. These differences provide valuable diagnostic characters for workers attempting to identify unknown bacteria. Samples of the organisms to be identified are placed on a variety of nutrient media and cultured at standard temperatures. By determining on which of the media the organisms will grow and on which they will not, and, when they grow, by comparing the color, texture, and other characteristics of the colony they produce with data established for known species, it is often possible to assign the unknown organisms to the proper group or even to the proper species (Figs. 21.14, 21.15).

Most bacteria, as we said, are heterotrophs, but some are either chemosynthetic or photosynthetic autotrophs. The chemosynthetic bacteria oxidize inorganic compounds, such as ammonia, nitrite, sulfur, hydrogen gas, or ferrous iron, and trap the released energy. The mechanisms of some of these oxidations were discussed on pp. 162–163. The bacteria that oxidize ammonia or nitrite are the nitrifying bacteria, important in the nitrogen cycle.

There are two basic kinds of photosynthetic bacteria, which differ in the type of bacteriochlorophyll they possess—the green bacteria and the purple bacteria. Neither group has chlorophyll *a*, the chief light-trapping pigment in higher plants. And unlike higher plants, neither group ever uses water as the ultimate electron donor in photosynthesis; hence neither produces molecular oxygen. Indeed, bacterial photosynthesis is a strictly anaerobic process; it cannot take place in the presence of O_2. Depending on the species of bacteria, the electron (and hydrogen) donor for reduction of NADP is molecular hydrogen, reduced sulfur compounds (such as H_2S), or organic compounds, but the oxidation of these substances is not a light-driven process. In fact, the photosynthetic bacteria possess only Photosystem I (and carry out only cyclic photophosphorylation); they have no Photosystem II.

The pigments and enzymes of the light-trapping process in photosynthetic bacteria are located in chromatophores, organelles composed of vesicles or paired membranous lamellae, which differ in structural details between the purple and the green bacteria. The chromatophores are not contained within any membrane-bounded structure that could be interpreted as a chloroplast. Hence the enzymes involved in the "dark" reactions of photosynthesis (i.e. carbon fixation via the Calvin cycle), which are in the stroma portion of plant chloroplasts, are assumed to be in the general cytoplasm of the bacterial cell.

In 1971 it was discovered that certain bacteria (genera *Halobacterium* and *Halococcus*) that live in extremely saline environments (salt lakes, brines) carry out a unique type of photosynthesis that does not use chlorophyll. When living under anaerobic conditions, these curious organisms synthesize a carotenoid pigment very similar to the rhodopsin of the vertebrate retina. This pigment, located in the plasma membrane, to which it imparts a purplish color, is bleached by light in a process that releases protons from the cell; the proton gradient thus established across the plasma membrane

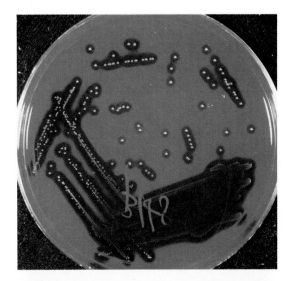

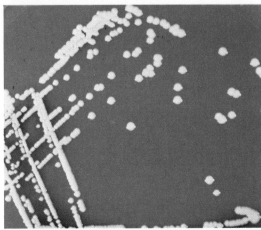

21.15 Different appearances of colonies of two kinds of bacteria growing on blood agar
Top: Colonies of *Streptococcus* are surrounded by black zones of hemolysis (breakdown of red blood cells).   Bottom: Colonies of *Sarcina lutea* have no surrounding zones of hemolysis.   [Courtesy Elliot Scientific Corp.]

provides energy for ATP synthesis by the chemiosmotic process (pp. 178–179).

The astonishing diversity of energy-yielding metabolic pathways seen in the bacteria—greater by far than in all other forms of life combined—is interpreted by many investigators as reflecting the variety of evolutionary "experiments" with metabolism during the long period when procaryotes were the only living things on the earth.

Bacteria as agents of disease Perhaps the bacteria best known to most people are the ones that cause diseases in human beings, domesticated animals, and cultivated plants. Most of the so-called germs are either bacteria or viruses (though a few are fungi, protozoans, or parasitic worms). The idea that bacteria can cause disease—often called the Germ Theory of Disease—was first developed by Louis Pasteur in the late nineteenth century. Initially scorned, the idea soon gained the support of such prominent scientists and physicians of the day as Joseph Lister, Robert Koch, Thomas Burrell, and Ferdinand Cohn. Lister, an English surgeon, was one of the first to realize the implications of Pasteur's discoveries for surgical procedures. He initiated use of antiseptic techniques in the operating room, using carbolic acid solution as a disinfectant. Koch, a German physician, showed that a bacillus was the cause of anthrax in horses, cows, sheep, and human beings, and he later demonstrated that another bacillus caused tuberculosis in humans. Burrell, an American botanist at the University of Illinois, showed that a plant disease—fire blight of pears—was caused by bacteria. Cohn, a German botanist who published a classic text on bacteria, is often considered the father of modern bacteriology.

In the course of his investigations of anthrax and tuberculosis, Robert Koch formulated the rules of procedure for proving that a particular microorganism is the cause of a particular disease. A slightly modified version of these rules, traditionally called *Koch's postulates,* is still used today:

1. It must be shown that the microorganism in question is always present in diseased hosts.
2. The microorganism must be isolated from the diseased host and grown in pure culture (i.e. in a culture containing only that one species of microorganism).
3. Microorganisms obtained from the pure culture, when injected into a healthy susceptible host, must produce the disease in that host.
4. Microorganisms must be isolated from the experimentally infected host, grown in pure culture, and compared with the microorganisms in the original culture.

The procedures outlined by Koch have been followed by hundreds of bacteriologists, and bacteria have been shown to cause a long list of human diseases, including bubonic plague, cholera, diphtheria, syphilis, gonorrhea, leprosy, scarlet fever, tetanus, tuberculosis, typhoid fever, whooping cough, bacterial pneumonia, bacterial dysentery, meningitis, strep throat, boils, and abscesses. To this list we can now add such diseases as typhus fever and Rocky Mountain spotted fever, which are caused by rickettsias and are spread to humans by the bites of arthropods (e.g. ticks, mites, and fleas), and also such diseases as ornithosis, lymphogranuloma, and trachoma, which are caused by the chlamydias. Both the rickettsias and the chlamydias are obligate intracellular parasites once thought to be intermediate between viruses and bacteria, but now known to be simply very small, specialized bacteria. Equally long lists could be compiled of bacterial diseases of other animals or of plants.

Microorganisms cause disease symptoms in a variety of ways. In some cases their immense numbers simply place such a tremendous material burden on the host's tissues that they interfere with normal function. In other cases the microorganisms actually destroy cells and tissues. In still other cases bacteria produce poisons, called *toxins.* These may be exotoxins, which are poisons released from the living bacterial cell into the host's tissues, as in diphtheria or tetanus, or they may be endotoxins, which are poisons retained in the cells of the bacteria that produce them and only released into the host when the bacteria die and disintegrate. There are even cases, as we mentioned in Chapter 16, where the disease symptoms are not caused directly by the microorganisms, but result from an excessive immune response to the microorganisms by the host's body.

We discussed in Chapter 16 the ways in which the human body resists attacks by pathogenic microorganisms. The first line of defense, once the pathogens have gotten into the body past the protective epidermal tissues, is phagocytic action by certain kinds of white blood cells; the pathogens are engulfed and destroyed. The second line of defense is production of antibodies that react with the antigens of the pathogens. Production of antibodies the first time a host individual is exposed to a particular antigen is a rather slow process, which may take days or even weeks. However, in most cases the immunity thus built up is relatively long-lasting. For example, a person who has once had whooping cough or chicken pox usually remains immune for life to further infection by the pathogens of those diseases. Immunity to some diseases, among them most respiratory infections, does not last for life, but may have a duration of several months.

Modern medicine often takes advantage of the body's antigen-antibody reaction to induce prophylactic immunity—immunity that prevents a first case of disease. The patient is inoculated with either a vaccine or an antiserum. A *vaccine* is material containing antigen from the pathogen. Sometimes the antigen consists of dead microorganisms; sometimes it consists of attenuated microorganisms, i.e. microorganisms that are alive, but have been treated in a manner that weakens them sufficiently to prevent their causing disease; and sometimes it consists either of a small amount of active bacterial toxin (enough to induce formation of antibodies, but not enough to produce disease) or of inactivated toxin, called toxoid, as in tetanus toxoid. Vaccines, whatever the kind, induce active immunity in the patient; i.e. they stimulate the patient to produce his own antibodies. They are therefore rather slow-acting, but their effects are long-lasting. By contrast, inoculation with *antiserum* produces almost immediate immunity, but the immunity lasts only a short time, because an antiserum contains presynthesized antibodies instead of antigens, and hence produces passive rather than active immunity. An antiserum is made by injecting antigen into some other animal, usually a horse, waiting until the animal has produced antibodies specific for that antigen, and then removing blood serum containing the antibodies from the animal.

The immunity that a newborn baby acquires from its mother is an example of a natural passive immunity. The milk produced in the first few days after birth consists of **colostrum,** a thin yellowish fluid that is rich in protein and contains numerous antibodies. These maternal antibodies provide the baby with an immediate short-term protection against certain microorganisms; babies fed exclusively on bottle formulas lack this protection.

Beneficial bacteria   Contrary to the popular impression, beneficial bacteria outnumber harmful ones. In Chapter 19 we mentioned the nitrogen-fixing bacteria and the role of bacteria as organisms of decay—a process that not only prevents the accumulation of dead bodies and metabolic wastes but also converts materials such as the nitrogen of proteins into a form usable by other living things. We have also mentioned the bacteria in the intestine that synthesize vitamins absorbed by the body and that aid in the digestion of certain materials. Anyone who has been given doses of antibiotics massive enough to exterminate his intestinal flora can testify to the ensuing disturbances in normal intestinal activity.

Bacteria are also of great importance in many industrial processes. Manufacturers often find it easier and cheaper to use cultured microorganisms in certain difficult syntheses than to try to perform the syntheses themselves. Among the many substances manufactured commercially by means of bacteria are acetic acid (vinegar), acetone, butanol, lactic acid, and several vitamins. Bacteria are also used in the retting of flax and hemp, a process that decomposes the pectin material holding the cellulose fibers together; the fibers, once freed, may be used in making linen, other textiles, and rope. Commercial preparation of skins for making leather goods often involves use of bacteria, as does the curing of tobacco.

Many branches of the food industry depend on bacteria. You are probably aware of the central role of bacteria in the making of dairy products, such as butter and the various kinds of cheeses; the characteristic flavor of Swiss cheese, for example, is due in large part to propionic acid produced by bacteria.

Many farmers depend on bacterial action in the making of silage for use as cattle feed. Also of considerable interest to farmers is the possibility that bacteria pathogenic for destructive insects may eventually be usable instead of insecticides.

Particularly interesting is the use of bacteria in the production of antibiotics that can help control other bacteria. Most of the antibiotic drugs in use today (but not penicillin) are produced by various species of bacteria of the Actinomycetes group or, if synthesized artificially, were discovered in these organisms. Among these drugs are streptomycin, Aureomycin, Terramycin, and neomycin.

CYANOPHYTA

The Cyanophyta, or blue-green algae, are procaryotic unicellular or filamentous photosynthetic organisms. The unicellular forms are either rods or spheres, which may occur singly or as colonies embedded in a gelatinous matrix (Fig. 21.16). The filamentous forms are multicellular in the sense that plasmodesmata penetrate their walls, interconnecting the cytoplasm of the cells, and that in some species there is division of labor among the cells,

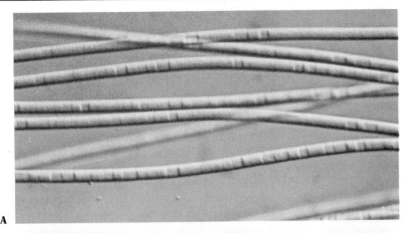

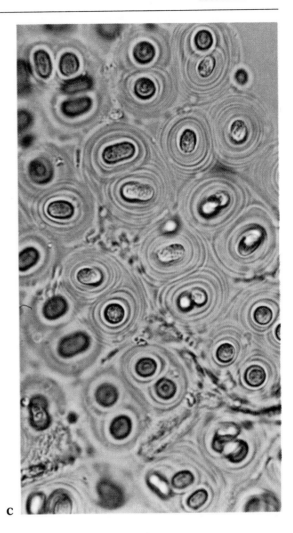

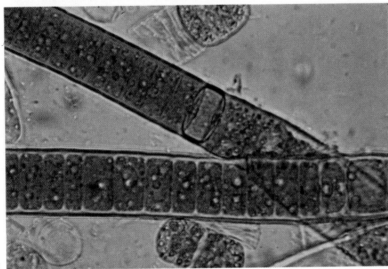

A

B

C

certain ones performing nitrogen fixation for all. Yet, for the most part, the individual cells remain the important units of function. The cells are typically larger than bacterial cells, but smaller than most eucaryotic cells, though the ranges overlap.

The cell walls of blue-green algae, like those of bacteria, contain murein. Outside the wall proper, there is often a sheath of gelatinous material composed of pectic materials and mucopolysaccharides (glycoproteins).

All blue-green algae possess photosynthetic pigments located in flattened membranous vesicles called thylakoids (see Fig. 20.5B, p. 909). These structures are similar to the chromatophores of the photosynthetic bacteria and, like the chromatophores, are not contained within chloroplasts.

The large cell vacuole characteristic of higher-plant cells is absent; also lacking are mitochondria, endoplasmic reticulum, Golgi apparatus, and a nuclear membrane. There are many ribosomes, however, and numerous proteinaceous granules and granules of a stored carbohydrate material called polyglucan, which is very similar to the glycogen used as a storage product in animal cells. Gas vacuoles and a few other membranous vesicles

21.16 Some representative blue-green algae
(A) *Oscillatoria*, a filamentous alga without a sheath. Members of this genus exhibit an odd oscillatory movement.   (B) *Scytonema*, another filamentous alga, has oval, nearly rectangular cells. The lighter colored cell in the diagonal filament is a heterocyst, a cell specialized for carrying out nitrogen fixation. (C) *Gloeothece*, a unicellular alga in which groups of nearly spherical cells are enclosed in layers of gelatinous material.   [Courtesy E. V. Gravé (A); J. M. Kingsbury, Cornell University (B, C).]

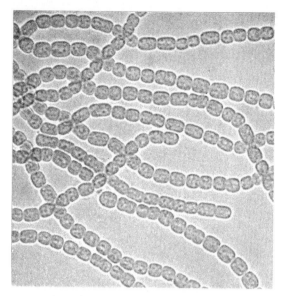

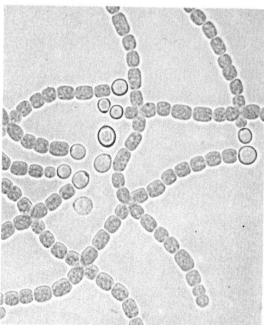

21.17 The effect of environmental nitrogen on heterocyst formation by a blue-green alga, *Anabaena* Top: When grown in a culture containing sodium nitrate, the filaments form no heterocysts.   Bottom: On a medium containing N$_2$ as the only nitrogen source, the filaments form numerous heterocysts.   [Courtesy Rosemarie Rippka, Institut Pasteur, Paris.]

probably derived from the thylakoids are also sometimes present. No blue-green algae ever possess flagella. The peculiar gliding motion exhibited by many species is like that seen in myxobacteria and, like it, remains unexplained.

The chlorophyll of the blue-green algae is chlorophyll *a*, the pigment also found in higher plants, rather than bacterial chlorophyll, and like the higher plants blue-green algae generate molecular oxygen as a by-product of their photosynthesis. As we saw in Chapter 20, it was probably the blue-green algae that initiated the oxygen revolution some 2.3 billion years ago.

In addition to chlorophyll and various carotenoids (two of which are known only in blue-green algae), these organisms contain **phycocyanin** (blue pigment) or sometimes **phycoerythrin** (red pigment). It is the presence of phycocyanin with the chlorophyll that gives these algae their characteristic blue-green color. However, not all "blue-green" algae are blue-green; black, brown, yellow, red, grass green, and other colors also occur. The periodic redness of the Red Sea is due to a species that contains a particularly large amount of phycoerythrin.

One particularly important property of many blue-green algae is their ability to fix atmospheric nitrogen (N$_2$). This process depends on an enzyme, nitrogenase, that functions only under anaerobic conditions. Indeed, O$_2$ does more than inhibit the activity of nitrogenase: It poisons the enzyme irreversibly. The species of blue-green algae most active in N$_2$ fixation are, with few exceptions, filamentous forms that produce a few highly specialized cells called **heterocysts**, where the N$_2$ fixation takes place (Fig. 21.16B).

A heterocyst is a cell with exceptionally thick walls whose nucleus has disappeared. Because O$_2$, it seems, is largely excluded from the interior of the heterocyst, nitrogenase can function there. An interesting adaptation for maintaining anaerobic conditions inside the heterocyst is its loss of Photosystem II, the oxygen-generating component of the photosynthetic apparatus. Equipped only with Photosystem I, the heterocyst can generate ATP in the light, but cannot manufacture carbohydrate and must therefore depend on neighboring cells to supply it with this essential material, presumably via plasmodesmata. It is presumably also via plasmodesmata that the heterocyst exports fixed nitrogen to the other cells of the filament. Heterocysts are rarely present when the algae are growing in an environment where the supply of already fixed nitrogen (e.g., sodium nitrate) is ample, but they form quickly if the fixed nitrogen becomes limiting (Fig. 21.17).

Nitrogen fixation by blue-green algae proceeds best when the heterocysts are in an atmosphere with less than 10 percent free O$_2$. At higher concentrations, some O$_2$ can leak into the heterocysts and begin to inhibit the nitrogenase. It is interesting that photosynthesis, too, is inhibited in blue-green algae when the O$_2$ concentration is above 10 percent. It has been suggested that the metabolism of blue-green algae functions best with low O$_2$ concentrations because these organisms are relicts of nearly two billion years of evolution under conditions of an oxygen-poor earthly atmosphere in pre-Silurian times. What lends credence to this idea is that even many species that do not form heterocysts and do not normally possess nitrogenase apparently have a gene that codes for nitrogenase; these species can synthesize the enzyme and fix N$_2$ when conditions are strictly anaerobic.

Cell division in blue-green algae, as in bacteria, is by binary fission (Fig. 21.18). Reproduction also frequently occurs by fragmentation of filaments.

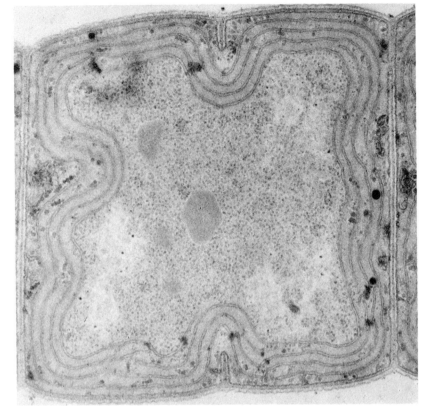

21.18 Cell division by a blue-green alga, *Plectonema boryanum*

Top: A cell in a very early stage of division. Walls have begun to grow inward at a point midway along the cell's length. Note the photosynthetic membranes lying near the periphery of the cell.   Bottom: The new walls have grown farther into the cell, but division is not yet complete. × 43,000.   [Courtesy R. Malcolm Brown, Jr., University of North Carolina.]

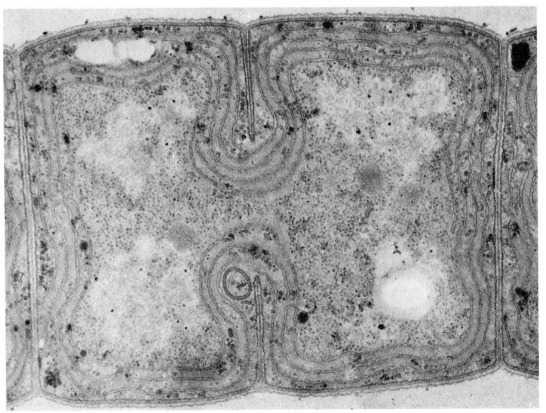

935

No form of genetic recombination has been unequivocally demonstrated in these organisms, but there are now several reports of what appears to be recombination by a process similar to bacterial transformation, i.e. the picking up of DNA from the surrounding medium. It has also been suggested, though not proved, that transduction by the viruses that attack blue-green algae may sometimes occur.

Blue-green algae flourish in numerous and varied habitats. Many live in freshwater, and a few are marine. Some species are very common on or in soil—among them species that live in the extremely nitrogen-poor soil often found in the tropics. There are also species that inhabit deserts, growing in microfissures just below the surface of rocks, where light can penetrate and tiny amounts of moisture are trapped. Other species are often found growing on the sides of damp rocks and flowerpots and on the bark of trees. Blue-green algae tend to be particularly abundant on wet cliffs and ledges. Ponds or lakes containing a rich supply of organic matter, particularly nitrogenous compounds, often develop huge populations ("blooms," as they are called) of blue-green algae, which may make the water so green that objects only a few centimeters below the surface are entirely invisible. Such blooms may give the water an objectionable odor, clog filters of water supplies, and even be toxic to livestock. Some species of blue-green algae live mutualistically with fungi in the compound plants called lichens. A few species (along with some bacteria) live in habitats that rank among the most inhospitable known—the hot springs that occur in various parts of the world; these species can grow well at temperatures as high as 74°C (133°F).

The occurrence of blue-green algae in some environments so bleak that no other organisms can inhabit them is probably due, in part, to the fact that the nitrogen-fixing species have the simplest nutritional requirements of any known organisms. In addition to N_2 and CO_2, which are readily available from the atmosphere, they need only water, light, and minerals.

PROCHLOROPHYTA

The discovery of what appears to be a new division (phylum) of moneran organisms was announced in 1976 by R. A. Lewin of the Scripps Institution of Oceanography. These bright green procaryotic unicells were first thought to be blue-green algae, but analysis of their pigment system revealed them to be quite different. They have both chlorophyll *a* and *b*, whereas blue-green algae have only chlorophyll *a*. And their accessory pigments are limited to the standard carotenoids found in higher plants, with no trace of the red or blue pigments so characteristic of the Cyanophyta. In short, these organisms, designated Prochlorophyta by Lewin, have a pigment system like that of the green algae and the higher land plants.

The Prochlorophyta are of considerable theoretical interest because, being essentially procaryotic green algae, they resemble eucaryotic chloroplasts much more closely than the Cyanophyta do. It has been a weakness of the endosymbiont hypothesis of the origin of eucaryotic cells that it postulated blue-green algae as the progenitors of chloroplasts, despite their very different pigment system; the Prochlorophyta would seem more likely candidates for the role of progenitors. It would be rash to embrace this possibility too strongly, however, until much more is learned about their distribution. At present, the only known representatives of the group live as extracellular associates of tunicates (a group of invertebrate marine chordates; see p. 1045), but it seems possible that some of the other supposed blue-green algae living in association with other marine organisms may also prove, on closer study, to be Prochlorophyta.

SUGGESTED READING

ADLER, J., 1976. "The sensing of chemicals by bacteria," *Scientific American*, April. (Offprint 1337)

BERG, H. C., 1975. "How bacteria swim," *Scientific American*, August.   *The structure and mode of action of bacterial flagella.*

BURKE, D. C., 1977. "The status of interferon," *Scientific American*, April. (Offprint 1356)   *Good update on interferon research.*

CAMPBELL, A. M., 1976. "How viruses insert their DNA into the DNA of the host cell," *Scientific American*, December. (Offprint 1347)

CLOWES, R. C., 1973. "The molecule of infectious drug resistance," *Scientific American*, April. (Offprint 1269)   *How bacteria pass the plasmid-borne genes for resistance to antibiotics from one cell to another.*

COSTERTON, J. W., G. G. GEESEY, and K.-J. CHENG, 1978. "How bacteria stick," *Scientific American*, January.   *On the surface molecules of bacteria that enable these to adhere to host cells—and on the potential for development of a new kind of antibiotic to attack those molecules.*

ECHLIN, P., 1966. "The blue-green algae," *Scientific American*, June.

HOLLAND, J. J., 1974. "Slow, inapparent, and recurrent viruses," *Scientific American*, February. (Offprint 1289)   *On viruses that cause degenerative diseases of humans without revealing their presence by the usual symptoms of infection.*

KAPLAN, M. M., and R. G. WEBSTER, 1977. "The epidemiology of influenza," *Scientific American*, December. (Offprint 1375)   *Recombination between human and animal strains of the influenza virus as a source of new medically important strains.*

LURIA, S. E., J. E. DARNELL, D. BALTIMORE, and A. CAMPBELL, 1978. *General virology*, 3rd ed. Wiley, New York.   *Excellent textbook on all aspects of the biology of viruses.*

RAFFERTY, K. A., 1973. "Herpes viruses and cancer," *Scientific American*, October.   *On the possibility that these common viruses are causal agents of some types of human cancer.*

SHARON, N., 1969. "The bacterial cell wall," *Scientific American*, May.   *On the peculiar structure of bacterial walls, and the way many antibiotics (e.g. penicillin) block their synthesis.*

STANIER, R. Y., E. A. ADELBERG, and J. L. INGRAHAM, 1976. *The microbial world*, 4th ed. Prentice-Hall, Englewood Cliffs, N.J.   *Excellent general microbiology text.*

STOECKENIUS, W., 1976. "The purple membrane of salt-loving bacteria," *Scientific American*, June. (Offprint 1340)   *Rhodopsin as the light-trapping pigment of a newly discovered kind of photosynthesis carried out by certain bacteria.*

WALSBY, A. E., 1977. "The gas vacuoles of blue-green algae," *Scientific American*, August. (Offprint 1367)   *How blue-green algae regulate their buoyancy.*

Chapter 22

THE PROTISTAN KINGDOM

The delimitation of a kingdom Protista has had a checkered history, as we found in Chapter 20 (see Table 20.2, p. 913). At times the kingdom has included both procaryotic and eucaryotic organisms; at other times it has been restricted to eucaryotic members. At times it has taken in both primarily unicellular groups (e.g. Protozoa, Euglenophyta) and groups primarily multicellular but lacking extensive tissue differentiation (e.g. Phaeophyta, Rhodophyta); at other times it has comprised only the unicellular ones. It has sometimes included the fungi and sometimes not. In short, the need has often been felt to set aside a kingdom for organisms at an evolutionary level offering little basis or justification for distinguishing between plant and animal; but viewpoints have varied as to where the dividing lines should be drawn.

Recent opinion has increasingly favored a definition of Protista as taking in primarily unicellular or colonial eucaryotic groups (or very simply multicellular or multinucleate ones). Though some of these groups tend to be plantlike, others funguslike, and still others animal-like, they share many characteristics. Thus all the plantlike protistan groups, though primarily photosynthetic, contain nonphotosynthetic members nutritionally similar to the animal-like protozoans. The slime molds, though funguslike at some stages of their life cycles, are remarkably like amoeboid protozoans at other stages. And some of the plantlike or funguslike protistans are very motile, while some of the animal-like ones are sedentary.

Even though the protistans will be treated in three sections here, according to the evolutionary thrusts—plant, fungal, or animal—they represent, it should be kept in mind, in view of their common characteristics, that such a division is in many ways artificial. Indeed, with the exception of two small groups of funguslike organisms (the chytrids and hyphochytrids), all the diverse groups we here assign to the Protista have historically been united by zoologists in a single phylum, the Protozoa.

ANIMAL-LIKE PROTISTA

Protozoans are usually said to be unicellular. However, as the great student of invertebrate animals Libbie Henrietta Hyman, of the American Museum of Natural History, pointed out, "Each protozoan is to be regarded not as equivalent to a cell of a more complex animal but as a complete organism with the same properties and characteristics as cellular animals." Although they "necessarily lack tissues and organs, since these are defined as aggregations of differentiated cells," many do exhibit "a remarkable degree of functional differentiation." Instead of organs, they have functionally equivalent subcellular structures called *organelles.* In recognition of the complexity of Protozoa, which often far exceeds that of other individual cells, Hyman and many other biologists prefer to call them *acellular* organisms, i.e. organisms whose bodies do not exhibit the usual construction of cells.

Protozoans occur in a great variety of habitats, including the sea, freshwater, soil, and the bodies of other organisms—in fact, wherever there is moisture. Most are solitary, but some are colonial. Many are free-living, but others are commensalistic, mutualistic, or parasitic. They are typically heterotrophic. Though some flagellated organisms that possess chlorophyll and are photosynthetic are placed in the Protozoa by many authors, we shall assign these plantlike forms to various protistan algal groups, to be discussed later in this chapter, or, in some cases, to the green algae, discussed in Chapter 23.

The heterotrophic protozoans usually digest food particles in food vacuoles (see p. 215 and Figs. 5.17 and 5.18). There are no special organelles for gas exchange, the general cell membrane serving as the exchange surface. Many species, particularly those living in hypotonic media such as freshwater, possess contractile vacuoles, which function primarily in osmoregulation (see pp. 319–320 and Figs. 8.7–8.9). Small amounts of nitrogenous waste may also be expelled by the contractile vacuoles, but most of it is released as ammonia by diffusion across the general cell surface. Locomotion is by formation of pseudopodia or by means of beating cilia or flagella (see pp. 485–486). Where a single individual has many cilia, their action is coordinated by a system of fibrils connecting their basal bodies; these fibrils seem to have conductile properties rather like those of nerves in multicellular animals. Reproduction is sometimes asexual and sometimes sexual. Most freshwater and parasitic protozoans can encyst when conditions are unfavorable; they secrete a thick resistant case around themselves and become dormant.

The Protozoa have been divided into five groups: Mastigophora, Sarcodina, Sporozoa, Cnidospora (which we shall not discuss), and Ciliata. Some of these are heterogeneous assemblages of structurally similar, but probably not closely related, organisms. The relationships of the five groups to one another are unclear. Biologists who choose to treat the Protozoa as a single phylum of unicellular animals usually list the five groups as classes; those who treat the protozoans as Protista usually assign them full phylum rank. We shall follow the latter system.

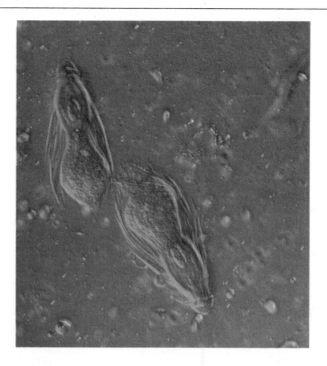

22.1   *Trichonympha,* a flagellate that inhabits the gut of termites
The flagellate helps digest the cellulose in the termite's diet.   [Courtesy E. V. Gravé.]

MASTIGOPHORA

The Mastigophora (also called Zooflagellata) are protozoans that possess flagella as their principal locomotor organelles. They appear to be the most primitive of all the Protozoa, and it seems likely that some (and possibly all) of the other protozoan groups arose from them. Many zoologists (though certainly not all) believe that the flagellate protozoans were also the ancestors of the multicellular animals; similarly, most botanists regard the flagellated photosynthetic protists as the ancestors of the multicellular plants. There is good reason to think, then, that flagellated unicells played a key role in the evolution of life on earth.

This is not to suggest that any flagellate species still living today was the ancestor of these other organisms, but simply that the ancestral flagellates from which modern flagellate protozoans and modern flagellate protistan algae are descended probably also gave rise to the other protozoans and to multicellular plants and animals. These relationships might be diagramed as follows:

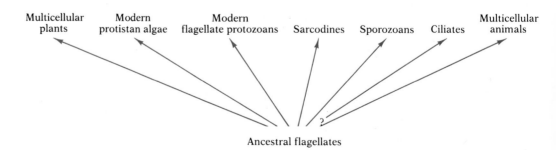

A few of the Mastigophora are free-living in salt- or freshwater, but most live as symbionts in the bodies of higher plants or animals. Several species, for instance, are found in the gut of termites, where they participate in the digestion of the cellulose consumed by the termite (Fig. 22.1); we mentioned this as an example of mutualism in Chapter 18.

Trypanosoma (Fig. 22.2) is a genus of parasitic zooflagellates that cause several severe diseases in human beings and domestic animals. *Trypanosoma gambiense*, for example, is the causative agent in African sleeping sickness.[1] The trypanosomes live in the blood of their host, where they multiply and release the poisonous by-product of their metabolism. In humans or domestic animals (but not in the native wild mammals of Africa), they eventually invade the nervous system, causing lethargy and finally death. The trypanosomes are spread from host to host by blood-sucking tsetse flies (genus *Glossina*). When a tsetse fly sucks blood from an infected animal, some of the trypanosomes are sucked into its intestine, where they multiply and undergo several developmental changes. They then migrate to the fly's salivary glands, where they undergo additional developmental changes and continue to multiply. If the fly now bites an uninfected vertebrate, some of the trypanosomes are injected from the salivary glands into the vertebrate host. Sleeping sickness, which makes large parts of Africa nearly uninhabitable for humans, is difficult to control, because the many wild animals serve as a constant reservoir of trypanosomes. It may eventually be eradicated by extermination of the tsetse flies—admittedly a huge undertaking.

SARCODINA

The Sarcodina are the amoeboid Protozoa. They are thought to be more closely related to the zooflagellates than to the other protozoan groups, because some zooflagellates undergo amoeboid phases, and, conversely, some Sarcodina have flagellated stages.

[1] African sleeping sickness should not be confused with ordinary sleeping sickness, or encephalitis, which is caused by a virus.

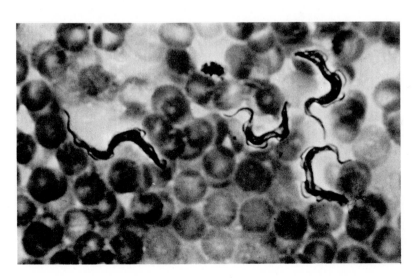

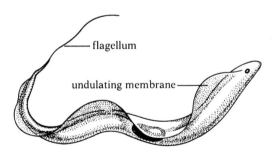

22.2 *Trypanosoma gambiense,* **the cause of African sleeping sickness**
Left: Photograph of the protozoans among red blood cells. × 1,500.   Right: Drawing showing the general structure of the organism.   [Left: Courtesy E. V. Gravé.]

22.4   A group of calcareous foraminiferan shells
[Oxford Scientific Films, Bruce Coleman Inc.]

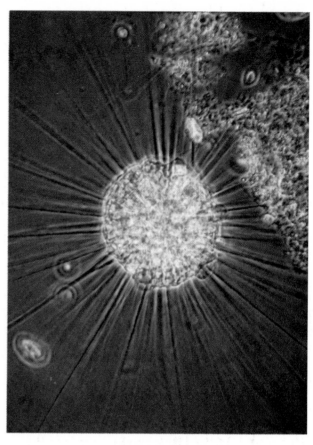

22.3   A sarcodine *(Actinosphaerium)* **with long pointed pseudopods**
[Courtesy E. V. Gravé.]

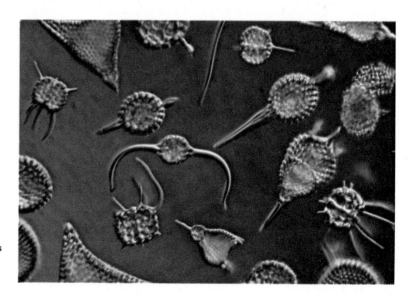

22.5   A group of siliceous radiolarian shells
[Courtesy E. V. Gravé.]

The most familiar sarcodines are the naked freshwater species of the genera *Amoeba* and *Pelomyxa* (see Fig. 1.7, p. 15), which have asymmetrical bodies that constantly change shape as new pseudopods are formed and old ones retracted. These pseudopods, which are large and have rounded or blunt ends, function both in locomotion (see p. 485 and Fig. 11.27) and in feeding by phagocytosis (see p. 93 and Fig. 3.15; also p. 214). The food consists of small algae, other protozoans, and even some small multicellular animals such as rotifers and nematode worms.

Also included in the Sarcodina are several groups of protozoans that secrete hard calcareous or siliceous shells around themselves. These shells, often quite elaborate and complex, can be used in species identification. The pseudopods of the shelled Sarcodina are usually thin and pointed (Fig. 22.3); in some forms they have no locomotory function, being exclusively feeding devices. This type of pseudopod does not flow around and engulf the prey, but instead functions as a trap. When prey organisms touch it, they become stuck in a mucoid adhesive secretion that coats the pseudopod surface. The secretion apparently contains proteolytic enzymes that initiate digestion of the prey, which is eventually enclosed in a food vacuole and drawn toward the interior of the cell.

Two groups of shelled sarcodines, the Foraminifera and the Radiolaria, have played major roles in the geologic history of the earth. Both groups are extremely abundant in the oceans, and when the individuals die their shells become important components of the bottom mud. The shells of foraminiferans (Fig. 22.4), which are calcareous, are especially prevalent in the mud at depths of 2,500 to 4,500 m; at greater depths they tend to dissolve, because of the increased carbon dioxide content of the water. The bottom ooze in deeper parts of the ocean is composed chiefly of the siliceous shells of radiolarians (Fig. 22.5). Both these groups have been abundant for a very long time; fossils of foraminiferans go back to the Cambrian period, and fossils of radiolarians occur in Precambrian rocks. Much of the limestone and chalk now present on the earth was formed from deposits of foraminiferan shells, and radiolarian shells have contributed to the formation of siliceous rocks such as chert.

SPOROZOA

The Sporozoa, as their name implies, usually have a sporelike infective cyst stage in their life cycles. They lack special locomotor organelles (except in the male gametes). All are internal parasites, and they usually have complex life cycles. They cause a variety of serious diseases in humans (e.g. coccidiosis and malaria), as well as in both domestic and wild animals.

As you probably know, malaria, which is caused by species of the genus *Plasmodium*, is transmitted from host to host by female *Anopheles* mosquitoes. When an anopheline mosquito bites a person and starts to suck blood, it releases saliva containing a chemical that prevents coagulation of the blood and often, too, *Plasmodium* cells of a stage called sporozoites (Fig. 22.6: 5). The sporozoites enter the person's bloodstream and are carried to the liver, where they enter liver cells and grow for 5–15 days. Then each sporozoite divides, producing a large number of new cells called merozoites. These are released from the liver cells and penetrate into red blood cells,

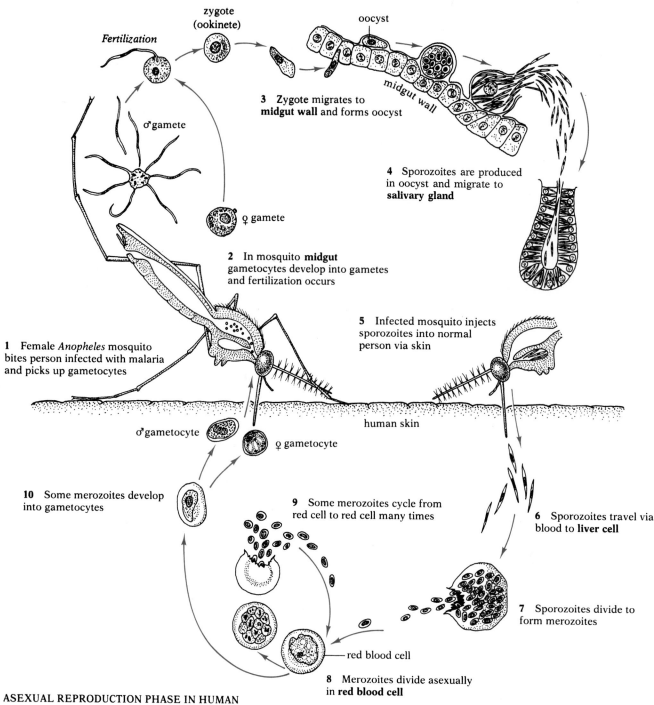

SEXUAL REPRODUCTION PHASE IN MOSQUITO

zygote
(ookinete)

Fertilization

oocyst

♂ gamete

midgut wall

3 Zygote migrates to
midgut wall and forms oocyst

4 Sporozoites are produced
in oocyst and migrate to
salivary gland

♀ gamete

2 In mosquito **midgut**
gametocytes develop into gametes
and fertilization occurs

5 Infected mosquito injects
sporozoites into normal
person via skin

1 Female *Anopheles* mosquito
bites person infected with malaria
and picks up gametocytes

human skin

♂ gametocyte

♀ gametocyte

10 Some merozoites develop
into gametocytes

9 Some merozoites cycle from
red cell to red cell many times

6 Sporozoites travel via
blood to **liver cell**

7 Sporozoites divide to
form merozoites

red blood cell

8 Merozoites divide asexually
in **red blood cell**

ASEXUAL REPRODUCTION PHASE IN HUMAN

**22.6 The life cycle of *Plasmodium*, the malarial
parasite**

where they reproduce asexually, producing additional merozoites. At regular intervals (48 hours in some types of malaria; in one type, 72 hours), all infected red cells burst, releasing the merozoites, which enter new blood cells and repeat the asexual reproductive process. Thus, in the most common form of malaria, hordes of merozoites are released into the bloodstream from ruptured red cells every 48 hours. The host experiences attacks of chills and fever each time such a release of merozoites occurs; these symptoms apparently stem from toxins discharged into the blood by the rupturing red cells.

Eventually some of the merozoites develop into special sexual cells capable of becoming either male or female gametes; they will not mature as gametes, however, as long as they remain in the blood of the human host. But if an anopheline mosquito sucks blood containing these cells, they complete their development in the midgut of the mosquito. The male gametes then fertilize the female gametes, and the zygote thus produced, which is amoeboid, works its way into the wall of the gut and encysts. Within the cyst, a series of divisions ultimately produce new sporozoites, which are released when the cyst ruptures. These sporozoites migrate to the salivary glands of the mosquito, from which they are discharged into a new vertebrate host when the mosquito feeds. Efforts at malarial control have largely been attempts to eradicate *Anopheles* mosquitoes, either directly by use of insecticides or indirectly by destroying their breeding places.

CILIATA

The phylum Ciliata is the largest of the four protozoan phyla and also the most homogeneous. It differs markedly from the other phyla, and its relationship to them is unclear. A possible derivation from the Mastigophora is suggested by a small group of internal symbionts of frogs (the class Opalinata, now usually placed in the Mastigophora) that shows a combination of some ciliate and some flagellate traits.

As their name implies, ciliates possess numerous cilia as locomotory organelles (Figs. 22.7, 22.8). In most species the cilia are present throughout life, but in a few (Suctoria) they are absent in the adult stages. The ciliates exhibit the greatest elaboration of subcellular organelles of any Protozoa (Fig. 22.9; see also Fig. 1.6, p. 15, and Fig. 8.7, p. 318), and it is to them that the term "acellular" applies best. As we saw with *Paramecium* (see Fig. 5.17, p. 214), they may have a special oral groove and cytopharynx into which food particles are drawn in currents produced by beating cilia, and they often have an anal "pore" through which indigestible wastes are expelled from food vacuoles. Conductile fibrils connect the bases of the cilia, and there may be a system of contractile fibers (sometimes striated) analogous to the muscular system of multicellular animals. Stiffened plates occasionally found in the pellicle of the cell together constitute a "skeleton." In some species there is a long stalk by which the individual may attach itself to the substratum (Figs. 22.7, 22.8C). A few species have tentacles for capture of prey. Some can discharge toxic threadlike darts called **trichocysts** that resemble the nematocysts of coelenterates (Fig. 22.9); these may function in defense against predators, in capturing prey, or in anchoring the organism to the substratum during feeding.

Colpoda cucullus

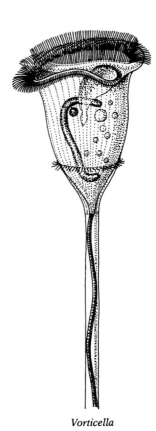

Vorticella

22.7 Representations of two ciliate protozoans
The drawings illustrate the remarkable complexity of these unicellular organisms, to which the term "acellular" applies especially well.

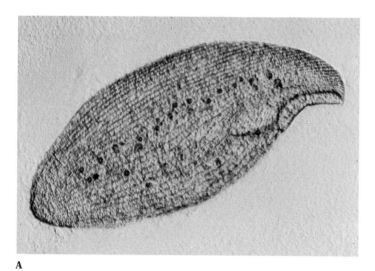

A

B

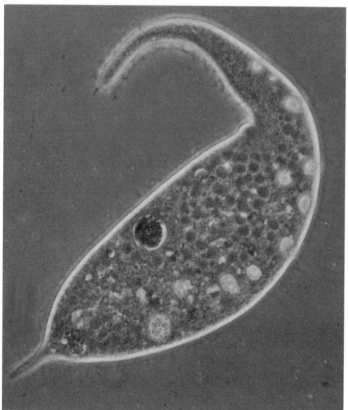

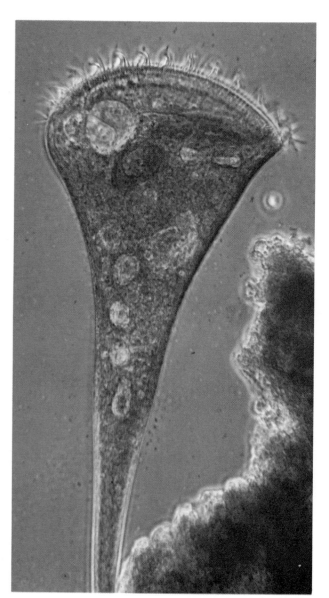

C

22.8 Photographs of three representative ciliates
(A) *Loxoda.* (B) *Dileptus.* (C) *Stentor.* [Courtesy
Hervé Chaumeton (A), Roman Vishniac (B), E. V.
Gravé (C).]

22.9   A living ciliate viewed with Nomarski optics
This technique permits study of internal detail without damage to the organism. The large nucleus is visible near the center of the cell. Below the nucleus, and to the left, are contractile vacuoles of the sort filled by small fusion vesicles, some of which can be seen around the periphery of the largest vacuole. The cell contains numerous food vacuoles of various shapes containing several kinds of algae, including diatoms and filamentous forms. An array of undischarged trichocysts can be seen just inside the plasma membrane at right.   [Courtesy Thomas Eisner, Cornell University.]

Ciliates differ from all other protozoans in having two types of nuclei: a large macronucleus and one or more small micronuclei (see Figs. 1.6 and 5.17). The macronucleus, which is polyploid (about 860-ploid in *Paramecium aurelia*!), controls the normal metabolism of the cell. The diploid micronuclei are concerned only with reproduction and with giving rise to the macronucleus. During asexual reproduction, which is usually by transverse binary fission, the micronuclei divide mitotically.[2] The macronucleus, however, divides amitotically; i.e. it forms no spindle and appears to divide by simple constriction. Since division in the macronucleus is preceded by DNA replication and each daughter nucleus receives a full set of genetic material, the division process must be more precise than it appears.

Many ciliates reproduce occasionally by a sexual process called ***conjugation.*** Two individuals of appropriate mating types come together and adhere in the oral region; there is some fusion in the area of contact (Fig. 22.10). Next, the micronuclei divide by meiosis. Then all but two of the resulting haploid micronuclei in each cell disintegrate; the macronucleus also usually disintegrates. One of the two nuclei in each cell remains stationary and functions as the female nucleus. The second moves into the other cell and fuses with that cell's female nucleus in the process of fertilization. Thus each cell acts as both male and female, donating a nucleus and receiving one in return, and when the two cells part each has a new recombinant diploid nucleus. This nucleus then undergoes one or more divisions, and some of the new nuclei thus produced develop into macronuclei. Following a variable number of cytoplasmic cleavages, the normal number of micro- and macronuclei per cell is restored.

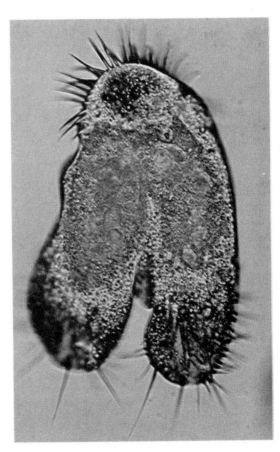

22.10   Two *Stylonychia* individuals in conjugation
[Manfred Kage, Peter Arnold, Inc.]

[2] The nuclear membrane of the micronuclei of ciliates does not disappear during mitosis, and the spindle forms inside the membrane-bounded nucleus.

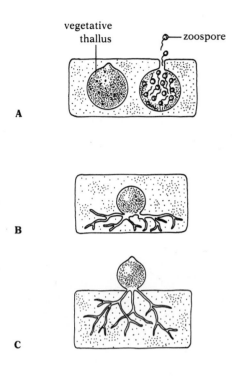

A

B

C

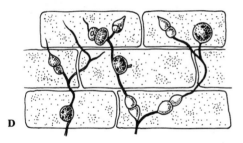

D

22.11   A variety of thallus forms of chytrids
(A) Some species have a simple saclike thallus (body)
located entirely within the host cell. When flagellated
zoospores are produced, a tube grows out to the
membrane of the host cell and then ruptures, the
zoospores being released to the exterior.   (B) Other
species that live entirely inside their host cells have
rootlike rhizoids for nutrient absorption.   (C) Still
other species, in which the main part of the thallus is
outside the host cell, send rhizoids into the host.
(D) Some species exhibit a simple form of
multicellularity, and their threadlike bodies exploit
several host cells simultaneously.

FUNGUSLIKE PROTISTA

Among the several major groups of fungal-type organisms that are at the
protistan level of complexity, we can, for convenience, recognize two categories: (1) those divisions[3] considered true fungi by more traditional classifications, but containing only unicellular or very simply multicellular organisms, and (2) those divisions traditionally set apart from the true fungi
and called slime molds.

PROTOMYCOTA (THE TRUE FUNGUSLIKE PROTISTS)

The two groups included here, the chytrids and hyphochytrids, are small
organisms of a more or less typical fungal type, but clearly at the protistan
level of complexity. Many are aquatic, living as parasites or saprophytes in
or on algae or other plants, but some are found in soil and a few are internal
parasites of such animals as mosquito larvae, nematode worms, or liver
flukes.

   The haploid bodies of these organisms may be simple sacs living entirely
inside a cell of the host (Fig. 22.11A), or sacs with rootlike nutrient-absorbing
protuberances called rhizoids (Fig. 22.11B, C), or sometimes filamentous
forms with saclike reproductive structures (Fig. 22.11D). Reproduction
begins when the sacs become multinucleate upon repeated mitotic division
of the nucleus without accompanying cytokinesis. Eventually the cytoplasm
is partitioned in such a way that each nucleus receives some. The newly
formed cells, each of which develops a single whiplash flagellum (located
posteriorly in the chytrids and anteriorly in the hyphochytrids), are then
released into the surrounding medium. Under some circumstances these
flagellated cells function as gametes in sexual reproduction. But more often
they function as asexual reproductive cells called **zoospores,** settling down in
a suitable location and developing into a new sac or filament, depending on
the species.

GYMNOMYCOTA (THE SLIME MOLDS)

The slime molds are curious organisms decidedly animal-like at some stages
in their life cycle and plantlike at others. Often placed among the Protozoa in
the animal kingdom by zoologists, they have been placed among the fungi in
the plant kingdom in traditional botanical classifications. When four or five
kingdoms are recognized, they seem to fit best in the Protista.

   The slime molds are generally found growing on damp soil, rotting logs,
leaf mold, or other decaying organic matter in moist woods, where they look
like glistening viscous masses of slime. They are sometimes white, but are
often colored red or yellow (Fig. 22.12).

   In the so-called **true slime molds** (division Myxomycota), the vegetative
phase of the life cycle is a large diploid multinucleate (coenocytic) amoeboid

---

[3] "Division" and "phylum" are equivalent terms. Whereas "phylum" is applied to animals and
animal-like organisms, "division" is applied to plants and plantlike organisms.

mass called a ***plasmodium,*** which moves about slowly and feeds on particles of organic material by phagocytosis. The behavior of the naked plasmodium is thus animal-like. Under certain conditions, however, the plasmodium becomes stationary and develops ***fruiting bodies,*** which may be either simple rounded masses or elaborate stalked organs; at this stage the appearance and behavior of the organism are plantlike. Meiosis occurs within the ***sporangia*** (spore-producing structures) of the fruiting bodies, and the haploid cells thus formed are released as spores, whose walls contain cellulose. When the spores germinate, they produce naked flagellated gametes. These fuse in pairs to form zygotes, which soon lose their flagella and become amoeboid. As this amoeboid form flows along the substratum, engulfing bacteria and other organic particles and digesting them in vacuoles, its diploid nucleus undergoes repeated mitotic divisions without accompanying cytokinesis. In this way the zygote develops into a multinucleate plasmodium, which may grow to a length of 25 cm (although 5–8 cm is a more common size). Some growth may also occur by fusion of the cytoplasms of two or more zygotes or young plasmodia. In summary, then, the life cycle proceeds from diploid amoeboid plasmodium, to stationary spore-producing plasmodium, to haploid spores, to flagellated gametes, to zygote, and back to amoeboid plasmodium (Fig. 22.13).

The life cycle of the ***cellular slime molds*** (division Acrasiomycota) is quite different from that of the true slime molds. Current evidence indicates that the two groups are not closely related and should be placed in separate divisions.[4] Their relationships to other organisms are unknown.

[4] Two other small groups, the Plasmodiophoromycota and the Labyrinthulomycota, are also usually put in the Gymnomycota, although they are probably not closely related to either the Myxomycota or the Acrasiomycota. Thus Gymnomycota is more a grouping of convenience than a true phylogenetic category.

Authors who wish to emphasize the animal characteristics of the slime molds often use the name "Mycetozoa."

22.12   A yellow slime mold growing on a log in Michigan
[L. West, Bruce Coleman Inc.]

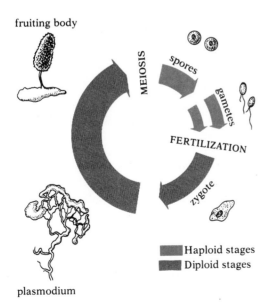

22.13   Life cycle of a true slime mold
See text for description.

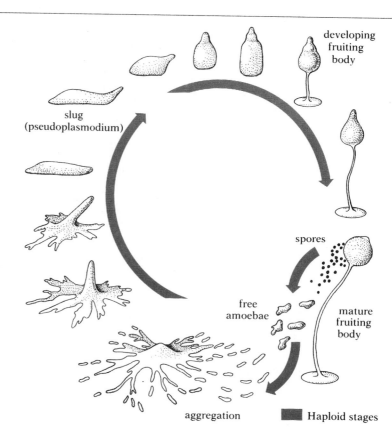

22.14 Life history of a cellular slime mold
See text for discussion.

In the cellular slime molds the spores do not develop into flagellated gametes, but instead give rise to free-living soil-inhabiting amoeboid cells, each with a single haploid nucleus (Fig. 22.14). The amoebae feed on bacteria and other organic matter. During this feeding stage the amoebae divide repeatedly (both mitosis and cytokinesis occur), producing independent uninucleate daughter cells. As the local food supply diminishes, the behavior of the amoebae suddenly changes; they cease feeding and begin to stream into central collecting points, where they clump together to form a sluglike *pseudoplasmodium.* The individual haploid cells retain their separate identities within the slug; they do not fuse. The pseudoplasmodium of the cellular slime molds is thus very different from the diploid multinucleate plasmodium of the true slime molds. The slug may move around as a unit for a while, but eventually it becomes sedentary and forms a stalked fruiting body in which new spores are produced. Notice that this life cycle does not include any sexual events, and that all stages are apparently haploid.

Cellular slime molds can easily be grown in the laboratory, and they have been used extensively in studies of the factors influencing development. It has been found that during the stage when the individual amoebae are free-living, they are nonpolar, i.e. they have no distinct anterior or posterior end (Fig. 22.15). However, as the food supply dwindles, they become polar, and some begin to secrete cyclic AMP from their posterior ends into the medium. Other polar amoebae, detecting the cAMP at their anterior ends, turn toward the signal and at the same time secrete cAMP from their poste-

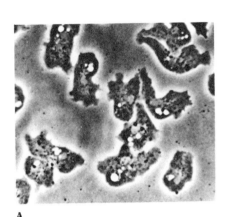

A

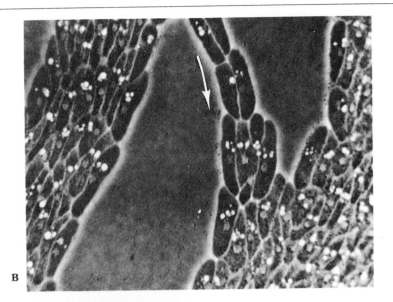

B

C

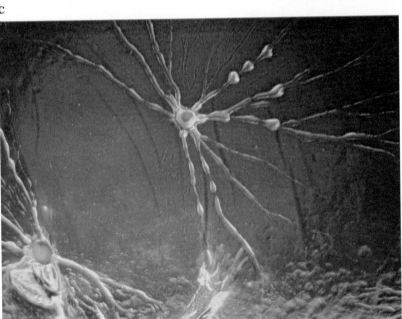

22.15 *Dictyostelium*, a cellular slime mold
(A) In the free-living stage each cell is an irregularly
shaped amoeba. × 450.   (B) When aggregation begins,
the cells take on an elongate shape, become polar (i.e.
have distinct anterior and posterior ends), and begin to
move in an oriented direction (indicated here by the
arrow). × 450.   (C) Two aggregations
(pseudoplasmodia) are in process of formation, with
long streams of amoebae funneling into the centers of
aggregation.   [(A–B) Courtesy K. B. Raper, *Proc. Am.
Philos. Soc.*, vol. 104, 1960. (C) Courtesy Roman
Vishniac.]

rior ends, thus stimulating still other amoebae to orient in the same direction.
The amoebae secrete the cAMP in pulses rather than in a steady stream, and
each time an oriented amoeba detects a pulse it moves forward for about 100
seconds. In this fashion more and more amoebae are recruited and move
toward the aggregation center, where the pseudoplasmodium is formed. It is
especially intriguing to biologists that the substance responsible for the
remarkable behavioral coordination in aggregating slime-mold amoebae
should be cAMP, which, as we saw, is involved in many sorts of intracellular
control in animals.

PLANTLIKE PROTISTA

Several groups of primarily unicellular, often flagellated, organisms have traditionally been regarded by botanists as algal members of the plant kingdom, because many of their members possess chlorophyll and many have cell walls. These same organisms have traditionally been placed in the Protozoa by zoologists. They exemplify, perhaps better than any other living organisms, the difficulty of distinguishing between plants and animals at the unicellular level. Their assignment to the kingdom Protista makes such distinction unnecessary.

EUGLENOPHYTA (THE EUGLENOIDS)

The euglenoids are unicellular organisms that show a combination of plant-like and animal-like characteristics. They are plantlike in that many species have chlorophyll and are photosynthetic; they are animal-like in lacking a cell wall and being highly motile, and the species that lack chlorophyll are heterotrophic, like animals. Although their pigmentation (chlorophylls *a* and *b* and carotenoids) is like that of the green algae and land plants, they seem to have no close relatives among the algae. Botanical classifications usually put them in a division by themselves. There are about 25 genera of euglenoids, containing approximately 450 species. Most live in freshwater, but a few are found in soil, on damp surfaces, or even in the digestive tracts of certain animals.

A representative genus is *Euglena*. A typical cell of *Euglena* is elongate ovoid, with a long flagellum emerging from an anterior invagination (Fig.

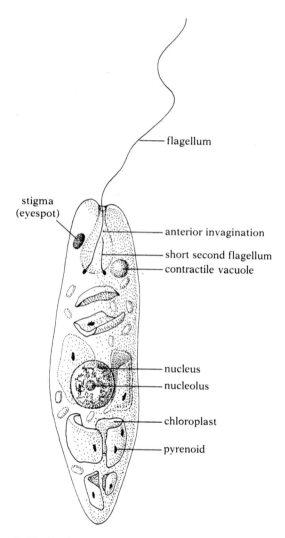

flagellum

stigma
(eyespot)

anterior invagination

short second flagellum
contractile vacuole

nucleus
nucleolus

chloroplast

pyrenoid

22.16 *Euglena*
Above: For a full description of the structures shown in this drawing, see text.   Right: Photograph of a group of *Euglena*.   [Courtesy E. V. Gravé]

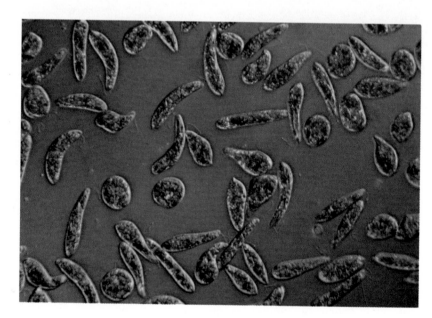

22.16).[5] Since the cell lacks a wall, it is fairly flexible, and its shape may change somewhat as it swims about; however, the pellicle, a proteinaceous layer just beneath the plasma membrane, prevents excessive alterations of its shape. The large nucleus contains a prominent nucleolus that is unusual in not disappearing during mitosis. The nuclear membrane too remains intact during mitosis, instead of disappearing as it does in most cells. An orange granule, called the *stigma* (or eyespot), is located near the anterior end of the cell and functions as a light detector leading to phototactic responses (there is no stigma in nonphotosynthetic euglenoid species). Most green euglenoids have a special organelle, the *pyrenoid,* that functions in the production of paramylum, a polymer of glucose that these organisms use as a storage product instead of starch or glycogen; paramylum is unique with the euglenoids. A large contractile vacuole lies near the anterior end of the cell.

The euglenoid species that lack chlorophyll are obligate heterotrophs. The species that have chlorophyll are facultative heterotrophs and can survive in the dark if they have a source of organic nutrients; indeed, even in the light they are not entirely autotrophic, since they require one or more vitamins in their diet. It is easy to destroy the chloroplasts of *Euglena* by treatment with streptomycin or heat (or by keeping the cells in the dark for a long time). A colorless strain of *Euglena* can be produced in this way, for chloroplasts are self-perpetuating organelles, which cannot be regained once they are lost. It is thus possible to make a lineage of organisms cross the traditional boundary (presence or absence of chlorophyll) between the plant and animal kingdoms.

Reproduction in euglenoids is by longitudinal mitotic cell division. Sexual reproduction has never been conclusively demonstrated in them.

22.17   A unicellular chrysophyte with characteristically unequal flagella

CHRYSOPHYTA (THE YELLOW-GREEN AND GOLDEN-BROWN ALGAE AND THE DIATOMS)

As the names of many algal divisions indicate, the earliest classifications of algae were based on color, which in turn depends on the sorts of pigments the cells contain. Fortunately, later study of other important characters—particularly the details of flagellation, the type of reserve materials produced, and the chemistry of the cell wall—showed that algae of like pigmentation usually share such characters as well and that the old color classification was still acceptable. Thus most of the species in the six classes of algae placed together in the division Chrysophyta are some shade of yellow or brown (caused in part by a predominance of carotenoids), and they also resemble each other in possessing chlorophylls *a* and *c*, but no *b*, and in using a polysaccharide called chrysolaminaran[6] instead of starch as their reserve material. The walls of many species are impregnated with silica or calcium. Two anteriorly attached flagella of unequal length are common (Fig. 22.17), but some species have no flagella, some have one, and some have two that are equal.

---

[5] *Euglena* possesses a second shorter flagellum that does not emerge from the anterior invagination. In some other euglenoids both flagella are emergent.

[6] Formerly called leucosin.

The majority of the Chrysophyta are unicellular or colonial, although a few have small simple multicellular bodies. Reproduction is usually asexual but occasionally sexual. Most of the yellow-green and golden-brown algae live in freshwater, but a few are marine. Diatoms are abundant in both fresh- and saltwater habitats.

The diatoms, which some authorities assign to a division of their own, separate from the other Chrysophyta, are of special interest for several reasons. They are unusual in that the vegetative cells are ordinarily diploid—not haploid, as might be expected in such simple and seemingly primitive plantlike organisms. Unlike most other Chrysophyta, diatoms lack flagella; some species, however, produce flagellated sperm cells. Silica-impregnated glasslike walls, composed of two pieces that fit together like a box with its lid (Fig. 22.18), often give the cells a jewel-like appearance; the different species exhibit a great variety of shapes and ornamentations (Fig. 22.19). The classification of the diatoms is based almost entirely on the characters of the walls, or shells, as they are commonly called.

When the cells of diatoms die, their shells sink to the bottom, where they may accumulate in large numbers, forming deposits of a material called diatomaceous earth. This material is used as an ingredient in many commercial preparations, including detergents, polishes, paint removers, decolorizing and deodorizing oils, and fertilizers. It is also extensively used as a filtering agent and as a component in insulating and soundproofing products.

The diatoms play an extremely important role in aquatic food webs, in both freshwater and marine habitats. They are the most abundant component of marine plankton, for example; it is not unusual for a gallon of seawater to contain as many as one or two million diatoms. *Plankton* consists, by definition, of small organisms floating or drifting near the surface. Planktonic organisms are generally divided into two groups—phytoplankton (plant plankton) and zooplankton (animal plankton). The organisms constituting phytoplankton are the principal photosynthetic producers in marine communities.

22.18 Scanning electron micrograph of a diatom (*Trinacria regina*)
The organism has a distinctly boxlike structure.
× 363. [Courtesy Turtox/Cambosco, Macmillan Science Co., Inc.]

22.19 Some representative diatoms
Left: *Synedra acus*, with a long rodlike shape.
× 234. Middle: *Triceratium pentacrinus*, a star-shaped diatom. × 520. Right: *Arachnoidiscus ehrenbergi*, a wheel-like organism. × 455. [Courtesy Turtox/Cambosco, Macmillan Science Co., Inc.]

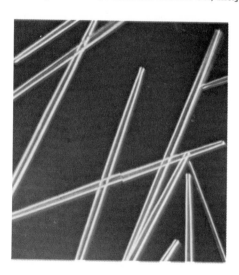

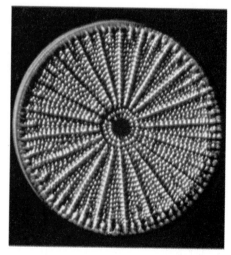

PYRROPHYTA (THE DINOFLAGELLATES)

The dinoflagellates are small, usually unicellular, organisms (Fig. 22.20). A cell wall may or may not be present; if present, it is composed largely of cellulose. Photosynthetic species possess chlorophylls *a* and *c;* they usually have a yellowish-green to brown color due to an abundance of carotenoids and xanthophylls, several of which are unique in these organisms. There are many colorless species; some feed on particulate organic matter, and others live as parasites in a variety of marine invertebrates. The reserve material is either starch or oils.

The nuclei of dinoflagellates are unlike those of any other organisms. The chromosomes lack centromeres and remain permanently in the condensed configuration, even during interphase. The nucleolus and nuclear membrane persist during cell division, and no spindle is formed. Long cytoplasmic channels, probably containing microtubules, intrude into the nucleus during division, and these may replace the spindle in the movement of the chromosomes.

Most species possess two very unequal flagella, which are attached laterally. One of these runs along a groove to the posterior end of the cell and extends beyond the cell like a tail; the other lies in a groove that encircles the midportion of the cell like a belt (Fig. 22.21).

Many dinoflagellates produce trichocysts similar to those of ciliate protozoans such as *Paramecium.* A few even produce nematocysts, which are so much like those of coelenterates that it was once thought they were acquired by the dinoflagellates from coelenterates.

Some species of dinoflagellates can produce light and are responsible for much of the luminescence often seen in ocean water at night.

Dinoflagellates are second only to the diatoms as primary producers of organic matter in the marine environment; they play a lesser role in freshwater. Not only are they important as food, but they also function as symbionts for an amazing variety of marine invertebrate animals that in some cases appear to be unable to survive without them. Some types of corals take as much as 60 percent of the carbon their dinoflagellate symbionts fix in photosynthesis; it is estimated that such corals would deposit calcium in their skeletons only one tenth as fast if they lacked the symbionts.

A number of species of dinoflagellates are poisonous. Some of these contain red pigments, and when they occur in great abundance they produce the so-called red tides that sometimes kill many millions of fish. Red tides are fairly common in the Gulf of Mexico off the coast of Florida.

SUGGESTED READING

Bold, H. C., and M. J. Wynne, 1978. *Introduction to the algae.* Prentice-Hall, Englewood Cliffs, N.J. *Comprehensive, rather technical text that includes the protistan algae.*

Bonner, J. T., 1969. "Hormones in social amoebae and mammals," *Scientific American,* June. (Offprint 1145) *The role of cyclic AMP in the communication system of the cellular slime molds.*

Grell, K. G., 1973. *Protozoology,* 2nd ed. Springer, Heidelberg. *Excellent comprehensive treatment of the Protozoa.*

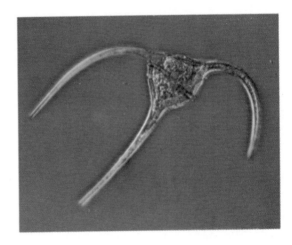

22.20 Photograph of *Ceratium,* a dinoflagellate
[Courtesy E. V. Gravé.]

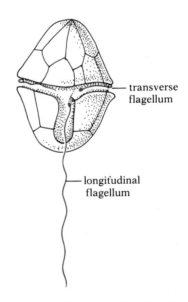

22.21 Drawing of a freshwater dinoflagellate (*Glenodinium cinctum*)
The two flagella lie in distinctive grooves on the surface of the cell.

transverse flagellum

longitudinal flagellum

Chapter 23

THE PLANT KINGDOM

The various divisions of the plant kingdom have traditionally been separated into two groups: the *Thallophyta* and the *Embryophyta.* Though not usually recognized as taxonomic categories in modern classifications, the groupings are useful in that they include plants at similar levels of structural complexity. The Thallophyta are the more primitive plants, the Embryophyta the more advanced. We can summarize the chief distinctions between them as follows:

1. Thallophytes usually show little if any tissue differentiation; there is thus no anatomical basis for distinguishing roots, stems, or leaves, the entire plant being known as a *thallus.* There is far more differentiation in embryophytes; the higher embryophytes have distinct roots, stems, and leaves.

2. The reproductive structures of thallophytes are often unicellular and, whether unicellular or multicellular, lack a protective wall or jacket of sterile (i.e. nonreproductive) cells.[1] The reproductive structures of embryophytes, by contrast, are multicellular and have a jacket of sterile cells.

3. The zygotes of thallophyte plants do not develop into embryos within the female reproductive organs where they are produced,[2] whereas the early stages of embryonic development in embryophytes take place while the embryo is still contained within the female reproductive organ (hence the name "embryophyte").

In older classification systems the Thallophyta included both the various groups called algae (the photosynthetic thallophytes) and the groups called fungi (the nonphotosynthetic thallophytes). In systems like the one in this text, in which the fungi have kingdom status and the various unicellular

---

[1] There are some seeming exceptions, where the reproductive structures appear superficially to have jacket cells.

[2] Except in some red algae.

algae are assigned to either the kingdom Protista or the kingdom Monera, only algal groups with many multicellular members remain in the thallophyte portion of the plant kingdom. A formal outline of the classification of the plant kingdom is found in the Appendix. The following list relates the traditional groupings to the formal classification:

Algal Thallophyta
Division Chlorophyta
Division Phaeophyta
Division Rhodophyta

Embryophyta
Division Bryophyta
Division Tracheophyta

CHLOROPHYTA (THE GREEN ALGAE)

The green algae are of particular interest, because they are generally regarded as the group from which the land plants arose. They are thus probably the only algal division that has not been a phylogenetic dead end. Like land plants, the green algae possess chlorophylls *a* and *b* and carotenoids; unlike many other algae, they have no unusual chlorophylls. The majority of green algae live in freshwater, but some live in moist places on land, and there are many marine species.

Many divergent evolutionary tendencies, all probably beginning with walled and flagellated unicellular organisms, can be traced in the Chlorophyta: (1) the evolution of motile colonies; (2) a change to nonmotile unicells and colonies; (3) the evolution of extensive tubelike bodies with numerous nuclei but without cellular partitions (coenocytic organisms); (4) the evolution of multicellular filaments and even three-dimensional leaflike thalluses.

These evolutionary trends can be studied especially well in the green algae, because many unicellular and rudimentarily multicellular members of the group are still extant. But for that very reason the green algae are difficult to place in a classification system that separates Protista from Plantae. It makes no evolutionary sense to assign the unicellular green algae to one kingdom and their multicellular relatives to another. Hence we shall treat the whole group as true plants whose unicellular representatives have remained at the protistan level.

Chlamydomonas as a representative unicellular green alga   *Chlamydomonas* is a genus of unicellular green algae that probably resemble the ancestral organisms from which the rest of the plant kingdom arose. Its many species are common in ditches, pools, and other bodies of freshwater and in soils. The individual organism is an oval haploid cell with a wall composed largely of glycoprotein; it differs from the walls of many other green algae in lacking cellulose.[3] There are two anterior flagellae of equal length and a single large cup-shaped chloroplast that fills from one half to two thirds of the basal portion of the cell (Fig. 23.1). Inside the chloroplast

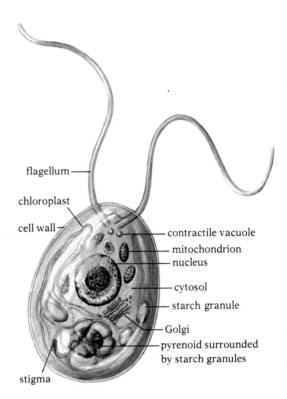

flagellum
chloroplast
cell wall
contractile vacuole
mitochondrion
nucleus
cytosol
starch granule
Golgi
pyrenoid surrounded by starch granules
stigma

23.1   Mature cell of *Chlamydomonas*

---

[3] In the absence of cellulose from its cell walls, *Chlamydomonas* probably differs from the organisms ancestral to the rest of the plant kingdom.

are numerous chlorophyll-bearing lamellae, often arranged in stacks rather like the grana of higher plants. A conspicuous pyrenoid in the basal portion of the chloroplast functions as the site of starch synthesis. A stigma, or eyespot, which helps mediate the organism's positive phototaxis (but is not essential for that behavior), is also located inside the large chloroplast. The cell has no large central vacuole such as is seen in mature cells of higher plants. Two small contractile vacuoles lying near the base of the flagella discharge alternately and rhythmically.[4]

Asexual reproduction is common in *Chlamydomonas* (Fig. 23.2). A vegetative cell resorbs its flagella; then mitotic division of the nucleus and cytokinesis take place. This process gives rise to two daughter cells, both of which lie within the wall of the original cell. In some species the two daughter cells are promptly released by breakdown of the wall; in other species the daughter cells themselves divide while still inside the wall of the parent cell, and a total of 4 (as in the figure), 8, 16, or more daughter cells are produced, depending on the species and conditions of growth. The daughter cells develop a wall and flagella just before they are released as free *zoospores,* which are motile asexual reproductive cells, i.e. motile reproductive cells not specialized as gametes. In *Chlamydomonas* the zoospores are smaller than mature vegetative cells, but otherwise indistinguishable from them; in many species of algae, however, there are noticeable morphological differences between the zoospores and the mature cells. The free zoospores soon grow to full size, completing the asexual reproductive cycle.

Under certain conditions, especially when the concentration of nitrogen in the medium is low, *Chlamydomonas* may reproduce sexually. A mature haploid vegetative cell divides mitotically to produce several gamete cells, which develop walls and flagella and are released from the parent cell. Gametes (usually of two different mating types) are attracted to each other and form large clusters. Eventually the clustered cells move apart in pairs. The members of a pair are positioned end to end, with their flagella, which bear species-specific and mating-type–specific attractant sites at their tips, in close contact. The cells then shed their walls, and their cytoplasms slowly fuse. Finally, their nuclei unite in the process of fertilization, which produces a single diploid cell, the zygote. The zygote sheds its flagella, sinks to the bottom, and develops a thick protective wall. It can withstand unfavorable environmental conditions, such as the drying up of the pond or the cold of winter. When conditions are again favorable, it germinates, dividing by meiosis to produce four (or eight) new flagellated haploid cells, which are released into the surrounding water. The new cells quickly mature, thus completing the sexual reproductive cycle.

Because sexual reproduction in most species of *Chlamydomonas* (e.g. *Chlamydomonas reinhardtii*) is at a very simple level, it can give us insight into the way sexuality probably arose. There are no separate male and female individuals. Furthermore, though the gametes usually differ in their mating-type–specific attractant sites, they are usually morphologically indistinguishable; i.e. they cannot be separated into male gametes (sperm) and female gametes (eggs). Such a condition, where all gametes are alike, is

[4] The descriptions of *Chlamydomonas* and other algae given in this chapter are based largely on material in J. M. Kingsbury, *Biology of the algae*, published privately, Ithaca, N.Y., 1963, and on discussions with Kingsbury.

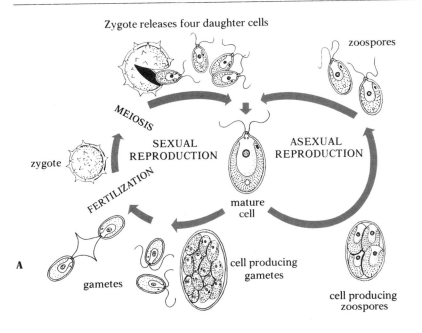

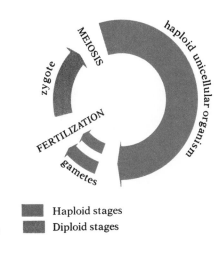

23.2 Life history of *Chlamydomonas*
(A) Diagram showing all stages of both the sexual and asexual cycles.   (B) Schematic representation of life cycle for comparison with those of other organisms. Note that the zygote is the only diploid stage. This type of life cycle was probably characteristic of the first sexually reproducing unicellular organisms, and it may thus be the type from which all other types arose.

called ***isogamy;*** it is probably the primitive (ancestral) condition in plants. The isogametes of *Chlamydomonas* are indistinguishable from vegetative cells; they may be viewed simply as small vegetative cells that tend to fuse and act as gametes under certain conditions.[5] This, too, is probably the primitive condition; the specialization of gametes as morphologically distinctive cells—a characteristic of most higher plants and animals—is surely a later evolutionary development.

Notice that the haploid stages of the life cycle of *Chlamydomonas* are the dominant ones; the only diploid stage is the zygote. Dominance of the haploid stages is characteristic of most very primitive plants, and it seems clear that this was the ancestral condition.

The volvocine series   As an example of one of the evolutionary tendencies that can be traced in the Chlorophyta, let us examine the so-called volvocine or motile-colony series. This is a series of genera showing a gradual progression from the unicellular condition of *Chlamydomonas* to an elaborate colonial organization.

Gonium may be taken as an example of the simplest colonial stage. Each colony of *Gonium* is made up of 4, 8, 16, or 32 cells (depending on the species), each of which is morphologically similar to *Chlamydomonas*. The cells are embedded in a mucilaginous matrix and are arranged in a flat or slightly curved plate. In some species delicate cytoplasmic strands run between the cells; these may provide a route for direct interaction and coordination between the cells of the colony. That some sort of coordination does indeed exist is shown by the organized fashion in which the flagella of all the cells beat together and thus enable the colony to swim as a unit.

[5] A few species of *Chlamydomonas* are anisogamous (one gamete is larger than the other), and a few are even oogamous (one of the gametes is a nonmotile egg cell).

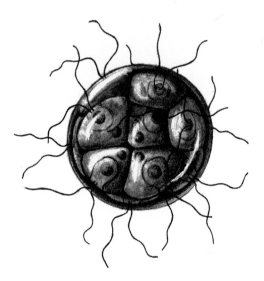

23.3   *Pandorina* colony
The cells are embedded in a gelatinous matrix.

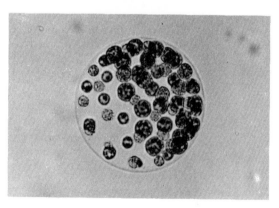

23.4   Photograph of a *Pleodorina* colony
[Courtesy J. M. Kingsbury, Cornell University.]

When asexual reproduction occurs in *Gonium,* all the cells in a colony divide simultaneously. When each cell of the parent colony has divided enough times to contain within its wall the same number of daughter cells as in the parent colony, its wall disintegrates and the daughter cells are released. However, the daughter cells from a single parent cell do not swim off independently as separate zoospores as in *Chlamydomonas.* Instead, they remain together in a common matrix and mature into a new *Gonium* colony. Thus each cell of the parent colony gives rise to a complete new colony.

Sexual reproduction in *Gonium* is similar to that in *Chlamydomonas.* Individual free-swimming cells are released from a colony and function as gametes, fusing in pairs to form zygotes. The gametes are isogamous. As in *Chlamydomonas,* the zygote is the only diploid stage in the life cycle.

Pandorina is a genus of colonial forms slightly more complex than *Gonium.* Each colony is a hollow sphere in which 8, 16, or 32 cells are arranged in a single layer about the periphery, their flagella oriented to the outside of the sphere (Fig. 23.3). Three main advances over *Gonium* are noticeable: (1) The colony shows some regional differentiation; it has definite anterior and posterior halves (detectable both by the orientation of the colony when it is swimming and by the larger size of the stigma in the anterior cells than in the posterior ones). (2) The vegetative cells of the colony are so dependent on one another that they cannot live apart from the colony, and the colony itself cannot survive if disrupted or broken. (3) Sexual reproduction is **heterogamous;** i.e. it involves two kinds of gametes. In *Pandorina* the male gametes are smaller than the female gametes, but both types of gametes have flagella and are free-swimming. This type of heterogamy, where the only morphological difference between the gametes is one of size, is called **anisogamy.**

Eudorina is a still more advanced genus. The spherical colonies contain 16 or 32 cells. The differences between the anterior and posterior portions of the colony are greater than in *Pandorina,* and the heterogamy is more pronounced in that the large female gametes are not released, but remain embedded in the matrix of the colony and are fertilized there by the much smaller free-swimming male gametes.

A still more advanced genus is *Pleodorina* (Fig. 23.4), whose spherical colonies are composed of 32 to 128 cells. These large colonies exhibit considerable division of labor. The anterior cells are vegetative, almost never participating in reproduction. The posterior cells, which function in both asexual and sexual reproduction, are much larger. Sexual reproduction is heterogamous and, as in *Eudorina,* the female gametes are retained within the parental colony; in fact, in some species the large female gametes lose their flagella and thus become true nonmotile egg cells. This type of advanced heterogamy, where only the male gamete is motile and the female gamete is a nonflagellated nonmotile egg cell, is called **oogamy.**

The culmination of the evolutionary series traced here is represented by the genus *Volvox* (Fig. 23.5). Its spherical colonies are very large, consisting of about 500–50,000 cells. Delicate cytoplasmic strands between cells make possible some intercellular communication (Fig. 23.6). Most of the cells are exclusively vegetative. A few cells (between two and 50), scattered in the posterior half of the colony, are much larger than the others and are specialized for reproduction. Each of the female reproductive cells can give rise to an entire new daughter colony (Figs. 23.5, 23.7). Sexual reproduction is always oogamous.

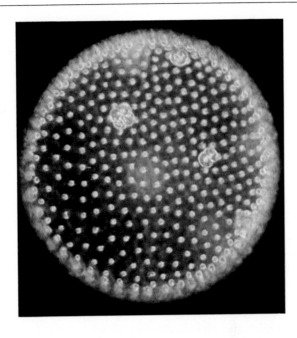

23.5   A *Volvox* colony
The colony is a sphere with a single layer of cells
embedded in gelatinous material; the interior of the
sphere is filled with a watery mucilage. Some small
daughter colonies can be seen developing in this
colony.   [Courtesy Roman Vishniac.]

**23.6   Cells of *Volvox* interconnected by protoplasmic
strands**
[Courtesy Roman Vishniac.]

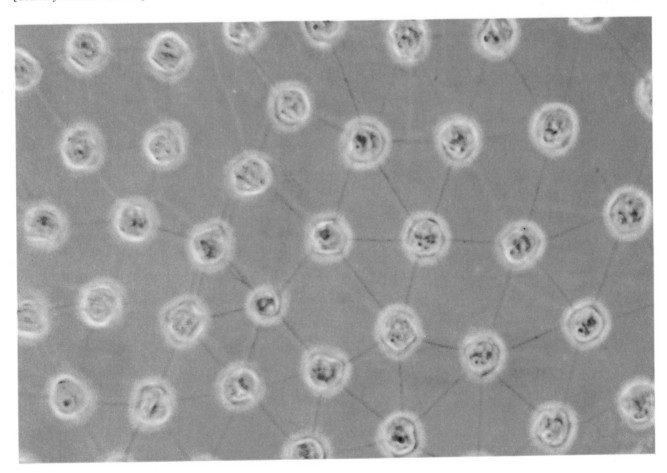

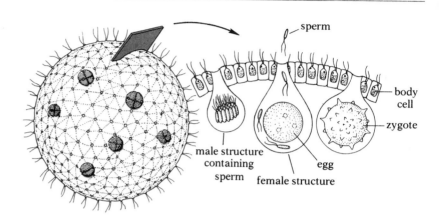

23.7   Reproduction in *Volvox*
Left: The colony is very large, containing 500 to 50,000 vegetative cells. Six daughter colonies (color) at various stages of development can be seen still embedded in the matrix of the parent colony.   Right: Section through the surface of a colony showing male and female reproductive structures. Sperm released by the male structure enter the female structure and fertilize the egg. After a period of inactivity, the zygote divides meiotically, and the haploid cells thus formed then divide mitotically, producing a new daughter colony, which is eventually released. The colony shown here is producing both male and female gametes; in some species or strains the sexes are separate, and a given colony produces only one kind of gamete.

We can summarize the major lines of evolutionary changes manifest in this series as follows: (1) a change from unicellular to colonial life, and a tendency for the number of cells in the colonies to increase; (2) increasing coordination of activity among the cells; (3) increasing interdependence among the vegetative cells, so that they cannot live apart from the colonies and the colonies cannot survive if disrupted; (4) increasing division of labor, particularly between vegetative and reproductive cells; (5) a gradual change from isogamy to anisogamy to oogamy. In both the oogamous algae and the oogamous higher plants the female gametes are characteristically located in a particular place (i.e. retained within the parental organism or colony), and the meeting of gametes becomes a less random process. Hence fewer female gametes need be produced, and more energy can be devoted to providing a large store of nutrients in those few.

In tracing the series *Chlamydomonas–Gonium–Pandorina–Eudorina–Pleodorina–Volvox*, we do not mean to imply that each genus evolved from the preceding one; the available evidence will not allow us to decide whether it did or not. But it does seem likely that each of these genera evolved from an ancestor that resembled in many important ways the modern genus placed just before it in this series, and therefore that the actual evolutionary progression from some unicellular ancestor to *Volvox* involved a series of stages similar to those represented by the modern genera discussed here. Study of this series suggests how complex colonial forms may have evolved, and indicates one possible way in which multicellularity may have arisen in plants. After all, it is largely an arbitrary decision whether one calls *Volvox* colonial or multicellular.

Although multicellular animals certainly did not evolve from *Volvox* or any of the other genera discussed here, a similar evolutionary series, beginning with a nonwalled unicellular organism, may have been the beginning of multicellularity in the animal kingdom.

Some multicellular green algae   Many green algae have a multicellular stage in their life cycle. In most cases this stage is a filamentous thallus, which may be either nonbranching or branching, depending on the species (Fig. 23.8). Let us take *Ulothrix* as a first example.

The species of *Ulothrix* are unbranched filamentous forms, most of which live in freshwater although a few are marine (Fig. 23.9). The filament of each plant is a very small threadlike structure attached to the substratum by a specialized cell called a *holdfast.* Except for the holdfast cell, all the cells of the filament are identical and are arranged end to end in a single series. The filament increases in length as its cells divide horizontally and as the new cells thus added to the chain grow to mature size. Adjacent cells have common end walls—a basic step in the evolution of multicellularity in algae. Each cell contains a single nucleus and a single large chloroplast.

Ulothrix may reproduce by fragmentation (each fragment growing into a complete plant), by asexually produced zoospores, or by sexual processes. In asexual reproduction any cell of the filament except the holdfast may act as a *sporangium,* producing zoospores, each of which has four flagella. After the zoospores are released, they swim about for a short while, and then settle down and give rise to a new filament.

Sexual reproduction is isogamous. The zygote (stage 4 in Fig. 23.9), formed by the union of two of the biflagellate gametes, develops a thick wall and functions as a resting stage capable of withstanding unfavorable environmental conditions. At germination the zygote divides by meiosis, producing haploid zoospores (stage 1), each of which will grow into a new filament (stage 2). The main difference, then, between this life cycle and that of *Chlamydomonas* is the addition of the haploid multicellular stage (stage 2).

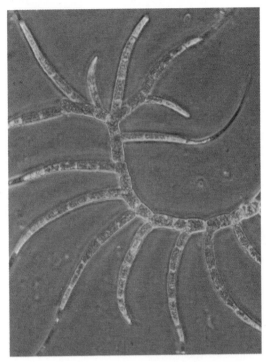

23.8 *Stigeoclonium,* **a branching filamentous green alga**
[Courtesy Roman Vishniac.]

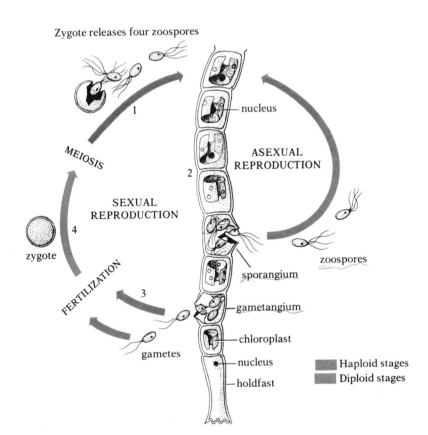

Zygote releases four zoospores

23.9 Life history of *Ulothrix*

The haploid plant may reproduce either asexually or sexually (though a single filament would never reproduce in both ways at once as shown here). Asexual reproduction is the more common; certain cells of the filament develop into sporangia (spore-producing structures) and produce zoospores, which settle down and develop into new filaments. Under certain environmental conditions the filament may cease reproducing asexually and begin reproducing sexually; a cell becomes specialized as a gametangium (gamete-producing structure) and produces isogametes. Two such gametes may fuse in fertilization, producing a zygote, which divides meiotically and releases zoospores.

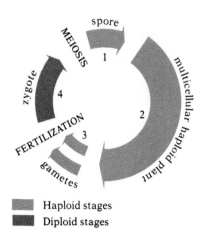

Haploid stages
Diploid stages

23.10 Life cycle characteristic of most multicellular green algae
Note that multicellularity is present only in the haploid phase.

As in *Chlamydomonas*, the only diploid stage is the zygote. This type of life cycle is diagramed in a more generalized form in Fig. 23.10.

Spirogyra (Fig. 23.11), a genus of rather odd filamentous green algae occurring in a great variety of freshwater habitats, is perhaps the most widely distributed member of the Chlorophyta. It has a sexual life cycle similar to that of *Ulothrix* except that the gamete cells are not flagellated and are not released from the plant that produces them. Instead, two filaments come to lie side by side; protuberances develop on the sides of the cells where they are in contact; the walls between the protuberances of each pair of cells disintegrate; and then one cell becomes amoeboid, moves through the conjugation tube, and fuses with the other cell, forming a zygote (Fig. 23.12). Relatively few green algae reproduce by this conjugation process.

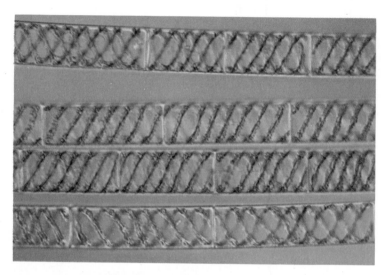

23.11 Photograph of living *Spirogyra* plants
[Courtesy E. V. Gravé.]

23.12 Diagrammatic representations of *Spirogyra*
(A) A single vegetative cell removed from the filament and partially sectioned. Note the unusual spiral chloroplast that runs the length of the cell (some species have more than one chloroplast); numerous pyrenoids are associated with the chloroplast. The cell has a large central vacuole, in which the nucleus is suspended by cytoplasmic threads; these threads connect to the peripheral cytosol, which forms a layer just inside the cell wall. (B) Conjugating filaments. The two filaments lie side by side, and a conjugation tube develops between each pair of cells. One cell acts as the sperm, moving through the tube to fuse with the other cell (see middle pair of cells). The zygote thus formed is the only diploid stage in the life cycle.

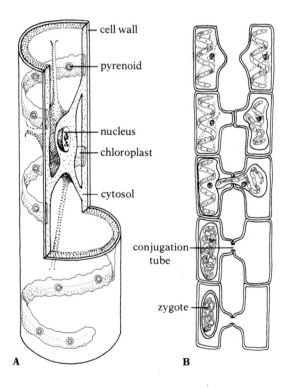

cell wall

pyrenoid

nucleus

chloroplast

cytosol

conjugation tube

zygote

A

B

Ulva, or sea lettuce, is an example of a green alga with an expanded leaflike thallus two cells thick (Figs. 23.13 and 1.10, p. 17). Its sexual life cycle is more complex than those of the other green algae discussed here in that it includes both multicellular haploid and multicellular diploid stages (stages 2 and 5 in Fig. 23.14). The entire cycle can be summarized as follows: Haploid zoospores (stage 1) divide mitotically to produce the haploid multicellular thalluses of stage 2. These may reproduce either asexually by means of zoospores or sexually by means of gametes (stage 3). Fusion of pairs of gametes (fertilization) produces diploid zygotes (stage 4). Upon germination, the zygotes divide mitotically (not meiotically as in the green algae previously discussed), producing diploid multicellular thalluses (stage 5). Eventually certain reproductive cells (sporangia) of these diploid plants divide by meiosis, producing haploid zoospores, which begin a new cycle. A life cycle of this type is said to exhibit **alternation of generations** in that a haploid multicellular phase alternates with a diploid multicellular phase. The haploid multicellular stage is customarily called a **gametophyte** (meaning that it is a plant that can produce gametes), and the diploid multicellular stage is called a **sporophyte** (meaning that it is a plant that can reproduce only by spores).

We have seen that multicellularity in plants arose first in the gametophyte, and that many green algae have no sporophyte stage. *Ulva* shows a more advanced life cycle in that both gametophyte and sporophyte stages are present. Furthermore, the two stages are equally prominent in *Ulva,* being nearly equal in duration and almost identical in appearance; in other words, the haploid portion of the life cycle is no longer dominant over the diploid.

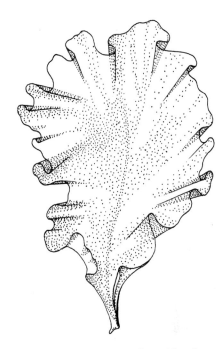

23.13   *Ulva,* a marine green alga with a three-dimensional leaflike thallus

PHAEOPHYTA (THE BROWN ALGAE)

The brown algae are almost exclusively marine, the few freshwater species being quite rare. Many of the plants called seaweeds are members of this division. They are most common along rocky coasts of the cooler parts of the oceans, where they normally grow attached to the bottom in the littoral

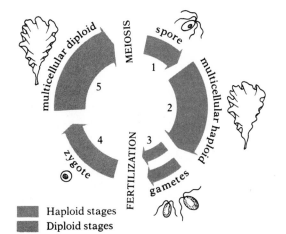

23.14   Life cycle of *Ulva*
The gametophyte (multicellular haploid) and sporophyte (multicellular diploid) stages are equally prominent. *Ulva* and its close relatives are unusual among the green algae in having a life cycle of this sort; most of the Chlorophyta have no alternation of generations, the sporophyte stage being absent.

23.15   Brown algae (mostly *Fucus*) growing on rocks exposed at low tide
[Anne Wertheim, Bruce Coleman Inc.]

23.16   A giant kelp (*Macrocystis pyrifera*)
At the base of each of the flattened blades is a gas-filled bladder that functions as a float.   [Courtesy Douglas Faulkner.]

(intertidal) and upper sublittoral zones. They may be seen in great abundance covering the rocks exposed at low tide along the New England coast (Fig. 23.15). A few species occur in warmer seas, and some of these differ from the majority of brown algae in being able to live and grow when detached from the substratum; e.g. some species of *Sargassum* form dense floating mats that cover much of the surface of the so-called Sargasso Sea, which occupies some six and a half million square kilometers of ocean between the West Indies and North Africa.

All brown algae are multicellular, and most are macroscopic, some growing as long as 45 m or more. The thallus (plant body) may be a filament, or it may be a large and rather complex three-dimensional structure (Figs. 23.16 and 1.8, p. 16). The latter type of thallus has apparently arisen several times independently; in some species it develops from interwoven and tightly compacted filaments, and in others it results from cell divisions in more than one plane. The individual cells have cell walls composed of cellulose and a gummy material called alginic acid. The walls sometimes have pits through which plasmodesmata pass. The cells usually contain a large vacuole, one or several plastids, and sometimes a pyrenoid. Unlike the cells of most higher land plants, they usually have centrioles.

Like all photosynthetic plants, the Phaeophyta possess chlorophyll *a*. However, they have chlorophyll *c* instead of the chlorophyll *b* found in green algae and in land plants. Large amounts of a xanthophyll carotenoid called **fucoxanthin,** which are also present, give the characteristic brownish color to these algae. A polysaccharide called laminaran is the principal storage product.

Asexual reproduction is most often by flagellated zoospores. Sexual reproduction often involves specialized multicellular sex organs called **gametangia.** In isogamous species all the gametangia are alike. But in heterogamous species they are of two kinds. When the heterogamy is of the oogamous type, the gametangia that produce sperm are called **antheridia** and those that produce eggs **oogonia.**

In brown algae the sex organs are ordinarily not enclosed by a protective layer of sterile jacket cells. The life cycle is usually characterized by an alternation of gametophyte (haploid) and sporophyte (diploid) multicellular generations. In many forms, such as *Ectocarpus*, the gametophyte and sporophyte (stages 2 and 5 in Fig. 23.17) are essentially similar in structure, and neither can be said to be dominant. In other forms, such as *Laminaria*, the haploid gametophyte is reduced and the diploid sporophyte is much larger and more prominent. In a few, such as *Fucus* (Fig. 23.18), reduction of the haploid stages has progressed so far that there is no longer any multicellular haploid gametophyte and the only haploid cells in the life cycle are the gametes (Fig. 23.19); such a life cycle, which is very rare in plants, is the sort seen in animals.

Let us look more closely at a few representative genera of brown algae. *Ectocarpus* has a branching filamentous thallus. The diploid sporophyte plants (stage 5 in Fig. 23.17) sometimes bear small unicellular sporangia, in which haploid zoospores (stage 1) are produced by meiosis. After swimming about for a while, the zoospores settle down and develop into haploid multicellular gametophyte plants (stage 2). These plants may bear multicellular gametangia, in which morphologically isogamous gametes (stage 3) are produced. Two gametes (from different plants) may fuse in fertilization to form a zygote (stage 4), which is motile at first, but soon settles down and germinates, giving rise to a new diploid multicellular sporophyte plant (stage 5), thus completing the cycle.

Laminaria belongs to the group of brown algae commonly called kelps.

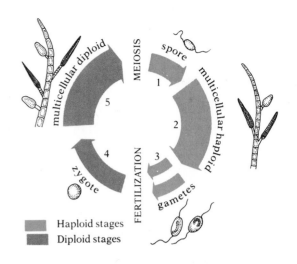

23.17 Life cycle of *Ectocarpus*
The gametophyte (multicellular haploid) and sporophyte (multicellular diploid) stages are equally prominent.

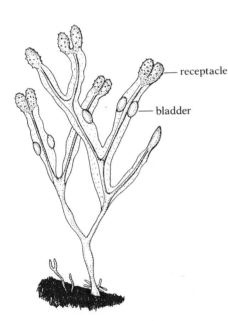

23.18 *Fucus* (often called rockweed), a brown alga common along northern coasts
Left: Each thallus is flattened and characterized by repeated dichotomous branching. Each younger axis consists of a midrib and thin paired wings. In some species (including the one shown here), there are bladders (floats) at intervals along the wings. The tips of fertile thalluses develop swollen reproductive structures called receptacles, whose surface is pocked by numerous tiny openings that lead into cavities (conceptacles) where the sex organs are located. In some species each individual has both male and female organs; in others the sexes are separate.   Right: The receptacles can be seen especially well here. [Right: R. P. Carr, Bruce Coleman Inc.]

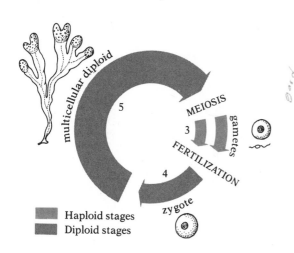

Haploid stages
Diploid stages

23.19 Life cycle of *Fucus*

Fucus and its close relatives have a very unusual life cycle. They are the only multicellular plants in which the multicellular haploid stage (the gametophyte) is completely absent and meiosis produces gametes directly (i.e. both stages 1 and 2 are absent). In this respect their life cycle is like that of animals.

The sporophyte thallus is large (about 2 m long in *Laminaria;* 45 m long or more in some kelps) and consists of a rootlike **holdfast,** a stemlike **stipe,** and an expanded leaflike **blade** (see Fig. 6.1, p. 238). Although thallophyte plants usually lack tissue differentiation, the stipe of some kelps has an outer surface tissue (epidermis), a middle tissue (cortex) containing many plastids, and a central core tissue (medulla); it may even have a meristematic layer similar to the cambium of higher vascular plants and, in a few genera, a phloemlike conductive tissue in the medulla. In short, these brown algae are complex plants that have convergently evolved many similarities to the vascular plants. However, none of them have a protective layer of sterile jacket cells around their reproductive organs; none develop multicellular embryos inside the oogonia; none have a cuticle; and none have xylem.

RHODOPHYTA (THE RED ALGAE)

The red algae are mostly marine seaweeds (Figs. 23.20 and 1.9, p. 16), but a few live in freshwater or on land. They often occur at greater depths than the brown algae. Most are multicellular and are attached to the substratum, but a few species are unicellular. No red algae attain the very large sizes often seen in brown algae.

The cell walls contain cellulose and also large quantities of mucilaginous material. The reserve product is not starch, but a polysaccharide similar to it called floridean starch. Red algae are an important source of commercial colloids—among others, agar used in culturing bacteria; suspending agents used in chocolate milk and puddings; stabilizers used in ice creams, some cheeses, and salad dressings; and moisture retainers used in icings, creams, and marshmallows.

In addition to chlorophyll *a,* which is found in all photosynthetic organisms except the photosynthetic bacteria, the Rhodophyta often possess chlorophyll *d,* which is not found in any other group of plants. They also contain phycocyanins and phycoerythrins. It is the phycoerythrins that give many of these algae their characteristic reddish color. It should be emphasized, however, that "red algae" are not always red; many are black, violet, brownish, yellow, or even green (Fig. 23.20).

The accessory pigments of the Rhodophyta play an important role in absorbing light for photosynthesis. The wavelengths preferentially absorbed by chlorophyll *a* for use in photosynthesis are among those at the ends of the visible spectrum (recall the discussion of the absorption spectrum of chlorophyll, p. 143). But those wavelengths do not penetrate to the depths at which the red algae grow, partly because of the imperfect transparency of the water, partly because they are selectively absorbed by pigmented phytoplankton. The wavelengths that do penetrate deep enough are mostly those of the central portion of the spectrum, which are not readily absorbed by chlorophyll *a.* But these wavelengths can be absorbed by the accessory pigments of the Rhodophyta, which then pass the energy to chlorophyll *a.* Thus the accessory pigments make it possible for red algae to live at depths where other algae, lacking these pigments, cannot survive.

The life cycles of red algae are usually very complex, and only a few have been worked out in detail. There is commonly some sort of alternation of generations. Flagellated cells never occur; even the sperm cells lack flagella and must be carried to the egg cells by water currents.

23.20 Two examples of red algae
Left: *Kallimenia reniformis*, a species with a flattened bladelike thallus.   Right: *Furcellaria fastigiata*, a branching alga that appears more brown than red.   [Courtesy Hervé Chaumeton.]

THE MOVEMENT ONTO LAND

Let us now turn to the Embryophyta, which have evolved numerous adaptations for life on land. We have seen that life probably arose in water, and that many plants, notably the algae, are still largely restricted to the aquatic environment; the few algae that live on land are not truly terrestrial, occurring, as they do, only in very moist places and actually living in a film of moisture. The evolutionary move from an aquatic mode of existence to a terrestrial one was not simple, for the terrestrial environment is in many ways hostile to life. Among the many problems faced by land plants are the following:

1. Obtaining enough water when fluid no longer bathes the entire surface of the plant body.
2. Transporting water and dissolved substances from restricted areas of intake to other parts of the plant body, and transporting the products of photosynthesis to those part of the plant that no longer carry out this process for themselves.
3. Preventing excessive loss of water by evaporation.
4. Maintaining a sufficiently extensive moist surface for gas exchange when the surrounding medium is air instead of liquid.
5. Supporting a large plant body against the pull of gravity when the buoyancy of an aqueous medium is no longer available.
6. Carrying out reproduction when there is little water through which flagellated sperm may swim and when the zygote and early embryo are in severe danger of desiccation.
7. Withstanding the extreme fluctuations in temperature, humidity, wind, light, and other environmental parameters to which terrestrial organisms are often subjected.

Much of the evolution of the embryophyte plants can best be understood in terms of adaptations that help solve these problems.

As we indicated earlier in this chapter, embryophyte plants characteristically have multicellular sex organs with an outer layer of sterile (nonreproductive) jacket cells that help protect the enclosed gametes from desiccation; male and female sex organs of this type are known as **antheridia** and **archegonia**.[6] The sporangia in embryophytes are also multicellular, and they too have a layer of jacket cells. All embryophytes are oogamous, and the egg cells are fertilized while they are still contained within the archegonia. Each zygote develops into a multicellular diploid embryo while still inside the archegonium. The embryo obtains some of its water and nutrients from the parent plant and is thus a parasite on it. This type of embryonic development, which is clearly an adaptation permitting the stages of development most susceptible to desiccation to take place in a favorably moist microenvironment, is strongly reminiscent of the internal gestation of mammals.

The surfaces of the aerial parts of the plant bodies of embryophytes are usually covered by a waxy cuticle, which waterproofs the epidermis and helps prevent excessive water loss.

The principal pigments in embryophytes are chlorophylls *a* and *b* plus carotenoids, and the reserve material is starch (Table 23.1). In other words, these plants are biochemically similar to the Chlorophyta, from which they almost certainly arose.

[6] Like "oogonium," "archegonium" denotes an organ producing female gametes, but most botanists restrict "archegonium" to the jacketed female reproductive organs of embryophytes, "oogonium" to the unjacketed ones of thallophytes.

TABLE 23.1 *A comparison of the major plant divisions*

Characteristics	Chlorophyta	Phaeophyta	Rhodophyta	Bryophyta	Tracheophyta
Usually, flagellated sperm	+	+	−	+	+ or −
Chlorophyll *b*	+	− (*c* instead)	− (*d* instead)	+	+
Principal reserve material usually starch	+	−	−	+	+
Sporophyte equal or dominant to gametophyte in most species	*	+	*	−	+
Usually, multicellular sex organs with jacket cells	−	−	−	+	+
Embryo development within archegonium	−	−	−	+	+
Cuticle usually present	−	−	−	+	+
Both xylem and phloem present in most species	−	−	−	−	+

* Among Chlorophyta and Rhodophyta species for which the full life cycle is known, some have a dominant gametophyte and some have a dominant sporophyte. Since many species have not yet been studied, it is not possible to say which condition is the more usual.

BRYOPHYTA (THE LIVERWORTS, HORNWORTS, AND MOSSES)

The bryophytes are relatively small plants that grow in moist places on land—on damp rocks and logs, on the forest floor, in swamps or marshes, or beside streams and pools. Some species can survive periods of drought, but only by becoming dormant and ceasing to grow. In short, the bryophytes live on land, but they have never become fully emancipated from their ancestral aquatic environment, and they have therefore never become a dominant group of plants. Their great dependence on a moist environment is linked to two characteristics: They retain flagellated sperm cells, which must swim to the egg cells in the archegonia; and most lack vascular tissues—and hence the means for efficient long-distance internal transport of fluids. A few bryophytes, however, possess cells that resemble the sieve cells of phloem, and some bryophytes possess elongated cells that sometimes transport water and are thus functional analogues of tracheids. The absence of true xylem cells with secondarily thickened walls, which function as major supportive elements in vascular plants, has probably also limited the size the bryophytes can attain.

The bryophytes are thought by some botanists to have arisen from filamentous green algae. Indeed, a very young moss plant, called a protonema (Fig. 23.21), often closely resembles a green algal filament. As the plant grows, it forms some branches (rhizoids) that enter the ground and function like roots, anchoring the plant and absorbing water and nutrients; different branches form upright shoots with stemlike and leaflike parts.

Other botanists hypothesize that the bryophytes arose from true vascular plants by "downgrade" evolution, i.e. by secondary evolutionary loss of structures. They point to the rudimentary vascular tissue of some bryophytes, which they regard as vestiges of tissues well developed in ancestral forms rather than as newly evolved structures, and they also point to the nonfunctional guard cells seen in some bryophytes, which may be indicative of the ancestral condition.

We noted earlier among the larger and more complex algae, most of which exhibit alternation of generations, an apparent evolutionary tendency, in many instances, toward reduction of the gametophyte (multicellular haploid) stage and increasing emphasis of the sporophyte (multicellular diploid) stage. In brown algae, for example, the sporophyte is at least as prominent as the phylogenetically older gametophyte, and it is often much more prominent. As we shall see later, the same evolutionary tendency is found in the vascular plants. But this tendency is not apparent in the bryophytes, where the haploid gametophyte (stage 2 in Fig. 23.22) is clearly the dominant stage in the life cycle. The "leafy" green moss plant or liverwort is the gametophyte. These plants bear antheridia and archegonia in which gametes (stage 3) are produced by mitosis. The flagellated male gametes (sperm) are released from the antheridia and swim through a film of moisture, such as rain or heavy dew. Responding to chemical attractants, they swim to archegonia, where they fertilize the egg cells, producing zygotes (stage 4). Each zygote then divides mitotically, producing a diploid sporophyte (stage 5).

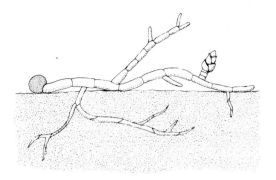

23.21 A young moss plant
The spore (color) gives rise to a filamentous plant (called a protonema) that strikingly resembles a green alga. The protonema develops into the mature moss plant.
[Modified from H. J. Fuller and O. Tippo, *College botany*, Holt, Rinehart & Winston, Inc., 1954.]

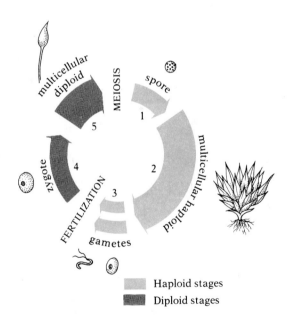

23.22 Life cycle of a bryophyte
Both gametophyte (stage 2) and sporophyte (stage 5) are present. The former is dominant.

In a moss this sporophyte is a relatively simple structure consisting of three parts: a foot embedded in the "leafy" green gametophyte, a stalk, and a distal capsule, or sporangium (Figs. 23.23, 23.24). The sporophyte has chloroplasts and carries out some photosynthesis, but it also obtains nutrients parasitically from the gametophyte to which it is attached. Meiosis occurs within the mature capsule of the sporophyte, producing haploid spores (stage 1), which are released. These spores, encased in highly specialized walls that are extremely resistant to degradation, may remain inactive for a long time (sometimes many years) if conditions are unfavorable. When they do germinate, they develop into protonemata and eventually into mature gametophyte plants (stage 2), thus completing the life cycle.

The gametophyte plants of some liverworts resemble mosses except that the "leaves" are scaly in appearance, and the "stem" is prostrate. Other liverworts grow as flat green structures lying on the substratum (see Fig. 1.11, p. 17). In some species the antheridia and archegonia are borne in receptacles located at the top of stalks that arise from the flat part of the plant body (Fig. 23.25); in other species there is no receptacle or stalk (sometimes only the stalk is missing), the reproductive structures being embedded in the upper portion of the prostrate "leaf." The life cycle is much like that of mosses, except that the sporophyte is even simpler (Fig. 23.26). Asexual reproduction sometimes occurs by production of special cells called gemmae, usually borne in cuplike structures located on the surface of the flat gametophyte (Fig. 23.27). When detached from the parent plant, the gemmae can grow into new gametophytes.

23.23   Gametophyte and sporophyte stages of a moss
The haploid gametophyte (color) is the lower "leafy" plant; the "leaves," except at their midrib, are only one cell thick. The diploid sporophyte plant (black), which consists of a foot (not visible here), a stalk, and a capsule, is attached to the gametophyte and is to some degree parasitic on it. (The capsule of the sporophyte is here covered by a cap—the calyptra—derived from the archegonium of the gametophyte; in time it will fall away, leaving the capsule fully exposed.)

23.24   Photograph of moss plants showing both the leafy green gametophytes and the red-stalked sporophytes
[David Overcash, Bruce Coleman Inc.]

23.25   Liverworts (*Marchantia*) with stalked receptacles bearing archegonia
[Courtesy E. S. Ross.]

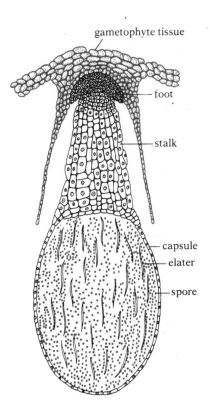

gametophyte tissue

foot

stalk

capsule

elater

spore

23.26   Sporophyte of *Marchantia*

The sporophyte of this liverwort is a small structure consisting of a foot, a short stalk, and a capsule. The foot remains embedded in the gametophyte plant, in the tissue of the undersurface of the umbrella-shaped receptacle (see Fig. 23.25). The mature capsule contains spores and elaters, which are elongate cells with spirally thickened walls. Eventually the wall of the capsule dries and bursts, releasing the spores. Ejection of the spores is aided by the elaters, which twist and jerk as they dry, thus throwing the spores from the capsule.

23.27   Gemma cups

Two cups can be seen here on the gametophyte of *Marchantia*. The gemma cups function in asexual reproduction.   [Courtesy V. N. Rockcastle, Cornell University.]

TRACHEOPHYTA (THE VASCULAR PLANTS)

Though most bryophytes live on land, in a sense they are not fully terrestrial. The tracheophytes, by contrast, have evolved a host of adaptations to the terrestrial environment that have enabled them to invade all but the most inhospitable land habitats. In the process, they have diverged sufficiently from one another for botanists to classify them in five subdivisions:

Division Tracheophyta
    Subdivision Psilopsida (psilopsids)
    Subdivision Lycopsida (club mosses)
    Subdivision Sphenopsida (horsetails)
    Subdivision Pteropsida (ferns)
    Subdivision Spermopsida (seed plants)

All members of this division (with a few minor exceptions) possess four important attributes lacking in even the most advanced and complex algae: a protective layer of sterile jacket cells around the reproductive organs; multicellular embryos retained within the archegonia; cuticles on the aerial parts; and xylem (see Table 23.1). All four are obviously fundamental adaptations for a terrestrial existence. Many other such adaptations, absent in the earliest tracheophytes, appear in more advanced members of the division; a history of the evolution of these adaptations is a history of the increasingly extensive exploitation of the terrestrial environment by vascular plants. Let us briefly trace the history of adaptation to life on land.

PSILOPSIDA

The oldest undisputed fossil representatives of the vascular plants can be placed late in the Silurian period, which means that they lived more than 395 million years ago (see Table 20.1, p. 906). They are classified in the subdivision Psilopsida, most of whose members lived during the Devonian period

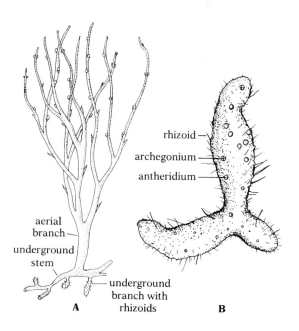

23.28 *Psilotum*
(A) Sporophyte. (B) Gametophyte: This stage is entirely subterranean. The two plants are not drawn to the same scale; the sporophyte is actually much larger than the gametophyte. (C) Photograph of a living *Psilotum.* [(C) Courtesy Carolina Biological Supply Co.]

and then became extinct. Two living genera, *Psilotum* and *Tmesipteris*, have traditionally been regarded as members of this ancient group, but recent evidence from embryology and from the morphology of the gametophyte— as D. W. Bierhorst of the University of Massachusetts has pointed out—suggests that they may actually be very primitive ferns. If this is so, then Psilopsida contains only extinct species. Whether *Psilotum* and *Tmesipteris* should be retained in the Psilopsida despite the differences between them and the ancient members of that class, from which they are separated by about 400 million years with no intervening fossils, or whether they should be transferred to the ferns, despite their lack of true roots and the presence of other primitive features, is still an open question.

The psilopsid sporophytes are simple dichotomously branching plants that lack leaves[7] and have no true roots, although they do have underground stems that bear unicellular rhizoids similar to root hairs (Fig. 23.28A). The aerial stems are green and carry out photosynthesis. There is no cambium, and hence no secondary growth. Sporangia develop at the tips of some of the aerial branches. Within the sporangia meiosis produces haploid spores.

In *Psilotum* and *Tmesipteris* the spores give rise to minute subterranean gametophytes (Fig. 23.28B). Each gametophyte bears both archegonia and antheridia and thus produces both eggs and sperm. When the gametes unite in fertilization, they form diploid zygotes that develop into the sporophyte plants described above, thus completing the life cycle. Note that although the diploid sporophyte (stage 5) is more prominent in the modern genera and hence may be said to be dominant, the haploid gametophyte (stage 2) is still relatively large. It seems possible that the earliest Psilopsida had an alternation of coequal generations.

Botanists, noting the resemblance between the Psilopsida and certain branching filamentous green algae, have assumed it was from such algae that these primitive vascular plants arose. Since it is the psilopsid sporophyte, not the gametophyte, that most resembles the green algae, it must be concluded that, if the vascular plants did indeed evolve from green algae, those algae probably had a prominent sporophyte stage, at least coequal with the gametophyte.

23.29 *Lycopodium* **plants with strobili**
[R. P. Carr, Bruce Coleman Inc.]

LYCOPSIDA (THE CLUB MOSSES)

The first representatives of the subdivision Lycopsida appeared in the middle of the Devonian period, almost 10 million years after the first Psilopsida. During the late Devonian and the Carboniferous periods these were among the dominant plants on land. Some of them were very large trees that formed the earth's first forests. Toward the end of the Paleozoic era, however, the group was displaced by more advanced types of vascular plants, and only five genera are alive today. One of these, *Lycopodium* (often called running pine or ground pine), is common in many parts of the United States and is frequently used in Christmas decorations (Fig. 23.29).[8]

Unlike the psilopsids, lycopsids have true roots. It is generally supposed that these arose from branches of the ancestral algae that penetrated the soil and branched underground. Lycopsids also have true leaves, which are

---

[7] Some Psilopsida have scalelike structures that superficially resemble leaves.

[8] The other living genera of lycopsids are *Phylloglossum*, *Selaginella*, *Isoetes*, and *Stylites*.

thought to have arisen as simple scalelike outgrowths (emergences) from the outer tissues of the stem. Certain of the leaves that have become specialized for reproduction bear sporangia on their surfaces. Such reproductive (fertile) leaves are called **sporophylls.** In many lycopsids the sporophylls are congregated on a short length of stem and form a conelike structure (strobilus) (Fig. 23.30). The cone is rather club-shaped; hence the name "club mosses" for the lycopsids (note, however, that lycopsids are not related to the true mosses, which are bryophytes).

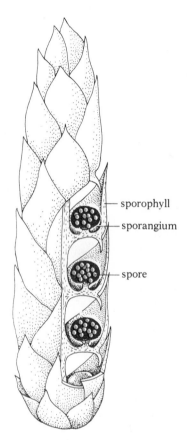

23.30   A strobilus of a lycopsid, partially sectioned to show the arrangement of sporangia on the sporophylls

23.31   **Carboniferous swamp forest**
Note the tree sphenopsids with their jointed stems and whorls of leaves. The trunk at right is a lycopsid. [Portion of group in Carnegie Museum, Pittsburgh. Used by permission.]

23.32 A fossil of a sphenopsid
A whorl of leaves is located at each joint of the
stem.   [Courtesy Field Museum of Natural History,
Chicago.]

The spores produced by *Lycopodium* are all alike, and each can give rise to
a gametophyte that will bear both archegonia and antheridia. However,
some lycopsids (e.g. *Selaginella*) have two types of sporangia, which produce
different kinds of spores. One type of sporangium produces very large
spores called ***megaspores,*** which develop into female gametophytes bearing
archegonia; the other type produces small spores called ***microspores,*** which
develop into male gametophytes bearing antheridia. Plants like *Lycopodium*
that produce only one kind of spore, and hence have only one kind of
gametophyte that bears both male and female organs, are said to be ***homosporous.*** Plants like *Selaginella* that produce both megaspores (female) and
microspores (male)—i.e. in which the sexes are separate in the gametophyte
generation—are said to be ***heterosporous.***

SPHENOPSIDA (THE HORSETAILS)

The sphenopsids first appear in the fossil record late in the Devonian period.
They became a major component of the land flora during the Carboniferous
period and then declined. Members of the one living genus, *Equisetum*, are
commonly called horsetails or scouring rushes. Though most of these are
small (less than one meter), some of the ancient sphenopsids were large trees
(Fig. 23.31). Much of the coal we use today was formed from the dead bodies
of these plants.

Like the lycopsids, sphenopsids possess true roots, stems, and leaves. The
stems are hollow and are jointed. Whorls of leaves occur at each joint (Fig.
23.32). Many of the extinct sphenopsids had cambium, and hence secondary
growth, but the modern species do not. Spores are borne in terminal cones
(strobili) (Fig. 23.33). In *Equisetum* all spores are alike (i.e. the plants are
homosporous) and give rise to small gametophytes that bear both archegonia and antheridia (i.e. the sexes are not separate).

23.33 *Equisetum*
At right is a stalk bearing a mature reproductive
cone.   [Courtesy Hervé Chaumeton.]

23.34 Bracken fern in autumn
The large leafy sporophyte plants are beginning to yellow. [R. P. Carr, Bruce Coleman Inc.]

23.36 Leaves of sensitive fern
The sterile leaves have expanded blades, but the fertile leaves (sporophylls) are spikes bearing grapelike clusters of reproductive organs. [Ray Simons, Photo Researchers, Inc.]

23.35 Undersurface of fertile leaf (sporophyll) of polypody fern
Each of the round dots is a sorus, which is a cluster of many tiny sporangia. [N. Fox-Davies, Bruce Coleman Inc.]

PTEROPSIDA (THE FERNS)

In the opinion of many botanists, the ferns evolved from the Psilopsida. They first appeared in the Devonian period and greatly increased in importance during the Carboniferous. Their decline late in the Paleozoic era was much less severe than that of the psilopsids, lycopsids, and sphenopsids, and, as you doubtless know, there are many modern species.

The ferns are fairly advanced plants with a very well developed vascular system and with true roots, stems, and leaves. The leaves are thought to have arisen in another way than those of the lycopsids. Instead of emergences, they are probably flattened and webbed branch stems; i.e. a group of small terminal branches probably became arranged in the same plane (planated), and the interstices filled with tissue.[9] Such leaves are larger, and provide a much greater surface area for photosynthesis, than the emergence leaves of the lycopsids and sphenopsids.

The leaves of ferns are sometimes simple, but more often they are compound, being divided into numerous leaflets that may give the plant a lacy appearance.[10] In a few ferns (e.g. the large tree ferns of the tropics), the stem is upright, forming a trunk. But in most modern ferns, especially those of temperate regions, the stems are prostrate on or in the soil, and the large leaves are the only parts normally seen.

The large leafy fern plant is the diploid sporophyte phase (Fig. 23.34). Spores are produced in sporangia located in clusters on the underside of some leaves (sporophylls) (Fig. 23.35). In some species the sporophylls are relatively little modified and look like the nonreproductive leaves. In other species the sporophylls look quite different from vegetative leaves; sometimes they are so highly modified that they do not look like leaves at all, forming spikelike structures instead (Fig. 23.36).

Most modern ferns are homosporous; i.e. all their spores are alike. After germination, the spores develop into gametophytes that bear both archegonia and antheridia (Fig. 23.37). These gametophytes are tiny (less than one centimeter wide), thin, and often more or less heart-shaped. Although most people are familiar with the sporophytes of ferns, few have ever seen a gametophyte, and even fewer would guess that it had anything to do with a fern. Small and obscure as it is, however, the fern gametophyte is an independent photosynthetic organism. Here, then, is a life cycle in which all five principal stages are present, but in which the multicellular haploid stage has been much reduced and the multicellular diploid stage emphasized (Fig. 23.38).

In some respects, the ferns (and also the three primitive groups of vascular plants discussed above) are no better adapted for life on land than the bryophytes. Their vascularized sporophytes can live in drier places and grow bigger, but for a number of reasons—because their nonvascularized free-living gametophytes can survive only in moist places, because their sperm are

[9] Leaves arising as emergences are called microphylls. Those arising as planated and webbed branch systems are called megaphylls.

[10] When a fern leaf, or frond, is divided into leaflets, the leaflets are called pinnae. The pinnae may themselves be subdivided into pinnules.

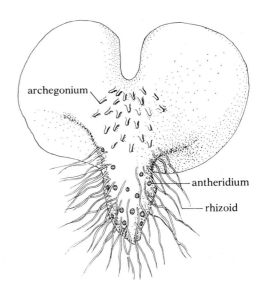

23.37   Fern gametophyte
This is a much-magnified view of the undersurface of the tiny heart-shaped organism.   [Modified from H. J. Fuller and O. Tippo, *College botany*, Holt, Rinehart & Winston, Inc., 1954.]

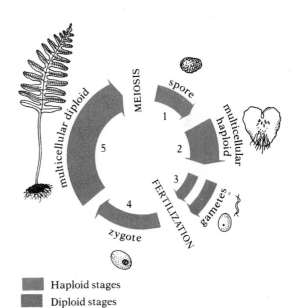

Haploid stages
Diploid stages

23.38   Life cycle of ferns
Both gametophyte (stage 2) and sporophyte (stage 5) are present; the latter is much the more prominent. Compare this life cycle with that of bryophytes, shown in Fig. 23.22.

THE EVOLUTION OF SPOROPHYTE DOMINANCE

An obvious question, in view of the tendency toward increasing dominance of the sporophyte generation in so many major groups of plants, is why this shift of emphasis in the life cycle should have occurred. Why should diploidy be adaptively superior to haploidy?

The question can be answered only tentatively. In haploid organisms there can be no such thing as dominance and recessiveness; the one dose of each gene must be phenotypically expressed. This means that selection is very rigorous; deleterious new mutations cannot easily be covered up, and the effects of beneficial new mutations are immediately felt. In diploid organisms, by contrast, recessive genes harmful or nearly neutral in their effects may be carried in populations for a long time and may, if conditions change, become more beneficial in the future; such recessive genes thus represent a pool of genetic potential that may allow greater flexibility of adaptational response. Another source of evolutionary potential is the heterozygosity that diploidy makes possible; as we have seen earlier, heterozygous phenotypes are often adaptively superior to either of the corresponding homozygous phenotypes.

There is reason to believe, then, that the superiority of diploidy is the greater evolutionary flexibility it confers in the face of predation, intense competition, unstable environmental conditions, and other situations that make for rigorous selection.

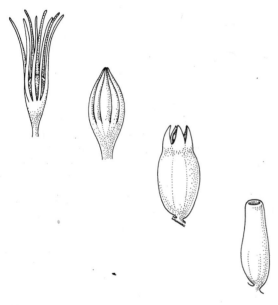

23.39   A model for the possible evolution of the seed
Shown here are the seeds of four species of extinct seed ferns (pteridosperms); note the progressive development of the integument. This sequence is thought to be similar to the one by which seeds of other plants evolved. [Redrawn from H. N. Andrews, *Science*, vol. 142, 1963. Copyright 1963 by the American Association for the Advancement of Science.]

flagellated and must have a film of moisture through which to swim to the egg cells in the archegonia, and because the young sporophyte develops directly from the zygote without passing through any protected seedlike stage—these plants are most successful in habitats where there is at least a moderate amount of moisture.

SPERMOPSIDA (THE SEED PLANTS)

The seed plants have been by far the most successful in fully exploiting the terrestrial environment. They first appeared in the late Devonian, and in the Carboniferous they soon replaced the lycopsids and sphenopsids as the dominant land plants, a position they still hold today. In these plants the gametophytes are even more reduced than in the ferns—they are not photosynthetic or free-living—and the sperm of most modern species are not independent free-swimming flagellated cells. In addition, the young embryo, together with a rich supply of nutrients, is enclosed within a desiccation-resistant seed coat and can remain dormant for extended periods if environmental conditions are unfavorable (Fig. 23.39). In short, the aspects of the reproductive process that are most vulnerable in more primitive vascular plants have been eliminated in the seed plants (Table 23.2).

The seed plants have traditionally been divided into two classes, the Gymnospermae and the Angiospermae. In recent years, however, it has become increasingly clear that the relationships between the five groups bracketed together as the gymnosperms are not particularly close and that these groups differ from one another at least as much as they differ from the angiosperms. Consequently many modern classifications recognize each of the gymnospermous groups as a separate class. We have adopted this procedure in the technical classification given in the Appendix and outlined below, but shall discuss the gymnospermous groups together.

Subdivision Spermopsida
    Class Pteridospermae ⎫
    Class Cycadae |
    Class Ginkgoae   ⎬ Gymnosperms
    Class Coniferae |
    Class Gneteae   ⎭
    Class Angiospermae

The gymnosperms   The first gymnosperms appear in the fossil record in the late Devonian, some 350 million years ago. Many of those first seed plants had bodies that closely resembled the ferns, and indeed for many years their fossils were thought to be fossils of ferns. Slowly, however, evidence accumulated that some of the "ferns" that were such important components of the coal-age forests produced seeds, not spores. Today these fossil plants—usually called the seed ferns—are grouped together as the class Pteridospermae of the subdivision Spermopsida. No members of this class survive today.

Another ancient group, the cycads and their relatives (class Cycadae), may have arisen from the seed ferns. These plants first appeared in the Permian period and became very abundant during the Mesozoic era. They had large palmlike leaves; the palmlike plants so often shown in pictures of the dinosaur age are usually cycads, not true palms. The cycads declined after the rise of the angiosperms in the Cretaceous period, but nine genera containing over a hundred species are in existence today (Fig. 23.40). They are generally called sago palms and are fairly common in some tropical regions. One genus (*Zamia*) occurs in Florida.

23.40 A living cycad
Though often called "sago palms," these plants are not really palms at all, but members of an ancient gymnosperm group.   [K. W. Fink, National Audubon Society, Photo Researchers, Inc.]

TABLE 23.2   *A comparison of the subdivisions of Tracheophyta*

Characteristics	Psilopsida	Lycopsida	Sphenopsida	Pteropsida	Spermopsida	
					Gymnosperms	Angiosperms
Vascular tissue	+	+	+	+	+	+
True roots and leaves	−	+	+	+	+	+
Megaphyllous leaves	−	−	−	+	+	+
Gametophyte retained in sporophyte tissue	−	−	−	−	+	+
Sperm cells without flagella	−	−	−	−	+ (− in primitive groups)	+
Production of seeds	−	−	−	−	+	+
Flowers and fruit	−	−	−	−	−	+

23.41   Branch of a larch tree in spring
The leaves are needlelike. At this time of year the new female cones have a pinkish coloration.   [Courtesy Roman Vishniac.]

23.42   Cross section of a pine needle
There are many stomata in the thick epidermal layer, particularly along the lower right edge. Six large resin ducts are just outside the prominent endodermis that bounds the stele.   [Courtesy Roman Vishniac.]

The Ginkgoae are still another once widespread group now nearly extinct. There is only one living species, the ginkgo or maidenhair tree, often planted as a lawn tree, but almost unknown in the wild.

By far the best-known group of gymnosperms is the conifers (class Coniferae), which include such common species as pines, spruces, firs, cedars, hemlocks, yews, and larches. The leaves of most of these plants are small evergreen needles or scales (Fig. 23.41), with an internal arrangement of tissues (Fig. 23.42) that differs somewhat from that in angiosperms, which we examined in earlier chapters. This group first arose in the Carboniferous period and was very common during the Mesozoic era. It remains an important part of the earth's flora.

Let us follow in some detail the life cycle of a pine tree as an example of the seed method of reproduction. The large pine tree is the diploid sporophyte stage (stage 5). This tree produces reproductive structures called **cones,** of which there are two kinds: large female cones, in whose sporangia meiosis gives rise to haploid megaspores (stage 2), and small male cones, in whose sporangia meiosis gives rise to haploid microspores (Fig. 23.43). (Production of distinctive male and female spores—heterospory—is characteristic of all seed plants, both gymnosperms and angiosperms.) In both kinds of cones the sporangia are produced by highly modified leaves (sporophylls).

Each scale of a female cone bears two sporangia on its upper (adaxial) surface (Fig. 23.44B). Each sporangium is encased in an integument with a small opening, the **micropyle,** at one end (Fig. 23.44C). Meiosis takes place inside the sporangium, producing four haploid megaspores, three of which soon disintegrate. Next, the single remaining megaspore gives rise, by repeated mitotic divisions, to a multicellular mass, which is the female game-

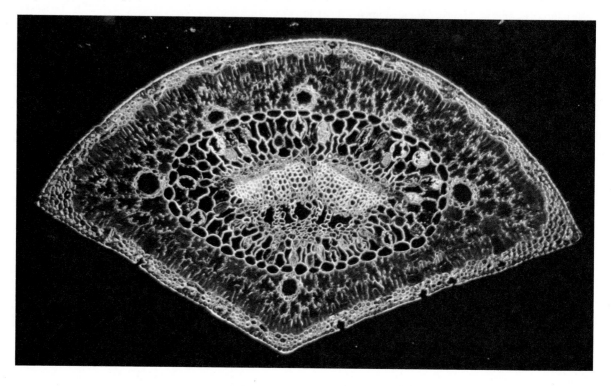

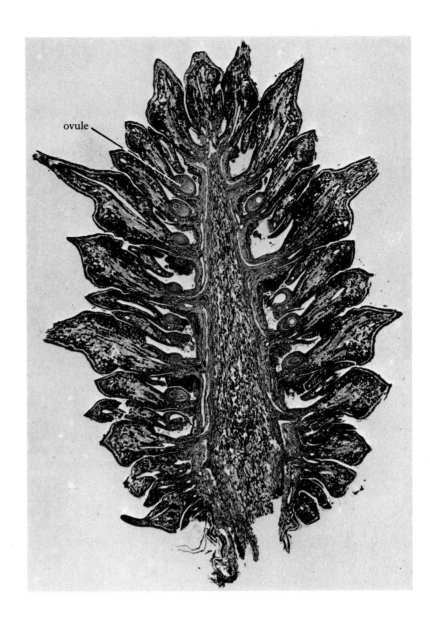

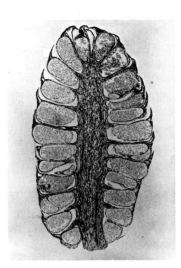

23.43 Sections of male and female pine cones
Left: Male cone. Each sporophyll (cone scale) bears a
large sporangium that becomes a pollen sac.   Right:
Female cone. Ovules can be seen on the surface of the
sporophylls near their base.   [Courtesy Thomas Eisner,
Cornell University (left); Carolina Biological Supply Co.
(right).]

23.44 Pollen grains and ovules of pine
(A) Pollen grain, composed of four cells, two of which
are degenerate.   (B) Scale from female cone. The two
ovules, each containing a sporangium, lie on the surface
of the scale near its point of attachment to the cone
axis.   (C) Section of an ovule.   [Modified from H. J.
Fuller and O. Tippo, *College botany*, Holt, Rinehart &
Winston, Inc., 1954.]

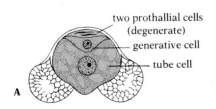

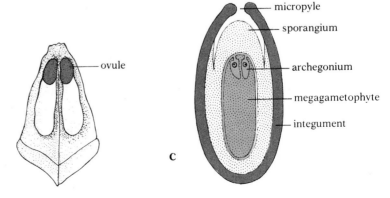

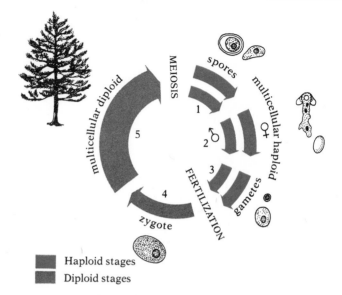

23.45 Life cycle of pine
The familiar tree is the sporophyte (multicellular diploid) stage. The gametophytes (multicellular haploid) are very tiny and cannot lead an independent existence. In all the haploid stages the sexes are separate.

Haploid stages
Diploid stages

tophyte (megagametophyte). When mature, the female gametophyte produces two to five tiny archegonia at its micropylar end. Egg cells develop in the archegonia. Note that the megaspore is never released from the sporangium, and that the female gametophyte derived from it remains embedded in the sporangium, which is still attached to the cone scale. The composite structure consisting of integument, sporangium, and female gametophyte is called an *ovule.*

Each of the many microspores produced by meiosis in a sporangium of a male cone becomes a *pollen grain.* It develops a thick coat, which is highly resistant to loss of water, and winglike structures on each side, which doubtless aid its dispersal by wind. Within the pollen grain the haploid nucleus divides mitotically several times, and walls develop around each nucleus. In this manner the pollen grain becomes four-celled (Fig. 23.44A). Two of the cells soon degenerate;[11] the two cells that remain are called the generative cell and the tube cell. The mature pollen grain is released from the cone when the sporangium bursts. A single male cone may release millions of tiny pollen grains, which may be carried many miles (sometimes as many as a hundred) by the wind. Note that the pollen grains are multicellular haploid structures (if four cells may be said to be "multi") and that they constitute the male gametophyte (microgametophyte) (stage 2 in Fig. 23.45).

Most of the millions of pollen grains released by a pine tree fail to reach a female cone. But of the few that sift down between the scales of a female cone, some land in a sticky secretion near the open micropylar end of an ovule. As this secretion dries, it is drawn through the micropyle, carrying the pollen grains with it. The arms of the integument around the micropyle then swell and close the opening. When a pollen grain comes in contact with the end of the sporangium just inside the micropyle, it develops a tubular out-

[11] The two cells that degenerate are called prothallial cells.

EVOLUTION OF PINE CONES

A male pine cone is a spiral cluster of cone scales, which are highly modified reproductive leaves (sporophylls) on a short section of stem (Fig. 23.43, left). The same description was generally thought to apply to the larger female cone, too, until the fossil evidence recently forced a revision of that assumption.

Apparently the earliest female cones were compound structures composed of a section of stem bearing a series of modified nonreproductive leaves called bracts (figure A). In the axil where each bract joined the stem was a very short bud-like lateral branch; a few of its spirally arranged scalelike leaves bore sporangia (i.e. they were sporophylls). In the course of evolution, all the dwarf lateral branches moved close together, and each was reduced to one to three sporophylls, which fused with several tiny sterile leaves to form a single compound structure called an ovuliferous scale. Each ovuliferous scale, in turn, partly fused with the bract in whose axil it developed (figure B). Thus each "scale" of a modern female pine cone consists of a bract (which is a modified leaf) and an ovuliferous scale (derived from a dwarf branch consisting of fused sporophylls and sterile leaves) (Fig. 23.43, right).

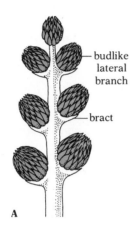
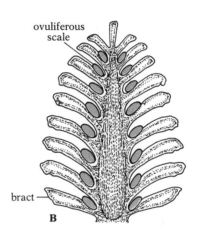

growth, the ***pollen tube.*** The nucleus of the tube cell enters the tube, followed by the generative cell. The generative cell then divides, and one of the daughter cells thus produced divides again, producing two sperm cells.[12] Thus a germinated pollen grain contains four active nuclei plus the two nuclei of the degenerate cells; this six-nucleate condition is as far as the male gametophyte of pine ever develops toward multicellularity.

The pollen tube grows down through the tissue of the sporangium and penetrates into one of the archegonia of the female gametophyte.[13] There it discharges its sperm cells, one of which fertilizes the egg cell. The resulting zygote (stage 4) then divides mitotically to produce a tiny embryo sporo-

---

[12] The two cells produced by division of the generative cell are the sterile cell and the spermatogenous cell. It is the latter that then divides to give rise to the two sperm cells.

[13] Since an ovule contains several archegonia, several embryos may begin development, but usually only one completes it.

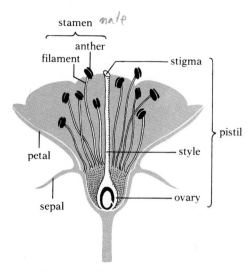

23.46   The parts of a flower

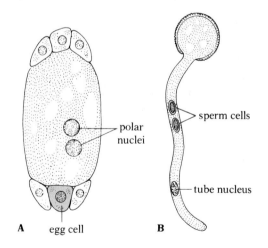

23.47   Gametophytes of an angiosperm
(A) Female gametophyte (embryo sac), which is composed of seven cells. One cell is much larger than the others and contains the two polar nuclei. (B) Male gametophyte (pollen grain and tube).

phyte consisting of a hypocotyl and an epicotyl. The embryo is still contained in the female gametophyte, which is itself contained in the sporangium. Finally, the entire ovule is shed from the cone as a *seed,* which consists of three main components: a seed coat derived from the old integument, stored food material derived from the tissue of the female gametophyte, and an embryo.[14]

According to the fossil evidence, the gymnosperms did not evolve from the ferns; the two groups appear to have arisen independently from an ancestral group, probably the long-extinct progymnosperms. It is nonetheless instructive to summarize the advances in the life cycle of pine as compared with that of a typical fern:

1. The sporophylls are more highly modified and less leaflike.
2. There are two types of sporangia, which produce two types of spores—microspores (male) and megaspores (female).
3. Two kinds of gametophytes are derived from the two types of spores (i.e. the sexes are separate in the gametophyte stage).
4. The gametophytes are much further reduced than those of ferns; they do not possess chlorophyll and are not free-living. A male gametophyte consists only of the six-nucleate pollen grain and tube. A female gametophyte is only a mass of haploid tissue that remains in the sporangium and is parasitic on it.
5. There are usually no flagellated sperm cells.[15]
6. The young embryo is contained within a seed.

The angiosperms   The first undisputed representatives of the angiosperms are from the early Cretaceous. The group underwent great expansion later in the Cretaceous, and these plants became the dominant land flora of the Cenozoic era, as they are today.

We have seen that the reproductive structures of gymnosperms are cones (or similar structures) and that the ovules, which later become the seeds, are borne naked on the surface of the cone scales. The reproductive structures of angiosperms, by contrast, are flowers, and the ovules are enclosed within modified leaves called carpels.

A flower is generally interpreted as a short length of stem with modified leaves attached to it. The modified leaves of a typical flower (Fig. 23.46) occur in four sets attached to the enlarged end (receptacle) of the flower stalk: (1) The *sepals* enclose and protect all the other floral parts during the bud stage. They are usually small, green, and leaflike, but in some species they are large and brightly colored. All the sepals together form the *calyx.* (2) Internal to the sepals are the *petals,* which together form the *corolla.*[16] In flowers pollinated by insects, birds, or other animals, the petals are usually quite showy, but in those pollinated by wind they are often reduced or even

[14] As a rule, the sporangium (or nucellus, as botanists generally call the sporangium in the ovule) eventually disintegrates and is not present in the seed, but in a few species it may be preserved, usually as the inner layer of the seed coat.

[15] The pollen tubes of primitive gymnosperms, such as ginkgoes and cycads, produce flagellated sperm cells, but these have only a very short distance to swim within the cytoplasm of the pollen tube to reach the egg cells, the pollen grains having already been carried to the ovules by the wind.

[16] The calyx and corolla together constitute the perianth.

absent. (3) Just inside the circle of the corolla are the ***stamens,*** which are the male reproductive organs; i.e. they are the sporophylls that produce the microspores.[17] Each stamen consists of a stalk, called a ***filament,*** and a terminal ovoid pollen-producing structure called an ***anther.*** (4) In the center of the flower is the female reproductive organ, the ***pistil*** (some species have more than one pistil per flower). Each pistil consists of an ***ovary*** at its base, a slender stalk (more than one in some species) called a ***style,*** which rises from the ovary, and an enlarged apex called a ***stigma.*** The pistil is derived from one or more sporophylls, which in flowers are called ***carpels.***[18] All four kinds of floral organs—sepals, petals, stamens, and pistils—are present in so-called complete flowers, but some flowers, which are said to be incomplete, lack one or more of them.[19]

Within the ovary are one or more (at least one for each carpel) sporangia, called ***ovules,*** which are attached by short stalks to the wall of the ovary. Meiosis occurs once in each ovule, with formation of four haploid megaspores, three of which usually soon disintegrate. The remaining megaspore then divides mitotically several times, producing, in most species, a structure composed of seven cells, one of which is much larger than the others and contains two nuclei, called polar nuclei (Fig. 23.47). This haploid seven-celled eight-nucleate structure is the much-reduced female gametophyte (often called an embryo sac).[20] One of the cells located near the micropylar end will act as the egg cell.

Each anther has four sporangia in each of which many cells undergo meiosis, producing numerous haploid microspores. The wall of each microspore thickens, and the nucleus divides mitotically, producing a generative nucleus and a tube nucleus. The resulting thick-walled two-nucleate structure is a pollen grain—a male gametophyte—which is released from the anther when the mature sporangium (Fig. 23.48) splits open.

A pollen grain germinates when it falls (or is deposited) on the stigma of a pistil, which is usually rough and sticky. A pollen tube begins to grow, and the two nuclei of the pollen grain move into it. The generative nucleus (which is surrounded by a plasma membrane and is thus technically a cell, though with virtually no cytoplasm) then divides, giving rise to two sperm cells (Fig. 23.47B).[21] The pollen tube grows down through the tissues of the

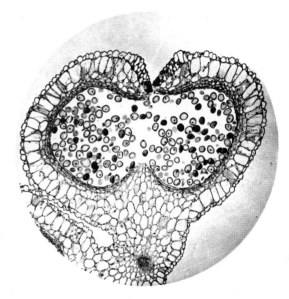

23.48 Pollen chamber of a lily anther
When the pollen chamber (derived from a sporangium) opens, the numerous pollen grains within it will be released.   [Courtesy Turtox/Cambosco, Macmillan Science Co., Inc.]

---

[17] It is not entirely certain that the stamens and carpels evolved directly from individual leaves. Some investigators consider them compound structures derived from the fusion of a bract leaf and a short reproductive shoot. Whether they are right, or whether, in accordance with the older, more widely held view, the stamens and carpels are highly modified single leaves, it is probably safe to regard the actual spore-producing parts as sporophylls.

[18] A simple pistil is composed of only one sporophyll, or carpel. A compound pistil is composed of several fused carpels.

[19] In some species, such as corn, willow, oak, and walnut, the stamens and pistils are in separate flowers. Incomplete flowers of this type, in which only one of the two kinds of reproductive structures is present, are called imperfect flowers. Flowers with both stamens and pistils (whether complete or incomplete) are called perfect flowers.

[20] The embryo sac of some species has more than eight nuclei, and that of a limited number of other species has fewer than eight. Furthermore, in some species no cytokinesis occurs, and all the nuclei lie in the same mass of cytoplasm. However, the most common sort of embryo sac is the seven-celled eight-nucleate type described in the text.

[21] The pollen grains in most species are released in the two-nucleate condition and the division of the generative nucleus does not take place until germination, but in some species this division occurs earlier and the pollen grains are released from the anthers in the three-nucleate condition.

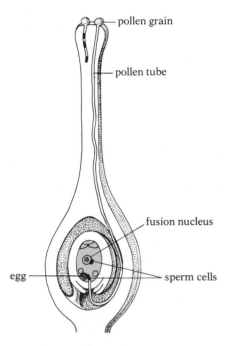

23.49 Fertilization of an angiosperm
Pollen grains land on the stigma and give rise to pollen tubes that grow downward through the style. One of the pollen tubes shown here has reached the ovule in the ovary and discharged its sperm cells into it. One sperm will fertilize the egg cell, and the other will unite with the diploid fusion nucleus (derived from the two polar nuclei) to form a triploid nucleus, which will give rise to endosperm.

23.50 Fruit of the dandelion
Each dandelion "seed" is actually a seed enclosed in a small dry fruit. On one end of the fruit is a parachute, which aids in dispersal by the wind. [Jane Burton, Bruce Coleman Inc.]

stigma and style and enters the ovary (Fig. 23.49). When the tip of the pollen tube reaches an ovule, it enters the micropyle and then discharges the two sperm cells into the female gametophyte (embryo sac). One of the sperm fertilizes the egg cell, and the zygote thus formed develops into an embryo sporophyte. By the time fertilization occurs, the two polar nuclei of the female gametophyte have combined to form a diploid *fusion nucleus,* with which the second sperm unites to form a triploid nucleus. This nucleus undergoes a series of divisions, and a triploid tissue, called *endosperm,* is formed. The endosperm functions in the seed as a source of stored food for the embryo.

After fertilization, the ovule matures into a seed, which, as in pine, consists of seed coat, stored food, and embryo. However, the angiosperm seed differs from that of pine in being enveloped by the ovary. It is the ovary that develops into the *fruit,* usually enlarging greatly in the process. Sometimes other structures associated with the ovary, such as the receptacle, are also incorporated into the fruit. The ripe fruit may burst, expelling the seeds, as in peas (where the pod is the fruit). Or the ripe fruit with the seeds still inside may fall from the plant, as in tomatoes, squash, cucumbers, apples, peaches, and acorns. The fruit not only helps protect the seeds from desiccation during their early development, before they have fully ripened, but often also facilitates their dispersal by various means—the wind, say (Fig. 23.50), or an animal, which, attracted by the fruit, carries it to other locations or eats both fruit and seeds and later releases the unharmed seeds in its feces.

The main features in which the angiosperm life cycle differs from that of gymnosperms can be summarized as follows:

1. The reproductive structures are flowers instead of cones. The sporophylls (stamens and pistils) of flowers are even less leaflike than those of cones.
2. The ovules are embedded in the tissues of the female sporophylls instead of lying bare on their surface.
3. The gametophytes are even more reduced than those of gymnosperms. The male gametophyte (pollen grain and tube) has only three nuclei. The female gametophyte usually has only eight nuclei.
4. In pollination the pollen grains are not deposited close to the opening of the ovule, as in gymnosperms, but are deposited on the stigma instead. The pollen tube thus has much farther to grow in angiosperms.
5. Angiosperms have "double fertilization," one sperm fertilizing the egg cell and the other uniting with the fusion nucleus to give rise to triploid endosperm.[22] Gymnosperms, by contrast, have single fertilization; one sperm fertilizes the egg, and the other soon deteriorates. The stored food in the seed of gymnosperms is the haploid tissue of the female gametophyte and is thus developmentally quite different from the triploid endosperm of the angiosperms.
6. The seeds of angiosperms are enclosed in fruits that develop from the ovaries and associated structures; gymnosperms have no fruit.

Angiosperms also differ from gymnosperms in other than reproductive characteristics—in the structure of the xylem, for example, which in angiosperms often contains vessels, but seldom does in gymnosperms.

The class Angiospermae is customarily divided into two subclasses, the Dicotyledoneae and the Monocotyledoneae. The dicots include oaks, maples, elms, willows, roses, beans, clover, tomatoes, asters, and dandelions. The monocots include grasses, corn, wheat, rye, daffodils, irises, lilies, and palms. There are certain basic differences between the two groups:

1. As the names imply, the embryos of dicots have two cotyledons, whereas those of monocots have only one.
2. Dicots often have vascular cambium and secondary growth; monocots usually do not.
3. The vascular bundles in the stems of young dicots are arranged in a circle or fused to form a tubular vascular cylinder; monocots have more scattered vascular bundles.
4. The leaves of dicots usually have net venation; those of monocots usually have parallel venation.
5. Dicot leaves generally have petioles; monocots generally do not. Many monocots can be recognized by the way the leaf base clasps the stem (as in corn).
6. The flower parts of dicots usually occur in fours or fives or multiples of these (e.g. four sepals, four petals, four stamens); those of monocots usually occur in threes or multiples of three (Fig. 23.51).

[22] In some angiosperms (e.g. lilies), the endosperm is pentaploid ($5n$).

23.51 Dicot and monocot flowers compared
Top: In a milkweed flower, a dicot, the main parts occur in fives.   Bottom: In a tulip flower, a monocot, the parts occur in threes or sixes.   [Courtesy E. S. Ross (top); Douglas Quine, Cornell University (bottom).]

SUGGESTED READING

BANKS, H. P., 1970. *Evolution and plants of the past.* Wadsworth, Belmont, Calif. *Excellent short book on fossil plants, with special emphasis on the evolutionary relationships of the tracheophyte groups.**

BOLD, H. C., and M. J. WYNNE, 1978. *Introduction to the algae.* Prentice-Hall, Englewood Cliffs, N.J. *Thorough, rather technical treatment of all the algal groups.*

GRANT, V., 1951. "The fertilization of flowers," *Scientific American,* June. (Offprint 12) *The special adaptations of flowers that help ensure their pollination.*

JENSEN, W. A., and F. B. SALISBURY, 1972. *Botany: An ecological approach.* Wadsworth, Belmont, Calif. *Very readable general text written from an evolutionary and ecological point of view.*

RAVEN, P. H., R. F. EVERT, and H. CURTIS, 1976. *Biology of plants,* 2nd ed. Worth, New York. *A well-written broad coverage of botany.*

* Available in paperback.

Chapter 24

THE FUNGAL KINGDOM

The several divisions of organisms that we here place in the kingdom Fungi were in the past often grouped together in a single division, the Eumycophyta, or true fungi.

The fungi are eucaryotic organisms, and they are primarily multicellular or multinucleate. Their multicellularity, however, differs from that of plants and animals in that partitions between nucleated compartments, or "cells," are generally either absent or only partial; thus the cytoplasm is continuous. Moreover, each compartment often has more than one nucleus. For many fungi, therefore, "multinucleate" is perhaps a more accurate description than "multicellular." The "cells" are usually organized into branched filaments called *hyphae,* which form a mass called a *mycelium.* In most fungi the basic component of the cell wall is not cellulose (though a small amount may be present in a few species), but chitin, a derivative polysaccharide containing nitrogen.

The fungi are parasitic or saprophytic, though a few are at times predatory (see Fig. 5.16, p. 212). Most saprophytic fungi secrete digestive enzymes onto their food material and absorb the products of the extracellular digestion by means of rootlike rhizoids or haustoria (see p. 211). Parasitic fungi may carry out extracellular digestion, or they may directly absorb materials produced by the body of their host.

Some fungi are parasitic on or in animals, including humans; many skin diseases, including "ringworm" and athlete's foot, are caused by fungi, and there are several serious fungal diseases of the lungs. Other fungi are parasitic on plants, and some of these cause annual losses of hundreds of millions of dollars when they attack agricultural crops. Still others cause spoilage of bread, fruit, vegetables, and other foodstuffs, and deterioration of leather goods, fabrics, paper, lumber, and other valuable products.

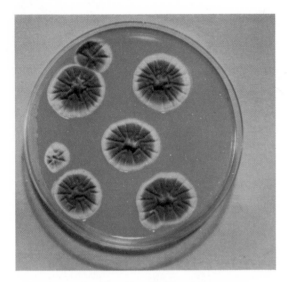

24.1   Seven colonies of *Penicillium chrysogenum* growing on culture medium in a Petri dish
Almost the entire world's supply of commercial penicillin is produced by this fungus in pharmaceutical laboratories.   [Courtesy Pfizer Inc.]

The numerous pathogenic or destructive fungi should not cause us to forget the many others that are beneficial, however. Yeasts are used extensively in the manufacture of alcoholic products and to make bread dough rise. The antibiotic penicillin is obtained from a fungus (Fig. 24.1). Fungi are important in the manufacture of many cheeses, and certain mushrooms are regularly used as food. And fungi, together with bacteria, decompose vast quantities of dead organic matter that would otherwise rapidly accumulate and make the earth uninhabitable.

Reproduction in the fungi may be either asexual or sexual, but in both cases the haploid stages are usually dominant (the Oomycota are an exception). The characteristics of sexual reproduction are especially important in distinguishing the four divisions of true fungi currently recognized.[1]

OOMYCOTA (THE WATER MOLDS)

As their common name "water molds" suggests, most of the Oomycota are aquatic. Some, however, live in the soil, and some are parasites of higher plants. Among the latter is the species *Phytophthora infestans*, the cause of late blight of potatoes, a historic scourge responsible for such devastations as the Irish potato famine of 1845–47, which did much to change the course of Irish history. Other members of this group of fungi—notably the agents of downy mildew of sugar beets and grapes—can cause severe economic damage.

There are a number of striking differences between the Oomycota and the other three groups of true fungi, which have led some authorities to postulate that they may represent an entirely separate evolutionary development: (1) They are the only true fungi that typically have flagellated zoospores. (2) Their sexual reproduction is oogamous. (3) Unlike other fungi, they are diploid during most of their life cycle. (4) They are unusual among the fungi in lacking chitin in their cell walls; though some cellulose is present, the principal component of their walls is another polysaccharide with linkages between adjacent sugar units different from those in cellulose. In most instances the hyphae lack cellular partitions (septa); the multinucleate cytoplasm is therefore a continuous (coenocytic) system.

ZYGOMYCOTA (THE CONJUGATION FUNGI)

The hyphae of the Zygomycota characteristically lack cross walls, although they contain many haploid nuclei (i.e. they are coenocytic). Cross walls appear only during the formation of reproductive structures. Neither the spores nor the gametes are motile. Sexual reproduction is accomplished by the conjugative fusion of morphologically indistinguishable cells from hyphae of two different mycelia (Fig. 24.2B).

As an example of a member of this division, let us examine the common black bread mold, *Rhizopus*. The hyphae of this mold form a whitish or grayish mycelium on the bread. If the mycelium is examined carefully, it can be seen to include three types of hyphae: hyphae (called stolons) that form a

---

[1] Other bases of classification are generally unsatisfactory. Hence it is often difficult to classify species for which sexual stages have not been found; such species are customarily assigned to a category called Fungi Imperfecti.

network on the surface of the bread; rootlike hyphae (called rhizoids) that penetrate into the bread and function both in anchoring the plant and in absorbing nutrients; and hyphae (called sporangiophores) that grow upright from the surface and bear globular sporangia on their ends (Fig. 24.2A; see also Fig. 1.13, p. 19). Thousands of asexual spores are produced in each sporangium. The spores, which have no flagella, are very tiny and light, and when liberated at maturity (by disintegration of the wall of the sporangium) they may passively be carried long distances by wind, rain, or animals. If a spore lands in a suitable location, where conditions are warm and moist, it germinates and soon gives rise to a new mass of hyphae, thus completing the asexual cycle.

Sexual reproduction in *Rhizopus* resembles that of the green alga *Spirogyra* (see Fig. 23.12, p. 964). Short branches from two hyphae (which must be of different mating types or sexes) contact each other at their tips (Fig. 24.2B). Cross walls soon form just back of the tips of these hyphal branches. The gamete cells thus delimited then fuse to form a zygote, which develops a thick protective wall and enters a period of dormancy usually lasting from one to three months. At germination the nucleus of the zygote undergoes meiosis, and a short hypha grows from the zygote. This haploid hypha promptly produces a sporangium, which releases asexual spores that grow into new mycelia. Note that the only diploid stage in the entire sexual cycle is the zygote (Fig. 24.3).

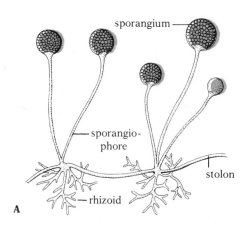

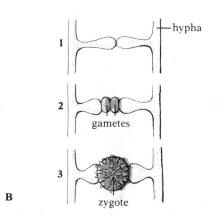

24.2 *Rhizopus*
(A) Hyphae with sporangia. (B) Sexual reproduction by conjugation: (1) Short branches from two different hyphae meet. (2) The tips of the branch hyphae are cut off as gametes. (3) The gametes fuse in fertilization to form a zygote with a thick spiny wall.

24.3 Life cycle of *Rhizopus*
See text for description.

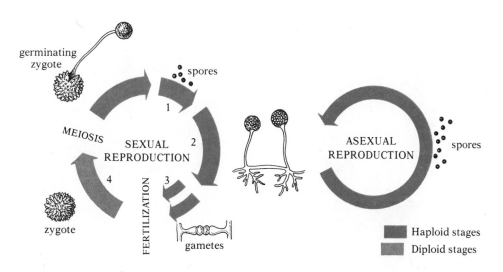

Members of the Zygomycota are widespread as saprophytes in soil and dung; they are thus mostly terrestrial, only a few species living in aquatic habitats. Some, however, are parasites on plants, animals, or other fungi.

ASCOMYCOTA (THE SAC FUNGI)

The members of this large division are very diverse, varying all the way from unicellular yeasts through powdery mildews and cottony molds to complex cup fungi. These last form a cup-shaped structure composed of many hyphae tightly packed together (Fig. 24.4). The vegetative hyphae of Ascomycota, unlike those of Zygomycota, are septate (i.e. they possess cross walls), but the septa are usually incomplete, having large holes in their centers. The cytoplasm of adjacent cells is thus continuous.

Though their vegetative structures differ, all Ascomycota resemble each other in forming a reproductive structure called an *ascus* during their sexual cycle. An ascus is a sac within which haploid spores (usually eight, but sometimes four) are produced; all the spores in an ascus are derived from a single parent cell (remember discussion of the ascus of the red bread mold, *Neurospora;* see p. 637 and Fig. 15.11). The events leading to the formation of a mature ascus are shown in Fig. 24.5.

24.4   The cup fungus *Peziza* growing on a log in a rain forest in Central America

[M. P. L. Fogden, Bruce Coleman Inc.]

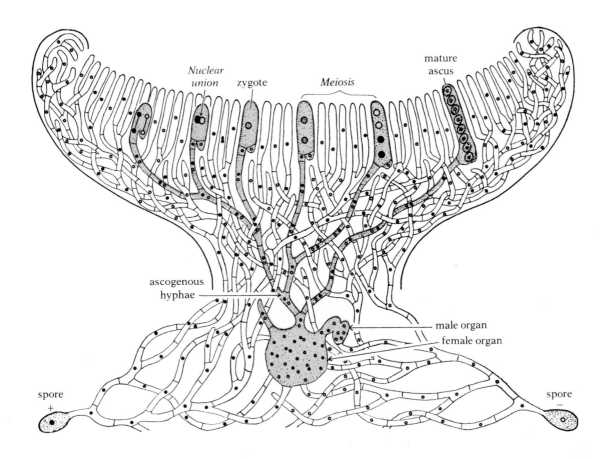

24.5 Diagram of a magnified section through a cup fungus

Hyphae of two haploid mycelia, one derived from a plus (female) spore and the other from a minus (male) spore, participate in forming the cup structure and in producing the spores. The plus mycelium bears a female organ (ascogonium), and the minus mycelium bears a male organ (antheridium). A tube grows from the female organ to the male organ, and then minus nuclei (small circles), acting as male nuclei, move into the female organ and become associated with plus nuclei (black dots). Next, hyphae grow from the female organ; each cell in these hyphae contains two associated nuclei, one minus and one plus. Such a cell, containing two nuclei, is said to be dikaryotic; when, as here, the nuclei are genetically different (heterokaryotic), the cells are effectively diploid even though the individual nuclei are haploid. The terminal cells of the dikaryotic hyphae eventually become elongate, and their nuclei unite, forming a zygote nucleus. The zygote nucleus promptly divides meiotically, and each of the four haploid nuclei thus formed then divides mitotically, producing a total of eight small spore cells, which are still contained within the wall of the old zygote cell, now called an ascus. When the mature ascus ruptures, the spores are released.   [From L.W. Sharp, *Fundamentals of cytology,* McGraw-Hill Book Co., copyright © 1943. Used by permission.]

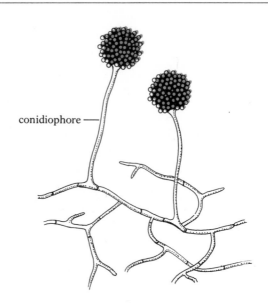

A

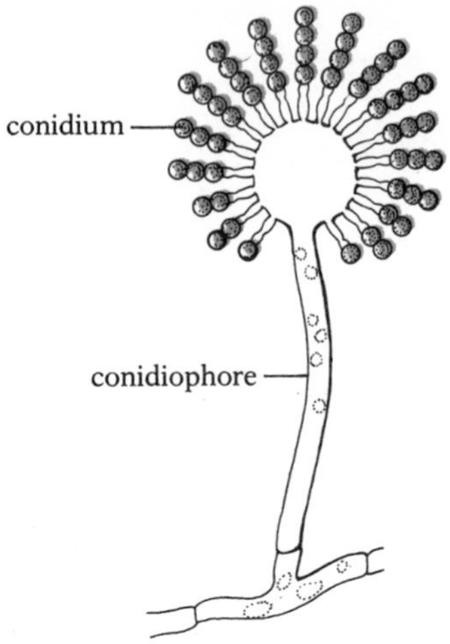

24.6   Asexual reproductive structures of *Aspergillus*, a member of the Ascomycota
(A) Two conidiophores arise from a mycelium.   (B) An enlarged section through a conidiophore shows that the structure bears numerous spores, called conidia, arranged in chains.

B

24.7   Budding in brewer's yeast
A small new cell is pinched off the larger parent cell.

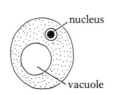

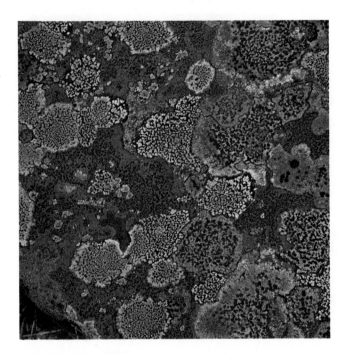

24.8   Two types of lichens
Left: Both *Lecidea atrata* (rust-red) and *Lecidea lithophila* (gray-black) grow as a crust, shown here on a boulder in the Cairngorms, Scotland.   Right: British soldiers, with their bright red tops, have a more upright growth.   [Jane Burton (left) and J. Shaw (right), Bruce Coleman Inc.]

Most Ascomycota also reproduce asexually by means of special spores called *conidia* (Fig. 24.6). Conidia are produced in chains at the end of conidiophore hyphae (but not inside sporangia). Each conidium can grow into a new fungal plant.

It may seem strange that yeasts are considered members of the Ascomycota. They are unicellular, and they reproduce asexually by **budding** (Fig. 24.7), not by conidia formation. However, under certain conditions a single yeast cell may function as an ascus, producing four spores. The spores are more resistant to unfavorable environmental conditions than vegetative cells, and they may enable yeasts to survive temperature extremes or periods of prolonged drying.

In previous chapters we mentioned the *lichens,* plants composed of a fungus and an alga growing together in a complex symbiotic relationship (Fig. 24.8). The fungal components of most lichens are Ascomycota, though in a few tropical forms the fungus is a member of the Basidiomycota, discussed below. The algal components are usually Chlorophyta, although they may be Cyanophyta.

24.9   Two representative Basidiomycota
Left: Fly agaric mushroom *(Amanita muscaria)*, a
deadly poisonous species. The spore-carrying gills on
the underside of the cap, which greatly enlarge the
spore-bearing surface, are clearly visible in the
specimen that has fallen on its side. Right: One of the
many kinds of edible mushrooms, *Hydnum repandum.*
[Jane Burton, Bruce Coleman Inc. (left); Hervé
Chaumeton (right).]

BASIDIOMYCOTA (THE CLUB FUNGI)

Many of the largest and most conspicuous fungi—puffballs, mushrooms,
toadstools, and bracket fungi—are Basidiomycota (Fig. 24.9). Though the
above-ground portion of these plants looks like a solid mass of tissue, and in
some is differentiated into a stalk and a prominent cap, it is nevertheless
composed of hyphae, as are all fungi. It should be emphasized that the
above-ground portion, or fruiting body, of many mushrooms is only a small
part of the total plant; there is an extensive mass of hyphae in the soil. The
hyphae are septate.

The members of this class are distinguished by their possession of club-
shaped reproductive structures called **basidia** (Fig. 24.10). The cells of the
hyphae that produce basidia are binucleate (dikaryotic), containing one
haploid male nucleus and one haploid female nucleus (usually designated as
minus and plus). Certain terminal cells of these hyphae, located in rows, or
"gills," on the undersurface of the cap of those Basidiomycota called gill
fungi (Fig. 24.10), become zygotes when their two nuclei fuse in fertilization.
The zygote then becomes a basidium. Its diploid nucleus divides by meiosis,
producing four new haploid nuclei. Four small protuberances develop on
the end of the basidium, and the haploid nuclei migrate into these. The tip of
each protuberance then becomes walled off as a spore, which is usually
ejected from the basidium. Each spore may give rise to a new mycelium.

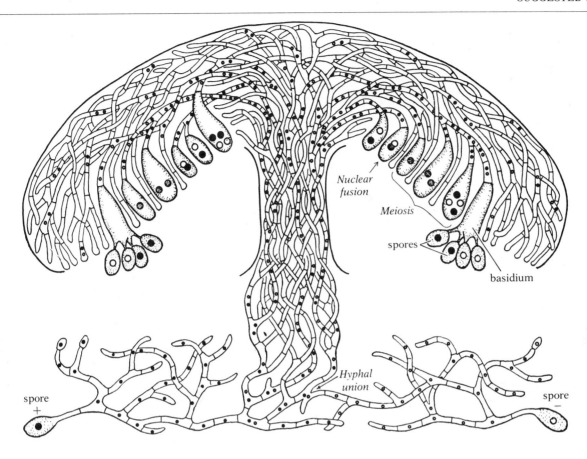

24.10 Diagram of section through a mushroom

Hyphae from two uninucleate mycelia—one of the plus, the other of the minus, mating type—unite and give rise to dikaryotic hyphae, which then develop into the aboveground part of the mushroom. The entire stalk and cap are composed of these hyphae tightly packed together. Spores are produced by basidia on the lower surface of the cap.   [From L. W. Sharp, *Fundamentals of cytology*, McGraw-Hill Book Co., copyright © 1943. Used by permission.]

SUGGESTED READING

ABMADJIAN, V., 1963. "The fungi of lichens," *Scientific American*, February.

COOKE, R. C., 1978. *Fungi, man and his environment.* Longman, London.   *Short but fascinating treatment of the biology of fungi, with emphasis on the many ways these organisms affect human beings.*

EMERSON, R., 1952. "Molds and men," *Scientific American*, January. (Offprint 115)   *The diversity of the fungi, and their many effects on human lives.*

LITTEN, W., 1975. "The most poisonous mushrooms," *Scientific American*, March.   *The members of the genus* Amanita *and the highly toxic compound they produce.*

RAVEN, P. H., R. F. EVERT, and H. CURTIS, 1976. *Biology of plants*, 2nd ed. Worth, New York.   *Includes an excellent treatment of the fungi.*

WEBSTER, J., 1977. *Introduction to fungi.* Cambridge University Press, New York.   *Rather technical treatment of the various groups of fungi.**

* Available in paperback.

Chapter 25

THE ANIMAL KINGDOM

You already have some familiarity with many phyla of the animal kingdom from our previous consideration of their vital processes—their nutrient procurement, gas exchange, internal transport, excretion, coordination, and development. In our earlier discussions we paid particular attention to representatives of the Coelenterata, Platyhelminthes (flatworms), Mollusca, Annelida (segmented worms), Arthropoda, Echinodermata, and Chordata (especially vertebrates). We shall again concentrate on these large and important groups, but shall also briefly mention some smaller phyla we previously ignored. Our aim here is to bring together the various phases of animal life discussed in other chapters and to suggest both the immense diversity within the kingdom and possible evolutionary patterns.

A formal outline of the classification of the animal kingdom used in this book is given in the Appendix. We must stress that this classification, though a widely used one, is not accepted by all biologists. Much is uncertain regarding the evolution of animal groups, and biologists vary in their interpretation of the data.

PORIFERA (THE SPONGES)

Sponges are aquatic, mostly marine animals (Fig. 25.1). Although the larvae are ciliated and free-swimming, the adults are always sessile and are usually attached to rocks or shells or other submerged objects. They are multicellular, but show few of the features ordinarily associated with multicellular animals. For example, they have no digestive system, no nervous system,[1] and no circulatory system. In fact, they have no organs of any kind, and even

25.1 Colonial orange sponges growing off Heron Island, Australia
Water, carrying microscopic food particles and oxygen, flows into the central body cavity of each of the sponges through the numerous pores in their body walls, and is discharged through the large oscula. [Courtesy Douglas Faulkner.]

---

[1] In some sponges a few cells of the body wall possess elongate processes that may have special conductile properties.

their tissues are not well defined. They thus represent a very low grade of organization.

The body of a sponge is rather like a perforated sac. Its wall is composed of three layers: an outer layer of flattened epidermal cells, a gelatinous middle layer with wandering amoeboid cells, and an inner layer of flagellated cells (Fig. 25.2). These last are unusual in that the base of the flagellum is encircled by a delicate collar; such cells are called *collar cells* (or choanocytes).

The wall of a sponge is perforated by numerous pores, each surrounded by a single pore cell. Water currents flow through the pores into the central cavity (spongocoel) and out through a larger opening (osculum) at the end of the body; the flow is enhanced by the beating of the flagella of the collar cells. Microscopic food particles brought in by the water currents adhere to the collar cells and are engulfed; the food may be digested in food vacuoles of the collar cells themselves or passed to the amoeboid cells for digestion. The water currents also bring oxygen to the cells and carry away carbon dioxide and nitrogenous wastes (largely ammonia).

Sponges characteristically possess an internal skeleton secreted by the amoeboid cells. This skeleton is composed of crystalline *spicules* or proteinaceous fibers or both. The spicules are made of calcium carbonate or siliceous material (chiefly silicic acid); their chemical composition and their shape are the basis for sponge classification. The fibrous skeletons of the bath sponges *(Spongia)* are cleaned and sold for many uses. A living bath sponge, which looks rather like a piece of raw liver, bears little resemblance to the familiar commercial object.

Among the free-living flagellated Protista are some collared organisms (Fig. 25.3) that closely resemble the collar cells of sponges—cells found in no other organisms. For this reason many biologists think that the Porifera evolved from collared flagellates. Other biologists disagree, pointing out that larval sponges have no collar cells, and suggest that sponges arose instead from a hollow free-swimming colonial flagellate. In any case, it seems likely that sponges arose from the Protista independently of the other multicellu-

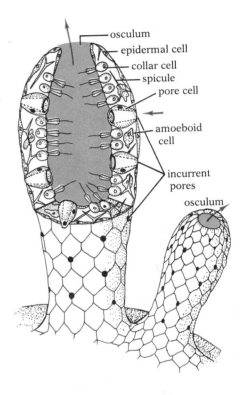

25.2 Detailed view of two colonial sponges
One sponge is shown partially sectioned, the other in exterior view. The color arrows indicate the path of water through the sponges. These sponges are of the asconoid type, with a simple tubular body. The majority of sponges (syconoid or leuconoid types) have complexly folded walls in which water passes through a network of channels, but their basic body plan may be considered an elaboration of the asconoid plan.

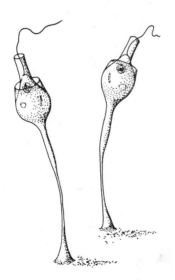

25.3 Two choanoflagellates *(Codosiga)*
These single-celled organisms strikingly resemble the collar cells of sponges, shown in Fig. 25.2. They have traditionally been classified as Protozoa by zoologists and as colorless members of the algal division Chrysophyta by botanists.

lar animals. The phylum Porifera would then stand as an evolutionary development entirely separate from the rest of the animal kingdom, and it must be concluded that multicellular animals evolved at least twice from the protozoans.

Because the Porifera differ so greatly from other multicellular animals, and probably arose independently, they are often regarded as constituting a separate subkingdom, the Parazoa.

THE RADIATE PHYLA

The two radiate phyla—Coelenterata and Ctenophora—comprise radially symmetrical animals with bodies at a relatively simple grade of construc-

25.4   A branching colonial hydrozoan (*Plumularia setacea*)
The colony is composed of a stalk with many lateral branches each bearing six to eight polyps.   [Courtesy Hervé Chaumeton.]

tion. These animals have definite tissue layers, but no distinct internal organs, no head, and no central nervous system, though they possess nerve nets (see Fig. 10.2, p. 401). There is a digestive cavity, but it has only one opening, which must serve as both mouth and anus; i.e. it is a gastrovascular cavity. Tentacles are usually present. There is no coelom or other internal space between the wall of the digestive cavity and the outer body wall.

It was once thought that the bodies of these animals consist of only two layers of cells—an outer epidermis (ectoderm) and an inner gastrodermis (endoderm)—but it is now known that a third layer called mesoglea (mesoderm) usually occurs between these two (see Fig. 5.19, p. 216), just as it does in higher multicellular animals. However, this mesodermal layer is not as well developed in the radiate phyla as in higher phyla. Usually gelatinous, it has a few scattered cells, which may be amoeboid or fibrous.

COELENTERATA

The phylum Coelenterata (also called Cnidaria) contains a variety of aquatic organisms, among them the hydras mentioned so frequently in earlier chapters, jellyfishes, sea anemones, and corals. The hydras live in freshwater, but most other coelenterates are marine.

The coelenterate body shows some cell specialization and division of labor. Thus the outer epidermis contains sensory–nerve cells, gland cells, special cells that produce nematocysts, small interstitial cells, and epitheliomuscular cells. These last are the main structural elements of the epidermis and consist of a columnar cell body with several contractile basal extensions.

The contractile elements of coelenterates, then, are parts of cells that also have other important functions; there are no cells specialized exclusively for contraction—no separate muscle cells. Note also that these contractile elements are ectodermal, not mesodermal as in most higher animals. The gastrodermis also contains contractile elements; these are basal extensions of cells whose cell bodies constitute the bulk of the lining of the gastrovascular cavity and function in digestion. Here again a single cell performs two functions that in higher animals are performed by separate elements. In short, there is some division of labor among cells in coelenterates, but it is never as complete as in most bilateral multicellular animals; and most functions performed by tissues derived from mesoderm in other animals are performed by ectodermal or endodermal cells in coelenterates.

The phylum Coelenterata is divided into three classes: Hydrozoa, Scyphozoa, and Anthozoa.

Class Hydrozoa   The best-known members of this class are the freshwater hydras. We have discussed their feeding (p. 215), gas exchange (p. 245), nervous control (p. 401), and locomotion (p. 467). Little more need be said about them here. In many ways, however, hydras are not typical members of their class. Many hydrozoans are colonial (Fig. 25.4) and have a complex life cycle, in which a sedentary hydralike *polyp* stage alternates with a free-swimming jellyfishlike *medusa* stage (Fig. 25.5). By contrast, hydras are solitary and have only a polyp stage (which is not completely sedentary).

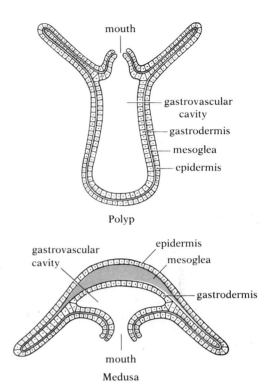

25.5 Diagram contrasting polyp and medusa
The basic structure of these two forms is the same. A medusa is like a flattened polyp turned upside down.

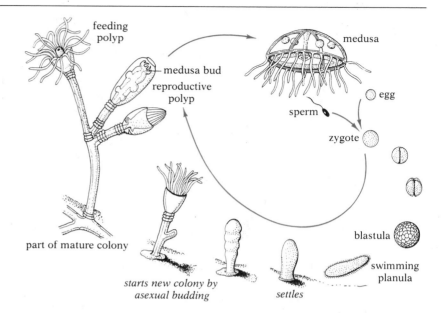

feeding
polyp

medusa bud
reproductive
polyp

medusa

egg

sperm

zygote

blastula

part of mature colony

*starts new colony by
asexual budding*

settles

swimming
planula

25.6   Life cycle of a colonial hydrozoan *(Obelia)*
Since the medusas are of separate sexes, the eggs and
sperm are produced by different individuals. See text
for description.

Let us examine *Obelia* as an example of a colonial hydrozoan (Fig. 25.6).
Much of the life of *Obelia* is passed as a sedentary branching colony of
polyps. The colony arises from an individual hydralike polyp by asexual
budding; the buds fail to separate, and the new polyps remain attached by
hollow stemlike connections. The gastrovascular cavities of all the polyps
are interconnected via the cavity in the stems. The cells lining the stem
cavity have long flagella that circulate the fluid in the cavity. Partly digested
food can be passed from one polyp to another in this moving fluid. Both the
stems and the polyps (mouth and tentacles excepted) are enclosed in a hard
chitinous case secreted by the ectoderm. Rings or joints in the case at
intervals along the stems permit some flexibility for the colony.

A mature *Obelia* colony consists of two kinds of polyps: feeding polyps
with tentacles and nematocysts; and reproductive polyps without tentacles,
which regularly bud off tiny transparent free-swimming medusas. The reproductive polyps are nourished with food captured by the feeding polyps
and passed to them through the common gastrovascular cavity in the connecting stems. In view of this division of labor, and of the structural continuity between the polyps, it might be argued that the so-called colony is not
really a colony, but a complex individual.

Each medusa is umbrella-shaped or bell-shaped, with numerous tentacles
hanging from the margin of the bell (Fig. 25.7). A tube with a mouth at its end

25.7   A hydrozoan medusa *(Gonionemus)*
Gonionemus spends most of its life as a medusa—a
weakly swimming bell-like creature with numerous
tentacles.   [Courtesy Carolina Biological Supply Co.]

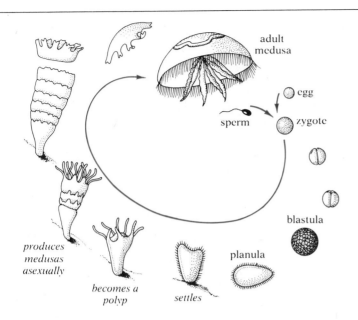

25.8 Life cycle of a jellyfish (*Aurelia*)
The polyps are shown much enlarged. Since the medusas are of separate sexes, the eggs and sperm are produced by different individuals. See text for description.

hangs from the middle of the undersurface of the bell (where the clapper of a real bell would be). The medusas are the dispersal and sexual stage in the life cycle. They swim feebly by alternately contracting and relaxing the contractile cells in the bell, but much of their movement is a matter of drifting with the water currents.

Certain cells in mature medusas undergo meiosis and give rise to either sperm or eggs, which are released into the surrounding water. Fertilization takes place, and the resulting zygote develops into a hollow blastula. Gastrulation does not take place by the process of invagination described in Chapter 17. Instead, endodermal cells proliferated by the wall (ectoderm) of the blastula wander into the blastocoel until they completely fill it. This solid gastrula then develops into an elongate ciliated larva called a ***planula.*** The planula eventually settles to the bottom, attaches by one end to some object, and develops a mouth and tentacles at the other end, becoming a polyp that gives rise to a new colony. The life cycle is thus completed. Note that the alternation of polyp and medusa stages in a coelenterate like *Obelia* differs from the alternation of generations in plants in that both polyp and medusa are diploid; as in all multicellular animals, the only haploid stage in the life cycle is the gametes.

Class Scyphozoa   The scyphozoans are the true jellyfishes. In these animals the medusa is the dominant and conspicuous stage in the life cycle, and the polyp stage is restricted to a small larva. This larva, which develops from the planula, promptly produces medusas by budding (Fig. 25.8). Scyphozoan medusas resemble the hydrozoan medusas already described except that they are usually much larger and have long oral arms (endodermal tentacles) arising from the margin of the mouth; the marginal tentacles may be much reduced, as in *Aurelia* (Fig. 25.8), or they may be quite long, as in *Pelagia* (Fig. 25.9).

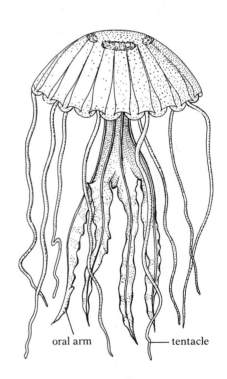

25.9 A jellyfish with oral arms and long marginal tentacles (*Pelagia*)

25.10 A sea anemone *(Stephanauge)*
The animal's mouth is clearly visible within the ring of tentacles. [Courtesy Hervé Chaumeton.]

25.11 A coral *(Eunicella verrucosa)*
Eunicella illustrates a type of coral in which the hard skeleton is elongate and branching. [Courtesy Hervé Chaumeton.]

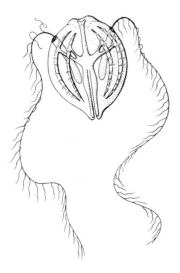

25.12 A common ctenophore *(Pleurobrachia)*
Note the rows of ciliary plates, as well as the very long antennae, which emerge from sheaths that extend deep into the body.

Class Anthozoa The class Anthozoa includes sedentary polypoid forms such as sea anemones (Figs. 25.10 and 1.14, p. 20), sea fans, and corals (Fig. 25.11). All are marine. There is no trace of a medusa stage in their life cycle. They are the most advanced members of the Coelenterata, and their body structure is much more complex than that of simple polyps like the hydras. They possess a tubular pharynx leading into the gastrovascular cavity, which is divided into numerous radiating compartments by longitudinal septa; their mesoderm (mesoglea) is much thicker than that of other coelenterates and is often elaborated into a fibrous connective tissue; and their muscles are much better developed.

The corals, anthozoans that secrete a hard limy skeleton, have played a very important role in the geologic history of the earth, particularly in tropical oceans. As their skeletons have accumulated over the ages, they have formed many reefs, atolls, and islands, especially in the South Pacific. The Great Barrier Reef, a coral ridge many kilometers wide that extends more than 1,600 km along the eastern coast of Australia, is a particularly impressive example of the way these lowly animals can change the face of the earth.

CTENOPHORA

Like the coelenterates, members of the phylum Ctenophora, known as comb jellies or sea walnuts, are radial animals with a saclike body composed of epidermis, mesoglea, and gastrodermis; and like the coelenterates, they have a digestive cavity of the gastrovascular type and lack a coelom and definitive organ systems. Unlike the coelenterates, however, they have independent mesodermal muscles, lack nematocysts (the tentacles, when present, may have adhesive cells instead), do not have the polymorphic life cycle so common in coelenterates, and characteristically have eight rows of ciliary plates (combs) that run across the surface of the transparent body from the upper pole to the lower pole like the lines of longitude on a globe (Fig. 25.12). The cilia in the eight rows beat in unison and enable the animal to swim feebly. Most ctenophores float near the surface of the sea, chiefly near shore. They are carried about by currents and tides, and large numbers of them may be blown into bays during storms or swept ashore by high waves.

ORIGIN OF THE METAZOA

The gastraea and planuloid hypotheses Speculation concerning the origin of the Metazoa (all multicellular animals except the Porifera, or sponges, and the Mesozoa, a group of extremely simple multicellular organisms that live as internal parasites of various invertebrates) has long centered on the radiate phyla just discussed—not only because the radiates, particularly the hydrozoan coelenterates, seem to be among the simplest metazoans, but also because their saclike, essentially two-layered bodies strikingly resemble embryonic gastrulas. Now, a spherical colonial flagellate like *Volvox* (see p. 960) certainly resembles an embryonic blastula. As long ago as 1874 Ernst Haeckel suggested that the ancestor of multicellular animals was a hollow-sphere colonial flagellate similar to *Volvox* (though presumably without cell walls or chlorophyll), and that the coelenterates arose from this hypothetical ancestor, called a **blastaea,** by a process of invaginating gastrulation. The higher animals would then have arisen from the early two-layered (diploblastic) **gastraea** ancestor of the modern coelenterates by assuming a creeping mode of life and slowly becoming bilateral (Fig. 25.13). According to this hypothesis, the blastula and gastrula stages in the embryonic development of higher animals are recapitulations of early steps in the evolution of these animals from colonial flagellates.

But gastrulation in coelenterates, it was soon pointed out, does not take place by invagination. Instead, as we have already seen, the endoderm arises by inwandering of cells produced at the inner surface of the ectoderm, and a solid gastrula, which develops into a planula larva, is formed. Both the hollowing out of the interior to form the gastrovascular cavity and the breaking-through of a mouth occur later, when the larva develops into a polyp. Thus it seems likely that the invaginative type of gastrulation was a later evolutionary development and not the original method of formation of endoderm. Consequently, although the idea of a hollow blastaea ancestor as the starting point was retained in later versions of Haeckel's hypothesis, the idea of a gastraea stage was abandoned and a **planuloid** stage hypothesized instead (Fig. 25.14).

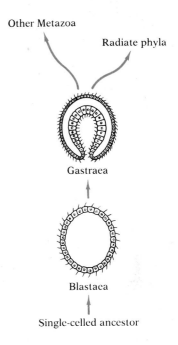

25.13 Diagram of the origin of the Metazoa according to Haeckel's blastaea-gastraea hypothesis

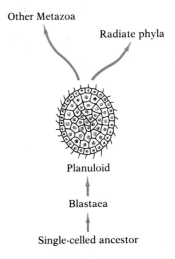

25.14 Diagram of the origin of the Metazoa according to the planuloid hypothesis

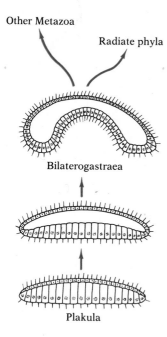

25.15 Diagram of the origin of the Metazoa according to Bütschli's plakula-bilaterogastraea hypothesis
In this interpretation the earliest form of the plakula was a two-layered organism with no internal space. The two layers gradually separated, and the organism became bilateral as it differentiated front and rear ends appropriate to its creeping mode of locomotion. [After K. G. Grell, University of Tübingen (1974).]

The plakula hypothesis and the rediscovery of *Trichoplax* Neither Haeckel's gastraea hypothesis nor its alternative, positing a planuloid stage, explains why gastrulation evolved in the first place, or why the endodermis became specialized for nutritional activities. To meet these difficulties, Otto Bütschli of the University of Heidelberg suggested in 1884 that the earliest metazoan may have been a ***plakula**,* a two-layered, flattened creature creeping about on the sea bottom (Fig. 25.15). Since the ventral cell layer of such an organism would be the layer that would come in contact with food particles most often, it would be reasonable to expect that it should evolve nutritive specializations, and that the dorsal cell layer should evolve protective and perhaps locomotory ones.

Now, if food items were captured beneath the plakula, and if the ventral epithelium was charged with digesting the food, it would have been advantageous, as Bütschli pointed out, for the organism to elevate part of its body and form a temporary digestive chamber from which neither the food nor the digestive enzymes secreted onto it could easily escape. In this manner, then, the plakula might sometimes have become a temporary gastraea. Since Bütschli regarded the plakula as a creeping rather than a swimming organism, he went one step further and suggested that bilateral symmetry[2] would have been more likely than radial symmetry; hence he envisioned a ***bilaterogastraea*** (Fig. 25.15). In other words, he thought that bilateral symmetry was a very early characteristic of the Metazoa and that the radial symmetry of such groups as the Coelenterata and Ctenophora evolved secondarily.

When Bütschli first put forward his ideas, he noted that just one year earlier a living organism had been found that corresponded in many ways with his hypothetical plakula. That organism was a tiny animal discovered in seawater aquaria by F. E. Schulze of the University of Graz in Austria. The organism, which Schulze named *Trichoplax adhaerens*, was a flattened, two-layered creature that crept about on the bottom. Despite its marked resemblance to Bütschli's hypothetical plakula, *Trichoplax* was soon forgotten, because it was presumed (on very questionable evidence) to be merely a larva of a coelenterate.

There the plakula hypothesis and *Trichoplax* rested, in oblivion, until 1969, when K. G. Grell of the University of Tübingen in Germany rediscovered *Trichoplax* on some algae sent to him from the Red Sea (Fig. 25.16). He soon found that, just as Bütschli had speculated eighty-five years earlier, *Trichoplax* does, in fact, often rear up to form a temporary digestive cavity (Fig. 25.17). Moreover, he showed that this simple organism, with only two principal tissue layers and no organs (Fig. 25.18), is not a larva, because it is capable of reproduction—asexual when conditions are not crowded, sexual when they are.

Grell has proposed a new phylum, Placozoa, for *Trichoplax.* In his estimation *Trichoplax* may be the most primitive multicellular animal known. If it is, then the evolution of the metazoans may indeed have followed the pat-

[2] Bilateral symmetry is the property of having two similar sides. A bilaterally symmetrical animal has definite dorsal (upper) and ventral (lower) surfaces and definite anterior (head) and posterior (tail) ends.

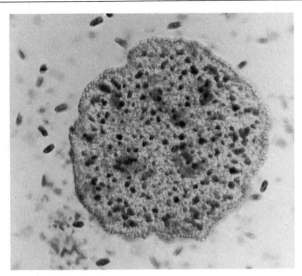
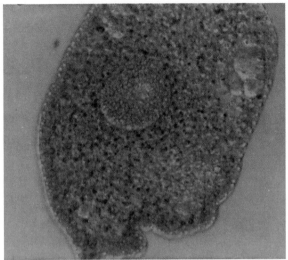

25.16 Photographs of *Trichoplax adhaerens*
Left: The animal has a flat platelike appearance. Of irregular shape, it has no constant symmetry. The many spherical particles in its surface layer are globules of a fatty material.   Right: An animal containing a large egg cell.   [Courtesy K. G. Grell, University of Tübingen.]

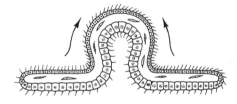

25.17 Formation of a temporary digestive chamber by *Trichoplax adhaerens*
The animal elevates part of its body to form a cavity. It thus becomes, in effect, a temporary gastraea.

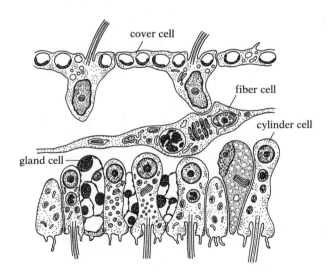

25.18 The histology of *Trichoplax adhaerens*
The thin dorsal epithelium, which is transparent, consists of cells of a single type, called cover cells by Grell; each bears a flagellum, and some incorporate large spherical vesicles containing fatty material. The much thicker ventral epithelium contains flagellated cylinder cells and nonflagellated gland cells. In the space between the two epithelia are some fiber cells, which probably function in locomotion.   [After K. G. Grell, University of Tübingen.]

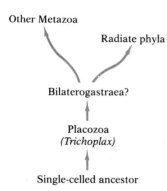

25.19   Diagram of the origin of the Metazoa according to Grell's Placozoa version of the plakula hypothesis
Grell suggests that the Placozoa may be ancestral to the sponges too, but since the evidence for such a filiation is not so convincing, this relationship is not diagramed here.

tern proposed by Bütschli so long ago (Fig. 25.19).[3] But because *Trichoplax* has no bilateral symmetry, Grell is uncertain whether the evolutionary stage following the plakula was a gastraea or a bilaterogastraea.

The ciliate hypothesis   The idea of metazoan origin from flagellates via blastaea and planuloid steps has been widely accepted among biologists, and Grell's new version of the plakula hypothesis is likely to win broad support as well. But other proposals have been put forward, the most notable of which is that multicellular animals arose from multinucleate ciliates, not from flagellates.

According to this so-called ciliate hypothesis, the bodies of the earliest metazoans were syncytial (coenocytic); i.e. they contained many nuclei, each controlling the cytoplasm around it, but had no cellular partitions. Formation of cellular membranes around each nucleus and its associated cytoplasm would have produced a typical multicellular animal, probably resembling some of the simplest flatworms (see Fig. 25.21). Note that, since this hypothesis starts out with a bilaterally symmetrical ciliate, it assumes, like Bütschli's, that the first metazoans were bilateral and that the radial symmetry of the Coelenterata and Ctenophora is secondary.

THE ACOELOMATE BILATERIA

There are two phyla—the Platyhelminthes and the Nemertina—that contain what biologists generally regard as the most primitive bilaterally symmetrical animals. In both phyla the body is composed of three well-developed tissue layers—ectoderm, mesoderm, and endoderm—and the mesoderm is a

[3] Whittaker's recent classification puts the phylum Placozoa and the phylum Mesozoa, whose members are internal parasites of various invertebrates, together in a separate subkingdom, Agnotozoa. Though we follow this practice for convenience in the Appendix, it seems likely that the Mesozoa are secondarily simple as parasites and are not related to the Placozoa.

25.20   An aquatic turbellarian (*Prosthecereus vittatus*)
[Courtesy Hervé Chaumeton.]

solid mass that fills what once was the embryonic blastocoel. In other words, there is no coelom—no cavity between the digestive tract and the body wall (see Fig. 25.32A). For this reason, these two phyla are known as the acoelomate bilateria.

PLATYHELMINTHES (THE FLATWORMS)

The flatworms, as their name implies, are dorsoventrally flattened, elongate animals.[4] Their digestive cavity (not always present) resembles that of coelenterates; i.e. it is a gastrovascular cavity—a cavity with a single opening that must serve as both mouth and anus. However, there is a muscular pharynx leading into the cavity, and the cavity itself is often profusely branched, especially in the free-living species (see p. 216 and Fig. 5.20). As in coelenterates, the amount of extracellular digestion is limited, most of the food particles being phagocytized and digested intracellularly by the cells of the wall of the gastrovascular cavity. Respiratory and circulatory systems are absent.[5] However, there is a flame-cell excretory system (see p. 320 and Fig. 8.10), and there are well-developed reproductive organs (usually both male and female in each individual).

That both the excretory system, with its flame bulbs and tubules, and the reproductive organs should be present signifies that the flatworms have advanced beyond the tissue level of construction seen in the radiate phyla to an organ level of construction. The more extensive development of mesoderm, leading to greater division of labor, was probably a major factor in making this advance possible. Mesodermal muscles are well developed. Several longitudinal nerve cords running the length of the body and a tiny "brain" ganglion located in the head constitute a central nervous system (see p. 402 and Fig. 10.3).

The phylum is divided into three classes: Turbellaria, Trematoda, and Cestoda. The last two are entirely parasitic.

Class Turbellaria The members of this class, of which the freshwater planarians often mentioned in earlier chapters are examples, are free-living flatworms ranging from microscopic size to a length of several centimeters. The body is clothed by an epidermal layer, which is usually ciliated (at least in part). Although a few turbellarians live on land, most are aquatic (the majority marine) (Fig. 25.20).

Turbellarians usually have a gastrovascular cavity, but most members of one small order, the Acoela, do not (Fig. 25.21). For a variety of reasons (not just the absence of a digestive cavity), some biologists have considered the Acoela the most primitive bilateral animals, and have suggested that a primitive *acoeloid* organism might well have arisen from a planuloid ancestor. In their view, both the more complex flatworms and the other metazoan phyla probably evolved from such an acoeloid organism. This version of the early evolution of the Metazoa is diagramed in Fig. 25.22.

---

[4] The term "worm" is applied to a great variety of unrelated animals. It is a descriptive, not a taxonomic, term that denotes possession of a slender elongate body, usually without legs or with very short ones.

[5] A so-called "lymphatic system," present in some flukes, may represent a primitive circulatory system.

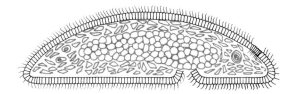

25.21 An acoel flatworm
There is a ventral mouth but no digestive cavity. The entire interior of the animal is filled with an almost solid mass of tissue (color).

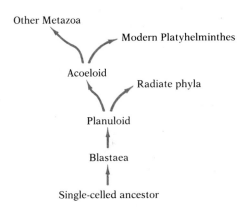

25.22 Diagram of the origin of the bilateral Metazoa according to the blastaea-planuloid-acoeloid hypothesis

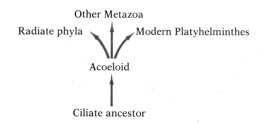

25.23 Diagram of the origin of the Metazoa according to the ciliate-acoeloid hypothesis

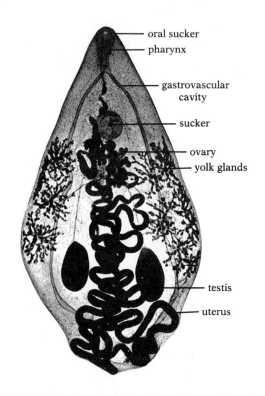

25.24 A fluke that parasitizes the oviducts of the domestic hen (*Prosthogonimus macrorchis*)
[Courtesy Turtox/Cambosco, Macmillan Science Co., Inc.]

Proponents of the ciliate hypothesis of metazoan origin also assume an acoeloid stage at the root of the Metazoa, but they derive the acoeloid directly from a multinucleate ciliate, not from a planuloid. They point out that acoels are about the same size as some ciliates, are ciliated, and sometimes have poorly developed cellular partitions. They think the primitive acoeloid organism was the ancestor of the radiate phyla as well as of the bilateral phyla. Their ideas are diagrammed in Fig. 25.23.

It is unclear how the acoeloid condition, if it is primitive, could be related to Grell's Placozoa hypothesis. However, some biologists regard the Acoela as secondarily simple, not primitive; if this view is correct, the Acoela pose no problem for Grell's hypothesis.

Class Trematoda (The flukes) The flukes are parasitic flatworms. They lack cilia, and in place of the cellular epidermis of their turbellarian ancestors they have a thick cuticle secreted by the cells below. This cuticle is highly resistant to enzyme action and is thus an important adaptation to a parasitic way of life. Flukes characteristically possess suckers, usually two or more, by which they attach themselves to their host (Fig. 25.24). They have a two-branched gastrovascular cavity, which does not ramify throughout the body like that of turbellarians. A large proportion of their bodies is occupied by reproductive organs, including two or more large testes, an ovary, a long much-coiled uterus in which eggs are stored prior to laying, and yolk glands.

The members of one order of flukes are ectoparasites (external parasites) on the gills or skin of freshwater and marine fishes. A few of these flukes sometimes wander into the body openings of their hosts, and it may have been from such a beginning that the endoparasitism (internal parasitism) typical of the members of the other two orders arose.

The endoparasitic flukes often have very complicated life cycles involving two to four different kinds of hosts. The blood fluke, *Schistosoma japonicum*, common in China, Japan, Taiwan, and the Philippines, is a species with two hosts.

The adult blood fluke inhabits blood vessels near the intestine of a human being. When ready to lay its eggs, it pushes its way into one of the very small blood vessels in the wall of the intestine. There it deposits so many eggs that the vessel ruptures, discharging the eggs into the intestinal cavity, whence they are carried to the exterior in the feces. If there is a modern sewage system, that is the end of the story.

But in many Asiatic countries, human feces are regularly used as fertilizer. Thus the eggs get into water in rice fields, irrigation canals, or rivers, where they hatch into tiny ciliated larvae. A larva swims about until it finds a snail of a certain species; it dies if it cannot soon locate the correct species. When it finds such a snail, it bores into the body of the snail and feeds on its tissues. It then reproduces asexually, and the new individuals thus produced leave the snail and swim about until they come in contact with the skin of a human being, such as a farmer wading in a rice paddy or a child swimming in a pond. They attach themselves to the skin and digest their way through it and into a blood vessel. Carried by the blood to the heart and lungs, they eventually reach the vessels of the intestine, where they settle down, mature, and lay eggs, thus initiating a new cycle.

Schistosomas in the body of a person cause a serious disease called ***schistosomiasis***, which is characterized initially by a cough, rash, and body pains,

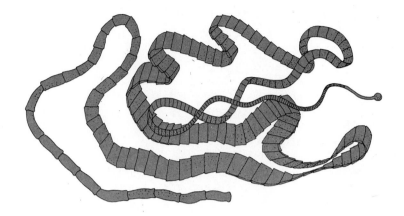

25.25 A tapeworm
The body is composed of a small head and neck, followed by a large number of segments called proglottids. As the proglottids ripen, they break off and pass with the host's feces to the outside. New proglottids are produced just back of the neck.

followed by severe dysentery and anemia. The disease so saps the strength of its victims that they become weak and emaciated and often die of other diseases to which their weakened condition makes them susceptible. Schistosomiasis is one of the most widespread and debilitating human diseases in the world today, but because it is confined to the warmer regions of the earth, most inhabitants of North America and Europe are hardly aware of its existence.

A fluke with three hosts is the Chinese liver fluke, *Clonorchis sinensis.* The adult lives in the liver of a human being, where it lays its eggs. The eggs pass into the intestine with the bile and are carried to the exterior in the feces. If they get into water and are eaten by a certain species of snail, they hatch, and the larvae bore into the lymph spaces of the snail, where they reproduce asexually. The new individuals thus produced live their entire lives in the snail, but they too reproduce asexually, and their progeny leave the snail, burrow through the skin of a fish, and encyst in the fish's muscles. If a person eats the raw or insufficiently cooked fish, his digestive enzymes weaken the walls of the cysts, and the flukes emerge in his intestine, migrate up the bile duct to the liver, and settle down to start a new cycle. All three hosts—snail, fish, and human—are necessary for completion of the reproductive cycle of this fluke.

Class Cestoda (The tapeworms)   Adult tapeworms (Fig. 25.25) live as internal parasites of vertebrates, almost always in the intestine. However, the life cycle usually involves one or two intermediate hosts, which may be invertebrate or vertebrate, depending on the species. The life cycle of the beef tapeworm, in which the intermediate host is a cow and the final host is a human being, was outlined on page 785.

Tapeworms exhibit many special adaptations for their parasitic way of life. Like the flukes, they have a resistant cuticle instead of the epidermis of their free-living ancestors. And they have neither mouth nor digestive tract. Bathed by the food in their host's intestine, they absorb predigested nutrients across their general body surface. Diffusion, probably augmented by active transport, suffices to provision all the cells, because none are far from the surface.

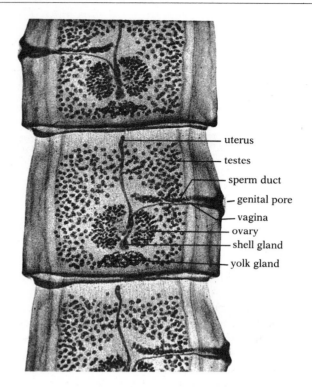

uterus
testes
sperm duct
genital pore
vagina
ovary
shell gland
yolk gland

**25.26   Scolex and proglottids of the dog tapeworm,
*Taenia pisiformis***
Left: Scolex. Note the hooks and suckers.
Right: Mature proglottids.   [Courtesy
Turtox/Cambosco, Macmillan Science Co., Inc.]

The head of a tapeworm is a small knoblike structure called a *scolex,*
which usually bears suckers, and often also hooks, by which the worm
attaches to the wall of the host's intestine (Fig. 25.26, left). Immediately
behind the scolex is a neck region, which is followed by a very long rib-
bonlike body (beef tapeworms occasionally grow to 23 m, fish tapeworms to
18 m, and pork tapeworms to 8 m). This long body is usually divided by
transverse constrictions into a series of segments called *proglottids* (Fig.
25.26, right).

Each proglottid is essentially a reproductive sac, containing both male
and female organs. Sperm cells, usually from a more anterior proglottid of
the same animal, enter the genital pore and fertilize the egg cells, which are
then combined with yolk from a yolk gland and enclosed in a shell. The
fertilized eggs, already undergoing development, are stored in a uterus,
which may become so engorged with eggs that it occupies most of the
volume of a mature proglottid. Eventually all the sexual organs except the
uterus degenerate, and the proglottid, now "ripe," detaches from the worm
and passes out of the host's body with the feces. As ripe proglottids are
released from the end of the worm, new ones are produced just back of the
neck. A single ripe proglottid may contain more than 100,000 eggs, and the
annual output of one worm may be more than 600 million.

If an appropriate intermediate host eats food contaminated with feces
containing tapeworm eggs, its enzymes digest the shells of the eggs. The
embryos thus released bore through the wall of the host's intestine, enter a
blood vessel, and are carried by the blood to the muscles, where they encyst.
If a person eats the raw or "rare" meat of this intermediate host (e.g. beef,

pork, or fish), the walls of the cysts are digested away; the young tapeworms attach to the intestinal wall and, nourished by an abundant supply of food, begin to grow and produce eggs, thus starting a new cycle.

NEMERTINA (THE PROBOSCIS WORMS)

The members of the phylum Nemertina are long slender worms (Fig. 25.27) characterized by a very long eversible muscular proboscis enclosed in a tubular cavity at the anterior end of the body. This proboscis, which is used in capturing prey and also in defense, is often two or more times the length of the worm's body and is somewhat coiled when enclosed in its sheath. The worms are common along both the Atlantic and Pacific shores of the United States. They are usually found sheltered under stones, shells, or seaweeds, or burrowing in the sand or mud in shallow water.

Nemertines resemble turbellarian flatworms (their probable ancestors) in their nervous systems and in many other ways, e.g. in having a ciliated epidermis, a solid mesoderm, and a flame-cell excretory system. But they differ from them in two important characteristics not encountered in the animals considered thus far. First, they have a *complete digestive system*—one that has two openings, a mouth and an anus. Such a system makes possible specialization of sequentially arranged chambers for different functions and thus permits an assembly-line processing of food, as we saw in Chapter 5. Second, they have a simple blood circulatory system, which presumably facilitates transport of materials from one part of the body to another.

25.27 A nemertine worm *(Lineus longissimus)*
[Courtesy Hervé Chaumeton.]

DIVERGENCE OF THE PROTOSTOMIA AND DEUTEROSTOMIA

We saw in Chapter 17 that the embryonic cavity called the archenteron, formed during gastrulation, becomes the digestive tract of an adult animal. But the archenteron has only one opening to the outside, the blastopore. In animals like coelenterates and flatworms, where the digestive tract is a gastrovascular cavity, the blastopore becomes the combined mouth and anus. Now, in nemertine worms and the other higher animals that have complete digestive systems, does the blastopore become the mouth, or does it become the anus? Embryologists have shown that in nemertines the site of the embryonic blastopore becomes the mouth and that the anus is an entirely new opening. This is also the case in many other animals, including nematode worms, molluscs, and annelids. But in a few phyla, among them two large and important ones—the Echinodermata and the Chordata—the situation is reversed: The embryonic blastopore becomes the anus, and the mouth is the new opening.

This fundamental difference in embryonic development suggests that a major split occurred in the animal kingdom soon after the origin of a bilateral ancestor. One evolutionary line led to all the phyla in which the blastopore becomes the mouth; these phyla are often called the *Protostomia* (from the Greek *protos*, first, and *stoma*, mouth). The other evolutionary line led to the phyla in which the blastopore becomes the anus and a new mouth is formed; these phyla are called the *Deuterostomia* (from the Greek *deuteros*, second, later, and *stoma*).

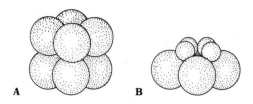

25.28 Radial and spiral cleavage patterns
(A) Radial cleavage, characteristic of deuterostomes.
The cells of the two layers are arranged directly above
each other.   (B) Spiral cleavage, characteristic of
protostomes. The cells in the upper layer are located in
the angles between the cells of the lower layer.

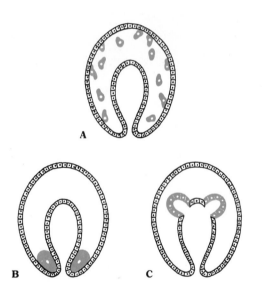

25.29 Different modes of origin of mesoderm
(A) In the radiate phyla mesoderm (colored cells) arises
from inwandering cells derived from the ectoderm. A
small amount of the mesoderm of the protostome
phyla arises in this way also.   (B) In most protostome
phyla the bulk of the mesoderm arises from initial cells
located near the blastopore, at the junction between
the ectoderm and the endoderm.   (C) In the
deuterostome phyla the mesoderm arises as pairs of
pouches from the endodermal wall of the archenteron.

As might be expected if the Protostomia and Deuterostomia diverged at a very early stage of their evolution, they differ in a number of other fundamental characters besides the mode of formation of mouth and anus. A further essential difference between them is that the early cleavage stages are usually determinate in protostomes and indeterminate in deuterostomes; i.e. the developmental fates of the first few cells of a protostome embryo are usually already at least partly determined, and if these cells are separated no one of them can form a complete individual, whereas the fates of the first few cells of a deuterostome embryo are not determined, and each cell, if separated, can develop into a normal individual (there can be identical twinning). Furthermore, the two groups exhibit strikingly different patterns of cleavage; the early cleavages in protostomes are usually oblique to the polar axis[6] of the embryo and thus give rise to a spiral arrangement of cells, whereas the early cleavages in deuterostomes are either parallel or at right angles to the polar axis and thus give rise to a so-called radial arrangement of cells (Fig. 25.28). The basic larval types are also different in the two groups, as we shall see later in this chapter.

Another fundamental difference between the protostome and deuterostome phyla is seen in the method of origin of the mesoderm in the embryo. The mesoderm of the radiate phyla arises from inwandering cells derived from the ectoderm. A small amount of mesoderm forms this way in the protostomes also, though not usually in the deuterostomes. Most of the mesoderm in protostomes and all of it in deuterostomes is derived from endoderm instead of ectoderm. However, in protostomes this mesoderm arises as a solid ingrowth of cells from a single initial cell located near the blastopore (Fig. 25.29B), whereas in deuterostomes (except vertebrates) it arises by a saclike outfolding of the gut wall, as we saw in Chapter 17.[7]

Still another difference, correlated with the preceding one, has to do with the method of formation of the coelom, if one is present. A true coelom is defined as a cavity enclosed entirely by mesoderm and located between the digestive tract and the body wall. In the coelomate protostomes this cavity usually arises as a split in the initially solid mass of mesoderm. In the deuterostomes, by contrast, the coelom arises as the cavity in the mesodermal sacs as they evaginate from the wall of the archenteron.[8]

To repeat, the Protostomia and the Deuterostomia differ most conspicuously in the fate of the embryonic blastopore, the determinateness and pattern of the initial cleavages, the mode of origin of the mesoderm and of the coelom (if one is present), and in type of larva. They also differ in many nondevelopmental traits. To give but one example, the visual receptor cells of the protostomes generally have a receptor organelle of specialized microvilli (forming a rhabdom), whereas the analogous organelle in the deuterostomes is composed of specialized cilia.

---

[6] The polar axis runs between the animal and vegetal poles.

[7] There are actually a variety of other ways in which mesoderm may arise from endoderm, but embryologists usually interpret them as variants of one or the other of the two processes described here.

[8] A coelom that arises as a split in an initially solid mass of mesoderm is called a schizocoelom. One that forms as the cavity in a pouch of mesoderm is called an enterocoelom. The coelomate protostomes are sometimes called the schizocoelous phyla, and the deuterostomes the enterocoelous phyla.

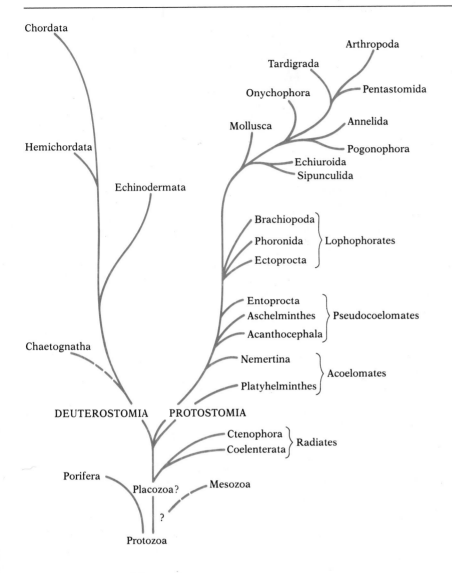

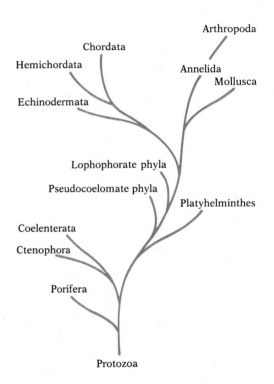

25.30 One interpretation of evolutionary relationships among the animal phyla

25.31 Another interpretation of evolutionary relationships among the animal phyla
[After R. D. Barnes.]

The differences, however, are not quite so clear-cut as we may have implied. Most of the contrasting characters are subject to exceptions, and some are far less distinct in the animals themselves than descriptions and diagrams tend to suggest. Indeed, some authors have seriously questioned whether the protostome-deuterostome dichotomy is real or imagined. With the gathering of more extensive comparative data on nonmorphological and nonembryological characters, ideas about the relationships of the animal phyla may undergo marked changes.

Figure 25.30, drawn in the traditional form of a phylogenetic tree, assumes the reality of the split between the protostomes and the deuterostomes; it shows one of many possible interpretations of evolutionary relationships among the animal phyla. Figure 25.31 shows another interpretation.

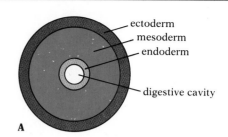

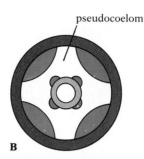

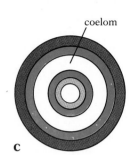

25.32 Diagrams of acoelomate, pseudocoelomate, and coelomate body types
(A) Acoelomate body. There is no body cavity; the entire space between the ectoderm and endoderm is filled by a solid mass of mesoderm.
(B) Pseudocoelomate body. There is a functional body cavity, but it is not entirely bounded by mesoderm.
(C) Coelomate body. The body cavity is completely bounded by mesoderm.

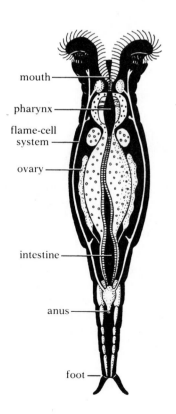

25.33 Section of a rotifer

THE PSEUDOCOELOMATE PROTOSTOMIA

In several protostome phyla the body cavity is functionally analogous to a coelom but differs from a true coelom, which is entirely enclosed by mesoderm, in being partly bounded by ectoderm and endoderm. Such a cavity is called a ***pseudocoelom*** (Fig. 25.32). It is actually only the remnant of the embryonic blastocoel.

Three phyla of pseudocoelomate protostomes are recognized by many biologists: Acanthocephala, Entoprocta, and Aschelminthes. The last will be discussed at greater length below.

Adult Acanthocephala are endoparasites in the digestive tracts of vertebrates; the larvae live in invertebrates. These animals are often called spiny-headed worms because they have a proboscis armed with rows of recurved hooks. They have no digestive tract.

The Entoprocta are tiny sessile, mostly marine, animals that live attached by a stalk to rocks, shells, pilings, or other animals such as crabs or sponges. Most are colonial. Their digestive tract is U-shaped, and both mouth and anus open inside a circle of ciliated tentacles. They feed on small plankton such as diatoms and protozoans.

ASCHELMINTHES

The largest and most important pseudocoelomate phylum is the Aschelminthes. This group includes an array of generally small, wormlike animals, without a definitely delimited head, that have a straight or slightly curved complete digestive tract, and a cuticle. There is no respiratory or circulatory system. A flame-cell excretory system occurs in most classes but not in nematodes, which have a special type of excretory system unique with them.

The phylum is divided into five classes, each recognized as a separate phylum by some biologists. We shall discuss only two: Rotifera and Nematoda.

Class Rotifera The rotifers—or wheel animalcules, as they are commonly called—are microscopic, usually free-living aquatic animals with a crown of cilia at the anterior end (Fig. 25.33). The cilia are generally arranged in a circle, and when beating they often give the appearance of a rotating wheel; hence the name of the class. When feeding, rotifers attach themselves by a tapering posterior "foot," and the beating cilia draw a current of water into the mouth. In this manner, very small protozoans and algae are swept into a complicated muscular pharynx, where they are ground up by seven hard jawlike structures.[9]

The freshwater rotifers are extremely abundant. Anyone examining a drop of water for Protozoa under a microscope is likely to see one or more of these interesting animals. In fact, most of them are no larger than protozoans, and it is often difficult, when encountering them for the first time, to realize that they are multicellular.

Class Nematoda (The roundworms) Nematode worms have round elongate bodies that usually taper nearly to a point at both ends (Fig. 25.34). Unlike flatworms, they have no cilia. The body is enclosed in a tough cuticle (Fig. 25.35). Just under the epidermal layer of the body wall are bundles of longitudinal muscles; there are no circular muscles. The lack of circular muscles and the stiff cuticle severely limit the types of movements possible for the worms, and they usually thrash about in what appears to be a random and inefficient manner.

The wall of the digestive tract consists of a single layer of endodermal cells; there is usually no muscle layer around the intestine, except sometimes at its posterior end. Between the intestine and the body wall is a fluid-filled cavity. As you can see from Fig. 25.35, this cavity is bounded internally by the endodermal wall of the intestine, and it is bounded externally in part by the bands of mesodermal muscle and in part (between the muscle bands) by the ectodermal layer of the body wall. Since the cavity is not entirely enclosed by mesoderm, it is not a coelom, but a pseudocoelom.

Nematodes are extremely abundant, and occur in almost every type of habitat. Of the many free-living in soil or water, most are very tiny, often microscopic. A single spadeful of garden soil may contain a million or more, and a bucket of water from a pond usually contains comparable numbers. Many other nematodes are internal parasites of both plants and animals; these also are often small, but some may attain a length of one meter. So numerous and widespread are nematodes that N. A. Cobb has written:

> If all the matter in the universe except the nematodes were swept away, our world would still be dimly recognizable, and if, as disembodied spirits, we could then investigate it, we should find its mountains, hills, vales, rivers, lakes, and oceans represented by a film of nematodes. The location of towns would be decipherable, since for every massing of human beings there would be a corre-

[9] This complex pharynx, a distinctive feature of the Rotifera, is called a mastax.

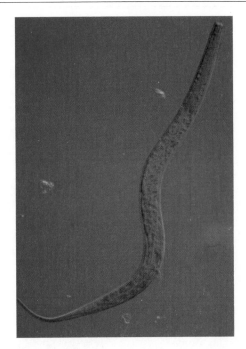

25.34 A living nematode worm viewed with Nomarski optics
[Courtesy Thomas Eisner, Cornell University.]

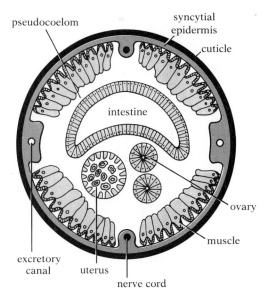

25.35 Diagrammatic cross section of a nematode worm

25.36   Photograph of *Trichinella spiralis* encysted in muscle

[Courtesy Bausch and Lomb, *Focus.*]

sponding massing of certain nematodes. Trees would still stand in ghostly rows representing our streets and highways. The location of the various plants and animals would still be decipherable, and, had we sufficient knowledge, in many cases even their species could be determined by an examination of their erstwhile nematode parasites.[10]

Nematodes parasitic on cultivated plants cause an annual loss of millions of dollars. Others parasitic on human beings cause some serious diseases. *Trichinella spiralis*, for example, causes the disease called **trichinosis,** often contracted by eating insufficiently cooked pork. Adult *Trichinella* worms inhabit the small intestine of numerous species of mammals, among them hogs. Impregnated females bore through the wall of the intestine and deposit young larvae (which hatched from the eggs while still inside the uterus of the female) in the lymphatic vessels of their host. The larvae are carried by the lymph and blood to all parts of the body. They then bore out of the vessels, eventually entering every organ and tissue. However, only those that bore into skeletal muscles (especially the muscles of the diaphragm, ribs, tongue, and eyes) survive. In the muscles they grow in size (to about one millimeter) and then curl up and encyst (Fig. 25.36); the thick wall of the cyst is formed by the host's tissues. If insufficiently cooked pork containing such cysts is eaten by a human being, the walls of the cysts are digested away and the worms complete their development in his intestine. The adult worms then deposit larvae in the lymph vessels in the wall of the intestine, and the larvae move through his body as they do through a hog's, eventually encysting in muscles.

Most of the damage of trichinosis occurs during the migration of the larvae, when half a billion or more may simultaneously bore through the body after one infection. Symptoms include excruciating muscular pains, fever, anemia, weakness, and sometimes localized swellings. Some victims die, and those that do not may sustain permanent muscular damage. Prevention of the disease is simple: Pork must be thoroughly cooked to kill the encysted larvae. It is well to remember that one ounce of infected pork in the center of a large chunk of meat where heat does not reach it may easily contain as many as 100,000 encysted worms, each of which, when mature, may produce 1,500 young larvae in the body of a new host.

Among other nematodes that parasitize humans are (1) *Ascaris,* a large worm (up to 30 cm long), which lives in the digestive tract, lays eggs that pass to the outside with the host's feces, and infects new hosts when vegetables grown in soil contaminated with feces are eaten without adequate washing; (2) hookworms, tiny worms, widespread in warm climates, that cause a severely debilitating disease usually gotten by going barefoot on soil contaminated with feces containing eggs; (3) pinworms, likewise tiny worms, common in schoolchildren, who usually get infected by putting unclean fingers with eggs on them in their mouths; (4) filaria worms, which are spread by the bite of certain mosquitoes in tropical and subtropical areas. Filaria worms live in the lymphatic system, where they may accumulate in such numbers that they block the flow of lymph, causing accumulation of fluid and often enormous swelling (elephantiasis) of the infected part of the body (see Fig. 7.33, p. 294).

[10] Cited in R. Buchsbaum, *Animals without backbones* (2nd ed.; University of Chicago Press, 1948), pp. 156–157.

25.37 **Phoronid worms** *(Phoronis hippocrepia)*
[Courtesy Hervé Chaumeton.]

THE COELOMATE PROTOSTOMIA

All the protostome phyla except the ones discussed above possess true coeloms that in most groups arise as a split in an initially solid mass of mesoderm. All have a complete digestive tract, and most have well-developed circulatory, excretory, and nervous systems.

THE LOPHOPHORATE PHYLA

There are three small phyla, Phoronida, Ectoprocta, and Brachiopoda, that resemble one another in having a *lophophore*—a fold, usually horseshoe-shaped, that encircles the mouth and bears numerous ciliated tentacles. The lophophore is a feeding device; its tentacular cilia create water currents that sweep plankton and tiny particles of detritus into a groove leading to the mouth.

All members of the lophophorate phyla are aquatic, and most are marine. Adults are usually sessile and secrete a protective case, tube, or shell around themselves, but the larvae are ciliated and free-swimming. The digestive tract is U-shaped in Phoronida and Ectoprocta and in some Brachiopoda; the anus lies outside the crown of tentacles.

The phylum Phoronida contains only about 15 species of wormlike animals that inhabit a tube of their own secretion (Fig. 25.37). They are found either buried in the sand or attached to rocks, shells, or other objects in shallow seas.

The Ectoprocta (or Bryozoa) are often called moss animals. They are very tiny (usually less than half a millimeter long), colonial, sessile animals enclosed in a case open only at the lophophore end (Fig. 25.38). Unlike most other coelomate protostomes, they lack both excretory and circulatory systems. They superficially resemble entoprocts, which were formerly included in the same phylum with them, but entoprocts have a pseudocoelom instead of a coelom and their anus is inside the ring of tentacles,[11] which is not a true lophophore.

[11] The name "Entoprocta" means internal anus (i.e. an anus inside the crown of tentacles) and the name "Ectoprocta" means external anus (i.e. outside the crown of tentacles).

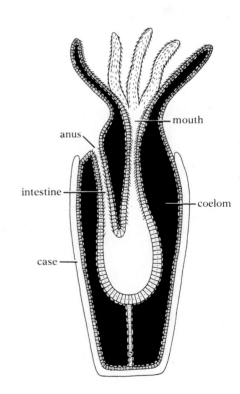

25.38 **Section through the body of an ectoproct, showing the U-shaped digestive tract**
The anus is located outside the ring of tentacles.

25.39 A brachiopod *(Terebratulina)*
The shape of the shell is reminiscent of old Roman
lamps. Brachiopods usually attach themselves to the
substratum by means of a stalk. [Courtesy Hervé
Chaumeton.]

25.40 A chiton
Note the series of plates composing the chiton's shell.
[Oxford Scientific Films.]

The Brachiopoda, shelled animals that superficially resemble molluscs,
are often called lamp shells because they are shaped rather like an old
Roman oil lamp (Fig. 25.39). They are usually permanently attached to the
ocean bottom by a fleshy stalk. There are only about 260 living species of
brachiopods, but more than 30,000 fossil species are known; they were
among the most common animals in the Paleozoic seas.

The relationships of the lophophorate phyla to the other phyla are very
poorly understood.

MOLLUSCA

The phylum Mollusca is the second-largest in the animal kingdom; it contains nearly 50,000 living species and 35,000 fossil species. Among the best-
known molluscs are snails and slugs, clams and oysters, squids and octopuses.

The various groups of molluscs may differ considerably in outward appearance, but most have fundamentally similar body plans.[12] The soft body
consists of three principal parts: (1) a large ventral muscular *foot*, which can
be extruded from the shell (if one is present) and functions in locomotion;
(2) a *visceral mass* above the foot, which contains the digestive system, the
excretory organs (nephridia), the heart, and other internal organs; and (3) a
heavy fold of tissue called the *mantle,* which covers the visceral mass and
which in most species contains glands that secrete a shell. The mantle often
overhangs the sides of the visceral mass, thus enclosing a *mantle cavity,*
which frequently accommodates gills (see Fig. 25.47).

Molluscs have an open circulatory system; i.e. during part of each circuit
the blood is in large open sinuses where it bathes the tissues directly. Blood
drains from the sinuses into vessels that run out into the gills, where the
blood is oxygenated. From the gills, the blood goes to the heart, which
pumps it into vessels that lead it back to the sinuses; a typical circuit, then, is
heart–sinuses–gills–heart.

Most marine molluscs pass through one or more ciliated free-swimming
larval stages, but freshwater and land snails complete the corresponding
developmental stages while still in the egg and hatch as miniature editions of
the adult.

The Mollusca are customarily divided into six classes: Amphineura, Monoplacophora, Gastropoda, Scaphopoda, Pelecypoda, and Cephalopoda.

Class Amphineura Exclusively marine animals, the amphineurans are
generally regarded as the most primitive living members of the Mollusca,
and are thus a suitable group with which to begin study of the basic molluscan body plan.

The best-known Amphineura are the chitons, which have an ovoid bilaterally symmetrical body with a shell consisting of eight serially arranged
dorsal plates (Fig. 25.40). They have an anterior mouth and a posterior anus

---

[12] The generalized body plan described here characterizes all living Mollusca except the solenogasters, which constitute the subclass Aplacophora of the class Amphineura. These very primitive
wormlike molluscs have no head, mantle, foot, shell, or nephridia. Some authorities (e.g. Hyman,
1967) elevate them to full class rank.

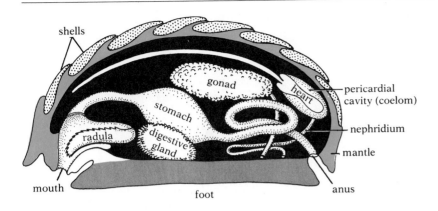

shells

gonad

heart

pericardial cavity (coelom)

stomach

nephridium

radula

digestive gland

mantle

mouth

foot

anus

25.41 Lateral view of a section of a chiton

25.42 A marine gastropod mollusc (Trivia monaca)
The animal moves along the substratum by means of the large muscular foot, here plainly seen. Notice also the head region, bearing prominent antennae, eyes, and a long siphon through which water is brought to the gills. [Courtesy Hervé Chaumeton.]

(Fig. 25.41). The coelom is reduced to a small cavity surrounding the heart.[13]

Chitons lead a sluggish, nearly sessile life. They creep about on the surface of rocks in shallow water, rasping off fragments of algae with a horny toothed organ called a *radula.* Their broad flat foot can develop tremendous suction, and when disturbed they clamp down so tenaciously to the rock that they can hardly be pried loose.

Class Monoplacophora Members of this class have long been known as fossils, but until 1952 it was thought that all had been extinct for about 350 million years. In 1952, however, ten living specimens (genus *Neopilina)* were dredged from a deep trench in the Pacific Ocean off the coast of Costa Rica.

These specimens sparked a lively debate on the ancestry of the Mollusca, because they show some internal segmentation, a characteristic seen in no other members of the phylum. Since it was already known that the early cleavage pattern and larval type of molluscs show striking similarities to the corresponding developmental stages in the segmented worms (Annelida), the segmentation of *Neopilina* led many biologists to conclude that the ancestral molluscs were segmented animals, perhaps primitive annelids. Many other biologists, however, are convinced that the segmentation of *Neopilina* is secondary, not primitive, and that the original molluscan body was unsegmented. Whichever view is correct, it seems clear that the Mollusca and the Annelida are fairly closely related.

Class Gastropoda (The snails and their relatives) Most gastropods have a coiled shell (Fig. 25.42). In some cases, however, the coiling is minimal. Some species—e.g. the nudibranchs—have lost the shell (Fig. 25.43).

The early larva in gastropods is bilateral, but, as it develops, the digestive tract bends downward and forward until the anus comes to lie close to the mouth. Then the entire visceral mass rotates through an angle of 180 degrees, coming to lie dorsal to the head in the anterior part of the body. Most

25.43 A nudibranch gastropod (Hermissenda crassicornis)
As the designation "nudibranch" suggests, the animal has no shell. [R. N. Mariscal, Bruce Coleman Inc.]

[13] The lumina of the gonads and nephridia are also thought to be remnants of the coelom.

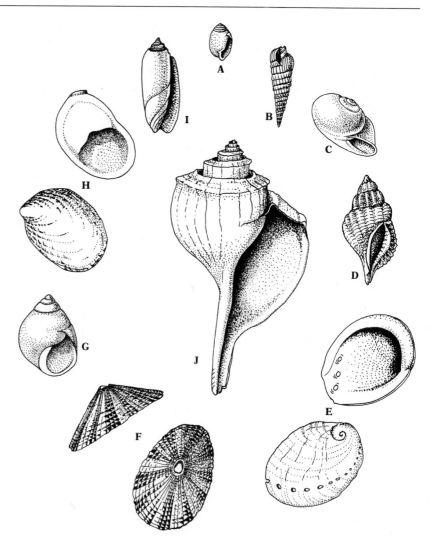

25.44 Some representative gastropods
(A) Salt-marsh snail *(Melampus).* (B) Auger shell
(Terebra). (C) Moon shell *(Polinices).* (D) Oyster drill
(Urosalpinx). (E) Abalone *(Haliotis),* dorsal and ventral
views. (F) Californian keyhole limpet *(Diodora),* lateral
and dorsal views. (G) Periwinkle *(Littorina).* (H) Boat
shell *(Crepidula),* dorsal and ventral views. (I) Olive
shell *(Oliva).* (J) Channeled conch *(Busycon).* [Based
in part on drawings by Louise G. Kingsbury.]

25.45 A scaphopod mollusc *(Dentalium senegalensis)*
The shell is a long tapering tube open at both ends.
[Courtesy Hervé Chaumeton.]

of the visceral organs on one side (usually the left) atrophy, and growth
proceeds asymmetrically, producing the characteristic spiral.

Except for the peculiar twisting and coiling of their bodies, gastropods are
thought to be rather like the ancestral molluscs. They have a distinct head
with well-developed sense organs, and most have a well-developed radula
and feed on bits of plant or animal tissue that they grate, rasp, or brush loose
with this organ.

Gastropods occur in a great variety of habitats. The majority are marine,
and their often large and decorative shells are among the most prized finds
on a beach (Fig. 25.44), but there are also·many freshwater species and some
that live on land. The land snails (see Fig. 1.17, p. 21) are among the few
groups of fully terrestrial invertebrates. In most of them, the gills have
disappeared, but the mantle cavity has become very highly vascularized and
functions as a lung. Such snails are said to be pulmonate. Some pulmonate
snails have secondarily returned to the water and must periodically come to
the surface to obtain air.

25.46 A scallop (Chlamys opercularis), a representative pelecypod
The shell is composed of two hinged valves. Note the numerous small eyes around the edges of the mantle. [Courtesy Hervé Chaumeton.]

Class Scaphopoda (The tusk shells)   Scaphopods have a long tubular shell, open at both ends (Fig. 25.45). One end is usually smaller than the other, and the shell thus has a tusklike or toothlike appearance. All scaphopods are marine, living buried in mud or sand. The living animals are seldom seen, but the shells of dead ones can sometimes be found washed up on a beach.

Class Pelecypoda (The bivalve molluscs)   As the term "bivalve" indicates, these animals have a two-part shell (Fig. 25.46). The two parts, or valves, are usually similar in shape and size and are hinged on one side (the animal's dorsum) (Fig. 25.47). The animals open and shut them by means of large muscles. Among the more common bivalves are clams, oysters, scallops, cockles, file shells, and mussels. Most lead rather sedentary lives as adults, though scallops sometimes swim about by rapidly opening and shutting the valves of their shells.

Pelecypods, which have no radula, are filter feeders, straining tiny food particles from the water flowing across their gills. The process is described on p. 219.

Class Cephalopoda (Squids, octopuses, and their relatives)   Many of the cephalopods bear little outward resemblance to other molluscs. Unlike their sedentary relatives, they are often specialized for rapid locomotion and a predatory way of life—for killing and eating large prey, such as fishes or crabs. Though fossil cephalopods often have large shells, these are much reduced or absent in most modern forms.[14] The body is elongate, with a large and well-developed head encircled by long tentacles.

Some species attain large size, often being several meters long. The giant squids (Architeuthis) of the North Atlantic are the largest living invertebrates; the biggest recorded individual was 17 m long (including the tentacles) and weighed approximately 2 tons. Octopuses (Fig. 25.48) never grow anywhere near this size (except in Hollywood).

[14] *Nautilus*, a modern form with a well-developed shell, is an exception.

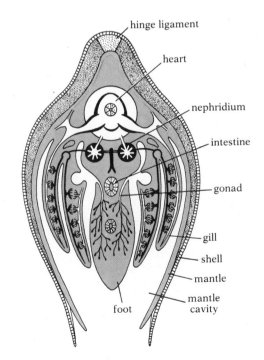

25.47   Cross section of a clam
[Modified from W. Stempell, *Zoologie im Grundriss*, Borntraeger, 1926.]

25.48   A swimming octopus *(Octopus vulgaris)*
Note the large suckers on the tentacles.   [Courtesy
Hervé Chaumeton.]

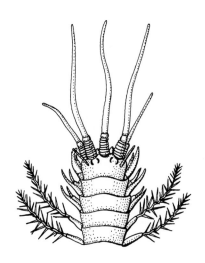

**25.49   Head and first two gill-bearing segments of a
polychaete worm *(Diopatra)***
[Modified from R. D. Barnes, *Invertebrate zoology,*
Saunders, 1963.]

Cephalopods, particularly squids (see Fig. 6.11, p. 246), have convergently
evolved many similarities to vertebrates. For example, squids have internal
cartilaginous supports analogous to the vertebrate skeleton, and they even
have a cartilaginous braincase rather like a skull. Furthermore, they have an
exceedingly well-developed nervous system with a large and complex brain.
Perhaps the most striking of all the squids' similarities to vertebrates are
their large camera-type image-forming eyes, which work exactly the way
ours do.

ANNELIDA

The Annelida, or segmented worms, have received attention in various connections throughout this book. We have considered their digestive system
(p. 217), gas exchange (p. 245), closed circulatory system (p. 278), nephridia
(p. 321), nervous system (pp. 403 and 452), and their hydrostatic skeleton,
muscle arrangement, and methods of locomotion (p. 467). The discussion
here will therefore be brief.

The phylum is usually divided into three classes: Polychaeta, Oligochaeta,
and Hirudinea.

Class Polychaeta   Polychaetes are marine annelids with a well-defined
head bearing eyes and antennae (Fig. 25.49). Each of the numerous serially
arranged body segments usually bears a pair of lateral appendages called
parapodia that function in both locomotion and gas exchange (see Fig. 6.10,
p. 245, and Fig. 11.3, p. 467). There are numerous stiff setae (bristles) on the
parapodia (the name "Polychaeta" means many chaetae, i.e. setae).

Some polychaetes swim or crawl about actively; others are more sedentary, usually living in tubes they construct in the mud or sand of the ocean

bottom. These tubes may be simple mucus-lined burrows, membranous structures, or elaborately constructed dwellings composed of sand grains cemented together. The tubes of some species are straight, while those of others are U-shaped and have two openings (Fig. 25.50). The beating parapodia keep water currents flowing through the tubes; these currents bring oxygen, and in some cases food particles, to the worm. Many of the tube dwellers are beautiful animals, often colored bright red, pink, or green; some are iridescent. Among the most beautiful are the fanworms and peacock worms, which have a crown of colorful, much-branched fanlike or feather-like processes that they wave in the water at the entrance to their tubes (Fig. 25.51).

All the segments of the body are usually much alike. The coelom of each is partly separated from the coeloms of adjacent segments by membranous intersegmental partitions; the partitions of many polychaetes are not complete, however, and in some species they have been entirely lost. Each segment generally has its own ventral ganglion and its own pair of nephridia.[15]

The sexes are separate in the majority of species. In primitive polychaetes most segments produce gametes, but in more advanced species gamete production is restricted to a few specialized segments. The gametes are usually shed into the coelom and leave the body through the nephridia. Fertilization is external. In many species development includes a ciliated free-swimming larval stage called a *trochophore* (see Fig. 25.76).

[15] In a few species of polychaetes, there is only one pair of nephridia for the whole animal.

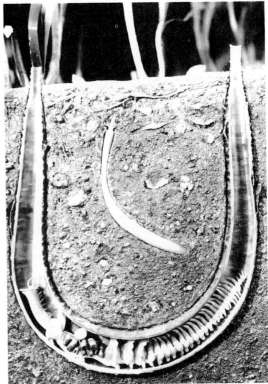

25.50 Polychaete and sipunculid
The parchment worm *(Chaetopterus pergamentacus)* is a polychaete that lives in a U-shaped tube (here shown in section). Between the arms of the tube is a sipunculid *(Phascolosoma gouldii);* the Sipunculida are a small phylum of worms related to the Annelida. [Courtesy American Museum of Natural History.]

25.51 Peacock worms *(Sabella pavonina)*
One worm has been removed from its tube, so that its segmented body can be seen. [Jane Burton, Bruce Coleman Inc.]

Class Oligochaeta The class Oligochaeta contains the earthworms and many freshwater species. They differ from polychaetes in that they lack a well-developed head and parapodia, have fewer setae (the name "Oligochaeta" means few setae), usually combine male and female organs in the same individual, and usually have more complete intersegmental partitions. We have described most of the important characteristics of earthworms in earlier chapters (see Figs. 5.21 and 5.22, p. 218; Fig. 7.19, p. 278; Fig. 8.11, p. 322; Fig. 9.37, p. 382; Fig. 10.4, p. 403).

Class Hirudinea (The leeches) The leeches, which probably evolved from oligochaetes, are the most specialized annelids. Their body is dorsoventrally flattened and often tapered at both ends. The first and last segments are modified to form suckers, of which the posterior one is much the larger. We have already described the way these suckers are used in locomotion (see Fig. 11.2, p. 466). Leeches show almost no internal segmentation; the intersegmental partitions have been completely lost except in a few very primitive species.

Some leeches are predaceous, capturing invertebrate prey such as worms, snails, and insect larvae and swallowing them whole. More familiar are the bloodsuckers, which attack a variety of vertebrate and invertebrate hosts (Fig. 25.52). When a leech of this type attacks a host, it selects a thin area of the host's integument, attaches itself by its posterior sucker, applies the anterior sucker very tightly to the skin, and either painlessly slits the skin with small bladelike jaws or dissolves an opening by means of enzymes. It then secretes into the wound a substance (hirudin) that prevents coagulation of the blood, and begins to suck the blood, usually consuming an enormous quantity at one feeding and then not feeding again for a fairly long time (some leeches have been known to go unfed for more than a year without apparent harm).

ONYCHOPHORA

There are only about 65 living species of this small phylum, all restricted to tropical regions or to the temperate parts of the Southern Hemisphere (Australia, New Zealand, South Africa, and the Andes). They are mostly confined to very moist habitats on land, living beneath leaves, logs, or stones in forests, and are active at night.

Looking rather like caterpillars, onychophorans have a segmented wormlike body with from 14 to 43 pairs of short unjointed legs (Fig. 25.53). These animals are of special interest because they have a combination of annelid and arthropod characters and are regarded as an early evolutionary offshoot from the line leading to the arthropods from an ancient annelidlike ancestor (see Fig. 25.56). They have a thin flexible permeable cuticle more like the cuticle of annelids than the exoskeleton of arthropods, and, like the annelids, they have a pair of nephridia in each segment. However, they resemble arthropods in having claws and an open circulatory system. The tracheal respiratory system of modern onychophorans probably evolved independently of the tracheae of terrestrial arthropods.

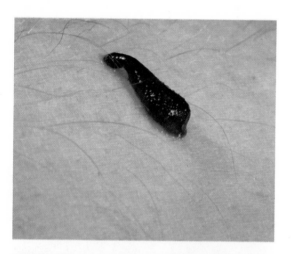

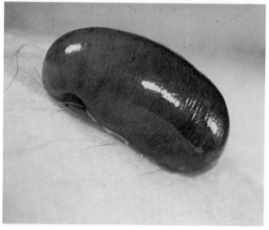

25.52 A terrestrial leech of Australia, sucking blood from a person's arm
Top: The leech has just begun feeding.
Bottom: Somewhat later its body has become engorged with blood. [Oxford Scientific Films.]

25.53   An onychophoran
Note the numerous short unjointed legs and the prominent antennae. [Peter Ward, Bruce Coleman Inc.]

ARTHROPODA

The phylum Arthropoda is by far the largest of the phyla. Nearly a million species have been described, and there are doubtless hundreds of thousands more yet to be discovered. Probably more than 80 percent of all animal species on earth belong to this phylum.

Arthropods are characterized by jointed chitinous exoskeletons and jointed legs. The exoskeleton, which is secreted by the epidermis, functions both as a point of attachment for muscles and as a protective armor, but it imposes limitations on growth and must be periodically molted if the animal is to undergo much increase in size (Fig. 25.54; see also Figs. 17.20 and 17.21, pp. 712–713, 714). The arthropod cuticle is not restricted to the exterior surface of the body; long rod-shaped processes that often project from the surface deep into the interior of the animal function as bases for muscle attachment, and both the anterior and the posterior portions of the digestive tract (and the tracheae of land arthropods) are lined with cuticle. These internal extensions of the exoskeleton are also shed at each molt.

Together with their elaborate exoskeleton, arthropods have evolved a complex musculature quite unlike that of most other invertebrates. It comprises not only longitudinal and circular bands, as in so many invertebrates, but also separate muscles that, running in myriad directions, make possible an extensive repertoire of movements. Most of the muscles are striated.

The nervous system is very well developed. Of a similar organization as the annelid nervous system, it consists of a dorsal brain and a ventral double nerve cord (see Fig. 10.4, p. 403). Primitively there were ganglia in each segment, but in many groups the ganglia have tended to move forward and fuse into larger ganglionic masses. Sensory organs are many and varied. Like the nervous system, discussed more extensively in Chapter 10 (p. 404), hormonal control, too, is well developed in arthropods (see p. 355).

As we have seen (p. 278, Fig. 7.20), arthropods have an open circulatory system. There is usually an elongate dorsal vessel called the heart, which pumps the blood forward into arteries (the extent of these arteries varies greatly among the various groups of arthropods). From the arteries, the

25.54   A molting centiped
Its new yellow-orange exoskeleton glistening, the animal is backing out of its old exoskeleton. [Jane Burton, Bruce Coleman Inc.]

25.55 A fossil of a trilobite
Note the two longitudinal furrows that partition the animal's body into a median lobe and two lateral lobes. It is this tripartite arrangement that suggested the name "Trilobita." [Courtesy N. F. Snyder, Cornell University.]

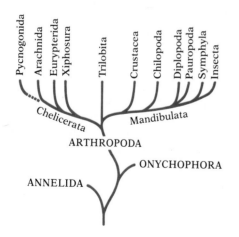

25.56 Diagram of possible relationships between the Annelida, the Onychophora, and the subphyla and classes of the Arthropoda

blood goes into open sinuses, where it bathes the tissues directly. Eventually the blood returns to the posterior portion of the heart.

The body spaces through which the blood moves constitute the **hemocoel**—not a true coelom but a cavity derived from the embryonic blastocoel. Although arthropods almost certainly descended from an annelidlike ancestor with well-developed coelomic cavities, and although such cavities develop in arthropod embryos, they are not retained as the functional body cavity in the adult.[16]

In most aquatic arthropods (excluding secondarily aquatic ones), excretion of nitrogenous wastes (primarily ammonia) is principally by way of the gills. Aquatic species usually also have special saclike glands, located near or in the head, that play a minor role in excretion; these glands (usually called coxal glands or green glands) have their own ducts leading to the outside. The excretory organs in most groups of terrestrial arthropods are Malpighian tubules (see Fig. 8.18, p. 330).

The sexes are usually separate. Fertilization is internal in all terrestrial and in most aquatic forms.

It is generally held that arthropods evolved either from a polychaete annelid or from the ancestor of the polychaetes. The arthropod body plan may be viewed as an elaboration and specialization of the segmented body of that annelid ancestor. The evidence indicates that the first arthropods had long wormlike bodies composed of many nearly identical segments, each bearing a pair of legs. All the legs were alike. Among the host of modifications of this ancestral body plan that have arisen in the various groups of arthropods during the millions of years of their evolution, four tendencies stand out: (1) reduction in the total number of segments; (2) grouping of segments into distinct body regions, such as a head and trunk, or a head, thorax, and abdomen; (3) increasing cephalization, i.e. incorporation of more segments into the head and concentration of nervous control and sensory perception in or just behind the head; (4) specialization of the legs of some segments for a variety of functions other than locomotion, and complete loss of legs from many other segments.

Subphylum Trilobita Arthropods were very abundant in the Paleozoic seas, and fossils from that era are plentiful. Particularly common in rocks of the first half of the Paleozoic are the fossils of an extinct group, the Trilobita (Fig. 25.55). The fossils show a usually oval and flattened shape and three body regions: a head, apparently composed of four fused segments, that bore a pair of slender antennae and, often, compound eyes; a thorax consisting of a variable number of separate segments; and an abdomen (pygidium), composed of several fused segments. It is not to this tripartite division, however, that the name "Trilobita" refers, but to a division of the body into a median lobe and two lateral lobes by two prominent longitudinal furrows running along the dorsum.

Trilobites, though they were surely different from the first arthropods, exhibiting specializations of their own (e.g. the longitudinal furrows and the fusion of the abdominal segments), nevertheless seem to approach the hypothetical arthropod ancestor more closely than any other known group. One primitive character stands out—the lack of specialization and structural

[16] The cavity of the gonads (and that of the excretory ducts in some arthropods) may be a remnant of the true coelom.

differentiation of the appendages. The fossils show that every segment bore a pair of legs, and that all these legs, including those of the four head segments, were nearly identical. There were thus no appendages specialized as mouthparts.

Trilobites are so different from all other arthropods that they are often regarded as a separate subphylum. Two other subphyla—Chelicerata and Mandibulata—are usually recognized. In both these groups the tendency toward specialization of some appendages and loss of others is quite evident; thus in both, the appendages of the most anterior segments have been modified as mouthparts and no longer function in locomotion.

Figure 25.56 shows one interpretation of the relationships among the arthropod subphyla and classes.

Subphylum Chelicerata   The chelicerate body is usually divided into two regions: a cephalothorax (prosoma) and an abdomen. There are no antennae. The appendages corresponding to the first pair of postoral legs in ancestral arthropods and trilobites are modified as mouthparts called *chelicerae,* which may be either pincerlike or fanglike. The cephalothorax usually bears five other pairs of appendages besides the chelicerae; in some groups these are all walking legs, while in others only the last four pairs are legs, the first pair being modified as feeding devices called *pedipalps,* which are often much longer than the chelicerae (Fig. 25.57). The legs of the abdominal segments have been either lost or modified into respiratory or sexual structures.

The subphylum Chelicerata includes four classes. One (Eurypterida) consists entirely of animals extinct since the Paleozoic era, and the members of another (Pycnogonida, the sea spiders) are very rare marine animals.

Members of a third class (Xiphosura) are familiar to anyone who has spent some time on the Atlantic beaches of North America (or the coast of Asia from Japan and Korea to Malaysia and Indonesia). These are the horseshoe crabs (Fig. 25.58), which are not really crabs at all but living relics of an ancient chelicerate class most members of which have been extinct for millions of years. The most common species is named *Limulus polyphemus.*

25.57   A whipscorpion *(Mastigoproctus giganteus)* Whipscorpions have six pairs of appendages: a pair of fanglike chelicerae (not visible in the photograph); a pair of stout toothed pincerlike pedipalps; and four pairs of legs (the long and slender first pair have a sensory-tactile function and are not used in walking). The posterior knob with its "whip" has slits through which the animal can spray a poisonous secretion. [M. P. L. Fogden, Bruce Coleman Inc.]

25.58   A horseshoe crab [C. B. Frith, Bruce Coleman Inc.]

25.59 A spider *(Araneus marmoreus)*
The four pairs of legs characteristic of the Arachnida are easy to make out. Note how the animal places its first pair of legs on strands of its web; it can detect vibrations when prey touch the web, and can often even distinguish, by the type of vibration, what sort of prey it is. [Courtesy Hervé Chaumeton.]

Members of the fourth class of chelicerates—Arachnida—are familiar to everyone. These are the spiders (Fig. 25.59), ticks, mites, daddy longlegs, scorpions, whipscorpions (Fig. 25.57), and their relatives. Though the various groups of arachnids differ structurally in many ways, most have two body regions: a cephalothorax and an abdomen (these are not distinguishable in ticks, mites, or daddy longlegs). There are often simple eyes on the cephalothorax (see Fig. 10.33, p. 437) but never any compound eyes or antennae. The cephalothorax bears six pairs of appendages: a pair of chelicerae, a pair of pedipalps, and four pairs of walking legs. In most groups prey is seized and torn apart by the pedipalps. The chelicerae may also function in manipulating prey, or they may be modified as poison fangs, as in spiders. The abdomen of arachnids may be long, as in scorpions, or short, as in spiders. In some (including spiders), the bases of one or two pairs of abdominal appendages are retained as much-modified book lungs; in others, no trace of abdominal appendages remains. In addition to the book lungs, many arachnids have tracheae, and some respire by means of tracheae only. Several arachnid groups possess glands that secrete silk.

25.60 A crab and a prawn, representatives of the larger, better-known crustaceans
Top: The crab *Macropipus depurator*. Bottom: A freshwater prawn. [Courtesy Hervé Chaumeton (top); Hans Reinhard, Bruce Coleman Inc. (bottom).]

A

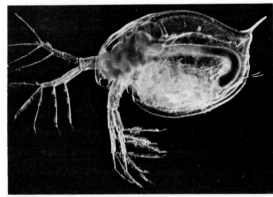

B

C

25.61   Some representatives of the smaller crustaceans
(A) a marine amphipod, *Gammarus.*   (B) The water flea, *Daphnia.*   (C) A marine isopod, *Idotea.*   (D) A group of sow bugs (also called wood lice or pill bugs), one of the very few terrestrial crustaceans.   [Courtesy Hervé Chaumeton (A, C); Oxford Scientific Films (B); G. R. Roberts (D).]

D

Subphylum Mandibulata   The members of this subphylum differ from chelicerates in having antennae and in having *mandibles* instead of chelicerae as their first pair of mouthparts. Mandibles are modified from the basal segment (coxa) of the ancestral legs, and function in biting and chewing (though in some species they are secondarily modified for piercing and sucking). They are never clawlike or pincerlike, as chelicerae often are. In most mandibulates there are two additional pairs of mouthparts called *maxillae.*

The subphylum comprises six classes. We shall briefly describe four of them here: Crustacea, Chilopoda, Diplopoda, and Insecta.

Class Crustacea   Some representatives of this class, such as crayfish, lobsters, shrimps, and crabs (Fig. 25.60), are well known to most people. But there are many other species of Crustacea, which bear little superficial resemblance to these familiar animals; among them are fairy shrimps, water fleas, brine shrimps, sand hoppers, barnacles, and sow bugs; many of these are very small odd-looking creatures (Fig. 25.61).

Crustacea characteristically have two pairs of antennae, a pair of mandibles, and two pairs of maxillae. But the rest of the appendages vary greatly from group to group, and whatever could be said about those of one group, such as crayfish and lobsters, would have little relevance to those of other groups. In fact, the Crustacea are an enormously diverse assemblage of animals that can hardly be characterized in any simple way. Some have a cephalothorax and an abdomen; others have a head and a trunk, or a head,

25.62   Goose barnacles *(Lepas)*
These animals are sedentary and secrete a protective shell. Most species of barnacles lack the long stalk so prominent in the goose barnacles.   [Oxford Scientific Films.]

25.63   A milliped
Each segment (except a few at the front and rear) bears two pairs of legs. [A. Cosmos Blank, National Audubon Society, Photo Researchers, Inc.]

thorax, and abdomen, or even a unified body. Most are free-living, but some are parasitic. Most are active swimmers, but some, like barnacles, secrete a shell and are sessile (Fig. 25.62). The majority are marine, but there are many freshwater species, and a few, such as sow bugs, are terrestrial and have a simple tracheal system. We could go on listing divergences, but the point of the amazing diversity of this group has been made. This is a class in which the basic arrangement of a segmented body with numerous jointed appendages has been modified and exploited in countless ways as the members of the class have diverged into different habitats and adopted different modes of life.

Class Chilopoda   Members of this class are called centipeds (or hundred-legged worms). Their body is divided into two regions, a head and a trunk (see Fig. 1.20, p. 23). The trunk is elongate and often somewhat flattened. The head bears a single pair of antennae and three pairs of mouthparts (mandibles and two pairs of maxillae). The animals are carnivorous, and the legs of the first trunk segment are modified as large poison claws. Each of the other trunk segments bears a single pair of walking legs. All centipeds are terrestrial and respire by means of tracheae. Excretion is by Malpighian tubules. The genital ducts open at the rear of the body.

Class Diplopoda   These animals are called millipeds (or thousand-legged worms). They superficially resemble centipeds and in fact were once placed with them (and with two other smaller groups) in a class called Myriapoda. However, it is now known that centipeds and millipeds are not closely related, and each is placed in a separate class.

The milliped body is divided into a head and a trunk (Fig. 25.63). The head bears a pair of antennae but only two pairs of mouthparts (mandibles, and a pair of maxillae fused to form a platelike underlip). The animals have no poison claws, and are not carnivorous, feeding largely on decaying organic matter of various types. Each of the first four or five trunk segments (depending on the species) bears a single pair of legs, but each of the other

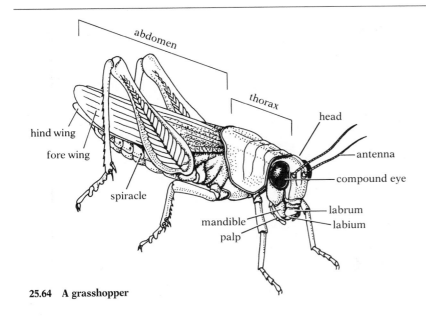

25.64 A grasshopper

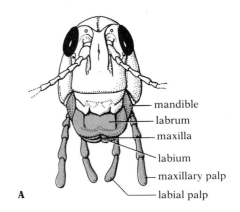

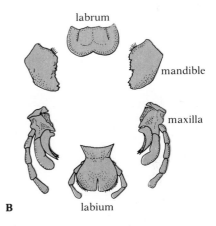

25.65 The mouthparts of a grasshopper
(A) Front view of head, with mouthparts in place.
(B) The mouthparts removed from the head, but kept
in their proper relative positions. The mandibles and
probably the labrum (upper lip) are derived from the
basal segments of ancestral legs; all the other segments
of those legs have been lost. The maxillae and labium
(lower lip) retain more of the segments of the legs from
which they are derived; the basal segments are
enlarged, but the distal segments form slender leglike
structures called palps, which bear many sensory
receptors.   [Modified from T. I. Storer and R. L.
Usinger, *General zoology*, McGraw-Hill Book Co.,
copyright © 1957. Used by permission.]

segments bears two pairs of legs (and also two pairs of spiracles); it is clear
that each of the double-legged segments is formed by the fusion of two
segments. Respiration is by tracheae, and excretion is by Malpighian tubules. The genital ducts open anteriorly, on the second segment. In most
milliped orders the legs (one or both pairs) of the seventh segment in the
males are highly modified and function as organs for inserting sperm into
the female reproductive tract (see Fig. 18.39, p. 799).

Class Insecta   This is an enormous group of diverse animals that occupy
almost every conceivable habitat on land and in freshwater. If numbers are
the criterion by which to judge biological success, then the insects are the
most successful group of animals that has ever lived; there are more species
of insects than of all other animal groups combined. But there is one restriction on their dominant role: They do not occur in the sea (although a
few species walk on the ocean surface or live in brackish water); the role
played by insects on land is played in the sea by Crustacea.

There are a few insect fossils from the Devonian, but it was in the Carboniferous and Permian periods that insects took their place as one of the
dominant groups of animals (see Table 20.1, p. 906). By the end of the
Paleozoic era, many of the modern orders had appeared, and the number of
species was enormous. A second great period of evolutionary radiation
began in the Cretaceous and continues to the present time; this second
radiation is correlated with the rise of flowering plants.

The insect body is divided into three regions: a head, a thorax, and an
abdomen (Fig. 25.64). The head segments are completely fused, and in adults
their boundaries cannot be distinguished. The head bears numerous sensory
receptors, usually including compound eyes; one pair of antennae; and
three pairs of mouthparts derived from ancestral legs. The mouthparts include a pair of mandibles, a pair of maxillae, and a lower lip, or *labium,*
formed by fusion of the two second maxillae (Fig. 25.65). The upper lip, or

25.66 A desert locust in flight
Like most flying insects except the flies, the locust—also called grasshopper—has two pairs of wings. The front wings are leathery and, when the animal is at rest, provide a protective covering for the fragile pleated rear wings, which are the ones important for flight. [Courtesy Stephen Dalton.]

labrum, which has not traditionally been classified as a mouthpart, may also be derived from ancestral legs.

The thorax is composed of three segments, each of which bears a pair of walking legs. In many insects (but not all), the second and third thoracic segments each bear a pair of wings, and the animals are vigorous fliers (Fig. 25.66).

The abdomen is composed of a variable number of segments (12 or fewer). Abdominal segments are devoid of legs, but highly modified remnants of the ancestral appendages may be present at the posterior end, where they function in mating and egg laying.[17]

We have already discussed the insects at considerable length in other chapters and need not repeat ourselves here. See the discussion of the insect tracheal system (p. 255, Figs. 6.23–6.25), the circulatory system (p. 278, Fig. 7.20), the Malpighian excretory organs (p. 329, Fig. 8.18), hormonal control (p. 355, Fig. 9.21), nervous control and sensory perception (p. 404, Fig. 10.4; p. 438, Figs. 10.34, 10.35; p. 445, Fig. 10.41), the exoskeleton and muscles (p. 468, Fig. 11.4B, C), behavior (Chapter 12), and development (p. 711, Figs. 17.16, 17.20, 17.21).

---

[17] A few primitive insects retain vestiges of appendages on many abdominal segments. These may have a sensory function.

The insects are classified in approximately 26 orders (the exact number depends on the authority cited). The following are among the more familiar:

THYSANURA   Bristletails and silverfish. Small, primitive, wingless; chewing mouthparts; long tail-like appendages on rear of abdomen. Incomplete metamorphosis. Common in houses, particularly in kitchens and bathrooms; sometimes damage books in libraries.

ODONATA   Dragonflies and damselflies (Fig. 25.67A). Medium to large, rapid-flying, predaceous on other insects. Two pairs of long membranous wings; chewing mouthparts; very large compound eyes. Immature stages (nymphs) in freshwater; incomplete metamorphosis.

ORTHOPTERA   Grasshoppers, crickets, walking sticks, mantids, cockroaches (Fig. 25.67B). Usually two pairs of wings—the coarse-textured fore wings, which are narrower than the hind wings, are not used in flight, but function (when animal is at rest) as covers for the folded fanlike hind wings; chewing mouthparts. Gradual metamorphosis.

ISOPTERA   Termites. Highly social insects. Wings briefly present only on members of the reproductive caste; usually chewing mouthparts. Gradual metamorphosis.

HEMIPTERA   True bugs (Fig. 25.67C). Usually two pairs of wings—basal half of fore wings thick and leathery, distal half membranous, hind wings membranous; piercing-sucking mouthparts. Gradual metamorphosis.

ANOPLURA   Sucking lice (Fig. 25.67D). External parasites. Wingless; piercing-sucking mouthparts; legs and claws adapted for clinging to host. Gradual metamorphosis.

COLEOPTERA   Beetles (Fig. 25.67E). Two pairs of wings—fore wings hard, meeting along middorsal line, forming a protective case for the folded membranous hind wings when animal is at rest; chewing mouthparts. Complete metamorphosis.

LEPIDOPTERA   Moths and butterflies. Two pairs of large scale-covered wings; chewing mouthparts in larvae, sucking (but not piercing) mouthparts in adults. Complete metamorphosis.

DIPTERA   True flies: mosquitoes, gnats, midges, houseflies, horseflies, etc. (Fig. 25.67F). One pair of membranous wings; highly modified hind wings acting as tiny balancing organs; piercing-sucking or sponging mouthparts. Complete metamorphosis.

SIPHONAPTERA   Fleas (Fig. 25.67G). Intermittent ectoparasites. Small, body laterally compressed; no wings; piercing-sucking mouthparts; long legs, adapted for jumping. Complete metamorphosis.

HYMENOPTERA   Sawflies, ants, bees, wasps (Fig. 25.67H). Usually two pairs of membranous wings, interlocked in flight; chewing or chewing-lapping mouthparts; thorax and abdomen connected by a very narrow waist. Complete metamorphosis.

A more complete listing of the insect orders is given in the Appendix.

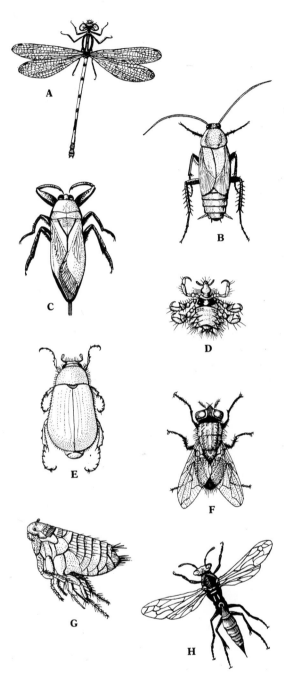

25.67 Some representatives of the major insect orders
(A) Damselfly (order Odonata). (B) Cockroach (Orthoptera). (C) Bug (Hemiptera). (D) Louse (Anoplura). (E) Beetle (Coleoptera). (F) Fly (Diptera). (G) Flea (Siphonaptera). (H) Wasp (Hymenoptera).

THE DEUTEROSTOMIA

There are only four phyla in the Deuterostomia, and only two of these—Echinodermata and Chordata—are major groups. We shall discuss these and a small phylum—Hemichordata—that is important from an evolutionary standpoint. The other phylum—Chaetognatha (arrow worms)—contains only a few, exclusively marine species and is not important for our purposes here.[18]

ECHINODERMATA

The echinoderms are exclusively marine, mostly bottom-dwelling animals. They are common in all seas and at all depths from the intertidal zone to the ocean deeps. Included in this distinctive phylum are the sea stars, brittle stars, sea urchins, sand dollars, sea cucumbers, and sea lilies.

The adults are radially symmetrical, but the larvae are bilateral, and it is generally held that echinoderms evolved from bilateral ancestors. The radial symmetry probably arose as an adaptation to a sessile way of life. Most of the modern echinoderms (with the exception of sea lilies) move about slowly and are thus not completely sessile, but the ancient echinoderms apparently were.

Almost all members of this phylum possess an internal skeleton composed of numerous calcareous plates embedded in the body wall. These plates may be separate, or they may be fused to form a rigid boxlike structure. The skeleton often bears many bumps or spines—they are particularly noticeable in sea urchins—that project from the surface of the animal (see Fig. 25.71). It is this characteristic that gives the animals the name "Echinodermata" (from the Greek *echino*, spiny, and *derma*, skin).

Echinoderms have a well-developed coelom in which the various internal organs are suspended. The complete digestive system is the most prominent of the organ systems. There is no special excretory system, and the blood circulatory system, though present, is poorly developed. The nervous system is radially organized, consisting of nerve networks that connect to ringlike ganglionated nerve cords running around the body of the animal (there are often three of these cords); there is no brain.

A characteristic unique in echinoderms is their ***water-vascular system.*** This is a system of tubes (usually called canals) filled with watery fluid. Water can enter the system through a sievelike plate, called a madreporite, on the surface of the animal. A tube from this plate leads to a ring canal that encircles the esophagus (Fig. 25.68). Five radial canals branch off the ring canal and run along symmetrically spaced grooves or bands on the surface of the animal. Many short side branches from the radial canals lead to hollow ***tube feet*** that project to the exterior. Each tube foot is a thin-walled hollow cylinder, with a sucker on its end. At the base of each tube foot is a muscular ampulla containing fluid. When the ampulla contracts, the fluid,

[18] Until recently, another small phylum of marine worms—Pogonophora (beard worms)—was placed in the Deuterostomia, but the discovery that part of their body (broken off all older specimens) is segmented and bears numerous setae has prompted the transfer of this phylum to the Protostomia near the Annelida.

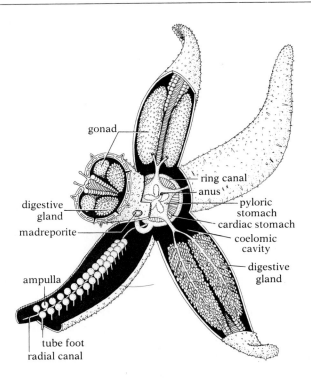

gonad

ring canal
anus
pyloric
stomach
cardiac stomach
coelomic
cavity
digestive
gland

digestive
gland

madreporite

ampulla

tube foot
radial canal

25.68   Dissection of a sea star (dorsal view)

prevented by a valve from flowing into the radial canal, is forced into the tube foot, which is thereby extended. The foot attaches to the substratum by its sucker, and then longitudinal muscles in its wall contract, shortening it and pulling the animal forward (while forcing the water back into the ampulla). This cycle of events, repeated rapidly by the many tube feet of an animal like a sea star, enables it to move slowly. The tube feet may also enable it to hold tightly to a rock or other object by applying suction, or to pull open the valves of the shell of a clam or oyster, on which the sea star will feed.

The sexes are usually separate. Eggs and sperm are shed into the surrounding water, where fertilization occurs. Cleavage is radial and indeterminate. The larva, which is ciliated and free-swimming, has a complete digestive tract (see Fig. 17.19, p. 711); the anus is derived from the embryonic blastopore, the mouth being a new opening.

Class Asteroidea (The sea stars)   The body of a sea star (starfish) consists of a central *disc* and usually of five rays, or *arms,*[19] each with a groove bearing rows of tube feet running along the middle of its lower surface. The outer surface of the animal is studded with many short spines and numerous tiny skin gills, which are thin fingerlike evaginations of the body wall that protrude to the outside between the plates of the endoskeleton (see Fig. 6.9, p. 245). The cavity of each skin gill is continuous with the general coelom. Scattered between the spines and skin gills are often numerous small jaw-like structures called pedicellariae, which are used for protection and for

[19] Though most species of sea stars have five arms, some have more.

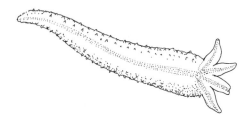

25.69   A detached arm of a sea star regenerating a new body
[Modified from L. H. Hyman, *The invertebrates,* McGraw-Hill Book Co., copyright © 1955. Used by permission.]

25.70   A brittle star (Ophiopholis aculeata) and a basket star (Gorgonocephalus eucnemis)
Left: The body disc of the brittle star is relatively small, and the arms are long and slender.   Right: The arms of the basket star branch repeatedly to produce a mass of coils resembling tentacles.   [Robert Dunne (left) and Charlie Ott (right), Photo Researchers, Inc.]

capturing very small animals. The madreporite is on the upper surface (but not in the center; in this respect the radial symmetry of the animal is not perfect).

The mouth is located in the center of the lower surface of the disc and the anus in the center of the upper surface (thus the lower surface is the morphological anterior end of the animal, and the upper surface is the morphological posterior end). The digestive tract of a sea star is straight and very short, consisting of a short esophagus, a broad stomach that fills most of the interior of the disc, and a very short intestine. The stomach is divided by a constriction into two parts: a large eversible part (cardiac stomach) at the esophageal (lower) end and a smaller noneversible part (pyloric stomach) at the intestinal (upper) end. Attached to the noneversible part are five pairs of large digestive glands; each pair of glands lies in the coelomic cavity of one of the arms (Fig. 25.68).

When the sea star feeds, it pushes the lower part of the stomach out through the mouth, turning it inside out and placing it over food material such as the soft body of a clam or oyster. The stomach secretes digestive enzymes onto the food, and digestion begins. The partly digested food is then taken into the upper part of the stomach and into the digestive glands, where digestion is completed and the products are absorbed.

Sea stars have amazing regenerative abilities. Even a single detached arm can regenerate an entire new animal (Fig. 25.69).

Other echinoderm classes   Although members of the four other echinoderm classes often show little superficial resemblance to sea stars, their structure is fundamentally similar. For example, sea urchins and sea cucumbers lack the five arms of sea stars, but they do have five bands of tube feet and thus show the same basic pentaradiate symmetry.

Class Ophiuroidea   These are the brittle stars, serpent stars, and basket stars (Fig. 25.70). They superficially resemble true sea stars (Asteroidea) in

25.71 A sea urchin
Sea urchins and their relatives have numerous large spines on their hard boxlike skeletons. [Courtesy Roman Vishniac.]

25.72 Sea cucumbers *(Cucumaria miniata)*
Notice the rows of tube feet. There are five of these rows, which correspond to the five arms of a sea star. [R. N. Mariscal, Bruce Coleman Inc.]

having five arms, but the arms are longer, much slenderer, more flexible, often branched, and grooveless. The body disc is relatively small. The tube feet have no ampullae and are not used in locomotion, which is by rapid lashing of the arms. There is a large stomach, but no intestine and no anus. Gas exchange is by invaginated pouches in the periphery of the disc.

Class Echinoidea These are the sea urchins, sand dollars, and heart urchins. They have no arms, but do have five bands of tube feet. The body is spherical, or flattened and oval, and is covered with long spines (Fig. 25.71). The plates of the endoskeleton are fused to form a rigid box or case. There is a complex chewing apparatus just inside the mouth, and the intestine is long and coiled. Gas exchange is by small but highly branched gills or by modified tube feet.

Class Holothuroidea The sea cucumbers (Fig. 25.72) differ from the other echinoderms in having a much reduced endoskeleton and a leathery body. Also unlike the members of the classes discussed above, they lie on their sides rather than on the oral surface. The mouth is surrounded by tentacles attached to the water-vascular system. There is a very long coiled intestine, and gas exchange is usually by complexly branched respiratory trees attached to the cloaca (see p. 248, Fig. 6.14).

25.73   A fossil sea lily
[Courtesy American Museum of Natural History.]

Class Crinoidea   This is the oldest and most primitive of the living classes of echinoderms. The sea lilies, as most Crinoidea are commonly called, are attached to the substratum by a long stalk and are thus sessile (Fig. 25.73). They have long feathery arms (often branched) around the mouth, which is on the upper side; the sea lilies thus differ from the Asteroidea, Ophiuroidea, and Echinoidea in that their morphological anterior end is directed upward and their morphological posterior end downward. Some modern crinoids—the feather stars—lack a stalk and are not sessile.

HEMICHORDATA

The hemichordates, many of which belong to a class called acorn worms (Fig. 25.74), are entirely marine. Often found living in U-shaped burrows in sand or mud along the coast,[20] acorn worms are fairly large, ranging from $6\frac{1}{2}$ to 43 cm in length. Their bodies consist of an anterior conical proboscis (thought by some to resemble an acorn—hence their name), a collar, and a long trunk (Fig. 25.75). The mouth is situated ventrally, at the junction between the proboscis and the collar.

A particularly important feature in the hemichordates is a series of *gill slits* in the wall of the pharynx. Water drawn into the mouth is forced back into the pharynx and out through these slits into branchial sacs, which open to the exterior via atrial pores. Oxygen is removed from the indrawn water,

---

[20] Some acorn worms live under rocks or shells instead of burrowing; and not all of the burrowers make U-shaped tubes. There is one class of hemichordates that are not acorn worms.

25.74   An acorn worm *(Balanoglossus)*
The short rounded proboscis is partially enveloped by the collar behind it. Only a part of the long trunk is shown.   [Courtesy Hervé Chaumeton.]

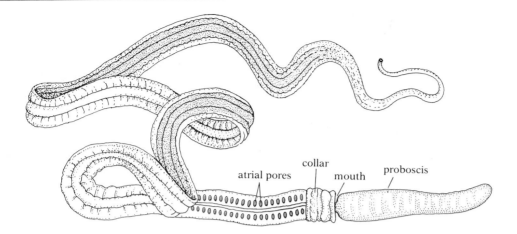

and carbon dioxide released into it, by blood in beds of capillaries in the septa between the slits. Food particles carried by the water into the pharynx do not pass through the gill slits, but instead move posteriorly into the esophagus. The pharynx, then, acts as a strainer or filter, separating food particles from the water.

Another important characteristic of hemichordates is the occurrence during development of a ciliated larval stage that strikingly resembles the larvae of some echinoderms.

THE RELATIONSHIPS BETWEEN ECHINODERMS, HEMICHORDATES, AND CHORDATES

It may seem strange that the Echinodermata form the major phylum generally considered most closely related to our own phylum, the Chordata. After all, sea stars, sea urchins, and sea cucumbers don't look like animals with which one would expect to claim kinship. But as we saw earlier when we discussed the differences between the Protostomia and Deuterostomia, certain characteristics seem to link echinoderms, hemichordates, and chordates and set them apart from all the protostome phyla. These characteristics include formation of the anus from the embryonic blastopore, radial and indeterminate cleavage, origin of the mesoderm as pouches, and formation of the coelom as the cavities in the mesodermal pouches.

The Hemichordata have long held special interest for zoologists because their apparent affinities to both the Echinodermata and the Chordata seem to provide additional evidence of a relationship between those two large and important phyla. The ciliated larvae of hemichordates are so much like those of some echinoderms that they were mistaken for echinoderms when first discovered. This larval type, sometimes called a *dipleurula* (Fig. 25.76), is found only in the echinoderms and hemichordates.[21] It has a band of cilia that forms a ring encircling the mouth. It thus differs from the trochophore larva found in many protostomes (including some turbellarian Platyhelminthes, the lophophorate phyla, Mollusca, and Annelida), which has a

25.75 An adult acorn worm
This particular genus *(Saccoglossus)* has a more elongate proboscis than *Balanoglossus*, shown in Fig. 25.74.

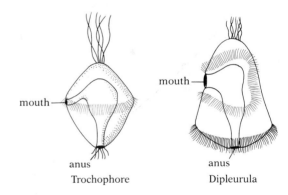

25.76 Trochophore and dipleurula larval types
The band of cilia of the trochophore is located anterior to the mouth, whereas the corresponding band of the dipleurula encircles the mouth.

---

[21] The larva of a hemichordate is called a tornaria, and the echinoderm larva it resembles is called a bipinnaria (Fig. 17.19, p. 711). Both are of the dipleurula type.

TABLE 25.1 A comparison of some of the major animal phyla

Phylum	Symmetry	Cleavage	Body cavity	Digestive tract	Circulatory system	Ciliated larva	Segmentation
Coelenterata	Radial	Determinate	None	Gastrovascular cavity	Absent	Planula	Absent
Platyhelminthes	Bilateral	Determinate	None	Gastrovascular cavity	Absent	Trochophorelike in some	Absent or correlated with reproduction
Aschelminthes	Bilateral	Determinate	Pseudocoelom	Complete, with mouth from blastopore	Absent	None or a unique type	Absent
Mollusca	Bilateral	Determinate	Coelom much reduced	Complete, with mouth from blastopore	Open	Trochophore	Absent (except in Neopilina)
Annelida	Bilateral	Determinate	Coelom	Complete, with mouth from blastopore	Closed	Trochophore	Present
Arthropoda	Bilateral	Determinate	Hemocoel (coelom, degenerate)	Complete, with mouth from blastopore	Open	None	Present
Echinodermata	Secondarily radial	Indeterminate	Coelom	Complete, with anus from blastopore	A special type; often poorly developed	Dipleurula	Absent
Hemichordata	Bilateral	Indeterminate	Coelom	Complete, with anus from blastopore	Open	Dipleurula	Absent
Chordata	Bilateral	Indeterminate	Coelom	Complete, with anus from blastopore	Closed (except in tunicates)	None	Present

band of cilia encircling the body anterior to the mouth. The similar larvae of hemichordates and echinoderms, as well as the similarities in early embryology mentioned above, indicate that these two groups must stem from a common ancestor. In view of the complicated metamorphosis that in echinoderms produces a radial adult from a bilateral larva, it seems likely that echinoderms have deviated greatly from the ancestral type and that hemichordates are probably nearer that ancestral type.

The most obvious resemblance of hemichordates to chordates is their possession of pharyngeal gill slits, which are found in all chordates but nowhere else in the animal kingdom. The hemichordates also have a dorsal nerve cord that is sometimes hollow and resembles the dorsal hollow nerve cord characteristic of chordates. Because of these resemblances, the hemichordates were regarded for many years as primitive members of the phylum Chordata. Though they are now generally regarded as a separate phylum, which may actually be closer to the echinoderms than to the chordates, recognition of their ties with both Chordata and Echinodermata has helped clarify the phylogenetic relationship between these two major groups. Note that there is no suggestion here that chordates evolved from echinoderms, but simply that the two groups diverged from a common ancestor at some remote time.

Some of the important characteristics of the major animal phyla are compared in Table 25.1.

INVERTEBRATE CHORDATA

Throughout this book we have used the terms "vertebrate" and "invertebrate," and have assigned all the animals discussed to one or the other of the categories they designate. But this division of the animal kingdom is in many respects an odd one, because neither category coincides with any phylum or group of phyla. Indeed, the term "vertebrate" designates only a part of one phylum; the rest of that phylum and all the other phyla then fall under the heading "invertebrate." The phylum that contains both invertebrate and vertebrate members is Chordata.

The phylum Chordata is customarily divided into three subphyla: Urochordata, Cephalochordata, and Vertebrata. These share three important characteristics: (1) All have, at least during embryonic development, a structure called a *notochord* (whence the name "Chordata"). This is a flexible supportive rod running longitudinally through the dorsum of the animal just ventral to the nerve cord. (2) All have pharyngeal gill slits (or pouches) at some stage in their development. (3) All have a dorsal hollow nerve cord.

The Urochordata and the Cephalochordata are both invertebrate; i.e. they have no backbone.

Subphylum Urochordata (The tunicates)   In the best-known class of tunicates (sometimes called sea squirts), the adults are sessile marine animals that little resemble other chordates except in having pharyngeal gill slits.[22] Water taken in through the mouth (also called the incurrent siphon) goes into a large pharynx, and then filters through the gill slits into a chamber

---

[22] Members of two smaller classes of tunicates are free-swimming planktonic organisms.

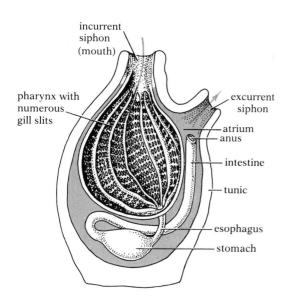

25.77   Cutaway diagram of an adult tunicate
The color arrows show the path of respiratory water, which is drawn into the pharynx through the incurrent siphon, passes through the gill slits into the atrium, and then exits through the excurrent siphon. Oxygen is absorbed from the water across the walls of the gill slits. Food particles drawn into the pharynx with the water do not pass through the gill slits, but instead move through the pharynx into the esophagus.

called the atrium, from which it passes to the exterior through the excurrent siphon (Fig. 25.77). The gill slits function in both gas exchange and feeding, acting as a strainer for removing small food particles from the water flowing through them. The food particles become caught in a layer of mucus in a ciliated groove of the pharynx, called the endostyle, and are carried by the mucus into the esophagus, which leads to the stomach. According to one hypothesis, the pharyngeal gill slits, which are so distinctive a trait of chordates, first evolved as an adaptation for this sort of filter feeding and only later came to function in gas exchange as well.

Larval tunicates, which are motile, show much more resemblance to the other chordates. With their elongate bilaterally symmetrical bodies and long tails, they look rather like tadpoles. They possess a well-developed dorsal hollow nerve cord and a notochord beneath it in the tail region (Fig. 25.78). When the larvae settle down and undergo metamorphosis to the adult form, the notochord and most of the nerve cord are lost.

Some biologists hold that the tunicates and vertebrates descended from a common ancestor that was free-swimming and resembled a modern tunicate larva. If this is so, then the sessile structure of modern adult tunicates is a later specialization. An alternative hypothesis is that the common ancestor was sessile, more like adult tunicates, and that vertebrates evolved from its motile larva; in other words, in the line leading to the vertebrates, the larval stage increased in importance and duration, until finally it could reproduce without undergoing metamorphosis, and the ancestral sessile stage dropped out of the life cycle entirely.

Subphylum Cephalochordata (The lancelets)   There are about 30 species of these small marine animals. Though capable of swimming, they spend most of their time buried tail down in sand in shallow water, with only their anterior end exposed. They are filter feeders, taking in water through the mouth and straining it in the pharynx. The water passes through pharyngeal gill slits into a large chamber, the atrium, and thence to the exterior through an atrial pore. Oxygen is removed from the water as it passes through the gill slits. Food particles do not pass through the gill slits, but move posteriorly into the digestive tract.

25.78   A larval tunicate
The arrows indicate the path of inflowing and outflowing water.

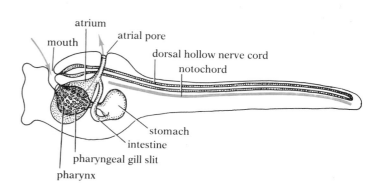

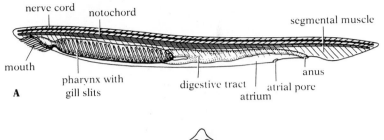

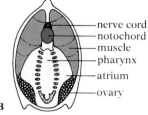

25.79 Diagrammatic drawing of an adult lancelet (amphioxus)
(A) Longitudinal view. (B) Cross section, showing relationship between the pharynx, the pharyngeal gill slits, and the atrium.

The genus of lancelets most commonly studied is *Branchiostoma,* more usually called amphioxus. The body of a typical specimen is about 5 cm long, translucent, and shaped rather like a fish (Fig. 25.79). Both the dorsal hollow nerve cord and the notochord are well developed and are retained through life. A feature not seen in tunicates but characteristic of both cephalochordates and vertebrates is segmentation. In lancelets this segmentation is most noticeable in the muscles, which are in V-shaped, segmentally arranged bundles.

VERTEBRATE CHORDATA

As the name "Vertebrata" implies, the animals in this group are characterized by an endoskeleton that includes a backbone composed of a series of vertebrae. The vertebrae develop around the notochord, which in most vertebrates is present in the embryo only. The serial arrangement of the vertebrae and the organization of the muscles are the principal tokens of segmentation.

We discussed the anatomy, physiology, behavior, and development of vertebrates at length in other parts of this book. Here we shall be concerned primarily with the evolutionary history of the group.

Class Agnatha The vertebrates are one of the few major animal groups not represented among the Cambrian fossils. The oldest vertebrate fossils are from the Ordovician period, which began some 500 million years ago (see Table 20.1, p. 906). Those first vertebrate fossils are of bizarre fishlike animals covered by thick plates of bony material. Though they had a skeleton, they lacked an important character found in all later vertebrates—jaws. Furthermore, most of them had no paired fins. These ancient fishes constitute the class Agnatha (which means jawless). Most were probably filter feeders, straining food material from mud and water flowing through their gill systems.

25.80   Evolution of the vertebrate classes

25.81   The head and pharyngeal region of a lamprey
The animal has a large oral sucker instead of jaws, and seven prominent external gill openings. [Oxford Scientific Films, Bruce Coleman Inc.]

The Agnatha continued as an important group through the Silurian period, sharing the seas with the already abundant sponges, coelenterates, brachiopods (which were far more numerous then than now), molluscs (particularly gastropods and cephalopods), trilobites and eurypterids, and echinoderms. But by the end of the Silurian the Agnatha had begun to decline, and they disappear from the fossil record by the end of the Devonian (Fig. 25.80).

A few peculiar species living today, the lampreys (Fig. 25.81) and the hagfishes, are generally classified as Agnatha, although they are quite unlike

25.82 Reconstruction of an extinct placoderm
Notice the hinged jaws. [Courtesy Field Museum of
Natural History, Chicago.]

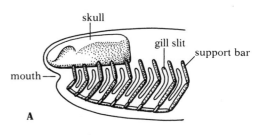

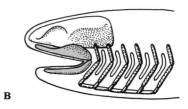

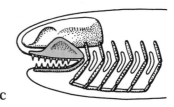

25.83 Evolution of the hinged jaws of vertebrates
(A) The earliest vertebrates had no jaws. The structures
(color) that in their descendants would become jaws
were gill support bars. (B) A pair of gill support bars
has been modified into weak jaws (the two most
anterior support bars, shown in A, were lost). (C) The
jaws have become larger and stronger. [Modified from
A. S. Romer, *The vertebrate body*, Saunders, 1949.]

the Paleozoic armored species.[23] They have a soft body without either armor
or scales; they have a cartilaginous skeleton, having lost all trace of bone;
and their jawless mouth is modified as a round sucker that is lined with
many horny teeth and accommodates a rasping tongue. They feed by attaching themselves by their sucker to other fishes, rasping a hole in the skin
of the prey and sucking blood and other body fluids. The lampreys have a
larval filter-feeding stage that strikingly resembles amphioxus.

Class Placodermi The decline of the ancient Agnatha coincided with the
rise of a second class of armored fishes—the Placodermi—which first appeared in the Silurian, having probably arisen from primitive agnaths. The
Placodermi were an important group during the Devonian, but most became
extinct by the end of that period; a few survived until the Permian, when
they too disappeared.

The Placodermi mark a notable advance in vertebrate evolution in their
possession of hinged jaws (Fig. 25.82). The acquisition of hinged jaws was
one of the most important events in the history of vertebrates, because it
made possible a revolution in the method of feeding and hence in the entire
mode of life of early fishes. They became more active and wide-ranging
animals, usually with paired fins. Many became ferocious predators. Even
those that remained mud feeders were evidently adaptively superior to the
ecologically similar agnaths, which they gradually replaced.

Anatomical and embryological studies have convinced biologists that the
hinged jaws of the placoderms developed from a set of gill support bars (Fig.
25.83). Notice that hinged jaws arose independently in two important animal
groups, the arthropods and the vertebrates, but that, although they are
functionally analogous structures, they arose in entirely different ways—in
the one case from ancestral legs and in the other from skeletal elements in
the wall of the pharyngeal region.

[23] Lampreys and hagfishes are so different from the Paleozoic Agnatha, called ostracoderms,
that some biologists erect a separate class for them.

25.84   A nurse shark (left) and two rays (right)
The rays have flattened bodies adapted for life on the
bottom. They swim by "flying" through the water,
using the thin lateral parts of their bodies as "wings."
[Oxford Scientific Films (left) and Hans Reinhard
(right), Bruce Coleman Inc.]

Though the Placodermi themselves have been extinct at least 230 million
years, two other classes that arose from them in the Devonian[24] are still
important elements of our fauna. These are the Chondrichthyes (cartilaginous fishes) and the Osteichthyes (bony fishes).

Class Chondrichthyes   The modern Chondrichthyes (sharks, skates, rays,
and their relatives) (Fig. 25.84) are distinguished by their cartilaginous skeletons; bone is unknown in the group. Though a cartilaginous skeleton might
at first be taken as a primitive trait, it is not thought to be one in Chondrichthyes. Their ancestors among the Placodermi probably had bony skeletons, and loss of the bone must be regarded as an evolutionary specialization.

Chondrichthyes have neither swim bladders nor lungs. Osmoregulation in
the subclass Elasmobranchii is unusual, involving retention of high concentrations of urea in the body fluids (see p. 317). Fertilization is internal, and
the eggs have tough leathery shells. Most species are predaceous, but a few
are plankton feeders.

Class Osteichthyes   The other class that arose from the Placodermi—the
Osteichthyes—includes most of the fishes familiar to you. This is a large
class, whose members are the dominant vertebrates in both freshwater and
the oceans, as they have been since the Devonian—the so-called Age of
Fishes. More than 17,000 species are known, and many remain to be discovered, particularly in the deeper parts of the oceans. According to some
biologists, the total number of living species may be as high as 40,000. A
tremendously varied lot, they range from organisms a centimeter long when
mature to giants more than 6 m long. They assume a host of different shapes,
many of them bizarre and grotesque to our eyes. Some are sluggish and
sedentary, while others can swim at speeds as great as 80 km an hour.
Almost every type of food is used by some species of fish.

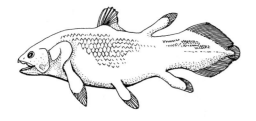

25.85   *Latimeria*, a modern lobe-finned fish

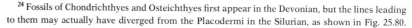

[24] Fossils of Chondrichthyes and Osteichthyes first appear in the Devonian, but the lines leading
to them may actually have diverged from the Placodermi in the Silurian, as shown in Fig. 25.80.

25.86 The movement of vertebrates onto land
Top: A Devonian lobe-finned fish *(Eusthenopteron),* which probably pulled itself out of the water onto mud flats and sandbars. Bottom: An early amphibian *(Diplovertebron).* Its legs were better suited for locomotion on land than the lobe fins of *Eusthenopteron,* but it too probably spent most of its time in the water. [Courtesy American Museum of Natural History.]

The earliest members of this class probably lived in freshwater. In addition to gills, they had lungs, which they probably used as supplementary gas-exchange devices when the water was stagnant and deficient in oxygen. As we saw in Chapter 6, the ventral lungs have been modified into a dorsal swim bladder in most modern bony fishes (see Fig. 6.22, p. 253), which rely for gas exchange almost exclusively on their gills (see Fig. 6.12, p. 247). But there are still a few living relict species with lungs.

Soon after the Osteichthyes arose from the Placodermi, the class split into two divergent groups. One underwent great evolutionary radiation, giving rise to nearly all the bony fishes alive today. The other radiated considerably in the late Paleozoic, but today is represented by only six relict species—five species of lungfishes (one in Australia, three in Africa, and one in South America) and one species of "lobe-fin" known only from deep waters off the southeast coast of Africa. Despite its rarity in the present fauna, this second group of Osteichthyes is of special evolutionary interest, because it is thought to have been ancestral to the land vertebrates.

Let us look more closely at the ancient lobe-fin fishes.[25] This group has long been known from fossils, but until 1939 it was thought to have been entirely extinct for some 75 million years. In that year a specimen was caught off the east coast of South Africa; since then, additional specimens of this living fossil, called coelacanths (genus name *Latimeria*), have been caught and studied (Fig. 25.85). The coelacanths are not the particular lobe-fins thought to be the ancestors of land vertebrates, but they resemble those ancestral forms in many ways.

In addition to lungs, the lobe-fins had another important preadaptation for life on land—the large fleshy bases of their paired pectoral and pelvic fins. At times, especially during droughts, lobe-fins living in freshwater probably used these leglike fins to pull themselves onto sandbars and mud flats (Fig. 25.86). They may even have managed, albeit with great difficulty, to crawl to a new pond or stream when the one they were in dried up.

Now, by the Devonian period the land had already been colonized by plants (see Table 20.1, p. 906), but was still nearly devoid of animal life (there is a fossil of what may have been a land scorpion from the Silurian, and the first insects and millipeds appeared in the Devonian, but they did not become common until the Carboniferous). Hence any animal that could survive on land would have had a whole new range of habitats open to it without competition. Any lobe-fin fishes that had appendages slightly better

[25] Known technically as Crossopterygii.

25.87 Two modern amphibians
Left: A gold-lined frog *(Rana erythraea)* from Malaya.
Right: A banded salamander *(Salamandra salamandra).*
[S. C. Bisserot (left) and Hans Reinhard (right), Bruce
Coleman Inc.]

suited for land locomotion than those of their fellows would have been able
to exploit these habitats more fully; through selection pressure exerted over
millions of years, the fins of these first vertebrates to walk (or rather crawl)
on land would slowly have evolved into legs. Thus by the end of the Devonian, with a host of other adaptations for life on land evolving at the same
time, one group of ancient lobe-fin fishes must have given rise to the first
amphibians.

Classes Amphibia and Reptilia   Numerous fossils indicate that, as would
be expected, the first amphibians were still quite fishlike (Fig. 25.86, bottom).
In fact, they probably spent most of their time in the water. But as they
progressively exploited the ecological opportunities open to them on land,
they slowly became a large and diverse group. So numerous were they
during the Carboniferous that that period is often called the Age of Amphibians, just as the period before it, the Devonian, is called the Age of
Fishes. The amphibians were still abundant in the Permian, but during that
period they slowly declined as the members of a new class, the Reptilia,
replaced them.

The end of the Permian, which also marked the end of the Paleozoic era,
was a time of great change, both geological and biological. The ancestral
Appalachian Mountains were built up; the last trilobites and the last placoderms disappeared; the once common brachiopods declined; and older
types of corals, molluscs, echinoderms, crustaceans, and fishes were replaced by more modern representatives of those groups. This so-called
Permo-Triassic crisis also witnessed the extinction of most groups of amphibians. By the end of the Triassic, the only members of this class that
survived were the immediate ancestors of the few small groups of modern
Amphibia—the salamanders (order Urodela), the apodes (order Apoda), and
the frogs and toads (order Anura) (Fig. 25.87).

The first reptiles had evolved from primitive amphibians by the late Carboniferous. The class expanded during the Permian, replacing its amphibian

predecessors, and became a huge and dominant group during the Mesozoic era, which is often called the Age of Reptiles.

One might well wonder why the reptiles were so effectively able to displace the once dominant amphibians. There were doubtless many reasons, but surely one of the most compelling was that the reptiles, unlike the amphibians, were terrestrial in the fullest sense of the word. Amphibians continued to use external fertilization and to lay fishlike eggs—eggs that had no amnion or shell and hence had to be deposited either in water or in very moist places on land, lest they dry up (Fig. 25.88). Larval development remained aquatic. Amphibians were thus bound to the ancestral freshwater environment by the necessities of their mode of reproduction. Furthermore, even adult amphibians probably had thin moist skin and were in danger of desiccation if conditions became very dry.[26] Reptiles, on the other hand, used internal fertilization, laid amniotic shelled eggs (Fig. 25.89), had no larval stage, and had dry, scaly, relatively impermeable skin. Evolution of the amniotic egg—often called the "land egg"—which provides a fluid-filled chamber in which the embryo may develop even when the egg itself is in a dry place, was an advance as important in the conquest of land as the evolution of legs by the Amphibia.

The Reptilia had many other characteristics that made them better suited for terrestrial life than the Amphibia. The legs of the ancient amphibians were small, weak, attached far up on the sides of the body, and oriented laterally; hence they were unable to support much weight, and the belly of the animal often dragged on the ground; walking was doubtless slow and labored, as it is in salamanders today. The legs of reptiles were usually larger and stronger and could thus support more weight and effect more rapid locomotion; in many (though not all) species they were also attached lower on the sides and oriented more vertically, so that the animal's body cleared

---

[26] All modern amphibians have thin moist skin that functions as a respiratory organ (in addition to the gills and/or lungs), but this may not have been true of all ancient amphibians.

25.88   Eggs of a spotted salamander *(Ambystoma maculatum)* attached to underwater twigs
[D. Lyons, Bruce Coleman Inc.]

25.89   Baby lizards hatching from their eggs
The shells of reptilian eggs are usually leathery, not brittle like birds' eggs, as can be seen here from the way the shells have buckled and bent.   [Hans Pfletschinger, Peter Arnold, Inc.]

A

B

25.90 Representatives of the main groups of modern reptiles
(A) A tuatara *(Sphenodon).*   (B) A lizard *(Cordylas).*
(C) A snake *(Vipera).*   (D) A turtle *(Pseudemys).*   (E) A
crocodile *(Crocodylus).*   [J. Markham (A), Alan Blank
(B), and Hans Reinhard (C, D), Bruce Coleman Inc.
John Moss, Photo Researchers, Inc. (E).]

D

C

E

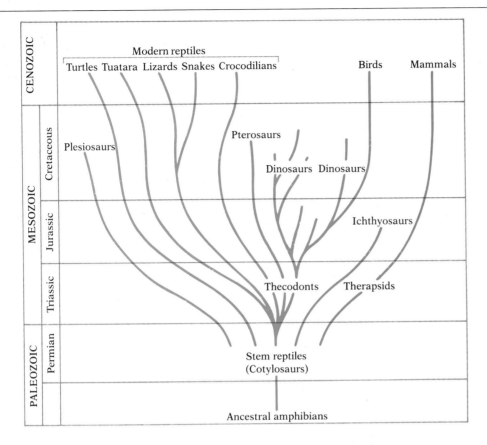

25.91   **Evolution of the reptilian groups**

the ground. Whereas the lungs of amphibians were poorly developed and inefficiently ventilated, those of reptiles were fairly well developed, and greater rib musculature made their ventilation more efficient. Whereas the amphibian heart was three-chambered (two atria and one ventricle), that of reptiles was four-chambered (though the partition between the ventricles was seldom complete); hence there was less chance of mixing oxygenated and deoxygenated blood.

The class Reptilia is represented in our modern fauna by members of four groups: turtles (order Chelonia), crocodiles and alligators (order Crocodylia), lizards and snakes (order Squamata), and the tuatara (order Rhynchocephalia) (Fig. 25.90). The tuatara *(Sphenodon punctatum)*, which is found only on a few islands off the coast of New Zealand, is the sole surviving member of its ancient order. Members of the other three orders are fairly abundant, totaling about 6,500 living species.

All the living reptiles except the crocodilians are directly descended from an important Permian group called the stem reptiles or root reptiles (cotylosaurs) (Fig. 25.91). This group also gave rise to several other lineages, including two (ichthyosaurs and plesiosaurs) that returned to the aquatic environment (Fig. 25.92), one (therapsids) (Fig. 25.93) that ultimately led to the mammals, and one (thecodonts) that in its turn gave rise to crocodilians, the flying reptiles called pterosaurs, the great assemblage of reptiles called dinosaurs, and the birds. The dinosaurs were extremely abundant and varied during the Jurassic and Cretaceous periods (Figs. 25.94–25.96).

25.92 Plesiosaurs (left) and ichthyosaurs (right)
[From a mural by C. R. Knight, courtesy Field Museum of Natural History, Chicago.]

25.93 Restoration of a therapsid reptile
Fossil evidence indicates that the mammals evolved from the therapsids. Some investigators think the advanced therapsids had hair (not shown here).
[Courtesy American Museum of Natural History.]

25.94 *Triceratops* and *Tyrannosaurus*, two ancient dinosaurs
Triceratops (left) was a herbivore. *Tyrannosaurus* was a giant carnivore, about 47 feet long and 19 feet high.
[From a mural by C. R. Knight, courtesy Field Museum of Natural History, Chicago.]

25.95 *Brontosaurus,* **a giant amphibious dinosaur**
Adults probably weighed as much as 25–35 tons.
[From a mural by C. R. Knight, courtesy Field Museum
of Natural History, Chicago.]

25.96 *Stegosaurus,* **an armored herbivorous dinosaur**
[From a mural by C. R. Knight, courtesy Field Museum
of Natural History, Chicago.]

25.97 A Cedar Waxwing feeding its young
Many baby birds have brightly colored mouth linings, which act as releasers of the parental feeding response. [Courtesy D. G. Allen.]

Recent research has led some investigators to propose a radical revision in the long-established conception of dinosaurs. They think that, unlike modern reptiles, which are cold-blooded, these animals may have been warm-blooded. Their bones sometimes show indications of the extensive vascularization characteristic of warm-blooded animals, and it appears that at least some of them may have had heat-conserving body coverings. For example, the particular dinosaurs from which the birds arose probably had feathers as insulation before the first true birds appeared. Other investigators, however, consider the idea that the dinosaurs were characteristically warm-blooded unconvincing.

By the end of the Cretaceous (which was also the end of the Mesozoic era), all the plesiosaurs and pterosaurs had disappeared. The dinosaurs, too, had disappeared, except for one group of specialized modern descendants, the birds. Only members of the four groups of modern reptiles remained as representatives of this once enormous class.

The decline of the dinosaurs was not as sudden as is often supposed; it took tens of millions of years. But it was a dramatic event in the history of life on earth nonetheless. Why previously successful animals should have died out on such a scale has never been satisfactorily explained. Extinction was not limited to reptiles; many invertebrates, such as a widespread and abundant group of shelled cephalopods (ammonites), also disappeared. Yet many other groups living in the same sorts of habitats not only did not become extinct but did not even undergo significant change.

Class Aves By the late Triassic or early Jurassic, at least two lineages of reptiles, descended from the theocodonts, had developed the power of flight. One of these lineages, the pterosaurs, included animals with wings consisting of a large membrane of skin stretched between the body and the enormously elongated arm and fourth finger; some species had wingspreads as great as 8 m. The pterosaurs were common for a time, but eventually became extinct. The other lineage developed wings of an entirely different sort, in which many long feathers, derived from scales, were attached to the modified forelimbs. This line eventually became sufficiently different from the other reptiles to be designated as a separate class—Aves—the birds. Not all authorities think the birds deserve a class of their own. Some would place them in the Reptilia as surviving dinosaurs; others would erect a new class that would include both the dinosaurs and the birds.

The oldest known fossil bird *(Archaeopteryx)*, from the middle Jurassic, still had many reptilian characters, e.g. teeth and a long jointed tail. Neither of these traits is present in modern birds, which have a beak instead of teeth and only a tiny remnant of the ancestral tail bones (the tail of a modern bird consists only of feathers).

Along with wings, birds evolved a host of other adaptations for their very active way of life. One of the most important was warm-bloodedness (homeothermy)—the ability to maintain a high and constant metabolic rate, and hence great activity, despite fluctuations in environmental temperature. An anatomical feature that helped make possible the metabolic efficiency necessary for homeothermy was the complete separation of the two ventricles of the heart; birds have completely four-chambered hearts. The insulation against heat loss provided by the body feathers plays an important role in temperature regulation; in modern birds all the scales except those of the feet are modified as feathers. Among other adaptations for flight are light hollow bones and an extensive system of air sacs attached to the lungs (Figs. 6.19, 6.20, p. 252). Birds also have very keen senses of vision, hearing, and equilibrium.

The newly hatched young of birds are usually not yet capable of complete temperature regulation, and they cannot fly. In many species, in fact, they are featherless, blind, and virtually helpless. Accordingly, most birds exhibit elaborate nest-building and parental-care behavior (Fig. 25.97; see also Fig. 12.59, p. 548).

Class Mammalia   Both birds and mammals evolved from reptiles that were probably at least partly homeothermic, and both became highly successful groups of organisms. But the two groups did not arise from the same ancestral reptilian stock. The line leading to the mammals split off from the stem reptiles early in the Permian (Fig. 25.91), while that leading to the birds probably diverged from the dinosaurs in the Jurassic.

The mammals themselves, it should be clearly understood, did not appear in the Permian. But the Permian saw the rise of the therapsid reptiles, some of which became very mammal-like (Fig. 25.93); they may even have had hair. Precisely at what point therapsids ceased and mammals began, it is impossible to say: There was no sudden transformation of reptile into mammal, no dramatic event to mark the appearance of the first member of our class. Hence no attempt is made in our review below of some of the characters that distinguish modern mammals from stem reptiles to specify when each of these characters appeared.

Mammals have a four-chambered heart and are homeothermic. They have a diaphragm, which increases breathing efficiency. There is increased separation (by the palate) of the respiratory and alimentary passages. The body is covered with an insulating layer of hair. The limbs are oriented ventrally and lift the body high off the ground. The lower jaw is composed of only one bone (compared with six or more in most reptiles), and the teeth are complexly differentiated for a variety of functions. There are three bones in the middle ear (compared with one in reptiles and birds). The brain, particularly the neocortex, is much larger than in reptiles, and behavior is more easily modifiable by experience. No eggs are laid (except in monotremes); embryonic development occurs in the uterus of the mother, and the young are born alive. After birth, the young are nourished on milk secreted by the mammary glands of the mother.

As indicated above, there is one small group of mammals—the monotremes—that are fundamentally different from all other members of the class. They lay eggs; yet they secrete milk. In many other ways they are a curious blend of reptilian traits, mammalian traits, and traits peculiar to themselves. It seems clear that they were a very early offshoot of the mammalian lineage and were not ancestral to the other mammals. Some biologists think they should be considered mammal-like reptiles rather than reptilelike mammals. The only living monotremes are the echidna (or spiny anteater) and the duck-billed platypus (Fig. 25.98); both are found in Australia, echidnas also occurring in New Guinea.

The main stem of mammalian evolution split into two parts very early, one

25.98   Duck-billed platypus, an egg-laying mammal
The platypus is well adapted for aquatic life.   [Taronga Zoo, Sydney (Tom McHugh, Photo Researchers, Inc.).]

25.99 A kangaroo with young in pouch
The young kangaroo (called a joey) seen here is hundreds of times larger than when it first entered its mother's pouch; it is no longer attached to a nipple, and it often comes out of the pouch for extended periods. [Courtesy Australian Information Service.]

leading to the marsupials and the other to the placentals. The characteristic difference between them is that marsupial embryos remain in the uterus for a relatively short time and then complete their development while attached to a nipple in an abdominal pouch of the mother (Fig. 25.99), whereas placental embryos complete their development in the uterus.

The living placental mammals are classified in approximately 16 orders, several of which contain species familiar to almost everyone. A few of the most important orders are listed below:

INSECTIVORA   Moles, shrews
CHIROPTERA   Bats
PRIMATES   Lemurs, monkeys, apes, humans
EDENTATA   Sloths, anteaters, armadillos
LAGOMORPHA   Rabbits, hares, pikas
RODENTIA   Rats, mice, squirrels, gophers, beavers, porcupines
CETACEA   Whales, dolphins, porpoises
CARNIVORA   Cats, dogs, bears, raccoons, weasels, skunks, minks, badgers, otters, hyenas, seals, walruses
PROBOSCIDEA   Elephants
PERISSODACTYLA   Odd-toed ungulates (hoofed animals): horses, zebras, tapirs, rhinoceroses
ARTIODACTYLA   Even-toed ungulates: pigs, hippopotamuses, camels, deer, giraffes, antelopes, cattle, sheep, goats, bison

The oldest fossils identified as placental mammalian ones are from the Jurassic. They are of small, probably secretive creatures that are thought to have fed primarily on insects. They remained a relatively unimportant part of the fauna until the end of the Mesozoic. Of the modern orders, Insectivora is closest to this ancient group. The great radiation from the insectivore ancestors dates from the beginning of the Cenozoic era, as the mammals rapidly filled the many niches left open by the demise of the dinosaurs. The Cenozoic, which includes the present, is aptly termed the Age of Mammals.

EVOLUTION OF THE PRIMATES

As members of the mammalian order Primates, we naturally have a special interest in its evolutionary history, and, in particular, in that part of its history that concerns the origin of human beings.

Fossil evidence indicates that the primates arose from an arboreal stock of small shrewlike insectivores very early in the Cenozoic (Fig. 25.100). The group soon split into several evolutionary lines that have had independent histories ever since. Though the modern representatives of these evolutionary lines are a rather heterogeneous lot, most of them share the following characteristics: (1) retention of the clavicle (collarbone), which is greatly reduced or lost in many other mammals; (2) development of a shoulder joint permitting relatively free movements in all directions, and an elbow joint permitting some rotational movement; (3) retention of five functional digits on each foot; (4) enhanced individual mobility of the digits, especially the first digits (thumb and big toe), which are usually apposable; (5) modification of the claws into flattened nails; (6) development of sensitive tactile pads on the digits; (7) abbreviation of the snout or muzzle; (8) elaboration of the visual apparatus and development of binocular vision; (9) expansion of the brain, particularly the cerebral cortex; (10) usually only two mammae; (11) usually only one young per pregnancy. Most of these traits are correlated with an aboreal way of life.

In quadrupedal terrestrial mammals the limbs function as props and as instruments of propulsion for running and galloping; they have tended to evolve toward greater stability at the expense of freedom of movement. Think of the forelimbs of a dog or a horse: The clavicles are greatly reduced or lost; the two limbs are positioned close together under the animal, and their movement is restricted largely to one plane (i.e. they can move easily back and forth, but cannot be spread far to the side like human arms). By contrast, in an animal leaping about in the branches of a tree, the limbs function in grasping, clasping, and swinging; mobility at the shoulder, elbow, and digit joints facilitates such activities, as does attachment of the limbs (braced by the clavicles) far apart at the sides of the body instead of underneath.

The eyes of many quadrupedal terrestrial mammals (e.g. horses, cows, dogs) are located on the sides of the head. As a result, they can survey a very wide total visual field, but the fields of the two eyes overlap only slightly; i.e. the animals have little binocular stereoscopic (three-dimensional) vision. But stereoscopic vision aids in localizing near objects, and an animal jumping from limb to limb obviously must be able to detect very accurately the position of the next limb. Hence the arboreal way of life of the early primates doubtless led to selection for stereoscopic vision and, consequently, for eyes directed forward rather than laterally. This change, in turn, would have led to the distinctive flattened, forward-directed face of most higher primates.

Now, hands capable of grasping the next limb and keen eyes with broadly overlapping fields of vision would not by themselves have met the requirements of an arboreal way of life. Essential, too, would have been neural and muscular mechanisms capable of very precise eye-hand coordination. This need was doubtless one of the factors that led to the early expansion of the primate brain.

We could continue in this manner, relating other characteristics of primates to the demands of arboreal life, but the point has been made: Many of the traits most important to us as human beings first evolved because our distant ancestors lived in trees.

The prosimians The living primates are usually classified in two suborders: Prosimii and Anthropoidea. The first, the prosimians ("pre-monkeys"), are a miscellaneous group of more or less primitive primates, including the lemurs, aye-ayes, lorises, pottos, galagos, and tarsiers.

The living lemurs and aye-ayes are found only on the island of Madagascar off the east coast of Africa. Their relatives, the lorises, pottos, and galagos, inhabit southern Asia and tropical Africa. Most lemurs are fairly small arboreal animals with a bushier coat than is usual among higher primates. They have fairly long foxlike snouts and bushy tails, and hardly resemble the higher primates (Fig 25.101). But they have apposable first digits, and the digits are usually provided with flattened nails.

25.100 A Malay tree shrew
The living tree shrews, which are intermediate in many of their traits between the Insectivora and the Primates, are thought to resemble the early ancestors of the modern Primates. They are not closely related to squirrels (members of the Rodentia), which they superficially resemble. [Courtesy American Museum of Natural History.]

25.101 A ring-tailed lemur
The animal has a long snout and a bushy tail—both uncharacteristic of the higher Primates—but, as can be seen here, its hands and feet have apposable first digits. [G. D. Dodge and D. R. Thompson, Bruce Coleman Inc.]

25.102   A tarsier
Note the distinct face and the large forward-directed eyes.   [A. G. Nelson, Animals Animals.]

The tarsier, which is a small crepuscular animal found in the Philippines and the East Indies, is a more advanced and specialized prosimian than the lemurs (Fig 25.102). In some respects it shows more superficial resemblance to monkeys, though it differs in many ways from all other primates. It has a much shorter muzzle than a lemur, and thus has a more distinct face. The eyes are enormous and are directed more completely forward than in lemurs. The hind limbs are long and specialized for leaping. The long tail is naked except at the end.

The monkeys   The first members of the suborder Anthropoidea had diverged from a prosimian stock by the Oligocene epoch. Actually, two lines of anthropoids probably arose at about the same time from closely related prosimians. One of these led to the New World monkeys (including the marmosets), and the other, which soon split, led to the Old World monkeys and to the apes. These hypothetical relationships are diagramed in Fig. 25.103.

The following is the formal classification within the anthropoid suborder:

Suborder Anthropoidea
  Superfamily Ceboidea
    Family Cebidae, New World monkeys
    Family Callithricidae, marmosets
  Superfamily Cercopithecoidea
    Family Cercopithecidae, Old World monkeys and baboons
  Superfamily Hominoidea
    Family Pongidae, apes
    Family Hominidae, humans

The New World and Old World monkeys differ in far more ways than we can mention here. Three differences, however, are easily seen even on a casual visit to a zoo: (1) Most New World monkeys have a prehensile tail that they use almost like another hand for grasping branches; the tail of Old World monkeys is not prehensile. (2) The nostrils of New World monkeys are separated by a wide partition and are thus oriented in a lateral direction; the nostrils of Old World monkeys are not widely separated and are directed forward and down. (3) New World monkeys lack the naked brightly colored areas on the buttocks (ischial callosities) so common in Old World monkeys.

25.103   Hypothetical phylogenetic relationships of the living groups of Primates

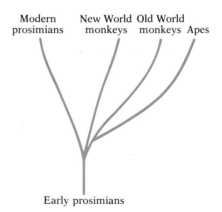

Among the best-known New World monkeys are capuchins (the traditional organ-grinders' monkeys), howlers (Fig 25.104), spider monkeys, and squirrel monkeys. Examples of Old World monkeys are macaques, mandrills, baboons, proboscis monkeys, mona monkeys, and the sacred hanuman monkeys of India. One of the macaques, commonly called the rhesus monkey, has been used extensively in physiological and psychological research; when physiologists or psychologists refer to "the monkey," this is usually the species they mean.

The apes The living great apes (Pongidae) fall into four groups: gibbons, orangutans, gorillas, and chimpanzees. All are fairly large animals that have no tail, a relatively large skull and brain, and very long arms. All have a tendency, when on the ground, to walk semi-erect.

The gibbons, of which several species are found in southeast Asia, represent a lineage that probably split from the others soon after the pongid line itself arose. They are the smallest of the apes (about 3 feet tall when standing). Their arms are exceedingly long, reaching the ground even when the animal is standing erect. The gibbons are amazing arboreal acrobats and spend almost all their time in trees (Fig. 25.105).

25.104 A howler monkey (Alouatta seniculus)
This is one of the New World monkeys with a prehensile tail. [Jan Lindblad, Photo Researchers, Inc.]

25.105 A gibbon (Hylobates moloch) from Sunda Island, Borneo
Notice the extremely long arms. [Jack Dermid, Bruce Coleman Inc.]

25.106 An orangutan *(Pongo pygmaeus)* **with young** [Monkey Jungle, Miami, Fla. R. P. Fontaine, Photo Researchers, Inc.]

The one living species of orangutan is native to Sumatra and Borneo (Fig. 25.106). Though the orangs are fairly large (males average about 165 pounds), and their movements slow and deliberate, they nevertheless spend most of their time in trees and only rarely descend to the ground.

Gorillas, of which there are two forms in Africa, are the largest of the apes (Fig. 25.107); wild adult males may weigh as much as 450 pounds (up to 600 pounds in zoos) and stand 6 feet tall. Their arms, while proportionately much longer than those of humans, are not as long as those of gibbons and orangs. Unlike gibbons and orangs, gorillas spend most of their time on the ground. Despite their fierce appearance, they are not usually aggressive.

Chimpanzees, which are native to tropical Africa, have been used extensively in psychological experiments. In general appearance they are the most human-looking of the living apes (see Fig. 25.110). They are about the same size as orangs, but their arms are shorter. Although they spend most of their time in trees, they descend to the ground more frequently than orangs, and sometimes even adopt a bipedal position (their usual locomotion, however, is quadrupedal, with the knuckles of the hand used for support). They are quite intelligent and can learn to perform a variety of tasks, such as opening doors and manipulating household gadgets. There has been considerable recent success in teaching them to communicate with sign language.

THE EVOLUTION OF HUMAN BEINGS

The earliest members of the family Hominidae (humans) probably arose from the same pongid stock that produced the gorillas and chimpanzees. Both paleontological evidence and biochemical and serological data indicate that gorillas, chimpanzees, and humans are more closely related, in terms of recentness of common ancestry, than any one of them is to orangutans or gibbons. Indeed, comparisons of the amino acid sequences of their

25.107 A male mountain gorilla *(Gorilla gorilla)* [George Holton, Photo Researchers, Inc.]

proteins have led some investigators to conclude that humans and chimps share about 99 percent of their genes. We can diagram the relationships as shown in Fig. 25.108.

Some investigators (especially V. Sarich and A. Wilson of the University of California, Berkeley) have interpreted the molecular evidence (from analysis of proteins) as indicating that humans have been separated from chimpanzees and gorillas for only about 4 million years, from orangutans about 7 million years, and from gibbons about 10 million years. From the fossil evidence it appears that these time intervals are probably incorrect—the separation from chimpanzees is more likely to have occurred between 10 and 15 million years ago—but the sequences deduced by the two methods (molecular and paleontological) are identical.

The current conception of the common ancestor of modern apes (at least of gorillas and chimpanzees) and humans is founded largely on fossils of several species assigned to the genus *Dryopithecus*, which first appeared some 25 million years ago (during the early Miocene) and ranged widely over Europe, Africa, and Asia.[27] These animals had a skull with a low rounded cranium, moderate supraorbital ridges, and moderate forward projection of face and jaws. The arms were only modestly specialized for brachiation—swinging from branch to branch—and the feet indicate some tendency toward bipedal posture.

Only a few of the many anatomical changes that occurred in the course of evolution from ape ancestor to modern humans can be mentioned here: (1) The jaw became shorter (making the muzzle shorter), and the teeth became smaller. (2) The point of attachment of the skull to the vertebral column shifted from the rear of the braincase to a position under the braincase, the skull thus becoming balanced more on top of the vertebral column (Fig. 25.109). (3) The braincase became much larger, and, as it did, a prominent vertical forehead developed. (4) The eyebrow ridges and other keels on the skull were reduced as the muscles that once attached to them became smaller. (5) The nose became more prominent, with a distinct bridge and tip. (6) The arms (though probably never as long as in the modern apes) became shorter. (7) The feet became flattened, and then an arch developed. (8) The big toe moved back into line with the other toes and ceased being apposable. The various fossil humans are intermediate in these characteristics.

One of the most distinctive traits of human beings is their bipedal locomotion and upright posture. How might this trait have evolved? One possibility immediately suggests itself. When our ancestors moved from the forest to the savanna, their forelimbs, adapted to an arboreal existence, may by that very fact have been preadapted for uses other than locomotion. Not that they didn't serve in locomotion on the savanna: They almost certainly did. But the locomotion may well have been of the kind in which the knuckles rather than the palms are on the ground—the method used by gorillas and chimpanzees today. Now, knuckle-walking enables chimpanzees to carry objects in their hands as they walk—an important advance in transport capabilities. But a hand that can carry objects can also manipulate objects as weapons. Chimpanzees, in fact, often use sticks when threatening, and they sometimes throw stones or other objects at such enemies as ba-

[27] Included in *Dryopithecus*, as used here, are the African forms sometimes called *Proconsul*.

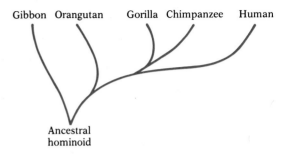

25.108 Hypothetical evolutionary relationships among the living apes and humans

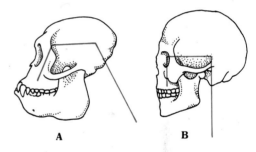

25.109 Human and monkey skulls compared with respect to their attachment to the vertebral column (A) The color line shows how the vertebral column attaches to the rear of the monkey skull, forming an obtuse angle with the horizontal axis of the cranium; note also the facial angle. (B) In the modern human, the vertebral column joins the skull more ventrally, forming approximately a right angle with the axis of the cranium; the face, too, is oriented nearly at a right angle to the cranial axis.

boons. When an animal is using a weapon, it cannot simultaneously use its forelimbs for support; it must adopt an upright posture. That posture not only facilitates manipulation of the weapon but also makes the animal appear larger as it directs a threat display at an enemy or an opponent. Greater use of weapons necessarily meant less reliance on forelimbs for support and locomotion.

The increased dependence on hunting as a means of obtaining food may also have contributed to freeing the forelimbs from locomotion. Though small objects could be carried in the hands during quadrupedal locomotion, the transportation of larger objects, such as recently killed prey, required bipedal locomotion.

Hands with apposable thumbs are preadapted not only for carrying, but also for manipulation of tools. This, too, can be observed in chimpanzees. Jane Goodall of the Gombe Stream Research Centre in Tanzania has reported a host of tool-using practices by these animals: the use of sticks for poking and prying, of blades of grass to collect and eat ants and termites (Fig. 25.110), of leaves for personal grooming, of stones to crack nuts. If such behavior became increasingly important to the early representatives of the hominid line, then there might have been selection for keeping the hands free, and hence for evolution of upright posture.

Another advantage of upright posture would have been the increased ease of maintaining surveillance—an important requirement for an animal living on the ground in the open country, if not for a chimpanzee living prevalently on the forest floor.

Fossil humans The first human fossil bones were found in 1856 in Germany by Johann Karl Fuhlrott. They excited lively debate, with Fuhlrott and his supporters maintaining that they were remnants of an ancient manlike organism quite different from modern humans, and his opponents objecting that they were simply the remains of a person who had suffered several deformities. It was many years before enough similar fossils were found to establish the validity of Fuhlrott's claim conclusively.

Many bones of ancient humans have been discovered since 1856. At first, the tendency of anthropologists was to erect both a new genus and a new species for each new find, without regard to biological criteria for erecting such categories. The result was a very long list of names that gave no indication of the probable relationships of the organisms they designated. More recently, however, the modern biological ideas concerning speciation and intraspecific variation have been increasingly applied to the study of fossil humans, and this, together with the discovery of many new fossils, particularly in Africa, has begun to improve our understanding of human evolution and to bring some order out of a growing jumble of information. But much is still unknown, and there is still considerable controversy over how the data should be interpreted; the brief sketch given here must be taken as only one of a number of possible interpretations.

The oldest fossils now assigned to the family Hominidae are of *Ramapithecus punjabicus*[28] from the late Miocene (about 12 to 14 million years ago). *Ramapithecus* appears to have coexisted for a time with *Dryopithecus*, from

25.110 A chimpanzee using a twig as a tool to get termites out of their nest
[Baron Hugo van Lawick, © National Geographic Society.]

---

[28] Included in *Ramapithecus punjabicus* are the fossils formerly called *Kenyapithecus*.

which it probably descended. It was an apelike creature, but one that exhibited early stages in reduction of the size of the incisors and canines—a change possibly correlated with increased use of the hands rather than the teeth in obtaining food. It was probably *Ramapithecus* that first moved the hominid line from the forest to the savanna.

The first truly manlike hominids are usually assigned to the genus *Australopithecus* (from *australis*, southern, and *pithecus*, ape—"southern," because the first specimens were found in South Africa).[29] The oldest known fossils of *Australopithecus*, which have usually been assigned to the species *Australopithecus africanus*, are about 3.8 million years old, and it is generally thought that the species must have originated from *Ramapithecus* at least 4 million (and perhaps as much as 6 million) years ago. It seems, then, that the emergence of humans, long associated with the Pleistocene, must now be relegated to the Pliocene (see Table 20.1, p. 906).

The australopithecine species, often called South African ape men, were apparently fully bipedal, though their stance was not as upright as that of modern humans (Fig. 25.111). They had large jaws, but almost no forehead or chin, and their cranial capacity was only about 450–550 cc, compared with 350–450 for normal chimpanzees and 1,200–1,600 (average about 1,360) for normal modern humans (Fig. 25.112). They probably used unworked stones as tools. Most of these ancient fossils are from eastern and southern Africa, but specimens have also been found in North Africa and Java. There were probably at least two species, which apparently lived contemporaneously for at least a million years. One, with smaller teeth and a more gracile form, which probably included meat in its diet, was *A. africanus*, the species mentioned earlier. The other, named *Australopithecus robustus*, had considerably larger teeth and larger bony crests on the skull, was about 40 cm

[29] *Australopithecus* includes the forms originally described in *Paranthropus*, *Meganthropus*, and *Zinjanthropus*.

25.111 A reconstruction of *Australopithecus africanus* standing
Although this primitive human was fully bipedal, his stance was not as erect as that of our own species. [Reproduced by permission of Rainbird Publishing Group from R. E. Leakey and R. Lewin, *Origins*, Dutton, 1977.]

25.112 Reconstruction of head of *Australopithecus africanus*
[Courtesy American Museum of Natural History.]

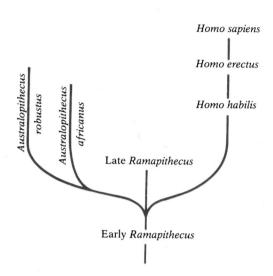

25.113 Hypothetical phylogenetic relationships among the known hominids

The scheme given here, with the *Australopithecus* and *Homo* lines branching separately from *Ramapithecus*, is one many investigators endorse. Others think *Homo* probably arose from an early species of *Australopithecus*, called *Australopithecus afarensis*, which also gave rise to *A. africanus*.

taller, and was probably primarily herbivorous.[30] *A. robustus* probably arose from *A. africanus*, but it is thought that the two species came to live sympatrically.

After the discovery of the australopithecines, the view long prevailed that *A. robustus* was an evolutionary dead end, that it had not given rise to any modern descendants, but that *A. africanus* was the likely ancestor of our own genus *Homo*. That interpretation came into serious question when remains of what clearly seemed to have been a form of *Homo* were recovered from deposits that also contained bones of both *A. africanus* and *A. robustus*. It seemed that the three forms must have been contemporaneous. Evidence has mounted in the last few years that the early representative of *Homo*, usually called *Homo habilis* (though there is some disagreement on whether this is the correct designation), may have originated almost as long ago as *Australopithecus*. In the view of many investigators today, *Homo habilis* probably arose directly from a *Ramapithecus* ancestor, and the genera *Homo* and *Australopithecus* thus represent separate evolutionary lines—one that eventually died out and one that continued until the present (Fig. 25.113). An alternative hypothesis, recently put forward by several investigators (but not generally accepted), holds that the oldest australopithecine fossils (2.9–3.8 million years old) represent a separate species, which they have named *Australopithecus afarensis* and that this species was the common ancestor of both the *Homo* line and the *A. africanus*–*A. robustus* line.

Homo habilis had a larger cranial capacity than the australopithecines, usually between 650 and 775 cc, and there is convincing evidence that *H. habilis* not only used stones as tools but also chipped and shaped them for various purposes.

A later stage in human evolution is represented by fossils that may be classified as *Homo erectus*[31] (originally described as *Pithecanthropus erectus*, often called Java man) (Fig 25.114). Specimens have been found in Asia, Africa, and Europe. This species, which almost certainly descended from *H. habilis*, first appeared about 1.5 million years ago. Its cranial capacity was considerably larger, averaging about 900 cc. However, the facial features remained primitive, with a projecting massive jaw, large teeth, almost no chin, a receding forehead, heavy bony eyebrow ridges, and a broad low-bridged nose. Not only did the members of this species make and use tools, but they also used fire. Casts of the interior of the skulls indicate the presence of the speech areas of the brain; of course we have no way of knowing whether or how language was used.

Modern humans are given the Latin name *Homo sapiens* (wise man). Early representatives of this species first appeared about 250,000 years ago. It seems highly likely that *H. sapiens* evolved from *H. erectus*, but it is uncertain where the early stages of this evolutionary transition occurred. The oldest fossils of *H. sapiens* are from England and continental Europe, but the species may well have migrated there from Africa or southern Asia.

During the period from about 100,000 to 40,000 years ago, a very distinctive form of *H. sapiens*, designated *Homo sapiens neanderthalensis*, Neanderthal

---

[30] Some investigators recognize a second large-toothed species, designated *Australopithecus boisei*, but most regard it as synonymous with *A. robustus*.

[31] *Homo erectus* includes the forms originally described in *Pithecanthropus*, *Sinanthropus*, *Telanthropus*, and *Atlanthropus*. Some authorities also prefer to regard as early *Homo erectus* the fossils we have called *Homo habilis*.

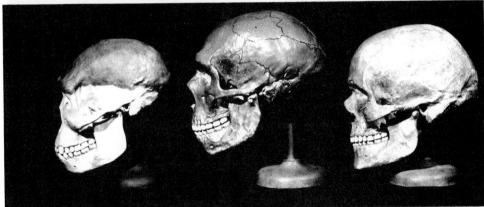

25.114 Prehistoric humans
Top: Restorations of (left to right) Java man, Neanderthal man, and Cro-Magnon man. Bottom: Lateral view of restored skulls of the same three types. Note differences in size of braincase, height of forehead, size of eyebrow ridge, length of jaw, shape of chin, size of teeth, etc. [Courtesy American Museum of Natural History.]

man, lived throughout most of Europe and also in parts of Asia and Africa.[32] This is the form to which the bones discovered by Fuhlrott belonged (Fig. 25.114). Neanderthals were 5–5½ feet tall, and had a receding forehead, prominent eyebrow ridges, and a receding chin, but their brain was as big as that of a modern human (perhaps a little bigger). They made many kinds of tools, and they buried their dead, which has been interpreted as a capacity on their part for abstract and religious thought. Neanderthals disappeared soon after the modern form, *Homo sapiens sapiens*, arrived in their range. This modern form, the earliest representatives of which are often called Cro-Magnon man (Fig. 25.114), is the only kind of human found on the earth today. Why the Neanderthals disappeared so rapidly after Cro-Magnon came on the scene is unclear. They may have been eliminated by combat or competition, or they may merely have been absorbed into *Homo sapiens sapiens* by interbreeding.

[32] A few workers still follow the older practice of regarding Neanderthal man as specifically distinct from *Homo sapiens*, and designating this form *Homo neanderthalensis*.

The human races   As we saw in Chapter 18, widespread species often tend to become subdivided into geographic races. Humans are no exception. *Homo sapiens* is an extremely variable species, and regional populations are often recognizably different (or were, prior to the great mobility of the last few centuries). Thus Scandinavians tend to have blue eyes and a fair complexion, while south Europeans tend to have brown eyes and a darker complexion. Eskimos look different from Mohawk Indians, and they in turn look different from Apaches. Pygmies of the Congo are obviously different from their taller neighbors. Many of the differences probably reflect adaptations to different environmental conditions. Thus, for example, the prevalence of darker skin in tropical and subtropical regions may be a protective adaptation against damaging ultraviolet solar radiation.

Now, races, by definition, are regional populations that differ genetically but have no effective intrinsic isolating mechanisms. There are seldom sharp boundaries between them, and they intergrade over wide areas. Designation of races in most species is thus an arbitrary matter, and there is no such thing as a "pure" race. An almost unlimited number of races of the snake *Coluber constrictor* could be erected, depending on which of the characters whose distributions are illustrated in Fig. 18.35 (p. 795) are chosen for emphasis. The same considerations apply to humans. Some authorities have chosen to recognize as many as 30 races, while others recognize only three: the traditional Caucasoid, Mongoloid, and Negroid. Another widely used system recognizes five: the traditional three, plus American Indians and Australian aborigines.

No one of these systems has any more biological validity than the others, since races, as categories, are human inventions. What is biologically real is the geographic variation within the species *Homo sapiens*, a variation that will surely tend to break down as people move about more and more. The main barriers to interbreeding in many parts of the world are now cultural or social rather than geographic, and it seems very unlikely that such barriers will even approach the effectiveness of the original geographic barriers. Hence, whatever races are recognized now, it seems probable that they will become less and less distinct as time goes on. This, too, is a phenomenon that has occurred countless times in other species.

The interaction of cultural and biological evolution   One of the most interesting discoveries of modern anthropologists is that early hominids used tools long before their brains were much larger than those of apes. Thus the old idea that a large brain and high intelligence were necessary prerequisites for the use of tools has been discredited. The early hominids' use of tools may, in fact, have been an important factor in leading to evolution of higher intelligence. Once the use and making of tools began, individuals that excelled in these endeavors would surely have had an advantage over their less talented fellows. There would thus have been strong selection for neural mechanisms making possible improved fashioning and use of tools. Thus perhaps, instead of considering culture the crown of the fully evolved human intelligence, we should regard early cultural development and increasing intelligence as two faces of the same coin; in a sense, the highly developed brain of modern human beings may be as much a consequence as a cause of culture.

Cultural evolution can proceed at a far more rapid pace than biological evolution. Words as units of inheritance are much more effective than genes in spreading new developments and in giving dominance to new approaches originating with a few talented individuals. But the two types of evolution continue to be interwoven just as they were in the use of tools; they may well be even more so in the future.

Human beings, by their unrivaled ability to alter their environment, are influencing in profound ways the evolution of all species with which they come in contact. Thus, as we have seen, new strains of bacteria have evolved in response to use of antibiotics; many species of insects have evolved new physiological and behavioral traits as a consequence of the intense selection resulting from use of insecticides; the clearing of forests for agricultural purposes has led to drastic decline in the population densities of some species and to increase in others; long-established balances between prey species and their predators and parasites have been destroyed, often with far-reaching consequences for the entire ecosystem.

While some human actions on the environment have been deliberate, many have been unintentional. But whether deliberate or unintentional, they have precipitated a period of wholesale and rapid change unmatched since life began. Since disruption of the ecosystem will unavoidably grow as civilization expands, the great challenge to our species is to use the resources of our knowledge and technology to guide the change in ways that will benefit both our own species and the other organisms around us. As Theodosius Dobzhansky has put it, "Creation is not an act but a process; it did not happen five or six thousand years ago but is going on before our eyes. Man is not compelled to be a mere spectator; he may become an assistant, a collaborator, a partner in the process of creation."[33]

Not only can we humans influence the evolution of other species, but we now also have the ability to alter deliberately some aspects of the future evolution of our own species. Thus modern medicine, by saving people with gross genetic defects that would once have been fatal, permits perpetuation of genes that natural selection would formerly have eliminated. Does this mean that we should practice eugenics—deliberately restrict, by law or by social pressure, the perpetuation of some genetic traits and encourage the perpetuation of others? Some thinkers, concerned over what they see as the inevitable physical "decline" of the species, have urged just that. Others have pointed out that, although there may indeed be an increase in traits that would once have been maladaptive, modern human beings live in an environment of their own design, in which those traits are no longer so deleterious. As Dobzhansky says, "man . . . adapts his environments to his genes more frequently and efficiently than his genes to his environments." One can argue, for example, that a decision to restrict reproduction by diabetics would focus too narrowly on a genetic defect whose symptomatic expression is now largely controllable. Individual diabetics might bear genes for musical ability, or artistic talent, or intellectual acuteness, or compassionate behavior. Would not these traits be more important to society than the diabetes?

---

[33] This and the following quotations are from "Evolution: Implications for religion," by T. Dobzhansky, in *Changing man: The threat and the promise*, ed. by K. Haselden and P. Hefner (Doubleday, 1968).

Our ability to control our own evolution may eventually no longer depend primarily on regulating reproduction. Now that the genetic code has been deciphered, the day will surely come when the DNA of genes can deliberately be altered in order to design, at least in part, new human beings. When that day comes, how do we decide what to design? What do we look for in human beings? We might all agree to rid our species of the genes for muscular dystrophy or sickle-cell anemia (at least in areas where there is no malaria), but once the techniques for achieving these apparently worthy ends are mastered, suggestions will surely be put forward for other alterations to which we cannot all agree. As Dobzhansky has asked, "shall we endeavor to breed a race of brawny athletes, or brainy intellectuals, or sensitive esthetes, or some combination of these qualities, or a population containing certain proportions of each kind?" Who shall decide? And who shall control the ones who decide?

More immediately pressing, perhaps, is the problem of regulating the size of human populations now that we have interfered with the action of the many former natural regulating factors. Already some people are asking whether we should abandon the campaigns to eradicate malaria, to cure cancer and heart disease, or to slow the aging process. They point out that the current population problem is the result of major advances in the technology of death control, which societies have generally been as eager to accept as they have been reluctant to accept compensatory birth control. Since no species can continue indefinitely with its birth and death rates unbalanced, and since there is little evidence that our species will consent soon enough to the efficacious population-control measures that become increasingly necessary as the traditional killers are overcome, perhaps the only solution is less death control, heartless as that may seem.

These complex and unnerving questions—at once biological, economic, political, and moral—must be faced, and soon. Human beings have already gone too far toward modifying biological evolution to pull back. The answers they give to these questions in the next few generations may well have as profound an influence on the future of life as anything that has happened since the first cells materialized in the primordial seas.

SUGGESTED READING

BAKKER, R. T., 1975. "Dinosaur renaissance," *Scientific American*, April. (Offprint 916) *Reasons for believing that the dinosaurs were warm-blooded and that the birds descended from them.*

BARNES, R. D., 1974. *Invertebrate zoology*, 3rd ed. Saunders, Philadelphia. *Rather technical, comprehensive treatment of all the invertebrate groups.*

BORROR, D. J., D. M. DeLONG, and C. A. TRIPLEHORN, 1976. *An introduction to the study of insects*, 4th ed. Holt, Rinehart & Winston, New York. *Thorough coverage of all aspects of insect biology.*

BUCHSBAUM, R., 1948. *Animals without backbones*, 2nd ed. University of Chicago Press, Chicago. *One of the most readable and fascinating discussions of the invertebrates ever written. Not technical.**

*Available in paperback.

BUCHSBAUM, R., and L. J. MILNE, 1960. *The lower animals: Living invertebrates of the world.* Doubleday, Garden City, N.Y.   *One of the "Living animals of the world" books. Like the others in the series—on amphibians, birds, fishes, insects, mammals, and reptiles—beautifully illustrated, well written, and nontechnical.*

COCHRAN, D. M., 1961. *Living amphibians of the world.* Doubleday, Garden City, N.Y.

DAWSON, T. G., 1977. "Kangaroos," *Scientific American,* August. (Offprint 1366)   *How their adaptive strategies resemble those of ungulates in other parts of the world.*

GILLIARD, E. T., 1958. *Living birds of the world.* Doubleday, Garden City, N.Y.

HERALD, E. S., 1961. *Living fishes of the world.* Doubleday, Garden City, N.Y.

KLOTS, A. B., and E. B. KLOTS, 1959. *Living insects of the world.* Doubleday, Garden City, N.Y.

LAPAN, E. A., and H. J. MOROWITZ, 1972. "The Mesozoa," *Scientific American,* December.   *An excellent account of these tiny invertebrates.*

LEAKEY, R. E., and R. LEWIN, 1977. *Origins.* Dutton, New York.   *An account, sometimes controversial, of human evolution. The Leakey family discovered many of the important human fossils.*

LEAKEY, R. E., and R. LEWIN, 1978. "The hominids of East Turkana," *Scientific American,* August.   *On some of the most recently discovered human fossils.*

OLDROYD, H., 1973. *Insects and their world,* 3rd ed. British Museum (Natural History), London.   *Short and well written.*

ROMER, A. S., 1959. *The vertebrate story,* 4th ed. University of Chicago Press, Chicago.   *Still fascinating, though a bit out of date in places.*

SANDERSON, I. T., 1955. *Living mammals of the world.* Doubleday, Garden City, N.Y.

SCHMIDT, K. P., and R. F. INGER, 1957. *Living reptiles of the world.* Doubleday, Garden City, N.Y.

SIMONS, E. L., 1977. "Ramapithecus," *Scientific American,* May. (Offprint 695)   *On the discovery of many new fossils of this earliest member of the hominid family, and the light they shed on human evolution.*

TELEKI, G., 1973. "The omnivorous chimpanzee," *Scientific American,* January. (Offprint 682)   *Evidence that chimpanzees, formerly thought entirely herbivorous, sometimes hunt and kill other mammals for food.*

TULLAR, R. M., 1977. *The human species: Its nature, evolution, and ecology.* McGraw-Hill, New York.   *Includes a clearly written, simple summary of current ideas about human evolution.*

VILLEE, C. A., W. F. WALKER, and R. D. BARNES, 1973. *General zoology,* 4th ed. Saunders, Philadelphia.   *Good textbook covering all animal groups.*

APPENDIX: A CLASSIFICATION
OF LIVING THINGS

The classification given here is one of many in current use. Some other systems recognize more or fewer divisions and phyla, and combine or divide classes in a variety of other ways; but compared with the large areas of agreement, the differences between the various classifications are minor. Chapters 20–25 discuss certain of the points at issue between advocates of different systems.

Botanists have traditionally used the term "division" for the major groups that zoologists have called phyla. In classifications recognizing only two or three kingdoms, this difference in terminology causes little difficulty, because usage can be consistent within each kingdom. But when a kingdom Protista is recognized, as it is here, consistency is achieved only at the expense of violating well-established usage. The Protista contain some plantlike or funguslike groups traditionally called divisions and some animal-like groups traditionally called phyla. These usages we have respected.

Most classes within a division or phylum are listed here, but where there is only one class it is not named. For some classes (e.g. Insecta and Mammalia) orders are given too. Except for a few extinct groups of particular evolutionary importance (e.g. Placodermi), only groups with living representatives are included. A few of the better-known genera are mentioned as examples in each of the taxons.

Whenever possible, an estimate (a very rough one) of the number of living species is provided for higher taxons.

KINGDOM MONERA

DIVISION SCHIZOMYCETES.   Bacteria* (1,400)

CLASS MYCOPLASMATA. *Mycoplasma, Acholeplasma*
CLASS RICKETTSIAE. *Rickettsia, Coxiella*
CLASS CHLAMYDIAE. *Chlamydia*
CLASS ACTINOMYCETES. *Streptococcus, Staphylococcus, Arthrobacter, Actinomyces, Streptomyces*
CLASS EUBACTERIA. *Escherichia, Rhizobium, Spirillum, Salmonella, Nitrosomonas, Serratia, Thiocystis*
CLASS MYXOBACTERIA. *Myxococcus, Chondromyces*
CLASS SPIROCHETES. *Leptospira, Spirocheta, Treponema*

DIVISION CYANOPHYTA.   Blue-green algae (1,800). *Gloeocapsa, Microcystis, Oscillatoria, Nostoc, Scytonema*

DIVISION PROCHLOROPHYTA.   *Prochloron*

KINGDOM PROTISTA

SECTION PROTOPHYTA: Algal protists

DIVISION EUGLENOPHYTA.   Euglenoids (800). *Euglena, Eutreptia, Phacus, Colacium*

DIVISION CHRYSOPHYTA

CLASS CHRYSOPHYCEAE. Golden-brown algae (650). *Chrysamoeba, Chromulina, Synura, Mallomonas*
CLASS HAPTOPHYCEAE (or Prymnesiophyceae). Haptophytes and coccolithophores. *Isochrysis, Prymnesium, Phaeocystis, Coccolithus, Hymenomonas*
CLASS XANTHOPHYCEAE. Yellow-green algae (360). *Botrydiopsis, Halosphaera, Tribonema, Botrydium*
CLASS EUSTIGMATOPHYCEAE. Eustigmatophytes. *Pleurochloris, Visheria, Pseudocharaciopsis*
CLASS CHLOROMONADOPHYCEAE. Chloromonads. *Gonyostomum, Reckertia*
CLASS BACILLARIOPHYCEAE. Diatoms (10,000). *Pinnularia, Arachnoidiscus, Triceratium, Pleurosigma*

DIVISION PYRROPHYTA.   Dinoflagellates (1,000). *Gonyaulax, Gymnodinium, Ceratium, Gloeodinium*

DIVISION CRYPTOPHYTA.   Cryptomonads. *Cryptomonas, Chroomonas, Chilomonas, Hemiselmis*

---

*There is no generally accepted classification for bacteria at the higher taxon level. One recent classification divides the bacteria into 17 distinct divisions, some without formal names. Another important classification assigns them to 19 "parts," most without formal names. The classification used here, recognizing seven classes, is a conservative one.

SECTION PROTOMYCOTA: Fungal protists

DIVISION HYPHOCHYTRIDIOMYCOTA.   Hyphochytrids (25). *Rhizidiomyces*

DIVISION CHYTRIDIOMYCOTA.   Chytrids (1,000). *Olpidium, Rhizophydium, Diplophlyctis, Cladochytrium*

SECTION GYMNOMYCOTA: Slime molds

DIVISION PLASMODIOPHOROMYCOTA.   Plasmodiophores (or endoparasitic slime molds). *Plasmodiophora, Spongospora, Woronina*

DIVISION LABYRINTHULOMYCOTA.   Net slime molds. *Labyrinthula*

DIVISION ACRASIOMYCOTA.   Cellular slime molds (26). *Dictyostelium, Polysphondylium*

DIVISION MYXOMYCOTA.   True slime molds (400). *Physarum, Hemitrichia, Stemonitis*

SECTION PROTOZOA: Animal-like protists

PHYLUM MASTIGOPHORA

CLASS ZOOFLAGELLATA (or Zoomastigina). "Animal" flagellates (5,000). *Trypanosoma, Calonympha, Chilomonas, Trichonympha*
CLASS OPALINATA.† Opalinids (200). *Opalina, Zelleriella*

PHYLUM SARCODINA.   Pseudopodal protozoans (11,500)

CLASS RHIZOPODEA. Naked and shelled amoebae, foraminiferans. *Amoeba, Pelomyxa, Entamoeba, Arcella, Globigerina, Textularia*
CLASS ACTINOPODEA. Radiolarians, heliozoans, acantharians. *Aulacantha, Acanthometron, Actinosphaerium, Actinophrys*

PHYLUM SPOROZOA.   Sporulation protozoans (3,600)

CLASS TELOSPOREA. *Monocystis, Gregarina, Eineria, Toxoplasma, Plasmodium*
CLASS PIROPLASMEA. *Babesia, Theileria*

PHYLUM CNIDOSPORA.   Cnidosporians (1,100)

CLASS MYXOSPOREA. *Myxobolus, Myxidium, Ceratomyxa*
CLASS MICROSPOREA. *Nosema, Thelohania, Pleistophora*

PHYLUM CILIATA.   Ciliates (6,000). *Paramecium, Stentor, Vorticella, Spirostomum*

---

† The opalinids are sometimes placed in the Ciliata, because they have cilia instead of flagella, but they lack the other diagnostic characters of Ciliata. It must be admitted, however, that they do not fit well in the Mastigophora either.

KINGDOM PLANTAE

DIVISION CHLOROPHYTA. Green algae (7,000). *Chlamydomonas, Volvox, Ulothrix, Spirogyra, Oedogonium, Ulva*

DIVISION CHAROPHYTA. Stoneworts (300). *Chara, Nitella, Tolypella*

DIVISION PHAEOPHYTA. Brown algae (1,500). *Sargassum, Ectocarpus, Fucus, Laminaria*

DIVISION RHODOPHYTA. Red algae (4,000). *Nemalion, Polysiphonia, Dasya, Chondrus, Batrachospermum*

DIVISION BRYOPHYTA (23,600)

CLASS HEPATICAE. Liverworts. *Marchantia, Conocephalum, Riccia, Porella*

CLASS ANTHOCEROTAE. Hornworts. *Anthoceros*

CLASS MUSCI. Mosses. *Polytrichum, Sphagnum, Mnium*

DIVISION TRACHEOPHYTA. Vascular plants

Subdivision Psilopsida. *Psilotum, Tmesipteris*

Subdivision Lycopsida. Club mosses (1,500). *Lycopodium, Phylloglossum, Selaginella, Isoetes, Stylites*

Subdivision Sphenopsida. Horsetails (25). *Equisetum*

Subdivision Pteropsida. Ferns (10,000). *Polypodium, Osmunda, Dryopteris, Botrychium, Pteridium*

Subdivision Spermopsida. Seed plants

CLASS PTERIDOSPERMAE. Seed ferns. No living representatives

CLASS CYCADAE. Cycads (100). *Zamia*

CLASS GINKGOAE (1). *Gingko*

CLASS CONIFERAE. Conifers (500). *Pinus, Tsuga, Taxus, Sequoia*

CLASS GNETEAE (70). *Gnetum, Ephedra, Welwitschia*

CLASS ANGIOSPERMAE. Flowering plants

SUBCLASS DICOTYLEDONEAE. Dicots (225,000). *Magnolia, Quercus, Acer, Pisum, Taraxacum, Rosa, Chrysanthemum, Aster, Primula, Ligustrum, Ranunculus*

SUBCLASS MONOCOTYLEDONEAE. Monocots (50,000). *Lilium, Tulipa, Poa, Elymus, Triticum, Zea, Ophyrys, Yucca, Sabal*

KINGDOM FUNGI

DIVISION OOMYCOTA. Water molds, white rusts, downy mildews (400). *Saprolegnia, Phytophthora, Albugo*

DIVISION ZYGOMYCOTA. Conjugation fungi (250)

CLASS ZYGOMYCETES. *Rhizopus, Mucor, Phycomyces, Choanephora, Entomophthora*

CLASS TRICHOMYCETES. *Stachylina*

DIVISION ASCOMYCOTA. Sac fungi (12,000)

CLASS HEMIASCOMYCETES. Yeasts and their relatives. *Saccharomyces, Schizosaccharomyces, Endomyces, Eremascus, Taphrina*

CLASS PLECTOMYCETES. Powdery mildews, fruit molds, etc. *Erysiphe, Podosphaera, Aspergillus, Penicillium, Ceratocystis*

CLASS PYRENOMYCETES. *Sordaria, Neurospora, Chaetomium, Xylaria, Hypoxylon*

CLASS DISCOMYCETES. *Sclerotinia, Trichoscyphella, Rhytisma, Xanthoria, Pyronema*

CLASS LABOULBENIOMYCETES. *Herpomyces, Laboulbenia*

CLASS LOCULOASCOMYCETES. *Cochliobolus, Pyrenophora, Leptosphaeria, Pleospora*

DIVISION BASIDIOMYCOTA. Club fungi (15,000)

CLASS HETEROBASIDIOMYCETES. Rusts and smuts. *Ustilago, Urocystis, Puccinia, Phragmidium, Melampsora*

CLASS HOMOBASIDIOMYCETES. Toadstools, bracket fungi, mushrooms, puffballs, stinkhorns, etc. *Coprinus, Marasmius, Amanita, Agaricus, Lycoperdon, Phallus*

KINGDOM ANIMALIA

SUBKINGDOM PARAZOA

PHYLUM PORIFERA. Sponges (5,000)

CLASS CALCAREA. Calcareous (chalky) sponges. *Scypha, Leucosolenia, Sycon, Grantia*

CLASS HEXACTINELLIDA. Glass sponges. *Euplectella, Hyalonema, Monoraphis*

CLASS DEMOSPONGIAE. *Spongilla, Euspongia, Axinella*

CLASS SCLEROSPONGIAE. Coralline sponges. *Ceratoporella, Stromatospongia*

SUBKINGDOM AGNOTOZOA

PHYLUM PLACOZOA (1). *Trichoplax*

PHYLUM MESOZOA (50)

CLASS DICYEMIDA. *Dicyema, Pseudicyema, Conocyema*

CLASS ORTHONECTIDA. *Rhopalura*

SUBKINGDOM METAZOA

SECTION RADIATA

PHYLUM COELENTERATA (or Cnidaria)

CLASS HYDROZOA. Hydrozoans (3,700). *Hydra, Obelia, Gonionemus, Physalia*

CLASS SCYPHOZOA. Jellyfishes (200). *Aurelia, Pelagia, Cyanea*

CLASS ANTHOZOA. Sea anemones and corals (6,100). *Metridium, Pennatula, Gorgonia, Astrangia*

PHYLUM CTENOPHORA. Comb jellies (90)

CLASS TENTACULATA. *Pleurobrachia, Mnemiopsis, Cestum, Velamen*

CLASS NUDA. *Beroe*

SECTION PROTOSTOMIA

PHYLUM PLATYHELMINTHES. Flatworms (10,000)

CLASS TURBELLARIA. Free-living flatworms. *Planaria, Dugesia, Leptoplana*

CLASS TREMATODA. Flukes. *Fasciola, Schistosoma, Prosthogonimus*

CLASS CESTODA. Tapeworms. *Taenia, Dipylidium, Mesocestoides*

PHYLUM GNATHOSTOMULIDA (100). *Gnathostomula, Haplognathia*

PHYLUM NEMERTINA (or Rhynchocoela). Proboscis worms (650)

CLASS ANOPLA. *Tubulanus, Cerebratulus*

CLASS ENOPLA. *Amphiporus, Prostoma, Malacobdella*

PHYLUM ACANTHOCEPHALA. Spiny-headed worms (500). *Echinorhynchus, Gigantorhynchus*

PHYLUM ASCHELMINTHES

CLASS ROTIFERA. Rotifers (1,700). *Asplanchna, Hydatina, Rotaria*

CLASS GASTROTRICHA (200). *Chaetonotus, Macrodasys*

CLASS KINORHYNCHA (or Echinodera) (100). *Echinoderes, Semnoderes*

CLASS NEMATODA. Round worms (12,000). *Ascaris, Trichinella, Necator, Enterobius, Ancylostoma, Heterodera*

CLASS NEMATOMORPHA. Horsehair worms (230). *Gordius, Paragordius, Nectonema*

PHYLUM ENTOPROCTA (60). *Urnatella, Loxosoma, Pedicellina*

PHYLUM PRIAPULIDA (8). *Priapulus, Halicryptus*

PHYLUM ECTOPROCTA (or Bryozoa). Bryozoans, moss animals (4,000)

CLASS GYMNOLAEMATA. *Paludicella, Bugula*

CLASS PHYLACTOLAEMATA. *Plumatella, Pectinatella*

PHYLUM PHORONIDA (15). *Phoronis, Phoronopsis*

PHYLUM BRACHIOPODA. Lamp shells (300)

CLASS INARTICULATA. *Lingula, Glottidia, Discina*

CLASS ARTICULATA. *Magellania, Neothyris, Terebratula*

PHYLUM MOLLUSCA. Molluscs

CLASS AMPHINEURA

SUBCLASS APLACOPHORA. Solenogasters (250). *Chaetoderma, Neomenia, Proneomenia*

SUBCLASS POLYPLACOPHORA. Chitons (600). *Chaetopleura, Ischnochiton, Lepidochiton, Amicula*

CLASS MONOPLACOPHORA (6). *Neopilina*

CLASS GASTROPODA. Snails and their allies (univalve molluscs) (40,000). *Helix, Busycon, Crepidula, Haliotis, Littorina, Doris, Limax*

CLASS SCAPHOPODA. Tusk shells (350). *Dentalium, Cadulus*

CLASS PELECYPODA. Bivalve molluscs (7,500). *Mytilus, Ostrea, Pecten, Mercenaria, Teredo, Tagelus, Unio, Anodonta*

CLASS CEPHALOPODA. Squids, octopuses, etc. (600). *Loligo, Octopus, Nautilus*

PHYLUM POGONOPHORA. Beard worms (100). *Siboglinum, Lamellisabella, Oligobrachia, Polybrachia*

PHYLUM SIPUNCULIDA (250). *Sipunculus, Phascolosoma, Dendrostomum*

PHYLUM ECHIUROIDA (80)

CLASS ECHIURIDA. *Echiurus, Urechis, Ikeda*

CLASS SACTOSOMATIDA. *Sactosoma*

PHYLUM ANNELIDA. Segmented worms

CLASS POLYCHAETA (including Archiannelida). Sandworms, tubeworms, etc. (5,400). *Nereis, Chaetopterus, Aphrodite, Diopatra, Arenicola, Hydroides, Sabella*

CLASS OLIGOCHAETA. Earthworms and many freshwater annelids (3,100). *Tubifex, Enchytraeus, Lumbricus, Dendrobaena*

CLASS HIRUDINEA. Leeches (300). *Trachelobdella, Hirudo, Macrobdella, Haemadipsa*

PHYLUM ONYCHOPHORA (65). *Peripatus. Peripatopsis*

PHYLUM TARDIGRADA. Water bears (300). *Echiniscus, Macrobiotus*

PHYLUM PENTASTOMIDA. Tongue worms (60). *Cephalobaena, Linguatula*

PHYLUM ARTHROPODA

Subphylum Trilobita. No living representatives

Subphylum Chelicerata

CLASS EURYPTERIDA. No living representatives

CLASS XIPHOSURA. Horseshoe crabs (4). *Limulus*

CLASS ARACHNIDA. Spiders, ticks, mites, scorpions, whipscorpions, daddy longlegs, etc. (55,000). *Archaearanea, Latrodectus, Argiope, Centruroides, Chelifer, Mastigoproctus, Phalangium, Ixodes*

CLASS PYCNOGONIDA. Sea spiders (500). *Nymphon, Ascorhynchus*

Subphylum Mandibulata

CLASS CRUSTACEA (26,000). *Homarus, Cancer, Daphnia, Artemia, Cyclops, Balanus, Porcellio*

CLASS CHILOPODA. Centipeds (3,000). *Scolopendra, Lithobius, Scutigera*

CLASS DIPLOPODA. Millipeds (8,000). *Narceus, Apheloria, Polydesmus, Julus, Glomeris*

CLASS PAUROPODA (300). *Pauropus*

CLASS SYMPHYLA (130). *Scutigerella*

CLASS INSECTA. Insects (900,000)

ORDER COLLEMBOLA. Springtails. *Isotoma, Achorutes, Neosminthurus, Sminthurus*

ORDER PROTURA. *Acerentulus, Eosentomon*

ORDER DIPLURA. *Campodea, Japyx*

ORDER THYSANURA. Bristletails, silverfish, firebrats. *Machilis, Lepisma, Thermobia*

ORDER EPHEMERIDA. Mayflies. *Hexagenia, Callibaetis, Ephemerella*

ORDER ODONATA. Dragonflies, damselflies. *Archilestes, Lestes, Aeshna, Gomphus*

ORDER ORTHOPTERA. Grasshoppers, crickets, walking sticks, mantids, cockroaches, etc. *Schistocerca, Romalea, Nemobiius, Megaphasma, Mantis, Blatta, Periplaneta*

ORDER ISOPTERA. Termites. *Reticulitermes, Kalotermes, Zootermopsis, Nasutitermes*

ORDER DERMAPTERA. Earwigs. *Labia, Forficula, Prolabia*

ORDER EMBIARIA (or Embiidina or Embioptera). *Oligotoma, Anisembia, Gynembia*

ORDER PLECOPTERA. Stoneflies. *Isoperla, Taeniopteryx, Capnia, Perla*

ORDER ZORAPTERA. *Zorotypus*

ORDER CORRODENTIA. Book lice. *Ectopsocus, Liposcelis, Trogium*

ORDER MALLOPHAGA. Chewing lice. *Cuclotogaster, Menacanthus, Menopon, Trichodectes*

ORDER ANOPLURA. Sucking lice. *Pediculus, Phthirius, Haematopinus*

ORDER THYSANOPTERA. Thrips. *Heliothrips, Frankliniella, Hercothrips*

ORDER HEMIPTERA. True bugs. *Belostoma, Lygaeus, Notonecta, Cimex, Lygus, Oncopeltus*

ORDER HOMOPTERA. Cicadas, aphids, leafhoppers, scale insects, etc. *Magicicada, Circulifer, Psylla, Aphis, Saissetia*

ORDER NEUROPTERA. Dobsonflies, alderflies, lacewings, mantispids, snakeflies, etc. *Corydalus, Hemerobius, Chrysopa, Mantispa, Agulla*

ORDER COLEOPTERA. Beetles, weevils. *Copris, Phyllophaga, Harpalus, Scolytus, Melanotus, Cicindela, Dermestes, Photinus, Coccinella; Tenebrio, Anthonomus, Conotrachelus*

ORDER HYMENOPTERA. Wasps, bees, ants, sawflies. *Cimbex, Vespa, Glypta, Scolia, Bembix, Formica, Bombus, Apis*

ORDER MECOPTERA. Scorpionflies. *Panorpa, Boreus, Bittacus*

ORDER SIPHONAPTERA. Fleas. *Pulex, Nosopsyllus, Xenopsylla, Ctenocephalides*

ORDER DIPTERA. True flies, mosquitoes. *Aedes, Asilus, Sarcophaga, Anthomyia, Musca, Chironomus, Tabanus, Tipula, Drosophila*

ORDER TRICHOPTERA. Caddisflies. *Limnephilus, Rhyacophila, Hydropsyche*

ORDER LEPIDOPTERA. Moths, butterflies. *Tinea, Pyrausta, Malacosoma, Sphinx, Samia, Bombyx, Heliothis, Papilio, Lycaena*

SECTION DEUTEROSTOMIA

PHYLUM CHAETOGNATHA. Arrow worms (60). *Sagitta, Spadella*

PHYLUM ECHINODERMATA

CLASS CRINOIDEA. Crinoids, sea lilies (630). *Antedon, Ptilocrinus, Comactinia*

CLASS ASTEROIDEA. Sea stars (1,600). *Asterias, Ctenodiscus, Luidia, Oreaster*

CLASS OPHIUROIDEA. Brittle stars, serpent stars, basket stars, etc. (2,000). *Asteronyx, Amphioplus, Ophiothrix, Ophioderma, Ophiura*

CLASS ECHINOIDEA. Sea urchins, sand dollars, heart urchins (860). *Cidaris, Arbacia, Strongylocentrotus, Echinanthus, Echinarachnius, Moira*

CLASS HOLOTHUROIDEA. Sea cucumbers (900). *Cucumaria, Thyone, Caudina, Synapta*

PHYLUM HEMICHORDATA (90)

CLASS ENTEROPNEUSTA. Acorn worms. *Saccoglossus, Balanoglossus, Glossobalanus*

CLASS PTEROBRANCHIA. *Rhabdopleura, Cephalodiscus*

PHYLUM CHORDATA. Chordates

Subphylum Urochordata (or Tunicata). Tunicates (2,000)

CLASS ASCIDIACEA. Ascidians or sea squirts. *Ciona, Clavelina, Molgula, Perophora*

CLASS THALIACEA. *Pyrosoma, Salpa, Doliolum*

CLASS LARVACEA. *Appendicularia, Oikopleura, Fritillaria*

Subphylum Cephalochordata. Lancelets, amphioxus (30). *Branchiostoma, Asymmetron*

Subphylum Vertebrata. Vertebrates

CLASS AGNATHA. Jawless fishes (50). *Cephalaspis,* Pteraspis,* Petromyzon, Entosphenus, Myxine, Eptatretus*

CLASS PLACODERMI. No living representatives

CLASS CHONDRICHTHYES. Cartilaginous fishes (625). *Squalus, Hyporion, Raja, Chimaera*

CLASS OSTEICHTHYES. Bony fishes (30,000)

SUBCLASS SARCOPTERYGII

ORDER CROSSOPTERYGII (or Coelacanthiformes). Lobe-fins. *Latimeria*

ORDER DIPNOI (or Dipteriformes). Lungfishes. *Neoceratodus, Protopterus, Lepidosiren*

SUBCLASS BRACHIOPTERYGII. Bichirs. *Polypterus*

SUBCLASS ACTINOPTERYGII. Higher bony fishes. *Amia, Cyprinus, Gadus, Perca, Salmo*

CLASS AMPHIBIA (2,600)

ORDER ANURA. Frogs and toads. *Rana, Hyla, Bufo*

ORDER URODELA. Salamanders. *Necturus, Triturus, Plethodon, Ambystoma*

ORDER APODA. *Ichthyophis, Typhlonectes*

CLASS REPTILIA (6,500)

ORDER CHELONIA. Turtles. *Chelydra, Kinosternon, Clemmys, Terrapene*

ORDER RHYNCHOCEPHALIA. Tuatara. *Sphenodon*

ORDER CROCODYLIA. Crocodiles and alligators. *Crocodylus, Alligator*

ORDER SQUAMATA. Snakes and lizards. *Iguana, Anolis, Sceloporus, Phrynosoma, Natrix, Elaphe, Coluber, Thamnophis, Crotalus*

CLASS AVES. Birds (8,600). *Anas, Larus, Columba, Gallus, Turdus, Dendroica, Sturnus, Passer, Melospiza*

*Extinct.

CLASS MAMMALIA. Mammals (4,100)

SUBCLASS PROTOTHERIA

ORDER MONOTREMATA. Egg-laying mammals. *Ornithorhynchus, Tachyglossus*

SUBCLASS THERIA. Marsupial and placental mammals

ORDER MARSUPIALIA. Marsupials. *Didelphis, Sarcophilus, Notoryctes, Macropus*

ORDER INSECTIVORA. Insectivores (moles, shrews, etc.). *Scalopus, Sorex, Erinaceus*

ORDER DERMOPTERA. Flying lemurs. *Galeopithecus*

ORDER CHIROPTERA. Bats. *Myotis, Eptesicus, Desmodus*

ORDER PRIMATES. Lemurs, monkeys, apes, humans. *Lemur, Tarsius, Cebus, Macacus, Cynocephalus, Pongo, Pan, Homo*

ORDER EDENTATA. Sloths, anteaters, armadillos. *Bradypus, Myrmecophagus, Dasypus*

ORDER PHOLIDOTA. Pangolin. *Manis*

ORDER LAGOMORPHA. Rabbits, hares, pikas. *Ochotona, Lepus, Sylvilagus, Oryctolagus*

ORDER RODENTIA. Rodents. *Sciurus, Marmota, Dipodomys, Microtus, Peromyscus, Rattus, Mus, Erethizon, Castor*

ORDER CETACEA. Whales, dolphins, porpoises. *Delphinus, Phocaena, Monodon, Balaena*

ORDER CARNIVORA. Carnivores. *Canis, Procyon, Ursus, Mustela, Mephitis, Felis, Hyaena, Eumetopias*

ORDER TUBULIDENTATA. Aardvark. *Orycteropus*

ORDER PROBOSCIDEA. Elephants. *Elephas, Loxodonta*

ORDER HYRACOIDEA. Coneys. *Procavia*

ORDER SIRENIA. Manatees. *Trichechus, Halicore*

ORDER PERISSODACTYLA. Odd-toed ungulates. *Equus, Tapirella, Tapirus, Rhinoceros*

ORDER ARTIODACTYLA. Even-toed ungulates. *Pecari, Sus, Hippopotamus, Camelus, Cervus, Odocoileus, Giraffa, Bison, Ovis, Bos*

GLOSSARY

The Glossary gives brief definitions of the most important recurrent terms used in the text, excluding taxonomic designations. For fuller definitions, consult the index, where italicized page numbers refer you to explanations of key terms in context.

Of the basic units of measurement, some are tabulated on p. A2, others have their own alphabetical entries.

Interalphabetized with the vocabulary are the main prefixes and combining forms used in biology. You will notice that, while they are generally of Greek or Latin origin, many of them have acquired a new meaning in biology (examples: *blasto-, -cyte, caryo-, -plasm*). Familiarity with these forms will make it easier for you to learn and remember the numerous terms in which they are incorporated.

TABLE 1 *Standard prefixes of the metric system*

kilo- (k)	1,000	10^3
deci- (d)	0.1	10^{-1}
centi- (c)	0.01	10^{-2}
milli- (m)	0.001	10^{-3}
micro- (μ)	0.000001	10^{-6}
nano- (n)	0.000000001	10^{-9}

TABLE 2 *Common units of length, weight, and liquid capacity*

kilometer (km)	1,000 m	0.62137 mile
meter (m)		39.37 inches
centimeter (cm)	0.01 m	0.39 inch
millimeter (mm)	0.001 m	0.039 inch
micrometer* (μm)	10^{-6} m	
nanometer (nm)	10^{-9} m	
angstrom† (Å)	10^{-10} m	
kilogram (kg)	1,000 g	2.2 pounds
gram (g)		0.035 ounce
milligram (mg)	0.001 g	
microgram (μg)	10^{-6} g	
liter (l)	1,000 cm³	1.057 quarts
milliliter (ml)	0.001 l	

*Formerly called micron.
†No longer used; nanometer used instead.

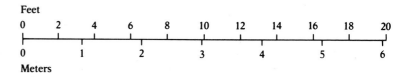

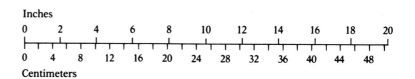

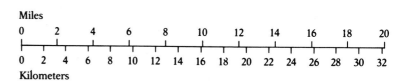

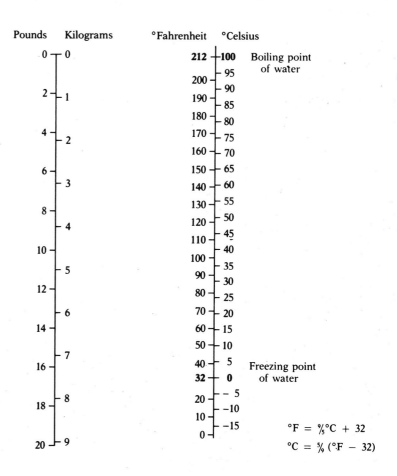

$$°F = \tfrac{9}{5}°C + 32$$

$$°C = \tfrac{5}{9}\,(°F - 32)$$

a- Without, lacking.

ab- Away from, off.

abdomen [L belly]   In mammals, the portion of the trunk posterior to the thorax, containing most of the viscera except heart and lungs. In other animals, the posterior portion of the body.

absolute zero   The temperature ($-273°C$) at which all thermal agitation ceases. The lowest possible temperature.

acellular   Not constructed on a cellular basis.

acid [L *acidus* sour]   A substance that increases the concentration of hydrogen ions when dissolved in water, that has a pH lower than 7.

ACTH   *See* adrenocorticotrophic hormone.

action potential   *See* potential.

active site   In an enzyme, the portion of the molecule that reacts with a substrate molecule.

active transport   Movement of a substance across a membrane by a process requiring expenditure of energy by the cell.

ad- Next to, at, toward.

adaptation   Any genetically controlled characteristic that increases an organism's fitness, usually by helping the organism to survive and reproduce in the environment it inhabits.

adenosine diphosphate (ADP)   A doubly phosphorylated organic compound that can be further phosphorylated to form ATP.

adenosine triphosphate (ATP)   A triply phosphorylated organic compound that functions as "energy currency" for organisms.

adipose [L *adeps* fat]   Fatty.

ADP   *See* adenosine diphosphate.

adrenal [L *renes* kidneys]   An endocrine gland of vertebrates located near the kidneys.

adrenalin   A hormone produced by the adrenal medulla that stimulates "fight-or-flight" reactions.

adrenocorticotrophic hormone (ACTH)   A hormone produced by the pituitary that stimulates the adrenal cortex.

adsorb [L *sorbēre* to suck up]   Hold on a surface.

advanced   New, unlike the ancestral condition.

aerobic [L *aer* air]   With oxygen.

agonistic [Gk *agōnistēs* combatant]   Having to do with attack, escape, or fear.

alcohol   Any of a class of organic compounds in which one or more OH groups are attached to a carbon backbone.

alkaline   Having a pH of more than 7. *See* base.

all-, allo- [Gk *allos* other]   Other, different.

allele   Any of several alternative gene forms at a given chromosomal locus.

allopatric [L *patria* homeland]   Having different ranges.

all-or-none   The property of responding maximally or not at all.

allosteric   Of an enzyme: one that can exist in two or more conformations.   *Allosteric control:* control of the activity of an allosteric enzyme by determination of the particular conformation it will assume.

alveolus [L little hollow]   A small cavity, especially one of the microscopic cavities that are the functional units of lungs.

amino acid   An organic acid carrying an amino group ($-NH_2$); the building-block compound of proteins.

amnion [Gk caul]   An extraembryonic membrane that forms a fluid-filled sac containing the embryo in reptiles, birds, and mammals.

amoeboid [Gk *amoibē* change]   Amoebalike in the tendency to change shape by protoplasmic flow.

amylase [L *amylum* starch]   A starch-digesting enzyme.

an- Without.

anabolism [Gk *ana-* upward; *metabolē* change]   The biosynthetic building-up aspects of metabolism.

anaerobic [L *aer* air]   Without oxygen.

analogous   Of characters in different organisms: similar in function and often in superficial structure but of different evolutionary origins.

angio-, -angium [Gk *angeion* vessel]   Container, receptacle.

anion   A negatively charged ion.

anterior   Toward the front end.

antheridium [Gk *anthos* flower]   Male reproductive organ of a plant; produces sperm cells.

antibody   A protein, produced by the immune system, that destroys or inactivates a particular antigen.

antigen   A foreign substance, usually a protein or polysaccharide, that stimulates an organism's immune system to produce antibodies against it. In rare cases (e.g. the antibodies for the A and B blood antigens), antibodies may be produced even without stimulation by the antigen.

anus [L ring]   Opening at the posterior end of the digestive tract, through which indigestible wastes are expelled.

aorta   The main artery of the systemic circulation.

apical   At, toward, or near the apex, or tip, of a structure such as a plant shoot.

apo- Away from.

apoplast   The network of cell walls and intercellular spaces within a plant body; permits extensive extracellular movement of water within the plant.

aposematic [Gk *sēma* sign]   Serving as a warning, with reference particularly to colors and structures that signal possession of defensive devices.

arch- [Gk *archein* to begin]   Primitive, original.

archegonium [Gk *archegonos* the first of a race]   Female reproductive organ of a higher plant; produces egg cells.

archenteron [Gk *enteron* intestine]   The cavity in an early embryo that becomes the digestive cavity.

arteriole   A small artery.

artery   A blood vessel that carries blood away from the heart.

artifact   A by-product of scientific manipulation rather than an inherent part of the thing observed.

ascus [Gk *askos* bag]   The elongate spore sac of a fungus of the Ascomycota group.

asexual   Without sex.

atmosphere (atm) (unit of pressure)   The normal pressure of air at sea level: 101,325 newtons per square meter (approx. 14.7 pounds per square inch).

atom [Gk *atomos* indivisible]   The smallest unit of an element, not divisible by ordinary chemical means.

atomic mass unit (amu)   *See* dalton.

atomic weight   The average weight of an atom of an element relative to C^{12}, an isotope of carbon with six neutrons in the nucleus. The atomic weight of C^{12} has arbitrarily been fixed as 12.

ATP   *See* adenosine triphosphate.

auto-   Self, same.

autonomic nervous system   A portion of the vertebrate nervous system, comprising motor neurons that innervate internal organs and are not normally under direct voluntary control.

autosome [Gk *sōma* body]   Any chromosome other than a sex chromosome.

autotrophic [Gk *trophē* food]   Capable of manufacturing organic nutrients from inorganic raw materials.

auxin [Gk *auxein* to grow]   Any of a class of plant hormones that promote cell elongation and can diffuse into a decapitated plant from an agar block, causing the plant to bend in the dark (Went test).

axon [Gk *axōn* axis]   A fiber of a nerve cell that conducts impulses away from the cell body and can release transmitter substance.

bacteriophage [Gk *phagein* to eat]   A virus that attacks bacteria, *abbrev.* phage.

basal   At, near, or toward the base (i.e. the point of attachment) of a structure such as a limb.

base (or alkali)   A substance that increases the concentration of hydroxyl ions when dissolved in water, that has a pH higher than 7.

basidium   The spore-bearing structure of Basidiomycota (club fungi).

bi-   Two.

bilateral symmetry   The property of having two similar sides, with definite upper and lower surfaces and definite anterior and posterior ends.

binary fission   Reproduction by the division of a cell into two essentially equal parts by a nonmitotic process.

bio- [Gk *bios* life]   Life, living.

biogenesis [Gk *genesis* source]   Origin of living organisms from other living organisms.

biological magnification   Increasing concentration of relatively stable chemicals as they are passed up a food chain from initial consumers to top predators.

biomass   The total weight of all the organisms, or of a designated group of organisms, in a given area.

biome   A major climax plant formation.

biotic   Pertaining to life.

blasto- [Gk *blastos* bud]   Embryo.

bastocoel [Gk *koilos* hollow]   The cavity of a blastula.

blastopore [Gk *poros* passage]   The opening from the cavity of the archenteron to the exterior in a gastrula.

blastula   An early embryonic stage in animals, preceding the delimitation of the three principal tissue layers; frequently spherical and hollow.

buffer   A substance that binds H^+ ions when their concentration rises and releases them when their concentration falls, thereby minimizing fluctuations in the pH of a solution.

caecum [L *caecus* blind]   A blind diverticulum of the digestive tract.

calorie [L *calor* heat]   The quantity of energy, in the form of heat, required to raise the temperature of one gram of pure water one degree from 14.5 to 15.5°C. The nutritionists' Calorie (capitalized) is 1,000 calories, or one kilocalorie.

cambium [L *cambiare* to exchange]   The principal lateral meristem of vascular plants; gives rise to most secondary tissue.

cAMP   *See* cyclic adenosine monophosphate.

capillarity [L *capillus* hair]   The tendency of aqueous liquids to rise in narrow tubes with hydrophilic surfaces.

capillary [L *capillus*]   A tiny blood vessel with walls one cell thick, across which exchange of materials between the blood and the tissues takes place; receives blood from arteries and carries it to veins. Also, a similar vessel of the lymphatic system.

carbohydrate   Any of a class of organic compounds composed of carbon, hydrogen, and oxygen in a ratio of about two hydrogens and one oxygen for each carbon. Examples: sugar, starch, cellulose.

carboxyl group   The —COOH group characteristic of organic acids.

cardiac [Gk *kardia* heart]   Pertaining to the heart.

carnivore [L *carnis* of flesh; *vorare* to devour]   An organism that feeds on animals.

carotenoid [L *carota* carrot]   Any of a group of red, orange, and yellow accessory pigments of plants, found in plastids.

carrying capacity   The maximum population that a given environment can support indefinitely.

cartilage A specialized type of dense fibrous connective tissue with a rubbery intercellular matrix.

caryo- [Gk *karyon* kernel] Nucleus.

Casparian strip A lignified and waterproofed thickening in the radial and end walls of endodermal cells of plants.

cata- Down.

catabolism [Gk *katabolē* a throwing down] The degradational breaking-down aspects of metabolism.

catalysis [Gk *katalyein* to dissolve] Acceleration of a chemical reaction by a substance that is not itself permanently changed by the reaction.

catalyst A substance that produces catalysis.

cation A positively charged ion.

caudal [L *cauda* tail] Pertaining to the tail.

cell cycle The cycle of cellular events from one mitosis through the next. Four stages are recognized, of which the last—distribution of genetic material to the two daughter nuclei—is mitosis proper.

cell sap *See* sap.

cellulose [L *cellula* cell] A complex polysaccharide that is a major constituent of most plant cell walls.

centi- [L *centum* hundred] One hundredth.

central nervous system A portion of the nervous system that contains interneurons and exerts some control over the rest of the nervous system. In vertebrates, the brain and spinal cord.

centri- [L *centrum* center] Center.

centrifugation [L *fugere* to flee] The spinning of a mixture at very high speeds to separate substances of different densities.

centriole A cylindrical cytoplasmic organelle located just outside the nucleus of animal cells and the cells of some lower plants; associated with the spindle during mitosis and meiosis.

centromere [Gk *meros* part] A special region on a chromosome that attaches to a spindle microtubule during mitosis or meiosis.

cephalization [Gk *kephalē* head] Localization of neural coordinating centers and sense organs at the anterior end of the body.

cerebellum [L small brain] A part of the hindbrain of vertebrates that controls muscular coordination.

cerebrum [L brain] Part of the forebrain of vertebrates, the chief coordination center of the nervous system.

character Any structure, functional attribute, behavioral trait, or other characteristic of an organism.

character displacement The rapid divergent evolution in sympatric species of characters that minimize competition and/or hybridization between them.

chemosynthesis Autotrophic synthesis of organic materials, energy for which is derived from inorganic molecules.

chitin [Gk *chitōn* tunic] Polysaccharide that forms part of the hard exoskeleton of insects, crustaceans, and other invertebrates; also occurs in the cell walls of fungi.

chlorophyll [Gk *chlōros* greenish yellow; *phyllon* leaf] The green pigment of plants necessary for photosynthesis.

chloroplast A plastid containing chlorophyll.

chrom-, -chrome [Gk *chrōma* color] Colored; pigment.

chromatid A single chromosomal strand.

chromatography Process of separating substances by adsorption on media for which they have different affinities.

chromosome [Gk *sōma* body] A filamentous structure in the cell nucleus (or nucleoid) along which the genes are located.

cilium [L eyelid] A short hairlike locomotory organelle on the surface of a cell (*pl.* cilia).

cisterna [L cistern] A cavity, sac, or other enclosed space serving as a reservoir.

cistron The genetic unit of function; synonymous with the gene as biochemically defined.

cleavage Division of a zygote or of the cells of an early embryo.

climax (ecological) A relatively stable stage reached in some ecological successions.

cline [Gk *klinein* to lean] Gradual variation, correlated with geography, in a character of a species.

cloaca [L sewer] Common chamber that receives materials from the digestive, excretory, and reproductive systems.

clone [Gk *klōn* twig] A group of cells or organisms derived asexually from a single ancestor and hence genetically identical.

co- With, together.

codon The unit of genetic coding, three nucleotides long.

coel-, -coel [Gk *koilos* hollow] Hollow, cavity; chamber.

coelom A body cavity bounded entirely by mesoderm.

coenocytic [Gk *koinos* common] Having more than one nucleus in a single mass of cytoplasm.

coenzyme A nonproteinaceous organic molecule that plays an accessory role, but a necessary one, in the catalytic action of an enzyme.

coleoptile [Gk *koleon* sheath; *ptilon* feather] A sheath around the young shoot of grasses.

collenchyma [Gk *kolla* glue] A supportive tissue in plants in which the cells usually have thickenings at the angles of the walls.

colloid [Gk *kolla*] A stable suspension of particles that, though larger than in a true solution, do not settle out.

colon The large intestine.

com- Together.

commensalism [L *mensa* table] A symbiosis in which one party is benefited and the other party receives neither benefit nor harm.

community In ecology, a unit composed of all the populations living in a given area.

competition   In ecology, utilization by two or more individuals, or by two or more populations, of the same limited resource; an interaction where both parties are harmed.

condensation reaction   A reaction joining two compounds with resultant formation of water.

conditioning   The associating, as a result of reinforcement, of a response with a stimulus with which it was not previously associated.

conformation (of a protein) [L *conformatio* symmetrical forming]   The three-dimensional pattern according to which the polypeptide chains of a protein coil (secondary structure), fold (tertiary structure), and—if there is more than one chain—fit together (quaternary structure).

conjugation [L *jugare* to join, marry]   Process of genetic recombination between two organisms (e.g. bacteria, algae) through a cytoplasmic bridge between them.

connective tissue   A type of animal tissue whose cells are embedded in an extensive intercellular matrix; connects, supports, or surrounds other tissues and organs.

contractile vacuole   An excretory and/or osmoregulatory vacuole in some cells, which, by contracting, ejects fluids from the cell.

cooperativity   The phenomenon of enhanced reactivity of the remaining binding sites of a protein as a result of the binding of substrate at one site.

cork [L *cortex* bark]   A waterproofed tissue, derived from the cork cambium, that forms at the outer surfaces of the older stems and roots of woody plants; the outer bark or periderm.

corpus luteum [L yellow body]   A yellowish structure in the ovary, formed from the follicle after ovulation, that secretes estrogen and progesterone (*pl.* corpora lutea).

cortex [L bark]   In plants, tissue between the epidermis and the vascular cylinder of stems and roots. In animals, the outer barklike tissue of some organs, as *cerebral cortex, adrenal cortex*, etc.

cotyledon [Gk *kotylē* cup]   A "seed leaf," a food-digesting and -storing part of a plant embryo.

covalent bond   A chemical bond resulting from the sharing of a pair of electrons.

crossing-over   Exchange of parts between two homologous chromosomes.

cross section   *See* section.

cryptic [Gk *kryptos* hidden]   Concealing.

cuticle [L *cutis* skin]   A waxy layer on the outer surface of leaves, insects, etc.

cyclic adenosine monophosphate (cyclic AMP or **cAMP)**   Compound, synthesized in living cells from ATP, that functions as an intracellular mediator of hormonal action; also plays a part in neural transmission and some other kinds of cellular control systems.

cyst [Gk *kystis* bladder, bag]   (1) A saclike abnormal growth.   (2) Capsule that certain organisms secrete around themselves and that protects them during resting stages.

-cyte, cyto- [Gk *kytos* container]   Cell.

cytochrome   Any of a group of iron-containing pigments important in the electron transport of oxidative phosphorylation and photophosphorylation.

cytokinesis [Gk *kinēsis* motion]   Division of the cytoplasm of a cell.

cytoplasm   All of a cell except the nucleus.

cytosol   The relatively fluid, less structured part of the cytoplasm of a cell, excluding organelles and membranous structures.

dalton   A unit of mass equal to one twelfth the atomic weight of C^{12}, or 1.66024×10^{-24} gram. Formerly called atomic mass unit (amu).

deamination   Removal of an amino group.

deciduous [L *decidere* to fall off]   Shedding leaves each year.

dehydration reaction   A condensation reaction.

deme [Gk *dēmos* population]   A local unit of population of any one species.

dendr-, dendro- [Gk *dendron* tree]   Tree; branching.

dendrite   A short unsheathed fiber of a nerve cell—often spiny, usually branched and tapering—that receives many synapses and leads impulses toward the cell body.

deoxyribonucleic acid (DNA)   A nucleic acid found especially in the cell nucleus—the genetic material.

-derm [Gk *derma* skin]   Skin, covering; tissue layer.

di-   Two.

dicot   A member of a subclass of the angiosperms, or flowering plants, distinguished mainly by the presence of two cotyledons in the embryo; *cf.* monocot.

differentiation   The process of developmental change from an immature to a mature form, especially in a cell.

diffusion   The movement of dissolved or suspended particles from one place to another as a result of their heat energy (thermal agitation).

digestion   Hydrolysis of complex nutrient compounds into their building-block units.

diploid [Gk *diploos* double]   Having two of each type of chromosome.

disaccharide   A double sugar, i.e. one composed of two simple sugars.

distal [L *distare* to stand apart]   Situated away from some reference point (usually the main part of the body).

diverticulum [L *devertere* to turn aside]   A blind sac branching off a cavity or canal.

DNA   *See* deoxyribonucleic acid.

dominant   (1) Of an allele: exerting its full phenotypic effect despite the presence of another allele of the same gene, whose phenotypic expression it blocks. *Dominant phenotype, dominant character:* one caused by a dominant allele.   (2) Of an individual: occupying a high position in the social hierarchy.

dormancy [L *dormire* to sleep]   The state of being inactive, quiescent. In plants, particularly seeds and buds, a period in which growth is arrested until environmental conditions become more favorable.

dorsal [L *dorsum* back]   Pertaining to the back.

duodenum [From a Latin phrase meaning 12 *(duodecim)* finger's-breadths long]   The first portion of the small intestine of vertebrates, into which ducts from the pancreas and gallbladder empty.

ecosystem [Gk *oikos* habitation]   The sum total of physical features and organisms occurring in a given area.

ecto-   Outside, external.

ectoderm   The outermost tissue layer of an animal embryo. Also, tissue derived from the embryonic ectoderm.

effector   The part of an organism that produces a response, e.g. muscle, cilium, flagellum.

egg   An egg cell or female gamete. Also a structure in which embryonic development takes place, especially in birds and reptiles; consists of an egg cell, various membranes, and often a shell.

electron   A negatively charged primary subatomic particle.

electronic charge unit   The charge of one electron, or 1.6021×10^{-19} coulomb.

elimination (or defecation)   The release of unabsorbed wastes from the digestive tract. *Cf.* excretion.

embryo   A plant or animal in an early stage of development; generally still contained within the seed, egg, or uterus.

emulsion [L *emulsus* milked out]   Suspension, usually as fine droplets, of one liquid in another.

-enchyma [Gk *parenchein* to pour in beside]   Tissue.

end-, endo-   Within, inside; requiring.

endergonic [Gk *ergon* work]   Energy-absorbing.

endocrine [Gk *krinein* to separate]   Pertaining to ductless glands that produce hormones.

endoderm   The innermost tissue layer of an animal embryo.

endodermis   A plant tissue, especially prominent in roots, that surrounds the vascular cylinder; all endodermal cells have Casparian strips.

endoplasmic reticulum [L *reticulum* network]   A system of membrane-bounded channels in the cytoplasm.

endoskeleton   An internal skeleton.

endosperm [Gk *sperma* seed]   A nutritive material in seeds.

entropy   Measure of the disorder of a system.

enzyme [Gk *zymē* leaven]   A protein that acts as a catalyst.

epi-   Upon, outer.

epicotyl   The portion of the axis of a plant embryo above the point of attachment of the cotyledons; forms most of the shoot.

epidermis [Gk *derma* skin]   The outermost portion of the skin or body wall of an animal.

episome [Gk *sōma* body]   Genetic element at times free in the cytoplasm, at other times integrated into a chromosome.

epithelium   An animal tissue that forms the covering or lining of all free body surfaces, both external and internal.

erythrocyte [Gk *erythros* red]   A red blood cell, i.e. a blood cell containing hemoglobin.

esophagus [Gk *phagein* to eat]   An anterior part of the digestive tract; in mammals it leads from the pharynx to the stomach.

estrogen [L *oestrus* frenzy]   Any of a group of vertebrate female sex hormones.

estrous cycles [L *oestrus*]   In female mammals, the higher primates excepted, a recurrent series of physiological and behavioral changes connected with reproduction.

eu- [Gk *eus* good]   Most typical, true.

eucaryotic cell   A cell containing a distinct membrane-bounded nucleus, characteristic of all organisms except bacteria, blue-green algae, and prochlorophytes.

evaginated [L *vagina* sheath]   Folded or protruded outward.

eversible [L *evertere* to turn out]   Capable of being turned inside out.

evolution [L *evolutio* unrolling]   Change in the genetic makeup of a population with time.

ex-, exo-   Out of, outside; producing.

excretion   Release of metabolic wastes and excess water. *Cf.* elimination.

exergonic [Gk *ergon* work]   Energy-releasing.

exoskeleton   An external skeleton.

extrinsic   External to, not a basic part of; as in *extrinsic isolating mechanism.*

fauna   The animals of a given area or period.

feces [L *faeces* dregs]   Indigestible wastes discharged from the digestive tract.

feedback   The process by which a control mechanism is regulated through the very effects it brings about.

fermentation   Anaerobic production of alcohol, lactic acid, or some similar compound from carbohydrate via the glycolytic pathway.

fertilization   Fusion of nuclei of egg and sperm.

fetus [L *fetus* pregnant]   An embryo in its later development, still in the egg or uterus.

fitness   The probable genetic contribution of an individual (or allele or genotype) to succeeding generations. *Inclusive fitness:* the sum of an individual's personal fitness plus the fitness of that individual's relatives devalued in proportion to their genetic distance from the individual.

fixation   (1) Conversion of a substance into a biologically more usable form, as the conversion of CO_2 into carbohydrate by photosynthetic plants or the incorporation of N_2 into more complex molecules by nitrogen-fixing bacteria.   (2) Process of treating living tissue for microscopic examination.

flagellum [L whip] A long hairlike locomotory organelle on the surface of a cell.

flora The plants of a given area or period.

follicle [L *follis* bag] A jacket of cells around an egg cell in an ovary.

follicle-stimulating homone (FSH) A gonadotrophic hormone of the anterior pituitary that stimulates growth of follicles in the ovaries of females and function of the seminiferous tubules in males.

food chain Sequence of organisms, including producers, consumers, and decomposers, through which energy and materials may move in a community.

foot-candle Unit of illumination; the illumination of a surface produced by one standard candle at a distance of one foot. *See also* lambert.

free energy Usable energy in a chemical system; energy available for producing change.

fruit A mature ovary or cluster of ovaries (sometimes with additional structures associated with the ovary).

fruiting body A spore-bearing structure (e.g. the aboveground portion of a mushroom).

FSH *See* follicle-stimulating hormone.

gamete [Gk *gametē(s)* wife, husband] A sexual reproductive cell that must usually fuse with another such cell before development begins; an egg or sperm.

gametophyte [Gk *phyton* plant] A haploid plant that can produce gametes.

ganglion [Gk tumor] A structure containing a group of cell bodies of neurons (*pl.* ganglia).

gastr-, gastro- [Gk *gastēr* belly] Stomach; ventral; resembling the stomach.

gastrovascular cavity An often branched digestive cavity, with only one opening to the outside, that conveys nutrients throughout the body; found only in animals without circulatory system.

gastrula A two-layered, later three-layered, animal embryonic stage.

gastrulation The process by which a blastula develops into a gastrula, usually by an involution of cells.

gel Colloid in which the suspended particles form a relatively orderly arrangement; *cf.* sol.

-gen; -geny [Gk *genos* birth, race] Producing; production, generation.

gene [Gk *genos*] The unit of inheritance; usually a portion of a DNA molecule that codes for some product such as a protein, tRNA, or rRNA.

gene flow The movement of genes from one part of a population to another, or from one population to another, via gametes.

gene pool The sum total of all the genes of all the individuals in a population.

generator potential *See* potential.

genetic drift Change in the gene pool purely as a result of chance, and not as a result of selection, mutation, or migration.

genome A haploid set of chromosomes, with the genes they bear. *Genomic region:* a portion of a chromosome bearing a subset of the organism's genes.

genotype The particular combination of genes present in the cells of an individual.

germ cell A sexual reproductive cell; an egg or sperm.

gibberellin A plant hormone one of whose effects is stem elongation in some dwarf plants.

gill An evaginated area of the body wall of an animal, specialized for gas exchange.

gizzard A chamber of an animal's digestive tract specialized for grinding food.

glucose [Gk *glykys* sweet] A six-carbon sugar; plays a central role in cellular metabolism.

glycogen [Gk *glykys*] A polysaccharide that serves as the principal storage form of carbohydrate in animals.

glycolysis [Gk *glykys*] Anaerobic catabolism of carbohydrates to pyruvic acid.

Golgi apparatus Membranous subcellular structure that plays a role in storage and modification particularly of secretory products.

gonadotrophic Stimulatory to the gonads.

gonadotrophin A hormone stimulatory to the gonads, i.e. a gonadotrophic hormone.

gonads [Gk *gonos* seed] The testes or ovaries.

gram molecule *See* mole.

granum [L grain] A stacklike grouping of photosynthetic membranes in a chloroplast (*pl.* grana).

guard cell A specialized epidermal cell that regulates the size of a stoma of a leaf.

habit [L *habitus* disposition] In biology, the characteristic form or mode of growth of an organism.

habitat [L it lives] The kind of place where a given organism normally lives.

haploid [Gk *haploos* single] Having only one of each type of chromosome.

hem-, hemat-, hemo- [Gk *haima* blood] Blood.

hematopoiesis [Gk *poiēsis* making] The formation of blood.

hemoglobin A red iron-containing pigment in the blood that functions in oxygen transport.

hepatic [Gk *hēpar* liver] Pertaining to the liver.

herbaceous [L *herbaceus* grassy] Having a stem that remains soft and succulent; not woody.

herbivore [L *herba* grass; *vorare* to devour] An animal that eats plants.

Hertz A unit of frequency (as of sound waves) equal to one cycle per second.

hetero- [Gk *heteros* other] Other, different.

heterogamy [Gk *gamos* marriage] The condition of producing gametes of two or more different types.

heterotrophic [Gk *trophē* food] Incapable of manufacturing organic compounds from inorganic raw materials, therefore requiring organic nutrients from the environment.

heterozygous [Gk *zygōtos* yoked] Having two different alleles of a given gene.

Hg [L *hydrargyrum* mercury] The symbol for mercury. Pressure is often expressed in *mm Hg*—the pressure exerted by a column of mercury whose height is measured in millimeters (at 0°C, 1 mm Hg = 133.3 newtons per square meter).

hilum Region where blood vessels, nerves, ducts, enter an organ.

hist- [Gk *histos* web] Tissue.

histology The structure and arrangement of the tissues of organisms; the study of these.

histone One of a class of basic proteins serving as structural elements of eucaryotic chromosomes.

homeo-, homo- [Gk *homoios* like] Like, similar.

homeostasis The tendency in an organism toward maintenance of physiological and psychological stability.

homeothermic [Gk *thermē* heat] Capable of self-regulation of body temperature; warm-blooded.

home range An area within which an animal tends to confine all or nearly all its activities for a long period of time.

homologous Of chromosomes: bearing genes for the same characters. Of characters in different organisms: inherited from a common ancestor.

homozygous [Gk *zygōtos* yoked] Having two doses of the same allele of a given gene.

hormone [Gk *horman* to set in motion] A control chemical secreted in one part of the body that affects other parts of the body.

hybrid In evolutionary biology, a cross between two species. In genetics, a cross between two genetic types.

hydr-, hydro- [Gk *hydōr* water] Water; fluid; hydrogen.

hydration Formation of a sphere of water around an electrically charged particle.

hydrocarbon Any compound containing only carbon and hydrogen.

hydrogen bond A weak chemical bond formed when a hydrogen atom is shared by two electronegative atoms, usually oxygen or nitrogen.

hydrolysis [Gk *lysis* loosing] Breaking apart of a molecule by addition of water.

hydrostatic [Gk *statikos* causing to stand] Pertaining to the pressure and equilibrium of fluids.

hydroxyl ion The OH⁻ ion.

hyper- Over, overmuch; more.

hypertonic Of a solution (or colloidal suspension): tending to gain water from some reference solution (or colloidal suspension) separated from it by a semipermeable membrane—usually because it has a higher osmotic concentration than the reference solution.

hypertrophy [Gk *trophē* food] Abnormal enlargement, excessive growth.

hypha [Gk *hyphē* web] A fungal filament.

hypo- Under, lower; less.

hypocotyl The portion of the axis of a plant embryo below the point of attachment of the cotyledons; forms the base of the shoot and the root.

hypothalamus [Gk *thalamos* inner chamber] Part of the posterior portion of the vertebrate forebrain, containing important centers of the autonomic nervous system and centers of emotion.

hypotonic Of a solution (or colloidal suspension): tending to lose water to some reference solution (or colloidal suspension) separated from it by a semipermeable membrane—usually because it has a lower osmotic concentration than the reference solution.

inducer In embryology, a substance that stimulates differentiation of cells or development of a particular structure. In genetics, a substance that activates particular genes.

inorganic compound A chemical compound not based on carbon.

in situ [L in place] In its natural or original position.

insulin [L *insula* island] A hormone produced by the β islet cells in the pancreas that helps regulate carbohydrate metabolism, especially conversion of glucose into glycogen.

integument [L *integere* to cover] A coat, skin, shell, rind, or other protective surface structure.

inter- Between (e.g. *interspecific*, between two or more different species).

intra- Within (e.g. *intraspecific*, within a single species).

intrinsic Inherent in, a basic part of; as in *intrinsic isolating mechanism*.

invaginated [L *vagina* sheath] Folded or protruded inward.

invertebrate [L *vertebra* joint] Lacking a backbone, hence an animal without bones.

in vitro [L in glass] Not in the living organism, in the laboratory.

in vivo [L in the living] In the living organism.

ion An electrically charged atom.

iso- Equal, uniform.

isogamy [Gk *gamos* marriage] The condition of producing gametes of only one type, no distinction existing between male and female.

isolating mechanism An obstacle to interbreeding, either extrinsic, such as a geographical barrier, or intrinsic, such as structural or behavioral incompatibility.

isotonic   Of a solution (or colloidal suspension): tending neither to gain nor to lose water when separated from some reference solution (or colloidal suspension) by a semipermeable membrane—usually because it has the same osmotic concentration as the reference solution.

isotope [Gk *topos* place]   An atom differing from another atom of the same element in the number of neutrons in its nucleus.

kilo-   A thousand.

kin-, kino- [Gk *kinēma* motion]   Motion, action.

lactic acid   A three-carbon organic acid produced in animals and some microorganisms by fermentation.

lambert   In metric system, unit of brightness of a light source; approximately equivalent to 929 foot-candles (q.v.).

lamella [L thin plate]   A thin platelike structure; a fairly straight intracellular membrane.

larva [L ghost, mask]   Immature form of some animals that undergo radical transformation to attain the adult form.

lateral   Pertaining to the side.

lenticel [L *lenticella* small lentil]   A porous region in the periderm of a woody stem through which gases can move.

leukocyte [Gk *leukos* white]   A white blood cell.

LH   *See* luteinizing hormone.

ligament [L *ligare* to bind]   A type of connective tissue linking two bones in a joint.

lignin [L *lignum* wood]   An organic compound in wood that makes cellulose harder and more brittle.

linkage   The location of two or more genes on the same chromosome, which, in the absence of crossing-over, causes the characters they control to be inherited together.

lip- [Gk *lipos* fat]   Fat or fatlike.

lipase   A fat-digesting enzyme.

lipid   Any of a variety of compounds insoluble in water but soluble in ethers and alcohols; includes fats, oils, waxes, phospholipids, and steroids.

locus [L place]   A particular location on a chromosome, hence often used synonymously with gene (*pl.* loci).

lumen [L light, opening]   The space or cavity within a tube or sac (*pl.* lumina).

lung   An internal chamber specialized for gas exchange in an animal.

luteinizing hormone (LH)   A gonadotrophic hormone of the pituitary that stimulates conversion of a follicle into a corpus luteum and secretion of progesterone by the corpus luteum; also stimulates secretion of sex hormone by the testes.

lymph [L *lympha* water]   A fluid derived from tissue fluid and transported in special lymph vessels to the blood.

lymphocyte   White blood cell that, upon stimulation by an antigen, gives rise to plasma cells, which produce antibody.

-lysis, lyso- [Gk *lysis* loosing]   Loosening, decomposition.

lysogenic   Of bacteria: carrying bacteriophage capable of lysing, i.e. destroying, other bacterial cells.

lysosome   A subcellular organelle that stores digestive enzymes.

macro-   Large.

Malpighian tubule   An excretory diverticulum of the digestive tract in insects and some other arthropods.

matrix [L *mater* mother]   A mass in which something is embedded, e.g. the intercellular substance of a tissue.

medulla [L marrow, innermost part]   (1) The inner portion of an organ, e.g. *adrenal medulla*.   (2) The *medulla oblongata*, a portion of the vertebrate hindbrain that connects with the spinal cord.

medusa [*after* Medusa, mythological monster with snaky locks]   The free-swimming stage in the life cycle of a coelenterate.

mega-   Large. Female.

megaspore   A spore that will germinate into a female plant.

meiosis [Gk *meiōsis* diminution]   A process of nuclear division in which the number of chromosomes is reduced by half.

meristematic tissue [Gk *meristos* divisible]   A plant tissue that functions primarily in production of new cells by mitosis.

meso-   Middle.

mesoderm   The middle tissue layer of an animal embryo.

mesophyll [Gk *phyllon* leaf]   The parenchymatous middle tissue layers of a leaf.

meta-   Posterior, later; change in.

metabolism [Gk *metabolē* change]   The sum of the chemical reactions within a cell (or a whole organism), including the energy-releasing breakdown of molecules (catabolism) and the synthesis of complex molecules and new protoplasm (anabolism).

metamorphosis [Gk *morphē* form]   Transformation of an immature animal into an adult. More generally, change in the form of an organ or structure.

micro-   Small. Male. In units of measurement, one millionth.

microorganism   A microscopic organism, especially a bacterium, virus, or protozoan.

microspore   A spore that will germinate into a male plant.

middle lamella   A layer of substance deposited between the walls of adjacent plant cells.

milli-   One thousandth.

mineral   In biology, any naturally occurring inorganic substance, excluding water.

mitochondrion [Gk *mitos* thread; *chondrion* small grain]   Subcellular organelle in which aerobic respiration takes place.

mitosis [Gk *mitos*]   Process of nuclear division in which complex movements of chromosomes along a spindle result in two new nuclei with the same number of chromosomes as the original nucleus.

modulator A control chemical that stabilizes an allosteric enzyme in one of its alternative conformations.

mold Any of many fungi that produce a cottony or furry growth.

mole The amount of a substance that has a weight in grams numerically equal to the molecular weight of the substance. One mole of a substance contains 6.023×10^{23} molecules of that substance; hence one mole of a substance will always contain the same number of molecules as a mole of any other substance.

molecular weight The weight of a molecule calculated as the sum of the atomic weights of its constituent atoms.

molecule A chemical unit consisting of two or more atoms bonded together.

mono- One.

monocot A member of a subclass of the angiosperms, or flowering plants, distinguished mainly by the presence of a single cotyledon in the embryo; *cf.* dicot.

-morph, morpho- [Gk *morphē* form] Form, structure.

morphogenesis The establishment of shape and pattern in an organism.

morphology The form and structure of organisms or parts of organisms; the study of these.

motivation The internal state of an animal that is the immediate cause of its behavior.

motor neuron A neuron leading from the central nervous system toward an effector.

mouthparts Structures or appendages near the mouth used in manipulating food.

mucosa Any membrane secreting mucus (a slimy protective substance), e.g. the membrane lining the stomach and intestine.

muscle [L *musculus* small mouse, muscle] A contractile tissue of animals.

mutation [L *mutatio* change] Any relatively stable heritable change in the genetic material.

mutualism A symbiosis in which both parties benefit.

mycelium [Gk *mykēs* fungus] A mass of hyphae forming the body of a fungus.

myo- [Gk *mys* mouse, muscle] Muscle.

NAD *See* nicotinamide adenine dinucleotide.

NADP *See* nicotinamide adenine dinucleotide phosphate.

nano- [L *nanus* dwarf] One billionth.

natural selection Differential reproduction in nature, leading to an increase in the frequency of some genes or gene combinations and to a decrease in the frequency of others.

navigation The initiation and/or maintenance of movement toward a goal by means other than recognition of landmarks.

nematocyst [Gk *nēma* thread; *kystis* bag] A specialized stinging cell in coelenterates; contains a hairlike structure that can be ejected.

neo- New.

neocortex Portion of the cerebral cortex in mammals, of relatively recent evolutionary origin; often greatly expanded in the higher primates and dominant over other parts of the brain.

nephr- [Gk *nephros* kidney] Kidney.

nephridium An excretory organ consisting of an open bulb and a tubule leading to the exterior; found in many invertebrates, e.g. segmented worms.

nephron The functional unit of a vertebrate kidney, consisting of Bowman's capsule, convoluted tubule, and loop of Henle.

nerve [L *nervus* sinew, nerve] A bundle of neuron fibers.

nerve net A nervous system without any central control, as in coelenterates.

neuron [Gk nerve, sinew] A nerve cell.

neutron An electrically neutral subatomic particle with approximately the same mass as a proton.

niche The functional role and position of an organism in the ecosystem.

nicotinamide adenine dinucleotide (NAD) An organic compound that functions as an electron acceptor, especially in respiration.

nicotinamide adenine dinucleotide phosphate (NADP) An organic compound that functions as an electron acceptor, especially in biosynthesis.

nitrogen fixation Incorporation of nitrogen from the atmosphere into substances more generally usable by organisms.

node (of plant) [L *nodus* knot] Point on a stem where a leaf or bud is (or was) attached.

notochord [Gk *nōtos* back; *chordē* string] In the lower chordates and in the embryos of the higher vertebrates, a flexible supportive rod running longitudinally through the back just ventral to the nerve cord.

nucleic acid Any of several organic acids that are polymers of nucleotides and function in transmission of hereditary traits, in protein synthesis, and in control of cellular activities.

nucleoid A region, not bounded by a membrane, where the chromosome is located in a procaryotic cell.

nucleolus A dense body within the nucleus, usually attached to one of the chromosomes; consists of multiple copies of the genes for rRNA.

nucleotide A chemical entity consisting of a five-carbon sugar with a phosphate group and a purine or pyrimidine attached; building-block unit of nucleic acids.

nucleus (of cell) [L kernel] A large membrane-bounded organelle containing the chromosomes.

nutrient [L *nutrire* to nourish] A food substance usable in metabolism as a source of energy or of building material.

nymph [Gk *nymphē* bride, nymph] Immature stage of insect that undergoes gradual metamorphosis.

olfaction [L *olfacere* to smell]   The sense of smell.

omnivorous [L *omnis* all; *vorare* to devour]   Eating a variety of foods, including both plants and animals.

ontogeny [Gk *ōn* being]   The course of development of an individual organism.

oo- [Gk *ōion* egg]   Egg.

oogamy   A type of heterogamy in which the female gametes are large nonmotile egg cells.

oogonium   Unjacketed female reproductive organ of a thallophyte plant.

oral [L *oris* of the mouth]   Relating to the mouth.

organ [Gk *organon* tool]   A body part usually composed of several tissues grouped together into a structural and functional unit.

organelle   A well-defined subcellular structure.

organic compound   A chemical compound containing carbon.

organism   An individual living thing.

orientation   The act of turning or moving in relation to some external feature, e.g. a source of light.

osmol   Measure of osmotic concentration; the total number of moles of osmotically active particles per liter of solvent.

osmoregulation   Regulation of the osmotic concentration of body fluids in such a manner as to keep them relatively constant despite changes in the external medium.

osmosis [Gk *ōsmos* thrust]   Movement of a solvent (usually water in biology) through a semipermeable membrane.

osmotic pressure   The pressure that must be exerted on a solution or colloid to keep it in equilibrium with pure water when it is separated from the water by a semipermeable membrane; hence a measure of the tendency of the solution or colloid to take in water.

ov-, ovi- [L *ovum* egg]   Egg.

ovary   Female reproductive organ in which egg cells are produced.

ovulation   Release of an egg from the ovary.

ovule   A plant structure, composed of an integument, sporangium, and megagametophyte, that develops into a seed after fertilization.

ovum   A mature egg cell (*pl.* ova).

oxidation   Energy-releasing process involving removal of electrons from a substance; in biological systems, generally by the removal of hydrogen (or sometimes the addition of oxygen).

pancreas   In vertebrates, a large glandular organ located near the stomach that secretes digestive enzymes into the duodenum and also produces hormones.

papilla [L nipple]   A small nipplelike protuberance.

para-   Alongside of.

parapodium [Gk *podion* little foot]   One of the paired segmentally arranged lateral flaplike protuberances of polychaete worms.

parasitism [Gk *parasitos* eating with another]   A symbiosis in which one party benefits at the expense of the other.

parasympathetic nervous system   One of the two parts of the autonomic nervous system.

parathyroids   Small endocrine glands of vertebrates located near the thyroid.

parenchyma   A plant tissue composed of thin-walled, loosely packed, relatively unspecialized cells.

parthenogenesis [Gk *parthenos* virgin]   Production of offspring without fertilization.

pathogen [Gk *pathos* suffering]   A disease-causing organism.

pectin   A complex polysaccharide that cross-links the cellulose fibrils in a plant cell wall and is a major constituent of the middle lamella.

pellicle [L *pellis* skin]   A thin skin or membrane.

pepsin [Gk *pepsis* digestion]   A protein-digesting enzyme of the stomach.

peptide bond   A bond between two amino acids resulting from a condensation reaction between the amino group of one acid and the acidic group of the other.

peri-   Surrounding.

pericycle   A layer of cells inside the endodermis but outside the phloem of roots and stems.

periderm   The corky outer bark of older stems and roots.

peristalsis [Gk *stalsis* contraction]   Alternating waves of contraction and relaxation passing along a tubular structure such as the digestive tract.

permeable [L *permeare* to go through]   Of a membrane: permitting other substances to pass through.

permease [L *permeare*]   A protein that functions as a carrier molecule in a membrane.

petiole [L *pediculus* small foot]   The stalk of a leaf.

PGAL   *See* phosphoglyceraldehyde.

pH   Symbol for the logarithm of the reciprocal of the hydrogen ion concentration; hence a measure of acidity. A pH of 7 is neutral; lower values are acidic, higher values alkaline (basic).

phage   *See* bacteriophage.

phagocytosis [Gk *phagein* to eat]   The active engulfing of particles by a cell.

pharynx   Part of the digestive tract between the oral cavity and the esophagus; in vertebrates, also part of the respiratory passage.

phenotype [Gk *phainein* to show]   The physical manifestation of a genetic trait.

pheromone [Gk *pherein* to carry + hormone]   A substance that, secreted by one organism, influences the behavior or physiology of other organisms of the same species.

phloem [Gk *phloios* bark]   A plant vascular tissue that transports organic materials; the inner bark.

-phore [Gk *pherein* to carry]   Carrier.

phosphoglyceraldehyde (PGAL)   A three-carbon phosphorylated carbohydrate, important in both photosynthesis and glycolysis.

phospholipid   A compound composed of glycerol, fatty acids, a phosphate group, and often a nitrogenous group.

phosphorylation   Addition of a phosphate group.

photo- [Gk *phōs* light]   Light.

photon   One of the discrete units or packets into which radiant energy is subdivided.

photoperiodism   A response by an organism to the duration and timing of the light and dark conditions.

photosynthesis   Autotrophic synthesis of organic materials in which the source of energy is light.

phototropism [Gk *tropos* turn]   A turning response to light.

-phyll [Gk *phyllon* leaf]   Leaf.

phylogeny [Gk *phylē* tribe]   Evolutionary history of an organism.

physiology [Gk *physis* nature]   The life processes and functions of organisms; the study of these.

-phyte, phyto- [Gk *phyton* plant]   Plant.

phytochrome   A protein pigment of plants sensitive to red and far-red light.

pinocytosis [Gk *pinein* to drink]   The active engulfing by cells of liquid or of very small particles.

pistil   The female reproductive organ of a flower, composed of one or more megasporophylls.

pith   A tissue (usually parenchyma) located in the center of a stem (rarely a root), internal to the xylem.

pituitary   An endocrine gland located near the brain of vertebrates; known as the master gland because it secretes hormones that regulate the action of other endocrine glands.

placenta [Gk *plax* flat surface]   An organ in mammals, made up of fetal and maternal components, that aids in exchange of materials between the fetus and the mother.

plasm-, plasmo-, -plasm [Gk *plasma* something formed or molded]   Formed material; plasma; cytoplasm.

plasma   Blood minus the cells and platelets.

plasma membrane   The outer membrane of a cell.

plasmid   A small circular piece of DNA free in the cytoplasm of a bacterial cell and replicated independently of the cell's chromosome.

plasmodesma [Gk *desma* bond]   A delicate cytoplasmic connection between adjacent plant cells (*pl.* plasmodesmata).

plasmolysis   Shrinkage of a plant cell away from its wall when in a hypertonic medium.

plastid   Relatively large organelle in plant cells that functions in photosynthesis and/or nutrient storage.

pleiotropic [Gk *pleiōn* more]   Of a gene: having more than one phenotypic effect.

poikilothermic [Gk *poikilos* various; *thermē* heat]   Incapable of precise self-regulation of body temperature, dependent on environmental temperature; cold-blooded.

pollen grain [L *pollen* flour dust]   A microgametophyte of a seed plant.

poly-   Many.

polymer [Gk *meros* part]   A large molecule consisting of a chain of small molecules bonded together by condensation reactions or similar reactions.

polymorphism [Gk *morphē* form]   The simultaneous occurrence of several discontinuous phenotypes in a population.

polyp [Gk *polypous* many-footed]   The sedentary stage in the life cycle of a coelenterate.

polypeptide chain   A chain of amino acids linked together by peptide bonds.

polyploid   Having more than one complete sets of chromosomes.

polysaccharide   Any carbohydrate that is a polymer of simple sugars.

population   In ecology, a group of individuals belonging to the same species.

portal system [L *porta* gate]   A blood circuit in which two beds of capillaries are connected by a vein (e.g. *hepatic portal system*).

posterior   Toward the hind end.

potential   Short for *potential difference:* the electrification of one point or structure relative to the electrification of some other point or structure.   *Resting p.:* a relatively steady potential difference across a cell membrane, particularly of a nonfiring nerve cell or a relaxed muscle cell.   *Action p.:* a sharp change in the potential difference across the membrane of a nerve or muscle cell that is propagated along the cell; in nerves, identified with the nerve impulse.   *Generator p.:* a change in the potential difference across the membrane of a sensory cell that, if it reaches a threshold level, may trigger an action potential along the associated neural pathway.

predation [L *praedatio* plundering]   The feeding of free-living organisms on other organisms.

primitive [L *primus* first]   Old, like the ancestral condition.

primordium [L *primus; ordiri* to begin]   Rudiment, earliest stage of development.

pro-   Before.

proboscis [Gk *boskein* to feed]   A long snout; an elephant's trunk. In invertebrates, an elongate, sometimes eversible process originating in or near the mouth that often serves in feeding.

procaryotic cell   A type of cell that lacks a membrane-bounded nucleus; found only in bacteria, blue-green algae, and prochlorophytes.

progesterone [L *gestare* to carry]   One of the principal female sex hormones of vertebrates.

prot-, proto-   First, primary.

protease   A protein-digesting enzyme.

protein   A long polypeptide chain.

proteolytic   Protein-digesting.

proton   A positively charged primary subatomic particle.

protoplasm   Living substance, the material of cells.

provirus   Viral nucleic acid integrated into the genetic material of a host cell.

proximal   Near some reference point (often the main part of the body).

pseudo-   False; temporary.

pseudocoelom   A functional body cavity not entirely enclosed by mesoderm.

pseudopod, pseudopodium [L *podium* foot]   A transitory cytoplasmic protrusion of an amoeba or an amoeboid cell.

pulmonary [L *pulmones* lungs]   Relating to the lungs.

purine   Any of several double-ringed nitrogenous bases important in nucleotides.

pyloric [Gk *pylōros* gatekeeper] · Referring to the junction between the stomach and the intestine.

pyrimidine   Any of several single-ringed nitrogenous bases important in nucleotides.

pyruvic acid   A three-carbon compound produced by glycolysis.

race   A subspecies.

radial symmetry   A type of symmetry in which the body parts are arranged regularly around a central line (in animals, running through the oral-anal axis) rather than on the two sides of a plane.

radiation   As an evolutionary phenomenon, divergence of members of a single lineage into different niches or adaptive zones.

recessive   Of an allele: not expressing its phenotype in the presence of another allele of the same gene, therefore expressing it only in homozygous individuals. *Recessive character, recessive phenotype:* one caused by a recessive allele.

rectum [L *rectus* straight]   The terminal portion of the intestine.

reduction   Energy-storing process involving addition of electrons to a substance; in biological systems, generally by the addition of hydrogen (or sometimes the removal of oxygen).

reflex [L *reflexus* bent back]   A functional unit of the nervous system, involving the entire pathway from receptor cell to effector.

reinforcement (psychological)   Reward for a particular behavior.

releaser   A structure, action, sound, etc., that gives rise to releasing stimuli, i.e. to stimuli particularly effective in triggering behavioral responses.

renal [L *renes* kidneys]   Pertaining to the kidney.

respiration [L *respiratio* breathing out]   (1) The release of energy by oxidation of fuel molecules.   (2) The taking in of O_2 and release of CO_2; breathing.

resting potential   *See* potential.

reticulum [L little net]   A network.

retina   The tissue in the rear of the eye that contains the sensory cells of vision.

rhizoid [Gk *rhiza* root]   Rootlike structure.

ribonucleic acid (RNA)   Any of several nucleic acids in which the sugar component is ribose and one of the nitrogenous bases is uracil.

ribosome   A small cytoplasmic organelle that functions in protein synthesis.

RNA   *See* ribonucleic acid.

salt   Any of a class of generally ionic compounds that may be formed by reaction of an acid and a base, e.g. table salt, NaCl.

sap   Water and dissolved materials moving in the xylem; less commonly, solutions moving in the phloem.   *Cell sap:* the fluid content of a plant-cell vacuole.

saprophyte [Gk *sapros* rotten]   A heterotrophic plant or bacterium that lives on dead organic material.

sarcomere [Gk *sarx* flesh; *meros* part]   The region of a skeletal-muscle myofibril extending from one Z line to the next; the functional unit of skeletal muscle contraction.

sclerenchyma [Gk *sclēros* hard]   A plant supportive tissue composed of cells with thick secondary walls.

section   *Cross* or *transverse s.:* section at right angles to the longest axis.   *Longitudinal s.:* section parallel to the longest axis.   *Radial s.:* longitudinal section along a radius.   *Sagittal s.:* vertical longitudinal section along the midline of a bilaterally symmetrical animal.

seed   A plant reproductive entity consisting of an embryo and stored food enclosed in a protective coat.

segmentation   The subdivision of an organism into more or less equivalent serially arranged units.

selection pressure   In a population, the force for genetic change resulting from natural selection.

semipermeable   Permeable only to solvent (usually water); less strictly: differentially permeable, i.e. permeable to some substances but not to others.

sensory neuron   A neuron leading from a receptor cell to the central nervous system.

septum [L barrier]   A partition or wall (*pl.* septa).

sessile [L *sessilis* of sitting, low]   Of animals, sedentary. Of plants, without a stalk.

sex-linked   Of genes: located on the X chromosome.

shoot   A stem with its leaves, flowers, etc.

sieve element   A conductile cell of the phloem.

sinus [L curve, hollow]   (1) A channel for the passage of blood lacking the characteristics of a true blood vessel.   (2) A hollow within bone or another tissue (e.g. the air-filled sinuses of some of the facial bones).

sol   Colloid in which the suspended particles are dispersed at random; *cf.* gel.

solute   Substance dissolved in another (the solvent).

solution [L *solutio* loosening]   A homogeneous molecular mixture of two or more substances.

solvent   Medium in which one or more substances (the solute) are dissolved.

-soma, somat-, -some [Gk *sōma* body]   Body, entity.

somatic   Pertaining to the body; to all cells except the germ cells; to the body wall. *Somatic nervous system:* a portion of the nervous system that is at least potentially under control of the will; its reflex arcs include one sensory and one motor neuron (*cf.* autonomic nervous system).

specialized   Adapted to a special, usually rather narrow, function or way of life.

speciation   The process of formation of new species.

species [L kind]   The largest unit of population within which effective gene flow occurs or could occur.

sperm [Gk *sperma* seed]   A male gamete.

sphincter [Gk *sphinktēr* band]   A ring-shaped muscle that can close a tubular structure by contracting.

spindle   A microtubular structure with which the chromosomes are associated in mitosis and meiosis.

sporangium   A plant structure that produces spores.

spore [Gk *spora* seed]   An asexual reproductive cell, often a resting stage adapted to resist unfavorable environmental conditions.

sporophyll [Gk *phyllon* leaf]   A modified leaf that bears spores.

sporophyte [Gk *phyton* plant]   A diploid plant that produces spores.

stamen [L thread]   A male sexual part of a flower; a microsporophyll of a flowering plant.

starch   A glucose polymer, the principal polysaccharide storage product of vascular plants.

stele [Gk *stēlē* upright slab]   The vascular cylinder in the center of a root or stem, bounded externally by the endodermis.

stereo- [Gk *stereos* solid]   Solid; three-dimensional.

steroid   Any of a number of complex, often biologically important compounds (e.g. some hormones and vitamins), composed of four interlocking rings of carbon atoms.

stimulus   Any environmental factor that is detected by a receptor.

stoma [Gk mouth]   An opening, regulated by guard cells, in the epidermis of a leaf or other plant part (*pl.* stomata).

stroma [Gk *strōma* bed, mattress]   The ground substance within such organelles as   oplasts and mitochondria.

subspecies   A genetically distinctive geographic subunit of a species.

substrate   (1) The base on which an organism lives, e.g. soil.   (2) In chemical reactions, a substance acted upon, as by an enzyme.

succession   In ecology, progressive change in the plant and animal life of an area.

sucrose   A double sugar composed of a unit of glucose and a unit of fructose; table sugar.

suspension   A heterogeneous mixture in which the particles of one substance are kept dispersed by agitation.

sym-, syn-   Together.

symbiosis [Gk *bios* life]   The living together of two organisms in an intimate relationship.

sympathetic nervous system   One of the two parts of the autonomic nervous system.

sympatric [L *patria* homeland]   Having the same range.

symplast   In a plant, the system constituted by the cytoplasm of cells interconnected by plasmodesmata.

synapse [Gk *haptein* to fasten]   A juncture between two neurons.

synapsis   The pairing of homologous chromosomes during meiosis.

synergistic [Gk *ergon* work]   Acting together with another substance or organ to achieve or enhance a given effect.

systemic circulation   The part of the circulatory system supplying body parts other than the gas-exchange surfaces.

-tactic   Referring to a taxis.

taxis   A simple continuously oriented movement in animals (e.g. phototaxis, geotaxis) (*pl.* taxes).

taxonomy [Gk *taxis* arrangement]   The classification of organisms on the basis of their evolutionary relationships.

tendon [L *tendere* to stretch]   A type of connective tissue attaching muscle to bone.

territory   A particular area defended by an individual against intrusion by other individuals, particularly of the same species.

testis   Primary male sex organ in which sperm are produced (*pl.* testes).

thalamus [Gk *thalamos* inner chamber]   Part of the rear portion of the vertebrate forebrain, a center for integration of sensory impulses.

thallus [Gk *thallos* young shoot]   A plant body exhibiting relatively little tissue differentiation and lacking true roots, stems, and leaves.

thorax [Gk *thōrax* breastplate]   In mammals, the part of the trunk anterior to the diaphragm, which partitions it from the abdomen. In insects, the body region between the head and the abdomen, bearing the walking legs and wings.

thymus [Gk *thymos* warty excrescence]   Glandular organ that plays an important role in the development of immunologic capabilities in vertebrates.

thyroid [Gk *thyreoeidēs* shield-shaped]   An endocrine gland of vertebrates located in the neck region.

thyroxin   A hormone, produced by the thyroid, that stimulates a speedup of metabolism.

tissue [L *texere* to weave]   An aggregate of cells, usually similar in both structure and function, that are bound together by intercellular material.

trachea   In vertebrates, the part of the respiratory system running from the pharynx into the thorax; the "windpipe." In land arthropods, an air duct running from an opening in the body wall to the tissues.

tracheid   An elongate thick-walled tapering conductile cell of the xylem.

trans-   Across; beyond.

transcription   In genetics, the synthesis of RNA along a DNA template.

transduction [L *ducere* to lead]   The transfer of genetic material from one host cell to another by a virus.

transformation   The incorporation by bacteria of fragments of DNA released into the medium from dead cells.

translation   In genetics, the synthesis of a polypeptide along an mRNA template.

translocation   In botany, the movement of organic materials from one place to another within the plant body, primarily through the phloem.

transpiration   Release of water vapor from the aerial parts of a plant, primarily through the stomata.

-trophic [Gk *trophē* food]   Nourishing; stimulatory.

tropism [Gk *tropos* turn]   A turning response to a stimulus, primarily by differential growth patterns in plants.

turgid [L *turgidus* swollen]   Swollen with fluid.

turgor pressure [L *turgēre* to be swollen]   The pressure exerted by the contents of a cell against the cell membrane or cell wall.

tympanic membrane [Gk *tympanon* drum]   A membrane of the ear that picks up vibrations from the air and transmits them to other parts of the ear; the eardrum.

unit membrane   A membrane composed of two layers of protein with two layers of lipid between them.

urea   The nitrogenous waste product of mammals and some other vertebrates, formed in the liver by combination of ammonia and carbon dioxide.

ureter   The duct carrying urine from the kidney to the bladder in higher vertebrates.

urethra   The duct leading from the bladder to the exterior in higher vertebrates.

uric acid   An insoluble nitrogenous waste product of most land arthropods, reptiles, and birds.

uterus   In mammals, the chamber of the female reproductive tract in which the embryo undergoes much of its development; the womb.

vaccine [L *vacca* cow]   Drug containing an antigen, administered to induce active immunity in the patient.

vacuole [L *vacuus* empty]   A membrane-bounded vesicle or chamber in a cell.

vascular tissue [L *vasculum* small vessel]   Tissue concerned with internal transport, such as xylem and phloem in plants and blood and lymph in animals.

vaso- [L *vas* vessel]   Blood vessel.

vector [L *vectus* carried]   Transmitter of pathogens.

vegetative   Of plant cells and organs: not specialized for reproduction. Of reproduction: asexual. Of bodily functions: involuntary.

vein [L *vena* blood vessel]   A blood vessel that transports blood toward the heart.

vena cava [L hollow vein]   One of the two large veins that return blood to the heart from the systemic circulation of vertebrates.

ventral [L *venter* belly]   Pertaining to the belly or underparts.

vessel cell   A highly specialized cell of the xylem, with thick secondary walls and extensively perforated end walls.

villus [L shaggy hair]   A highly vascularized fingerlike process from the intestinal lining or from the surface of some other structure (e.g. a chorionic villus of the placenta) (*pl.* villi).

virion   A free virus particle.

virus [L slime, poison]   A submicroscopic noncellular, obligatorily parasitic particle, composed of a protein shell and a nucleic acid core, that exhibits some properties normally associated with living organisms, including the ability to mutate and to evolve.

viscera [L]   The internal organs, especially those of the great central body cavity.

vitamin [L *vita* life]   An organic compound, necessary in small quantities, that a given organism cannot synthesize for itself and must obtain prefabricated in the diet.

X chromosome   The female sex chromosome.

xylem [Gk *xylon* wood]   A vascular tissue that transports water and dissolved minerals upward through the plant body.

Y chromosome   The male sex chromosome.

yolk   Stored food material in an egg.

zoo- [Gk *zōion* animal]   Animal; motile.

zoospore   A ciliated or flagellated plant spore.

zygote [Gk *zygōtos* yoked]   A fertilized egg cell.

zymogen [Gk *zymē* leaven]   An inactive precursor of a digestive enzyme.

INDEX

Boldface numbers refer to pages that carry illustrations, generally in addition to text, on the subject in question. *Italicized* page numbers identify definitions in context or the main treatment of a subject mentioned in several parts of the book.